# MOLECULAR SPECTRA

*and*

# MOLECULAR STRUCTURE

## III. Electronic Spectra and Electronic Structure of Polyatomic Molecules

GERHARD HERZBERG
*National Research Council of Canada*

VNR VAN NOSTRAND REINHOLD COMPANY
NEW YORK CINCINNATI TORONTO LONDON MELBOURNE

VAN NOSTRAND REINHOLD COMPANY Regional Offices:
New York Cincinnati Chicago

VAN NOSTRAND REINHOLD COMPANY Foreign Offices:
London Toronto Melbourne



ISBN: 0-442-03387-7



Manufactured in the United States of America.

Published by VAN NOSTRAND REINHOLD COMPANY
135 West 50th Street, New York, N.Y. 10020

Published simultaneously in Canada by
VAN NOSTRAND REINHOLD COMPANY LTD.

15 14 13 12 11 10

# PREFACE

The present volume is the third and concluding volume in the series Molecular Spectra and Molecular Structure, which I started in 1939 with the publication of Volume I, Spectra of Diatomic Molecules, and continued in 1945 with the publication of Volume II, Infrared and Raman Spectra of Polyatomic Molecules. A revised edition of Volume I was published in 1950.

The present third volume deals with Electronic Spectra and Electronic Structure of Polyatomic Molecules, a field that has greatly developed in the last two decades. In contrast to diatomic molecules, the spectroscopic study of polyatomic molecules was for a long time carried out principally by work in the infrared region and by the study of the Raman effect. However, the recent considerable improvement in vacuum ultraviolet techniques and the development of new techniques for the study of free radicals have made possible a rapid development of the study of electronic spectra. The object of this book is to present the results of this new development without, of course, neglecting the significant older work on the subject.

The book is aimed at the serious student of molecular structure. Without going into long mathematical developments, a good deal of the theory of the subject is treated, and emphasis is laid on the comparison with experiment. As in the two preceding volumes, numerous spectrograms are intended to give a feeling for the experimental basis of the subject, while a large number of energy level diagrams and other illustrations should help to make the theoretical developments more easily visualizable.

Although in many instances cross references to the earlier volume are given, the material in this book should be understandable without having the other volumes constantly at hand. The basic information on rotational and vibrational energy levels has been repeated and where necessary, brought up to date in this volume.

For the convenience of the reader and in order not to interrupt the text, certain tables of symmetry types and characters of the more important point groups, of spin functions, direct products, and resolution of species when going to lower symmetry, have been collected in Appendices. Here are also extensive tables of molecular constants of most of the polyatomic molecules (up to 12 atoms) for which discrete absorption or emission spectra have been analysed. The data on the ground states of these molecules have been brought up to date and replace those given in Volume II. An attempt was made to make these tables as up-to-date as possible and the results of important papers up to the end of 1965 have been included. However, with the prolifieration of scientific journals and the immense amount of material published annually, it is impossible not to overlook some important contributions. My apologies go to those authors whose work has not been properly quoted or has been overlooked.

The manuscript of the book was completed in June 1965, and only a few later developments could be included in the proofs.

As in the other two volumes, the detailed subject index of the book includes all symbols and quantum numbers as well as molecules discussed in the book.

G. Herzberg

Ottawa, Canada
July, 1966

## Acknowledgements

It gives me great pleasure to acknowledge the help and cooperation I have received from many friends and associates during the preparation of this book. I am particularly indebted to Drs. J. T. Hougen and A. E. Douglas who read the whole manuscript and made innumerable suggestions for the improvement of the manuscript, as well as clarified by discussion many problems that arose in the writing of the manuscript. I am equally grateful to Professor H. C. Longuet-Higgins, F.R.S., for invaluable help with some of the theoretical problems that arose, especially in connection with double groups. I am also indebted to Professor D. M. Bishop (University of Ottawa) for a critical reading of the manuscript of Chapter III.

In the proofreading I had the enthusiastic and critical help of Drs. J. H. Callomon, K. P. Huber and J. W. C. Johns, each of whom went over the whole galley proof with the greatest care, pointing out many errors and inconsistencies, and making most valuable suggestions for improvements. Several colleagues have given me invaluable help in the preparation of the tables of molecular constants by supplying the latest information about molecules in which they were particularly expert.

In the preparation of the numerous illustrations, I am happy to acknowledge the invaluable aid of Mrs. M. Selwood, Mr. R. O'Kell and Mr. J. Shoosmith. Mrs. Selwood also gave constant help in many other of the numerous details connected with the preparation of such an extensive manuscript.

Various drafts of the manuscript were typed by Miss P. Hanson, Miss K. Assing and Miss O. B. Maddock. Miss Hanson in addition also prepared the author index and gave extensive help with the proofreading. I am most grateful to all of them for their efficient and devoted help.

Finally, I must acknowledge my debt to those authors who agreed to the use of their illustrations and, in the case of spectrograms, supplied original prints for this purpose. They are:

Prof. R. B. Bernstein (University of Wisconsin): Fig. 175
Dr. J. H. Callomon (University College, London): Fig. 199
Dr. C. A. Coulson (Oxford University): Fig. 119c
Prof. F. W. Dalby (University of British Columbia): Fig. 109
Dr. R. N. Dixon (Sheffield University): Figs. 46, 47, 74, 75
Dr. A. E. Douglas (National Research Council, Ottawa): Fig. 98, 101, 192
Prof. H. Eyring (University of Utah): Fig. 142
Dr. K. P. Huber (National Research Council, Ottawa): Fig. 183
Prof. K. K. Innes (Vanderbilt University): Figs. 105, 200
Dr. J. W. C. Johns (National Research Council, Ottawa): Figs. 78, 112
Dr. B. Kleman (Research Institute of National Defence, Stockholm): Fig. 84
Prof. R. S. Mulliken (University of Chicago): Fig. 197
Dr. D. A. Ramsay (National Research Council, Ottawa): Figs. 95, 177
Dr. Y. Tanaka (Air Force Cambridge Research Laboratories): Fig. 189
Prof. K. Watanabe (University of Hawaii): Figs. 184, 188, 191
Dr. J. K. G. Watson (University of Reading): Fig. 83

The permission to use these figures was readily granted by the editors of the following societies or journals:

The Canadian Journal of Physics (N.R.C.): Figs. 78, 84, 101, 108, 109, 112, 177, 189, 192
The Faraday Society: Figs. 46, 47, 98, 199
Journal of Chemical Physics: Figs. 175, 188, 191, 197

Journal of Molecular Spectroscopy (Academic Press, Inc.): Fig. 200
Journal of the Optical Society: Fig. 184
The Royal Society: Figs. 74, 75, 92, 95, 97, 185, 193
Spectrochimica Acta (Pergamon): Fig. 105

G. Herzberg

# CONTENTS

# TABLES

# INTRODUCTION

**General remarks.** In the two preceding volumes* the spectra of diatomic molecules and the infrared and Raman spectra of polyatomic molecules have been discussed. In the present volume the electronic spectra of polyatomic molecules will be treated. As in the volume on diatomic molecules, both chemically stable and unstable molecules (free radicals) will be included.

Almost all electronic transitions of polyatomic molecules give rise to spectra in the visible and ultraviolet regions; only very few extend into the infrared region. The study of these spectra gives information about the various electronic states and their rotational and vibrational energy levels. For stable molecules, information about the rotational and vibrational levels of the electronic ground state is in general obtainable more readily and with greater precision from a study of infrared, microwave and Raman spectra. The importance of the investigation of electronic spectra of stable molecules lies in the information they supply about the electronic structure of these molecules, about the geometrical structure in excited electronic states, and about their ionization potentials, dissociation energies, etc. But for chemically unstable molecules (free radicals) for which infrared and Raman spectra are not readily obtained, the study of electronic spectra is also at present the only source of information about the rotational and vibrational structure of the ground state and therefore of the geometrical structure in this state.

Most electronic spectra of polyatomic molecules have been studied in absorption, but a few have been obtained in emission in electric discharges, in fluorescence or phosphorescence, and in flames. The absorption spectra of polyatomic free radicals have been obtained mostly by the flash photolysis technique, that is, by taking the absorption spectrum during the photodecomposition of an appropriate parent compound by means of a bright flash [see, e.g., Porter (1006)[1] and Ramsay (1039)]. The spectra of some free radicals have also been obtained in high temperature chemical equilibria [see, e.g., Wieland (1295), Garton (407) and Kleman (679)] both in absorption and in thermal emission. Various types of electric discharges have been described for obtaining spectra of free radicals and polyatomic ions in emission, especially by Schüler and his collaborators (1109) (1114) [see also Callomon (172)]. A few such spectra have been observed in various types of flames, e.g., the $NH_2$ bands in an oxy-ammonia flame, the ethylene flame bands, believed to be due to HCO, in many hydrocarbon flames, and a few others [see Pearse and Gaydon (32)].

* G. Herzberg, Molecular Spectra and Molecular Structure: Volume I: Spectra of Diatomic Molecules, 2nd ed.; Volume II: Infrared and Raman Spectra of Polyatomic Molecules (Van Nostrand, Princeton, 1950 and 1945). These volumes are cited in this book as Volume I and Volume II, respectively.

[1] Numbers in ( ) refer to the bibliography, p. 671.

In the present volume, as in the preceding one, we shall limit ourselves to the consideration of the spectra of *simple molecules* with not more than 12 atoms as observed in the *gaseous state*. For the spectra of more complicated molecules and for spectra in the liquid and solid states, the reader is referred to Förster (15), West (42), Platt (990)(991), Jaffe and Orchin (24), Murrell (27), Mason (799a) and Sandorfy (34).

**Symmetry of polyatomic molecules.** As in the discussion of infrared and Raman spectra, symmetry considerations are of fundamental importance for an understanding of electronic spectra. For the convenience of the reader we summarize here briefly the discussion of this topic given previously in Volume II, pp. 1–12.

The nuclear framework of a molecule (like any geometrical figure) may have one or more of the following *symmetry elements*:

(1) a plane of symmetry usually designated by $\sigma$;

(2) a center of symmetry usually designated by $i$;

(3) a $p$-fold axis of symmetry usually designated by $C_p$, where $p = 1, 2, 3, \ldots$; $C_1$ means no rotational symmetry, $C_2$ a two-fold axis of symmetry, etc.; $C_\infty$ means an axis of cylindrical symmetry as in a linear or diatomic molecule;

(4) a $p$-fold rotation-reflection axis usually designated by $S_p$; a molecule having such an element of symmetry will be transformed into itself by a rotation through an angle $360°/p$ followed by a reflection at a plane perpendicular to the axis;

(5) the identity designated by $I$ (or $E$).

In general a molecule has several of these symmetry elements. By combining more and more symmetry elements, systems of higher and higher symmetry are obtained. However not all combinations of symmetry elements, but only certain ones, are possible. These possible combinations of symmetry operations are called *point groups*. A molecule in a given *conformation* (that is, in a given relative arrangement of the nuclei) must belong to one of these point groups. The following point groups are of importance for polyatomic molecules.

*The point groups* $\boldsymbol{C_p}$ have only a $p$-fold axis of symmetry $C_p$ and no other element of symmetry apart from the identity $I$. $\boldsymbol{C_1}$ is a point group without symmetry. The only "symmetry" element is the identity $I$. $\boldsymbol{C_2}$ is a system having only a two-fold axis of symmetry.

*The point groups* $\boldsymbol{S_p}$ have only a $p$-fold rotation-reflection axis which, however, implies the presence of certain other elements of symmetry. Point groups $\boldsymbol{S_p}$ are defined only for even $p$. The symmetry element $S_2$ is equivalent to $i$ and the point group $\boldsymbol{S_2}$ is usually called $\boldsymbol{C_i}$. The symmetry element $S_4$ always implies $C_2$ and therefore the point group $\boldsymbol{S_4}$ has the symmetry elements $I$, $S_4$ and $C_2$.

*The point groups* $\boldsymbol{C_{pv}}$ have a $p$-fold axis $C_p$ and $p$ planes of symmetry $\sigma_v$ through the axis (the subscript $v$ is meant to indicate "vertical" assuming that the symmetry axis is placed vertically). The point group $\boldsymbol{C}_{1v}$ has only a plane of symmetry and is usually written $\boldsymbol{C}_s$. The point group $\boldsymbol{C}_{2v}$ has the symmetry elements $C_2$ and two $\sigma_v$. The point group $\boldsymbol{C}_{\infty v}$ has an $\infty$-fold axis and an infinite number of planes through the axis. It is exemplified by a non-symmetrical linear or diatomic molecule.

*The point groups* $\boldsymbol{D}_p$ (dihedral groups) have a $p$-fold axis of symmetry, $C_p$, and $p$ two-fold axes $C_2$ perpendicular to $C_p$. $\boldsymbol{D}_1$ is identical with $\boldsymbol{C}_2$ and is not considered as belonging to the groups $\boldsymbol{D}_p$. $\boldsymbol{D}_2$ is often called $\boldsymbol{V}$ (from the German "Vierergruppe"). It has three mutually perpendicular two-fold axes and no other symmetry elements.

*The point groups* $\boldsymbol{C}_{ph}$ have a $p$-fold axis $C_p$ and a (horizontal) plane $\sigma_h$ perpendicular to it. $\boldsymbol{C}_{1h}$ is identical with $\boldsymbol{C}_s$; it has only one plane of symmetry. In the point group $\boldsymbol{C}_{2h}$ there is in addition to $C_2$ and $\sigma_h$ a center of symmetry $i$.

*The point groups* $\boldsymbol{D}_{pd}$ have a $p$-fold axis $C_p$, $p$ two-fold axes perpendicular to it and $p$ planes of symmetry, $\sigma_d$, bisecting the angles between two successive two-fold axes going through $C_p$. $\boldsymbol{D}_{2d}$ is often called $\boldsymbol{V}_d$.

*The point groups* $\boldsymbol{D}_{ph}$ have a $p$-fold axis $C_p$, $p$ vertical planes of symmetry, $\sigma_v$, and a horizontal plane of symmetry, $\sigma_h$, perpendicular to $C_p$. A number of other symmetry elements are a necessary consequence of those given. The point group $\boldsymbol{D}_{2h}$ is often called $\boldsymbol{V}_h$. It has three mutually perpendicular two-fold axes, three mutually perpendicular planes of symmetry and as a consequence a center of symmetry. The point group $\boldsymbol{D}_{3h}$ has a three-fold axis $C_3$, three $C_2$ at right angles to the former and three planes $\sigma_v$ through it, as well as one plane, $\sigma_h$, perpendicular to it, but no center of symmetry. The point group $\boldsymbol{D}_{\infty h}$ has a $\infty$-fold axis $C_\infty$, an infinite number of $C_2$ perpendicular to the $C_\infty$, an infinite number of planes through the $C_\infty$ and a plane of symmetry perpendicular to the $C_\infty$ which implies a center of symmetry $i$. This is the symmetry of symmetrical linear polyatomic molecules.

*The point group* $\boldsymbol{T}_d$ has three mutually perpendicular two-fold axes and four three-fold axes as well as planes of symmetry, $\sigma_d$, through each pair of three-fold axes. This is the symmetry of the regular tetrahedron.

*The point group* $\boldsymbol{O}_h$ has three mutually perpendicular four-fold axes $C_4$ and four three-fold axes $C_3$, as well as a center of symmetry $i$ and as a consequence numerous other elements of symmetry (see Table 1). The regular octahedron has this symmetry.

*The point group* $\boldsymbol{I}_h$ (icosahedral group) has six five-fold axes, ten three-fold axes and fifteen two-fold axes; in addition, there is a center of symmetry (absent in the point group $\boldsymbol{I}$) which entails a number of planes of symmetry and of rotation-reflection axes (see Table 1). This is the symmetry of the regular icosahedron and of the pentagondodecahedron.

*The point group* $\boldsymbol{K}_h$ (continuous rotation-inversion group) has an infinite number of infinite axes of symmetry ($C_\infty$) all going through the same point which is also a center of symmetry. The presence of this center of symmetry implies in addition an infinite number of planes of symmetry. All atoms belong to this

TABLE 1. SYMMETRY ELEMENTS AND EXAMPLES OF THE MORE IMPORTANT POINT GROUPS[a]

| Point group | Symmetry elements[b] | Examples |
|---|---|---|
| $\boldsymbol{C}_1$ | No symmetry | CHFClBr |
| $\boldsymbol{C}_2$ | One $C_2$ | $H_2O_2$, $N_2H_4$, HClC=C=CHCl |
| $\boldsymbol{C}_3$ | One $C_3$ | |
| $\boldsymbol{C}_i$ ($\equiv \boldsymbol{S}_2$) | $i$ ($\equiv S_2$) | trans ClBrHC—CHBrCl |
| $\boldsymbol{C}_s$ ($\equiv \boldsymbol{C}_{1v} \equiv \boldsymbol{C}_{1h}$) | One $\sigma$ | NOCl, $N_3H$ |
| $\boldsymbol{C}_{2v}$ | One $C_2$, two $\sigma_v$ | $H_2O$, $H_2CO$, $CH_2Cl_2$, $H_2C{=}CCl_2$ |
| $\boldsymbol{C}_{3v}$ | One $C_3$, three $\sigma_v$ | $NH_3$, $CH_3Cl$, $H_3C{-}CCl_3$ |
| $\boldsymbol{C}_{4v}$ | One $C_4$, one $C_2$ (coincident with $C_4$), four $\sigma_v$ | $SF_5Cl$, $B_5H_9$[c], $XeOF_4$, $BrF_5$ |
| $\boldsymbol{C}_{6v}$ | One $C_6$, one $C_3$, one $C_2$ (both coincident with $C_6$), six $\sigma_v$ | |
| $\boldsymbol{C}_{\infty v}$ | One $C_\infty$, any $C_p$, infinite number of $\sigma_v$ | CN, HCN, COS, HC≡CCl |
| $\boldsymbol{D}_2$ ($\equiv \boldsymbol{V}$) | Three $C_2$ (mutually perpendicular) | |
| $\boldsymbol{D}_4$ | One $C_4$, one $C_2$ (coincident with $C_4$), four $C_2$ ($\perp$ to $C_4$) | cyclo-$C_8H_8$ |
| $\boldsymbol{C}_{2h}$ | One $C_2$, one $\sigma_h$, $i \equiv S_2$ | trans HClC=CClH, trans $C_6H_2Cl_2Br_2$, $C_2H_2O_2$ (glyoxal) |
| $\boldsymbol{C}_{3h}$ | One $C_3$, one $\sigma_h$, one $S_3$ (coincident with $C_3$) | $B(OH)_3$ |
| $\boldsymbol{D}_{2d}$ ($\equiv \boldsymbol{V}_d$) | Three $C_2$ (mutually $\perp$), one $S_4$ (coincident with one $C_2$), two $\sigma_d$ (through $S_4$) | $H_2C{=}C{=}CH_2$, |
| $\boldsymbol{D}_{3d}$ ($\equiv \boldsymbol{S}_{6v}$) | One $C_3$, three $C_2$ ($\perp$ to $C_3$), $S_6$ (coincident with $C_3$), $i$, three $\sigma_d$ | $C_2H_6$ (staggered), $C_2Cl_6$, cyclo-$C_6H_{12}$ (chairform) |
| $\boldsymbol{D}_{4d}$ ($\equiv \boldsymbol{S}_{8v}$) | One $C_4$, four $C_2$ ($\perp$ to $C_4$), $S_8$ (coincident with $C_4$), $C_2$ (coincident with $C_4$), four $\sigma_d$ | puckered octagon form of $S_8$ (sulfur) |
| $\boldsymbol{D}_{5d}$ ($\equiv \boldsymbol{S}_{10v}$) | One $C_5$, five $C_2$ ($\perp$ to $C_5$), $S_{10}$ (coincident with $C_5$), $i$, five $\sigma_d$ | $Fe(C_5H_5)_2$ (ferrocene) |
| $\boldsymbol{D}_{2h}$ ($\equiv \boldsymbol{V}_h$) | Three $C_2$ (mutually $\perp$), three $\sigma$ (mutually $\perp$), $i$ | $C_2H_4$, $N_2O_4$, $C_4N_2H_4$ (pyrazine), $B_2H_6$ |
| $\boldsymbol{D}_{3h}$ | One $C_3$, three $C_2$ ($\perp$ to $C_3$), three $\sigma_v$, one $\sigma_h$ | $BCl_3$, $CH_3$, 1, 3, 5-$C_6H_3Cl_3$, $B_3N_3H_3$ |
| $\boldsymbol{D}_{4h}$ | One $C_4$, four $C_2$ ($\perp$ to $C_4$), four $\sigma_v$, one $\sigma_h$, one $C_2$, one $S_4$ (both coincident with $C_4$), $i$ | $C_4H_8$ (cyclobutane) |
| $\boldsymbol{D}_{5h}$ | One $C_5$, five $C_2$ ($\perp$ to $C_5$), five $\sigma_v$, one $\sigma_h$ | $IF_7$[d] |
| $\boldsymbol{D}_{6h}$ | One $C_6$, six $C_2$ ($\perp$ to $C_6$), six $\sigma_v$, one $\sigma_h$, one $C_2$, $C_3$, $S_6$ (each coincident with $C_6$), $i$ | $C_6H_6$, $C_6Cl_6$ |
| $\boldsymbol{D}_{7h}$ | One $C_7$, seven $C_2$ ($\perp$ to $C_7$), seven $\sigma_v$, one $\sigma_h$ | $C_7H_7$ (tropyl radical), $C_7H_7^+$ |
| $\boldsymbol{D}_{\infty h}$ | $C_\infty$, infinite number of $C_2$ ($\perp$ to $C_\infty$) and of $\sigma_v$, one $\sigma_h$, and $C_p$ and $S_p$ (coincident with $C_\infty$), $i$ | $O_2$, $CO_2$, $CH_2$, $C_2H_2$, $C_4H_2$ |
| $\boldsymbol{T}_d$ | Three $C_2$ (mutually $\perp$), four $C_3$, six $\sigma$, three $S_4$ (coincident with the $C_2$) | $CH_4$, $CCl_4$, $P_4$, $OsO_4$ |

TABLE 1—*continued*

| Point group | Symmetry elements | Examples |
|---|---|---|
| $\boldsymbol{O}_h$ | Three $C_4$ (mutually $\perp$), four $C_3$, $i$, three $S_4$ and $C_2$ (coincident with $C_4$), six $C_2$, nine $\sigma$, four $S_6$ (coincident with $C_3$) | $SF_6$, $UF_6$ |
| $\boldsymbol{I}_h$ | Six $C_5$, ten $C_3$, fifteen $C_2$, $i$, six $S_{10}$ (coincident with $C_5$), ten $S_6$ (coincident with $C_3$), fifteen $\sigma$ | $(B_{12}H_{12})^{--}$ |
| $\boldsymbol{K}_h$ | An infinite number of $C_\infty$ and $\sigma$, $i$ | all atoms |

[a] For other point groups see Vol. II.
[b] The element $I$ (identity), which is contained in every point group, has been omitted.
[c] See Hrostowski and Myers (582).
[d] See Lohr and Lipscomb (759).

group but no molecules. Without the center of symmetry we have the continuous rotation group $\boldsymbol{K}$.

In Table 1 the symmetry elements of all the important point groups are listed and in each case one or two examples of molecules or radicals belonging to these point groups are given. Point groups not likely to be encountered in polyatomic molecules have been omitted. For illustrations of some of the point groups see Figs. 1, 2 and 3 of Volume II [see also Cotton (6)].

It may be noted that in polyatomic molecules all symmetry elements of the point groups other than infinite axes ($C_\infty$) are caused by the presence of identical nuclei. The corresponding (geometrical) symmetry operations can therefore be replaced by suitable *permutations* of these identical nuclei or by permutations combined with an inversion. These permutations however do not constitute the complete *permutation-inversion group* of $n$ nuclei, which, except for $n = 2$ or 3, has many more elements (namely $2 \times n!$) than any of the geometrical point groups with $n$ identical atoms. This is because in the geometrical point groups only those permutations are included that can be brought about by rigid rotations and by reflections.

A molecule with $n$ identical nuclei has in general more than one equilibrium conformation. Each of the permutations of the nuclei which is not equivalent to a rotation or reflection leads to a new equilibrium conformation in the $3N$–6 dimensional space of the internal coordinates (see Vol. II, pp. 26 and 220). In addition, every non-planar molecule has a left and a right conformation corresponding to inversion. All these equilibrium conformations are completely equivalent and only one of them need be considered (and is usually considered) as long as they are separated by high and practically unsurmountable potential barriers. There are, however, molecules for which some of these barriers are not high, that is, for which some of the permutations or permutation-inversions are "*feasible*" (i.e., may actually occur within the time of an experiment). Well-known examples are the inversion of $NH_3$, the internal torsion of $C_2H_6$ about the C—C axis leading to permutations of the H atoms not obtainable by rigid rotations, the internal torsions of the $CH_3$ groups in acetone, and others. Examples of permutations or permutation-inversions which for all practical purposes are *not* feasible, are the permutation of two H atoms in $CH_4$, the permutation of the two C atoms in $C_2H_6$, the inversion of most non-planar molecules (other than $NH_3$ and some of its derivatives).

Since in almost all non-rigid molecules only a few permutations or permutation-inversions are feasible we need in general not use, for a discussion of their symmetry properties, the full permutation-inversion group with its large number of symmetry elements ($2 \times n_1! \times n_2! \times \cdots$ if there are $n_1$ identical nuclei of one kind, $n_2$ of a second kind, etc.). A sub-group of this group must be used which includes the symmetry operations of the standard point group (assuming the molecule to be rigid) and those permutations and permutation-inversions that are feasible. In simple cases this sub-group may be isomorphous with one of the standard point groups; for example for $NH_3$ if inversion is included as a feasible operation a group isomorphous with $\boldsymbol{D}_{3h}$ is obtained. In going from one equilibrium conformation of $NH_3$ to the inverted one, one passes through the planar conformation of point group $\boldsymbol{D}_{3h}$. Longuet–Higgins (767) and Hougen (575) have recently considered in detail the molecular symmetry groups appropriate to $CH_3BF_2$, $N_2H_4$, $C_4H_6$ (dimethyl acetylene) when internal rotations and inversions are taken into account. We refer to their papers for a more rigorous treatment of *non-rigid molecules*. For all rigid molecules (rigid in the sense that all identical potential minima are separated by unsurmountable potential barriers) we can use the standard point groups discussed earlier and can ignore the existence of all but one equilibrium conformation. In electronic spectra non-rigidity (in this sense) leads to observable effects only in a few simple cases.

CHAPTER I

# ELECTRONIC STATES

## 1. Electronic Energy, Electronic Eigenfunctions, Potential Surfaces

As in diatomic molecules (Vol. I, p. 146) the nuclei in a polyatomic molecule are held together by the electrons. Depending on the quantum numbers of the electrons we have different electronic states of the molecule. This dependence will be discussed in detail in Chapter III. In the present chapter we shall simply accept the fact that there are many different electronic states; we shall classify them and discuss their vibrational and rotational energy levels.

**Total energy and electronic energy; potential surfaces.** The total energy of the molecule (neglecting spin and magnetic interactions) consists of the potential and kinetic energies of the electrons and the potential and kinetic energies of the nuclei. If we imagine the nuclei to be fixed, the electronic energy (i.e., the sum of potential and kinetic energy) will have a constant value; but this energy value will change if we change the positions of the nuclei.

On account of the smallness of the mass of the electrons compared to that of the nuclei the electrons move much more rapidly than the nuclei; and therefore the electronic energy, when the nuclei are no longer fixed, assumes the value corresponding to the momentary positions of the nuclei. Thus in order to change the positions of the nuclei, not only must work be done against the Coulomb repulsion of the nuclei, but also work must be supplied for the necessary change of electronic energy. In other words, the sum of the electronic energy and the Coulomb potential of the nuclei acts as the potential energy under whose influence the nuclei carry out their vibrations.

In a molecule with $N$ nuclei there are $3N - 6$ (or, for linear molecules, $3N - 5$) relative coordinates (see Vol. II, p. 61) which describe completely the motions of the nuclei. Therefore the potential energy under which the nuclei move in a given electronic state can be represented by a $3N - 6$ (or $3N - 5$) dimensional hypersurface in a $3N - 5$ (or $3N - 4$) dimensional space. Every electronic state of a polyatomic molecule is characterized by such a potential surface. There may be large differences in the shape of these potential surfaces for different electronic states. If the potential surface has no minimum the electronic state is unstable; if the potential surface has at least one minimum the electronic state is stable. Potential surfaces with several minima occur not infrequently (sometimes corresponding to different isomers). In general the potential minima of different electronic states occur at different values of the internuclear distances; often the symmetry of the molecule (its point group) is different in different electronic states.

Even in the simplest case, that of a triatomic molecule, the potential surface is three (or four) dimensional in a four (or five) dimensional space. Only by keeping certain coordinates fixed can the potential surface be represented in three dimensional space. Examples of potential surfaces will be given in Chapter IV (compare also Fig. 66 of Vol. II).

**Total and electronic eigenfunction.** In wave mechanics the fact that the total energy is constant is represented by the Schrödinger equation

$$H\psi = E\psi. \tag{I, 1}$$

Here $\psi$ is the wave function and $H$ the Hamiltonian operator

$$H = T + V;$$

$V$ is the potential energy and $T$ the kinetic energy, given by

$$T = \frac{1}{2m}\sum_i p_i^2 + \frac{1}{2}\sum_k \frac{1}{M_k} p_k^2$$

where the first term is the kinetic energy of the electrons (mass $m$), the second that of the nuclei (mass $M_k$). If the linear momenta $p_i$ and $p_k$ are replaced by the corresponding operators, one obtains the Schrödinger equation in the more conventional form

$$\frac{1}{m}\sum_i\left(\frac{\partial^2\psi}{\partial x_i^2} + \frac{\partial^2\psi}{\partial y_i^2} + \frac{\partial^2\psi}{\partial z_i^2}\right) + \sum_k \frac{1}{M_k}\left(\frac{\partial^2\psi}{\partial x_k^2} + \frac{\partial^2\psi}{\partial y_k^2} + \frac{\partial^2\psi}{\partial z_k^2}\right) + \frac{8\pi^2}{h^2}(E - V)\psi = 0, \tag{I, 2}$$

where $x_i, y_i, z_i$ are the coordinates of the electrons, and $x_k, y_k, z_k$ those of the nuclei. The potential energy $V$ is the sum of an electronic and nuclear contribution

$$V = V_e + V_n \tag{I, 3}$$

where $V_e$ includes the mutual potential energy of the electrons as well as the potential energy of the electrons with respect to the nuclei and $V_n$ is the mutual potential energy of the nuclei.

In a first approximation the solution of (I, 2) can be written

$$\psi = \psi_e(\ldots, x_i, y_i, z_i, \ldots)\,\psi_{vr}(\ldots, x_k, y_k, z_k, \ldots) \tag{I, 4}$$

where $\psi_e$ and $\psi_{vr}$ are solutions of the equations

$$\sum_i\left(\frac{\partial^2\psi_e}{\partial x_i^2} + \frac{\partial^2\psi_e}{\partial y_i^2} + \frac{\partial^2\psi_e}{\partial z_i^2}\right) + \frac{8\pi^2 m}{h^2}(E^{el} - V_e)\psi_e = 0 \tag{I, 5}$$

and

$$\sum_k \frac{1}{M_k}\left(\frac{\partial^2\psi_{vr}}{\partial x_k^2} + \frac{\partial^2\psi_{vr}}{\partial y_k^2} + \frac{\partial^2\psi_{vr}}{\partial z_k^2}\right) + \frac{8\pi^2}{h^2}(E - E^{el} - V_n)\psi_{vr} = 0 \tag{I, 6}$$

respectively. That (I, 4) is an approximate solution of (I, 2) can be seen immediately if it is substituted in (I, 2) and eqs. (I, 5) and (I, 6) are taken into account. One finds that (I, 2) is satisfied if

$$\sum_k \frac{2}{M_k}\left\{\frac{\partial\psi_e}{\partial x_k}\frac{\partial\psi_{vr}}{\partial x_k} + \frac{\partial\psi_e}{\partial y_k}\frac{\partial\psi_{vr}}{\partial y_k} + \frac{\partial\psi_e}{\partial z_k}\frac{\partial\psi_{vr}}{\partial z_k} + \psi_{vr}\left(\frac{\partial^2\psi_e}{\partial x_k{}^2} + \frac{\partial^2\psi_e}{\partial y_k{}^2} + \frac{\partial^2\psi_e}{\partial z_k{}^2}\right)\right\}$$

can be neglected, that is, if the variation of $\psi_e$ with the nuclear coordinates is sufficiently slow so that its first and second derivatives $\partial\psi_e/\partial x_k, \ldots, \partial^2\psi_e/\partial x_k{}^2, \ldots$ can be neglected. That this condition is usually fulfilled to a satisfactory approximation has been shown in detail by Born and Oppenheimer (131).

The equation (I, 5) for $\psi_e$ is the Schrödinger equation of the electrons moving in the field of the fixed nuclei and having a potential energy $V_e$ (which is a function of the electronic coordinates $x_i$, $y_i$, $z_i$). For different nuclear positions, $V_e$ is different, and therefore the eigenfunctions $\psi_e$ and eigenvalues $E^{el}$ of this equation depend on the nuclear coordinates as parameters.

The second equation (I, 6) is the Schrödinger equation of the nuclei moving under the action of the potential $E^{el} + V_n$. Thus in this approximation the motion can be separated, into an electronic motion in a more or less fixed nuclear frame characterized by the function $\psi_e$, and a nuclear motion (vibration plus rotation) in a potential field $E^{el} + V_n$, characterized by the function $\psi_{vr}$.

As for diatomic molecules the minimum value of $E^{el} + V_n$, that is, the minimum of the potential surface of a given stable electronic state is considered as *the* electronic energy of this state and designated $E_e$. The total energy may then be written, in the above approximation,

$$E = E_e + E_{vr} \qquad \text{(I, 7)}$$

where $E_{vr}$ is the vibrational-rotational energy derived from eq. (I, 6).

**Classification of electronic states.** In the Schrödinger equation for the electronic motion (I, 5), $V_e$ is the potential energy of the electrons in the field of the (fixed) nuclei. As mentioned before, in a first approximation (which is in general a good approximation), we may use for the discussion of the electronic motion the equilibrium positions of the nuclei. Therefore $V_e$ has the symmetry of the molecule in the particular electronic state. Thus if a symmetry operation is carried out the Schrödinger equation for the electronic motion is unchanged. As a consequence, the electronic eigenfunction for non-degenerate states can only be symmetric or antisymmetric for each of the symmetry operations permitted by the symmetry of the molecule in the equilibrium position, i.e., can only remain unchanged or merely change sign. For degenerate states[1], the eigenfunction can only change into a linear combination of the two (or more) degenerate eigenfunctions such that the square of the eigenfunction, which represents the electron density, remains unchanged. Different eigenfunctions may behave differently with respect to the various symmetry operations of a given point group; but since in general not all

[1] For separably degenerate states which occur for a few not very important point groups the eigenfunction is multiplied by a complex factor ($e^{\pm i\beta}$, $\beta = 2\pi/p$) upon rotation about a $p$-fold axis (see Vol. II, p. 99).

symmetry elements of a point group are independent of one another, only certain combinations of behavior of the eigenfunctions with regard to the symmetry operations are possible. Such combinations of symmetry properties are called *symmetry types* or *species* (see Vol. II, p. 104). In group-theoretical language, they are the *irreducible representations* of the point group considered. Each electronic eigenfunction, and therefore each electronic state, belongs to one of the possible symmetry types (species, representations) of the point group of the molecule in the equilibrium position. All this is precisely the same as for the vibrational and rotational eigenfunctions (see Vol. II).

If there is only one element of symmetry (as in the point groups $\boldsymbol{C}_2$ and $\boldsymbol{C}_s$), there are only two types of electronic states, those whose eigenfunctions are symmetric and those whose eigenfunctions are antisymmetric with respect to that element of symmetry. These symmetry types (species) are designated $A$ and $B$ for $\boldsymbol{C}_2$, and $A'$ and $A''$ for $\boldsymbol{C}_s$. Here it must be emphasized that while in a non-linear triatomic XYZ molecule there can only be normal vibrations and vibrational levels of species $A'$, there can be electronic states of both types, $A'$ and $A''$.

In point groups with two independent elements of symmetry (but no higher than two-fold axes) we have four types of electronic states which may be designated $++$, $+-$, $-+$ and $--$ if the first sign indicates the behavior with respect to the first element of symmetry, the second sign that with respect to the second. This applies for example to the point groups $\boldsymbol{C}_{2v}$ and $\boldsymbol{C}_{2h}$ for which the species are actually designated $A_1$, $A_2$, $B_1$, $B_2$ and $A_g$, $A_u$, $B_u$, $B_g$ respectively. Again, in a molecule $XY_2$ of point group $\boldsymbol{C}_{2v}$, while only $A_1$ and $B_2$ vibrations and vibrational levels can occur, all four types of electronic states will occur.

If there are more-than-two-fold axes, *degenerate species* arise. For example, for the point group $\boldsymbol{C}_{3h}$ the eigenfunctions may be symmetric or antisymmetric with respect to the plane of symmetry $\sigma_h$ and they may be symmetric or "degenerate" with respect to the three-fold axis $C_3$. Thus there are four species which might be described by $++$, $-+$, $+d$, $-d$ where the first symbol gives the behavior with respect to $\sigma_h$, the second with respect to $C_3$. Actually the designations of these four species are $A'$, $A''$, $E'$, $E''$. Here $E$ is the designation for degenerate symmetry types (from the German "entartet").

It must be noted that in all these examples there is only a finite, usually small, number of species. However, this number increases with increasing number of symmetry elements, but only in the case of infinity-fold axes, i.e., for linear molecules, is there an infinite number of species, which are identical with those of diatomic molecules. For example, for the point group $\boldsymbol{C}_{\infty v}$ there are two non-degenerate species $\Sigma^+$ and $\Sigma^-$ and an infinite number of degenerate species $\Pi, \Delta, \Phi, \ldots$ corresponding to the values $\Lambda = 1, 2, 3, \ldots$ of the electronic orbital angular momentum about the symmetry axis.

In Tables 48–55 of Appendix I, the symmetry types of the more important point groups are summarized in a condensed form. More details may be found in Volume II, pp. 104–123. The parts of the tables above and to the left of the broken lines are identical with corresponding tables of Volume II. The remainder will be explained below. The numbers given under the various symmetry operations for each species are the so-called *characters* which for non-degenerate species are simply +1 or −1, depending on whether an eigenfunction of this species is symmetric or antisymmetric with respect to the particular operation, while for degenerate species it is the "trace" of the transformation matrix. There are several point groups (e.g., $\boldsymbol{D}_{2h}$, $\boldsymbol{D}_{3d}$, $\boldsymbol{D}_{4h}, \ldots, \boldsymbol{O}_h, \boldsymbol{I}_h$) which differ from corresponding simpler groups ($\boldsymbol{D}_2, \boldsymbol{D}_3, \boldsymbol{D}_4, \ldots, \boldsymbol{O}, \boldsymbol{I}$) by the addition of a center of symmetry. Their character tables are not given here explicitly (see, however, Vol. II, pp. 108f) since they can be obtained easily by multiplication of the symmetry elements of the simpler groups by $i\,[I \times i = i, C_2 \times i = \sigma, C_3 \times i = S_6, C_4 \times i = S_4, \ldots]$ thus obtaining the additional symmetry elements, and by multiplying the characters by +1 or −1, thus obtaining an even ($g$) and an odd ($u$) species for each species of the simpler group. This procedure is represented symbolically by the equations $\boldsymbol{D}_{2h} = \boldsymbol{D}_2 \times \boldsymbol{C}_i$, $\boldsymbol{D}_{3d} = \boldsymbol{D}_3 \times \boldsymbol{C}_i, \ldots$ written under the character tables for $\boldsymbol{D}_2, \boldsymbol{D}_3, \ldots$.

As in Volume II, the species of the translations $T_x$, $T_y$, $T_z$ and the rotations $R_x$, $R_y$, $R_z$ are indicated in the Tables 48–55 at the right. This information will be important in the discussion of the selection rules (see Chap. II).

The classification of the electronic states of a polyatomic molecule according to the species of the various point groups is initially based on the assumption that the nuclei are fixed in their equilibrium positions (see p. 9). If the nuclei were fixed in positions other than their equilibrium positions and if the symmetry of the displaced positions is different from that of the equilibrium positions, the species of the electronic eigenfunctions would be different. However, since there must clearly be a one-to-one correspondence between the electronic eigenfunctions in the two conformations, we can, at least for small displacements (vibrations), classify the electronic eigenfunctions according to the species of the equilibrium conformation. We must note, however, that in degenerate electronic states a splitting of the potential surface for certain displacements from the equilibrium conformation may arise, since in the displaced conformation the symmetry may be lower and degenerate species may not exist (see section 2 of this chapter). The problem of the correlation of the species of different point groups will be taken up in Chapter III, section 1.

In different electronic states the equilibrium conformation of a polyatomic molecule may belong to different point groups and thus the species of several point groups may be represented among the manifold of electronic states. Apart from that, there are, of course, as for diatomic molecules differences in the magnitudes of the internuclear distances in different electronic states.

In the preceding discussion and in the tables of Appendix I we have considered the classification of electronic states only for the standard (geometrical) point groups. For molecules in which the transition from one equilibrium conformation to another is "feasible" (non-rigid molecules, see p. 5) other symmetry groups of higher order must be considered. The species of a number of these groups have been considered by Myers and Wilson (922), Longuet–Higgins (767), Hougen (575) and Stone (1169). We need not consider them here since, at least up to now, in electronic spectra of polyatomic molecules only those non-rigid molecules have been sufficiently studied for which the appropriate symmetry group is isomorphous with one of the standard point groups. An important example is $NH_3$ for which, as already mentioned, if inversion is taken into account, the point group is isomorphous with $\boldsymbol{D}_{3h}$, that is, the vibronic states (see p. 20) can be classified according to the species of this point group.

If we consider a degenerate electronic state in a conformation of high symmetry it is clear that on lowering the symmetry sufficiently the potential function (i.e., the purely electronic energy) will split, since for the conformation of lower symmetry the reason for degeneracy no longer exists. However, such a splitting does not arise for the vibrational energy levels, not even for the lowest one with $v_i = 0$, since the degeneracy with regard to rotation about a symmetry axis becomes transformed into degeneracy with regard to appropriate permutations, corresponding to the equivalent equilibrium conformations that arise when the symmetry is lowered. We shall come back to this point in the discussion of vibrational levels.

**Electronic angular momentum.** Let us consider the form of the electronic eigenfunction of a *linear molecule*, assuming the nuclei to be fixed on the $z$ axis. This electronic eigenfunction must be a function of the $3n$ coordinates of the $n$ electrons. We choose cylindrical coordinates $z_i$, $\rho_i$, $\varphi_i$ where $\rho_i$ is the distance from the axis and $\varphi_i$ the azimuthal angle. Clearly one of the angles, say $\varphi_1$, is arbitrary. Therefore, we introduce new azimuthal angles $\varphi_i' = \varphi_i - \varphi_1$. The Schrödinger equation must be invariant with respect to a rotation of the whole system about the $z$ axis, that is, with respect to $\varphi_1$, and therefore the electronic eigenfunction must have the form

$$\psi_e = \chi^{\pm}(\ldots, z_i, \rho_i, \varphi_i', \ldots)e^{\pm i\Lambda\varphi_1} \tag{I, 8}$$

where $\Lambda$ is a positive integer and where the function $\chi^-$ is identical with $\chi^+$ except that $\varphi_i'$ is replaced by $-\varphi_i'$. The angular momentum of the electrons about the $z$ axis, $P_z^{(e)}$, is obtained in quantum mechanics (see Vol. I, p. 16) by operating with $(h/2\pi i)\, \partial/\partial\varphi_1$ on the wave function. Thus, one obtains the well-known result

$$P_z^{(e)} = \pm\Lambda\frac{h}{2\pi}. \tag{I, 9}$$

The two functions (I, 8) correspond to a circulation of the electrons in one or the opposite sense about the $z$ axis with angular momentum $\Lambda(h/2\pi)$. They belong, for the same $\chi$, to one and the same energy value, that is, to a doubly degenerate electronic state, viz., to a $\Pi, \Delta, \Phi, \ldots$ state for $\Lambda = 1, 2, 3, \ldots$. If in (I, 8) we put $\Lambda = 0$, there are also two solutions, but they have no longer the same energy; one solution $\chi^+ + \chi^-$ corresponds to a $\Sigma^+$ state, the other $\chi^+ - \chi^-$ to a $\Sigma^-$ state. Each of them is non-degenerate.

For *non-linear molecules* in non-degenerate states, just as for linear molecules in $\Sigma$ states, the electronic angular momentum is zero. In degenerate states the eigenfunctions are similar in form to (I, 8) except that now $\varphi_1$ appears also in $\chi^{\pm}$

since there is no longer cylindrical symmetry. As a result, the angular momentum is not simply $\Lambda(h/2\pi)$, but is smaller, depending on the extent to which the presence of the off-axis nuclei impedes the orbital motion of the electrons. We can therefore write for the electronic angular momentum in degenerate electronic states of axial molecules

$$P_z^{(e)} = \zeta_e \frac{h}{2\pi} \tag{I, 10}$$

where $\zeta_e$ $(= \int \psi_e^* L_z \psi_e d\tau_e)$ is in general non-integral and may be positive or negative. If the contributions to the electronic angular momentum are produced by atoms lying on the symmetry axis, $\zeta_e$ may be close to an integer, $\pm\Lambda$. For example, in $CH_3F$ or $CH_3CCF$ the fluorine end of the molecule is quasi-linear and will contribute to $\zeta_e$ just as in a linear molecule.

It may be noted that in such a case $\zeta_e$ values near $\pm 3$, $\pm 6, \ldots$ would not occur since states with $\Lambda = 3, 6, \ldots$ of the linear conformation go over into $A_1$, $A_2$ pairs of a $\boldsymbol{C}_{3v}$ conformation (see Chap. III, section 1).

For the cubic point groups, no electronic angular momentum arises for the doubly degenerate ($E$) states but such an angular momentum does arise for triply degenerate ($F$) states. As for the axial point groups, it is non-integral. The component in any molecule-fixed direction is given by $+\zeta_e(h/2\pi)$, 0 or $-\zeta_e(h/2\pi)$, where $\zeta_e$ is a constant for a given state. As for the axial point groups the value of $\zeta_e$ is not limited.

**Multiplicity.** Up to now, in the classification of electronic states we have disregarded the effects of electron spin. The electronic eigenfunction $\psi_e$ has been considered to be a function of the positional coordinates of the electrons only, and the symmetry types (species) refer to symmetry properties of these orbital wave functions. The full electronic eigenfunctions must take account of the fact that each electron has a spin $s = \frac{1}{2}$ which can orient itself parallel or antiparallel to any preferred direction. As long as the coupling of the individual spins with the orbital motion is small, the spins of the individual electrons form a resultant $\boldsymbol{S}$ which is half-integral for an odd number, and integral for an even number of electrons, just as for atoms and diatomic molecules. The resultant spin $\boldsymbol{S}$ is characteristic for each electronic state[2]. Any of the (single-valued) symmetry types discussed in the preceding pages may occur with any of the $S$ values compatible with the number of electrons present.

As for atoms and diatomic molecules, the coupling of the spin $\boldsymbol{S}$ with the orbital motion may lead to a splitting of the molecular electronic state into $2S + 1$ components. This multiplicity is written as a superscript in front of the symbol representing the symmetry type. Thus, for $S = 0$ we have ${}^1A_1$, ${}^1B_2$, ${}^1E', \ldots$ states, for $S = \frac{1}{2}$ we have ${}^2A_1$, ${}^2B_2$, ${}^2E', \ldots$ states, for $S = 1$ we have ${}^3A_1$, ${}^3B_2$, ${}^3E', \ldots$ states and so on. Actually, an observable splitting does not always arise

[2] Here we are following the usual practice of designating angular momentum vectors ($\boldsymbol{S}, \boldsymbol{J}, \boldsymbol{K}\ldots$) by bold-face letters, while the corresponding quantum numbers ($S, J, K\ldots$) are designated by light-face letters (see Vol. I, p. 16).

because the spin is not influenced directly by an electric field but only by a magnetic field. According to elementary concepts of classical and quantum mechanics, a magnetic moment arises whenever there is an electronic angular momentum. As we have seen above, all degenerate states of axial molecules have in general an electronic angular momentum, and therefore there is a fairly strong magnetic field that can orient the spin $\boldsymbol{S}$: we expect a fairly large multiplet splitting except for the lightest molecules.

On the other hand, in non-degenerate states there is no electronic angular momentum and thus no magnetic moment. Therefore, as long as rotation of the molecule is neglected, no appreciable multiplet splitting arises if $S \neq 0$. This is exactly similar, and due to the same reasons, as for the $S$ states of atoms and $\Sigma$ states of diatomic (and linear polyatomic) molecules. As for atoms and diatomic molecules, the states are classified by the value of $2S + 1$, the multiplicity (or spin degeneracy). Even if no splitting is present, the spin degeneracy can be demonstrated through an external magnetic field.

In non-degenerate states a very small magnetic field is produced by the rotation of the molecule, and this will give rise to a very small splitting into $2S + 1$ components. For degenerate electronic states of linear molecules the splitting pattern without rotation is entirely similar to diatomic molecules (Vol. I, p. 215). Very little is known about the splitting in degenerate states of non-linear molecules.

As for atoms and diatomic molecules, the magnitude of the multiplet splitting would be expected to increase with increasing nuclear charge. It is in general small for the lighter polyatomic molecules, while for the heavier ones for which it may be quite large, it has not been studied in great detail.

**Spin functions.** As long as the multiplet splitting (i.e., the spin-orbit interaction) is small, the total electronic eigenfunction $\psi_{es}$ may be written as a product of an orbital function $\psi_e$ and a spin function $\beta$ (see Vol. I, p. 217)[3]

$$\psi_{es} = \psi_e\beta. \tag{I, 11}$$

For the spin function we may use a space-fixed or a molecule-fixed coordinate system. The former is particularly appropriate when spin-orbit interaction is very small. In that case the spin function is not affected by any of the symmetry operations permitted by the point group of the molecule (it is totally symmetric), and the species of the total electronic eigenfunction is the same as that of the orbital function. In diatomic molecules this situation corresponds to Hund's case (b) in which multiplet levels of a given $N$ have the same symmetry properties.

If spin-orbit interaction is not negligibly small, it is more appropriate to use molecule-fixed spin functions which are affected by symmetry operations and therefore must belong to one of the species of the point group of the molecule. In order to determine the species of the spin function, let us first consider the symmetry properties of the spin functions of a free atom (point group $\boldsymbol{K}_h$). Wigner (44) has shown that for integral spin (i.e., for an even number of electrons) the spin

[3] Although strictly speaking this resolution is possible only for one or two electrons (see Chap. III, section 2c), the results discussed here are independent of this restriction.

function belongs to one of the even species of the group $\boldsymbol{K}_h$, viz., $D_{0g}$, $D_{1g}$, $D_{2g}, \ldots$ depending on whether $S = 0, 1, 2, \ldots$ respectively (see Table 55 of Appendix I). For example, for $S = 1$ we have the triply degenerate species $D_{1g}$ (corresponding to the orbital species $P_g$). One set of three eigenfunctions would be

$$\beta_{+1} = \rho e^{+i\varphi}, \quad \beta_0 = \rho e^{0\varphi}, \quad \beta_{-1} = \rho e^{-i\varphi}, \tag{I, 12}$$

which transform, on rotating by an angle $\varphi'$ according to

$$\beta_{+1}{}^{\text{tr.}} = \beta_{+1}e^{i\varphi'}, \quad \beta_0{}^{\text{tr.}} = \beta_0, \quad \beta_{-1}{}^{\text{tr.}} = \beta_{-1}e^{-i\varphi'}$$

and therefore the character of this operation is

$$\chi = 1 + e^{i\varphi'} + e^{-i\varphi'} = 1 + 2\cos\varphi' \tag{I, 13}$$

as listed in Table 55.

When the spin $S$ is half-integral, a formal application of the transformation rules (i.e., multiplication by $e^{im_S\varphi'}$ for rotation by $\varphi'$; $m_S = S, S-1, \ldots, -S$) leads to the result that the spin function changes sign after rotation by $2\pi$ (since $e^{im_S(\varphi'+2\pi)} = -e^{im_S\varphi'}$ for $S$ half-integral). These eigenfunctions are therefore *two-valued* (or *double-valued*) and do not belong to any of the species to which the ordinary single-valued eigenfunctions belong. For their classification it is necessary to extend the normal point groups. These *extended point groups* (or *double groups*) are actually the ones for which the species and characters are given in Appendix I. For the continuous rotation group $\boldsymbol{K}$ one obtains upon extension the additional species $D_{\frac{1}{2}}$, $D_{\frac{3}{2}}$, $D_{\frac{5}{2}}, \ldots$ corresponding to $S$ (or $J$) $= \frac{1}{2}, \frac{3}{2}, \frac{5}{2}$, respectively.

In order to fit the extended point groups into the general group-theoretical scheme and to find two-valued representations (species) in point groups of lower symmetry than $\boldsymbol{K}$, it is necessary to add somewhat fictitious symmetry elements, as was first done by Bethe (116) [see also Landau and Lifshitz (26)]. One assumes that a rotation by $2\pi$ does not return the system to itself but that this is done only by a rotation by $4\pi$. The rotation by $2\pi$ is a new symmetry element called $R$ with respect to which an eigenfunction may be symmetric or antisymmetric. As a consequence there are new symmetry elements $RC_2$, $R\sigma$, $RC_3, \ldots$ if $C_2$, $\sigma$, $C_3, \ldots$ are the original symmetry elements. For two-fold axes ($C_2$) and planes of symmetry ($\sigma$) these new elements ($RC_2$, $R\sigma$) belong to the same classes and merely cause a doubling of the order of the class; but for all more-than-two-fold axes and for the center of symmetry they cause a doubling of the number of classes; for example, while in the ordinary point group $\boldsymbol{C}_{3v}$ we have the two elements, $C_3$ and $C_3{}^2$, in the class designated $2C_3$ we have now, in the extended point group, the four elements $C_3$, $C_3{}^2$, $RC_3 \equiv C_3{}^4$, and $RC_3{}^2 \equiv C_3{}^5$ which form the two classes designated by $2C_3$ and $2C_3{}^2$ and contain the elements $C_3$, $RC_3{}^2$ and $C_3{}^2$, $RC_3$ respectively, and similarly for other point groups. These differences arise because the rotation by $2\pi \pm \varphi'$ is no longer equivalent to the rotation by $\pm\varphi'$. For the species $D_0$, $D_1$, $D_2$, $D_3, \ldots$ of the continuous point group $\boldsymbol{K}$, and similarly for all single-valued species of the point groups of lower symmetry, the characters of the new symmetry elements ($R$, $C_3{}^4, \ldots, iR$) are the same as those of the corresponding old elements ($I, C_3, \ldots, i$), but for the double-valued species the characters have the opposite sign (see the Tables in Appendix I).

Let us consider in more detail the characters of species $D_{\frac{3}{2}}$ of $\boldsymbol{K}$. This species is quadruply degenerate; there are four linearly independent eigenfunctions for each state of this species. Therefore, the character for the identity operation is 4. The four eigenfunctions can be written

$$\psi_M = \rho\, e^{iM\varphi} \quad \text{with} \quad M = +\tfrac{3}{2}, +\tfrac{1}{2}, -\tfrac{1}{2}, -\tfrac{3}{2}. \tag{I, 14}$$

Therefore, the character for a rotation by an angle $\varphi'$ is, similar to (I, 13),

$$\begin{aligned} \chi_{\varphi'} &= e^{i\frac{3}{2}\varphi'} + e^{i\frac{1}{2}\varphi'} + e^{-i\frac{1}{2}\varphi'} + e^{-i\frac{3}{2}\varphi'} \\ &= 2\cos\tfrac{1}{2}\varphi' + 2\cos\tfrac{3}{2}\varphi'. \end{aligned} \tag{I, 15}$$

Similarly for a rotation by $2\varphi'$

$$\chi_{2\varphi'} = 2\cos\varphi' + 2\cos 3\varphi'. \tag{I, 16}$$

The characters for the elements $R$, $RC_\infty^\varphi, \ldots$ are the negative of those just given because of the property of the eigenfunction that they change sign upon rotation by $2\pi$.

In the group $\boldsymbol{K}_h$ there are, in addition, the center of symmetry and the two-fold axes and planes of symmetry following from it. The character of $i$ is $+1$ or $-1$ yielding a $g$ and a $u$ species for each of the single- and the double-valued species, while the characters for $C_2$ and $\sigma$ are 0 except for $D_0$.

In a similar way the characters of the species of the other extended point groups in Appendix I have been obtained [see Bethe (116), Jahn (616), Opechowsky (949), Satten (1099) and Judd (654)]. The number of species of an extended group, just as that of a simple group, equals the number of classes of symmetry elements (number of columns in the Tables). The degree of degeneracy $d_i$ of the species is readily obtained from the condition that

$$\sum d_i^2 = g \tag{I, 17}$$

where $g$ is the order of the group which equals the total number of symmetry elements. [Here separably degenerate species (see Vol. II, pp. 99 and 120) must be counted as two non-degenerate species.]

The designation of the double-valued species has not yet been standardized. We adopt here a designation which is an extension of that used by Jahn (616) for $\boldsymbol{D}_p$ and $\boldsymbol{C}_{pv}$ with even $p$ only, that is, we use $E_{\frac{1}{2}}, E_{\frac{3}{2}}, E_{\frac{5}{2}}, \ldots$ for the doubly degenerate representations (species) corresponding to $M = \pm\frac{1}{2}, \pm\frac{3}{2}, \pm\frac{5}{2}, \ldots$ of a free atom (point group $\boldsymbol{K}_h$) and we use $G_{\frac{3}{2}}$ and $I_{\frac{5}{2}}$ for the quadruply and sextuply degenerate representations corresponding to $J = \frac{3}{2}$ and $\frac{5}{2}$ respectively. For the point groups of lowest symmetry $\boldsymbol{C}_1$, $\boldsymbol{C}_i$, $\boldsymbol{C}_3$, there are double-valued representations of unit dimension ($B_{\frac{1}{2}}, B_{\frac{1}{2}g}, B_{\frac{1}{2}u}, B_{\frac{3}{2}}$). A state of such a species is however not non-degenerate since it can be shown that there is always another state of the same species at the same energy as long as no magnetic field is applied. Two coinciding states of species $B_{\frac{1}{2}}$ are entirely similar to a separably degenerate $E_{\frac{1}{2}}$ state except that the former have identical transformation properties while the two components of an $E_{\frac{1}{2}}$ state differ (they can be different because of the higher symmetry). Both the two $B_{\frac{1}{2}}$ states and the two components of an $E_{\frac{1}{2}}$ state can be separated by a magnetic field.

For free atoms and diatomic as well as linear polyatomic molecules a clear understanding of the effect of the electron spin on the energy levels is much more readily obtained by means of the vector model than by means of the preceding group-theoretical considerations. However, for molecules belonging to finite point groups, that is, for non-linear molecules (as well as for atoms in crystals), the vector model cannot be used since the number of species is finite (and often very small) and therefore there is no one-to-one correspondence between the various $S$ values and the species, as there is for atoms and diatomic and linear molecules. We must therefore establish the species of the spin function for various $S$ values for all the important point groups. This can now easily be done since we know the species for the point group $\boldsymbol{K}_h$; we only need to establish into which species the species of $\boldsymbol{K}_h$ resolve in the point groups of lower symmetry. The result is given in Table 56 of Appendix II.

As an example for the derivation of Table 56 let us consider point group $\boldsymbol{C}_{3v}$. To find the behavior of the spin function with respect to $C_3$, we put $\varphi' = 120°$. In this way we see that $D_{\frac{1}{2}g}$, the species corresponding to $S = \frac{1}{2}$ in point group $\boldsymbol{K}_h$, goes into $E_{\frac{1}{2}}$; $D_{1g}$ corresponding to $S = 1$ leads to a reducible representation in $\boldsymbol{C}_{3v}$ with characters 3, 0, . . . which on reduction yields $E + A_2$; similarly, $D_{\frac{3}{2}g}$ corresponding to $S = \frac{3}{2}$ leads to a representation with characters $4, -1, \ldots, -4, +1, \ldots$ which yields on reduction $E_{\frac{1}{2}} + E_{\frac{3}{2}}$, and so on. The spin value $S = \frac{5}{2}$ corresponds to $E_{\frac{1}{2}} + E_{\frac{3}{2}} + E_{\frac{1}{2}}$ since $E_{\frac{5}{2}}$ does not exist for $\boldsymbol{C}_{3v}$. It does exist for several point groups of higher symmetry; for these, $S = \frac{5}{2}$ corresponds to $E_{\frac{1}{2}} + E_{\frac{3}{2}} + E_{\frac{5}{2}}$, for example, for $\boldsymbol{D}_{3h}$. But only for the point groups $\boldsymbol{C}_{\infty v}$ and $\boldsymbol{D}_{\infty h}$ does this regularity continue to any $S$ value, integral or half-integral; only for them is $M_s$ defined for any $S$ value and thus only for them can the vector model replace the group-theoretical considerations.

It may be noted that on going to point groups of lower and lower symmetry, the spin functions for integral spin are eventually resolved into $2S + 1$ non-degenerate functions, corresponding to $2S + 1$ states of (slightly) different energy. However, the spin functions for half-integral spin are at most resolved into functions which are still doubly degenerate (account being taken of the degeneracy of species $B_{\frac{1}{2}}$, $B_{\frac{1}{2}g}$, $B_{\frac{1}{2}u}$ and $B_{\frac{3}{2}}$ mentioned earlier). This remaining degeneracy, as was first recognized by Kramers, is caused by the fact that as long as no magnetic field is present there is in all atomic systems an additional symmetry element: time reversal, that is, the wave equation is invariant if the time $t$ is replaced by $-t$ [see Wigner (44) or Landau and Lifshitz (26)]. This degeneracy caused by time reversal is now usually called *Kramers degeneracy*, and pairs of states like the two coinciding $B_{\frac{1}{2}}$ (or $B_{\frac{3}{2}}$) states or the two components of an $E_{\frac{1}{2}}$ (or $E_{\frac{3}{2}}$, $E_{\frac{5}{2}}$) state are called *Kramers doublets.*

**Species of multiplet components.** We are now ready to determine what are the species of the individual multiplet components when the multiplet splitting, even for zero rotation, is not negligibly small. For this purpose, that is, in order to find the species of $\psi_{es}$, we must form the direct product of the species of the spin function $\beta$ and the species of the coordinate function $\psi_e$. This multiplication is done in the same way as the multiplication of vibrational species discussed in Volume II (pp. 124f): one multiplies the characters and reduces the set so obtained to the sum of characters of the resulting species, which are uniquely determined in this way [see eq. (II, 87) of Vol. II].

For example in a $^3\Pi$ state of a linear molecule, the spin function ($S = 1$) according to Appendix II has the species $\Sigma^-$ and $\Pi$, while the coordinate function is $\Pi$. The product $\Sigma^- \times \Pi$ has the characters (see Appendix I): $+2$, $\cos 2\varphi, \ldots, 0$ and therefore yields directly a single $\Pi$ state, while $\Pi \times \Pi$ has the characters $4, 4\cos^2\varphi, 4\cos^2 2\varphi, \ldots, 0$ which can be reduced to the sum of the characters of $\Sigma^+ + \Sigma^- + \Delta$. Thus, we see that $^3\Pi$ is split into four states $\Sigma^+, \Sigma^-, \Pi$, and $\Delta$. The same result is obtained by means of the vector model: $^3\Pi \rightarrow {}^3\Pi_{0^+}, {}^3\Pi_{0^-}, {}^3\Pi_1$ and $^3\Pi_2$. The general species designations $\Sigma^+, \Sigma^-, \Pi, \Delta$ indicate now the resultant electronic angular momentum $\boldsymbol{\Omega} = \boldsymbol{\Lambda} + \boldsymbol{S}$.

Similarly, for a $^3E$ state of a non-linear molecule, for example of point group $C_{3v}$, the spin function has species $A_2$ and $E$ (see Appendix II). Upon multiplication by the species ($E$) of the coordinate function we obtain for the total function $\psi_{es}$ the species $A_2 \times E + E \times E$. The characters of the product $A_2 \times E$ are $2, -1, 0$, that is, those of $E$, while the characters of $E \times E$ are $4, 1, 0$ which can only be obtained from $A_1 + A_2 + E$. Thus, the $^3E$ state splits into four states $A_1 + A_2 + E + E$. The first two correspond to $\Sigma^+$ and $\Sigma^-$ of the linear case, but there is no longer a difference in symmetry between $\Omega = 1$ and $\Omega = 2$; we obtain simply two $E$ states. The vector model cannot be applied.

In a similar way we find for a $^2\Pi$ state of a linear molecule from the product $E_{\frac{1}{2}} \times \Pi$ the components $E_{\frac{1}{2}} + E_{\frac{3}{2}}$ corresponding to $^2\Pi_{\frac{1}{2}}$ and $^2\Pi_{\frac{3}{2}}$; or for a $^2E$ state of a $C_{3v}$ molecule from $E_{\frac{1}{2}} \times E$ the components $E_{\frac{1}{2}} + E_{\frac{3}{2}}$. Always the total electronic eigenfunctions of systems with an odd number of electrons, just as the spin functions alone, belong to two-valued species in contrast to systems with an even number of electrons which have single-valued species.

For convenience we reproduce in Appendix III the multiplication tables, including those for double-valued species, for all the important point groups. Only point groups without center of symmetry are included. For corresponding point groups *with* a center of symmetry the same multiplication tables apply with the addition of the $(g, u)$ rule, that is, $g \times g = g$, $g \times u = u$, $u \times u = g$. The tables will be useful not only for evaluating the effect of electron spin, but later also for the determination of vibronic species, the determination of the manifolds of states arising from open electron configurations and for the correlation of molecular electronic states.

The direct product of two identical degenerate species is the sum of the *symmetric* and the *antisymmetric product* [see Landau and Lifshitz (26)]. The symmetric product (which we shall not define here) determines, for example, the species of the vibrational levels $2\nu_i$ of a degenerate vibration $\nu_i$ or the species of the electronic states resulting from two equivalent electrons (unless the Pauli principle introduces still further restrictions). In Appendix III the species forming the antisymmetric product are put in square brackets.

The product representation (I, 11) of the total electronic eigenfunction is, of course, only a first approximation. In a higher approximation we must write

$$\psi_{es} = \psi_e\beta + \chi_{es}, \tag{I, 18}$$

where $\chi_{es}$ contains both spin and space coordinates. However, the species of $\chi_{es}$ must be the same as that of $\psi_e\beta$ since from general principles $\psi_{es}$ must belong to one of the species of the point group concerned, and there can be no change of species when the spin-orbit interaction is gradually increased.

**Magnitude of multiplet splitting.** The magnitude of the splitting produced by spin-orbit interaction is readily calculated only for linear molecules. Just as for diatomic molecules, if spin-orbit interaction is relatively small (Russell–Saunders coupling), we have for the electronic energy of the multiplet components

$$T_e = T_0 + A\Lambda\Sigma \tag{I, 19}$$

where $\Sigma$ is the component of $S$ along the internuclear axis; in other words, in this approximation the multiplet components are equidistant and for $^3\Pi$ the separation of the two states with $\Omega = 0$ ($\Sigma^+$ and $\Sigma^-$) is zero. The splitting constant $A$ increases rapidly with the atomic number of the atoms forming the molecule. The splitting is produced by the magnetic field in the direction of the internuclear axis which is caused by the orbital motion of the electrons ($\Lambda \neq 0$). The constant $A$ may be positive or negative corresponding to regular and inverted states (see Vol. I, p. 216). For non-degenerate states ($\Lambda = 0$, $\Sigma$ states) there is in this approximation no splitting.

As for diatomic molecules, when the spin-orbit coupling is very large, eq. (I, 19) no longer applies and $\mathbf{\Lambda}$ is no longer defined. Only $\mathbf{\Omega}$, the total electronic angular momentum about the internuclear axis, remains well defined, corresponding to the well-defined species discussed above.

For non-linear molecules belonging to axial point groups, formulae similar to (I, 19) have not been developed, but one may expect a similar relation except that $A\Lambda$ is replaced by $A\zeta_e$, that is, if $\zeta_e$ is small, the multiplet splitting is small. The theory of spin-orbit interaction in aromatic molecules and ions has been developed by McClure (804) and McConnell (807).

For molecules belonging to cubic point groups in triply degenerate ($F$) states the splitting formula according to Van Vleck (1239) [see also Griffith (16)], is essentially that for free atoms as long as Russell–Saunders coupling applies, viz.,

$$T_e = T_0 + \tfrac{1}{2}A\zeta_e J'(J' + 1) \tag{I, 20}$$

where $J'$ is a pseudo-inner quantum number which would represent the total electronic angular momentum if $\zeta_e$ were equal to 1, that is,

$$J' = L + S, L + S - 1, \ldots, |L - S|$$

where here $L = 1$. For doubly degenerate states of molecules belonging to cubic point groups just as for non-degenerate states no first-order multiplet splitting is expected.

**Mutual perturbation of electronic states.** When in a certain approximation two electronic states lie fairly close together, they may perturb each other, that is, "repel" each other and adopt each other's properties on account of mixing of the eigenfunctions. If spin-orbit coupling is very small, such perturbations can only occur between states of the same (orbital) species and the same multiplicity. It is of course in general not easy to establish the presence of such perturbations but they are readily recognized when a state belonging to a Rydberg series lies close to a state of a different electron configuration but of the same species. There will then be a deviation from the normal Rydberg formula and, if there is enough mixing, an extra member of the Rydberg series.

If spin-orbit interaction is not negligible, perturbations can also occur between states of different orbital species as long as the over-all electronic species is the same. For example, if a $^1A_1$ state of a $C_{3v}$ molecule lies close to a $^3E$ state, spin-orbit induced perturbation can occur for one of the component states of $^3E$, viz., $A_1$; or if a $^3A_1$ state lies close to $^3E$, two components of the latter can perturb the two components $A_2 + E$ of the former.

Additional perturbations between electronic states can arise, when the molecule is rotating, on account of Coriolis interaction or, when it is vibrating, on account of vibronic interactions. These perturbations which affect only the higher rotational or vibrational levels of either electronic state will be discussed in sections 2 and 3 of this chapter.

## 2. Vibrational Structure of Electronic States, Interaction of Vibration and Electronic Motion

In a given electronic state a polyatomic molecule may be excited to any of the possible vibrational levels. Each of these levels may be called a *vibrational-electronic* state or, for short, following Mulliken, a *vibronic* state.

### (a) Non-degenerate electronic states

**Vibronic energies.** As long as the electronic state is non-degenerate, the vibrational levels are of the same type as discussed in Volume II for the electronic ground states (which were implicitly assumed to be non-degenerate). The vibronic energy is then simply the sum of the electronic energy $E_e$ corresponding to the minimum of the potential function and the vibrational energy $E_v$,

$$E_{ev} = E_e + E_v, \qquad (I, 21)$$

or, using term values,

$$T_{ev} = T_e + G(v_1, v_2, \ldots). \qquad (I, 22)$$

Here the vibrational contribution according to eq. (II, 281) of Volume II is given by

$$G(v_1, v_2, v_3, \ldots) = \sum_i \omega_i\left(v_i + \frac{d_i}{2}\right) + \sum_i \sum_{k \geqslant i} x_{ik}\left(v_i + \frac{d_i}{2}\right)\left(v_k + \frac{d_k}{2}\right) + \sum_i \sum_{k \geqslant i} g_{ik} l_i l_k + \cdots. \qquad (I, 23)$$

In this equation $v_i$, $v_k$ are the vibrational quantum numbers, $\omega_i$ are the vibrational frequencies for infinitesimal amplitudes, $x_{ik}$ and $g_{ik}$ are anharmonicity constants,

$d_i$, $d_k$ are the degeneracies (1, 2 or 3) of the vibrations $i$, $k$, and $l_i$ and $l_k$ are angular momentum quantum numbers of degenerate vibrations,

$$l_i = v_i, v_i - 2, v_i - 4, \ldots, 1 \text{ or } 0 \tag{I, 24}$$

and similarly for $l_k$. For non-degenerate vibrations, $l_i = 0$ and $g_{ik} = 0$.

Frequently it is more convenient to refer the vibronic energy levels to the lowest one in a given electronic state, that is, to write

$$T_{ev} = T_0 + G_0(v_1, v_2, \ldots) \tag{I, 25}$$

where

$$G_0(v_1, v_2, \ldots) = \sum_i \omega_i{}^0 v_i + \sum_i \sum_{k \geqslant i} x_{ik}{}^0 v_i v_k + \sum_i \sum_{k \geqslant i} g_{ik} l_i l_k + \cdots. \tag{I, 26}$$

For the relation between $\omega_i$ and $\omega_i{}^0$ and the observed fundamentals $\nu_i$ see Volume II, p. 211. In (I, 23) and (I, 26) the existence of Fermi resonances has been neglected as has been the finer interaction of degenerate vibrations with each other, which leads to small splittings of levels in which two or more degenerate vibrations are excited (see Vol. II, pp. 212f).

The interaction of vibration and electronic motion is taken into account in the preceding formulae largely by the fact that the $\omega_i$ and $x_{ik}$ correspond to the potential function of the electronic state considered. However, the interaction of the electronic state considered with other electronic states of different species on account of vibronic interaction is not included (see below).

**Vibronic eigenfunction and vibronic species.** In a first approximation the vibronic eigenfunction, which describes the electronic and vibrational motions in the molecule, can be written as a product

$$\psi_{ev} = \psi_e(q, 0)\, \psi_v(Q) \tag{I, 27}$$

of the electronic eigenfunction $\psi_e(q, 0)$ for the equilibrium position and the vibrational eigenfunction $\psi_v(Q)$. Here $q$ and $Q$ stand for all the electronic and nuclear (normal) coordinates respectively. Both $\psi_e$ and $\psi_v$ have symmetry properties in accord with one of the species of the point group of the equilibrium conformation. Therefore the vibronic eigenfunction must also belong to one of these species even if the product resolution (I, 27) no longer applies. The vibronic species is therefore simply obtained by "multiplication" of the species of electronic and vibrational eigenfunctions. This multiplication, or, in group-theoretical language, the formation of the direct product, has already been discussed in the preceding section, and the results for all important cases can be found in Appendix III. This Appendix can also be used in order to find the species of the vibrational eigenfunction from the species of the normal vibrations (see Vol. II, pp. 126 and 129). Sometimes left superscripts $e$, $v$, $ev$ are added to the species symbol whenever it is not clear from the content whether an electronic, a vibrational or a vibronic species is meant.

As an example, consider an electronic state of species $B_{1g}$ (or ${}^eB_{1g}$) of an ethylene-type molecule (point group $\boldsymbol{D}_{2h}$). If the vibrations $\nu_4(a_u)$, $\nu_7(b_{1u})$, $\nu_9(b_{2u})$ are each excited by one quantum (or by an odd number of quanta), that is, if the vibrational eigenfunction has species $B_{3u}$ (or ${}^{v}B_{3u}$), the total eigenfunction will have species $B_{2u}$ (also written ${}^{ev}B_{2u}$). Figures 2 and 10 below will present further examples of species of vibronic levels in non-degenerate as well as degenerate electronic states.

When spin-orbit interaction is large, we must use instead of $\psi_e$, the eigenfunction including spin $\psi_{es}$ in (I, 27). Correspondingly, $\psi_{ev}$ becomes $\psi_{esv}$, the vibronic eigenfunction including spin. The species of $\psi_{esv}$ are obtained from $\psi_{ev}$ in the same way as the species of $\psi_{es}$ was previously (p. 17) obtained from the species of $\psi_e$ and the species of the spin function (Appendix II). For half-integral spin we have two-valued vibronic species.

In a higher approximation, on the right-hand side of (I, 27) a term $\chi(Q, q)$ has to be added which depends both on the positions of the nuclei and the positions of the electrons but in such a way that electronic and nuclear coordinates ($q$ and $Q$) cannot be separated. However, the previous results about the species of the vibronic eigenfunction are not changed, since there must be a one-to-one correlation of the energy levels for a gradual change of the strength of vibronic interaction and since, for any interaction, $\psi_{ev}$ (or $\psi_{esv}$) must belong to one of the species of the point group of the equilibrium conformation of the molecule. The reason for this conclusion is that the potential energy (and therefore the wave equation) is symmetric with respect to all symmetry operations of this point group.

**Correlation of vibronic levels for planar and non-planar equilibrium conformation.** It happens fairly frequently that a molecule that is planar in one electronic state is non-planar in another one. It is therefore of interest to consider the way in which the energy levels change in a given electronic state when we imagine the equilibrium conformation changed from planar to non-planar. If the molecule is non-planar, there are two identical potential minima, and therefore all vibronic levels are double on account of inversion doubling, one component level being symmetric, the other antisymmetric with respect to inversion. Since the potential function of the non-planar molecule has the same symmetry as that of the planar molecule, we can use the same species to describe the vibronic levels, if in the non-planar case inversion doubling is taken into account.

As an example, in Fig. 1a we show a vibrational energy level diagram for a non-planar (at left) and a planar (at right) $XYZ_2$ molecule in a totally symmetric electronic state of point group $\boldsymbol{C}_s$ and $\boldsymbol{C}_{2v}$, respectively. The levels shown in the planar case correspond to the out-of-plane bending vibration[4] $\nu_4(b_1)$ and have alternately the vibronic species ${}^{ev}A_1$, ${}^{ev}B_1$, ${}^{ev}A_1, \ldots$. In the non-planar case they correspond to the bending vibration $\nu_4(a')$, that tends to open (or close) the pyramid and, if inversion doubling is disregarded, would all be ${}^{ev}A'$ levels. But the individual inversion components are symmetric and antisymmetric with respect to the plane of the planar conformation, that is, are $A_1$ and $B_1$, and can be correlated

[4] Note that the orientation of the axes and therefore the designation of $B_1$ and $B_2$ has been changed compared to Volume II in order to conform to the more recent international recommendations [Mulliken (912)].

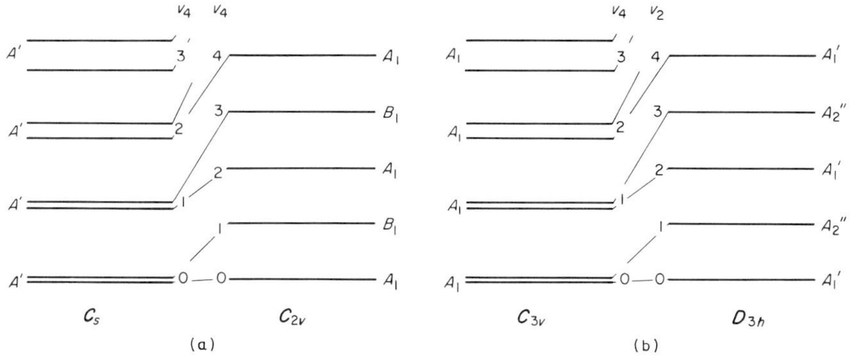


FIG. 1. **Correlation of vibrational energy levels for non-planar and planar molecules in totally symmetric electronic states.** (*a*) Non-planar and planar $XYZ_2$ (point groups $\boldsymbol{C_s}$ and $\boldsymbol{C_{2v}}$ respectively). (*b*) Non-planar and planar $XY_3$ (point groups $\boldsymbol{C_{3v}}$ and $\boldsymbol{D_{3h}}$ respectively). It is assumed that, as in $NH_3$, the inversion doubling in the non-planar conformation increases rapidly with $v$. When the doubling is resolved the inversion components of a given $v$ are often distinguished as $v^+$ and $v^-$ (e.g., $0^+, 0^-, 1^+, 1^-, \cdots$) or by the species symbols corresponding to the planar conformation.

with those of the planar molecule in the way shown. Figure 1b shows a similar correlation between planar and non-planar $XY_3$ molecules of point groups $D_{3h}$ and $C_{3v}$, respectively. In order to find a corresponding correlation for electronic states of species different from $A_1$ or $A_1'$, one has only to multiply all vibronic species in the figure by the electronic species (even if it is degenerate). It must be noted that, in both examples, levels that have the same vibrational quantum number in the non-planar conformation are correlated with levels of different vibrational quantum number in the planar conformation. In intermediate cases in which the inversion doubling is resolved but small, successive levels are often numbered $0^+, 0^-, 1^+, 1^-, \cdots$.

A quantitative calculation of the energy levels for various heights of the potential barrier separating the two equivalent non-planar conformations has been given by Somorjai and Hornig (1142) [see also Chan, Zinn, Fernandez and Gwinn (186)].

The correlation between a *linear and non-linear* conformation is more complicated, since a vibrational degree of freedom goes over into a rotational degree of freedom. There is no inversion doubling. This case will be considered later (p. 120f).

## (b) Degenerate electronic states: linear molecules

**Vibronic species.** The procedure for obtaining the vibronic species is exactly the same for degenerate electronic states as for non-degenerate ones, that is, we have to "multiply" the vibrational and electronic species. However, now when degenerate vibrations are excited, in general several vibronic species result from a given combination of electronic and vibrational species, and therefore, several vibronic states result which in a zero approximation have the same energy.

For example, if a bending vibration (species $\Pi$) is singly excited in a $\Pi$ electronic state of a linear molecule, the vibronic species are $\Pi \times \Pi = \Sigma^+ + \Sigma^- + \Delta$ (see Appendix III). If a degenerate vibration is doubly excited, the vibrational species are $\Sigma^+$ and $\Delta$ (see Table 32, p. 127 of Vol. II, or the "symmetrical" product in

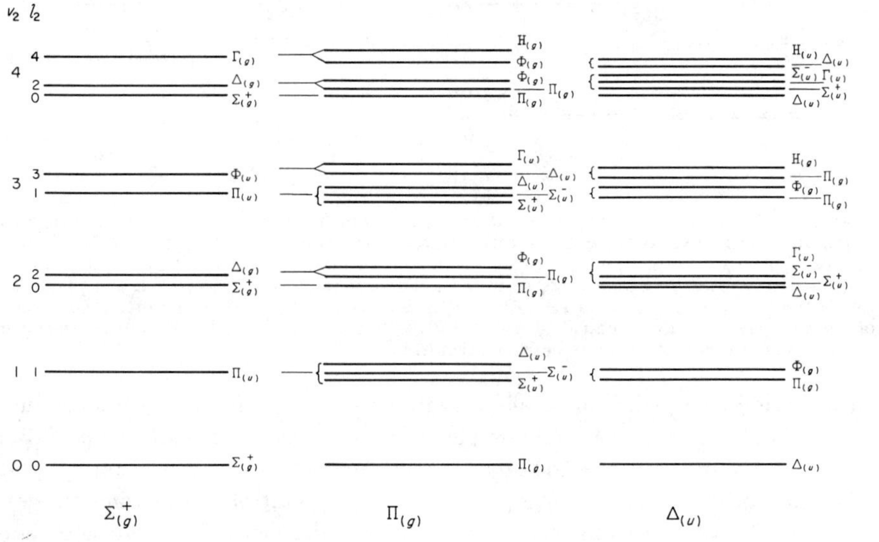


FIG. 2. **Vibronic species of the vibrational levels in $\Sigma^+$, $\Pi$ and $\Delta$ electronic states of linear molecules.** The subscripts $g$ or $u$ added in brackets give the species designation for $\Sigma_g^+$, $\Pi_g$ and $\Delta_u$ electronic states of symmetrical linear molecules (point group $\boldsymbol{D}_{\infty h}$).

Appendix III of this volume) and consequently, the vibronic species are $(\Sigma^+ + \Delta) \times \Pi = \Pi + \Pi + \Phi$. In a similar way one finds for the higher vibrational levels of $\Pi$ and $\Delta$ electronic states the vibronic species given in the second and third columns of Fig. 2. In the first column, for comparison, the vibronic species in a non-degenerate (totally symmetric) electronic state ($\Sigma^+$) are given.

**Vibronic angular momentum.** In degenerate vibrations and therefore in degenerate vibrational levels there is a vibrational angular momentum in the direction of the symmetry axis (Vol. II, pp. 75 and 402f). In degenerate electronic states this vibrational angular momentum is coupled with the electronic angular momentum and the two form a resultant which may be called *vibronic angular momentum.*

For linear molecules each degenerate vibration $\nu_i$ has a vibrational angular momentum about the internuclear axis of magnitude $l_i(h/2\pi)$ where $l_i = v_i$, $v_i - 2, \ldots, 1$ or 0. The total vibrational angular momentum is $l(h/2\pi)$, where

$$l = \left| \sum_i (\pm l_i) \right|. \tag{I, 28}$$

Here the $\pm$ sign takes account of the two possible orientations with respect to the internuclear axis. The values $l = 0, 1, 2, \ldots$ correspond to $\Sigma, \Pi, \Delta, \ldots$ vibrational levels. The electronic orbital angular momentum about the internuclear axis is $\Lambda(h/2\pi)$ with $\Lambda = 0, 1, 2, \ldots$ corresponding to $\Sigma, \Pi, \Delta, \ldots$ electronic states. The resultant vibronic angular momentum is $K(h/2\pi)$, where

$$K = |\pm\Lambda \pm l|. \tag{I, 29}$$

Here $K = 0, 1, 2, \ldots$ corresponds to $\Sigma, \Pi, \Delta, \ldots$ vibronic states. Equation (I, 29) represents the result of applying the vector model to the vibrational and electronic motion. The results are the same as already given in Fig. 2 on the basis of group-theoretical multiplication of species. Note that here $K$ (just as $\Lambda$ and $l$) in accordance with spectroscopic practice [see Mulliken (912)] has been taken to be an unsigned quantity since the two directions along the internuclear axis are entirely equivalent. However, in theoretical work $K$ as well as $\Lambda$ and $l$ are usually used *with* sign so that (I, 29), is written

$$K = \Lambda + l. \tag{I, 29a}$$

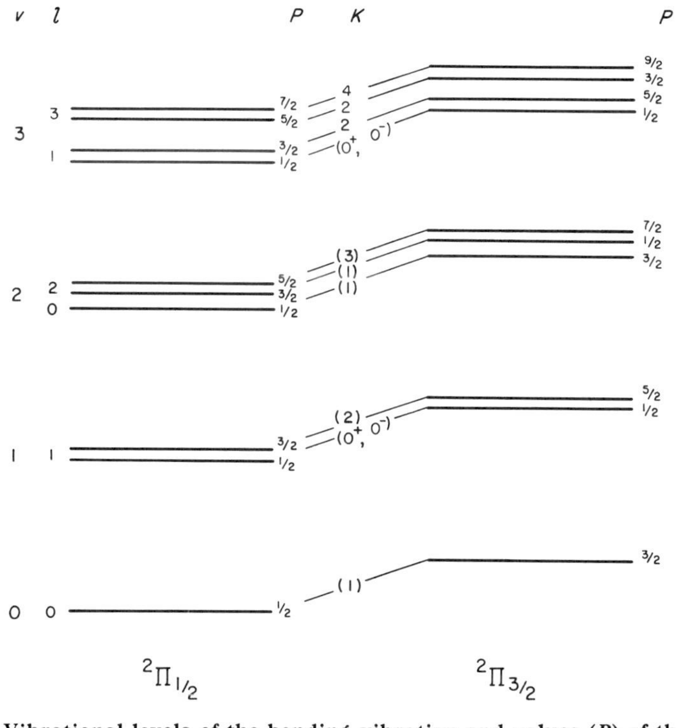


FIG. 3. **Vibrational levels of the bending vibration and values (*P*) of the resultant vibronic angular momentum including spin in a $^2\Pi$ electronic state of a linear molecule with large spin splitting.** The values of $K$ are indicated in brackets. $K$ is not a "good" quantum number when the spin splitting is large. Vibronic (Renner–Teller) interaction has been neglected: see Fig. 8.

If the electron spin is strongly coupled with the orbital angular momentum, we have, as for diatomic molecules, the *total electronic angular momentum* $\Omega(h/2\pi)$, where $\Omega = |\Lambda + \Sigma|$, and $\Sigma(h/2\pi)$ is the component of the spin along the internuclear axis. One has then to form the resultant of $\mathbf{\Omega}$ and $\boldsymbol{l}$ which is called $\boldsymbol{P}$, the *resultant vibronic angular momentum including spin.* We have for the corresponding quantum number

$$P = |\pm\Omega \pm l|. \tag{I, 30}$$

For example, in a $^2\Pi$ electronic state of a linear molecule in which a bending vibration is singly excited ($l = 1$), we have the $P$ values $\frac{5}{2}, \frac{1}{2}, \frac{3}{2}, \frac{1}{2}$. Figure 3 gives an energy level diagram for this and a few higher levels of the bending vibration of a $^2\Pi$ state when the spin splitting is large. The same results can also be obtained from group theory by using double-valued representations for the electronic and therefore also for the vibronic eigenfunctions. The vibronic species are $E_{\frac{1}{2}}, E_{\frac{3}{2}}, \ldots$ for $P = \frac{1}{2}, \frac{3}{2}, \ldots$ respectively (see Appendix I).

In most observed $^2\Pi$ states the spin splitting is small or intermediate in magnitude, and therefore usually the classification of the vibronic levels according to $K$ is sufficient; however, when one wants to distinguish the individual doublet components, the classification by $P$ is necessary.

**Vibronic interaction (Renner–Teller effect) in singlet electronic states.** In a zero approximation the vibronic levels for a given set of $l_i$ values (or, in a triatomic molecule, for a given $l$ value) coincide. However there are no reasons of symmetry for the coincidence of these levels, and therefore in a sufficiently high approximation they must have slightly different energies and there will then be as many different component levels as there are vibronic species for the particular set of $l_i$ values.

Formally we may consider these splittings as produced by the interaction of the electronic and vibrational angular momenta similar to the interaction between the orbital angular momenta of two electrons in a diatomic molecule (see Vol. I, p. 334) which produces the energy difference of states of the same electron configuration (for example, the difference between $\Pi$ and $\Phi$ states resulting from the configuration $\pi\delta$). In the present case the interaction of the angular momenta is the result of the forces that correspond to those terms in the wave equation that are neglected when the product resolution $\psi_e \psi_v$ is made (see p. 21). In general there are several values of the vibrational angular momentum ($l = v_2, v_2 - 2, \ldots, 1$ or $0$) at the same zero-approximation energy and therefore most resultant angular momenta occur in pairs except those with $K = |v_2 + \Lambda|, |v_2 + \Lambda - 2|, \ldots, |v_2 - \Lambda + 2|$ (that is, for $\Pi$ states only the one with $K = |v_2 + \Lambda|$); it is the "repulsion" between the members of such pairs which dominates the splitting pattern (see below).

In order to get a quantitative relation for the splittings we must consider the variation of the potential energy with the bending coordinate. As was first recognized by Teller (542) and subsequently worked out in detail by Renner (1069), the potential function in a degenerate electronic state splits into two when the molecule is bent. This is because in the bent position of a linear molecule, just as for a bent molecule, there are no degenerate electronic states since there is no more-than-two-fold axis. In Figs. 4b and c at the top the potential energy is shown schematically as a function of the bending angle for a Π and a Δ state and

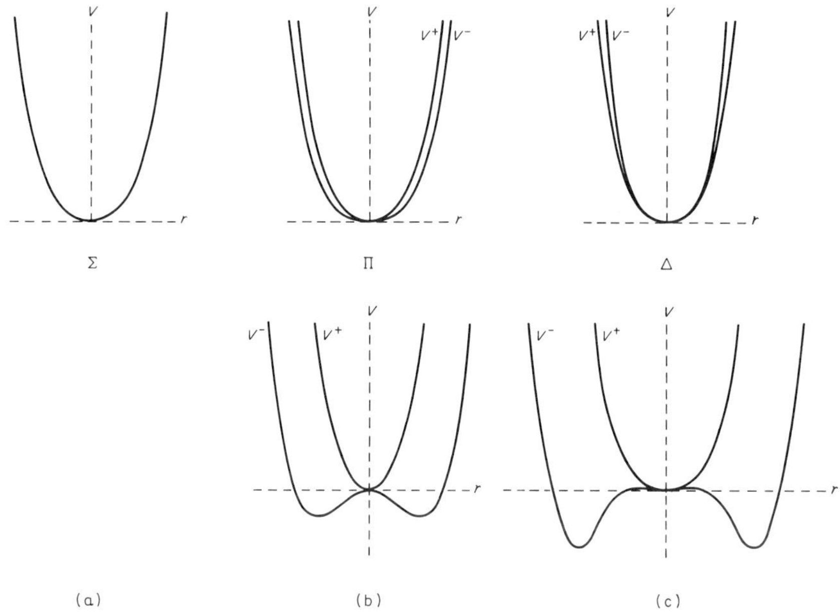


FIG. 4. **Potential functions for the bending vibration in Σ, Π and Δ electronic states of linear molecules.** The abscissa is the bending coordinate. The two diagrams at the top in (*b*) and (*c*) correspond to small vibronic interaction, those at the bottom to large vibronic interaction.

compared in Fig. 4a with that for a Σ state. For the Π and Δ states we call the upper (inner) potential function $V^+$, the lower (outer) one $V^-$. Since the potential functions for reasons of symmetry must be even functions of the bending coordinate $r$, the zero-order potential function can be written

$$V^0 = ar^2 + br^4 + \cdots \tag{I, 31}$$

and the splitting of the potential function in Π, Δ, ... states must also have the same form, viz.

$$V^+ - V^- = \alpha r^2 + \beta r^4 + \cdots. \tag{I, 32}$$

It is clear that if the vibronic interaction is so large that $\frac{1}{2}\alpha > a$, the $V^-$ curve will not have a minimum but a maximum at $r = 0$, and will in general have a minimum at some non-zero value of $r$. Such a case is illustrated by the lower diagram of Fig. 4b. Here we have therefore a splitting of the Π state into one component state in which the molecule is linear and one in which it is bent.

It can be shown [see Pople and Longuet–Higgins (1002)] that for Δ electronic states the quantity $\alpha$ in (I, 32) vanishes. Therefore, the splitting becomes noticeable only at larger $r$ values. Even if the vibronic interaction is very large, there will always be a minimum at $r = 0$ in both curves as long as in the unperturbed function $V^0$ the quadratic term is the dominant one. But there may then be a second minimum in the lower curve at a non-zero $r$-value as shown in the lower diagram of Fig. 4c.

In a triatomic linear molecule there is only one way of bending the molecule. The bent conformation has the symmetry $\boldsymbol{C_s}$ for an unsymmetric (XYZ) molecule and $\boldsymbol{C_{2v}}$ for a symmetric ($XY_2$) molecule. In the former case, all degenerate electronic states, $\Pi, \Delta, \ldots$, split into one $A'$ and one $A''$ state each, for $r \neq 0$. In the latter case, the species in the bent conformation differ for different types of degenerate states. As will be shown in more detail in Chapter III, section 1, a $\Pi_g$ electronic state splits into $A_2$ and $B_2$, $\Pi_u$ into $A_1 + B_1$, $\Delta_g$ into $A_1 + B_1$, $\Delta_u$ into $A_2 + B_2$. In each case, the electronic eigenfunction of one component state is symmetric, that of the other is antisymmetric with respect to the molecular plane. The appropriate species designations $A'$, $A''$, or $A_1$, $B_1$, etc. could be ascribed to the two potential functions $V^+$ and $V^-$. However it is not possible to say generally whether $V^+$ is $A'$, $V^-$ is $A''$ or whether the reverse correlation applies. Sometimes the two component states are distinguished as $\Pi^{(+)}$, $\Pi^{(-)}$, or $\Delta^{(+)}$, $\Delta^{(-)}$, etc. corresponding to $V^+$ and $V^-$. This designation should not be confused with $\Pi^+, \Pi^-, \Delta^+, \Delta^-, \ldots$ used to distinguish the two $l$- or $\Lambda$-type components of a $\Pi, \Delta, \ldots$ state.

For a four-atomic linear molecule, there are an infinite number of ways of bending it. For a symmetrical $X_2Y_2$ molecule there are two symmetrical ways of bending it leading to a $\boldsymbol{C_{2v}}$ or a $\boldsymbol{C_{2h}}$ symmetry. For $\boldsymbol{C_{2v}}$, the symmetries of the electronic eigenfunctions are the same as for $XY_2$ molecules; for $\boldsymbol{C_{2h}}$ we obtain $A_g + B_g$ from $\Pi_g, \Delta_g, \ldots$ and $A_u + B_u$ from $\Pi_u, \Delta_u, \ldots$.

We must now investigate the effect of the splitting of the potential function on the *vibrational energy levels*. This is not a simple problem because the energy levels are not simply the energy levels belonging to two independent potential functions $V^+$ and $V^-$. Classically the reason for this complication is the fact that the potential surfaces are in contact and that therefore the molecule may change over from one to the other surface unless symmetry requirements prevent it from doing so. The quantum-theoretical treatment of Renner (1069) for $\Pi$ states of linear $XY_2$ molecules has shown that only the $\Sigma$ vibronic levels (see Fig. 2) can be rigorously assigned to one or the other potential function, the $\Sigma^+$ levels to $V^+$, the $\Sigma^-$ levels to $V^-$, or conversely, the $\Sigma^+$ levels to $V^-$ and the $\Sigma^-$ levels to $V^+$. The $\Pi, \Delta, \ldots$ vibronic levels belong strictly speaking to both potential functions. However in a certain very rough approximation the higher one of a pair of $\Pi$ vibronic levels may be assigned to the upper potential curve ($V^+$), the lower one, as well as the single $\Pi$ level at $v = 0$, to the lower curve ($V^-$); but there is no reason of symmetry that this should be so, and it is therefore far from a rigorous assignment. Similar considerations apply to $\Delta$ vibronic levels which also occur in pairs except for $v_2 = 1$ for which only a single $\Delta$ level occurs (see Fig. 2).

If we use as a zero-approximation the product form (I, 27) for the vibronic eigenfunction, we must substitute for $\psi_e$ in a linear molecule according to (I, 8)

$$\psi_e = \chi_e^{\pm} e^{\pm i\Lambda\nu} \tag{I, 33}$$

and for $\psi_v$ (see eq. II, 58 of Vol. II)

$$\psi_v = \rho_{v_2 l}(r)\ e^{\pm il\varphi}. \tag{I, 34}$$

Here $\nu$ is the azimuth of one of the electrons ($\equiv \varphi_1$ of I, 8) and $\chi_e^{\pm}$ gives the dependence on the other coordinates of this electron and all the coordinates of the other electrons; $\varphi$ is the azimuth of the molecular plane and $\rho_{v_2 l}(r)$ a function of the displacement $r$ from the equilibrium position. The double sign in front of $i\Lambda\nu$ and $il\varphi$ corresponds to the electronic and vibrational degeneracy.

The coupling of vibration and electronic motion depends on $r$ and on the azimuth $\nu - \varphi$ of the first electron with respect to the molecular plane. More specifically, according to Pople and Longuet–Higgins (1002), the perturbation function for small coupling is

$$H' = fr^2 \cos 2(\nu - \varphi). \tag{I, 35}$$

For a $\Pi$ electronic state ($\Lambda = 1$), the following combinations of the degenerate zero-approximation eigenfunctions must be used with this perturbation:

for $K=0,\ l=1,\ v_2=1, 3, \ldots,$

$$\begin{aligned} \psi_0^+ &= \chi_e^+ \rho_{v_2 1}(r)\, e^{i(\nu-\varphi)} + \chi_e^- \rho_{v_2 1}(r)\, e^{-i(\nu-\varphi)} \\ \psi_0^- &= \chi_e^+ \rho_{v_2 1}(r)\, e^{i(\nu-\varphi)} - \chi_e^- \rho_{v_2 1}(r)\, e^{-i(\nu-\varphi)} \end{aligned} \tag{I, 36}$$

for $K = 1,\ l = 0, 2,\ v_2 = 2, 4, \ldots$

$$\begin{aligned} \psi_1^+ &= \chi_e^+ \rho_{v_2 0}(r)\, e^{i\nu} + \chi_e^- \rho_{v_2 2}(r)\, e^{-i(\nu-2\varphi)} \\ \psi_{-1}^+ &= \chi_e^- \rho_{v_2 0}(r)\, e^{-i\nu} + \chi_e^+ \rho_{v_2 2}(r) e^{+i(\nu-2\varphi)} \\ \psi_1^- &= \chi_e^+ \rho_{v_2 0}(r)\, e^{i\nu} - \chi_e^- \rho_{v_2 2}(r)\, e^{-i(\nu-2\varphi)} \\ \psi_{-1}^- &= \chi_e^- \rho_{v_2 0}(r)\, e^{-i\nu} - \chi_e^+ \rho_{v_2 2}(r)\, e^{+i(\nu-2\varphi)}. \end{aligned} \tag{I, 37}$$

The two functions for $K = 0$ have the species $\Sigma^+$ and $\Sigma^-$, respectively, and retain these species for any strength of the vibrational-electronic interaction. It can be shown that the $\psi_0^+$ functions even in higher approximation are dependent only on the potential function $V^+$ (or only on $V^-$) and entirely independent of $V^-$ (or $V^+$), and conversely, that $\psi_0^-$ depends only on $V^-$ (or only on $V^+$). It is tempting to assume that in an analogous way for $K = 1$ the first pair of functions $\psi_1^+$ and $\psi_{-1}^+$ belong to $V^+$, and the second pair $\psi_1^-$ and $\psi_{-1}^-$ belong to $V^-$ (or vice versa). But while each pair does represent a complete vibronic $\Pi$ state ($K = 1$), it is not either symmetric or antisymmetric with respect to reflection at a plane through the internuclear axis, i.e., to simultaneous reversal of the sign of $\nu$ and $\varphi$ and therefore does not belong wholly to one or the other potential function. Rather, there are non-vanishing matrix elements of the perturbation (I, 35) between $\psi_1^+$ for a given $v_2$ and $\psi_1^-$ for another $v_2$, that is, every $\Pi$ vibronic level depends on both $V^+$ and $V^-$. Similar conclusions apply to $\Delta, \Phi, \ldots$ vibronic levels. Nevertheless, in a very rough first approximation it is often possible to assign the $\psi_K^+$ functions to the one potential function (say $V^+$) and $\psi_K^-$ to the other (say $V^-$).

On the basis of the preceding considerations Renner (1069) obtained, for small and medium interaction of vibration and electronic motion but disregarding the quartic term in (I, 31), the following vibronic term values in an electronic $\Pi$ state of a linear $XY_2$ molecule:

for $K = 0$ ($\Sigma$ vibronic states)

$$G^{\pm}(v_2, 0) = \omega_2\sqrt{1 \pm \epsilon}\,(v_2 + 1), \quad v_2 = 1, 3, 5, \ldots; \tag{I, 38}$$

for $K \neq 0$, $v_2 = K - 1$ (lowest single vibronic level of species $\Pi, \Delta, \ldots$)

$$G(v_2, K) = \omega_2[(v_2 + 1) - \tfrac{1}{8}\epsilon^2 K(K + 1)] \tag{I, 39}$$

and for $K \neq 0$, $v_2 > K - 1$ (remaining $\Pi, \Delta, \ldots$ vibronic states, occurring in pairs)

$$G^{\pm}(v_2, K) = \omega_2(1 - \tfrac{1}{8}\epsilon^2)(v_2 + 1) \pm \tfrac{1}{2}\omega_2\epsilon\sqrt{(v_2 + 1)^2 - K^2}. \tag{I, 40}$$

Here $\omega_2$ is the bending frequency (in $cm^{-1}$ if the vibronic energy $G(v_2, K)$ is to be obtained in $cm^{-1}$) and $\epsilon$ is the Renner parameter: $\epsilon = \alpha/2a$ where $a$ and $\alpha$ are the coefficients of $r^2$ in eqs. (I, 31) and (I, 32). To obtain the total vibronic energy, the quantity $G(v_2, K)$ must be substituted in (I, 22) and (I, 23) for the part depending on $v_2$, that is, for $\omega_2(v_2+1) + g_2 l_2^{\,2}$. The anharmonic terms $x_{ik}$ are not considered in this approximation. Such terms have been discussed by Hougen and Jesson (579).

In Fig. 5 the energy levels obtained from these formulae are shown for $K = 0$, 1 and 2 as a function of $\epsilon$. For $K = 0$ the formula (I, 38) is exact when the vibrations are strictly harmonic and the coupling purely quadratic. For increasing coupling ($\epsilon$) the lower components of the higher vibrational levels cross the upper components of the lower vibrational levels, and eventually tend to $G = 0$ for $\epsilon = 1$, that is, when the lower potential curve has flattened to a horizontal line. The $\Pi$ and $\Delta$ vibronic levels behave similarly, but here the formulae are not exact for higher $\epsilon$ values and do not give the convergence of the lower components to $G = 0$ at $\epsilon = 1$ which is assumed in Fig. 5. In addition, the formulae give crossings of the energy curves (broken lines) which in higher approximation are avoided as shown by the full lines. In Fig. 6 an energy level diagram for small coupling ($\epsilon = 0.1$) is given which shows all the vibronic levels in their correct relative position (neglecting anharmonic terms) up to $v = 6$ and $K = 4$.

Pople and Longuet Higgins (1002) have carried out calculations for the case (lower part of Fig. 4b) in which the lower of the two potential functions has a maximum for $r = 0$ and a minimum for a non-zero $r$ value, that is, corresponds to a non-linear equilibrium conformation. In order to consider this case, it is necessary to include higher power (anharmonic) terms in the potential energy and in the coupling energy. For the vibronic levels corresponding to the upper potential function (in which the molecule remains linear) Pople and Longuet–Higgins found the following formula

$$G^{+}_{vK} = \omega_2\{(v_2 + 1) - \tfrac{1}{2}\bar{\alpha}[(v_2 + 1) - \sqrt{(v_2 + 1)^2 - K^2}] + \tfrac{1}{2}\bar{b}[3(v_2 + 1)^2 - K^2] + \tfrac{3}{4}\bar{\beta}(v_2 + 1)\sqrt{(v_2 + 1)^2 - K^2}\} \tag{I, 41}$$

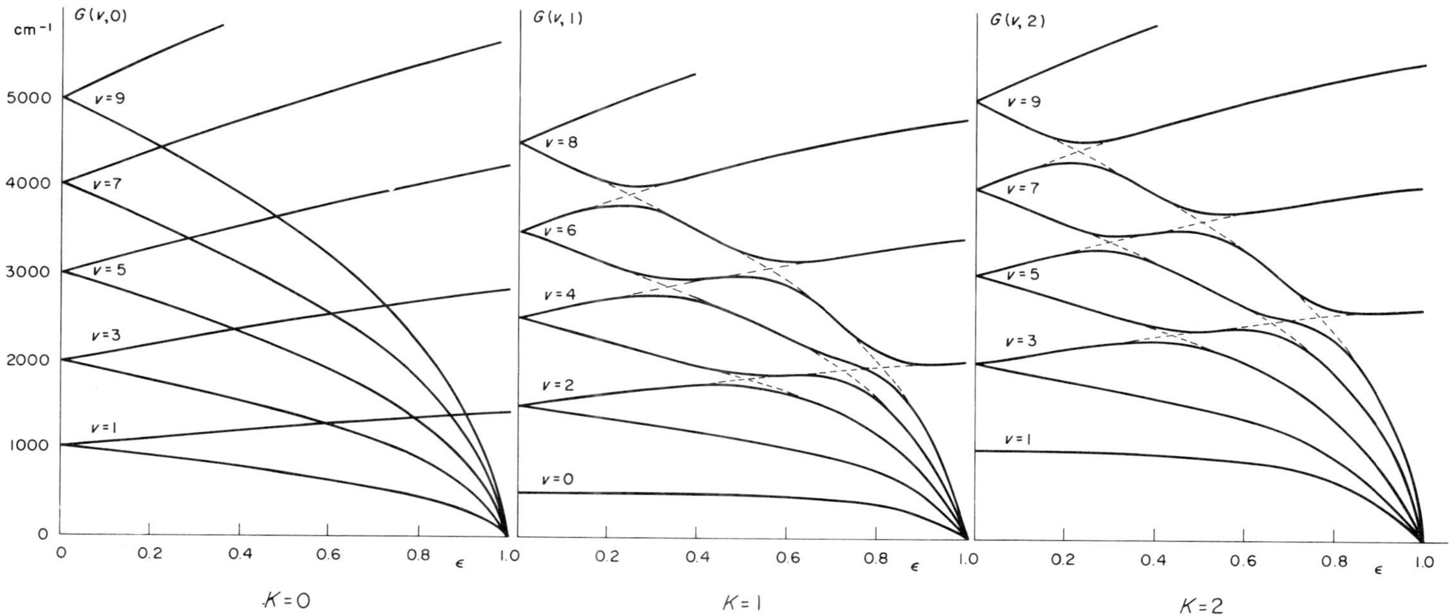


FIG. 5. **Variation of the energies of the Σ, Π, Δ vibronic levels of a Π electronic state as a function of the Renner parameter ε.** A bending frequency of 500 cm$^{-1}$ is assumed; the lowest level for $\epsilon = 0$ occurs therefore at an energy 500 cm$^{-1}$ and corresponds to $K = 1$ (central diagram). The broken-line correlations correspond to the approximation in which the formulae (I, 39) and (I, 40) hold.

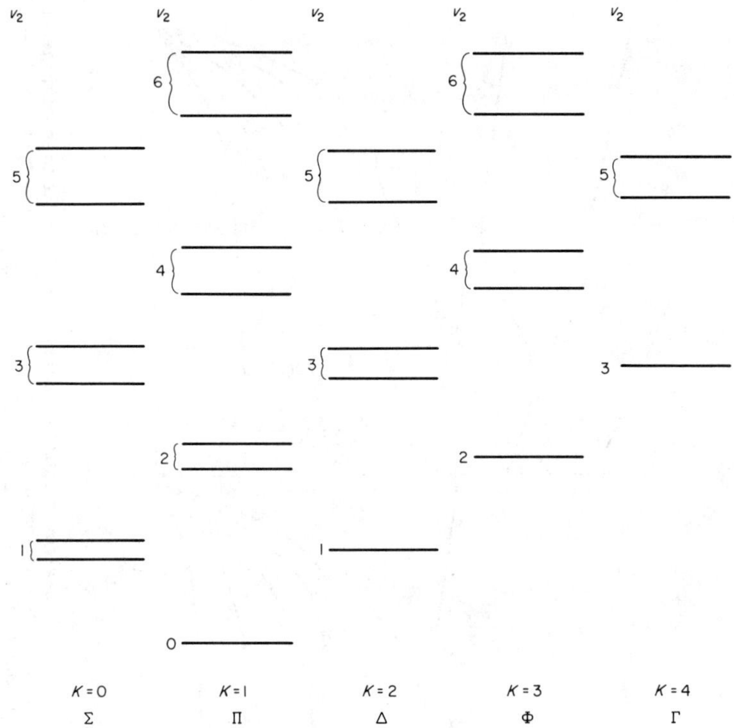


FIG. 6. **Diagram of vibronic levels in a Π electronic state for ε = 0.1.** All levels except those with $K = v_2 + 1$ occur in pairs on account of vibronic interaction. Anharmonicity is neglected.

where $\bar{\alpha} = \alpha/(2a + \alpha)$, $\bar{\beta} = \beta/(2a + \alpha)$, $\bar{b} = b/(2a + \alpha)$ and $a, b, \alpha, \beta$ are the potential constants defined in (I, 31) and (I, 32). If $K$ is small compared to $v$, eq. (I, 41) can be approximated by

$$G^+_{vK} = \omega_2\left\{(v_2 + 1) + \frac{3}{4}(2\bar{b} + \bar{\beta})(v_2 + 1)^2 - \frac{1}{8}\left(\frac{2\bar{\alpha}}{(v_2 + 1)} + 4\bar{b} + 3\bar{\beta}\right)K^2\right\}. \quad \text{(I, 42)}$$

In the derivation of (I, 41) it is assumed that the eigenfunctions $\psi_1^+$ can be assigned to the levels belonging to the upper curve $V^+$. The expression (I, 42) has the same form as the standard expression for the vibrational energy of a bending vibration without regard for vibrational-electronic coupling, that is,

$$G(v_2, l_2) = \omega_2(v_2 + 1) + x_{22}(v_2 + 1)^2 + g_{22}l_2^2 \quad \text{(I, 43)}$$

except that $l_2$ is replaced by $K$ and the coefficient of $K^2$ is in general much larger than $g_{22}$ because it contains a quadratic potential constant in addition to higher ones. For negligible vibronic coupling $\bar{\alpha} \to 0$, $\bar{\beta} \to 0$, and therefore $x_{22} \to \frac{3}{2}\omega_2\bar{b}$, $g_{22} \to -\frac{1}{2}\omega_2\bar{b}$ if other anharmonic constants are neglected, or if $\omega_2 \ll \omega_1, \omega_3$; thus in this approximation $g_{22} \to -\frac{1}{3}x_{22}$. Conversely, a large deviation of the observed

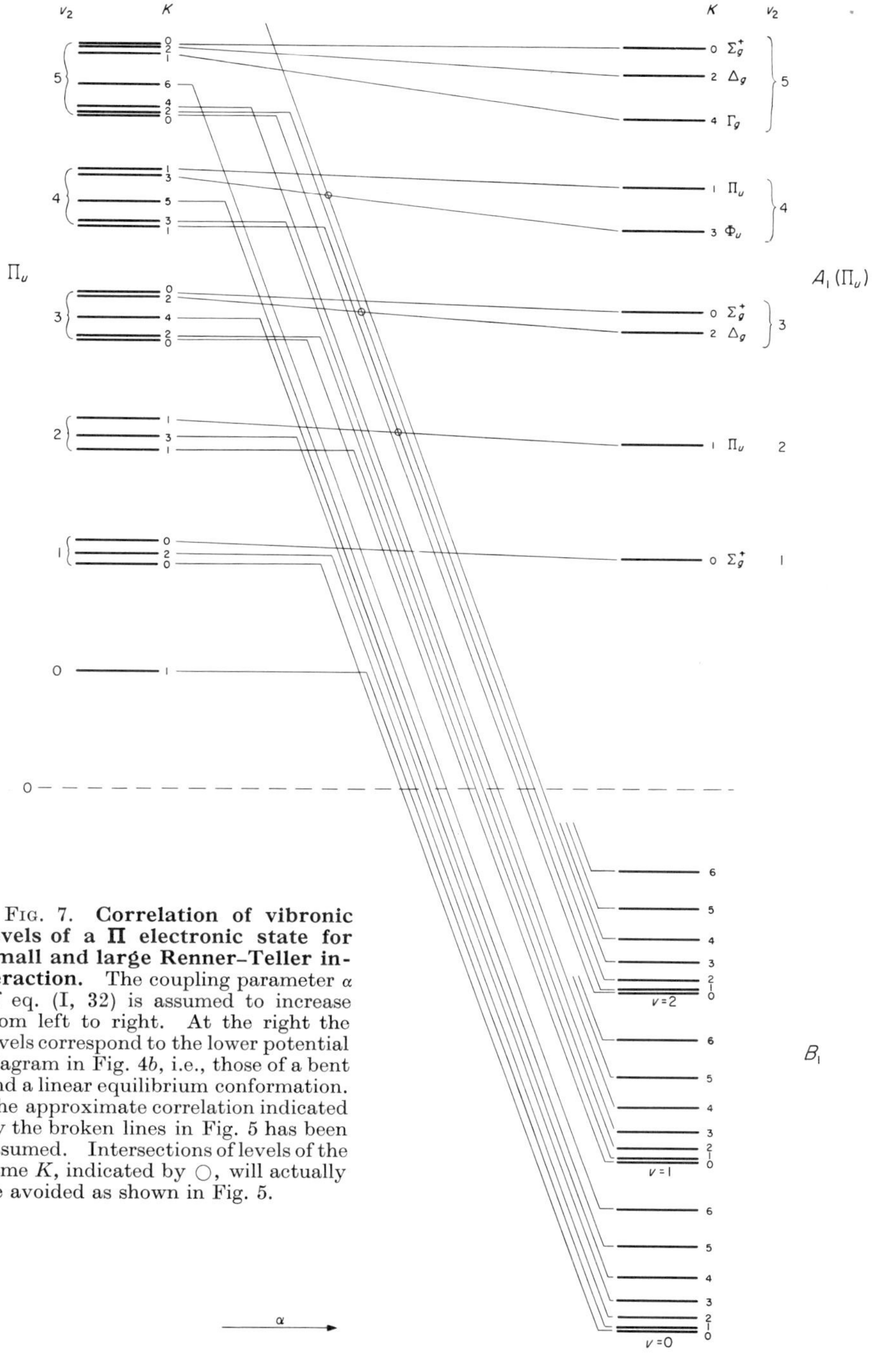


FIG. 7. **Correlation of vibronic levels of a Π electronic state for small and large Renner–Teller interaction.** The coupling parameter $\alpha$ of eq. (I, 32) is assumed to increase from left to right. At the right the levels correspond to the lower potential diagram in Fig. 4*b*, i.e., those of a bent and a linear equilibrium conformation. The approximate correlation indicated by the broken lines in Fig. 5 has been assumed. Intersections of levels of the same $K$, indicated by ○, will actually be avoided as shown in Fig. 5.

coefficient of $K^2$ from the value $-x_{22}/3$ can serve to indicate the presence of a case in which a Π state of the linear conformation is split into one state with linear and one with bent equilibrium conformation on account of strong Renner–Teller coupling (see the lower diagram of Fig. 4b and the discussion of $NH_2$ and $CH_2$ in Chap. II).

For Δ states, no formulae similar to (I, 38–40) and applicable to small Renner–Teller coupling have been developed;[4a] but Longuet–Higgins (768) has derived a formula for those levels that in a certain approximation may be assigned to the upper potential function of a widely split Δ state when the lower function has minima at $r \neq 0$. We shall not reproduce this formula but only mention that according to it for Δ states unlike Π states, the coefficient of $K^2$ is not anomalously large, but that there is an appreciable quartic term in $K$, particularly for small $v$, whose presence in an observed spectrum may be used to establish the case illustrated by the lower diagram of Fig. 4c.

No special formula for the vibronic levels of the lower potential function $V^-$ has been worked out for either Π or Δ electronic states; but in the approximation in which (I, 42) holds, one may expect the usual formula for the vibrational levels of a bent molecule to hold. Some idea of the actual energy levels may be obtained by a consideration of the diagram Fig. 7 in which for a Π electronic state schematically the levels for small interaction are correlated with those for very large interaction. In this diagram there are a few intersection points of vibronic states of the same $K \neq 0$ (indicated by small circles). At each of these intersections the situation is similar to that shown in Fig. 5b and c, i.e., the "intersection" does not actually take place.

The theory of Fermi resonance in Π electronic states has been discussed by Hougen (570). No detailed discussion of the Renner–Teller splittings for linear molecules other than $XY_2$ and XYZ has yet been given. For molecules like linear $X_2Y_2$ further complications are likely to arise on account of the presence of more than one bending vibration.

**Vibronic interaction in doublet states.** The first case of a Renner–Teller splitting was observed in a $^2\Pi$ state, and even now more examples of such splittings in $^2\Pi$ states than in $^1\Pi$ states are known. It is therefore important to consider the effect of the presence of a non-zero electron spin on the vibronic interaction or, conversely, the effect of vibronic interaction on the multiplet structure. This was first done by Pople (1001). The results of his calculations are illustrated in Fig. 8 which gives the correlation of the actual energy levels of a $^2\Pi$ electronic state shown in the center with those for vanishing Renner–Teller interaction, shown at the left, and with those for vanishing spin-orbit interaction, shown at the right. The levels at the left are taken from Fig. 3 but they show a single level for each set of coinciding levels; the levels at right are taken from Fig. 6 but symbols corresponding to the presence of spin are added, even though spin-orbit coupling is assumed to be zero. The component levels with $K = v_2 + 1$ at the right, split

[4a] Since this was written, Merer and Travis (826) have developed such formulae.

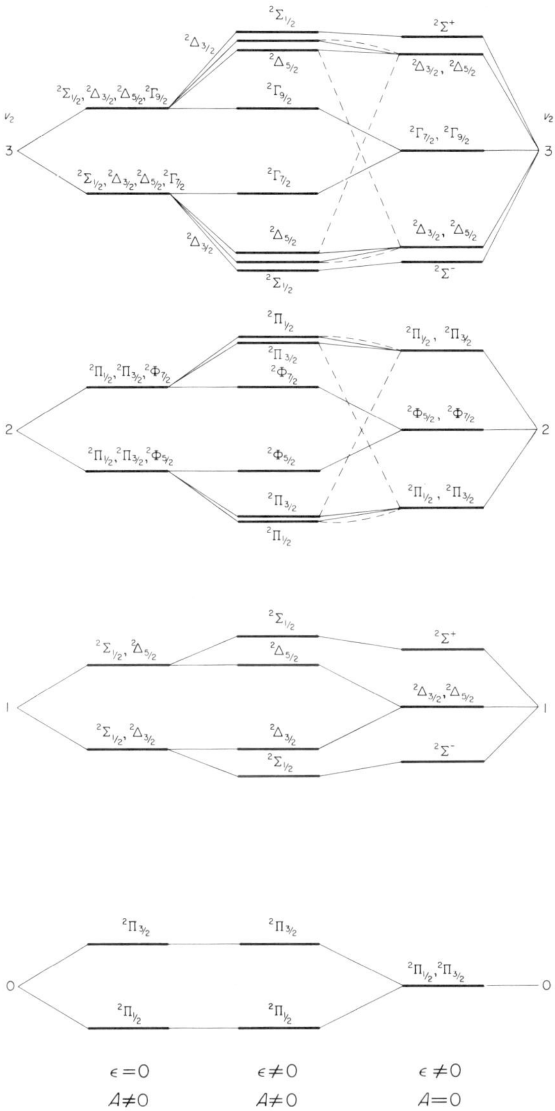


FIG. 8. **Correlation of the vibronic levels of a $^2\Pi$ electronic state for zero vibronic (left) and zero spin-orbit (right) interaction with those for which both interactions are non-zero (center).** Energy levels that coincide in a given approximation are shown as one level with the term symbols indicated. Note that at the left (and in the center) the quantum number $K$ (indicated by the symbols $\Sigma$, $\Pi$, $\Delta$, . . .) is not a good quantum number, but $P$ is a good quantum number throughout.

upon introduction of spin-orbit coupling in a first approximation in the same way as if there were no vibronic interaction, i.e., instead of (I, 39) we have

$$G(v_2, K, \pm\tfrac{1}{2}) = \omega_2(v_2 + 1) \pm \tfrac{1}{2}A - \tfrac{1}{8}\epsilon^2\omega_2 K(K + 1), \qquad \text{(I, 44)}$$

where $\pm\frac{1}{2}$ stands for the spin quantum number $\Sigma$, which represents the component of $S$ along the internuclear axis and where $A$ is the usual spin-orbit coupling constant. In this approximation the spin splitting for the component $K = v_2 + 1$ is

$$\Delta\nu_{K=v_2+1} = A, \qquad \text{(I, 45)}$$

the same as without vibronic interaction. In a higher approximation, as was shown by Hougen (568), one finds a slight dependence on $K$:

$$\Delta\nu_{K=v_2+1} = A[1 - \tfrac{1}{8}\epsilon^2 K(K + 1)]. \qquad \text{(I, 46)}$$

For vibronic levels with $K < v_2 + 1$, there is a considerable change. Pople (1001) [as corrected by Hougen (568)] gives

$$\begin{aligned} G^+(v_2, K, \pm\tfrac{1}{2}) &= \omega_2(1 - \tfrac{1}{8}\epsilon^2)(v_2 + 1) + \tfrac{1}{2}A^*_{v_2,K} \mp \frac{\epsilon^2 A\omega_2 K(v_2 + 1)}{8A^*_{v_2,K}} \\ G^-(v_2, K, \pm\tfrac{1}{2}) &= \omega_2(1 - \tfrac{1}{8}\epsilon^2)(v_2 + 1) - \tfrac{1}{2}A^*_{v_2,K} \pm \frac{\epsilon^2 A\omega_2 K(v_2 + 1)}{8A^*_{v_2,K}}, \end{aligned} \qquad \text{(I, 47)}$$

where

$$A^*_{v_2,K} = \sqrt{A^2 + \epsilon^2\omega_2{}^2[(v_2 + 1)^2 - K^2]} \qquad \text{(I, 48)}$$

is an effective spin-splitting constant modified by vibronic interaction. Equation (I, 47) should be compared with (I, 40) into which it goes over when $A \to 0$. If $A$ is small compared to $\epsilon\omega_2$, (I, 47) may be simplified to

$$\begin{aligned} G^+(v_2, K, \pm\tfrac{1}{2}) &= \omega_2(1 - \tfrac{1}{8}\epsilon^2)(v_2 + 1) + \tfrac{1}{2}\epsilon\omega_2\sqrt{(v_2 + 1)^2 - K^2} \\ &\quad \mp \frac{\epsilon AK(v_2 + 1)}{8\sqrt{(v_2 + 1)^2 - K^2}} \\ G^-(v_2, K, \pm\tfrac{1}{2}) &= \omega_2(1 - \tfrac{1}{8}\epsilon^2)(v_2 + 1) - \tfrac{1}{2}\epsilon\omega_2\sqrt{(v_2 + 1)^2 - K^2} \\ &\quad \pm \frac{\epsilon AK(v_2 + 1)}{8\sqrt{(v_2 + 1)^2 - K^2}} \end{aligned} \qquad \text{(I, 49)}$$

Coming from the left-hand side of Fig. 8 we see that the doublet splitting of the states with $0 < K < v_2 + 1$, e.g., of the two vibronic $^2\Pi$ states of $v_2 = 2$ or of the two vibronic $^2\Delta$ states of $v_2 = 3$, is increased by the introduction of Renner–Teller coupling from the value $A$ to

$$\Delta\nu_{0<K<v_2+1} = A^*_{v_2,K}. \qquad \text{(I, 50)}$$

We may consider this as a result of the mutual "repulsion" of states of the same $P$ which is greatest for the smallest $P$ value. (There is no repulsion for the two central components since they have different $P$ values.) The center of one doublet is shifted down, that of the other up by the amount $\epsilon^2 A\omega_2 K(v_2 + 1)/(8A^*_{v_2,K})$; one doublet is normal, the other is inverted.

On the other hand, coming from the right of Fig. 8 two correlations are possible: either each pair of states coinciding for $A = 0$ splits into one state of the upper and one of the lower group in the center (broken-line correlation) with the splitting given by (I, 50), or the upper pair is correlated with the two levels of the upper group, the lower pair with the two levels of the lower group in the center (full-line correlation). For this correlation the splitting is much smaller, namely, from (I, 47)

$$\Delta\nu_{0<K<v+1} = \frac{\epsilon^2 A\omega_2 K(v_2 + 1)}{4A^*_{v_2K}}. \qquad \text{(I, 51)}$$

For the vibronic levels with $K = 0$ ($^2\Sigma$), which occur for odd $v_2$, there is no splitting. But as $A$ increases, the two $^2\Sigma$ states of a given $v_2$ are pushed apart by the combined effect of spin-orbit and vibronic interactions (see Fig. 8). The vibronic energy is again given by eq. (I, 47) if we put $K = 0$. The separation of the two $^2\Sigma$ states is given by (I, 50) with $K = 0$. While at the right of Fig. 8, for vanishing spin-orbit coupling, one of the $^2\Sigma$ states is $^2\Sigma^-$, the other $^2\Sigma^+$, in the center of Fig. 8 as well as at the left when spin-orbit coupling is large, the +, − character is no longer well defined: the $^2\Sigma$ states at the left are more nearly like $\frac{1}{2}$ states in the sense used for Hund's case (c) of diatomic molecules (see Vol. I, p. 236).

**Vibronic interaction in triplet states.** Hougen (569) has also considered the Renner–Teller splittings for $^3\Pi$ states of $XY_2$ molecules. Figure 9 shows the results in a way similar to Fig. 8. Every vibronic state at the right, upon introduction of spin-orbit interaction, splits into three component states (which differ in the value of $P = |K + \Sigma|$) except the $^3\Sigma$ vibronic states which split into two states, a 0 and a 1 state as in Hund's case (c) of diatomic molecules ($0^+$ from $^3\Sigma^-$ and $0^-$ from $^3\Sigma^+$). In a first approximation the component states with $P = K$ have the same energy as without spin-orbit interaction, while the energy of the states with $P = K \pm 1$ is given by (I, 47) if $A$ is replaced by $2A$. Again the vibronic states with $K = v_2 + 1$ have the same spin splitting as without vibronic interaction (see Fig. 9).

## (c) Degenerate electronic states: non-linear molecules

**Vibronic species.** Just as for linear molecules, several vibronic species and therefore several vibronic levels result for a given excitation of degenerate vibrations in a degenerate electronic state of a non-linear molecule (in this sub-section, "non-linear" refers not to "bent" molecules but to molecules that have at least one $C_p$ with $p > 2$). For example, if in a planar $XY_3$ molecule (point group $\boldsymbol{D}_{3h}$) a degenerate $e'$ vibration is singly excited in an $E''$ electronic state, we obtain from Table 57 (Appendix III) the vibronic species $^vE' \times {}^eE'' = {}^{ev}A_1'' + {}^{ev}A_2'' + {}^{ev}E''$. If the same vibration is doubly excited, i.e., if the vibrational species is $A_1' + E'$ (see Table 32, p. 127 of Vol. II), we obtain the vibronic species $({}^vA_1' + {}^vE') \times {}^eE'' = {}^{ev}E'' + {}^{ev}A_1'' + {}^{ev}A_2'' + {}^{ev}E''$. In Fig. 10a and b the vibronic species in several

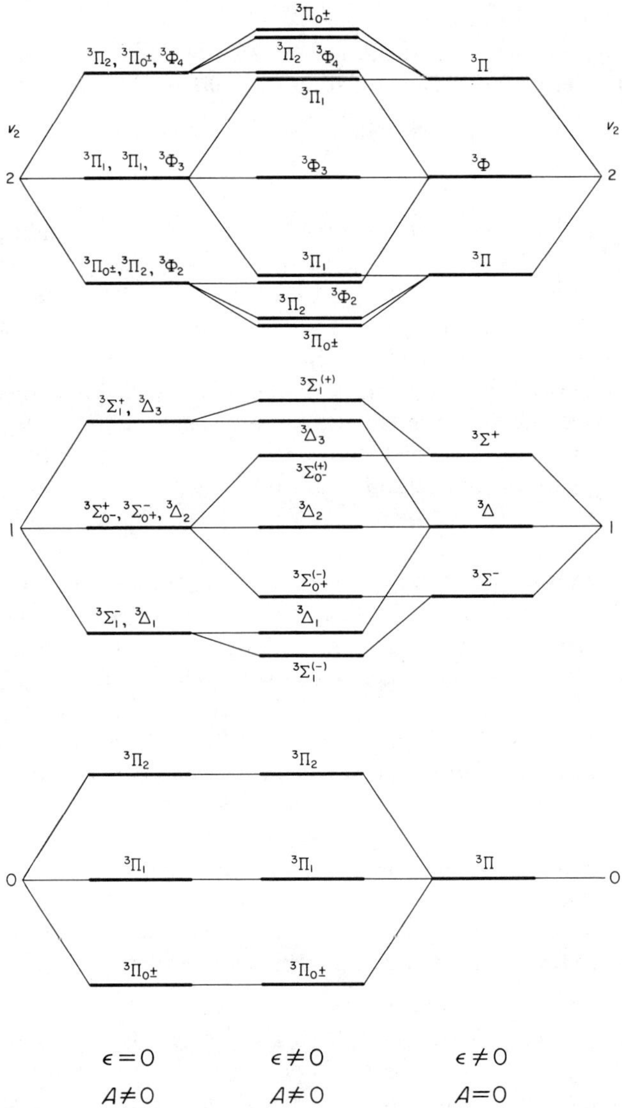


FIG. 9. **Correlation of the vibronic levels of a $^3\Pi$ electronic state for zero vibronic (left) and zero spin-orbit (right) interaction with those for which both interactions are non-zero (center) [after Hougen (569)].** See legend of Fig. 8.

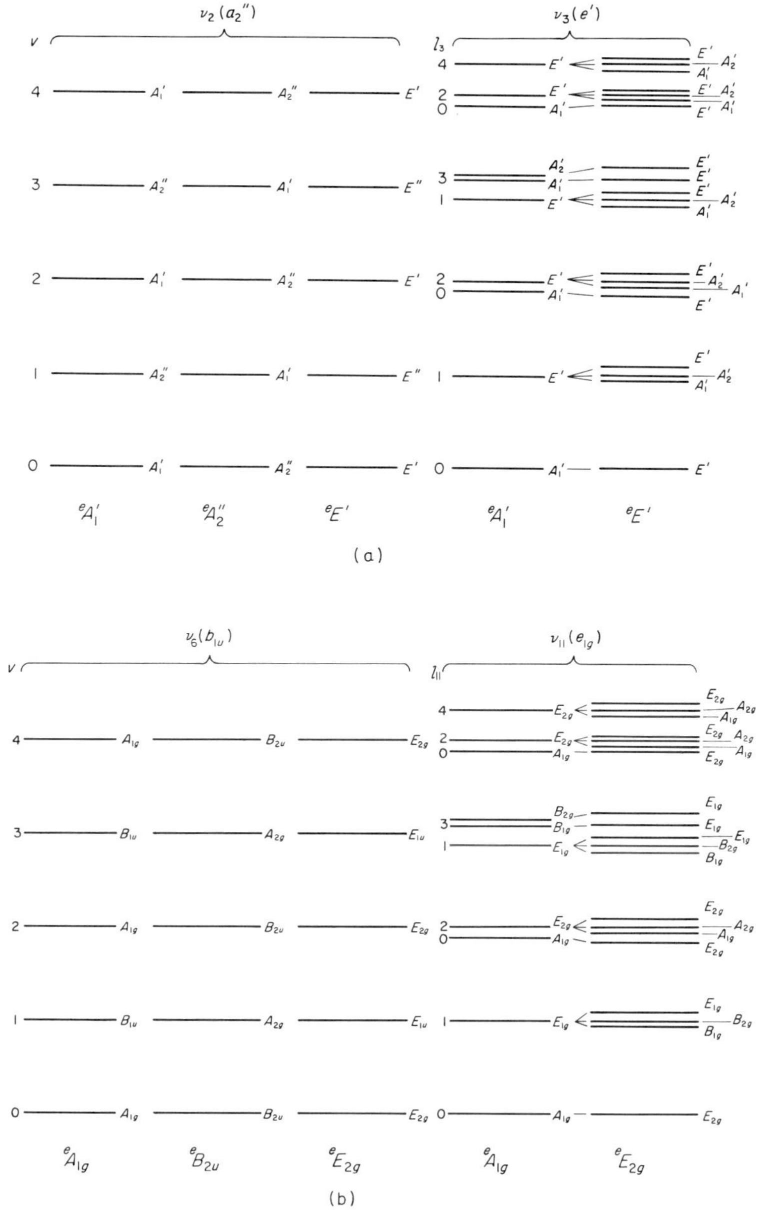

FIG. 10. **Vibronic species of the vibrational levels of non-degenerate and degenerate vibrations in various types of electronic states of** (*a*) $\boldsymbol{D_{3h}}$ **molecules and** (*b*) $\boldsymbol{D_{6h}}$ **molecules.** The species in the body of the figure are vibronic species. The prefix *ev* has been omitted in accordance with common practice. The same diagrams can be used for $\boldsymbol{C_{3v}}$ and $\boldsymbol{C_{6v}}$ molecules respectively if ′ and ″ and $g$ and $u$ are disregarded.

degenerate and non-degenerate electronic states of $\boldsymbol{D}_{3h}$ and $\boldsymbol{D}_{6h}$ molecules are shown for $v \leq 4$. On the basis of these diagrams and the rules given above, it is a simple matter to derive similar diagrams for other electronic species and other point groups. If vibronic interaction is taken into account, there will be as many different vibronic levels as there are vibronic species for each vibrational level. But the degenerate ($E$, $F$, . . .) vibronic levels are not split by the vibronic interactions. In particular, the lowest vibrational level of a degenerate electronic state always remains a single vibronic level of a degeneracy equal to that of the electronic state. This holds even when large amplitude motions are possible as in the case of several potential minima separated by low (surmountable) barriers (see p. 5). Only the interaction with rotation can produce a splitting (see section 3).

**The Jahn–Teller theorem.** Before we can evaluate the magnitude of the splittings between the different vibronic levels obtained in the above described manner, we must, just as for linear molecules, consider the splitting of the potential function for non-totally symmetric displacements of the nuclei. Qualitatively the reason for the splitting of the potential function is the same as for linear molecules: in the displaced position of the nuclei the symmetry is lower and, in general, all electronic states are non-degenerate; instead of one doubly degenerate electronic state, we obtain for the displaced position of the nuclei two non-degenerate electronic states of slightly different energy. Similarly, for triply degenerate electronic states we obtain three non-degenerate or one non-degenerate and one doubly degenerate state depending on the type of the displacement.

For some of the non-totally symmetric displacements we have in a first approximation simply two parabolic potential functions which run together (osculate) at the symmetrical position where they both have minima (when vibronic interaction is small), just as for linear molecules (Fig. 4b and c). However, Jahn and Teller (618) have shown that in a non-linear molecule, unlike a linear molecule, there is always at least one non-totally symmetric normal coordinate that causes a splitting of the potential function such that *the potential minima are not in the symmetrical position.* Rather they are at a certain distance from this symmetrical position which is greater the larger the vibrational-electronic interaction. A one-dimensional cross-section through the potential function in a simple case is shown in Fig. 11. The two components of the potential function cross each other at the "original" equilibrium position with a non-zero angle. Thus, the symmetrical conformation is *not* the position of minimum energy if vibronic interaction is taken into account; rather, several (equivalent) minima of potential energy arise for certain unsymmetrical conformations.

For example, for a molecule like $CH_3I$ in a degenerate electronic state, if vibronic interaction is not negligible, the I atom will not be in equilibrium on the axis of symmetry but will have three symmetrical equilibrium positions off the axis. The potential function still has $\boldsymbol{C}_{3v}$ symmetry, but there is no potential minimum when the I atom is on the axis. If the vibronic interaction is

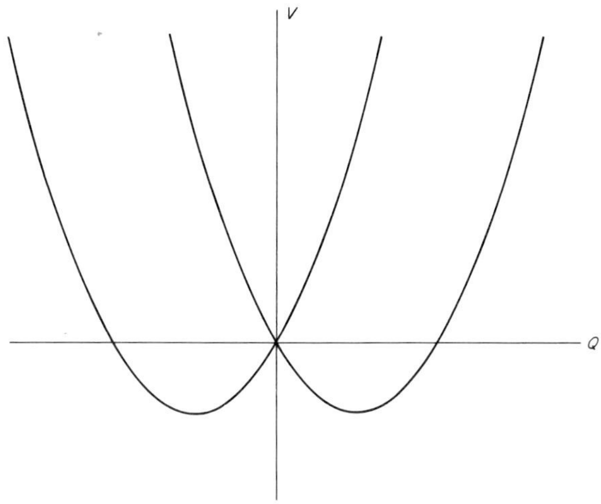


FIG. 11. **Cross section through the potential function of a non-linear molecule in a degenerate electronic state when vibronic interaction is large.** $Q$ is a non-totally symmetric normal coordinate that gives rise to strong Jahn–Teller interaction.

very strong, the minima are very far off axis, and very large amounts of vibrational energy would be required to bring the molecule from one of the (equivalent) potential minima to another. In such a case it would be better to consider the molecule as unsymmetrical (point group $\boldsymbol{C}_s$). However, if the vibronic interaction is weak, small amounts of vibrational energy suffice to make the system go from one minimum to another, i.e., to cover the whole symmetrical potential surface. In that case it is better to consider the molecule as symmetrical and study the modifications of the energy levels of such a symmetrical system by the vibronic (Jahn–Teller) interactions treated as a perturbation.

Qualitatively, one can demonstrate the validity of the Jahn–Teller theorem most easily for an $X_4$ molecule of point group $\boldsymbol{D}_{4h}$, i.e., when the four X atoms are at the corners of a square [see Sponer and Teller (1155)]. In a displaced position of the antisymmetric vibration $\nu_2(b_{1g})$, as shown in Fig. 12a, the conformation of the nuclei is no longer square. In this conformation for a degenerate electronic state (say $E_g$) there is a splitting of the degeneracy, i.e., there are two electronic states of species $B_{2g}$ and $B_{3g}$ of the point group $\boldsymbol{D}_{2h}$ (see Table 60, Appendix IV); $B_{2g}$ is antisymmetric with respect to the $yz$ plane, $B_{3g}$ with respect to $xz$. Let the energies be $W_{2g}(Q_2)$ and $W_{3g}(Q_2)$ where $Q_2$ stands for the normal coordinate, and let $W_{2g}(Q_2) > W_{3g}(Q_2)$. If we now reverse the displacements (i.e., consider $-Q_2$) as in Fig. 12b, we have again a splitting. Let the energies be $W_{2g}(-Q_2)$ and $W_{3g}(-Q_2)$ where as before $W_{2g}$ and $W_{3g}$ belong to the electronic functions that are antisymmetric with respect to the $yz$ and $xz$ planes, respectively. Since Fig. 12b can be obtained from Fig. 12a by a simple rotation by 90°, it follows that

$$W_{2g}(-Q_2) = W_{3g}(Q_2) \quad \text{and} \quad W_{3g}(-Q_2) = W_{2g}(Q_2) \qquad \text{(I, 52)}$$

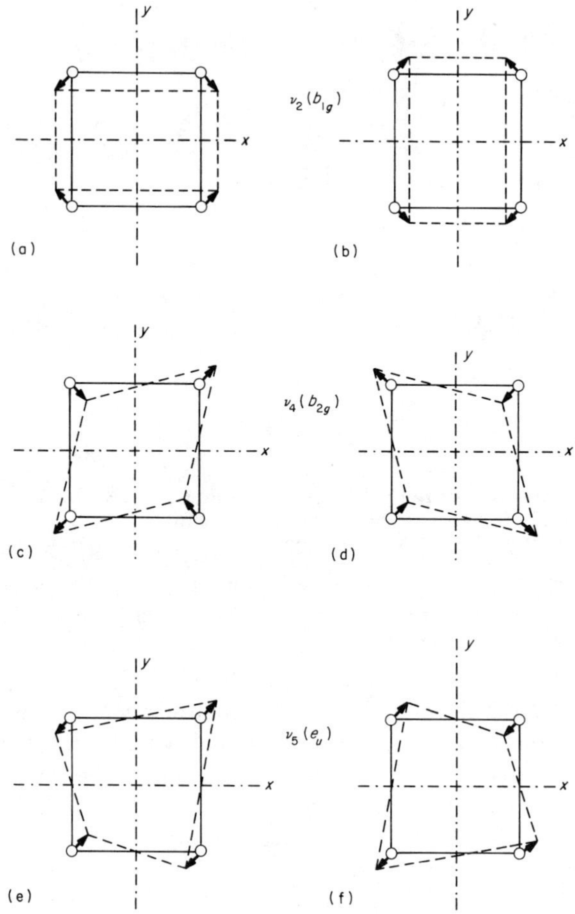


FIG. 12. **Non-totally symmetric normal modes of a planar square $X_4$ molecule.** The broken-line quadrangles represent the displaced conformations. For each vibration, $\nu_2$, $\nu_4$, $\nu_5$, two opposite phases are shown.

and therefore, as shown by Fig. 13a, the two energy curves $W_{2g}$ and $W_{3g}$ must cross each other. They do so at $Q_2 = 0$ where by assumption we have an electronic degeneracy $[W_{2g}(0) = W_{3g}(0)]$. Thus neither $W_{2g}$ nor $W_{3g}$ need have a minimum at $Q_2 = 0$ and therefore in general each of them will vary linearly with $Q_2$ near $Q_2 = 0$. A minimum, if it occurs at all, will occur at a non-zero $Q_2$ value in accordance with the Jahn–Teller theorem.

In exactly the same way we see that the $b_{2g}$ bending vibration $\nu_4$ of $X_4$, shown in Fig. 12c and d, gives the energy curves as a function of $Q_4$ in Fig. 13b which also cross, and for exactly the same reasons as those for $Q_2$. It should be noted that the splitting of the energy as a function of $Q_4$ is in general different from that as a function of $Q_2$. As a consequence, if we plot the potential energy as a function

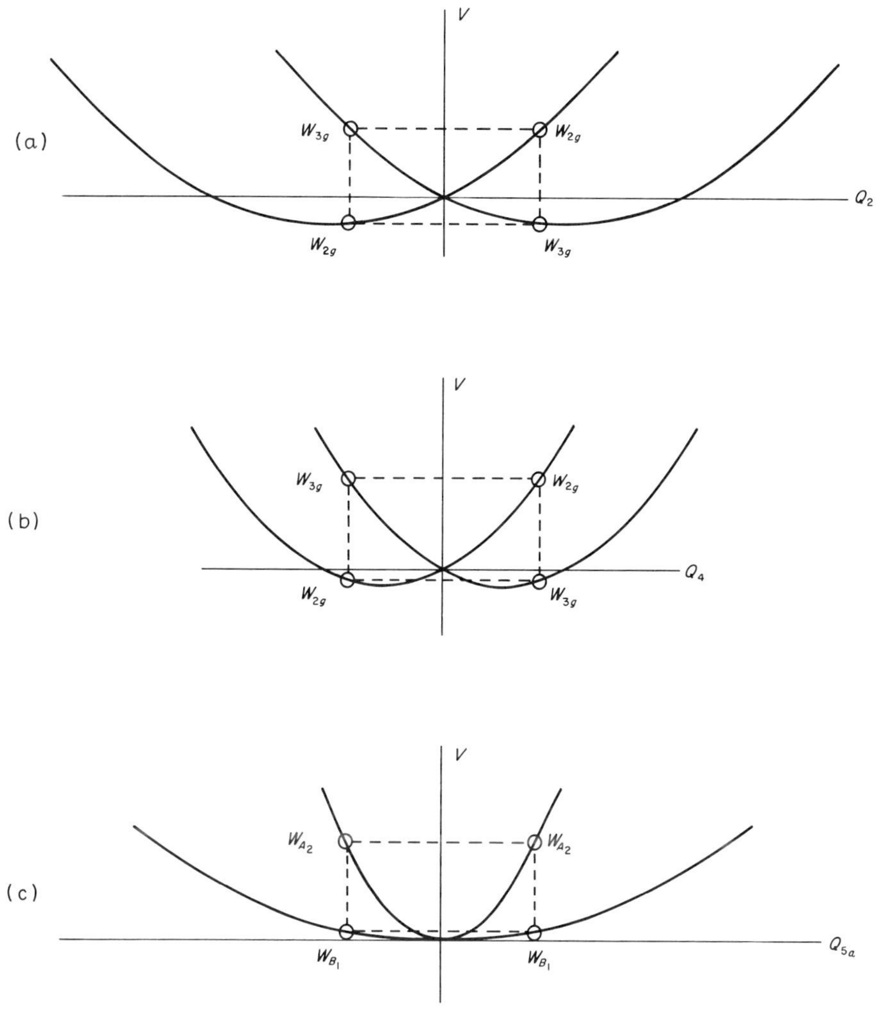


FIG. 13. **Variation of the potential energy in a degenerate electronic state of a planar square $X_4$ molecule as a function of** (*a*) **the normal coordinate $Q_2$ corresponding to $\nu_2(b_{1g})$,** (*b*) **the normal coordinate $Q_4$ corresponding to $\nu_4(b_{2g})$, and** (*c*) **the normal coordinate $Q_{5a}$ corresponding to one of the components of $\nu_5(e_u)$ (compare Fig. 12).**

of the two normal coordinates $Q_2$ and $Q_4$ (which for each X atom are at right angles to each other, see Fig. 12), we obtain a surface with two minima and two saddle points as illustrated by the contour diagram Fig. 14.

If we apply the same procedure to one component of the degenerate bending vibration $\nu_{5a}$ (see Fig. 12e and f), the displaced conformation has symmetry $\boldsymbol{C}_{2v}$ (with a diagonal of the square as the $C_2$ axis) and the degenerate electronic state splits into an $A_2$ and a $B_1$ state (see Table 59). Suppose that $W_{A_2}(Q_{5a}) > W_{B_1}(Q_{5a})$ for positive $Q_{5a}$. If we now reverse the displacements (Fig. 12f), since the two

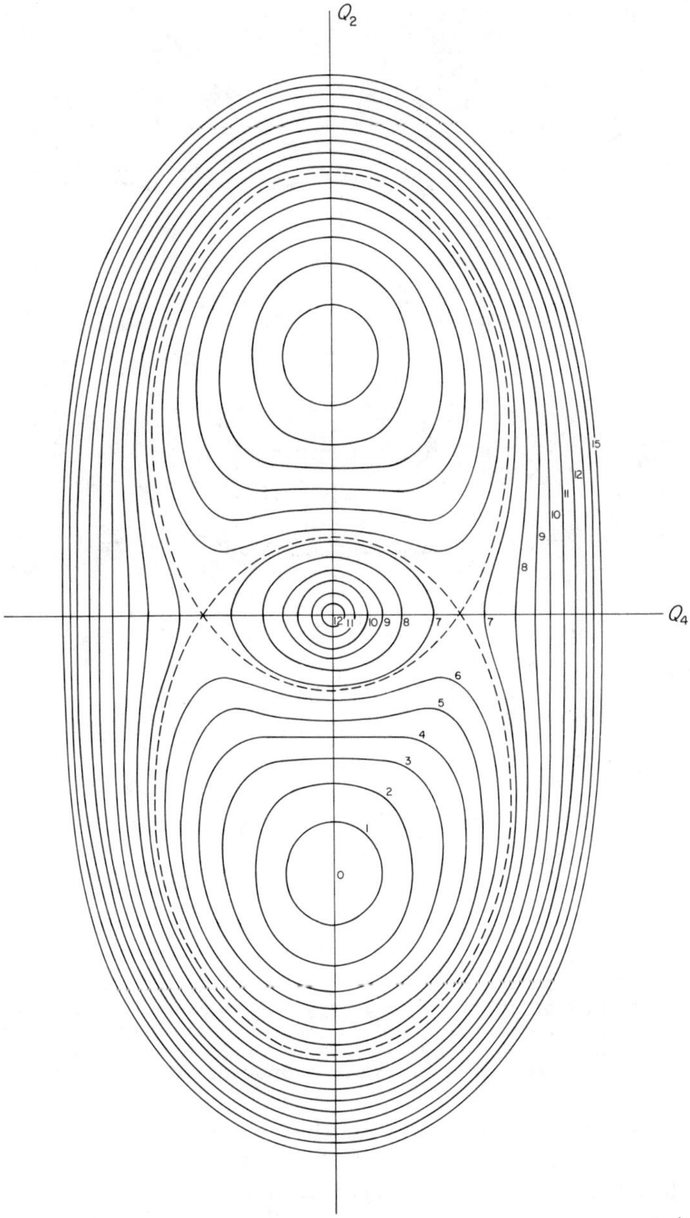


FIG. 14. **Contour diagram of the potential surface of a planar square $X_4$ molecule in a degenerate electronic state as a function of the two normal coordinates $Q_2$ and $Q_4$.** There is a conical peak in the center ($Q_2 = Q_4 = 0$) and two bowls above and below the abscissa (compare the numbers on the contour lines which give relative values of the potential energy).

diagrams differ only by a rotation by 180° which does not change the species of the two electronic component states, we see that here

$$W_{A_2}(-Q_{5a}) = W_{A_2}(Q_{5a}) \quad \text{and} \quad W_{B_1}(-Q_{5a}) = W_{B_1}(Q_{5a})$$

and therefore the two energy curves do not cross over; rather, they both have a minimum at $Q_{5a} = 0$ where they coincide: there is no Jahn–Teller instability. The same applies, of course, to the other component $\nu_{5b}$ of the degenerate vibration (see Vol. II, p. 92, Fig. 37), and therefore the potential surface in the space of the two mutually degenerate normal coordinates $Q_{5a}$ and $Q_{5b}$ splits into two coaxial paraboloids which coincide at $Q_5 = 0$. This is exactly the same behavior as in degenerate states of linear molecules: we have a Renner–Teller but not a Jahn–Teller splitting. [For more details see Hougen (577).]

For molecules with three-, five-, six-fold axes, the proof of the Jahn–Teller theorem cannot be obtained in such a simple qualitative way as just presented for molecules with four-fold axes. We consider here only the proof for an $X_3$ molecule of point group $\boldsymbol{D}_{3h}$ following Moffitt and Liehr (869) who discuss a more general case.

Let the two degenerate normal coordinates of the vibration $\nu_2$ be $Q_{2a}$ and $Q_{2b}$ as shown in Fig. 15, the first being symmetric, the second antisymmetric with respect to the $yz$ plane. Let $q$ represent the coordinates of the electrons and let $\psi_{ea}$ and $\psi_{eb}$ be the two electronic eigenfunctions belonging to a degenerate electronic state (species $E'$). They are eigenfunctions of the electronic Hamiltonian $H_e = T_e + V_e$ (see p. 8) which depends on the normal coordinates as parameters.

It is convenient to introduce the complex normal coordinates and electronic eigenfunctions

$$\begin{aligned} Q_2^+ &= Q_{2a} + iQ_{2b} = re^{+i\varphi} \\ Q_2^- &= Q_{2a} - iQ_{2b} = re^{-i\varphi} \\ \psi_e^+ &= \psi_{ea} + i\psi_{eb} \\ \psi_e^- &= \psi_{ea} - i\psi_{eb}. \end{aligned} \tag{I, 53}$$

The electronic energy of the system is determined by the secular equation

$$\begin{vmatrix} H_{++} - E & H_{+-} \\ H_{-+} & H_{--} - E \end{vmatrix} = 0. \tag{I, 54}$$

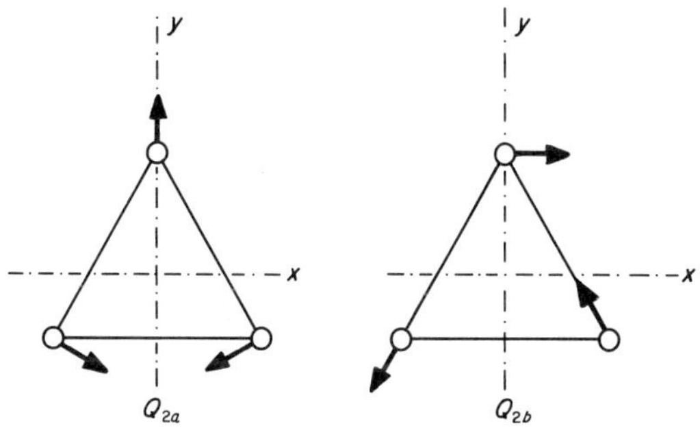


FIG. 15. **The two components of the degenerate mode of a symmetrical $X_3$ molecule.**

where the matrix elements are given by

$$\begin{aligned} H_{++} &= \int \psi_e^{+*} H_e \psi_e^{+} dq, \qquad H_{--} = \int \psi_e^{-*} H_e \psi_e^{-} dq \\ H_{+-} &= \int \psi_e^{+*} H_e \psi_e^{-} dq, \qquad H_{-+} = \int \psi_e^{-*} H_e \psi_e^{+} dq. \end{aligned} \tag{I, 55}$$

The Hamiltonian $H_e$ can be developed in a power series of the normal coordinates. We only consider the dependence on the degenerate normal coordinates $Q_2^+$ and $Q_2^-$ and write

$$H_e = H_0 + H_1^+ Q_2^- + H_1^- Q_2^+ + H_2^+ Q_2^{-2} + H_2^- Q_2^{+2} + \cdots. \tag{I, 56}$$

Substituting into the matrix elements (I, 55) we obtain

$$\begin{aligned} H_{++} = \int \psi_e^{+*} H_0 \psi_e^{+} dq &+ Q_2^- \int \psi_e^{+*} H_1^+ \psi_e^{+} dq + Q_2^+ \int \psi_e^{+*} H_1^- \psi_e^{+} dq \\ &+ Q_2^{-2} \int \psi_e^{+*} H_2^+ \psi_e^{+} dq + Q_2^{+2} \int \psi_e^{+*} H_2^- \psi_e^{+} dq + \cdots \end{aligned} \tag{I, 57}$$

$$\begin{aligned} H_{+-} = \int \psi_e^{+*} H_0 \psi_e^{-} dq &+ Q_2^- \int \psi_e^{+*} H_1^+ \psi_e^{-} dq + Q_2^+ \int \psi_e^{+*} H_1^- \psi_e^{-} dq \\ &+ Q_2^{-2} \int \psi_e^{+*} H_2^+ \psi_e^{-} dq + Q_2^{+2} \int \psi_e^{+*} H_2^- \psi_e^{-} dq \end{aligned} \tag{I, 58}$$

and similarly for $H_{--}$ and $H_{-+}$.

For a rotation by 120° about the axis of symmetry, $Q_2^+$, $\psi_e^+$ and $\psi_e^{-*}$ are multiplied by $\omega = e^{+2\pi i/3}$ while $Q_2^-$, $\psi_e^-$ and $\psi_e^{+*}$ are multiplied by $\bar{\omega} = e^{-2\pi i/3}$. Since the Hamiltonian must be totally symmetric, it follows that $H_1^+$ and $H_1^-$ for a rotation by 120° are multiplied by $\omega$ and $\bar{\omega}$, respectively, while $H_2^+$ and $H_2^-$ are multiplied by $\omega^2$ and $\bar{\omega}^2$, respectively. The integrals in (I, 57) and (I, 58) are different from zero only if the integrands are invariant under all symmetry operations permitted by the point group, in particular under the operation $C_3$ (rotation by 120°). If this is kept in mind, it is readily seen that in $H_{++}$, and similarly $H_{--}$, the linear terms in $Q_2$ vanish. On the other hand, in $H_{+-}$ and $H_{-+}$ the first term (independent of $Q_2$) vanishes while one of the linear terms, $Q_2^+$ for the $H_{+-}$ and $Q_2^-$ for $H_{-+}$, does *not* ordinarily vanish. Therefore, if we neglect the quadratic terms, we can write

$$H_{++} = W_0 = H_{--}, \quad H_{+-} = CQ_2^+, \quad H_{-+} = CQ_2^-.$$

Substituting into (I, 54) we obtain

$$\begin{vmatrix} W_0 - E & CQ_2^+ \\ CQ_2^- & W_0 - E \end{vmatrix} = 0 \tag{I, 59}$$

and therefore

$$E = W_0 \pm C\sqrt{Q_2^+ \cdot Q_2^-}$$

or, with (I, 53)

$$E = W_0 \pm Cr, \tag{I, 60}$$

where $r = \sqrt{Q_{2a}^2 + Q_{2b}^2}$ is the magnitude of the normal coordinate.

Equation (I, 60) shows that in a first approximation the electronic energy varies linearly with the displacement $r$, increasing for one component state, decreasing for the other. The potential minimum is not at $r = 0$. This is the statement of the Jahn–Teller theorem for an $X_3$ molecule of point group $\boldsymbol{D}_{3h}$. As long as quadratic and higher terms in the Hamiltonian are neglected, the two potential surfaces depend only on the magnitude $r$ of the degenerate normal coordinate, and thus we obtain in the $Q_{2a}$, $Q_{2b}$ space a surface of rotational symmetry described by rotating the diagram Fig. 11 about the $z$ axis. This is shown in Fig. 16 and, in the form of a contour diagram, in Fig. 17a. There is a circular trough of constant depth around the origin.

If the quadratic terms in the matrix elements $H_{+-}$ and $H_{-+}$ are not neglected, one finds that as a function of $\varphi$ there are three minima for each value of $r$. Such

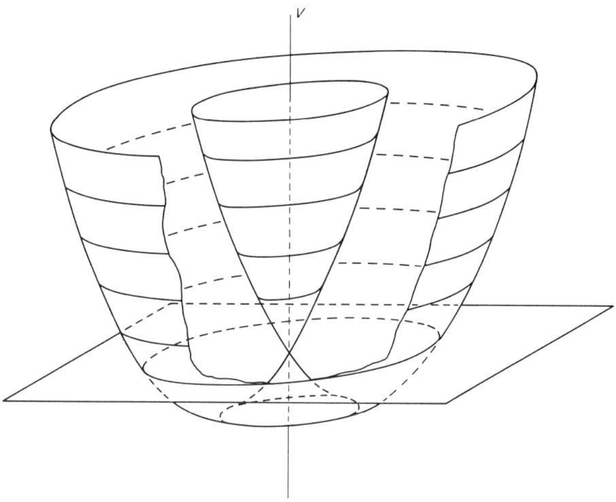


FIG. 16. **Potential energy surface of a non-linear molecule in a degenerate electronic state as a function of the two components of a degenerate normal coordinate (first approximation).** The horizontal plane shown gives the energy of the minimum of potential energy before vibronic interaction is introduced.

a surface is shown by contour lines in the $Q_{2a}$, $Q_{2b}$ plane in Fig. 17b. There are three bowls separated by saddle points, all of which are joined together in the center by a conical peak. The contour diagrams show only the lower of the two surfaces. To help in visualizing the upper surface, Fig. 18 shows a cross section through the $yz$ plane. The cross sections through the other two minima and the $z$ axis are, of course, identical with Fig. 18. A cross section through the $xz$ axis would give a symmetrical picture like Fig. 11, but the minima of the potential curves in this cross section do not correspond to the bottom of the two bowls.

The three potential minima of Fig. 17b correspond to the three conformations of the molecule shown in Fig. 19. Each nucleus has three potential minima for its motion, but, of course, the motions of the three nuclei are not independent of one another. A simpler case is the motion of the central X atom in a planar $XY_3$ molecule when the Y atoms are heavy (as, e.g., in $BCl_3$). The potential surface Fig. 17b applies then directly to the motion of X in real space (rather than in normal-coordinate space). Similarly, as already mentioned, in an $XYZ_3$ molecule the motion of the X atom is given by a potential surface essentially like Fig. 17b.

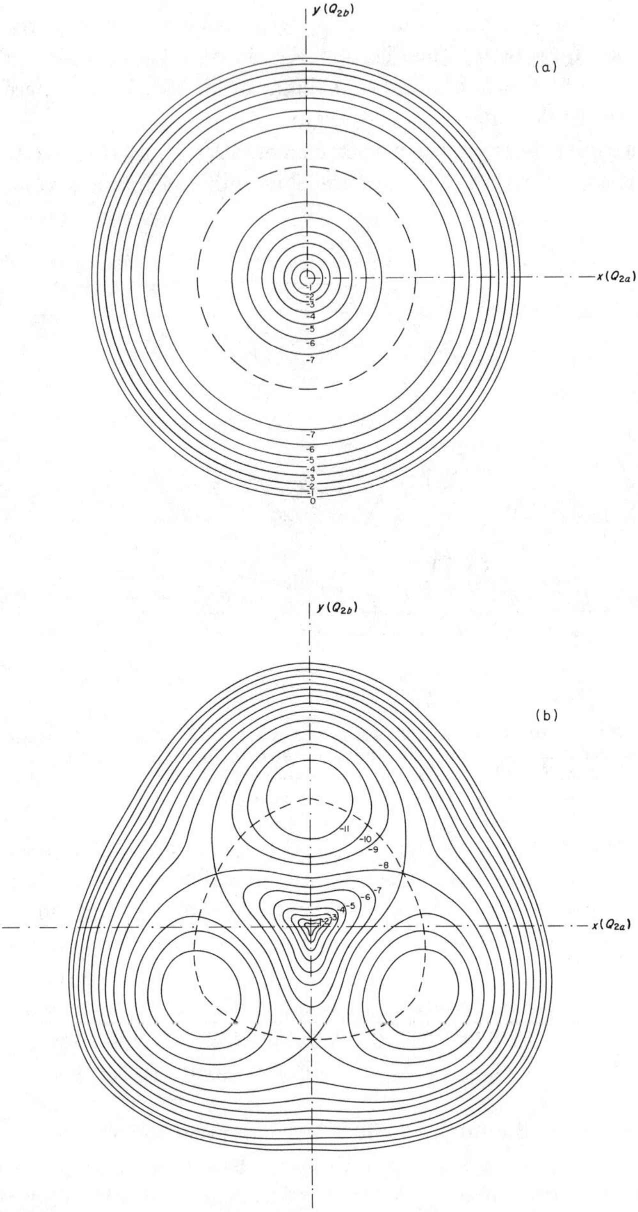


FIG. 17. **Contour diagrams of the lower part of the potential surface of a $C_{3v}$ (or $D_{3h}$) molecule in a degenerate electronic state** (*a*) **in a first approximation** (*b*) **in a higher approximation when quadratic and higher terms in the vibronic interaction are taken into account.** Both figures have a central conical peak. The trough (broken line) going around this peak has uniform depth in (*a*) but has three minima in (*b*). Only the part of the surface below the horizontal plane indicated in Fig. 16 is shown, that is, the part below the minimum for zero vibronic interaction. Therefore the relative $V$ values given are all negative.

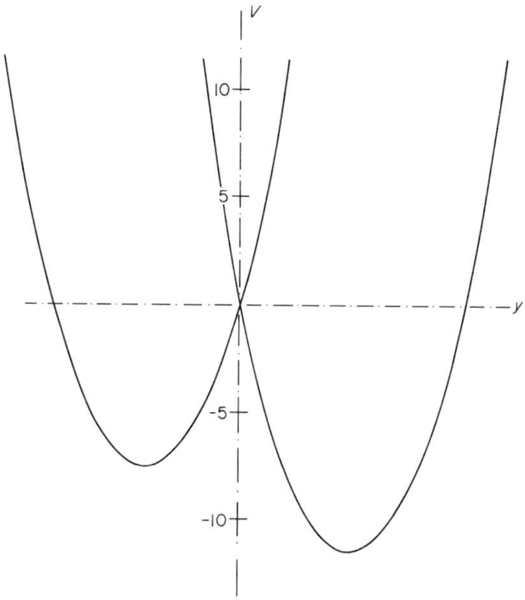


FIG. 18. **Cross section of the potential function Fig. 17b in the $yz$ plane.** The scale given on the $V$ axis corresponds to that indicated by the numbers on the contour lines in Fig. 17b.

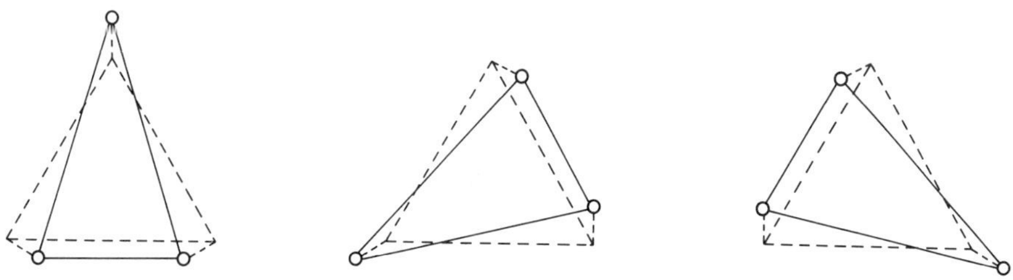

FIG. 19. **Three equilibrium conformations of a symmetrical $X_3$ molecule in a degenerate electronic state when Jahn–Teller interaction is strong as in Fig. 17b.** The broken-line triangle in each diagram gives the original equilibrium conformation before vibronic interaction is introduced.

The treatment of the Jahn–Teller theorem for other more complicated cases is similar [see Jahn and Teller (618), Van Vleck (1237), Moffitt and Liehr (869), Moffitt and Thorson (870) and Liehr (747)]. A linear term in the matrix elements $H_{+-}$ and $H_{-+}$ and therefore an instability of the symmetrical configuration will always arise when integrals of the type $\int \psi_e^{+*} H_1^+ \psi_e^- dq$ do not vanish, and they will not vanish when the product $\psi_e^{+*}\psi_e^-$ has the same species as a non-totally symmetric vibration (since $H_1^+$ has the opposite symmetry, see above). If $G$ is the (degenerate) species of $\psi_e$, then the species of $\psi_e^{+*}\psi_e^-$ is that of $(G)^2$, the symmetrical product of $G$ with itself. These products are included in Table 57, Appendix III. The symmetrical product always includes the totally symmetric species, but a vibration of this species can, of course, not lead to a splitting of the degeneracy [see also Table 32 (p. 127) of Vol. II].

TABLE 2. SPECIES OF NORMAL COORDINATES THAT WILL PRODUCE JAHN–TELLER INSTABILITY IN DEGENERATE ELECTRONIC STATES OF NON-LINEAR MOLECULES

| Point group | Degenerate electronic state | Species of normal coordinates causing instability |
|---|---|---|
| $\mathbf{D}_{3h}$,[$\mathbf{C}_{3v}$, $\mathbf{D}_3$, $\mathbf{C}_{3h}$, $\mathbf{C}_3$][a] | $E'$<br>$E''$ | $e'$<br>$e'$ |
| $\mathbf{D}_{4h}$,[$\mathbf{C}_{4v}$, $\mathbf{D}_4$, $\mathbf{D}_{2d}$, $\mathbf{C}_{4h}$, $\mathbf{C}_4$, $\mathbf{S}_4$][b] | $E_g$<br>$E_u$ | $b_{1g}, b_{2g}$<br>$b_{1g}, b_{2g}$ |
| $\mathbf{D}_{5h}$,[$\mathbf{C}_{5v}$, $\mathbf{D}_5$, $\mathbf{C}_{5h}$, $\mathbf{C}_5$][c] | $E_1'$<br>$E_1''$<br>$E_2'$<br>$E_2''$ | $e_2'$<br>$e_2'$<br>$e_1'$<br>$e_1'$ |
| $\mathbf{C}_{6v}$,[$\mathbf{C}_6$] | $E_1$<br>$E_2$ | $e_2$<br>$e_2$ |
| $\mathbf{D}_{6h}$,$\mathbf{C}_{6h}$,[$\mathbf{D}_6$, $\mathbf{D}_{3d}$, $\mathbf{S}_6$][d] | $E_{1g}$<br>$E_{1u}$<br>$E_{2g}$<br>$E_{2u}$ | $e_{2g}$<br>$e_{2g}$<br>$e_{2g}$<br>$e_{2g}$ |
| $\mathbf{D}_{4d}$, $\mathbf{C}_{8v}$, $\mathbf{D}_8$ | $E_1$<br>$E_2$<br>$E_3$ | $e_2$<br>$b_1, b_2$<br>$e_2$ |
| $\mathbf{T}_d$, $\mathbf{O}$,[$\mathbf{T}$][e] | $E$<br>$F_1$<br>$F_2$ | $e$<br>$e, f_2$<br>$e, f_2$ |
| $\mathbf{O}_h$ | $E_g$<br>$E_u$<br>$F_{1g}$<br>$F_{1u}$<br>$F_{2g}$<br>$F_{2u}$ | $e_g$<br>$e_g$<br>$e_g, f_{2g}$<br>$e_g, f_{2g}$<br>$e_g, f_{2g}$<br>$e_g, f_{2g}$ |
| $\mathbf{I}_h$ | $F_{1g}$<br>$F_{1u}$<br>$G_g$<br>$G_u$<br>$H_g$<br>$H_u$ | $h_g$<br>$h_g$<br>$g_g, h_g$<br>$g_g, h_g$<br>$g_g, h_g$<br>$g_g, h_g$ |

[a] For $\mathbf{C}_{3v}$, $\mathbf{D}_3$, and $\mathbf{C}_3$ the ′ and ″ should be omitted.

[b] For $\mathbf{C}_{4v}$, $\mathbf{D}_4$, $\mathbf{D}_{2d}$ ($\equiv \mathbf{V}_d$) the subscripts $g$ and $u$, for $\mathbf{C}_{4h}$ the subscripts 1 and 2, and for $\mathbf{C}_4$ and $\mathbf{S}_4$ all subscripts should be omitted.

[c] For $\mathbf{C}_{5v}$, $\mathbf{D}_5$, and $\mathbf{C}_5$ the ′ and ″ should be omitted.

[d] For $\mathbf{D}_6$ the subscripts $g$ and $u$ should be omitted. For $\mathbf{D}_{3d}$ the subscripts 1 and 2 of $E$ and $e$ should be omitted.

[e] For $\mathbf{T}$ the subscripts 1 and 2 should be omitted.

For the convenience of the reader, in Table 2 the species of all vibrations that can cause Jahn–Teller instability are given for the degenerate electronic states of all important point groups.

For a molecule with a five-fold axis of symmetry in a degenerate electronic state the potential surface as a function of the two components of an appropriate normal coordinate (see Table 2) would have five minima symmetrically placed around the original equilibrium position, entirely similar to Fig. 17b for a three-fold axis; for a seven-fold axis, there would be seven minima. However, for molecules with a four-fold axis, as we have seen, only two potential minima appear, and similarly for six- and eight-fold axes only three and four minima, respectively. The difference between odd and even axes is due to the fact that for the latter the vibrations that cause instability are symmetric with respect to rotation by 180° about the axis.

In all cases (both odd and even axes), the potential energy as a function of the coordinates of all nuclei has the full symmetry of the "original" point group no matter how large the interaction of vibration and electronic motion, no matter how much the minima resulting from Jahn–Teller instability deviate from the axis of symmetry. Figure 19 illustrates this for an $X_3$ molecule since it gives all the (Jahn–Teller distorted) equilibrium positions of all the nuclei: the full potential function obtained by superimposing the three diagrams of Fig. 19 has $\boldsymbol{D}_{3h}$ symmetry. The corresponding diagram for an $X_4$ molecule (Fig. 12) has only two equilibrium positions for each nucleus, but the potential function is invariant under all $\boldsymbol{D}_{4h}$ symmetry operations as is readily seen by superimposing diagrams similar to those in Fig. 19. If a central atom is present as in an $XY_4$ molecule this atom (X) does not take part in $b_{1g}$ or $b_{2g}$ vibrations, and therefore, Jahn–Teller instability does not affect it: it has only one equilibrium position on the axis. The symmetry of the potential function is again $\boldsymbol{D}_{4h}$ no matter how large the effect of Jahn–Teller instability is on the Y atoms.

The degenerate states of molecules which cannot be described by the standard point groups (non-rigid molecules, see p. 5f) are also subject to Jahn–Teller interaction but no detailed discussion of such cases has yet been given. Only the case of molecules with inversion symmetry is easily treated: for example for $NH_3$ the symmetry group is effectively $\boldsymbol{D}_{3h}$, and in a degenerate electronic state the H atoms have now six instead of two equivalent equilibrium positions.

The magnitude of the Jahn–Teller interaction in a given degenerate electronic state is very difficult to predict. It depends on the way in which the various filled molecular orbitals (see Chap. III) affect the vibrational motion. Coulson and Strauss (243) have attempted a prediction for a few simple cases ($CH_4^+$, $CF_4^+$, $NH_3^+$ and $NH_3$), but the experimental data are as yet quite insufficient for a satisfactory comparison.

An interesting discussion of the Jahn–Teller theorem in terms of forces rather than potential functions, using the Hellman–Feynman theorem, has been given by Clinton and Rice (210).

**Effect of electron spin on Jahn–Teller instability.** The preceding considerations apply to orbital degeneracy. Jahn (616) and Mulliken and Teller (917) have considered the effect of electron spin. The general result is that spin-orbit interaction tends to reduce the effects of Jahn–Teller instability.

For molecules with an even number of electrons, the spin function and thus the spin-orbit function has only one-valued representations just as the orbital function, and therefore for these molecules the general theorem is unchanged: any state with a degenerate spin-orbit function is unstable in the symmetrical conformation since there is always a non-totally symmetric normal coordinate on which the potential energy depends linearly (see Table 2). For example, in a $^3E$ state of a $C_{3v}$ molecule, which is orbitally unstable, the spin causes a splitting into the states $A_1 + A_2 + E + E$ (see p. 18) of which only the last two will be unstable;

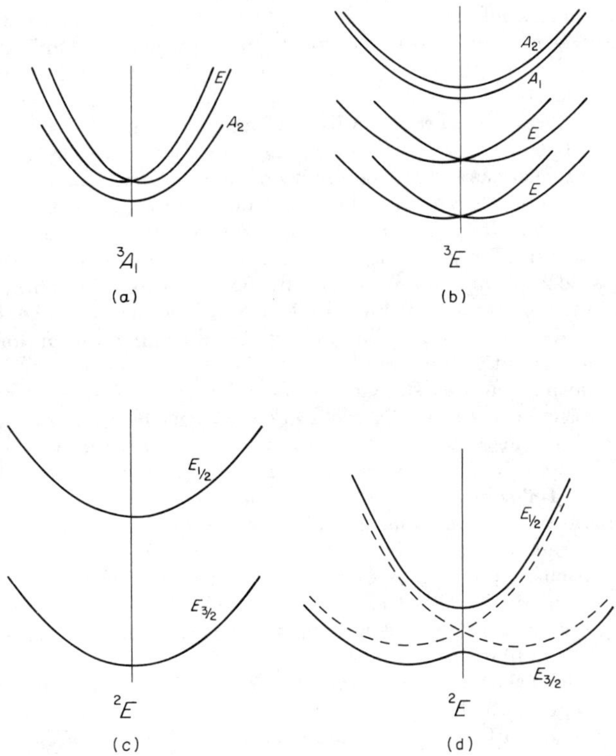


FIG. 20. **Cross sections through potential surfaces in triplet and doublet states when Jahn–Teller interaction is present.** (*a*) **$^3A_1$, small spin-orbit interaction;** (*b*) **$^3E$, large spin-orbit interaction;** (*c*) **$^2E$, large spin-orbit interaction;** (*d*) **$^2E$, small spin-orbit interaction.** The broken line curves in (*d*) give the potential function for zero spin-orbit interaction. All diagrams are qualitative.

the first two component states have a symmetrical equilibrium conformation. On the other hand, for a $^3A_1$ state, which is orbitally stable, we have a splitting into $A_2 + E$, of which only the first is stable, the second unstable. However, the $E$ component is only very slightly unstable (i.e., the equilibrium position is very nearly the symmetric one) because in an orbitally non-degenerate state spin-orbit coupling is always very small. A one-dimensional plot of the potential function for the two examples is shown in Fig. 20a and b. The splitting for $^3A_1$ is exaggerated by a large factor.

The situation is different for an odd number of electrons. The electron spin is not directly affected by an electric field, and therefore the coupling of vibration and electronic motion cannot remove the degeneracy caused by a single spin, the so-called Kramers degeneracy (see p. 17). As we have seen previously, for the description of spin functions corresponding to half-integral $S$ values the extended point groups (double groups) have to be used. Jahn (616) has shown that it is now the antisymmetric (rather than the symmetric) product of the species of the spin-orbit function with itself which must have the same species as one of the non-totally symmetric normal vibrations in order to make Jahn–Teller instability possible.

As can be seen from Table 57 (Appendix III), for all axial point groups the antisymmetric product of any doubly degenerate two-valued representation with itself is totally symmetric; that is, $E_{\frac{1}{2}}$, $E_{\frac{3}{2}}$, . . . states cannot be split by vibronic interaction in accordance with Kramers' theorem that a two-fold spin degeneracy cannot be split by any non-magnetic interactions. Thus, for all axial point groups when spin-orbit interaction is large, there is no Jahn–Teller instability. For example, a $^2A_1$ state of a $C_{3v}$ molecule assumes species $E_{\frac{1}{2}}$ when spin-orbit coupling is large, but vibronic coupling cannot remove the degeneracy. There is no Jahn–Teller instability. Only a magnetic field such as that connected with rotation can remove the degeneracy.

In a $^2E$ state the orbital part of the degeneracy will lead to Jahn–Teller instability if spin-orbit interaction is small. For large spin-orbit coupling $^2E$ will split into two states of species $E_{\frac{1}{2}}$ and $E_{\frac{3}{2}}$ (see Fig. 20c). The introduction of vibronic coupling produces no further splitting; each of the two doublet components remains doubly degenerate for any displacement of the nuclei. The potential minima are in the symmetrical conformation. For intermediate strength of spin-orbit coupling, when the effects of vibronic instability of the orbital state and of spin splitting are of similar magnitude, we obtain again two potential functions for $^2E$, one with a single minimum in the symmetrical conformation, the other with two minima off axis as shown schematically in Fig. 20d. For comparison the potential functions that would arise for zero spin-orbit interaction are shown as broken lines. If spin-orbit interaction is included (full-line curves), there are for small displacements two states $E_{\frac{1}{2}}$ and $E_{\frac{3}{2}}$, each of which is doubly degenerate but does not split and which for large displacements go over into the two Jahn–Teller states. Thus, one state with a minimum and one with a maximum on the symmetry axis is formed. In other words, spin-orbit interaction for half-integral spin removes near the axis the point of intersection of the potential functions, that is, reduces the instability produced by orbital degeneracy.

For cubic and icosahedral point groups, there are for an odd number of electrons two-valued representations with a dimension higher than two, and these electronic degeneracies can be split by vibronic interaction. For example, for a tetrahedral or octahedral molecule there are, for half-integral spin, four-fold degenerate electronic states of species $G_{\frac{3}{2}}$ (or $G_{\frac{3}{2}g}$, $G_{\frac{3}{2}u}$). The antisymmetric product of this species with itself (see Table 57) contains the non-totally symmetric species $E$ and $F_2$ (or $E_g$ and $F_{2g}$ for $\boldsymbol{O}_h$). Vibrations of these species will produce Jahn–Teller instability in $G_{\frac{3}{2}}$ states, i.e., a splitting of the potential function into two, just as in Fig. 11. But each component function still corresponds to a spin-doublet (Kramers' doublet) which cannot be split by vibronic interaction.

**Vibronic energy levels.** The discussion up to now has dealt with the effect of the vibronic interactions on the potential functions in degenerate electronic states of non-linear molecules. This effect is sometimes referred to as *static Jahn–Teller effect.* We must now consider the energy levels that arise when Jahn–Teller instability is present, that is, consider the *dynamic Jahn–Teller effect.* For this purpose we must solve the wave equation when potential functions of the type discussed above (Figs. 14, 16, 17) are substituted into it. This was done by Moffitt and Liehr (869), Moffitt and Thorson (870)(871), and Longuet–Higgins, Öpik, Pryce and Sack (769) [see also the summaries by Longuet–Higgins (766) and Child and Longuet–Higgins (194)]. We shall only discuss the results.

The general result is (as was to be expected) that each vibrational level of given $v$ splits into as many different vibronic levels as there are species resulting from the multiplication of electronic and vibrational species, for example for $\boldsymbol{D}_{3h}$ and $\boldsymbol{D}_{6h}$ into as many levels as indicated in Fig. 10a and b. This splitting may be called *the Jahn–Teller splitting.* Here it may be remembered that even in a non-degenerate electronic state the higher levels of a degenerate vibration are split by anharmonicity (see Fig. 10). In a degenerate electronic state the potential function, as we have seen, is clearly not harmonic, and it is therefore not surprising that further splittings occur. It must, however, be emphasized that all genuine vibronic degeneracies remain, i.e., $E$ vibronic levels are not split, no matter how strong the vibronic interaction. This is because the permutation-inversion symmetry is unchanged and therefore the potential function always has the full "original" symmetry even if the symmetrical conformation no longer corresponds to a potential minimum (see p. 51).

The calculations so far made of the vibronic energy levels in doubly degenerate electronic states have all introduced a simplification of the potential function by assuming that the quadratic terms in (I, 56) are negligible and therefore that the potential function has rotational symmetry as in Fig. 16, i.e., has no separate minima at the bottom of the "moat". For this reason the pairs of levels $A_1$, $A_2$ are not split but behave like other degenerate vibronic levels. Even with this neglect, the solution of the wave equation is quite cumbersome, and explicit formulae for the energy levels can be given only for the limiting cases of very small and very large Jahn–Teller coupling. For very small coupling the authors mentioned [see also Child (193)] obtained the following formula for the vibronic levels in $X_3$ molecules

$$G(v_2, l) = \omega_2(v_2 + 1) \mp 2D\omega_2(l \pm 1). \tag{I, 61}$$

Here $D$ is a coupling parameter defined in such a way that $D\omega_2$ is the depth of the bottom of the "moat" below the "original" equilibrium position. For each value of $l$ ($=v_2, v_2-2, \ldots, 1$ or 0), there are two energy values corresponding to the upper and lower signs, except for $l = 0$, for which only the upper sign holds. According to Child (193) eq. (I, 61) holds for $D < 0.05$. Moffitt and Thorson (870) and Child (190)(193) have given corresponding formulae for $\boldsymbol{D}_{4h}$, $\boldsymbol{T}_d$ and $\boldsymbol{O}_h$ molecules in degenerate states.

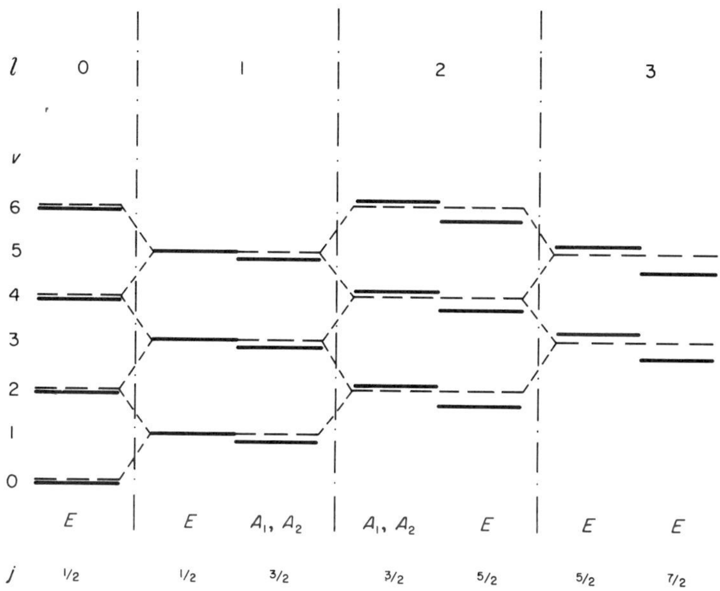


Fig. 21. **Energy levels of a degenerate bending vibration in a degenerate electronic state of a $C_{3v}$ (or $D_{3h}$) molecule for very small vibronic (Jahn–Teller) interaction, compared with the corresponding levels for zero vibronic interaction.** The heavy horizontal lines represent the actual energy levels, the broken horizontal lines the levels without vibronic interaction. Levels of the same $j$ but different $l$ are connected by sloping broken lines which are intended to indicate that these levels interact with one another when vibronic interaction is introduced. The $A_1$ and $A_2$ pairs are not split in this approximation.

In Fig. 21 the energy levels given by (I, 61) (indicated by heavy lines) are compared with the levels for zero vibronic interaction (assuming a strictly harmonic vibration). The two energy levels, into which a state of given $v_2$ and $l \neq 0$ is split, are the same two (degenerate) vibronic levels which were previously discussed in connection with Fig. 10. Here the further splitting of some levels into $A_1$ and $A_2$ is neglected. In this approximation eq. (I, 61) gives for a given $v$ a splitting into $v+1$ equidistant energy levels of spacing $4D\omega_2$.

When the coupling constant $D$ is not very small compared to 1, the energy levels can no longer be represented by an explicit formula. Rather elaborate numerical calculations are necessary. Longuet–Higgins, Öpik, Pryce and Sack (769) and Moffitt and Thorson (871) have given tables of these energy values for a number of $D$ values[5]. In Fig. 22 the energy levels are plotted for three $D$ values,

[5] Longuet–Higgins *et al.* use $k^2 = 2D$ as parameter.

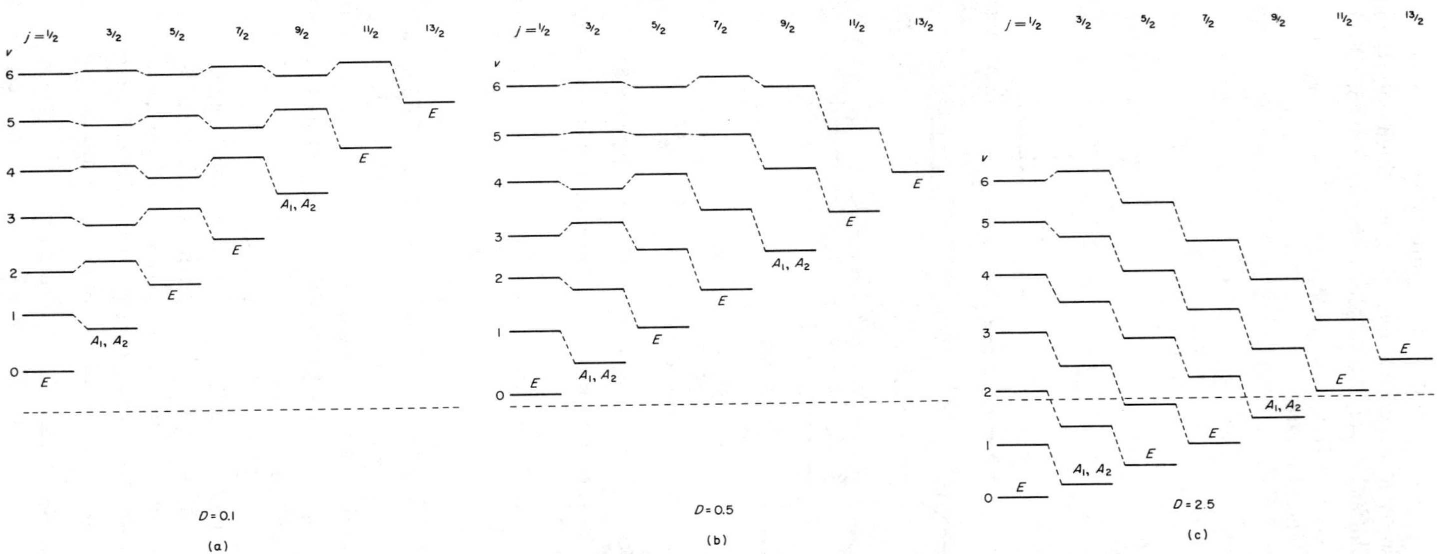


FIG. 22. **Energy levels of a degenerate bending vibration in a degenerate electronic state of a $C_{3v}$ (or $D_{3h}$) molecule for three values of the Jahn–Teller parameter** (*a*) $\boldsymbol{D = 0.1}$, (*b*) $\boldsymbol{D = 0.5}$, (*c*) $\boldsymbol{D = 2.5}$. Levels belonging to the same $v$ value are connected by broken lines. $l$ is no longer well defined and is omitted here. The horizontal broken line gives the energy of the minimum of the potential function for zero vibronic interaction. The vibronic species is the same for all levels of a given column and is indicated only for the lowest one.

0.1, 0.5 and 2.5. For $D = 0.1$, as can be seen, the levels can still be represented at least approximately by the formula (I, 61); for a given $v$ they are nearly equidistant. For higher $D$ values eq. (I, 61) is not even approximately valid. For these cases the quantum number $l$ of the vibration, because of the coupling of vibration and electronic motion, is no longer a good quantum number. Instead, a new quantum number $j$ has been introduced[6] by Longuet–Higgins *et al.* which takes the values $\frac{1}{2}, \frac{3}{2}, \ldots, v + \frac{1}{2}$ (or $l \pm \frac{1}{2}$) and which distinguishes the different levels of a given $v$ (but with the proviso that $A_1$ and $A_2$ still coincide in this approximation). For point group $\boldsymbol{D}_{3h}$ or $\boldsymbol{C}_{3v}$, the levels with $j = \frac{1}{2}, \frac{5}{2}, \frac{7}{2}, \frac{11}{2}, \frac{13}{2}, \ldots$, are $E$, those with $j = \frac{3}{2}, \frac{9}{2}, \frac{15}{2}, \ldots$ are $A_1, A_2$. The relation between the $j$ and $l$ values is indicated in Fig. 21.

In Fig. 23 the lowest vibronic levels are shown in their relation to the potential functions for the same $D$ values as in Fig. 22. For $D = 0.1$ and 0.5 all energy levels lie above the central hump, for $D = 2.5$ eight levels lie below it.

For large Jahn–Teller coupling, that is, for a very deep moat, the three (or more) potential minima similar to Fig. 17b would be expected to be very pronounced. But, neglecting them, Longuet–Higgins, Öpik, Pryce and Sack (769) have obtained a very simple formula for the energy levels which may be written

$$G(u, j) = \omega(u + \tfrac{1}{2}) + A^{\dagger}j^2. \qquad \text{(I, 62)}$$

Here the energy is referred to the bottom of the trough as zero; $\omega$ is the frequency and $u$ the quantum number of radial vibrations while

$$A^{\dagger} = \frac{h}{8\pi^2 c\mu r^2} \qquad \text{(I, 63)}$$

where $\mu$ is the reduced mass and $r$ the distance of the minimum from the symmetry axis of the potential surface. In other words, the vibronic energy is the sum of the vibrational energy of an oscillator vibrating radially toward and away from the symmetrical configuration and of a (two-dimensional) rotator moving in the potential trough around the symmetry axis. The latter motion corresponds, of course, in the case of an $X_3$ molecule to the circular motion of each X nucleus (see Fig. 19). The angular momentum of this motion for the assumed large Jahn–Teller coupling is $j(h/2\pi)$ (see below).

No detailed calculations of the effect of the three potential minima (Fig. 17b) on the energy levels have yet been made. If the minima are very deep, we have in effect a non-symmetrical molecule, and regular vibrational levels are again obtained. The molecule can be treated without considering the Jahn–Teller theorem. But for an intermediate case, there will be, in addition to a splitting of the $A_1$, $A_2$ degeneracy, further apparently irregular shifts of the vibronic $E$ levels compared to those in Fig. 22.

For linear molecules, as we have seen, the vibronic levels can be assigned, at least approximately, to one or the other branch of the potential function. This is not possible for non-linear molecules because the two potential functions are much

[6] Moffitt and Thorson (871) have instead introduced $\overline{\Lambda} = 2j$.

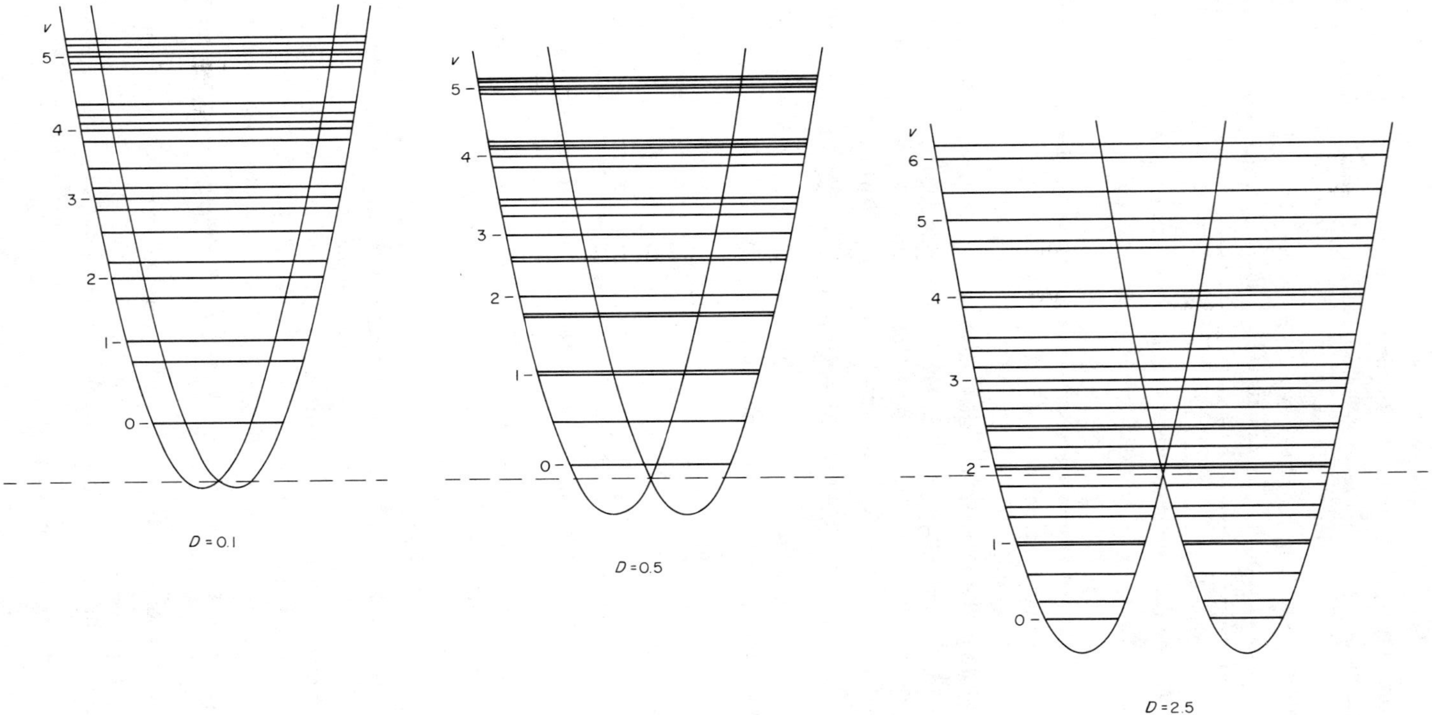


FIG. 23. **Energy levels as in Fig. 22 in their relation to the Jahn–Teller distorted potential functions.** Each potential diagram in the present approximation has rotational symmetry about a vertical axis through the point of intersection of the two curves. As a guide for comparison with Fig. 22 the levels with $j = \frac{1}{2}$ are marked.

more closely connected: classically, starting from a point on the upper (inner) cone (see Fig. 16) the image point must necessarily cross the vertex to the lower cone, i.e., to the other surface. This interconnectedness is the intrinsic reason for the greater complication of the vibronic levels.

It should be noted that Renner–Teller type interactions (produced by even-powered terms in the potential function) can also occur in non-linear molecules. They represent the effects of the potential minima referred to above as well as the effects of degenerate normal coordinates that do not cause Jahn–Teller instability [see, e.g., Hougen (577)]. On the other hand, Jahn–Teller interactions (involving odd-powered terms in the potential function) cannot occur in linear molecules.

**Correlation of vibronic levels.** It is instructive to consider how the vibronic levels for small Jahn–Teller interaction are correlated with those for very large interaction when the off-axis potential minima are fairly deep. It is also instructive to compare this correlation with the corresponding one for a non-degenerate electronic state. This comparison is made in Fig. 24a and b for a degenerate bending vibration, say $\nu_b$, of a $C_{3v}$ molecule in a non-degenerate and a degenerate electronic state respectively. At the extreme left of each diagram the vibronic levels are shown for zero or small Jahn–Teller interaction; at the extreme right they are shown for a corresponding molecule of symmetry $C_s$. We assume from left to right a gradual deformation of the potential function from one of $C_{3v}$ to one of $C_s$ symmetry brought about by a hypothetical deformation of the electronic orbitals (Fig. 24a) or by increasing vibronic interaction (Fig. 24b). If the potential minimum for the $C_s$ conformation is sufficiently deep, the vibrational levels result from the excitation of the two bending vibrations, $\nu_{b_1}$ and $\nu_{b_2}$, which correspond to the single degenerate bending vibration of the symmetric conformation. The positions of these levels are given in a first approximation (neglecting anharmonicity) by

$$T = T_0 + \omega_{b_1}(v_{b_1} + \tfrac{1}{2}) + \omega_{b_2}(v_{b_2} + \tfrac{1}{2}).$$

The levels have the species $A'$ or $A''$ depending on whether the antisymmetric one of the two vibrations, say $\nu_{b_2}$, is excited by an even or odd number of quanta. Since there are necessarily three identical minima each level is really triply degenerate. This degeneracy, which is usually disregarded for rigid molecules, cannot be neglected when the correlation to the symmetrical case is to be considered.

The levels at the left in Fig. 24a are given by (see eq. I, 43)

$$T = T_s + \omega_b(v_b + 1) + g_{bb}l_b^2$$

while those at the left in Fig. 24b are given by eq. (I, 61). If the coupling conditions are gradually changed so that the three potential minima are less and less deep each of the triply degenerate levels at the right will split into a doubly degenerate and a non-degenerate level. The latter will be $A_1$ or $A_2$ depending on whether without splitting the level is $A'$ or $A''$. The reason for this particular splitting will become clear in Chapter III; it is similar to the splitting of torsional

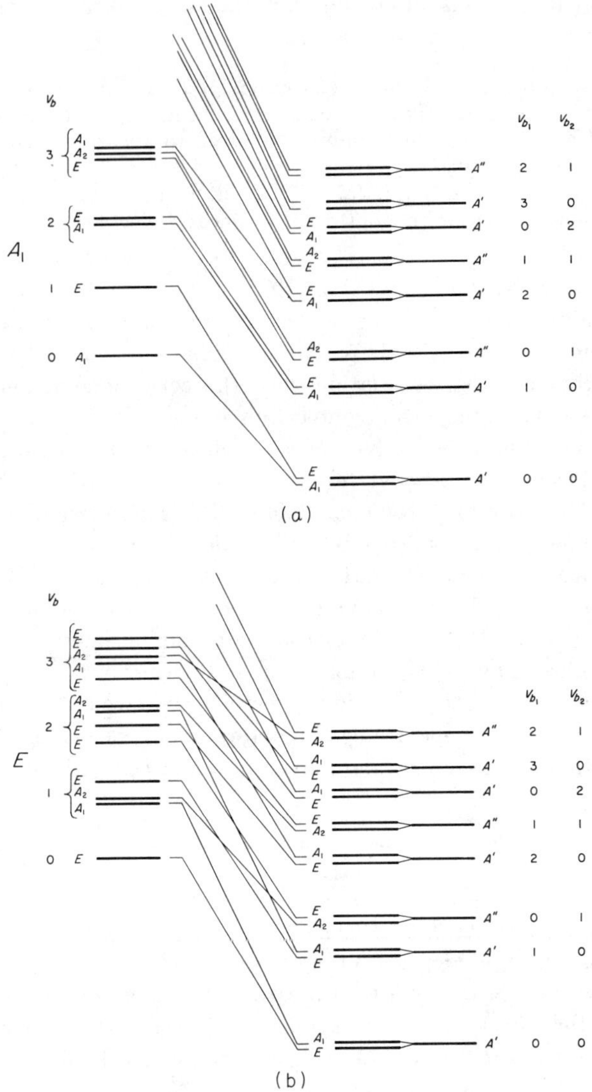


FIG. 24. **Correlation of the vibronic levels of a degenerate bending vibration of a $C_{3v}$ molecule with those of the corresponding two non-degenerate bending vibrations of a deformed molecule of symmetry $C_s$.** (*a*) **For a non-degenerate electronic state of the $C_{3v}$ molecule;** (*b*) **for a degenerate electronic state of the $C_{3v}$ molecule: transition from small to very large Jahn–Teller interaction.** At the extreme right the levels are designated according to $C_s$ symmetry, assuming only one potential minimum. The splitting produced by the existence of two other identical minima is shown in the second column of levels from the right.

levels in an ethane-type molecule (see Vol. II, pp. 225 and 495). The correlation from left to right in Fig. 24 has now simply to be made so that levels of the same species are connected without intersections. In consequence, a doubly degenerate level remains doubly degenerate for any intermediate coupling condition and similarly a non-degenerate level remains non-degenerate.

The degeneracies at left are normally said to be vibrational (Fig. 24a) or vibronic (Fig. 24b) in character. In both cases they are due to the presence of a three-fold axis of symmetry of a "rigid" molecule. At the right in both Figs. 24a and b the degeneracies are normally said to be due to the presence of three identical potential minima and the resulting permutation symmetry. Since the point group $C_{3v}$ used at left is a sub-group of the permutation-inversion group that must be used at the right when all three potential minima are considered, there is no splitting of the degeneracy of any vibronic level in going from a symmetrical equilibrium conformation to an unsymmetrical one of the same molecule (and the same applies to molecules of other point groups with degenerate species). On the contrary, as shown by Fig. 24 it is the levels of the unsymmetrical molecule that split as one changes to a more symmetrical arrangement (such a change is of course possible only if there is permutation symmetry in the unsymmetrical conformation). For example in both Figs. 24a and b the lowest level ($v_{b_1} = 0$, $v_{b_2} = 0$) at right splits and goes over into the $v_b = 0$ level and one of the $v_b = 1$ levels at left.

A splitting of the *electronic* degeneracy does occur when the electronic motion is considered in the field of the *fixed nuclei* as indicated by the splitting of the potential function (static Jahn–Teller effect, see Fig. 16). But when vibration is included and vibronic levels are considered no degeneracies are split upon lowering the symmetry of the equilibrium conformation [see also Watson (1278b)].

It must be emphasized that the actual positions of the energy levels in an intermediate case of Jahn–Teller interaction cannot be obtained by simple interpolation between the extreme cases as assumed in Fig. 24b. The actual energy of each of the vibronic levels goes through a number of strong oscillations similar to those of the energy curves of the $K \neq 0$ vibronic levels of a $\Pi$ electronic state of a linear molecule (see Fig. 5).

As can be seen from Fig. 24 the main difference between the correlation of vibronic levels for a non-degenerate and the correlation for a degenerate electronic state is that in the latter case there are twice as many levels to be correlated. This may not be obvious in the right part of the diagram, but there is actually a second electronic state which corresponds to the inner conical part of the potential surface in Fig. 16. Since by assumption at the right the potential minima are very deep the upper electronic state lies very high and is not shown.

Strictly speaking there is an additional doubling of all levels in Fig. 24 on account of the possibility of inversion. Only for $NH_3$ type molecules does this lead to an observable doubling. If on the $C_3$ axis there were two identical atoms as in methyl cyanide ($CH_3CN$) a further doubling would arise but is in all conceivable cases entirely negligible because this permutation is not a "feasible" operation (see p. 5).

**Vibronic eigenfunctions.** Just as for degenerate electronic states of linear molecules when Renner–Teller interaction is not negligible so here for non-negligible Jahn–Teller interaction the eigenfunction of the system can no longer be expressed as a simple product of an electronic and a vibrational function. Rather,

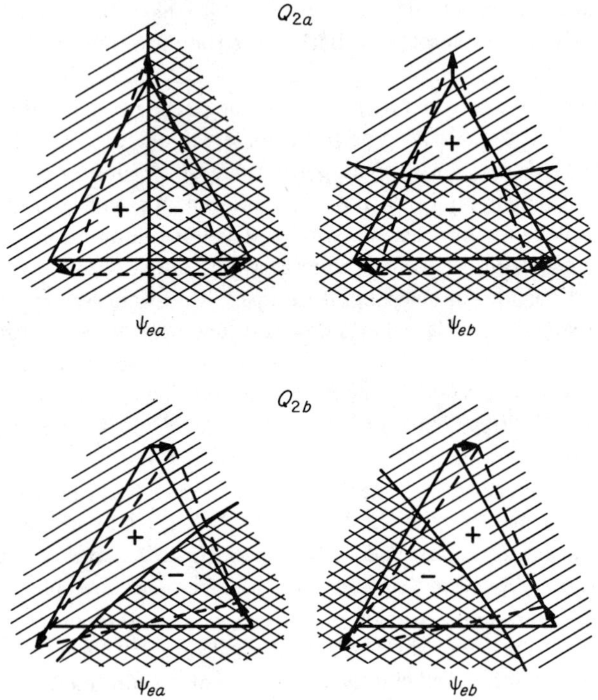


FIG. 25. **Forms of electronic eigenfunctions in a degenerate electronic state ($E$) of an $X_3$ molecule of symmetry $D_{3h}$ as adapted to the two components $Q_{2a}$ and $Q_{2b}$ of the degenerate vibration.** Hatching and cross-hatching indicate regions of positive and negative $\psi_e$. The exact shape and position of the nodal surfaces except in the first diagram is not determined by symmetry.

the two components of the electronic function and the two components of the vibrational function are intimately mixed: according to Moffitt and Liehr (869), Longuet–Higgins (766), and others, the vibronic wave function can be written in a good approximation as

$$\psi = \chi_{2a}(Q)\,\psi_{ea}(q, Q) + \chi_{2b}(Q)\,\psi_{eb}(q, Q). \tag{I, 64}$$

Here $\psi_{ea}$ and $\psi_{eb}$ are the two electronic eigenfunctions that are degenerate in the symmetrical conformation; they depend on the nuclear coordinates $Q$ as parameters and on the electronic coordinates $q$; $\chi_{2a}$ and $\chi_{2b}$ are the two components of the vibrational eigenfunction which depend on $Q$ only. In the undisplaced position (all $Q = 0$), $\psi_{ea}$ and $\psi_{eb}$ are two functions with two mutually perpendicular nodal surfaces through the axis of symmetry. In the displaced position of each of the degenerate normal coordinates, the positions of the nodal surfaces are adapted to the nuclear displacements as shown schematically in Fig. 25 for an $X_3$ molecule.

An important property of the electronic components $\psi_{ea}$ and $\psi_{eb}$ is that upon rotation of the displacement vector $Q$ by $\pi$, $\psi_{ea}$ goes into $\psi_{eb}$, and $\psi_{eb}$ goes into $-\psi_{ea}$; as a result, for a full rotation by $2\pi$ (i.e., going once around the "moat") both $\psi_{ea}$ and $\psi_{eb}$ go over into their negatives [Longuet–Higgins (766)]. Therefore, in order that the total eigenfunction $\psi$ of equation (I, 64) be single-valued, as of course it must be, the vibrational factors $\chi_{2a}(Q)$ and $\chi_{2b}(Q)$ must also change sign for such a rotation. This will be the case if the complex conjugate form of these vibrational functions $\chi_{2a} \pm i\chi_{2b}$ has the factor $e^{ij\varphi}$ with half-integral $j$. This quantum number $j$ represents for small vibronic interaction an angular momentum (see below). One finds that $j = l \pm \frac{1}{2}$. The vibrational part of the vibronic eigenfunction is different from a simple two-dimensional oscillator function characterized by $v$ and $l$ since it depends on the quantum number $j$ which, in a sense, replaces $l$. It is the mixing of levels with the same $j$ (connected by oblique lines in Fig. 21) that produces the differences from the energy levels of a two-dimensional oscillator. For a more detailed discussion of the vibronic eigenfunctions we refer to Longuet–Higgins (766) and Child and Longuet–Higgins (194).

It should be emphasized again that in all detailed treatments published so far, the potential minima at the bottom of the moat which must surely be present have been neglected in order to make the calculations possible. Neither wave functions nor energy levels for the case with minima have as yet been evaluated.

**Vibronic angular momentum.** We have seen previously that, when we consider the electronic motion in a polyatomic molecule independent of vibration and rotation (fixed center system), in a degenerate electronic state there will be in general an electronic angular momentum, $\zeta_e(h/2\pi)$, about the symmetry axis where $\zeta_e$ is usually not integral.

In the free molecule, of course, the total angular momentum about the symmetry axis must be integral $[K(h/2\pi)]$, that is, there is always a certain amount of rotational angular momentum to compensate for the non-integral magnitude of the electronic angular momentum. (In a linear molecule where rotation about the symmetry axis is impossible, the electronic (orbital) angular momentum must be integral, viz., $\Lambda(h/2\pi)$.) The excitation of non-degenerate vibrations has no effect on the angular momentum about the symmetry axis, but when degenerate vibrations are excited, they contribute a vibrational angular momentum about the symmetry axis. As we have seen in Vol. II, p. 405, the vibrational angular momentum when a degenerate vibration $\nu_i$ is singly excited is

$$\zeta_i \frac{h}{2\pi} \qquad \text{(I, 65)}$$

where the Coriolis parameter $\zeta_i$ is a number between $-1$ and $+1$ which in principle can be calculated from the potential constants of the molecule [see, e.g., Nielsen

(936)]. If several (doubly) degenerate vibrations are multiply excited, the vibrational angular momentum is

$$\zeta_v \frac{h}{2\pi} = \sum (\pm l_i \zeta_i) \frac{h}{2\pi}. \qquad \text{(I, 66)}$$

In a first approximation, for very small vibrational-electronic interaction the total internal angular momentum about the symmetry axis is the sum or difference of electronic and vibrational angular momentum

$$|\zeta_t| = |\zeta_e \pm \zeta_v|. \qquad \text{(I, 67)}$$

However, when vibronic coupling is no longer very small, there is, as we have seen, a mixing of the vibrational and electronic wave functions, and as a consequence, the total internal angular momentum is no longer given by such a simple expression. Child and Longuet–Higgins (194) have shown that for an $X_3$ molecule the total internal angular momentum (vibronic angular momentum) is given by[7]

$$|\zeta_t| = |(\zeta_e + \tfrac{1}{2}\zeta_2)d - j\zeta_2|. \qquad \text{(I, 68)}$$

Here $\zeta_2$ refers to the only doubly degenerate vibration $\nu_2$ of $X_3$, and $d$ is a "quenching parameter" which varies from $\pm 1$ to 0 as one goes from very weak to very strong coupling. It is positive when $j = l + \frac{1}{2}$ and negative when $j = l - \frac{1}{2}$.

In the limit of very small coupling, the expression (I, 68) gives the same answer as (I, 67):

$$\begin{aligned}
\text{for } v_2 &= 0, l = 0, j = \tfrac{1}{2}, d = +1: & \zeta_t &= \zeta_e \\
\text{for } v_2 &= 1, l = 1, j = \tfrac{1}{2}, d = -1: & \zeta_t &= -\zeta_e - \zeta_2 \\
& \qquad\qquad\quad\; j = \tfrac{3}{2}, d = +1: & \zeta_t &= -\zeta_e + \zeta_2 \\
\text{for } v_2 &= 2, l = 2, j = \tfrac{3}{2}, d = -1: & \zeta_t &= -\zeta_e - 2\zeta_2 \\
& \qquad\qquad\quad\; j = \tfrac{5}{2}, d = +1: & \zeta_t &= -\zeta_e + 2\zeta_2 \\
& \qquad\; l = 0, j = \tfrac{1}{2}, d = +1: & \zeta_t &= \zeta_e.
\end{aligned}$$

Here the correct signs for the $\zeta_t$ values have been included as derived from the general discussion of Mills (855). We shall not go through the rather involved reasoning that leads to these signs even though the correct choice of signs is very important for an understanding of the band structure of transitions to or from levels with $\zeta_t \neq 0$ [see Chapter II, section 3(b)].

When the coupling is not very small, considerable deviations from the simple values above occur. According to Child (193) when $D < 0.005$ the quenching parameter is given by

$$d = 1 - 4(v_2+1)D \quad \text{for} \quad j = l + \tfrac{1}{2},$$

and

$$d = -1 + 4(v_2+1)D \quad \text{for} \quad j = l - \tfrac{1}{2}.$$

For larger values of $D$, Child (191) has given tables of $d$ for $j = \frac{1}{2}$ and $\frac{3}{2}$. Even

[7] Here for the sake of consistency with the choice of sign of $\zeta_v$ of Boyd and Longuet–Higgins (132) the sign of $\zeta_2$ is the opposite of that used by Child and Longuet–Higgins (194). Personal discussion with Longuet–Higgins has confirmed the need for this change.

when $D$ is as small as 0.125 the $d$ values in the above example are 0.663, $-0.213$, 0.535, 0.046, ([7a]) and $-0.005$ leading to substantial changes of the $\zeta_t$ values.

For very strong coupling, the electronic angular momentum is completely quenched ($d \approx 0$), and the resulting angular momentum is, according to (I, 68), given by

$$\zeta = j\zeta_2, \qquad \text{(I, 69)}$$

that is, in the lowest vibrational level it is $\frac{1}{2}\zeta_2$ instead of $\zeta_e$. However, the formulae (I, 68) and (I, 69) neglect the minima in the potential trough (see Fig. 17). This is clearly not a good approximation when the coupling is strong[8]. While no detailed calculations have yet been made for the trough with minima, one may expect that in such a case both the electronic and the vibrational angular momenta will be quenched.

Vibronic angular momenta for axial molecules with more than one degenerate vibration causing Jahn–Teller instability and for molecules of the cubic point groups have not yet been discussed in detail.

The vibronic angular momentum $\zeta$ has no direct effect on the vibronic energy levels. It does, however, greatly affect the rotational energy levels as will be discussed in section 3 of this chapter. It also determines the Zeeman splittings in a magnetic field.

### (d) Vibronic interactions of different electronic states.

**General remarks.** In the two preceding sub-sections (b) and (c), we have seen that the simple product representation (I, 27) of the molecular wave function breaks down when two electronic states have the same energy, i.e., are degenerate with each other. The two electronic eigenfunctions are then thoroughly mixed with the vibrational eigenfunctions. A *partial mixing* may also arise when the two (or more) electronic states are well separated from each other. We must now briefly consider the conditions for such a mixing and its consequences.

**Interaction of electronic states of the same species.** All calculated electronic states of the same species mutually interact since there are always some terms in the wave equation for the electronic motion which are neglected in a first or higher approximation and which when introduced would lead to a slight mixing of states of the same species. The resultant shifts of the electronic energy levels or changes of the potential function are not always easy to establish unless the first approximation on which the unperturbed energy levels are based is a very rough one, or unless one has a Rydberg series of electronic states in which it is easy to establish a deviation from the Rydberg formula produced by the presence of another state of the same species but not belonging to the series (just as in atomic spectra; see ref. (21), p. 170). Interaction between non-Rydberg states of the same species is of great importance for an understanding of valence and the

[7a] not available.

[8] Note, for example, that an $A_1$, $A_2$ pair (for example, the $j = \frac{3}{2}$ levels) is degenerate only when the depths of the minima are zero.

stability of electronic states (see Chap. III). It has, however, little effect on electronic transitions except on their over-all intensities. The preceding considerations also apply to multiplet states with a given orbital species when spin-orbit interaction is small, or to individual multiplet components of the same species of the spin-orbit function when spin-orbit interaction is large.

**Interaction of electronic states of different species.** In a polyatomic molecule, unlike a diatomic molecule, a mixing (interaction) of electronic states of different species can be brought about by the interaction of vibration and electronic motion. This is because it is now the vibronic species that must be the same if the two states are to interact with each other. Identity of the vibronic species is possible for suitable vibrational levels of the two electronic states of different species. In such cases one expects to find shifts of the vibrational levels of each of the two electronic states from their "normal" position in the sense of a mutual "repulsion": one has *vibronic perturbations*. The magnitude of these perturbations depends inversely on the separation of the unperturbed levels. At the same time each of the vibronic states which perturb each other assumes properties of the other electronic state, and this leads to the occurrence of forbidden transitions (see Chap. II).

The reason for the occurrence of vibrational perturbations between electronic states of different species may also be stated in the following way: when the nuclei are in a displaced position of a non-totally symmetric vibration the molecule has lower symmetry than in the equilibrium position; electronic states that have different species in the equilibrium position may have the same species in the point group of lower symmetry of the displaced position and may therefore perturb each other. We obtain therefore a mutual distortion of the potential functions of the two electronic states; but unlike the interaction of two electronic states of the same species, here the distortion is not a general one but occurs only as a function of certain normal coordinates. One may consider it as a consequence of these distortions of the potential functions that certain vibronic levels of the two electronic states are shifted (perturbed). The presence of such perturbations is not easily established unambiguously since Fermi resonances within each electronic state may lead to similar shifts. Indeed, the perturbations here considered are the vibronic analogue of Fermi resonances.

**Selection rules for vibronic perturbations.** Just as for perturbations in diatomic molecules, vibronic perturbations in polyatomic molecules will be largest when the two potential surfaces of the two electronic states intersect each other (or come very close to each other). For the levels in the neighborhood of the region of intersection, the overlap of the eigenfunctions is most favorable and thus for these levels the perturbations are likely to be large both for interactions of electronic states of the same and of different species. Intersections of potential surfaces of states of the same species are in general prohibited for diatomic molecules (non-crossing rule, see Vol. I, p. 295), but for polyatomic molecules they can under certain circumstances occur, as was first shown by Teller (1197) (see also Chap. IV). For this reason perturbations between vibrational levels of electronic states of the same species are in many ways similar to those between states of different species except that for the former even totally symmetric vibrational levels can perturb one another.

While in principle two electronic states of any species can perturb each other if suitable vibrations are excited, we shall show below that in general this perturbation will be very weak unless the following selection rule holds: *the species of the two electronic states must not differ by more than the species of one of the normal vibrations.* In other words, the product of the species of a normal vibration and that of one of the electronic states must be equal to the species of the other electronic state. Consider, for example, a $\Pi_g$ and a $\Sigma_g^+$ electronic state of a linear molecule. Only a $\pi_g$ vibration transforms $\Sigma_g^+$ into $\Pi_g$. Such a vibration does not occur in a triatomic linear molecule, but does in more-than-triatomic molecules. Therefore, according to the above selection rule, $\Pi_g - \Sigma_g^+$ perturbations do not occur for triatomic linear molecules but do occur for four-atomic linear molecules. On the other hand, $\Pi_u - \Sigma_g^+$ or $\Pi_g - \Sigma_u^+$ or $\Pi_g - \Pi_u$ perturbations do occur even in linear triatomic molecules.

In a similar way, one finds for a molecule of point group $C_{2v}$ that the pairs of electronic states $A_1 - A_2$, $A_1 - B_1$, $A_1 - B_2$, $A_2 - B_1$, $A_2 - B_2$ and $B_1 - B_2$ may perturb each other strongly if and only if there are normal vibrations of type $A_2$, $B_1$, $B_2$, $B_2$, $B_1$ and $A_2$, respectively. Thus, for a triatomic molecule of point group $C_{2v}$, like $H_2O$, which has only normal vibrations of type $A_1$ and $B_2$, only the pairs of states $A_1 - B_2$ and $A_2 - B_1$ can perturb each other (in addition to pairs of the same species, of course). In this case, since there are no vibrational levels of species other than $A_1$ and $B_2$ the selection rule for perturbations is rigorous. For a molecule like $H_2CO$, which has normal vibrations of types $A_1$, $B_1$ and $B_2$, all but the pairs $A_1 - A_2$ and $B_1 - B_2$ can perturb each other. Finally, when, as for a molecule like $CH_2Cl_2$ and others, all four types of normal vibrations occur, all pairs of electronic states can perturb each other.

As a third example, let us consider molecules of point group $C_{3v}$ which may have electronic states $A_1$, $A_2$ and $E$. According to the above rule, the pairs of states $A_1 - A_2$, $A_1 - E$ and $A_2 - E$ may perturb each other if there are normal vibrations of type $A_2$, $E$ and $E$, respectively. For a molecule such as $CH_3Cl$ which does not have $A_2$ vibrations (but, of course, does have $A_2$ electronic states), only the pairs $A_1 - E$ and $A_2 - E$ can perturb each other in this approximation.

The proof of the above rule for perturbations is fairly simple [Herzberg and Teller (542)]. In zero approximation the electronic eigenfunction is considered to be the same during the whole vibrational motion; we shall designate it by $\psi_e(Q_0)$ where $Q_0$ represents the coordinates of the nuclei in the equilibrium position. In a higher approximation we have to take account of the dependence of $\psi_e$ on the displacements, that is, we have to use $\psi_e(Q)$ where the coordinates $Q$ vary during the vibration. We can resolve $\psi_e(Q)$ into a series

$$\psi_e(Q) = \sum_i a_i(Q)\psi_e^i(Q_0) \tag{I, 70}$$

where the $\psi_e{}^i(Q_0)$ are the different electronic eigenfunctions in the equilibrium position. The coefficients $a_i(Q)$ are all small except $a_1(Q)$, which is the coefficient of $\psi_e^1(Q_0)$, the zero approximation eigenfunction of the electronic state we are considering. In the equilibrium position $a_1(Q) = a_1(Q_0) = 1$ while all the other $a_i(Q_0)$ are zero.

The perturbed field of force in which the electrons move is that of the displaced nuclei. This field can be considered as arising from the unperturbed field (corresponding to the equilibrium positions of the nuclei) by neutralizing the nuclear charges in the equilibrium position by suitable negative charges and adding at the same time equal positive charges at the displaced positions, in other words by adding the field of small electric dipoles to the original field. There is one such dipole for each of the nuclei, its direction being that of the displacement and its magnitude being equal to the product of the displacement and the nuclear charge (or the effective nuclear charge). If the perturbation energy produced by these dipoles is $H_I(Q)$, then according to standard perturbation theory, in first approximation, the coefficients $a_i(Q)$ are given by

$$a_i(Q) = \int \frac{\psi_e^{1*}(Q_0)\, H_I(Q)\, \psi_e^i(Q_0)}{E_1(Q_0) - E_i(Q_0)}\, d\tau_e \tag{I, 71}$$

where $E_1(Q_0)$ and $E_i(Q_0)$ are the energy values in the equilibrium position of the two electronic states 1 and $i$.

The perturbation energy $H_I$ is in a first approximation a linear function of the displacements and has the symmetry of these displacements. Therefore $H_I$ can be resolved into a sum of terms, each of which depends linearly on only one normal coordinate, $\xi_1, \xi_2, \ldots$ that is,

$$H_I = H_I^1\xi_1 + H_I^2\xi_2 + \cdots. \tag{I, 72}$$

Therefore the expression for $a_i(Q)$ will be equal to a sum of integrals corresponding to the different normal coordinates. In order that a term corresponding to a particular normal coordinate in this sum be different from zero, the integrand must be invariant under all symmetry operations. Therefore the sum, that is $a_i(Q)$, is different from zero only if there is at least one normal coordinate which by multiplication with $\psi_e^{1*}(Q_0)\psi_e^i(Q_0)$ gives a totally symmetric product, or in other words if there is at least one normal coordinate whose species when multiplied with that of $\psi_e{}^1$ gives the species of $\psi_e{}^i$. This is the above rule.

Since $E_1(Q_0) - E_i(Q_0)$ appears in the denominator of eq. (I, 71), it is clear that the perturbation is stronger the closer the two electronic states are. A rough estimate shows that the perturbation may be considerable if the separation of the two electronic states is less than 1 eV.

A detailed theoretical treatment of vibronic coupling between two electronic states has recently been given by Fulton and Gouterman (402).

## 3. Rotational Structure of Electronic States; Coupling of Rotation with Vibration and Electronic Motion

In a first approximation the rotation in each electronic state of a polyatomic molecule can be considered independent of vibration and electronic motion, that is, can be dealt with in the same way as has been done for the ground electronic state in the discussion of pure rotation spectra in Chapter I of Volume II. However, in polyatomic even more than in diatomic molecules, the interactions of rotation with vibration and electronic motion are very important and may produce drastic changes of the rotational levels (here sometimes called rovibronic levels). These effects are very different in linear, symmetric top and asymmetric top molecules. We shall consider separately each group of molecules.

### (a) Linear molecules

**Rotational levels in $^1\Sigma$ electronic states.** The simplest case is that of the rotational levels in $^1\Sigma$ electronic states of linear molecules. Here the interaction of rotation and electronic motion can for most purposes be entirely neglected. We have for a vibrational level $[v] \equiv v_1, v_2, \ldots$, the rotational term values (in cm$^{-1}$)

$$F_{[v]}(J) = B_{[v]}J(J+1) - D_{[v]}J^2(J+1)^2 + \cdots \quad \text{(I, 73)}$$

where $J$ is the quantum number of the total angular momentum and $B_{[v]}$ is the rotational constant given by

$$B_{[v]} = B_e - \sum_i \alpha_i(v_i + \tfrac{1}{2}d_i). \quad \text{(I, 74)}$$

The summation is over all normal vibrations; $d_i = 1$ for non-degenerate, $d_i = 2$ for doubly degenerate vibrations. The constants $\alpha_i$ are small compared to $B$ and represent the interaction of rotation and vibration. $B_e$ is the equilibrium rotational constant defined by

$$B_e = \frac{h}{8\pi^2 c I_B^{\,e}} \quad \text{(I, 75)}$$

where $I_B^{\,e}$ is the equilibrium moment of inertia,

$$I_B^{\,e} = \sum m_i r_i^2, \quad \text{(I, 76)}$$

and $r_i$ is the distance of the nucleus $i$ from the center of mass. Equation (I, 73) may have to be very slightly modified if there is strong $l$-uncoupling as may happen especially for Rydberg states. In (I, 76) one can use in most cases for $m_i$ the atomic masses. In this way the contributions of the electrons to the moment of inertia are partially taken into account.

The rotational constant $D_{[v]}$ represents the effect of centrifugal stretching. It is almost always less than $10^{-5}B_{[v]}$ and can often be neglected except for very precise evaluations. One has, similar to (I, 74)

$$D_{[v]} = D_e + \sum \beta_i(v_i + \tfrac{1}{2}d_i). \quad \text{(I, 77)}$$

The $\beta_i$ are small compared to $D_e$ and can almost always be disregarded in polyatomic molecules.

When one or more degenerate (bending) vibrations are singly or multiply excited, degenerate vibrational levels arise of species $\Pi, \Delta, \ldots$ all of which are doubly degenerate. The molecule is then really a symmetric top and the energy levels are given by

$$F_{[v]}(J) = B_{[v]}[J(J+1) - l^2] + A_{[v]}l^2 - \cdots \quad \text{(I, 78)}$$

where $l$ is the quantum number of the resultant vibrational angular momentum about the internuclear axis. For most purposes (I, 78) can be replaced by (I, 73) since $l$ is fixed for a given vibrational level and therefore $(A_{[v]} - B_{[v]})l^2$ can be taken into the vibrational formula. But now each rotational level is doubly degenerate. The interaction of rotation and vibration removes this degeneracy and leads to *l-type doubling*. The magnitude of the splitting for $\Pi$ vibrational levels is given in a first approximation by

$$\Delta\nu = qJ(J + 1). \tag{I, 79}$$

For a symmetrical triatomic molecule

$$q_2 = \frac{B_e^{\,2}}{\omega_2}\left(1 + \frac{4\omega_2^{\,2}}{\omega_3^{\,2} - \omega_2^{\,2}}\right)(v_2 + 1). \tag{I, 80}$$

For an unsymmetrical linear triatomic molecule, or for more-than-triatomic molecules, one finds [Nielsen (936)]

$$q_i = \frac{B_e^{\,2}}{\omega_i}\left(1 + 4\sum_k \frac{\zeta_{ik}^2\omega_i^{\,2}}{\omega_k^{\,2} - \omega_i^{\,2}}\right)(v_i + 1) \tag{I, 81}$$

where $\omega_i$ is the frequency of the bending vibration considered, $\omega_k$ that of a stretching vibration that can have Coriolis interaction with $\omega_i$, and $\zeta_{ik}$ a Coriolis coefficient that depends in a fairly complicated way on the masses and potential constants[9].

Another way of stating the result of $l$-type doubling is to say that the $B$ values of the two component levels are smaller and larger by $\frac{1}{2}q$ than that of the "original" $\Pi$ vibrational level. The $l$-type doubling constant $q_i$ is closely related to $\alpha_i$ in (I, 74) for the same vibration: $\alpha_i$ differs from $q_i$ only by an additive term due to anharmonicity (see Vol. II, p. 376). In a higher approximation, terms in $J^2(J + 1)^2$ have to be added to (I, 79) and terms dependent on $v_3$ and $v_1$ have to be included in (I, 80).

For $\Delta$ vibrational levels the $l$-type doubling is in a first approximation given by [Nielsen (936)]

$$\Delta\nu = q_\Delta J^2(J + 1)^2 \tag{I, 82}$$

where $q_\Delta$ is very much smaller than $q$ for $\Pi$ vibrational levels. It is in general of the order of, but usually smaller than, the rotational constant $D$. The presence of $l$-type doubling in a $\Delta$ vibrational level therefore means that the two components have different effective $D$ values but the same $B$ value.

The *symmetry properties* of the rotational levels are the same as those discussed in Volume II for infrared and Raman spectra. We have *positive* (+) and *negative* (−) *rotational levels*, depending on whether the total eigenfunction remains unchanged or changes sign upon inversion. For symmetrical linear molecules we have, in addition, *symmetric* (*s*) and *antisymmetric* (*a*) *rotational levels* depending on whether the eigenfunction remains unchanged or changes sign upon exchange of the nuclei on one side of the center with those on the other.

In Fig. 26 the rotational levels of $\Sigma^+$, $\Sigma^-$ and $\Pi$ vibrational levels of a singlet

[9] Note that Nielsen's $q$ is $\frac{1}{2}$ of our $q$.

electronic state are represented including their symmetry properties. In Π vibrational levels, and similarly in Δ, Φ, . . . levels, there is a positive and a negative rotational level for each $J$ value; these are split by $l$-type doubling in such a way that alternately the upper and the lower component is +. If the Π level corresponds to single excitation of a bending vibration in a $\Sigma^+$ electronic state, the

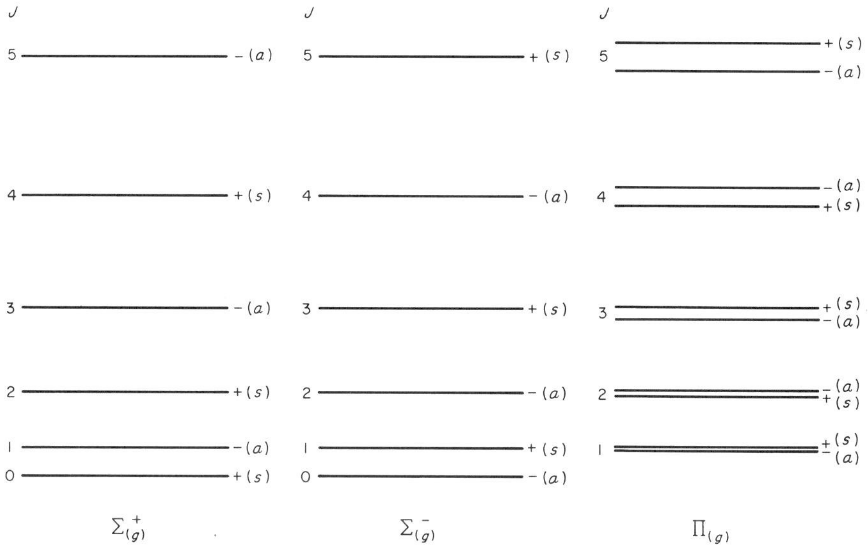


FIG. 26. **Symmetry properties of rotational levels of $\Sigma^+$, $\Sigma^-$ and Π vibronic levels of a singlet electronic state of a linear ($C_{\infty v}$) molecule.** The designations in brackets refer to a symmetric linear molecule assuming "even" ($g$) vibronic levels. For "odd" ($u$) vibronic levels $s$ and $a$ have to be exchanged throughout.

upper components of the levels with even $J$ are −, those with odd $J$ are +, while in a $\Sigma^-$ electronic state the reverse order applies. For $\Sigma_g$, $\Pi_g$, $\Delta_g$, . . . vibronic levels of symmetrical molecules, that is, for levels whose vibrational-electronic eigenfunction remains unchanged upon reflection at the center, the "positive" levels are symmetric, the "negative" antisymmetric (see the symbols in brackets in Fig. 26), and the converse is the case for $\Sigma_u$, $\Pi_u$, $\Delta_u$, . . . vibronic levels. The ratio of the statistical weights of symmetric and antisymmetric levels is determined by the ratio of the two expressions (I, 9) and (I, 8) of Volume II. (It is $(I + 1)/I$ if there is only one nucleus with non-zero spin ($I$) on each side of the center of symmetry and if this nucleus follows Bose statistics.)

*Rotational perturbations* may arise just as for diatomic molecules if two vibrational levels of two different electronic states have nearly the same energy, and if at the same time the rotational term curves intersect or come very close together (see Vol. I, p. 284, Fig. 135). But for polyatomic molecules perturbations can also arise between different vibrational levels of the same electronic state. In both cases the two mutually interacting vibrational levels must either have the same vibronic species (i.e., the same $l$ value and for $l = 0$ must both be $\Sigma^+$ or both $\Sigma^-$), or can differ in $l$ only by one.

The first kind of perturbations ($\Delta l = 0$), also referred to as *homogeneous perturbations*, is simply a special case of vibrational perturbations of the Fermi resonance type. The rotational energy aids in making the resonance very close, and therefore even when the matrix element of the perturbation function is very small, an appreciable perturbation can arise. The usual selection rules for perturbations, $\Delta J = 0$ and $+ \nleftrightarrow -$, must, of course, also be fulfilled. The second kind of rotational perturbations ($\Delta l = \pm 1$), also referred to as *heterogeneous perturbations*, is made possible solely by the Coriolis forces which arise on account of the rotation. This kind of perturbation increases with $J$. The same Coriolis forces are also at least partly responsible for $K$-type doubling.

**Rotational levels in $^1\Pi$, $^1\Delta, \ldots$ electronic states.** In $^1\Pi$, $^1\Delta, \ldots$ electronic states there is an electronic orbital angular momentum $\boldsymbol{\Lambda}$ (see p. 12) and therefore an electronic double degeneracy. This degeneracy is split when the molecule rotates, in exactly the same way and for the same reasons as $\Lambda$-type doubling in diatomic molecules (see Vol. I, p. 226). For $\Pi$ electronic states, the $\Lambda$ doubling is again given by (I, 79), but the splitting constant $q$ is now approximately given by

$$q_e \approx \frac{2B_e^2}{\nu_e} \tag{I, 83}$$

where $\nu_e$ is the separation to the nearest electronic state of appropriate symmetry ($\Sigma^+$ or $\Sigma^-$). Since $\nu_e$ is usually much larger than $\omega_i$, it is seen, by comparison with (I, 80), that $q_e$ is in general much smaller than $q$ for $l$-type doubling.

If now a bending vibration is excited, we have, in addition to $\Lambda$, a vibrational angular momentum $\boldsymbol{l}$ about the internuclear axis, and the two angular momenta interact strongly and form a resultant $\boldsymbol{K}$ (Renner–Teller interaction, see p. 25f). Similar to (I, 78), the rotational energy is now given by

$$F_{[v]}(J) = B_{[v]}[J(J + 1) - K^2] - D_{[v]}[J(J + 1) - K^2]^2. \tag{I, 84}$$

Again for most purposes the terms in $K^2$, since they are constant in a given vibronic state, can be taken into the vibrational energy formula, and one can use, as for $^1\Sigma$ states, eq. (I, 73) except that here the smallest value of $J$ is $K$ and that each level with $K \neq 0$ is doubly degenerate.

The double degeneracy of states with $K \neq 0$ is split by the interaction of rotation and electronic motion as well as by that of rotation and vibration. This splitting may be designated *K-type doubling*. The magnitude of the splitting is again proportional to $J(J + 1)$ for $\Pi$ vibronic states, to $J^2(J + 1)^2$ for $\Delta$ vibronic states, etc. Johns (633) has studied this effect in more detail for the $\Pi$ vibronic states of an electronic $\Pi$ state and has shown that in a first approximation for $v_2 \neq 0$

$$q_{ev} = \tfrac{1}{2}q_e \pm q_v$$

where the + sign applies to the upper of the two Π states that arise for a given $v_2$, the − sign to the lower.

It should perhaps be emphasized that because of the strong interaction of **Λ** and ***l***, we never see an $l$-type doubling superposed on a Λ-type doubling. The electronic degeneracy is split by the much stronger interaction with vibration whenever $l \neq 0$ (Renner–Teller interaction), and we have to deal with separate vibronic states (of different $K$) rather than components of an (electronic) Λ doublet on which the coupling of ***l*** with rotation acts.

The symmetry properties (+, −, $s$, $a$) of the rotational levels of $\Sigma^+$, $\Sigma^-$, Π, Δ, . . . vibronic levels of $^1\Pi$, $^1\Delta$, . . . ,electronic states are entirely similar to those of the corresponding vibronic levels of $^1\Sigma$ electronic states (see Fig. 26).

**Rotational levels in $^2\Sigma$ and $^3\Sigma$ electronic states.** When the electronic orbital angular momentum Λ about the internuclear axis is zero (Σ states) the electron spin is coupled to the rotation of the molecule (Hund's case (b), see Vol. I, p. 221). Since this coupling is very weak the rotational energy is, in a first approximation, given by

$$F_{[v]}(N) = B_{[v]}N(N + 1) - D_{[v]}N^2(N + 1)^2 \tag{I, 85}$$

where $N$ is the *quantum number of the total angular momentum apart from spin*[10].

If the coupling of the spin with the rotation is introduced one obtains just as for diatomic molecules for the two component levels of a $^2\Sigma$ state (neglecting the centrifugal stretching term)

$$\begin{aligned} F_1(N) &= B_{[v]}N(N + 1) + \tfrac{1}{2}\gamma_{[v]}N \\ F_2(N) &= B_{[v]}N(N + 1) - \tfrac{1}{2}\gamma_{[v]}(N + 1). \end{aligned} \tag{I, 86}$$

Here $F_1$ and $F_2$ refer to the components with $J = N + \frac{1}{2}$ and $J = N - \frac{1}{2}$ respectively, which coincide for negligible coupling. The splitting is given by $\gamma_{[v]}(N + \frac{1}{2})$. The splitting constant $\gamma$ is usually very small compared to $B$, and often the splitting even for large $N$ values is too small to be resolved.

For $^3\Sigma$ states each rotational level except the one with $N = 0$ is split into three component levels $F_1$, $F_2$ and $F_3$ corresponding to $J = N + 1, N, N - 1$, respectively. The formulae for the levels are somewhat more complicated than for $^2\Sigma$ states because of the presence of spin-spin interaction in addition to spin-rotation interaction. According to Schlapp (1103) one has

$$\begin{aligned} F_1(N) &= B_{[v]}N(N + 1) + (2N + 3)B_{[v]} - \lambda \\ &\qquad - \sqrt{(2N + 3)^2B_v{}^2 + \lambda^2 - 2\lambda B_v} + \gamma(N + 1) \\ F_2(N) &= B_{[v]}N(N + 1) \\ F_3(N) &= B_{[v]}N(N + 1) - (2N - 1)B_{[v]} - \lambda \\ &\qquad + \sqrt{(2N - 1)^2B_v{}^2 + \lambda^2 - 2\lambda B_v} - \gamma N \end{aligned} \tag{I, 87}$$

[10] Note that since the publication of Volume I by international agreement the designation of the quantum number of the total angular momentum apart from spin has been changed from $K$ to $N$ (see Mulliken (912)).

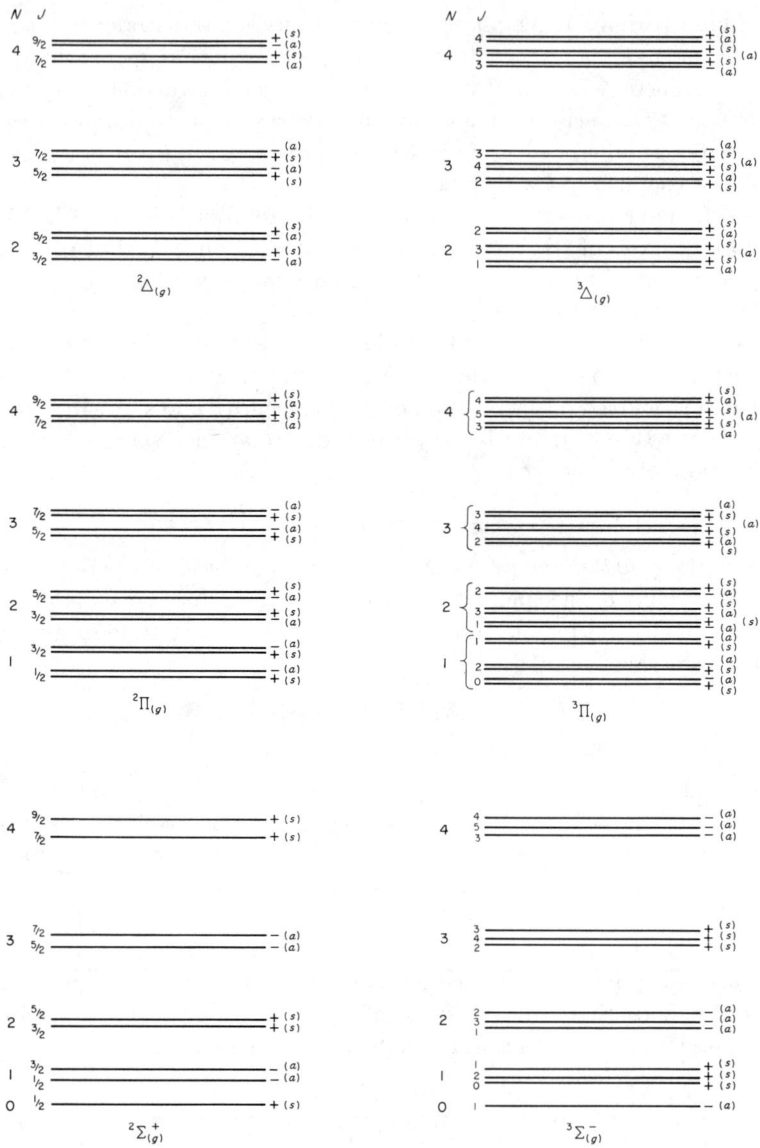

FIG. 27. **Rotational energy levels and their quantum numbers and symmetry properties in $^2\Sigma$, $^2\Pi$, $^2\Delta$, $^3\Sigma$, $^3\Pi$, $^3\Delta$ vibronic levels in $^2\Sigma$ and $^3\Sigma$ electronic states of linear ($C_{\infty v}$) molecules.** The same diagrams apply to electronic states other than $^2\Sigma$ or $^3\Sigma$ as long as spin-orbit interaction is weak [Hund's case (b)]. The designations in brackets refer to a symmetric linear molecule ($D_{\infty h}$) assuming "even" ($g$) vibronic levels. For "odd" ($u$) vibronic levels $s$ and $a$ have to be exchanged throughout.

where $\lambda$ and $\gamma$ are constants characterizing the spin-spin and spin-rotation coupling (see Vol. I, p. 223). Refinements of Schlapp's formulae have been given by Miller and Townes (849) and Tinkham and Strandberg (1221).

Strictly speaking, the formulae just given apply only to $^2\Sigma$ and $^3\Sigma$ vibronic levels of electronic $^2\Sigma$ and $^3\Sigma$ states. If degenerate vibrations are excited, we have, in addition, $^2\Pi$, $^2\Delta, \ldots$, $^3\Pi$, $^3\Delta, \ldots$ vibronic levels. However, for these vibronic levels $K \equiv l$, that is, $\boldsymbol{K}$ is entirely due to the vibrational angular momentum. Unlike $\boldsymbol{\Lambda}$, the vibrational angular momentum $\boldsymbol{l}$ does not produce a large magnetic moment in the internuclear axis; rather, the magnetic moment produced by $\boldsymbol{l}$ is of the same order as that produced by $\boldsymbol{N}$. Therefore, at least for large $N$, the spin remains coupled to the over-all rotation just as in $^2\Sigma$, $^3\Sigma, \ldots$ vibronic states. Hund's case (b) applies to all these vibronic levels. Thus, rotational energy formulae for $^2\Pi$, $^3\Pi$ and $^2\Delta$, $^3\Delta$ vibronic levels are the same as for $^2\Sigma$, $^3\Sigma$ except that $N(N + 1)$ has to be replaced by $N(N + 1) - l^2$ and that $N \geq l$. Therefore, the formulae (I, 86) and (I, 87) can be used if the constant term $-B_{[v]}l^2$ is incorporated in the vibrational term values. A more detailed theoretical consideration of this case has not yet been given.

The two or three component levels for a given $N$ have the same symmetry properties ($+$, $-$, $s$, $a$). The dependence of these symmetry properties on $N$ is the same as that on $J$ for the corresponding singlet states. Figure 27 shows rotational energy level diagrams for a few examples of vibronic levels of $^2\Sigma$ and $^3\Sigma$ electronic states (compare the discussion of case (b) states in Vol. I, p. 239).

**Rotational levels in $^2\Pi$ electronic states.** In $\Pi$ electronic states there is always a magnetic moment along the internuclear axis produced by the electronic orbital angular momentum $\boldsymbol{\Lambda}$. Therefore, at least for sufficiently small rotation, the spin is coupled to the axis [Hund's case (a)]. Uncoupling arises for higher rotational levels. As long as no degenerate vibrations are excited, the rotational levels are, for any degree of uncoupling, given by the same formulae which were derived by Hill and Van Vleck for diatomic molecules, namely,

$$\begin{aligned} F_1(J) &= B_{[v]}[(J + \tfrac{1}{2})^2 - \Lambda^2 - \tfrac{1}{2}\sqrt{4(J + \tfrac{1}{2})^2 + Y(Y - 4)\Lambda^2}] \\ F_2(J) &= B_{[v]}[(J + \tfrac{1}{2})^2 - \Lambda^2 + \tfrac{1}{2}\sqrt{4(J + \tfrac{1}{2})^2 + Y(Y - 4)\Lambda^2}], \end{aligned} \qquad \text{(I, 88)}$$

where $Y = A/B_{[v]}$ and $A$ is the coupling constant in the spin-orbit coupling term $A\boldsymbol{L}\cdot\boldsymbol{S}$ of the Hamiltonian. In the present case $\Lambda = 1$. $F_1(J)$ is the term series that for large rotation [case (b)] forms the levels with $J = N + \frac{1}{2}$, while $F_2(J)$ forms those with $J = N - \frac{1}{2}$, corresponding for a regular $^2\Pi$ state (positive $A$) to $^2\Pi_{\frac{1}{2}}$ and $^2\Pi_{\frac{3}{2}}$, respectively, but corresponding for an inverted $^2\Pi$ state (negative $A$) to $^2\Pi_{\frac{3}{2}}$ and $^2\Pi_{\frac{1}{2}}$, respectively. It is readily seen that the separation of the first level $J = \frac{1}{2}$ of $^2\Pi_{\frac{1}{2}}$ from the first level $J = \frac{3}{2}$ of $^2\Pi_{\frac{3}{2}}$ is very nearly $A$. For large $|A|$ the formulae (I, 88) can be approximated in the way indicated in eqs. (V, 29) and (V, 30) of volume I (p. 233).

In the formulae (I, 88) the effects of centrifugal stretching and of $\Lambda$-type doubling are neglected. The former can be taken care of by the addition of a term

$$-D_v[(J - \tfrac{1}{2})(J + \tfrac{1}{2})^2(J + \tfrac{3}{2}) + 1]$$

to both equations [Almy and Horsfall (66), Dixon (281)]. The latter can be taken into account by the addition of a small term $\varphi_i(J)$ which is different for the two $\Lambda$ doubling components. Unlike the case of a $^1\Pi$ state (see p. 72) one finds when $|A|$ is large [Hund's case (a)] for $^2\Pi_{\frac{1}{2}}$

$$\varphi_i = \pm\tfrac{1}{2}p(J + \tfrac{1}{2}) \tag{I, 89}$$

and for $^2\Pi_{\frac{3}{2}}$

$$\varphi_i = \pm\tfrac{1}{2}s(J - \tfrac{1}{2})(J + \tfrac{1}{2})(J + \tfrac{3}{2}) \approx \pm\tfrac{1}{2}s(J + \tfrac{1}{2})^3 \tag{I, 90}$$

where $s \approx p/Y^2$, that is, in general $s \ll p$. Except for small $|A|$ only the doubling of $^2\Pi_{\frac{1}{2}}$ is resolved. If it is, it establishes immediately which of the observed spin-doublet components is $^2\Pi_{\frac{1}{2}}$ and which $^2\Pi_{\frac{3}{2}}$.

As an illustration of spin doubling, in Fig. 28 the rotational energy levels in the $^2\Pi$ ground state of NCO are plotted (a) as a function of $J$, and (b) as a function of $N$ after subtraction of the purely rotational terms $B_0J(J + 1)$ and $B_0N(N + 1)$, respectively.

If degenerate vibrations are excited, for each set of vibrational quantum numbers, in general, several vibronic levels arise, as discussed in section 2 (see particularly Fig. 2 and 6). The splitting between these levels (Renner–Teller splitting) is effected by spin-orbit coupling in the way previously described (p. 34f).

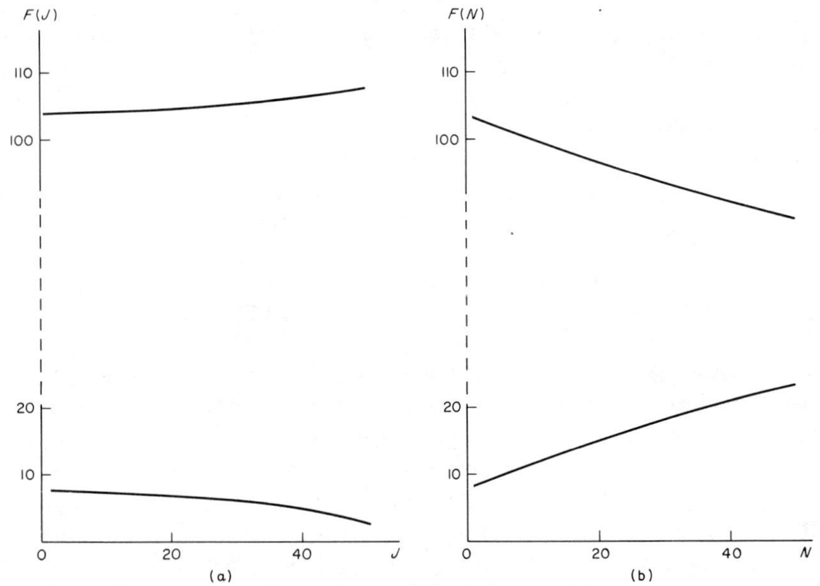


FIG. 28. **Spin doubling in the $^2\Pi$ ground state of NCO** (*a*) **as a function of** $J$, (*b*) **as a function of** $N$. The purely rotational part of the energy, $BJ(J + 1)$ and $BN(N + 1)$ respectively, has been subtracted.

As we have seen, the resultant vibronic levels are determined by the parameters $A$ and $\epsilon$ and the vibrational quantum number according to eq. (I, 47). We must now consider the rotational levels in such a case. Hougen (568) has discussed in detail the effect of the vibronic coupling on the rotational levels of a $^2\Pi$ electronic state of a triatomic molecule.

We consider first Hougen's results for the $^2\Sigma^{(+)}$ and $^2\Sigma^{(-)}$ vibronic states which arise for odd values of $v_2$ (see Fig. 8). Here the effect of the interaction of rotation and vibronic motion is particularly pronounced since for these levels $K = 0$, even though $\Lambda = 1$. The electronic orbital angular momentum ($\Lambda = 1$) by itself would cause a large spin splitting as it actually does for $v_2 = 0$ but, because for $\Sigma$ vibronic states it is compensated by the vibrational angular momentum ($K = \Lambda - l = 0$), no splitting can arise for zero rotation. However, with increasing rotation the strong coupling of the electronic orbital angular momentum $\boldsymbol{L}$ to the linear axis breaks down slightly allowing a component to develop in the direction of the axis of rotation, and thus producing a much stronger than normal coupling of the spin to the axis of rotation. Hougen finds for the rotational energies

$$\begin{aligned}
{}^2\Sigma^{(+)}\text{:}\quad & F_1(N) = B_{[v]}^{(+)}N(N+1) + \tfrac{1}{2}\gamma_{[v]}^{(+)}(N+1) \\
& F_2(N) = B_{[v]}^{(+)}N(N+1) - \tfrac{1}{2}\gamma_{[v]}^{(+)}N \\
{}^2\Sigma^{(-)}\text{:}\quad & F_1(N) = B_{[v]}^{(-)}N(N+1) + \tfrac{1}{2}\gamma_{[v]}^{(-)}(N+1) \\
& F_2(N) = B_{[v]}^{(-)}N(N+1) - \tfrac{1}{2}\gamma_{[v]}^{(-)}N.
\end{aligned} \qquad \text{(I, 91)}$$

Here the normal $B_{[v]}$ values have been replaced by effective values

$$B_{[v]}^{(\pm)} = B_{[v]}\left(1 \pm \frac{A^2 B_{[v]}}{(A^*_{v_2,0})^3}\right) \qquad \text{(I, 92)}$$

and the splitting constants $\gamma$ are given by

$$\gamma_{[v]}^{(\pm)} = 2B_{[v]}\left(1 - \left|\frac{(v_2+1)\epsilon\omega_2}{A^*_{v_2,0}}\right| \pm \frac{A^2 B_{[v]}}{(A^*_{v_2,0})^3}\right). \qquad \text{(I, 93)}$$

The quantity $A^*_{v_2,0}$ is given by the previous equation (I, 48) putting $K = 0$. The second term in the bracket in (I, 92) is small compared to 1 as long as $A^*$ is not too small, and therefore the effective $B$ values do not differ much from $B_{[v]}$. More important is the result that the splitting constants $\gamma$ have an appreciable magnitude. Only when the spin-orbit coupling constant $A$ is small compared to $(v_2 + 1)\epsilon\omega_2$ [see eq. (I, 48)], would $\gamma$ be small. Even when the spin-orbit coupling constant is only moderately large the second term in the bracket of (I, 93) will be substantially smaller than 1, and therefore the spin splitting of the $^2\Sigma$ vibronic levels will be large; the constants $\gamma$ may then reach values of the order of $B_{[v]}$, i.e., will be very much greater than in $^2\Sigma$ electronic states.

In Fig. 29, as examples, the observed rotational levels in the 010 $^2\Sigma^{(+)}$ vibronic level of the $^2\Pi$ ground state of NCO and in the 010 $^2\Sigma_g^{(+)}$ vibronic level of the first excited $^2\Pi_u$ state of $BO_2$ are shown to scale. In the first example $\gamma$ is not large (but still much larger than in a normal $^2\Sigma$ electronic state), while in the second example it is so large that the $^2\Sigma^{(+)}$ state might be more properly considered as a $P = \frac{1}{2}$ state (see p. 26) with a large $P$-type doubling (similar to $\Omega$-type doubling for diatomic molecules; see Vol. I, p. 229). This interpretation is in accordance

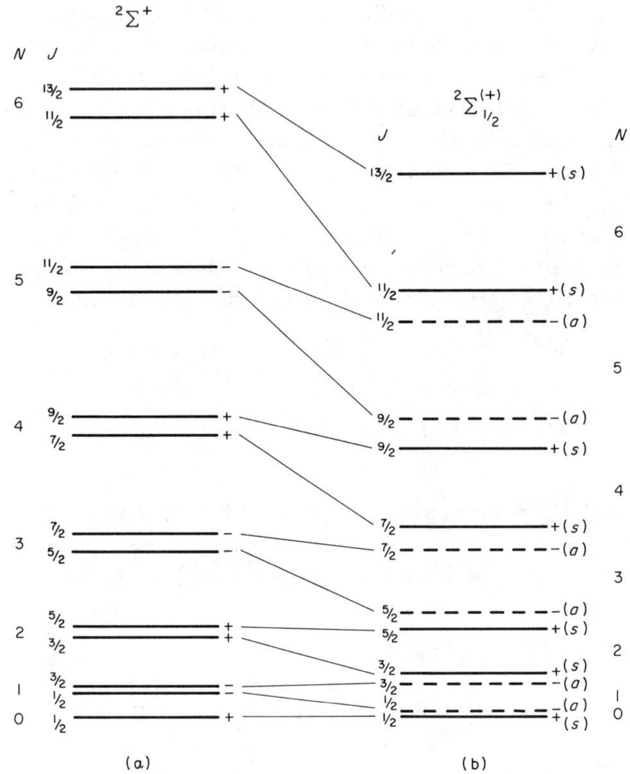


FIG. 29. **Lowest observed rotational levels in the 010 $^2\Sigma^+$ vibronic level of** (*a*) **the $^2\Pi$ ground state of NCO after Dixon (281) and** (*b*) **the first excited $^2\Pi$ state of $BO_2$ after Johns (630).** The missing levels in $BO_2$ are indicated by broken lines. The correlation lines between (*a*) and (*b*) indicate the transition from small to large $\gamma$.

with the previous conclusion that the $\Sigma^+$, $\Sigma^-$ character for $^2\Sigma$ vibronic levels is no longer rigorously defined. We have a transition to Hund's case (c) in which the quantum number $K$ no longer exists, i.e., no $\Sigma^+$ and $\Sigma^-$ states exist. It should be noted that in Fig. 29 the symmetry properties of the rotational levels do not change in going from left to right.

In Table 3 the observed magnitudes of $\gamma$ are compared with those calculated from eq. (I, 93) for $BO_2$ [Johns (630)] and NCO [Dixon (281)]. The agreement is very satisfactory. It is seen that in the $\tilde{A}\ ^2\Pi_u$ excited state of $BO_2$ the $\gamma$ value is larger than the $B_{[v]}$ value.

TABLE 3. SPLITTING CONSTANTS $\gamma$ OF VIBRONIC $^2\Sigma$ STATES IN ELECTRONIC $^2\Pi$ STATES

| Molecule | Electronic state | Vibronic state | $B_{[v]}$ cm$^{-1}$ | $\gamma$ obs. (cm$^{-1}$) | $\gamma$ cal. (cm$^{-1}$) |
|---|---|---|---|---|---|
| NCO | $\tilde{X}\ ^2\Pi$ | 010 $^2\Sigma^{(+)}$ | 0.3912 | 0.118 | 0.10 |
| $BO_2$ | $\tilde{X}\ ^2\Pi_g$ | 010 $^2\Sigma_u^{(+)}$ | 0.3308 | 0.163 | 0.15 |
| | $\tilde{A}\ ^2\Pi_u$ | 010 $^2\Sigma_g^{(+)}$ | 0.3108 | 0.474 | 0.47 |
| | $\tilde{A}\ ^2\Pi_u$ | 010 $^2\Sigma_g^{(-)}$ | 0.3125 | 0.444 | 0.47 |

If the $^2\Sigma$ vibronic levels are considered on the basis of Hund's case (c) Hougen (568) finds the following approximate energy formulae

$$\begin{aligned} {}^2\Sigma_{\text{(upper)}}:\quad & F_e(J) = B_{[v]}^{(+)}J(J+1) + \tfrac{1}{2}p_v(J+\tfrac{1}{2}) \\ & F_f(J) = B_{[v]}^{(+)}J(J+1) - \tfrac{1}{2}p_v(J+\tfrac{1}{2}) \\ {}^2\Sigma_{\text{(lower)}}:\quad & F_e(J) = B_{[v]}^{(-)}J(J+1) + \tfrac{1}{2}p_v(J+\tfrac{1}{2}) \\ & F_f(J) = B_{[v]}^{(-)}J(J+1) - \tfrac{1}{2}p_v(J+\tfrac{1}{2}) \end{aligned} \qquad \text{(I, 94)}$$

where $e$ and $f$ distinguish the two $P$-doubling components. $B_{[v]}^{(+)}$ and $B_{[v]}^{(-)}$ are given by (I, 92) while

$$p_v = 2B\,\frac{(v_2+1)\epsilon\omega_2}{A^*_{v_2,0}}. \qquad \text{(I, 95)}$$

As before, the upper and lower $^2\Sigma$ state are separated by

$$\Delta\nu = A^*_{v_2,0} = \sqrt{A^2 + (v_2+1)^2\epsilon^2{\omega_2}^2}. \qquad \text{(I, 96)}$$

When the Renner parameter $\epsilon$ is small, the $P$-type doubling, according to (I, 94) and (I, 95), is small, while the spin doubling in case (b) (also called $\rho$-type doubling) according to (I, 91) and (I, 93) is large. But the two descriptions are equivalent: it is easily seen that for small $\epsilon$ (i.e., $\gamma \approx 2B$) the resultant large splitting according to (I, 91) and (I, 93) just brings levels with the same $J$ into coincidence as they should be according to (I, 94) (compare Fig. 29).

Both formulae (I, 91) and (I, 94) were indeed obtained as expansions of the following more general formulae:

$$\begin{aligned} {}^2\Sigma_{\text{(upper)}}:\quad T_{[v]}^{(+)}(J) &= G(v_1, v_2, v_3) + B_{[v]}(J+\tfrac{1}{2})^2 \\ &\quad + \tfrac{1}{2}\sqrt{(\Delta\nu)^2 \pm 4B_{[v]}(J+\tfrac{1}{2})(v_2+1)\epsilon\omega_2 + 4B_{[v]}^2(J+\tfrac{1}{2})^2} \\ {}^2\Sigma_{\text{(lower)}}:\quad T_{[v]}^{(-)}(J) &= G(v_1, v_2, v_3) + B_{[v]}(J+\tfrac{1}{2})^2 \\ &\quad - \tfrac{1}{2}\sqrt{(\Delta\nu)^2 \pm 4B_{[v]}(J+\tfrac{1}{2})(v_2+1)\epsilon\omega_2 + 4B_{[v]}^2(J+\tfrac{1}{2})^2}. \end{aligned} \qquad \text{(I, 97)}$$

These formulae, unlike (I, 91) and (I, 94), give the total rovibronic energy including the over-all splitting $\Delta\nu$ between the two $^2\Sigma$ states. $G(v_1, v_2, v_3)$ is essentially the ordinary vibrational energy with two very small correction terms [see Hougen (568)]. The upper sign in each of the two equations (I, 97) gives the set of levels $-, +, -, \ldots$ the lower sign the set $+, -, +, \ldots$ each with $J = \frac{1}{2}, \frac{3}{2}, \frac{5}{2}, \ldots$ respectively.

Equation (I, 97) holds even when $\Delta\nu$ is small compared to $2BJ$ that is, when eqs. (I, 91) and (I, 94) are not applicable. We shall not give expansions for this case (they are in fact very similar to eqs. (V, 28a) of Vol. I, p. 233). But we give in Fig. 30 the correlation of the levels for $\Delta\nu \ll 2BJ$ and small $\epsilon$ (at the extreme left) to those of medium spin-orbit coupling and small $\epsilon$ (in the center). In the same figure at the right is given the correlation to the case when the vibronic coupling ($\epsilon$) is large compared to the

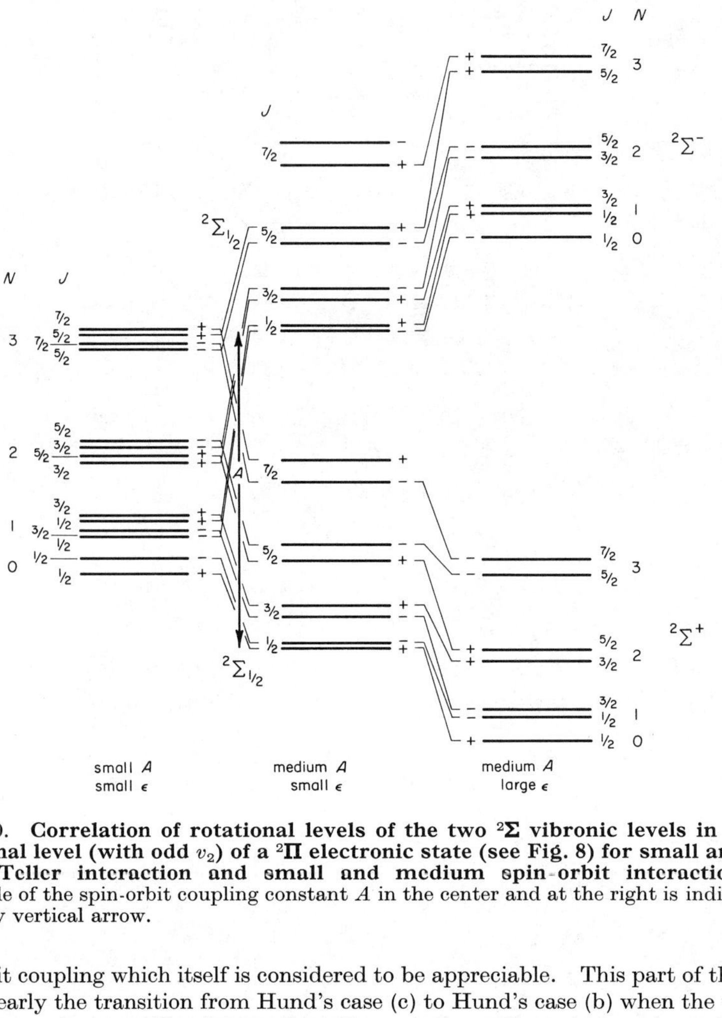


FIG. 30. **Correlation of rotational levels of the two $^2\Sigma$ vibronic levels in a given vibrational level (with odd $v_2$) of a $^2\Pi$ electronic state (see Fig. 8) for small and large Renner–Teller interaction and small and medium spin-orbit interaction.** The magnitude of the spin-orbit coupling constant $A$ in the center and at the right is indicated by the heavy vertical arrow.

spin-orbit coupling which itself is considered to be appreciable. This part of the figure shows clearly the transition from Hund's case (c) to Hund's case (b) when the vibronic coupling goes from small to large values [from $\epsilon\omega_2(v_2 + 1) \ll A$ to $\epsilon\omega_2(v_2 + 1) \gg A$]. It explains why, in the intermediate region, when spin-orbit and vibronic coupling are of the same order, a large doubling arises in each $^2\Sigma$ state, which may equally be considered as $P$-type and as $\rho$-type doubling.

In most practical cases it is convenient to use the classification $^2\Sigma^+$, $^2\Sigma^-$, and to apply the formula (I, 91) even if the splitting is large.

For vibronic levels other than $^2\Sigma$ belonging to a $^2\Pi$ electronic state, only slight modifications of the Hill and Van Vleck formulae arise:

For vibronic levels with $K = v_2 + 1$, the highest $K$ value of a given set (e.g., the $\Delta$ state arising for $v_2 = 1$; see Fig. 6) Hougen (568) found that the rotational levels are given by the Hill and Van Vleck formulae (I, 88) except that $Y(Y - 4)\Lambda^2$ must be replaced by $Y(Y\Lambda - 4K)\Lambda$ where $Y = A_{\text{eff.}}/B$ is determined by an effective $A$ value

$$A_{\text{eff.}} = A\left[1 - \frac{\epsilon^2}{8} K(K + 1)\right]. \qquad \text{(I, 98)}$$

For large spin-orbit coupling one may also describe the rotational levels of each doublet component by an effective $B$ value

$$B_{\text{eff.}} = B\left(1 \pm \frac{B}{|A_{\text{eff.}}|}\right) \qquad \text{(I, 99)}$$

similar to the practice for diatomic molecules with large spin doubling. The upper sign refers to the doublet component of higher energy. As mentioned in section 2, without rotation the doublet splitting of the vibronic level with $K = v_2 + 1$ is the same as for the pure electronic state ($^2\Pi$).

For vibronic levels of intermediate $K$ values, which always occur in pairs (see Fig. 8), Hougen (568) has given explicit formulae which we shall not reproduce. However, in the limiting case of large splitting we can again express the rotational energy levels by effective $B$ values. Hougen gives for the two spin components of the upper and lower vibronic component of a pair of states with $0 < K < v_2 + 1$;

$$\begin{aligned} B_{\text{eff.}}^{(\text{upper})} &= B\left\{1 + \frac{BA^2}{(A^*_{v_2,K})^3} \pm \frac{4B\omega_2[(v_2 + 1)^2 - K^2]}{(v_2 + 1)|A|A^*_{v_2,K}}\right\} \\ B_{\text{eff.}}^{(\text{lower})} &= B\left\{1 - \frac{BA^2}{(A^*_{v_2,K})^3} \pm \frac{4B\omega_2[(v_2 + 1)^2 - K^2]}{(v_2 + 1)|A|A^*_{v_2,K}}\right\}. \end{aligned} \qquad \text{(I, 100)}$$

Here it must be realized that the pairing of the four states of given $K$ and $P = K \pm \frac{1}{2}$ is somewhat arbitrary as indicated by the two correlations in Fig. 8 but, irrespective of that, $B_{\text{eff.}}^{(\text{upper})}$ in (I, 100) refers to the upper pair of levels, $B_{\text{eff.}}^{(\text{lower})}$ to the lower. For a regular $^2\Pi$ electronic state the highest and lowest of the four levels have $P = K - \frac{1}{2}$ while the two intermediate ones have $P = K + \frac{1}{2}$. The reverse is the case for an inverted $^2\Pi$ state. Each of the four levels resulting from (I, 100) is still doubly degenerate. This degeneracy may be split by the finer interaction between vibronic motion and rotation ($K$-type or $P$-type doubling).

Hougen's formulae refer to linear triatomic molecules. It is very likely that they will also hold for linear four-, five-, ... atomic molecules except that the parameter $\omega_2\epsilon(v_2 + 1)$ will have to be replaced by another appropriate quantity.

**Rotational levels in other multiplet states.** No detailed discussion of doublet electronic states other than $^2\Pi$ states has yet been given in the literature. One may expect, however, that for $^2\Delta$, $^2\Phi$, ... electronic states similar formulae as given above for $^2\Pi$ will apply except that the relation (I, 96) for $\Delta\nu$ will have to be replaced by

$$\Delta\nu = \sqrt{A^2\Lambda^2 + (v_2 + 1)^2\epsilon^2\omega_2{}^2}. \qquad \text{(I, 101)}$$

For $^3\Pi$ electronic states Hougen (569) has developed formulae similar to those for $^2\Pi$ given above. As we have seen in section 2, if spin-orbit interaction is large, the $^3\Sigma^+$ and $^3\Sigma^-$ vibronic levels of a $^3\Pi$ state are each split widely (even without rotation) into $^3\Sigma_1$ and $^3\Sigma_{0^-}$ or $^3\Sigma_{0^+}$ levels. The $^3\Sigma_1$ vibronic levels split with increasing rotation on account of $P$-type doubling ($P = 1$). The $^3\Sigma_{0^-}$ and $^3\Sigma_{0^+}$ vibronic levels remain single, and their + and − characteristic remains well defined. When spin-orbit coupling is small (case (b)) and the three components of each $^3\Sigma$ vibronic state are close together for zero rotation, a splitting linear in $N$ will arise, and, just as for $^2\Sigma$ vibronic states, the splitting constants $\gamma$ are anomalously large. For the behavior of vibronic levels having $K \neq 0$ we refer to Hougen (569).

The symmetry properties of the rotational levels in the various vibronic levels are again exactly analogous to those of diatomic molecules (Vol. I, p. 239).

For $^3\Delta$ electronic states considerations similar to those for $^3\Pi$ apply, but no detailed discussion has yet been published.

### (b) Symmetric top molecules

**Rotational levels in non-degenerate vibrational levels of non-degenerate singlet electronic states.** For the simplest case of non-degenerate singlet electronic states the rotational levels of symmetric top molecules have been discussed in detail in Volume II. We summarize here only the results. The *rotational term values of a prolate top*, assuming that there are no vibrational (or electronic) degeneracies, are given by

$$F_{[v]}(J, K) = B_{[v]}J(J+1) + (A_{[v]} - B_{[v]})K^2 - D_J J^2(J+1)^2 - D_{JK}J(J+1)K^2 - D_K K^4. \quad \text{(I, 102)}$$

The principal terms, the only ones occurring for a rigid symmetric top, are the first two. The last three terms in (I, 102) are *centrifugal stretching* terms which are always small compared to the first two except at very high $J$ and $K$ values. Their variation with $v_i$ can almost always be neglected. However, the dependence of the rotational constants $B_{[v]}$ and $A_{[v]}$ on the $v_i$ is important. We have

$$\begin{aligned} B_{[v]} &= B_e - \sum \alpha_i^B(v_i + \tfrac{1}{2}d_i) + \cdots \\ A_{[v]} &= A_e - \sum \alpha_i^A(v_i + \tfrac{1}{2}d_i) + \cdots \end{aligned} \quad \text{(I, 103)}$$

where the summation is over all normal vibrations of the molecule. The $\alpha_i^A$ and $\alpha_i^B$ are analogous to the $\alpha_i$ in linear molecules; they are small compared to $A$ and $B$ respectively, but by no means negligibly small. The equilibrium rotational constants $A_e$ and $B_e$ are related to the equilibrium moments of inertia $I_A^e$ (about the top axis) and $I_B^e$ (about an axis perpendicular to the top axis and through the center of mass), by the relations

$$A_e = \frac{h}{8\pi^2 c I_A^e}, \qquad B_e = \frac{h}{8\pi^2 c I_B^e}. \quad \text{(I, 104)}$$

The rotational quantum number $K$ corresponds to the angular momentum $\boldsymbol{K}$ about the top axis whose magnitude is $K(h/2\pi)$. It is the component of the total angular momentum $\boldsymbol{J}$. While for linear molecules $K$ can be different from zero only when there is an electronic or vibrational angular momentum, here, even without such a contribution, $K$ may be different from zero, i.e., we may have a pure (rigid) rotation about the top axis without contributions from electronic and vibrational motions. But $\boldsymbol{K}$ may also include such contributions (see below). It must be noted that always $J \geq K$.

In Fig. 31a a rotational energy level diagram of a prolate symmetric top in a non-degenerate state is shown. For each $K$ value there are a series of rotational levels with $J = K, K + 1, K + 2, \ldots$.

According to the international nomenclature [Mulliken (912)], the rotational constants of an asymmetric top are designated in such a way that

$$A > B > C. \tag{I, 105}$$

The corresponding axes are called $a$-axis, $b$-axis, $c$-axis. For the *prolate* symmetric top we have $B = C$; the top axis is the $a$-axis. On the other hand, for the *oblate* symmetric top we have $A = B$; the top axis is the $c$-axis, that is, the moment of inertia about the top axis is the largest and the corresponding rotational constant must be designated $C$. If it is substituted in place of $A$ in (I, 102) we

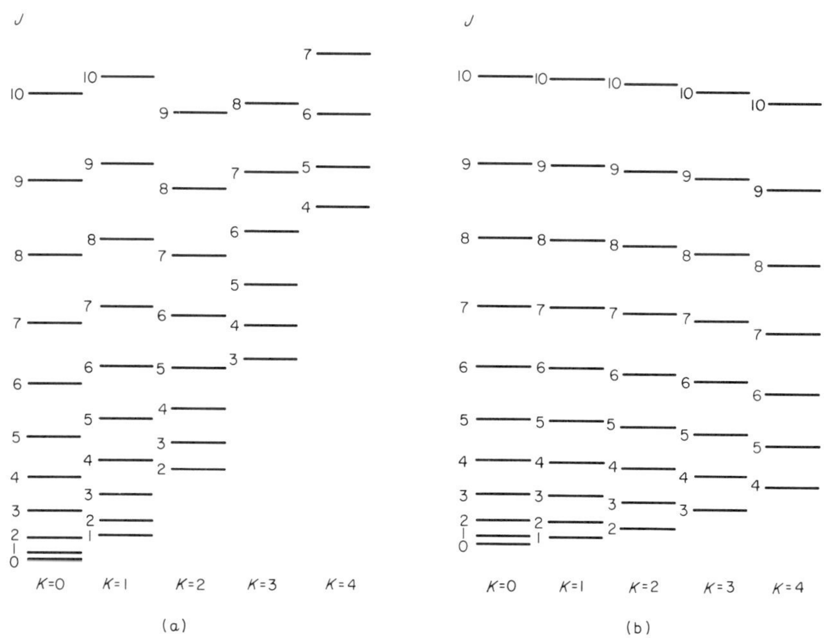


Fig. 31. **Rotational energy levels of** (*a*) **a prolate** (*b*) **an oblate symmetric top in a non-degenerate vibronic state.** The levels are arranged in different columns according to their $K$ values.

have for the *oblate symmetric top*, neglecting centrifugal stretching terms,

$$F_{[v]}(J, K) = B_{[v]}J(J + 1) - (B_{[v]} - C_{[v]})K^2 \tag{I, 106}$$

where

$$C_{[v]} = C_e - \sum \alpha_i^C(v_i + \tfrac{1}{2}d_i), \qquad C_e = \frac{h}{8\pi^2 c I_C^e}. \tag{I, 107}$$

Levels with the same $J$ now shift downwards with increasing $K$ instead of upwards, as shown in Fig. 31b.

The formulae given here hold both for molecules that are symmetric tops on account of their symmetry as well as those that have accidentally two equal moments of inertia. *Planar symmetric top molecules* are always oblate tops. For them

$$I_C{}^e = 2I_B{}^e \tag{I, 108}$$

and therefore

$$C_e = \tfrac{1}{2}B_e. \tag{I, 109}$$

If such a planar molecule is a symmetric top on account of symmetry, there is also a relation between the three centrifugal distortion constants, as was shown by Dowling (304), viz.,

$$D_{JK} = -\tfrac{2}{3}(D_J + 2D_K). \tag{I, 110}$$

**Rotational levels in degenerate vibrational levels of non-degenerate singlet electronic states.** In degenerate vibrational levels (which occur for all genuine symmetric top molecules) the Coriolis forces in the rotating molecule, according to Teller and Tisza (1198) and Teller (1196), produce a splitting of the degeneracy which, in a first approximation, increases linearly with the quantum number $K$ (see Vol. II, p. 401). The reason for this splitting is the fact that now the angular momentum about the top axis, $Kh/2\pi$, is the sum of a rotational and a vibrational contribution. The latter is $\zeta_v h/2\pi$ (see p. 63) and therefore the rotational contribution is $(K \mp \zeta_v)h/2\pi$ where the minus sign applies when the vibrational angular momentum is parallel, the plus sign when it is antiparallel to $\boldsymbol{K}$. Correspondingly in the rotational energy formulae (I, 102) and (I, 106) we have to replace $AK^2$ by $A(K \mp \zeta_v)^2$ and $CK^2$ by $C(K \mp \zeta_v)^2$ respectively. This substitution means that we have to add to eq. (I, 102) for the prolate top the term

$$\mp 2A_{[v]}\zeta_v K + A_{[v]}\zeta_v{}^2 \tag{I, 111}$$

and to eq. (I, 106) for the oblate top the term

$$\mp 2C_{[v]}\zeta_v K + C_{[v]}\zeta_v{}^2. \tag{I, 112}$$

Usually the second term in (I, 111) or (I, 112) is omitted since it is independent of rotation and may be incorporated in the vibrational term value. In the simplest case, when a single degenerate vibration is singly excited, $\zeta_v = \zeta_i$ is a quantity whose magnitude is determined by the form of the two components of the normal mode and is characteristic of the degenerate vibration (see Vol. II, p. 404). The $\zeta_i$ values of degenerate vibrations lie between $+1$ and $-1$. Except when there is only one degenerate vibration of a given kind the $\zeta_i$ depend in a fairly complicated way on the potential constants but the sums of the $\zeta_i$ for a given species follow very simple formulae: see Volume II, pp. 404 and 405 [11], Boyd and Longuet-Higgins (132), Lord and Merrifield (771), Meal and Polo (816), Mills and Duncan (856).

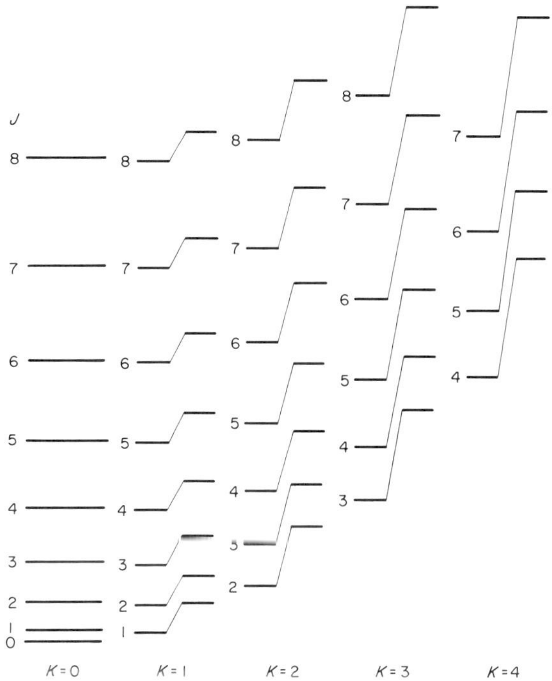


FIG. 32. **Rotational energy levels of a prolate symmetric top in a degenerate vibrational (or vibronic) state with $\zeta_v = 0.4$.** The pairs of levels with $K \neq 0$ connected by sloping lines arise from single levels in Fig. 31; they are split by Coriolis interaction.

If several degenerate vibrations are singly or multiply excited, the total vibrational angular momentum $\zeta_v$ is given by $\sum(\pm\zeta_i l_i)$ [see eq. (I, 66)]. Thus we have for the general rotational energy formula for prolate symmetric top molecules in non-degenerate electronic states (omitting the term independent of $K$ and $J$)

$$F_{[v]}(J, K) = B_{[v]}J(J+1) + (A_{[v]} - B_{[v]})K^2 - 2A_{[v]}\left(\sum \pm l_i\zeta_i\right)K \quad \text{(I, 113)}$$

and similarly for oblate symmetric top molecules

$$F_{[v]}(J, K) = B_{[v]}J(J+1) - (B_{[v]} - C_{[v]})K^2 - 2C_{[v]}\left(\sum \pm l_i\zeta_i\right)K. \quad \text{(I, 114)}$$

Here the centrifugal stretching terms have been omitted. In Fig. 32 the rotational

[11] Note that eq. (IV, 49) of Volume II contains an error: it should be

$$\zeta_{10} + \zeta_{11} + \zeta_{12} = \frac{B}{2A} \quad \text{(not 0).}$$

levels in a degenerate vibrational level with $\zeta_v = \sum l_i\zeta_i = 0.4$ are shown. Figure 33 shows the variation of the energy with $K$ for a given $J$ for three $\zeta_v$ values. It should be noted that for large $\zeta_v[> \frac{1}{2}(1 - B/A)]$ one component of the $K = 1$ level, and for still larger $\zeta_v$, of the $K = 2, 3, \ldots$ levels lies below the $K = 0$ level. For

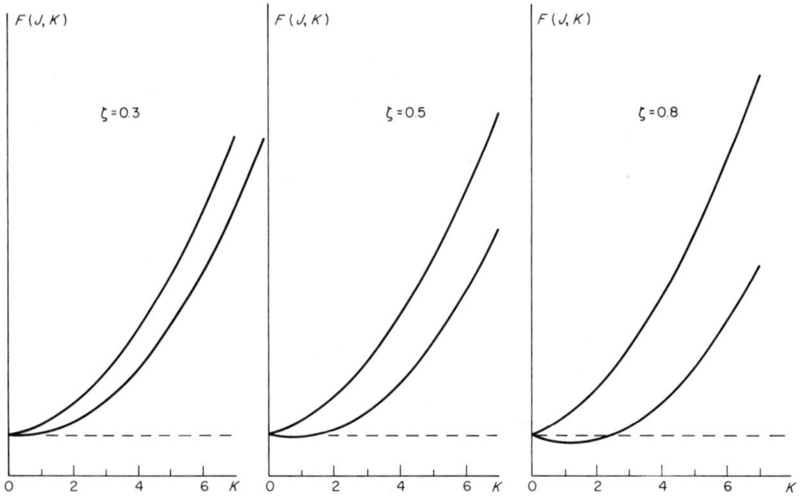


Fig. 33. **Variation with $K$ of the rotational energy of a prolate symmetric top in a degenerate vibronic state for three different $\zeta_v$ values.** The same curves hold for each $J$ value; but note that $J \geqslant K$.

an oblate top, since $C < B$, the energy decreases with $K$ except that for sufficiently large $\zeta_v$ one component of the $K = 1, 2, \ldots$ levels may be above the $K = 0$ level.

Garing, Nielsen and Rao (406) have given centrifugal stretching terms depending on the $\zeta_i$ which in a higher approximation have to be added to (I, 113) or (I, 114) in addition to the usual centrifugal stretching terms given in (I, 102). These additional terms are

$$+2\left(\sum \pm l_i\zeta_i\right) K\,[(2D_J + D_{JK})J(J+1) + 2(D_K + D_{JK})K^2]. \qquad \text{(I, 115)}$$

**Rotational levels in degenerate singlet electronic states.** As we have seen in section 1, there is in general an electronic angular momentum $\zeta_e(h/2\pi)$ in degenerate electronic states. The interaction of the electronic and rotational angular momentum, just as the interaction of vibrational and rotational angular momentum, leads to a splitting of the degeneracy expressed by a term

$$\mp 2A_{[v]}\zeta_e K \quad \text{or} \quad \mp 2C_{[v]}\zeta_e K \qquad \text{(I, 116)}$$

which is to be added to the rotational energy expression of the prolate and oblate top respectively. The energy levels are therefore entirely similar to those shown in Fig. 32 or 33. Only the cause of the splitting is different.

In degenerate *vibronic* levels of degenerate electronic states it is the resultant vibronic angular momentum $\zeta_t(h/2\pi)$ that determines the splitting, that is, we have the general energy formula (neglecting centrifugal stretching terms)

$$F_{[v]}(J, K) = B_{[v]}J(J + 1) + (A_{[v]} - B_{[v]})K^2 \mp 2A_{[v]}\zeta_t K. \qquad \text{(I, 117)}$$

As discussed in section 2, $\zeta_t$ is, for small vibronic interaction, simply the sum or difference of the vibrational and electronic angular momentum

$$|\zeta_t| = |\zeta_e \pm \zeta_v|. \qquad \text{(I, 118)}$$

For the sign of $\zeta_t$ see the discussion on p. 64. For stronger vibronic interaction more complicated formulae apply such as (I, 68) for an $X_3$ molecule (neglecting the minima in the potential trough).

For molecules with three- and six-fold axes ($\boldsymbol{C}_{3v}$, $\boldsymbol{D}_{3h}$, . . ., $\boldsymbol{C}_{6v}$, $\boldsymbol{D}_{6h}$, . . .), the combination of a degenerate vibrational and a degenerate electronic state always leads to one degenerate ($E$) vibronic state and a pair ($A_1$, $A_2$) of non-degenerate vibronic states (see p. 39). One of the two $\zeta_t$ values of (I, 118) corresponds to the former, the other to the latter. However, if the splitting between $A_1$ and $A_2$ is large, the vibrational angular momentum is completely suppressed and the energy levels are given by (I, 102) or (I, 106). For intermediate or small splitting between $A_1$ and $A_2$ Hougen (572) has derived the formula (for prolate tops)

$$F_{[v]}(J, K) = B_{[v]}J(J + 1) + (A_{[v]} - B_{[v]})K^2 \mp \sqrt{\tfrac{1}{4}(\Delta\nu)^2 + 4A_{[v]}^2K^2\zeta_t^2}, \qquad \text{(I, 119)}$$

where $\Delta\nu$ is the $A_1$, $A_2$ splitting for zero rotation. It is immediately seen that for very small $\Delta\nu$ this formula goes over into (I, 117), while for very large $\Delta\nu$ it goes over into (I, 102) for each component $A_1$ and $A_2$. If $\Delta\nu$ is not too large, development of (I, 119) leads to a slightly different effective $A_{[v]}$ value for the two components, viz.,

$$A_{[v]}\left(1 \pm \frac{4A_{[v]}\zeta_t^2}{|\Delta\nu|}\right) \qquad \text{(I, 120)}$$

where the upper sign goes with the upper component ($+\frac{1}{2}|\Delta\nu|$), the lower sign with the lower component ($-\frac{1}{2}|\Delta\nu|$).

**Rotational levels in multiplet electronic states.** In orbitally non-degenerate electronic states, spin-orbit coupling is usually very small, just as in $\Sigma$ electronic states of linear or diatomic molecules [Hund's case (b)], but it will increase with increasing $J$ and $K$. As for linear molecules, we introduce the quantum number $N$ of the total angular momentum apart from spin which now replaces $J$ in all the previous symmetric top formulae. The spin $\boldsymbol{S}$ is added to $\boldsymbol{N}$ to give the total angular momentum

$$\boldsymbol{J} = \boldsymbol{N} + \boldsymbol{S} \qquad \text{(I, 121)}$$

and therefore we have for the quantum numbers (see Vol. I, p. 25)

$$J = N + S, N + S - 1, \ldots, |N - S|. \qquad \text{(I, 122)}$$

Thus, as for diatomic and linear polyatomic molecules, each level of given $N$ is split into $2S + 1$ components except when $N < S$. As examples, we present in

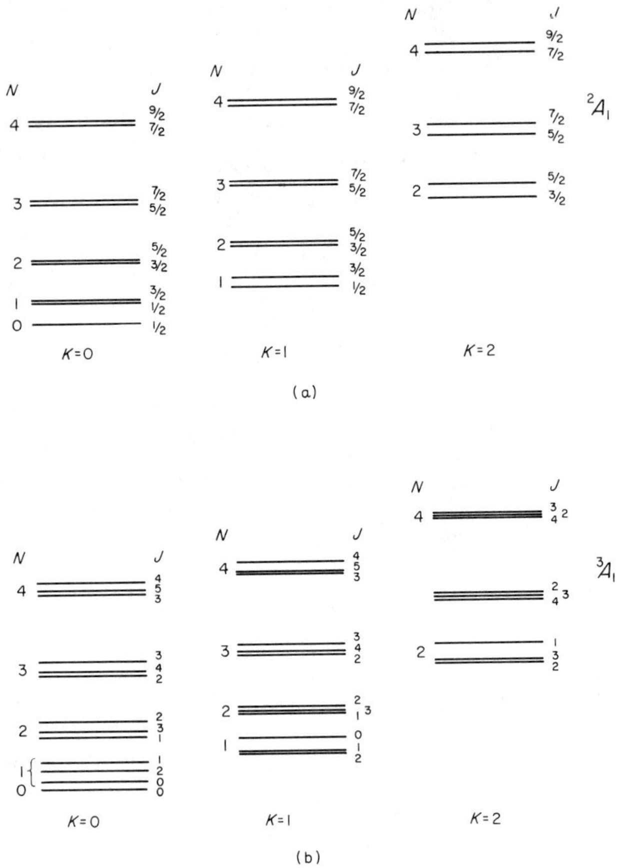


FIG. 34. **Rotational energy levels** (*a*) **in a** $^2A_1$ **and** (*b*) **in a** $^3A_1$ **vibronic state of a** $C_{3v}$ **molecule.** The magnitude of the doublet and triplet splitting is greatly exaggerated but roughly drawn to scale in different levels according to the splitting formulae (I, 123) and (I, 124).

Fig. 34 the lowest rotational energy levels of a $^2A_1$ and a $^3A_1$ vibronic state of a $C_{3v}$ molecule.

Splitting formulae can readily be obtained from the general formulae for asymmetric top molecules derived by Henderson (493) [see also Raynes (1059)]. One finds for doublet states

$$F_1(N, K) = F_0(N, K) + \frac{1}{2}\left(\kappa \frac{K^2}{N(N+1)} + \mu\right)N$$

$$F_2(N, K) = F_0(N, K) - \frac{1}{2}\left(\kappa \frac{K^2}{N(N+1)} + \mu\right)(N+1). \qquad \text{(I, 123)}$$

Here $F_0(N, K)$ is the rotational term value without spin, while $F_1$ and $F_2$ are the term values for $J = N + \frac{1}{2}$ and $J = N - \frac{1}{2}$ respectively; $\kappa$ and $\mu$ are spin-rotation coupling constants which depend in a rather complicated way on the moments of

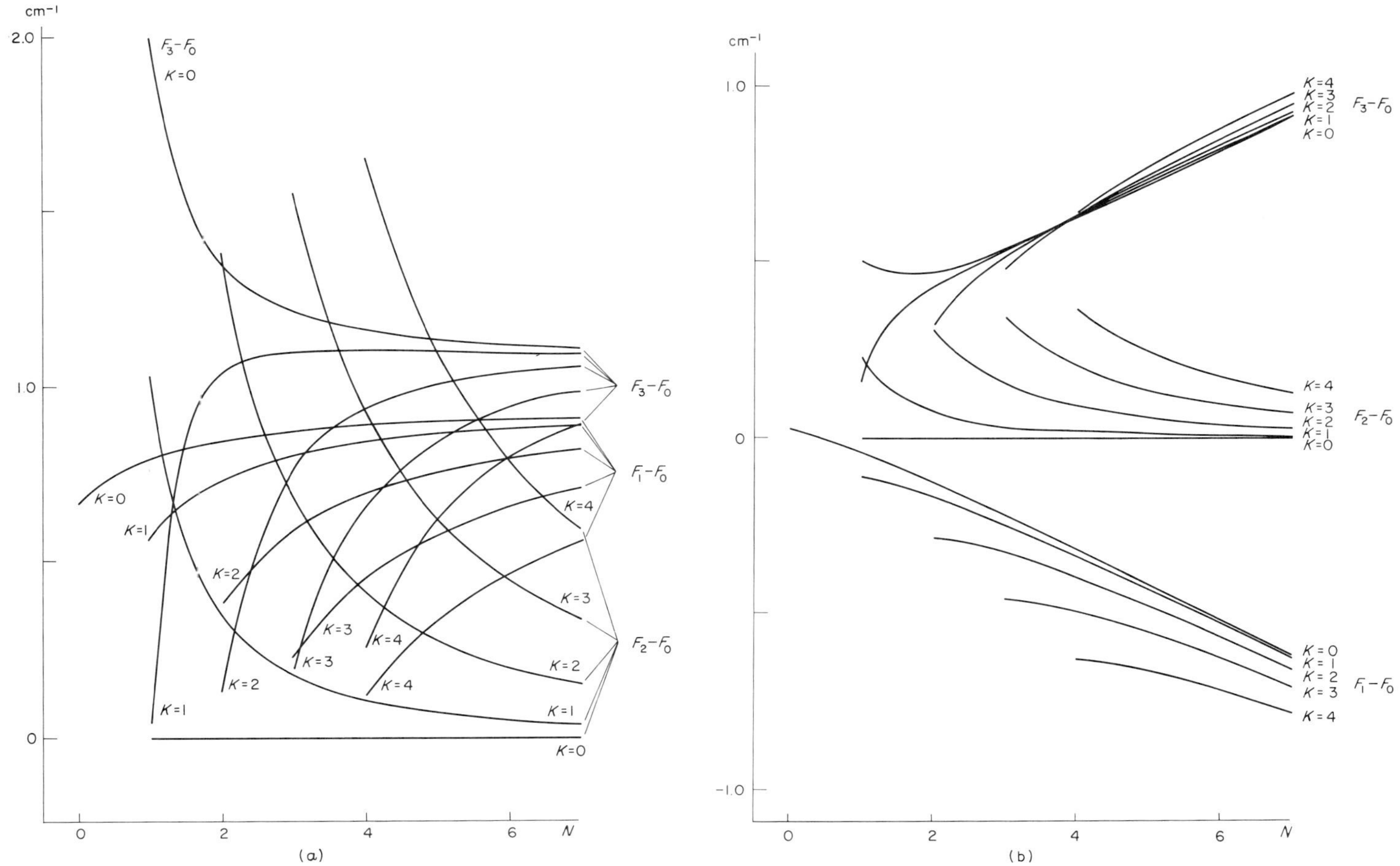


**FIG. 35. Variation of triplet splitting with $N$ in a non-degenerate electronic state of a symmetric top molecule.** (*a*) For $\lambda = -1.0$, $\mu = -0.004$, $\kappa = -0.06$; (*b*) for $\lambda = -0.2$, $\mu = -0.1$, $\kappa = -0.06$. The quantities $F_i - F_0$ as obtained from eqs. (I, 124) are plotted against $N$. To obtain the actual energies of the levels $F_0(N, K)$ must be added.

inertia and the average distances between the nuclei and the unpaired electron. The coupling constants $a_0$ and $a$ used by Van Vleck (1238) and Raynes (1059) are related to those used here by $\kappa = -3a$, $\mu = a - a_0$. If $\kappa$ is small, the dependence of the splitting on $N$ is the same as for $^2\Sigma$ states of linear molecules. However, in general $\kappa$ is larger than $\mu$ and therefore for a given $K$ the splitting decreases rather than increases with $N$ (see Fig. 34a).

For triplet states an additional term arises on account of spin-spin interaction. One finds for the component levels [see Henderson (493) and Raynes (1059)]

$$\begin{aligned} F_1(N, K) &= F_0(N, K) + \left(\kappa \frac{K^2}{(N+1)^2} + \mu\right)(N+1) - 2\lambda \frac{(N+1)^2 - K^2}{(N+1)(2N+3)} \\ F_2(N, K) &= F_0(N, K) - \kappa \frac{K^2}{N(N+1)} - 2\lambda \frac{K^2}{N(N+1)} \\ F_3(N, K) &= F_0(N, K) - \left(\kappa \frac{K^2}{N^2} + \mu\right)N - 2\lambda \frac{N^2 - K^2}{N(2N-1)}. \end{aligned} \qquad \text{(I, 124)}$$

Here $\lambda$ is the spin-spin coupling constant similar to the $\lambda$ occurring for $^3\Sigma$ states of linear molecules (corresponding to Van Vleck's $\alpha$ which is $\frac{2}{3}\lambda$). Indeed for $K = 0$ the formulae (I, 124) go over into Kramers' formulae for $^3\Sigma$ (see Vol. I, p. 223) which give a first approximation to the Schlapp formulae (I, 87) when $\lambda$ is small compared to $B$. When $\lambda$ is not small compared to $B$ additional terms have to be added to (I, 124) which we shall not discuss, but which are fully dealt with by Raynes (1059).

As an illustration, in Fig. 35a and b, the variation of the three triplet components with $N$ according to (I, 124) is shown [after subtraction of $F_0(N, K)$] for two sets of values of the constants, one with large $|\lambda|$ and small $|\mu|$, and the other with smaller $|\lambda|$ but much larger $|\mu|$. In both examples $\kappa$ has been assumed to have the same value ($\kappa = -0.06$).

In degenerate electronic states, since in general $\zeta_e$ is non-zero, there is an orbital magnetic moment in the direction of the symmetry axis and therefore a fairly large spin splitting is expected similar to that for $\Pi, \Delta, \ldots$ electronic states of linear molecules. Thus far, no detailed theoretical discussion of this case has been given. One would expect that for large multiplet splitting the rotational energy levels in each multiplet component can be represented by effective rotational constants which are slightly different for the different multiplet components.

**Symmetry properties of rotational levels.** The symmetry properties of the rotational eigenfunctions of symmetric top molecules in non-degenerate electronic states have been discussed in detail in Volume II. At the time of writing Volume II, it was generally considered that for the classification of the rotational and over-all (*rovibronic*) functions only the rotational sub-group need and should be used. Since then, it has been shown by Hougen (571)(573) that (except for linear molecules) a classification according to the full symmetry group of the molecule is also possible and sometimes necessary. In Fig. 36 we reproduce the diagram of rotational levels from Volume II (Fig. 118) with the completed

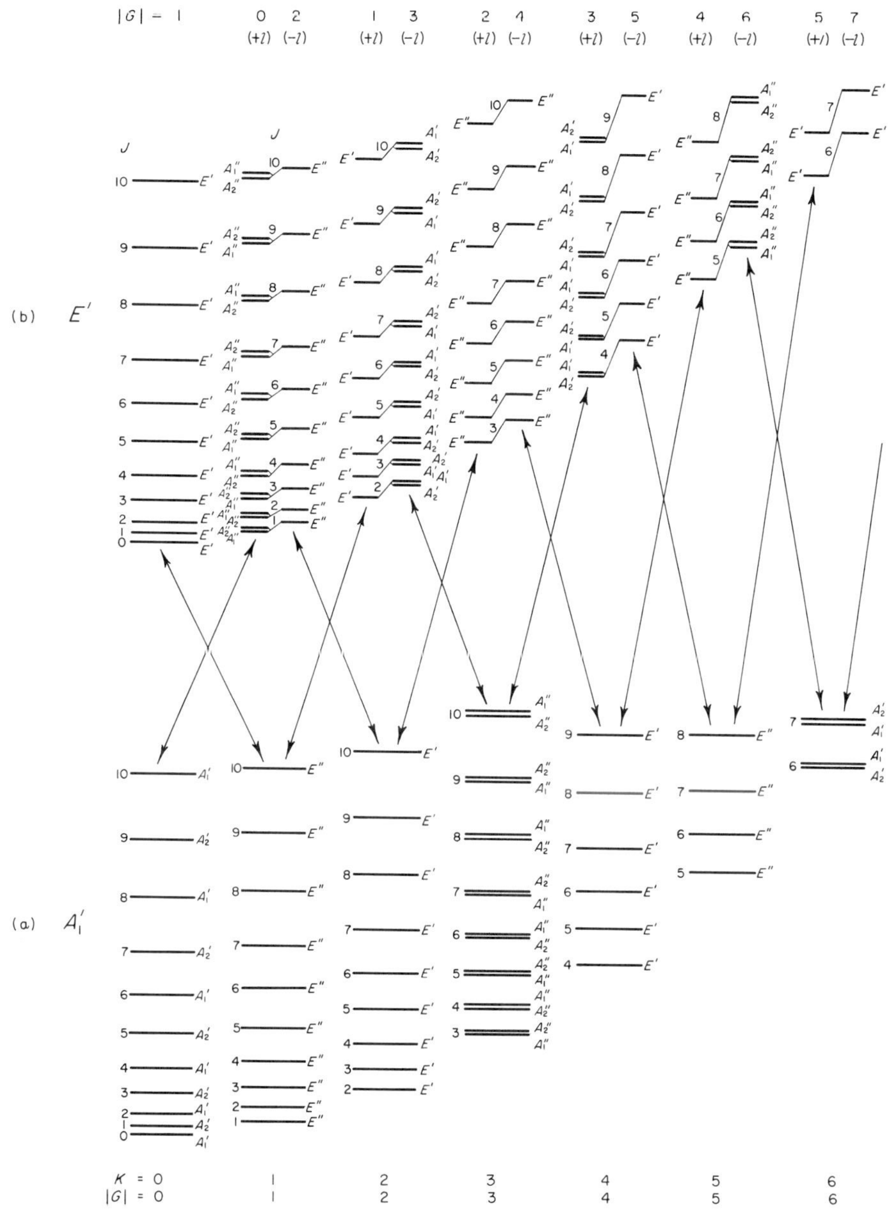

FIG. 36. **Rotational energy levels in $A_1'$ and $E'$ vibronic states of $D_{3h}$ molecules with rovibronic species designations.** At top and bottom Hougen's quantum number $G$ is given. The oblique arrows give the transitions for a $\perp$ band (see p. 230f). By dropping all ′ and ″ the corresponding diagram for a $D_3$ or $C_{3v}$ molecule is obtained.

species designations. These designations apply to a molecule of point group $\boldsymbol{D}_{3h}$, while the original figure giving the species for the rotational sub-group $\boldsymbol{D}_3$ can also be used for the full group $\boldsymbol{C}_{3v}$ (obtained from Fig. 36 by dropping all $'$ and $''$).

For a totally symmetric (non-degenerate) electronic state ($A_1'$; Fig. 36a) the rotational eigenfunctions and therefore the rovibronic eigenfunctions for $K = 3, 6, 9, \cdots$ are non-degenerate; but there are always two levels, $A_1$ and $A_2$, for each $J$ value because of the $K$ degeneracy; for $K = 6, 12, \ldots$ they are $A_1'$ and $A_2'$, for $K = 3, 9, \ldots$ they are $A_1''$ and $A_2''$. For $K = 0$ there is only one function for each $J$ value, which is alternately $A_1'$ and $A_2'$. All other rotational levels have species $E$, that is, have a degeneracy that cannot be split by any higher order interactions; the species is $E'$ for even $K$, $E''$ for odd $K$. In an $A_2'$ electronic state the subscripts 1 and 2 in Fig. 36a must be interchanged, while for $A_1''$ and $A_2''$ electronic states the $'$ and $''$ must be interchanged. It may be noted that for axial point groups with a four-, five- or six-fold axis, the species will run through a cycle of 4, 5 or 6 members as a function of $K$. If there is a plane of symmetry $\sigma_h$ the length of the cycle is doubled for three-, five-, . . ., fold axes as in the example of Fig. 36.

In a degenerate electronic (or vibronic) state, the rovibronic species are the same as for a degenerate vibrational level as discussed in Volume II. Figure 36b shows this for an $E'$ vibronic state of a $\boldsymbol{D}_{3h}$ molecule. Here all the levels with $K = 0$ are doubly degenerate ($E'$). For $K \neq 0$ we have the Coriolis splitting into two levels which for $K = 3, 6, 9, \ldots$ are both $E'', E', E'', \ldots$ respectively, while for intermediate $K$ values, one Coriolis component consists of two non-degenerate levels as shown. Again $'$ and $''$ alternate as a function of $K$.

In Volume II the levels represented by the upper sign in (I, 111) and (I, 112) were called $+l$ levels, those corresponding to the lower sign were called $-l$ levels, they are the lower and upper component levels respectively for positive $\zeta_v$ or $\zeta_t$. We shall very slightly modify this designation and, following Mills (855), write $(+l)$ and $(-l)$ levels in order to counteract the impression that the two levels correspond to positive and negative $l$. For large vibronic interaction, $l$ is no longer a good quantum number and the two sets of levels of eq. (I, 117) will then be called $(+j)$ and $(-j)$ levels since they are now characterized by the quantum number $j$ defined in section 2(c).

Hougen (571) has introduced in place of the designation $(+l)$ and $(-l)$ levels a new quantum number $G$ which is given by

$$G = G_e + G_v \mp K \qquad \text{(I, 125)}$$

where in the present example ($\boldsymbol{D}_{3h}$ molecule) $G_e = 0$ for non-degenerate and $G_e = \pm 1$ for degenerate electronic states, where $G_v = \sum(\pm l_i)$ and where $G$ is only defined modulo 3. The values of $|G|$ are added to the levels in Fig. 36. For a discussion of the $G$ values in other point groups see Hougen (571). The quantum number $G$ is convenient for the formulation of the selection rule for $(\pm l)$ levels (see Chap. II).

For linear (and diatomic) molecules, the over-all symmetry property $+$ and $-$ giving the behavior of the rovibronic eigenfunction with regard to inversion is of great importance. For axial molecules it is not as important unless the potential barrier opposing inversion is not very high. If it is high, then every level thus far considered really consists of two very nearly coinciding levels, one that is symmetric, and one that is antisymmetric with regard to inversion (i.e., a + and a − level). For most purposes we can disregard the fact that there are two levels rather than one level. But if the barrier is low, as, for example, in $NH_3$, we have a doubling, the inversion doubling, and we must take account of the + and − character of the levels (see Fig. 120, p. 412 of Vol. II). If the molecule is planar, each level is *either* + *or* −, but this character is then already implicitly taken into account in the other symmetry properties, at least when the full symmetry group is used to describe the rotational or rovibrational levels.

For a discussion of the statistical weights of the rotational levels we only need the behavior of the wave functions with regard to the rotational sub-group as discussed in detail in Volume II, p. 409f. In an axial molecule with a $p$-fold axis, there must be at least one set of $p$ identical nuclei. In our example of a $\boldsymbol{D}_{3h}$ molecule, if the identical nuclei have zero spin (and therefore follow Bose statistics), only the levels of species $A_1$ of the rotational sub-group, i.e., $A_1'$ and $A_1''$ of the full symmetry group, actually occur, that is, there would be for an $E'$ vibronic state (see Fig. 36b) no levels with $K = 0, 3, 6, \ldots$ and alternately only the lower $(+l)$ and upper $(-l)$ components of the levels with $K = 1, 2, 4, 5, \ldots$. For a non-degenerate vibronic state ($A_1'$ or $A_2'$ or $A_1''$ or $A_2''$), only the levels with $K = 0, 3, 6, \ldots$ would be present (see Fig. 36a).

If the identical nuclei have spin $I = \frac{1}{2}$ (and follow Fermi statistics), both $A_2$ and $E$ rotational levels (i.e., $A_2'$, $A_2''$, $E'$ and $E''$ of the full symmetry group) occur but not $A_1$ (i.e., not $A_1'$ and $A_1''$). The nuclear spin contribution to the statistical weight of the $A_2$ levels is 4, to that of the $E$ levels it is 2. If the identical nuclei have $I = 1$ (and follow Bose statistics), all three species of rotational levels $A_1$, $A_2$, $E$ occur with statistical weights 10:1:8; and if the identical nuclei have spin $I = \frac{3}{2}$ (Fermi statistics) the weights are 4:20:20. Thus we have an alternation of statistical weights as a function of $K$ in a non-degenerate vibronic state: it is 4:2:2:4:2: . . . for $I = \frac{1}{2}$, it is 11:8:8:11:8: . . . for $I = 1$, and it is 24:20:20:24:20: . . . for $I = \frac{3}{2}$. For degenerate vibronic states the same alternation occurs for both the lower and the upper Coriolis component, but they are out of step (see Fig. 36b): the higher weight occurs for $K = 1, 4, 7, \ldots$ for the $(+l)$ levels and for $K = 2, 5, 8, \ldots$ for the $(-l)$ levels.

Because of the difference in statistical weight for $A_1$ and $A_2$ rotational levels, there is an alternation of weights as a function of $J$ for $K = 0$ of a non-degenerate vibronic state. For $A_1'$ and $A_1''$ vibronic states the weight ratio of even and odd rotational levels is 1:0 for $I = 0$, it is 0:4 for $I = \frac{1}{2}$, it is 10:1 for $I = 1$ and 4:20 for $I = \frac{3}{2}$, while for $A_2'$ and $A_2''$ vibronic states it is the reverse. The same weight ratios apply to the pairs of $A_1$, $A_2$ levels for each $J$ of $K = 3, 6, \ldots$ in a non-degenerate state, and similarly for the $A_1$, $A_2$ components of $(+l)$ or $(-l)$ levels with $K = 1, 2, 4, 5, \ldots$ of degenerate vibronic states (see Fig. 36).

For molecules of point group $C_{3v}$ the rotational sub-group has only species

$A$ and $E$, that is, $A_1$ and $A_2$ levels of Fig. 36 have the same weight which is equal to the sum of the weights for $\boldsymbol{D}_{3h}$, that is, 1, 4, 11, 24 for $I = 0, \frac{1}{2}, 1, \frac{3}{2}$ respectively. The alternation of weights as a function of $K$ is the same as for $\boldsymbol{D}_{3h}$, but there is no alternation for $K = 0$ as a function of $J$.

As mentioned earlier, all the levels of $\boldsymbol{C}_{3v}$ molecules have a double degeneracy due to the possibility of inversion. If this degeneracy is split (inversion doubling), then the distinction between $A_1$ and $A_2$ must be made since then the molecule can be considered as belonging to point group $\boldsymbol{D}_{3h}$. Thus, for $K = 0$ there is again an alternation of weights for each inversion doubling component.

TABLE 4. ALTERNATION OF STATISTICAL WEIGHTS IN AXIAL MOLECULES[a]

| Molecule | Point group | Nuclear spin of identical atoms | Alternation in $K$[b] | Alternation in $J$ for $K = 0$[c] |
|---|---|---|---|---|
| $XY_3$, $XYZ_3$ | $\boldsymbol{C}_{3v}$ | $I = 0$ | 1:0:0:1 | — |
| | | $I = \frac{1}{2}$ | 4:2:2:4 | — |
| | | $I = 1$ | 11:8:8:11 | — |
| | | $I = \frac{3}{2}$ | 24:20:20:24 | — |
| $XY_3$, $X_3$ | $\boldsymbol{D}_{3h}$ | $I = 0$ | 1:0:0:1 | 1:0 |
| | | $I = \frac{1}{2}$ | 4:2:2:4 | 0:4 |
| | | $I = 1$ | 11:8:8:11 | 10:1 |
| | | $I = \frac{3}{2}$ | 24:20:20:24 | 4:20 |
| cyclo $C_3H_6$ | $\boldsymbol{D}_{3h}$ | $I = \frac{1}{2}$ | 24:20:20:24 | 8:16 |
| $XY_4$ | $\boldsymbol{C}_{4v}$ | $I = 0$ | 1:0:0:0:1 | — |
| | | $I = \frac{1}{2}$ | 4:3:6:3:4 | — |
| | | $I = 1$ | 24:18:21:18:24 | — |
| $XY_4$ | $\boldsymbol{D}_{4h}$ | $I = 0$ | 1:0:0:0:1 | 1:0 |
| | | $I = \frac{1}{2}$ | 4:3:6:3:4 | 1:3 |
| | | $I = 1$ | 24:18:21:18:24 | 21:3 |
| $X_2Y_6$ ($C_2H_6$)[d] | $\boldsymbol{D}_{3d}$ | $I_X = 0, I_Y = \frac{1}{2}$ | 24:20:20:24 | 8:16 |
| | | $I_X = 0, I_Y = 1$ | 249:240:240:249 | 138:111 |
| $C_6H_6$ | $\boldsymbol{D}_{6h}$ | $I = \frac{1}{2}$ | 10:11:9:14:9:11:10 | 7:3 |
| $C_3H_4$ (allene) | $\boldsymbol{D}_{2d}$ | $I = \frac{1}{2}$ | 10:6:10:6:10 | 7:3 |

[a] The alternations given here hold for a totally symmetric vibronic state.
[b] The statistical weights for $K = p, p + 1, \ldots, 2p$ are given where $p$ is the order of the symmetry axis.
[c] The ratio of statistical weights for even and odd levels is given which is also the ratio of the $A_1$, $A_2$ pair for $K \neq 0$.
[d] Assuming a high barrier.

The symmetry properties and statistical weights of the rotational levels for molecules of several other point groups have been discussed by Placzek and Teller (988), Wilson (1305)(1306), Schäfer (1100) and Mizushima (862). Although their discussion is based on the rotational sub-group, it would be easy to translate their results to the full symmetry group with the help of Hougen's paper (571). In Table 4 the results of the preceding discussion with regard to the statistical weights are summarized, including a number of cases not treated above.

**$l$-type doubling.** In addition to the first order effect of Coriolis interaction which causes the splitting into $(+l)$ and $(-l)$ [or $(+j)$ and $(-j)$] levels, Coriolis forces produce further higher order splittings and shifts of energy levels. Similar splittings and shifts are also produced by anharmonic terms in the potential energy and by the fact that in the displaced positions of non-totally symmetric vibrations the molecule is no longer an exact symmetric top. The most important of these splittings is the analogue of the $l$-type doubling of linear molecules. It was first recognized by Nielsen (934) in the microwave spectra of $CH_3CN$ and $CH_3NC$ and by Garing, Nielsen and Rao (406) in the infrared spectrum of $NH_3$.

Without considering higher order effects, the $A_1$ and $A_2$ levels for a given $J$ and $K$ would have the same energy, just as have the two component levels of given $J$ of a $\Pi$ vibronic state of a linear molecule. When a degenerate vibration $\nu_i$ is excited, Coriolis interaction or simply rotation-vibration interaction will produce a splitting into two component levels which is called $l$-type doubling even though for symmetric top molecules unlike linear molecules the (vibrational) angular momentum is not $l_i(h/2\pi)$ but $l_i\zeta_i(h/2\pi)$ (see p. 64). Garing, Rao and Nielsen (406) have shown that, just as for linear molecules, for $K = 1$ the doubling is in a first good approximation given by

$$\Delta\nu = q_v J(J + 1). \tag{I, 126}$$

Here it must be emphasized that only the $(+l)$ component is split (see Fig. 36b); the $(-l)$ component has species $E$ and therefore cannot be split in any approximation. It must also be noted that in $XY_3$ if Y has a nuclear spin $\frac{1}{2}$ (e.g., when Y = H) alternately the upper and lower level of the $l$-type doublets is missing since only the $A_2$ levels occur (see Fig. 36). In principle, the $(-l)$ levels for $K = 2$, the $(+l)$ levels for $K = 4$, etc. will also be split, but just as the $l$-type doubling of $\Delta$, $\Phi$, ... states of linear molecules, here the doubling comes out only in a much higher approximation and is proportional to a correspondingly higher power of $J(J + 1)$. The same applies to the $A_1A_2$ doublets for $K = 3$ of a non-degenerate vibrational level (see Fig. 36a).

Again, as for the linear molecules, the splitting constants $q$ can be resolved into two components

$$q_v = q_v^{\text{Cor.}} + q_v^{\text{as.}}, \tag{I, 127}$$

one due to Coriolis interaction with another vibrational level, and one due to the asymmetry of the displaced position. Garing, Nielsen and Rao (406) give[12] for $\nu_4$ of non-planar $XY_3$

$$q_v^{\text{Cor.}} = \frac{4\zeta_{24}{}^2 B_v{}^2 \omega_4}{\omega_4{}^2 - \omega_2{}^2}, \qquad q_v^{\text{as.}} = \frac{2B_v{}^2 a'}{\omega_4}, \tag{I, 128}$$

where $\zeta_{24}$ is a Coriolis interaction coefficient of magnitude $\leq 1$ (see Vol. II, p. 404) and $a'$ is a coefficient which is not given explicitly. Similar expressions would apply to other axial molecules.

It is readily seen that in an axial molecule when a degenerate vibration is

[12] Garing, Nielsen and Rao give for the first term

$$\frac{8\zeta_{24}{}^2 B_e{}^2}{8\pi^2 c^2(\omega_4{}^2 - \omega_2{}^2)}$$

which is dimensionally incorrect and was presumably intended to be the quantity given in the text.

excited, rotation about an axis perpendicular to the symmetry axis will give rise to a Coriolis force which tends to excite a non-degenerate vibration (compare the similar discussion in Vol. II, p. 374). Thus, since in $NH_3$ $\nu_4(e)$ and $\nu_2(a_1)$ are fairly close together, there will be Coriolis interaction between them, giving rise to $q_v^{\text{Cor.}}$ above. Similarly, as we shall discuss in more detail below, a slight deviation from a symmetric top gives rise to a $K$-type doubling which for $K = 1$ is proportional to $J(J + 1)$, thus accounting for the term $q_v^{\text{as.}}$; it is independent of the presence of other interacting vibrations.

The magnitude of $q_v$, according to (I, 128), depends essentially on $B^2/\omega_i$ just as for $l$-type doubling in linear molecules. For $NH_3$, $B = 9.94$ cm$^{-1}$ and therefore, with $\omega_4 = 1628$ cm$^{-1}$, $q_v$ is of the order of 0.06 cm$^{-1}$, while for $CO_2$, with $B = 0.391$ and $\omega_2 = 664$, $q_v$ is of the order 0.00023. Thus, on account of the much higher $B$ value the $l$-type doubling in $NH_3$ is much larger than in $CO_2$. Since the cause of $l$-type doubling is precisely the same in both cases we prefer not to adopt Garing, Nielsen and Rao's designation of "giant" $l$-type doubling for the $NH_3$ case. The magnitude of the $l$-type doubling, according to (I, 128), depends also on the proximity of other vibrations with which the degenerate vibration considered can interact. In (I, 128) only the most important term for $XY_3$ is given; in principle, other similar terms corresponding to interaction with other vibrations have to be added to $q_v^{\text{Cor.}}$, for example, for $XY_3$ the term $4\zeta_{14}{}^2B_v{}^2\omega_4/(\omega_4{}^2 - \omega_1{}^2)$, but these terms are in general small. For a recent more detailed discussion of $l$-type doubling see Weber (1281).

**$j$-type doubling.** As we have seen above, in degenerate *electronic* states the rotational energy levels are entirely similar to those for degenerate vibrational levels except that now $\zeta$ is either simply the electronic $\zeta_e$, or when degenerate vibrations are excited, the sum or difference of electronic and vibronic $\zeta$'s. Again we expect to find a splitting of the $A_1$, $A_2$ doublets for $K = 1$. This splitting will be called $j$-type doubling since it is now the quantum number $j$ that characterizes the vibronic levels (see p. 63f); the quantum number $l$ is zero for the lowest level of a degenerate electronic state, and yet there will be a doubling, namely, of $j$-type. Moreover for higher vibrational levels $l$ may not be well defined if the Jahn–Teller interaction is strong.

Child (191) has studied the rotational levels in a degenerate electronic state and has shown that the $(+j)$ levels with $K = 1$ of a $j = \frac{1}{2}$ vibronic state are split according to the formula (I, 126); however, here, particularly in the $v = 0$ level, the splitting does not originate from Coriolis interaction of different vibrations but is entirely due to vibronic-rotational interaction within the vibronic state considered. It vanishes when Jahn–Teller interaction is zero and is therefore not an analogue of $\Lambda$-type doubling resulting from pure electronic-rotational coupling. It may be thought of as being produced by the fact that in the equilibrium position of the degenerate electronic state the molecule is not symmetrical if Jahn–Teller interaction is non-zero, and that therefore we have a slightly asymmetric top which will exhibit a doubling in the $K = 1$ rotational levels similar to the asymmetry component of $l$-type doubling. Conversely, the observation of such a doubling in a vibronic level with $l = 0$ would represent a direct proof for an asymmetric equilibrium position in a degenerate electronic state as required by the Jahn–Teller theorem.

Child (191) gives the following formula for the coupling constant $q_v$ in a degenerate electronic state of an $X_3$ molecule

$$q_v = 2aB\sqrt{\frac{B}{\omega_2}} \qquad \text{(I, 129)}$$

where $a$ is a constant of order 1 which depends on the Jahn–Teller parameter $D$ [$a = 0$ for $D = 0$; $a = 0.832$ for $D = 0.125$ and $v_2 = 0$; $a = 0.394$ for $D = 0.125$ and $v_2 = 1$; for other values see Child (191)]. It is significant that the dependence of the $j$-type doubling constant on $B$ and $\omega_2$ is somewhat different from that in the asymmetry term $q_v^{\text{as.}}$ of $l$-type doubling (see eq. I, 128). If the two parameters $a$ and $a'$ were equal, $j$-type doubling would be larger by a factor $\sqrt{\omega_2/B}$ than the asymmetry component of $l$-type doubling. In view of the difficulty of determining the Jahn–Teller parameter $D$ experimentally, it may be pointed out that if $q_v$ is experimentally determined for a $j = \frac{1}{2}$ state in a degenerate electronic state, one can obtain from it directly this parameter $D$, if the frequency of the degenerate vibration causing Jahn–Teller instability is known, since the constant $a$ in (I, 129) is a function of $D$ (see the tables given by Child). To be sure, Child's formulae hold, strictly speaking, only for an $X_3$ molecule but could probably be fairly easily generalized.

As for $l$-type doubling, the $j$-type doubling for the $(-j)$ levels of $K = 2$, the $(+j)$ levels of $K = 4$, etc. is very much smaller than that of the $(+j)$ levels of $K = 1$ and varies with the fourth or a higher power of $J$. On the scale used in Fig. 36b it was not possible to show this variation: the magnitude of the splitting shown in this figure is much exaggerated. For vibronic levels with $j > \frac{1}{2}$ splittings of the levels with $K = 1$ are not expected in the approximation considered by Child. It is, however, likely, although not explicitly stated by Child, that for higher $j$ values when Jahn–Teller interaction is small, $l$-type doubling will arise as long as $l$ is approximately defined.

In addition to the cause of $j$-type doubling considered above, one may expect that when other electronic states are nearby or even other suitable vibronic states of the same electronic state, Coriolis interactions will add to the magnitude of the splitting, but this has not yet been discussed in detail.

**Other effects of higher order rovibronic interactions.** The same interactions that produce $l$-type doubling also produce contributions to the rotational constants $\alpha_i^B$. These contributions are of opposite sign for the two vibrations $\nu_i$ and $\nu_k$ between which Coriolis interaction takes place. For example, in an $XY_3$ molecule, according to Garing, Nielsen and Rao (406), $B_{v_2=1}$ is smaller by $q_v^{\text{Cor.}}$, $B_{v_4=1}$ is larger by $\frac{1}{2}q_v^{\text{Cor.}}$ than it would have been without this interaction. This modification of the $B_{[v]}$ values is taken care of by using appropriate $\alpha_i^B$ and need not be further discussed [compare the explicit formulae for the $\alpha_i$ of $XY_3$ and $XYZ_3$ molecules given by Silver and Shaffer (1129) and Shaffer (1117)(1118)]. There is also a lowering of $\zeta_i$ by an amount $q_v^{\text{Cor.}}/4C_v$ which, in a way similar to the corrections on $B_v$, may be considered as a contribution to $\alpha_i^\zeta$.

While there is, in the present approximation, no $l$-type doubling for $K > 1$, there is an effect on the separation of the $(+l)$ and $(-l)$ levels when $q_v$ is large. Without the higher order effects here considered, i.e., according to eq. (I, 113 or 114), the separation of corresponding $(+l)$ and $(-l)$ levels for a given $K$ is independent of $J$ (see Fig. 36b). According to Garing, Nielsen and Rao (406), the higher order effects produce an additional term in the energy formula which is quartic in $J$. This term is[13] for oblate $XY_3$ molecules

$$\pm q_v^2 \frac{[J(J+1) - K(K \mp 1)][J(J+1) - (K \mp 1)(K \mp 2)]}{16(K \mp 1)[(1-\zeta_v)C_e - B_e]} \tag{I, 130}$$

where the upper signs refer to the $(+l)$ levels, the lower to the $(-l)$ levels. For positive $\zeta_v$, since $C_v < B_v$, the term (I, 130) leads to an increase in the $(+l) - (-l)$ splitting, for sufficiently negative $\zeta_v$ it leads to a decrease of that splitting.

The term (I, 130) must also be added to (I, 114) for $K = 0$ and for the $(-l)$ levels of $K = 1$ (which are not affected by $l$-type doubling). For $K = 0$ the $\mp$ in the denominator cancels the $\pm$ sign in front of $q_v^2$, and thus only a single value for each $J$ is obtained as required since the $K = 0$ levels are not split. For the $(-l)$ levels of $K = 1$ the lower sign has to be used throughout. In all these cases the term (I, 130) has the same effect as the term in $D_J$ and is very difficult to separate from it in the analysis of observed spectra.

As an example, Fig. 37a shows the observed separations for the $(+l)$ and $(-l)$ levels of the $\nu_4$ vibrational level in the ground state of $NH_3$ for $K = 1$ and $K = 2$.

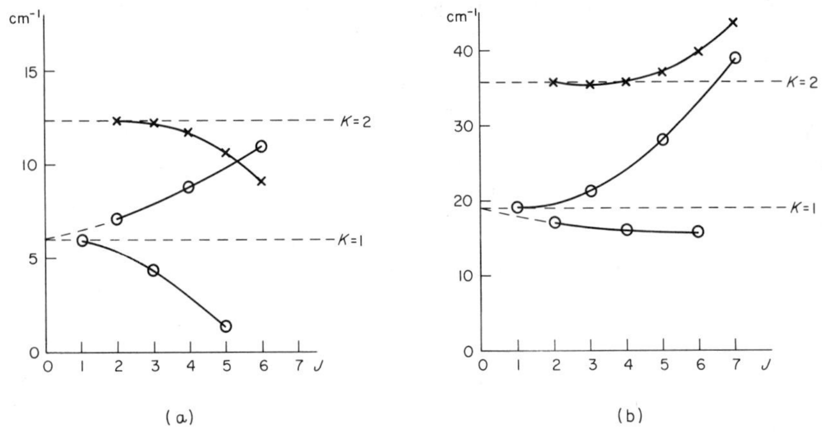


FIG. 37. **Observed first order Coriolis splitting** (*a*) **in the 0001 level of the ground state (vibronic species *E*) and** (*b*) **the 0100 level of the *E*″ upper state of the 1600 Å bands of $NH_3$.** The broken lines indicate what the splitting would be without higher order effects. The $K = 1$ levels show a marked $l$-type (or $j$-type) doubling of the $(+l)$ levels; therefore two curves result for the separations of corresponding $(+l)$ and $(-l)$ levels. For $NH_3$, for one set only even for the other only odd $J$ values occur.

In both cases the points would fall on horizontal straight lines if the higher order terms were not present. For $K = 1$ there are two curves instead, corresponding to the $l$-type doubling in the $(+l)$ component; the average of the two curves drops slightly at high $J$; but for $K = 2$ there is an appreciable drop, i.e., an appreciable reduction of the splitting between $(+l)$ and $(-l)$ levels for high $J$ in accordance with (I, 130) ($\zeta_4$ is sufficiently negative).

A term very similar to (I, 130) must be added according to Child (191) to the rotational energy formula in a $j = \frac{1}{2}$ vibronic state of a degenerate electronic state,

[13] Including the correction of sign given by Rao, Brim, Hoffman, Jones and McDowell (1051).

when Jahn–Teller interaction is present (even if a degenerate vibration is *not* excited). But now $q_v$ is given by (I, 129), and therefore the shifts are liable to be much larger unless the Jahn–Teller parameter $D$ is very small. A probable example has been observed in the excited $E''$ state of $NH_3$ by Douglas and Hollas (295). Figure 37b shows the $(+j) - (-j)$ splittings for $K = 1$ and 2 for the state in which the non-degenerate vibration $\nu_2$ is singly excited. If the much smaller scale is noted, the greater $j$-type doubling and the greater deviation of the $(+j) - (-j)$ splitting from a constant value is clearly seen. To be sure, both the doubling and the shifts may be partly (or wholly) due to interaction with the nearby $A_2''$ state.

It is hardly necessary to emphasize that local perturbations for a small range of $J$ values can occur for symmetric top molecules just as for linear molecules and for the same reasons, i.e., on account of Fermi and Coriolis interactions between different vibrational levels of given electronic states, or between different vibronic levels of different electronic states.

## (c) Spherical top molecules

Since in spherical top molecules all three principal moments of inertia are the same, the zero-approximation formula for the rotational energy levels is even simpler than for symmetric top molecules, namely,

$$F^0_{[v]}(J) = B_{[v]}J(J + 1) - D_{[v]}J^2(J + 1)^2 \qquad \text{(I, 131)}$$

where as before

$$B_{[v]} = B_e - \sum \alpha_i(v_i + \tfrac{1}{2}d_i). \qquad \text{(I, 132)}$$

The only difference from symmetric top molecules is that $d_i$ can now have the values 1, 2 and 3, that is, we may have triply degenerate as well as doubly degenerate and non-degenerate vibrations if the molecule is a spherical top on account of symmetry.

Considering that here levels with different $K$ coincide, it is clear that each level of given $J$ has a $(2J + 1)$-fold degeneracy (in addition to the $(2J + 1)$-fold space degeneracy). This degeneracy is partially removed, that is, we have a splitting into a number of sub-levels on account of the finer interaction of vibration and rotation to which now the interaction of electronic motion and rotation has to be added.

For a discussion of these splittings it is necessary to consider first the symmetry types of the rotational eigenfunctions.

**Symmetry properties of rotational levels.** In Volume II, p. 449, the rotational levels of the spherical top were classified according to the rotational sub-group of the point group considered. According to Hougen (573), just as for symmetric top molecules, it is possible and for some purposes necessary to classify them according to the full symmetry of the point group. Hougen has shown that the rotational eigenfunctions of a spherical top behave like the even species $D_{Jg}$ of the continuous rotation-inversion group $\boldsymbol{K_h}$ (see Table 55 of Appendix I). These species are $(2J + 1)$-fold degenerate. They must be resolved into the species of the point group of the molecule considered. We shall only discuss tetrahedral molecules of point group $\boldsymbol{T_d}$ which has the species $A_1$, $A_2$, $E$, $F_1$, $F_2$. These are

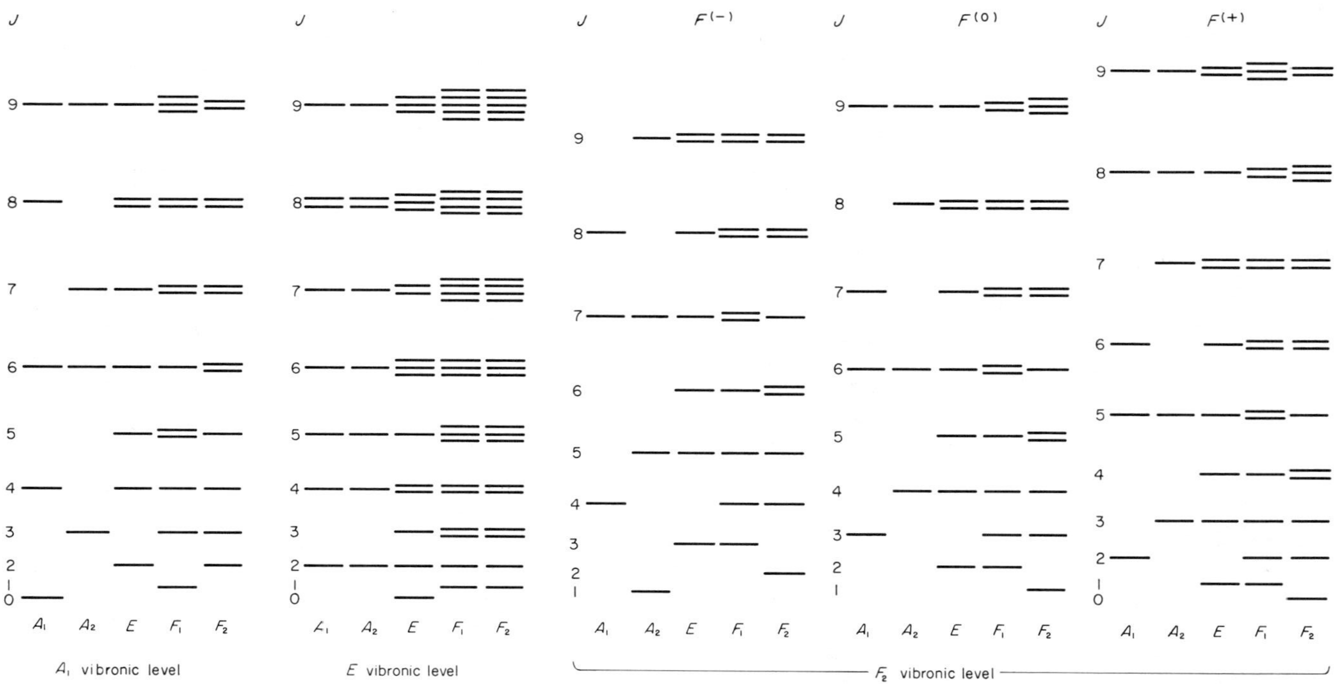


**FIG. 38. Number and rovibronic species of rotational levels of tetrahedral molecules in $A_1$, $E$ and $F_2$ vibronic levels.** For $A_2$ and $F_1$ vibronic levels the same rotational levels result as for $A_1$ and $F_2$ vibronic levels, respectively, except that the rovibronic species $A_1$ and $A_2$ and $F_1$ and $F_2$ have to be exchanged throughout. For the $F_2$ vibronic levels three sets of rotational levels arise called $F^{(-)}$, $F^{(0)}$ and $F^{(+)}$. The spacings of sub-levels for a given $J$ are not to scale. Note that Jahn (617) and Hecht (485) exchange $_1$ and $_2$ for odd $J$ values.

the possible species of the rotational levels. For the lowest $J$ values the reduction of $D_{Jg}$ into the species of $\boldsymbol{T}_d$ is given in Table 58 of Appendix IV. The lowest level $J = 0$ has species $A_1$; the next level $J = 1$ has species $F_1$, i.e., neither of these levels can be split in any approximation. For $J = 2$ we obtain $E + F_2$, for $J = 3$ we obtain $A_2 + F_1 + F_2$, i.e., here splittings are possible (see below).

In Fig. 38 the species of the rotational levels are given up to $J = 9$ for $A_1$, $E$ and $F_2$ vibronic states (compare Fig. 138, p. 450 of Vol. II). The species of the rotational levels of the $E$ vibronic state have been obtained from those of the $A_1$ state simply by multiplication of all species by $E$, and similarly for $F_2$. Corresponding diagrams for $A_2$ and $F_1$ vibronic states differ from those for $A_1$ and $F_2$ by an exchange of the subscripts $_1$ and $_2$ throughout. It should be noted that Jahn (617) and Hecht (485) reverse the designations $A_1$, $A_2$ and $F_1$, $F_2$ *for odd $J$ values*. Figure 38 applies also to molecules of point group $\boldsymbol{O}$ and can be readily extended to octahedral molecules (point group $\boldsymbol{O}_h$) if $g$ and $u$ are added to the species, depending on whether the vibronic state is even or odd.

Because of the possibility of inversion, each one of the levels shown in Fig. 38 is actually double, one being $+$, the other $-$ with respect to inversion. But this doubling is entirely negligible in all cases thus far studied.

If the identical nuclei in a tetrahedral molecule have spin zero only the $A$ rovibronic levels (i.e., $A_1$ and $A_2$) of Fig. 38 exist. However, if $I \neq 0$ all five types of rotational levels occur but with different statistical weights. One finds (see Vol. II, p. 451) that for $I = \frac{1}{2}$ the weights are 5, 2 and 3 for $A$, $E$ and $F$ respectively, while for $I = 1$ they are 15, 12 and 18 respectively (the weights for $A_1$ are the same as for $A_2$, those for $F_1$ the same as for $F_2$).

**Fine structure in non-degenerate vibronic states.** Rotational levels of different species but the same $J$ of a given vibronic level are differently influenced by Coriolis interactions with the rotational levels of other vibronic levels or by centrifugal distortion and other higher order interactions. For this reason in a sufficiently high approximation there is a splitting into as many levels as indicated by the number of horizontal lines in Fig. 38. In other words, when the molecule is distorted by centrifugal forces or by non-totally symmetric vibrations, it is no longer an exact spherical top, and the reason for the $(2J + 1)$-fold degeneracy no longer applies: to an extent compatible with symmetry requirements the degeneracy is resolved. Jahn (617) and more recently Hecht (485) have discussed in detail the resulting splittings. Unfortunately these splittings cannot be represented by explicit formulae but depend on the matrix elements of the various perturbing terms.

Hecht has shown that in a good approximation the *relative* spacings for a set of levels of given $J$ are independent of the magnitude of the vibration-rotation (and presumably electronic-rotation) interactions. Most of the effect of these interactions can therefore be taken into account by adding to (I, 131) the term

$$\delta F_{[v]}(J) = D_{[v]}^{(t)} f(J, \kappa) \qquad \text{(I, 133)}$$

where $D^{(t)}_{[v]}$ is a constant, similar in magnitude to $D_{[v]}$, and characteristic for the particular vibronic level considered, and $f(J, \kappa)$ is a splitting function, the same for all levels, which depends on $J$ and the particular sub-level $\kappa$. Since it is not possible to give an explicit form for $f(J, \kappa)$, we give in Table 5 its numerical values for $J$ values up to 8. It will be noticed that the values of $f(J, \kappa)$ increase roughly as $J^2(J + 1)^2$. For the ground states of tetrahedral molecules, the term

Table 5. Splitting function $f(J, \kappa)$ for tetrahedral molecules [after Hecht (485)][a]

| $J$ | $\kappa$ | $f(J, \kappa)$ | $J$ | $\kappa$ | $f(J, \kappa)$ |
|---|---|---|---|---|---|
| 0 | $A_1$ | 0 | 6 | $A_1$ | 3780 |
| 1 | $F_1$ | 0 | | $F_1$ | 2880 |
| 2 | $F_2$ | 24 | | $F_2$ | 1862.2 |
| | $E$ | −36 | | $A_2$ | −1980 |
| 3 | $A_2$ | 360 | | $F_2$ | −3062.2 |
| | $F_2$ | 60 | | $E$ | −3420 |
| | $F_1$ | −180 | 7 | $F_1$ | 4762.5 |
| 4 | $F_2$ | 780 | | $E$ | 4224 |
| | $E$ | −120 | | $F_2$ | 2967 |
| | $F_1$ | −420 | | $A_2$ | 624 |
| | $A_1$ | −840 | | $F_2$ | −5676 |
| 5 | $F_1$ | 1661.4 | | $F_1$ | −7882.5 |
| | $E$ | 1260 | 8 | $F_2$ | 10038.9 |
| | $F_2$ | −840 | | $E$ | 8909.6 |
| | $F_1$ | −1661.4 | | $F_1$ | 5142.0 |
| | | | | $F_2$ | 881.1 |
| | | | | $E$ | −10589.6 |
| | | | | $F_1$ | −11022.0 |
| | | | | $A_1$ | −11760 |

[a] Note that for odd $J$ the designations $A_1$ and $A_2$ as well as $F_1$ and $F_2$ are exchanged compared to Hecht.

(I, 133) is in general exceedingly small compared to the rotational energy (I, 131). For example, for $CH_4$ in its ground state, $D^{(t)}_{[0]} = 4.0 \times 10^{-6}$ cm$^{-1}$, and therefore, using the $f(J, \kappa)$ of Table 13 the over-all splitting for $J = 8$, for example, is only 0.087 cm$^{-1}$.

The value of the splitting constant $D^{(t)}_{[v]}$ for a tetrahedral $XY_4$ molecule is, according to Hecht (485), in a first approximation, neglecting the dependence on $v$, given by

$$D^{(t)}_{[v]} = \frac{1}{5} B_e{}^3\left(\frac{1}{\omega_2{}^2} - \frac{\zeta_{23}{}^2}{\omega_3{}^2} - \frac{\zeta_{24}{}^2}{\omega_4{}^2}\right) \tag{I, 134}$$

where $\zeta_{23}$ and $\zeta_{24}$ are Coriolis-coupling coefficients similar to those defined in Volume II, p. 375. $(\zeta_{23}{}^2 + \zeta_{24}{}^2 = 1)$. The expression for $D^{(t)}_{[v]}$ may be compared with that for the ordinary centrifugal stretching constant $D_{[v]}$ for which Hecht obtains

$$D_{[v]} = 4B_e{}^3\left(\frac{2}{3\omega_1{}^2} + \frac{2}{15\omega_2{}^2} + \frac{\zeta_{23}{}^2}{5\omega_3{}^2} + \frac{\zeta_{24}{}^2}{5\omega_4{}^2}\right). \tag{I, 135}$$

For the $CH_4$ ground state this quantity is $1.0 \times 10^{-4}$ cm$^{-1}$. Thus in this case, the splittings are small compared to the shifts by centrifugal distortion. However, in excited (non-degenerate) electronic states the splittings are likely to be much larger, since other electronic states are in general much nearer and will cause larger interactions. In such cases the approximation on which (I, 133) is based may not be adequate.

**Fine structure in degenerate vibronic states.** In a triply degenerate state the Coriolis interaction produces a first order splitting of the degeneracy (see Vol. II, p. 447). According to Teller (1196), the rotational levels are given in a first approximation by

$$\begin{aligned} F^{(+)}_{[v]}(J) &= B_{[v]}J(J+1) + 2B_{[v]}\zeta(J+1) \\ F^{(0)}_{[v]}(J) &= B_{[v]}J(J+1) \\ F^{(-)}_{[v]}(J) &= B_{[v]}J(J+1) - 2B_{[v]}\zeta J. \end{aligned} \quad \text{(I, 136)}$$

Here $\zeta$ is, just as for symmetric top molecules, either a purely vibrational angular momentum (when the electronic state is non-degenerate) or a purely electronic angular momentum (when the electronic state is degenerate but no degenerate vibrations are excited) or a vibronic angular momentum (when degenerate vibrations are excited in a degenerate electronic state). As before, $\zeta$ is not integral, and its magnitude in simple cases is less than 1. In Fig. 39, as an example, the rotational levels in a degenerate state with $\zeta = 0.2$ are plotted. For $J = 0$ only the $F^{(+)}$ level exists since a total angular momentum $J = 0$ can only be obtained in one way, i.e., when the purely rotational angular momentum is opposite and equal to the vibronic angular momentum ($\zeta$). Note that Jahn–Teller instability in a triply degenerate electronic state does not remove the triple degeneracy of the vibronic levels but Coriolis interaction does in the rotating molecule.

Some authors [see Hecht (485)] introduce a quantum number $R$ corresponding to the purely rotational angular momentum. If $\zeta$ were equal to 1, one would have $J = R + 1$, $R$, $R - 1$. Since the rotational energy is $B_{[v]}R(R+1)$, one obtains on substitution of $R = J - 1, J, J + 1$ the three eqs. (I, 136) with $\zeta = 1$, confirming that $F^{(+)}(J)$ corresponds to levels in which rotational and vibronic angular momentum are opposite to each other.

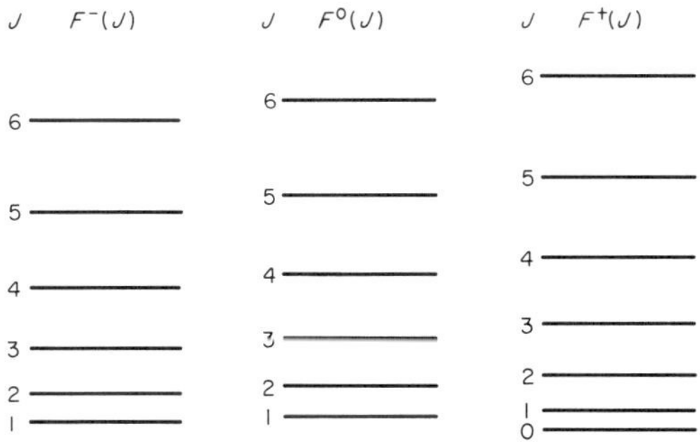


FIG. 39. **Rotational levels in a triply degenerate vibronic level of a tetrahedral or octahedral molecule for $\zeta = 0.2$.** Note that a level with $J = 0$ exists only for the $F^+(J)$ series. It has $F_2$ rovibronic symmetry, i.e., is triply degenerate.

When the finer interaction of rotation and vibronic motion is taken into account, each rotational level of Fig. 39 is split into as many sub-levels as indicated by the number of horizontal lines in the right-hand part of Fig. 38. However, now the splittings are in general much larger than for a non-degenerate state. In a second approximation one finds, according to Hecht (485), the following terms that have to be added to the expressions (I, 136):

$$\begin{aligned}\delta F_{[v]}^{(+)}(J) &= \alpha^* J(J+1) + 2\alpha^*\zeta(J+1) - D_{[v]}^{(+)}(J)\, f(J+1, \kappa)\\ \delta F_{[v]}^{(0)}(J) &= \alpha_0 J(J+1) + D_{[v]}^{(0)}(J)\, f(J, K)\\ \delta F_{[v]}^{(-)}(J) &= \alpha^* J(J+1) - 2\alpha^*\zeta J - D_{[v]}^{(-)}(J)\, f(J-1, \kappa).\end{aligned} \qquad \text{(I, 137)}$$

Here the quantities $\alpha^*$ and $\alpha_0$, which are of the order of ordinary $\alpha$'s, represent a slight change of the effective $B_{[v]}$ values, different for the central component from what it is for the two others. (In terms of the quantities defined by Hecht for the level $\nu_3$, $\alpha^* = -\frac{1}{3}Z_{3s} + 2F_{3s}$, $\alpha_0 = +\frac{2}{3}Z_{3s} + F_{3s}$.) The splitting function $f(J, \kappa)$ in the present approximation (neglect of non-diagonal terms) is the same as for non-degenerate states, but the coefficients $D_{[v]}^{(+)}(J)$, $D_{[v]}^{(0)}(J)$ and $D_{[v]}^{(-)}(J)$, and therefore the magnitudes of the splittings, are very much larger. Moreover, these coefficients depend on $J$ and in general decrease with increasing $J$. While the relative spacings in the splitting pattern of the $F_{[v]}^{(0)}(J)$ set are approximately the same as those of a non-degenerate state of equal $J$ as given in Table 5, eq. (I, 137) shows that for the $F_{[v]}^{(+)}(J)$ set the agreement in the splitting pattern exists between $F_{[v]}^{(+)}(J-1)$ and $f(J, \kappa)$, and for the $F_{[v]}^{(-)}(J)$ set between $F_{[v]}^{(-)}(J+1)$ and $f(J, \kappa)$ except for the sign. This agreement is also in accord with the resolution into the various species shown in Fig. 38. Note, however, that for an $F_2$ vibronic level to which Fig. 38 applies, the $A_1$, $A_2$ and similarly the $F_1$, $F_2$ designations are exchanged compared to an $A_1$ vibronic level. For large splittings and large $J$, non-diagonal matrix elements become important and produce additional shifts not taken into account in (I, 137).

The only example for which the formulae (I, 137) have been compared with experiment is $CH_4$; see Hecht (485) and Herranz and Stoicheff (514).

No detailed discussion of the rotational levels in doubly degenerate vibronic levels of tetrahedral molecules has yet been given. Nor has the effect of the electron spin been discussed. The rotational levels of octahedral molecules (as well as tetrahedral molecules) in non-degenerate and triply degenerate vibrational states have been discussed in detail by Moret–Bailly (878).

### (d) Asymmetric top molecules

**General rotational energy formulae.** As discussed in more detail in Volume II the rotational energy levels of an asymmetric top molecule in a given vibronic state (neglecting centrifugal distortion corrections) can be represented by

$$F_{[v]}(J_\tau) = \tfrac{1}{2}(B_{[v]} + C_{[v]})J(J+1) + [A_{[v]} - \tfrac{1}{2}(B_{[v]} + C_{[v]})]W_{J_\tau}^{[v]} \qquad \text{(I, 138)}$$

where $\tau = J, J-1, \ldots, -J$ numbers the $2J+1$ levels of a given $J$ and where $W_{J_\tau}^{[v]}$ is a dimensionless quantity determined by certain algebraic equations which depend on the effective rotational constants for the vibrational level $[v]$:

$$A_{[v]} = A_e - \sum \alpha_i^A(v_i + \tfrac{1}{2}), \qquad A_e = \frac{h}{8\pi^2 c I_A^e}$$
$$B_{[v]} = B_e - \sum \alpha_i^B(v_i + \tfrac{1}{2}), \qquad B_e = \frac{h}{8\pi^2 c I_B^e} \tag{I, 139}$$
$$C_{[v]} = C_e - \sum \alpha_i^C(v_i + \tfrac{1}{2}), \qquad C_e = \frac{h}{8\pi^2 c I_C^e}.$$

The equations for $W_{J_\tau}^{[v]}$ up to $J = 6$, in terms of the asymmetry parameter

$$b = \frac{C_{[v]} - B_{[v]}}{2[A_{[v]} - \tfrac{1}{2}(B_{[v]} + C_{[v]})]} \tag{I, 140}$$

are given in Volume II, p. 46; equations for higher $J$ values may be found in the papers by Nielsen (933), Randall, Dennison, Ginsburg and Weber (1046) and Ginsburg (424) [see also Allen and Cross (1)].

An alternative form of the energy formula is

$$F_{[v]}(J_\tau) = \tfrac{1}{2}(A_{[v]} + C_{[v]})J(J+1) + \tfrac{1}{2}(A_{[v]} - C_{[v]})E_{J_\tau} \tag{I, 141}$$

where $E_{J_\tau}$ is expressed in terms of the parameter

$$\kappa = \frac{2[B_{[v]} - \tfrac{1}{2}(A_{[v]} + C_{[v]})]}{A_{[v]} - C_{[v]}} = -\frac{1 + 3b}{1 - b} \tag{I, 142}$$

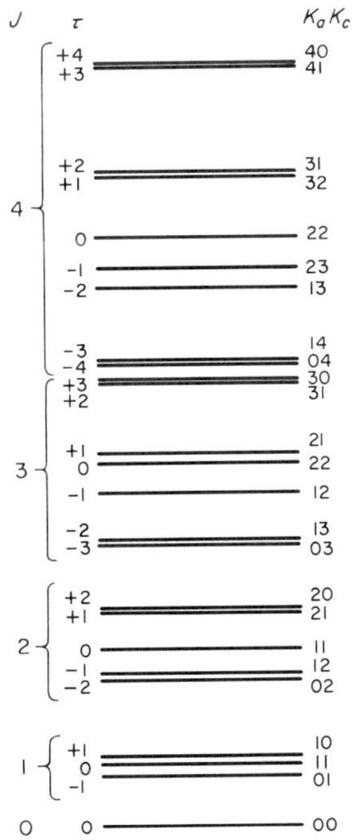


FIG. 40. **Rotational levels of an asymmetric top with $\kappa = -0.2$.** The $\tau$ values are given at left, the (equivalent) $K_a K_c$ values at right.

which takes the value $-1$ for a prolate symmetric top and $+1$ for an oblate symmetric top (while $b$ is 0 and $-1$ respectively for these limiting cases). Extensive tables for $E_{J_\tau}$ for $\kappa = -1.0, -0.9, \ldots, 0$ have been given by King, Hainer and Cross (670) to $J = 12$ and by Erlandsson (359) to $J = 40$. Still more finely subdivided tables for a limited range of $\kappa$ near the limiting symmetric top have been prepared by Fike (381) and Januzzi and Porto (623), and up to $J = 9$ for the whole range of $\kappa$ values by Blaker, Sidran and Kaercher (123). The usefulness of all these tables is extended by the fact that

$$E_{J_\tau}(\kappa) = -E_{J_{-\tau}}(-\kappa) \tag{I, 143}$$

In most of the tables the levels of a given $J$ are distinguished not by $\tau$ but by $K_a$ and $K_c$ (also designated $K_{-1}$ and $K_{+1}$), the $K$ values that the levels would have in the limiting prolate and oblate symmetric tops: the lowest level has $K_a = 0$, the two next levels $K_a = 1$, and so on, while the highest level has $K_c = 0$, the two next highest $K_c = 1$, and so on. Figure 40 shows the levels up to $J = 4$ in a fairly asymmetric case ($\kappa = -0.2$) with both the $J_\tau$ and the $J_{K_aK_c}$ designation. It is easily seen that

$$\tau = K_a - K_c. \tag{I, 144}$$

**Approximations for nearly symmetric tops.** When the asymmetric top is not too far removed from the limiting case of the symmetric top, the energy formula (I, 138) can be expanded into a power series in $K^2$ and $J(J+1)$. Equation (I, 138) goes over into the energy formula for the (rigid) prolate symmetric top by putting $W_{J_\tau}^{[v]} = K^2$ and $B = C$. For slight asymmetry, $W_{J_\tau}^{[v]}$ may be expressed in terms of the asymmetry parameter $b$ as follows [see Townes and Schawlow (40) and Davis and Beam (269)]:

$$W_{J_\tau}^{[v]} = K^2 + C_1 b + C_2 b^2 + C_3 b^3 + \cdots. \tag{I, 145}$$

Polo (999) has shown that this formula may be further developed into a series in terms of increasing powers of $b$ and of $J(J+1)$. His results can be written

$$\begin{aligned} F_{[v]}(J, K_a) = {} & \frac{1}{2}(B_{[v]} + C_{[v]})J(J+1) + \left[A_{[v]} - \frac{1}{2}(B_{[v]} + C_{[v]})\right] \\ & \times \left(1 - \frac{3}{8}b^2 - \frac{51}{512}b^4 - \cdots\right)K^2 + \Delta B_{\text{eff.}}^K J(J+1) \\ & + \Delta D_{\text{eff.}}^K J^2(J+1)^2 + \Delta H_{\text{eff.}}^K J^3(J+1)^3 + \cdots \end{aligned} \tag{I, 146}$$

where the coefficients $\Delta B_{\text{eff.}}^K$, $\Delta D_{\text{eff.}}^K$ and $\Delta H_{\text{eff.}}^K$ are given by,

$$\begin{aligned} \text{for } K = 0\text{:} \quad \Delta B_{\text{eff.}}^0 &= \left[A_{[v]} - \frac{1}{2}(B_{[v]} + C_{[v]})\right]\left(\frac{1}{4}b^2 + \frac{9}{128}b^4 + \cdots\right) \\ \Delta D_{\text{eff.}}^0 &= \left[A_{[v]} - \frac{1}{2}(B_{[v]} + C_{[v]})\right]\left(-\frac{1}{8}b^2 - \frac{19}{512}b^4 + \cdots\right) \\ \Delta H_{\text{eff.}}^0 &= \left[A_{[v]} - \frac{1}{2}(B_{[v]} + C_{[v]})\right]\left(-\frac{3}{512}b^4 + \cdots\right) \end{aligned} \tag{I, 147}$$

$$\text{for } K = 1: \quad \Delta B^1_{\text{eff.}} = \left[A_{[v]} - \frac{1}{2}(B_{[v]} + C_{[v]})\right] \times \left[\pm\frac{1}{2}b + \frac{1}{4}b^2 \mp \frac{3}{128}b^3 + \frac{9}{128}b^4 \mp \cdots\right]$$

$$\Delta D^1_{\text{eff.}} = \left[A_{[v]} - \frac{1}{2}(B_{[v]} + C_{[v]})\right]\left[-\frac{1}{32}b^2 \pm \frac{1}{64}b^3 - \frac{35}{3072}b^4 \pm \cdots\right]$$

$$\Delta H^1_{\text{eff.}} = \left[A_{[v]} - \frac{1}{2}(B_{[v]} + C_{[v]})\right]\left[\mp\frac{1}{512}b^3 + \frac{1}{1536}b^4 \mp \cdots\right] \tag{I, 148}$$

$$\text{for } K = 2: \quad \Delta B^2_{\text{eff.}} = \left[A_{[v]} - \frac{1}{2}(B_{[v]} + C_{[v]})\right]\left[\frac{1}{8}(2 \pm 1)b^2 + \frac{1}{128}(9 \pm 4)b^4 + \cdots\right]$$

$$\Delta D^2_{\text{eff.}} = \left[A_{[v]} - \frac{1}{2}(B_{[v]} + C_{[v]})\right] \times \left[\frac{1}{48}(2 \mp 3)b^2 + \frac{1}{2048}(23 \mp 32)b^4 + \cdots\right]$$

$$\Delta H^2_{\text{eff.}} = \left[A_{[v]} - \frac{1}{2}(B_{[v]} + C_{[v]})\right]\left[\frac{1}{18432}(57 \mp 64)b^4 + \cdots\right] \tag{I, 149}$$

$$\text{for } K = 3: \quad \Delta B^3_{\text{eff.}} = \left[A_{[v]} - \frac{1}{2}(B_{[v]} + C_{[v]})\right]\left[\frac{1}{4}b^2 \pm \frac{3}{128}b^3 + \frac{9}{128}b^4 \pm \cdots\right]$$

$$\Delta D^3_{\text{eff.}} = \left[A_{[v]} - \frac{1}{2}(B_{[v]} + C_{[v]})\right]\left[\frac{1}{64}b^2 \mp \frac{1}{64}b^3 + \frac{61}{10240}b^4 \mp \cdots\right]$$

$$\Delta H^3_{\text{eff.}} = \left[A_{[v]} - \frac{1}{2}(B_{[v]} + C_{[v]})\right]\left[\pm\frac{1}{512}b^3 - \frac{3}{5120}b^4 \pm \cdots\right] \tag{I, 150}$$

$$\text{for } K = 4: \quad \Delta B^4_{\text{eff.}} = \left[A_{[v]} - \frac{1}{2}(B_{[v]} + C_{[v]})\right]\left[\frac{1}{4}b^2 + \frac{1}{256}(18 \pm 1)b^4 + \cdots\right]$$

$$\Delta D^4_{\text{eff.}} = \left[A_{[v]} - \frac{1}{2}(B_{[v]} + C_{[v]})\right]\left[\frac{1}{120}b^2 + \frac{1}{15360}(46 \mp 45)b^4 + \cdots\right]$$

$$\Delta H^4_{\text{eff.}} = \left[A_{[v]} - \frac{1}{2}(B_{[v]} + C_{[v]})\right]\left[\frac{1}{46080}(-6 \pm 25)b^4 + \cdots\right] \tag{I, 151}$$

$$\text{for } K = 5: \quad \Delta B^5_{\text{eff.}} = \left[A_{[v]} - \frac{1}{2}(B_{[v]} + C_{[v]})\right]\left[\frac{1}{4}b^2 + \frac{9}{128}b^4 + \cdots\right]$$

$$\Delta D^5_{\text{eff.}} = \left[A_{[v]} - \frac{1}{2}(B_{[v]} + C_{[v]})\right]\left[\frac{1}{768}b^2 + \frac{79}{43008}b^4 + \cdots\right]$$

$$\Delta H^5_{\text{eff.}} = \left[A_{[v]} - \frac{1}{2}(B_{[v]} + C_{[v]})\right]\left[-\frac{1}{21504}b^4 + \cdots\right]. \tag{I, 152}$$

The $\pm$ signs in the preceding formulae indicate the $K$-type doubling caused by the deviation from a symmetric top. This doubling is largest for the $K = 1$ levels. For them, by substituting $b$ from (I, 140) and neglecting higher terms than the first power of $b$, one finds the effective $B$ values,

$$B^1_{\text{eff.}} = \tfrac{1}{2}(B_{[v]} + C_{[v]}) \pm \tfrac{1}{4}(C_{[v]} - B_{[v]}), \tag{I, 153}$$

that is, for the lower component ($d$)

$$B^{1}_{\text{eff.}}{}^{d} = \tfrac{1}{4}(B_{[v]} + 3C_{[v]}) \tag{I, 154}$$

and for the upper component ($c$)

$$B^{1}_{\text{eff.}}{}^{c} = \tfrac{1}{4}(3B_{[v]} + C_{[v]}), \tag{I, 155}$$

or in other words, the splitting of the $K = 1$ levels is given by

$$\Delta\nu_{cd}^{K=1}(J) = \tfrac{1}{2}(B_{[v]} - C_{[v]})J(J + 1). \tag{I, 156}$$

Similarly one finds for the splitting of the $K = 2$ levels, neglecting higher terms than the second power in $b$,

$$\Delta\nu_{cd}^{K=2}(J) = \frac{(B_{[v]} - C_{[v]})^2}{32[A_{[v]} - \frac{1}{2}(B_{[v]} + C_{[v]})]}(J - 1)J(J + 1)(J + 2). \tag{I, 157}$$

It may be noted that while the $K$-type splitting for $K = 1$ goes with the first power of $b$, that for $K = 2$ goes with the second power, that for $K = 3$ with the third power. For $K$ values higher than 4, as long as $|b|$ is not too large the $K$-type splitting is entirely negligible. Also the terms in $J^2(J + 1)^2$ and $J^3(J + 1)^3$ become very small and only a $K$ independent correction to the effective $B$ value remains, namely $\Delta B^{K}_{\text{eff.}} = [A_{[v]} - \frac{1}{2}(B_{[v]} + C_{[v]})](\frac{1}{4}b^2 + \frac{9}{128}b^4 + \cdots)$.

The formulae (I, 146–157) can be applied to nearly oblate symmetric tops if everywhere (including in $b$) $C$ and $A$ are exchanged. It must be emphasized that these formulae give good approximations to the energy of the near prolate or oblate top only when the asymmetry parameter $b$ (or $b^*$, see Vol. II, p. 49) is small and $J$ is not too large. At large $J$ values even for fairly small $b$ values many terms in the expansion must be included before convergence is reached.

Another method of approximating the energy levels of an asymmetric top is by the use of Mathieu functions (Golden (428), Gora (438a)). This method converges well for high $J$ and low $K$ values [see Innes (605) for an application of this method to the first excited state of $C_2H_2$]. Still another method using continued fractions has been described by Wait and Barnett (1254). With modern computing machines calculation of the energy levels by a direct solution of the original determinantal equations for the energy is entirely feasible and, as programs become available, is becoming preferable to the approximate methods even for small asymmetries [see, for example, Bennett, Ross and Wells (110)].

**Centrifugal distortion.** The effect of centrifugal distortions on the energy levels of asymmetric top molecules was first discussed in some detail by Kivelson and Wilson (676). Explicit formulae have been given by Polo (999) and Erlandsson (360)(361). There are here six distortion constants instead of the three ($D_J$, $D_{JK}$, $D_K$) for symmetric top molecules. The three additional ones were originally designated $\delta_J$, $R_5$ and $R_6$, but, following Nielsen (936), we designate them $D_1$, $D_2$ and $D_3$, respectively. In the limiting case of a symmetric top $D_1 = D_2 = D_3 = 0$.

We shall not consider the rather cumbersome formulae for the general case [Kivelson and Wilson (676); see also Allen and Cross (1)] but only present Polo's (999) expansion for small asymmetry. He finds for the rotational term values of a near-prolate asymmetric top

$$\begin{aligned} F_{[v]}(J, K) = [F_{[v]}(J_\tau)]_{\text{rigid}} &- D_K K^4 + 10 D_3 K^2 + 5 b D_2 K^2 (K^2 + 2) \\ &- [(D_{JK} + \tfrac{3}{2} b D_1 + 6 b D_2) K^2 + 4 D_3 + 4 b D_2] J(J + 1) \\ &- \left(D_J - b D_1 - b D_2 \frac{K^2 - 2}{K^2 - 1}\right) J^2 (J + 1)^2 + \frac{b D_1}{2(K^2 - 1)} J^3 (J + 1)^3. \end{aligned} \quad (\text{I, } 158)$$

This formula holds for $K = 0$ and $K > 3$ when the $K$-type doubling can be neglected. For $K = 1$, $K = 2$ and $K = 3$ additional terms have to be added which have opposite signs for the two components: For the lower component levels with $K = 1$ (i.e., with $K_c = J$) one has to add

$$-(\tfrac{3}{2} b D_3 + 2 D_2) J(J + 1) + (D_1 + b D_3) J^2 (J + 1)^2 - \tfrac{1}{8} D_3 b J^3 (J + 1)^3 \quad (\text{I, } 159)$$

while for the upper component levels ($K_c = J - 1$) the same quantity has to be subtracted; in addition in (I, 158) the term $b D_2 (K^2 - 2)/(K^2 - 1)$ (which would be infinite for $K = 1$) has to be replaced by $\frac{5}{4} b D_2$, and $b D_1 / 2(K^2 - 1)$ has to be replaced by $-\frac{1}{8} b D$. For $K = 2$ the additional terms are (added for the lower, subtracted for the upper component)

$$+2(D_3 - D_2 b) J(J + 1) - (D_3 - D_2 b - \tfrac{1}{2} D_1 b) J^2 (J + 1)^2 - \tfrac{1}{4} D_1 b J^3 (J + 1)^3 \quad (\text{I, } 160)$$

and for $K = 3$

$$+\tfrac{3}{2} D_3 b J(J + 1) - 2 D_3 b J^2 (J + 1)^2 + \tfrac{1}{8} D_3 b J^3 (J + 1)^3. \quad (\text{I, } 161)$$

These formulae are approximations and hold only for small asymmetry. Exact energy values can be obtained by substituting appropriate matrix elements [see Allen and Olson (57)] into the secular equation.

The distortion constants $D_J$, $D_{JK}$, $D_K$, $D_1$, $D_2$, $D_3$ depend on the potential constants of the molecule [see Kivelson and Wilson (676) and Nielsen (936)]. For planar molecules only four of the six distortion constants are independent of one another [Dowling (304); see also Oka and Morino (947), Hill and Edwards (550) and Parker (962)]. Like the potential constants, the centrifugal distortion constants may have quite different values in different electronic states. Even in a given electronic state, there will be a (slight) dependence of all the $D$ on the vibrational quantum numbers.

**Symmetry properties of rotational levels.** The asymmetric rotor functions are distinguished by their behavior with respect to rotation by 180°, $C_2{}^a$, $C_2{}^b$ and $C_2{}^c$, about the three principal axes. Since one of these operations is equivalent to the other two carried out in succession it is sufficient to indicate the behavior with respect to two of them; usually $C_2{}^c$ and $C_2{}^a$ are chosen. We have thus the four different types of rotational levels $++$, $+-$, $-+$, $--$ where the first sign refers to the behavior of the corresponding wave functions with respect to $C_2{}^c$, the second to the behavior with respect to $C_2{}^a$ [see Vol. II, p. 51].

If the molecule has symmetry, additional symmetry properties arise. In Volume II (p. 462) the symmetry properties of the over-all eigenfunctions with

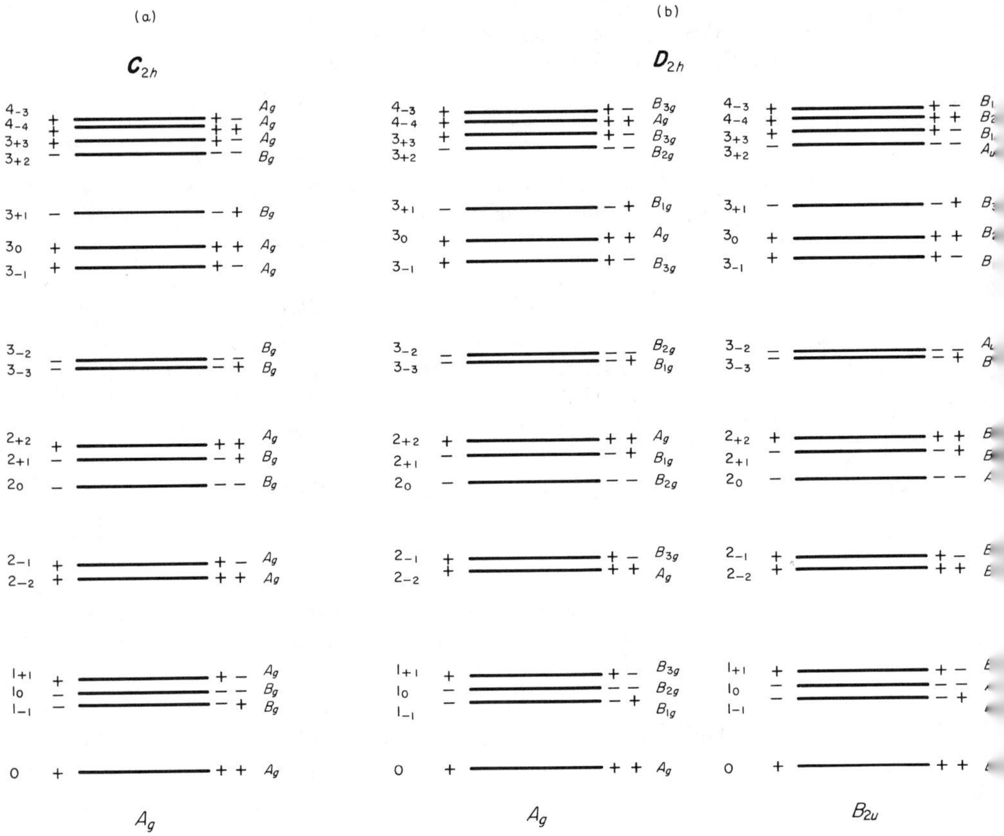


FIG. 41. **Asymmetric top species (+ −) and rovibronic (over-all) species of the lowest rotational levels, (*a*) of a $C_{2h}$ molecule in an $A_g$ vibronic state, (*b*) of a $D_{2h}$ molecule in $A_g$ and $B_{2u}$ vibronic states.** At the left of the levels the $J_\tau$ designations and the parities (for planar molecules) are given. It is assumed that in (*a*) the two-fold axis of symmetry is the *c* axis, in (*b*) the *x* and *z* axes are the *c* and *a* axes respectively. Note the difference in rovibronic species from Figs. 143–145 of Vol. II because of the different choice of axes.

respect to the rotational sub-group of the point group of the molecule have been described. We must now briefly discuss their behavior with respect to the full symmetry group. For asymmetric top molecules with a center of symmetry (point groups $C_i$, $C_{2h}$, $D_{2h}$) the full *rovibronic* (over-all) species is obtained simply from the species in the rotational sub-group by adding the subscript *g* or *u*, depending on whether the vibronic state is *g* or *u*. The reason for this simple rule is the fact that the rotational wave function of an asymmetric top remains unchanged for an inversion. As examples, we give in Fig. 41 the symmetry properties of the lowest rotational levels (a) in an $A_g$ vibronic state of a $C_{2h}$ molecule for which the $C_2$ is the $c$ axis and (b) in $A_g$ and $B_{2u}$ vibronic states of a $\boldsymbol{D}_{2h}$ molecule if the $x$ and $z$ axes are the $c$ and $a$ axes.

For asymmetric top molecules *without* a center of symmetry (i.e., point groups $\boldsymbol{C}_1$, $\boldsymbol{C}_2$, $\boldsymbol{C}_s$, $\boldsymbol{C}_{2v}$, $\boldsymbol{D}_2$) the relations are somewhat less simple. For a totally symmetric vibronic level we obtain the rovibronic species by establishing the behavior of the asymmetric rotator functions as described by the $++$, $+-$, $-+$, $--$ symbols with respect to the symmetry elements of the point group. For this purpose, as was shown by Hougen (573), reflection at a plane of symmetry is equivalent to a two-fold rotation about an axis perpendicular to that plane. Thus, for a molecule of point group $\boldsymbol{C}_s$ if the $c$ axis is perpendicular to the (only) plane of symmetry, remembering that the first sign of the $+-$ designation refers to $C_2^c$, the $++$ and $+-$ rotational levels will be $A'$, the $-+$ and $--$ levels will be $A''$ in an $A'$ vibronic state; if the $a$ axis is perpendicular to the plane of symmetry, the $++$ and $-+$ levels would be $A'$, the $+-$ and $--$ levels $A''$; and if the $b$ axis is perpendicular, the $++$ and $--$ levels would be $A'$, the $+-$ and $-+$ levels $A''$. For an $A''$ vibronic state the rovibronic species are reversed.

For a $\boldsymbol{C}_{2v}$ molecule, if the $a$ axis is the $C_2(z)$ axis and the $c$ axis the $x$ axis, the behavior of the asymmetric top functions with respect to reflection at the $yz$ plane is determined by rotation about the $c$ axis while that with respect to $C_2$ is determined by rotation about the $a$ axis. Therefore, in an $A_1$ vibronic state the $++$ levels are $A_1$, the $+-$ levels $B_2$, the $-+$ levels $A_2$ and the $--$ levels $B_1$. For other species of vibronic states the rovibronic species are obtained by multiplication of the vibronic with the rotational species. For other orientations of the axes, different results are obtained. Figure 42 gives energy level diagrams for a few examples and Table 6 gives the correlation of the rovibronic species with the species of the asymmetric rotor functions for all cases of interest.

In addition to the rovibronic species, we have to consider the parity $+$ or $-$ of each rotational level, that is, the behavior of the eigenfunctions with respect to inversion (irrespective of whether or not the molecule has a center of symmetry). As previously, for non-planar molecules each rotational level is double, consisting of a $+$ and a $-$ sub-level which correspond to the possibility of inversion. These two sub-levels are separated by a measurable amount only if the barrier separating the two equilibrium positions is small. Except in this fairly rare case, the parity classification is unimportant for non-planar molecules. For planar molecules, an inversion doubling does not arise: a rotational level is either "positive" or "negative". In a totally symmetric vibronic state the $++$ and $+-$ rotational levels are "positive", the $-+$ and $--$ levels are "negative" (see Vol. II, p. 465). These parities ($+$ or $-$) are indicated at the left of Figs. 41 and 42 for $A_g$, $A_1$ and $A'$ vibronic levels. The same parities apply also to those non-totally symmetric vibronic levels that are symmetric with respect to the plane of the molecule, e.g., $B_{2u}$ of $\boldsymbol{D}_{2h}$ (Fig. 41), $B_2$ of $\boldsymbol{C}_{2v}$ if the plane of the molecule is $yz$, and $B_u$ of $\boldsymbol{C}_{2h}$. The parities are reversed for those vibronic states whose eigenfunctions are anti-symmetric with respect to the plane of the molecule, e.g., $A_2$ and $B_1$ of $\boldsymbol{C}_{2v}$, $A_u$ and $B_g$ of $\boldsymbol{C}_{2h}$. Note that Figs. 41 and 42 apply to non-planar as well as planar

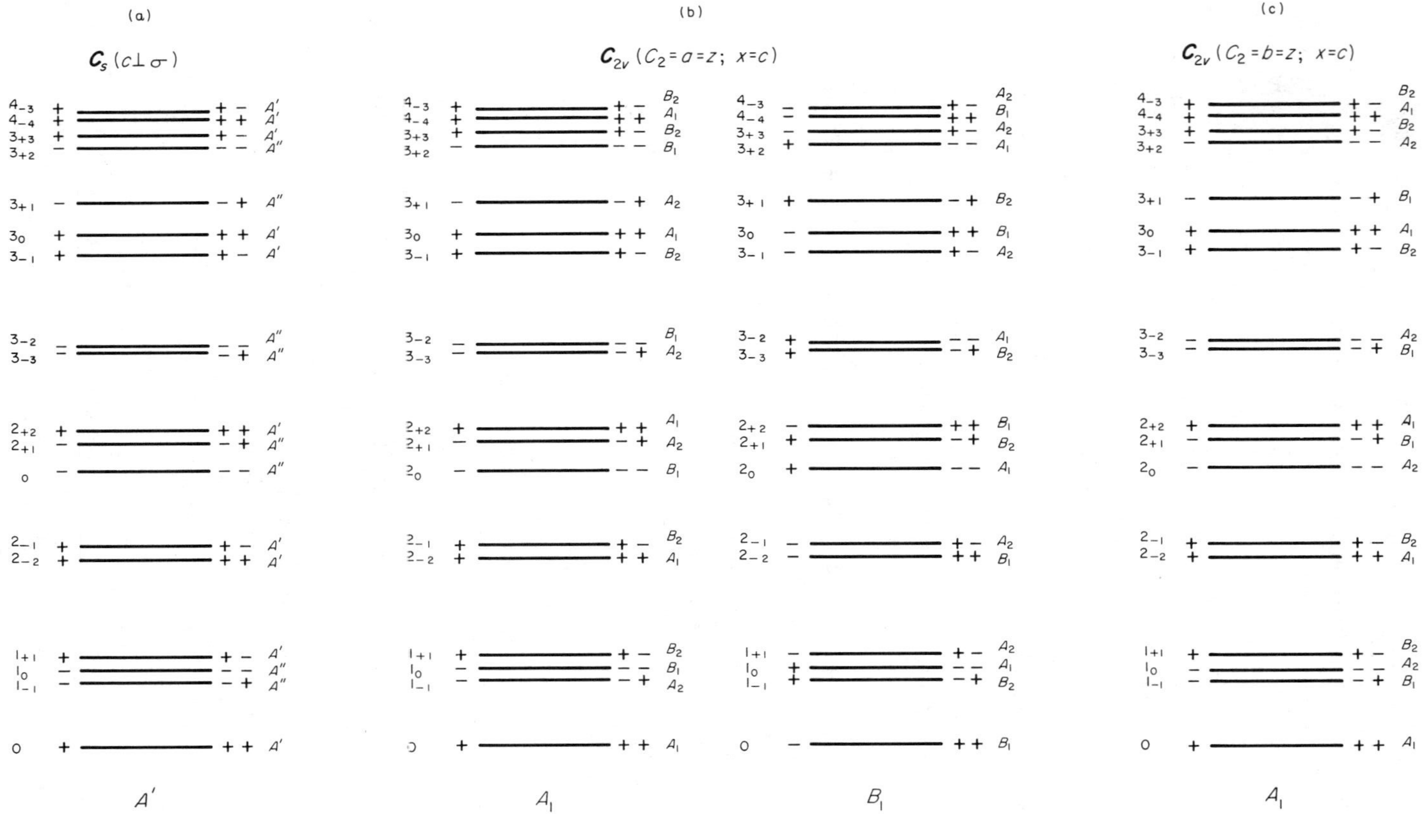


FIG. 42. **Asymmetric top species (+ −) and rovibronic (over-all) species of the lowest rotational levels of a $C_s$ and a $C_{2v}$ molecule in different vibronic states.** At the left the $J_\tau$ designations and parities (for planar molecules) are given. Note the orientations of

TABLE 6. ROVIBRONIC (OVER-ALL) SPECIES OF ASYMMETRIC TOP MOLECULES

| Asymmetric rotator designation | Point group and vibronic species | | | | | | | | | | | | | |
|---|---|---|---|---|---|---|---|---|---|---|---|---|---|---|
| | $\boldsymbol{C_s}$ | | | | | | $\boldsymbol{C_i}$ | | $\boldsymbol{C_2}$ | | | |
| | $(c \perp \sigma)$ | | $(a \perp \sigma)$ | | $(b \perp \sigma)$ | | | | $(C_2 = c)$ | | $(C_2 = a)$ | |
| | $A'$ | $A''$ | $A'$ | $A''$ | $A'$ | $A''$ | $A_g$ | $A_u$ | $A$ | $B$ | $A$ | $B$ |
| $++$ | $A'$ | $A''$ | $A'$ | $A''$ | $A'$ | $A''$ | $A_g$ | $A_u$ | $A$ | $B$ | $A$ | $B$ |
| $+-$ | $A'$ | $A''$ | $A''$ | $A'$ | $A''$ | $A'$ | $A_g$ | $A_u$ | $A$ | $B$ | $B$ | $A$ |
| $-+$ | $A''$ | $A'$ | $A'$ | $A''$ | $A''$ | $A'$ | $A_g$ | $A_u$ | $B$ | $A$ | $A$ | $B$ |
| $--$ | $A''$ | $A'$ | $A''$ | $A'$ | $A'$ | $A''$ | $A_g$ | $A_u$ | $B$ | $A$ | $B$ | $A$ |

| Asymmetric rotator designation | Point group and vibronic species | | | | | | | | | | | | | | | | | | | | | | | |
|---|---|---|---|---|---|---|---|---|---|---|---|---|---|---|---|---|---|---|---|---|---|---|---|---|
| | $\boldsymbol{C_{2v}}$ | | | | | | | | | | | | $\boldsymbol{C_{2h}}$ | | | | | | | | | | | |
| | $(C_2 = a = z,\ x = c)$[a] | | | | $(C_2 = b = z,\ x = c)$[b] | | | | $(C_2 = c = z,\ x = a)$[a] | | | | $(C_2 = z = c)$ | | | | $(C_2 = z = b)$ | | | | $C_2 = z = a)$ | | | |
| | $A_1$ | $A_2$ | $B_1$ | $B_2$ | $A_1$ | $A_2$ | $B_1$ | $B_2$ | $A_1$ | $A_2$ | $B_1$ | $B_2$ | $A_g$ | $A_u$ | $B_g$ | $B_u$ | $A_g$ | $A_u$ | $B_g$ | $B_u$ | $A_g$ | $A_u$ | $B_g$ | $B_u$ |
| $++$ | $A_1$ | $A_2$ | $B_1$ | $B_2$ | $A_1$ | $A_2$ | $B_1$ | $B_2$ | $A_1$ | $A_2$ | $B_1$ | $B_2$ | $A_g$ | $A_u$ | $B_g$ | $B_u$ | $A_g$ | $A_u$ | $B_g$ | $B_u$ | $A_g$ | $A_u$ | $B_g$ | $B_u$ |
| $+-$ | $B_2$ | $B_1$ | $A_2$ | $A_1$ | $B_2$ | $B_1$ | $A_2$ | $A_1$ | $A_2$ | $A_1$ | $B_2$ | $B_1$ | $A_g$ | $A_u$ | $B_g$ | $B_u$ | $B_g$ | $B_u$ | $A_g$ | $A_u$ | $B_g$ | $B_u$ | $A_g$ | $A_u$ |
| $-+$ | $A_2$ | $A_1$ | $B_2$ | $B_1$ | $B_1$ | $B_2$ | $A_1$ | $A_2$ | $B_2$ | $B_1$ | $A_2$ | $A_1$ | $B_g$ | $B_u$ | $A_g$ | $A_u$ | $B_g$ | $B_u$ | $A_g$ | $A_u$ | $A_g$ | $A_u$ | $B_g$ | $B_u$ |
| $--$ | $B_1$ | $B_2$ | $A_1$ | $A_2$ | $A_2$ | $A_1$ | $B_2$ | $B_1$ | $B_1$ | $B_2$ | $A_1$ | $A_2$ | $B_g$ | $B_u$ | $A_g$ | $A_u$ | $A_g$ | $A_u$ | $B_g$ | $B_u$ | $B_g$ | $B_u$ | $A_g$ | $A_u$ |

| Asymmetric rotator designation | Point group and vibronic species | | | | | | | | | | | | | | | |
|---|---|---|---|---|---|---|---|---|---|---|---|---|---|---|---|---|
| | $\boldsymbol{D_{2h}}(\equiv \boldsymbol{V_h})$[c] | | | | | | | | | | | | | | | |
| | $(z = a,\ x = c)$[d] | | | | | | | | $(z = b,\ x = c)$[e] | | | | $(z = c,\ x = b)$[e] | | | |
| | $A_g$ | $A_u$ | $B_{1g}$ | $B_{1u}$ | $B_{2g}$ | $B_{2u}$ | $B_{3g}$ | $B_{3u}$ | $A_g$ | $B_{1g}$ | $B_{2g}$ | $B_{3g}$ | $A_g$ | $B_{1g}$ | $B_{2g}$ | $B_{3g}$ |
| $++$ | $A_g$ | $A_u$ | $B_{1g}$ | $B_{1u}$ | $B_{2g}$ | $B_{2u}$ | $B_{3g}$ | $B_{3u}$ | $A_g$ | $B_{1g}$ | $B_{2g}$ | $B_{3g}$ | $A_g$ | $B_{1g}$ | $B_{2g}$ | $B_{3g}$ |
| $+-$ | $B_{3g}$ | $B_{3u}$ | $B_{2g}$ | $B_{2u}$ | $B_{1g}$ | $B_{1u}$ | $A_g$ | $A_u$ | $B_{3g}$ | $B_{2g}$ | $B_{1g}$ | $A_g$ | $B_{1g}$ | $A_g$ | $B_{3g}$ | $B_{2g}$ |
| $-+$ | $B_{1g}$ | $B_{1u}$ | $A_g$ | $A_u$ | $B_{3g}$ | $B_{3u}$ | $B_{2g}$ | $B_{2u}$ | $B_{2g}$ | $B_{3g}$ | $A_g$ | $B_{1g}$ | $B_{2g}$ | $B_{3g}$ | $A_g$ | $B_{1g}$ |
| $--$ | $B_{2g}$ | $B_{2u}$ | $B_{3g}$ | $B_{3u}$ | $A_g$ | $A_u$ | $B_{1g}$ | $B_{1u}$ | $B_{1g}$ | $A_g$ | $B_{3g}$ | $B_{2g}$ | $B_{3g}$ | $B_{2g}$ | $B_{1g}$ | $A_g$ |

[a] For $x = b$ exchange $B_1$ and $B_2$.
[b] For $x = a$ exchange $B_1$ and $B_2$.
[c] For $D_2$ omit $g$ and $u$.
[d] For $x = b$ exchange $B_2$ and $B_3$.
[e] For $x = a$ exchange $B_2$ and $B_3$; for odd vibronic species all rovibronic species are odd, otherwise they are the same as for the corresponding even species.

molecules except that for the former always both $+$ and $-$ parity arise for each level shown.

It will be seen from Figs. 41 and 42 that in planar molecules levels with different parity also have different rovibronic species. Therefore, parity does not give any additional information not already contained in the species when the full symmetry group is used. However, when the rotational levels of planar molecules are classified according to the rotational sub-group (as was done in Vol. II), the parity designation does add new information. In other words, for planar molecules the use of the species of the rotational sub-group *and* the parity is equivalent to the use of the species of the full symmetry group.

In order to ascertain the statistical weights of the rotational levels of different over-all (rovibronic) species, it is again sufficient to consider the rotational sub-group just as for symmetric top molecules. Rotational levels of those rovibronic species that differ only on account of the use of the full symmetry group have the same statistical weight. Thus, if the identical nuclei in the molecule have zero spin ($I = 0$) and follow Bose statistics, only the $A$ rotational levels occur. The levels $A'$ and $A''$ of point group $\boldsymbol{C}_s$ have the same weight and both occur for $I = 0$, and similarly $A_g$, $A_u$ of point groups $\boldsymbol{C}_i$, $\boldsymbol{C}_{2h}$ and $\boldsymbol{D}_{2h}$, $A_1$, $A_2$ of point group $\boldsymbol{C}_{2v}$. If the nuclear spin $I$ is different from zero, both $A$ and $B$ levels occur. For the point groups $\boldsymbol{C}_2$, $\boldsymbol{C}_{2h}$ and $\boldsymbol{C}_{2v}$ with a single axis of symmetry, the ratio of the statistical weights of the $A$ and $B$ levels is determined in the same way as for corresponding linear molecules (see Vol. II, p. 54). For example, in molecules with a single pair of identical nuclei which obey Fermi statistics and have spin $I = \frac{1}{2}$, like $H_2O$, $H_2CO$, $H_2O_2, \ldots$ the $B$ levels have three times the weight of the $A$ levels, while for $I = 1$ (Bose statistics), the $B$ levels have one-half of the weight of the $A$ levels.

For the point groups $\boldsymbol{D}_2$ and $\boldsymbol{D}_{2h}$ with three mutually perpendicular axes of symmetry, there are four species of the rotational sub-group: $A$, $B_1$, $B_2$, $B_3$ whose statistical weights depend on the spins of the identical nuclei present (see Vol. II, p. 54). Again if all nuclei have $I = 0$ only the $A$ levels (that is, $A_g$ and $A_u$ in $\boldsymbol{D}_{2h}$) occur. If there are four nuclei of spin $I = \frac{1}{2}$ as in $C_2H_4$, all levels occur, the weight ratio being 7:3:3:3 for $A$, $B_1$, $B_2$, $B_3$. If in addition there is one pair of nuclei with $I = \frac{1}{2}$ on the $z$ axis as in $C_2^{13}H_4$, the ratio is 16:24:12:12. For other examples, see Table 11, p. 54, of Vol. II. (Note, however, the different choice of axes adopted internationally since publication of Vol. II; see Mulliken (912).)

The effect of the nuclear spin on the relative statistical weights of the two nearly coinciding levels of non-planar molecules, which correspond to inversion, has been discussed in Volume II, p. 466.

**Sum rules.** As was mentioned in Volume II, p. 49, the average of all asymmetric top levels of a given $J$ follows the formula of the simple rotator with an average rotational constant. This conclusion holds even when centrifugal distortion is included. One finds

$$\frac{\sum F(J_\tau)}{2J+1} = \frac{1}{3}\left(A + B + C + \frac{1}{5}D_K - 2D_3\right)J(J+1) - \left(D_J + \frac{1}{3}D_{JK} + \frac{1}{5}D_K\right)J^2(J+1)^2. \quad \text{(I, 162)}$$

For each type of rotational level $(++, +-, -+, --)$ separate sum rules hold for each $J$ value which are important for the evaluation both of the rotational constants $A$, $B$ and $C$ and of the centrifugal stretching constants. These sum rules, first formulated by Mecke (818) for the rigid asymmetric top, have recently been given explicitly by Allen and Olson (57) for the non-rigid asymmetric top [see also Erlandsson (360) and Hill and Edwards (551) who have introduced into

TABLE 7. SUM RULES FOR ASYMMETRIC TOP MOLECULES

| Rotational species | Designation of levels to be added | Sum |
|---|---|---|
| + + | $F(0_0)$ | $0$ |
| | $F(2_{-2}) + F(2_{+2})$ | $4(A + B + C) - 72D_J - 24D_{JK} - 16D_K + 16D_3$ |
| | $F(3_0)$ | $4(A + B + C) - 144D_J - 48D_{JK} - 16D_K - 128D_3$ |
| | $F(4_{-4}) + F(4_0) + F(4_{+4})$ | $20(A + B + C) - 1200D_J - 400D_{JK} - 272D_K + 320D_3$ |
| | $F(5_{-2}) + F(5_{+2})$ | $20(A + B + C) - 1800D_J - 600D_{JK} - 272D_K - 880D_3$ |
| | $F(6_{-6}) + F(6_{-2}) + F(6_{+2}) + F(6_{+6})$ | $56(A + B + C) - 7056D_J - 2352D_{JK} - 1568D_K + 1568D_3$ |
| + − | $F(1_{+1})$ | $(A + B) - 4D_J - 2D_{JK} - D_K - 4D_1 + 4D_2 + 2D_3$ |
| | $F(2_{-1})$ | $4C + (A + B) - 36D_J - 6D_{JK} - D_K + 36D_1 - 12D_2 - 14D_3$ |
| | $F(3_{-1}) + F(3_{+3})$ | $4C + 10(A + B) - 288D_J - 120D_{JK} - 82D_K - 144D_1 + 24D_2 + 4D_3$ |
| | $F(4_{-3}) + F(4_{+1})$ | $20C + 10(A + B) - 800D_J - 200D_{JK} - 82D_K + 400D_1 - 40D_2 - 60D_3$ |
| | $F(5_{-3}) + F(5_{+1}) + F(5_{+5})$ | $20C + 35(A + B) - 2700D_J - 1050D_{JK} - 707D_K - 900D_1 + 60D_2 - 10D_3$ |
| | $F(6_{-5}) + F(6_{-1}) + F(6_{+3})$ | $56C + 35(A + B) - 5292D_J - 1470D_{JK} - 707D_K + 1764D_1 - 84D_2 - 154D_3$ |
| − + | $F(1_{-1})$ | $(B + C) - 4D_J - 8D_3$ |
| | $F(2_{+1})$ | $4A + (B + C) - 36D_J - 24D_{JK} - 16D_K - 8D_3$ |
| | $F(3_{-3}) + F(3_{+1})$ | $4A + 10(B + C) - 288D_J - 48D_{JK} - 16D_K + 64D_3$ |
| | $F(4_{-1}) + F(4_{+3})$ | $20A + 10(B + C) - 800D_J - 400D_{JK} - 272D_K - 320D_3$ |
| | $F(5_{-5}) + F(5_{-1}) + F(5_{+3})$ | $20A + 35(B + C) - 2700D_J - 600D_{JK} - 272D_K + 680D_3$ |
| | $F(6_{-3}) + F(6_{+1}) + F(6_{+5})$ | $56A + 35(B + C) - 5292D_J - 2352D_{JK} - 1568D_K - 1624D_3$ |
| − − | $F(1_0)$ | $(A + C) - 4D_J - 2D_{JK} - D_K + 4D_1 - 4D_2 + 2D_3$ |
| | $F(2_0)$ | $4B + (A + C) - 36D_J - 6D_{JK} - D_K - 36D_1 + 12D_2 - 14D_3$ |
| | $F(3_{-2}) + F(3_{+2})$ | $4B + 10(A + C) - 288D_J - 120D_{JK} - 82D_K + 144D_1 - 24D_2 + 4D_3$ |
| | $F(4_{-2}) + F(4_{+2})$ | $20B + 10(A + C) - 800D_J - 200D_{JK} - 82D_K - 400D_1 + 40D_2 - 60D_3$ |
| | $F(5_{-4}) + F(5_0) + F(5_{+4})$ | $20B + 35(A + C) - 2700D_J - 1050D_{JK} - 707D_K + 900D_1 - 60D_2 - 10D_3$ |
| | $F(6_{-4}) + F(6_0) + F(6_{+4})$ | $56B + 35(A + C) - 5292D_J - 1470D_{JK} - 707D_K - 1764D_1 + 84D_2 - 154D_3$ |

the sum rules the relation between the centrifugal stretching constants in planar molecules]. We shall not reproduce the general formulae here but give instead in Table 7 the sums in terms of the rotational constants for levels up to $J = 6$ for each rotational species. This table is an extension of Table 8 of Volume II to include centrifugal corrections. The sum rules are valid for any degree of asymmetry.

As can be seen from Table 7 in order to determine the rotational constants from the observed levels one must form sums of term values, and differences of the sums, for each rotational species. From the explicit formulae of Allen and Olson (57) one finds

$$\begin{aligned}
&\left|\sum F^{++}(J) - \sum F^{+-}(J) + \sum F^{-+}(J) - \sum F^{--}(J)\right| \\
&\qquad = (A + D_K + 6D_3)J(J+1) - (D_K + D_J + D_{JK})J^2(J+1)^2 \\
&\left|\sum F^{++}(J) - \sum F^{+-}(J) - \sum F^{-+}(J) + \sum F^{--}(J)\right| \\
&\qquad = (B + 4D_2 - 8D_3)J(J+1) - (D_J + 2D_1 - 2D_3)J^2(J+1)^2 \\
&\left|\sum F^{++}(J) + \sum F^{+-}(J) - \sum F^{-+}(J) - \sum F^{--}(J)\right| \\
&\qquad = (C - 4D_2 - 8D_3)J(J+1) - (D_J - 2D_1 - 2D_3)J^2(J+1)^2.
\end{aligned} \tag{I, 163}$$

Here the sums are over all levels of given symmetry (indicated by the superscript) and given $J$, i.e., the same sums as in Table 7. It is seen that in each case the dependence on $J$ is like that for a diatomic molecule, the coefficient of $J(J + 1)$ being essentially $A$, $B$ or $C$ and the coefficient of $J^2(J + 1)^2$ being a combination of centrifugal stretching constants. In order to apply either (I, 162) or (I, 163) it is necessary to know all the levels of a given $J$.

**Spin splitting.** Unlike linear and symmetric (or spherical) top molecules, an asymmetric top molecule cannot have an electronic orbital angular momentum and therefore in general the splitting of levels due to non-zero electron spin is small. In a direct way such a splitting can only be caused by the interaction of the spin with the very small magnetic moment arising from the rotation of the molecule as a whole. However, there is also an indirect effect of the coupling of the spin $\boldsymbol{S}$ with the orbital angular momentum $\boldsymbol{L}$ even though the latter is on the average equal to zero (i.e., even though the diagonal elements of $\boldsymbol{L}$ vanish).

A detailed theoretical discussion of the spin splitting in asymmetric top molecules has been given by Henderson (493) and Raynes (1059) [see also Lin (752)]. Their formulae adapted to symmetric top molecules have already been used for the discussion of the spin splitting in these molecules (p. 88f). For asymmetric top molecules, additional terms arise in the splitting formulae.

Using, as previously, $\boldsymbol{N}$ to designate the angular momentum of the molecule apart from spin, and assuming that in the previous energy formulae $J$ is replaced by $N$, we have again for the two components of a spin doublet, i.e., for $S = \frac{1}{2}$

$$\begin{aligned}
F_1(N_\tau) &= F_0(N_\tau) + \tfrac{1}{2}\gamma N \\
F_2(N_\tau) &= F_0(N_\tau) - \tfrac{1}{2}\gamma(N + 1)
\end{aligned} \tag{I, 164}$$

where $F_1$ refers to the component with $J = N + \frac{1}{2}$ and $F_2$ to that with $J = N - \frac{1}{2}$.

As before, $F_0(N_\tau)$ is the energy without spin splitting given by (I, 138) except that $J_\tau$ is replaced by $N_\tau$. The splitting constant $\gamma$ is in a first approximation given by the same formula as for symmetric top molecules except that for the $K = 1$ levels ($c$ and $d$) which are now usually fairly widely separated the splitting constant differs slightly. Raynes (1059) gives as a first approximation

$$\gamma = \kappa \frac{K^2}{N(N+1)} + \mu \pm \tfrac{1}{2}\eta_K. \qquad \text{(I, 165)}$$

Here $K$ is $K_a$ or $K_c$ depending on whether the molecule is nearer a prolate or nearer an oblate symmetric top; $\kappa$ and $\mu$ are the same spin-rotation constants as previously;[14] $\eta_K$ is negligible for $K > 1$, $\eta_1$ is directly equal to Van Vleck's (1238) spin-rotation constant $b$ which depends on the difference of spin rotation coupling in the $b$ and $c$ axes. Henderson (493) has given formulae for the difference in splitting constants for levels with $K > 1$ but this difference decreases rapidly with $K$ and will not be considered here. For a more elaborate treatment including a consideration of hyperfine structure see Lin (752) and Curl and Kinsey (260).

As an example, we give in Fig. 43 the observed doublet splitting in the ground state of $NH_2$ as a function of $N$ from the data of Dressler and Ramsay (308). The broken line curves correspond to $\kappa = 0.335$, $\mu = 0$, $\eta_1 = 0$, i.e., to the symmetric top approximation [see Herzberg (523)]. The full curves correspond to

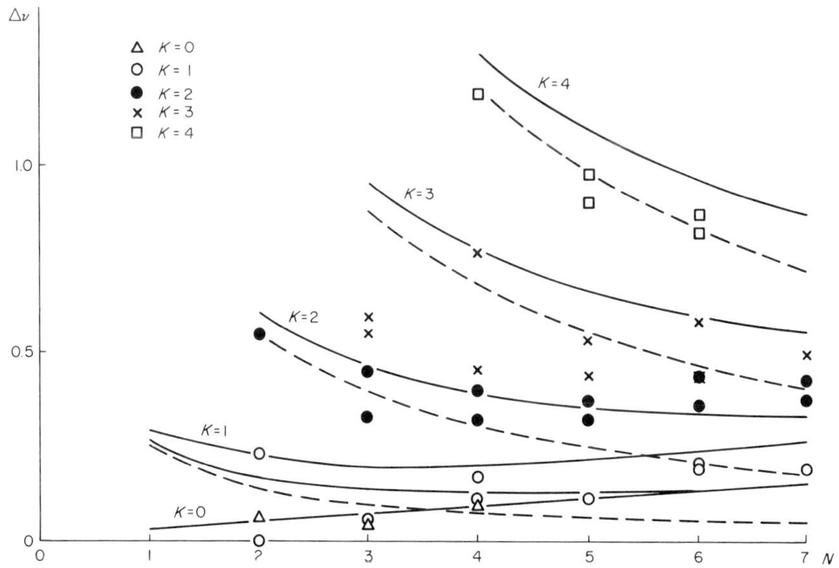


FIG. 43. **Comparison of observed and calculated doublet splittings in the ground state of $NH_2$ as a function of $N$ and $K$.** The broken line curves are calculated with $\kappa = 0.335$, $\mu = 0$, $\eta_1 = 0$; the full-line curves with $\kappa = 0.335$, $\mu = 0.02$, $\eta_1 = 0.016$.

[14] The spin-rotation constant $\kappa$ should not be confused with the asymmetry parameter $\kappa$ defined by eq. (I, 142).

$\kappa = 0.335$, $\mu = 0.02$, $\eta_1 = 0.016$. It is seen that introduction of the asymmetry term gives a slightly better representation of the observed data; but the fairly large difference in splitting in $K$-type doublets with $K > 1$ is not represented. For a discussion of the doublet splittings in $NO_2$ and $ClO_2$ see Raynes (1059).

For triplet states of asymmetric top molecules, just as for symmetric top molecules, an additional term arises in the splitting formulae [see eq. (I, 124)]; but now this additional term depends on which $K$-doubling component is considered. We shall not discuss the extended formulae here since no cases have yet been observed in which it was necessary to introduce these refinements. The only triplet states of asymmetric top molecules for which the triplet splittings have been resolved are the $^3A_2$ excited state of $H_2CO$ [Di Giorgio and Robinson (278) (1077), Raynes (1059)] and the $^3B_1$ excited state of $SO_2$ [Merer (822)]. In both cases the evaluation has been made in terms of the symmetric top approximation (eq. I, 124). The term in $\lambda$ is found to be much larger than the terms in $\kappa$ and $\mu$. In agreement with expectation the splittings are small compared to the rotational energy.

**Perturbations.** In addition to regular perturbations which lead to systematic deviations of levels from their "normal" positions or to splittings of near degeneracies, there are also here, as for linear molecules, irregular perturbations which affect only a few rotational levels of a single vibrational level. They may be of the *homogeneous* type, that is, be caused by interaction between two vibronic states of the same species having accidentally nearly the same energy for a small range of $J$ values (*Fermi interaction*), or they may be of the *heterogeneous* type, that is, be caused by the interaction of two vibronic states of different species (*Coriolis interaction*). The difference from the otherwise similar situation in rotation-vibration spectra (cf. Vol. II, p. 466) is that now the two interacting states may belong to different electronic states. While the homogeneous perturbations are then caused by vibronic interaction, the heterogeneous perturbations are caused by the interaction of rotation and electronic (or vibronic) motion. The Coriolis forces produced by the rotation cause an interaction between vibronic states whose species differ by the species of a rotation. Because of the lower symmetry of asymmetric top molecules, such perturbations are liable to be more frequent for them than for more symmetrical molecules. On the other hand, they are harder to detect since the rotational energy formulae are more complex. Very few examples of this type are known.

### (e) Other types of molecules

**Molecules with free internal rotation.** If there is free internal rotation of one part of a molecule against the other, there is an additional term in the rotational energy formula corresponding to this internal rotation. In the simplest case of a symmetric top molecule, when two equal parts can rotate relative to each other (e.g., dimethylacetylene), this additional term is

$$AK_i^2 \qquad \text{(I, 166)}$$

where $A$ is the over-all rotational constant for rotation about the symmetry axis and $K_i$ is the quantum number of internal rotation which can assume the values 0, 2, 4, . . . for even $K$ and 1, 3, 5, . . . for odd $K$. A few less symmetrical molecules have been discussed in Volume II, p. 492. For further details see Wilson, Lin and Lide (1307), Burkhard and Irvin (164) and the review article of Lin and Swalen (753). The energy formulae become much more involved when degenerate vibrations are excited. The rotational (and vibronic) levels must then be classified according to the more general point groups mentioned earlier (p. 11) rather than the standard point groups. The case of dimethylacetylene has been discussed on this basis by Bunker and Longuet–Higgins (162) and Hougen (575). We shall not consider these complications further since clear-cut examples in electronic spectra have not yet been found. One must, however, be prepared to find excited states of molecules in which free internal rotation exists even though there is no free rotation in the ground state.

**Molecules with hindered rotation.** When the internal rotation is not free, that is, when there are several potential minima as the two parts of the molecule are rotated with respect to each other, in a first approximation the molecule can be considered as a rigid molecule of the type discussed in the preceding subsections, with one fairly low vibrational frequency corresponding to a torsional vibration. However, as the amplitude of the torsional vibrations is increased or the barrier separating the minima is lowered, a gradual transition to the levels for free internal rotation takes place. This transition has been described in some detail in Volume II, p. 494. We shall not discuss it further here, but only refer to Wilkinson and Mulliken's (1303) discussion of the transition from hindered to free rotation for some of the excited states of $C_2H_4$, and the review of Lin and Swalen (753) [see also Hecht and Dennison (486) and Evett (368)].

**Molecules with inversion doubling.** When a small inversion doubling is present in a non-planar molecule, each inversion component has rotational levels of the same type as discussed in the preceding subsections. However, for larger inversion doublings, or for very precise measurements of molecules with small inversion doubling, the mutual interaction of inversion and rotation must be taken into account. In a first approximation one finds that the two components have very slightly different values of the rotational constants. For all practical cases of electronic spectra this is a sufficient approximation. For the interpretation of microwave spectra, higher approximations are required [see Townes and Schawlow (40)]. There is also an effect of this interaction on the $l$-type doubling in symmetric top molecules which has been observed in the microwave spectrum of $NH_3$ where it is proportional to the sixth power of $J$ [Nielsen and Dennison (937), Costain (232)]. Again, these effects are not likely to be observed in electronic spectra.

**Quasi-linear molecules.** When a molecule is only slightly bent, with increasing amplitude of the bending vibration, a transition of the rotational levels from those of a nearly symmetric top to those of a linear molecule takes place. Such a transition has already been briefly discussed in connection with the transition from small to very large Renner–Teller interaction (see Fig. 7). However, it occurs also when no Renner–Teller interaction is involved, that is, when in the linear conformation the molecule is in a $\Sigma$ state. A detailed discussion of the energy levels has been given by Thorson and Nakagawa (1216) and Dixon (285). The former authors introduced the name quasi-linear molecules for molecules for which a transition from a bent to a linear conformation can easily occur. The potential energy as a function of the bending coordinate is of the form given in Fig. 44. It is characteristic of quasi-linear molecules that the central maximum is not very high and that the two minima occur for comparatively small bending angles.

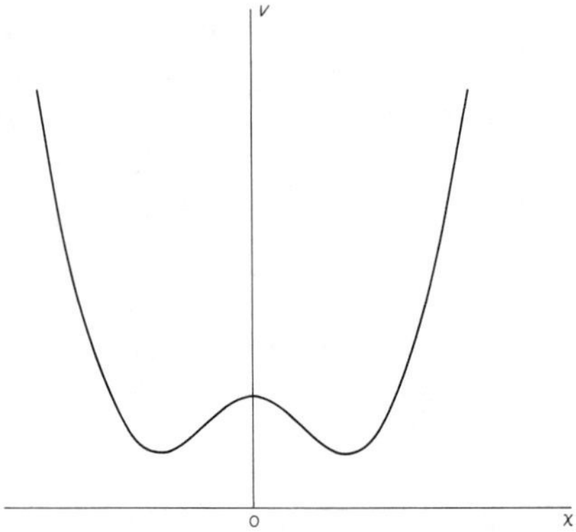


FIG. 44. **Potential energy as a function of the bending coordinate ($x$) in a quasi-linear molecule.**

Thorson and Nakagawa (1216) have used the expression

$$V = \tfrac{1}{2}kx^2 + \frac{K_B}{c^2 + x^2} \qquad \text{(I, 167)}$$

to represent a potential function of the type of Fig. 44 while Dixon (285) has used [see also Chan and Stelman (185)]

$$V = \tfrac{1}{2}kx^2 + \alpha e^{-\beta x^2}. \qquad \text{(I, 168)}$$

Here $x$ is the bending coordinate (often called $r$) and $K_B$, $c$, $\alpha$, $\beta$ are constants. It must of course be realized that the two minima of Fig. 44 are not separate minima since there is axial symmetry about the axis of the linear molecule. Figure 44 must be rotated about the ordinate axis to represent fully the potential function.

In Fig. 45 is shown schematically the correlation of energy levels for zero and finite height of the central hump. The quantum number $l$ at left corresponds to $K$ at the right and must be conserved in the correlation. As a consequence components of different vibrational levels at left become components of a single vibrational level at the right and vice versa.

Thorson and Nakagawa (1216) have given detailed graphs showing the way in which the levels change as a function of the barrier height, that is, have given a quantitative formulation of the schematic diagram Fig. 45. What is of even greater interest for the interpretation of actual spectra is the way in which the intervals between successive vibrational levels (the $\Delta G$ values) vary for a given barrier as one goes from the lower levels below the barrier to the higher levels above the barrier. This question has been studied by Dixon (285). We present in Fig.

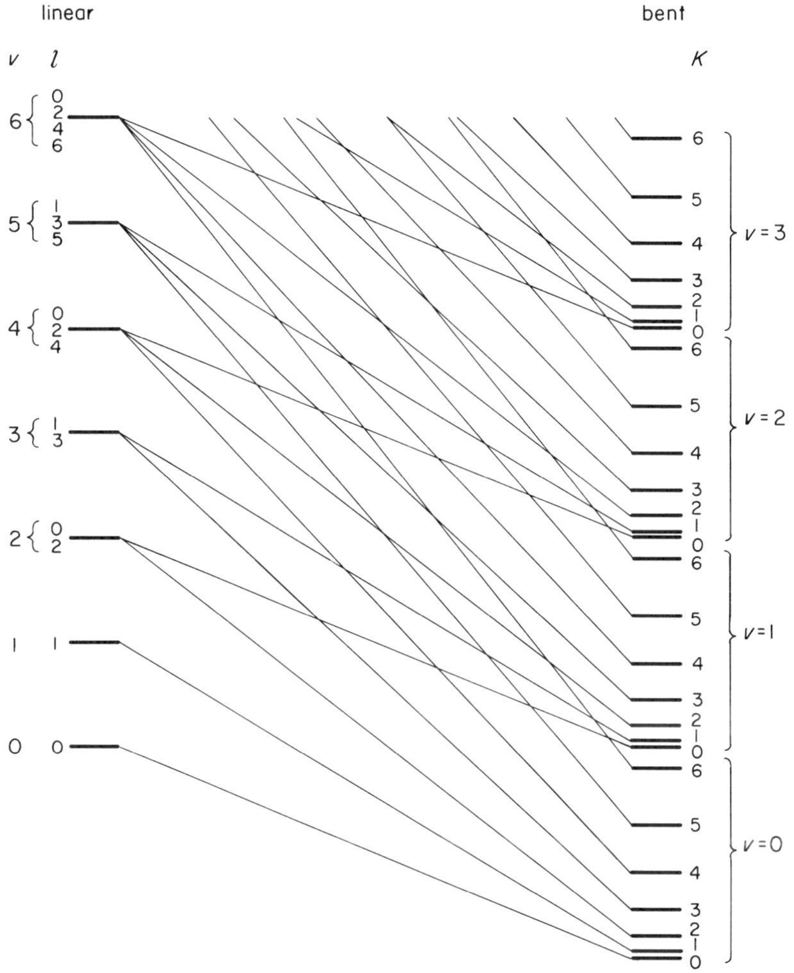


FIG. 45. **Correlation of energy levels of linear and bent molecules in a non-degenerate electronic state.** The height of the barrier increases from left to right; the energy curves are only qualitatively correct.

46 his results for the $\Delta G$ values for several values of $K$ and a barrier that is 20 times the vibrational frequency that would arise if the term $\alpha e^{-\beta x^2}$ in (I, 168) were disregarded. The interesting result is that there is a minimum in the $\Delta G(v, K)$ values near the top of the barrier, a minimum that is most pronounced for the lowest $K$ values. Such a behavior of the $\Delta G$ values was first observed by Ramsay (1042) for $PH_2$ and $PD_2$ and is interpreted by Dixon as produced by a potential hump in the upper state. On this basis the angle of the molecule in this state is estimated to be 120°.

In Fig. 47, following Dixon, an energy level diagram is shown for the same case as Fig. 46. The change-over of the energy levels from those of a nearly symmetric

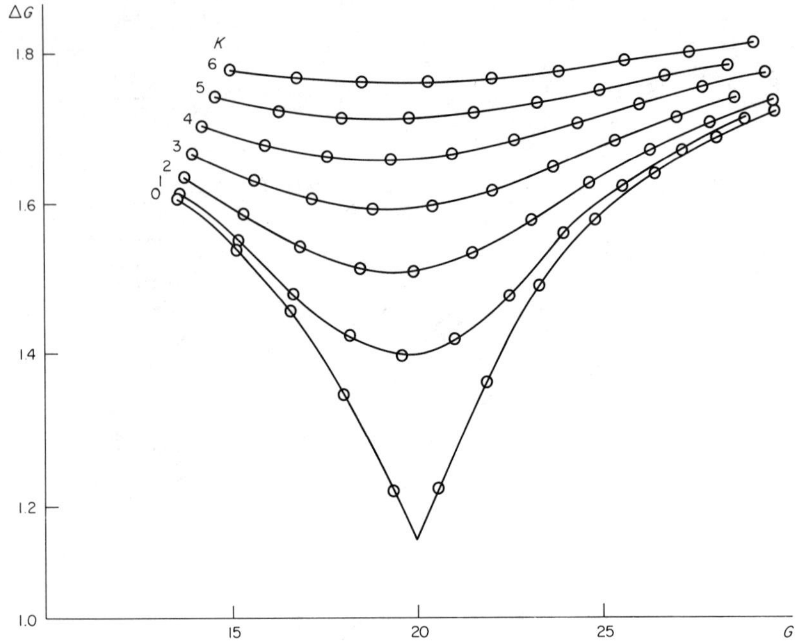


FIG. 46. **Variation of vibrational quantum $\Delta G$ as a function of vibrational energy $G$ in a quasi-linear molecule for various values of $K$ near the top of the barrier, after Dixon (285).** The average energy and the $\Delta G$ values are given in units of the vibrational frequency $\omega_0$ which would apply if there were no barrier. The barrier height is assumed to be $20\omega_0$. Note that above the top of the barrier the vibrational spacing for a given $K$ (i.e., $l$) tends to $2\omega_0$ since even and odd $l$ values alternate; below the top the spacing tends to $2\omega_0$ because the potential function is twice as narrow.

top at the bottom to those of a linear molecule at the top of the diagram is clearly seen.

## 4. Effects of Magnetic and Electric Fields on the Energy Levels

**Zeeman effect in non-degenerate electronic states.** Just as for atoms and diatomic molecules, if a polyatomic molecule is brought into a magnetic field, only certain orientations of the total angular momentum $\boldsymbol{J}$ are possible such that the component in the field direction is $M(h/2\pi)$ where

$$M = J, J-1, \ldots, -J \tag{I, 169}$$

(see Fig. 3, p. 17, of Vol. I). If the molecule has a magnetic moment $\mu$, the $2J+1$ states of different $M$ will have slightly different energies. We have

$$W = W_0 - \bar{\mu}_H H \tag{I, 170}$$

where $W_0$ is the energy of the molecule in the absence of a field, $\bar{\mu}_H$ is the mean value of the component of the magnetic moment in the field direction, and $H$ is the magnetic field intensity.

For molecules in non-degenerate singlet electronic states no permanent magnetic

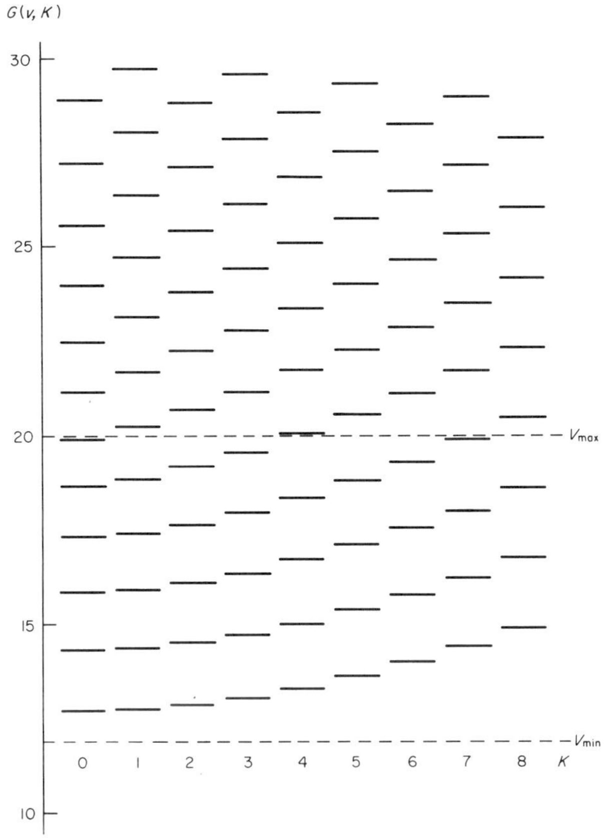


FIG. 47. **Energy levels of the bending vibration of a quasi-linear molecule near the top of the barrier, after Dixon (285).** The two horizontal broken lines give the energies of the minima and of the maximum of the potential energy. Note that $G = 0$ in this and the preceding figure does not correspond to the minimum of potential energy for the assumed barrier but does so only for the hypothetical case that $\alpha = 0$ in eq. (I, 168).

moment is produced by the electronic motion. Only the rotational motion produces a magnetic moment which consists of a positive contribution by the nuclei of the order $\sqrt{J(J+1)}$ nuclear magnetons and a negative contribution of the same order of magnitude by the electrons. One finds just as for diatomic molecules (Vol. I, p. 299) for the magnetic moment of the molecule in the direction of $\boldsymbol{J}$

$$\mu_J = g_r\sqrt{J(J+1)}\,\mu_{0n} \tag{I, 171}$$

where $\mu_{0n}$ is the nuclear magneton [$= (e/2m_pc)h/2\pi$; $m_p$ = mass of proton], and $g_r$, the Landé $g$ factor, is of the order of 1. Since the angle between $\boldsymbol{J}$ and the field direction is given by

$$\cos(\boldsymbol{J}, H) = \frac{M}{\sqrt{J(J+1)}}, \tag{I, 172}$$

and since $\bar{\mu}_H = \mu_J \cos(\boldsymbol{J}, H)$, we obtain, by substituting into (I, 170), for the energy in the field

$$W = W_0 - g_r\mu_{0n}HM. \tag{I, 173}$$

Thus, each rotational level of given $J$ is split into $2J + 1$ equidistant components. The magnitude of $g_r$ depends inversely on the separation of the electronic state considered from other electronic states [see Townes and Schawlow (40)]. In exceptional cases, if a suitable electronic state is close by, $g_r$ may become much larger than 1.

For linear molecules, just as for diatomic molecules, $g_r$ is a single constant for a given non-degenerate electronic state (since $\boldsymbol{J}$ is always perpendicular to the internuclear axis). Therefore the magnetic splitting between adjacent component levels is the same for different rotational levels of this state. However, for non-linear molecules, $g_r$ is different for different orientations of $\boldsymbol{J}$ with respect to the molecular frame, and therefore the splitting is different for different rotational levels of one and the same electronic state. Indeed, for the general polyatomic molecule, $g_r$ is a tensor whose axes do not necessarily coincide with the principal axes of inertia. However, they do coincide with them when the molecule has symmetry such that the axes are determined by it. This is the only case we shall consider. For an asymmetric top we have then three $g$ values $g_{aa}$, $g_{bb}$, $g_{cc}$ corresponding to the rotational magnetic moments in the three inertial axes. For a prolate symmetric top, $g_{bb} = g_{cc}$, and one finds for $g_r$ [see Townes and Schawlow (40)]

$$g_r = g_{bb} + (g_{aa} - g_{bb})\frac{K^2}{J(J+1)}. \qquad \text{(I, 174)}$$

If $K = 0$, $\boldsymbol{J}$ is perpendicular to the top axis and $g_r = g_{bb}$, that is, $g_r$ is determined entirely by the $g$ value for the $b$ axis. For an oblate symmetric top $g_{aa} = g_{bb}$ and $g_{aa}$ in (I, 174) must be replaced by $g_{cc}$.

For asymmetric top molecules, the three quantities $g_{aa}$, $g_{bb}$, $g_{cc}$ are all different, and the expression for $g_r$ is more complicated. Schwartz [as quoted by Burrus (167)] gives

$$\begin{aligned} g_r = \frac{1}{2}(g_{aa} + g_{cc}) &+ \frac{E_{J_\tau}}{2J(J+1)}(g_{aa} - g_{cc}) \\ &- \frac{1}{2J(J+1)}\frac{\partial E_{J_\tau}}{\partial \kappa}[(1+\kappa)g_{aa} - 2g_{bb} + (1-\kappa)g_{cc}] \end{aligned} \qquad \text{(I, 175)}$$

where $E_{J_\tau}$ and $\kappa$ are the quantities introduced in (I, 141) and (I, 142). It is readily seen that (I, 175) goes over into (I, 174) when $\kappa = -1, E_{J_\tau} = 2K^2 - J(J+1)$ and $g_{cc} = g_{bb}$. According to (I, 175), $g_r$ is different for every rotational level of an asymmetric top.

In Table 8 we give the observed values of the $g$ factors for a few molecules in their ground states. No values for any excited singlet electronic states are as yet available. It should be emphasized that (I, 175) does not apply to molecules (like HDO) for which the inertial axes are not determined by symmetry. For this case we refer to the discussion in Townes and Schawlow (40).

The Zeeman splittings in multiplet states ($S \neq 0$) are quite different from those in singlet states. In an orbitally non-degenerate state, the resultant spin $\boldsymbol{S}$ is always *weakly* coupled to the axis of rotation (because the magnetic moment associated with rotation is so small) and therefore it is readily uncoupled by an external magnetic field. Therefore, if the field is sufficiently large, we have a splitting into $2S + 1$ components characterized by the quantum number $M_S) = S, S-1, \ldots, -S)$. The energy formula is in first approximation

$$W = W_0 - 2M_S\mu_0 H \qquad \text{(I, 176)}$$

where $\mu_0$ is now the Bohr magneton, not the nuclear magneton[15]. The splitting in

[15] Strictly speaking, the factor 2 in this formula must be replaced by 2.00229 corresponding to the anomalous magnetic moment of the electron. But (I, 176) is not accurate to this degree anyway.

TABLE 8. LANDÉ $g$ FACTORS FOR THE ELECTRONIC GROUND STATES OF SOME MOLECULES

| Molecule | $g$ factor | References | Molecule | $g$ factor | References |
|---|---|---|---|---|---|
| $N_2O$ | $\pm 0.086$ | (625) | $H_2O$ | $g_{aa} = \pm 0.585$ | (40) |
| $CO_2$ | $(-)0.05508$ | (184) | | $g_{bb} = \pm 0.742$ | |
| $CS_2$ | $(-)0.02274$ | (184) | | $g_{cc} = \pm 0.666$ | |
| OCS | $-0.02889$ | (625)(184) | $H_2S$ | $g_{aa} = \pm 0.355$ | (167) |
| $C_2H_2$ | $-0.04903$ | (184) | | $g_{bb} = \pm 0.195$ | |
| $NH_3$ | $g_{cc} = 0.484$ | (362) | | $g_{cc} = \pm 0.209$ | |
| | $g_{bb} = g_{aa} = 0.560$ | | $SO_2$ | $g_{aa} = \pm 0.606$ | (167) |
| $CH_3CCH$ | $g_{aa} = \pm 0.31$ | (250) | | $g_{bb} = \pm 0.123$ | |
| | $\lvert g_{bb}\rvert \ll \lvert g_{aa}\rvert$ | | | $g_{cc} = \pm 0.074$ | |
| $CF_4$ | 0.031 | (184) | $O_3$ | $g_{aa} = \pm 2.960$ | (167) |
| $Ni(CO)_4$ | 0.0179 | (184) | | $g_{bb} = \pm 0.227$ | |
| $Fe(CO)_5$ | 0.0210 | (184) | | $g_{cc} = \pm 0.086$ | |
| | | | $H_2CO$ | $g_{aa} = -2.90$ | (686) |
| | | | | $g_{bb} = \pm 0.35$ | (687) |
| | | | | $g_{cc} = \pm 0.02$ | |

this case is more than 1000 times as great as without the presence of a spin and in a first approximation is independent of $J$ and $K$.

For small fields the energy formula becomes much more complicated. Each level of given $J$ splits into $2J + 1$ components with different $g$-factors for different $J$. While for very small fields this splitting would be linear with $H$, for intermediate fields it is no longer linear with $H$. With further increase of the field, the levels of a given spin multiplet form $2S + 1$ groups of $2N + 1$ levels each (where $N$ is the quantum number of the total angular momentum without spin). The centers of these groups are given by (I, 176). For a detailed discussion of the whole transition from small to large fields for a $^3\Sigma$ state of a diatomic molecule we refer to Tinkham and Strandberg (1222). Their treatment can be applied directly to $^3\Sigma$ states of linear molecules. Except for a brief discussion of the Zeeman effect in the microwave spectrum of $ClO_2$ by Tolles, Kinsey, Curl and Heidelberg (1223), no general discussion for non-linear molecules has as yet been given. For a specific case, the first excited $^3A_2$ state of $CS_2$ (in which the molecule is non-linear), Douglas and Milton (298) and Hougen (576) have given a detailed discussion. Here the situation is complicated by the fact that the spin splitting is very large and only one triplet component $B_2$ is observed. This component shows a Zeeman splitting increasing rapidly with $J$ and reaching values comparable to those given by (I, 176) at high $J$, when the three triplet components are strongly mixed.

**Zeeman effect in orbitally degenerate electronic states.** If in a linear molecule an electronic orbital angular momentum $\boldsymbol{\Lambda}$ is present, there will be a magnetic moment $\mu_\Lambda = \Lambda\mu_0$ in the internuclear axis. The component of this magnetic moment in the direction of $\boldsymbol{J}$ is

$$\mu_J = \mu_\Lambda \frac{\Lambda}{\sqrt{J(J+1)}}$$

and therefore we find, taking account of (I, 172), for the average component in the field direction

$$\bar{\mu}_H = \frac{\Lambda^2 M}{J(J+1)}\mu_0 \qquad \text{(I, 177)}$$

and thus for the energy in the field

$$W = W_0 - \frac{\Lambda^2}{J(J + 1)} \mu_0 HM. \tag{I, 178}$$

We see that the over-all splitting ($M = \pm J$) decreases inversely as $J+1$. For very low $J$ values it is of the order of the normal Zeeman splitting.

If in addition the spin $S$ is different from zero, we have a magnetic moment $(\Lambda + 2\Sigma)\mu_0$ in the internuclear axis as long as the spin is strongly coupled to the axis such that its component $\Sigma$ is $S, S-1, \ldots, -S$ [Hund's case (a)]. Here again the over-all splitting is inversely proportional to $J+1$ for each multiplet component (see Vol. I, p. 301). On the other hand, in Hund's case (b) when the spin is coupled to the axis of rotation we have, at least for large $J$, a splitting into $2S+1$ components given by (I, 176) modified by the effect of the orbital magnetic moment which again decreases with increasing $J$ (see Vol. I, p. 303).

Similar Zeeman splittings arise if a vibrational angular momentum is present as well as an electronic orbital angular momentum except when the resultant angular momentum $\boldsymbol{K}$ about the linear axis is zero, that is, for $\Sigma$ vibronic levels. For such levels we must go back to eq. (I, 173) for the splittings. However, one may expect that $g_r$ will be anomalously large for the same reasons that the spin-splitting constant $\gamma$ is large in $^2\Sigma$ or $^3\Sigma$ vibronic states of $^2\Pi$ or $^3\Pi$ electronic states (see p. 77f). No examples to check this prediction have as yet been studied.

No general discussion of the Zeeman splittings of orbitally degenerate states of free *non-linear molecules* has yet been given. Child and Longuet–Higgins (194) state that the splitting will be directly proportional to the quenching parameter $d$ (see p. 64) when Jahn–Teller interaction is small. One expects therefore that the energy formula is

$$W = W_0 - \frac{d\zeta_e K}{J(J + 1)} \mu_0 HM, \tag{I, 179}$$

that is, that the splitting increases with $K$ but decreases in the same manner as before with $J$.

Zeeman splittings in orbitally degenerate *multiplet* states of non-linear molecules have not yet been considered in the literature.

**Stark effect.** In an electric field also, a splitting of most molecular energy levels arises. Similar to the magnetic case we have for the energy in a field of intensity $E$

$$W = W_0 - \bar{\mu}_E E \tag{I, 180}$$

where $\bar{\mu}_E$ is now the mean component of the *electric* moment $\mu$ in the field direction.

As was shown in more detail in Volume I, a linear (first order) Stark effect arises only if the molecule has a permanent electric dipole moment and if in addition the angular momentum, $\boldsymbol{K}$, about the symmetry axis is not zero. Assuming that the permanent dipole moment $\mu$ lies in the direction of the top axis (as is necessarily the case for genuine symmetric tops) we see immediately that, just as in the magnetic case,

$$\mu_J = \mu \frac{K}{\sqrt{J(J + 1)}}$$

and thus

$$\bar{\mu}_E = \mu_J \frac{M}{\sqrt{J(J + 1)}} = \mu \frac{KM}{J(J + 1)}$$

and therefore

$$W = W_0 - \mu \frac{KM}{J(J + 1)} E. \tag{I, 181}$$

We have a splitting into $2J+1$ component levels whose spacing increases linearly with $E$ just as in the magnetic case. The proportionality factor, $\mu KM/[J(J+1)]$, is given directly by the dipole moment and the quantum numbers of the state considered. An experimental determination of the splitting allows therefore a simple and direct determination of the dipole moment provided the quantum numbers are known.

In the approximation of eq. (I, 181) no splitting arises for the levels with $K = 0$ because for them the dipole moment is perpendicular to the axis of rotation. For the same reason, for linear molecules in $^1\Sigma$ states no splitting arises. But in the next approximation when the interaction with neighboring levels is taken into account a Stark splitting does arise for $K = 0$; however, it is proportional to the square of the field intensity (*quadratic Stark effect*) and is in general very much smaller than the linear effect discussed above except at very high fields. A quadratic effect also arises for the levels with $K \neq 0$ which show a linear effect. According to Townes and Schawlow (40) the energy is given to second order by

$$W = W_0 - \mu \frac{KM}{J(J+1)} E + \frac{\mu^2 E^2}{2Bhc} \times \left\{ \frac{(J^2 - K^2)(J^2 - M^2)}{J^3(2J-1)(2J+1)} - \frac{[(J+1)^2 - K^2][(J+1)^2 - M^2]}{(J+1)^3(2J+1)(2J+3)} \right\}. \quad \text{(I, 182)}$$

Here $B$ is the rotational constant for the (prolate or oblate) symmetric top. For $K = 0$ eq. (I, 182) simplifies to (V, 97) of Volume I. Higher order terms as well as numerical tables and graphical representations of the Stark energy levels of symmetric tops have been given by Shirley (1127), and a graphical presentation of the Stark levels in asymmetric tops by West and Mizushima (1290).

For molecules without a permanent dipole moment the Stark splittings in a given electronic state depend on perturbations by other electronic states and are therefore very much smaller than even the second order Stark splittings of molecules with a dipole moment.

# CHAPTER II

# ELECTRONIC TRANSITIONS

In Chapter I we have discussed the various types of electronic states and their associated vibrational and rotational levels for various classes of polyatomic molecules. For a comparison of these theoretical results with experiment it is necessary now to consider the transitions between these levels. Just as for diatomic molecules, transitions from the vibrational and rotational levels of one electronic state to those of another give rise to a *band system*, except that now, in general, the structure of such a band system is much more complicated. Most band systems of polyatomic molecules have been observed in absorption only, but a few have been observed in emission.

## 1. Types of Electronic Transitions

As for diatomic molecules we distinguish *allowed* and *forbidden* electronic transitions. Allowed electronic transitions give rise to the most intense absorption spectra of polyatomic molecules. But forbidden electronic transitions do occur weakly and are, in fact, much more important for polyatomic than for diatomic molecules.

### (a) Allowed electronic transitions

We consider an electronic transition as allowed if it can occur as an electric dipole transition according to the selection rules to be discussed below, without taking account of the interaction of vibration or rotation with the electronic motion. In other words, we consider an electronic transition as allowed if it can occur for fixed nuclei.

**General selection rule.** Let $\psi_e'$ and $\psi_e''$ be the electronic eigenfunctions of the upper and lower state respectively of an electronic transition, and let both states be non-degenerate. The transition will be associated with an electric dipole moment, that is, it will be allowed, if and only if the matrix element

$$R_{e'e''} = \int \psi_e'^* \boldsymbol{M} \psi_e'' d\tau_e \quad \textit{is different from zero.} \tag{II, 1}$$

Here $\boldsymbol{M}$ is the dipole moment vector whose components are $\sum e_i x_i$, $\sum e_i y_i$ and $\sum e_i z_i$ (see Vol. I, p. 19). In other words, an electronic transition between non-degenerate states is allowed *if the product*

$$\psi_e'^* \boldsymbol{M} \psi_e'' \tag{II, 2}$$

*is totally symmetric* for at least one orientation of $\boldsymbol{M}$, that is, if the product $\psi_e'\psi_e''$ belongs to the same species as one of the components of $\boldsymbol{M}$.

If one or both states are degenerate the product (II, 2) will in general not be totally symmetric even for an allowed transition; but, if the transition is allowed, linear combinations of the mutually degenerate eigenfunctions and of the dipole components that are transformed into one another can be found which make the product totally symmetrical. It can be shown by group theory that a totally symmetric product is obtainable and therefore the transition is allowed if the direct product of the species ($\Gamma$) of $\psi_e'$, $\psi_e''$ and $\boldsymbol{M}$ has a totally symmetric component, i.e.,

$$\Gamma(\psi_e') \times \Gamma(\psi_e'') \times \Gamma(\boldsymbol{M}) = \text{totally symmetric} \tag{II, 3}$$

or, in other words, if $\Gamma(\psi_e') \times \Gamma(\psi_e'')$ belongs to the same species as one of the components of $\boldsymbol{M}$ (just as for non-degenerate states). Thus the general selection rule is exactly the same as for infrared spectra (see Vol. II, p. 252) except that the vibrational eigenfunctions have to be replaced by the electronic eigenfunctions.

As for the infrared spectrum, the general selection rule can usually be still further simplified for absorption spectra since for most stable molecules the ground state is totally symmetric. In order for an allowed transition to occur in absorption in such a case, the rule is simply that the upper state must have the species of one of the components of the dipole moment.

The selection rule (II, 1) applies strictly only for fixed nuclei. Actually the nuclei are not fixed and therefore we must consider the total eigenfunction which includes the nuclear coordinates. Neglecting the rotational motion for the time being (see below), or, in other words referring the motions of electrons and nuclei to a rotating, molecule-fixed coordinate system we can write in a good approximation (Born–Oppenheimer approximation) for the total eigenfunction

$$\psi_{ev} = \psi_e(q, Q)\,\psi_v(Q) \tag{II, 4}$$

where $q$ stands for all the electronic coordinates and $Q$ for all the nuclear coordinates. In writing $\psi_{ev}$ as a single product of $\psi_e$ and $\psi_v$, we are neglecting the finer interaction of electronic motion and vibration but are retaining that part of this interaction that can be expressed by the dependence of $\psi_e$ on $Q$. For convenience in discussing the subject we shall call the latter part *type (a) vibronic interaction*, the former part *type (b) vibronic interaction*.

Let us now resolve the electric dipole moment into two parts, one due to the electrons, the other due to the nuclei:

$$\boldsymbol{M} = \boldsymbol{M}_e + \boldsymbol{M}_n. \tag{II, 5}$$

The transition moment in this general case (considering only non-degenerate states) is then given by

$$\begin{aligned} R_{e'v'e''v''} &= \int \psi_{ev}'^{*}\boldsymbol{M}\psi_{ev}''d\tau_{ev} \\ &= \int \psi_v'^{*}\psi_v''d\tau_v \int \psi_e'^{*}\boldsymbol{M}_e\psi_e''d\tau_e + \int \psi_v'^{*}\boldsymbol{M}_n\psi_v''d\tau_v \int \psi_e'^{*}\psi_e''d\tau_e. \end{aligned} \tag{II, 6}$$

As long as the product resolution (II, 4) is valid, that is, as long as the finer interaction of electronic motion with vibration [*type* (*b*)] is neglected we have

$$\int \psi_e'^* \psi_e'' d\tau_e = 0, \tag{II, 7}$$

since the electronic eigenfunctions for a given position $Q$ of the nuclei are mutually orthogonal. Therefore we have

$$R_{e'v'e''v''} = \int \psi_v'^* \psi_v'' d\tau_v \int \psi_e'^* \boldsymbol{M}_e \psi_e'' d\tau_e, \tag{II, 8}$$

The second integral in (II, 8)

$$R_{e'e''}(Q) = \int \psi_e'^*(q, Q) M_e \psi_e''(q, Q) d\tau_e \tag{II, 9}$$

is the matrix element of the electric dipole moment for a given nuclear configuration ($Q$); it varies only slightly with $Q$. If we neglect the dependence of $\psi_e$ on the nuclear coordinates (that is, neglect both type (a) and type (b) vibronic interactions), we obtain

$$R_{e'e''}(Q_0) = \int \psi_e'^*(q, Q_0) M_e \psi_e''(q, Q_0) d\tau_e \tag{II, 10}$$

where $Q_0$ corresponds to a configuration of the nuclei near the equilibrium position of one of the two electronic states. In this approximation (II, 8) becomes

$$R_{e'e''v'v''} = R_{e'e''}(Q_0) \int \psi_v'^* \psi_v'' d\tau_v, \tag{II, 11}$$

that is, the transition probability, which is proportional to $R^2$, can be resolved into a factor depending on the nuclear motion alone and a factor depending on the electronic motion alone. In this approximation we see that the general electronic selection rule (II, 1) for fixed nuclei holds also for the vibrating molecule. The particular vibrational transitions which make the total transition moment $R_{e'v'e''v''}$ different from zero are those that make the second factor in (II, 11)

$$\int \psi_v'^* \psi_v'' d\tau_v$$

different from zero (see section 2).

Thus far in the preceding discussion it has been implicitly assumed that the molecule has the same symmetry in both the upper and lower electronic state. If that is so, for an allowed transition the product $\psi_e'^* \boldsymbol{M}_e \psi_e''$ must be symmetric with respect to all symmetry elements of the point group of the molecule. If however, as frequently happens, the equilibrium conformations of the molecule have different symmetry (belong to different point groups) in the two states then only the common symmetry elements must be considered. In consequence there are fewer restrictions than when the equilibrium conformations have the same high symmetry. For example, for a molecule $XY_3$ which is planar and symmetrical (point group $\boldsymbol{D}_{3h}$) in one state and non-planar (point group $\boldsymbol{C}_{3v}$) in another the question whether a transition is allowed or forbidden must be decided on the basis of the behavior of $\psi_e'^* \boldsymbol{M}_e \psi_e''$ with respect to the symmetry elements of point group $\boldsymbol{C}_{3v}$, not $\boldsymbol{D}_{3h}$.

If rotation is to be considered we must include a rotational eigenfunction and write

$$\psi = \psi_e \psi_v \psi_r \tag{II, 12}$$

where $\psi_r$ is a function of the Eulerian angles which refer the rotating axes to space-fixed axes. The dipole moment $\boldsymbol{M}^f$ with respect to space-fixed axes has the components

$$M_x^f = M_x \cos \alpha_x + M_y \cos \alpha_y + M_z \cos \alpha_z \tag{II, 13}$$

and similarly for $M_y^f$ and $M_z^f$. Here $\alpha_x$, $\alpha_y$, $\alpha_z$ are the angles between the molecule-fixed $x$, $y$, $z$ axes and the space-fixed $x$ axis (and similarly $\beta_x$, $\beta_y$, $\beta_z$ for the $y$ axis and $\gamma_x$, $\gamma_y$, $\gamma_z$ for the $z$ axis).

The $x$ component of the transition moment with respect to space-fixed axes is now

$$R^{(x)}_{e'v'r'e''v''r''} = \int \psi'^* M_x^f \psi'' d\tau. \tag{II, 14}$$

Substituting (II, 12), (II, 13) and (II, 5) we obtain

$$\begin{aligned} R^{(x)}_{e'v'r'e''v''r''} = & \int \psi_e'^* M_{ex} \psi_e'' d\tau_e \int \psi_v'^* \psi_v'' d\tau_v \int \psi_r'^* \cos \alpha_x \psi_r'' d\tau_r \\ & + \cdots + \int \psi_e'^* M_{ez} \psi_e'' \, d\tau_e \int \psi_v'^* \psi_v'' d\tau_v \int \psi_r'^* \cos \alpha_z \psi_r'' d\tau_r \\ & + \int \psi_e'^* \psi_e'' d\tau_e \int \psi_v'^* M_{nx} \psi_v'' d\tau_v \int \psi_r'^* \cos \alpha_x \psi_r'' d\tau_r \\ & + \cdots + \int \psi_e'^* \psi_e'' d\tau_e \int \psi_v'^* M_{nz} \psi_v'' d\tau_v \int \psi_r'^* \cos \alpha_z \psi_r'' d\tau_r \end{aligned} \tag{II, 15}$$

and similarly for the $y$ and $z$ components. The last three terms in (II, 15) vanish because of (II, 7) as long as the product resolution (II, 12) is valid. The integrals $\int \psi_v'^* \psi_v'' d\tau_v$, $\int \psi_r'^* \cos \alpha_x \psi_r'' d\tau_r$, etc., vanish unless certain relations between the vibrational and rotational quantum numbers of the upper and lower state are fulfilled, i.e., they determine the vibrational and rotational selection rules (see below). Assuming that these selection rules are fulfilled we see that the transition moment depends as before on the three components of the electronic transition moment (II, 9).

**Spin selection rule.** As we have seen in Chapter I, section 1, as long as spin-orbit interaction is small, the electronic eigenfunction including spin can be written as a product of an orbital and a spin function[1]

$$\psi_{es} = \psi_e \beta. \tag{II, 16}$$

Therefore we have for the transition moment including spin

$$\int \psi_{es}'^* \boldsymbol{M}_e \psi_{es}'' d\tau_e d\sigma = \int \psi_e'^* \boldsymbol{M}_e \psi_e'' d\tau_e \int \beta'^* \beta'' d\sigma. \tag{II, 17}$$

The second integral at the right vanishes for states of different spin since the spin functions corresponding to different $S$ values are orthogonal to one another. Therefore, as for atoms and diatomic molecules, we have the selection rule

$$\Delta S = 0 \tag{II, 18}$$

that is, *only states of the same multiplicity* (i.e., the same spin $S$) *combine with each other*; in other words, intercombinations are forbidden.

For higher atomic numbers of the constituent atoms the resolution (II, 16) is less and less rigorous. A multiplet splitting arises and at the same time the rule (II, 18) is no longer strictly valid: intercombinations do occur weakly.

[1] As mentioned in Chapter I this resolution is strictly possible only for one and two electron systems; but for more electrons it still gives the correct symmetry properties.

**Application to the more important point groups.** If a molecule has no symmetry (point group $C_1$), all electronic states can combine with one another except those of different multiplicity. If the molecule has one element of symmetry (as in point groups $C_s$, $C_i$, $C_2$), that is, if there are two species of electronic states, then three types of transitions are conceivable which may not all be allowed. For example for point group $C_i$ with species $A_g$ and $A_u$ all three components of $\boldsymbol{M}$ behave as $A_u$ and therefore according to the general selection rule only $A_g - A_u$ transitions can occur and not $A_g - A_g$ or $A_u - A_u$. On the other hand, for point group $C_s$ with species $A'$ and $A''$, the components $M_x$, $M_y$ have species $A'$, while $M_z$ has $A''$ (cf. the species of $T_x$, $T_y$, $T_z$ in Table 48 of Appendix I). Therefore $A' - A'$ and $A'' - A''$ transitions are possible with $M_x$ and $M_y$, while $A' - A''$ transitions are possible with $M_z$. There are no forbidden transitions (except of

TABLE 9. TRANSITION MOMENTS OF ELECTRONIC TRANSITIONS OF MOLECULES BELONGING TO VARIOUS POINT GROUPS

The species of one state participating in the transition is given in the top row of each sub-table, that of the other state in the column at the right. In the body of each sub-table $M_x$, $M_y$, $M_z$ give the orientation for the dipole moment for the particular transition. $M_{x,y}$ indicates that $M_x$ and $M_y$ are equivalent, $M_{x,y,z}$ that $M_x$, $M_y$ and $M_z$ are equivalent; *f.* refers to a forbidden transition.

| $C_s$ | $A'$ | $A''$ | |
|---|---|---|---|
| | $M_x, M_y$ | $M_z$ | $A'$ |
| | | $M_x, M_y$ | $A''$ |

| $C_i$ | $A_g$ | $A_u$ | |
|---|---|---|---|
| | *f.* | $M_x, M_y, M_z$ | $A_g$ |
| | | *f.* | $A_u$ |

| $C_2$ | $A$ | $B$ | |
|---|---|---|---|
| $(C_{2h})$[a] | $M_z$ | $M_x, M_y$ | $A$ |
| | | $M_z$ | $B$ |

| $C_{2v}$ | $A_1$ | $A_2$ | $B_1$ | $B_2$ | |
|---|---|---|---|---|---|
| | $M_z$ | *f.* | $M_x$ | $M_y$ | $A_1$ |
| | | $M_z$ | $M_y$ | $M_x$ | $A_2$ |
| | | | $M_z$ | *f.* | $B_1$ |
| | | | | $M_z$ | $B_2$ |

| $D_2$ | $A$ | $B_1$ | $B_2$ | $B_3$ | |
|---|---|---|---|---|---|
| $(D_{2h})$[a] | *f.* | $M_z$ | $M_y$ | $M_x$ | $A$ |
| | | *f.* | $M_x$ | $M_y$ | $B_1$ |
| | | | *f.* | $M_z$ | $B_2$ |
| | | | | *f.* | $B_3$ |

| $C_{3v}$ | $A_1$ | $A_2$ | $E$ | |
|---|---|---|---|---|
| | $M_z$ | *f.* | $M_{x,y}$ | $A_1$ |
| | | $M_z$ | $M_{x,y}$ | $A_2$ |
| | | | $M_z, M_{x,y}$ | $E$ |

| $D_3$ | $A_1$ | $A_2$ | $E$ | |
|---|---|---|---|---|
| $(D_{3d})$[a] | *f.* | $M_z$ | $M_{x,y}$ | $A_1$ |
| | | *f.* | $M_{x,y}$ | $A_2$ |
| | | | $M_z, M_{x,y}$ | $E$ |

| $D_{3h}$ | $A_1'$ | $A_2'$ | $A_1''$ | $A_2''$ | $E'$ | $E''$ | |
|---|---|---|---|---|---|---|---|
| $(C_{3h})$[b] | *f.* | *f.* | *f.* | $M_z$ | $M_{x,y}$ | *f.* | $A_1'$ |
| | | *f.* | $M_z$ | *f.* | $M_{x,y}$ | *f.* | $A_2'$ |
| | | | *f.* | *f.* | *f.* | $M_{x,y}$ | $A_1''$ |
| | | | | *f.* | *f.* | $M_{x,y}$ | $A_2''$ |
| | | | | | $M_{x,y}$ | $M_z$ | $E'$ |
| | | | | | | $M_{x,y}$ | $E''$ |

TABLE 9 (CONT'D)

| $\boldsymbol{D}_6$ $(\boldsymbol{D}_{6h})$[a] | $A_1$ | $A_2$ | $B_1$ | $B_2$ | $E_1$ | $E_2$ | |
|---|---|---|---|---|---|---|---|
| | f. | $M_z$ | f. | f. | $M_{x,y}$ | f. | $A_1$ |
| | | f. | f. | f. | $M_{x,y}$ | f. | $A_2$ |
| | | | f. | $M_z$ | f. | $M_{x,y}$ | $B_1$ |
| | | | | f. | f. | $M_{x,y}$ | $B_2$ |
| | | | | | $M_z$ | $M_{x,y}$ | $E_1$ |
| | | | | | | $M_z$ | $E_2$ |

| $\boldsymbol{D}_{2d}$ | $A_1$ | $A_2$ | $B_1$ | $B_2$ | $E$ | |
|---|---|---|---|---|---|---|
| | f. | f. | f. | $M_z$ | $M_{x,y}$ | $A_1$ |
| | | f. | $M_z$ | f. | $M_{x,y}$ | $A_2$ |
| | | | f. | f. | $M_{x,y}$ | $B_1$ |
| | | | | f. | $M_{x,y}$ | $B_2$ |
| | | | | | $M_z$ | $E$ |

| $\boldsymbol{C}_{\infty v}$ $(\boldsymbol{D}_{\infty h})$[a] | $\Sigma^+$ | $\Sigma^-$ | $\Pi$ | $\Delta$ | $\Phi$ | ... | |
|---|---|---|---|---|---|---|---|
| | $M_z$ | f. | $M_{x,y}$ | f. | f. | ... | $\Sigma^+$ |
| | | $M_z$ | $M_{x,y}$ | f. | f. | ... | $\Sigma^-$ |
| | | | $M_z$ | $M_{x,y}$ | f. | ... | $\Pi$ |
| | | | | $M_z$ | $M_{x,y}$ | ... | $\Delta$ |
| | | | | | $M_z$ | ... | $\Phi$ |
| | | | | | | ... | ⋮ |

| $\boldsymbol{T}_d$ | $A_1$ | $A_2$ | $E$ | $F_1$ | $F_2$ | |
|---|---|---|---|---|---|---|
| | f. | f. | f. | f. | $M_{x,y,z}$ | $A_1$ |
| | | f. | f. | $M_{x,y,z}$ | f. | $A_2$ |
| | | | f. | $M_{x,y,z}$ | $M_{x,y,z}$ | $E$ |
| | | | | $M_{x,y,z}$ | $M_{x,y,z}$ | $F_1$ |
| | | | | | $M_{x,y,z}$ | $F_2$ |

| $\boldsymbol{O}$ $(\boldsymbol{O}_h)$[a] | $A_1$ | $A_2$ | $E$ | $F_1$ | $F_2$ | |
|---|---|---|---|---|---|---|
| | f. | f. | f. | $M_{x,y,z}$ | f. | $A_1$ |
| | | f. | f. | f. | $M_{x,y,z}$ | $A_2$ |
| | | | f. | $M_{x,y,z}$ | $M_{x,y,z}$ | $E$ |
| | | | | $M_{x,y,z}$ | $M_{x,y,z}$ | $F_1$ |
| | | | | | $M_{x,y,z}$ | $F_2$ |

[a] The table applies to this point group if the $(g, u)$ rule is added, that is, $g \leftrightarrow u$, $g \not\leftrightarrow g$, $u \not\leftrightarrow u$.
[b] The table applies to this point group if subscripts 1 and 2 are omitted.

course those forbidden by the spin selection rule). Similarly for point group $\boldsymbol{C}_2$ one has $A-A$ and $B-B$ transitions with $M_z$ and $A-B$ transitions with $M_x$, $M_y$.

For linear non-symmetrical molecules (point group $\boldsymbol{C}_{\infty v}$) the dipole moment has species $\Sigma^+$ or $\Pi$. Therefore $\Sigma^+-\Sigma^+$, $\Sigma^--\Sigma^-$, $\Pi-\Pi$, $\Delta-\Delta$ transitions are possible with $M_z$ while $\Sigma^\pm-\Pi$, $\Pi-\Delta, \ldots$ transitions are possible with either $M_x$ or $M_y$ or any linear combination of them for which we write $M_{x,y}$. However, $\Sigma^+-\Sigma^-$, $\Sigma-\Delta$, $\Pi-\Phi, \ldots$ transitions are forbidden. For symmetrical linear molecules (point group $\boldsymbol{D}_{\infty h}$) the $(g, u)$ rule has to be added, that is, $g \leftrightarrow u$, $g \not\leftrightarrow g$, $u \not\leftrightarrow u$.

In a similar manner other point groups can be dealt with. All one has to do is to establish the species of the components of the dipole moment and to see whether one (or more) of them agree with the species resulting from the direct product of the species of the two states considered (see Appendix III). The species of the dipole moment can be found in Table 55 of Volume II or from the character tables in Appendix I of this volume. For the convenience of the reader we give in Table 9 the results for all possible transitions in the more important point groups.

## (b) Forbidden electronic transitions

Electronic transitions that are not in harmony with the selection rules discussed in the preceding sub-section are in general weak. However, some of these forbidden transitions may have appreciable intensities and therefore a more detailed discussion is necessary, the more so since some authors do not designate certain transitions as forbidden transitions which according to the notation used here must clearly be considered as such. Also it happens quite frequently that a transition which would be classified as a forbidden transition if the molecule had the same high symmetry in both states is actually an allowed transition because in one state the molecule has lower symmetry.

Several types of forbidden transitions may be distinguished depending on the cause for their appearance in violation of the previous selection rules.

### (α) *Magnetic dipole and electric quadrupole transitions*

Classically an oscillating magnetic dipole moment or electric quadrupole moment also leads to the (weak) emission or absorption of radiation. In quantum theory the transition probability produced by magnetic dipole or electric quadrupole radiation is obtained if the magnetic dipole moment or electric quadrupole moment is substituted in place of the electric dipole moment in the transition moment (II, 1). A non-zero value for this transition probability is obtained if the product $\psi_e'^*\psi_e''$ has the same species as one of the components of the magnetic dipole or electric quadrupole moments.

Since the magnetic dipole moment is an axial vector, its components behave in the same way as the three rotations $R_x$, $R_y$, $R_z$ and its species may be found in the Tables of Appendix I as the species of these rotations. The electric quadrupole moment is a tensor whose components behave in the same way as the components of the polarizability, that is, as the product of two translations. Therefore, the species given in Table 55 of Volume II (p. 252) for $\alpha_{xx}, \alpha_{xy}, \ldots$ apply. For example, for symmetrical linear molecules (point group $\boldsymbol{D}_{\infty h}$) the components of the magnetic dipole moment have the species $\Sigma_g^-$ and $\Pi_g$, those of the electric quadrupole moment $\Sigma_g^+$, $\Pi_g$, $\Delta_g$. Therefore, for a given transition to be allowed as magnetic dipole radiation the product of the electronic eigenfunctions of upper and lower states must have the species $\Sigma_g^-$ or $\Pi_g$. Thus in absorption from a totally symmetric ground state $\Sigma_g^- - \Sigma_g^+$, $\Pi_g - \Sigma_g^+$ transitions can occur. Similarly for transitions allowed as electric quadrupole radiation the product of the eigenfunctions must have one of the species $\Sigma_g^+$, $\Pi_g$ or $\Delta_g$, and in absorption from a totally symmetric ground state $\Sigma_g^+ - \Sigma_g^+$, $\Pi_g - \Sigma_g^+$ and $\Delta_g - \Sigma_g^+$ transitions can occur.

Similar considerations apply to other point groups. Table 10 lists all the forbidden transitions made possible in this way for the more important point groups assuming the lower state to be totally symmetric. For other lower state species, say $\Gamma$, the upper state species given have to be multiplied by $\Gamma$. For the visible region the intensity of magnetic dipole and electric quadrupole transitions is about $10^{-5}$ and $10^{-8}$ respectively of the intensity of strong electric dipole transitions; in the ultraviolet the intensity ratio is larger. In several point groups (without a center of symmetry) some transitions allowed as electric dipole radiation can also occur as magnetic dipole or quadrupole radiation. Such transitions are not listed in Table 10 since the electric dipole moment component then usually predominates by such a large factor.

The only observed case of a magnetic dipole spectrum of a polyatomic molecule is the || component of the near ultraviolet absorption system of $H_2CO$. Such an interpretation of the observed spectrum was first suggested by Sidman (1132) and

TABLE 10. FORBIDDEN TRANSITIONS MADE POSSIBLE BY MAGNETIC DIPOLE AND ELECTRIC QUADRUPOLE RADIATION FOR THE MORE IMPORTANT POINT GROUPS

Only transitions with totally symmetric lower states are listed. Others can be obtained from those given by "multiplication" with the species of the lower state.

| Point group | Magnetic dipole transitions | Electric quadrupole transitions |
|---|---|---|
| $\boldsymbol{C}_i$ | $A_g - A_g$ | $A_g - A_g$ |
| $\boldsymbol{C}_{2v}$ | $A_2 - A_1$ | $A_2 - A_1$ |
| $\boldsymbol{C}_{3v}$ | $A_2 - A_1$ | |
| $\boldsymbol{C}_{\infty v}$ | $\Sigma^- - \Sigma^+$ | $\Delta - \Sigma^+$ |
| $\boldsymbol{C}_{2h}$ | $A_g - A_g$, $B_g - A_g$ | $A_g - A_g$, $B_g - A_g$ |
| $\boldsymbol{C}_{3h}$ | $A' - A'$, $E'' - A'$ | $A' - A'$, $E'' - A'$ |
| $\boldsymbol{D}_{2h}$ | $B_{1g} - A_g$, $B_{2g} - A_g$, $B_{3g} - A_g$ | $A_g - A_g$, $B_{1g} - A_g$, $B_{2g} - A_g$, $B_{3g} - A_g$ |
| $\boldsymbol{D}_{3h}$ | $A_2' - A_1'$, $E'' - A_1'$ | $A_1' - A_1'$, $E'' - A_1'$ |
| $\boldsymbol{D}_{6h}$ | $A_{2g} - A_{1g}$, $E_{1g} - A_{1g}$ | $A_{1g} - A_{1g}$, $E_{1g} - A_{1g}$, $E_{2g} - A_{1g}$ |
| $\boldsymbol{D}_{\infty h}$ | $\Sigma_g^- - \Sigma_g^+$, $\Pi_g - \Sigma_g^+$ | $\Sigma_g^+ - \Sigma_g^+$, $\Pi_g - \Sigma_g^+$, $\Delta_g - \Sigma_g^+$ |
| $\boldsymbol{D}_{2d}$ | $A_2 - A_1$ | $A_1 - A_1$, $B_1 - A_1$ |
| $\boldsymbol{D}_{3d}$ | $A_{2g} - A_{1g}$, $E_g - A_{1g}$ | $A_{1g} - A_{1g}$, $E_g - A_{1g}$ |
| $\boldsymbol{T}_d$ | $F_1 - A_1$ | $A_1 - A_1$, $E - A_1$ |
| $\boldsymbol{O}_h$ | $F_{1g} - A_{1g}$ | $A_{1g} - A_{1g}$, $E_g - A_{1g}$, $F_{2g} - A_{1g}$ |

has recently been definitely established by Callomon and Innes (178) (see also p. 270). While for diatomic molecules forbidden electronic transitions have been observed which are entirely caused by magnetic dipole or quadrupole radiation (see Vol. I, p. 277), for polyatomic molecules forbidden electronic transitions caused in this way can also appear as electric dipole radiation if certain vibrations are excited (see below); these vibronically induced transitions are always much stronger than those vibrational transitions of the same electronic transition which can occur only as quadrupole or magnetic dipole radiation (e.g., the 0–0 band). Thus, as in the case of $H_2CO$, magnetic dipole (or quadrupole) transitions occur only as weak components of an electronic transition whose principal components while forbidden are much stronger on account of vibronic interaction[2]. For this reason magnetic dipole and electric quadrupole transitions are observed only very rarely in polyatomic molecules.

[2] While in the upper state of the near ultraviolet $H_2CO$ bands the molecule is non-planar and the transition is therefore formally allowed ($^1A'' - {}^1A'$), actually, since the inversion doubling is very large, the transitions of the individual components follow the rule for a planar-planar $^1A_2 - {}^1A_1$ transition and only for this reason could the magnetic dipole nature of the transition be established (see p. 521).

Magnetic dipole transitions are of great importance indirectly since in molecules of appropriate symmetry a transition with parallel components of electric and magnetic dipole transition moments gives rise to circular dichroism with an associated optical rotatory dispersion, that is, to the familiar optical activity of dissymmetric molecules [for a recent review, see Mason (799b)].

### (β) *Intercombinations*

If the multiplet splitting is not negligibly small the eigenfunction including spin can no longer be characterized by a fixed value of $S$, and therefore transitions violating the rule $\Delta S = 0$ may occur weakly. In such cases it is necessary to use the symmetry of the complete eigenfunction $\psi_{es}$ for a discussion of the selection rules, that is, in applying Table 9 one must use the species of the product of $\psi_e$ and the spin function (see Appendix II).

The most important and simplest case is that of a singlet-triplet intercombination. Most stable molecules have a singlet ground state and have a triplet state as the first excited state. For example for a linear molecule we may have a ${}^1\Sigma_g^+$ ground state and a ${}^3\Pi_u$ excited state. Since the spin function for a triplet state of a linear molecule has the species $\Sigma_g^- + \Pi_g$ the resultant species of the multiplet components of the ${}^3\Pi_u$ state are $\Pi_u + \Sigma_u^+ + \Sigma_u^- + \Delta_u$, of which according to Table 9 only $\Pi_u$ and $\Sigma_u^+$ can combine with the $\Sigma_g^+$ ground state. (This corresponds to the well-known fact, in diatomic molecular spectroscopy, that ${}^3\Pi_{0^+}$ and ${}^3\Pi_1$ combine with ${}^1\Sigma$ but not ${}^3\Pi_{0^-}$ and ${}^3\Pi_2$ if the multiplet splitting is large.) Similarly for a non-planar $XY_3$ molecule (point group $\boldsymbol{C}_{3v}$) the spin function of a triplet state has species $A_2 + E$. Therefore a ${}^3A_2$ state will split into one of species $A_1$ and one of species $E$, both of which can combine with a ${}^1A_1$ ground state, while without spin-orbit interaction the transition is forbidden by symmetry ($A_2 \not\leftrightarrow A_1$). As a last example let us consider a ${}^3F_2 - {}^1A_1$ transition of a tetrahedral molecule. Since here the spin function has species $F_1$, the ${}^3F_2$ state splits into four states $A_2$, $E$, $F_1$ and $F_2$ (see Appendix III) of which only the last can combine with the ${}^1A_1$ state. This corresponds to the fact that in atomic spectra in a ${}^3P - {}^1S$ transition only the ${}^3P_1$ component (which like $F_2$ is triply degenerate) can combine with ${}^1S_0$.

For doublet-quartet transitions the extended point groups must be used. Consider for example a ${}^4E'' - {}^2A_2''$ transition of a planar $XY_3$ molecule. The spin functions of quartet and doublet states are respectively $E_{\frac{1}{2}} + E_{\frac{3}{2}}$ and $E_{\frac{1}{2}}$ (see Appendix II). Therefore, taking account of spin-orbit interaction, ${}^4E''$ splits into $E_{\frac{1}{2}} + E_{\frac{3}{2}} + E_{\frac{1}{2}} + E_{\frac{5}{2}}$ while ${}^2A_2''$ transforms into $E_{\frac{5}{2}}$ (see Appendix III). The transitions $E_{\frac{1}{2}} - E_{\frac{5}{2}}$ can occur both with $M_z$ and $M_{x,y}$ as is readily seen from the direct products while $E_{\frac{3}{2}} - E_{\frac{5}{2}}$ can occur with $M_{x,y}$ only, and $E_{\frac{5}{2}} - E_{\frac{5}{2}}$ is forbidden.

No examples of doublet-quartet transitions have as yet been observed[2a], but

[2a] This statement applies only to free molecules in the gaseous phase. For ions of transition metals or rare earths in crystals or solutions such intercombinations have been observed. The best-known example is the line 6943 Å of ruby crystals, the first transition for which optical maser action was observed.

many examples of singlet-triplet transitions are known (see Chap. V). Lewis and his collaborators have interpreted the observed phosphorescence of many complicated polyatomic molecules as due to transitions from the first excited (triplet) electronic state to the (singlet) ground state. For simpler molecules only a few such intercombinations have been observed. One clear case in which the nature of an intercombination has been established by the triplet splitting is the $^3A_2 - {}^1A_1$ transition of formaldehyde [Di Giorgio and Robinson (278), Raynes (1060)]. A detailed theoretical treatment of the magnitude of spin-orbit coupling in $H_2CO$ and the intensity of the intercombination has been given by Sidman (1131) and Raynes (1060). As in atoms and diatomic molecules, it is the mixing of the triplet state with a singlet state (which combines strongly with the ground state) that makes it possible for the triplet-singlet transition to occur. For $H_2CO$ it is a $^1A_1$ state lying fairly high above $^3A_2$ which gives the main contribution to the transition with an intensity which is in rough agreement with the experimental value[3]. Sidman has also shown why triplet-singlet transitions in aromatic molecules are in general very much weaker, by about a factor 1000 [see also Hameka and Oosterhoff (467a) and Albrecht (54)].

Douglas (293) has used the Zeeman effect in order to establish whether or not a given transition observed in absorption from the ground state is an intercombination (see p. 272). In this way he has shown that the 3600 Å bands of $CS_2$ and the 3800 Å bands of $SO_2$ are intercombinations. Herzberg and Verma (523) (545) have attempted to identify intercombinations in HSiCl, HSiBr and symmetrical pyrazine by anomalies in the rotational structure (see below).

### (γ) *Transitions due to vibronic interactions*

The neglect of type (a) and type (b) vibronic interactions made in the derivation of the general electronic selection rule (II, 1) is often a serious one, and therefore frequently transitions forbidden by the electronic selection rule occur weakly on account of these vibronic interactions. As stated previously, the general selection rule valid when vibronic interactions are *not* neglected (but rotation is disregarded; see below) is

$$R_{e'v'e''v''} = \int \psi'^{*}_{ev} \boldsymbol{M} \psi''_{ev} d\tau_{ev} \tag{II, 19}$$

where it is now no longer possible to make the resolution expressed in eq. (II, 6) which leads to eq. (II, 11). In (II, 19) $\boldsymbol{M}$ is the dipole moment vector in the molecule-fixed coordinate system. Thus the general selection rule that takes account of vibronic interaction is that $R_{e'v'e''v''}$ *must be different from zero* or, in other words, in order that a given transition may occur, the product of the *vibronic* species of

[3] Sidman's statement that the selection rule $\Delta K = 0$ applies to $^3A_2 - {}^1A_1$ transitions produced by interaction of $^3A_2$ with $^1A_1$ while $\Delta K = \pm 1$ applies when the transition is produced by $^3A_2 - {}^1B_{1,2}$ interaction is incorrect as pointed out by Hougen (574): the selection rule is $\Delta K = 0, \pm 2$ independent of the nature of the singlet state that causes the intercombination to occur. Transitions with $\Delta K = \pm 1$ are forbidden by the symmetry rules in $\boldsymbol{C}_{2v}$ molecules (see p. 268f).

the two states involved must contain the species of at least one component of the electric dipole moment. Here it must be remembered that the vibronic species can always be obtained from the product form $\psi_{ev} = \psi_e\psi_v$ even when $\psi_e\psi_v$ is only a very rough approximation to $\psi_{ev}$ (see p. 22).

If rotation is included we must write

$$\psi = \psi_{ev}\psi_r \tag{II, 20}$$

and obtain instead of eq. (II, 15) now for the $x$ component of the transition moment

$$R^{(x)}_{e'v'r'e''v''r''} = \int \psi^*_{e'v'} M_x \psi_{e''v''} d\tau_{ev} \int \psi_r'^* \cos\alpha_x \psi_r'' d\tau_r + \cdots + \int \psi^*_{e'v'} M_z \psi_{e''v''} d\tau_{ev} \int \psi_r'^* \cos\alpha_z \psi_r'' d\tau_r \tag{II, 21}$$

and similarly for the $y$ and $z$ components. The integrals $\int \psi_r'^* \cos\alpha_x \psi_r'' d\tau_r$ etc., as before, determine the rotational selection rules (see section 3 of this chapter). If these selection rules are fulfilled, that is, if these integrals are different from zero, we see from (II, 21) that the transition moment is determined by eq. (II, 19); that is, the general selection rule in this approximation is not affected by rotation.

For an allowed electronic transition $\int \psi_e'^* \boldsymbol{M}_e \psi_e'' d\tau_e$ is different from zero, and if the vibronic functions $\psi_{ev}$ have the same species as the electronic function (as they do for all totally symmetric vibrational levels), the general selection rule is also fulfilled (see section 2b$\alpha$). But for a forbidden electronic transition we have

$$R_{e'e''}(Q_0) = \int \psi_e'^*(q, Q_0) \boldsymbol{M}_e \psi_e''(q, Q_0) d\tau_e = 0. \tag{II, 22}$$

Therefore, if $R_{e'v'e''v''}$ is to be different from zero, the vibronic species must be different from the electronic species and thus the vibrational transitions in such a forbidden electronic transition are different from those in an allowed electronic transition (see below). It is therefore clear that this type of forbidden transition has no analogue in diatomic molecules because for diatomic molecules the vibrations are always totally symmetrical and therefore the vibronic species is always the same as the electronic species. The reason for the existence of a different kind of forbidden transitions in polyatomic molecules is that there can be antisymmetric or degenerate vibrations which take the molecule to conformations of a symmetry that is lower than that of the equilibrium conformation and in which therefore the (electronic) selection rules are less restrictive.

If the definition (II, 22) of a forbidden electronic transition (viz., forbidden by symmetry) is fulfilled for one nuclear conformation $Q_0$, it is clearly fulfilled for any other conformation $Q$ of the same symmetry. We can therefore choose in $\psi_e'$ the equilibrium conformation $Q_0'$ of the upper state, and in $\psi_e''$ its equilibrium conformation $Q_0''$ and thus write the definition of a forbidden transition

$$\int \psi_e'(q, Q_0') \boldsymbol{M}_e \psi_e''(q, Q_0'') d\tau_e = 0. \tag{II, 22a}$$

On the other hand, when antisymmetric or degenerate vibrations are excited, nuclear configurations $Q_a$ occur which do not have the full symmetry of the assumed point group, and for these configurations $R_{ee}^{\prime\prime\prime}$ is no longer exactly zero since $\psi_e$ varies slightly with $Q$. Thus for certain vibrational transitions involving antisymmetric vibrations (see section 2) the electronic forbiddenness is not rigorous.

As an example, consider a ${}^1\Sigma_g{}^+ - {}^1\Sigma_g{}^+$ electronic transition of a $\boldsymbol{D}_{\infty h}$ molecule. In symmetric diatomic molecules this transition could occur only as quadrupole radiation and is therefore extremely weak. The same is true for the 0–0 band and generally transitions between totally symmetric vibrational levels of a linear polyatomic molecule of $\boldsymbol{D}_{\infty h}$ symmetry. However, there are $\Sigma_u^+$ and $\Pi_u$ vibronic levels in each ${}^1\Sigma_g^+$ state, and according to the general selection rule (II, 19) they can combine with the $\Sigma_g^+$ vibronic levels of the other electronic state with an intensity that depends on the strength of the vibronic interaction. For zero vibronic interaction these transitions would have vanishing intensity since then (II, 22) is accurately fulfilled, and this, together with (II, 11), leads immediately to $R_{e'v'e''v''} = 0$.

Similarly, while an $A_2 - A_1$ electronic transition is forbidden for $XYZ_2$ molecules of point group $\boldsymbol{C}_{2v}$, there are $B_1$ and $B_2$ vibronic levels in the $A_2$ electronic state and according to the general selection rule these vibronic levels can combine with the lowest vibrational level of the lower $A_1$ state as well as with higher $A_1$ vibrational levels of this state. Conversely the lowest vibrational level of the upper ($A_2$) electronic state (or other totally symmetric vibrational levels) can combine with the $B_1$ and $B_2$ vibronic levels of the lower ($A_1$) electronic state.

In the same way *forbidden components of allowed electronic transitions* may occur. For example, an $A_1 - A_1$ electronic transition in a $\boldsymbol{C}_{2v}$ molecule is allowed with $M_z$ (i.e., the transition moment is in the symmetry axis). The dipole components $M_x$ and $M_y$ do not give rise to an allowed transition. But there are $B_1$ and $B_2$ vibronic levels in either electronic state which according to (II, 19) may combine with the $A_1$ vibronic levels of the other state if the direction of the dipole moment is perpendicular to the symmetry axis. Similarly, a $\Sigma - \Pi$ electronic transition of a linear ($\boldsymbol{C}_{\infty v}$) molecule is allowed with the dipole moment perpendicular to the internuclear axis only ($M_{x,y}$). However, there are $\Pi$ vibronic levels in the $\Sigma$ electronic state which, according to (II, 19), can combine with the $\Pi$ vibronic levels of the $\Pi$ state with a dipole moment in the direction of the internuclear axis ($M_z$). Again all these forbidden components would have zero intensity if there were no interaction between vibration and electronic motion.

The strength of the vibronic interaction and therefore the intensity of the forbidden transitions (or forbidden components) between two electronic states $\tilde{X}$ and $\tilde{Y}$ depends on the proximity of a suitable third electronic state $\tilde{Z}$ to one of the two states $\tilde{X}$ or $\tilde{Y}$ (see Fig. 48). If the transition $\tilde{Z} - \tilde{X}$ is allowed with a dipole component $M_\mu$, and if $\tilde{Z}$ is not too far from $\tilde{Y}$, then the transition $\tilde{Y} - \tilde{X}$ can "borrow" intensity from $\tilde{Z} - \tilde{X}$ because of a mixing of the electronic eigenfunctions of $\tilde{Y}$ and $\tilde{Z}$. This mixing depends on the vibronic interaction: only states of the same vibronic species mix. Thus, only those vibrational levels of state $\tilde{Y}$ (Fig. 48) which have the same vibronic species as $\tilde{Z}$ will mix with $\tilde{Z}$ and will therefore combine with the ground state $\tilde{X}$ with the same dipole component ($M_\mu$) as that of the allowed $\tilde{Z} - \tilde{X}$ transition.

In a first approximation, as was shown by Herzberg and Teller (542), only those electronic states will mix whose species differ by no more than the species of a normal vibration. Under this condition and if in addition the two mutually perturbing states ($\tilde{Y}$ and $\tilde{Z}$) are less than 1 eV apart, one finds that the forbidden transition ($\tilde{Y}-\tilde{X}$) will have an appreciable intensity, comparable to that of a weak allowed transition.

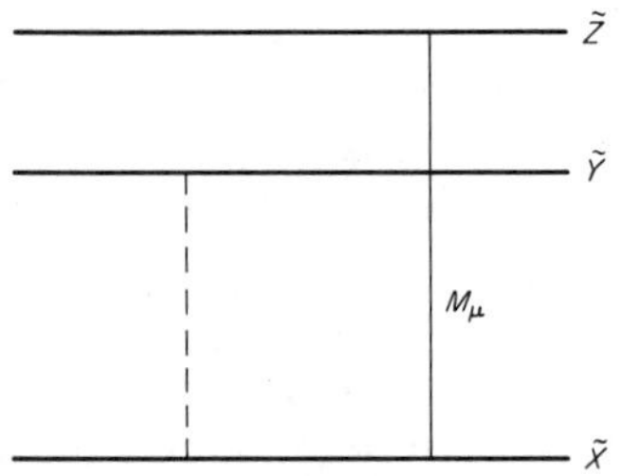


FIG. 48. **Energy level diagram explaining the occurrence of forbidden transitions.** The transition $\tilde{Z}-\tilde{X}$ is allowed with dipole component $M_\mu$. Because of (vibronic) mixing between $\tilde{Z}$ and $\tilde{Y}$ the forbidden transition $\tilde{Y}-\tilde{X}$ can "borrow" intensity from $\tilde{Z}-\tilde{X}$.

As an example, consider a planar $\boldsymbol{C}_{2v}$ molecule with the electronic states $A_1$, $A_2$, $B_2$ corresponding to $\tilde{X}$, $\tilde{Y}$, $\tilde{Z}$, respectively, in Fig. 48. The transition $B_2-A_1$ is allowed with the dipole component $M_y$ (in-plane). The states $A_2$ and $B_2$ can mix if there is a normal vibration of species $B_1$. This is the case for $H_2CO$ (out-of-plane bending) but not for $H_2O$. In $H_2CO$ therefore an $A_2-A_1$ transition could occur weakly with $M_y$. Actually the observed near ultraviolet absorption bands of $H_2CO$ were at one time thought to represent such a forbidden $A_2-A_1$ transition [Pople and Sidman (1003)] but it has been found that in the excited state the molecule is non-planar [see Chapter V, section 2b] and that the transition is therefore not electronically forbidden. Nevertheless, since the barrier against inversion in the excited state is low, in a certain approximation this transition can be treated as if it were a forbidden $A_2-A_1$ transition of a planar $\boldsymbol{C}_{2v}$ molecule.

Perhaps a better example is provided by the near ultraviolet absorption system of benzene. This system has been shown to be a forbidden ${}^1B_{2u}-{}^1A_{1g}$ transition (point group $\boldsymbol{D}_{6h}$). Such a transition is made possible by the interaction of ${}^1B_{2u}$ with the lowest ${}^1E_{1u}$ state (6.8 eV). These two states differ by the species $E_{2g}$ and, since there are (four) normal vibrations of this species in $C_6H_6$, an appreciable borrowing of intensity can occur. Murrell and Pople (921) have carried out a quantitative calculation of the interaction and have derived an intensity which is 8.5 times the observed intensity [see also Albrecht (53)].

Similar considerations apply to forbidden components of allowed electronic transitions. In such cases the transition $\tilde{Y}-\tilde{X}$ (Fig. 48) is allowed with a dipole component different from that of $\tilde{Z}-\tilde{X}$; but due to the vibronic mixing of $\tilde{Z}$ and $\tilde{Y}$, a forbidden component of $\tilde{Y}-\tilde{X}$ with the same dipole component as $\tilde{Z}-\tilde{X}$ can occur. An example for a linear molecule is provided by the ${}^2\Sigma-{}^2\Pi$ electronic transition of NCO in which a weak ∥ component (of type $\Pi - \Pi$) has been observed (see Chap. V, section 1c).

If the allowed component of $\tilde{Y} - \tilde{X}$ is weak and vibronic mixing of $\tilde{Y}$ and $\tilde{Z}$ is strong, it can happen that the forbidden component of $\tilde{Y} - \tilde{X}$ is stronger than the allowed component. This situation arises in the ${}^1B_{2u} - {}^1A_{1g}$ absorption system of naphthalene near 3200 Å in which, as shown by Craig, Hollas, Redies and Wait (253), the forbidden $M_z$ component is ten times stronger than the allowed $M_y$ component ($y$ and $z$ are the long and short axes in the plane of the molecule). For further theoretical discussion of forbidden components in allowed electronic transitions, see Albrecht (52).

It is possible that for an allowed transition the integral (II, 22) vanishes by coincidence even though this is not required by symmetry. In such a case the 0–0 band would also be very weak or absent just as for true forbidden transitions. If $\tilde{Y} - \tilde{X}$ in Fig. 48 is such a transition, and if $Z - \tilde{X}$ is a transition of the same kind for which the accidental cancellation of the electronic transition moment does not occur, then the excitation of totally symmetric vibrations in $\tilde{Y}$ can lead to vibronic mixing of $\tilde{Y}$ and $\tilde{Z}$ and the higher members of the progressions in the totally symmetric vibrations will occur in $\tilde{Y} - \tilde{X}$, "borrowing" intensity from $\tilde{Z} - \tilde{X}$. Such a case has recently been described by Craig and Gordon (251a) in the near ultraviolet spectrum of phenanthrene.

### (δ) *Transitions due to rotational-electronic interactions*

Rotational-electronic interaction is in general very weak, that is, the resolution (II, 20) does represent a very good approximation. However, when no other stronger interactions are present, the rotational-electronic interaction may lead to the very weak occurrence of forbidden transitions. For this to happen there must be in the neighborhood of one of the states ($\tilde{X}$ or $\tilde{Y}$ in Fig. 48) a state ($\tilde{Z}$) which differs from the former by the species of a rotation. The intensity of the forbidden transition will then depend strongly on the rotational quantum number corresponding to this rotation.

For example, for a $C_{2v}$ molecule an $A_2 - A_1$ electronic transition could appear on account of rotational-electronic interaction if there is another $A_1$ state in the neighborhood of the $A_2$ state (or an $A_2$ state in the neighborhood of $A_1$). Since the rotation about the symmetry axis has species $A_2$, an increasing interaction between the $A_1$ and the $A_2$ state will take place with increasing rotation, that is, at high $K$ or $J$, bands with a dipole component parallel to the symmetry axis ($M_z$) would appear. According to Pople and Sidman (1003) the intensity should go up with $K^2$ if the $a$ axis is the symmetry axis. Since the weak ∥ bands observed in $H_2CO$ do not show such a dependence they cannot be accounted for by rotational-electronic interaction; actually they have been assigned to magnetic dipole radiation [see sub-section (α)]. In the ultraviolet absorption spectrum of $CH_2N_2$ Merer (823) has found, in a ${}^1B - {}^1A_1$ electronic transition, forbidden components of vibronic type $A_2 - A_1$ which are in all probability produced by rotational-electronic (Coriolis) interaction.

### ($\epsilon$) *Enforced dipole transitions*

As has been discussed for diatomic molecules in Volume I, p. 280, transitions forbidden rigorously as dipole radiation may be made allowed by the application of strong electric fields, that is, may occur as enforced dipole radiation. The selection rules for enforced dipole radiation are similar to those for quadrupole radiation but will not be discussed in detail here. Instead of an external field, the intermolecular field in gases at high pressure or in liquids and solids may serve to "enforce" the transitions. However, thus far no such transitions have been definitely identified for free polyatomic molecules.

## 2. Vibrational Structure of Electronic Transitions

### (a) Structure of a band system: unsymmetrical molecules

The transitions from the various vibrational (and rotational) levels of one electronic state to those of another in a polyatomic molecule give rise to a *band system* just as for diatomic molecules. However, since there are several vibrations instead of one, the vibrational structure is in general very much more complicated than for diatomic molecules. This is particularly so in an unsymmetrical molecule for which no restrictions of the vibrational transitions exist other than those imposed by the Franck–Condon principle (see below).

**General formulae.** If the vibrational term values in the upper and lower states are $G'(v_1', v_2', v_3', \ldots)$ and $G''(v_1'', v_2'', v_3'', \ldots)$ respectively and, if the electronic term values are $T_e'$ and $T_e''$, the wave numbers of all possible vibrational transitions of a given electronic transition (i.e., of a band system) are represented by

$$\nu = \nu_e + G'(v_1', v_2', \ldots) - G''(v_1'', v_2'', \ldots) \tag{II, 23}$$

where

$$\nu_e = T_e' - T_e'' \tag{II, 23a}$$

is a constant for a given band system and is also called the *origin* of the band system.

Substituting $G(v_1, v_2, v_3, \ldots)$ from (I, 23) we obtain

$$\begin{aligned} \nu = \nu_e &+ \sum_i \omega_i'(v_i' + \tfrac{1}{2}) + \sum_i \sum_{k \geq i} x_{ik}'(v_i' + \tfrac{1}{2})(v_k' + \tfrac{1}{2}) \\ &- \sum_i \omega_i''(v_i'' + \tfrac{1}{2}) - \sum_i \sum_{k \geq i} x_{ik}''(v_i'' + \tfrac{1}{2})(v_k'' + \tfrac{1}{2}) \end{aligned} \tag{II, 24}$$

assuming that all vibrations are non-degenerate ($d_i = 1, l_i = 0$) as would always be the case for an unsymmetrical molecule.

For practical purposes it is often much more convenient to refer the energies to the lowest vibrational level in each state and to write

$$\begin{aligned} \nu = \nu_{00} &+ \sum_i \omega_i^{0\prime} v_i' + \sum_i \sum_{k \geq i} x_{ik}^{0\prime} v_i' v_k' + \cdots \\ &- \sum_i \omega_i^{0\prime\prime} v_i'' - \sum_i \sum_{k \geq i} x_{ik}^{0\prime\prime} v_i'' v_k'' - \cdots \end{aligned} \tag{II, 25}$$

where

$$\omega_i^0 = \omega_i + x_{ii} + \tfrac{1}{2}\sum_{i \neq k} x_{ik} \tag{II, 25a}$$

$$\nu_{00} = \nu_e + \tfrac{1}{2}\sum_i \omega_i' + \tfrac{1}{4}\sum_i \sum_{k \geq i} x_{ik}' - \tfrac{1}{2}\sum_i \omega_i'' - \tfrac{1}{4}\sum_i \sum_{k \geq i} x_{ik}'' \tag{II, 25b}$$

and where the $x_{ik}^0$ differ from the $x_{ik}$ only when higher terms $(y_{ijk}v_iv_jv_k)$ are included (see Vol. II, p. 208). $\nu_{00}$ is the wave number of the $0-0$ band, that is, the transition between the two lowest vibrational levels. The frequencies (in $\text{cm}^{-1}$) of the fundamentals (as observed for the ground state in the infrared or Raman spectrum) are related to the constants in (II, 25) and (II, 24) by

$$\nu_i = \omega_i^0 + x_{ii} = \omega_i + 2x_{ii} + \tfrac{1}{2}\sum_{k \neq i} x_{ik} + \cdots. \tag{II, 26}$$

It must be emphasized that the general formulae (II, 24) and (II, 25) apply to an idealized case in which no Fermi resonances occur. Actually, in many cases the occurrence of such resonances spoils the simple regularity expressed in the formulae: only the average of two (or more) bands corresponding to two (or more) interacting levels is correctly represented by the formulae.

**Progressions.** In Fig. 49 the vibrational levels in two electronic states are shown assuming that there are only two normal vibrations. While this assumption does not correspond to any actual case of an unsymmetrical molecule since the minimum number of vibrations is three, it does show more clearly the essential facts.

Starting with any level of the upper electronic state, e.g., $v_1' = 1$, $v_2' = 2$, and considering the possible transitions to the lower state, we do not just get a single $v_2''$ progression of bands as for diatomic molecules but a whole progression of progressions, one corresponding to $v_1'' = 0$, one to $v_1'' = 1$, one to $v_1'' = 2$, and so on, as indicated in Fig. 49a. In other words each member of the first progression is the starting point of a progression in the second vibration. If a third vibration were present, each member of each of the progressions in the second vibration would be the starting point of a progression in the third vibration, and so on.

In the same way, in absorption, starting from any vibrational level of the lower state we obtain not just one progression but a progression of progressions. This applies of course also to the lowest vibrational level ($v_1 = 0$, $v_2 = 0$), that is, to absorption at low temperature. Figure 49b shows the transitions in this case and Fig. 50 gives a schematic spectrum.

For a band system of a diatomic molecule all the bands can be conveniently written down in a Deslandres table. From what has been said above it is clear that now the transitions from each vibrational level of one electronic state, say the lower one, fill a whole square array corresponding to the various values of $v_1'$ and $v_2'$ and therefore the bands of the whole band system can only be written down in the form of a Deslandres table of Deslandres tables, one for each pair of values of $v_1''$, $v_2''$.

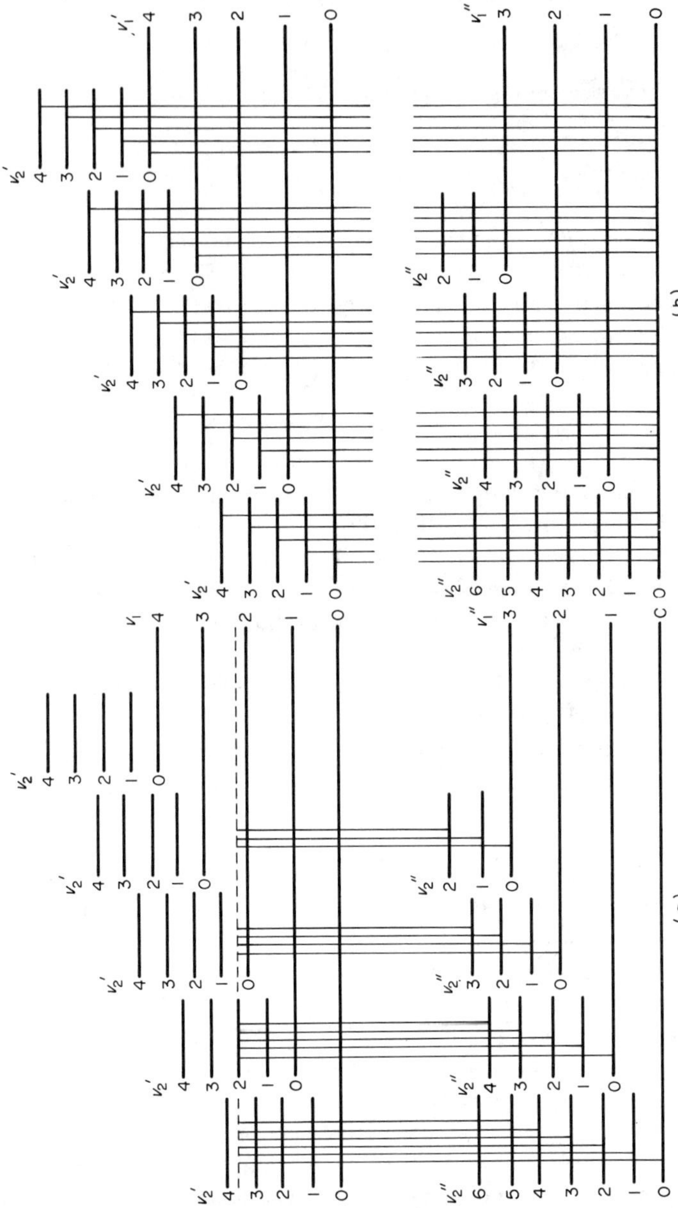


FIG. 49. **Energy level diagrams showing the origin of progressions (*a*) in emission, and (*b*) in absorption from a single initial level.** Only vibrational levels corresponding to two vibrations $\nu_1$ and $\nu_2$ are shown.

Alternatively one can write this *super-Deslandres table* in terms of $v_1'$, $v_1''$ inserting at each place the Deslandres table corresponding to $v_2'$, $v_2''$. This is shown schematically in Fig. 51. It is readily seen how this scheme would have to be expanded if there are more than two vibrations.

If the principal progression $v_1',0 \leftarrow 0,0$ is observed in absorption (see Fig. 49b) one obtains from it immediately the vibrational levels corresponding to excitation

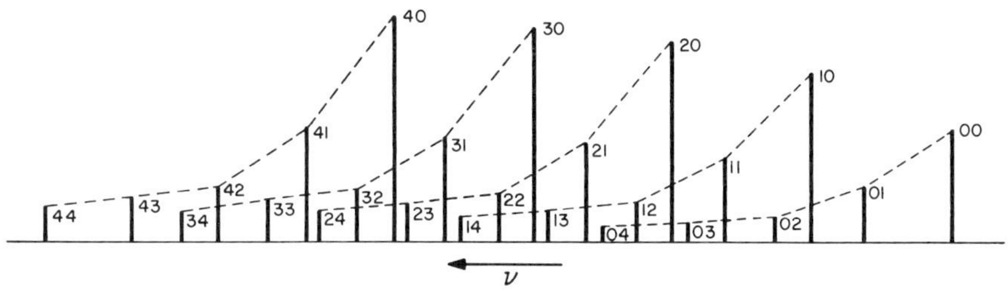


FIG. 50. **Schematic low temperature absorption spectrum showing progressions in two upper state vibrations.** This figure corresponds to the energy level diagram Fig. 49b. The numbers attached to the bands are the $v_1'v_2'$ values. All bands have $v_1'' = 0$, $v_2'' = 0$.

of $\nu_1'$ by various quanta in the upper state and therefore also the quantities $\omega_1^{0\prime}$, $x_{11}^{0\prime}$ and possibly higher terms. Similarly from the progression $0, 0 - v_1'', 0$ (see Fig. 49a) the lower state vibrational levels corresponding to $\nu_1''$ and the constants $\omega_1^{0\prime\prime}$, $x_{11}^{0\prime\prime}, \ldots$, are obtained. The same levels and constants are obtained from other progressions, $v_1', 0 \leftarrow a, b$ and $c, d - v_1'', 0$ where $a, b$ and $c, d$ are the vibrational quantum numbers of the fixed lower or upper state of these progressions. Apart from very small rotational effects the difference of corresponding bands in the two

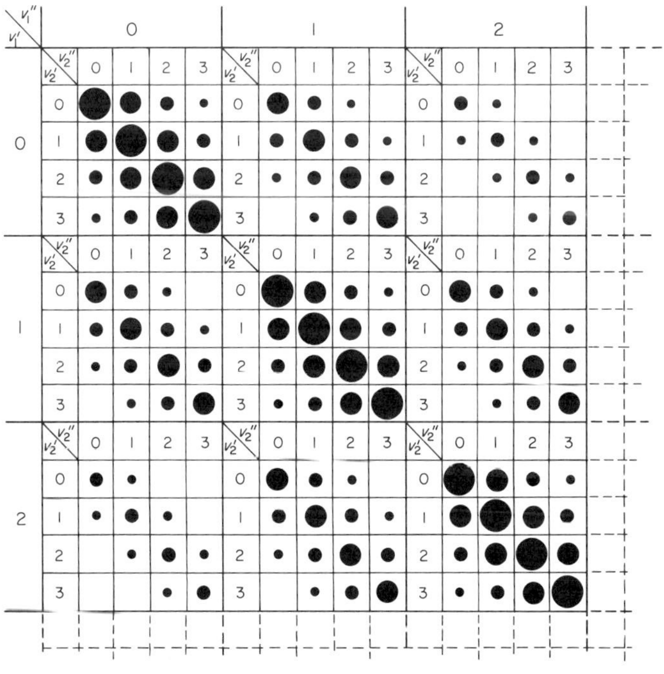


FIG. 51. **Form of Deslandres table for a band system of a polyatomic molecule assuming that only two vibrations are affected by the transition.** The size of the black dots is intended to indicate the intensity of the particular transition assuming only a slight change of molecular dimensions in the transition and assuming nearly equal population of vibrational levels in the initial state.

progressions $v_1', 0-0, 0$ and $v_1', 0-a, b$ and similarly $0, 0-v_1'', 0$ and $c, d-v_1'', 0$ must be exactly equal; this constant difference is directly equal to $G_0''(a, b)$ and $G_0(c, d)$ respectively. Such combination relations give a good check on the correctness of the assignments of the bands. They apply even if there are perturbations (Fermi

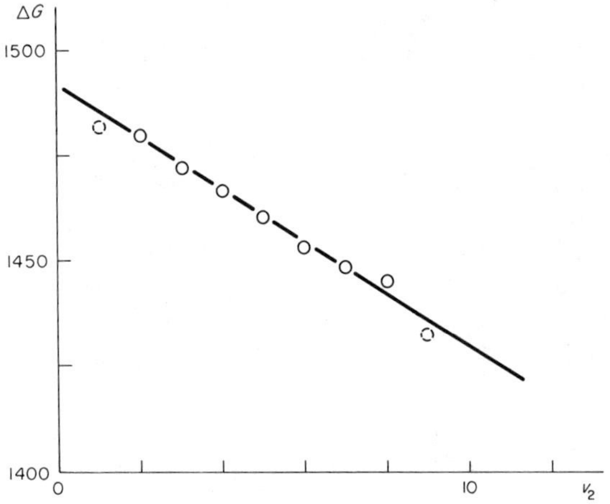


FIG. 52. **Separations $\Delta G$ of corresponding bands in the two progressions** $1v_2'0-000$ and $0v_2'0-000$ **of the $\tilde{A}-\tilde{X}$ systems of DCN as a function of** $v_2'$.

resonances). The same considerations apply to any other vibration that gives rise to a progression.

On the other hand if one compares a progression $v_1'0 \leftarrow 00$ with a progression $v_1'a \leftarrow 00$ one finds the difference of corresponding bands to vary slightly with $v_1'$ because of the term $x_{12}v_1'v_2'$ in the energy formula. In a first approximation this variation should be linear in $v_1$ and should of course be quite small since $x_{12} \ll \omega_1$. The anharmonic constant $x_{12}$ can be determined in this way. As an example in Fig. 52 the difference between the two progressions $1v_2'0 - 000$ and $0v_2'0 - 000$ of the $\tilde{A} - \tilde{X}$ system[3a] of DCN is plotted. The linear variation is clearly shown.

**Sequences.** For a diatomic molecule a band system can be considered as a series of sequences with $\Delta v = 0, \pm 1, \pm 2, \ldots$ corresponding to the diagonal and the lines parallel to the diagonal in the Deslandres table. These sequences form very characteristic groups in the band system if the vibrational frequencies in the upper and lower state are of similar magnitude and if a number of vibrational levels are excited in the initial state. Similar sequences occur in a band system of a polyatomic molecule except that there are now several parameters that distinguish one sequence from another, viz., $\Delta v_1 (= v_1' - v_2'')$, $\Delta v_2$, $\Delta v_3, \ldots$. Again we consider only the case of two normal vibrations. For each value of $\Delta v_1$ all possible values of $\Delta v_2$ can occur, and for a given pair of values $\Delta v_1, \Delta v_2$ both $v_1$ and $v_2$ can go

[3a] For a discussion of the short-hand designations $\tilde{X}, \tilde{A}, \tilde{B}, \ldots \tilde{a}, \tilde{b}, \ldots$ of observed states see p. 488.

through their whole range of values. Thus for $\Delta v_1 = 0, \Delta v_2 = 0$ corresponding to the principal diagonal of the super-Deslandres table (Fig. 51), there is a series of sequences in $v_2$, one for each $v_1$. The full vertical lines in the energy level diagram Fig. 53 represent these transitions. Similarly, a series of sequences arises from $\Delta v_1 = +1, \Delta v_2 = 0$, as well as from $\Delta v_1 = -1, \Delta v_2 = 0$ (oblique broken lines in

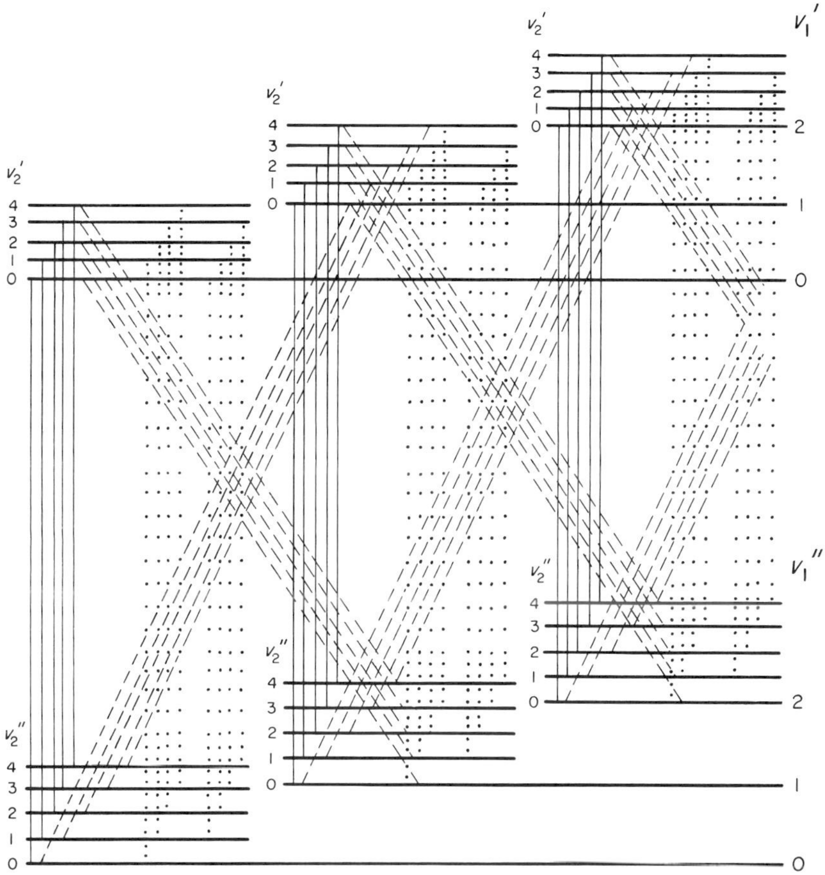


FIG. 53. **Energy level diagram showing the origin of sequences in a band system of a polyatomic molecule assuming that only two vibrations $\nu_1$ and $\nu_2$ are active.** The full-line transitions correspond to $\Delta v_1 = 0, \Delta v_2 = 0$, the broken-line transitions to $\Delta v_1 = \pm 1, \Delta v_2 = 0$, and the dotted transitions to $\Delta v_1 = 0, \Delta v_2 = \pm 1$. Sequences with higher $\Delta v_i$ values are not shown.

Fig. 53) and from $\Delta v_1 = 0, \Delta v_2 = +1$, as well as from $\Delta v_1 = 0, \Delta v_2 = -1$ (vertical dotted lines in Fig. 53), and so on for other pairs of $\Delta v_1, \Delta v_2$ values.

If the vibrational frequencies in the upper and lower electronic state were the same, then all bands of a given set of sequences (a given $\Delta v_1, \Delta v_2$) would coincide. If the vibrational frequencies are slightly different, then each sub-sequence of a given $\Delta v_1, \Delta v_2$ and $v_1$ forms a group of close-lying and approximately equidistant bands of spacing $\Delta\nu_2$ ($\approx \omega_2' - \omega_2''$) and the sub-sequences of different $v_1$ form equidistant groups of spacing $\Delta\nu_1$ ($\approx \omega_1' - \omega_1''$). This is shown schematically for $\Delta v_1 = 0, \Delta v_2 = 0$ in Fig. 54. Similar diagrams apply to the other pairs of $\Delta v_1, \Delta v_2$ values. If $\Delta\nu_2$ were larger than $\Delta\nu_1$, the sub-sequences in $v_1$ would form the closest groups and the separation of the groups would be $\Delta\nu_2$, opposite to the case

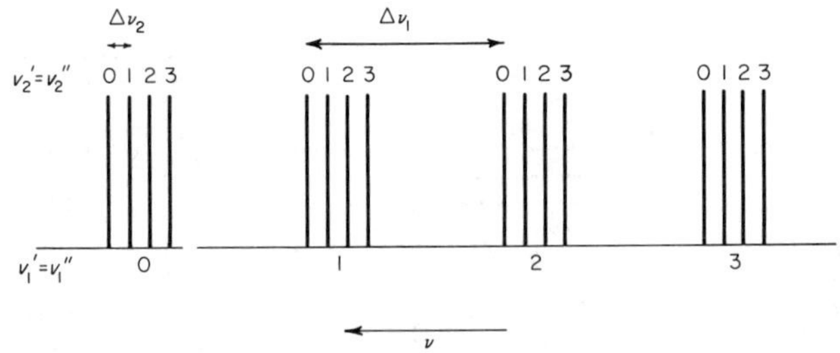


FIG. 54. **Schematic spectrum showing sequences with $\Delta v_1 = 0, \Delta v_2 = 0$ in a band system in which only two vibrations are active.** Both $\Delta\nu_1$ and $\Delta\nu_2$ have been assumed to be negative, i.e., $\nu_1' < \nu_1''$, $\nu_2' < \nu_2''$. Higher sequences with $\Delta v_i \neq 0$ are much weaker than those shown as long as $\Delta\nu_i$ is small.

illustrated in Fig. 54. It is easy to extend the diagrams shown to the case of three or four normal vibrations.

If the differences $\Delta\nu_i$ of the vibrational frequencies in the upper and lower state are large, different sequences overlap one another and a complicated pattern arises. In such a case the concept of sequences may not be of much help in analyzing the spectrum.

In absorption at sufficiently low temperature when no vibrational levels with $v_i > 0$ are thermally excited in the lower electronic state, no sequence structure exists since no more than one band of any sequence can occur. However, at moderate temperatures when one or more low-frequency vibrations are thermally excited, the sequences become important for these vibrations. When $\Delta\nu_i$ is relatively large, it is no longer obvious whether or not a particular band group is a sequence or rather a progression. In that event, other evidence to decide this question has to be sought. One difference between a sequence and a progression is that in the former a quadratic term may be (but need not be) considerable while for the latter it is usually quite small. A sequence may even show a reversal, that is, a "head of heads" may occur (see Vol. I, p. 160). Well-developed sequences are often prominent in emission spectra, for example, in the spectrum of NCN (see p. 499), and they are also observed frequently in absorption spectra of heavier molecules with low-frequency fundamentals such as benzene (see Fig. 70 and p. 557) and naphthalene [see Craig *et al.* (253)].

**Intensity distribution, Franck–Condon principle.** Even though in an unsymmetric molecule there are no sharp selection rules for the vibrational quantum numbers, there are enormous differences in the intensities of different vibrational transitions. In order to understand these differences we must apply, as for diatomic molecules, the Franck–Condon principle. As we have seen in section 1, as long as vibrational and electronic motion can be separated the transition probability can be resolved into one factor $R_{e'e''}$ depending essentially on the electronic motion, one factor $R_{v'v''}$ depending only on vibration and one factor $R_{r'r''}$ depending only on rotation. Neglecting rotation we can write according to eq. (II, 11) for the transition moment

$$R_{e'v'e''v''} = R_{e'e''}R_{v'v''} \tag{II, 27}$$

where $R_{e'e''}$ is given by (II, 9) and $R_{v'v''}$, the contribution of vibration to the transition moment, is given by

$$R_{v'v''} = \int \psi_v'^* \psi_v'' d\tau_v. \tag{II, 28}$$

$R_{e'e''}$, is in general very nearly the same for all vibrational transitions of a given electronic transition. It is therefore the second factor, $R_{v'v''}$, the *overlap integral* of the vibrational eigenfunctions of upper and lower states, that determines the relative intensities of the bands in a band system, just as in diatomic molecules.

If the potential functions in the two electronic states are nearly alike and therefore the internuclear distances and vibrational frequencies are nearly the same, then the vibrational eigenfunctions with different vibrational quantum numbers in the upper and lower state are very nearly orthogonal with respect to each other, and therefore $R_{v'v''}$ will be different from zero only if none of the vibrational quantum numbers change. This situation corresponds to the semi-classical consideration of Franck that after the "electron jump", the nuclei are in the same relative position as before and, since the relative potential energy has not changed, will remain at rest if they were at rest before, or will retain the same kinetic energy they had before. We have therefore $\Delta v_i = 0$ for all vibrational quantum numbers. In cold absorption there will be only one band of outstanding intensity, the $0-0$ band, and at somewhat higher temperatures bands of the principal sequences ($\Delta v_i = 0$) will occur with intensities determined mainly by the appropriate Boltzmann factor. This type of intensity distribution is found in many Rydberg transitions of polyatomic molecules, e.g., $H_2O$, $CH_2$, $CH_3$, . . . indicating that the ion has a potential function that is very similar to that of the neutral molecule.

If on the other hand the equilibrium position of the nuclei changes in going from the one electronic state to the other, and if this change is most nearly represented by the vibrational mode $\nu_i$, bands with $\Delta v_i \neq 0$ will occur strongly. If the change of equilibrium position is large, the maximum of intensity will no longer be at $v_i' = 0$. In such a case, if the molecule is initially in the equilibrium position of one state, then after the "jump" it is more or less high on the slope of the potential surface of the other state and, classically, will immediately start oscillating. Often this oscillation is not a simple motion but a Lissajous motion corresponding to a super-position of two or more normal vibrations of the excited state. The intensity maxima in the corresponding progressions will in general be at non-zero $v_i'$ values.

All this is very similar to what was discussed for diatomic molecules in Volume I, p. 194f, except that we have now a multi-dimensional potential surface (see Chap. IV and Vol. II, p. 202). No detailed calculations have been carried out on this basis with regard to the relative intensities for any unsymmetrical molecule. But one general formula may be mentioned for the case that the equilibrium position remains closely the same in the electronic transition but the vibrational frequency of a normal vibration changes from $\omega_i''$ to $\omega_i'$. One finds for the ratio of the intensity of the $0-0$ band to the sum of the intensities of all $v_i-0$ bands (incl. $0-0$) or, in other words, for the fractional intensity of the $0-0$ band

$$\frac{I_{0-0}}{\sum\limits_{v_i} I_{v_i-0}} = \frac{\sqrt{\omega_i'\omega_i''}}{\frac{1}{2}(\omega_i' + \omega_i'')}. \tag{II, 29}$$

Unless $\omega_i'$ is very different from $\omega_i''$ this ratio is very close to 1, that is, the $0-0$ band has overwhelming intensity compared to all the others. Even for $\omega_i' = \frac{1}{2}\omega_i''$ the above ratio is $\frac{2}{3}\sqrt{2} = 0.9426$.

### (b) Structure of a band system: symmetrical molecules

The structure of a band system of a molecule with one or more elements of symmetry is similar to that of an unsymmetrical molecule except that there are now specific selection rules which eliminate certain bands entirely and restrict the direction of the transition moment in those that do occur, thus simplifying their rotational structure. In addition, for molecules with degenerate vibrations, the vibrational energy formulae have to be changed appropriately. Finally, as we have seen, for symmetrical molecules certain electronic transitions are forbidden, but can occur weakly on account of vibronic interactions. The vibrational structure of these forbidden transitions is different from those of the allowed transitions and will be treated separately.

#### (α) *Allowed electronic transitions*

**General selection rule.** In an allowed electronic transition the electronic transition moment $R_{e'e''}$ is different from zero. Whether or not a transition from a certain vibrational level of the upper state ($v_i'$) to a certain vibrational level of the lower state ($v_i''$) occurs depends on the vibrational overlap integral (II, 28). In a symmetrical molecule, in order that this integral be different from zero, the integrand ($\psi_v'^*\psi_v''$) must be symmetric with respect to all symmetry operations permitted by the point group to which the molecule belongs, that is

$$\psi_v'^*\psi_v'' \textit{ must be totally symmetric} \tag{II, 30}$$

or, for degenerate vibrational levels a product of appropriate linear combinations of the mutually degenerate vibrational eigenfunctions must be totally symmetric. In either case (degenerate or non-degenerate) this conclusion is equivalent to the selection rule: *only vibrational levels of the same vibrational species in the upper as in the lower state can combine with each other.* This is the general vibrational selection rule in allowed electronic transitions. The difference of this selection rule from that for the pure vibration spectrum in the infrared should be noted: the dipole moment $\boldsymbol{M}$ does not enter (II, 28) and (II, 30) since it is already contained in the electronic transition moment $R_{e'e''}$.

According to the general vibrational selection rule (II, 30), in cold absorption, from the vibrationless ground state only totally symmetric vibrational levels of the excited state can be reached. Similarly in emission from the lowest vibrational level of the excited state only totally symmetric vibrational levels of the ground state can be reached.

The general selection rule (II, 30) is derived on the assumption of the validity of the product resolution (II, 4), that is, of negligible vibronic interaction of type (a). If this assumption is not fulfilled we must use for the transition moment the general expression

$$R_{e'v'e''v''} = \int \psi^*_{e'v'} \boldsymbol{M} \psi_{e''v''} d\tau_{ev} \qquad \text{(II, 19)}$$

that is, all those transitions may occur for which the product $\psi^*_{e'v'}\psi_{e''v''}$ has the same species as $\boldsymbol{M}$. In an allowed electronic transition we must choose for $\boldsymbol{M}$ that component that gives a non-vanishing $R_{e'e''}$ according to (II, 1) and therefore we are led back to (II, 30). Other components of $\boldsymbol{M}$ may also give non-vanishing $R_{e'v'e''v''}$ but they correspond by our definition (p. 138) to forbidden components of the electronic transition (see subsection $\beta$).

**Transitions between non-degenerate electronic states.** Since the higher vibrational levels of a totally symmetric vibration are totally symmetric it is clear that according to the general selection rule, progressions of bands corresponding to totally symmetric vibrations occur in symmetrical molecules, entirely similar to the progressions discussed earlier for unsymmetrical molecules. For example, if there are two totally symmetric vibrations we have the same double Deslandres table as discussed in the preceding subsection (Fig. 51). The intensity maxima in each progression are determined as before by the relative positions of the potential surfaces of the upper and lower states (see below). Totally symmetric in this connection means symmetric with respect to all symmetry operations that apply to both the equilibrium position of the one and that of the other electronic state (see p. 130).

An anomalous intensity distribution in the totally symmetric progressions will arise if the electronic transition moment happens to be very small and therefore the $0-0$ band is very weak. As mentioned earlier, higher members of the progressions may then be much stronger through vibronic interaction of one or more totally symmetric vibrations if there is near the upper state another electronic state of the same species which combines strongly with the lower state.

The higher vibrational levels of an antisymmetric vibration $\nu_k$ (i.e., a vibration that is antisymmetric with respect to at least one element of symmetry) are

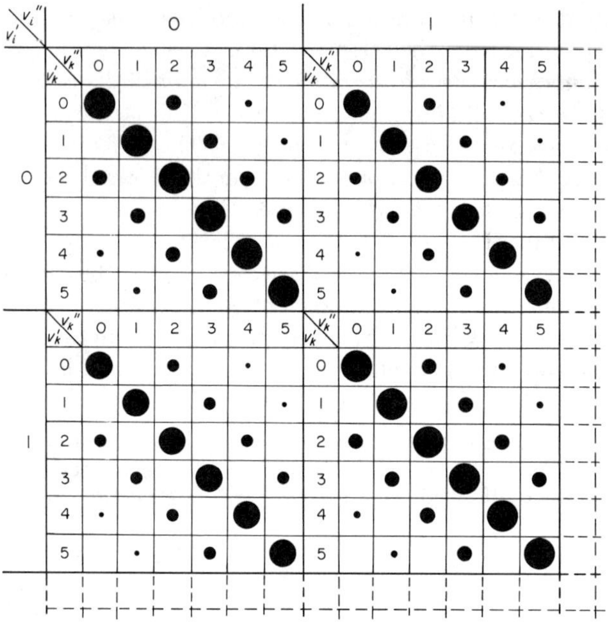

FIG. 55. **Form of Deslandres table for one symmetric ($\nu_i$) and one antisymmetric ($\nu_k$) vibration.** Compare Fig. 51. Note the absence of the odd sequences in $v_k$. The drop in intensity with increasing $|\Delta v_k|$ is usually much greater than represented by the size of the black circles.

symmetric for even $v_k$, antisymmetric for odd $v_k$. Therefore, according to the general selection rule, $v_k$ can only change by an even number, that is

$$\Delta v_k = 0, \pm 2, \pm 4, \ldots. \tag{II, 31}$$

If there is one totally symmetric vibration $\nu_i$ and one antisymmetric vibration $\nu_k$ then in the "double" Deslandres table all sequences with $\Delta v_k = \pm 1, \pm 3, \pm 5, \ldots$ are missing as shown schematically in Fig. 55.

The relative intensities of the transitions with $\Delta v_k = 0, \pm 2, \pm 4, \ldots$ which do occur can readily be predicted on the basis of the Franck–Condon principle. One finds that the transitions with $\Delta v_k = 0$ are always by far the most intense, that is, the bands in the main diagonals of the Deslandres sub-tables are the most intense. The reason for this conclusion is immediately seen when it is realized that as long as the symmetry is the same in upper and lower state the potential minimum occurs at the same value of the antisymmetric coordinate (viz., $\xi_k = 0$) for both states irrespective of any contraction or expansion of the molecule that conserves the symmetry. Just as for a symmetric vibration for which the potential minima have the same position in the two states, here an appreciable intensity of the bands with $\Delta v_k \neq 0$ can arise only when the frequency of the antisymmetric vibration in the two states is very different; but even when the frequency changes by as much as a factor 2 (which is very rare) the $\Delta v_k = 0$ transitions contain 94.4 per cent of the intensity [see eq. (II, 29)]. Thus in cold absorption, when only one symmetric and one antisymmetric vibration are present, only one progression of bands, corresponding to $0-0, 1-0, 2-0, \ldots$ in the symmetric vibration $\nu_i$, appears with appreciable intensity. All other bands, with $v_k \neq 0$, are very weak. For hot absorption instead of a progression of single bands we obtain a progression of sequences with $\Delta v_k = 0$ if $\nu_k$ is small enough to be thermally excited.

If several antisymmetric vibrations $\nu_k$, $\nu_l$, are present, transitions with $\Delta v_k = \pm 1, \pm 3, \ldots$ or $\Delta v_l = \pm 1, \pm 3 \ldots$, may appear (in addition to $\Delta v_k = 0, \pm 2, \ldots, \Delta v_l = 0, \pm 2, \ldots$) as long as for each symmetry element

$$\sum \Delta v_a = 0, \pm 2, \pm 4 \tag{II, 32}$$

where $v_a$ is the vibrational quantum number of a vibration that is antisymmetric with respect to that particular symmetry element for which $\sum \Delta v_a$ is formed. This is because only then will the resultant vibrational species be the same in the upper and lower state. As an example, consider a transition in symmetrical $X_2Y_4$ (point group $\boldsymbol{D}_{2h}$) in which, apart from totally symmetric vibrations, in the upper state $\nu_4(a_u)$, $\nu_5(b_{1g})$ and $\nu_8(b_{2g})$ and in the lower state $\nu_{12}(b_{3u})$ are singly excited. This transition is allowed by the general selection rule (II, 30) as is readily seen since $a_u \times b_{1g} \times b_{2g} \times b_{3u} = A_g$. At the same time $\Delta v_4 = +1$, $\Delta v_5 = +1$, $\Delta v_8 = +1$, $\Delta v_{12} = -1$, but for the vibrations antisymmetric to any one of the planes of symmetry $\sum \Delta v_a = +2$.

If a degenerate vibration is singly excited, the vibrational species is degenerate and therefore, according to the general selection rule, such a level cannot combine with one in which no degenerate vibration is excited, that is, a $1-0$ or $0-1$ transition of a degenerate vibration is always forbidden no matter what other non-degenerate vibrations are excited in the upper and lower state.

If a degenerate vibration $\nu_k$ is multiply excited we have to consider the quantum number $l_k = v_k, v_k - 2, \ldots, 1$ or $0$ which characterizes the different sub-levels that occur (see p. 21). It is easily seen (see Longuet–Higgins, Öpik, Pryce and Sack (769)) that for this quantum number we have the selection rule

$$\Delta l_k = 0 \tag{II, 33}$$

and therefore since $l_k$ is even or odd when $v_k$ is even or odd, we see that again

$$\Delta v_k = 0, \pm 2, \pm 4, \ldots . \tag{II, 31}$$

For several point groups the selection rule (II, 31) is supported by symmetry considerations, namely for those point groups, like $\boldsymbol{D}_{2d}$, $\boldsymbol{D}_{4h}$, $\boldsymbol{D}_{6h}$, $\boldsymbol{D}_{\infty h}$, for which only the even overtones of a degenerate vibration have a totally symmetric component and therefore only the even or only the odd vibrational levels can combine with a given level of the other electronic state. In such cases the selection rule (II, 31) is rigorous even if finer interactions are taken into account. (We do not consider here forbidden components of allowed electronic transitions, see subsection 2b($\beta$).) In other point groups (e.g., $\boldsymbol{D}_{3h}$, $\boldsymbol{C}_{3v}$, $\boldsymbol{T}_d, \ldots$) all overtones of the degenerate vibrations contain at least one totally symmetric component (see Table 32 of

Vol. II, p. 127) and therefore as far as symmetry is concerned a combination with a given totally symmetric level of the other electronic state is possible for even as well as odd $v_k$, that is, the rule (II, 31) is not rigorous. But in all cases the $1-0$ (or $0-1$) transition of a degenerate vibration is forbidden by symmetry and the rule (II, 31) holds at least to a good approximation. For degenerate vibrations, as for antisymmetric vibrations, the intensities of all transitions with $\Delta v_k \neq 0$ are

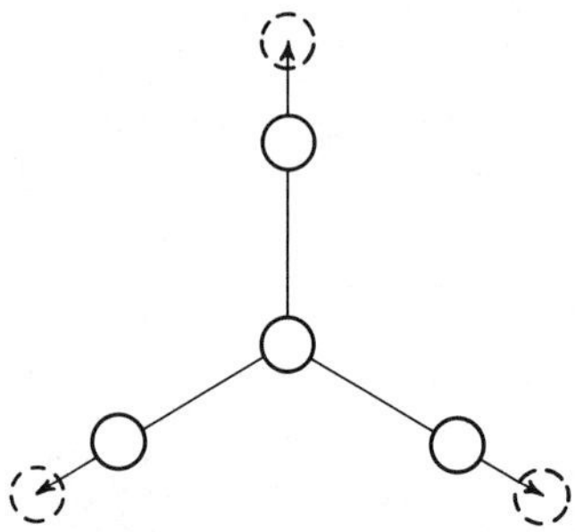

FIG. 56. **Application of the Franck–Condon principle to a planar symmetrical $XY_3$ molecule if the symmetry of the equilibrium conformation is the same in upper and lower state.** The full-line circles represent the equilibrium conformation of the lower state, the broken-line circles that of the upper state. The arrows indicate the motion started according to the Franck–Condon principle immediately after the electronic transition.

very weak compared to those with $\Delta v_k = 0$, even for a fairly large change of the vibrational frequency.

If several degenerate vibrations are present their simultaneous excitation in an electronic transition may take place whenever the resultant vibrational species (see Appendix III of this Volume and Table 32 of Vol. II) in the upper and lower states are the same. But again all these transitions have very small intensities compared to those with $\Delta v_k = 0$.

As an example consider a linear molecule of point group $\boldsymbol{D}_{\infty h}$ (e.g., $C_2H_2$) in which in the lower state the $\pi_g$ bending vibration is triply excited and the $\pi_u$ vibration is singly excited while in the excited state $\nu(\pi_g)$ is doubly, $\nu(\pi_u)$ triply excited. The resultant vibrational species of the lower state are $\Sigma_u^+$, $\Sigma_u^-$, $\Delta_u$, $\Delta_u$, $\Gamma_u$ while those of the upper are $\Pi_u$, $\Pi_u$, $\Pi_u$, $\Phi_u$, $\Phi_u$, $H_u$. Thus there are no equal species and the transitions from any sublevel of the upper to any sublevel of the lower state are forbidden even though for the $\pi_u$ vibration $\Delta v_k = 2$. This conclusion holds for any allowed electronic transition as long as vibronic coupling can be neglected.

Most of the preceding conclusions established here by rigorous symmetry considerations can also be derived qualitatively from an application of the elementary form of the Franck–Condon principle. For example, if a planar symmetrical $XY_3$ molecule has the same symmetry in an excited state as in the ground state but with a different XY distance (see Fig. 56), then immediately after the electron jump the nuclei will be symmetrically displaced from the new equilibrium position and will start oscillating in the symmetrical "breathing" vibration, the only totally symmetric vibration. This is the only vibration that can be strongly excited by light absorption or emission (if there is a change of equilibrium XY distance). This qualitative consideration illustrates clearly why antisymmetric or degenerate vibrations are not (or only very weakly) excited, i.e., why $\Delta v_k = 0$ gives the strongest transitions.

For symmetrical just as for unsymmetrical molecules, the positions of the intensity maxima in progressions of totally symmetric vibrations depend on the change of the positions of the potential minima as functions of the corresponding (totally symmetric) normal coordinates, that is, of the bond distances and angles involved in the particular vibrations in the upper and lower states. In general, for absorption from the lowest vibrational level of the ground state a single intensity maximum appears in a given progression but, just as for diatomic molecules, when the molecule is vibrationally excited in the ground state, progressions with two or even more intensity maxima may arise. If only one totally symmetric vibration is excited in the ground state, there will be two principal intensity maxima in the progression, corresponding to the two classical turning points of the vibrational motion. If several ($n$) totally symmetric vibrations are excited, the classical motion in the lower state is a Lissajous motion which has in general $2^n$ turning points. Therefore on a classical basis $2^n$ intensity maxima would be expected in a progression arising from such a lower state. Moreover, additional maxima may be expected which correspond to constructive interference of the wave functions of the upper and lower states in regions between the turning points, similar to the minor intensity maxima in progressions of diatomic molecules which sometimes occur in addition to the main maxima. Some quantitative formulae for the intensity ratios of bands in a progression as a function of the changes of the internuclear distances have recently been given by Coon, DeWames and Loyd (227) who have suggested their use for the converse operation, namely the determination of the changes of geometric parameters in an electronic transition.

The intensity distribution in sequences is quite different from that in progressions. In absorption it is determined by (a) the Boltzmann factor ($e^{-G(v'')/kT}$) and (b) the overlap integral (II, 28). For the members of the $\Delta v = 0$ sequence the overlap integral does in general not vary greatly; it is in a first approximation constant (independent of $v$) when there is no change in the equilibrium positions of the nuclei. As we have seen this condition is necessarily fulfilled for antisymmetric or degenerate vibrations. While therefore for the $\Delta v_k = 0$ sequences in these vibrations the overlap integral is always independent of $v$, for the $\Delta v_k \neq 0$ sequences, which are always very weak compared to the $\Delta v_k = 0$ sequences, a strong dependence on $v$ arises. In particular one finds if $|\nu_k' - \nu_k''|$ is not too large that for the $\Delta v_k = \pm 2$ sequences in an antisymmetric (non-degenerate) vibration the overlap integral increases with $(v_k + 1)$ and therefore the intensity increases with $(v_k'' + 1)^2$ in the $\Delta v_k = +2$ sequence, and with $(v_k' + 1)^2$ in the $\Delta v_k = -2$ sequence [see Sponer and Teller (1155)]. Similarly for the $\Delta v_k = \pm 4$ sequences the intensity would increase as $(v_k + 1)^4$. These increases have, of course, to be combined with the Boltzmann decrease. In Fig. 57 the theoretical relative intensities in these sequences are plotted for $T = 725°K$ and $\nu'_k = 200$, $\nu''_k = 250$ cm$^{-1}$, but the intensities of the sequences $\Delta v_k = \pm 2$ and $\Delta v_k = +4$ are multiplied by a factor 15 and 225 respectively in order to make it possible to plot them in a single diagram. For low temperature, of course, only the $0-0, 2-0, 4-0, \ldots$ bands would remain.

If degenerate vibrations are excited in the lower state the situation is similar except that a third factor comes in: the statistical weight of the lower state which

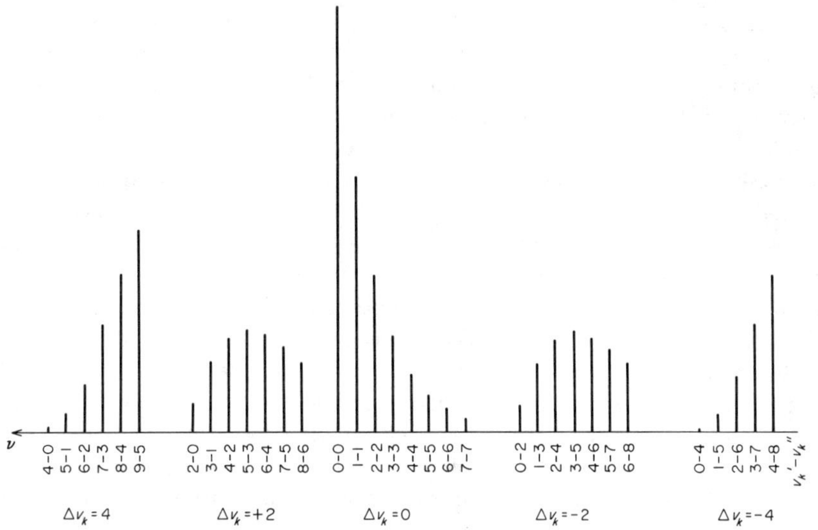


FIG. 57. **Intensity distribution in sequences of different $\Delta v_k$ values.** The separations of the bands are plotted to scale for $\nu'_k = 200$, $\nu''_k = 250$ cm$^{-1}$. The relative intensities are plotted for $T = 725°$ K. The actual intensities of the $\Delta v_k = \pm 2$ and $\pm 4$ sequences are $\frac{1}{15}$ and $\frac{1}{225}$ respectively of those plotted.

for doubly degenerate vibrations increases as $(v + 1)$ (disregarding the anharmonicity splitting; see p. 21). The intensities in the sequences (see Fig. 57) have to be multiplied by this factor.

The preceding considerations show that, while symmetry simplifies the spectrum considerably, even for a symmetrical molecule usually a rather complicated vibrational structure will result except for the simplest molecules. Often it is imperative, in order to obtain a vibrational analysis, to simplify the spectrum by studying the absorption at low temperature or by studying the emission under conditions under which only one vibrational level of the excited state is excited.

A relatively simple example is provided by the absorption spectrum of $HgCl_2$ near 1700 Å, first studied by Wehrli (1285) and analyzed by Sponer and Teller (1154)(1155). Figure 58 gives a reproduction of three spectrograms taken at three different pressures. At low pressure there is a single main progression in a frequency of 289 cm$^{-1}$ which in all probability corresponds to a stretching vibration in the excited state. Since the molecule is known to be linear in the ground state

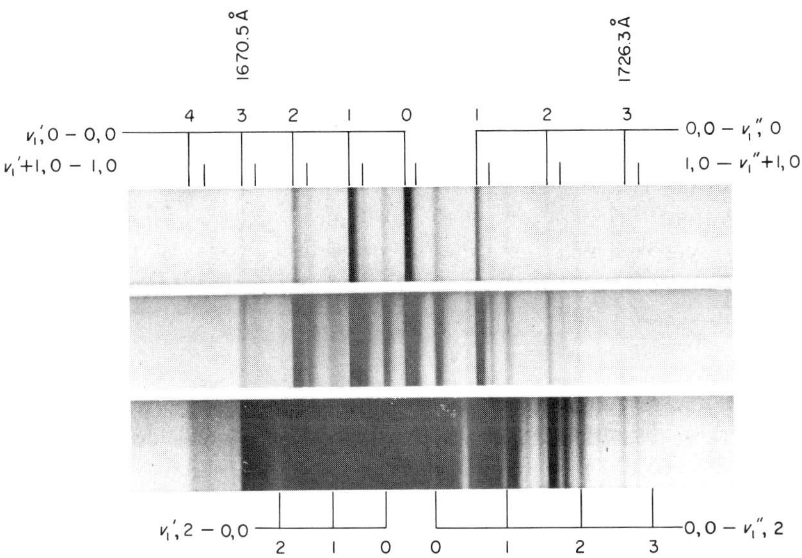


FIG. 58. **Absorption spectrum of $HgCl_2$ near 1700 Å.** In order to show both the stronger and the weaker bands three spectra taken at three different pressures are reproduced. The assignments of the bands (in the form $v_1'v_2' - v_1''v_2''$) are given above and below the spectrograms.

the observation in absorption of a single main progression in a stretching vibration can be understood only by the assumption that the molecule is also linear in the excited state since then only one totally symmetric vibration exists in the excited state as in the ground state and thus at low temperature only one progression can occur in absorption. Actually since $kT$ at the temperature used was not small compared to $\Delta G$, a (much weaker) progression was found extending to longer wavelengths and corresponding to the thermal excitation of the symmetrical stretching vibration in the ground state. In addition, very weak bands on either side of the $0-0$ band are present which must be assigned as $0-2$ and $2-0$ bands corresponding to the bending vibration $\nu_2$. At higher pressures, similar bands accompanying other main bands are visible as well as second members of the sequences of which the main bands are the first members. These are sequences in $\nu_1$; sequences in $\nu_2$ starting from the main bands are not resolved. There is, however, evidence that higher members of the sequences with $\Delta v_2 = \pm 2$ are stronger than the lower ones (see Fig. 57) and this accounts for certain shifts of these unresolved groups.

**Transitions between electronic states at least one of which is degenerate.** If the interaction of vibration and electronic motion is weak, the vibrational structure of electronic transitions involving degenerate electronic states is the same as of those involving only non-degenerate states. Again the vibrational transitions are determined by the general selection rule (II, 30) and the rules (II, 31) and (II, 32) derived from it. However, if the interaction of vibration and electronic motion is not negligibly small, we must consider the effect of the vibronic splittings (Renner–Teller and Jahn–Teller splittings) on the vibrational structure of the band system. Conversely, the complication of the vibrational structure produced by vibronic interactions may be used to recognize whether or not the upper and lower state of a band system is degenerate even if the rotational structure is not resolved.

In addition to the splitting of certain vibrational transitions, further transitions will occur which for zero vibronic interaction are strictly forbidden by the selection rule (II, 30). If vibronic interaction is strong the product resolution (II, 4) of the wave function is not valid and we must use for the transition moment the earlier expression

$$R_{e'v'e''v''} = \int \psi^*_{e'v'} \boldsymbol{M} \psi_{e''v''} d\tau_{ev} \tag{II, 19}$$

that is, all those transitions can occur for which the product $\psi^*_{e'v'}\psi_{e''v''}$ contains the species of the dipole moment characterizing the transition. While for non-degenerate electronic states this more general selection rule leads to the same transitions as the rule (II, 30), for degenerate electronic states this is no longer so when degenerate vibrations are excited, since then several vibronic states $\psi_{ev}$ arise from a given $\psi_e$ and of these one or more may give the right symmetry of $\psi^*_{e'v'}\psi_{e''v''}$ even if (II, 30) is not fulfilled. [Note that we are not considering here forbidden components of the electronic transition, for which (II, 19) is fulfilled because a dipole component different from that for the $0-0$ band is used; see subsection ($\beta$).]

In Fig. 59a and b the vibrational transitions in a $^1\Pi - ^1\Sigma$ electronic transition of a *linear molecule* are shown without and with vibronic splittings. Only the levels corresponding to excitation of a bending vibration are shown. Under the energy level diagrams schematic spectra are drawn. The transitions form three sequences $\Delta v_k = +2, 0, -2$ of which the one with $\Delta v_k = 0$ is by far the strongest. At low temperature only the $2-0$ and $0-0$ bands would occur. The effect of the vibronic splitting is, as seen in Fig. 59b, that certain transitions that are single in Fig. 59a are split into two or three components. The magnitude of the splitting, as we have seen in Chapter I, section 2, may be quite considerable, often much larger than the separation of bands in a sequence. Some transitions, indicated by broken lines in Fig. 59, are forbidden in the limiting case of zero vibronic interaction by the rule (II, 30) but can now occur according to the more general rule (II, 19) (see p. 137). For example the transitions to the extreme left in Fig. 59b would go from a $\Delta$ vibrational level to a $\Sigma$ vibrational level and is therefore forbidden according to (II, 30). However, vibronically the upper level is $\Pi$ and it can combine with the lower $\Sigma$ level with the same dipole component as the pure electronic transition according to the selection rule (II, 19) if vibronic interaction is present.

Splittings of the type illustrated by Fig. 59b have been observed for $C_3$ [Gausset, Herzberg, Lagerqvist and Rosen (411)] and $C_2H_2$ [Herzberg (523)]. For these molecules the *g*, *u* symmetry has to be included but does not change the

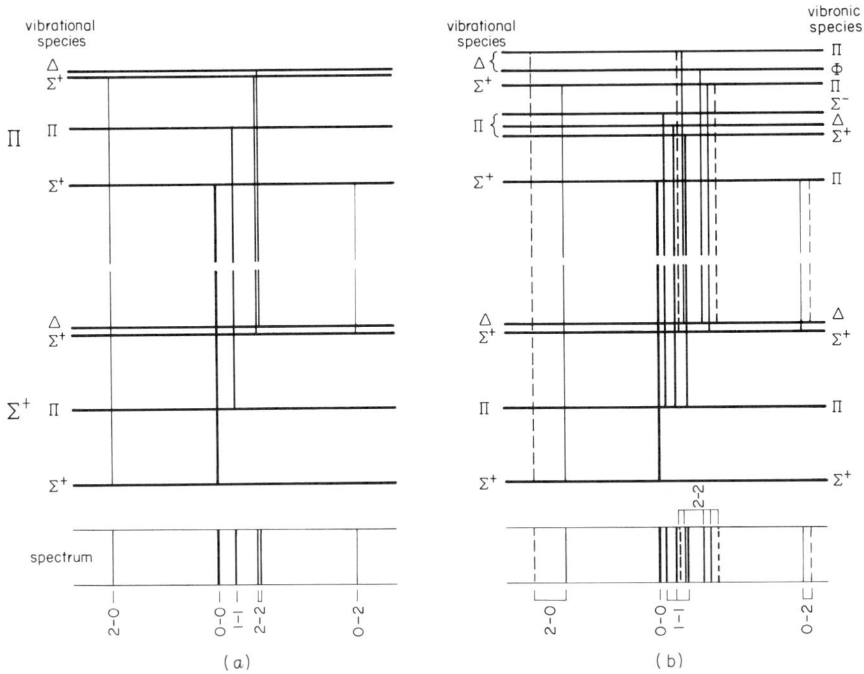


FIG. 59. **Vibrational transitions in the bending vibration for a $^1\Pi - {}^1\Sigma$ electronic transition of a linear (triatomic) molecule** (*a*) **without and** (*b*) **with vibronic (Renner–Teller) splitting.** The weights of the lines representing the transitions and of the "lines" in the schematic spectra indicate the relative intensities. Transitions occurring only when vibronic interaction is introduced are represented by broken lines.

spectrum except that the $\Delta v_k = \pm 1$ transitions are rigorously forbidden even for large vibronic coupling.

It is readily seen that if higher vibrational levels of the perpendicular vibration of a linear molecule come into play, e.g., in emission or in absorption at a high temperature, an extremely complicated vibrational structure may arise when the vibronic interaction is large. This was the reason that the vibrational analysis of the 4050 Å group of $C_3$, which is a $^1\Pi - {}^1\Sigma$ transition, presented such difficulties even though the molecule is linear in both upper and lower state. Not only is the vibronic interaction very large, but also the frequency of $\nu_2$ is very small in the ground state (63.5 cm$^{-1}$) while it has a much larger value (307 cm$^{-1}$) in the upper state. For these reasons the normally weak transitions with $\Delta v_k = \pm 2, \pm 4$ are relatively strong, and in addition, hot bands are very strong even at room temperature.

Additional complications arise when there is an appreciable multiplet splitting as is the case for NCO and $BO_2$ since, as we have seen in Chapter I, the spin splitting is affected by the bending vibration. Nevertheless, the analysis of the doublet transitions in NCO [Dixon (281)(282)] and $BO_2$ [Johns (630)] actually preceded the analysis of the singlet transitions in $C_3$ and $C_2H_2$. Figure 60 shows

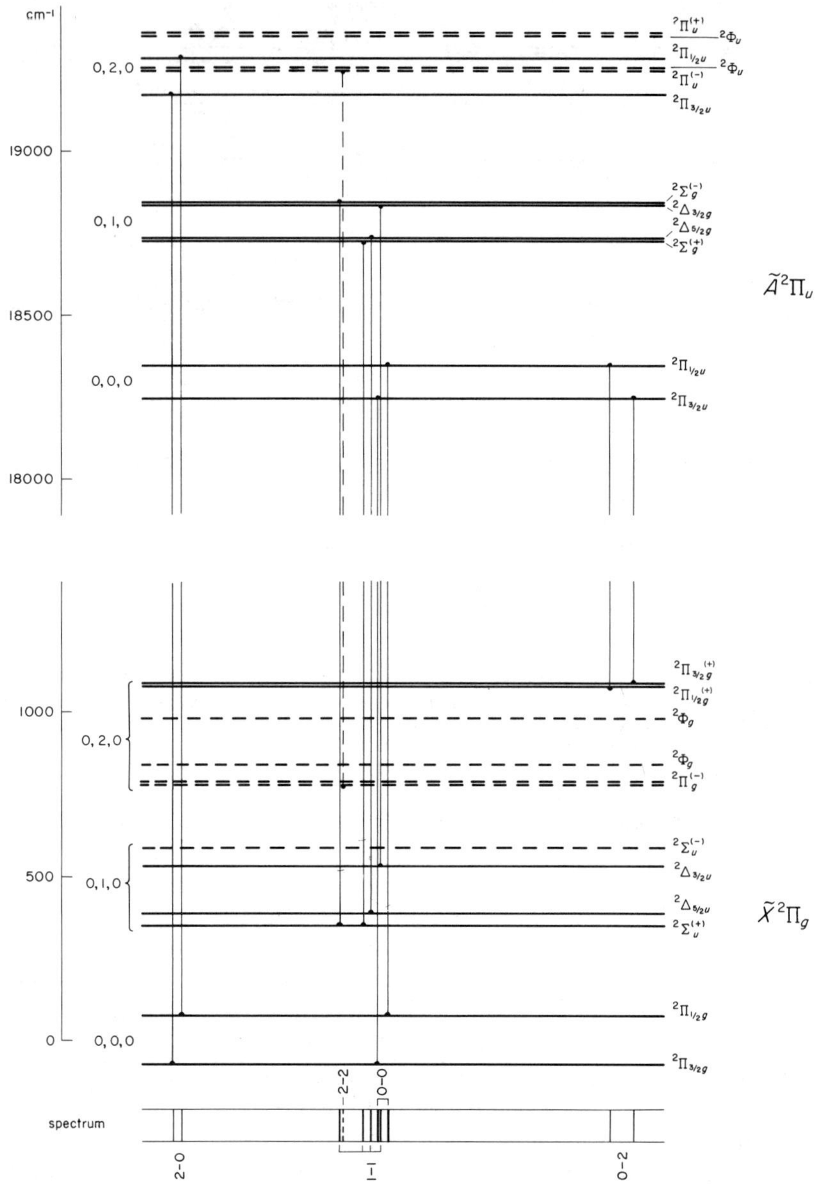


FIG. 60. **Observed vibronic levels and transitions for the $^2\Pi - {^2\Pi}$ system of $BO_2$ after Johns (630).** Only the levels and transitions involving the bending vibration $\nu_2$ are shown. Levels that have not been observed directly are indicated by broken lines. Both in the upper and lower state the $^2\Pi_u$ 020 vibronic levels are shifted on account of Fermi interaction with the $^2\Pi$ 100 level.

as an example part of the observed vibrational energy level diagram for the $^2\Pi - {}^2\Pi$ bands of $BO_2$ according to Johns, with a schematic spectrum below. The origin of the various sub-bands of the $1-1$, $2-2$, ... vibrational bands is illustrated by this figure. The positions of the component levels of 020 in upper and lower state

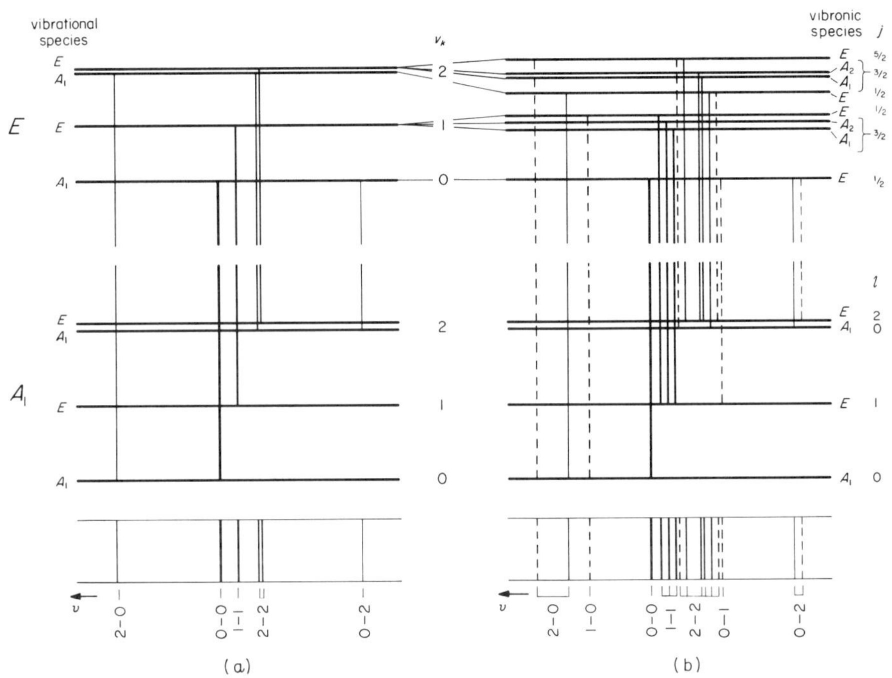


FIG. 61. **Vibrational transitions in a degenerate vibration $\nu_k$ for a $^1E - {}^1A_1$ electronic transition of a $C_{3v}$ molecule (*a*) without and (*b*) with vibronic (Jahn–Teller) splitting.** Compare the caption of Fig. 59. The quantum number $j$ at the right is well defined only if the three identical minima in the excited state can be combined into a trough of nearly cylindrical symmetry.

are slightly modified here compared to expectation (Fig. 8) by Fermi interaction with the 100 level.

An example of a $^3\Pi - {}^3\Sigma^-$ transition is provided by the band group at 3290 Å of the NCN molecule recently analyzed by Herzberg and Travis (543). Here both the triplet splitting and the vibronic splitting is fairly large ($A = -37.56$, $\epsilon\omega_2 = -85.7$ cm$^{-1}$). $^3\Pi - {}^3\Sigma^-$ transitions are also present but have not yet been fully analyzed in the vacuum ultraviolet spectrum of $CH_2$ [Herzberg (521)].

For *non-linear molecules* a similar complication of the vibrational structure occurs in $E-A$ or $E-E$ electronic transitions on account of vibronic interaction (Jahn–Teller effect). Figure 61 presents an energy level diagram, similar to Fig. 59, for a degenerate vibration of a $C_{3v}$ molecule, both for zero and for non-zero vibronic coupling. Again several bands that would be single without vibronic interaction are split on account of this interaction, in particular the $1-1$ and $2-2$

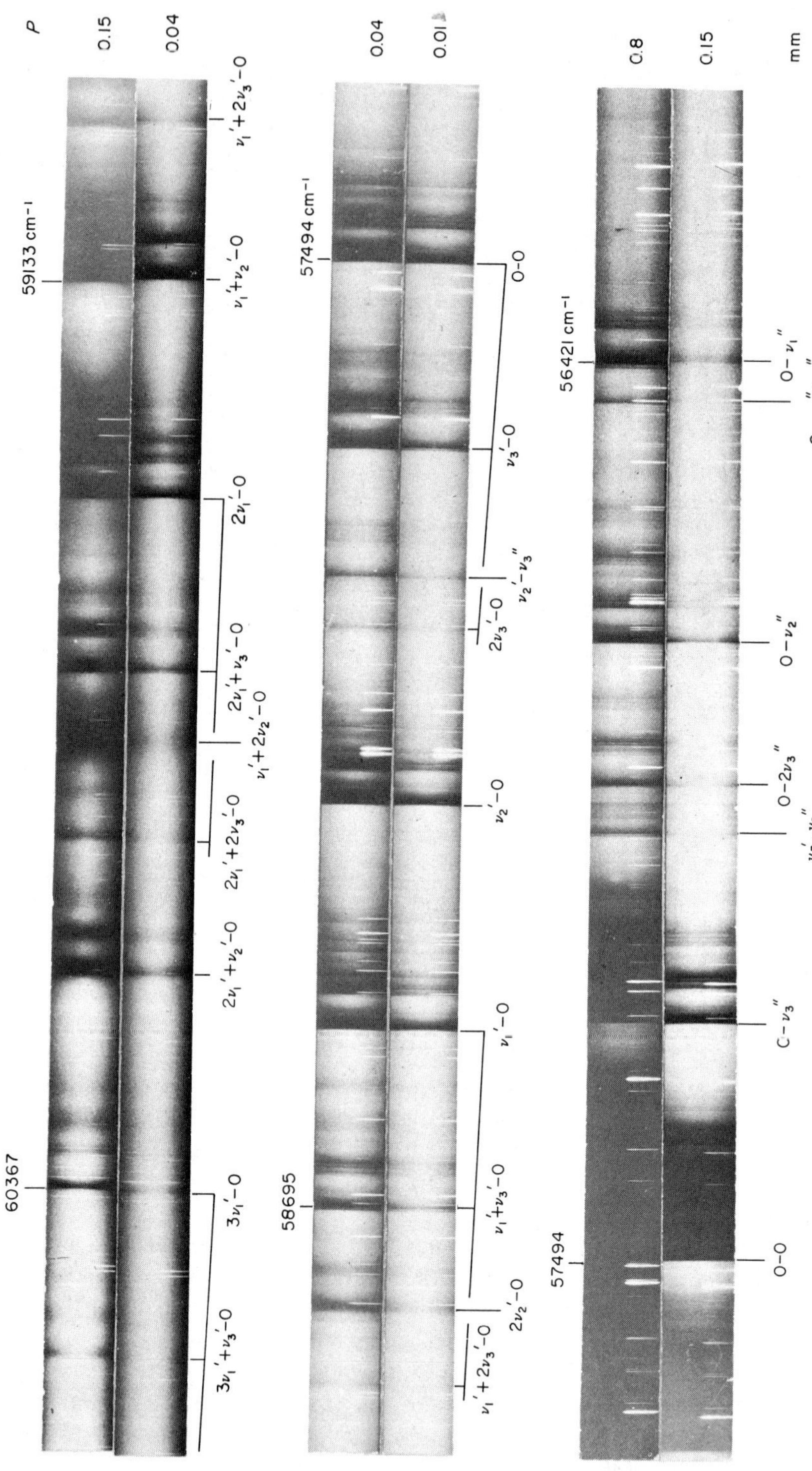


FIG. 62. **Absorption spectrum of $CF_3I$ in the region 1775 to 1650 Å.** For each of the successive sections, spectrograms at two different pressures are given in order to show clearly both the weak and the strong bands. These pressures are given at the right of each strip. The path length was 65 cm. The assignments of the principal bands are indicated. Each of these bands is joined by sequences extending to longer wavelengths.

bands in the degenerate vibration. Moreover several transitions (indicated by broken lines) would be forbidden for vanishing vibronic interaction according to the selection rule (II, 30) but do occur on the basis of the more general rule (II, 19). Among these "forbidden" vibrational transitions are in particular $1-0$ and $0-1$ in the degenerate bending vibration (see p. 153).

An example is supplied by the strong absorption system of $CF_3I$ extending from 1775 to 1650 Å which is reproduced in Fig. 62. At low pressure three progressions are immediately evident which extend from the $0-0$ band at 57494 $cm^{-1}$ to higher wave numbers, with spacings 231, 682 and 969 $cm^{-1}$. These three spacings must clearly correspond to the three totally symmetric vibrations $\nu_3'$, $\nu_2'$, $\nu_1'$ respectively of the molecule in the excited state. Combinations of these vibrations like $\nu_1' + \nu_3'$, $\nu_1' + 2\nu_3'$, etc. account for most of the other strong bands. At high pressure hot bands appear at the long wavelength side of the $0-0$ band, three of which are separated from the $0-0$ band by $286._5$, $743._3$ and $1075._4$ $cm^{-1}$. These numbers correspond exactly to the known totally symmetric infrared fundamentals $\nu_3''$, $\nu_2''$, $\nu_1''$ respectively of $CF_3I$ in the ground state. The agreement with the infrared data is quite striking and represents a nice confirmation of the correctness of the interpretation of the spectrum.

Each of the main absorption bands of $CF_3I$ is accompanied by weaker bands forming sequences which are clearly hot bands corresponding to the excitation of the two lowest vibrations $\nu_3$ and $\nu_6$ with 1, 2 and 3 quanta in the upper and lower states of the main bands. Up to this point the whole structure of the band system would be equally compatible with that of an $A_1-A_1$ and an $E-A_1$ electronic transition. It is only when one considers the detailed structure of the sequences just mentioned that a decision between these two alternatives can be reached.

In Fig. 63 the $0-0$ sequence and the $0-\nu_3''$ sequence are shown under larger magnification. It is seen that the second member in each of these sequences consists of at least three component bands. The longward component in the $0-0$ sequence has a separation of $56._3$ $cm^{-1}$ from the main band which agrees as closely as can be expected with the difference $\nu_3''-\nu_3' = 286._5-231._3 = 55._2$ $cm^{-1}$. This component must, therefore, be assigned as the one for which $\nu_3$ is excited in both upper and lower state. The fact that there are two other components of similar intensity, i.e., similar Boltzmann factor when only one other low lying fundamental is available ($\nu_6$), can be accounted for if it is assumed that the upper electronic state is degenerate (i.e., has species $E$) and that the vibronic interaction in this state produces a Jahn–Teller splitting of the level $\nu_6'$. As shown by Fig. 61 there are then three close lying levels ($E$, $A_1$, $A_2$) of which two ($A_1$ and $A_2$) may not be resolved. Thus the combination of the single lower level ($\nu_6''$) of species $E$ with the upper levels $E$ and $A_1$, $A_2$ would give rise to at least two component bands in the spectrum as observed.

It is significant that the splitting for the second member of the sequence starting with the $0-\nu_3$ band is identical with that for the second member of the $0-0$ sequence. These two members have the same upper states. Similar splittings are observed in other sequences, but these are not as clear, sometimes exhibiting more than three components. In the upper states of these other sequence members, in addition to $\nu_3'$ or $\nu_6'$, other vibrations $\nu_1'$, $\nu_2'$, $\nu_3'$ and their overtones are excited and Fermi resonances may easily occur which complicate the situation.

According to Fig. 61 the $2\nu_6' - 2\nu_6''$ bands of the $0-0$ and the $0-\nu_3''$ sequences should consist of six components, of which three may not be resolved. In other

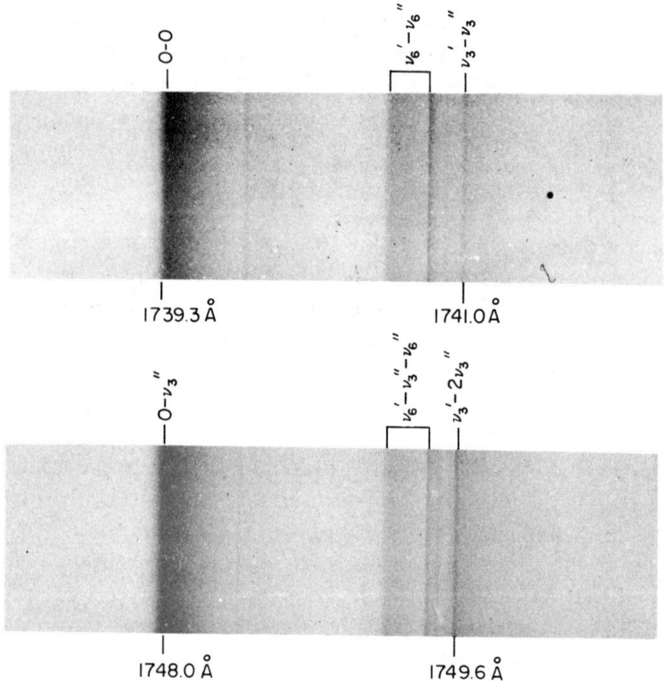


FIG. 63. **Spectrograms of the $0-0$ and $0-\nu_3''$ sequences of the $\tilde{B}-\tilde{X}$ bands of $CF_3I$.** These spectrograms represent two small sections of Fig. 62 taken with higher resolution. The two heads indicated by ⌐¬ correspond in each case to the $1-1$ band of the sequence in $\nu_6$ (see p. 161).

words, there would be four or more components plus one component due to $2\nu_3' - 2\nu_3''$. Four or five components are indeed observed, but a detailed interpretation is not yet possible because no theoretical estimates of the relative intensities are available.

It may be noted that the conclusion that the $CF_3I$ spectrum here discussed has a degenerate upper electronic state has so far been entirely based on the discussion of hot bands and their splittings. Additional confirmation of this conclusion has been obtained by the observation of the weak forbidden $\nu_6' - 0$ band and of its rotational structure (see below).

Further differences of $E-A$ or $E-E$ electronic transitions from an $A-A$ electronic transition (involving non-degenerate states only) arise when the vibronic interaction is large. As already mentioned, with increasing coupling of vibration and electronic motion the quantum number $l_k$ becomes less and less well defined and therefore the selection rule (II, 33) becomes less and less applicable. As a consequence the selection rule (II, 31) remains valid only for those degenerate vibrations that are antisymmetric with respect to one of the elements of symmetry of the system (if such an element is present). The reason for the breakdown of (II, 33) is that now we can no longer separate the electronic and vibrational part of the eigenfunction and must base our considerations of selection rules entirely on the vibronic eigenfunctions of the levels. Thus for a $\boldsymbol{C}_{3v}$ molecule (see Fig. 61b) a totally symmetric ($A_1$) lower vibrational level can combine with every vibrational level of an $e$ vibration of the $E$ upper electronic state since each of these levels has at least one $E$ vibronic component. But for a $\boldsymbol{D}_{3h}$ (or $\boldsymbol{D}_{3d}$) molecule in an $E'-A_1'$ (or $E_u-A_{1g}$) electronic transition the combination with the odd overtones of an $e''$ (or $e_u$) vibration would still be forbidden (that is (II, 31) is still valid) no matter how strong the vibronic interaction is, since the $E$ vibronic sublevels of the odd vibrational levels have $E''$ (or $E_g$) symmetry and according to (II, 19) cannot combine with the $A_1'$ (or $A_{1g}$) ground state.

For strong vibronic coupling in the degenerate electronic state we have instead of the quantum number $l_k$ the quantum number $j_k = l_k \pm \frac{1}{2}$ (see Chap. I, section 2c) for which according to Longuet-Higgins, Öpik, Pryce and Sack (769) the selection rule is

$$j_k' - l_k'' = \pm\tfrac{1}{2} \qquad \text{(II, 34)}$$

where $l_k''$ is the $l_k$ value in the lower, non-degenerate state. This rule establishes which of the "forbidden" vibrational transitions made possible by vibronic interactions [that is, of those that are possible according to (II, 33) but not according to (II, 30); see the broken-line transitions in Fig. 61b] may occur fairly strongly in a first approximation. It implies, for example, that if there are several $E$ vibronic levels for a given $v_k'$ (e.g. $v_k' = 2$, see Fig. 61b) only one of them can combine with a given $A_1$ vibrational level ($l_k'' = 0$) of the lower electronic state. It should, however, be emphasized that the introduction of the quantum number $j_k$ is based on the assumption of a trough-like potential function (Fig. 17a) without separate minima. To what extent the results here discussed would be modified by dropping the assumption is not known.

Longuet-Higgins, Öpik, Pryce and Sack have calculated for various strengths of vibronic interaction the intensity distribution in the $v_k'-0$ progression in absorption and the $0-v_k''$ progression in emission. We reproduce in Fig. 64 their results only for $D = 2.5$. Here it has been assumed that the vibrational frequency $\omega_k$ in the upper state is the same as in the lower. Therefore without vibronic interaction ($D = 0$) we would observe only a single band, the $0-0$ band in either progression. (For $\omega_k' \neq \omega_k''$ and zero vibronic interaction we would have two progressions of bands with very rapidly decreasing intensity and with spacings $2\omega_k'$ and $2\omega_k''$, since $\Delta v_k = 0, \pm 2, \ldots$.) Instead we have now an absorption progression ($E \leftarrow A_1$) with a somewhat irregular spacing of about $\omega_k'$ (see Fig. 22c) and an emission progression ($E \rightarrow A_1$) with a regular spacing $\omega_k''$ ($= \omega_k'$). The former has two intensity maxima, the latter has one. These maxima are not at the 0–0 band but at a non-zero $v_k'$ or $v_k''$ value which increases with increasing $D$ value. The difference from the case of zero vibronic interaction is very pronounced.

The reason for this difference is qualitatively apparent when one applies the Franck–Condon principle to the potential function of an $E$ (upper) state with

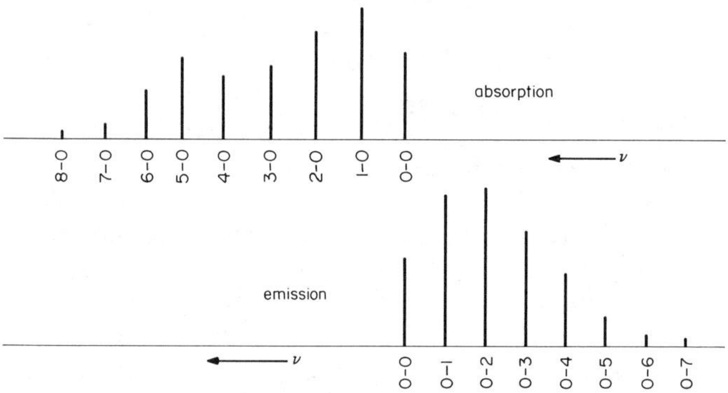


Fig. 64. **Intensity distribution in a progression in a degenerate vibration for an E – A electronic transition of a $C_{3v}$ molecule in emission and absorption when Jahn–Teller interaction is large ($D$ = 2.5)** [after Longuet–Higgins, Öpik, Pryce and Sack (769)]. It has been assumed that $\omega_k' = \omega_k''$. Therefore without vibronic interaction only the $0-0$ band would occur in either progression.

$D = 2.5$ (see Fig. 23c) and a normal potential function of a non-degenerate (lower) state with its minimum at the origin: transitions from the minimum of the upper curve to the lower give clearly a progression with intensity maximum at $v_k'' \neq 0$ while transitions from the minimum of the lower curve to the upper give a progression with an intensity maximum at $v_k' \neq 0$. Since there are two branches of the upper potential function it is not surprising that actually two intensity maxima arise corresponding to points above and below the cusp. It is interesting to compare the intensity distribution in the present case with that when the excited state is non-degenerate and has a non-symmetrical equilibrium position. Clearly the intensity distribution would be very similar except that there would be only one intensity maximum in the progression observed in absorption.

For very small vibronic interaction, as is readily seen from Fig. 21 or 22a, the progression in $\nu_k'$ will have alternately smaller and larger intervals (since only the levels with $j' = \frac{1}{2}$ can combine with the lowest vibrational level of the lower state). However in this case the intensity will very rapidly decrease from the $0-0$ band on. If there is a substantial difference between $\omega_k'$ and $\omega_k''$, alternate bands, with even $\Delta v_k$, would be stronger since they would be the only ones remaining for zero vibronic interaction.

No clear example of an intensity distribution as in Fig. 64 for an $E-A_1$ electronic transition has as yet been observed. However, the occurrence of a weak $1-0$ band in $\nu_6$ for both $CF_3I$ [Herzberg (523)] and $CH_3I$ [Mulliken and Teller (917)] indicates the beginning of such an anomalous intensity distribution. Since in these examples $D$ is very small, the intensity maximum is still at $v_6' = 0$, but the $1-0$ band (and similarly the $0-1$ band) could not occur at all without vibronic interaction[3b].

It is interesting to note that considerable changes in the infrared and Raman spectra are to be expected for molecules that have a degenerate electronic ground state, if vibronic coupling is not negligible. For example, it is immediately clear from Fig. 59b that a linear molecule in a $\Pi$ ground state would show in the infrared instead of one band corresponding to a bending frequency $\nu_k$, three bands corresponding to the three Renner–Teller components $\Sigma^+$, $\Sigma^-$ and $\Delta$. Similarly, according to Fig. 61b, in a $\boldsymbol{C}_{3v}$ molecule with an $E$ ground state instead of a single $\perp$ band corresponding to a degenerate fundamental $\nu_k$, three such bands would arise corresponding to the Jahn–Teller components $E$, $A_1$, $A_2$, each of which can combine with the lowest vibrational level which has species $E$. Another peculiarity first established by Child and Longuet–Higgins (194) is that in such molecules certain vibrations which are infrared-inactive in normal molecules can become infrared-active. The breathing vibration $\nu_1(a_1')$ of a planar symmetrical $XY_3$ molecule is an example. A general discussion for orbitally degenerate electronic states has been given by Child and Longuet–Higgins and for spin-degenerate states by Child (192).

No simple examples have as yet been found experimentally but Weinstock and Goodman (1286) have reported several interesting effects of Jahn–Teller interactions in the infrared spectra of a number of hexafluoride molecules of octahedral symmetry.

**Transitions between states with different symmetry of the equilibrium conformation.** Up to now we have assumed that the molecule has the same symmetry in the equilibrium positions of both upper and lower state. This assumption has been shown to be invalid for many observed spectra. Let us, therefore, consider the effect of a change of symmetry on the vibrational structure of a band system. As emphasized before we must in such a case apply the selection rules that correspond to the common elements of symmetry. If that is done, the same rules apply as before. However, it is useful to consider a few examples in order to see what precisely the effect of the difference in symmetry is on the vibrational structure and how, conversely, from the observed vibrational structure conclusions can be drawn about the difference of symmetry of the molecule in the two equilibrium positions.

A very striking example is provided by the HCN molecule. In the first absorption system ($\tilde{A}-\tilde{X}$) which begins at 1910 Å several long progressions are

[3b] It may be noted that there are two ways in which $1-0$ and $0-1$ bands in a degenerate vibration can arise, either by Jahn–Teller interaction (i.e., vibronic interaction within one and the same electronic state) as assumed in the preceding discussion, or by vibronic interaction with another electronic state of different symmetry. In the latter case the $1-0$ and $0-1$ bands in the $E-A_1$ electronic transition would have $\parallel$ structure, that is, would be $A_1-A_1$ vibronic transitions. The observed $\perp$ structure of the $CH_3I$ and $CF_3I$ bands in question eliminates this latter possibility.

observed, as shown schematically in Fig. 65. These progressions have intensity maxima at a considerable distance from their first bands. The separations of successive bands in each progression are about 940 cm$^{-1}$ while different progressions are separated by 1496 cm$^{-1}$. Thus two vibrations, 940 and 1496 cm$^{-1}$, are excited in the upper state. If HCN were linear in the upper state ($\tilde{A}$) as it is in the ground state, only the totally symmetric vibrations $\nu_3$(C—N) and $\nu_1$(C—H) could be excited with appreciable intensity and one would have to assign the

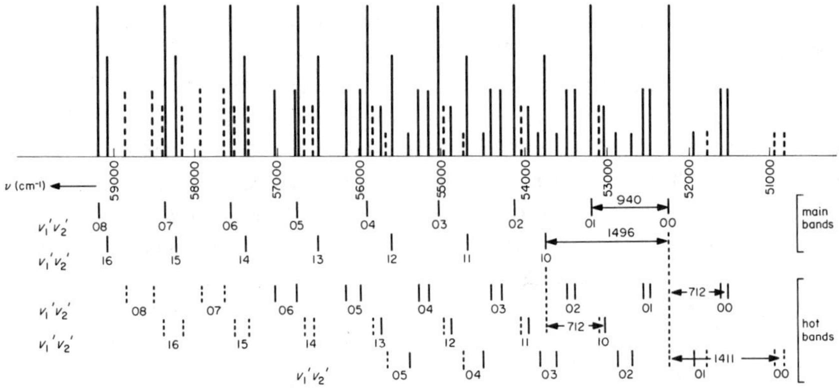


Fig. 65. **Observed progressions in the $\tilde{A}-\tilde{X}$ system of HCN.** A few bands that have not been observed but are useful in showing the regularities have been added as broken lines. The progressions are shown separately in the lower part of the diagram. For a discussion of the hot bands see p. 203f.

observed frequencies, 940 and 1496 cm$^{-1}$, to these two vibrations. Since in the ground state these frequencies are 2089 and 3312 cm$^{-1}$ respectively, they would have to drop by more than a factor 2 in the excited state if one wanted to retain the assumption that the molecule is linear in the excited state. This remote possibility is definitely excluded by the investigation of DCN which also shows the excitation of two vibrations in the excited $\tilde{A}$ state with frequencies 735 cm$^{-1}$ and 1500 cm$^{-1}$. The lack of an isotope shift between HCN and DCN for the higher of the two vibrations shows that it cannot be the C—H stretching vibration but must be the C—N stretching vibration (which in the ground state is 2089 cm$^{-1}$). Therefore the lower vibration (which shows a large isotope effect) must be assigned to the bending vibration $\nu_2$ which in the ground state is 712 cm$^{-1}$ for HCN and 569 cm$^{-1}$ for DCN[4]. The existence of long progressions in the bending vibration $\nu_2$ proves that the molecule is bent in the excited state, since for a linear-linear transition bands with $\Delta v_2 \neq 0$ would be very weak. The non-linearity in the excited state is fully confirmed by the rotational analysis.

It is interesting to consider what would be the structure of the *fluorescence spectrum* of HCN which would result from a transition from the lowest vibrational

[4] The alternative that the lower frequency of the excited state corresponds to the C—H stretching vibration which has the frequency 3312 cm$^{-1}$ in the ground state is most unlikely to be correct but is difficult to eliminate unambiguously on vibrational evidence alone.

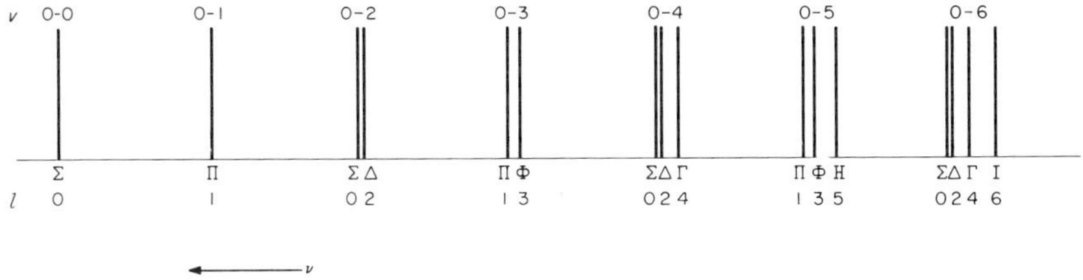


FIG. 66. **Hypothetical fluorescence spectrum of HCN corresponding to the $\tilde{A}-\tilde{X}$ absorption system.** Only one upper vibrational level is assumed to be excited and only the progression in $\nu_2''$, the bending vibration of the lower state, is shown. Note the alternation of sub-bands with even and odd $l$.

level of the excited state to the various vibrational levels of the ground state. (In spite of considerable effort such a fluorescence spectrum has not been observed, presumably on account of predissociation in the excited state.) One would expect to find a strong excitation of both the bending vibration ($\nu_2 = 712$ cm$^{-1}$) and the C≡N stretching vibration ($\nu_3 = 2089$ cm$^{-1}$) in the ground state. However, in this case, different from the absorption spectrum, the progression in the bending vibration would show with increasing $v_2$ a splitting into an increasing number of sub-bands corresponding to the different possible values of $l$. In Fig. 66 the structure of such a progression is shown schematically. The sub-bands are marked $\Sigma, \Pi, \Delta, \ldots$ corresponding to the value of $l$ in the lower state. It should be noticed that even and odd $l$ values alternate in succeeding members of the progression.

Absorption spectra similar to that of HCN have been found for $C_2H_2$ [Ingold and King (600)(601), Innes (605)] and $CS_2$ [Kleman (680)]. Here also the vibrational structure leads to the definite conclusion that these molecules which are linear in their ground states are non-linear in their first excited states.

A striking example of the opposite type is provided by the absorption spectrum of $NH_2$ in the visible region [Dressler and Ramsay (308)]. Here also a long progression of absorption bands is found, with a spacing of 622 cm$^{-1}$. From the rotational structure it is known that in the ground state the molecule is strongly bent with an angle of 103°. The frequency of 622 cm$^{-1}$ in the excited state can only correspond to a bending vibration and the fact that there is a long progression in this bending vibration therefore implies that there is a large change in the angle of bend of the molecule.

Actually a more detailed investigation shows that there is a series of sub-bands for each member of the progression similar to the sub-bands mentioned for the fluorescence spectrum of HCN, and the rotational analysis shows that the $l$ values of these sub-bands alternate between even and odd values in succeeding bands of the progression. Therefore at least in the region of the observed transitions the molecule must be considered as linear. Again, as before, $C_{2v}$ is the symmetry that the equilibrium positions of the two states have in common, and the selection rules for this symmetry, but not those for $\boldsymbol{D}_{\infty h}$ (or $\boldsymbol{C}_{\infty v}$), allow the occurrence of a long progression in $\nu_2$. Dixon (286a) has shown by a comparison of the observed separations of the bands of lowest $v$ with those expected for a quasi-linear molecule (see Chap. I, section 3c) that $NH_2$ in its first excited state is not strictly linear but rather that there is a small hump in the potential function when plotted against the angle of bend (see Fig. 44).

A second example of the same type, but not as completely resolved, is the absorption system of $NO_2$ in the region 1650–1350 Å studied by Ritchie and Walsh (1071). The alternation in each $v_2'$ progression between even and odd $l_2'$ values is clearly marked in that the intervals between the first three sub-bands are alternately in the ratio 4:12 and 8:16 as expected because of the terms $g_{22}l_2^2$ and $(A - \bar{B})K^2$ in the energy formulae for the upper and lower state respectively (see p. 211).

Several examples of *planar–non-planar transitions* are provided by the $NH_3$ molecule. A number of absorption band systems have been observed in the ultraviolet, starting at 2100 Å, and each of these at low temperature consists of a single fairly long progression of bands with a spacing of about 900 $cm^{-1}$ for $NH_3$ and 650 $cm^{-1}$ for $ND_3$ and an intensity maximum between $v' = 4$ and 6. The frequency of the upper state vibration suggests strongly that it corresponds to the symmetrical bending vibration $\nu_2$ which is indeed one of the two vibrations ($\nu_1$ and $\nu_2$) that can occur in long progressions in a $\boldsymbol{C}_{3v}$ molecule. The strong excitation of $\nu_2$ shows that a considerable change of the pyramidal angle (i.e., angle of the NH bond with the symmetry axis) occurs in these electronic transitions. It could be a decrease or an increase of this angle but actually, as we shall see, the angle increases so much that the molecule is planar in all known excited electronic states.

Both for a non-planar and a planar upper state (point group $\boldsymbol{D}_{3h}$) the $\boldsymbol{C}_{3v}$ selection rules apply since the lower state has $\boldsymbol{C}_{3v}$ symmetry. However, if inversion doubling is not negligible an important difference arises between a planar–non-planar and a non-planar–non-planar transition. This is because one of the inversion doublet components is symmetric, the other antisymmetric, with respect to a plane at right angles to the symmetry axis or, in other words, the effective point group of a non-planar molecule with non-zero inversion doubling is $\boldsymbol{D}_{3h}$ (since the potential field has this symmetry) and therefore the selection rules are those of $\boldsymbol{D}_{3h}$: only those transitions can occur for which the vibrational eigenfunction has the same symmetry with respect to the plane perpendicular to the symmetry axis in the upper and lower state [cf. the + and − signs in Fig. 67 and the selection rule (II, 30)]. As a consequence, as shown by Fig. 67a, in a non-planar–non-planar transition all bands of a $v_2'$ progression are double while in a planar–non-planar transition (Fig. 67b) they are single but because of the inversion splitting in the lower state alternately shifted to longer and shorter wavelengths. Figure 67b is drawn under the assumption that the upper electronic state is $A_2''$ and therefore the even vibrational levels of this state combine with the lower inversion components of the lower state, the odd with the upper components. The opposite transitions would arise for an $A_1'$ upper electronic state. Both $A_2'' - A_1$ and $A_1' - A_1$ transitions are allowed when the symmetry changes from $\boldsymbol{D}_{3h}$ to $\boldsymbol{C}_{3v}$.

In $NH_3$ no splitting is observed in any of the progressions originating from the lowest vibrational level of the ground state. Since in this state the inversion splitting is only 0.66 $cm^{-1}$, that is, is negligible if only the vibrational structure is considered, the absence of band splittings can be accounted for either by assuming that the inversion splitting in the excited state is also too small to be observed (i.e., the pyramidal angle is small and the barrier to inversion is high) or by assuming

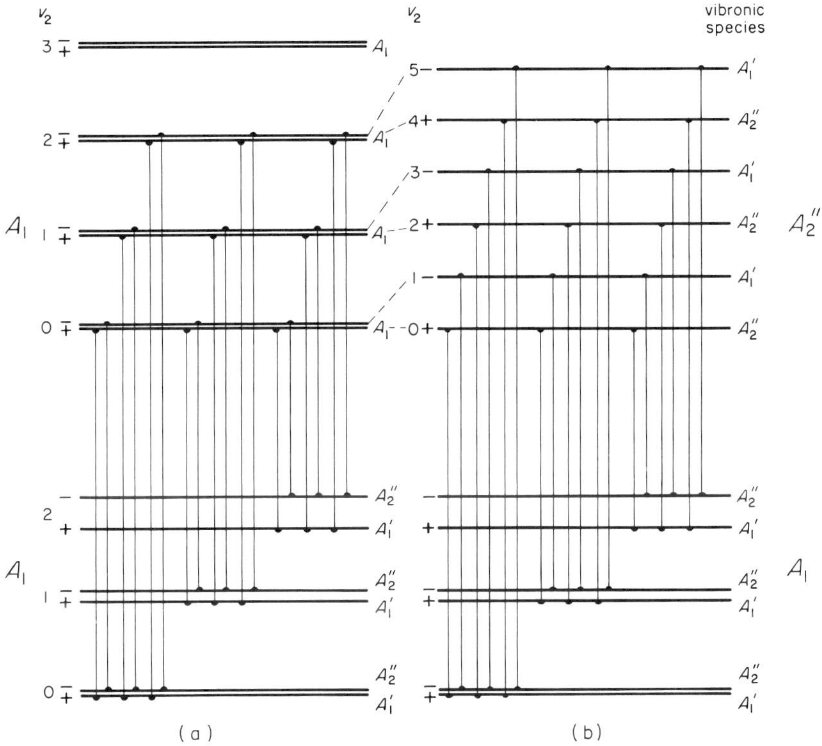


FIG. 67. **Energy level diagrams for** (*a*) **a non-planar – non-planar and** (*b*) **a planar – non-planar transition of a $C_{3v}$ molecule.** In the lower state ($A_1$) the inversion doubling increases rapidly with $v_2$; in the excited state it is assumed to be small for all $v_2$ in (*a*) while in (*b*), because of the planarity, no doubling arises. The correlation of the levels from the non-planar to the planar case is shown by the broken lines.

that the molecule is planar in the excited state in which case according to Fig. 67b no splittings are expected and the "staggering" in the progression would not be noticeable because of the smallness of the splitting in the lower state. However, the inversion splitting in the $v_2'' = 1$ and 2 levels of the lower state is quite considerable (36 and 313 $cm^{-1}$ respectively), and therefore a doubling or a "staggering" is expected for the hot bands (see Fig. 67). In fact, a staggering of the right magnitude has been observed in the $\tilde{A}-\tilde{X}$ system [Walsh and Warsop (1270)] proving the planar structure in the first excited state. To be sure, only a few hot bands are sufficiently free from overlapping to establish the staggering, but the analysis of the rotational structure (see section 3) has confirmed the conclusion in a very definite and convincing way.

In many other cases the study of the vibrational structure has led to definite conclusions about the symmetry of the molecule considered in its excited states (see Chap. V). The number of molecules which change their symmetry in some of their excited states is surprisingly large.

In general the previous remarks about the intensity distribution (p. 151f) apply also to the cases in which the symmetry of the molecule differs in the upper and lower state of a transition. However, a peculiarity arises when the electronic transition would be forbidden if the symmetry did not change, but is allowed when the symmetry does change, a situation that occurs not infrequently.

While strictly speaking in such a case the product resolution of $R$ into $R_{e'e''}R_{v'v''}$ can no longer be applied, in a first approximation the effect of vibronic interaction can be taken into account by considering the dependence of the electronic transition moment $R_{e'e''}$ on the nuclear coordinates. As an example Fig. 68 shows a cross section through the potential functions of the upper and lower states for a non-linear – linear transition. If the transition were allowed for the linear conformation the maximum of intensity would correspond to the "vertical"

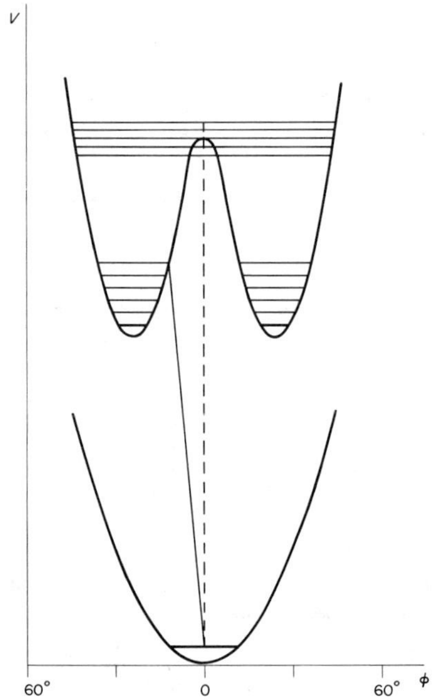


FIG. 68. **Potential curves for the bending vibration in a non-linear – linear transition.** The broken vertical line corresponds to the maximum of the Franck–Condon region in a transition that is allowed in the linear conformation, the full oblique line corresponds to the intensity maximum of a transition that is allowed only when the molecule is non-linear in one state.

transition indicated by the broken line, that is, one would expect transitions to levels in the neighborhood of the potential maximum of the upper state to be the strongest. But if the vertical transition is forbidden while the transition from minimum to minimum is allowed, one would expect an intensity maximum somewhere between the 0–0 band and the bands corresponding to the potential maximum because in the product $R_{e'e''} \times R_{v'v''}$ the first factor increases rapidly with increasing $\varphi$ (the angle of bending) while the second factor decreases.

The $\tilde{A}-\tilde{X}$ bands of HCN represent a striking example of such a situation. The excited state, in which the molecule is bent, according to Herzberg and Innes (527) is derived from a $^1\Delta$ state of the linear conformation which cannot combine with the $^1\Sigma^+$ ground state. Since in the bent conformation the electronic transition is allowed, the intensity maximum in the $v_2'$ progression occurs well below the potential maximum, at about $v_2' = 7$. As a consequence the band spacings in the progression are perfectly normal, decreasing very regularly with increasing $v_2'$ except at the shortward end where the $\Delta G$ curve flattens out in agreement with expectation (see p. 121f).

A very similar situation has been encountered in the first absorption system of $CS_2$ which also is non-linear in the excited state while it is linear in the ground state. Here the excited state is the $B_2$ component of a $^3A_2$ state (see Chap. V, section 1c) which in the linear conformation goes over into $^3\Delta_u$. The latter cannot combine with the ground state ($^1\Sigma_g{}^+$). Therefore again the vertical transition does not give the position of maximum intensity in the $v_2'$ progressions; these progressions are quite regular since they do not reach the point of maximum potential energy corresponding to the linear conformation.

If the electronic transition probability as a function of the bending angle $\varphi$ varies as strongly as in the cases just described, it is clear that the intensities of the hot bands are no longer determined simply by the Boltzmann factors and the overlap integral but depend also on the value of $R_{e'e''}$ that corresponds to the particular vibrational transition. Indeed, often the first strong suggestion that an observed electronic transition may involve a change of molecular symmetry comes from anomalously high intensities of higher members of a sequence in some low-frequency bending vibration.

### (β) *Forbidden electronic transitions*

If the occurrence of a forbidden electronic transition is determined solely by the electronic eigenfunctions, as it is for magnetic dipole and quadrupole transitions [section 1b ($\alpha$)] or intercombinations [section 1b ($\beta$)], the separation of the transition probability into one part, $R_{e'e''}$, depending only on the electronic eigenfunctions and one part, $R_{v'v''}$, depending only on the vibrational eigenfunctions remains possible as before. The vibrational part $R_{v'v''}$ is exactly the same as for allowed electronic transitions and therefore the vibrational structure of these forbidden electronic transitions is also exactly the same. However, the situation is radically different for those forbidden electronic transitions that are made possible by vibrational-electronic interactions [section 1b ($\gamma$)], and they are therefore the only ones which we need consider here separately.

**Selection rules.** Forbidden electronic transitions are defined as those for which

$$R_{e'e''}(Q_0) = \int \psi_e'^*(q, Q_0)\boldsymbol{M}_e\psi_e''(q, Q_0)d\tau_e = 0 \qquad \text{(II, 22)}$$

If this condition is not fulfilled for one component of the dipole moment $\boldsymbol{M}_e$ we have an allowed transition. The forbidden components of this allowed transition, that is, the components for which (II, 22) is fulfilled, can be treated in the same way as genuine forbidden transitions for which all three components of the matrix element vanish. In either case, as we have already seen in section 1b($\gamma$), the general selection rule is that the *vibronic* wave functions must obey the inequality

$$R_{e'v'e''v''} \equiv \int \psi_{ev}'^*\boldsymbol{M}\psi_{ev}''d\tau_{ev} \neq 0. \qquad \text{(II, 35)}$$

Thus we can immediately find which vibrational transitions in a forbidden electronic transition (or a forbidden component of an allowed electronic transition) may occur by ascertaining the vibronic species of the upper and lower state and establishing with the help of Table 9, in the same way as for electronic transitions, whether a dipole component exists which can make the integral in (II, 35) different from zero. If the lower state is totally symmetric, the transition is allowed when $\psi_{ev}'$ has the same species as one of the components of the dipole moment $\boldsymbol{M}$, just as for pure electronic transitions.

It is important to note that as long as vibronic interaction of type (a) or (b) is negligible, the condition (II, 35) cannot be fulfilled for any vibrational transition of a (symmetry) forbidden electronic transition because then $R_{e'v'e''v''}$ can be expressed by (II, 11) and therefore, since for a forbidden transition by definition $R_{e'e''}(Q_0)$ is zero,

$$R_{e'v'e''v''} = R_{e'e''}(Q_0) \int \psi_v'^*\psi_v''d\tau_v = 0 \qquad \text{(II, 36)}$$

for any $\psi_v'$ and $\psi_v''$. However, if vibronic interaction of type (a) is introduced, that is, the variation of $\psi_e$ and therefore of $R_{e'e''}$ with $Q$ [see eq. (II, 9)], we must go back to the expression (II, 8) for $R_{e'v'e''v''}$ which does in general not vanish for a forbidden transition because $\int \psi_e'^*\boldsymbol{M}_e\psi_e''d\tau_e$ does not vanish by symmetry for non-totally symmetric nuclear configurations even if $R_{e'e''}(Q_0)$ vanishes. If vibronic interaction of type (b) is introduced, the resolution (II, 4) can no longer be made and therefore even (II, 8) can no longer be applied; thus we must use the form (II, 35) for $R_{e'v'e''v''}$, and it may or may not be zero depending on the symmetry of the vibronic functions.

Nevertheless, as we have seen earlier, for a consideration of the symmetry properties of $\psi_{ev}$ it is sufficient to use the resolution (II, 4). For a forbidden transition (or a forbidden component of an allowed transition) it follows immediately from (II, 22) that the integrand

$$\psi_e'^*\boldsymbol{M}_e\psi_e'' \textit{ is non-totally symmetrical} \qquad \text{(II, 37)}$$

(otherwise the integral would not necessarily be zero, the transition would be allowed). Therefore, in order to make

$$\int \psi_{ev}'^{*} \boldsymbol{M} \psi_{ev}'' d\tau_{ev} = \int \psi_{e}'^{*} \psi_{e}'' \boldsymbol{M} \psi_{v}'^{*} \psi_{v}'' d\tau_{ev} \tag{II, 38}$$

different from zero we conclude that

> $\psi_v'^*\psi_v''$ *must be non-totally symmetrical for the same symmetry operations for which* $\psi_e'^*\boldsymbol{M}_e\psi_e''$ (or one of its components) *is non-totally symmetrical.* (II, 39)

In other words, those vibrational transitions can occur in a forbidden electronic transition (or a forbidden component of an allowed transition) for which $\psi_v'^*\psi_v''$ has the same symmetry as $\psi_e'^*M_e\psi_e''$ (or one of its components). This formulation of the general selection rule (II, 35) is sometimes more convenient than the direct use of (II, 35).

It follows immediately from the rule (II, 39) by comparison with (II, 30), that in a forbidden electronic transition none of the vibrational transitions of an allowed electronic transition can occur; but by no means all vibrational transitions forbidden in an allowed electronic transition can occur in a forbidden electronic transition. Indeed according to (II, 35) all transitions for which for all three components of $\boldsymbol{M}$

$$\psi_e'^*\psi_v'^*\boldsymbol{M}\psi_e''\psi_v''$$

is non-totally symmetric, are strictly forbidden for dipole radiation no matter how strong the interaction of vibration and electronic motion is. The only way in which they might occur as dipole radiation would be through interaction of rotation and vibronic motion, but this is quite rare (see p. 243).

For forbidden components of allowed electronic transitions we may state the general selection rule in still another way: A vibrational transition for which $\psi_v'^*\psi_v''$ is non-totally symmetric, and which is therefore forbidden by the vibrational selection rules for allowed electronic transitions, may yet occur weakly if there is a component $M_\mu$ of the dipole moment such that $\psi_e'^*\boldsymbol{M}_\mu\psi_e''$ has the same symmetry as $\psi_v'^*\psi_v''$, where of course $\boldsymbol{M}_\mu$ is not the dipole component of the allowed transition.

**Forbidden transitions between non-degenerate electronic states.** It follows immediately from the general selection rule that in all forbidden electronic transitions made possible by vibronic interactions *the* $0-0$ *band is absent*, just as are all the other vibrational transitions which would be allowed in an allowed electronic transition. According to the preceding considerations, the absence of the $0-0$ band in symmetry-forbidden electronic transitions is rigorous for electric dipole radiation as long as rotational-electronic interaction is neglected (i.e., it is rigorous for zero rotation)[4a].

[4a] The absence of the $0-0$ band in symmetry-forbidden transitions may also be understood from the simple consideration that, when there is no vibration, vibronic interaction is ineffective in producing a forbidden electronic transition to appear. Zero-point vibrations, since the corresponding wave functions are totally symmetric, do not contribute to vibronic interaction.

If non-degenerate vibrations are excited, the symmetry selection rule (II, 39) leads immediately to the rule

$$\sum \Delta v_a = \pm 1, \pm 3, \ldots \qquad \text{(II, 40)}$$

where the summation is over all those vibrations that are antisymmetric with respect to any symmetry operation with respect to which $\psi_e'^* \boldsymbol{M}_e \psi_e''$ is antisymmetric. The rule (II, 40) replaces the rule (II, 32) which holds for allowed electronic transitions. If only a single antisymmetric vibration is affected in a forbidden electronic transition we have according to (II, 40)

$$\Delta v_a = \pm 1, \pm 3, \ldots.$$

But it is by no means necessary for each antisymmetric vibration to change its quantum number by an odd number, only the sum of the changes must be odd, that is, at least one of the antisymmetric vibrations must change by an odd number.

As mentioned before, according to the Franck–Condon principle the intensity distribution in a progression of an antisymmetric vibration corresponds to that of a diatomic molecule for which the equilibrium position in the upper and lower electronic state is the same, that is, the intensity has its peak for the $0-0$ band and falls off the more rapidly the closer $\omega'$ is to $\omega''$. This is the reason that in allowed electronic transitions the totally symmetric vibrations predominate and antisymmetric vibrations are rarely observed except in sequences; for the same reason in forbidden electronic transitions the vibrational transitions in which only one antisymmetric vibration has

$$\Delta v_a = \pm 1$$

are in general the most prominent, and all others are quite weak. This holds both for forbidden electronic transitions and forbidden components of allowed transitions.

As an example consider an electronic transition $A_2 - A_1$ of a $\boldsymbol{C}_{2v}$ molecule such as $H_2O$ or $H_2CO$. As indicated in Table 9 this is a forbidden transition since there is no component of $\boldsymbol{M}$ for which $\psi_e'^* \boldsymbol{M} \psi_e''$ is totally symmetric [$\psi_e'^* M_z \psi_e''$ is antisymmetric with respect to both planes of symmetry, $\psi_e'^* M_x \psi_e''$ with respect to $\sigma_v(xz)$ and $\psi_e'^* M_y \psi_e''$ with respect to $\sigma_v(yz)$]. But on account of vibronic interactions this transition may occur weakly if $\psi_v'^* \psi_v''$ is antisymmetric with respect to both planes of symmetry (has species $A_2$) or if it is antisymmetric with respect to $\sigma_v(xz)$ (has species $B_2$) or if it is antisymmetric with respect to $\sigma_v(yz)$ (has species $B_1$). Therefore in general all non-totally symmetric vibrational levels of the upper state can combine with the totally symmetric vibrational levels of the lower state (and vice versa), that is, in this case all vibrational transitions which are forbidden in an allowed electronic transition can occur, but the three species of the product $\psi_v'^* \psi_v''$ ($A_2$, $B_2$, $B_1$) occur with different orientations of the dipole moment ($M_z$, $M_x$, $M_y$). In a molecule like $H_2O$ only $A_1$ and $B_2$ vibrational levels occur and therefore only one orientation of the dipole moment ($M_x$) can occur in a forbidden $A_2 - A_1$ electronic transition. In a molecule like (planar) $H_2CO$ all four types of vibrational levels occur and therefore all three orientations of the dipole moment ($M_x$, $M_y$ and $M_z$) are possible in a forbidden transition.

In Fig. 69 we present a schematic spectrum representing the vibrational structure of an $A_2 - A_1$ transition of a $\boldsymbol{C}_{2v}$ molecule. The vibrational transitions which would occur in a corresponding allowed electronic transition are indicated by dotted lines, assuming that only one totally symmetric vibration, $\nu_1$, is strongly

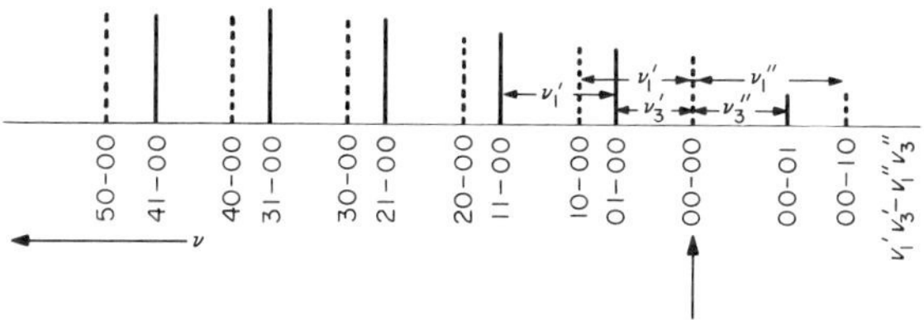


FIG. 69. **Schematic spectrum showing the vibrational structure of a forbidden electronic transition occurring on account of vibronic interaction.** The full lines represent the vibronically allowed transitions. The broken lines represent the transitions that would occur if the electronic transition were allowed.

excited. In such an allowed transition the first band at low temperature is the $0-0$ band (followed by the progression in $\nu_1$). In contrast in the forbidden transition the first band at low temperature is one in which an antisymmetric vibration, say $\nu_3$, is excited and the progression in $\nu_1$ joins onto it (full lines in the schematic spectrum). The first hot bands other than those belonging to sequences are also shown. For an allowed transition it is the $0-1$ band in the totally symmetric vibration, $\nu_1$; for a forbidden transition it can only be the $0-1$ band in the same antisymmetric vibration, $\nu_3$, that occurs in the main progression. The overall appearance of a forbidden band system is thus much like that of an allowed system except that it is built up on the $1-0$ and $0-1$ bands in the antisymmetric vibration instead of the $0-0$ band. Some authors who (contrary to the international nomenclature) call the $0-0$ band the origin of the band system, refer to the $1-0$ and $0-1$ bands as false origins (or vibronic origins).

It is important to note that in contrast to the allowed electronic transition, for the forbidden transition the separation of the first hot band from the first main band does not correspond to a vibrational frequency of the ground state but to the sum of the frequencies of the antisymmetric vibration in the upper and lower state ($\nu_3'' + \nu_3'$). Conversely the observation of a disagreement of the frequency difference of first main band and first hot band with any of the fundamentals in the ground state gives a strong indication that the transition under consideration is electronically forbidden. It may be mentioned that the dotted-line transitions in Fig. 69, while rigorously forbidden in an $A_2 - A_1$ electric dipole transition, may occur (extremely weakly) for a magnetic dipole transition of this type (compare the case of $H_2CO$ discussed on p. 270).

As a second example, consider a $B_{2u} - A_g$ transition of a $\boldsymbol{D}_{2h}$ molecule (e.g., ethylene or naphthalene). This transition is allowed with the dipole component

$M_y$. The other two components ($M_x$ and $M_z$) can occur as forbidden components of the same electronic transition if $\psi_v'^* \psi_v''$ has species $B_{1g}$ and $B_{3g}$ respectively, since $\psi_e'^* M_x \psi_e''$ and $\psi_e'^* M_z \psi_e''$ have these species. Thus, in addition to the main bands corresponding to totally symmetric upper vibrational levels (assuming absorption from the lowest vibrational level of the ground state) a $b_{1g}$ or $b_{3g}$ vibration may be singly excited quite weakly with dipole components ($M_x$ or $M_z$) different from that of the main transition ($M_y$). For naphthalene such a case has been observed [Craig, Hollas, Redies and Wait (253)] with $b_{3g}$. In this molecule the allowed transition has a rather small intrinsic intensity so that here the forbidden vibrational transitions have an intensity comparable to (or even somewhat larger than) the main allowed bands.

Finally, consider an electronic transition of type $B_{2u}-A_{1g}$ of a molecule of point group $\boldsymbol{D}_{6h}$. According to Table 9 such a transition is forbidden by the electronic selection rules. It may nevertheless be caused to appear as a forbidden electronic transition by vibronic interactions when vibrations of type $b_{1g}$ or $e_{2g}$ are excited since $\psi_e'^* M_z \psi_e''$ and $\psi_e'^* M_{x,y} \psi_e''$ have species $B_{1g}$ and $E_{2g}$ respectively.

The well-known absorption bands of $C_6H_6$ near 2600 Å have been shown to be due to such a $B_{2u}-A_{1g}$ electronic transition [see Sponer and Teller (1155)]. In $C_6H_6$ there are no $b_{1g}$ vibrations (see Vol. II, p. 363) but there are four $e_{2g}$ vibrations. Any one of the latter if singly excited could cause the appearance of the forbidden transition. The spectrum, shown in Fig. 70a, consists of a progression of main bands in the frequency 925 cm$^{-1}$. This frequency in all probability corresponds to the totally symmetric breathing vibration of the excited state (which is 995.4 cm$^{-1}$ in the ground state). There is a faint band at the long wavelength side separated by 1130 cm$^{-1}$ from the first main band. Its intensity increases with temperature, it is a "hot" band, but no fundamental of frequency 1130 cm$^{-1}$ is known. However, if the electronic transition is forbidden, then 1130 cm$^{-1}$ must be equal to the sum of the frequencies of an $e_{2g}$ vibration in the upper and the lower state in the same way as in Fig. 69 for an $A_2-A_1$ transition of a $C_{2v}$ molecule. The lowest $e_{2g}$ vibration $\nu_{18}$ in the ground state is at 608.0 cm$^{-1}$ which would make the corresponding frequency in the upper state 522.4 cm$^{-1}$, a reasonable value. In other words, the two bands under discussion are assigned as the $0-1$ and $1-0$ bands in $\nu_{18}$. Indeed in solid $C_6H_6$ at low temperature, where the selection rules are no longer as strict, the $0-0$ band has been observed at a distance of 522 cm$^{-1}$ from the first strong absorption band.

As a further confirmation of this assignment one finds that in fluorescence the $0-1$ band in $\nu_{18}$ is the band of shortest wavelength. This would be expected, if,

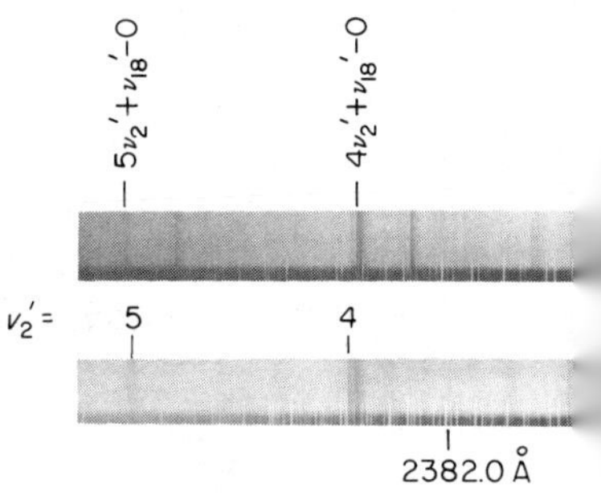


FIG. 70. **Absorption spectrum of $C_6H_6$ and $C_6D_6$ in the region 2600–2300Å.** The main progression (in $\nu_2$) is marked by solid leading lines, the $0-\nu_{18}$ hot band by a dotted leading line. The position of the $0-0$ band which does not occur is marked by a small arrow. Most (but not all) of the other bands are sequence bands joining onto each of the main bands.

as is usual, all molecules are brought by collisions into the lowest vibrational level of the excited electronic state before fluorescence emission. The other absorption and emission bands, accompanying the main bands, are readily understood as sequences in several of the other vibrations. In each case the vibration $\nu_{18}$ is excited in the upper state by one quantum more or less than in the lower state. The isotope shifts observed when going from $C_6H_6$ to $C_6D_6$ (see Fig. 70b) confirm in a convincing way the interpretation of the spectrum [see Sponer (1147) and Sponer and Teller (1155)].

**Forbidden transitions between electronic states at least one of which is degenerate.** When one or both of the two combining electronic states are degenerate we must take account of the Renner–Teller or Jahn–Teller splittings that may be present. If the occurrence of the forbidden transition is made possible by a non-degenerate antisymmetric vibration the situation is entirely similar to that for non-degenerate electronic states as illustrated by Fig. 69. For example for a ${}^1\Pi_g - {}^1\Sigma_g^+$ transition of a symmetrical linear molecule an excitation of a $\sigma_u^+$ vibration ($\nu_3$ in linear $XY_2$) can make the transition possible with dipole component $M_{x,y}$. As before, every observable $\perp$ band must have a $\sigma_u^+$ vibration at least singly excited in either the upper or lower state. Similarly in an $E'' - A_1'$ transition (or $E' - A_2''$) of a $\boldsymbol{D}_{3h}$ molecule (e.g., planar $XY_3$), only excitation of an $a_2''$ vibration ($\nu_2$ of planar $XY_3$) can make the transition possible with a "perpendicular" dipole moment. In these two cases vibronic splittings will only arise in hot bands and they have to be treated in a way entirely similar to that previously given for allowed electronic transitions (see p. 158f).

When, however, the forbidden transition is made possible by a degenerate vibration the situation is somewhat different because of the effect of Renner–Teller and Jahn–Teller splittings. According to the general selection rule only certain vibronic components of the degenerate electronic state can combine with the other electronic state (ground state). Figure 71 illustrates this for two examples: ${}^1\Pi_g - {}^1\Sigma_g^+$ of a $\boldsymbol{D}_{\infty h}$ molecule and $E'' - A_1'$ of a $\boldsymbol{D}_{3h}$ molecule. In the first case when a $\pi_u$ vibration (say $\nu_2$) is singly excited in the $\Pi_g$ electronic state three vibronic states arise of which only $\Sigma_u^+$ can combine with the $\Sigma_g^+$ lower state. When other totally symmetric vibrations are excited in addition it is always the $\Sigma_u^+$ component alone that can be reached from the lower state. The separation of the first strong band, the $1-0$ band in the bending vibration, from the missing $0-0$ band is no longer equal to the frequency of the bending vibration in the upper state but is higher or lower on account of the Renner–Teller splitting.

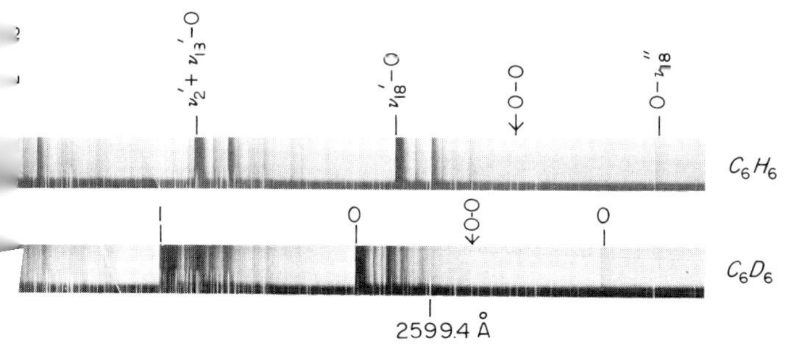

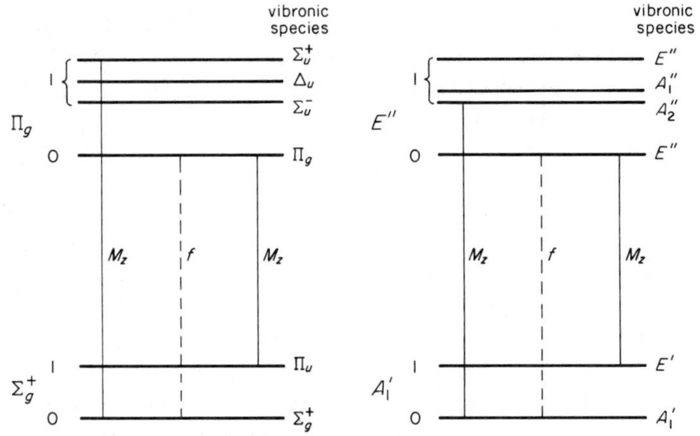


FIG. 71. **Vibronic transitions in a forbidden electronic transition with degenerate upper state made possible by the interaction with a degenerate vibration** (*a*) **for a** $^1\Pi_g - {}^1\Sigma_g^+$ **transition of a** $D_{\infty h}$ **molecule;** (*b*) **for a** $^1E'' - {}^1A_1'$ **transition of a** $D_{3h}$ **molecule.** Only the levels with $v = 0$ and 1 are shown.

In a similar way in an $E'' - A_1'$ transition of $D_{3h}$ the first strong band is the transition to the $A_2''$ vibronic component of the state in which an $e'$ vibration, say $\nu_3$, is singly excited and its separation from the $0-0$ band is different from $\nu_3(e')$ on account of the Jahn–Teller splitting.

Alternatively the excitation of a degenerate vibration in the non-degenerate *lower* state may also lead to the appearance of the forbidden electronic transition (see Fig. 71) giving rise in absorption to a hot band which is displaced from the (missing) $0-0$ band by the vibrational frequency of the degenerate vibration in the ground state.

If for a $^1\Pi_g - {}^1\Sigma_g^+$ transition of a symmetrical linear $XY_2$ molecule the Renner–Teller splitting in the $\Pi_g$ state is large, transitions with $\Delta v_2 = \pm 3, +5, \ldots$ which are very weak for small Renner–Teller splitting, will acquire appreciable intensity, that is, there will be progressions in addition to those whose first members are shown in Fig. 71. But transitions with $\Delta v_2 = 0, \pm 2, \pm 4$ remain absent since only for odd $v_2$ are there $\Sigma_u^+$ vibronic states in a $\Pi_g$ electronic state (see Fig. 2).

The situation is different for large Jahn–Teller splittings in an $E'' - A_1'$ electronic transition of a $D_{3h}$ molecule. Here all levels with $v_3 \neq 0$ of the $E''$ state have an $A_2''$ vibronic component and therefore transitions with $\Delta v_3 = \pm 2, \pm 3, \ldots$ will appear in addition to $\Delta v_3 = \pm 1$. This is analogous to the behavior in an allowed transition (e.g., $E' - A_1'$) where all $\Delta v_3$ values are possible for a large Jahn–Teller parameter $D$ (see p. 165f).

A somewhat different example would be a $\Delta - \Sigma^+$ transition of a $C_{\infty v}$ molecule. If this forbidden electronic transition occurs with a perpendicular dipole component $(M_{x,y})$ everything is as before, i.e., only transitions with $\Delta v_2 = \pm 1, \pm 3, \ldots$ occur where $v_2$ is again the quantum number of a bending vibration. As before transitions with $\Delta v_2 = \pm 1$ will be by far the most prominent except when Renner–Teller interaction is large. However, if the transition occurs with a parallel dipole component ($M_z$, $\Delta K = 0$) only $\Delta v_2 = \pm 2, \pm 4, \ldots$ is possible since only then can the $K$ values be the same in the upper and lower state (see Fig. 2). Thus the first strong band has $v_2' = 2$, that is, it is separated by $2\omega_2'$ from the $0-0$ band which is strictly forbidden. Hot bands can occur even with $\Delta v_2 = 0$, e.g., the $1-1$ band of type $\Pi - \Pi$ which would be near the forbidden $0-0$ band. The first fluorescence band from the lowest vibrational level of the upper state (vibronic species $\Delta$) would be the $0-2$ band of type $\Delta - \Delta$ which would lie longward of the $0-0$ band by $2\omega_2''$. However, it must be realized that a $\Delta - \Sigma$ transition with a dipole component $M_z$ can occur only if the $\Delta$ state is perturbed by a $\Sigma$ state (or the $\Sigma$ state by a $\Delta$ state) and such a perturbation is bound to be weak since a $\Delta$ and a $\Sigma$ state differ by more than the species of a single normal vibration (see pp. 66f and 140). Indeed no example has yet been found.

## (c) Isotope effects

Just as for diatomic molecules, a study of isotope shifts in electronic spectra of polyatomic molecules may be of help in ascertaining which molecule (or radical) is responsible for a given spectrum and it may also aid in the vibrational assignments in the spectrum. In addition, just as in infrared spectra of polyatomic molecules the vibrational frequencies of isotopic molecules in ground and excited states are needed if a complete set of force constants in either of these states is to be obtained. Finally, sometimes even when the rotational structure is not resolved conclusions about the geometrical structure of the molecule can be derived from vibrational isotope shifts.

**General formulae.** In general, the potential functions of two isotopic molecules in a given electronic state are very nearly identical. Exceptions to this rule are expected only if the electronic state is very close to another electronic state. In all but these exceptional cases the origin of a band system is therefore the same for the two isotopic molecules. Thus, we have, when the formula for the bands of the "normal" molecule is

$$\begin{aligned}\nu = \nu_e &+ \sum_i \omega_i'(v_i' + \tfrac{1}{2}) + \sum_i \sum_{k>i} x_{ik}'(v_i' + \tfrac{1}{2})(v_k' + \tfrac{1}{2}) + \cdots \\ &- \sum_i \omega_i''(v_i'' + \tfrac{1}{2}) - \sum_i \sum_{k>i} x_{ik}''(v_i'' + \tfrac{1}{2})(v_k'' + \tfrac{1}{2}) - \cdots,\end{aligned} \tag{II, 41}$$

for those of the isotopic molecule

$$\begin{aligned}\nu = \nu_e &+ \sum_i \omega_i^{*\prime}(v_i' + \tfrac{1}{2}) + \sum_i \sum_{k>i} x_{ik}^{*\prime}(v_i + \tfrac{1}{2})(v_k + \tfrac{1}{2}) + \cdots \\ &- \sum_i \omega_i^{*\prime\prime}(v_i'' + \tfrac{1}{2}) - \sum_i \sum_{k>i} x_{ik}^{*\prime\prime}(v_i + \tfrac{1}{2})(v_k + \tfrac{1}{2}) - \cdots.\end{aligned} \tag{II, 42}$$

Here while $\nu_e$ is the same for the bands of the two systems, the isotopic $\omega_i^*$ and $x_{ik}^*$ differ from $\omega_i$ and $x_{ik}$ but are determined by the same potential function.

Unfortunately, for polyatomic molecules it is not possible to give explicit formulae for the $\omega_i^*$ and $x_{ik}^*$, as it is for diatomic molecules, except when only one normal vibration of a given species exists. If this is the case, as for example for each of the vibrations of linear symmetric $XY_2$ molecules, we have

$$\omega_i^* = \rho\omega_i \tag{II, 43}$$

where $\rho$ is a factor depending only on the masses of the nuclei and the geometry of the molecule. If there are several vibrations $\omega_1, \omega_2, \ldots, \omega_f$ of a given species then according to the Teller–Redlich product rule (Vol. II, p. 231) it is only for the product of the vibrational frequencies that such a simple relation exists, viz.

$$\omega_1^*\omega_2^* \ldots \omega_f^* = \rho\omega_1\omega_2 \ldots \omega_f. \tag{II, 44}$$

The general formula for $\rho$ has been given in Volume II (p. 232) where also several examples have been discussed. Additional relations between the frequencies of isotopic molecules in a given electronic state have been discussed by Decius and Wilson (270), Sverdlov (1181), Brodersen and Langseth (155) and Heicklen (487). Similar relations between the $x_{ik}$ of isotopic molecules have been discussed by Dennison (272a) and Sverdlov (1182), but all of them are based on conjecture rather than rigorous derivation. In general, the $x_{ik}$ of a heavier isotopic molecule are smaller than those of the lighter species.

**Examples.** A good example of isotope shifts in linear symmetric $XY_2$ molecules is provided by the spectrum of $BO_2$ of which a small section is reproduced in Fig. 78, p. 192. All the main bands due to $B^{11}O_2$ are accompanied by weaker bands of $B^{10}O_2$. In the harmonic approximation the symmetric vibration $\nu_1$ should not show any shift while both $\nu_2$ and $\nu_3$ should be modified by the factor

$$\rho = \sqrt{\frac{(m_X{}^* + 2m_Y)m_X}{(m_X + 2m_Y)m_X{}^*}} = 1.0363.$$

This is indeed found to be the case to a good approximation [Johns (630)]. The isotope shift of the $000-000$ band shown in Fig. 78 is particularly important since it confirms unambiguously the conclusion from the assignments of some of the weaker bands, that the antisymmetric stretching vibration in the lower state $\nu_3''$ is unusually low (1322 cm$^{-1}$) while in the excited state it has the "normal" value $\nu_3' = 2357$ cm$^{-1}$. The calculated shift is $+19.5$ cm$^{-1}$ while $+21.5$ cm$^{-1}$ is observed. Considering that anharmonic terms have been disregarded, this agreement must be considered as very satisfactory.

An example of the use of the isotope effect for establishing the vibrational numbering is provided by the spectrum of HCO. Here a long progression of bands in $\nu_2'$ is observed, but its numbering is not obvious. The corresponding progression in DCO has, of course, a smaller spacing corresponding to the reduced value of $\nu_2'$. The $000-000$ bands of the two band systems must be close to each other since the isotope shift for this band is expected to be small. However, one finds two pairs of (extrapolated) bands, which could be considered as $000-000$ bands, one at 8489 and 8523 cm$^{-1}$ for HCO and DCO respectively and the other at 9294 and

9161 cm$^{-1}$. A decision between these two alternatives would be easy if all vibrational frequencies in the upper and lower state were known for both isotopes. This is not the case: only $\nu_1'$, $\nu_2'$, $\nu_3'$, $\nu_2''$, $\nu_3''$ are known for HCO and only $\nu_1'$, $\nu_2'$, $\nu_2''$, for DCO [Johns, Priddle and Ramsay (638)]. But by the use of the product rule in the upper and lower state and the assumption that the CO vibration $\nu_3''$ in DCO is nearly the same as in HCO it is possible from the observed shifts of the supposed 000 – 000 band to get an approximate value for the CH stretching vibration $\nu_1''$. Johns, Priddle and Ramsay (638) found 4000 and 2700 cm$^{-1}$ for the two alternative sets of 000 – 000 bands. Since a value of 4000 cm$^{-1}$ seems impossibly high, the first alternative is eliminated. This elimination is important because it establishes the nature of the upper electronic state: with the new numbering, even $K$ values in the upper state (in which the molecule is linear) go with odd $v_2'$. This result implies that $K$ differs from $l$ by an odd number, that is, in all probability $K - l = 1$. Thus the excited electronic state must be a $\Pi$ state [see Chap. I, section 2(b)].

There are many cases of free radical spectra for which only by the study of the isotope shifts a definite assignment to a specific radical has been possible. For example, in this way, it was shown that the molecule responsible for the so-called 4050 Å group (first observed in comets) cannot contain hydrogen since no shifts occurred by substitution of deuterium in the discharge producing the spectrum [Monfils and Rosen (874)]. The observation by Douglas (292) of six heads instead of one when a mixture of $C^{12}$ and $C^{13}$ was used established that three C atoms are present in the molecule responsible and indeed the fine structure of the bands shows that no other atoms are present, i.e. that the molecule responsible is the free $C_3$ radical.

In a similar way the spectrum of $CH_3$ was definitely identified by the observation of four isotopic bands in a 50:50 mixture of hydrogen and deuterium and of an isotope shift with $C^{13}$ [Herzberg (521)]. The identification of the spectra of $CH_2$, $NH_2$, HNCN and others was based on similar isotopic studies.

## 3. Rotational Structure of Electronic Transitions

Each of the bands produced by the vibrational transitions discussed in section 2 has a fine structure which is due to the various possible rotational transitions, just as for diatomic molecules. Frequently this fine structure is not resolved, either because it is so narrow that a resolution with the available means is not possible or because the line width is greatly increased on account of predissociation. Both reasons apply more frequently for electronic spectra of polyatomic molecules than for those of diatomic molecules since the moments of inertia are often larger and since the possibilities of predissociation are greater (see Chap. IV). Even if there is no predissociation, for heavy molecules the line width due to Doppler motion may be larger than the spacing of individual rotational lines and in that case, of course, a resolution is impossible.

The rotational fine structures of the bands in electronic transitions of polyatomic molecules are in general, particularly for non-linear molecules, much more complicated than those of diatomic molecules; moreover, often the electronic bands of a polyatomic molecule are more complicated than the infrared bands of the same molecule, since in the latter the rotational constants differ very little in upper and lower state while in the former, large differences may arise. Because of all these difficulties until recently the rotational fine structures of only comparatively few electronic transitions of polyatomic molecules have been completely resolved and analyzed, but rapid progress in this field has been made in the last few years.

There is one advantage in the study of electronic spectra compared to that of infrared and Raman spectra: In the latter frequently some features are obscured by the close superposition of hot bands usually corresponding to the thermal population of low-frequency bending vibrations. Just these vibrations often undergo large changes in frequency on electronic excitation, and therefore the hot bands of electronic transitions are more widely separated from the main bands so that both the main and the hot bands can be studied more easily and information can be gathered even about the ground state which is sometimes difficult to obtain from infrared spectra.

## (a) Linear Molecules

In this section we shall consider the band structures for those molecules which have a linear equilibrium conformation in at least one of the two combining states.

### (α) *Linear–linear transitions*

**Singlet bands.** The simplest case is that in which the molecule is linear in both upper and lower state and in which both states are singlet states. In such cases the band structure is entirely similar to that for diatomic molecules. This similarity exists between the vibronic symmetry of the polyatomic molecule and the corresponding electronic symmetry of a diatomic molecule: a $\Pi - \Sigma$ vibronic band has the same structure as a $\Pi - \Sigma$ electronic band of a diatomic molecule.

As always for dipole radiation the *selection rule* for the quantum number $J$ of the total angular momentum is

$$\Delta J = 0, \pm 1 \quad (J = 0 \not\leftrightarrow J = 0) \tag{II, 45}$$

and for the symmetry properties $+$, $-$, $s$, $a$, we have

$$+ \leftrightarrow -, \quad + \not\leftrightarrow +, \quad - \not\leftrightarrow - \tag{II, 46}$$

$$s \leftrightarrow s, \quad a \leftrightarrow a, \quad s \not\leftrightarrow a. \tag{II, 47}$$

For a proof of these rules see Volume II, p. 414. The rule (II, 45) leads to the three branches $P$, $Q$ and $R$ in the rotational structure of a band. There are three types of singlet bands: those without a $Q$ branch, ${}^1\Sigma - {}^1\Sigma$ bands; those with only a weak $Q$ branch, ${}^1\Pi - {}^1\Pi$, ${}^1\Delta - {}^1\Delta, \ldots$ bands and those with a strong $Q$ branch, ${}^1\Pi - {}^1\Sigma$, ${}^1\Sigma - {}^1\Pi$, ${}^1\Delta - {}^1\Pi, \ldots$ bands. These three types correspond to $\Delta K = 0$ with $K = 0$, $\Delta K = 0$ with $K \neq 0$, and $\Delta K = \pm 1$ respectively where $K$ is the quantum number of the angular momentum about the internuclear axis (see p.

25), that is, the sum of the electronic and vibrational angular momentum ($K = |\Lambda \pm l|$).

In *allowed electronic transitions* only states of the same vibrational species combine, that is, $l$ is the same in the upper and lower state, and therefore the band type depends only on the electronic species, that is, on $\Delta\Lambda (= \Delta K)$ as for diatomic molecules. If $\Delta\Lambda = \pm 1$ all allowed bands of a band system have strong $Q$ branches, while if $\Delta\Lambda = 0$ all allowed bands have either no $Q$ branch (if $K = 0$) or a very weak one (if $K \neq 0$). In a $\Sigma - \Sigma$ electronic transition, for example, in absorption from the vibrationless ground state only totally symmetric upper vibrational levels with $l = 0$ (species $\Sigma$) are reached and therefore all the "cold" absorption bands will be of type $\Sigma - \Sigma$, i.e. will have no $Q$ branch. If in the lower state a bending vibration is singly excited, that is if $l'' = 1$, then also in the upper state a bending vibration must be excited with $l' = 1$, the resulting band is still a $\|$ band but has a weak $Q$ branch. But in no case can an allowed vibrational band in this $\Sigma - \Sigma$ electronic transition have a strong $Q$ branch. Similar considerations apply to a $\Pi - \Sigma$ electronic transition in which all allowed vibrational bands have strong $Q$ branches.

However, when a suitable third electronic state perturbs one of the two states under consideration, *forbidden components* of the allowed electronic transition may occur, which for a $\Sigma - \Sigma$ electronic transition would have perpendicular structure (i.e., have a strong $Q$ branch) or for a $\Pi - \Sigma$ electronic transition have parallel structure (i.e., have no or only a weak $Q$ branch).

The formulae for the wave numbers of the lines in the branches are of course exactly the same here as for electronic bands of diatomic or infrared bands of polyatomic molecules. One has for the $P$ and $R$ branches a single formula

$$\begin{aligned}\nu_{P,R} = \nu_0 &+ (B'_v + B''_v)m + (B'_v - B''_v - D'_v + D''_v)m^2 \\ &- 2(D'_v + D''_v)m^3 - (D'_v - D''_v)m^4\end{aligned} \qquad \text{(II, 48)}$$

where $m = -J$ for the $P$ branch and $m = J + 1$ for the $R$ branch; $\nu_0$ is the *band origin*. For the $Q$ branch one finds

$$\nu_Q = \nu_0 + (B'_v - B''_v)J(J + 1) - (D'_v - D''_v)J^2(J + 1)^2. \qquad \text{(II, 49)}$$

Here, since $D \ll B$, the cubic and quartic terms can often be neglected for small $J$. Band head formation and the shading of bands are determined in the same way as for diatomic molecules. For $\boldsymbol{D}_{\infty h}$ molecules an intensity alternation arises (or alternate lines are missing) just as for infrared spectra. For further details see Volume II, p. 382.

As has been discussed in Chapter I, section 3a, whenever $K$ is different from zero a $K$ type doubling arises. In $\Pi - \Sigma$ bands this $K$-type doubling leads to a combination defect between the $P$ and $R$ branches on the one hand and the $Q$ branch on the other in exactly the same way as $\Lambda$-type doubling in diatomic molecular spectra or $l$-type doubling in infrared spectra of polyatomic molecules (see Fig. 119 of Vol. I and the accompanying discussion). As a consequence the $B'_v$ and $D'_v$ values in (II, 48) and in (II, 49) are not exactly the same. For $^1\Pi - {}^1\Pi$

and $^1\Delta - {}^1\Pi$ bands the $K$-type doubling causes an actual doubling of all the lines. The magnitude of the doubling (or the combination defect for $\Pi - \Sigma$ transitions) is larger than in corresponding diatomic cases whenever there is a contribution of the vibrational angular momentum to $K$ (see Chap. I, p. 72).

The technique of analyzing the bands of linear molecules is in most respects similar to that of diatomic molecules described in Volume I, Chapters IV and V.

Thus far only two examples of singlet linear – linear transitions have been fully analyzed: the 4050 Å group of the $C_3$ molecule [Gausset, Herzberg, Lagerqvist and Rosen (411)] and the 1240 Å system of $C_2H_2$ [Herzberg (524)], both of which are $^1\Pi - {}^1\Sigma$ electronic transitions. Figure 72 shows a spectrogram of the 0 – 0 band of $C_3$. The three branches $P$, $Q$, $R$ are clearly visible. Alternate lines are missing as can be concluded from the fact that the $P$ branch is apparently not a continuation of the $R$ branch. A combination defect exists between $P$, $R$ and $Q$ branches indicating a moderate $\Lambda$-type doubling in the $\Pi$ state ($q = 0.0004$ cm$^{-1}$). Examples of $^1\Delta - {}^1\Pi$ vibronic transitions have also been found in this system of $C_3$. They show the predicted doubling in all three branches. This doubling is here entirely due to $l$-type doubling in the lower state; it is unusually large because the bending frequency in the lower state is extremely low (63 cm$^{-1}$). Very few of the singlet electronic transitions of other well-known linear molecules like $CO_2$, $CS_2$, HCN, are linear – linear transitions and those that are have not yet been studied with sufficient resolution.

**Doublet bands.** As for diatomic molecules the band structure of doublet transitions depends on the coupling case (Hund's case a, b, c) to which the electronic states belong. However, when a bending vibration is excited, additional complications arise on account of vibronic interactions which cannot occur in diatomic molecules.

If the doublet splitting is small in both electronic states, that is, if both belong to Hund's case (b), the band structure is of course entirely similar to that of a corresponding singlet transition except for a small doubling of all branches which increases with $N$, the quantum number of total angular momentum apart from spin. This applies always to $^2\Sigma - {}^2\Sigma$ electronic transitions and would also hold for $^2\Pi - {}^2\Pi$, $^2\Delta - {}^2\Delta$ vibronic transitions belonging to $^2\Sigma - {}^2\Sigma$ electronic transitions since for them Renner–Teller splittings do not arise and the doublet splittings of $^2\Pi$, $^2\Delta$ vibronic states of $^2\Sigma$ electronic states are necessarily small.

For $^2\Pi - {}^2\Sigma$ or $^2\Sigma - {}^2\Pi$ electronic transitions, Renner–Teller splittings do arise leading to several sub-bands when $l \neq 0$ in the $\Pi$ state; but if the $^2\Pi$ state

FIG. 72. **Spectrogram of the 0 – 0 band of the 4050 Å group of the $C_3$ molecule after Gausset, Herzberg, Lagerqvist and Rosen (411).** The lines of the three branches $P$, $Q$ and $R$ are marked.

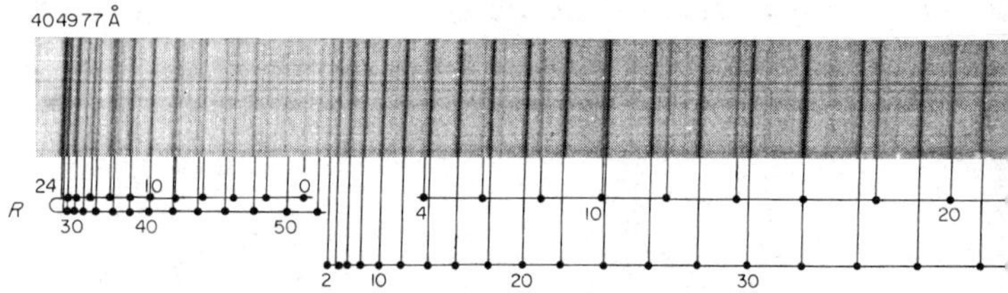

belongs to case (b) each of these sub-bands has a normal singlet-type structure. If the $^2\Pi$ state belongs to Hund's case (a) the band structure is different: the $0-0$ band has now 12 branches, six each for the $^2\Pi_{\frac{1}{2}}$ and $^2\Pi_{\frac{3}{2}}$ components just as for the corresponding bands of diatomic molecules (see Vol. I, p. 258f). Eight of these branches very nearly coincide in pairs, if as usual the splitting of the $^2\Sigma$ state is small, and therefore we have the characteristic eight-branch structure with four characteristic heads illustrated in Figs. 23 and 124c of Volume I by the well-known red CN bands (see also Fig. 74 below). The analysis of such bands proceeds in entirely the same way as for diatomic molecules and need not be discussed here. However by no means all the bands of a $^2\Pi-{}^2\Sigma$ or $^2\Sigma-{}^2\Pi$ electronic transition of a linear polyatomic molecule have this structure, since the vibronic species may differ from the electronic species.

For example the $1-1$ band of the bending vibration consists of three sub-bands (see Fig. 59b): a $^2\Sigma^+-{}^2\Pi$, a $^2\Sigma^--{}^2\Pi$ and a $^2\Delta-{}^2\Pi$ vibronic band for a $^2\Pi-{}^2\Sigma$ electronic transition (and similarly $^2\Pi-{}^2\Sigma^+$, $^2\Pi-{}^2\Sigma^-$, and $^2\Pi-{}^2\Delta$ for a $^2\Sigma-{}^2\Pi$ electronic transition). In the $^2\Sigma^+-{}^2\Pi$ and $^2\Sigma^--{}^2\Pi$ vibronic sub-bands the lower vibronic ($^2\Pi$) state, since it belongs to the $^2\Sigma$ electronic state, i.e. has no electronic angular momentum ($\Lambda = 0$), also belongs to case (b), while the $^2\Sigma^+$ and $^2\Sigma^-$ states, since they belong to a case (a) $^2\Pi$ electronic state, have a large $\rho$-type doubling, that is, are between case (a) and (c) (see Chap. I, section 3(a)). This gives rise to a band structure not normally encountered in diatomic molecules, consisting of six branches (two $P$, two $Q$, two $R$) similar to a $^2\Pi(b)-{}^2\Sigma$ band (see Fig. 124b of Vol. I) but with rapidly increasing splitting. Figure 73 gives as an example a Fortrat diagram of such a band of NCO for which however upper and lower states are reversed: the electronic transition is $^2\Sigma^+-{}^2\Pi$. The $^2\Delta-{}^2\Pi$ (or $^2\Pi-{}^2\Delta$) vibronic sub-band of the same vibrational $(1-1)$ transition has the normal structure of a $^2\Delta(a)-{}^2\Pi(b)$ [or $^2\Pi(b)-{}^2\Delta(a)$] band since the $^2\Delta$ state has the same splitting as the $^2\Pi$ electronic state.

For the $2-2$ band of a bending vibration in a $^2\Pi(a)-{}^2\Sigma$ electronic transition we have five vibronic sub-bands, one $^2\Phi-{}^2\Delta$, two $^2\Pi-{}^2\Delta$ and two $^2\Pi-{}^2\Sigma$ bands (see Fig. 59b). The only "normal" band of this group is $^2\Phi-{}^2\Delta$ since the $^2\Phi$ vibronic state has the same spin splitting as the $^2\Pi$ electronic state (see Fig. 8) and the $^2\Delta$ state has a very small spin splitting since it belongs to the $^2\Sigma$ electronic state. The four other sub-bands are somewhat anomalous since the splittings of the two $^2\Pi$ vibronic states are not simply like those of two independent $^2\Pi$ electronic states. As we have seen in Chapter I, section 3(a), the two vibronic $^2\Pi$ states are interacting strongly with each other and therefore they can only formally be separated into two states. If the Renner–Teller splitting is large we have in effect two fairly widely separated $^2\Pi-{}^2\Delta$ bands and two fairly widely separated $^2\Pi-{}^2\Sigma$ bands with different effective $B'$ values, one higher and the other lower than that

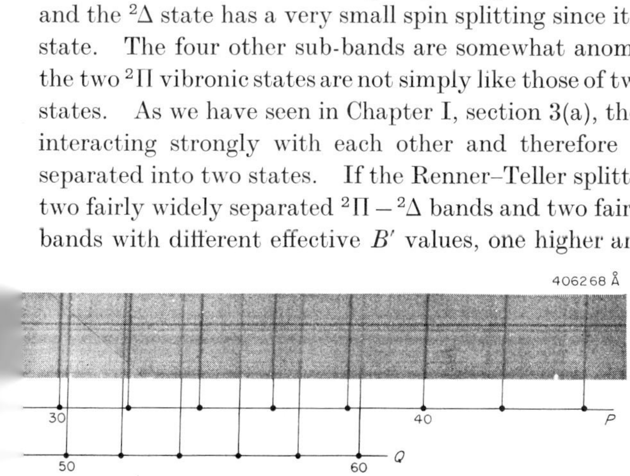

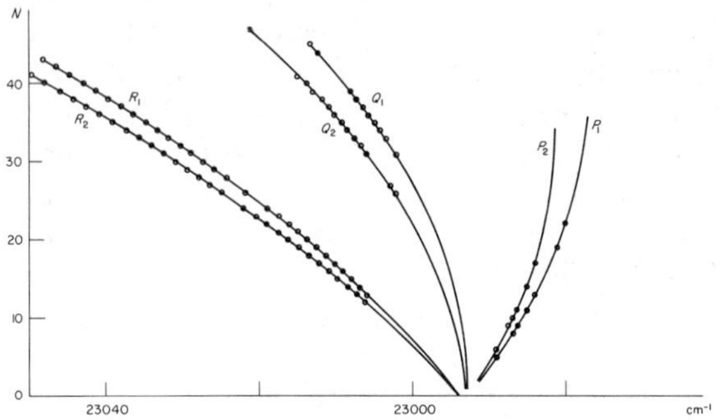


FIG. 73. **Fortrat diagram of the 010 – 010 $^2\Pi - ^2\Sigma^+$ vibronic band in the $^2\Sigma - ^2\Pi$ electronic transition of NCO at 4348 Å.** The circles represent the measured band lines, the curves are calculated. The gaps in the measurements are due to blending.

of the $^2\Phi$ state. In the two $^2\Pi$ states the spin splitting is quite different from that of the $^2\Phi$ state and has opposite direction. While the type of band structure is similar to the corresponding diatomic one ($^2\Pi_r - ^2\Delta$ and $^2\Pi_i - ^2\Delta$) and can be analyzed by the use of effective $B$ values, for precise $B$ determinations, particularly when spin-orbit and Renner–Teller interactions are of comparable magnitudes, it is necessary to analyze the two bands together since only the average of the four levels of a given $J$ (disregarding $K$-type doubling) will follow a simple formula.

As an example of a $^2\Sigma - ^2\Pi$(a) electronic transition in Fig. 74 the fine structure of the 0 – 0 band of NCO at 4400 Å is shown as observed by Dixon (281). All eight branches are clearly visible. In Fig. 75 under smaller magnification the 010 – 010 bands accompanying the 0 – 0 band of NCO are shown. All three sub-bands, $^2\Pi - ^2\Sigma^+$, $^2\Pi - ^2\Delta$, $^2\Pi - ^2\Sigma^-$ are visible although the last one only as a fragment. The $^2\Pi - ^2\Delta$ sub-band has much the same appearance as the 0 – 0 band while the $^2\Pi - ^2\Sigma^+$ sub-band has an entirely different appearance, showing only one strong $Q$ head and two $P$ heads. The doubling of the $P$ heads is due to the large $\rho$-type splitting of the $^2\Sigma^+$ vibronic state as is shown in more detail in the Fortrat diagram Fig. 73. 2 – 2 bands in the bending vibration have not been observed. For NCO in addition to other allowed vibrational bands a number of forbidden vibrational transitions with $\Delta K = 0$ corresponding to a forbidden component ($M_z$) of the dipole moment have been observed.

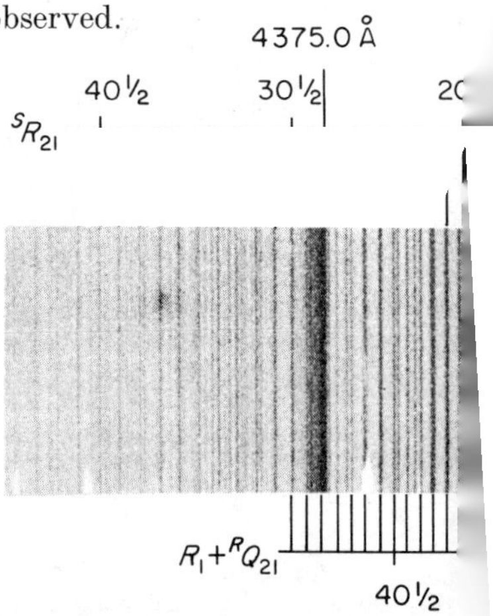


FIG. 74. **Spectrogram of the 0 – 0 band of the $^2\Sigma - ^2\Pi$(a) system of NCO at 4400 Å after Dixon (281).** The eight main branches are marked. Four of these form heads. The head at 4375.0 Å does not belong to the 0 – 0 band (see Fig. 75).

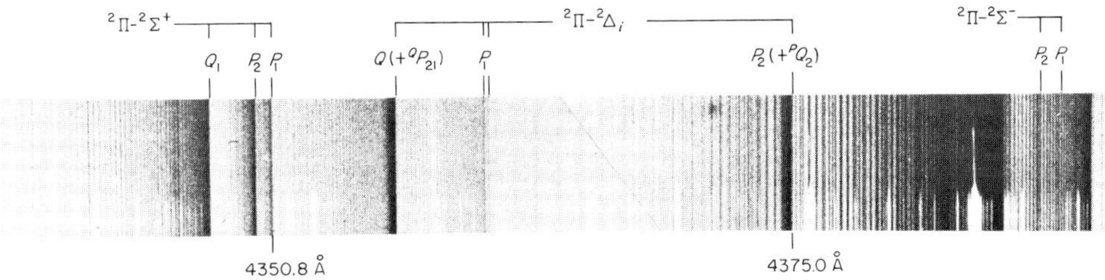


FIG. 75. **Spectrogram of the 010 – 010 bands of the $^2\Sigma-{}^2\Pi$(a) system of NCO near 4400 Å after Dixon (281).** The species symbols at the top are vibronic species. Compare the energy level diagram Fig. 59 showing how the three component bands originate. The two strong unmarked heads at the right belong to the 0 – 0 band (see Fig. 74).

Other examples of $^2\Sigma-{}^2\Pi$(a) electronic bands have been found in the spectra of $BO_2$ [Johns (630)], $CO_2^+$ [Bueso–Sanllehi (160)], $N_2O^+$ [Callomon (173), Callomon and Creutzberg (174)], $N_3$ [Douglas and Jones (297)] and $CS_2^+$ [Callomon (172)]. No examples of $^2\Pi-{}^2\Sigma$ electronic bands are known.

$^2\Pi-{}^2\Pi$ electronic bands of linear molecules again are entirely similar to those of diatomic molecules if no bending vibrations are excited. If both $^2\Pi$ states belong to case (b) even the vibronic bands corresponding to excitation of the bending vibrations have the same structure as corresponding electronic bands of diatomic molecules except of course that for each vibrational transition, on account of the Renner–Teller splittings several sub-bands appear instead of one. If however both the spin splitting and Renner–Teller splitting are appreciable in at least one of the two $^2\Pi$ states, somewhat different band structures appear for the vibronic bands. We shall consider only the case in which for both $^2\Pi$ states both interactions are appreciable, i.e. a $^2\Pi(a)-{}^2\Pi(a)$ transition with a non-zero value of $\epsilon$ in both states. The 0 – 0 band in such a transition is normal, that is, has two sub-bands $^2\Pi_{\frac{1}{2}}-{}^2\Pi_{\frac{1}{2}}$ and $^2\Pi_{\frac{3}{2}}-{}^2\Pi_{\frac{3}{2}}$, each with a strong $P$ and $R$ branch and a weak $Q$ branch; each of the branches is double if Λ-doubling is resolved. Because of the weakness of the $Q$ branches only two heads (rather than four, as for $^2\Sigma-{}^2\Pi$ transitions) are prominent.

In a 1 – 1 vibrational transition of the bending vibration as shown by Fig. 60 (p. 160), we have five sub-bands: $^2\Delta-{}^2\Delta$, $^2\Sigma^+-{}^2\Sigma^+$, $^2\Sigma^--{}^2\Sigma^-$, $^2\Sigma^+-{}^2\Sigma^-$ and

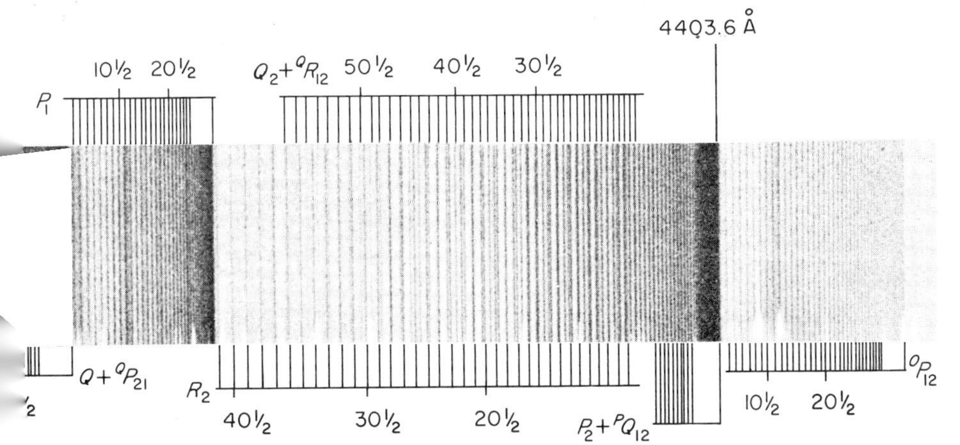

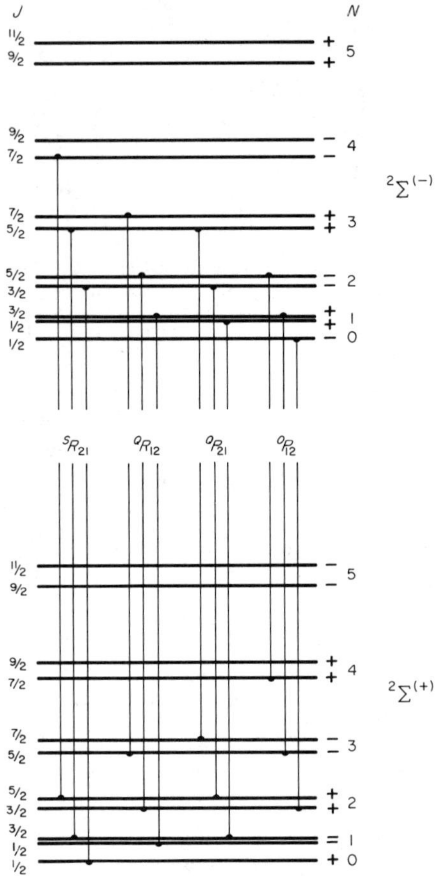


FIG. 76. **Energy level diagram for a $^2\Sigma^{(-)}-{}^2\Sigma^{(+)}$ sub-band of a 1 − 1 transition in the bending vibration of a $^2\Pi-{}^2\Pi$ electronic transition of a linear molecule.**

$^2\Sigma^- - {}^2\Sigma^+$. Of these five sub-bands only $^2\Delta-{}^2\Delta$ has a normal structure [except that in the Hill–Van Vleck formula $Y(Y - 4)\Lambda^2$ is replaced by $Y(Y\Lambda - 4K)\Lambda$; see p. 81]. The $^2\Sigma^+ - {}^2\Sigma^+$ and $^2\Sigma^- - {}^2\Sigma^-$ vibronic sub-bands are similar to ordinary $^2\Sigma-{}^2\Sigma$ bands of diatomic molecules but the doublet splitting is very much larger and for low $N$ the satellite branches $^RQ_{21}$ and $^PQ_{12}$ (see Fig. 117 of Vol. I) become noticeable. On account of the large doublet splitting two heads (rather than one for ordinary $^2\Sigma-{}^2\Sigma$ bands) appear.

As we have seen in Chapter I, section 3(a), the + or − character of $^2\Sigma$ is no longer well defined when the spin splitting is appreciable: $^2\Sigma$ then tends to case (c), that is, is better described by $^2\Sigma_{\frac{1}{2}}$ or simply $\frac{1}{2}$. This is the reason that $^2\Sigma^+ - {}^2\Sigma^-$ or $^2\Sigma^- - {}^2\Sigma^+$ transitions become possible; in case (c) they would be $^2\Sigma_{\frac{1}{2}} - {}^2\Sigma_{\frac{1}{2}}$ (or simply $\frac{1}{2}-\frac{1}{2}$) transitions.

Figure 76 shows an energy level diagram for a $^2\Sigma^- - {}^2\Sigma^+$ sub-band. The transitions permitted by the selection rules (II, 45) and (II, 46) are indicated. It is seen that four branches arise, two $Q$- form branches, one $S$- and one $O$- form branch. In Fig. 77 the Fortrat diagram of a $^2\Sigma^- - {}^2\Sigma^+$ vibronic band of $BO_2$ is compared with that of a $^2\Sigma^+ - {}^2\Sigma^+$ band belonging to the same $1-1$ transition.

In a $2-2$ transition of a bending vibration in a $^2\Pi - {}^2\Pi$ electronic transition, again five vibronic sub-bands can appear: one $^2\Phi - {}^2\Phi$ and four $^2\Pi - {}^2\Pi$ (see Fig. 60).

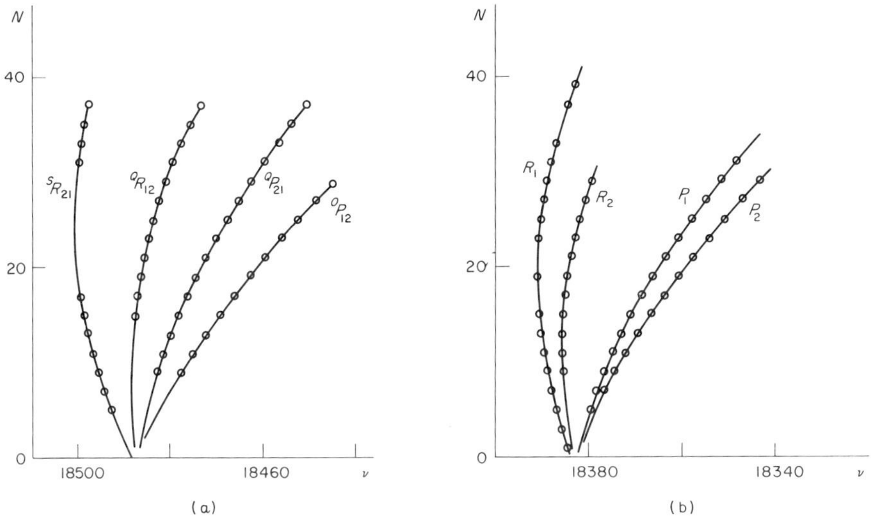


**Fig. 77. Fortrat diagrams of (*a*) the $^2\Sigma^- - {}^2\Sigma^+$ band and (*b*) the $^2\Sigma^+ - {}^2\Sigma^+$ band at 5404 and 5429 Å of the $1-1$ transition in the $\tilde{A}\,^2\Pi - \tilde{X}\,^2\Pi$ system of $BO_2$ after the data of Johns (630).**

Of these the first is normal, while the other four are affected by the mutual interaction between each pair of $^2\Pi$ states: the effective $B$ values of the two members of each pair of $^2\Pi$ states are different (see eq. I, 100) and other minor deviations from the normal formulae for the rotational levels occur (see Hougen (568)). Similar differences occur for the two $^2\Pi - {}^2\Pi$ vibronic sub-bands of a $2-0$ transition in the bending vibration.

An example of a $^2\Pi - {}^2\Pi$ electronic transition is provided by the green bands of $BO_2$ which have been analyzed in detail by Johns (630). Several but not all of the sub-bands of the $1-1$ transition have been found including a $^2\Sigma^- - {}^2\Sigma^+$ band. Figure 78 shows sections of a spectrogram of the $0-0$ band and some of the hot bands. Both the bands of $B^{11}O_2$ and $B^{10}O_2$ are visible. Since alternate rotational levels are missing in $BO_2$, the $\Lambda$ doubling leads to a "staggering" (instead of an actual doubling) in the branches of the $0-0$ band, while in the $^2\Sigma^+ - {}^2\Sigma^+$ hot band alternate lines are missing, so that it has a much more open structure than the $0-0$ band. Other examples of $^2\Pi - {}^2\Pi$ bands are the near ultraviolet $CO_2^+$ bands analyzed by Mrozowski (882)(883) before the development of the more recent theory, the near ultraviolet bands of NCO [Dixon (282)] and Schüler's T

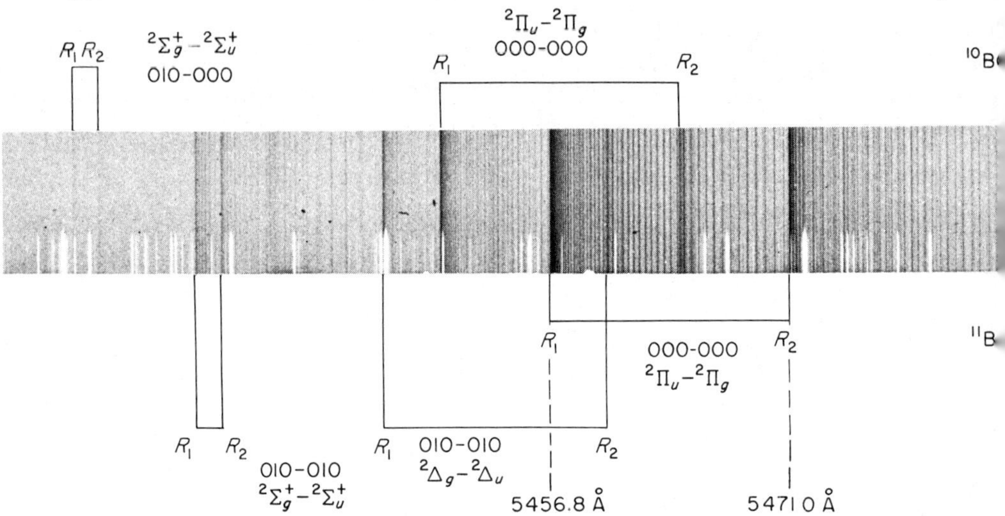


FIG. 78. **Spectrogram of the 0 – 0 and 1 – 1 bands of the $^2\Pi - ^2\Pi$ transition of $BO_2$ after Johns (630).** The heads of $B^{10}O_2$ bands are marked at the top, those of $B^{11}O_2$ (which are the stronger ones) at the bottom.

spectrum, extending from 5000 Å to 5900 Å which Callomon (171) showed to be due to ionized diacetylene, $C_4H_2{}^+$.

**Triplet bands.** The rotational structure of $^3\Sigma - ^3\Sigma$, $^3\Pi - ^3\Sigma$, $^3\Sigma - ^3\Pi$, $^3\Pi - ^3\Pi$ bands is in most respects analogous to those of the corresponding doublet bands. $^3\Sigma - ^3\Sigma$ electronic transitions have in general small triplet splittings just as corresponding bands of diatomic molecules and again this applies to all vibronic transitions ($^3\Pi - ^3\Pi$, $^3\Delta - ^3\Delta, \ldots$) belonging to $^3\Sigma - ^3\Sigma$ electronic transitions. Renner–Teller splittings do not arise. An example of a $^3\Sigma_u^- - ^3\Sigma_g^-$ transition is the 1415 Å band of $CH_2$ whose triplet structure has, however, not been resolved [Herzberg (521)].

In $^3\Pi(b) - ^3\Sigma$ electronic transitions the band structure is similar to that of $^1\Pi - ^1\Sigma$ transitions except that all lines are triple. In $^3\Pi(a) - ^3\Sigma$ electronic transitions the 0 – 0 band has the same structure as a $^3\Pi(a) - ^3\Sigma$ band of a diatomic molecule: there are 27 branches, nine for each of the components $^3\Pi_2 - ^3\Sigma$, $^3\Pi_1 - ^3\Sigma$, $^3\Pi_0 - ^3\Sigma$. The 1 – 1 band in the bending vibration would consist of three vibronic sub-bands $^3\Delta - ^3\Pi$, $^3\Sigma^+ - ^3\Pi$ and $^3\Sigma^- - ^3\Pi$. Irrespective of the strength of vibronic interaction the $^3\Delta - ^3\Pi$ sub-band has a normal structure with a triplet splitting in the $^3\Delta$ upper vibronic level equal to the spin splitting of the electronic $^3\Pi$ state while the $^3\Pi$ lower vibronic level has only a small spin splitting since it belongs to a $^3\Sigma$ electronic state. Because of the assumed large spin-orbit coupling in the $^3\Pi$ state each of the two $^3\Sigma$ vibronic levels is split into two components $^3\Sigma_1$ and $^3\Sigma_0$ (see p. 37) and as a result the two vibronic sub-bands $^3\Sigma^+ - ^3\Pi$ and $^3\Sigma^- - ^3\Pi$ each are split into two components with different origins. Both vibronic sub-bands would have 27 branches of which 9 would form the $^3\Sigma_0 - ^3\Pi$ component.

On the basis of the energy formulae of Hougen (569) it would be easy to derive formulae for all these branches.

The only example of a $^3\Pi(a) - {}^3\Sigma$ transition thus far studied in detail is a band system of the NCN radical near 3290 Å [Herzberg and Travis (543)]. Most of the predicted branches in the 0 – 0 band have been identified and the splitting of the $^3\Sigma - {}^3\Pi$ vibronic sub-bands of the 010 – 010 vibrational transitions has been observed.

**Intercombinations.** Triplet – singlet electronic transitions of linear polyatomic molecules are also closely related to corresponding diatomic transitions. For $^3\Sigma - {}^1\Sigma$ transitions we expect, when no bending vibrations are excited, bands with four branches (see Vol. I, p. 275). Such a band structure has recently been observed by Callomon and Davey (175) in the longest wavelength absorption system of $C_2N_2$ near 3000 Å. No hot bands corresponding to excitation of the bending vibration have yet been resolved but their structure could easily be predicted.

For $^3\Pi(a) - {}^1\Sigma$ transitions five branches would be expected, two for $^3\Pi_0 - {}^1\Sigma$ and three for $^3\Pi_1 - {}^1\Sigma$. Hot bands in the bending vibrations would be similar to those of $^3\Pi - {}^3\Sigma$ transitions except that several branches would be missing, e.g., in the $^3\Delta - {}^1\Pi$ vibronic sub-band the component $^3\Delta_3 - {}^1\Pi$ would be absent or weak. No examples of such transitions have as yet been identified.

### (β) *Bent – linear and linear – bent transitions*

For a number of molecules that are linear in their ground states, excited states have been found in which they are bent (i.e. have a bent equilibrium conformation), and conversely for some molecules which are bent in the ground state excited states exist in which they are linear.

**Main bands of bent – linear transitions.** If the molecule is non-linear in an excited state it is of course an asymmetric top in this state and we shall have to consider the transitions between these asymmetric top levels and the rotational levels of a linear molecule. We discuss first the case in which in the excited state the molecule, while strictly speaking an asymmetric top, is close to a prolate symmetric top and where therefore the quantum number $K$ of the angular momentum about the figure axis is well defined. The rotational energy levels can then be expressed by the symmetric-top-like expression (I, 146). In the lower state only the electronic and vibrational angular momenta contribute to $K$, that is, $K'' = |l'' \pm \Lambda''|$ and if in the ground state $\Lambda = 0$ we have $K'' = l''$.

The *selection rules* for the rotational transitions, as long as the upper state is close to a symmetric top, are the symmetric top selection rules (see Vol. II, p. 414)

$$\Delta K = K' - K'' = 0 \quad \text{for } M_z \quad (\| \text{ band}) \tag{II, 50}$$

and

$$\Delta K = K' - K'' = \pm 1 \quad \text{for } M_x, M_y \quad (\perp \text{ band}) \tag{II, 51}$$

combined with the usual selection rule for $J$.

Let us consider first, on the basis of these selection rules, the case of cold absorption illustrated at the left in the energy level diagram Fig. 79. In this case, for a $^1\Sigma$ electronic ground state we have $K'' = l'' = 0$ and therefore in the upper state we can reach only levels with $K = 0$ when the electronic transition moment is in the direction of the figure axis (broken vertical lines in Fig. 79), or only levels with $K = 1$ when the electronic transition moment is perpendicular to the figure axis (full vertical lines), or both $K = 0$ and $K = 1$ when the transition moment has both parallel and perpendicular components. Thus we expect to find a $\Sigma - \Sigma$ type band structure with a single $P$ and $R$ branch, or a $\Pi - \Sigma$ type band structure with $P$, $Q$ and $R$ branches, or a superposition of the two, separated by the interval between the $K = 0$ and $K = 1$ levels in the excited state [i.e. by $A' - \frac{1}{2}(B' + C')$].

Except in the last case, the structure of the absorption bands of a linear molecule which is non-linear in the excited state is therefore very similar to the structure of the absorption bands when both states are linear. This is particularly so since the dependence of the rotational levels on $J$ is of exactly the same form for a nearly symmetric top as for a linear molecule. Also, for a symmetric linear molecule the intensity alternation in a bent – linear transition is exactly the same as for a linear – linear transition.

However, there are certain slight but definite differences between linear – linear and bent – linear bands which at least in favorable cases make it possible to distinguish between them unambiguously. According to eqs. (I, 146) to (I, 152) the quartic term in $J^2(J + 1)^2$, apart from the normal centrifugal stretching contribution, contains an asymmetry contribution $\Delta D_{\text{eff.}}$ which is largest for $K = 0$ and increases with the square of the asymmetry parameter $b$. Thus the $\Sigma - \Sigma$ type bands of a bent – linear transition upon analysis carried out in the usual way will show an anomalously large effective $D'$ value and the same applies, to a less degree, to the $\Pi - \Sigma$ type bands ($\Delta D^1_{\text{eff.}} \approx \frac{1}{4}\Delta D^0_{\text{eff.}}$). Moreover the magnitude of this effect depends in a different way on isotope substitution than does the centrifugal stretching effect.

In $\Pi - \Sigma$ type bands of a bent – linear transition there is in addition to the $\Delta D$ effect a $K$-type doubling caused by the fact that the molecule is an asymmetric top in the bent conformation. This $K$-type doubling is much larger than the normal $K$-type doubling in a linear molecule where it can only be caused by $l$-type or $\Lambda$-type doubling. The magnitude of the asymmetry doubling for $K = 1$, according to eq. (I, 156), neglecting terms of higher powers, is

$$\Delta\nu = \tfrac{1}{2}(B - C)J(J + 1), \qquad \text{(II, 52)}$$

that is, it is larger the more the molecule deviates from a symmetric top in the excited state. For $\Pi - \Sigma$ type bands this doubling does not lead to a line doubling but only to a combination defect between $Q$ branch on the one hand and $P$ and $R$ branches on the other, just as does $l$-type and $\Lambda$-type doubling, that is, we have

$$\begin{aligned}[R(J) - Q(J)] - [Q(J + 1) - P(J + 1)] &= \Delta\nu_{J+1} + \Delta\nu_J \\ &= (B - C)(J + 1)^2. \qquad \text{(II, 53)}\end{aligned}$$

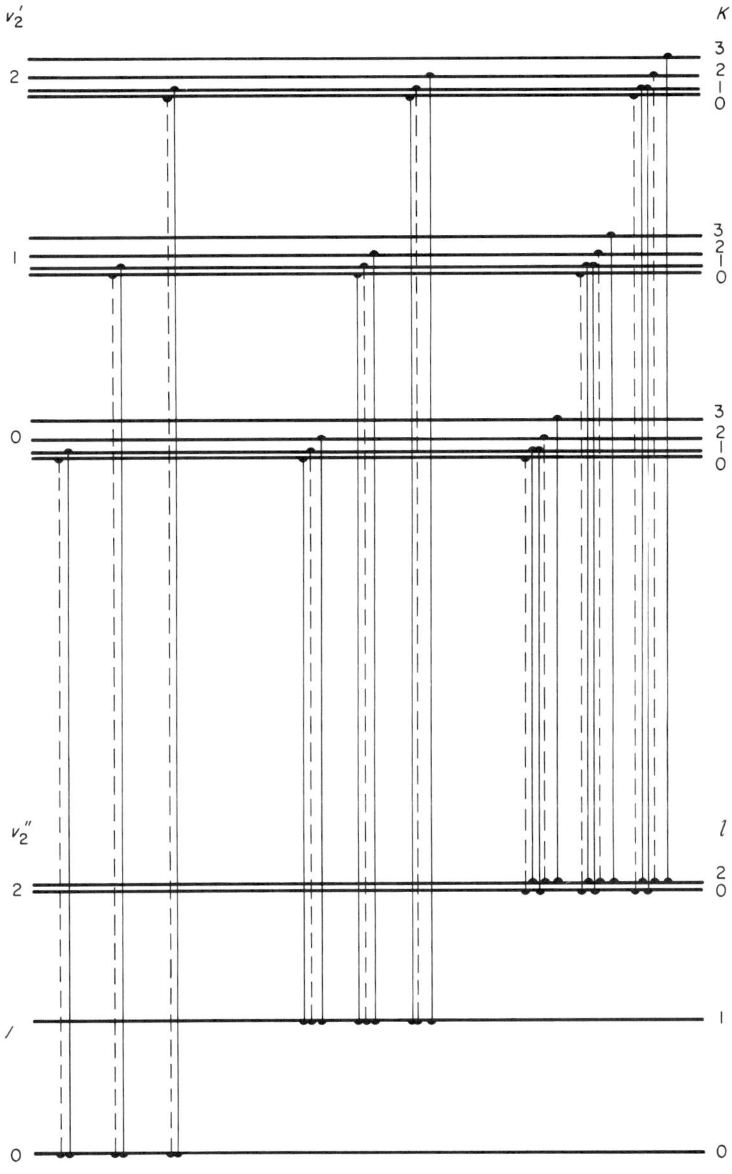


Fig. 79. **Energy level diagram showing origin of sub-bands in a bent–linear transition.** Only the vibrational levels of the bending vibration $\nu_2$ are shown. The full-line transitions correspond to a transition moment perpendicular to the plane of the molecule in the excited state ($K' - l'' = \pm 1$); the broken-line transitions correspond to the case when the transition moment is parallel to the $a$ axis in the excited state ($K' - l'' = 0$). Both full- and broken-line transitions occur when the transition moment is at a non-zero angle to the $a$-axis in the plane of the molecule.

This combination defect is usually large enough to be easily detectable whenever the band structure is resolved and its observation represents fairly strong proof of non-linearity in the excited state unless it can be shown that there are other reasons for an anomalously large $\Lambda$-doubling in the linear conformation. Figure 80

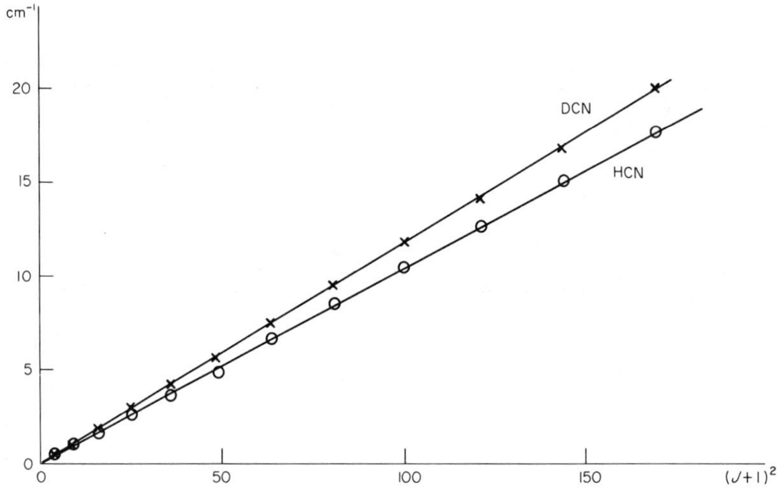


FIG. 80. **Combination defect in the 030 – 000 bands of the $\tilde{A} - \tilde{X}$ system of HCN and DCN.** The observed values of $R(J) - Q(J) - [Q(J + 1) - P(J + 1)]$ are plotted against $(J + 1)^2$.

shows as an example the combination defect observed in two corresponding bands of HCN and DCN in the $\tilde{A} - \tilde{X}$ system near 1900 Å. It is seen that eq. (II, 53) is well fulfilled. The fact that the combination defect is greater for DCN than for HCN confirms that it is produced by asymmetry and not by $\Lambda$- or $l$-type doubling.

The magnitude of the $K$-type doubling may also be obtained by evaluating separately the effective $B$ values $B^{PR}$ and $B^{Q}$ for the $P$, $R$ and $Q$ branches by means of the combination differences $\Delta_2F'(J)$ in the way described on p. 254 of Volume I and then taking the difference $B^{PR} - B^{Q}$ which would give the doubling according to

$$\Delta\nu = (B^{PR} - B^{Q})J(J + 1). \qquad \text{(II, 54)}$$

The combination defect as defined in (II, 53) and (II, 54) is positive when the upper levels of the $Q$ lines are the lower components of the $K$-type doublets and the upper levels of the $P$ and $R$ lines the upper components. The sign of the combination defect gives an important piece of information about the nature of the excited electronic state. Additional information results for symmetrical molecules from the "sign" of the intensity alternation, that is, whether even or odd lines are strong in the various branches.

The point group of a bent triatomic molecule arising from an unsymmetrical linear molecule is $C_s$, that of a bent triatomic molecule arising from a symmetrical linear molecule is $C_{2v}$ with the $C_2$ in the plane of the bent molecule. For bent four, five, ... atomic molecules derived from symmetric linear molecules the point group may also be $C_{2h}$ or $C_2$ or $C_1$. We shall consider in more detail only the three cases $C_{2v}$, $C_{2h}$ and $C_s$. Figure 81 shows the first rotational levels for the four different types of bent–linear transitions which arise if in the upper state the molecule belongs to point group $C_{2v}$ while in the lower state it is linear (point group $D_{\infty h}$) and is in a $^1\Sigma_g^+$ state. The symmetry properties of the rotational levels are given for the four species of vibronic levels of point group $C_{2v}$ and, in brackets, for the corresponding species of $C_{2h}$ assuming that in the former point group the $C_2$ is in the direction of the $b$-axis, in the latter point group in the direction of the $c$-axis. The symmetry classification of the rotational levels used here is that according to the rotational sub-group, not the full symmetry group (see p. 110). For point group $C_s$ the two diagrams at left correspond to $A'$, those at right to $A''$; moreover, for this point group the rotational sub-group has no symmetry and therefore the species symbols $A$ and $B$ of the rotational levels should be disregarded. In the lower state for which only the lowest vibrational level is shown ($l = 0$) the over-all species $s$ and $a$ are, of course, defined only in symmetric molecules. In addition to the over-all species also the parities + and − are indicated according to the rules given in Chapter I, section 3(a) and (d).

The selection rules applicable here are

$$+ \leftrightarrow -, \quad + \not\leftrightarrow +, \quad - \not\leftrightarrow - \qquad \text{(II, 55)}$$

$$A \leftrightarrow s, \quad B \leftrightarrow a, \quad A \not\leftrightarrow a, \quad B \not\leftrightarrow s \qquad \text{(II, 56)}$$

$$\begin{aligned} \Delta J &= 0, \pm 1 \text{ for } \Delta K = \pm 1 \text{ and for } \Delta K = 0 \text{ when } K \neq 0 \\ \Delta J &= \pm 1 \quad \text{for } \Delta K = 0, \text{ when } K = 0. \end{aligned} \qquad \text{(II, 57)}$$

Transitions forbidden by the parity rule (II, 55) are forbidden in all three cases ($C_s$, $C_{2v}$, $C_{2h}$) since the parities of the various levels are the same in all of them. These are the transitions to $K' = 0$ of the $A_2$ and $B_1$ vibronic states of $C_{2v}$, of the $A_u$ and $B_g$ vibronic states of $C_{2h}$ and of the $A''$ vibronic states of $C_s$. The allowed transitions for $C_{2v}$ (indicated by full and dot-dash lines) are those to $K = 1$ of $A_1$ and $B_1$ and to $K = 0$ of $B_2$ while for $C_{2h}$ they are those to $K = 1$ of $B_u$ and $A_u$ and to $K = 0$ of $B_u$ (indicated by full and broken lines). For $C_s$ all transitions except to $K = 0$ of $A''$ occur, (i.e. full, dashed, dot-dashed and dotted transitions in Fig. 81).

It is important to note that, on account of the parity rule, for an $A_1$ upper state the $Q$ lines go to the upper $K$-doubling components of $K = 1$ while for a $B_1$ upper state they go to the lower ones. The same difference occurs between $A'$ and $A''$ of $C_s$ and between $B_u$ and $A_u$ of $C_{2h}$. (This is because the parities of the rotational levels are reversed when the vibronic eigenfunction is antisymmetric with respect to the plane of the molecule.) Thus on the basis of the observed branches it is immediately possible to distinguish between the different vibronic

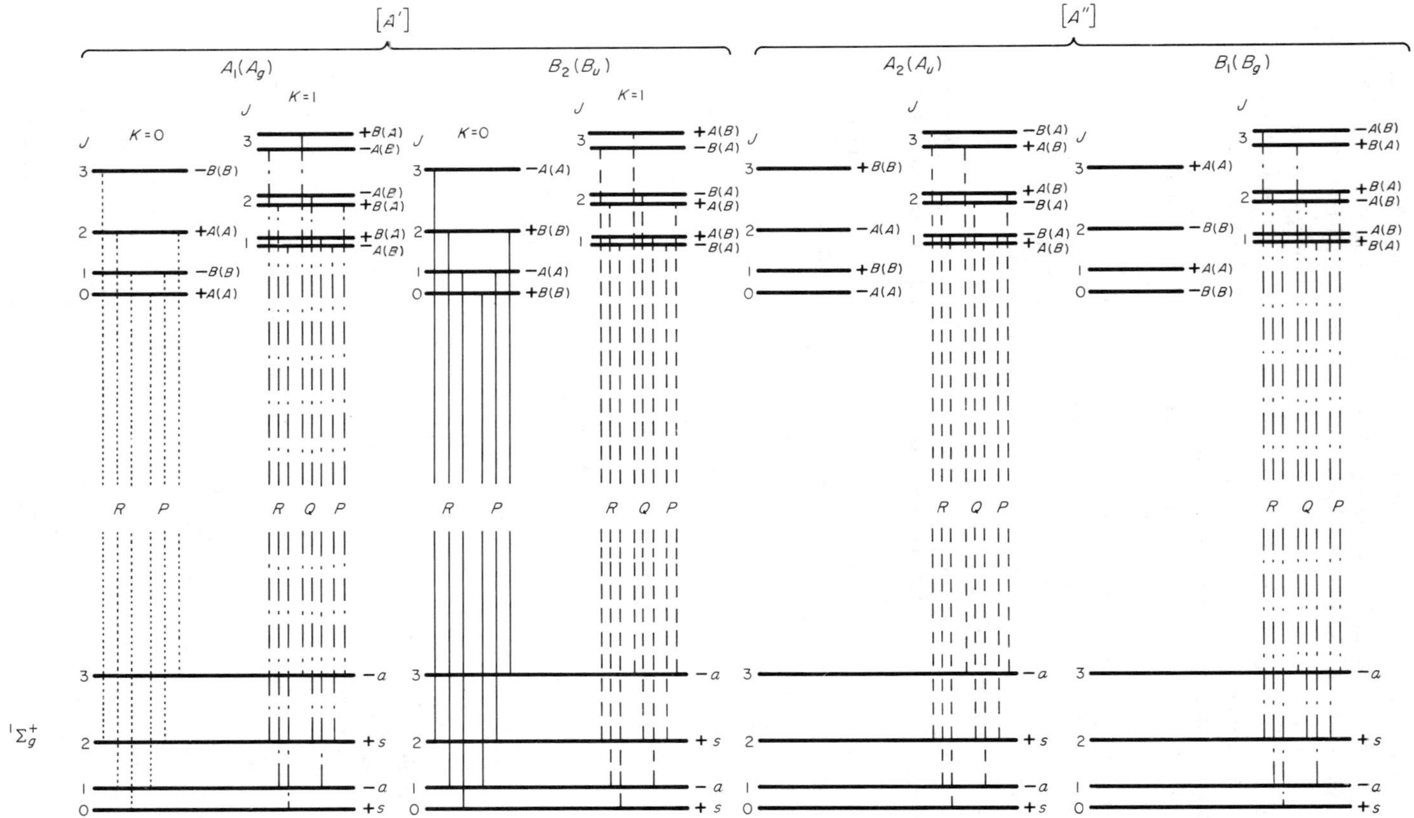


FIG. 81. **Energy level diagrams for bent–linear transitions assuming that in the excited state the molecule has $C_{2v}$ (or $C_{2h}$ or $C_s$) symmetry.** Separate diagrams are given for the four types of upper vibronic (electronic) states $A_1$, $B_2$, $A_2$, $B_1$, the lower state being $^1\Sigma_g^+$ of the linear conformation. Only levels with $K = 0$ and 1 in the upper state are shown since in the lower state only $l = 0$ occurs and transitions to the $K = 2$ upper levels are either forbidden or very weak (see p. 207). The symmetry species in ( ) refer to the case that $C_{2h}$ is the symmetry in the upper state; the symmetry species in [ ] to the case that it is $C_s$. The transitions indicated by full lines are allowed for all three symmetries, those indicated by dot-dash lines are allowed for $C_{2v}$ and $C_s$ but not for $C_{2h}$, while those indicated by broken lines are allowed for $C_{2h}$ and $C_s$ but not for $C_{2v}$; those indicated by dotted lines are allowed only for $C_s$.

species if the point group is known. A $B_2$ upper state is distinguished from $A_1$ and $B_1$ by the fact that only $K' = 0$ occurs in transitions from a totally symmetric ground state. Whether the molecule belongs to point group $\boldsymbol{C}_s$ or $\boldsymbol{C}_{2v}$ (or $\boldsymbol{C}_{2h}$) in the excited state follows usually from the point group in the ground state (see p. 197). But it is not possible a priori to decide between $\boldsymbol{C}_{2v}$ and $\boldsymbol{C}_{2h}$ for four- or more-atomic molecules. However, such a decision can be made from the band structure, if the alternation of the statistical weights in the $K' = 0$ and $K' = 1$ levels can be compared, as follows:

If the $Q$ lines of a $\perp$ band ($K' = 1$) of a bent–linear transition involve the *upper* $K$-doubling components the excited state is either $A_1$ of $\boldsymbol{C}_{2v}$ or $B_u$ of $\boldsymbol{C}_{2h}$. If it is the latter, there can also be, as shown by Fig. 81, a $\|$ component (with $K' = 0$) and its $P$ and $R$ branches should show an intensity alternation of the same sign as the $P$ and $R$ branches of the $\perp$ component. But if the excited state is $A_1$ of $\boldsymbol{C}_{2v}$, no $\|$ component arises; yet the sign of the alternation of weights in the $K' = 0$ levels can be established from hot bands (see below) and would be the same as that of the $Q$ levels (not the $P$, $R$ levels) with $K' = 1$. This is because, in this case, as shown in Fig. 81, the $Q$ levels of $K = 1$ have the same overall species ($A$ or $B$) as the $K = 0$ levels while the reverse is the case for $B_u$ of $\boldsymbol{C}_{2h}$.

If the $Q$ lines of a $\perp$ band ($K' = 1$) involve the *lower* $K$-doubling components, the excited state is either $B_1$ of $\boldsymbol{C}_{2v}$ or $A_u$ of $\boldsymbol{C}_{2h}$. Here the $Q$ levels of $K' = 1$ have opposite species to the $K = 0$ levels for $B_1$ of $\boldsymbol{C}_{2v}$ and the same species for $A_u$ of $\boldsymbol{C}_{2h}$ and therefore again these two cases can be distinguished by a comparison of the intensity alternation of the main bands with that in hot bands with $K' = 0$ (since no $\|$ components of the main bands arise).

Summarizing we have the following conclusions: If it can be established whether the $Q$ lines in a $\perp$ (main) band of a bent–linear transition go to the upper or lower $K$-doubling component, it follows immediately that the upper vibronic state is $A'$ or $A''$ respectively of $\boldsymbol{C}_s$, or $A_1$ or $B_1$ respectively of $\boldsymbol{C}_{2v}$, or $B_u$ or $A_u$ respectively of $\boldsymbol{C}_{2h}$. In symmetrical molecules the observed intensity alternation (whether it yields the same or the opposite sign of the alternation of statistical weights for the $K = 0$ and $K = 1$ levels) allows one to distinguish unambiguously between $\boldsymbol{C}_{2v}$ and $\boldsymbol{C}_{2h}$. It may be noted that pure $\|$ bands ($\Delta K = 0$) can occur only if the upper state is $B_2$ of $\boldsymbol{C}_{2v}$.

The analysis of the main bands of bent–linear transitions proceeds in the same way as for linear–linear transitions. For the upper state (in which the molecule is not linear) one obtains effective $B$ values which for a $\Pi - \Sigma$ type band are in a first approximation $\frac{1}{4}(3B_v + C_v)$ and $\frac{1}{4}(B_v + 3C_v)$ for the two components of the upper state [see eq. (I, 154 and I, 155)]. Thus approximate values for $B_v$ and $C_v$ are obtained. A value for $A_v$ in the upper state can be obtained directly from the main bands if the transition is not purely of the $\perp$ type but has a $\|$ component. The separation of the origins of corresponding $\Pi - \Sigma$ and $\Sigma - \Sigma$ type sub-bands gives directly $A_v$. For pure $\perp$ type transitions the study of the hot bands leads to a determination of $A_v$ as shown below. Improved values for $B_v$ and $C_v$ can then be obtained on the basis of the formulae (I, 147) and (I, 148) for slightly asymmetric top molecules by substituting the asymmetry parameter obtained from the approximate rotational constants.

A number of examples of bent–linear transitions have been observed and studied in detail. The main bands of the $\tilde{A}-\tilde{X}$ absorption system of HCN extending from 2000 to 1600 Å show the simple $P$, $Q$, $R$ structure described above. This is illustrated by the 030–000 band reproduced in Fig. 82. The considerable

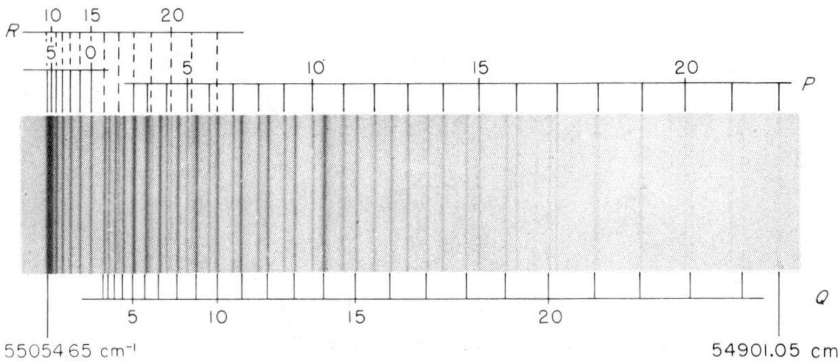


FIG. 82. **Spectrogram of the 030–000 band of the $\tilde{A}-\tilde{X}$ system of HCN at 1816 Å showing simple $P$, $Q$, $R$ branch structure [after Herzberg and Innes (527)].** The transition is of the type $^1A''-{}^1\Sigma^+$ corresponding to the two right-hand diagrams in Fig. 81.

combination defect already shown in Fig. 80 proves that the molecule is bent in the upper state while the sign of the defect shows that the upper state is an $A''$ state, that is, that the transition moment is perpendicular to the plane of the molecule. This conclusion is confirmed by the absence of any $\|$ component. An example of an $A'-{}^1\Sigma^+$ transition with both $\|$ and $\perp$ components is the $\tilde{C}-\tilde{X}$ system of HCN at 1500 Å. Even though here a complete resolution is not possible on account of predissociation, the main bands show clearly three heads which can be readily accounted for as the two heads of a $\Pi-\Sigma$ sub-band plus the single head of the $\Sigma-\Sigma$ component (see Fig. 88 below). This interpretation is confirmed by the hot bands (see below and Chap. V, section 1b).

A transition in a symmetrical molecule, similar to the $\tilde{A}-\tilde{X}$ system of HCN, is the near ultraviolet system of $C_2H_2$ in the region 2500–1900 Å. Figure 83 shows the fine structure of one of the main bands. The intensity alternation is clearly visible. Again the large combination defect shows that the molecule is bent in the excited state and the sign of this defect shows unambiguously that the transition moment is perpendicular to the plane of the molecule, that is, that the species of the upper state is either $A_u$ of $\boldsymbol{C}_{2h}$ or $B_1$ of $\boldsymbol{C}_{2v}$. While the study of the band shown in Fig. 83 by itself cannot lead to a distinction between these two alternatives the intensity alternation in hot bands with $K' = 0$ of the same upper vibronic state does lead to a decision. For $K' = 0$ the lines with odd $J'$ are found to be strong just as for the $Q$ levels of $K' = 1$: therefore the upper state must be $A_u$ of $\boldsymbol{C}_{2h}$ not $B_1$ of $\boldsymbol{C}_{2v}$, that is, $C_2H_2$ in the excited state has a staggered (trans) conformation.

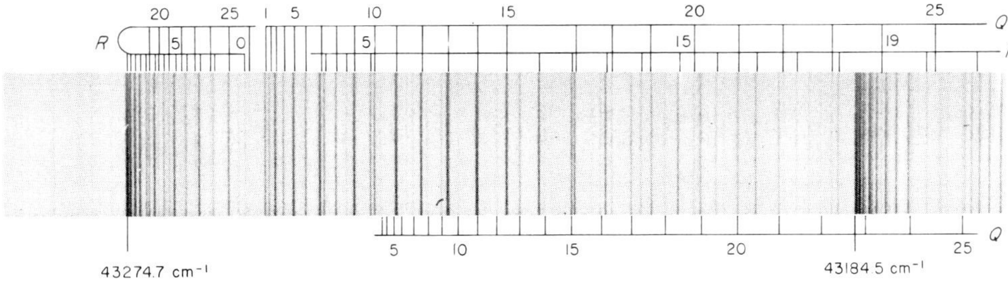


FIG. 83. **Spectrogram of the 00100 − 00000 band of the $\tilde{A}-\tilde{X}$ system of $C_2H_2$ at 2310 Å showing simple *P*, *Q*, *R* branch structure after Watson (1279).** The electronic transition is of the type ${}^1A_u - {}^1\Sigma_g^+$ corresponding to the third diagram in Fig. 81. The three branches of the main sub-band (of type $\Pi - \Sigma$) are marked at the top. At the bottom the lines of a weak, forbidden sub-band (of type $\Sigma^- - \Sigma^+$) are marked. It arises on account of "axis-switching" and consists solely of a $Q$ branch. It is very weak at low $J$ but becomes comparable in intensity to the main sub-band for high $J$ (see p. 209). The head at the right belongs to another band.

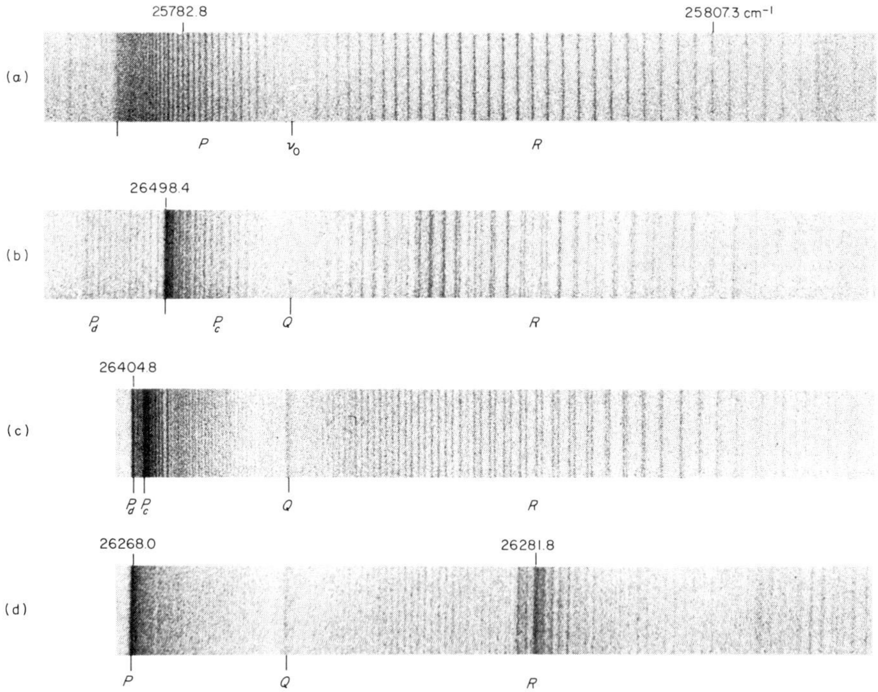


FIG. 84. **Spectrograms of several absorption bands of the $\tilde{a}-\tilde{X}$ system of $CS_2$: (*a*) a $\Sigma - \Sigma$ type band (040 − 040) at 3878 Å, (*b*) a $\Pi - \Pi$ type band (050 − 030) at 3773 Å, (*c*) a $\Delta - \Delta$ type band (060 − 040) at 3786 Å, (*d*) a $\Phi - \Phi$ type band (040 − 030) at 3806 Å** [after **Kleman (680)**]. Although the electronic transition is now known to be ${}^3A_2 - {}^1\Sigma_g^+$, only the $B_2 - {}^1\Sigma_g^+$ component is observed. Therefore, for the first band the second diagram in Fig. 81 applies, while for the other bands similar diagrams with $l = 1, 2, 3$ in the lower state apply. The second component of the $\Pi - \Pi$ band is clearly indicated by the branch extending to the left from the $P$ head of the first component. It does, however, not form a visible head. Note the increasing intensity of the line-like $Q$ branch in going from $\Pi - \Pi$ to $\Delta - \Delta$ to $\Phi - \Phi$. The $K$-type doubling in $\Delta - \Delta$ is clearly shown. It is just visible in $\Phi - \Phi$. It should be noted that in this figure, unlike all other spectrograms, the wavelengths increase from right to left.

An example of a purely parallel type transition is provided by the near ultraviolet absorption bands of $CS_2$ in the region 3800–3300 Å. Figure 84a shows one of the main bands with its simple $P$ and $R$ branch. The zero gap is one and a half times (not twice) the spacing of the lines, showing that alternate lines are missing (just as in the main infrared bands) in accordance with the zero nuclear spin of the sulphur nuclei. If the molecule were linear in the excited state (as it is in the ground state) the electronic transition would be $\Sigma_u^+ - \Sigma_g^+$; if it were bent (point group $C_{2v}$) it would be $B_2 - \Sigma_g^+$, which according to Fig. 81 is the only possibility for a $\Sigma - \Sigma$ type transition. From the structure of the main bands alone a decision with regard to the equilibrium conformation of the excited state is not readily possible. But the hot bands do show conclusively that the molecule in the excited state is bent (see below).

**Hot bands of bent–linear transitions.** At room temperature and even more at elevated temperatures the bending vibrations of linear molecules are often excited with one, two or even more quanta. Absorption from such levels gives rise to absorption bands (hot bands) of a somewhat different structure from that of the bands just discussed. The central and right-hand parts of Fig. 79 show the origin of the various sub-bands ($K$ structure) but the levels with different $J$ are not shown (note that $J \geq K$ in the upper and $J \geq l$ in the lower state).

In Fig. 79 the transitions corresponding to a parallel orientation of the transition moment (i.e. $\|$ to the top axis) are shown as broken lines, those corresponding to a perpendicular orientation as full lines. According to the selection rule (II, 50), for a *parallel transition* the $v_2 = 1$, $\Pi$ level of the lower state, combines only with the $K = 1$ sub-levels of the upper state, the $v_2 = 2$, $\Sigma$ level only with the $K = 0$ sub-levels, the $v_2 = 2$, $\Delta$ level only with the $K = 2$ sub-levels, and so on. The $\Pi - \Pi$ type bands corresponding to $v_2 = 1$ have a structure similar to $^1\Pi - {}^1\Pi$ bands of diatomic molecules (see Vol. I, p. 266) with very weak $Q$ branches. But unlike the diatomic case, here the $K$-type doubling (asymmetry splitting) in the upper state is large while in the lower state the $l$-type doubling is small and therefore the pairs of branches $P_c$, $P_d$, $Q_c$, $Q_d$, $R_c$, $R_d$ split fairly widely with increasing $J$. Figure 84b shows a spectrogram of such a band observed in the band system of $CS_2$ already mentioned. Figure 85 shows the corresponding Fortrat diagram. It is seen that the two $P$ heads are widely separated quite unlike ordinary $\Pi - \Pi$ bands. Conversely from the observation of such a large splitting for $CS_2$ it can immediately be concluded that in the excited state the molecule is not linear, and indeed from the magnitude of the splitting according to (II, 52) the difference $B - C$ is obtained which, together with the $\bar{B} = \frac{1}{2}(B + C)$ values, gives $B$ and $C$ separately and therefore the angle of bend. For $CS_2$ Kleman (680) found in this way an angle of 136°.

For the $\Delta - \Delta$ and $\Phi - \Phi$ type bands the splitting is much smaller but still noticeable as shown by the $CS_2$ bands reproduced in Fig. 84c and d. At the same time the intensity of the $Q$ branch increases with $K$ in essentially the same way as in the sub-bands of a $\|$ infrared band of a symmetric top (see p. 422 of Vol. II).

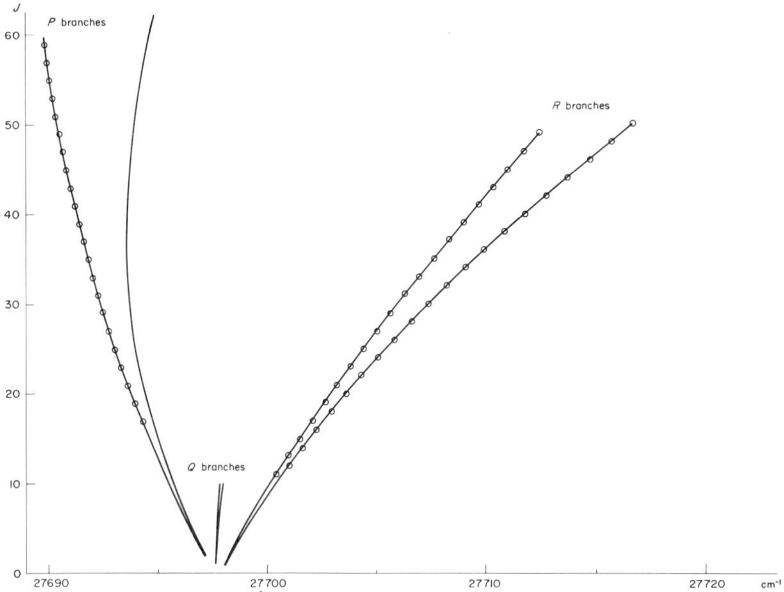


FIG. 85. **Fortrat diagram for the $CS_2$ band at 3610 Å.** This diagram corresponds to the spectrum in Fig. 84b. One of the *P* branches and the two *Q* branches have not been resolved. The curves represent the calculated positions.

It should be noted that in $CS_2$ (and similarly in other molecules with identical nuclei of zero spin) alternately the upper and lower component of the *K*-type doublet levels is missing, that is, as long as the *K*-type doubling is small the band structure may also be described as consisting of single *P*, *R* and *Q* branches each with "staggering".

For a *perpendicular transition*, according to the selection rule (II, 51), the $v_2 = 1, \Pi$ lower state can combine with both $K = 0$ and $K = 2$ in the upper state giving rise to two sub-bands of type $\Sigma - \Pi$ and $\Delta - \Pi$ (see Fig. 79). From the formula for the energy levels of a nearly symmetric top (I, 146) we obtain for the separation of the two sub-bands $4[A - \frac{1}{2}(B + C)]$. Since for a slightly bent molecule $A$ is large, the separation of the sub-bands is comparatively large. In Fig. 86 is given a Fortrat diagram of such a pair of sub-bands as observed in HCN. Figure 87 shows the corresponding spectrum. While the $\Sigma - \Pi$ band consists of single *P*, *R* and *Q* branches, the $\Delta - \Pi$ band consists of doublet branches because of the *K*-type doubling in the upper state and the *l*-type doubling in the lower state. The latter doubling leads to a combination defect in the $\Sigma - \Pi$ band which is in general much smaller than the defect in the main $\Pi - \Sigma$ bands (caused by the *K*-type doubling in the $K = 1$ upper state).

According to the Franck–Condon principle, because of the bent structure in the excited state there will be a long progression of such pairs of sub-bands ($\Sigma - \Pi$ and $\Delta - \Pi$) running parallel to the main progression ($\Pi - \Sigma$). It is readily seen

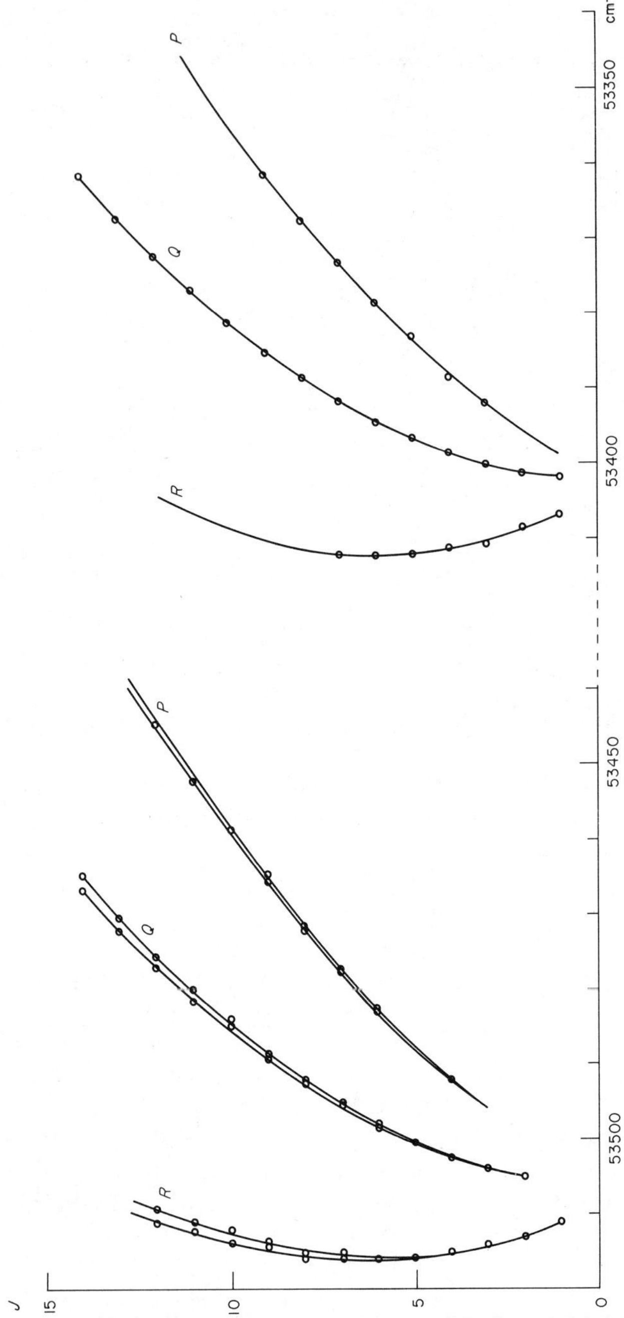


FIG. 86. **Fortrat diagrams for the two sub-bands of the 020 − 010 band of the $\tilde{A} - \tilde{X}$ system of HCN.** To the left is the $\Delta - \Pi$, to the right is the $\Sigma - \Pi$ sub-band. The corresponding spectrum is shown in Fig. 87.

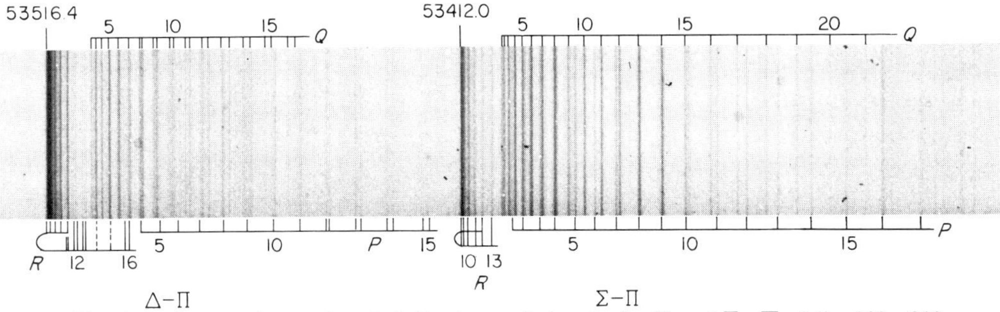


FIG. 87. **Spectrogram showing the two sub-bands Δ − Π and Σ − Π of the 020 − 010 band of HCN.** The *P*, *Q* and *R* branches have been marked. Note the doubling of the branches at high *J* in the Δ − Π sub-band (see also Fig. 86).

from Fig. 79 that the separation of neither hot band from the corresponding main band is exactly the vibrational frequency $\nu_2''$ of the bending vibration in the ground state. Rather the separation of the Σ − Π sub-band from the corresponding main band is somewhat greater than $\nu_2''$, since its upper state ($K = 0$) is slightly lower than the upper state ($K = 1$) of the main band, while the separation of the Δ − Π sub-band from the corresponding main band is somewhat smaller than $\nu_2''$. The actual amounts of these displacements give the spacings of the $K = 0, 1, 2$ levels in the upper state: viz. Σ − Π is separated by

$$\nu_2'' + [A - \tfrac{1}{2}(B + C)](1^2)$$

from Π − Σ while Δ − Π is separated from it by

$$\nu_2'' - [A - \tfrac{1}{2}(B + C)](2^2 - 1^2).$$

The magnitudes of these separations change slightly from band to band since $A$, $B$, $C$ vary slightly in going to the higher vibrational levels of the upper state.

According to the preceding discussion the observation in the absorption spectrum of a linear molecule of a double series of hot bands accompanying the main bands such that one series has a smaller the other a larger separation than $\nu_2''$ from the main series is clear evidence that the molecule is bent in the excited state, and such evidence can be obtained even if the fine structure is not resolved. Double series of hot bands have been found for $C_2H_2$ and $C_2D_2$ [Ingold and King (600), Innes (605)] as well as for HCN and DCN [Herzberg and Innes (527)]. The previous Fig. 65 (p. 168) shows schematically the relation of the hot bands to the main bands for HCN.

According to Fig. 79, for a $\perp$ band three sub-bands arise from the $v_2 = 2$ level of the lower state, one of type Π − Σ from the Σ vibrational sublevel and two of types Π − Δ and Φ − Δ from the Δ sublevel. The Π − Σ band is in all respects similar in structure to the main bands. The Π − Δ type sub-band is similar to the Δ − Π type arising from $v = 1$, but the $K$-type doubling in the upper state is much larger and therefore the splitting of the branches is greater. The $l$-type doubling in the lower state is negligibly small for $l = 2$. The Φ − Δ sub-bands are similar to the Π − Δ except that the $K$-type doubling is much smaller. A few sub-bands with $v = 2$ have been observed and analyzed for $C_2H_2$ and HCN. An extension of these considerations to $v = 3$ of the lower state will not be given here; it can very easily be made by the reader.

It is clear that from the separations of the sub-bands the rotational constant $A$ can be obtained in a number of ways if $\bar{B} = \frac{1}{2}(B + C)$ has been determined from the $J$ structure of the sub-bands. However, actually there are two constants to be determined because the centrifugal stretching constant $D_K$ is often relatively large for a slightly bent molecule and must be taken into account in the evaluation. Thus for the interval between the two levels with $K = K_2$ and $K = K_1$ of a given upper vibronic state we must use

$$\nu(K_2) - \nu(K_1) = (A - \bar{B})(K_2^2 - K_1^2) - D_K(K_2^4 - K_1^4). \qquad \text{(II, 58)}$$

As an example Table 11 gives the observed intervals in HCN for the upper states $\tilde{A}(000)$ to $\tilde{A}(050)$. It should be noted that in each case $\frac{1}{4}[\nu(2) - \nu(0)]$ is smaller than $[\nu(1) - \nu(0)]$. This is the effect of $D_K$. The resulting values of $A$ as well as of $D_K$ show a fairly rapid increase with increasing amplitude of the bending vibration in the excited state.

TABLE 11. SUB-BAND INTERVALS AND ROTATIONAL CONSTANTS $A - \bar{B}$ AND $D_K$ IN THE $\tilde{A} - \tilde{X}$ SYSTEM OF HCN

| Level | $\nu_{\Delta-\Pi} - \nu_{\Sigma-\Pi}$ $= \nu(2) - \nu(0)$ | $\nu_{\Pi-\Pi}{}^{a} - \nu_{\Sigma-\Pi}$ $= \nu(1) - \nu(0)$ | $A - \bar{B}$ | $D_K$ |
|---|---|---|---|---|
| $\tilde{A}(000)$ | 83.20 | 21.41 | 21.61 | 0.20 |
| $\tilde{A}(010)$ | 92.55 | 23.47 | 23.58 | 0.11 |
| $\tilde{A}(020)$ | 104.05 | 26.75 | 27.00 | 0.25 |
| $\tilde{A}(030)$ | 118.89 | 30.72 | 31.05 | 0.33 |
| $\tilde{A}(040)$ | 142.45 | 37.57 | 38.22 | 0.65 |
| $\tilde{A}(050)$ | 181.69 | 52.99 | 55.51 | 2.52 |

[a] $\Pi - \Pi$ sub-bands are not directly observed but their wave numbers are readily obtained from the observed $\Pi - \Sigma$ main bands by subtraction of $\nu_2''$.

As pointed out earlier, in addition to $\|$ and $\perp$ bands we may also have *hybrid bands* when there is both a $\|$ and a $\perp$ component of the transition moment. The relative intensities of the two component bands depend on the ratio of the squares of the components of the transition moment. For molecules that are planar in the excited state a hybrid band will in general arise when the transition moment lies in the plane of the molecule and when in the excited state the molecule belongs to point group $\boldsymbol{C}_s$ or $\boldsymbol{C}_{2h}$ since in that case the direction of the transition moment is not determined by symmetry. A hybrid band cannot arise when the molecule belongs to point group $\boldsymbol{C}_{2v}$ since according to Table 9 the transition moment for a given electronic transition can only be in the $x$ or $y$ or $z$ direction and these directions are either parallel or perpendicular to the top axis.

As mentioned before, the $\tilde{C} - \tilde{X}$ bands of HCN represent an example of such a hybrid transition. This is confirmed by the hot-band structure: there are three (rather than two) sub-bands with $v_2'' = 1$ corresponding to every main band, as shown by the spectrogram in Fig. 88. As expected the shortest and longest wavelength sub-bands of such a group have two heads each ($R$ and $Q$) since they are $\Delta - \Pi$ and $\Sigma - \Pi$ transitions while the middle band has only one head since it is a $\Pi - \Pi$ transition. The intervals between the bands are roughly in the ratio 1:3 and from them $A - \frac{1}{2}(B + C)$ can be determined even though the fine structure

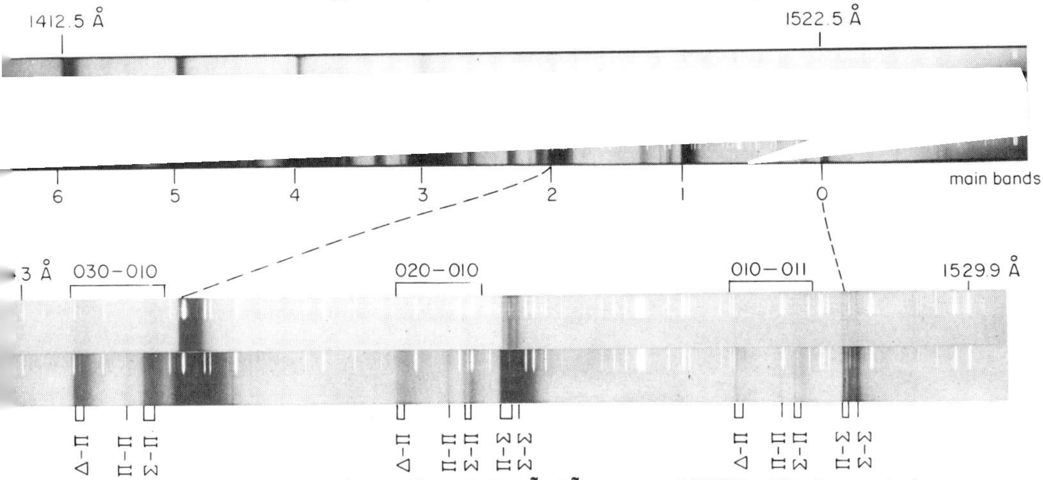


Fig. 88. **Spectrogram of a section of the $\tilde{C} - \tilde{X}$ system of HCN.** The top part gives a general view of the absorption spectrum of HCN in the region 1600 to 1300 Å. At the bottom the first few bands are shown at higher resolution to show the sub-band structure. The bands of the main progression show three heads, two corresponding to the $\perp$ component ($\Pi - \Sigma$) and one corresponding to the $\parallel$ component ($\Sigma - \Sigma$); see p. 200. The hot bands have three sub-bands, two with double heads ($R$, $Q$) and one with a single head ($R$).

of the bands is not resolved. It is interesting to note that the $\Pi - \Pi$ sub-band rapidly decreases in relative intensity compared to $\Delta - \Pi$ and $\Sigma - \Pi$ with increasing $v_2'$. This indicates that with increasing amplitude of the bending vibration the transition moment becomes more and more nearly perpendicular to the top axis. Presumably this change is closely related to the rapid increase of $A$ with $v_2$ which indicates that the $a$-axis approaches more and more the direction of the C—N bond (or in other words, in spite of the increasing amplitude of vibration the molecule on the average is less and less bent).

Hot hybrid bands with $v_2'' = 2$ would have five sub-bands: $\Sigma - \Sigma$, $\Pi - \Sigma$, $\Pi - \Delta$, $\Delta - \Delta$, $\Phi - \Delta$. No example has as yet been observed.

Thus far we have assumed that in the bent conformation of the excited state the molecule is nearly a symmetric top, that is, that the asymmetry parameter $b$ is small. If this is no longer the case we can still designate the rotational levels according to $K$, the quantum number of rotation about the $a$-axis, but the $K$-type doubling is now very large and $K$ is no longer a "good" quantum number. Therefore transitions violating the selection rule $\Delta K = 0, \pm 1$ may occur. Thus for example, from the ground state ($l'' = 0$) not only $K = 0$ and $K = 1$ in the upper state can be reached but also $K = 2, 3, \ldots$. It is easily seen by considering the over-all species of the rotational levels that when in the nearly symmetric case only $\Delta K = 0$ or only $\Delta K = \pm 1$ occurs, in the asymmetric case only even or only odd

$\Delta K$ values respectively (not both, as in hybrid bands) can occur. However, even for a large asymmetry the transitions with $\Delta K = 0, \pm 1$ still represent the strongest transitions (see section 3d($\gamma$) below). The intensity drops rapidly with increasing $|\Delta K|$ the more so since for one of the combining states $K$ is still exactly defined.

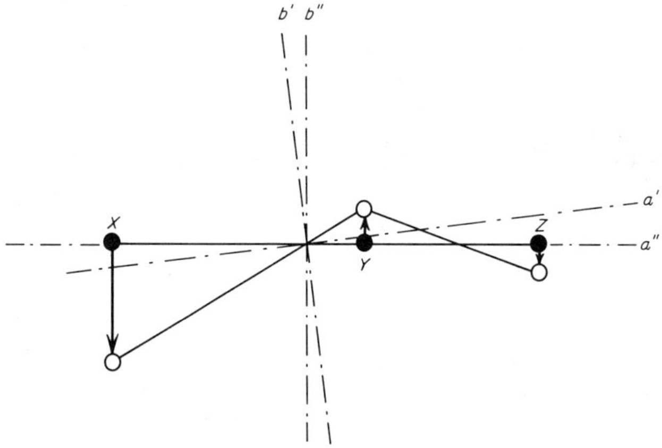


FIG. 89. **Change of axis system in a bent – linear transition of an XYZ molecule.** The black circles indicate the equilibrium positions of the nuclei in the lower, the open circles the equilibrium positions in the upper state. The arrows indicate the initial motions of the nuclei immediately after the quantum jump.

Another cause for the occurrence of transitions with unorthodox $\Delta K$ values has recently been discussed by Hougen and Watson (580). It is briefly referred to as *axis switching*: According to the elementary form of the Franck–Condon principle, if the molecule goes from a state with linear equilibrium conformation to one with a bent conformation it will start out to oscillate from the linear conformation about the new equilibrium position. This is shown schematically in Fig. 89. It is clear however that the principal axes of inertia in the upper state ($a'$, $b'$, $c'$) do not in general "coincide" with those in the lower state ($a''$, $b''$, $c''$). In the example of Fig. 89, if the amplitude of the bending vibration in the bent form (open circles) is increased until the turning points of the motion for the three atoms are on a straight line, there is no reason of symmetry why this line should coincide with the inertial axis $a'$ of the bent form. But the straight line in question according to Franck's principle must correspond to the initial conformation immediately after the "quantum jump" from the lower (linear) state. Only if there is sufficient symmetry will the two axis systems coincide, for example, for a molecule that belongs to point group $\boldsymbol{D}_{\infty h}$ in the ground state and $\boldsymbol{C}_{2v}$ in the excited state. But in all unsymmetrical cases as for example, for bent – linear transitions of an XYZ molecule (see Fig. 89) or even of a symmetrical $X_2Y_2$ molecule if the point group of the excited state is $\boldsymbol{C}_{2h}$, axis switching takes place. The angle between the two axis systems is usually quite small, even in extreme cases less than 10°. However, because of the difference in the axis system for the rotational eigenfunctions the matrix elements for transitions with $\Delta K \neq \pm 1$ for a $\perp$ band and $\Delta K \neq 0$ for a $\parallel$ band are no longer zero even if the molecule in the excited state is very close to a symmetric top. Thus forbidden sub-bands with unorthodox $\Delta K$ values may be expected to occur. More detailed calculations show that the intensity formulae for these forbidden sub-bands contain in addition to the terms of the normal (Hönl–London) formulae a factor $J^2$ if $\Delta K$ differs by 1 from the orthodox values and a factor $J^4$ if the difference is 2.

Examples for the effect of axis switching have been found by Watson (1279) in the near ultraviolet absorption spectrum of $C_2H_2$. Already in the early work of Ingold and King (601) and Innes (605) a few sub-bands with $\Delta K = 0$ and $+2$ were found while most of the bands have $\Delta K = \pm 1$. Watson established that these additional sub-bands have very small intensity at low $J$ as required by the axis switching mechanism. In addition he observed some transitions with $\Delta K = 3$ ($\Phi - \Sigma$ type sub-bands) in which the dependence on $J$ is still more pronounced. Watson has also confirmed the prediction that in the forbidden $\Delta K = 0$ sub-bands of a $\perp$ transition the $P$ and $R$ branches are weak and have zero intensity in a $\Sigma - \Sigma$ type sub-band, i.e., only the $Q$ branch occurs (as Fig. 81 shows, the $P$ and $R$ branches would violate the parity rule). In Fig. 83, which shows the $1-0$ main band of $C_2H_2$, the $Q$ branch of the forbidden $\Sigma - \Sigma$ type sub-band is visible; it is weak for low $J$ but comparable in intensity to the $P$, $Q$, $R$ branches of the main band for high $J$. Here, as in all other forbidden sub-bands, the observed relative intensities fit in well with theoretical expectation. The angle between the two axis systems in $C_2H_2$ is only 2.5°. The non-observation of similar transitions in HCN is accounted for, since the angle is smaller (1.5°) and since the relative intensity depends on the square of this angle.

**Linear – bent transitions.** The difference between transitions in which the molecule is linear in the lower state and those in which it is linear in the upper state (and non-linear in the lower) is mainly that in the latter, even for cold absorption, in the lower state a number of levels of different $K$ are populated while in the former only one such level $K = l = 0$ is present. Since according to the Franck–Condon principle, in linear – bent transitions in the excited state the higher vibrational levels of the bending vibration (each with several $l'$ values) are excited, each vibrational band will consist of a number of sub-bands even at low temperature. The energy level diagram Fig. 90a illustrates this situation for a $\parallel$ transition, Fig. 90b for a $\perp$ transition, assuming that the excited state is a $\Sigma$ electronic state. In the excited state only the levels $v' = 5, 6, 7$ are shown. There is an alternation of odd and even $l$ values for odd and even $v$. As is seen directly from the diagram, the number of sub-bands of a $\parallel$ band equals the number of $l'$ values for each $v'$, (in the example of Fig. 90a, three for $5-0$, four for $6-0$ and $7-0$). For a $\perp$ band the number of sub-bands is twice as large since $\Delta K$ can be both $+1$ and $-1$ (except when $K = 0$ in the ground state); thus the $5-0$ band consists of six, the $6-0$ band of seven, the $7-0$ band of eight sub-bands.

In the excited electronic state the vibrational levels of a bending vibration

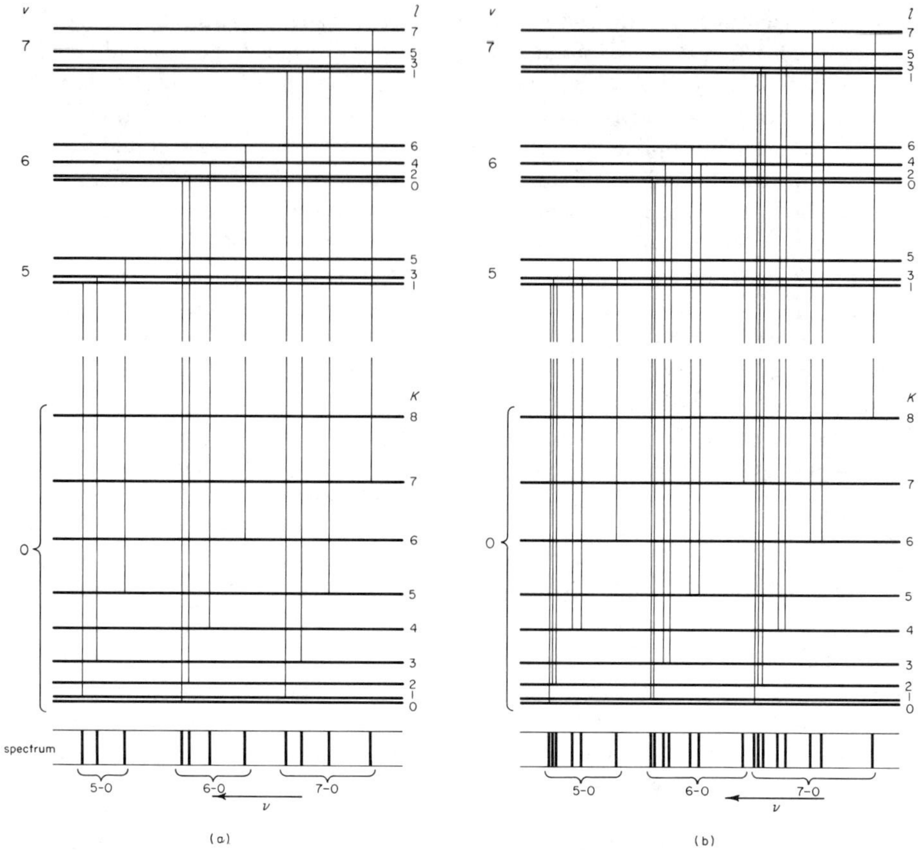


FIG. 90. **Energy level diagram for a progression in the bending vibration of a linear – bent transition (*a*) for a ∥ transition, (*b*) for a ⊥ transition.** Only one vibrational level ($v = 0$) of the lower state is shown (cold absorption). At the bottom schematic spectra are drawn.

are normally represented by

$$\omega_i(v_i + 1) + x_{ii}(v_i + 1)^2 + g_{ii}l^2.$$

In other words for a given $v_i$ the energy depends quadratically on $l$, just as in the ground state it depends quadratically on $K$. Therefore the coarse structure of the resulting bands will be very similar to that of ∥ or ⊥ bands of a symmetric top (see the next subsection). However here in general the coefficients of the quadratic terms in the upper and lower state, $g_{ii}$ and $A - \frac{1}{2}(B + C)$, are quite different in magnitude and consequently there will be in general a very strong divergence in the series of sub-bands for each $v'$ (see the schematic spectra at the bottom of Fig. 90). In addition, unlike the case of the symmetric top, the number of sub-bands is strictly limited by the $v'$ value in the upper state and may be used to determine this $v'$ value. Furthermore sub-bands with even or odd $K''$ values are missing for odd or even $v'$ respectively if $\Delta K = 0$, and for even or odd $v'$ respectively if $\Delta K = \pm 1$. As a result of this alternation we expect an apparent staggering in a progression of the bending vibration because the first sub-bands in successive vibrational transitions come alternately from $K = 0$ and $K = 1$. For hybrid bands the two band structures ($\|$ and $\perp$) are simply superimposed.

If the deviation from a symmetric top (now in the lower state) is large, transitions with $\Delta K = \pm 2, \pm 4, \ldots$ may occur very weakly in $\|$ bands, and transitions with $\Delta K = \pm 3, \pm 5$ in $\perp$ bands in addition to the main transitions with $\Delta K = 0$ and $\pm 1$. Even for a nearly symmetric top, just as for bent–linear transitions, sub-bands with unorthodox $\Delta K$ values may also be produced by axis switching if the location of the principal axes in the lower state is not determined by symmetry alone (see p. 208).

The fine structure of the sub-bands is determined by the selection rules (II, 55) to (II, 57). It is entirely similar to that of bent–linear transitions. However, here higher $K$ and $l$ values occur and therefore more and more lines are missing at the beginning of the branches, since $J \geq K$. Conversely from the number of missing lines the $K$ values can be ascertained and therefore it can be established whether or not alternate sub-bands are missing, that is, whether or not the molecule is linear in the excited state. At the same time the values of the vibrational quantum number $v$ in the excited state can be determined from the $K$ values that occur.

While no example of a linear–bent transition with a $\Sigma$ upper state has yet been completely resolved, Ritchie and Walsh (1071) have resolved the $K$ structure of such a transition in the ultraviolet absorption spectrum of $NO_2$. There are several progressions in the bending vibration ($\nu_2' = 600\ \text{cm}^{-1}$) all of which show a pronounced staggering. Each band consists of three main peaks but the spacings between these peaks are not the same in successive bands but alternate between 21 and 64 $\text{cm}^{-1}$ for one group, and 42 and 85 $\text{cm}^{-1}$ for the next group, that is, alternate in the ratio 4:12 and 8:16 respectively. This is precisely what is expected in a case like that shown in Fig. 90a, assuming that these groups correspond alternately to $K = 0, 2, 4$ and $K = 1, 3, 5$. We can conclude therefore that the molecule is linear in the excited state and that the transition is ${}^2\Sigma^+ - {}^2A_1$. The constant factor in the ratios, i.e. 5.3 $\text{cm}^{-1}$, gives the difference between $A - \bar{B}$ of the lower and $g_{22}$ of the upper state. An example of a $\perp$ transition as in Fig. 90b has not yet been observed.

If the upper electronic state is a $\Pi$ or $\Delta$ state we must take vibronic interactions and the resulting Renner–Teller splittings into account. As we have seen in Chapter I, section 2b, each single level of the upper state in Fig. 90 is then split into two levels with $K = |l \pm \Lambda|$, except when $l = \Lambda$ or $l = 0$ in which case there are three or one component levels respectively. If the upper state is $\Pi_u$ (of $\boldsymbol{D}_{\infty h}$) and the lower state $A_1$ (of $\boldsymbol{C}_{2v}$) we have a $\perp$ transition with $\Delta K = \pm 1$. In Fig. 91 an energy level diagram for this case is shown assuming very small Renner–Teller splitting ($\epsilon = 0.02$). For such a small splitting the effect of anharmonicity cannot be neglected and has been taken into account following Hougen and Jesson (579).

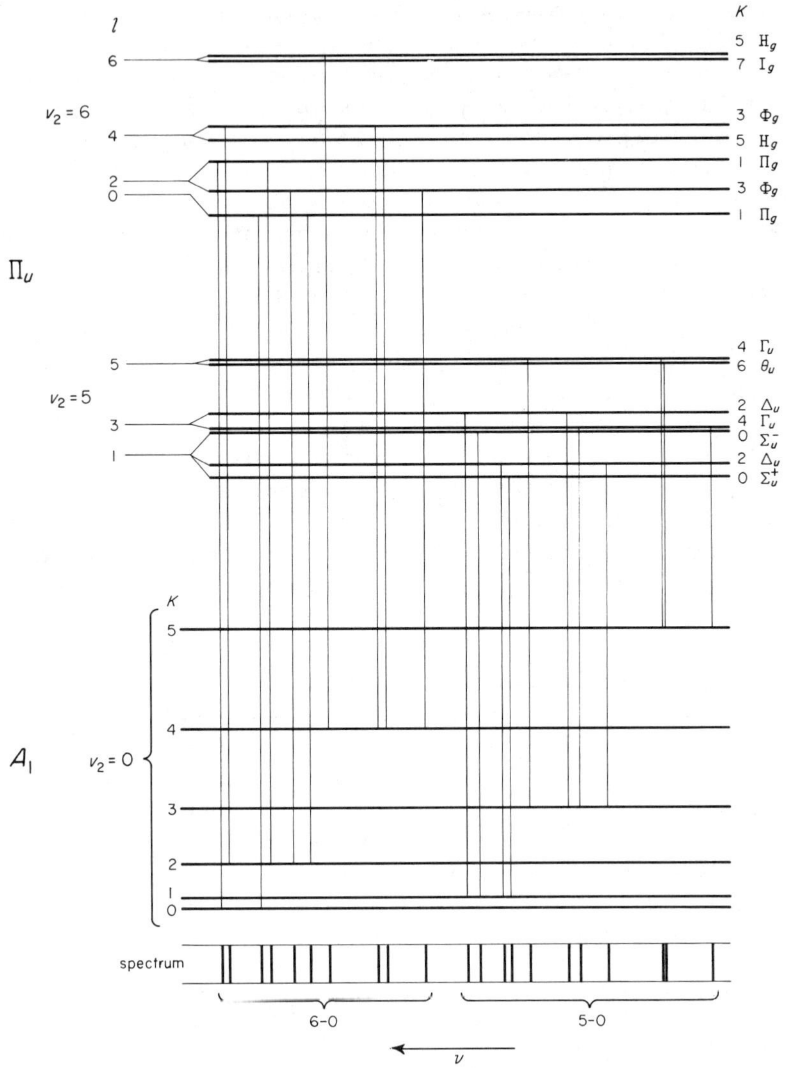


FIG. 91. **Energy level diagram for a linear – bent transition when the upper state is an electronic Π state with small but non-zero Renner–Teller splitting.** The anharmonicity constant $g_{22}$ has been assumed to be fairly large in the upper state. The Renner–Teller splitting is smaller than the anharmonicity splitting. The formulae of Hougen and Jesson (579) were used for the levels.

There are now almost twice as many sub-bands as for a Σ upper state (Fig. 90b). As before, sub-bands with even or odd $K''$ are missing depending on whether $v'$ is odd or even. No examples of this type have yet been studied.

The evaluation of the rotational constants in the upper and lower states of linear – bent transitions proceeds of course in much the same way as for bent –

linear transitions. Thus the effective $B$ value for the lower state gives essentially $\frac{1}{2}(B + C)$ while the $K$-type doubling for $K'' = 1$ gives $\frac{1}{2}(B - C)$ with appropriate corrections when the molecule is a strongly asymmetric top [see Chap. I, section 3d]. Since all $K''$ values are represented in each lower vibrational level the rotational constants $A''_v$ can be somewhat more readily determined than for bent – linear transitions observed in absorption; we only need to take appropriate differences of sub-band origins $\nu_0(K' - K'')$. For example (see Fig. 90b), neglecting centrifugal stretching and higher asymmetry terms,

$$\nu_0(1-0) - \nu_0(1-2) = F''(K = 2) - F''(K = 0) = 4[A'' - \tfrac{1}{2}(B'' + C'')]$$
$$\nu_0(2-1) - \nu_0(2-3) = F''(K = 3) - F''(K = 1) = 8[A'' - \tfrac{1}{2}(B'' + C'')].$$

Here the second relation, because of the limitation of the $K$ values in the upper state, cannot be formed from the same band as the first.

For the quartic terms in the energy formulae, asymmetry terms become very important even for slightly asymmetric tops. A determination of centrifugal stretching constants is therefore quite difficult. However, this difficulty does not greatly affect the evaluation of the main rotational constants.

**Linear – bent transitions between Renner–Teller states.** As has been mentioned in Chapter I, section 2b, several cases of "linear" molecules are known where the Renner–Teller splitting is so large that in effect two separate electronic states result: a higher one in which the equilibrium conformation is linear (or nearly linear) and a lower one in which it is bent (see Fig. 4). Transitions between these two Renner–Teller states can occur and are of the linear – bent type. The transition moment for these transitions is necessarily perpendicular to the plane of the molecule in the lower state. They have the same coarse structure as described earlier for linear – bent transitions with upper $\Sigma$ state (Fig. 90b), except that now $v'$ is one or two less than the highest $K'$ in each band (see Fig. 7), depending on whether the "original" state from which both upper and lower state are derived is $\Pi$ or $\Delta$ (since $K'_{\max} = |l' + \Lambda|$).

An especially simple example is the red absorption system of HCO [Herzberg and Ramsay (538)]. Here all the vibronic levels with $K' > 0$ are strongly predissociated and only the sub-bands with $K' = 0$ are observed to have a discrete structure. Figure 92 shows one of these bands with its simple $P, Q$ and $R$ branches. The observed bands form a simple progression but, if this is indeed a linear – bent transition, they represent only alternate bands since only alternate vibrational levels have $K' = 0$, i.e. are of the $\Sigma$ type (see Fig. 7). Actually intermediate bands with $K' = 1$ have been found but they are very diffuse [see Johns, Priddle and Ramsay (638)]. The simple structure of the discrete bands might lead one to conclude that the molecule is linear in both excited and ground state but the considerable combination defect between $Q$ and $P$, $R$ branches, which is the same for all bands of the progression and is larger for DCO than for HCO, proves beyond doubt that the molecule is non-linear in the ground state.

A more general example is provided by the red absorption spectrum of $NH_2$ (the so-called $\alpha$-bands of ammonia). Here the sub-band structure is not at all obvious because the moments of inertia are small, the rotational constants large, and because in the lower state the molecule is a strongly asymmetric top. Therefore the spectrum has the appearance of a many-line spectrum and only after considerable effort was it finally analyzed in detail by Ramsay (1041) and Dressler and Ramsay (308).

A very similar spectrum was later found and analyzed for $CH_2$ by Herzberg (521) and Herzberg and Johns (530). Since this spectrum does not have the

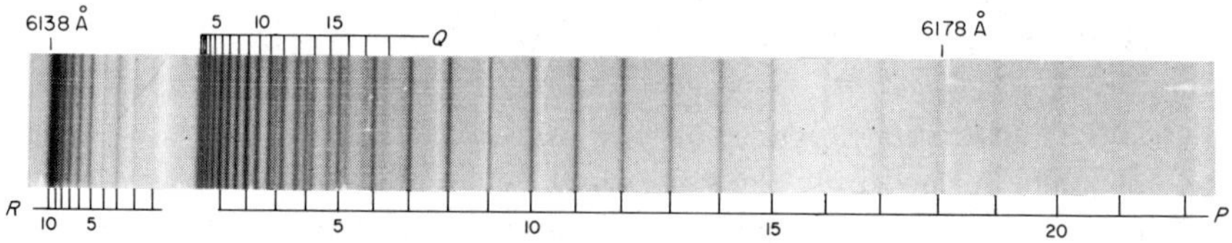


FIG. 92. **Spectrogram of the 090 − 000 band of the $\tilde{A}-\tilde{X}$ system of HCO at 6138 Å.** The $P$, $Q$ and $R$ branches are marked. They are numbered according to the $N$ values in the lower state (with $K = 1$). Each line is an unresolved doublet.

complication of spin-doublet structure exhibited by $NH_2$, we use it for further illustration. Figure 93 shows the most intense part of the 14 − 0 band. Using a diagram similar to Fig. 90 but taking account of Fig. 7 in the upper state and applying the selection rule $\Delta K = \pm 1$ we find that the complete 14 − 0 band would have eleven main sub-bands, five with $\Delta K = +1$ and six with $\Delta K = -1$. In addition there could be much weaker sub-bands with $\Delta K = \pm 3$ and possibly higher $\Delta K$ values. Actually, thus far only five sub-bands, two with $\Delta K = +1$ and three with $\Delta K = -1$, have been found. Of these, three are to be seen in Fig. 93. Each sub-band, except the one with $K' = 0$, has six branches, two $P$, two $R$ and two $Q$ branches (on account of asymmetry doubling) while the sub-band with $K' = 0$ has only three, one $P$, one $Q$ and one $R$ branch. As an illustration, in Table 12 the wave numbers of the lines in the observed branches of the 14 − 0 band are given. The designation of the branches follows the usual rules: the left superscripts of the symbols $P$, $Q$, $R$ give the $\Delta K$ value: $p$ means $\Delta K = -1$, $r$ means $\Delta K = +1$. The first subscript gives the $K_a$ value of the lower state, the second subscript the $K_c$ value (see p. 106). The latter distinguishes the two $K$-doubling components for a given $K_a$. Thus the three branches of the band with $K' = 0$ and $K'' = 1$ are ${}^pR_{1,J-1}(J)$, ${}^pQ_{1,J}(J)$ and ${}^pP_{1,J-1}(J)$, the six branches for the transition $K' = 2 \leftarrow K'' = 1$ are ${}^rR_{1,J-1}(J)$, ${}^rR_{1,J}(J)$, ${}^rQ_{1,J-1}(J)$, ${}^rQ_{1,J}(J)$, ${}^rP_{1,J-1}(J)$, ${}^rP_{1,J}(J)$ and similarly for other sub-bands (see Figs. 81 and 107). Alternatively, the individual lines are simply described by the $J_{K_aK_c}$ values of upper and lower state, e.g., the line ${}^pQ_{1,J}(3)$ would be described as $3_{03} - 3_{13}$.

The difficulties in the analysis of the bands of both $NH_2$ and $CH_2$ have been: (1) the branches are short since the rotational constants are large, (2) the $K$-type splitting for $K'' = 1$ is so large (since the molecule in the lower state is a strongly asymmetric top) that pairs of branches distinguished only by the $K$-doublet component to which they belong are difficult to recognize (see Fig. 93), (3) the six

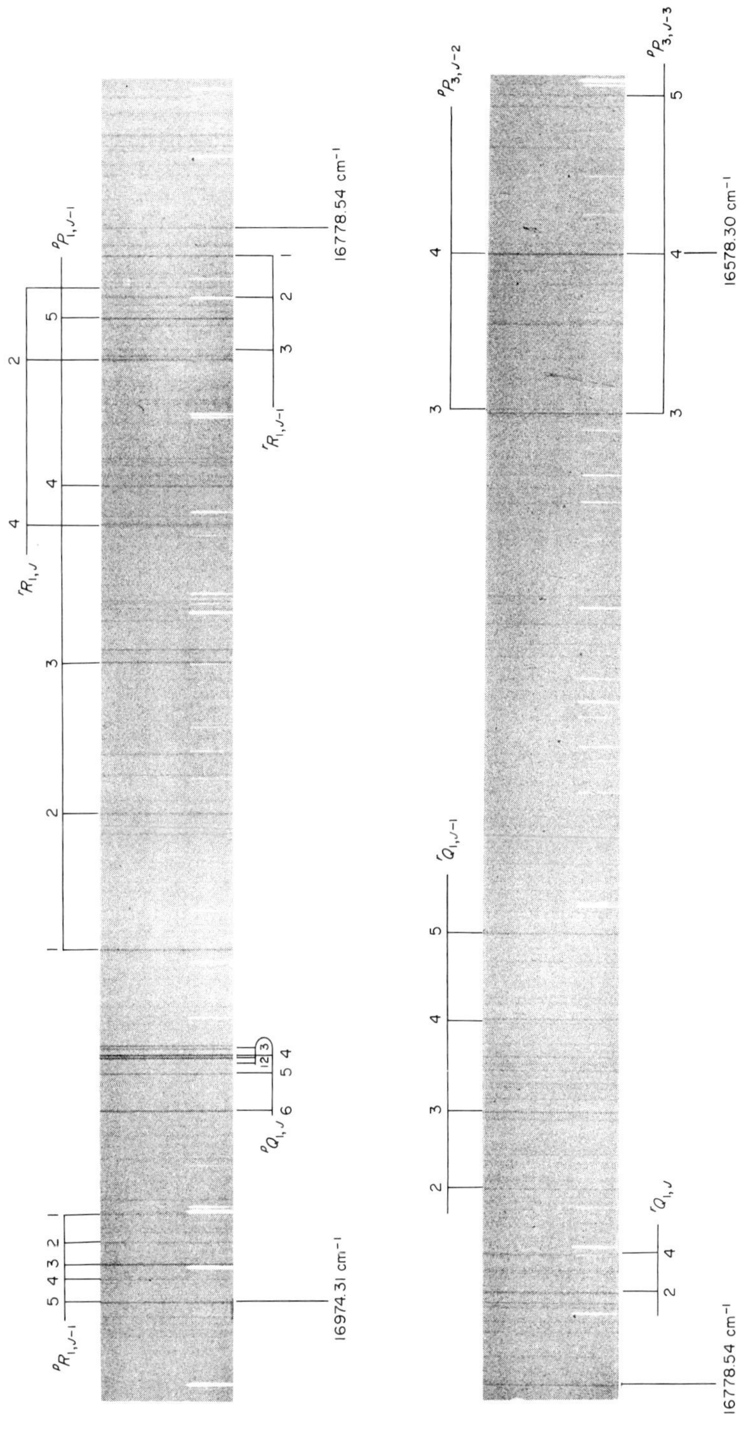


FIG. 93. **Spectrogram of the 0 1 4 0 − 0 0 0 band of the red $CH_2$ system at 5900 Å.** Only branches of the principal sub-bands of type $\Sigma-\Pi$, $\Delta-\Pi$ and $\Delta-\Phi$ are shown. In the branch symbols the superscripts $p$ and $r$ give $l' - K''_a$, the first subscript gives $K''_a$.

branches of a given sub-band may have very different intensities, in particular the branches for which $\Delta K$ and $\Delta J$ have opposite sign are in general very weak and often not observed (see Table 12), (4) because of the deviations from a symmetric top the cubic and quartic terms in the branch formulae may be relatively large thus making it difficult to recognize the branches, a difficulty that is increased because of the small number of lines in each branch.

TABLE 12. FINE STRUCTURE OF THE 0,14,0 – 0,0,0 BAND OF THE RED SYSTEM OF $CH_2$

| | 0 – 1 sub-band | | | 2 – 1 sub-band | | | |
|---|---|---|---|---|---|---|---|
| $J$ | ${}^pR_{1,J-1}$ | ${}^pQ_{1,J}$ | ${}^pP_{1,J-1}$ | ${}^rR_{1,J-1}$ | ${}^rR_{1,J}$ | ${}^rQ_{1,J-1}$ | ${}^rQ_{1,J}$ |
| 0 | | | | | | | |
| 1 | 16957.69 | 16930.37 | 16909.39 | 16783.57 | 16786.76 | | |
| 2 | 963.20 | 929.33 | 884.47 | 790.88 | 802.24 | 16743.52 | 16755.27 |
| 3 | 967.25 | 927.65 | 856.96 | 800.38 | 812.82 | 729.77 | |
| 4 | 970.00 | 928.79 | 825.05 | | 831.98 | 710.21 | 761.97 |
| 5 | 974.31 | 932.19 | 794.80 | | | 697.98 | |
| 6 | | 939.01 | 767.31 | | | | |
| 7 | | | 743.52 | | | | |

| | 2 – 3 sub-band | | | | 4 – 3 sub-band | | | |
|---|---|---|---|---|---|---|---|---|
| | ${}^pQ_{3,J-3}$ | ${}^pQ_{3,J-2}$ | ${}^pP_{3,J-3}$ | ${}^pP_{3,J-2}$ | ${}^rR_{3,J-3}$ | ${}^rR_{3,J-2}$ | ${}^rQ_{3,J-3}$ | ${}^rQ_{3,J-2}$ |
| 0 | | | | | | | | |
| 1 | | | | | | | | |
| 2 | | | | | | | | |
| 3 | 16652.93 | | 16606.10 | 16605.38 | 16338.36 | 16338.64 | | |
| 4 | | 16648.85 | | 578.30 | | 354.31 | | 16263.64 |
| 5 | | | 550.71 | | | | 16256.21 | |
| 6 | | | | 514.30 | | | | |

In addition, two lines of the 4 – 5 sub-band have been identified.

In spite of these difficulties, by a systematic application of the method of constant differences it has been possible to analyze a fairly large number of $NH_2$ and $CH_2$ bands. There are two types of combination relations that are important for the analysis of such complicated bands (see Fig. 107 below):

(1) Combination relations within a given band or even a given sub-band. For example within the $K' = 2 \leftarrow K'' = 1$ sub-band we have (assuming that the doubling for the $K' = 2$ levels is negligible)

$$ {}^rR_{1,J-1}(J) - {}^rQ_{1,J-1}(J+1) = {}^rQ_{1,J-1}(J) - {}^rP_{1,J-1}(J+1), \qquad \text{(II, 59)} $$

and similar relations for the other component of the $K'' = 1$ level as well as corresponding relations in other sub-bands. Different sub-bands with one state in common, e.g. $K' = 2 \leftarrow K'' = 1$ and $K' = 2 \leftarrow K'' = 3$, must obey combination relations such as

$$ {}^rR_{1,J-1}(J) - {}^rQ_{1,J-1}(J) = {}^pQ_{3,J-2}(J+1) - {}^pP_{3,J-2}(J+1), \qquad \text{(II, 60)} $$

which may also be written

$$^{r}R_{1,J-1}(J) - {}^{p}Q_{3,J-2}(J+1) = {}^{r}Q_{1,J-1}(J) - {}^{p}P_{3,J-2}(J+1) \qquad \text{(II, 61)}$$

and similar relations between other pairs of sub-bands.

(2) Combination relations between different bands. For example, the values of the differences (II, 59) and (II, 61) must, of course, recur in other bands with the same lower state while differences like (II, 60) must recur in other bands with the same upper state. In addition, there are differences like

$$^{r}R_{1,J}(J-1) - {}^{p}P_{3,J-2}(J+1)$$

and

$$^{r}R_{1,J-1}(J-1) - {}^{p}P_{3,J-3}(J+1)$$

which must recur in all bands with the same lower state.

Table 13 gives a few examples of the agreement of combination differences in the red $CH_2$ bands. Using these and other similar differences it was established that a given band contains only either even or only odd $K'$ values and that even and odd $K'$ values alternate in successive bands of the main progression. This observation, according to the discussion above, represents a clear demonstration that the $CH_2$ molecule must be linear or nearly linear in the upper state of the red bands while the presence of several successive $K''$ values in the lower state and the large $K$-type doubling particularly for $K'' = 1$ prove that the molecule is strongly bent in the lower state.

For a $\Sigma$ electronic state as upper state the contribution of $l_2$ to the energy in a triatomic molecule would be

$$g_{22}l_2{}^2$$

where $g_{22}$ is of the order of the anharmonic constants (e.g. $g_{22} = +1.7$ cm$^{-1}$ for the ground state of $CO_2$) and is usually positive. For $CH_2$ the value of the coefficient of $K'^2$ (which would be that of $l'^2$ if the upper state were a $\Sigma$ state) is found to be $-16$ cm$^{-1}$. Such a large negative value of this coefficient seems impossible for a $\Sigma$ electronic state but is precisely what one would expect from the theory of the Renner–Teller effect (see p. 30f) for the upper component of a $\Pi$ or $\Delta$ electronic state of a linear molecule. The present observations on $CH_2$ are insufficient to decide whether the upper (and lower) state is derived from a $^1\Pi$ or a $^1\Delta$ state, but it is readily seen from the electron configuration (see Chap. III) that only a $^1\Delta$ state is available. For an experimental proof of this conclusion it would be necessary to establish what the $K'$ values are for the lowest $v'$ values.

A more detailed study of the lower vibrational levels of the excited electronic states of both $NH_2$ [Dixon (286a)] and $CH_2$ [Herzberg and Johns (530)] shows that they do not follow exactly a simple vibrational formula of the type

$$\omega_2(v_2+1) + x_{22}(v_2+1)^2 + g_{22}l^2$$

but that deviations occur of the type expected for molecules with a slight potential hump at the linear conformation (see Chap. I, section 3(e)): If the $\Delta G$ values of the

Σ vibronic levels are plotted as a function of the vibrational energy they are found to have a minimum at low $v_2$ values as expected for quasi-linear molecules (see Fig. 46, p. 122). In addition the vibronic Π levels for low $v_2$ approach the Σ levels rather than lie halfway between successive Σ levels (see Fig. 47). Dixon has estimated that the equilibrium angle for $NH_2$ is about 144° and the barrier

TABLE 13. COMBINATION DIFFERENCES IN THE RED $CH_2$ BANDS [AFTER HERZBERG AND JOHNS (530)]

| $J$ | ${}^pR_{1,J-1}(J) - {}^pQ_{1,J}(J+1)$ | | ${}^pQ_{1,J}(J) - {}^pP_{1,J-1}(J+1)$ | |
|---|---|---|---|---|
| | 080 – 000 | 0140 – 000 | 080 – 000 | 0140 – 000 |
| 1 | 28.21* | 28.36 | 44.65 | 45.90* |
| 2 | 35.50 | 35.55 | 72.48* | 72.37 |
| 3 | 38.41* | 38.46 | 102.51 | 102.60 |
| 4 | 37.89* | 37.81 | 134.01* | 133.99 |
| 5 | 35.40* | 35.30 | 164.91* | 164.88 |
| 6 | | | 195.83* | 195.49 |
| $J$ | ${}^rR_{1,J-1}(J-1) - {}^pP_{3,J-3}(J+1)$ | | ${}^rR_{1,J}(J-1) - {}^pP_{3,J-2}(J+1)$ | |
| | 080 – 000 | 0140 – 000 | 080 – 000 | 0140 – 000 |
| 2 | 177.45 | 177.47 | 181.50 | 181.38 |
| 3 | 212.48 | | 224.00 | 223.94 |
| 4 | 249.66 | 249.67 | 271.61 | |
| 5 | | | 317.66 | 317.68 |

* An asterisk indicates the use of a blended line for the particular combination difference.

of the order of 800 $cm^{-1}$. For the higher vibrational levels of the excited state both $CH_2$ and $NH_2$ behave as if they were linear and therefore it is justified to consider the spectra as linear – bent transitions. The spectrum of $BH_2$ recently observed by Herzberg and Johns (531) represents a purer case of a linear – bent transition in which no potential hump is detectable in the excited state.

**Emission spectra.** In the preceding discussion of bent – linear and linear – bent transitions we have assumed them to be observed in absorption. The analogue of cold absorption is emission from the lowest vibrational level of the excited state. Here, when the molecule is non-linear in the upper state the band structure is entirely similar to that in the absorption spectrum when the molecule is non-linear in the lower state, and, conversely, the emission from an excited state in which the molecule is linear and in which it is in the lowest vibrational level, gives rise to a spectrum that is similar in every way to the absorption spectrum when the molecule is linear in the lower state. If in emission several vibrational levels are excited in the upper state the band structures are similar to the corresponding case of hot absorption.

Two band systems of the type here under consideration have been observed in emission: the $NH_2$ bands discussed earlier occur in emission in various flames as well as discharges and have also been observed in the spectra of comets. The only difference from the absorption spectrum lies in the occurrence of bands for which in the lower state one or more vibrations are singly or multiply excited. The second band system is that of the carbon monoxide flame bands which have been a puzzle for many decades, but which have recently been shown by Dixon (283) to be due to a bent – linear transition of $CO_2$. Here all observed transitions come from the lowest two vibrational levels of an excited ($B_2$) state in which the molecule is strongly bent ($\theta = 122°$) while the lower levels are highly excited vibrational levels of the ground state in which the molecule is linear. The characteristic alternation of even and odd sub-bands in successive bands of a progression in $v_2$ appears, but the vibrational structure is complicated by the presence of Fermi resonance. The transition is of the $\|$ type (see Fig. 90a), that is, $K' = l''$ and bands with $l''$ values up to 4 have been assigned. The determination of $A - \bar{B}$ in the excited state, just as for the $CS_2$ absorption spectrum, cannot be made directly (since $\Delta K = 0$) but requires a knowledge of the energy differences between levels with different $l$ in the lower state. For $CO_2$ these energy differences can be extrapolated from infrared data [Courtoy (246)]. The resulting rotational constants of the upper state are given in Table 64 of Appendix VI.

**Multiplet splitting.** If the resultant spin of the electrons in one or both of the electronic states is not zero and if spin-orbit coupling is not negligibly small, a multiplet structure of each of the rotational lines of the preceding treatment will occur. As we have seen in Chapter I, section 3, for non-linear conformations of a polyatomic molecule, the multiplet splitting is in general small and is of the same type as in Hund's case (b) for diatomic molecules. For linear conformations the multiplet splitting may be large or small depending on whether Hund's case (a) or case (b) respectively applies (we disregard case (c) for the present discussion). Thus for bent – linear and linear – bent transitions we may have case (b) – case (a) and case (a) – case (b) transitions or case (b) – case (b) transitions. If in the linear conformation case (a) applies, we have to consider separately the combination of each multiplet component of this state with the non-linear state according to the selection rules (II, 55) to (II, 57). Since no example of such a case is known we shall not discuss it further.

For case (b) – case (b) transitions the quantum number $N$ of the angular momentum apart from spin is well defined and for it the selection rule is (see Vol. I, p. 244)[5]

$$\Delta N = 0, \pm 1 \qquad \text{(II, 62)}$$

with the restriction [see (eq. II, 57)]

$$\Delta N = 0 \text{ is forbidden for } K' = K'' = 0. \qquad \text{(II, 62a)}$$

In other words, if the multiplet splitting is neglected we can use $N$ in place of $J$ and everything is the same as for singlets. If one wants to consider the multiplet

[5] Note that in Volume I, $K$ was used instead of $N$.

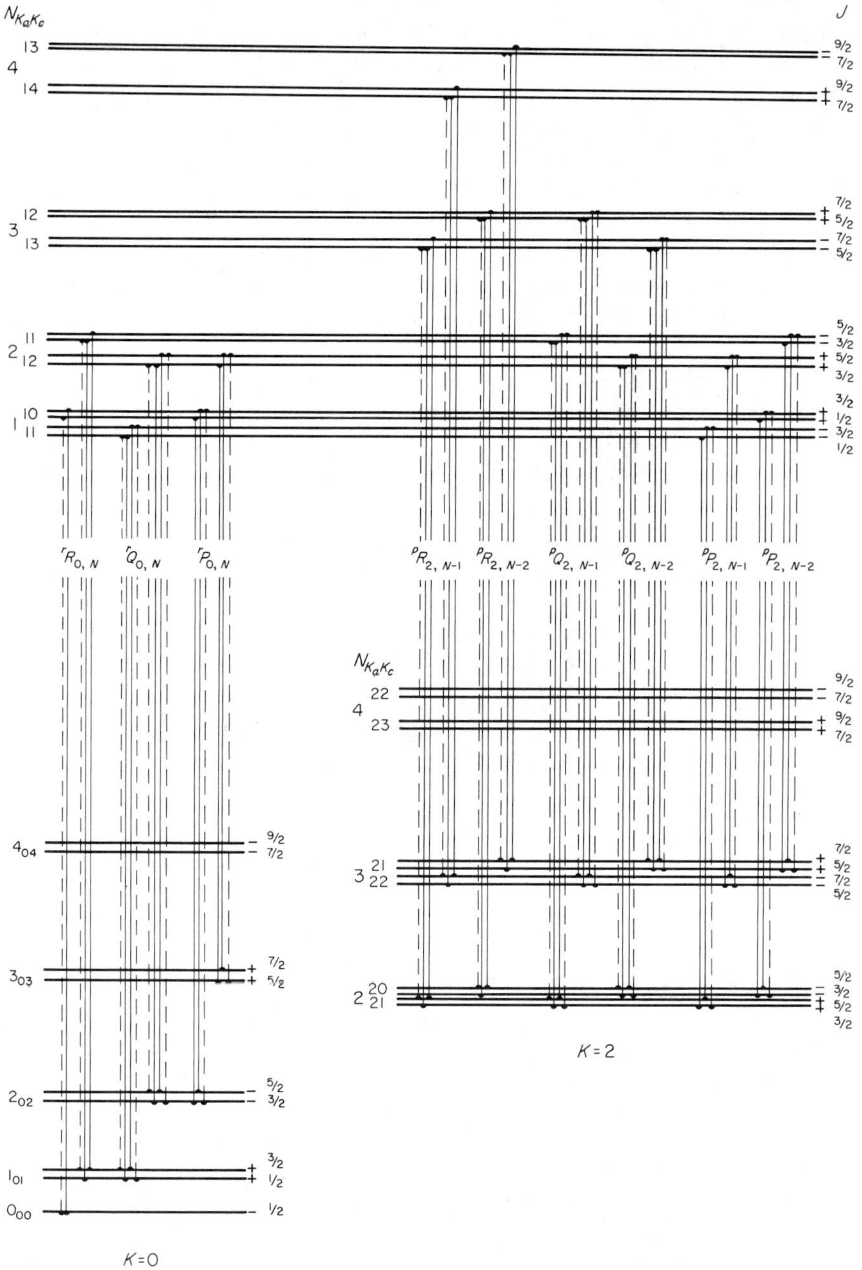


FIG. 94. **Energy level diagram showing spin doubling in the K = 1 ↔ K = 0 and K = 1 ↔ K = 2 sub-bands of a linear – bent transition.** In the excited state $K_a$ and $K_c$ values are formally assigned to the levels ($K_a = K$, $K_c = N$, $N - 1$). Branches with $\Delta J = \Delta N$ (main branches) are indicated by solid lines, branches with $\Delta J \neq \Delta N$ (satellite branches) are indicated by broken lines.

splitting one has to apply the selection rule (II, 45) for $J$ as well as (II, 62) for $N$. As for diatomic molecules the strongest branches are those for which $\Delta J = \Delta N$ and therefore in a doublet or triplet transition in place of each single line of a singlet transition there are two or three strong components respectively. The splitting of these doublets or triplets is the difference or sum of the level splittings in the upper and lower states. The level splittings can be determined separately only if at least some lines with $\Delta J \neq \Delta N$ are observed.

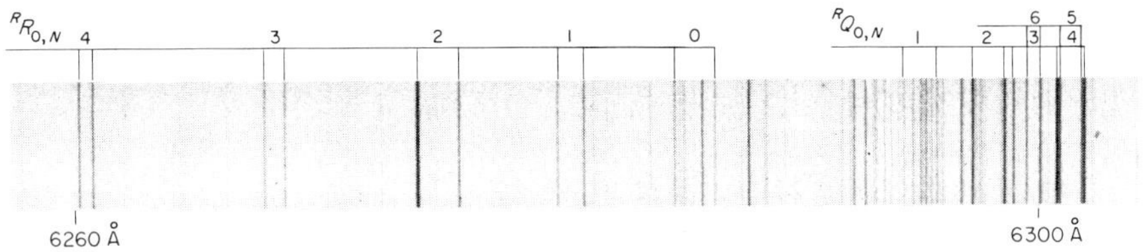


FIG. 95. **Spectrogram of the K = 1 ← K = 0 sub-band of the 6300 Å band of $NH_2$ showing spin doubling [after Dressler and Ramsay (308)].** Only the main branches ($\Delta N = \Delta J$) are marked. For each $N$ value there are two lines whose spacing corresponds to the difference of the spin doubling in upper and lower state.

As an example Fig. 94 gives an energy level diagram for the $K = 1 \leftrightarrow K = 0$ and $K = 1 \leftrightarrow K = 2$ sub-bands of a linear–bent transition with spin doubling. The origin of the main branches ($\Delta J = \Delta N$, solid lines) and of the satellite branches ($\Delta J \neq \Delta N$, broken lines) is clearly shown. The figure is drawn approximately to scale for the $NH_2$ 080 ← 000 band at 6300 Å. Figure 95 shows a spectrogram of the $K = 1 \leftarrow K = 0$ sub-band, in which the spin doublets are well resolved. It is from this and similar spectra that the spin splittings of the lower state represented in Fig. 43 have been obtained.

**Forbidden transitions.** For triplet–singlet transitions (or generally *intercombinations*) the spin fine structure of a bent–linear or linear–bent transition would be expected to be similar to corresponding linear–linear transitions. Also the Zeeman splitting in a magnetic field would be expected to be similar, that is, unlike singlet–singlet transitions it would be of the order of the "normal" Zeeman splitting. However, it has been recognized recently [Herzberg (523), Hougen (574)] that the selection rule for $K$ [corresponding now to the component of $\mathbf{N}$, not $\mathbf{J}$, in the top axis] is changed to

$$\Delta K = 0, \pm 1, \pm 2 \qquad \text{(II, 63)}$$

with certain restrictions (see section 3b) and thus additional sub-bands with unorthodox $\Delta K$ values may occur. In addition, in Hund's case (b), the selection rule for $N$ is changed to

$$\Delta N = 0, \pm 1, \pm 2 \qquad \text{(II, 64)}$$

making possible the occurrence of $S$- and $O$-form branches in the sub-bands[6].

[6] For doublet–quartet transitions the selection rules would be $\Delta K = 0, \pm 1, \pm 2, \pm 3$ and $\Delta N = 0, \pm 1, \pm 2, \pm 3$.

There are thus the following criteria for recognizing triplet – singlet transitions of the bent – linear or linear – bent type:

(a) a fine structure of individual "lines"
(b) a large Zeeman splitting in a magnetic field
(c) the occurrence of unorthodox sub-bands with $\Delta K$ values not normally found for $\parallel$ or $\perp$ bands
(d) the occurrence of branches with $\Delta N = \pm 2$ [assuming Hund's case (b)].

Thus far no case has been found in which all four criteria have been observed.

The only example of a triplet – singlet bent – linear transition known at present is provided by the near ultra violet absorption bands of $CS_2$. The simple structure of these bands was for a long time considered as evidence that they represent a singlet – singlet transition until Douglas (293) showed that there is an appreciable splitting in a magnetic field. Such a splitting can only be accounted for if the upper state is triplet. The apparent absence of a line splitting without magnetic field has recently been explained by Douglas and Milton (298) and Hougen (576) as due to a fairly large triplet splitting of the excited $^3A_2$ state of which only the $B_2$ component is observed. The low intensity of the transitions to the other two components ($A_1$ and $B_1$) at high $J$ is understandable on the basis of Hougen's theory.

Forbidden electronic transitions which occur on account of vibronic interaction have the same rotational structure as corresponding allowed transitions. For example, in a $^1A_2 - {}^1\Sigma_g^+$ bent – linear transition of an $XY_2$ molecule, which is forbidden by the dipole selection rules, vibronic transitions are possible from the 000 level of the ground state to those upper vibrational levels in which the antisymmetric stretching vibration is excited by odd quanta; since the vibronic species of these upper states is $^1B_1$ the corresponding bands would have the same fine structure as the bands of $^1B_1 - {}^1\Sigma_g^+$ electronic transitions.

A $^1A_2 - {}^1\Sigma_g^+$ electronic transition could also occur as a magnetic dipole transition (see p. 134f). The band structure would be entirely similar to that of a $^1A_1 - {}^1\Sigma_g^+$ electric dipole transition, that is, it would include the 0 – 0 band; but, as emphasized before, the intensity of such magnetic dipole transitions would in general be much smaller than that of the electric dipole transitions induced by vibronic interactions in the same electronic transition. No actual example of either type has been observed for linear – bent or bent – linear transitions.

### (b) Symmetric top molecules

We shall consider in this section the rotational structure of bands of only such molecules which at least in one of the two combining states are symmetric tops on account of symmetry. If in the second state the molecule is somewhat asymmetric and therefore not a genuine symmetric top, we shall assume in this sub-section that it is still sufficiently close to a symmetric top that the effect of asymmetry terms can be disregarded.

**Selection rules.** The selection rules for the rotational quantum number in electronic transitions of symmetric top molecules are the same as for infrared bands since according to (II, 15) they are determined by the same matrix elements of the direction cosines

$$\int \psi_r'^* \cos \alpha_x \psi_r'' d\tau_r, \quad \text{etc.}$$

as for infrared spectra [see eq. (IV, 56) of Vol. II]. Thus, when the transition moment of the electronic transition is parallel to the top axis (|| bands) we have

$$\begin{aligned} \Delta K &= 0, \quad \Delta J = 0, \pm 1 \quad \text{if } K \neq 0 \\ \Delta K &= 0, \quad \Delta J = \pm 1 \quad \text{if } K = 0. \end{aligned} \tag{II, 65}$$

On the other hand, when the transition moment is perpendicular to the top axis (⊥ bands) we have

$$\Delta K = \pm 1, \qquad \Delta J = 0, \pm 1. \tag{II, 66}$$

In addition, just as for linear molecules we have selection rules for the over-all symmetry properties. The selection rule for the +, − symmetry (symmetry or antisymmetry with regard to inversion of the over-all function) is the same as for linear molecules, i.e.

$$+ \leftrightarrow -, \quad + \not\leftrightarrow +, \quad - \not\leftrightarrow -. \tag{II, 67}$$

This rule is important only for planar molecules and those non-planar molecules for which inversion doubling is resolved. For all other molecules, since there are for each rotational level a + and a − sub-level of the same energy this rule can be disregarded.

Furthermore we have the selection rules for the over-all (rovibronic) species which are analogous to and just as strong as the $s \not\leftrightarrow a$ rule of linear molecules. If the over-all species is taken with regard to the rotational sub-group we have the rule, just as for infrared and Raman spectra (see Vol. II, p. 415), that the over-all species does not change, that is

$$A \leftrightarrow A, \; E \leftrightarrow E, \ldots, \; A \not\leftrightarrow E, \ldots. \tag{II, 68}$$

For allowed electronic transitions of symmetric top molecules this rule does not introduce any restrictions beyond those introduced by the other selection rules.

However, when the over-all species is taken with respect to the full symmetry group (see Chap. I, section 3) the selection rule is different and does introduce further restrictions. Hougen (571) finds that only those (rovibronic) transitions can occur for which the direct product of the over-all species of upper and lower state contains the species of the product

$$T_z R_z$$

of a translation along the symmetry axis and a rotation about this axis. For the point groups $\boldsymbol{C}_{3v}$, $\boldsymbol{C}_{4v}$, ... this product has species $A_2$, for $\boldsymbol{C}_{3h}$, $\boldsymbol{C}_{5h}$ ... species $A''$, for $\boldsymbol{D}_{3h}$, $\boldsymbol{D}_{5h}$, ... species $A_1''$, for $\boldsymbol{D}_{4h}$, $\boldsymbol{D}_{6h}$, ... and $\boldsymbol{D}_{3d}$, $\boldsymbol{D}_{5d}$, ... species $A_{1u}$ and for

$\boldsymbol{D}_{2d}$, $\boldsymbol{D}_{4d}$, ... species $B_1$. Thus for example for a $\boldsymbol{C}_{3v}$ molecule the selection rules for the over-all species are

$$A_1 \leftrightarrow A_2; \quad E \leftrightarrow E; \quad A_1, A_2 \not\leftrightarrow E; \quad A_1 \not\leftrightarrow A_1; \quad A_2 \not\leftrightarrow A_2. \tag{II, 69}$$

Similarly for $\boldsymbol{D}_{3h}$

$$\begin{gathered} A_1' \leftrightarrow A_1''; \quad A_2' \leftrightarrow A_2''; \quad E' \leftrightarrow E''; \quad A_1' \not\leftrightarrow A_1', A_2', A_2''; \\ A_1'' \not\leftrightarrow A_1'', A_2', A_2''; \quad E' \not\leftrightarrow E'; \quad E'' \not\leftrightarrow E''. \end{gathered} \tag{II, 70}$$

Finally we must consider the selection rule for the component levels of the first order Coriolis splitting that arises in every degenerate vibronic state [$(+l)$ and $(-l)$ levels, see Chap. I, section 3b]. Selection rules for special cases have long been known for infrared transitions [see Vol. II, p. 415] and to some extent also for electronic transitions [see Mulliken and Teller (917)]. Hougen (571) has recently derived general selection rules in terms of the new quantum number $G$ introduced by him (see Chap. I, section 3b). He finds

$$\Delta G = 0 \bmod n \tag{II, 71}$$

for a molecule with an $n$-fold axis of symmetry. If there is an $m$-fold rotation-reflection axis with $m = 2n$ the selection rule is

$$\Delta G = \tfrac{1}{2}m \bmod m. \tag{II, 72}$$

We need not go into the operation of these selection rules for $G$ in detail since one can express them also and perhaps more conveniently in terms of the $(+l)$ and $(-l)$ levels. Mills (855) has recently derived the selection rules for $(+l)$ and $(-l)$ levels in a very simple and elegant way, but we shall only give the results which agree with those obtained by Hougen's method.

For the simplest case of an $E - A_1$ (or $E - A_2$) transition the selection rules are (as in the infrared)

$$\begin{aligned} \Delta K &= +1 \text{ occurs only for } (+l) \leftrightarrow (0) \\ \Delta K &= -1 \text{ occurs only for } (-l) \leftrightarrow (0). \end{aligned} \tag{II, 73}$$

Here as usual the upper state is put first. The symbol (0) is used to indicate the single level of a given $K$ in the non-degenerate state. The same selection rule applies for $E_1 - A_1$ and $E_1 - A_2$ transitions in point groups that distinguish $E_1$, $E_2$, ... degenerate states. For the inverse transitions $A_{1,2} - E$ or $A_{1,2} - E_1$ the selection rules are

$$\begin{aligned} \Delta K &= +1 \text{ occurs only for } (0) \leftrightarrow (-l) \\ \Delta K &= -1 \text{ occurs only for } (0) \leftrightarrow (+l). \end{aligned} \tag{II, 74}$$

In point groups with an "even" axis of symmetry we have $E - B_1$, $E - B_2$, $E_2 - B_1$ or $E_2 - B_2$ transitions. For them the selection rules (II, 73) are reversed. Similarly for $B_{1,2} - E$, $B_{1,2} - E_2$ the rules (II, 74) are reversed.

For the $\parallel$ components of $E - E$ transitions of all axial point groups *except* $\boldsymbol{D}_{2d}$, $\boldsymbol{D}_{4d}$, ... one finds the selection rule

$$\Delta K = 0: \quad (+l) \leftrightarrow (+l) \quad \text{and} \quad (-l) \leftrightarrow (-l) \tag{II, 75}$$

while for the point groups $\boldsymbol{D}_{2d}$, $\boldsymbol{D}_{4d}$, . . . the opposite selection rule applies, viz.

$$\Delta K = 0: \quad (+l) \leftrightarrow (-l) \quad \text{and} \quad (-l) \leftrightarrow (+l). \tag{II, 76}$$

For the $\perp$ components of $E - E$ transitions of the point groups $\boldsymbol{C}_3$, $\boldsymbol{C}_{3h}$, $\boldsymbol{C}_{3v}$, $\boldsymbol{D}_{3h}$, $\boldsymbol{D}_{3d}$ one finds the selection rule[7]

$$\begin{aligned} \Delta K &= +1: \quad (-l) \leftrightarrow (+l) \\ \Delta K &= -1: \quad (+l) \leftrightarrow (-l). \end{aligned} \tag{II, 77}$$

For axial point groups with a more than four-fold axis the selection rule (II, 77) holds only for $E - E$ transitions of the same index ($E_1 - E_1$, $E_2 - E_2$, . . .) if such transitions can occur with a $\perp$ component of the transition moment. For other transitions $E_i - E_k$ ($i \neq k$) the results are given in Table 14 which also contains a summary of the selection rules just discussed.

TABLE 14. SELECTION RULES FOR $(+l)$ AND $(-l)$ LEVELS IN ELECTRONIC TRANSITIONS INVOLVING DEGENERATE ELECTRONIC STATES

| Point groups | Transition | | Selection rules | | |
|---|---|---|---|---|---|
| | Upper state | Lower state | $\Delta K = 0$ | $\Delta K = +1$ | $\Delta K = -1$ |
| $\boldsymbol{C}_3$, $\boldsymbol{C}_4$, $\boldsymbol{C}_{3h}$, $\boldsymbol{C}_{4h}$, $\boldsymbol{C}_{3v}$, $\boldsymbol{C}_{4v}$, $\boldsymbol{D}_{3h}$, $\boldsymbol{D}_{4h}$, $\boldsymbol{D}_{3d}$, $[\boldsymbol{D}_{2d}]$[a] | $E$ | $A_1$ or $A_2$ | — | $(+l) \leftrightarrow (0)$ | $(-l) \leftrightarrow (0)$ |
| | $A_1$ or $A_2$ | $E$ | — | $(0) \leftrightarrow (-l)$ | $(0) \leftrightarrow (+l)$ |
| | $E$ | $B_1$ or $B_2$ | — | $(-l) \leftrightarrow (0)$ | $(+l) \leftrightarrow (0)$ |
| | $B_1$ or $B_2$ | $E$ | — | $(0) \leftrightarrow (+l)$ | $(0) \leftrightarrow (-l)$ |
| | $E$ | $E$ | $(+l) \leftrightarrow (+l)$, $(-l) \leftrightarrow (-l)$ | $(-l) \leftrightarrow (+l)$ | $(+l) \leftrightarrow (-l)$ |
| $\boldsymbol{C}_5$, $\boldsymbol{C}_6$, . . ., $\boldsymbol{C}_{5h}$, $\boldsymbol{C}_{6h}$, . . . $\boldsymbol{C}_{5v}$, $\boldsymbol{C}_{6v}$, . . ., $\boldsymbol{D}_{5h}$, $\boldsymbol{D}_{6h}$, . . . $[\boldsymbol{D}_{4d}]$[c], $\boldsymbol{D}_{5d}$, $[\boldsymbol{D}_{6d}]$[c], . . . | $E_1$ | $A_1$ or $A_2$ | — | $(+l) \leftrightarrow (0)$ | $(-l) \leftrightarrow (0)$ |
| | $A_1$ or $A_2$ | $E_1$ | — | $(0) \leftrightarrow (-l)$ | $(0) \leftrightarrow (+l)$ |
| | $E_2$[b] | $B_1$ or $B_2$ | — | $(-l) \leftrightarrow (0)$ | $(+l) \leftrightarrow (0)$ |
| | $B_1$ or $B_2$ | $E_2$ | — | $(0) \leftrightarrow (+l)$ | $(0) \leftrightarrow (-l)$ |
| | $E_i$ | $E_i$[d] | $(+l) \leftrightarrow (+l)$, $(-l) \leftrightarrow (-l)$ | $(-l) \leftrightarrow (+l)$ | $(+l) \leftrightarrow (-l)$ |
| | $E_{i+1}$ | $E_i$[d] | — | $(+l) \leftrightarrow (+l)$ | $(-l) \leftrightarrow (-l)$ |
| | $E_i$ | $E_{i+1}$[d] | — | $(-l) \leftrightarrow (-l)$ | $(+l) \leftrightarrow (+l)$ |

[a] For $\boldsymbol{D}_{2d}$ the selection rule for the $\parallel$ component of $E - E$ is reversed, i.e. $(+l) \leftrightarrow (-l)$, $(-l) \leftrightarrow (+l)$.
[b] For point groups with an 8-, 10- . . . , fold axis $E_2$ must be replaced by $E_3$, $E_4$, . . . . But $\boldsymbol{D}_{5d}$ which has an $S_{10}$ counts here as a point group with a five-fold axis since $\boldsymbol{D}_{5d} \equiv \boldsymbol{D}_5 \times \boldsymbol{C}_i$; there are no $B$ states.
[c] For $\boldsymbol{D}_{4d}$ the selection rule for $E_2 - E_2$, $E_3 - E_1$ and $E_1 - E_3$ which occur with a $\parallel$ component only is $(+l) \leftrightarrow (-l)$, $(-l) \leftrightarrow (+l)$; for $\boldsymbol{D}_{6d}$ the same selection rule applies to $E_3 - E_3$, $E_2 - E_4$, $E_4 - E_2$, $E_1 - E_5$ and $E_5 - E_1$.
[d] $i = 1, 2, 3, \ldots$.

**Transitions between non-degenerate states ($\parallel$ bands).** An allowed electronic transition between non-degenerate states of a molecule that is a symmetric top on account of symmetry, is necessarily a parallel transition, that is, one for which only $M_z$ is different from zero. Therefore all allowed bands in such a band system would be $\parallel$ bands obeying the selection rule (II, 65). If the rotational constants $A$ and $B$ are not very different in the upper and lower electronic state the band structure will be identical with that of $\parallel$ infrared bands as discussed in detail in

[7] Note that $E - E$ transitions of $\boldsymbol{C}_4$, $\boldsymbol{C}_{4h}$, $\boldsymbol{C}_{4v}$, $\boldsymbol{D}_{4h}$, $\boldsymbol{D}_{4d}$ have no $\perp$ components.

Volume II, p. 416f: we have a band with a $P$, $Q$ and $R$ branch with little shading. In this band each "line" consists of a number of component lines with different $K$ values ($K = 0, 1, 2, \ldots, J^*$ where $J^*$ is the smaller of the $J$ values in the upper and lower state). For an electronic transition, unlike an infrared band, the rotational constants in the upper and lower state may be very different from one another and, if that happens, the individual sub-bands will no longer reinforce one another in the way they do in the infrared; as a consequence, a fairly complicated structure arises (see Fig. 97).

As in the infrared (see Vol. II, p. 421) the intensity distribution in an absorption band is given by the relation

$$I_{KJ} = C\nu A_{KJ} g_{KJ} e^{-F(K,J)/kT} \tag{II, 78}$$

where $A_{KJ}$ is the line strength, $g_{KJ}$ the statistical weight of the lower state, $\nu$ the wavenumber of the line and $k$ Boltzmann's constant in cm$^{-1}$/degree. The line strengths are given by the Hönl–London formulae, as in the infrared, which are for a $\parallel$ band

$$\text{for the } R \text{ branches } (\Delta J = +1): \quad A_{KJ} = \frac{(J+1)^2 - K^2}{(J+1)(2J+1)}$$

$$\text{for the } Q \text{ branches } (\Delta J = 0): \quad A_{KJ} = \frac{K^2}{J(J+1)} \tag{II, 79}$$

and

$$\text{for the } P \text{ branches } (\Delta J = -1): \quad A_{KJ} = \frac{J^2 - K^2}{J(2J+1)}.$$

Here it may be noted that the $Q$ branch has zero intensity for $K = 0$, and that in all $Q$ branches the intensity decreases rapidly with $J$.

As we have seen in Chapter I, for a molecule that is a symmetric top on account of symmetry there is an alternation of statistical weights ($g_{KJ}$) as a function of $K$ and therefore there is an alternation of intensities of the sub-bands of different $K$; for example, for a three-fold axis of symmetry (e.g. point groups $\boldsymbol{C}_{3v}$, $\boldsymbol{C}_{3h}$, $\boldsymbol{D}_{3h}$, $\boldsymbol{D}_{3d}$) and zero nuclear spin of the off-axis atoms the sub-bands with $K \neq 3n$ are missing while for non-zero nuclear spin these sub-bands have lower intensity than the sub-bands with $K = 3n$. Thus we have, as in the infrared, the characteristic alternation: strong, weak, weak, strong, .... If there are only three off-axis atoms of nuclear spin $\frac{1}{2}$ the intensity alternation is in the ratio 2:1; for spin 1 it is in the ratio 11:8.

For molecules of point groups $\boldsymbol{D}_{3h}$ and $\boldsymbol{D}_{3d}$ there is in addition an intensity alternation in $J$ but only in the sub-band with $K = 0$. For planar $XY_3$ molecules lines with even or odd $J$ are missing in the $K = 0$ sub-band if the nuclear spin of the Y atoms is $I = 0$ or $\frac{1}{2}$ while there is an intensity alternation in the ratio 10:1 if $I = 1$. Whether lines with even or odd $J$ are the preferred ones depends on the nature of the electronic or vibronic state and on the nuclear statistics. For example for an $A_2'' - A_1'$ transition the even lines (numbered according to the $J$ value of the lower $A_1'$ state) are strong for Bose statistics (e.g. if the identical nuclei are deuterons) while the odd lines are strong for Fermi statistics and are the only ones that occur if the three identical nuclei have $I = \frac{1}{2}$ (e.g. are H nuclei). Conversely if an intensity alternation in $J$ is observed in a molecule with a three-fold axis it establishes immediately that the molecule must have $\boldsymbol{D}_{3h}$ or $\boldsymbol{D}_{3d}$ symmetry in at least one of the two combining states, and the sense of the intensity alternation gives information about the species of the electronic (or vibronic) states.

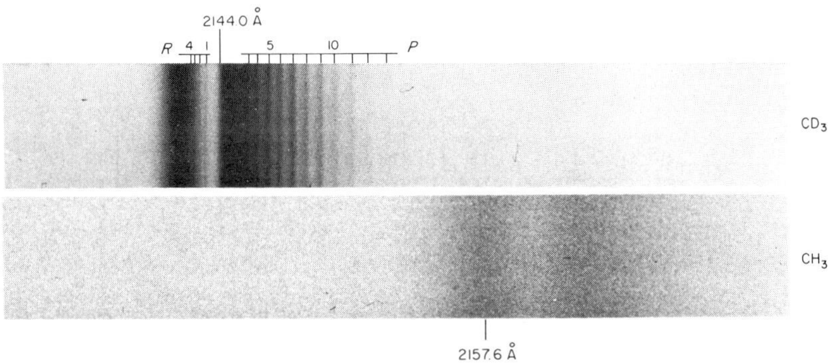


FIG. 96. **Spectrograms showing the parallel bands of $CD_3$ and $CH_3$ at 2144 Å and 2160 Å, respectively.** The transition involved is ${}^2A_1' - {}^2A_2''$. The $R$ and $P$ branches of the $CD_3$ band are marked (compare Fig. 97). The $Q$ branch forms the head at 2144.0 Å. In $CH_3$ only two diffuse maxima are visible, corresponding to the $R$ and $Q + P$ branches.

Examples of $\parallel$ bands are found in the absorption spectrum of $CH_3$ and $CD_3$. In Fig. 96a the 2140 Å band of $CD_3$ is reproduced. Unfortunately the lines are somewhat broad on account of predissociation, but the $\parallel$ structure of the band is clearly shown. Figure 97 shows diagrammatically the sub-bands and their superposition; in particular it shows how on account of the strong intensity alternation in the $K = 0$ sub-band a weak intensity alternation in the resulting band arises if the molecule is planar in one of the states. Conversely the observations of a just detectable intensity alternation in the tail of the $P$ branch and a strong alternation at the beginning of the $R$ branch for $CD_3$ (the line $R(0)$ is missing or very weak in the spectrum Fig. 96a) show unambiguously that the molecule is planar in at least one state. That it is planar or nearly planar also in the other state then follows from the fact that only one strong band occurs in the band system, showing according to the Franck–Condon principle that the molecular conformation must be nearly the same in both states. The statement that the molecule is planar is meant to include the case in which the molecule is so slightly non-planar that the inversion doubling is so large that only one of the component levels is observed at room temperature.

The fact that in the $CD_3$ spectrum of Fig. 96a the even lines are observed to be weaker than the odd ones shows that the electronic transition is either ${}^2A_1' - {}^2A_2''$ or ${}^2A_1'' - {}^2A_2'$ of which only the former is compatible with the electron configuration of the ground state of $CH_3$ (see Chapter III).

Another example is the first ultraviolet absorption system of $NH_3$ and $ND_3$ which extends from 2200 to 1700 Å. Here also predissociation makes the $NH_3$

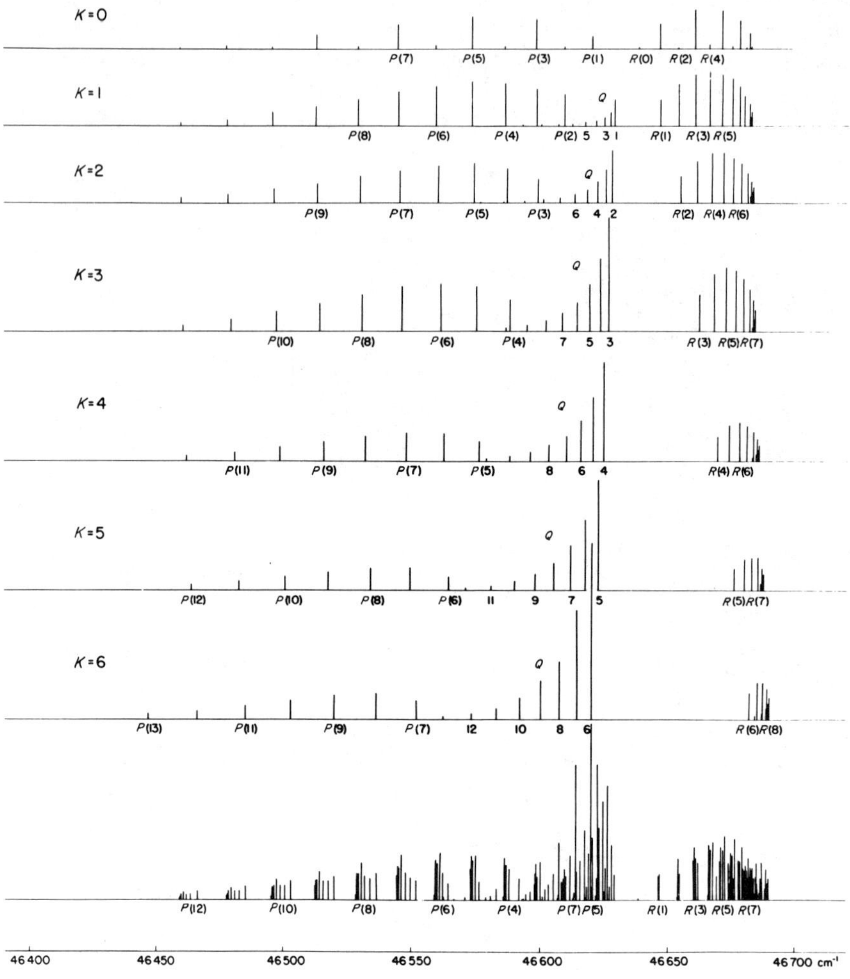


FIG. 97. **Sub-bands of the 2140 Å CD$_3$ band (schematic).** The heights of the lines indicate the intensities calculated with the help of the Hönl–London formulae. At the bottom the superposition of the sub-bands is shown. Only the $K = 0$ sub-bands show an intensity alternation leading to a very slight alternation of the unresolved lines of the $P$ branch (see Fig. 96). Note that in this figure the direction of increasing $\lambda$ is opposite to the normal one.

bands almost completely diffuse while some of the bands of $ND_3$ show a clear fine structure. But unlike $CH_3$ the ground state of $NH_3$ is well known to be non-planar (point group $C_{3v}$) and to have electronic species $A_1$. In the excited state, however, the molecule is planar and the electronic species is $A_2''$. The planarity in the excited state follows, just as for $CD_3$, from the fact that alternate lines are very weak in the $K = 0$ sub-bands (intensity alternation 1:10 for $ND_3$). Moreover, in successive bands of the main progression, which corresponds to the bending vibration $\nu_2(a_2'')$ in the excited state, alternately the even and odd lines in the $K = 0$

sub-bands are weak since the upper vibrational levels are alternately $A_1'$ and $A_2''$. Correspondingly in the lower state alternately the upper and the lower inversion doubling component combines with the upper state (see section 2b($\alpha$)). In Fig. 98 a spectrogram of the $1-0$ band of $ND_3$ is shown. Because of the greater difference between $B'$, $C'$ and $B''$, $C''$, the band structure is rather more complicated than for $CD_3$ (Fig. 96a) and the $Q$ branch instead of forming a line-like feature is fairly widely spaced. The lines with $J = K$, corresponding to the lowest $J$ for a given $K$, form the most prominent series in this $Q$ branch (see also Fig. 97). The beginning of

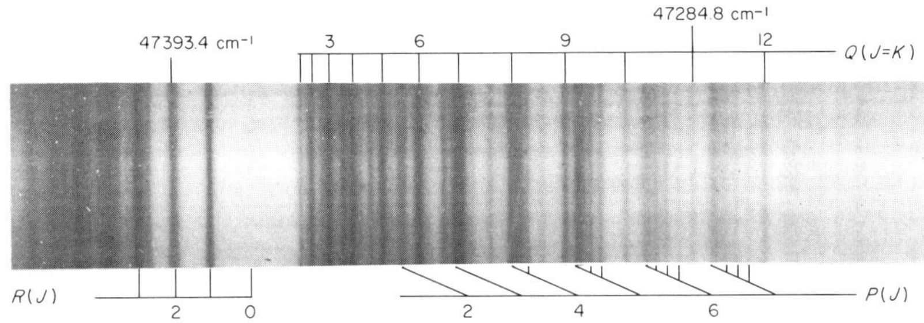


FIG. 98. **Spectrogram of the $1-0$ band of the $\tilde{A}-\tilde{X}$ system of $ND_3$ at 2111 Å after Douglas (294).** In the $Q$ branch only the lines with $J = K$ are marked; all other lines, with $K < J$, are much weaker. Only for the higher lines in the $P$ branch is the $K$ structure well developed as indicated. The strongest lines in the $R$ and $P$ branches are those with $K = 0$. The corresponding band of $NH_3$ is much more diffuse on account of predissociation.

the $R$ branch is relatively free from blending and shows clearly that $R(0)$ is very weak as is the $K = 0$ component of $R(2)$, showing that the upper state is vibronically $A_1'$ as expected for an $A_2''$ electronic state with one quantum of an $a_2''$ vibration excited.

The two examples just discussed are spectra of oblate symmetric tops. No resolved $\parallel$ bands of prolate tops have yet been observed. For a calculated band structure in such a case with a large difference between $A'-B'$ and $A''-B''$ see Volume II, Fig. 122c, p. 418.

The rotational constants $B'$ and $B''$ can be determined from the sub-bands of $\parallel$ bands in exactly the same way as for linear molecules. Even if the sub-bands are not resolved as for $CD_3$ a determination of $B'$ and $B''$ is possible although the values obtained are apt to be much less accurate since they are affected by the fact that the unresolved $K$ structure is different for the lower $J$ values than for the higher ones. If the $K$ structure is only partially resolved (as for $ND_3$ in Fig. 98) and if the ground state rotational constants are known, it is best to evaluate the upper state energies by adding the ground state rotational energy to the wavenumber of each line and then determine the upper state rotational constants by plotting the upper state energies against $K^2$ for a given $J$ and against $J(J + 1)$ for each $K$. The constants $A'$ and $A''$ (or $C'$ and $C''$) cannot be determined from a $\parallel$ band of a prolate (or oblate) symmetric top; only their difference $(A'-A'')$ can be determined if the $K$ structure is resolved.

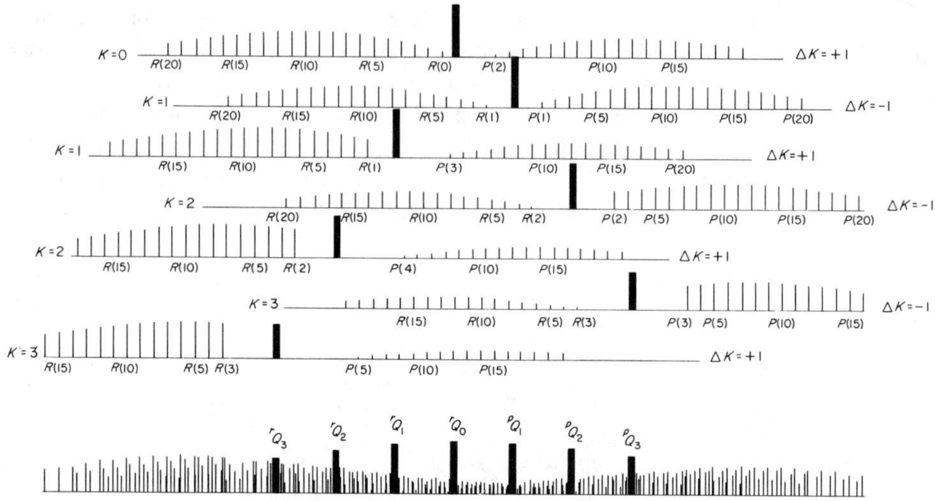


FIG. 99. **Calculated structure of a ⊥ band when $B' \approx B''$, $A' \approx A''$.** In the upper part seven sub-bands are shown separately. At bottom the sub-bands are superimposed. The $Q$ branches in which many lines coincide are shown as black rectangles. The height of these rectangles and of the lines representing the $P$ and $R$ branches indicate the intensities. No intensity alternation is indicated.

**Transitions between a degenerate and a non-degenerate state (⊥ bands).** For transitions between a degenerate and a non-degenerate electronic state the transition moment is at right angles to the symmetry axis; all allowed vibrational transitions are ⊥ bands for which the selection rule (II, 66) for $K$ and $J$ and the selection rules (II, 73 and 74) for the $(+l)$ and $(-l)$ levels apply (see also Table 14).

The band structure obtained by applying the selection rules is identical with that of perpendicular bands in the infrared (see Vol. II, p. 428), except that $B'$ and $B''$ as well as $A'$ and $A''$ may now be very different. In Fig. 36 we have already given an energy level diagram for such a transition for a $\boldsymbol{D}_{3h}$ molecule. The transitions according to the selection rules (II, 66 and 73) are indicated by the oblique arrows. The resulting sub-bands (each with $P$, $Q$ and $R$ branch in $J$) form two branches, $r$ and $p$, as shown schematically in Fig. 99 (see also Fig. 128 of Vol. I). We have for the origins ($J = 0$) of the $r$ sub-bands using (I, 117) for the upper state and (I, 102) for the lower state and taking account of the selection rule for the $(+l)$ and $(-l)$ levels

$$\begin{aligned} \nu_r^0 &= \nu_0 + F'[0, K+1, (+l)] - F''(0, K); \\ &= \nu_0 + (A' - B')(K+1)^2 - 2A'\zeta(K+1) - (A'' - B'')K^2 \\ &= \nu_0 + [A'(1 - 2\zeta) - B'] + 2[A'(1 - \zeta) - B']K \\ &\quad + [(A' - B') - (A'' - B'')]K^2 \end{aligned} \tag{II, 80}$$

and for the origins of the $p$ sub-bands

$$\begin{aligned} \nu_p^0 &= \nu_0 + F'[0, K-1, (-l)] - F''(0, K); \\ &= \nu_0 + (A' - B')(K-1)^2 + 2A'\zeta(K-1) - (A'' - B'')K^2 \\ &= \nu_0 + [A'(1-2\zeta) - B'] - 2[A'(1-\zeta) - B']K \\ &\quad + [(A' - B') - (A'' - B'')]K^2. \end{aligned} \tag{II, 81}$$

Since $K = 0, 1, 2, \ldots$ for the $r$ branch and $K = 1, 2, 3, \ldots$ for the $p$ branch it is clear that one has a single series of sub-bands without a zero gap. Superficially this is similar to the structure of a $\parallel$ band. However, while in the latter, when there is no change of geometry ($A' = A''$, $B' = B''$), all sub-bands coincide, in a $\perp$ band a series of equidistant sub-bands arises with a separation $2[A(1-\zeta) - B]$. An even more important difference is that in a $\perp$ band there are always two sub-bands for each $K$ value except $K = 0$, and this can be ascertained, if the sub-bands are resolved, by the number of missing lines near the origins of the sub-bands, since $J \geq K$. At the same time there are combination differences $\nu_r(K) - \nu_p(K)$ and $\nu_r(K-1) - \nu_p(K+1)$ which must be exactly the same in different bands with the same upper and lower state respectively. Such differences, of course, do not exist in $\parallel$ bands. Thus while it is quite possible that an electronic $\parallel$ band may have a similar coarse structure to that of a $\perp$ band since in electronic bands $A'$, $B'$ and $A''$, $B''$ may be very different, a decision about the true nature of a given band is in principle always possible.

In electronic bands with a degenerate upper (or lower) state the quantity $\zeta$ in (II, 80 and 81) is the electronic or the vibronic angular momentum (see Chap. I, section 2). If for the present we consider only the case of very weak vibronic interaction we have simply [see eq. (I, 67)]

$$|\zeta| = |\zeta_e \pm \zeta_v|. \tag{II, 82}$$

For the sign of $\zeta$, which is important in evaluating (II, 80 and 81), see p.64. The second term in (II, 82) vanishes if no degenerate vibrations are excited. In absorption this condition is fulfilled for all the strong bands of an allowed electronic transition since, as we have seen, almost exclusively the totally symmetric vibrations are excited. Thus for the strong bands the band structure is determined by $\zeta_e$, and if, as is often the case, $\zeta_e$ is close to 1 the spacing $2[A(1-\zeta)-B]$ of the sub-bands is small or in other words the $\perp$ band will look like a $\parallel$ band, even if there is no change of geometry. The inherent reason for this conclusion is simply that the electrons do not contribute to the rotational energy but do contribute to the angular momentum about the symmetry axis: almost all of $\Delta K = \pm 1$ is provided by the electrons.

If $\zeta$ is small the spacing $2[A(1-\zeta)-B]$ of the sub-bands is positive for prolate and negative for oblate symmetric tops, that is, in the prolate case the $r$ sub-bands are on the shortward side of the band origin, the $p$ sub-bands on the longward side, while in the oblate case this order is reversed. But even in the prolate case such a reversal may arise when $\zeta$ is close to 1.

The intensity distribution in the rotational fine structure, as for the $\parallel$ bands, is determined by the Hönl–London formulae which are here

$$\text{for the } R \text{ branches } (\Delta J = +1): \quad A_{KJ} = \frac{(J + 2 \pm K)(J + 1 \pm K)}{(J + 1)(2J + 1)} \qquad \text{(II, 83a)}$$

$$\text{for the } Q \text{ branches } (\Delta J = 0): \quad A_{KJ} = \frac{(J + 1 \pm K)(J \mp K)}{J(J + 1)} \qquad \text{(II, 83b)}$$

$$\text{for the } P \text{ branches } (\Delta J = -1): \quad A_{KJ} = \frac{(J - 1 \mp K)(J \mp K)}{J(2J + 1)} \qquad \text{(II, 83c)}$$

where the upper sign refers to the $r$ sub-bands ($\Delta K = +1$), the lower to the $p$ sub-bands ($\Delta K = -1$). For $K = 0$ and $\Delta K = +1$ the values given by (II, 83a) have to be multiplied by 2. A graphical representation of the intensities is given in Fig. 99 (see also Fig. 128 of Vol. II). It should be noted that according to (II, 83) the $R$ branches of the $r$ sub-bands gain increasingly in intensity over the $P$ branches as $K$ increases while the $P$ branches gain over the $R$ branches in the $p$ sub-bands. As long as the spacing $2[A(1 - \zeta) - B]$ is positive this means that strong ${}^pP$ and ${}^rR$ branches are present in the longward and shortward wings of the bands while when $2[A(1 - \zeta) - B]$ is negative they crowd into the centre of the band and only the much weaker ${}^rP$ and ${}^pR$ branches are in the wings.

The intensity distribution in the bands is further governed by the statistical weights $g_{KJ}$ [see eq. (II, 78)]. The alternation of weights as a function of $K$ is much more obvious in $\perp$ bands than in $\parallel$ bands since the $Q$ branches of the sub-bands are in general well separated. The alternation of $g_{KJ}$ in $J$ for $\boldsymbol{D}_{3h}$, $\boldsymbol{D}_{4h}$, ... molecules will cause an intensity alternation only in the $1-0$ and $0-1$ sub-bands.

Thus far we have disregarded $j$-type (or $l$-type) doubling (p. 94), that is, the splitting between $A_1$ and $A_2$ rotational levels of equal $J$ and $K$. As we have seen in Chapter I this splitting has in general both an electronic and a vibrational component which for large vibronic interaction cannot be separated. For small interaction, when no degenerate vibration is excited the splitting is mainly electronic in origin. At any rate whether electronic or vibrational in origin it is appreciable only for the $(+j)$ [or $(+l)$] levels of $K = 1$ in the degenerate electronic state. As shown by Fig. 36 this splitting enters only into the $r$ sub-band with $K = 0$. Because of the symmetry selection rules (II, 69 and 70) the level splitting does not lead to a line splitting but only to a combination defect between $P$, $R$ and $Q$ branches of this sub-band, the $Q$ lines having one component level, the $P$ and $R$ lines the other as their upper states. For planar $XY_3$ molecules (point group $\boldsymbol{D}_{3h}$) if the nuclear spin of the Y atoms $I = 0$ only the $A_1$ rotational levels exist, for $I = \frac{1}{2}$ only the $A_2$ levels, and therefore alternate lines are missing in the three branches of the $1-0$ sub-band; no such alternation exists in any of the other sub-bands. For an $E' - A_1'$ transition as in Fig. 36, if $I = \frac{1}{2}$ (as for example in $CH_3$, $NH_3$, ...) the even lines are missing in the $P$, $Q$, $R$ branches of the $1-0$ sub-band. For an $E'' - A_2''$ transition (obtained from Fig. 36 by interchanging $'$ and $''$ throughout and, in the lower state, interchanging the subscripts 1 and 2) the odd lines would be missing if $I = \frac{1}{2}$.

If the $XY_3$ molecule is non-planar in the lower state the $A_1'$ and $A_2''$ vibrational levels will move together in pairs (see Fig. 67) but only one of them will combine with a given upper vibronic level, say $E'$ if the molecule is planar in the upper electronic state. There will now be a long progression of bands in the out-of-plane bending vibration $\nu_2(a_2'')$. The upper vibronic levels are alternately $E'$ and $E''$ and therefore combine alternately with the lower or upper inversion component of the ground state as in Fig. 67. Since alternate bands are of type $E'-A_1'$ and $E''-A_2''$ there will be a characteristic alternation of the intensity alternation: for even $v_2'$ the even lines, for odd $v_2'$ the odd lines are missing in the $1-0$ sub-bands assuming that the upper state is electronically $E'$. The reverse alternation applies for an electronic $E''$ state.

The rotational constants $B$ and $D$ for the upper and lower state of a $\perp$ band can be obtained from each sub-band in the same way as for linear and diatomic molecules except that a slight variation of the resulting effective $B$ values with $K$ may occur on account of the presence of a term in $D_{JK}$. We have for the upper state using (I, 102) with the additional term (I, 116)

$$\begin{aligned} R(J, K) - P(J, K) &= \Delta_2^J F'(J, K') \\ &= F'[J+1, K', (\pm l)] - F'[J-1, K', (\pm l)] \\ &= 4(B' - \tfrac{3}{2}D_J' - D_{JK}'K'^2)(J+\tfrac{1}{2}) - 8D_J'(J+\tfrac{1}{2})^3 \end{aligned} \qquad \text{(II, 84)}$$

where $K' = K \pm 1$ depending on whether an $r$ or $p$ sub-band is considered. An entirely similar relation holds for the lower state combination difference

$$R(J-1, K) - P(J+1, K) = \Delta_2^J F''(J, K).$$

Thus, neglecting the very small $D_J$ terms, one obtains initially the quantities $B - D_{JK}K^2$ (two for each $K$ value except $K = 0$, one from the $r$ sub-band and one from the corresponding $p$ sub-band). If these quantities are plotted against $K^2$ one obtains $D_{JK}$ from the slope of the resulting straight line and $B$ from the intercept at $K = 0$. Figure 100 shows such a plot for the $v_2' = 2$ level of the upper state of the $E'' - A_1$ bands of $NH_3$ (see below).

Just as from $\perp$ bands in the infrared, it is not possible to determine the constants $A'$, $A''$ and $\zeta'$ from $\perp$ bands in the ultraviolet. This is because there are no $p$ and $r$ sub-bands that have upper states in common (see Fig. 36). Only if one of the three quantities $A'$, $A''$, $\zeta'$ is known from other evidence, for example if $A''$ is known from the infrared and Raman spectrum, can the other two be determined. It is easily seen that the combination differences between sub-band origins ($J = 0$) are given by (remembering the selection rule for $(+l)$ and $(-l)$ levels)

$$\begin{aligned} \nu_r^0(K) - \nu_p^0(K) &= F'[0, K+1, (+l)] - F'[0, K-1, (-l)] \\ &= 4[A' - B']K - 8D_K'K(K^2+1) - 4A'\zeta'K, \end{aligned} \qquad \text{(II, 85a)}$$

$$\begin{aligned} \nu_r^0(K-1) - \nu_p^0(K+1) &= F''(0, K+1) - F''(0, K-1) - 4A'\zeta'K \\ &= 4[A'' - B'']K - 8D_K''K(K^2+1) - 4A'\zeta'K. \end{aligned} \qquad \text{(II, 85b)}$$

For $J \neq 0$ the term $-D_{JK}J(J+1)$ must be added in the square bracket at the

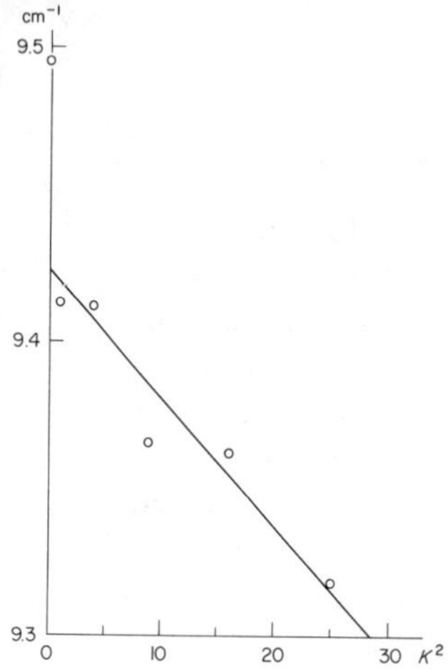


FIG. 100. **Observed effective B values as a function of $K^2$ for the $v_2' = 2$ level of the $\tilde{C}E''$ state of $NH_3$.** The circles correspond to the slopes of the $\Delta_2^J F'(J, K')$ curves according to eq. (II, 84).

right. If the rotational constants in the lower state are known the second equation allows one to determine $A'\zeta'$ and with that the first equation yields $A' - B'$ and $D_K'$.

If, as seems to happen fairly frequently, the Coriolis interaction is strong, the splitting of the $(+l)$ and $(-l)$ levels is no longer given by the simple expression $4A'\zeta'K$ but the additional term (I, 130) has to be considered. It is the actual splitting that is determined from (II, 85b) and must be substituted into (II, 85a) in order to determine the rotational constants of the upper state.

An alternative way of determining the rotational constants of the upper state, which is particularly suitable when the various branches have not all been observed to the same degree of completeness, is simply to add to each of the observed lines the appropriate rotational term values of the lower state if they are known from the infrared, Raman or microwave data. In this way the rotational term values of the upper state are obtained and can then be expressed in terms of the rotational constants $A'$, $B'$, $\zeta'$, $D_K'$, $D_{JK}'$, $D_J'$.

A fully resolved system of perpendicular bands of the $E - A$ type is the 1600 Å absorption system of $NH_3$ studied by Douglas and Hollas (295). The lower state is a $^1A_1$ state in which the molecule is non-planar, the upper state is an $E''$ state in which the molecule is planar. The main progression is one in the out-of-plane bending vibration $\nu_2'$. The proof for the planarity of the molecule in the excited state follows easily by applying the preceding discussion: the $r$ subbands with $K' - 1$, $K'' - 0$, unlike the other sub bands, have alternate lines missing in the $J$ structure and the lines that are present have alternately even and odd $J$ in successive bands of the main progression. The fact that for the 0–0

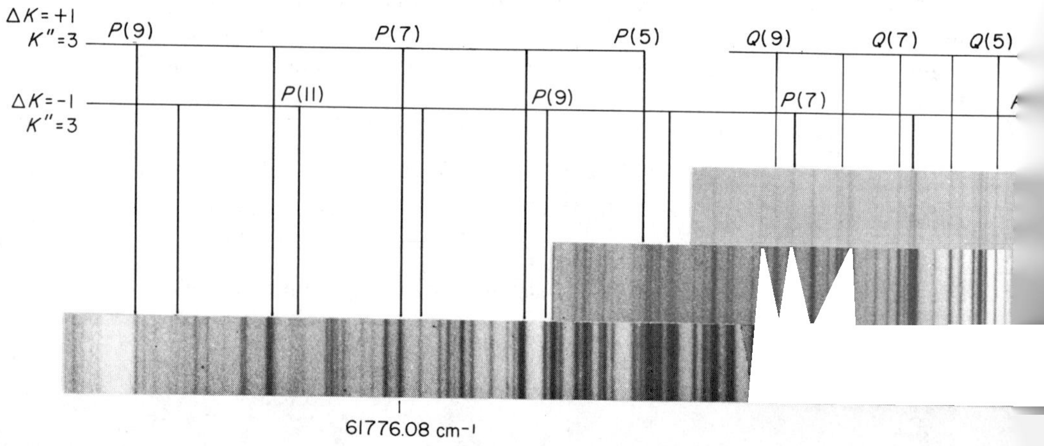

band the even lines are present shows that the upper state is $E''$ and not $E'$. The alternation in successive bands shows that the vibration that is excited must be antisymmetric with respect to the plane of the molecule (species $a_2''$). Figure 101 shows a spectrogram of one of the bands of this system. Only the two sub-bands with $K'' = 3$ are marked. It is seen how in this case since the molecule is an oblate symmetric top the strong branches of the sub-bands are all crowded into the center of the band. The characteristic intensity alternation: strong, weak, weak, strong, ... of the sub-bands is present but not readily discernible because the $Q$ branches are not line-like, and are rather close together. The value of $\zeta$ in the excited state is about 0.88 (for $v_2' = 0$). The $j$-type doubling in the $K' = 1$ levels is very pronounced leading to a large discrepancy between the positions of the $Q$ lines as calculated from the $P$ and $R$ lines and the actual positions of the $Q$ lines. The $j$-type doubling decreases rapidly in the higher vibrational levels ($q = 0.758$ for $v_2' = 1$ and $q = 0.170$ for $v_2' = 8$).

An example of a band system consisting of pseudo-parallel bands which are actually perpendicular bands is provided by the $\tilde{B}-\tilde{X}$ and $\tilde{C}-\tilde{X}$ absorption systems of $CH_3I$ near 2000 and 1800 Å respectively. The main bands of these systems consist of very narrow strong absorption peaks giving the impression of unresolved $\|$ bands; but from a consideration of the electron configurations and of the structure of some of the weaker bands it follows that the upper states are $E$ states, that is, that the bands are really $\perp$ bands whose sub-bands are telescoped together since $\zeta_e \approx 1$ [see Chap. V, section (3)]. The corresponding bands of $CF_3I$ are quite similar.

In most of the preceding discussion we have used $E-A$ transitions of $\boldsymbol{C}_{3v}$ or $\boldsymbol{D}_{3h}$ molecules as examples (see Fig. 36). For other point groups $E-A$ transitions are entirely similar except that the over-all (rovibronic) species of the rotational levels are different. Therefore the band structure differs from bands of $\boldsymbol{C}_{3v}$ or $\boldsymbol{D}_{3h}$ molecules only by the intensity alternation (see Table 4, p. 94). For example, for $XY_4$ molecules (point group $\boldsymbol{C}_{4v}$ or $\boldsymbol{D}_{4h}$) if the nuclear spin of Y is zero, only sub-bands with $K = 0, 4, 8, \ldots$ occur while for non-zero nuclear spin all

FIG. 101. **Spectrogram of the 3–0 band of the $\tilde{B}-\tilde{X}$ system of $NH_3$ at 1613 Å after Douglas and Hollas (295).** Only two sub-bands with $K'' = 3$ and $K' = 2$ and 4 are marked.

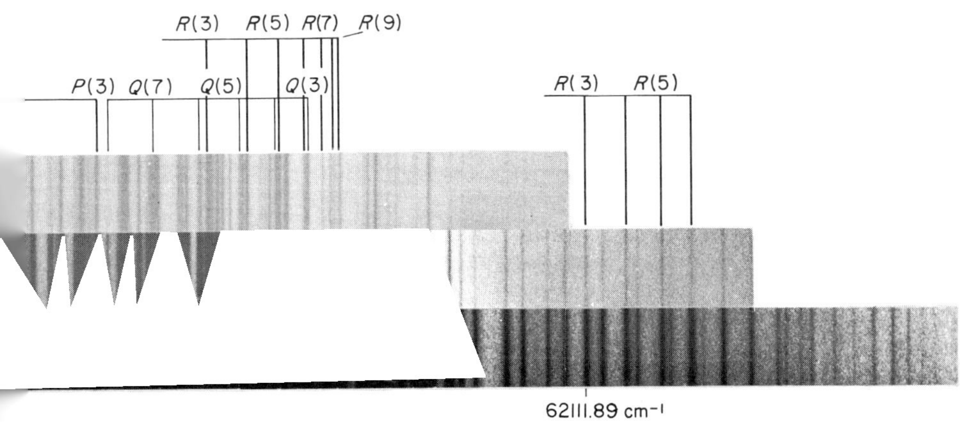

sub-bands occur but with intensities which for $I = \frac{1}{2}$ are 4, 3, 6, 3, 4, . . . for $K = 0, 1, 2, 3, 4, \ldots$ respectively. Similarly for a molecule like $C_6H_6$ with a six-fold axis the intensities would be 10, 11, 9, 14, 9, 11, 10, . . . for $K = 0, 1, 2, 3, 4, 5, 6, \ldots$. No examples of electronic bands have been sufficiently resolved to exhibit these alternations.

Thus far we have assumed that the vibronic interaction in the degenerate electronic state (i.e. the Jahn–Teller effect) is very small. If this interaction is no longer negligible certain vibronic transitions may occur which are forbidden by (II, 31) in the absence of this interaction, e.g. the $1-0$ and $0-1$ bands corresponding to a degenerate vibration ($\nu_k$). These transitions could have both $\|$ and $\perp$ components (see Fig. 61) but only the $\perp$ components will appear with appreciable intensity since they can "borrow" intensity from the main $\perp$ bands. Thus the band structure of these $1-0$ and $0-1$ bands is of the $\perp$ type but differs from that of the main bands because the effective $\zeta$ is different, as was first shown for $CH_3I$ by Mulliken and Teller (917).

In the $E$ vibronic sub-level of the $v_k = 1$ state, as we have seen in Chapter I, for small vibronic interaction

$$\zeta = -(\zeta_e' + \zeta_v'). \qquad \text{(II, 86)}$$

Therefore the spacing of the sub-bands will be (assuming $B' \approx B''$, $A' \approx A''$)

$$2[A(1 + \zeta_e' + \zeta_v') - B]$$

rather than $2[A(1 - \zeta_e') - B]$ for the main bands. In other words if $\zeta_e' \approx 1$ and $A \gg B$ the spacing of the sub-bands is more than twice the spacing in corresponding infrared bands and many times that in the main bands. The same conclusion applies to the $0-1$ band in the degenerate vibration; for this band both upper and lower state are degenerate, i.e. have a non-zero $\zeta$ value, $\zeta_e'$ in the upper and $\zeta_v''$ in the lower state. As we shall see below, the spacing of the sub-bands is now (again assuming $B' \approx B''$, $A' \approx A''$)

$$2[A(1 + \zeta_e' + \zeta_v'') - B].$$

For comparison it is interesting to note that in the $E-E$ component of the $1-1$ band (which is an allowed vibrational transition), as we shall see below, the spacing is

$$2[A(1 - \zeta_e' - \zeta_v' + \zeta_v'') - B],$$

that is, similar to that in the main bands.

Striking examples of perpendicular bands with a wide spacing of the sub-bands have been observed for $CH_3I$ by Scheibe, Povenz and Linström (1102) and Henrici and Grieneisen (505), and their correct interpretation was first found by Mulliken and Teller (917). Figure 102 reproduces a section of the $CH_3I$ spectrum showing both three narrow main bands and three of the $\perp$ bands with wide spacing and a clear intensity alternation: strong, weak, weak, strong, . . . . The broad "lines" are the unresolved $Q$ branches of the $\perp$ band. The rotational constants $A$ and $B$

are nearly the same in the upper and lower state so that all the lines in a given $Q$ branch nearly coincide and so that there is very little convergence in the series of $Q$ branches. The spacing of the $Q$ branches in two of the bands is about 20 cm$^{-1}$, almost three times the spacing in the infrared band $\nu_6(e)$. The vibration $\nu_6'$ is excited in the band at 1978 Å shown in Fig. 102. If the upper electronic state were

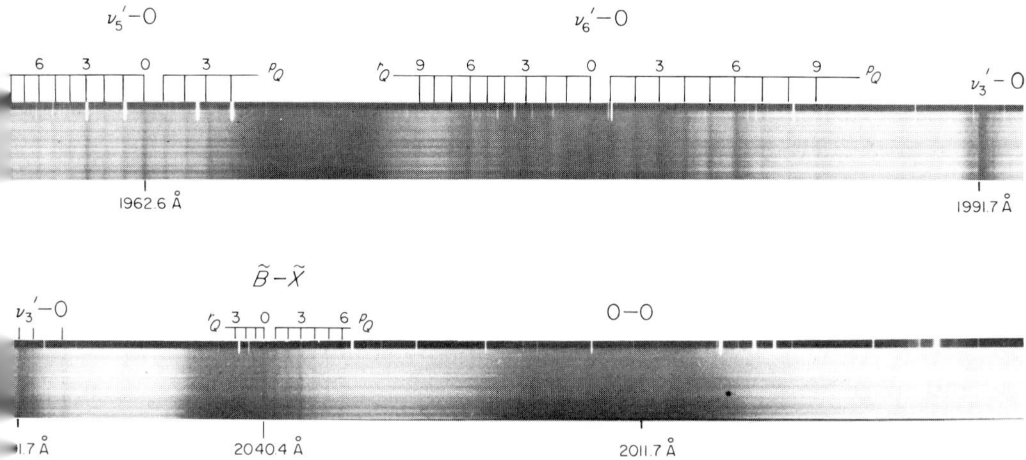


FIG. 102. **Absorption spectrum of $CH_3I$ between 2020 and 1955 Å showing bands with narrow and wide K structure [after Dunn and Herzberg (326)].** All but one of the bands shown belong to the $\tilde{C}-\tilde{X}$ system (see Chapter V, section 3b). All bands are $\perp$ bands (of vibronic type $E-A_1$), but three are unresolved because $\zeta_{\text{eff}} \approx 1$, while two show a large spacing of the $Q$ branches because $\zeta_{\text{eff}} \approx -\zeta_e - \zeta_{v_i}$.

an $A_1$ state as was suggested by the apparent $\parallel$ type structure of the main bands (see Fig. 102) a spacing of 20 cm$^{-1}$ could not be accounted for; but if the upper state is an $E$ state and $\zeta_e \approx 1$ then both the appearance of the main bands and the spacing in the bands with fine structure can be understood immediately. The corresponding $0-1$ band has also been observed; it has a similar structure but is not shown in Fig. 102. From it, since the ground state values of $A''$, $B''$, $\zeta''$ are known, one obtains directly (no longer neglecting the difference between $A'$ and $A''$, $B'$ and $B''$) $A_0' = 4.78$ cm$^{-1}$ and $\zeta_e' = 0.99_8$ (close to 1, as expected), and with these values one obtains from the $1-0$ band $A_1' = 4.75$ cm$^{-1}$ and $\zeta_6' = 0.34$.

Two very similar bands have been observed recently in the corresponding band system of $CF_3I$ [Herzberg (523)]. They also show the characteristic intensity alternation[3b].

For large vibronic interaction the formula (II, 86) for $\zeta$ has to be replaced by (I, 68) or its analogue in more general cases. In addition, the equilibrium conformation in the $E$ state shifts more and more away from the symmetrical one and therefore asymmetry effects will arise. The detailed nature of these effects has not yet been discussed; nor has any spectrum of this type been found and analyzed.

[3b] See footnote p. 167.

**Transitions between two degenerate states.** When both the upper and lower electronic state is degenerate, the transition moment for some point groups (e.g. $\boldsymbol{C}_{3v}$) can be both $\|$ and $\perp$ to the symmetry axis (see Table 9), that is, we have in general hybrid bands; in other point groups (e.g. $\boldsymbol{D}_{3h}$, $\boldsymbol{D}_{6h}$) only either a $\|$ or a $\perp$ component arises. $E-E$ vibronic transitions in $A-A$ or $E-A$ electronic transitions are similar and are included in the following discussion.

Because of the presence of first order Coriolis splitting in upper and lower state both the $\|$ and the $\perp$ components of an $E-E$ transition differ from ordinary $\|$ and $\perp$ bands. Figure 103a gives an energy level diagram for an $E'-E''$ transition of a $\boldsymbol{D}_{3h}$ molecule which has only a $\|$ component. The same diagram represents the $\|$ component of an $E-E$ transition of a $\boldsymbol{C}_{3v}$ molecule if $'$ and $''$ are omitted throughout. It is seen from this diagram how the Coriolis splitting in the upper and lower state causes a splitting of each sub-band with $K > 0$ into two (not four) components, one corresponding to the $(+l)$ levels, the other corresponding to the $(-l)$ levels. The two other conceivable components are excluded by the selection rule (II, 75).

From the energy formula (I, 117) we obtain for the sub-band origins

$$\nu_q^0 = \nu_0 \mp 2(A'\zeta' - A''\zeta'')K + [(A' - A'') - (B' - B'')]K^2. \qquad \text{(II, 87)}$$

The magnitude of the splitting is $4(A'\zeta' - A''\zeta'')K$ which may be considerable if the Coriolis interaction is different in the upper and lower state. Each sub-band consists of a $P$, $Q$ and $R$ branch corresponding to $\Delta J = -1, 0,$ and $+1$. However, in the $(+l)$, $K = 1$ sub-band a splitting of each branch, increasing with $J(J+1)$ arises on account of $l$- or $j$-type doubling, i.e. on account of the splitting between the $A_1$ and $A_2$ sub-levels. Here we must apply the selection rule (II, 70) for the over-all species. As a result only two (not four) branches of each type arise just as in $\Pi-\Pi$ or $\Delta-\Pi$ bands of linear molecules. The line splitting is the difference of the level splittings in the upper and lower state [given by (I, 126)] for the $Q$ lines and the sum of the level splittings for the $R$ and $P$ lines. The reverse would be the case for a $\boldsymbol{C}_{3v}$ molecule since here the rovibronic selection rule is (II, 69). In principle there would be similar splittings in the $(-l)$, $K = 2$; $(+l)$, $K = 4$, etc. sub-bands but this splitting is always negligibly small.

For $\boldsymbol{D}_{3h}$ and similar molecules with a plane of symmetry perpendicular to the three- or higher-fold axis an intensity alternation arises in the individual branches of the sub-band $(+l)$, $K = 1$ since the levels $A_1$ and $A_2$ have different statistical weights depending on the nuclear spin of the identical nuclei. But the alternation is opposite in the two branches of each pair arising from $l$-type doubling so that no intensity alternation will be noticeable unless the $l$-type doubling is resolved. If alternate lines are missing the apparent single branch resulting from the two components of a pair will show "staggering" similar to that in $\Pi-\Pi$ bands of symmetric linear molecules with zero nuclear spin. The magnitude of the intensity alternation depends in the same way on the number and spin of identical nuclei as in the previous discussion on the $K = 0$ sub-bands of $A-A$ transitions (p. 226).

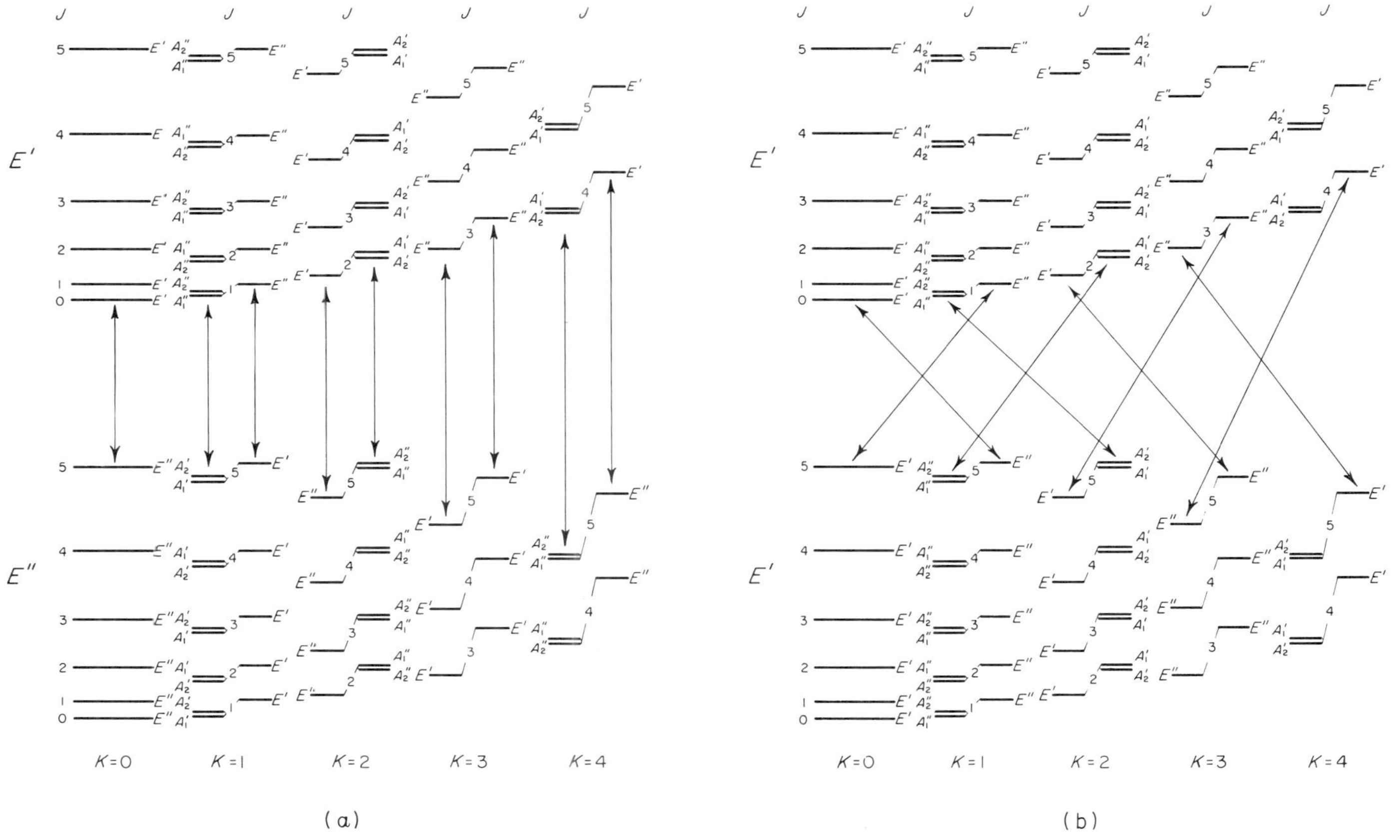


FIG. 103. **Energy level diagrams for $E - E$ transitions of $D_{3h}$ molecules,** (*a*) **for $E' - E''$ (∥ band),** (*b*) **for $E' - E'$ (⊥ band).** In both upper and lower state a positive $\zeta$ has been assumed, i.e., the $(+l)$ levels are below the $(-l)$ levels. The $l$-type doubling is only qualitatively indicated. Its dependence on $J$ and $K$ is not shown. The same figures apply to $E'' - E'$ and $E'' - E''$ respectively, if primes and double primes are everywhere exchanged. For $C_{3v}$ molecules the primes and double primes must be omitted and therefore both ∥ and ⊥ transition may occur in one and the same band. The transition arrows refer to the *sets* of levels of given $K$. The selection rule $\Delta J = 0, \pm 1$ applies as usual but is not expressed by the arrows.

For the $\perp$ component of an $E-E$ electronic (or vibronic) transition of molecules of point groups $\boldsymbol{C}_3$, $\boldsymbol{C}_{3h}$, $\boldsymbol{C}_{3v}$, $\boldsymbol{D}_{3h}$, $\boldsymbol{D}_{3d}$, the selection rule (II, 77) applies to the $(+l)$ and $(-l)$ levels and as a result, just as for an $E-A$ band only two branches of sub-bands, $r$ and $p$, arise; there is no doubling of the number of sub-bands. For $\boldsymbol{D}_{3h}$ molecules, $E'-E'$ and $E''-E''$ transitions have only $\perp$ components while $E'-E''$ transitions have only $\parallel$ components. In Fig. 103b the transitions possible on the basis of the selection rule (II, 77) are shown for an $E'-E'$ transition (for an $E''-E''$ transition one would simply have to exchange all primes and double primes). The same figure can also be used for $E-E$ transitions of $\boldsymbol{C}_{3v}$ molecules and for $E_u-E_g$ (or $E_g-E_u$) transitions of $\boldsymbol{D}_{3d}$ molecules if all primes and double primes are omitted and, for $\boldsymbol{D}_{3d}$, $u$ and $g$ added to the rovibronic species symbols. For both $\boldsymbol{C}_{3v}$ and $\boldsymbol{D}_{3d}$ there is a $\parallel$ component for the same electronic transition (see Fig. 103a).

From the energy formula (I, 117) we obtain for the $r$ sub-bands[8] assuming a prolate symmetric top and taking account of the selection rule (II, 77)

$$\begin{aligned}\nu_r^0 = \nu_0 &+ [A'(1 + 2\zeta') - B'] + 2[A'(1 + \zeta') + A''\zeta'' - B']K \\ &+ [(A' - B') - (A'' - B'')]K^2 \end{aligned} \qquad \text{(II, 88)}$$

and for the $p$ sub-bands

$$\begin{aligned}\nu_p^0 = \nu_0 &+ [A'(1 + 2\zeta') - B'] - 2[A'(1 + \zeta') + A''\zeta'' - B']K \\ &+ [(A' - B') - (A'' - B'')]K^2. \end{aligned} \qquad \text{(II, 89)}$$

Here $\zeta'$ and $\zeta''$ are the effective $\zeta$ values which for an $E-E$ electronic transition when no degenerate vibrations are excited are simply $\zeta_e'$ and $\zeta_e''$. If the rotational constants $A$ and $B$ in the upper and lower state are nearly the same, we see that the spacing of successive sub-bands is $2[A(1 + \zeta' + \zeta'') - B]$ which may be compared with $2[A(1 - \zeta') - B]$ for an ordinary $(E - A)$ $\perp$ band. If $\zeta' = \zeta_e'$ and $\zeta'' = \zeta_e''$ and if both $\zeta_e$ are close to 1 the spacing of the sub-bands will be about $6A - 2B$.

The similarity between the formulae (II, 88 and 89) and (II, 87) for the $\perp$ and $\parallel$ component of an $E-E$ transition may be noted. The linear term is in general much smaller for the $\parallel$ component as long as $\zeta'$ and $\zeta''$ have the same sign. The $K = 0$ sub-bands of the two components are separated by $A'(1 + 2\zeta') - B'$, and that of the $\parallel$ component has no $Q$ branch.

For an oblate symmetric top the same formulae (II, 87, 88 and 89) apply if $A'$ and $A''$ are replaced by $C'$ and $C''$ respectively. For axial point groups with four-fold axes ($\boldsymbol{C}_4$, $\boldsymbol{C}_{4v}$, $\boldsymbol{C}_{4h}$, $\boldsymbol{D}_{4h}$) $E-E$ transitions do not have $\perp$ components. For axial point groups with more than four-fold axes we must take account of the different $(\pm l)$ selection rules given in Table 14. For transitions between $E$ states of different index ($E_1-E_2$, $E_2-E_3$, ...) these selection rules lead to certain changes of sign in the terms in $\zeta$ in eqs. (II, 88 and 89).

In the fine structure of the sub-bands of the $\perp$ component of an $E-E$ transition just as in the $\parallel$ component we have single $P$, $Q$, $R$ branches except in the

[8] It should be noted that in the similar formulae for infrared bands reproduced in Volume II, p. 433, the sign in front of $A''\zeta''$ is incorrectly given.

sub-bands involving $K' = 1$ or $K'' = 1$, that is, in the $1-2$ and $2-1$ sub-bands. Because of the $l$- or $j$-type doubling in the $K = 1$ levels the lines in these sub-bands are doubled. Again we must apply the rovibronic selection rule (see p. 223f) to establish which sub-levels combine and again each of the sub-branches shows an intensity alternation (or alternate missing lines). Since the $l$-type doubling in the $K = 2$ levels is in general negligible the line-doublings in the $1-2$ and $2-1$ sub-bands give directly the $l$-type splittings of $K = 1$ in the upper and lower state respectively.

No examples of $E-E$ electronic transitions have as yet been sufficiently resolved to test all the predictions in the preceding paragraphs. However, as already mentioned, in $CH_3I$ a $0-1$ band in a degenerate vibration ($\nu_6''$) has been observed in an $E-A_1$ electronic transition. This vibronic transition is of the $E-E$ type and since the electronic transition is $E-A_1$ only the $\perp$ component is expected with a spacing of sub-bands of $2[A(1 + \zeta_e' + \zeta_v'') - B]$. Indeed the observed band has a spacing more than twice that in the corresponding infrared band (see also Chap. V, section 3b).

**Multiplet transitions.** If the two electronic states involved in a transition are not singlets, as has been assumed thus far, but doublets, triplets or higher multiplets, line splittings are expected just as for linear and diatomic molecules. As discussed in Chapter I, section 3b, the level splittings produced by spin-orbit interaction are small for all non-degenerate states. They are expected to be much larger for degenerate states but have not yet been discussed in detail.

As we have seen in Chapter I, for small spin-orbit coupling the levels are classified by a quantum number $N$ representing the total angular momentum apart from spin which assumes the role that $J$ has for singlet states while $J$ represents now the total angular momentum including spin. For $N$ we have as for linear molecules the selection rule

$$\Delta N = 0, \pm 1. \qquad \text{(II, 90)}$$

As a result the band structure is the same as for singlet – singlet transitions as long as the spin splittings are small. The fine structure levels then combine according to the rule

$$\Delta J = 0, \pm 1$$

together with the various symmetry rules. Since up to now no doublet – doublet, triplet – triplet, ... transitions of symmetric top molecules have been sufficiently well resolved for a fine structure analysis, we shall not consider the multiplet structure in more detail.

**Forbidden transitions.** In forbidden electronic transitions occurring on account of vibronic interactions certain vibrational transitions may occur weakly if they are in accordance with the vibronic selection rules. Similarly in allowed electronic transitions forbidden vibrational components may occur weakly if the vibronic selection rules are satisfied. The structure of these bands is entirely determined by the vibronic species of the upper and lower state; for example an $A_2''-A_1'$ vibronic transition in an $E''-A_1'$ forbidden electronic transition of point group $\boldsymbol{D}_{3h}$ has the structure of an $A_2''-A_1'$ electronic band as described above; an $E-A_1$ vibronic transition in an $A_1-A_1$ electronic transition of point group $\boldsymbol{C}_{3v}$ has the same rotational structure as an $E-A_1$ electronic transition; and similarly in other cases, except of course that the $\zeta$ values are different. A well-known example are the near ultraviolet absorption bands of $C_6H_6$ which represent a $B_{2u}-A_{1g}$ (forbidden) electronic transition. All the principal bands correspond to single excitation of $\nu_{18}(e_{2g})$ either in the upper or the lower state. They are therefore of type $E_{1u}-A_{1g}$ or $B_{2u}-E_{2g}$ and differ from allowed purely electronic transitions of this type mainly in that the Coriolis $\zeta$'s are entirely of vibrational origin. But since the excited state is a $B$ state, $\zeta_{\text{eff.}}$ in the formulae for the sub-band origins must be put equal to $-\zeta_{18}'$ for the $1-0$ and $+\zeta_{18}''$ for the $0-1$ bands. There are weaker bands corresponding to sequences in $\nu_{18}$ as well as combinations of several degenerate vibrations of the required species. The $\zeta$ values in these cases have been discussed by Callomon, Dunn and Mills (176) who have performed a partial rotational analysis of the $C_6H_6$ and $C_6D_6$ bands on this basis.

In forbidden electronic transitions produced by (weak) *spin-orbit interaction* the selection rules for $K$ and $N$ are changed, as has recently been recognized by Hougen (574) [see also Herzberg (523)]. Just as for linear (and diatomic) molecules spin-orbit interaction can mix states differing in $\Lambda$ by $\Delta S$, so in symmetric top molecules it can mix states differing in $K$ by $\Delta S$. For example in a triplet state a level with given $K$ can mix with $K+1$, $K$ and $K-1$ of a neighboring singlet state. For that reason in a triplet$-$singlet transition the selection rule for $K$ is

$$\Delta K = 0, \pm 1, \pm 2. \tag{II, 91}$$

For similar reasons the selection rule for $N$ is

$$\Delta N = 0, \pm 1, \pm 2 \tag{II, 92}$$

while of course the selection rule for $J$ remains the same as usual $[\Delta J = 0, \pm 1]$[9].

As a result of (II, 91), in the most general case a given vibrational transition of a triplet$-$singlet electronic transition would have five branches of sub-bands corresponding to the five $\Delta K$ values. But usually, for reasons of symmetry, not all five can occur. As one can readily see by considering the selection rule for the rovibronic species (which are the same for each set of levels of given $N$ and $K$), for example, for a ${}^3A_1-{}^1A_1$ transition of $\boldsymbol{C}_{3v}$ or ${}^3A_2''-{}^1A_1'$ of $\boldsymbol{D}_{3h}$ only $\Delta K = 0$ is possible. On the other hand for a ${}^3E-{}^1A_1$ transition of $\boldsymbol{C}_{3v}$ as well as ${}^3E'-{}^1A_1'$ of $\boldsymbol{D}_{3h}$ both $\Delta K = \pm 1$ and $\Delta K = \pm 2$ can occur. For ${}^3E-{}^1E$ transitions of $\boldsymbol{C}_{3v}$ all five values of $\Delta K$ are possible, while, for a ${}^3E''-{}^1E'$ (or ${}^3E'-{}^1E''$) transition of $\boldsymbol{D}_{3h}$ only $\Delta K = 0$ and $\pm 2$ are possible, for ${}^3E'-{}^1E'$ (or ${}^3E''-{}^1E''$) only $\Delta K = \pm 1$.

[9] For the corresponding selection rules for quartet$-$doublet transitions see footnote 6, p. 221.

In all these examples the selection rule for the $(+l)$, $(-l)$ levels is the same as for corresponding singlet–singlet transitions if $\Delta K = 0$ or $\pm 1$. But the $\Delta K = \pm 2$ transitions have the opposite $(+l)$, $(-l)$ selection rule.

No example of a triplet–singlet transition in a genuine symmetric top molecule has as yet been fully analyzed although the $CH_3I$ system previously discussed in all probability represents one triplet component of a $^3E - {}^1A_1$ transition (see p. 529). For examples among nearly symmetric tops see below.

The band structure in forbidden transitions occurring because of *magnetic dipole radiation* is entirely similar to that of ordinary allowed electric dipole transitions (just as for linear molecules). This is because the selection rules for $J$ and $K$ are the same while the selection rule for the rovibronic species is changed ($A_i \leftrightarrow A_i$ instead of $A_1 \leftrightarrow A_2$ for $C_{3v}$) just as is that for the electronic species.

Finally forbidden transitions caused by *Coriolis interaction* (i.e. by rovibronic interaction) must be mentioned. With increasing rotation vibronic levels that differ by the species of a rotation can mix and as a consequence if a suitable third vibronic state is close to the upper or lower state of a forbidden transition this transition will occur weakly for higher $J$ or $K$ values. For example, an $A_1 - A_2$ electronic (or vibronic) transition in a $C_{3v}$ molecule which is forbidden by the electric dipole selection rules can yet occur weakly with increasing $K$ since rotation about the top axis is of species $A_2$ and can mix the $A_2$ vibronic state with a nearby $A_1$ state and therefore the $A_1 - A_2$ transition can "borrow" intensity from the $A_1 - A_1$ transition. The mixing occurs for rotational levels of the same $J$ and the same rovibronic species and the rovibronic selection rules are the same as for allowed transitions.

No examples of magnetic dipole or Coriolis induced transitions have as yet been found for genuine symmetric top molecules.

### (c) Spherical top molecules

For a spherical top all three moments of inertia are equal and therefore in a first approximation the energy formula is very simple, and in fact identical with that of a linear molecule (see eq. I, 131). Naturally in this approximation a very simple band structure would arise. The actual band structure is greatly complicated by Coriolis interactions. We shall consider here only an $F_2 - A_1$ electronic transition of a $T_d$ (tetrahedral) molecule, the only type of transition that is allowed in absorption from a totally symmetric ($A_1$) ground state (see Table 9).

We shall disregard for the present considerations the effect of Jahn–Teller instability on the rotational energy levels. As we have seen in Chapter I, section 3c, first order Coriolis interaction splits the rotational levels of the $F_2$ state into three sets $F_2^{(+)}(J)$, $F_2^{(0)}(J)$ and $F_2^{(-)}(J)$ whose energies are given by eq. (I, 136). Just as for the infrared spectrum (see Vol. II, p. 453) there is a selection rule for the $F_2^{(+)}$, $F_2^{(0)}$, $F_2^{(-)}$ levels somewhat analogous to the selection rule for $(+l)$,

$(-l)$ levels for symmetric top molecules. Teller (1196) showed that only the following transitions can occur

$$\begin{aligned} &F_2^{(-)}-A_1 \quad \text{for } \Delta J = +1, \\ &F_2^{(0)}-A_1 \quad \text{for } \Delta J = 0, \\ &F_2^{(+)}-A_1 \quad \text{for } \Delta J = -1. \end{aligned} \tag{II, 93}$$

Thus instead of three branches of each type there is only one $P$, one $Q$ and one $R$ branch. Formulae for these branches have been given in Volume II, p. 454. If the rotational constants in the upper and lower state are nearly the same the $Q$ branch will be line-like while the $P$ and $R$ branches will consist of nearly equidistant lines with a spacing $2B(1 - \zeta)$. The Coriolis parameter $\zeta$ is similar to that for symmetric top molecules (see p. 230f).

If finer effects of Coriolis interaction are taken into account each rotational level is split into a number of component levels which can be characterized by over-all (rovibronic) species designations $A_1$, $A_2$, $E$, $F_1$, $F_2$ (see Fig. 38). The transitions between these levels are determined by the selection rule that the product of the rovibronic species must contain the species $A_2$ [see Hougen (573)] that is, that

$$A_1 \leftrightarrow A_2, \quad E \leftrightarrow E, \quad F_1 \leftrightarrow F_2. \tag{II, 94}$$

Since no electronic bands with discrete fine structure of any spherical top molecule have yet been resolved we shall not discuss the expected structure further but merely refer to three recent analyses of similar infrared bands by Hecht (485), Herranz (513) and Fox (395). The difference in notation for the rovibronic species in these papers from that used here (see p. 101) should be noted[10].

## (d) Asymmetric top molecules

It remains to discuss the rotational structure of electronic transitions between two states in both of which the molecule is an asymmetric top.

### (α) *General selection rules*

For the asymmetric top as for all other atomic systems we have the selection rule

$$\Delta J = 0, \pm 1 \quad (\textit{with the restriction } J = 0 \not\leftrightarrow J = 0). \tag{II, 95}$$

For multiplet states with small multiplet splitting when $N$ is the total angular momentum apart from spin we have in addition

$$\Delta N = 0, \pm 1. \tag{II, 96}$$

This selection rule replaces the first when the multiplet splitting is negligibly small.

As we have seen in Chapter I, section 3d, for every value of $J$ (or $N$) there are in a given vibronic state of an asymmetric top $2J + 1$ (or $2N + 1$) distinct rotational levels. Even if the molecule has no symmetry there are additional restric-

[10] In the notation used by the authors mentioned the rovibronic selection rule cannot be stated in such a simple way as in (II, 94).

tions for the combinations of the levels of a given $J$ value of one electronic state with those of a given $J$ value of another electronic state. These restrictions are related to the symmetry properties $++$, $+-$, $-+$, $--$ of the asymmetric top rotational eigenfunctions (see p. 109). The asymmetric top species are rigorously defined as long as rotation can be separated from vibronic motion and are for most purposes sufficiently well defined even if the interaction of rotation and vibronic motion is not vanishingly small. We have then, just as for infrared spectra, the following selection rules:

If the transition moment is in the direction of the axis of smallest moment of inertia (*a-axis*) only the transitions

$$++ \leftrightarrow -+ \quad \text{and} \quad +- \leftrightarrow -- \tag{II, 97}$$

can occur. If the transition moment is in the direction of the *b-axis* only the transitions

$$++ \leftrightarrow -- \quad \text{and} \quad +- \leftrightarrow -+ \tag{II, 98}$$

can occur and if the transition moment is in the direction of the *c-axis* only the transitions

$$++ \leftrightarrow +- \quad \text{and} \quad -+ \leftrightarrow -- \tag{II, 99}$$

can occur. If the transition moment does not lie in the direction of one of the principal axes the allowed transitions depend on which components of the transition moment are different from zero. But even in the most general case when all three components along the principal axes are different from zero, transitions between levels of the same symmetry are still forbidden, i.e.

$$++ \not\leftrightarrow ++, \quad +- \not\leftrightarrow +-, \quad -+ \not\leftrightarrow -+, \quad -- \not\leftrightarrow --. \tag{II, 100}$$

Energy level diagrams showing the operation of the selection rules (II, 97–99) have been given in Figs. 149, 154 and 160 of Volume II.

An alternative way of stating the selection rules just given is by means of $K_a$ and $K_c$, the quantum numbers of rotation about the $a$ and $c$ axes in the limiting cases of the corresponding prolate and oblate symmetric tops. As we have seen, each level of the asymmetric top is completely described by $J_{K_aK_c}$. For $K_a$ and $K_c$ one finds the selection rules [see Cross, Hainer and King (257)]:

if the transition moment is in the *a-axis*

$$\Delta K_a = 0, \pm 2, \ldots \qquad \Delta K_c = \pm 1, \pm 3, \ldots, \tag{II, 101}$$

if it is in the *b-axis*

$$\Delta K_a - \pm 1, \pm 3, \ldots \qquad \Delta K_c = \pm 1, \pm 3, \ldots, \tag{II, 102}$$

and if it is in the *c-axis*

$$\Delta K_a = \pm 1, \pm 3, \ldots \qquad \Delta K_c = 0, \pm 2, \ldots. \tag{II, 103}$$

It is to be understood that in these selection rules for a given $\Delta K_a$ the possible values of $\Delta K_c$ are not entirely unrestricted because the sum $K_a + K_c$ is either equal to $J$ ("even" levels according to Cross, Hainer and King) or $J + 1$ ("odd" levels); therefore, for example, for $\Delta K_a = 0$ and $\Delta J = 0$, $\Delta K_c$ can only have the values $\pm 1$, while for $\Delta K_a = 0$ and $\Delta J = +1$ it can only have the value $\Delta K_c = +1$ and for $\Delta K_a = 0$, $\Delta J = -1$ only $\Delta K_c = -1$; for $\Delta K_a = +1$ and $\Delta J = 0$ only $\Delta K_c = -1$, 0 or $+1$ is possible, and so on.

Although for unsymmetrical molecules the selection rules (II, 97–99) and (II, 101–103) hold only to the extent that the rotational motion can be separated from the vibronic motion and would be violated for example when there is strong Coriolis interaction, some (or all) of the selection rules are maintained irrespective of separability *if the molecule has symmetry*.

In principle the potential energy of all molecules has inversion symmetry (see p. 93) and therefore we have the parity rule for "positive" and "negative" rotational levels

$$+ \leftrightarrow -, \quad + \not\leftrightarrow +, \quad - \not\leftrightarrow -. \tag{II, 104}$$

As pointed out in Volume II, p. 415, this rule is of consequence for non-planar molecules only when the inversion doubling is not negligibly small. For planar molecules, for which the $c$-axis is always perpendicular to the plane of the molecule, the parity rule does introduce restrictions but these restrictions do not go as far as those introduced by (II, 97–99) although unlike the latter they are rigorous for electric dipole radiation. For example, remembering that the first sign in the $+-$ symbol gives the parity if the vibronic state is symmetric with respect to the plane of the molecule, we see that both the $a$-type and $b$-type transitions represented by eqs. (II, 97 and 98) are in accord with the parity rule and no others can occur between states that are symmetric with respect to the plane. The $c$-type transitions are also seen to be in accord with the parity rule if it is remembered that in a vibronic state that is antisymmetric with respect to the plane of the molecule the parities are opposite to the first sign in the $+-$ symbol.

More important are the selection rules for the over-all (rovibronic) species, although these too do not introduce any additional restrictions beyond (II, 97–99) but only make some (or all) of them rigorous. We have for the over-all species based on the rotational sub-group the simple rule that it does not change in a transition, i.e.

$$A \leftrightarrow A,\ B \leftrightarrow B, \ldots. \tag{II, 105}$$

This is an absolute rule for zero nuclear spin of the identical atoms; but even for non-zero nuclear spin it is still extremely strong (just like the ortho $\not\leftrightarrow$ para rule of diatomic molecules) since the coupling of nuclear spin with rotation is so very small. The rule holds not only for electric dipole radiation but also for magnetic dipole and quadrupole radiation, as well as for collisions, and is the basis for the intensity alternation (or alternate missing lines) in the various branches of bands of symmetrical molecules.

For the over-all species based on the full symmetry group we have again the selection rule (see p. 223) that the *product of the over-all species of upper and lower* *state must be that of the product $T_zR_z$ of a translation and a rotation.* This rule holds for electric dipole radiation only. In Table 15 the species of $T_zR_z$ is given for all asymmetric top point groups and, derived from it, the allowed rovibronic transitions are listed. It may be noted that by dropping $g$, $u$ for $\boldsymbol{C}_i$, $\boldsymbol{C}_{2h}$, $\boldsymbol{D}_{2h}$, prime and double-prime for $\boldsymbol{C}_s$ and subscripts 1 and 2 for $\boldsymbol{C}_{2v}$ the selection rules (II, 105) are obtained.

TABLE 15. ROVIBRONIC SELECTION RULES FOR ASYMMETRIC TOP POINT GROUPS

| Point group | Species of $T_zR_z$ | Allowed rovibronic transitions |
|---|---|---|
| $\boldsymbol{C}_1$ | $A$ | $A-A$ |
| $\boldsymbol{C}_i$ | $A_u$ | $A_u-A_g$ |
| $\boldsymbol{C}_s$ | $A''$ | $A''-A'$ |
| $\boldsymbol{C}_2$ | $A$ | $A-A,\ B-B$ |
| $\boldsymbol{C}_{2h}$ | $A_u$ | $A_u-A_g,\ B_u-B_g$ |
| $\boldsymbol{C}_{2v}$ | $A_2$ | $A_1-A_2,\ B_1-B_2$ |
| $\boldsymbol{D}_2$ | $A$ | $A-A,\ B_1-B_1,\ B_2-B_2,\ B_3-B_3$ |
| $\boldsymbol{D}_{2h}$ | $A_u$ | $A_u-A_g,\ B_{1u}-B_{1g},\ B_{2u}-B_{2g},\ B_{3u}-B_{3g}$ |

As an example of the operation of these selection rules let us consider an $A_1-A_1$ vibronic transition of a $\boldsymbol{C}_{2v}$ molecule in which the $a$-axis lies in the symmetry axis (e.g. $H_2CO$). As we have seen in Chapter I, Table 6, the correlation of $+-$ symmetry and rovibronic species for an $A_1$ vibronic state is

$$\begin{array}{cccc} ++ & -+ & -- & +- \\ A_1 & A_2 & B_1 & B_2 \end{array}$$

Therefore the selection rule (Table 15) that only $A_1-A_2$ and $B_1-B_2$ rovibronic transitions occur means that only $++ \leftrightarrow -+$ and $-- \leftrightarrow +-$ transitions occur in agreement with the selection rule (II, 97). In this case as well as all other vibronic transitions of point groups $\boldsymbol{C}_{2v}$, $\boldsymbol{D}_2$ and $\boldsymbol{D}_{2h}$ the selection rules of Table 15 lead to the same restrictions as the selection rules (II, 97–99) but they hold, unlike (II, 97–99), also when rovibronic (Coriolis) interaction is large and when as a result forbidden vibronic transitions occur. For the point groups of lower symmetry $\boldsymbol{C}_i$, $\boldsymbol{C}_s$, $\boldsymbol{C}_2$, $\boldsymbol{C}_{2h}$ the selection rules of Table 15, while in agreement with (II, 97–99) are less restrictive and therefore for normal allowed transitions the rules (II, 97–99) should be used; but when rovibronic interaction is large, i.e. in forbidden vibronic transitions only the rules of Table 15 remain valid.

### (β) *Slightly asymmetric tops*

If the molecule is only slightly asymmetric, the band structure can be most readily understood by starting out from the nearest symmetric top and then introducing the $K$-type doubling due to the asymmetry. Since this asymmetry doubling decreases rapidly with $K$, for high $K$ even fairly strongly asymmetric tops may be treated as nearly symmetric tops but for low $K$ the asymmetry must be very small for this to be possible.

For slightly asymmetric tops the general selection rules (II, 101–103) are replaced by the symmetric top selection rules

$$\Delta K_a = 0, \pm 1 \quad \text{or} \quad \Delta K_c = 0, \pm 1 \tag{II, 106}$$

depending on whether the molecule approaches a prolate or an oblate symmetric top. In either case $\Delta K = 0$ corresponds to the $\parallel$ component, $\Delta K = \pm 1$ to the $\perp$ component of the transition moment, that is, to the component $\parallel$ or $\perp$ to the top axis. Whether one or the other or both types of rotational transitions occur depends on the point group, the relation of the top axis to the symmetry elements, and the type of vibronic transition. These relations for electronic (vibronic) transitions are entirely similar to infrared transitions and are summarized in Table 16. Hybrid bands arise whenever the transition moment has components both in the top axis and perpendicular to it, $\parallel$ bands when the transition moment lies in the top axis, and $\perp$ bands when it is perpendicular to the top axis. The rigorous selection rules for the rovibronic species (Table 15) and the selection rules (II, 97–99) are valid of course for slightly as well as for strongly asymmetric tops.

TABLE 16. BAND TYPES FOR VARIOUS VIBRONIC TRANSITIONS IN SLIGHTLY ASYMMETRIC TOP MOLECULES

| Point group | Vibronic transitions | Band type[a] | | |
|---|---|---|---|---|
| | | Top axis $\parallel z$[b] | Top axis $\parallel y$ | Top axis $\parallel x$ |
| $\boldsymbol{C}_1$ | $A-A$ | hybrid ($A, B, C$) | hybrid ($A, B, C$) | hybrid ($A, B, C$) |
| $\boldsymbol{C}_i$ | $A_u-A_g$ | hybrid ($A, B, C$) | hybrid ($A, B, C$) | hybrid ($A, B, C$) |
| $\boldsymbol{C}_s$ | $A'-A'$, $A''-A''$ | $\perp$ band ($B, C$) | hybrid ($A, B$) | hybrid ($A, B$) |
| | $A'-A''$ | $\parallel$ band ($A$) | $\perp$ band ($B$ or $C$) | $\perp$ band ($B$ or $C$) |
| $\boldsymbol{C}_2$, $[\boldsymbol{C}_{2h}]$[c] | $A-A$, $B-B$ | $\parallel$ band ($A$) | $\perp$ band ($B$ or $C$) | $\perp$ band ($B$ or $C$) |
| | $A-B$ | $\perp$ band ($B, C$) | hybrid ($A, B$) | hybrid ($A, B$) |
| $\boldsymbol{C}_{2v}$ | $A_i-A_i$, $B_i-B_i$ | $\parallel$ band ($A$) | $\perp$ band ($B$ or $C$) | $\perp$ band ($B$ or $C$) |
| | $A_1-B_1$, $A_2-B_2$ | $\perp$ band ($B$ or $C$) | $\perp$ band ($B$ or $C$) | $\parallel$ band ($A$) |
| | $A_1-B_2$, $A_2-B_1$ | $\perp$ band ($B$ or $C$) | $\parallel$ band ($A$) | $\perp$ band ($B$ or $C$) |
| $\boldsymbol{D}_2$, $[\boldsymbol{D}_{2h}]$[c] | $A-B_1$, $B_2-B_3$ | $\parallel$ band ($A$) | $\perp$ band ($B$ or $C$) | $\perp$ band ($B$ or $C$) |
| | $A-B_2$, $B_1-B_3$ | $\perp$ band ($B$ or $C$) | $\parallel$ band ($A$) | $\perp$ band ($B$ or $C$) |
| | $A-B_3$, $B_1-B_2$ | $\perp$ band ($B$ or $C$) | $\perp$ band ($B$ or $C$) | $\parallel$ band ($A$) |

[a] The symbols $A$, $B$, $C$ in brackets indicate whether the transition moment has components in the $a$, $b$ or $c$ axis for a near-prolate top (top axis = $a$ axis); for a near-oblate top (top axis = $c$ axis) $A$ and $C$ must be exchanged.
[b] The $z$ axis is assumed to be $\perp$ to $\sigma$ of $\boldsymbol{C}_s$, but identical with $C_2(z)$ in $\boldsymbol{C}_2$, $\boldsymbol{C}_{2h}$, $\boldsymbol{C}_{2v}$, $\boldsymbol{D}_2$.
[c] For $\boldsymbol{C}_{2h}$ and $\boldsymbol{D}_{2h}$ add the $g$, $u$ rule; it does not affect the band type.

**Parallel bands.** The structure of $\parallel$ bands of slightly asymmetric tops is entirely similar to that of $\parallel$ bands of symmetric tops except that there is a doubling in all three branches of all sub-bands with $K > 0$ on account of the asymmetry doubling. Figure 104 shows an energy level diagram for the $1-1$ sub-band indicating how the branches originate. In order to see that a splitting into only two and not four lines arises, it is necessary to take account of the symmetry rules

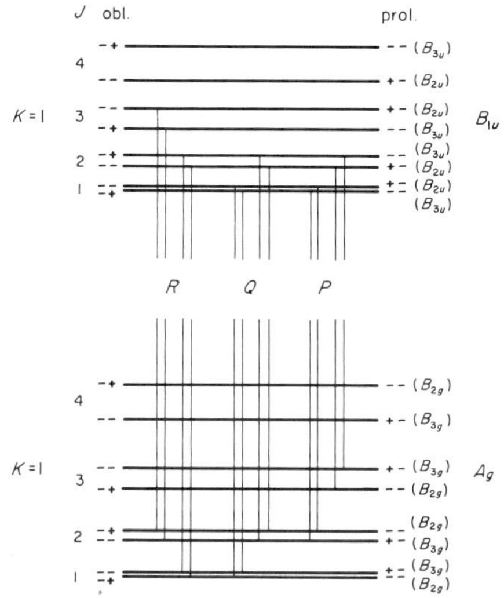


FIG. 104. **Energy level diagram showing the origin of the six branches in a 1 – 1 sub-band of a ∥ band of a slightly asymmetric top.** At the left the asymmetric top species (+ –) for an oblate, at the right for a prolate top are given. In addition at the right in brackets the rovibronic species for a $B_{1u} - A_g$ vibronic transition of a $\boldsymbol{D}_{2h}$ molecule are given. The top axis is assumed to be the $z$-axis.

(II, 97) for a prolate or (II, 99) for an oblate top. The asymmetric top species for the two alternatives are given at the right and left of the energy levels in Fig. 104. The result is the same for both: there are only two possible combinations for each transition that is single without asymmetry splitting; the line splitting is the sum of the level splittings in the $Q$ branch, the difference in the $P$ and $R$ branches. This result is independent of $K$ and holds for both prolate and oblate tops. The same results can also be obtained (and more rigorously) from the rovibronic species if the symmetry is high enough. As an example in Fig. 104 at the right the rovibronic species for a $B_{1u} - A_g$ vibronic transition of a $\boldsymbol{D}_{2h}$ molecule are indicated. It is seen that the selection rules of Table 15 are fulfilled and no other transitions than those shown are possible. If the symmetry is low ($\boldsymbol{C}_1$, $\boldsymbol{C}_i$, $\boldsymbol{C}_s$, $\boldsymbol{C}_2$) and if there is strong Coriolis interaction the (approximate) selection rules (II, 97) and (II, 99) break down and in principle a total of four lines can occur for each single line of the symmetric top, but no such case has as yet been observed. As we have seen in Chapter I the magnitude of the asymmetry splitting for low $J$ is largest for $K = 1$ and decreases rapidly with increasing $K$. For $K = 1$ it increases with $J(J + 1)$, for $K = 2$ with $J^2(J + 1)^2$, and so on.

For those sub-bands for which the $K$-doubling is resolved the combination differences $\Delta_2^J F(K, J)$ formed between appropriate components of $P$ and $R$ "lines" give immediately the effective $B$ values of the two component levels in the upper and lower state. Alternatively, combination differences $\Delta_1^J F(J, K)$ between $P$ and $Q$ or between $Q$ and $R$ lines can be used for the same purpose. The resulting effective $B$ values are related to the true $A_{[v]}$, $B_{[v]}$ and $C_{[v]}$ values by Polo's formulae (I, 146–152); in particular for $K = 1$ we have for a near-prolate top in a first approximation

$$B^1_{\text{eff.}}{}^{(c)} = \tfrac{1}{4}(3B_{[v]} + C_{[v]}), \qquad B^1_{\text{eff.}}{}^{(d)} = \tfrac{1}{4}(B_{[v]} + 3C_{[v]}) \tag{II, 107}$$

where ($c$) and ($d$) stand for the upper and lower components of the asymmetry doublets. From the observed $B_{\text{eff.}}$ values according to (II, 107) the rotational constants $B_{[v]}$ and $C_{[v]}$ can be determined both for the upper and lower electronic state, possibly after correction with the asymmetry terms of eq. (I, 148) and centrifugal distortion terms. However $A_{[v]}$ cannot be accurately determined from a $\|$ band since it enters only into the higher correction terms of the effective $B$ values. Only the difference $A'_{[v]} - A''_{[v]}$ is readily obtained. According to (I, 146) the sub-band origins are given in a first approximation by

$$\nu_0^{\text{sub}} = \nu_0 + \{[A'_{[v]} - \tfrac{1}{2}(B'_{[v]} + C'_{[v]})] - [A''_{[v]} - \tfrac{1}{2}(B''_{[v]} + C''_{[v]})]\}K^2. \tag{II, 108}$$

Thus if the $B_{[v]}$ and $C_{[v]}$ values have been determined $A'_{[v]} - A''_{[v]}$ is immediately found from a plot of $\nu_0^{\text{sub}}$ against $K^2$. Not infrequently $A''_{[v]}$ is known for the lower (ground) state from infrared, Raman or microwave spectra. Then a reliable value for $A'_{[v]}$ is immediately obtained from a $\|$ electronic band.

For planar molecules, if neither $A'_{[v]}$ nor $A''_{[v]}$ is known from other data, an approximate value can be obtained indirectly from $B_{[v]}$ and $C_{[v]}$ under the assumption that the inertial defect is zero (i.e. $I_C = I_A + I_B$). This assumption is in general fairly well fulfilled for the lowest vibrational level, that is for $A_{[0]}$, $B_{[0]}$, $C_{[0]}$.

When the asymmetry doubling is not resolved one obtains from the analysis of the $J$ structure only $\bar{B}_{[v]} = \tfrac{1}{2}(B_{[v]} + C_{[v]})$. Again for planar molecules if $A_{[v]}$ is known from other data $B_{[v]}$ and $C_{[v]}$ can be obtained separately on the assumption of zero inertial defect according to the relations (see Vol. II, p. 437)

$$B = -(A - \bar{B}) + \sqrt{A^2 + \bar{B}^2}, \qquad C = +(A + \bar{B}) - \sqrt{A^2 + \bar{B}^2}. \tag{II, 109}$$

Several examples of $\|$ bands of slightly asymmetric tops have been wholly or partially analysed both in allowed and forbidden electronic transitions. Among the former are the main bands of the ultraviolet system of $NO_2$ extending from 1650 to 1350 Å recently studied by Ritchie, Walsh and Warsop (1072), the main bands of the 3200 Å systems of the diazines [Innes and his collaborators (608) (609) (610) (831)], and of the 5600 Å system of tetrazine [Mason (799)]. The forbidden components of the 3900 Å bands of HC≡CCHO show clear examples of $\|$ bands [Brand, Callomon and Watson (141)]. The forbidden (magnetic dipole) components of the near ultraviolet $H_2CO$ system represent a fully resolved example of the $\|$ type. In $NO_2$, unlike the other cases, there is a strong divergence of the sub-bands and what are mainly observed are the $R$-heads of the sub-bands. These $R$ heads follow the formula (II, 108) and yield a value of $(A' - \bar{B}') - (A'' - \bar{B}'')$. The $J$ structure has been partially resolved but an approximate value

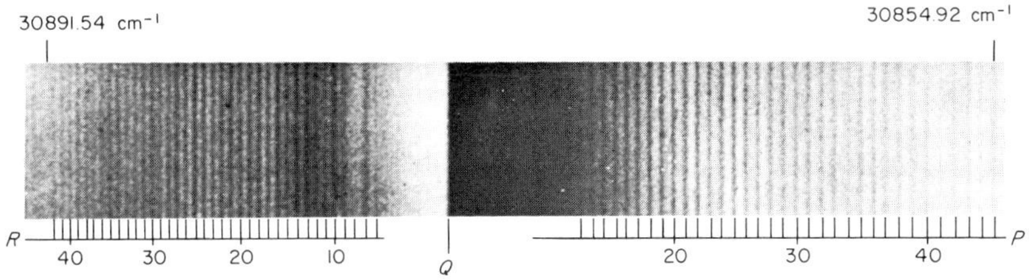


FIG. 105. **Spectrogram of the 0 – 0 band of the $\tilde{A} - \tilde{X}$ system of symmetrical diazine (pyrazine) at 3239 Å after Merritt and Innes (831).** The $K$ structure of the "lines" of the $P$ and $R$ branches is not resolved.

for $\bar{B}'$ could only be obtained by a stochastic method, i.e. by calculation of model bands with different values of $\bar{B}'$ using the value of $\bar{B}''$ known from microwave studies.

The diazines and tetrazines unlike the other molecules are oblate tops. Even though here the asymmetry parameter is rather large, these molecules can be considered under the heading of slightly asymmetric tops because the strongest lines have fairly high $K$ values for which the asymmetry splitting is very small. Since in addition apparently the moments of inertia or at least $C - \frac{1}{2}(A + B)$ in upper and lower state are so nearly alike that sub-bands of different $K_c$ are almost exactly superimposed a simple $P, Q, R$ structure results just as for infrared $\|$ bands. Figure 105 shows as an example the structure of the $0-0$ band of symmetrical diazine. It does not differ in any noticeable way from a $\|$ band of a genuine symmetric top.

**Perpendicular bands.** For $\perp$ bands of slightly asymmetric top molecules we have the selection rule $\Delta K = \pm 1$. In addition the symmetry rules (II, 97–99) and the rovibronic selection rules given in Table 15 must be observed. In Fig. 106 the detailed structure of a $\perp$ band is shown in a way similar to Fig. 99 for a symmetric top. For simplicity, $A' = A''$, $B' = B''$ and $C' = C''$ has been assumed. Exact rigid asymmetric top levels with $\kappa = -0.95$ were used in the construction of this figure. The relative intensities are from the tables of Cross, Hainer and King (257) assuming a temperature of 300°K. It is seen, by comparison with Fig. 99, that the appearance of the coarse structure ($K$ structure) is the same as for a genuine symmetric top: if as assumed the rotational constants in the upper and lower state are the same there is a series of equidistant sub-bands; if the rotational constants differ there is a diverging series of such sub-bands. For low resolution the $Q$ branches of these sub-bands even though no longer as line-like as for a symmetric top form the most prominent feature of the band. As before the series of sub-bands consists of two branches, an $r$ and a $p$ branch corresponding to $\Delta K = +1$ and $-1$ respectively which join together without a break.

From (I, 146) and the selection rule $\Delta K = \pm 1$, neglecting all correction terms, we obtain immediately for the sub-band origins

$$\nu_0^{sub.} = \nu_0 + [A' - \tfrac{1}{2}(B' + C')] \pm 2[A' - \tfrac{1}{2}(B' + C')]K + \{[A' - \tfrac{1}{2}(B' + C')] - [A'' - \tfrac{1}{2}(B'' + C'')]\}K^2. \qquad \text{(II, 110)}$$

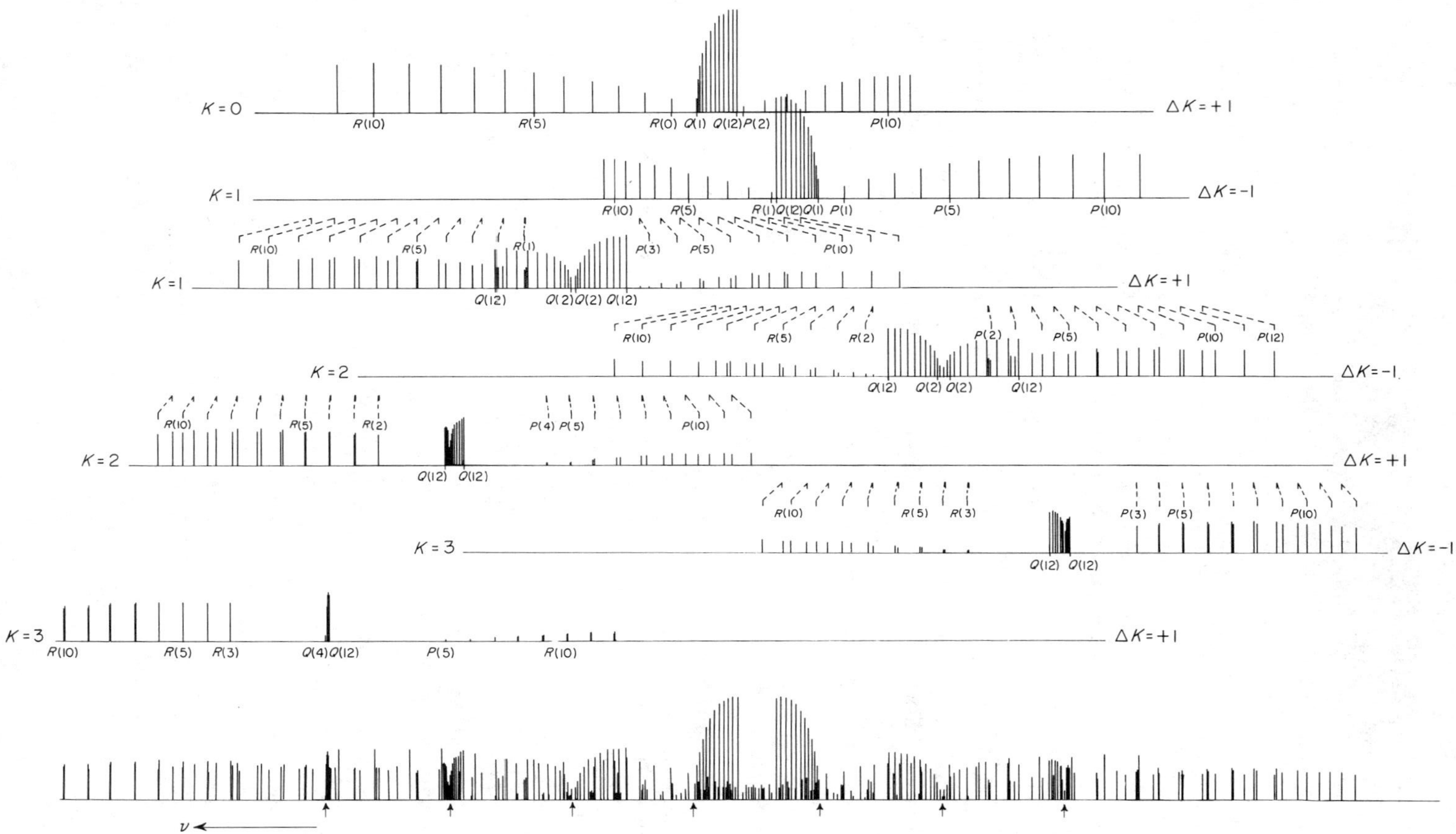


FIG. 106. **Calculated structure of a ⊥ band of a slightly asymmetric top molecule ($\kappa = 0.95$) assuming $A' \approx A''$, $B' \approx B''$, $C' \approx C''$.** In this spectrum

If the quadratic term is small (i.e. if there is little change in the rotational constants), we have a spacing of $2[A - \frac{1}{2}(B + C)]$ which should be compared with $2[A(1 - \zeta) - B]$ for the genuine symmetric top. The complication introduced by the Coriolis parameter $\zeta$ is not present here.

However, it happens not infrequently, especially in very nearly prolate top molecules, that there are pairs of skeletal deformation vibrations in which displacements are at right angles to each other (e.g., species $B_1$ and $B_2$ in planar $C_{2v}$) and not too different in frequency. The rovibronic levels of these modes may then be strongly coupled through $K$-dependent Coriolis forces, and as a consequence eq. (II, 110) ceases to apply. Rather, the levels have to be analysed in pairs with the help of degenerate perturbation theory [see Nielsen (933a)]; the energies and hence the sub-band origins again contain first order terms in $2A\zeta K$ as for symmetric tops. Examples of such analyses have been published for formaldehyde [Brand (138)] and propynal [Brand, Callomon and Watson (141)].

Small but significant differences between perpendicular bands of genuine and nearly symmetric tops arise in the $J$ structure on account of the asymmetry doubling for $K > 0$ and of the symmetry selection rules. For a prolate top we must distinguish three cases:

(1) the transition moment lies in the $c$-axis;
(2) the transition moment lies in the $b$-axis;
(3) the transition moment has components in both $b$- and $c$-axes.

Corresponding cases for an oblate top are obtained by replacing $c$ by $a$.

In Fig. 107 the possible transitions for case (1) are indicated by full lines in an energy level diagram assuming a prolate top of point group $C_{2v}$ [with $C_2(z)$ as $a$-axis and the $x$-axis as $c$-axis]. The spectrum in Fig. 106 is drawn for this case. The most important difference from a genuine symmetric top is the line doubling in all three branches of the $2-1$ and $1-2$ sub-bands, the much smaller doubling in the $3-2$ and $2-3$ sub-bands and the still smaller doublings in higher sub-bands. This doubling arises solely on account of the asymmetry of the top and will increase with increasing deviation from the symmetric top. According to the symmetry selection rule (II, 99), in the sub-bands of case (1) the splitting of the $Q$ lines is the sum, that of the $P$ and $R$ lines the difference of the level splittings in upper and lower state (see Fig. 107).

Since the level splittings for $K = 1$ are proportional to $J(J + 1)$ we see that, even though we assumed $A' = A''$, $B' = B''$, $C' = C''$ in Fig. 106 the two component $Q$ branches of the $2-1$ and $1-2$ sub-bands are not line-like but shaded in opposite directions. A similar difference for the $3-2$ and $2-3$ sub-bands is much less pronounced since the level splittings for $K = 2$ are much smaller. Because of the symmetry selection rule (see Fig. 107) the three branches of the $1-0$ and $0-1$ sub-bands remain single; the $Q$ lines go only to the lower, the $P$ and $R$ lines only to the upper component levels of the asymmetry doublets. As a result as shown in Fig. 106 the single $Q$ branch of the $1-0$ sub-band is shaded to the red, that of the $0-1$ sub-band is shaded to the violet while the $P$ and $R$ branches are

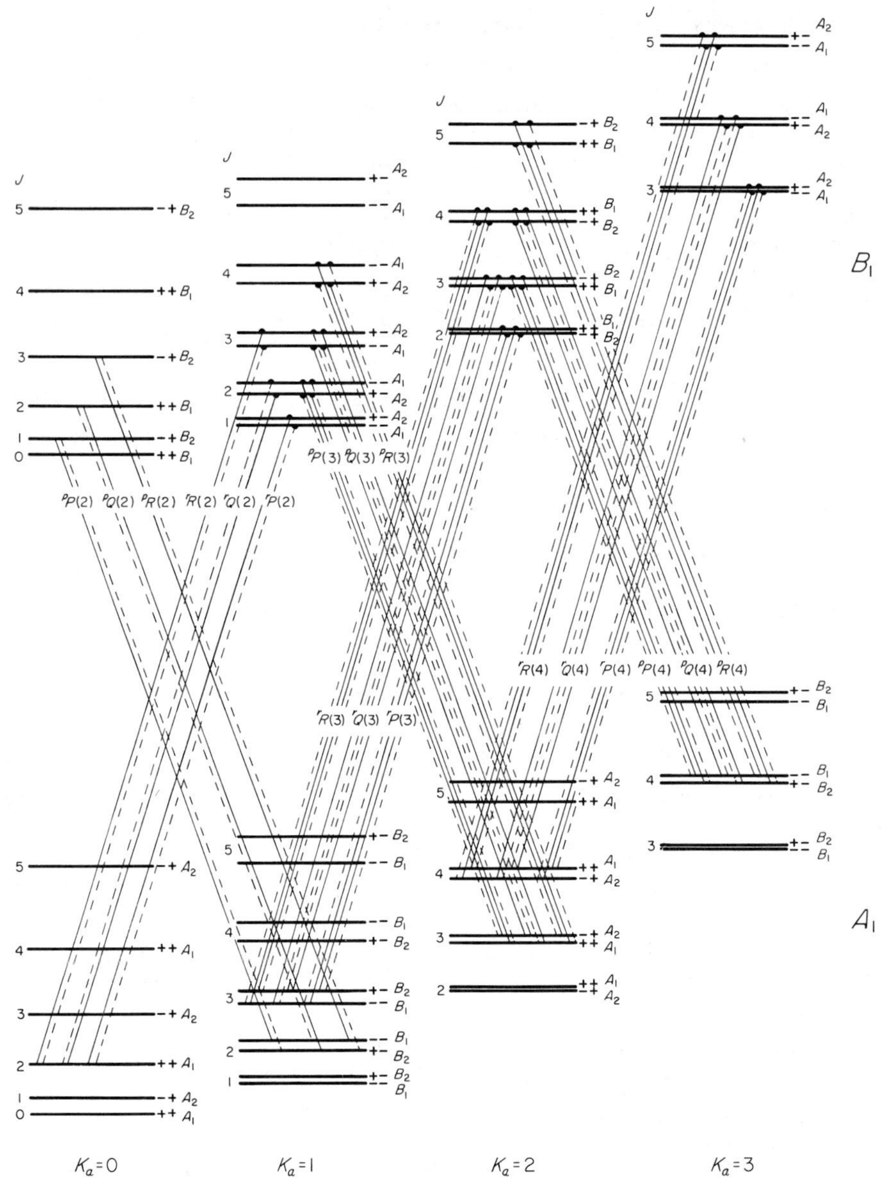


FIG. 107. **Energy level diagram for a $\perp$ band of a slightly asymmetric (prolate) top molecule.** The full-line transitions correspond to case (1), i.e. the transition moment in the $c$-axis; the broken-line transitions correspond to case (2), i.e., the transition moment in the $b$-axis. The rovibronic species added at the right of the levels correspond to a $B_1 - A_1$ transition of a $\mathbf{C}_{2v}$ molecule with the $C_2$ being the $b$-axis. Only one member of each branch is shown.

shaded in the opposite direction. Of course, when, as usually happens, the assumption $A' = A''$, $B' = B''$, $C' = C''$ underlying Fig. 106 is not fulfilled both $Q$ branches will in general be shaded in the same direction but one more so than the other and similarly for the $P$ and $R$ branches.

In case (2) (i.e. when the transition moment is in the $b$-axis) we must use the selection rule (II, 98) in place of (II, 99). The corresponding transitions are indicated by broken lines in Fig. 107. Now the $Q$ lines in the $1-0$ and $0-1$ sub-bands go to the upper rather than the lower component levels of $K = 1$ and therefore the shading of the $Q$ branches and of the $P$ and $R$ branches for $A' = A''$, $B' = B''$, $C' = C''$ is the reverse of that shown in Fig. 106. Similarly in the $2-1$ and $1-2$ (and higher) sub-bands the two components of the $Q$ lines are separated by the difference rather than the sum of the level splittings in the upper and lower state. From the sign of the combination defect in the $1-0$ and $0-1$ sub-bands or simply from the direction of shading of the $Q$ branches it is easy to decide whether the $Q$ lines correspond to the upper or lower component levels in the $K = 1$ set and therefore it is easy to decide whether the transition moment is in the $b$- or $c$-axis. In this way the nature of the vibronic transition can be established.

When the transition moment has both a component in the $b$ and one in the $c$-axis [case (3) above] both the transitions indicated by solid lines and those indicated by broken lines in Fig. 107 can occur. Thus there will be four component lines for all $P$, $Q$ and $R$ "lines" of the $2-1$, $1-2$, $3-2$, $2-3, \ldots$ sub-bands and two component lines for those of the $0-1$ and $1-0$ sub-bands. However, the two sets of components, those corresponding to the $b$-component and those corresponding to the $c$-component, of the transition moment will in general have different intensities unless by accident the transition moment is at 45° to the $b$- and $c$-axes. Case (3) can occur only for the point groups, $\boldsymbol{C}_1$, $\boldsymbol{C}_2$, $\boldsymbol{C}_i$, $\boldsymbol{C}_s$, $\boldsymbol{C}_{2h}$ for which not all three directions of the top axes are determined by symmetry.

While approximate values of the rotational constants can be estimated simply from the spacing of the sub-bands and of the lines within the sub-bands, for a more precise evaluation it is best to use the method of combination differences. For an evaluation of $A - \frac{1}{2}(B + C)$ and $D_K$ in the upper and lower state we use as before the sub-band origins. Since now $\zeta = 0$ we have instead of (II, 85) the simpler equations

$$\begin{aligned} \nu_r^0(K) - \nu_p^0(K) &= F'(0, K+1) - F'(0, K-1) = \Delta_2^K F'(0, K) \\ &= 4[A' - \tfrac{1}{2}(B' + C')]K - 8D_K'K(K^2+1) \\ \nu_r^0(K-1) - \nu_p^0(K+1) &= F''(0, K+1) - F''(0, K-1) = \Delta_2^K F''(0, K) \\ &= 4[A'' - \tfrac{1}{2}(B'' + C'')]K - 8D_K''K(K^2+1). \end{aligned} \qquad \text{(II, 111)}$$

From these relations $A - \frac{1}{2}(B + C)$ and $D_K$ can be determined either by the method of least squares or by one of the graphical methods described in Volumes I and II.

Strictly speaking according to (I, 146) it is not $A - \frac{1}{2}(B + C)$ but

$$[A - \tfrac{1}{2}(B + C)](1 - \tfrac{3}{8}b^2 + \cdots)$$

that is obtained. However, a fairly strong asymmetry is required to make the second bracket sufficiently different from 1 to affect the $A$ value within its accuracy.

Instead of using the sub-band origins in (II, 111) one may also take differences of corresponding lines in the sub-bands. If one uses averages of asymmetry doublets for this purpose one can see readily from (I, 146) and (I, 158) that only the terms

$$[\Delta B_{\text{eff.}}^{K+1} - \Delta B_{\text{eff.}}^{K-1} - 4D_{JK}K]J(J+1) + (\Delta D_{\text{eff.}}^{K+1} - \Delta D_{\text{eff.}}^{K-1})J^2(J+1)^2 + \cdots \quad \text{(II, 112)}$$

have to be added to (II, 111) if the asymmetry is assumed to be small, so that the effect of the constants $D_1$, $D_2$, $D_3$ can be neglected. Only the first term of (II, 112) is important.

The rotational constant $\bar{B} = \frac{1}{2}(B + C)$ is determined (just as $B$ for a genuine symmetric top) from combination differences in the $J$ structure. Similar to (II, 84) we have here

$$\begin{aligned}\Delta_2^J F(J, K) = 4[\tfrac{1}{2}(B + C) - \tfrac{3}{2}D_J - \tfrac{3}{2}\Delta D_{\text{eff.}}^K + \Delta B_{\text{eff.}}^K - D_{JK}K^2](J + \tfrac{1}{2})\\ - 8(D_J - \Delta D_{\text{eff.}}^K)(J + \tfrac{1}{2})^3 + \cdots. \quad \text{(II, 113)}\end{aligned}$$

Higher terms must be included unless the asymmetry is small. The observed $\Delta_2^J F(J, K)$ values are as usual obtained from $R_K(J - 1) - P_K(J + 1)$ for the lower and $R_K(J) - P_K(J)$ for the upper state.

By a least squares evaluation or the usual graphical methods we obtain the coefficients of $4(J + \frac{1}{2})$ and $8(J + \frac{1}{2})^3$ in (II, 113) which represent effective $B$ and $D$ values. The effective $B$ value is $\frac{1}{2}(B + C)$ except for the small asymmetry correction $\Delta B_{\text{eff.}}^K$ and the term $-D_{JK}K^2$. If the $\Delta_2^J F(J, K)$ have been obtained for several $K$ values, the slope of the plot of the effective $B$ against $K^2$ will give the centrifugal stretching constant $D_{JK}$. Its value must, of course, agree with the value obtained above from the $\Delta_2^K F$. The correction term $\Delta B_{\text{eff.}}^K$ is independent of $K$ if averages of the asymmetry doublets are used [see eqs. (I, 147–152)]. To a first good approximation it is given by

$$\Delta B_{\text{eff.}}^K = \frac{1}{16} \frac{(C - B)^2}{A - \frac{1}{2}(B + C)}. \quad \text{(II, 114)}$$

For a slightly asymmetric top this term is clearly very small compared to $\frac{1}{2}(B + C)$ and can often be neglected. In that event $\frac{1}{2}(B + C)$ is obtained as the intercept in the plot of the effective $B$ versus $K^2$.

Frequently not all three branches of all sub-bands are well resolved and only very few $\Delta_2^J F(J, K)$ may be determined. One can then also use the $\Delta_1^J F(J, K)$ values if the $Q$ branch and one of the other branches is resolved. However, this is possible only for those sub-bands for which the asymmetry doubling is small. If one uses the average of each doublet one finds neglecting asymmetry terms

$$\begin{aligned}\Delta_1^J F(J, K) &= F(J + 1, K) - F(J, K)\\ &= 2[\tfrac{1}{2}(B + C) - D_{JK}K^2](J + 1) - 4D_J(J + 1)^3. \quad \text{(II, 115)}\end{aligned}$$

The observed $\Delta_1^J F(J, K)$ values are obtained in the usual way (see p. 216) from $P$ and $Q$ or $Q$ and $R$ branches.

For the $1-0$ and $0-1$ sub-bands if only one $P$, one $Q$ and one $R$ branch are present [case (1) and (2) above, see Fig. 107], the $\Delta_1^J F(J, K)$ values obtained from $R$ and $Q$ do not agree with those obtained from $Q$ and $P$ branches. As is readily seen from Fig. 107 and as has already been discussed for linear – bent transitions of "linear" molecules the differences between the two $\Delta_1^J F(J, K)$ values give immediately the sum of the asymmetry doublings in two successive rotational levels. This sum is given by (see eq. II, 53)

$$\Delta\nu_{cd}^{K=1}(J+1) + \Delta\nu_{cd}^{K=1}(J) = (B - C)(J+1)^2. \qquad \text{(II, 116)}$$

Higher order corrections depend on $b^3$ and are usually negligible except for very high $J$. If a few values of the sum have been obtained from the observed spectrum, a very precise value of $B - C$ can be derived from the relation (II, 116) and therefore with $\frac{1}{2}(B + C)$ from $\Delta_2 F$ or $\Delta_1 F$ the constants $B$ and $C$ can be determined separately.

It may be noted that the sum (II, 116) can also be determined from $2-1$ or $1-2$ sub-bands if care is taken in forming the $\Delta_1 F(J, K)$ values from appropriate components of the $P$, $R$ and $Q$ doublets. In a similar way also the splittings of the $K = 2$ levels can be obtained either from $2-1$ and $1-2$ or from $3-2$ and $2-3$ sub-bands. Here, again the difference of the $\Delta_1 F(J, K)$ values gives the sum of the splittings of two successive rotational levels which is according to (I, 157)

$$\Delta\nu_{cd}^{K=2}(J+1) + \Delta\nu_{cd}^{K=2}(J) = \frac{1}{16}\frac{(B - C)^2}{[A - \frac{1}{2}(B + C)]} J(J+1)^2(J+2). \qquad \text{(II, 117)}$$

Thus, if for some reason the $1-0$ and $0-1$ sub-bands cannot be measured the value of $(B - C)$ can be determined, though less accurately, from the line splittings in $2-1$ and $1-2$ or $3-2$ and $2-3$ sub-bands, assuming that $A - \frac{1}{2}(B + C)$ has first been determined from the sub-band spacings.

Formulae similar to (II, 116) and (II, 117) hold for a nearly oblate symmetric top except that everywhere $A$ is replaced by $C$ and $C$ by $A$.

From the combination differences $\Delta_1^J F(J, K)$ it is easy to establish the sign of the combination defect, that is whether the $Q$ lines in the $1-0$ and $0-1$ sub-bands go to the lower or upper component level of the $K = 1$ set, and similarly whether the splitting of the $Q$ lines of $2-1$ and $1-2$ sub-bands is the sum or difference of the level splittings in upper and lower state (i.e. is larger or smaller than the splitting in corresponding $P$ and $R$ lines). If the first alternative applies (full lines in Fig. 107) it follows that the transition moment is in the $c$ axis; if the second alternative applies, it is in the $b$ axis (and similarly in the $a$ and $b$ axes respectively for a near oblate symmetric top). Thus important information about the nature of the electronic transition is obtained from the band structure. For example, if for a planar molecule of point group $C_s$ a near prolate $\perp$ band is observed in which the first of the above alternatives applies, that is, if the transition moment is in the $c$ axis it follows immediately that the transition is $A''-A'$ or $A'-A''$ and not $A'-A'$ or $A''-A''$. If a similar observation is made for a $C_{2v}$ molecule for which it is known from other evidence that the $a$ axis (top axis) is in the $C_2(z)$

axis, it follows immediately that the transition is $B_1-A_1$ (or $B_2-A_2$) and not $B_2-A_1$ (or $B_1-A_2$).

Further information about the nature of the electronic states giving rise to the observed transition can be obtained if an intensity alternation is observed in successive lines of the branches or in the series of sub-bands. An intensity alternation in $K$ immediately establishes that the top axis ($a$ or $c$) is an axis of symmetry and the sign of the intensity alternation (i.e. whether even or odd lines are strong) establishes whether the vibronic wave functions of the upper and lower state are symmetric or antisymmetric with respect to a rotation by 180° about that axis. On the other hand, if an intensity alternation in $J$ is observed, it establishes the existence of a two-fold axis of symmetry at right angles to the top axis and the sign of this intensity alternation depends on the symmetry of the wave functions with respect to rotation by 180° about the axis. Both types of intensity alternation occur in molecules of point group $\boldsymbol{D}_{2h}$.

In Fig. 107 the rovibronic species are added for a $B_1-A_1$ vibronic transition of a $\boldsymbol{C}_{2v}$ molecule with $C_2(z)$ as $a$ axis, corresponding to the full-line transitions. It is seen that the levels of a given $K$ are either all $A$ or all $B$ and that $A$ and $B$ alternate for successive $K$ values. Therefore only an intensity alternation in $K$ arises, none in $J$. If the identical atoms causing the two-fold axis of symmetry are H atoms (as in $H_2CO$) the $B$ levels are the strong ones and therefore in the example the sub-bands with odd $K''$ are strong, those with even $K''$ weak. The reverse is the case if the H atoms are replaced by D atoms (or if there are two sets of H atoms). Other cases are easily worked out according to the rules discussed earlier (p. 114).

Several examples of $\perp$ bands of slightly asymmetric top molecules have been studied in detail. A particularly clear example of a $\perp$ band is provided by the HNCN radical and is reproduced together with the corresponding band of DNCN in Fig. 108. Here the rotational constants in the upper and lower state are very nearly the same. Because of this and because of the very small asymmetry a band structure arises which is very similar to the schematic spectrum of a symmetric top in Fig. 99: a nearly equidistant series of line-like $Q$ branches with intermediate structure due to $P$ and $R$ branches. These absorption bands are typical $\perp$ bands just like infrared $\perp$ bands. The very wide spacing of the $Q$ branches ($\sim 40$ cm$^{-1}$) and the halving of this spacing for the deuterated compound shows that the small moment of inertia $I_A$ must be due almost entirely to

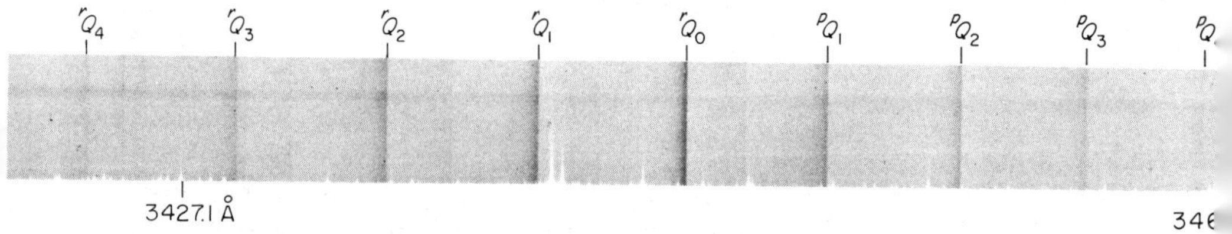


FIG. 108. **Spectrogram of the 0 – 0 band of the $\tilde{A}-\tilde{X}$ system of HNCN at 3440 Å after Herzberg and Warsop (547).** The heads of the $Q$ branches of the sub-bands are marked. The $P$ heads are visible to the right of the $Q$ heads of several sub-bands.

the H atom which accordingly must be assumed to be off the axis of a linear NCN chain. By the use of higher resolution several of the sub-bands have been fairly completely resolved and in the way outlined above all three rotational constants $A$, $B$, $C$ have been determined in both upper and lower state. The asymmetry parameter $b$ in the ground state is only $-1.798 \times 10^{-4}$. From the sign of the

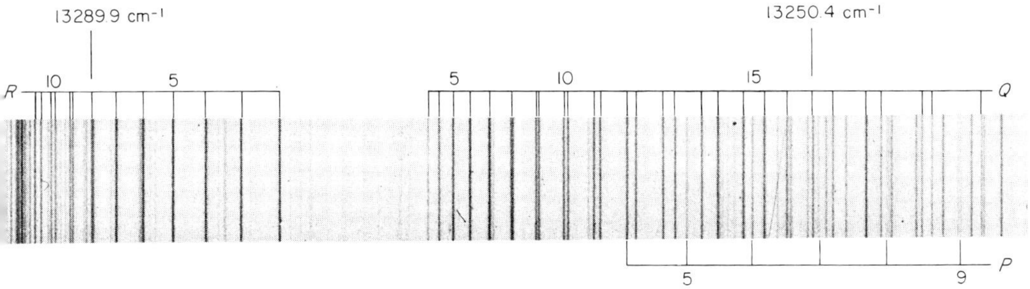


FIG. 109. **Spectrogram of the 3 – 2 sub-band of the 0 – 0 band of the Ã – X̃ system of HNO at 7534 Å after Dalby (264).** Note the $K$-type doubling at higher $J$ values.

combination defect in the 1 – 0 and 0 – 1 bands it has been established that the transition is $A''-A'$ (or $A'-A''$) that is, type $C$.

Another similar example is provided by the absorption system of HNO in the red region first observed by Dalby (264). Here, because of the smaller moments of inertia $I_B$ and $I_C$ a very complete resolution of the fine structure is possible. Figure 109 shows the 3 – 2 sub-band of the 0 – 0 band. The $P$ branch is much weaker than $Q$ and $R$ branch in agreement with the intensity formulae (II, 83). For the higher sub-bands the $P$ branch is not observed. The asymmetry splitting of lines with higher $J$ values into two components of equal intensity is clearly shown in each branch of the 3 – 2 sub-band. It is also to be noted that this splitting for a given $J$ is larger in the $Q$ branch than in $P$ and $R$ branch showing that this band corresponds to case (1) above, i.e. that the transition moment is in the $c$ axis and therefore that the electronic transition is $A''-A'$ (or $A'-A''$). It should be noted also that the $P$, $Q$, $R$ branches start with $J = 4$, 3 and 2 respectively in agreement with expectation for a 3 – 2 sub-band. Conversely it was in this way that the $K$ numbering was established. A corresponding spectrum of DNO has also been obtained and analyzed. For each isotope, from the combination differences, precise values of all three rotational constants have been obtained and from them the geometrical parameters $r_0$(NO), $r_0$(NH), ∢HNO have been determined (see Chap. V, section 1b).

A further important example, in a molecule with somewhat higher symmetry, is found in the near ultraviolet absorption system of $H_2CO$ first analyzed by Dieke and Kistiakowsky (277) as early as 1934 (indeed this was the first successful rotational analysis of an electronic spectrum of a polyatomic molecule). The analysis was extended by Dyne (332), Robinson (1075) and Callomon and Innes (178). This spectrum also shows the typical structure of ⊥ bands. There is a clear intensity alternation in $K$: sub-bands with odd $K''$ are strong showing that the lower electronic state is an $A_1$ or $A_2$ state. While the band structure does not allow one to distinguish between $A_1$ and $A_2$, the electron configuration of $H_2CO$ (see Chap. III) demands that it be $A_1$. Since in the upper state the even $K$ levels are strong, this state must be a $B$ (vibronic) state. The fact that the $Q$ lines of the main bands show a smaller asymmetry splitting than the $P$ and $R$ lines shows according to Fig. 107 that the transition moment lies in the $b$ axis (type $B$ bands), i.e. in the plane of the molecule. Thus the upper state is vibronically $B_2$ not $B_1$. There are also a few weaker bands in which the reverse splitting ratio of $Q$ and $P$, $R$ lines is observed (type $C$ bands) and for which therefore the upper state must be vibronically $B_1$. The complications which arise in the interpretation of the electronic and vibrational structure from the fact that the molecule is non-planar in the excited state will be discussed in Chapter V, section 2b.

Several other examples of $\perp$ bands of slightly asymmetric top molecules are described in the recent literature. King (667) and Paldus and Ramsay (957) have analyzed the green absorption bands of glyoxal, $(HCO)_2$, and have shown them to represent an $A_u - A_g$ transition of point group $C_{2h}$. Brand, Callomon and Watson (140)(141) have studied the near ultraviolet absorption bands of HC≡C—CHO (propynal) under high resolution finding typical $\perp$ bands with well-resolved sub-bands. The electronic transition is ${}^1A'' - {}^1A'$. A similar transition is observed in $C_2H_3CHO$ (acrolein) near 3860 Å; the $\perp$ structure of the bands has recently been firmly established and the $K$ structure analyzed by Hollas (565).

**Hybrid bands.** As indicated in Table 16, for point groups $C_1$, $C_i$, $C_s$, $C_2$ and $C_{2h}$, hybrid bands can appear, that is, for them both the rotational transitions described for a $\parallel$ band and those for a $\perp$ band can occur in one and the same vibronic transition. The relative intensity of the $\parallel$ and $\perp$ components depends on the orientation of the transition moment with respect to the top axes. It is easily seen from Table 16 that except for point groups $C_1$ (no symmetry) and $C_i$ the $\perp$ components of hybrid bands are single components following only one of the three selection rules (II, 97–99), that is the $P$, $Q$, $R$ branches with $K > 0$ have each only two not four components. However, the bands of $C_1$ and $C_i$ molecules are complete hybrids, that is, unless accidentally the transition moment lies in one of the principal axes all three components, type $A$, type $B$ and type $C$ are present. Good examples of hybrid bands have been found in the forbidden components of the 3800 Å system of propynal [Brand, Callomon and Watson (141)]. Unlike the main bands which are strictly $\perp$ bands (type $C$), the forbidden component consists of $A' - A'$ vibronic transitions which have both $\parallel$ and $\perp$ dipole components; in some of these bands the $K$ structure is well resolved and shows about equal intensities of the $\Delta K = 0$ and $\Delta K = \pm 1$ sub-bands (type $A$ and type $B$).

**Unresolved bands.** Only too frequently, for heavier molecules, the band structure cannot be resolved. Sometimes conclusions about the nature of the upper state can be derived from unresolved bands if the structure of the molecule in the ground state is known from microwave, infrared or Raman studies. Metropolis (833) has studied in some detail the band envelopes to be expected for near prolate $XY_2$ molecules. He has in particular discussed the question of the shading of the $K$ structure and the $J$ structure and the conclusions that can be drawn from it about the changes of internuclear distance and angle in the electronic transitions and has applied them to the near ultraviolet spectra of $SO_2$ and $ClO_2$. More recently Parkin (962a) has developed a computer program for obtaining directly the contours of asymmetric top bands, and Parkin and Innes (964) have applied it to a determination of the rotational constants of the upper state of the 2482 Å band of HFCO.

### (γ) *Strongly asymmetric tops*

When the molecule is a strongly asymmetric top the selection rule $\Delta K = 0, \pm 1$ is no longer valid since $K$ is no longer a good quantum number. Rather we must now apply the general selection rules (II, 101–103) for $K_a$ and $K_c$. As a consequence, many more sub-bands and branches will occur. In addition, since now the asymmetry splitting is large, even for fairly large $K_a$ or $K_c$ the resulting pairs of branches which were very close to each other in the slightly asymmetric top are now widely separated and cannot immediately be recognized as belonging together.

Just as in infrared spectra, depending on which one of the three selection rules (II, 101), (II, 102) or (II, 103) [or equivalently (II, 97), (II, 98) or (II, 99)] applies we have type $A$, type $B$ and type $C$ bands. In Volume II, pp. 470, 475 and 481, energy level diagrams for infrared transitions of the three types have been given. They are identical with corresponding diagrams for electronic transitions which we need therefore not reproduce here (but see Fig. 107 for a type $C$ band of a slightly asymmetric top). However, in order to give some idea of the structure of bands of strongly asymmetric top molecules we do present in Figs. 110 and 111 diagrams showing the various sub-bands of an $A$ type and a $C$ type band of a strongly asymmetric top assuming $A = 20.145$, $B = 11.185$, $C = 7.065$ cm$^{-1}$ in both upper and lower state. In the designation of the branches two superscripts are used preceding the symbol $P$ or $Q$ or $R$ (which gives the $\Delta J$ (or $\Delta N$) value); the first of these superscripts indicates the $\Delta K_a$, the second the $\Delta K_c$ value. Thus one has ${}^{qr}R$, ${}^{qt}R$, ${}^{qp}R$, ${}^{qn}R$, ${}^{sr}R$, ${}^{or}R$, . . . branches and similar $Q$ and $P$ branches in type $A$ bands and corresponding branches in the other band types. Unfortunately it was not possible to include the superscripts in the designations in Figs. 110 and 111.

The intensities of the lines in Figs. 110 and 111, indicated by their heights, have again been taken from the tables of Cross, Hainer and King (257). Since the example is closer to a prolate than to an oblate symmetric top, the lines following the prolate selection rules $\Delta K_a = 0$ for Fig. 110 and $\Delta K_a = \pm 1$ for Fig. 111 are the strongest and of these again those with the lowest $\Delta K_c$ are strongest. The branches with higher $\Delta K_a$ values are all very weak; in order to be able to plot them a larger intensity scale has been used than for the other branches. For slightly asymmetric tops all these branches have vanishing intensities.

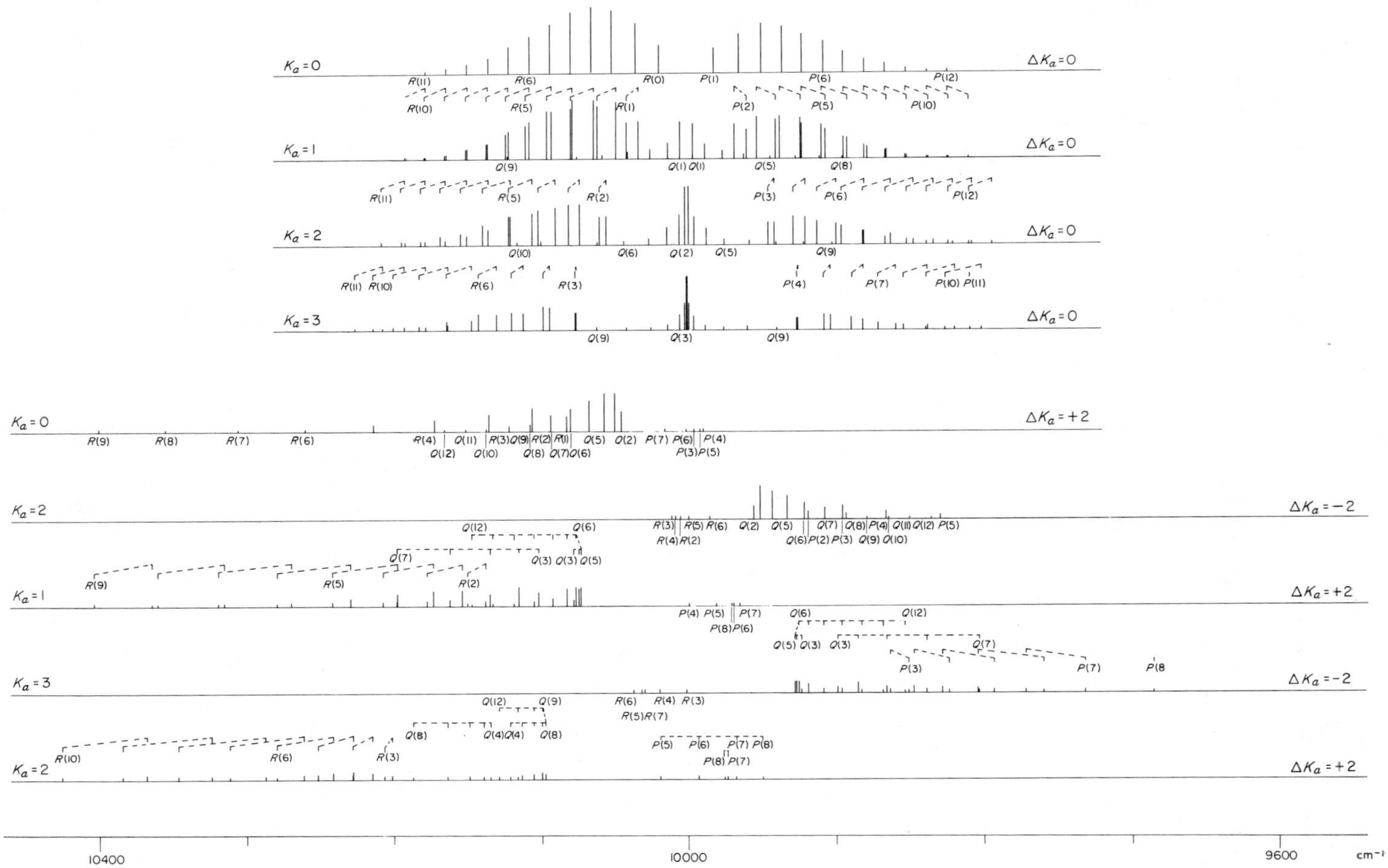


FIG. 110. **Calculated structures of the sub-bands of an *A*-type band of a strongly asymmetric top.** The assumed rotational constants are $A' = A'' = 20.145$, $B' = B'' = 11.185$, $C' - C'' = 7.065$ cm$^{-1}$. The band origin has been arbitrarily placed at 10000 cm$^{-1}$. The main sub-bands with $\Delta K_a = 0$ are in the upper part, the weaker sub-bands with $\Delta K_a = \pm 2$ in the lower part. The intensity scale in the lower part is 10 times greater than in the upper part. *K*-type (asymmetry) doublets are indicated by the oblique broken-line brackets.

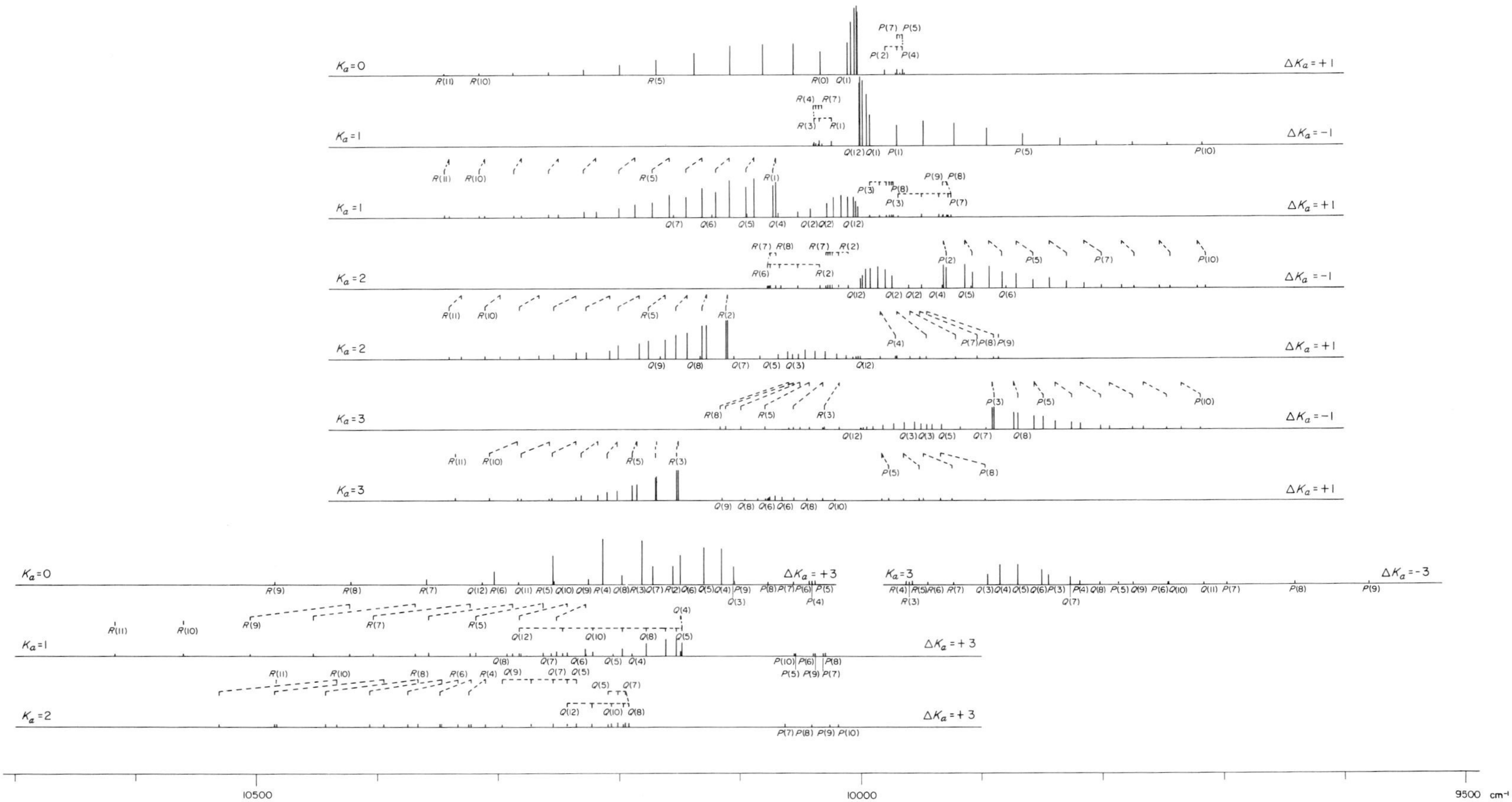


FIG. 111. **Calculated structures of the sub-bands of a *C*-type band of a strongly asymmetric top.** See the legend of Fig. 110. In the upper part are the main sub-bands with $\Delta K_a = \pm 1$; a few much weaker sub-bands with $\Delta K_a = \pm 3$ are given in the lower part at a tenfold intensity scale.

The intensity tables of Cross, Hainer and King (257) originally intended for the study of infrared bands are based on the assumption that the rotational constants in upper and lower state are the same. In electronic transitions there are often great differences between $A'$, $B'$, $C'$ and $A''$, $B''$, $C''$ and consequently large deviations from the intensities of Cross, Hainer and King must be expected. The recently developed computer programs for asymmetric top bands allow fairly readily an extension to include intensity calculations, and where they have been carried out, they bear out the expectation of considerable changes of the Cross, Hainer, King intensities [see Birss and Ramsay (120a)].

It can be seen in Figs. 110 and 111 that with increasing $K_a$ the $Q$ branches contract so that for asymmetric top molecules for which higher $K$ values are important there is an approach to the symmetric top structure. To get a true picture of the structure of an asymmetric top band the various sub-bands of each of the two diagrams must of course be superimposed. It is clear from Figs. 110 and 111 that then a spectrum arises in which individual branches are not easily recognized.

The analysis of electronic bands of strongly asymmetric tops proceeds of course in the same way as for infrared bands, and we can refer to the discussion in Volume II, p. 484f. Many more details about the various methods of analysis may be found in Allen and Cross's book (1). The same remarks apply to the methods of determining the rotational constants in the upper and lower state. If all levels for several $J$ values have been evaluated the best way of determining the rotational constants $A$, $B$, $C$ as well as some of the centrifugal stretching constants is by way of the relations (I, 163) of Chapter I.

In recent years with the increasing availability of electronic computers, it has become much easier to analyze asymmetric top bands and determine the rotational constants by a stochastic method, i.e. by calculating the whole band structure with approximate rotational constants and then repeating the procedure with slightly varied constants until a fit within the accuracy of the measurements is obtained [see Birss and Ramsay (120a)]. This method presumes that no perturbations occur in the excited state, a condition that is less often fulfilled in electronic than in infrared transitions.

The only examples of electronic bands of strongly asymmetric tops (and not involving a state in which the molecule is linear or nearly linear as in $NH_2$ and $CH_2$) which have been fully resolved and analyzed are bands near 1250 Å of $H_2O$ and $D_2O$ (Johns (631)). They are reproduced in Fig. 112. While in $H_2O$ the resolution is limited by predissociation (see Chap. IV) in $D_2O$ a fairly complete resolution is obtained. The analysis of these bands was greatly simplified by the knowledge of the lower state rotational levels from the infrared spectrum. Several sub-bands are indicated in Fig. 112. The fact that $\Delta K_c$ is even for all sub-bands, while $\Delta K_a$ is odd, shows that the band is a type $C$ band: the transition moment is perpendicular to the plane of the molecule. None of the infrared bands of $H_2O$ are of this type.

Bands similar to those of $H_2O$ have been found for $H_2S$ and $D_2S$ near 1390 and 1265 Å. However for $H_2S$ (and $D_2S$) the rotational constants $A$ and $B$ are

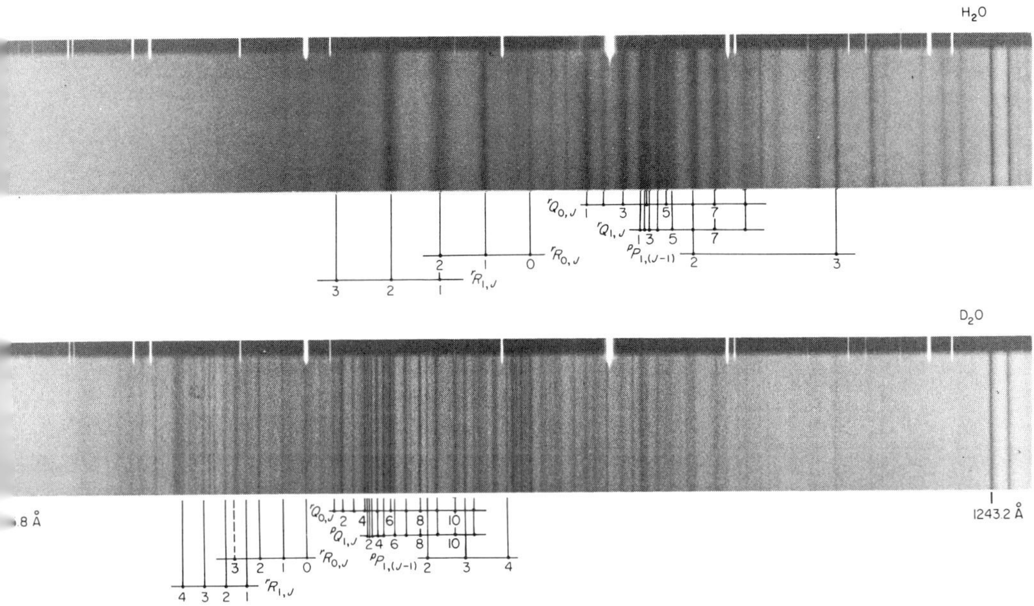


FIG. 112. **Spectrograms of the 0–0 bands of the $\tilde{C}-\tilde{X}$ systems of $H_2O$ and $D_2O$ after Johns (631).** The assignments of the lines of the sub-bands with $K_a = 0$ and 1 are given. Note the rapid increase of diffuseness with $J$ in $H_2O$.

not very different while $C \approx \frac{1}{2}A$. Therefore these bands are again nearly symmetric top bands but corresponding to an oblate top.

Only one example of a doublet–doublet transition in a strongly asymmetric top has as yet been analyzed [Dixon, Duxbury and Ramsay (286b)]: the red absorption bands of $PH_2$ which represent a ${}^2A_1-{}^2B_1$ transition (compare also the doublet structure in the linear–bent transition of $NH_2$, Fig. 95). No examples of triplet–triplet or higher multiplet transitions in asymmetric top molecules have been resolved and analyzed.

### (δ) *Forbidden transitions*

Just as for symmetric top molecules, the band structure of asymmetric top molecules in those forbidden electronic transitions that are made possible by vibronic interaction is the same as in corresponding allowed transitions: It is the vibronic symmetry of upper and lower state that determines the direction of the transition moment and thus the band structure.

**Transitions induced by Coriolis interaction.** The band structure is different in forbidden electronic (or vibronic) transitions that are caused to appear by the interaction with rotation, i.e. by Coriolis interaction. Consider for example the 00..–00.. band of an $A_2-A_1$ electronic transition (or any $A_2-A_1$ vibronic transition) of a $C_{2v}$ molecule. Such a transition is rigorously forbidden in the non-rotating molecule. However, if the $A_2$ state is near a third state that can combine with $A_1$ the state $A_2$ may mix with this third state with increasing rotation and will therefore be enabled in its higher rotational levels to combine with $A_1$. (Alternatively if there is a third state near the lower state $A_1$ which can combine with $A_2$, the $A_2-A_1$ transition can be made possible by Coriolis interaction.) If the third state interacting with $A_2$ is an $A_1$ state it is rotation about the $z$-axis that can cause mixing; if it is a $B_1$ or $B_2$ state it is rotation about the $x$- or $y$-axis respectively. If the $z$-axis is the $a$-axis, in the first case ($A_1-A_2$ Coriolis interaction) levels with the same $K_a$ can perturb one another, in the other two cases levels differing by $\pm 1$ in $K_a$ can perturb one another. Figure 113a shows an energy level diagram for the first case, Fig. 113b for the second case ($B_1-A_2$ Coriolis interaction).

Coriolis perturbations can take place only between states *of the same $J$ and the same rovibronic species*. Therefore, as Fig. 113a shows, in the first case ($A_1-A_2$ Coriolis interaction) a band with $\Delta K_a = 0$, similar to an ordinary $\|$ band, arises in which, however, the sub-band with $K_a = 0$ is missing and in which for $K_a \neq 0$ the opposite components of the asymmetry doublets appear as in a normal $A_1-A_1$ (or $A_2-A_2$) $\|$ band. The sub-bands with $K_a = 1, 2, \ldots$ made possible by Coriolis interaction are forbidden by any of the selection rules (II, 97–99) for the asymmetric top species which, as pointed out earlier, are not rigorous; the sub-bands are, however, in accordance with the selection rule for the rovibronic species (Table 15).

On the other hand, for the second case (Fig. 113b) the $A_2$ vibronic levels with $K_a = 0, 1, 2,$ mix with the $B_1$ levels with $K_a = 1, 0$ and $2, 1$ and $3$, etc. and therefore transitions with $\Delta K_a = 0$ and $\pm 2$ arise, i.e. $q$, $s$ and $o$ branches of sub-bands including a sub-band with $K_a' = K_a'' = 0$ with a strong $Q$ but no $P$ and $R$ branch. The other sub-bands have the usual $P$, $Q$, $R$ branches but since in this case the Coriolis interaction is due to rotation about the $x$-axis the transition becomes increasingly allowed with increasing $J$, not $K$, and the intensity distribution in the branches is therefore anomalous: the line strengths as given in the tables of Cross, Hainer and King (257) have to be multiplied by a factor $J(J + 1)$, i.e. lines of low $J$ are very weak.

In the third case ($B_2-A_2$ perturbation) everything is similar to the second ($B_1-A_2$) shown in Fig. 113b, except that the rovibronic species in the $B_2$ state are reversed compared to $B_1$ (i.e. $B_2$ is replaced by $B_1$, $B_1$ by $B_2$ and $A_2$ by $A_1$, $A_1$ by $A_2$).

For forbidden $B_2-B_1$ transitions caused to appear by Coriolis interaction the band structure is entirely similar to $A_2-A_1$ transitions: only the rovibronic species in Fig. 113 have to be appropriately changed (see Table 6, p. 113).

Similar considerations apply to $\boldsymbol{D}_{2h}$ molecules. Here it must be noted that forbidden transitions of the $g-g$ or $u-u$ type cannot be caused to appear by Coriolis interaction. Only $A_g-A_u$ and $B_{ig}-B_{iu}$ transitions can be produced in this way.

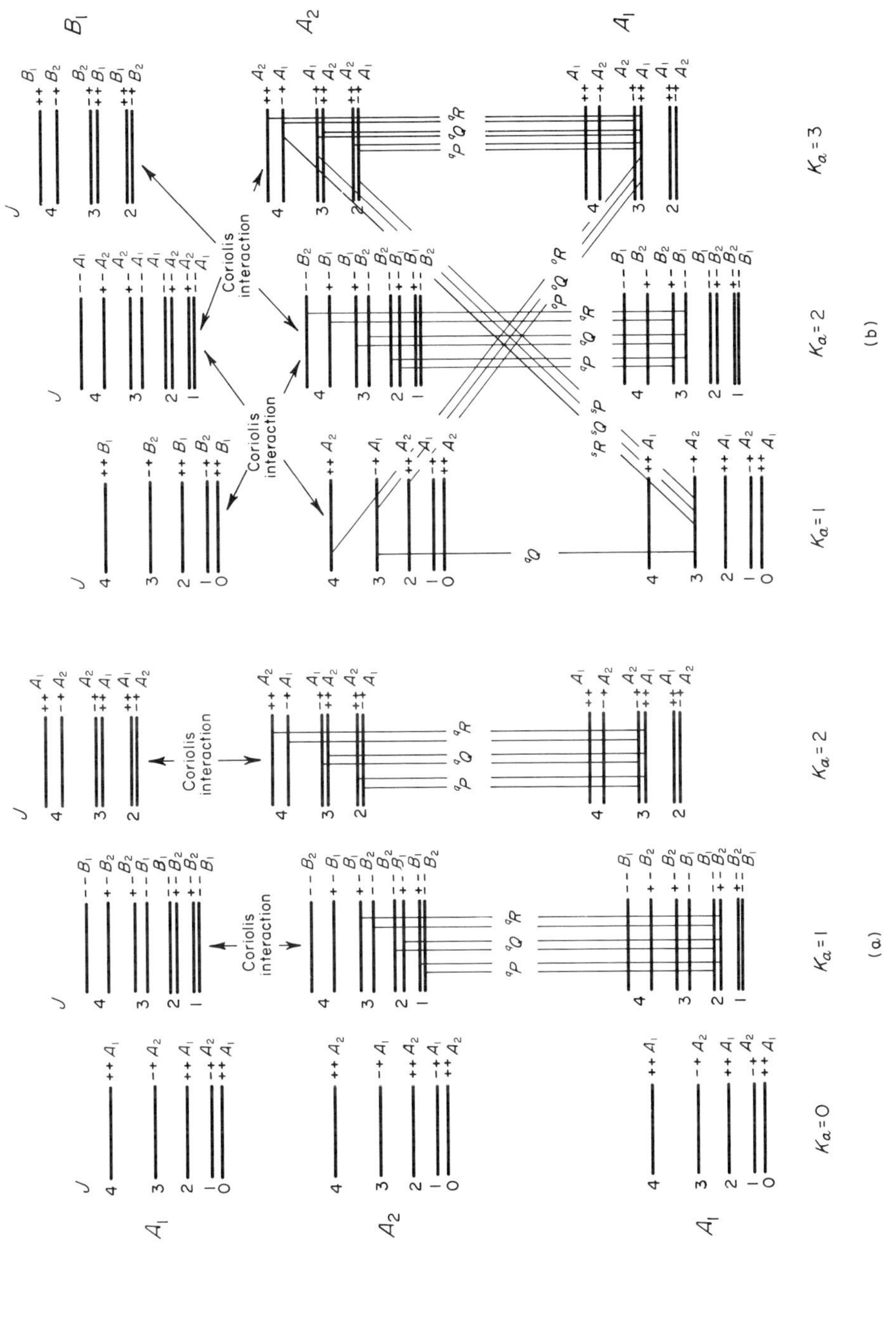


FIG. 113. **Rotational transitions made possible by Coriolis interaction in a forbidden ($A_2 - A_1$) electronic transition of a $C_{2v}$ molecule.** (*a*) **The state interacting with $A_2$ is an $A_1$ state,** (*b*) **the state interacting with $A_2$ is a $B_1$ state.** It is assumed that the molecule is near prolate and that the *a*-axis is in the *z*-axis ($C_2$-axis). The interacting state is shown at the top. The arrows indicate which rotational levels can interact with each other. Only one line of each branch is shown in the lower part.

For point groups of lower symmetry there are no forbidden electronic transitions other than $g-g$, $u-u$ transitions. Here Coriolis interaction can cause the weak occurrence of forbidden sub-bands in allowed electronic transitions. For example, in an $A''-A'$ transition of a nearly symmetric top molecule of point group $\boldsymbol{C}_s$, which normally has only transitions with $\Delta K = \pm 1$, transitions with $\Delta K = 0$ and $\pm 2$ become possible with increasing $J$ (as in Fig. 113b) if an $A'$ state is close to the $A''$ state and if that $A'$ state combines strongly with the $A'$ lower state of the $A''-A'$ transition.

Forbidden sub-bands can also be caused by "*axis switching*". Just as for bent–linear transitions (section 3a($\beta$)) axis switching will occur in asymmetric top transitions if the principal axes are not entirely determined by symmetry in one or both of the states involved. An example would be a bent XYZ molecule with a different angle in the upper and lower state. In such a case just as for linear–bent transitions the selection rule for $K$ is

$$\Delta K = 0, \pm 1, \pm 2, \ldots \qquad \text{(II, 118)}$$

where $K$ may be $K_a$ or $K_c$. Thus forbidden sub-bands [forbidden by the selection rules (II, 101–103)] may occur. Just as for forbidden sub-bands produced by Coriolis interaction the intensity factors in these forbidden sub-bands have an additional factor $J^2$ if $\Delta K_a$ or $\Delta K_c$ is one removed from the normal value. If $\Delta K_a$ or $\Delta K_c$ differs by 2 from a normal value the sub-bands can also appear on account of asymmetry but the part of the intensity due to axis switching has an additional factor $J^4$.

It may be noted that axis switching will produce forbidden sub-bands independent of any other nearby electronic state while Coriolis interaction can do so only if a suitable third electronic state is in the neighborhood of the upper or lower electronic state of the transition. The only type of forbidden electronic or vibronic transition possible in molecules of sufficiently low symmetry (i.e., low enough for axis switching to occur) are $g-g$ and $u-u$ transitions. Axis switching cannot help to produce such transitions.

**Triplet–singlet transitions.** As we have seen earlier (p. 242), spin-orbit interaction between two states of different spin $S$ mixes the eigenfunctions of levels of different $K$: for $\Delta S = 1$ we have $\Delta K = 0, \pm 1$ for the interacting levels. Therefore in triplet–singlet transitions of slightly asymmetric top molecules, the selection rule $\Delta K = 0, \pm 1$ of singlet–singlet transitions is replaced by

$$\Delta K = 0, \pm 1, \pm 2. \qquad \text{(II, 119)}$$

In addition, one finds for the quantum number $N$ of the total angular momentum apart from spin

$$\Delta N = 0, \pm 1, \pm 2. \qquad \text{(II, 120)}$$

At the same time of course the selection rules for the rovibronic species (Table 15) remain valid. For strongly asymmetric tops, sub-bands with $\Delta K_a = \pm 2$ or $\Delta K_c = \pm 2$ are allowed even in singlet–singlet transitions [see the general selection rules (II, 101–103)] but these sub-bands are usually quite weak. However, in triplet–singlet transitions they will have intensities of the same order as the $\Delta K_{a,c} = 0, \pm 1$ sub-bands.

For some molecules of higher symmetry, because of the rovibronic selection rules, not all $\Delta K$ in (II, 119) occur in a given triplet–singlet transition. For example for a $\boldsymbol{D}_{2h}$ molecule (assuming the $z$ axis to be the top axis), in $^3A_u - {}^1A_g$, $^3B_{iu} - {}^1B_{ig}$, $^3B_{1u} - {}^1A_g$ and $^3B_{3u} - {}^1B_{2g}$ bands only transitions with $\Delta K = 0, \pm 2$ can occur while in $^3B_{2u} - {}^1A_g$, $^3B_{3u} - {}^1A_g$, $^3B_{3u} - {}^1B_{1g}$, $^3B_{2u} - {}^1B_{1g}$ bands only transitions with $\Delta K = \pm 1$ can occur, as can easily be seen by writing down the rovibronic species for the rotational levels involved. In the transitions given, $g$ and $u$ and triplet and singlet can be exchanged and the order of the states can be reversed.

As the symmetry of the molecule is reduced there are more and more transitions for which both $\Delta K = 0, \pm 2$ and $\Delta K = \pm 1$ can occur. For the point groups $\boldsymbol{C}_2$, $\boldsymbol{C}_{2h}$ and $\boldsymbol{C}_s$, only when the top axis coincides with the $C_2$ or is perpendicular to the plane of symmetry is there a restriction: $A - A$ or $B - B$ transitions of $\boldsymbol{C}_2$ and $\boldsymbol{C}_{2h}$ and $A'' - A'$ of $\boldsymbol{C}_s$ have only transitions with $\Delta K = 0, \pm 2$ while $A - B$ and $A' - A'$, $A'' - A''$ transitions have only $\Delta K = \pm 1$. On the other hand, when the top axis is at right angles to the $C_2$ or in the plane $\sigma$, all electronic (or vibronic) transitions may have both $\Delta K = 0, \pm 2$ and $\Delta K = \pm 1$.

A detailed theoretical discussion of triplet – singlet transitions in slightly asymmetric top molecules has been given by Hougen (574). He has also derived intensity factors for all the branches of the various types of sub-bands.

It is interesting to compare the selection rules for triplet – singlet transitions with those for transitions in which Coriolis interaction or axis switching is important. Since the selection rule for $K$ is the same in all three cases the same branches of sub-bands arise and therefore when such anomalous sub-bands are observed without resolution of the $J$ structure it is not immediately obvious which of the three causes is responsible for the appearance of the anomalous sub-bands. However, in general both Coriolis interaction and axis switching can produce only a relatively weak occurrence of anomalous sub-bands; in order to produce a non-negligible intensity the former requires the presence of a second suitable electronic state nearby, the latter a substantial change of shape (i.e. of axes) in the transition. Even if these conditions are fulfilled the intensity of the anomalous sub-bands vanishes for low $J$ but it increases rapidly with $J$. On the other hand in triplet – singlet transitions the intensity distribution in the branches is normal and even at low $J$ the intensity of the anomalous sub-bands may be (but need not be) of the same order as that of the normal sub-bands. Furthermore it is only for triplet – singlet transitions that branches with $\Delta N = \pm 2$ may occur, and it is only for triplet – singlet transitions that a Zeeman splitting in a magnetic field is expected.

Several examples of triplet – singlet transitions in asymmetric top molecules are known: the 3900 Å system of $SO_2$ [see Merer (822)], the 3970 Å system of $H_2CO$ [Robinson and Di Giorgio (1078, 278), Raynes (1059, 1060)], the 4200 Å system of propynal [Brand, Callomon and Watson (140)] and the 4120 Å system of acrolein [Brand and Williamson (145), Hollas (565)]. In none of these cases have the anomalous branches been observed. A further example may be provided by the absorption (and fluorescence) bands of HSiCl and HSiBr in the visible region. They do show in addition to strong sub-bands with $\Delta K = \pm 1$ somewhat weaker sub-bands with $\Delta K = 0$ and $\pm 2$ [Herzberg and Verma (545)]. These can be readily accounted for if the electronic transition is assumed to be $^3A''-^1A'$. However, no triplet splitting has been resolved nor have branches with $\Delta N = \pm 2$ been observed. By assuming suitable parameters in Hougen's formulae the weakness of the $\Delta N = \pm 2$ branches in the sub-bands with $\Delta K = \pm 1$ could be understood, and these sub-bands are the only ones in which the absence of branches with $\Delta N = \pm 2$ has been well established. However, the possibility that the transition is singlet–singlet and the anomalous sub-bands are caused by axis switching cannot be entirely excluded. The possibility of Coriolis interaction as the cause for these sub-bands is excluded by the observation of $^qQ(K = 0)$ branches which could not be caused in this way.

Still another example is the 3700 Å absorption system of symmetrical diazine (pyrazine). Here Innes and Giddings (607) observed $Q$ branches with $\Delta K = \pm 2$ in addition to the strong $\Delta K = 0$ transitions. It was therefore suggested by Herzberg (523) that the transition is triplet–singlet and this conclusion was strikingly confirmed by the observation of a large Zeeman splitting by Douglas and Milton (299).

**Magnetic dipole transitions.** As pointed out in section 1 of this Chapter, magnetic dipole (and quadrupole) radiation can cause the occurrence of certain electronic transitions forbidden for electric dipole radiation. This statement applies also to vibronic transitions, i.e. when the interaction of vibration and electronic motion has been taken into account. Thus, for example, $A_2-A_1$ vibronic transitions of $C_{2v}$ or $A_g-A_g$ vibronic transitions of $C_{2h}$ which are rigorously forbidden for electric dipole radiation can occur as magnetic dipole radiation (see Table 10). The selection rules for $J$ and $K$ are the same as for electric dipole radiation but the rovibronic selection rule is reversed. Therefore, as is immediately seen from Fig. 113, in a magnetic dipole $A_2-A_1$ transition the same sub-bands and branches occur as in an electric dipole $A_1-A_1$ transition; in particular, in the $K' = 0 \leftrightarrow K'' = 0$ sub-band only $P$ and $R$ branches appear. According to the preceding discussion an $A_2-A_1$ vibronic transition can also be made to occur by Coriolis interaction or by spin-orbit interaction (that is, if it is $^3A_2-^1A_1$) but in both these cases the rovibronic selection rule is opposite to that for magnetic dipole radiation. As a consequence the opposite rotational transitions occur as in an $A_1-A_1$ transition; in particular the $K = 0 \leftrightarrow K = 0$ sub-band cannot occur at all by Coriolis interaction and if produced by spin-orbit interaction it would have a strong $Q$ branch and weak $O$ and $S$ branches ($\Delta N = 0, \pm 2$) but no $P$ and $R$ branches.

The only example of a magnetic dipole transition thus far known for polyatomic molecules is observed in the near ultraviolet bands of formaldehyde. These bands represent a $^1A_2-^1A_1$ electronic transition (disregarding here the fact that the molecule is slightly off-planar in the excited state). The main bands are $B_2-A_1$ vibronic bands; the 0–0 band is strictly forbidden as electric dipole radiation since it would be of vibronic type $A_2-A_1$. Its actual occurrence with a normal || structure can, as was shown in detail by Callomon and Innes (178), only be accounted for by the assumption that this band and a few similar ones represent magnetic dipole radiation.

## 4. Zeeman and Stark Effects

Very little experimental work has been done on the Zeeman and Stark effects in electronic spectra of polyatomic molecules. On the other hand, a large amount of work has been done on Zeeman and Stark effects in microwave spectra and in molecular beam spectroscopy but this as well as the fields of nuclear magnetic and paramagnetic resonance are outside the scope of this book. We shall therefore give here only a brief discussion of the expected splittings of band lines in magnetic and electric fields based on the previous discussion of the splittings of the energy levels, and mention the few cases in which these splittings have been observed.

**Selection rules.** In an electric or magnetic field the component levels are characterized by the magnetic quantum number $M$ (Chap. I, section 4). The selection rule for this quantum number is, just as for atoms and diatomic molecules (see Vol. I, p. 299),

$$\Delta M = 0, \pm 1 \qquad \text{(II, 121)}$$

with the restriction

$$M = 0 \nleftrightarrow M = 0 \quad \text{for } \Delta J = 0. \qquad \text{(II, 122)}$$

This selection rule holds for both electric and magnetic dipole radiation. However, while for electric dipole radiation the transitions with $\Delta M = 0$ are polarized || to the field and those with $\Delta M = \pm 1$ are polarized $\perp$ to the field the reverse polarization applies to magnetic dipole radiation.

If the resultant spin $\boldsymbol{S}$ is different from zero but not strongly coupled to the rotational angular momentum the selection rules (II, 121 and 122) apply also to the magnetic quantum number $M_N$ of the total angular momentum apart from spin ($\boldsymbol{N}$). In addition we have for the magnetic quantum number $M_S$ of the spin:

$$\Delta M_S = 0. \qquad \text{(II, 123)}$$

This rule holds strictly only when the interaction of spin and rotation is very small compared to the interaction of spin and magnetic field. It does not apply to intercombinations for which $S$ is different in upper and lower state and which are made possible by the interaction of the spin with the other angular momenta. For these intercombinations

$$\Delta M_S = 0, \pm 1. \qquad \text{(II, 124)}$$

**Zeeman patterns.** Splitting patterns for the lines of diatomic molecules were presented in Volume I, p. 302. The behavior of the lines of linear polyatomic molecules is the same in every detail as long as vibronic interactions can be neglected. Thus in $^1\Sigma - {}^1\Sigma$ transitions a splitting into $3(2J - 1)$ lines arises for every line of the $P$ branch and into $3(2J + 1)$ lines for every line of the $R$ branch. However in general this splitting is extremely small since according to (I, 173) the over-all splitting for high $J$ (between $M = +J$ and $M = -J$) is given by

$$\Delta W = hc\Delta\nu = 2(g_r' - g_r'')\mu_{0n}HJ$$

and since $g_r$ is usually less than 1. It has not yet been observed in a single case. (For $H = 10000$ gauss, $J = 20$ and $g_r' - g_r'' = 1$ the splitting would be 0.01017 $\text{cm}^{-1}$.)

In contrast, the Zeeman splittings expected in $^1\Pi - {}^1\Sigma$ (or $^1\Sigma - {}^1\Pi$) electronic transitions are much larger for low $J$ since the level splittings in the $\Pi$ state according to eq. (I, 178) are of the order of Bohr magnetons rather than nuclear magnetons; but the splittings decrease inversely as $J + 1$, while the number of components increases as before. Even this much larger Zeeman splitting has not yet been observed in a single case. According to (I, 176) the spin produces a large magnetic splitting of the levels; yet because of the selection rule (II, 123) the Zeeman patterns of multiplet $\Sigma - \Sigma$, $\Pi - \Sigma, \ldots$ transitions are the same as those of corresponding singlet transitions if both states belong to case (b). It is only when the coupling case or the multiplicity changes in the transition that the level splitting corresponding to the spin is directly observed in the spectrum. No examples have as yet been observed in linear–linear transitions of polyatomic molecules.

For non-linear molecules in transitions between non-degenerate states the situation is similar to that for $^1\Sigma - {}^1\Sigma$ transitions of linear molecules. The line splitting is extremely small except at very high $J$ values, but the dependence on $J$ is more complicated than for linear molecules [cf. eqs. (I, 174 and 175)]. Similarly for multiplet transitions between orbitally non-degenerate states, as long as the multiplicity is the same in the upper and lower state, only very small line splittings arise because of the selection rule (II, 123). However, a large splitting arises in intercombinations if the spin is only loosely coupled to the rotation. For such transitions, according to the selection rule (II, 124) combined with the splitting formula (I, 176), we obtain a splitting of each line into three components separated by twice the "normal" Zeeman splitting ($\mu_0 H$). Each component is again split into $3(2N - 1)$ or $3(2N + 1)$ components but this secondary splitting is of the order of $\frac{1}{1000}$ of the normal splitting.

An example of a large Zeeman splitting, approaching for high $J$ the normal Zeeman splitting, has been found by Douglas and Milton (298) in the near ultraviolet absorption bands of $CS_2$. They have shown that these $CS_2$ bands represent a $^3A_2 - {}^1\Sigma_g^+$ transition, the molecule having a $\boldsymbol{C}_{2v}$ symmetry in the upper state (see p. 202). However, here a complication arises because the triplet splitting in the upper state is fairly large and only one of the triplet components ($B_2$) combines strongly with the ground state. With increasing $N$ this triplet component becomes more and more mixed with the other triplet components and therefore an increasing

Zeeman splitting arises which approaches but does not reach the simple case described above.

If one or both of the electronic states are degenerate, in general, on account of the energy formula (I, 179) for the splittings of the levels of a degenerate state, a large line splitting arises which is of the order of the normal Zeeman splitting. However, just as for $^1\Pi - {}^1\Sigma$ transitions of linear molecules the number of components increases with $2J + 1$ and only for the lowest $J$ value can resolved Zeeman patterns be expected. The expected patterns for $K' - K''$ sub-bands would be similar to corresponding patterns of $\Lambda' - \Lambda''$ transitions of linear molecules (see Fig. 146 of Vol. I) except that on account of the factor $d\zeta_e$ in (I, 179) the scale of the pattern is different. No such patterns have been resolved but Douglas (294) has observed broadening of low $J$ lines in the $E'' - A_1$ transition of $NH_3$ near 1600 Å, with an effective $g$ value of about 0.6.

**Magnetic rotation spectra.** An alternative way of studying the Zeeman effect of polyatomic molecules is by means of magnetic rotation spectra. As explained in more detail in Volume I, p. 306, these are spectra of the light transmitted by an absorption tube between two crossed nicols (polaroids) when a magnetic field is applied. Transmission will occur for those wavelengths for which a rotation of the plane of polarization has taken place, and a rotation of the plane of polarization will take place in the immediate neighborhood of magnetically sensitive lines, in the region of anomalous dispersion corresponding to these lines, if the applied magnetic field is sufficiently strong.

According to the preceding discussion of Zeeman effects, electronic transitions between non-degenerate singlet states should not exhibit a magnetic rotation spectrum (except at extremely high fields). On the other hand, in transitions between a degenerate and a non-degenerate electronic state ($^1\Pi - {}^1\Sigma$, $^1E - {}^1A$, etc.) a strong magnetic rotation spectrum should appear corresponding to lines of low $J$. Because of the restriction to low $J$, a considerable simplification of the spectrum compared to the absorption spectrum should result. While such a simplification has been found for some diatomic molecules (see Vol. I, p. 306) no good example is known for polyatomic molecules.

For transitions between non-degenerate doublet states a magnetic rotation spectrum may be expected (not restricted to low $J$) if the uncoupling of the spin from the rotation proceeds at different rates as a function of $N$ in upper and lower state. (For complete uncoupling, because of the selection rule $\Delta M_S = 0$, no Zeeman splitting and therefore no magnetic rotation spectrum is expected.) An example has been found in the visible absorption spectrum of $NO_2$, first by Wood and Dieke (1318). Recently this spectrum has been studied under high resolution by Douglas (294a). The simplification of the spectrum is not as great as expected and a full analysis has not yet been achieved. For another doublet transition, the visible and near ultraviolet $ClO_2$ bands, no magnetic rotation spectrum was found [Eberhardt and Renner (340)] presumably because the spin is largely uncoupled in both upper and lower state.

The greatest interest attaches to the magnetic rotation spectra corresponding to singlet – triplet transitions. For these transitions always a large Zeeman splitting arises and therefore a strong magnetic rotation spectrum is expected. The first such magnetic rotation spectrum was that of $CS_2$ studied 25 years ago by Kusch and Loomis (702) although at that time it was not recognized as a triplet – singlet transition (see p. 272). More recently Eberhardt and Renner (340) have discussed several cases of such transitions; for example, they confirmed in this way that the formaldehyde absorption bands in the region 3970–3600 Å represent triplet – singlet transitions. The main singlet – singlet transition at 3500–3000 Å shows magnetic rotation only in one band, probably on account of a perturbation caused in the singlet upper state by the triplet state. Another good example is glyoxal (see p. 539). Kusch (701) has recently rediscussed the old results of Kusch and Loomis on $CS_2$ and found them to be in agreement with the assumption that a magnetic moment develops through rotational distortion so that maximum magnetic rotation occurs for fairly high $J$ values in $P$ and $R$ branch giving rise, under low dispersion, to characteristic doublets. This explanation is in accord with Douglas and Milton's (298) and Hougen's (576) interpretation of the Zeeman effect of $CS_2$ quoted earlier.

One important condition for the appearance of a magnetic rotation spectrum, emphasized by Eberhardt and Renner (340), is that the absorption lines must be sharp enough so that they can be split by the magnetic field used. No magnetic rotation spectrum is therefore expected for slightly predissociated absorption bands even if they fulfil the other conditions. This seems to be the reason that no magnetic rotation spectrum corresponding to the ultraviolet absorption system of $NO_2$ has been found.

The theory of magnetic rotation spectra, particularly of the triplet – singlet kind, has recently been developed by Hameka (467).

**Stark patterns.** As we have seen in Chapter I, section 4, if a symmetric top molecule has a permanent electric dipole moment the splittings of the energy levels in an electric field in a first approximation are entirely similar to those in a magnetic field and therefore, since the selection rules are the same, the Stark patterns of the lines in an electric field are also similar: we expect a splitting into $3(2J + 1)$, $3(2J)$ and $3(2J - 1)$ components for $R$, $Q$ and $P$ lines respectively. The over-all splitting except for the lowest $J$ values is given by

$$\Delta W = hc\Delta\nu = \frac{2K}{J + 1}(\mu' - \mu'')E$$

where $\mu'$ and $\mu''$ are the dipole moments in the upper and lower states. As in the Zeeman patterns the over-all splitting decreases with increasing $J$ but increases with $K$.

The intensity distribution within each Stark pattern is again similar to that in corresponding Zeeman patterns (see Fig. 146 of Vol. I). It is different for the light that is polarized parallel and for the light that is polarized perpendicular to the field direction. For the $\|$ component of $Q$ lines and the $\perp$ components of $R$ and $P$ lines there is zero intensity in the center and a maximum of intensity at the edges of the patterns and therefore even though a complete resolution may not be possible a separation into two "lines" is more easily achieved for these components. Such a resolution was recently obtained by Freeman and Klemperer (399) for several lines of $H_2CO$. Although this molecule is not rigorously a symmetric top, for the lines of higher $K$ values investigated by Freeman and Klemperer it is sufficiently close to a symmetric top that the simple formulae given above and in Chapter I can be applied. From the magnitude of the splitting one obtains directly $\mu' - \mu''$ and therefore since $\mu''$ is known from the microwave spectrum, the dipole moment $\mu'$ in the excited state is obtained. For $H_2CO$ Freeman and Klemperer find $\mu' = 1.48 \pm 0.07$ debye compared to $\mu'' = 2.34 \pm 0.02$ debye for the ground state. Similar studies for other molecules promise to give interesting information about the distribution of charges in its dependence on the electron configuration.

The preceding considerations apply also to linear molecules with non-zero dipole moments as long as in at least one state $\Lambda \neq 0$. For $\Sigma - \Sigma$ transitions, just as for $K' = 0 \rightarrow K'' = 0$ sub-bands of symmetric tops, no first order Stark splitting arises. For $E - A$ or $E - E$ transitions of symmetric tops the effect of first order Coriolis interaction on the Stark patterns must be considered. This has not yet been done. Nor have Stark patterns for electronic bands of asymmetric tops been considered.

As in magnetic rotation spectra, if the absorption tube is placed between two crossed nicols, upon application of an electric field, transmission will occur in the neighborhood of those lines that are sensitive to the electric field: we have a Kerr effect spectrum. Such an effect has been observed for $H_2CO$ by Freeman and Klemperer for the same lines for which they observed Stark splittings.

Buckingham and Ramsay (159a) have recently used the Stark effect to modulate lines of high $K$ and low $J$ in the absorption spectrum of $H_2CO$. By using different directions of polarization it is possible to distinguish $Q$ lines from $P$ and $R$ lines.

## CHAPTER III

# BUILDING-UP PRINCIPLES, ELECTRON CONFIGURATIONS AND STABILITY OF ELECTRONIC STATES

In the first two chapters we have taken the existence of various stable electronic states of a polyatomic molecule for granted and have considered the types of electronic states, their vibrational and rotational levels as well as the structure of the spectra arising from transitions between them. We shall now study the problem of what particular electronic states a given molecule is expected to have from theory. In other words we shall try to understand, on the basis of quantum theory, the manifold of electronic states, their relative positions and their stability in a way similar to that attempted in Chapter VI of Volume I for diatomic molecules.

As for diatomic molecules, we can determine the manifold of electronic states of a polyatomic molecule by making use of building-up principles:

(1) we may build up the molecule from the *separated atoms* or *groups of atoms*, or

(2) we may build up the molecule by starting out from the *united atom or molecule*, or

(3) we may build up the molecule by adding the electrons one by one to the fixed nuclear frame obtaining various *electron-configurations*.

For polyatomic molecules, unlike diatomic molecules there is an additional way of deriving the term manifold:

(4) we may form the molecule in its actual conformation by starting out from a conformation of different (higher or lower) symmetry.

After we have in one of these ways determined the manifold of the electronic states we must investigate the question which of these states are stable and what is their relative order. This will be done in the last section of this chapter.

## 1. Correlation of Electronic States

The building-up principles (1), (2) and (4) consist essentially in establishing the correlation rules for electronic states of polyatomic molecules. This matter is somewhat complicated compared to diatomic molecules because now there are several ways in which the molecule can be separated into atoms or groups of atoms and also several ways in which the atoms of the molecule can be brought together to a united atom or molecule. For example, the HCN molecule may be built up from H + CN or HC + N or H + C + N. The united atom is Si (with 14 electrons) but we may also consider the united molecule AlH (by uniting only C and N) or $N_2$ (by uniting only H and C).

## (a) Derivation of molecular electronic states from those of the united atom or molecule

The united atom or molecule has in general a higher symmetry than the molecule considered (e.g. $H_2CO \rightarrow O_2 \rightarrow S$). In order to obtain the molecular electronic states corresponding to a given state of the united atom (or molecule), we must therefore resolve the species of the latter (of point group $\boldsymbol{P}$) into the species of the point group ($\boldsymbol{Q}$) to which the molecule belongs. This resolution is readily done by finding for the species in question in the character table of point group $\boldsymbol{P}$ (see Appendix I) the characters for the symmetry elements of point group $\boldsymbol{Q}$. These characters belong either to a certain species of $\boldsymbol{Q}$ or to a sum of certain species of $\boldsymbol{Q}$ which are thus determined (see Vol. II, p. 236). In Table 58 of Appendix IV this resolution is given for the first ten species of the spherical point group of free atoms (viz., the point group $\boldsymbol{K}_h$, see Table 55 of Appendix I) into those of point groups $\boldsymbol{O}_h$, $\boldsymbol{T}_d$, $\boldsymbol{D}_{\infty h}$, $\boldsymbol{D}_{6h}$, $\boldsymbol{D}_{4h}$, $\boldsymbol{D}_{3h}$, $\boldsymbol{C}_{3v}$, $\boldsymbol{D}_{2d}$, $\boldsymbol{D}_{2h}$, $\boldsymbol{C}_{2v}$ and $\boldsymbol{C}_s$. Similarly Table 59 of Appendix IV gives the resolution of the first twelve species of point group $\boldsymbol{D}_{\infty h}$ into those of $\boldsymbol{D}_{6h}$, $\boldsymbol{C}_{6v}$, $\boldsymbol{D}_{4h}$, $\boldsymbol{D}_{3h}$, $\boldsymbol{C}_{3v}$, $\boldsymbol{D}_{2d}$, $\boldsymbol{D}_{2h}$, $\boldsymbol{C}_{2v}$, $\boldsymbol{C}_{2h}$ and $\boldsymbol{C}_s$. For $\boldsymbol{D}_{2h}$, $\boldsymbol{C}_{2v}$, $\boldsymbol{C}_{2h}$ and $\boldsymbol{C}_s$ the correlation is given for several possible orientations of the symmetry elements of $\boldsymbol{D}_{\infty h}$ with respect to those of the point group considered.

The correlation between the united atom and the point group $\boldsymbol{D}_{\infty h}$ is of course the same as that for diatomic molecules as first derived by Wigner and Witmer (1298) (see Vol. I, p. 322). The other correlations were first derived by Bethe (116) and Mulliken (890).

The correlations of Table 59 are implicitly contained in Table 58 but are given explicitly for the convenience of the reader. Similarly Table 60 of Appendix IV gives correlations for a few other point groups all of which are implicitly contained in Table 58 and indeed have been given in Table 53 of Volume II (p. 237).

Tables 59 and 60 can also serve for the correlations with a united molecule of any of the point groups given and can easily be extended to those point groups which are not given explicitly. However, care must be taken concerning the relative orientation of the symmetry axes of the point groups involved. Correlation with $\boldsymbol{C}_{\infty v}$ is obtained from that with $\boldsymbol{D}_{\infty h}$ by simply omitting the subscripts $g$ and $u$; similarly correlation with $\boldsymbol{D}_3$ is obtained from that with $\boldsymbol{D}_{3h}$ by omitting the $'$ and $''$, etc. (see the footnotes of the tables).

As long as spin-orbit coupling is small, the spin is conserved in the correlation between molecule and united atom, that is, e.g. a triplet state of the molecule can only be correlated with a triplet state of the united atom (or united molecule).

If the spin-orbit coupling is not small, then we must correlate the species of the total eigenfunction including spin. The determination of the species of the total eigenfunction has been discussed in Chapter I, p. 17f. The correlation of the two-valued species characteristic of half-integral spin values is contained in Table 56 of Appendix II. The species $D_{\frac{1}{2}}$ of atoms ($S = \frac{1}{2}$) correlates with $E_{\frac{1}{2}}$ or $E_{\frac{1}{2}g}$ or $2B_{\frac{1}{2}}$ while $D_{\frac{3}{2}}$ correlates with $G_{\frac{3}{2}g}$ of the cubic point groups or $E_{\frac{1}{2}} + E_{\frac{3}{2}}$ or $E_{\frac{1}{2}g} + E_{\frac{3}{2}g}$ or $E_{\frac{1}{2}} + 2B_{\frac{3}{2}}$ or $E_{\frac{1}{2}} + E_{\frac{1}{2}}$, or $E_{\frac{1}{2}g} + E_{\frac{1}{2}g}$ or $2B_{\frac{1}{2}g} + 2B_{\frac{1}{2}g}$ of point groups of lower symmetry.

In order to illustrate the application of the correlation rules let us consider a few examples:

The united atom of $CH_4$ is Ne. Assuming tetrahedral symmetry for $CH_4$ (point group $\boldsymbol{T_d}$), we obtain from the lowest states of Ne according to Table 58 the molecular states in the second column of Table 17. On this basis one expects the

TABLE 17. CORRELATION OF THE LOWEST STATES OF Ne WITH THOSE OF $CH_4$, $NH_3$, $H_2O$

| United atom: Ne | | $CH_4$ | $NH_3$ [a] | $H_2O$ [a] |
|---|---|---|---|---|
| $1s^22s^22p^6$ | $^1S_g$ | $^1A_1$ | $^1A_1$ | $^1A_1$ |
| $1s^22s^22p^53s$ | $^3P_u$ | $^3F_2$ | $^3A_1 + {^3E}$ | $^3A_1 + {^3B_1} + {^3B_2}$ |
| | $^1P_u$ | $^1F_2$ | $^1A_1 + {^1E}$ | $^1A_1 + {^1B_1} + {^1B_2}$ |
| $1s^22s^22p^53p$ | $^3D_g$ | $^3E + {^3F_2}$ | $^3A_1 + 2\,{^3E}$ | $2\,{^3A_1} + {^3A_2} + {^3B_1} + {^3B_2}$ |
| | $^1D_g$ | $^1E + {^1F_2}$ | $^1A_1 + 2\,{^1E}$ | $2\,{^1A_1} + {^1A_2} + {^1B_1} + {^1B_2}$ |
| | $^3P_g$ | $^3F_1$ | $^3A_2 + {^3E}$ | $^3A_2 + {^3B_1} + {^3B_2}$ |
| | $^1P_g$ | $^1F_1$ | $^1A_2 + {^1E}$ | $^1A_2 + {^1B_1} + {^1B_2}$ |
| | $^3S_g$ | $^3A_1$ | $^3A_1$ | $^3A_1$ |
| | $^1S_g$ | $^1A_1$ | $^1A_1$ | $^1A_1$ |

[a] The numbers in front of the species symbols indicate the number of times a given species occurs if that number is greater than 1.

ground state of $CH_4$ to be a $^1A_1$ state. The first excited state $^3P$ of Ne gives a $^3F_2$ state of $CH_4$, and so on. In a similar way the second and third columns of Table 17 give the electronic states of $NH_3$ and $H_2O$ corresponding to those of Ne (which is again the united atom) assuming point groups $\boldsymbol{C_{3v}}$ and $\boldsymbol{C_{2v}}$, respectively, for these molecules. For $H_2O$, for example, the $^1D_g$ state of Ne gives rise to five electronic states $^1A_1$, $^1A_1$, $^1A_2$, $^1B_1$, $^1B_2$; no degeneracies remain.

The united atom of $CH_3$ and $NH_2$ is F. Table 18 which is similar to Table 17 gives the electronic states of $CH_3$ and $NH_2$ that result from the lowest states of

TABLE 18. CORRELATION OF THE LOWEST STATES OF THE F ATOM WITH THOSE OF $CH_3$ AND $NH_2$

| United atom: F | | $CH_3$ $\boldsymbol{D_{3h}}$ | $CH_3$ $\boldsymbol{C_{3v}}$ | $NH_2$ $\boldsymbol{D_{\infty h}}$ | $NH_2$ $\boldsymbol{C_{2v}}$ |
|---|---|---|---|---|---|
| $1s^22s^22p^5$ | $^2P_u$ | $^2A_2'' + {^2E'}$ | $^2A_1 + {^2E}$ | $^2\Sigma_u^+ + {^2\Pi_u}$ | $^2A_1 + {^2B_1} + {^2B_2}$ |
| $1s^22s^22p^43s$ | $^4P_g$ | $^4A_2' + {^4E''}$ | $^4A_2 + {^4E}$ | $^4\Sigma_g^- + {^4\Pi_g}$ | $^4A_2 + {^4B_1} + {^4B_2}$ |
| | $^2P_g$ | $^2A_2' + {^2E''}$ | $^2A_2 + {^2E}$ | $^2\Sigma_g^- + {^2\Pi_g}$ | $^2A_2 + {^2B_1} + {^2B_2}$ |
| $1s^22s^22p^43p$ | $^4D_u$ | $^4A_1'' + {^4E'} + {^4E''}$ | $^4A_2 + 2\,{^4E}$ | $^4\Sigma_u^- + {^4\Pi_u} + {^4\Delta_u}$ | $^4A_1 + 2\,{^4A_2} + {^4B_1} +$ |
| | $^2D_u$ | $^2A_1'' + {^2E'} + {^2E''}$ | $^2A_2 + 2\,{^2E}$ | $^2\Sigma_u^- + {^2\Pi_u} + {^2\Delta_u}$ | $^2A_1 + 2\,{^2A_2} + {^2B_1} +$ |
| | $^4S_u$ | $^4A_1''$ | $^4A_2$ | $^4\Sigma_u^-$ | $^4A_2$ |
| | $^2S_u$ | $^2A_1''$ | $^2A_2$ | $^2\Sigma_u^-$ | $^2A_2$ |

fluorine under the assumption that $CH_3$ is planar (point group $\boldsymbol{D}_{3h}$) or non-planar (point group $\boldsymbol{C}_{3v}$) and that $NH_2$ is linear (point group $\boldsymbol{D}_{\infty h}$) or non-linear (point group $\boldsymbol{C}_{2v}$). Here the ground state of the united atom gives rise to several molecular electronic states, one of which would normally be the ground state of the molecule. The others are likely to be low-lying states of the molecule.

As a third example we give in Table 19 the molecular electronic states of $CH_2$ that result from the low-lying states of the united atom, O, both for a linear and non-linear conformation of the molecule.

TABLE 19. CORRELATION OF THE LOWEST STATES OF THE O ATOM WITH THOSE OF $CH_2$

| O | | $CH_2$ $\boldsymbol{D}_{\infty h}$ | $CH_2$ $\boldsymbol{C}_{2v}$ |
|---|---|---|---|
| $1s^22s^22p^4$ | $^3P_g$ | $^3\Sigma_g^- + {}^3\Pi_g$ | $^3A_2 + {}^3B_1 + {}^3B_2$ |
| | $^1D_g$ | $^1\Sigma_g^+ + {}^1\Pi_g + {}^1\Delta_g$ | $2\,{}^1A_1 + {}^1A_2 + {}^1B_1 + {}^1B_2$ |
| | $^1S_g$ | $^1\Sigma_g^+$ | $^1A_1$ |
| $1s^22s^22p^33s$ | $^5S_u$ | $^5\Sigma_u^-$ | $^5A_2$ |
| | $^3S_u$ | $^3\Sigma_u^-$ | $^3A_2$ |
| $3p$ | $^5P_g$ | $^5\Sigma_g^- + {}^5\Pi_g$ | $^5A_2 + {}^5B_1 + {}^5B_2$ |

For molecules containing more than one heavy atom, it is often of greater interest to consider the correlation to a united molecule than to the united atom. The united molecule of both $H_2CO$ and $C_2H_4$ is the $O_2$ molecule. In Table 20 on the basis of Table 59 of Appendix IV, the correlation is given for the lowest states of $O_2$, assuming $H_2CO$ and $C_2H_4$ to belong to the point groups $\boldsymbol{C}_{2v}$ and $\boldsymbol{D}_{2h}$, respectively. Since the first two excited states of $O_2$ lie very low, there is no reason to expect the ground states of $O_2$ ($^3\Sigma_g^-$) to correspond to the ground states of $H_2CO$ and $C_2H_4$ and indeed the observed ground states are $^1A_1$ and $^1A_g$ respectively which cannot arise from $^3\Sigma_g^-$ of the united molecule.

TABLE 20. MOLECULAR ELECTRONIC STATES OF $H_2CO$ AND $C_2H_4$ ARISING FROM THE LOWEST STATES OF $O_2$

| $O_2$ | $H_2CO$ ($\boldsymbol{C}_{2v}$) | $C_2H_4$ ($\boldsymbol{D}_{2h}$) |
|---|---|---|
| $^3\Sigma_g^-$ | $^3A_2$ | $^3B_{1g}$ |
| $^1\Delta_g$ | $^1A_1 + {}^1A_2$ | $^1A_g + {}^1B_{1g}$ |
| $^1\Sigma_g^+$ | $^1A_1$ | $^1A_g$ |
| $^3\Sigma_u^+$ | $^3A_1$ | $^3B_{1u}$ |
| $^1\Sigma_u^-$ | $^1A_2$ | $^1A_u$ |
| $^3\Delta_u$ | $^3A_1 + {}^3A_2$ | $^3A_u + {}^3B_{1u}$ |

### (b) Correlation of electronic states for different conformations of a given molecule

It is frequently of interest to know in which way the species of a molecular electronic state changes when the molecule is distorted into shapes belonging to different point groups. One may, for example, ask into which states the states of normal planar $C_2H_4$ (point group $\boldsymbol{D}_{2h}$) go over when the two $CH_2$ groups are rotated about the C=C axis so that they are in planes at right angles to each other (point group $\boldsymbol{D}_{2d}$), or similarly how the states of staggered $C_2H_6$ (point group $\boldsymbol{D}_{3d}$) are correlated with those of eclipsed $C_2H_6$ (point group $\boldsymbol{D}_{3h}$). Such questions can readily be answered by recourse to Tables 59 and 60 of Appendix IV. They are closely related to the question of the correlation of vibrations of isotopic molecules of higher and lower symmetry, a question that was considered in Volume II, Chapter II, 6. In particular, Table 53 of Volume II (p. 237) gives a number of such correlations (see also Table 60, p. 577). However, for the study of electronic structures some additional correlations, which are of no interest in vibrational problems and were therefore not included in Volume II, are now of particular interest. A few such correlations are presented in Table 21, including the above mentioned cases of $C_2H_4$ and $C_2H_6$.

TABLE 21. CORRELATION OF SPECIES OF DIFFERENT POINT GROUPS CORRESPONDING TO DIFFERENT CONFORMATIONS OF A GIVEN MOLECULE

| $\boldsymbol{D}_{2h}$ | $\rightarrow \boldsymbol{D}_2 \rightarrow$ | $\boldsymbol{D}_{2d}$ |
|---|---|---|
| $A_g$ | $A$ | $A_1$ or $B_1$ |
| $A_u$ | $A$ | |
| $B_{1g}$ | $B_1$ | $A_2$ or $B_2$ |
| $B_{1u}$ | $B_1$ | |
| $B_{2g}$ | $B_2$ | $E$ |
| $B_{2u}$ | $B_2$ | |
| $B_{3g}$ | $B_3$ | $E$ |
| $B_{3u}$ | $B_3$ | |

| $\boldsymbol{D}_{3d}$ | $\rightarrow \boldsymbol{D}_3 \rightarrow$ | $\boldsymbol{D}_{3h}$ |
|---|---|---|
| $A_{1g}$ | $A_1$ | $A_1'$ or $A_1''$ |
| $A_{1u}$ | $A_1$ | |
| $A_{2g}$ | $A_2$ | $A_2'$ or $A_2''$ |
| $A_{2u}$ | $A_2$ | |
| $E_g$ | $E$ | $E'$ or $E''$ |
| $E_u$ | $E$ | |

| $\boldsymbol{C}_{2v}$ | $\rightarrow \boldsymbol{C}_2 \rightarrow$ | $\boldsymbol{C}_{2h}$ |
|---|---|---|
| $A_1$ | $A$ | $A_g$ or $A_u$ |
| $A_2$ | $A$ | |
| $B_1$ | $B$ | $B_g$ or $B_u$ |
| $B_2$ | $B$ | |

There is one important difference between the Tables 59 and 60 (as well as Table 53 of Vol. II) on the one hand and Table 21 on the other. When the species of a point group of higher symmetry $\boldsymbol{P}$ are resolved into those of a point group of lower symmetry $\boldsymbol{Q}$ (but such that all symmetry elements of $\boldsymbol{Q}$ are also symmetry elements of $\boldsymbol{P}$), the correlations are always unambiguous. But the reverse correlations are not always unambiguous. For example, an $A_1$ state in point group $\boldsymbol{C}_{3v}$ may result from an $S_g$, or $P_u$, or $D_g$, or $F_g$, ... state of a united atom, or from an $A_1$ or $F_2$ state of a $\boldsymbol{T}_d$ molecule. There is no way of telling from the correlation rules which is correct. The same ambiguity arises when one wants to correlate the species of two point groups of different but equally high symmetry. Thus in the example of $C_2H_4$ (or $C_2H_6$) when the two $CH_2$ (or $CH_3$) groups are twisted with respect to each other, first a conformation of lower symmetry is formed ($\boldsymbol{D}_2$ for $C_2H_4$, $\boldsymbol{D}_3$ for $C_2H_6$) and the correlation from this point group of lower symmetry to the final point group of higher symmetry is ambiguous. Thus the $A_g$ state of planar $C_2H_4$ can go into an $A_1$ or a $B_1$ state of perpendicular $C_2H_4$, or conversely, an $A_1$ state of perpendicular $C_2H_4$ can go over into an $A_g$ or an $A_u$ state of planar $C_2H_4$, and similarly in other cases. In Table 21 the species of the intermediate point groups are included.

In Tables 58 and 59, the columns for the point groups $\boldsymbol{D}_{2h}$ and $\boldsymbol{D}_{2d}$ or for $\boldsymbol{D}_{3h}$ and $\boldsymbol{D}_{3d}$ seem to have a one-to-one correlation, but this is only apparent since these tables only give the resolution of the species of a point group of high symmetry into $\boldsymbol{D}_{2h}$, $\boldsymbol{D}_{2d}$, $\boldsymbol{D}_{3h}$, $\boldsymbol{D}_{3d}$ but not the actual correlation between $\boldsymbol{D}_{2h}$ and $\boldsymbol{D}_{2d}$ or between $\boldsymbol{D}_{3h}$ and $\boldsymbol{D}_{3d}$, which depends on the path chosen. On the other hand, if one asks, for example, which states in non-linear $C_2H_2$ of $\boldsymbol{C}_{2v}$ or $\boldsymbol{C}_{2h}$ symmetry arise from a $\Sigma_g^+$ state of linear $C_2H_2$, it is clear that only the states $A_1$ and $A_g$ respectively arise, that is, $A_1$ of $\boldsymbol{C}_{2v}$ goes over into $A_g$ of $\boldsymbol{C}_{2h}$ if one goes via the linear conformation and knows that one is dealing with a $\Sigma_g^+$ state. But if one starts with $A_1$ of $\boldsymbol{C}_{2v}$ without knowing the corresponding state in the linear conformation, then one can only conclude from the correlation rules that the corresponding linear state is $\Sigma_g^+$ or $\Pi_u$ or $\Delta_g$ or $\Phi_u$, etc., and therefore that the corresponding state of the $\boldsymbol{C}_{2h}$ conformation is $A_g$ or $A_u$ or $B_g$ or $B_u$. The correlation via the $\boldsymbol{C}_2$ form, that is, by twisting the C—H bonds around the C—C bond, gives a less ambiguous result showing that only $A_g$ or $A_u$ can result from $A_1$ of $\boldsymbol{C}_{2v}$ (Table 21).

Again, in all these correlations there must be spin conservation as long as spin-orbit coupling is small, that is, singlets correlate with singlets, doublets with doublets, etc.

## (c) Derivation of molecular electronic states from those of the separated atoms or groups of atoms

Like the correlation with the united atom or molecule, the correlation between the electronic states of a polyatomic molecule and those of the separated atoms or groups of atoms can be obtained by an appropriate generalization of the Wigner–Witmer rules for diatomic molecules (Vol. I, p. 315f).

### (α) *Linear molecules*

**Unsymmetrical molecules (point group $\boldsymbol{C}_{\infty v}$).** For unsymmetrical linear molecules we can use the vector model in order to derive the molecular electronic states from those of the separated atoms or groups of atoms. As for diatomic molecules we obtain the possible $\Lambda$ values by algebraical addition of the $M_{L_i}$ values of all atoms (or groups of atoms).

$$\Lambda = \left| \sum M_{L_i} \right| \tag{III, 1}$$

where $M_{L_i}$ for each atom takes the values $L_i$, $L_i - 1, \ldots, -L_i$. The possible total spin values $\boldsymbol{S}$ are obtained by vectorial addition of the individual spin values $\boldsymbol{S}_i$, that is

$$\boldsymbol{S} = \sum \boldsymbol{S}_i. \tag{III, 2}$$

For the quantum number $S$ this means that we have to form partial resultants according to the usual rule:

$$S_{ik} = S_i + S_k, S_i + S_k - 1, \ldots, |S_i - S_k| \tag{III, 3}$$

which are then added according to the same rule. For example, for four atoms with spins $S_1$, $S_2$, $S_3$, $S_4$ we have for the quantum number of the resultant spin

$$S = S_{12} + S_{34}, S_{12} + S_{34} - 1, \ldots, |S_{12} - S_{34}|$$

where

$$S_{12} = S_1 + S_2, S_1 + S_2 - 1, \ldots, |S_1 - S_2|,$$
$$S_{34} = S_3 + S_4, S_3 + S_4 - 1, \ldots, |S_3 - S_4|.$$

The $\Sigma$ states resulting from (III, 1) occur in pairs if at least two $M_{L_i}$ are different from zero. The members of a pair differ by a reversal of the signs of all $M_{L_i}$ values. One of each pair is $\Sigma^+$, the other $\Sigma^-$. The one $\Sigma$ state which is not a member of a pair is that state which arises when all $M_{L_i} = 0$. This state, as for diatomic molecules, is $\Sigma^+$ or $\Sigma^-$ depending on whether the sum

$$L_1 + L_2 + L_3 + \cdots + \sum l_{i_1} + \sum l_{i_2} + \sum l_{i_3} + \cdots$$

is even or odd, respectively (see also below). Here the sums $\sum l_i$ are extended over all the electrons in each atom; even or odd $\sum l_i$ corresponds to even or odd *parity* ($g$ or $u$).

As an example, consider the building-up of the HCN molecule from its atoms in their respective ground states, i.e., $^2S_g$, $^3P_g$ and $^4S_u$. Since only one $L_i$ is different from 0, the value of $\Lambda$ is determined by this $L_i$ alone according to (III, 1), that is, $\Lambda = 1$ or 0; in other words we have a $\Pi$ or a $\Sigma$ state. The addition of the spins gives for the resultant spin the values $S = 3, 2, 2, 1, 1, 0$. Thus we have the molecular states: $^1\Sigma$, $^1\Pi$, two $^3\Sigma$, two $^3\Pi$, two $^5\Sigma$, two $^5\Pi$, $^7\Sigma$, $^7\Pi$. Here the $\Sigma$ states according to the above rule are all $\Sigma^+$.

As a second less simple example, consider the states of FCN that arise from the atomic states $^2P_u$, $^3P_g$, $^2D_u$ (that is, we assume the F and C atoms to be in their ground states but the N atom to be in its first excited state). The formation of the resulting $\Lambda$ values according to (III, 1) gives the states $\Sigma(9)$, $\Pi(8)$, $\Delta(6)$, $\Phi(3)$, $\Gamma$, where the numbers in brackets give the number of times a given species occurs (if it is $>1$); the vector addition of the $S_i$ gives $S = 2, 1, 1, 0$. Of the nine $\Sigma$ states five are $\Sigma^+$, four are $\Sigma^-$. Thus, we obtain the states: $^1\Sigma^+(5)$, $^1\Sigma^-(4)$, $^1\Pi(8)$, $^1\Delta(6)$, $^1\Phi(3)$, $^1\Gamma$, $^3\Sigma^+(10)$, $^3\Sigma^-(8)$, $^3\Pi(16)$, $^3\Delta(12)$, $^3\Phi(6)$, $^3\Gamma(2)$, $^5\Sigma^+(5)$, $^5\Sigma^-(4)$, $^5\Pi(8)$, $^5\Delta(6)$, $^5\Phi(3)$, $^5\Gamma$.

It is clear from these examples that this type of correlation does not greatly restrict the number of possibilities and is consequently not very useful. A more useful correlation is obtained if we consider the building-up from smaller molecules. If an atom with orbital angular momentum $\boldsymbol{L}$ and a diatomic molecule with orbital angular momentum $\boldsymbol{\Lambda}_d$ about the internuclear axis are brought together to form a linear molecule, the resultant $\boldsymbol{\Lambda}$ is given by

$$\Lambda = |M_{L_1} + M_{L_2}| \tag{III, 1a}$$

where $M_{L_1} = L, L - 1, \ldots, -L$ and $M_{L_2} = \pm \Lambda_d$, while the resultant spin is as before given by (III, 2). Thus if HCN is formed from CH in its ground state ($^2\Pi$) and N in its ground state ($^4S$), we obtain only one $\Lambda$ value, viz. $\Lambda = 1$ while the resultant spin is $S = 2$ or 1; in other words, we obtain HCN in a $^5\Pi$ or a $^3\Pi$ state. If the N atom instead is in its first excited state $^2D$ more $\Lambda$ values arise, viz. $\Lambda = 3, 2, 1, 1, 0, 0$ which are now to be combined with the spin values 0 and 1. Thus we obtain the states $^1\Sigma^+$, $^1\Sigma^-$, $^1\Pi(2)$, $^1\Delta$, $^1\Phi$, $^3\Sigma^+$, $^3\Sigma^-$, $^3\Pi(2)$, $^3\Delta$, $^3\Phi$ of HCN.

These examples and several others are collected in Table 22, where, however, the multiplicities are omitted since they can always be easily derived from (III, 2). Each of the resultant states given occurs with each of the possible multiplicities. Naturally the states arising from a given set of say three atoms in certain states must be the same as those obtained by first bringing two atoms together and then combining the resulting states of this diatomic molecule with the state of the third atom. This rule can easily be verified in the example of HCN.

TABLE 22. ELECTRONIC STATES OF LINEAR MOLECULES RESULTING FROM THE STATES OF UNEQUAL SEPARATED GROUPS

| States of separated groups | Resulting molecular states |
|---|---|
| $S_g + \Sigma^+$ or $S_u + \Sigma^-$ | $\Sigma^+$ |
| $S_g + \Sigma^-$ or $S_u + \Sigma^+$ | $\Sigma^-$ |
| $S_g + \Pi$ or $S_u + \Pi$ | $\Pi$ |
| $S_g + \Delta$ or $S_u + \Delta$ | $\Delta$ |
| $P_g + \Sigma^+$ or $P_u + \Sigma^-$ | $\Sigma^-, \Pi$ |
| $P_g + \Sigma^-$ or $P_u + \Sigma^+$ | $\Sigma^+, \Pi$ |
| $P_g + \Pi$ or $P_u + \Pi$ | $\Sigma^+, \Sigma^-, \Pi, \Delta$ |
| $P_g + \Delta$ or $P_u + \Delta$ | $\Pi, \Delta, \Phi$ |
| $D_g + \Sigma^+$ or $D_u + \Sigma^-$ | $\Sigma^+, \Pi, \Delta$ |
| $D_g + \Sigma^-$ or $D_u + \Sigma^+$ | $\Sigma^-, \Pi, \Delta$ |
| $D_g + \Pi$ or $D_u + \Pi$ | $\Sigma^+, \Sigma^-, \Pi, \Pi, \Delta, \Phi$ |
| $D_g + \Delta$ or $D_u + \Delta$ | $\Sigma^+, \Sigma^-, \Pi, \Delta, \Phi, \Gamma$ |
| $\Sigma^+ + \Sigma^+$ or $\Sigma^- + \Sigma^-$ | $\Sigma^+$ |
| $\Sigma^+ + \Sigma^-$ | $\Sigma^-$ |
| $\Sigma^+ + \Pi$ or $\Sigma^- + \Pi$ | $\Pi$ |
| $\Sigma^+ + \Delta$ or $\Sigma^- + \Delta$ | $\Delta$ |
| $\Pi + \Pi$ | $\Sigma^+, \Sigma^-, \Delta$ |
| $\Pi + \Delta$ | $\Pi, \Phi$ |
| $\Delta + \Delta$ | $\Sigma^+, \Sigma^-, \Gamma$ |

If two diatomic or linear polyatomic groups are brought together to form a linear molecule the resultant $\Lambda$ is clearly given by the sum of the $M_{L_i}$ ($= \pm\Lambda_i$) values while the multiplicities are obtained as before. The lower part of Table 22 gives the most important examples of this type. It can easily be extended to include higher $\Lambda_i$ values.

The question of whether the resultant $\Sigma$ states in such cases are $\Sigma^+$ or $\Sigma^-$ is easily answered since in a first approximation the wave function is a product of the wave functions of the component systems, that is the resultant species are obtained from the direct product of the component species (see Table 57 of Appendix III). Thus $\Sigma^+ + \Sigma^+$ as well as $\Sigma^- + \Sigma^-$ gives $\Sigma^+$, while $\Sigma^+ + \Sigma^-$ gives $\Sigma^-$. If one component system is an atom, we must first resolve the atomic species according to Table 58 into those of point group $C_{\infty v}$ and then form the direct product. In this way we see that $P_g + \Sigma^+$ gives $\Sigma^-$ and $\Pi$ because $P_g$ is resolved into $\Sigma^-$ and $\Pi$, and similarly in other cases.

As an example consider the molecule HCCF built up from CH + CF. Both parts have $^2\Pi$ ground states, and therefore from Table 22 we see that the states $^1\Sigma^+$, $^1\Sigma^-$, $^1\Delta$, $^3\Sigma^+$, $^3\Sigma^-$, $^3\Delta$ will arise.

In the previous example of HCN the combination of the ground states of CH ($^2\Pi$) and N ($^4S$) does not give rise to a $^1\Sigma$ state. On the other hand the infrared spectrum shows the ground state to be $^1\Sigma$. Thus we conclude that HCN in its ground state does not dissociate adiabatically into the ground states of CH and N even though it does dissociate into the ground states of H ($^2S$) and CN ($^2\Sigma$). In a similar way one finds that the $N_2O$ molecule in its $^1\Sigma$ ground state cannot dissociate into $N_2$ + O in their ground states ($^1\Sigma + {}^3P$) nor into N + NO in their ground states ($^4S + {}^2\Pi$). Such conclusions are of importance for a discussion of the unimolecular decomposition of these and other molecules (see Chap. IV).

**Symmetrical molecules (point group $D_{\infty h}$).** If two identical linear groups are brought together to form a linear molecule the resulting states are also easily derived. If the identical groups are in different states we can use the previous methods (lower part of Table 22) except that every one state of the previous treatment now occurs twice, once as a $g$ state and once as a $u$ state. This is because the same states arise when the excitation energy of the two parts is exchanged and the resolution of this resonance degeneracy leads to a splitting into a $g$ and a $u$ state just as for diatomic molecules formed from identical atoms in different states. For example a CH radical in the $^2\Pi$ ground state and another in the $^2\Delta$ excited state give rise to the states $^1\Pi_g$, $^1\Pi_u$, $^1\Phi_g$, $^1\Phi_u$, $^3\Pi_g$, $^3\Pi_u$, $^3\Phi_g$ and $^3\Phi_u$ of $C_2H_2$ (see Table 22).

If the two identical groups are in the same states, there is no resonance degeneracy and therefore only as many states arise as for unequal groups (Table 22) but some of these states are $g$ and some are $u$, and this symmetry character alternates for the different possible spin values in the same way as for diatomic molecules. Table 23 shows the results of a group-theoretical consideration for the most important cases [see Kotani (689)]. One sees, for example, that if two CH

TABLE 23. ELECTRONIC STATES OF SYMMETRICAL LINEAR MOLECULES ($D_{\infty h}$) RESULTING FROM IDENTICAL STATES OF THE SEPARATED EQUAL GROUPS

| States of separated groups | Resulting molecular states |
|---|---|
| $^1\Sigma^+ + {}^1\Sigma^+$ or $^1\Sigma^- + {}^1\Sigma^-$ | $^1\Sigma_g^+$ |
| $^2\Sigma^+ + {}^2\Sigma^+$ or $^2\Sigma^- + {}^2\Sigma^-$ | $^1\Sigma_g^+$, $^3\Sigma_u^+$ |
| $^3\Sigma^+ + {}^3\Sigma^+$ or $^3\Sigma^- + {}^3\Sigma^-$ | $^1\Sigma_g^+$, $^3\Sigma_u^+$, $^5\Sigma_g^+$ |
| $^1\Pi + {}^1\Pi$ | $^1\Sigma_g^+$, $^1\Sigma_u^-$, $^1\Delta_g$ |
| $^2\Pi + {}^2\Pi$ | $^1\Sigma_g^+$, $^1\Sigma_u^-$, $^1\Delta_g$, $^3\Sigma_u^+$, $^3\Sigma_g^-$, $^3\Delta_u$ |
| $^3\Pi + {}^3\Pi$ | $^1\Sigma_g^+$, $^1\Sigma_u^-$, $^1\Delta_g$, $^3\Sigma_u^+$, $^3\Sigma_g^-$, $^3\Delta_u$, $^5\Sigma_g^+$, $^5\Sigma_u^-$, $^5\Delta_g$ |
| $^1\Delta + {}^1\Delta$ | $^1\Sigma_g^+$, $^1\Sigma_u^-$, $^1\Gamma_g$ |
| $^2\Delta + {}^2\Delta$ | $^1\Sigma_g^+$, $^1\Sigma_u^-$, $^1\Gamma_g$, $^3\Sigma_u^+$, $^3\Sigma_g^-$, $^3\Gamma_u$ |
| $^3\Delta + {}^3\Delta$ | $^1\Sigma_g^+$, $^1\Sigma_u^-$, $^1\Gamma_g$, $^3\Sigma_u^+$, $^3\Sigma_g^-$, $^3\Gamma_u$, $^5\Sigma_g^+$, $^5\Sigma_u^-$, $^5\Gamma_g$ |

groups in their $^2\Pi$ ground states are brought together, only the states $^1\Sigma_g^+$, $^1\Sigma_u^-$, $^1\Delta_g$, $^3\Sigma_u^+$, $^3\Sigma_g^-$, $^3\Delta_u$ of $C_2H_2$ result.

In order to determine the electronic states from the separated atoms, we can again proceed in much the same way as for unsymmetrical molecules, except that we have to treat each pair of identical atoms together and determine, according to the Wigner–Witmer rules, the molecular states resulting from it. The resulting states for the various pairs of identical atoms are then combined by the same rules as before.

As an example, let us consider the formation of linear $CH_2$ from its three atoms. The two H atoms in their $^2S$ ground states give the molecular states $^1\Sigma_g^+$ and $^3\Sigma_u^+$ (as for $H_2$) and the carbon atom in its $^3P_g$ ground state gives according to Table 58 in a field of $D_{\infty h}$ symmetry the molecular states $^3\Sigma_g^-$ and $^3\Pi_g$. Combining the states of H + H and C then gives the following states of linear $CH_2$: $^1\Sigma_u^-$, $^1\Pi_u$, $^3\Sigma_g^-$, $^3\Sigma_u^-$, $^3\Pi_g$, $^3\Pi_u$, $^5\Sigma_u^-$, $^5\Pi_u$. In Table 24 a number of similar examples are collected including the examples of unsymmetrical molecules previously discussed.

It remains to consider the building-up of a symmetrical linear molecule from unlike groups, for example, forming a symmetrical $XY_2$ molecule from XY + Y. Clearly we must proceed in the same way as for unsymmetrical molecules obtaining both the $\Lambda$ and $S$ values of the resultant states. However, we do not obtain in this way the $g$, $u$ symmetry of these states. The reason for this lack of specificity of the correlation rules lies in the fact that we start out from an unsymmetrical conformation of point group $C_{\infty v}$ in which $g$ and $u$ are not defined, while in forming the molecule from equal parts at each stage the symmetry is $D_{\infty h}$ and $g$ and $u$ are defined. Thus, even for symmetrical molecules if we start out from separated unlike groups, Table 22 contains all the information (except for the spin) that we can obtain directly.

In some cases it is possible to obtain information about the $g$, $u$ symmetry indirectly by comparison with the result of building up the molecule in a symmetrical way. As an example consider the linear $CH_2$ molecule. If we bring

Table 24. Examples of molecular electronic states of linear molecules resulting from certain states of the separated atoms[1]

| Molecule: Type | Molecule: Prototype | Atomic states | Molecular states |
|---|---|---|---|
| XYZ | HCN | X $^2S_g$ + Y $^3P_g$ + Z $^4S_u$ | $^1\Sigma^+$, $^1\Pi$, $^3\Sigma^+(2)$, $^3\Pi(2)$, $^5\Sigma^+(2)$, $^5\Pi(2)$, $^7\Sigma^+$, $^7\Pi$ |
| | | X $^2S_g$ + Y $^5S_u$ + Z $^4S_u$ | $^1\Sigma^+$, $^3\Sigma^+(2)$, $^5\Sigma^+(2)$, $^7\Sigma^+(2)$, $^9\Sigma^+$ |
| | | X $^2S_g$ + Y $^3P_g$ + Z $^2D_u$ | $^1\Sigma^+(2)$, $^1\Sigma^-$, $^1\Pi(3)$, $^1\Delta(2)$, $^1\Phi$, $^3\Sigma^+(4)$, $^3\Sigma^-(2)$, $^3\Pi(6)$, $^3\Delta(4)$, $^3\Phi(2)$, $^5\Sigma^+(2)$, $^5\Sigma^-$, $^5\Pi(3)$, $^5\Delta(2)$, $^5\Phi$ |
| | FCN | X $^2P_u$ + Y $^3P_g$ + Z $^2D_u$ | $^1\Sigma^+(5)$, $^1\Sigma^-(4)$, $^1\Pi(8)$, $^1\Delta(6)$, $^1\Phi(3)$, $^1\Gamma$, $^3\Sigma^+(10)$, $^3\Sigma^-(8)$, $^3\Pi(16)$, $^3\Delta(12)$, $^3\Phi(6)$, $^3\Gamma(2)$, $^5\Sigma^+(5)$, $^5\Sigma^-(4)$, $^5\Pi(8)$, $^5\Delta(6)$, $^5\Phi(3)$, $^5\Gamma$ |
| $X_3$ | $H_3$ | X $^2S_g$ + X $^2S_g$ + X $^2S_g$ | $^2\Sigma_g^+$, $^2\Sigma_u^+$, $^4\Sigma_u^+$ |
| | $H_3^+$ | $X$ $^2S_g$ + $X^+$ $^1S_g$ + $X$ $^2S_g$ | $^1\Sigma_g^+(2)$, $^1\Sigma_u^+$, $^3\Sigma_g^+$, $^3\Sigma_u^+(2)$ |
| $XY_2$ | $BeH_2$ | Y $^2S_g$ + X $^1S_g$ + Y $^2S_g$ | $^1\Sigma_g^+$, $^3\Sigma_u^+$ |
| | $BH_2$ | Y $^2S_g$ + X $^2P_u$ + Y $^2S_g$ | $^2\Sigma_g^+$, $^2\Sigma_u^+$, $^2\Pi_g$, $^2\Pi_u$, $^4\Sigma_g^+$, $^4\Pi_g$ |
| | $CH_2$ | Y $^2S_g$ + X $^3P_g$ + Y $^2S_g$ | $^1\Sigma_u^-$, $^1\Pi_u$, $^3\Sigma_g^-$, $^3\Sigma_u^-$, $^3\Pi_g$, $^3\Pi_u$, $^5\Sigma_u^-$, $^5\Pi_u$ |
| | | Y $^2S_g$ + X $^5S_u$ + Y $^2S_g$ | $^3\Sigma_g^-$, $^5\Sigma_g^-$, $^5\Sigma_u^-$, $^7\Sigma_g^-$ |
| | | Y $^2S_g$ + X $^1D_g$ + Y $^2S_g$ | $^1\Sigma_g^+$, $^1\Pi_g$, $^1\Delta_g$, $^3\Sigma_u^+$, $^3\Pi_u$, $^3\Delta_u$ |
| | $CO_2$ | Y $^3P_g$ + X $^3P_g$ + Y $^3P_g$ | $^1\Sigma_g^+(2)$, $^1\Sigma_u^+$, $^1\Sigma_g^-$, $^1\Sigma_u^-(3)$, $^1\Pi_g(2)$, $^1\Pi_u(4)$, $^1\Delta_g$, $^1\Delta_u(2)$, $^1\Phi_u$, $^3\Sigma_g^+(4)$, $^3\Sigma_u^+(5)$, $^3\Sigma_g^-(7)$, $^3\Sigma_u^-(5)$, $^3\Pi_g(10)$, $^3\Pi_u(8)$, $^3\Delta_g(5)$, $^3\Delta_u(4)$, $^3\Phi_g(2)$, $^3\Phi_u$, $^5\Sigma_g^+(3)$, $^5\Sigma_u^+(3)$, $^5\Sigma_g^-(4)$, $^5\Sigma_u^-(4)$, $^5\Pi_g(6)$, $^5\Pi_u(6)$, $^5\Delta_g(3)$, $^5\Delta_u(3)$, $^5\Phi_g$, $^5\Phi_u$, $^7\Sigma_g^+$, $^7\Sigma_u^+(2)$, $^7\Sigma_g^-(3)$, $^7\Sigma_u^-$, $^7\Pi_g(4)$, $^7\Pi_u(2)$, $^7\Delta_g(2)$, $^7\Delta_u$, $^7\Phi_g$ |

[1] Numbers in ( ) following the species symbols indicate the number of times the particular species arises if this number is larger than one.

together $H(^2S) + CH(^2\Pi)$ in their ground states, according to Table 22 and the spin rule (III, 2) we obtain the molecular states $^1\Pi$ and $^3\Pi$; but it is not possible to say whether these states are $g$ or $u$. If, however, we bring the atoms together symmetrically in their ground states we get the molecular states given in Table 24. Among these is only a single $^1\Pi$ state which is $^1\Pi_u$, and, since the $^2\Pi$ ground state of CH is formed from normal atoms, the $^1\Pi$ state obtained from $H(^2S) + CH(^2\Pi)$ must be this $^1\Pi_u$. Similar information about the $^3\Pi$ state cannot be obtained since, according to Table 24, there are two $^3\Pi$ states, $^3\Pi_g$ and $^3\Pi_u$ which result from $H(^2S) + C(^3P) + H(^2S)$ and it is not possible to say unambiguously which of these two corresponds to the $^3\Pi$ state resulting from $H(^2S) + CH(^2\Pi)$.

Actually, from energy considerations and by comparing with the states resulting from H + C + H, an answer can be found in the present case, viz., that the $^3\Pi$ state from H + CH is $^3\Pi_g$. However this result cannot be generalized. For example for $C_2H_2$ formed from $H(^2S) + C_2H(^2\Pi)$ the two $\Pi$ states are $^1\Pi_g$ and $^3\Pi_u$.

### (β) *Non-linear molecules*

**Building-up from individual atoms.** If we want to study the question as to what molecular electronic states of a non-linear molecule arise from given states of the separated atoms, we must proceed in a way similar to that used for linear molecules, that is, we must, following Kotani (689), first determine the states of each set of identical atoms in the point group of the complete molecule and then compound the results of all the sets in order to get the states of the molecule.

Consider the formation of a non-linear XYZ molecule (point group $C_s$) from X, Y and Z. We must resolve the species of X, Y and Z into those of the point group $C_s$ using Table 58 (Appendix IV). For example, if the atomic states are $^2S_g$, $^4S_u$ and $^3P_g$ as in HNO, we find $^2A'$, $^4A''$ and $^3A' + {}^3A'' + {}^3A''$, respectively. According to the rules for the direct product (Table 57 of Appendix III), we find for the species of the resulting molecular states $A''$, $A'$ and $A'$. The multiplicity, as in the linear case, is obtained by the vector addition of the $\boldsymbol{S}$ vectors (eq. III, 2) and this yields in the present case the resultant spins 3, 2, 2, 1, 1, 0, that is, we obtain the molecular states $^1A'(2)$, $^1A''$, $^3A'(4)$, $^3A''(2)$, $^5A'(4)$, $^5A''(2)$, $^7A'(2)$, $^7A''$.

In the formation of a non-linear $XY_2$ molecule (point group $C_{2v}$) from X and 2Y, we must consider the two Y atoms together as a diatomic molecule and then resolve the species of $D_{\infty h}$ obtained into those of $C_{2v}$ taking account of the fact that the two-fold axis of $C_{2v}$ is at right angles to the $C_\infty$ of $D_{\infty h}$. As an example, consider that the Y atoms are in a $^2S_g$ state and the X atom in a $^3P_g$ state as in the formation of $H_2O$ (or non-linear $CH_2$) from ground state atoms. The two Y atoms, according to the Wigner–Witmer rules, give the states $^1\Sigma_g^+$ and $^3\Sigma_u^+$ of $Y_2$ which, when resolved into $C_{2v}$, give, according to Table 59, $^1A_1$ and $^3B_2$. The $^3P_g$ state of X gives on resolution (Table 58) $^3A_2 + {}^3B_1 + {}^3B_2$. Combining the states of $Y_2$ and of X (i.e. forming the direct product) we obtain

$$^1A_1 \times (^3A_2 + {}^3B_1 + {}^3B_2) = {}^3A_2 + {}^3B_1 + {}^3B_2$$

and

$$^3B_2 \times (^3A_2 + {}^3B_1 + {}^3B_2) = {}^{5,3,1}B_1 + {}^{5,3,1}A_2 + {}^{5,3,1}A_1.$$

In other words we obtain the states $^1A_1$, $^1A_2$, $^1B_1$, $^3A_1$, $^3A_2(2)$, $^3B_1(2)$, $^3B_2$, $^5A_1$, $^5A_2$, $^5B_1$. If one of the Y atoms were in an excited $^2S_g$ state, then the four states $^1\Sigma_g^+$, $^1\Sigma_u^+$, $^3\Sigma_g^+$, $^3\Sigma_u^+$ of $Y_2$ would arise which give $^1A_1$, $^1B_2$, $^3A_1$, $^3B_2$ in $\boldsymbol{C}_{2v}$ symmetry. Combined with $^3P_g$ $(= {}^3A_2 + {}^3B_1 + {}^3B_2)$ they give the following states of $XY_2$: $^1A_1$, $^1A_2(2)$, $^1B_1(2)$, $^1B_2$, $^3A_1(2)$, $^3A_2(4)$, $^3B_1(4)$, $^3B_2(2)$, $^5A_1$, $^5A_2(2)$, $^5B_1(2)$, $^5B_2$.

The situation is less simple if we want to bring together three identical atoms X to form a triangular molecule of point group $\boldsymbol{D}_{3h}$. Suppose the three atoms are in identical $^2S_g$ states and let us first bring the molecule together unsymmetrically, that is, let one X atom approach the other two in a conformation of point group $\boldsymbol{C}_{2v}$. The two latter atoms considered together form, according to the above, the states $^1A_1$ and $^3B_2$, while the single atom forms a $^2A_1$ state. The product of these species yields the states $^2A_1$, $^2B_2$ and $^4B_2$ of $X_3$ in the conformation of an isosceles but not equilateral triangle. The same states (with different orientation of the $z$ axis) arise if we bring any of the other X atoms up to the remaining two. If we now consider the symmetrical ($\boldsymbol{D}_{3h}$) configuration, it is clear on the basis of Table 60 (Appendix IV) that the $^4B_2$ state, since it is the only quartet state, must go over into $^4A_2'$. The two doublet states $^2A_1$ and $^2B_2$ can either go to $^2A_1'$ and $^2A_2'$ or form one degenerate $^2E'$ state. It can be shown from group theory that actually the latter correlation is the correct one [see Kotani (689)]. Qualitatively this result is due to the equivalence of the three (unsymmetrical) ways in which $X_3$ can be brought together. Thus only the two states $^2E'$ and $^4A_2'$ result from three X atoms in $^2S_g$ states if they form an equilateral $X_3$ triangle.

It may at first sight seem surprising that three X atoms in totally symmetric $S_g$ states give anything but $A_1'$ states of $X_3$. However, it must be realized that because of the possibility of exchange of electrons there are actually the six spatial eigenfunctions

$$\psi_a(1)\psi_b(2)\psi_c(3), \quad \psi_a(2)\psi_b(3)\psi_c(1), \quad \psi_a(3)\psi_b(1)\psi_c(2)$$
$$\psi_a(3)\psi_b(2)\psi_c(1), \quad \psi_a(2)\psi_b(1)\psi_c(3), \quad \psi_a(1)\psi_b(3)\psi_c(2)$$

These functions can be combined to form one totally symmetric, one antisymmetric and two degenerate functions (i.e., symmetric, antisymmetric, etc. with respect to exchange of electrons). Since the spin functions of three atoms of spin $\frac{1}{2}$ are either totally symmetric (four functions) or degenerate (two pairs of functions) as shown in the similar situation for three nuclei of $I = \frac{1}{2}$ in Volume II, p. 410, the over-all eigenfunction can be antisymmetric, as required by the Pauli principle, only if either the antisymmetric spatial function is combined with the four symmetric spin functions (yielding $^4A_2'$) or the degenerate spatial functions with the degenerate spin function yielding $^2E'$.

In Table 25 the examples used in the preceding discussion as well as a few others are collected.

The treatment of four-atomic molecules proceeds in a similar way. We consider only a few examples. Suppose that, in an $XYZ_2$ molecule of point

TABLE 25. ELECTRONIC STATES OF NON-LINEAR TRIATOMIC MOLECULES RESULTING FROM GIVEN STATES OF THE SEPARATED ATOMS

| Molecule | | | Atomic states | Molecular states |
|---|---|---|---|---|
| Type | Point group | Prototype | | |
| XYZ | $C_s$ | | $^2S_g + {}^4S_u + {}^2S_g$ | $^2A''$, $^4A''(2)$, $^6A''$ |
| | | HNO | $^2S_g + {}^4S_u + {}^3P_g$ | $^1A'(2)$, $^1A''$, $^3A'(4)$, $^3A''(2)$, $^5A'(4)$, $^5A''(2)$, $^7A'(2)$, $^7A''$ |
| $XY_2$ | $C_{2v}$ | $H_2O$, $CH_2$ | $^3P_g + {}^2S_g + {}^2S_g$ | $^1A_1$, $^1A_2$, $^1B_1$, $^3A_1$, $^3A_2(2)$, $^3B_1(2)$, $^3B_2$, $^5A_1$, $^5A_2$, $^5B_1$ |
| | | $NH_2$ | $^4S_u + {}^2S_g + {}^2S_g$ | $^2B_1$, $^4A_2$, $^4B_1$, $^6B_1$ |
| | | $CF_2$ | $^3P_g + {}^2P_u + {}^2P_u$ | $^1A_1(8)$, $^1A_2(7)$, $^1B_1(7)$, $^1B_2(5)$, $^3A_1(13)$, $^3A_2(14)$, $^3B_1(14)$, $^3B_2(13)$, $^5A_1(8)$, $^5A_2(7)$, $^5B_1(7)$, $^5B_2(5)$ |
| | | $NF_2$ | $^4S_u + {}^2P_u + {}^2P_u$ | $^2A_1(2)$, $^2A_2$, $^2B_1(4)$, $^2B_2(2)$, $^4A_1(4)$, $^4A_2(5)$, $^4B_1(5)$, $^4B_2(4)$, $^6A_1(2)$, $^6A_2$, $^6B_1(4)$, $^6B_2(2)$ |
| | | | $^2S_g + {}^2S_g + {}^2S_g$ | $^2A_1$, $^2B_2$, $^4B_2$ |
| $X_3$ | $D_{3h}$ | $H_3$ | $^2S_g + {}^2S_g + {}^2S_g$ | $^2E'$, $^4A_2'$ |
| | | | $^2S_u + {}^2S_u + {}^2S_u$ | $^2E''$, $^4A_2''$ |
| | | $H_3^+$ | $^2S_g + {}^1S_g + {}^2S_g$ | $^1A_1'$, $^1E'$, $^3A_2'$, $^3E'$ |
| | | $F_3$ | $^2P_u + {}^2P_u + {}^2P_u$ | $^2A_1'(4)$, $^2A_1''(4)$, $^2A_2'(4)$, $^2A_2''(4)$, $^2E'(10)$, $^2E''(9)$, $^4A_1'(3)$, $^4A_1''(4)$, $^4A_2'(3)$, $^4A_2''$, $^4E'(4)$, $^4E''(4)$ |
| | | $O_3$ | $^3P_g + {}^3P_g + {}^3P_g$ | $^1A_1'(4)$, $^1A_1''(3)$, $^1A_2'$, $^1A_2''(3)$, $^1E'(4)$, $^1E''(4)$, $^3A_1'(5)$, $^3A_1''(7)$, $^3A_2'(8)$, $^3A_2''(7)$, $^3E'(13)$, $^3E''(14)$, $^5A_1'(4)$, $^5A_1''(4)$, $^5A_2'(4)$, $^5A_2''(4)$, $^5E'(9)$, $^5E''(10)$, $^7A_1'$, $^7A_1''(3)$, $^7A_2'(4)$, $^7A_2''(3)$, $^7E'(4)$, $^7E''(4)$ |
| | | $N_3$ | $^4S_u + {}^4S_u + {}^4S_u$ | $^2E''$, $^4A_1''$, $^4A_2''$, $^4E''$, $^6A_2''$, $^6E''$, $^8E''$, $^{10}A_2''$ |

group $C_{2v}$, Z is in a $^2S_g$, and X as well as Y in a $^3P_g$ state as in $H_2CO$ if formed from normal atoms. The states of the non-linear $CH_2$ group have already been derived (see above and Table 25). We have only to multiply by $^3P_g$ of O resolved into the species of $C_{2v}$, that is, by $^3A_2 + {}^3B_1 + {}^3B_2$. A large number of states arise of which the singlet states are $^1A_1(5)$, $^1A_2(4)$, $^1B_1(4)$, $^1B_2(5)$.

If a planar $XY_3$ molecule of point group $D_{3h}$ is formed from an X atom in a $^3P_g$ state and three Y atoms in $^2S_g$ states (as for $CH_3$) we obtain the resulting states by multiplying the species of $Y_3$ formed from $^2S_g$ (see Table 25), with the species into which $^3P_g$ is resolved in point group $D_{3h}$, i.e. $^3A_2' + {}^3E''$. This multiplication yields $^2A_1'$, $^2A_1''$, $^2A_2''$, $^2E'$, $^2E''(2)$, $^4A_1'$, $^4A_1''$, $^4A_2''$, $^4E'$, $^4E''(2)$, $^6A_1'$, $^6E''$. If the point group is instead $C_{3v}$, that is, if the molecule is non-planar we merely have to take account of the correlation between $D_{3h}$ and $C_{3v}$ in Table 60 (Appendix IV) and obtain the states $^2A_1(2)$, $^2A_2$, $^2E(3)$, $^4A_1(2)$, $^4A_2$, $^4E(3)$, $^6A_1$, $^6E$. If the X atom is in a $^4S_u$ state, as for the formation of $NH_3$ from ground state atoms, we find directly, if we resolve the species of $Y_3$ into those of $C_{3v}$ (obtaining $^2E + {}^4A_2$)

and observe that ${}^4S_u$ goes over into ${}^4A_2$, the following states: ${}^1A_1$, ${}^3A_1$, ${}^3E$, ${}^5A_1$, ${}^5E$, ${}^7A_1$. Thus the observed ground states ${}^1A_1$ and ${}^2A_2''$ of $NH_3$ and $CH_3$ respectively can be formed from normal atoms.

For a molecule $X_4$ consisting of four identical atoms a number of different symmetrical arrangements are possible. If the symmetry is that of a regular tetrahedron (point group $\boldsymbol{T_d}$) and if all four X atoms are in the same ${}^2S_g$ state, we obtain according to Kotani the three states ${}^1E$, ${}^3F_1$ and ${}^5A_2$. It is readily checked that on lowering the symmetry to $\boldsymbol{C_{3v}}$ the same states result as would be obtained by forming directly a $\boldsymbol{C_{3v}}$ molecule according to the methods of the preceding paragraph (see also Table 26).

If the tetrahedron is somewhat distorted so that the $X_4$ molecule belongs to point group $\boldsymbol{D_{2d}}$, the resulting states are immediately derived from those for $\boldsymbol{T_d}$

TABLE 26. ELECTRONIC STATES OF NON-LINEAR FOUR-, FIVE- AND SIX-ATOMIC MOLECULES RESULTING FROM GIVEN STATES OF THE SEPARATED ATOMS

| Molecule | | | Atomic states | Molecular states |
|---|---|---|---|---|
| Type | Point group | Prototype | | |
| XYZW | $\boldsymbol{C_s}$ | HNCN | ${}^2S_g+{}^4S_u+{}^3P_g+{}^4S_u$ | ${}^2A'(4)$, ${}^2A''(8)$, ${}^4A'(6)$, ${}^4A''(12)$, ${}^6A'(5)$, ${}^6A''(10)$, ${}^8A'(3)$, ${}^8A''(6)$, ${}^{10}A'$, ${}^{10}A''(2)$ |
| $XYZ_2$ | $\boldsymbol{C_{2v}}$ | $H_2CO$ | ${}^3P_g + {}^3P_g + 2\,{}^2S_g$ | ${}^1A_1(5)$, ${}^1A_2(4)$, ${}^1B_1(4)$, ${}^1B_2(5)$, ${}^3A_1(9)$, ${}^3A_2(8)$, ${}^3B_1(8)$, ${}^3B_2(11)$, ${}^5A_1(7)$, ${}^5A_2(6)$, ${}^5B_1(6)$, ${}^5B_2(8)$, ${}^7A_1(2)$, ${}^7A_2(2)$, ${}^7B_1(2)$, ${}^7B_2(3)$ |
| $XY_3$ | $\boldsymbol{D_{3h}}$ | $CH_3$ | ${}^3P_g + 3\,{}^2S_g$ | ${}^2A_1'$, ${}^2A_1''$, ${}^2A_2''$, ${}^2E'$, ${}^2E''(2)$, ${}^4A_1'$, ${}^4A_1''$, ${}^4A_2''$, ${}^4E'$, ${}^4E''(2)$, ${}^6A_1'$, ${}^6E''$ |
| | $\boldsymbol{C_{3v}}$ | $CH_3$ | ${}^3P_g + 3\,{}^2S_g$ | ${}^2A_1(2)$, ${}^2A_2$, ${}^2E(3)$, ${}^4A_1(2)$, ${}^4A_2$, ${}^4E(3)$, ${}^6A_1$, ${}^6E$ |
| | | $NH_3$ | ${}^4S_u + 3\,{}^2S_g$ | ${}^1A_1$, ${}^3A_1$, ${}^3E$, ${}^5A_1$, ${}^5E$, ${}^7A_1$ |
| $X_4$ | $\boldsymbol{C_{3v}}$ | $H_4$ | $4\,{}^2S_g$ | ${}^1E$, ${}^3A_2$, ${}^3E$, ${}^5A_2$ |
| | $\boldsymbol{T_d}$ | $H_4$ | $4\,{}^2S_g$ | ${}^1E$, ${}^3F_1$, ${}^5A_2$ |
| | | $P_4$ | $4\,{}^4S_u$ | ${}^1A_1$, ${}^1A_2$, ${}^1E$, ${}^3F_1(2)$, ${}^3F_2$, ${}^5A_2$, ${}^5E(2)$, ${}^5F_1$, ${}^5F_2$, ${}^7A_2$, ${}^7F_1(2)$, ${}^7F_2$, ${}^9A_2$, ${}^9E$, ${}^9F_1$, ${}^{11}F_1$, ${}^{13}A_2$ |
| | $\boldsymbol{D_{2d}}$ | $H_4$ | $4\,{}^2S_g$ | ${}^1A_1$, ${}^1B_1$, ${}^3A_2$, ${}^3E$, ${}^5B_1$ |
| | $\boldsymbol{D_{4h}}$ | $H_4$ | $4\,{}^2S_g$ | ${}^1A_{1g}$, ${}^1B_{1g}$, ${}^3A_{2g}$, ${}^3E_u$, ${}^5B_{1g}$ |
| | $\boldsymbol{D_{2h}}$ | $H_4$ [a] | $4\,{}^2S_g$ | ${}^1A_g(2)$, ${}^3B_{1g}$, ${}^3B_{2u}$, ${}^3B_{3u}$, ${}^5A_g$ |
| $XYZ_3$ | $\boldsymbol{C_{3v}}$ | $NaCH_3$ | ${}^2S_g + {}^3P_g + 3\,{}^2S_g$ | ${}^1A_1(2)$, ${}^1A_2$, ${}^1E(3)$, ${}^3A_1(4)$, ${}^3A_2(2)$, ${}^3E(6)$, ${}^5A_1(3)$, ${}^5A_2$, ${}^5E(4)$, ${}^7A_1$, ${}^7E$ |
| | | $FCH_3$ | ${}^2P_u + {}^3P_g + 3\,{}^2S_g$ | ${}^1A_1(5)$, ${}^1A_2(4)$, ${}^1E(9)$, ${}^3A_1(10)$, ${}^3A_2(8)$, ${}^3E(18)$, ${}^5A_1(7)$, ${}^5A_2(5)$, ${}^5E(12)$, ${}^7A_1(2)$, ${}^7A_2$, ${}^7E(3)$ |
| $XY_4$ | $\boldsymbol{T_d}$ | $CH_4$ | ${}^3P_g + 4\,{}^2S_g$ | ${}^1A_1$, ${}^1E$, ${}^1F_1$, ${}^1F_2$, ${}^3A_1$, ${}^3E$, ${}^3F_1(2)$, ${}^3F_2(3)$, ${}^5A_1$, ${}^5E$, ${}^5F_1$, ${}^5F_2(2)$, ${}^7F_2$ |
| $X_6$ | $\boldsymbol{D_{6h}}$ | $H_6$ | $6\,{}^2S_g$ | ${}^1A_{1g}(2)$, ${}^1B_{2u}$, ${}^1E_{2g}$, ${}^3A_{2g}$, ${}^3B_{1u}(2)$, ${}^3E_{1u}(2)$, ${}^3E_{2g}$, ${}^5A_{1g}$, ${}^5E_{1u}$, ${}^5E_{2g}$, ${}^7B_{1u}$ |

[a] Assuming the four H atoms to be in the $xy$-plane.

by means of Table 60 (Appendix IV). One finds $^1A_1$, $^1B_1$, $^3A_2$, $^3E$, $^5B_1$. If instead the four X atoms form a square (point group $\boldsymbol{D}_{4h}$), one finds from Kotani's paper the states $^1A_{1g}$, $^1B_{1g}$, $^3A_{2g}$, $^3E_u$, $^5B_{1g}$ and if the four X atoms form a rectangle (point group $\boldsymbol{D}_{2h}$), the corresponding states are, according to Table 60, $^1A_g(2)$, $^3B_{1g}$, $^3B_{2u}$, $^3B_{3u}$, $^5A_g$.

For five-atomic molecules $XYZ_3$ of point group $\boldsymbol{C}_{3v}$ one proceeds in the same way as for $XY_3$ molecules. There is an additional atom on the axis whose resolved species have to be multiplied with those of $YZ_3$.

For building up a tetrahedral $XY_4$ molecule we first resolve the atomic state of the central atom X according to point group $\boldsymbol{T}_d$ and then multiply the resulting species by those of the $Y_4$ group given above. In this way one obtains for $CH_4$ formed from a C atom in its $^3P_g$ ground state and four H atoms in their ground states the states $^1A_1$, $^1E$, $^1F_1$, $^1F_2$, $^3A_1$, $^3E$, $^3F_1(2)$, $^3F_2(3)$, $^5A_1$, $^5E$, $^5F_1$, $^5F_2(2)$, $^7F_2$. This group of states includes the $^1A_1$ ground state.

Finally we consider as an example of a six-atomic molecule, the hexagonal $X_6$ molecule (point group $\boldsymbol{D}_{6h}$) formed from six X atoms in $^2S_g$ states. From Kotani's tables one finds in this case the states $^1A_{1g}(2)$, $^1B_{2u}$, $^1E_{2g}$, $^3A_{2g}$, $^3B_{1u}(2)$, $^3E_{1u}(2)$, $^3E_{2g}$, $^5A_{1g}$, $^5E_{1u}$, $^5E_{2g}$, $^7B_{1u}$.

The examples of four-, five- and six-atomic molecules discussed above are collected in Table 26. A few additional ones have been included.

**Building-up from unlike groups.** The formation of non-linear polyatomic molecules from individual atoms leads in general to a very large number of molecular states and consequently the knowledge of these states is only of modest value. Much more specific information can be obtained by building up the molecule from two (or at the most three) parts. We consider first the case in which the two parts are unequal.

If neither of the parts into which we separate the molecule has a lower symmetry than the complete molecule and if during the building-up, at all stages, the molecule retains its full symmetry then the problem under consideration is readily treated in the same way as the problem of building-up the molecule from separated atoms. This is illustrated by the following two examples.

If the molecule $XYZ_2$ is to be built up from X + $YZ_2$ by moving the X atom along the symmetry axis of the $YZ_2$ group, we only have to resolve the state of X into the species of point group $\boldsymbol{C}_{2v}$ and multiply these species with that of the $YZ_2$ group. For example if $YZ_2$ is $CH_2$ in its lowest singlet state $^1A_1$ and X is O in its $^3P_g$ ground state we know from Table 58 that the latter state gives $^3A_2 + {}^3B_1 + {}^3B_2$ which combined with $^1A_1$ of $CH_2$ gives the three states $^3A_2$, $^3B_1$ and $^3B_2$ of $H_2CO$. (This result, incidentally, implies that the $^1A_1$ ground state cannot arise in this way.) In a similar way we can also consider the case in which $YZ_2$ is a linear molecule which is combined with X put on a two-fold axis of $YZ_2$ to give $XYZ_2$ of point group $\boldsymbol{C}_{2v}$. We simply have to resolve the species of linear $YZ_2$ into that of point group $\boldsymbol{C}_{2v}$ taking account of the change of $z$ axis. For example, if $CH_2$ were in its $^3\Sigma_g^-$ ground state it would yield $^3B_1$ in $\boldsymbol{C}_{2v}$ symmetry (i.e. in its bent form) and combined with O $^3P_g$ would yield the states $^1A_1$, $^1A_2$, $^1B_2$, $^3A_1$, $^3A_2$, $^3B_2$, $^5A_1$, $^5A_2$, $^5B_2$ of the $H_2CO$ molecule.

Similarly, if the molecule $XYZ_3$ is built up from X and $YZ_3$ by moving the X atom along the symmetry axis of $YZ_3$, we must resolve the state of X and if necessary the state of $YZ_3$ into the species of point group $\boldsymbol{C_{3v}}$. For example, if $YZ_3$ is $CH_3$ in its $^2A_2''$ ground state and X is F in its $^2P_u$ ground state, the species in $\boldsymbol{C_{3v}}$ are $^2A_1$ and $^2A_1 + {}^2E$ respectively which by multiplication give the states $^1A_1$, $^1E$, $^3A_1$, $^3E$ of $CH_3F$. If $CH_3$ is in its first excited state $^2E'$ we get the states $^1A_1$, $^1A_2$, $^1E(2)$, $^3A_1$, $^3A_2$, $^3E(2)$. In a similar way one can build up an $XY_3$ molecule from the states of X and $Y_3$ as has, in fact, already been done in deriving the results of Table 26 except that in this derivation all the states of $Y_3$ resulting from a given state of Y were considered together. In Table 27 several other examples in addition to those discussed in detail here are collected.

TABLE 27. ELECTRONIC STATES OF NON-LINEAR POLYATOMIC MOLECULES RESULTING FROM GIVEN STATES OF THE SEPARATED UNLIKE PARTS

| Molecule | | | States of separated parts | Resulting molecular states |
|---|---|---|---|---|
| Type | Point group | Proto-type | | |
| $XY_2$ | $\boldsymbol{C_{2v}}$ | $H_2O$ | $X(^3P_g) + Y_2(^1\Sigma_g^+)$ | $^3A_2$, $^3B_1$, $^3B_2$ |
| | | | $X(^1D_g) + Y_2(^1\Sigma_g^+)$ | $^1A_1(2)$, $^1A_2$, $^1B_1$, $^1B_2$ |
| | | | $X(^3P_g) + Y_2(^3\Sigma_u^+)$ | $^1A_1$, $^1A_2$, $^1B_1$, $^3A_1$, $^3A_2$, $^3B_1$, $^5A_1$, $^5A_2$, $^5B_1$ |
| | | | $XY(^2\Pi) + Y(^2S_g)$ | $^1A_1$, $^1A_2$ or $^1B_1$, $^3A_1$ or $^3B_2$, $^3A_2$ or $^3B_1$ |
| XYZ | $\boldsymbol{C_s}$ | HCO | $X(^2S_g) + YZ(^1\Sigma^+)$ | $^2A'$ |
| | | | $X(^2S_g) + YZ(^3\Pi)$ | $^2A'$, $^2A''$, $^4A'$, $^4A''$ |
| | | | $XY(^2\Pi) + Z(^3P_g)$ | $^2A'(3)$, $^2A''(3)$, $^4A'(3)$, $^4A''(3)$ |
| $XY_3$ | $\boldsymbol{C_{3v}}$ | $NH_3$ | $X(^4S_u) + Y_3(^2E')$ | $^3E$, $^5E$ |
| | | | $X(^4S_u) + Y_3(^4A_2')$ | $^1A_1$, $^3A_1$, $^5A_1$, $^7A_1$ |
| | $\boldsymbol{D_{3h}}$ | $CH_3$ | $Y(^2S_g) + XY_2(^1A_1)$ | $^2A_1'$ or $^2E'$ |
| | | | $Y(^2S_g) + XY_2(^3\Sigma_g^-)$ | $^2A_2''$ or $^2E''$, $^4A_2''$ or $^4E''$ |
| $XYZ_2$ | $\boldsymbol{C_{2v}}$ | $H_2CO$ | $X(^3P_g) + YZ_2(^1A_1)$ | $^3A_2$, $^3B_1$, $^3B_2$ |
| | | | $X(^3P_g) + YZ_2(^3\Sigma_g^-)$ | $^1A_1$, $^1A_2$, $^1B_2$, $^3A_1$, $^3A_2$, $^3B_2$, $^5A_1$, $^5A_2$, $^5B_2$ |
| | | | $X(^1D_g) + YZ_2(^1A_1)$ | $^1A_1(2)$, $^1A_2$, $^1B_1$, $^1B_2$ |
| $XYZ_3$ | $\boldsymbol{C_{3v}}$ | $FCH_3$ | $X(^2P_u) + YZ_3(^2A_2'')$ | $^1A_1$, $^1E$, $^3A_1$, $^3E$ |
| | | | $X(^2P_u) + YZ_3(^2E')$ | $^1A_1$, $^1A_2$, $^1E(2)$, $^3A_1$, $^3A_2$, $^3E(2)$ |
| | | $LiCH_3$ | $X(^2S_g) + YZ_3(^2A_2'')$ | $^1A_1$, $^3A_1$ |
| | | | $X(^2S_g) + YZ_3(^2E')$ | $^1E$, $^3E$ |
| | | $FSiH_3$ | $X(^2P_u) + YZ_3(^2A_1)$ | $^1A_1$, $^1E$, $^3A_1$, $^3E$ |
| $XY_4$ | $\boldsymbol{T_d}$ | $CH_4$ | $Y(^2S_g) + XY_3(^2A_2'')$ | $^1A_1$ (or $^1F_2$), $^3A_1$ (or $^3F_2$) |
| | | | $X(^3P_g) + Y_4(^1E)$ | $^3F_1$, $^3F_2$ |
| | | | $X(^3P_g) + Y_4(^3F_1)$ | $^1A_1$, $^1E$, $^1F_1$, $^1F_2$, $^3A_1$, $^3E$, $^3F_1$, $^3F_2$, $^5A_1$, $^5E$, $^5F_1$, $^5F_2$ |
| | | | $X(^3P_g) + Y_4(^5A_2)$ | $^3F_2$, $^5F_2$, $^7F_2$ |
| $X_2Y_4$ | $\boldsymbol{D_{2h}}$ | $C_2H_4$ | $X_2Y_3(^2A') + Y(^2S_g)$ | $^1A'$, $^3A'$ ($\equiv A_g$ or $B_{1u}$ or $B_{2u}$ or $B_{3g}$) |
| | | | $X_2Y_3(^2A'') + Y(^2S_g)$ | $^1A''$, $^3A''$ ($\equiv A_u$ or $B_{1g}$ or $B_{2g}$ or $B_{3u}$) |

It is interesting to see what happens in a case like $CH_3I$ where the iodine atom has a large doublet splitting ($\Delta\nu$ is almost 1 eV) and where therefore the individual doublet components must be correlated. As we have seen in Chapter I, section 1, the $^2A_1$ state of (deformed) $CH_3$ goes over into an $E_{\frac{1}{2}}$ state when the spin function is included. Similarly the $^2P_u = {}^2A_1 + {}^2E$ state of the I atom becomes $E_{\frac{1}{2}} + E_{\frac{1}{2}} + E_{\frac{3}{2}}$. Here the (lower) $^2P_{\frac{3}{2}}$ component corresponds to $E_{\frac{1}{2}} + E_{\frac{3}{2}}$ while the upper $^2P_{\frac{1}{2}}$ component corresponds to $E_{\frac{1}{2}}$. If now the species of I and $CH_3$ are multiplied one gets from $I(^2P_{\frac{1}{2}})$ the states $A_1$, $A_2$, $E$ of $CH_3I$ and from $I(^2P_{\frac{3}{2}})$ the states $A_1$, $A_2$, $E$, $E$, $E$. One can easily check that the states given previously without regard for spin-orbit coupling go over into those just given if account is taken of the fact that the spin function for $S = 1$ (triplets) is $A_2 + E$. Considering the non-crossing rule it must be concluded that the observed ground state of $CH_3I$ is the $A_1$ state arising from the lower doublet component $^2P_{\frac{3}{2}}$ of the I atom.

Let us now consider the case in which the symmetry of at least one of the parts (point group $\boldsymbol{Q}$) is less than that of the complete molecule (point group $\boldsymbol{P}$) and where during the building-up process the conformation of the nuclei has the lower symmetry ($\boldsymbol{Q}$) of one of the parts. Consider for example the formation of $CH_3$ (point group $\boldsymbol{D}_{3h}$) from $CH_2 + H$. During the formation the symmetry is at most $\boldsymbol{C}_{2v}$. In the way indicated in the preceding paragraphs we can immediately obtain the resulting states in terms of the lower symmetry: if $CH_2$ is in the $^1A_1$ and H is in the $^2S_g$ state, we find the single state $^2A_1$ of $CH_3$, or if $CH_2$ is in the $^3\Sigma_g^-$ state, we find the states $^2B_1$ and $^4B_1$ of $CH_3$. If now the third H atom is brought into a symmetrical position with regard to the remainder of the molecule, that is, if the molecule is brought to $\boldsymbol{D}_{3h}$ symmetry, we must use the correlation given in Table 60 (Appendix IV) to find the states of the completed molecule that correspond to those obtained in the deformed conformation of symmetry $\boldsymbol{C}_{2v}$. We see that the $^2A_1$ state from $CH_2(^1A_1) + H\ {}^2S_g$ can give either $^2A_1'$ or one component of $^2E'$ of $\boldsymbol{D}_{3h}$ while the $^2B_1$ and $^4B_1$ states resulting from $CH_2(^3\Sigma_g^-)$ $+ H\ {}^2S_g$ can give either $^2A_2''$ and $^4A_2''$ or one component of $^2E''$ and $^4E''$. It is not possible to say from symmetry considerations alone which of these two alternatives in each case applies. Just as for linear molecules there are sometimes indirect means of finding a unique correlation, but in general one must be satisfied with the correlation given by the lower symmetry. We refer to Table 27 for the correlation between the states of $H + CH_3$ and those of $CH_4$, and other similar cases.

Finally we consider the case in which both parts have a higher symmetry than the molecule to be built up but in which during the formation process the symmetry is less than that of the molecule. An example is a non-linear $XY_2$ molecule formed from $Y + XY$ when the Y atom is not on the XY-axis. In the general position the point group is $\boldsymbol{C}_s$ and therefore we must resolve both the state of Y and that of XY into the species of $\boldsymbol{C}_s$. If for example Y is in a $^2S_g$ and XY in a $^2\Pi$ state, as for the ground state of $H + OH$, we obtain a resolution into $^2A'$ and $^2A' + {}^2A''$ respectively, and therefore the molecular states are $^1A'$, $^1A''$, $^3A'$ and $^3A''$. Again from Table 60 we cannot tell whether the $^1A'$ state gives $^1A_1$ or $^1B_2$ of the symmetrical $XY_2$ molecule and similarly for the others. In this instance we see from Table 25 that $^3P_g + 2\,^2S_g$ does not give a $^1B_2$ state and therefore, since it is known that the $^2\Pi$ ground state of OH is formed from ground state atoms, the $^1A'$ state from $^2S_g + {}^2\Pi$ must correspond to $^1A_1$, but similar conclusions are not possible for the other pairs. Naturally for unsymmetrical molecules like HNO and HCO this ambiguity does not arise.

**Building-up from like groups.** When the two parts from which the molecule is formed are alike it is in general possible to determine unambiguously the resulting states even though the symmetry of the parts is lower than that of the molecule. This is because even at large separation of the two parts the full symmetry of the molecule may exist. The situation is very similar to that discussed previously for linear molecules.

If the two like groups are in different states, a resonance exists in that either the one or the other group can be in the higher state and therefore a splitting into two states arises one of which is symmetrical while the other is antisymmetrical with regard to the plane of symmetry or center of symmetry that arises when the two equal parts are put in a symmetrical position with regard to each other (as in the molecule to be formed).

Consider as an example the formation of $C_2H_4$ (point group $\boldsymbol{D}_{2h}$) from a $CH_2$ group in a $^1A_1$ state and one in a $^1B_1$ state. If we had two unequal groups, e.g. $CF_2$ and $CH_2$, forming a molecule of point group $\boldsymbol{C}_{2v}$ (e.g. $F_2C$—$CH_2$) a single $^1B_1$ state would arise if one group is excited, and another $^1B_1$ state of different energy if the other group is excited. But if the two groups are identical the two $^1B_1$ states will have the same energy as long as the groups are far apart. Because of this resonance, the two states mix and as the two identical groups come closer together two new states result corresponding to the sum and difference of the two original wave functions: one of these states is symmetric the other is antisymmetric with respect to the new element of symmetry introduced by bringing the two groups together, i.e. the center of symmetry or plane of symmetry. In the present case, when the resulting molecule has $\boldsymbol{D}_{2h}$ symmetry, the two states are $^1B_{2g}$ and $^1B_{3u}$ (see also Table 60). Other examples of this type are given in the upper part of Table 28.

If the two equal parts are in the same state, there is no resonance and no doubling of the number of states arises. We must then decide which of the resulting states are $g$, which are $u$, or which are $'$ and which $''$. Even though no general treatment of this question has been given in the literature we can obtain an answer in all cases of interest by using the correlation with corresponding diatomic (or linear polyatomic) molecules for which the results have been given by Wigner and Witmer (see Vol. I and section 1c($\alpha$) above).

For example, if in forming $C_2H_4$ the two $CH_2$ groups are in different $^1A_1$ states, we would according to the previous discussion obtain two states of $C_2H_4$ viz., $^1A_g$ and $^1B_{1u}$. But if the $^1A_1$ states are identical, only one state of $C_2H_4$ arises and this state must be $^1A_g$ since, in forming the united molecule $O_2$, we obtain from two $^1S_g$ states of O (which in $CH_2$ correspond to $^1A_1$) one single state

TABLE 28. ELECTRONIC STATES OF NON-LINEAR POLYATOMIC MOLECULES RESULTING FROM GIVEN STATES OF THE SEPARATED EQUAL PARTS

| Molecule | | | States of separated equal parts | Resulting molecular states |
|---|---|---|---|---|
| Type | Point group | Prototype | | |
| $X_2Y_2$ | $\boldsymbol{C_2}$ | $H_2O_2$ | $^2\Pi + {}^2\Sigma^+$ | $^1A(2)$, $^1B(2)$, $^3A(2)$, $^3B(2)$ |
| | $\boldsymbol{C_{2h}}$ | exc. $C_2H_2$ | $^2\Pi + {}^2\Sigma^-$ | $^1A_g$, $^1A_u$, $^1B_g$, $^1B_u$, $^3A_g$, $^3A_u$, $^3B_g$, $^3B_u$ |
| $X_2Y_4$ | $\boldsymbol{D_{2h}}$ | $C_2H_4$ | $^1A_1 + {}^1B_1$ | $^1B_{2g}$, $^1B_{3u}$ |
| | | | $^3\Sigma_g^- + {}^1A_1$ | $^3B_{2g}$, $^3B_{3u}$ |
| $X_2Y_6$ | $\boldsymbol{D_{3d}}$ | $C_2H_6$ | $^2A_2'' + {}^2E'$ | $^1E_g$, $^1E_u$, $^3E_g$, $^3E_u$ |
| | | | $^2A_2'' + {}^2A_1'$ | $^1A_{1g}$, $^1A_{2u}$, $^3A_{1g}$, $^3A_{2u}$ |
| | $\boldsymbol{D_{3h}}$ | | $^2A_2'' + {}^2E'$ | $^1E'$, $^1E''$, $^3E'$, $^3E''$ |
| | | | $^2A_2'' + {}^2A_1'$ | $^1A_1'$, $^1A_2''$, $^3A_1'$, $^3A_2''$ |
| $X_2Y_2$ | $\boldsymbol{C_2}$ | $H_2O_2$ | $^2\Sigma^+ + {}^2\Sigma^+$ (or $^2\Sigma^- + {}^2\Sigma^-$) | $^1A$, $^3B$ |
| | | | $^2\Pi + {}^2\Pi$ (or $^2\Delta + {}^2\Delta$) | $^1A(3)$, $^1B$, $^3A$, $^3B(3)$ |
| | $\boldsymbol{C_{2h}}$ | exc. $C_2H_2$ | $^2\Sigma^+ + {}^2\Sigma^+$ (or $^2\Sigma^- + {}^2\Sigma^-$) | $^1A_g$, $^3B_u$ |
| | | | $^2\Pi + {}^2\Pi$ (or $^2\Delta + {}^2\Delta$) | $^1A_g(2)$, $^1A_u$, $^1B_g$, $^3A_u$, $^3B_g$, $^3B_u(2)$ |
| $X_2Y_4$ | $\boldsymbol{D_{2h}}$ | $C_2H_4$ | $^1A_1 + {}^1A_1$, $^1A_2 + {}^1A_2$, $^1B_1 + {}^1B_1$, $^1B_2 + {}^1B_2$ | $^1A_g$ |
| | | | $^3A_1 + {}^3A_1, \ldots, {}^3\Sigma_g^- + {}^3\Sigma_g^-, \ldots$ | $^1A_g$, $^3B_{1u}$, $^5A_g$ |
| | | $N_2O_4$ | $^2A_1 + {}^2A_1$, $^2A_2 + {}^2A_2$, $^2B_1 + {}^2B_1$, $^2B_2 + {}^2B_2$ | $^1A_g$, $^3B_{1u}$ |
| $X_2Y_6$ | $\boldsymbol{D_{3d}}$ | $C_2H_6$ | $^2A_1' + {}^2A_1'$, $^2A_2'' + {}^2A_2''$ (or $^2A_1 + {}^2A_1$, $^2A_2 + {}^2A_2$) | $^1A_{1g}$, $^3A_{2u}$ |
| | | | $^2E' + {}^2E'$ or $^2E'' + {}^2E''$ | $^1A_{1g}$, $^1A_{1u}$, $^1E_g$, $^3A_{2g}$, $^3A_{2u}$, $^3E_u$ |
| | | | $^1A_1 + {}^1A_1$ or $^1A_2 + {}^1A_2$ | $^1A_{1g}$ |
| | | | $^3A_1 + {}^3A_1$ or $^3A_2 + {}^3A_2$ | $^1A_{1g}$, $^3A_{2u}$, $^5A_{1g}$ |
| | | | $^1E + {}^1E$ | $^1A_{1g}$, $^1A_{1u}$, $^1E_g$ |
| $X_2Y_6$ | $\boldsymbol{D_{3h}}$ | | $^2A_1 + {}^2A_1$ or $^2A_2 + {}^2A_2$ | $^1A_1'$, $^3A_2''$ |
| | | | $^2E + {}^2E$ | $^1A_1'$, $^1A_1''$, $^1E'$, $^3A_2'$, $^3A_2''$, $^3E''$ |
| | | | $^1A_1 + {}^1A_1$ or $^1A_2 + {}^1A_2$ | $^1A_1'$ |
| | | | $^3A_1 + {}^3A_1$ or $^3A_2 + {}^3A_2$ | $^1A_1'$, $^3A_2''$, $^5A_1'$ |
| | | | $^1E + {}^1E$ | $^1A_1'$, $^1A_1''$, $^1E'$ |

$^1\Sigma_g^+$ which according to Table 59 corresponds to the state $^1A_g$ of $C_2H_4$. If we form the $C_2H_4$ molecule from two $CH_2$ groups in their $^3\Sigma_g^-$ ground states, we must first resolve $^3\Sigma_g^-$ into a species of point group $\boldsymbol{C}_{2v}$ with the $z$-axis perpendicular to the internuclear axis of $CH_2$ obtaining $^3B_1$. Combining the two parts in identical $B_1$ states we would obtain a single state $A_1$ of the whole molecule if it had $\boldsymbol{C}_{2v}$ symmetry, with spin values 0, 1, 2. In $\boldsymbol{D}_{2h}$ symmetry, according to Table 60, $A_1$ correlates either with $A_g$ or $B_{1u}$. Since no resonance occurs, only one of these arises for each multiplicity. Again by comparing with the united molecule (i.e. by forming $O_2$ from two O atoms) we see that the states $^1A_g$, $^3B_{1u}$, $^5A_g$ result. These and several other examples are collected in the lower part of Table 28.

## 2. Electron Configurations

The correlations between united atom, molecule and separated atoms or groups of atoms discussed in the preceding section give an idea of the manifold of all the electronic states of a molecule. These correlations are based mainly on symmetry considerations and, as is often the case in such situations, while giving an exact idea of the possible states they give very little idea as to which of them will be stable and even less as to the nuclear conformations of maximum stability, i.e., as to molecular geometries. The third method for building-up a molecule, mentioned at the beginning of this Chapter, the method of electron configurations, has turned out to be more useful and important. We shall now discuss it in more detail. This method gives not only the manifold of electronic states, but, in addition, gives without detailed calculation some information about the order of the electronic states and, particularly, allows predictions about the nature of the ground state and other low lying states. It forms the basis of the molecular orbital method of understanding the *stability* of electronic states which is to be discussed in section 3.

### (a) Classification of orbitals

Just as for atoms and diatomic molecules, the motions of the electrons in a polyatomic molecule can in a first very rough approximation be considered as independent. In other words, we may consider separately the motion of each electron in the field of the nuclei and of the average field of the other electrons. In wave mechanics the motion of an electron $i$ is characterized by a wave function $\psi_i$ which is substantially different from zero only in the neighborhood of the nuclei and which vanishes at infinity. Following Mulliken (888) such one-electron wave functions and the corresponding electron motions are called *orbitals*. For atoms with a single electron these orbitals are the familiar wave functions of the H atom and H-like ions. For atoms with several electrons they are similar, but more complicated functions, *atomic orbitals*, which have the same symmetry properties as the wave functions of one-electron atoms. They are designated as $s, p, d, \ldots$ orbitals depending on the value $l = 0, 1, 2, \ldots.$ of the quantum number of the orbital angular momentum. For diatomic molecules we have corresponding *molecular orbitals* designated $\sigma, \pi, \delta, \ldots$ orbitals depending on the value $\lambda = 0, 1, 2, \ldots$ of the quantum number of the component of the orbital angular momentum along the internuclear axis (see Vol. I, Chap. VI, 3). The orbitals for linear polyatomic molecules are entirely similar. If there is a center of symmetry (point group $\boldsymbol{D}_{\infty h}$), the orbitals can only be symmetric or antisymmetric with respect to this center, we have $\sigma_g, \sigma_u, \pi_g, \pi_u, \ldots$ orbitals. Qualitatively the form of these orbitals has been illustrated in Fig. 155 of Volume I (p. 326).

It is clear that also for a non-linear conformation of the nuclei the eigenfunction (orbital) corresponding to the motion of a single electron must have symmetry properties corresponding to one of the species of the point group of the nuclear conformation. This follows from the same reasoning as given in Chapter I, section 1, for the over-all electronic eigenfunction. If there are several electrons, we consider each electron moving in the combined field of the nuclei and the other electrons. In general, this field does not have at every instant the full symmetry of the point group of the nuclear conformation, but if we average the field of the other electrons in a suitable manner, we can again obtain a field that has this symmetry. In general this suitably averaged field represents a good approximation, but it is well to bear in mind that only with this assumption can the orbitals be classified in the same way as the electronic states. One uses small letters for the species symbols of orbitals corresponding to the capital letters used for the general species designation, just as for atoms and diatomic molecules.

It must be noted that not necessarily all species of a point group occur as species of orbitals. Only atomic orbitals $s_g, p_u, d_g, f_u, g_g, \ldots$ occur (since electronic states of atoms are $g$ or $u$ depending on whether $\sum l_i$ is even or odd, that is, for individual electrons depending on whether $l_i$ is even or odd), and therefore usually the subscripts $g$ and $u$ are omitted. As a consequence (see Table 58, Appendix IV), for diatomic or linear polyatomic molecules (point groups $\boldsymbol{C}_{\infty v}$ and $\boldsymbol{D}_{\infty h}$) orbitals of species $\Sigma^-$ do not occur and therefore in referring to $\sigma$ electrons the superscript $+$ is usually omitted. In all other point groups, orbitals of all species may occur even though some types of orbitals occur only for a rather high $l$ value of the corresponding atomic orbital. For example the lowest $l$ value for which $a_2$ orbitals arise in point group $\boldsymbol{T}_d$ is $l = 6$ (not included in Table 58); similarly $a_{2g}$ orbitals of $\boldsymbol{O}_h$ arise first for $l = 6$ and $a_{1u}$ first for $l = 9$.

No properties of the orbitals other than the behavior with regard to the symmetry operations of the point group of the nuclear framework are rigorously defined. Since for a given molecule there are in general several orbitals of each species it is now customary to distinguish them by a number preceding the species symbol, which indicates whether it is the first, second, third, ... orbital of its kind if arranged according to energy. Thus one has $1\sigma, 2\sigma, 3\sigma, \ldots$ $1\pi, 2\pi, 3\pi, \ldots$ orbitals of $\boldsymbol{C}_{\infty v}$ or $1a_1, 2a_1, 3a_1, \ldots, 1a_2, 2a_2, \ldots$ $1b_1, 2b_1, \ldots,$ orbitals of $\boldsymbol{C}_{2v}$, and similarly in other cases. The number here introduced should not be confused with the principal quantum number. As we have seen in Volume I the principal and the azimuthal quantum numbers of the electron in the united atom or in the separated atoms are often added to the symbol for an orbital in diatomic

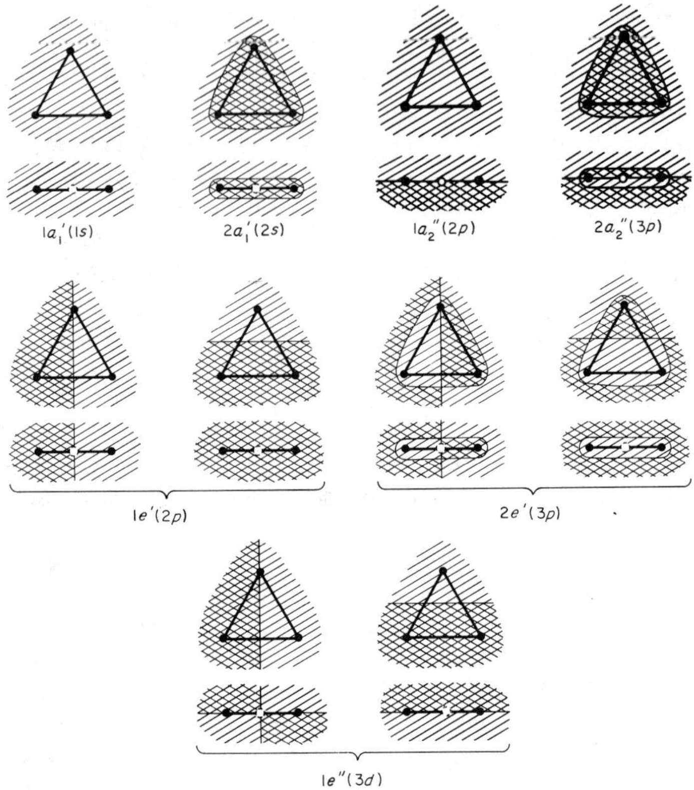


FIG. 114. **Nodal surfaces of the orbital functions of an $X_3$ molecule of point group $\boldsymbol{D_{3h}}$.** For each orbital function two projections (|| and $\perp$ to the $X_3$ plane) are shown. The sign of the wave functions is indicated by simple (+) and cross (−) hatching. The nodal surfaces are indicated by thin solid lines separating areas of simple and cross hatching. The orbitals $3a_1'$, $4a_1'$, . . ., $3a_2''$, $4a_2''$, . . ., $3e'$, . . . , $2e''$, . . . . (not shown) have more and more nodal surfaces.

molecules. For polyatomic molecules such designations can also be used, particularly for the more highly excited (Rydberg) orbitals. They will be described in more detail in the next subsection.

As an example, we illustrate in Fig. 114 the lowest orbitals of an $X_3$ molecule of point group $\boldsymbol{D_{3h}}$. Only the nodal surfaces and the relative signs of the orbital wave functions on each side of these surfaces are indicated. The same diagrams apply also to planar $XY_3$ except that there may now also be nodal surfaces inside the $Y_3$ triangle. Similar diagrams for other point groups would be easy to prepare.

**Mathematical formulation: molecular (symmetry) orbitals.** The wave equation of an electron $i$ moving in the field of the (fixed) nuclei plus the field of the other electrons is

$$\nabla^2\psi_i + \frac{8\pi^2 m}{h^2}\left(\epsilon_i - V_{ni} - \sum_{k \neq i} V_{ik}\right)\psi_i = 0. \qquad \text{(III, 4)}$$

Here $V_{ni}$ is the potential produced by the nuclei on $i$ while $V_{ik}$ is the potential produced by electron $k$ on $i$ and $\epsilon_i$ is the energy. It is not possible to give a general solution of this equation. However, for linear molecules, just as for diatomic molecules, it is easy to show (see Vol. I, p. 328) that the solutions must be of the form

$$\psi_i = \chi_i(z_i, \rho_i)e^{\pm i\lambda_i\varphi_i} \tag{III, 5}$$

where $\lambda_i = 0, 1, 2, \ldots$ represents the orbital angular momentum about the molecular axis in units $h/2\pi$ and corresponds to the designation $\sigma, \pi, \delta, \ldots$ of the electron. $\varphi_i$ is the azimuthal angle and $\chi_i$ is a function of the $z$-coordinate of the electron and its perpendicular distance $\rho_i$ from the axis. When the molecule is non-linear not even the general form of $\psi_i$ can readily be given except that if $\sum V_{ik}$ is suitably averaged, $\psi_i$ will have symmetry properties corresponding to one of the species of the point group.

It is, however, possible to obtain an explicit representation of the molecular orbitals in the limiting case of the united atom or the separated atoms. One-electron eigenfunctions of atoms (atomic orbitals) are known exactly for hydrogen-like systems and good approximations can be given for many-electron systems.

Approximate orbitals that are often used are *Slater orbitals*. They are similar to hydrogen-like orbitals except that the radial part has no nodes but is expressed in the form

$$r^n e^{-\mu r}$$

where $n$ and $\mu$ are adjustable parameters. For example, in considering the orbitals in non-linear $CH_2$ we can start out from the united atom O for which, in its ground state, the Slater orbitals, written in terms of $C_{2v}$ symmetry, are:

$$\begin{aligned}
1s \to a_1{:}\quad \psi_1 &= \sqrt{\frac{\mu_1^{\,3}}{\pi}}\, e^{-\mu_1 r_O} \\
2s \to a_1{:}\quad \psi_2 &= \sqrt{\frac{\mu_2^{\,5}}{3\pi}}\, r_O\, e^{-\mu_2 r_O} \\
2p_z \to a_1{:}\quad \psi_3 &= \sqrt{\frac{\mu_3^{\,5}}{\pi}}\, r_O \cos\vartheta\, e^{-\mu_3 r_O} \\
2p_y \to b_2{:}\quad \psi_5 &= \sqrt{\frac{\mu_5^{\,5}}{\pi}}\, r_O \sin\vartheta \sin\varphi\, e^{-\mu_5 r_O} \\
2p_x \to b_1{:}\quad \psi_7 &= \sqrt{\frac{\mu_7^{\,5}}{\pi}}\, r_O \sin\vartheta \cos\varphi\, e^{-\mu_7 r_O}
\end{aligned} \tag{III, 6}$$

Here $r_O$ is the distance from the O nucleus (in Bohr radii), $\vartheta$ is the angle with the $z$-axis, $\varphi$ the azimuthal angle in the $xy$-plane and the constants $\mu$ according to Slater are for oxygen $\mu_1 = 7.7$ and $\mu_2 = \mu_3 = \mu_5 = \mu_7 = 2.275$. It is easily checked that $\psi_7$ and $\psi_5$ have indeed the $b_1$ and $b_2$ symmetry, that is, they are approximations to the molecular orbitals of $CH_2$. The same orbitals with different values of $\mu$ can be used for $H_2O$.

If on the other hand we start out from the separated atoms, we can use the same orbitals but we must now use Slater parameters $\mu_i$ corresponding to the central atom rather than the united atom and we must add to the orbitals (III, 6) the orbitals provided by the two hydrogen atoms. The wave function of a hydrogen atom in its ground state is

$$\varphi(1s_{\mathrm{H}}) = \frac{1}{\sqrt{\pi}} e^{-r_{\mathrm{H}}} \tag{III, 7}$$

where $r_{\mathrm{H}}$ is the distance from the H nucleus (in Bohr radii). The wave function of the two H atoms symmetrically placed with regard to C or O can only be symmetric or antisymmetric with respect to the plane of symmetry (which is perpendicular to the plane of the molecule) that is, we have the two molecular orbitals

$$1s_{\mathrm{H}'}, 1s_{\mathrm{H}''}\begin{cases} a_1: & \psi_4 = \dfrac{1}{\sqrt{2\pi}}(e^{-r_{\mathrm{H}'}} + e^{-r_{\mathrm{H}''}}) \\ b_2: & \psi_6 = \dfrac{1}{\sqrt{2\pi}}(e^{-r_{\mathrm{H}'}} - e^{-r_{\mathrm{H}''}}). \end{cases} \tag{III, 8}$$

The seven functions $\psi_1, \psi_2, \ldots, \psi_7$ are the orbitals of the $CH_2$ or $H_2O$ molecule which arise from the ground state configurations of C or O and H at large separations of the atoms. They are also called symmetry orbitals. In Fig. 115 these orbitals are illustrated schematically. Their symmetry properties are apparent from these diagrams.

When the three atoms are no longer very far apart from one another the orbitals will change. In a first approximation according to standard methods of quantum mechanics the new orbital wave functions are linear combinations of the old, but such that only functions of the same species can be combined. In this way we obtain the seven molecular orbital wave functions (MO's):

$$\begin{aligned} \tilde{\psi}_1(1a_1) &= c_{11}\psi_1 + c_{12}\psi_2 + c_{13}\psi_3 + c_{14}\psi_4 \\ \tilde{\psi}_2(2a_1) &= c_{21}\psi_1 + c_{22}\psi_2 + c_{23}\psi_3 + c_{24}\psi_4 \\ \tilde{\psi}_3(3a_1) &= c_{31}\psi_1 + c_{32}\psi_2 + c_{33}\psi_3 + c_{34}\psi_4 \\ \tilde{\psi}_4(4a_1) &= c_{41}\psi_1 + c_{42}\psi_2 + c_{43}\psi_3 + c_{44}\psi_4 \\ \tilde{\psi}_5(1b_2) &= c_{55}\psi_5 + c_{56}\psi_6 \\ \tilde{\psi}_6(2b_2) &= c_{65}\psi_5 + c_{66}\psi_6 \\ \tilde{\psi}_7(1b_1) &= \psi_7. \end{aligned} \tag{III, 9}$$

The *molecular orbitals* thus obtained as *linear combinations of atomic orbitals* are in the literature often referred to as LCAO MO's. Of the four $a_1$ orbitals numbered $1a_1$, $2a_1$, $3a_1$, $4a_1$ in the order of their energy, the first one is very similar to the carbon or oxygen atom $1s$ orbital, that is, $\psi_1$ greatly predominates in the linear combination $\tilde{\psi}_1$; the orbital $\tilde{\psi}_7$ is identical with one of the carbon (or oxygen) $2p$ orbitals ($\psi_7$); all the other orbitals are real mixtures of the original atomic orbitals. For the approximate shapes of the orbitals $\tilde{\psi}_2$ to $\tilde{\psi}_7$ see Fig. 124 farther below.

An actual calculation of the coefficients $c_{ik}$ has been carried out by Ellison and Shull (353) for the ground state of $H_2O$ by means of a variation calculation [see also McWeeny and Ohno (815)]. The result depends, of course, on the assumed H—O—H angle. For an angle of 105° the result is given in Table 41, p. 388. It must be emphasized that in a higher approximation it is necessary to

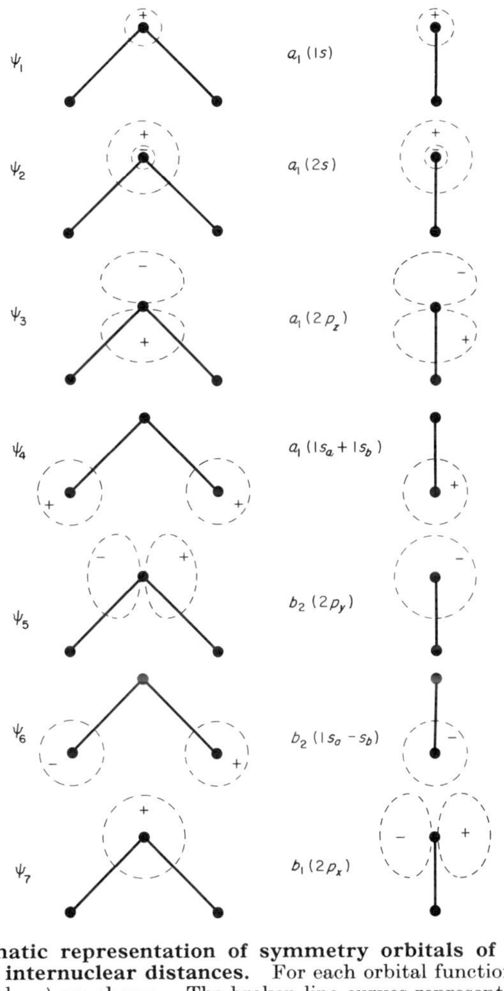


FIG. 115. **Schematic representation of symmetry orbitals of a non-linear $XH_2$ molecule for large internuclear distances.** For each orbital function two projections (|| and ⊥ to the $XH_2$ plane) are shown. The broken line curves represent the points at which the orbital function has a certain small magnitude, say 1/10 of the maximum magnitude.

take account of higher atomic orbitals and to introduce them into the appropriate linear combinations. However, fortunately, since the higher orbitals are separated from the lower ones by a considerable energy interval the effect of the higher orbitals is usually fairly small.

Because of the importance of the concept of molecular orbitals we consider a second example namely the planar $CH_3$ molecule. The orbitals of the C atom resolved into those of point group $\boldsymbol{D}_{3h}$ (see Table 58) are in the Slater approximation

$$
\begin{array}{lll}
1s_C \rightarrow a_1' & \psi_1 = \sqrt{\dfrac{\mu_1{}^3}{\pi}}\, e^{-\mu_1 r_C} & \\
2s_C \rightarrow a_1' & \psi_2 = \sqrt{\dfrac{\mu_2{}^5}{3\pi}}\, r_C\, e^{-\mu_2 r_C} & \\
2p_z \rightarrow a_2'' & \psi_4 = \sqrt{\dfrac{\mu_2{}^5}{\pi}}\, r_C \cos\vartheta\, e^{-\mu_2 r_C} & \\
\left.\begin{array}{l} 2p_y \\ \\ 2p_x \end{array}\right\} \rightarrow e' & \left\{\begin{array}{l} \psi_5 = \sqrt{\dfrac{\mu_2{}^5}{\pi}}\, r_C \sin\vartheta \sin\varphi\, e^{-\mu_2 r_C} \\ \psi_7 = \sqrt{\dfrac{\mu_2{}^5}{\pi}}\, r_C \sin\vartheta \cos\varphi\, e^{-\mu_2 r_C} \end{array}\right. &
\end{array}
\qquad \text{(III, 10)}
$$

where $\mu_1 = 5.7$ and $\mu_2 = 1.56$ [see Duncanson and Coulson (324)] and $r_C$ is the distance from the C nucleus. The functions $\psi_5$ and $\psi_7$ form now a degenerate pair of species $E'$.

Each of the three hydrogen atom orbitals is again given by (III, 7). However, we must now combine them in such a way that they form orbitals (symmetry orbitals) belonging to the point group $\boldsymbol{D}_{3h}$. As explained in more detail below (p. 303), group theory shows that in point group $\boldsymbol{D}_{3h}$ the three H orbitals form one $a_1'$ and one $e'$ molecular orbital. In a first approximation they can be represented by

$$
1s_{H'}, 1s_{H''}, 1s_{H'''} \rightarrow \left\{\begin{array}{ll}
a_1' & \psi_3 = \dfrac{1}{\sqrt{3\pi}}\left(e^{-r_{H'}} + e^{-r_{H''}} + e^{-r_{H'''}}\right) \\
e' & \left\{\begin{array}{l} \psi_6 = \dfrac{1}{\sqrt{6\pi}}\left(2e^{-r_{H'}} - e^{-r_{H''}} - e^{-r_{H'''}}\right) \\ \psi_8 = \dfrac{1}{\sqrt{2\pi}}\left(e^{-r_{H''}} - e^{-r_{H'''}}\right). \end{array}\right.
\end{array}\right.
\qquad \text{(III, 11)}
$$

It is readily seen that, upon rotation by 120°, $\psi_6$ goes over into a linear combination of $\psi_6$ and $\psi_8$ as it should do for a degenerate pair.

If now the four atoms C + 3H are brought together in a symmetrical way, the resulting molecular orbitals are in a first approximation linear combinations of those $\psi_i$ that have the same species, that is, we obtain the six molecular orbitals

$$
\begin{aligned}
\tilde{\psi}_1(1a_1') &= c_{11}\psi_1 + c_{12}\psi_2 + c_{13}\psi_3 \\
\tilde{\psi}_2(2a_1') &= c_{21}\psi_1 + c_{22}\psi_2 + c_{23}\psi_3 \\
\tilde{\psi}_3(3a_1') &= c_{31}\psi_1 + c_{32}\psi_2 + c_{33}\psi_3 \\
\tilde{\psi}_4(1a_2'') &= \psi_4 \\
\tilde{\psi}_5(1e') &= c_{55}\psi_5 + c_{56}\psi_6 \\
\tilde{\psi}_6(2e') &= c_{65}\psi_5 + c_{66}\psi_6.
\end{aligned}
\qquad \text{(III, 12)}
$$

The functions $\psi_7$ and $\psi_8$ give the same linear combinations as $\psi_5$ and $\psi_6$ (with the same coefficients) representing the other component of the $1e'$ and $2e'$ orbitals.

Of the three orbitals $1a_1'$, $2a_1'$ and $3a_1'$ the lowest, $1a_1'$, represents again essentially the carbon $1s$ orbital.

In order to determine the actual form of the molecular orbitals, that is, to determine the coefficients $c_{jl}$ in (III, 9), (III, 12) or similar equations, it is necessary to solve eq. (III, 4) for each of the orbitals $\tilde{\psi}_i$. Such a solution is made difficult by the presence of the term $-\sum V_{ik}$ which represents the effect of the other electrons on electron $i$. More specifically, since $|\tilde{\psi}_k|^2$ is the charge density produced by electron $k$, its effect on $i$ is given by

$$V_{ik} = e^2 \int \frac{|\tilde{\psi}_k|^2}{r_{ik}} d\tau_k \qquad \text{(III, 13)}$$

where $e$ is the electronic charge and $r_{ik}$ the distance of electron $i$ from the charge element $e|\tilde{\psi}_k|^2 d\tau_k$ of electron $k$. Thus each of the equations (III, 4) contains the orbital functions $\tilde{\psi}_i$, $\tilde{\psi}_k$ of all the $N$ electrons present; we have a system of $N$ simultaneous differential equations. They can be solved only by a method of iteration and successive approximation. The solution is called the *self-consistent field approximation without exchange* (or *Hartree-approximation*): The wave functions of the electrons are self-consistent if upon calculation of all the $V_{ik}$ according to (III, 13) with these functions and subsequent solution of (III, 4) the same functions are obtained back again. The energies $\epsilon_i$ obtained in this way are the ionization potentials of the molecule corresponding to removal of electron $i$ from the orbital $\tilde{\psi}_i$.

In the transition from atomic orbitals to molecular orbitals by the LCAO method we frequently find [see eqs. (III, 9) and (III, 12)] that a molecular orbital is a mixture (*a hybrid*) of different atomic orbitals originating in the same atom. This mixing is sometimes called *hybridization*, although usually this designation is reserved for the corresponding phenomenon in valence-bond theory (see section 3a). The mixing of atomic $s$ and $p$ orbitals of the same principal quantum number ($s$-$p$ hybridization) is particularly important. In the process of molecule formation from the separated atoms (or groups of atoms) or from the united atom (or molecule) the degree of mixing in general changes greatly. In the free atom there is no $s$-$p$ mixing since $s$ and $p$ orbitals have different symmetry. In other words, the mixing arises only because the symmetry is lowered in the molecule and because therefore orbitals of the same molecular species arise from atomic orbitals of different species.

**Species of orbitals formed from equivalent atoms (group orbitals).** The species of the molecular orbitals formed from given atomic orbitals of a number of equivalent atoms in a symmetrical molecule is readily obtained by simple group-theoretical considerations. For example, in the case of $CH_3$ discussed above, the three original $1s$ orbital functions $\varphi_H'$, $\varphi_H''$, $\varphi_H'''$ if attached to three H nuclei of a $\boldsymbol{D}_{3h}$ conformation will have the following transformation matrices for the symmetry operations $I$, $C_3$, $\sigma_v$ and $\sigma_h$

$$I:\ \begin{matrix}1&0&0\\0&1&0\\0&0&1\end{matrix};\qquad C_3:\ \begin{matrix}0&1&0\\0&0&1\\1&0&0\end{matrix};\qquad \sigma_v:\ \begin{matrix}0&0&1\\0&1&0\\1&0&0\end{matrix};\qquad \sigma_h:\ \begin{matrix}1&0&0\\0&1&0\\0&0&1\end{matrix}$$

that is, the characters (the sums of the diagonal elements) are 3, 0, 1, 3 respectively. This representation can be immediately reduced to that of $A_1' + E'$ since the characters of $A_1'$ and $E'$ are 1, 1, 1, 1 and 2, $-1$, 0, 2 respectively (see Table 51, Appendix I) which add up to 3, 0, 1, 3. Thus we obtain an $a_1'$ and an $e'$ orbital.

If the H atoms, instead, contribute each a $2p$ electron (or if we consider generally an $XY_3$ molecule where the Y atoms have $2p$ (or $3p$) electrons), we can deal with the

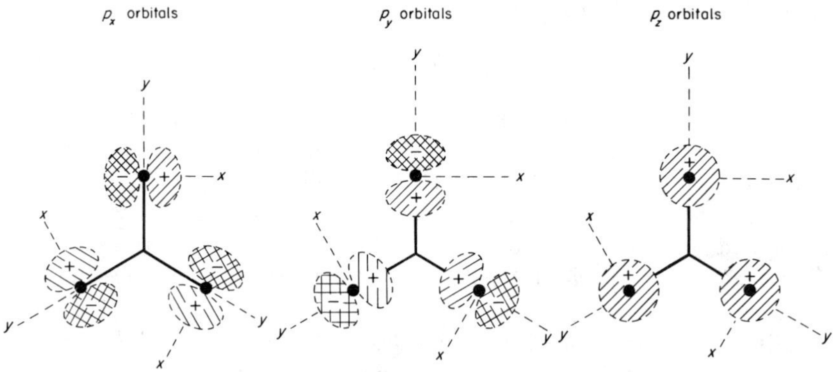


FIG. 116. **Transformation properties of molecular orbitals of a $D_{3h}$ molecule resulting from $p_x$, $p_y$, $p_z$ orbitals of three equivalent atoms.** Compare the transformation matrices below. Each configuration shown is degenerate with all those that arise from it by any of the symmetry operations.

$p_x$, $p_y$, $p_z$ components separately if we place the coordinate axes symmetrically for each H atom as in Fig. 116. We find for the transformation matrices of $p_x$, $p_y$, $p_z$, as is readily verified from Fig. 116

$$
\begin{array}{ccccc}
 & I & C_3 & \sigma_v & \sigma_h \\
p_x & \begin{matrix} 1 & 0 & 0 \\ 0 & 1 & 0 \\ 0 & 0 & 1 \end{matrix} & \begin{matrix} 0 & 1 & 0 \\ 0 & 0 & 1 \\ 1 & 0 & 0 \end{matrix} & \begin{matrix} 0 & 0 & -1 \\ 0 & -1 & 0 \\ -1 & 0 & 0 \end{matrix} & \begin{matrix} 1 & 0 & 0 \\ 0 & 1 & 0 \\ 0 & 0 & 1 \end{matrix} \\
p_y & \begin{matrix} 1 & 0 & 0 \\ 0 & 1 & 0 \\ 0 & 0 & 1 \end{matrix} & \begin{matrix} 0 & 1 & 0 \\ 0 & 0 & 1 \\ 1 & 0 & 0 \end{matrix} & \begin{matrix} 0 & 0 & 1 \\ 0 & 1 & 0 \\ 1 & 0 & 0 \end{matrix} & \begin{matrix} 1 & 0 & 0 \\ 0 & 1 & 0 \\ 0 & 0 & 1 \end{matrix} \\
p_z & \begin{matrix} 1 & 0 & 0 \\ 0 & 1 & 0 \\ 0 & 0 & 1 \end{matrix} & \begin{matrix} 0 & 1 & 0 \\ 0 & 0 & 1 \\ 1 & 0 & 0 \end{matrix} & \begin{matrix} 0 & 0 & 1 \\ 0 & 1 & 0 \\ 1 & 0 & 0 \end{matrix} & \begin{matrix} -1 & 0 & 0 \\ 0 & -1 & 0 \\ 0 & 0 & -1. \end{matrix}
\end{array}
$$

Therefore the characters are for $p_x$, 3 0 $-1$ 3, for $p_y$, 3 0 1 3 and $p_z$, 3 0 1 $-3$ leading to the orbitals $a_2' + e'$ for $p_x$, $a_1' + e'$ for $p_y$ and $a_2'' + e''$ for $p_z$.

The examples just given, together with similar examples for all important point groups, are collected in Table 61, Appendix V. The forms of the orbital wave functions are similar to (III, 8) and (III, 11). Explicit forms for some of the more complicated cases have been given by Kotani, Ohno and Kayama (690) and Hoffmann and Gouterman (561). In the so-called ligand field theory (see p. 405) molecular orbitals of the type discussed here are called ligand orbitals and in crystal field theory they are called crystal field molecular orbitals. A more neutral name would be *group orbitals.*

**Localized and equivalent orbitals.** For the purposes of one form of valence theory, the valence-bond (or electron pair bond) theory, it is necessary to introduce orbitals that are quite different from the molecular orbitals described so far. The latter are non-localized orbitals which extend over the whole molecule (see Fig. 114); they are not adapted to a simple explanation of the existence of directed chemical bonds although they are an excellent tool for an understanding of the

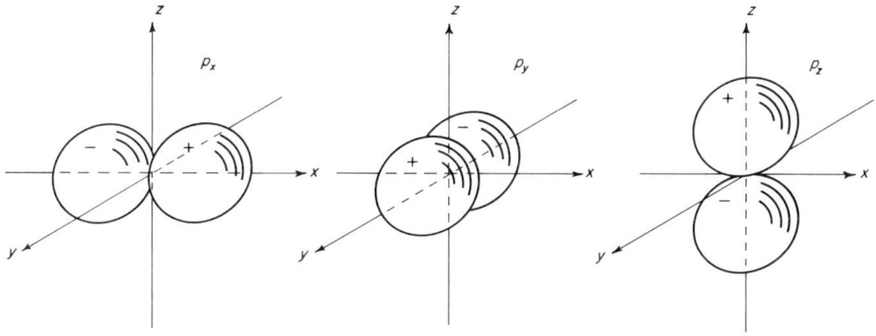


FIG. 117. **Schematic representations of the three components of a $p$ orbital.** The spherically shaped surfaces contain the region in which the orbital function is larger in magnitude (+ or −) than a certain small quantity. They are shown in isometric projection. Strictly speaking, the lobes are not spherical but have the slightly flattened shape shown in Fig. 119a.

excited states of a molecule (see further below). If we form localized orbitals, we give up at least temporarily the requirement that molecular wave functions must have symmetry properties (species) corresponding to the symmetry of the molecule. Rather we neglect in a first approximation the presence of all but the two atoms between which the bond is to be formed and consider what are essentially diatomic orbitals of axial symmetry formed from the atomic orbitals of the two atoms adjacent to the bond.

The $p$ orbitals of an atom (X) are triply degenerate. We can choose as the three component functions the functions $p_x$, $p_y$, $p_z$ given in (III, 6) which we can also write

$$\begin{aligned} \psi(p_x) &= x\,f(r) \\ \psi(p_y) &= y\,f(r) \\ \psi(p_z) &= z\,f(r) \end{aligned} \qquad \text{(III, 14)}$$

where $f(r)$ is a function of $r$ which for Slater orbitals takes the simple form

$$f(r) = Ce^{-\mu r}. \qquad \text{(III, 15)}$$

The functions (III, 14) are represented schematically in Fig. 117. This figure shows clearly that $\psi(p_x)$ is localized around the $x$-axis, $\psi(p_y)$ around the $y$-axis and $\psi(p_z)$ around the $z$-axis. (For a more precise representation of one of these functions see Fig. 119a.) If another atom, say Y, is brought along one of the axes, say $x$, its orbitals are similarly oriented with respect to the axis except for $s$

orbitals which are spherically symmetric. The bond orbitals of the bond X–Y are in a first approximation linear combinations of the X and Y atomic orbitals.

As an example consider the $H_2O$ molecule. Let the $x$- and $y$-axes be in the direction of the two OH bonds. If the eigenfunction of the $1s$ electron of the first H atom ($H_a$) is $\psi(H_a;\, 1s)$, that of the second $\psi(H_b;\, 1s)$, we have, together with the oxygen orbitals $\psi(O;\, 2p_x)$ and $\psi(O;\, 2p_y)$, four atomic orbitals which in the molecule go over into the linear combinations

$$\begin{aligned}\tilde{\psi}_1 &= c_{11}\psi(H_a;\, 1s) + c_{12}\psi(O;\, 2p_x)\\ \tilde{\psi}_2 &= c_{21}\psi(H_a;\, 1s) + c_{22}\psi(O;\, 2p_x)\\ \tilde{\psi}_3 &= c_{33}\psi(H_b;\, 1s) + c_{34}\psi(O;\, 2p_y)\\ \tilde{\psi}_4 &= c_{43}\psi(H_b;\, 1s) + c_{44}\psi(O;\, 2p_y).\end{aligned} \qquad \text{(III, 16)}$$

The first pair of orbitals is localized around the $H_a$—O bond, the second pair around the $H_b$—O bond, in other words these orbitals are molecular orbitals with regard to two nuclei only. The presence of a third nucleus (and any electrons in the other bond) is considered as a small perturbation which in this approximation is neglected. In one of each pair of orbitals, say $\tilde{\psi}_1$ and $\tilde{\psi}_3$, the coefficients have the same sign, i.e. the positive lobe of the $2p$ function overlaps a positive H atom function and as a result the energy is lower than in the free atoms; in the other pair, $\tilde{\psi}_2$ and $\tilde{\psi}_4$, the coefficients have opposite signs, i.e. the negative lobe of the $2p$ function overlaps a positive H atom function and the energy is higher than in the free atoms. The former are bonding, the latter anti-bonding orbitals (see below).

The orbitals (III, 16) do not transform according to one of the species of the point group $C_{2v}$. They have however the property of *equivalence*, that is, the first two are with regard to the $H_a$—O bond the exact equivalent of what the other two are with regard to the $H_b$—O bond. Such *equivalent orbitals* were first explicitly introduced by Lennard–Jones (738). From these orbitals one can construct proper molecular orbitals having the correct symmetry properties simply by forming $\tilde{\psi}_1 + \tilde{\psi}_3$ or $\tilde{\psi}_2 + \tilde{\psi}_4$ which belong to species $a_1$ or by forming $\tilde{\psi}_1 - \tilde{\psi}_3$ or $\tilde{\psi}_2 - \tilde{\psi}_4$ which belong to species $b_2$.

If *three* atoms Y are brought up to a central atom X, there are three equivalent directions and therefore three equivalent localized orbitals entirely similar to (III, 16) but now in the three directions $x$, $y$, $z$. Such a situation corresponds to that in $NH_3$. The transformation to proper molecular orbitals is not quite as simple but is similar to the determination of the resultant states of three $^2S$ atoms (see p. 288).

It is not possible to form *four* equivalent localized orbitals from only the $p$ orbitals of the central atom if these orbitals are to be orthogonal to one another (that is, linearly independent). It is however possible, as was first shown by Pauling (969), to obtain four equivalent localized orbitals if the energy difference between $2s$ and $2p$ is neglected, since under this condition any linear combination of the one $s$ and the three $p$-orbital functions is also a solution of the wave equation for the same energy value and the orthogonality condition can be met. Four suitable mutually orthogonal orbitals obtained by such *hybridization* of $s$ and $p$ orbitals are

$$\begin{aligned}\psi_1 &= \tfrac{1}{2}\psi(2s) + \tfrac{1}{2}\sqrt{2}\,\psi(2p_x) + \tfrac{1}{2}\psi(2p_z)\\ \psi_2 &= \tfrac{1}{2}\psi(2s) - \tfrac{1}{2}\sqrt{2}\,\psi(2p_x) + \tfrac{1}{2}\psi(2p_z)\\ \psi_3 &= \tfrac{1}{2}\psi(2s) + \tfrac{1}{2}\sqrt{2}\,\psi(2p_y) - \tfrac{1}{2}\psi(2p_z)\\ \psi_4 &= \tfrac{1}{2}\psi(2s) - \tfrac{1}{2}\sqrt{2}\,\psi(2p_y) - \tfrac{1}{2}\psi(2p_z),\end{aligned} \qquad \text{(III, 17)}$$

where $\psi(2s)$, $\psi(2p_x)$, $\psi(2p_y)$, $\psi(2p_z)$ are the atomic orbitals previously given (eq. III, 6, or III, 14). In Fig. 118 the orientation of the axes used here is indicated with respect to a regular tetrahedron.

If $a$, $b$, $c$, $d$ designate the corners of the tetrahedron the atomic orbital functions (III, 17) can also be written

$$\begin{aligned}\psi_1 &= \tfrac{1}{2}\psi(2s) + \tfrac{1}{2}\sqrt{3}\,\psi(2p_a)\\ \psi_2 &= \tfrac{1}{2}\psi(2s) + \tfrac{1}{2}\sqrt{3}\,\psi(2p_b)\\ \psi_3 &= \tfrac{1}{2}\psi(2s) + \tfrac{1}{2}\sqrt{3}\,\psi(2p_c)\\ \psi_4 &= \tfrac{1}{2}\psi(2s) + \tfrac{1}{2}\sqrt{3}\,\psi(2p_d),\end{aligned} \qquad \text{(III, 18)}$$

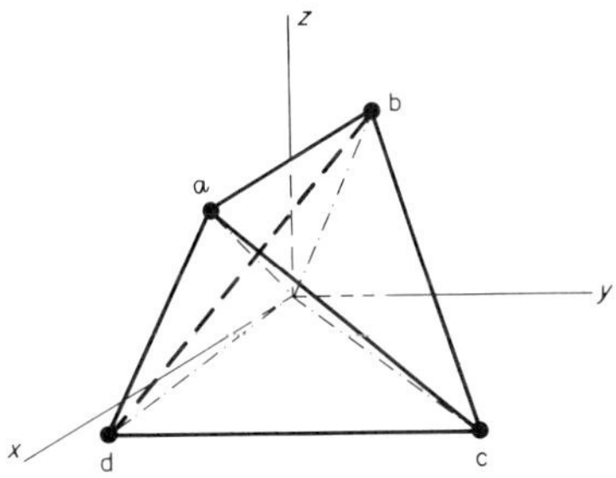

FIG. 118. **Orientation of axes to describe tetrahedral orbitals.**

where $\psi(2p_a)$ is of the form of a $2p_z$ function if the $z$-direction were in the direction from the central atom to $a$, and similarly for $\psi(2p_b)$, $\psi(2p_c)$ and $\psi(2p_d)$. [Note that $\psi(2p_a)$, $\psi(2p_b)$, . . . are not mutually orthogonal.] From the form (III, 17) it can immediately be seen that the four orbitals $\psi_1$, $\psi_2$, $\psi_3$, $\psi_4$, are mutually orthogonal [since $\psi(2s)$, $\psi(2p_x)$, $\psi(2p_y)$, $\psi(2p_z)$ are orthonormal]. The form (III, 18) makes it obvious that the four orbitals are equivalent.

It is interesting to compare one of the hybrid orbitals (III, 18), say $\psi_1$, with the orbital $\psi(2p_a)$. Figure 119a and b show the form of $\psi(2p_a)$ and $\psi_1$ respectively by means of contour diagrams using the Slater forms of eq. (III, 6). Here it must be noted that Slater's approximation neglects the radial node of $\psi(2s)$. Figure 119c gives $\psi_1$ using the more rigorous (self-consistent) forms of $\psi(2s)$ and $\psi(2p)$ following Moffitt and Coulson (868). In either approximation it is seen that the wave function $\psi_1$ is concentrated much more in the direction toward $a$ than is $\psi(2p_a)$ which is equally concentrated on either side of a plane through C perpendicular to the C–$a$ axis (see Fig. 119a).

If now an H nucleus (or a similar nucleus) is brought to $a$, a new orbital $\psi(\mathrm{H}_a, 1s)$ becomes available to any electron that may be present (this is apart from excited orbitals which are not under discussion here). If the energy of this H atom orbital is not too different from that of $\psi_1$ it will mix with $\psi_1$ and as a result we have the two orbitals

$$\begin{aligned}\tilde{\psi}_1 &= c_{11}\psi_1(\mathrm{C};\, 2s,\, 2p_a) + c_{12}\psi(\mathrm{H}_a;\, 1s)\\ \tilde{\psi}_2 &= c_{21}\psi_1(\mathrm{C};\, 2s,\, 2p_a) + c_{22}\psi(\mathrm{H}_a;\, 1s),\end{aligned} \qquad \text{(III, 19)}$$

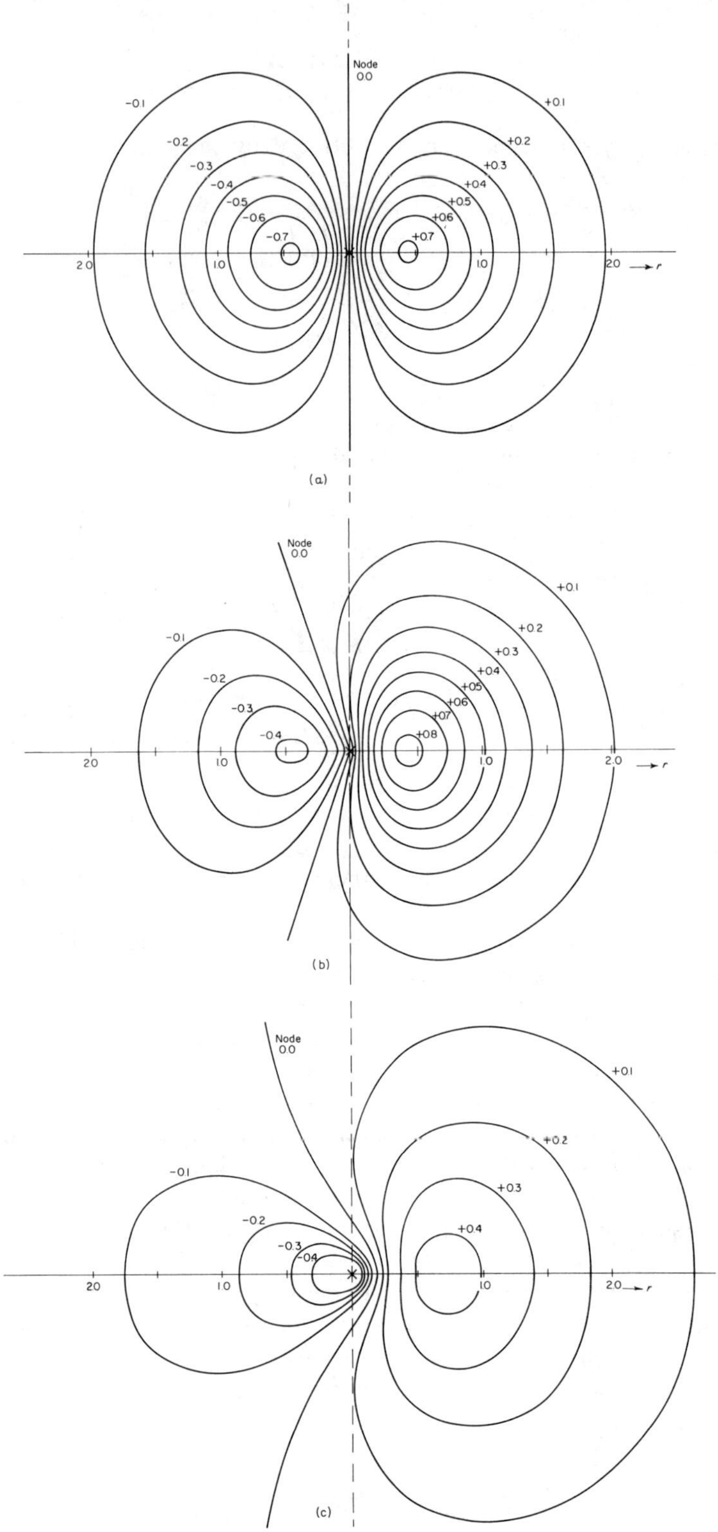
Node
0.0
-0.1
-0.2
-0.3
-0.4
-0.5
-0.6
-0.7
+0.1
+0.2
+0.3
+0.4
+0.5
+0.6
+0.7
2.0
1.0
1.0
2.0
r
(a)
Node
0.0
-0.1
-0.2
-0.3
-0.4
+0.1
+0.2
+0.3
+0.4
+0.5
+0.6
+0.7
+0.8
2.0
1.0
1.0
2.0
r
(b)
Node
0.0
-0.1
-0.2
-0.3
-0.4
+0.1
+0.2
+0.3
+0.4
2.0
1.0
1.0
2.0
r
(c)

one of lower and one of higher energy than the energy for large separation of $H_a$ and C. There are similar pairs of orbitals equivalent to (III, 19) for each of the other tetrahedral directions ($b$, $c$, $d$). None of these eight localized orbitals (also called *tetrahedral hybrids*) belongs to any of the species of point group $\boldsymbol{T_d}$; but there are, as before for $H_2O$, linear combinations of these localized orbitals which are no longer localized and which do belong to the species of $\boldsymbol{T_d}$.

If instead of mixing (hybridizing) $2s$ with $2p_x$, $2p_y$, $2p_z$ we mix it only with $2p_x$ and $2p_y$ and leave $2p_z$ unchanged ($sp^2$ hybridization) we can obtain *three* equivalent (normalized) orbitals which are localized in the $x,y$-plane and point to the corners $a$, $b$, $c$ of an equilateral triangle. Similar to (III, 18) they are given by

$$\begin{aligned}\psi_1 &= \frac{1}{\sqrt{3}}\,\psi(2s) + \sqrt{\frac{2}{3}}\,\psi(2p_a)\\ \psi_2 &= \frac{1}{\sqrt{3}}\,\psi(2s) + \sqrt{\frac{2}{3}}\,\psi(2p_b)\\ \psi_3 &= \frac{1}{\sqrt{3}}\,\psi(2s) + \sqrt{\frac{2}{3}}\,\psi(2p_c).\end{aligned} \qquad \text{(III, 20)}$$

These functions must again be combined with the orbitals of the attached atoms as indicated by (III, 19) to give two localized orbitals for each bond. These orbitals are called *trigonal hybrids*. They are of importance in all cases in which three equivalent bonds lie in one plane.

If we mix $2s$ with say $2p_z$ only and leave $2p_x$ and $2p_y$ unchanged ($sp$ hybridization), we can obtain two equivalent orbitals which are localized in the positive and negative $z$ direction and are given by

$$\begin{aligned}\psi_1 &= \frac{1}{\sqrt{2}}\,[\psi(2s) + \psi(2p_a)]\\ \psi_2 &= \frac{1}{\sqrt{2}}\,[\psi(2s) + \psi(2p_b)].\end{aligned} \qquad \text{(III, 21)}$$

These orbitals, or their combinations with those of the attached atoms, are called *digonal hybrids*.

It can be easily shown [see Lennard-Jones (738)] that *six* equivalent orbitals localized in the directions to the corners of a regular octahedron cannot be obtained by any combination of $s$ and $p$ atomic orbitals but that at least two $d$ orbitals must be included in the hybridization. For the mathematical form of these orbitals see, e.g. Coulson (7). The various types of hybridization and the atomic orbitals

---

FIG. 119. **Cross sections of $2p$ and of tetrahedral (hybrid) orbitals of the C atom.** ($a$) **Slater $2p$ orbital ($\mu_3$ = 2.275),** ($b$) **Slater tetrahedral orbital** $[\frac{1}{2}\psi(2s) + \frac{1}{2}\sqrt{3}\psi(2p_a)]$, ($c$) **tetrahedral orbital built from self-consistent $2s$ and $2p$ orbitals [after Moffitt and Coulson (868)].** The cross sections are represented by contour lines along which the orbital function has a fixed value. The abscissa axis represents the direction from the central (carbon) atom (marked **x**) to one of the corners of the tetrahedron (scale in Bohr radii). Note the difference in sign of $\psi$ at left and right.

of the central atom needed to form equivalent localized orbitals are summarized in Table 29.

TABLE 29. TYPES OF HYBRIDIZATION TO FORM EQUIVALENT LOCALIZED ORBITALS, ACCORDING TO KIMBALL (666a) AND MACEK AND DUFFEY (786)

| Number of equivalent orbitals | Symmetry of molecule | Orbitals used for hybridization |
|---|---|---|
| 2 | linear $XY_2$, $\boldsymbol{D}_{\infty h}$ | $sp$, $pd$ |
| 3 | planar $XY_3$, $\boldsymbol{D}_{3h}$ | $sp^2$, $dp^2$, $sd^2$, $d^3$ |
| 3 | non-planar $XY_3$, $\boldsymbol{C}_{3v}$ | $p^3$, $pd^2$ |
| 4 | tetrahedral $XY_4$, $\boldsymbol{T}_d$ | $sp^3$, $sd^3$ |
| 4 | square $XY_4$, $\boldsymbol{D}_{4h}$ | $sp^2d$, $p^2d^2$ |
| 5 | bipyramidal $XY_5$, $\boldsymbol{D}_{3h}$ | $sp^3d$, $spd^3$ |
| 6 | octahedral $XY_6$, $\boldsymbol{O}_h$ | $sp^3d^2$ |
| 12 | icosahedral $XY_{12}$, $\boldsymbol{I}_h$ | $sp^3d^5f^3$ |

In general it is easy to derive which atomic orbitals must be hybridized in order to obtain a given number of equivalent orbitals. One must establish what the transformation properties of the equivalent orbitals are, reduce the resulting representation into its irreducible components and see from which atomic orbitals these components can arise[2]. For example for four tetrahedral orbitals $\psi_1, \psi_2, \psi_3, \psi_4$, we see immediately that they transform for the operations $I$, $C_3$ and $\sigma_d$ as

$$\begin{array}{ccccc} & I & C_3 & \sigma_d \\ \psi_1 & 1\,0\,0\,0 & 1\,0\,0\,0 & 1\,0\,0\,0 \\ \psi_2 & 0\,1\,0\,0 & 0\,0\,1\,0 & 0\,1\,0\,0 \\ \psi_3 & 0\,0\,1\,0 & 0\,0\,0\,1 & 0\,0\,0\,1 \\ \psi_4 & 0\,0\,0\,1 & 0\,1\,0\,0 & 0\,0\,1\,0. \end{array}$$

Therefore the characters are 4, 1, 2. According to Table 53 (Appendix I) these characters can be obtained only by adding those of species $A_1 + F_2$. The species $A_1$ according to Table 58 (Appendix IV) can only be obtained from $s, f, g, \ldots$ atomic orbitals while $F_2$ can only be obtained from $p, d, f, \ldots$ orbitals. The lowest combination of orbitals to give four tetrahedral localized orbitals is therefore $sp^3$, the next lowest $sd^3$. The combinations $sf^3$, $p^3f$, $d^3f$, $f^4, \ldots$ would also give tetrahedral orbitals but in practice they are not of great importance since the $f$ orbitals for the lighter elements have rather high energies.

For some purposes, we may use $s$-$p$ hybridized orbitals instead of the pure $p_x$ and $p_y$ localized orbitals previously described even in an $XY_2$ molecule like $H_2O$; that is, in place of $\psi(O; 2p_x)$ and $\psi(O; 2p_y)$ in (III, 16) we may use (omitting the normalization factor $\sqrt{1 + \lambda^2}$)

$$\begin{aligned} \psi_a(O; 2s, 2p_a) &= \lambda\psi(O; 2s) + \psi(O; 2p_a) \\ \psi_b(O; 2s, 2p_b) &= \lambda\psi(O; 2s) + \psi(O; 2p_b). \end{aligned} \qquad \text{(III, 22)}$$

[2] Compare the similar procedure for determining which molecular orbitals can arise from a number of equivalent atomic orbitals in equivalent atoms (p. 303f).

Here $\lambda$ is the hybridization parameter. When it is zero we have pure $p$ localized orbitals oriented at 90° to each other and in that case $\psi(\mathrm{O}; 2p_a) = \psi(\mathrm{O}; 2p_x)$ and $\psi(\mathrm{O}; 2p_b) = \psi(\mathrm{O}; 2p_y)$. For different degrees of hybridization the angle between the two directions to $a$ and $b$ varies; or, in other words, if we want to obtain an angle $\vartheta$ different from 90° between the two localized orbitals we must introduce hybridization between $2s$ and $2p$. A simple calculation [see Coulson (7)] shows that

$$\lambda^2 = -\cos\vartheta. \qquad \text{(III, 23)}$$

For $\lambda = 1/\sqrt{3}$ we obtain $\vartheta = 109°\,28'$, i.e. tetrahedral orbitals, for $\lambda = 1/\sqrt{2}$ we obtain $\vartheta = 120°$, i.e. trigonal orbitals. In $H_2O$ where empirically $\vartheta = 104°\,30'$ we would need $\lambda = 0.50$ in order that the localized orbitals point in the direction of the two H atoms. While these two orbitals are equivalent the third orbital arising from the atomic orbitals $sp^2$ is not equivalent with them but is given by

$$\psi_c(\mathrm{O}; 2s, 2p_c) = \mu\psi(\mathrm{O}; 2s) + \psi(\mathrm{O}; 2p_c), \qquad \text{(III, 24)}$$

if $c$ is the direction bisecting the YXY angle. In order that $\psi_c$ be orthogonal to $\psi_a$ and $\psi_b$ we must have

$$\mu = \frac{\cos\frac{1}{2}\vartheta}{\lambda} \qquad \text{(III, 25)}$$

which for $\vartheta = 104°\,30'$ yields $\mu = 1.22$. $\psi_c$ is the so-called lone-pair orbital of $H_2O$. According to eq. (III, 24 and 25) it is mainly a $2s$ atomic orbital of the O atom but with some admixture of $2p$.

In a similar way, in a molecule like $NH_3$, $s$-$p$ hybridization can be introduced. The resulting orbitals are of the form (III, 18) except that the coefficients are different, viz.

$$\begin{aligned}
\psi_a(\mathrm{N}; 2s2p_a) &= \lambda\psi(\mathrm{N}; 2s) + \psi(\mathrm{N}; 2p_a)\\
\psi_b(\mathrm{N}; 2s2p_b) &= \lambda\psi(\mathrm{N}; 2s) + \psi(\mathrm{N}; 2p_b)\\
\psi_c(\mathrm{N}; 2s2p_c) &= \lambda\psi(\mathrm{N}; 2s) + \psi(\mathrm{N}; 2p_c)\\
\psi_d(\mathrm{N}; 2s2p_d) &= \mu\psi(\mathrm{N}; 2s) + \psi(\mathrm{N}; 2p_d)
\end{aligned} \qquad \text{(III, 26)}$$

where as before

$$\lambda^2 = -\cos\vartheta, \qquad \text{(III, 27)}$$

$\vartheta$ being the angle between the NH bonds. In addition

$$\mu = \frac{\cos\beta}{\lambda} \qquad \text{(III, 28)}$$

where $\beta$ is the angle of an NH bond with the symmetry axis $(\cos\beta = \sqrt{\frac{1}{3} + \frac{2}{3}\cos\vartheta})$. The functions $\psi_a$, $\psi_b$, $\psi_c$ are hybridized equivalent orbitals corresponding to $p_x$, $p_y$, $p_z$ before hybridization; $\psi_d$ is the lone-pair orbital. Only when $\vartheta$ is the tetrahedral angle, i.e. $\cos\vartheta = -\frac{1}{3}$, are all four orbitals equivalent. Upon molecule formation the three H atom orbitals will mix with the three equivalent orbitals of the N atom yielding two orbitals of the type (III, 19) for each one of the equivalent orbitals.

**Spin-orbitals.** All the orbitals treated so far, atomic orbitals, molecular orbitals, group orbitals, localized orbitals, equivalent orbitals, are functions of the spatial coordinates of the electrons only. As long as spin-orbit coupling is small the spin dependence of one-electron wave functions can be taken into account by multiplying the orbital wave function $\psi$ by a spin function $\varphi$ which depends on the spin variable $\sigma$. Thus we have for the complete *spin-orbital wave function* or, for short, *spin-orbital*,

$$\Psi = \psi(x, y, z)\varphi(\sigma). \tag{III, 29}$$

The spin variable $\sigma$ can assume only the two values $+\frac{1}{2}$ and $-\frac{1}{2}$ which are the components of the spin in a magnetic field. Thus the spin introduces a degeneracy (spin-degeneracy) such that for each energy value there are two eigenfunctions

$$\Psi^+ = \psi\,\varphi(+\tfrac{1}{2}) \quad \text{and} \quad \Psi^- = \psi\,\varphi(-\tfrac{1}{2}). \tag{III, 30}$$

In a magnetic field, there will be a splitting into two levels of slightly different energy. The symmetry of the spin functions for various point groups and for various values of the resultant spin has already been discussed in Chapter I, p. 14f.

### (b) Order and correlation of orbitals

In introducing the concept of molecular orbitals in the preceding subsection we have already considered the order of the orbitals in the limiting cases of the united atom (or united molecule) and the separated atoms (or groups of atoms). For diatomic molecules, from the correlation of the orbitals in the limiting cases the energetical order in intermediate cases can be estimated, as considered in detail in Volume I. The same method can be applied to polyatomic molecules even though the quantitative uncertainties of this method are even larger here. Moreover, a separate correlation diagram must be drawn for each type of molecule.

**Linear $XY_2$ molecules.** Let us first consider a linear $XH_2$ molecule. The energies of the orbitals of the united atom with $n = 2$ and $n = 3$ are shown at the extreme left in Fig. 120, while those of the orbitals of the separated atoms are shown at the extreme right. Here it is assumed that the $1s$ orbital of H has an energy slightly below that of $2p$ of X, as for example in $CH_2$ or $BH_2$. For the X atom, as for the united atom, only the orbitals with $n = 2$ and 3 are shown. Each orbital of hydrogen is double since there are two H atoms. For $NH_2$ the order of the $2p_X$ and $1s_H$ orbitals at the right would have to be reversed. Next to the atomic orbitals at the left are shown the molecular orbitals into which they go over in a linear conformation (see the previous discussion, p. 297). Next to the extreme right the molecular orbital energies corresponding to the orbitals of the separated atoms are shown. The $1s$ orbitals of the two H atoms give the two molecular orbitals $\sigma_g$ and $\sigma_u$ since they can be combined as

$$\psi(1s_{H'}) + \psi(1s_{H''}) \quad \text{and} \quad \psi(1s_{H'}) - \psi(1s_{H''}). \tag{III, 31}$$

The molecular orbitals are numbered in the order of their energy $2\sigma_g$, $3\sigma_g$, $4\sigma_g$, . . .,

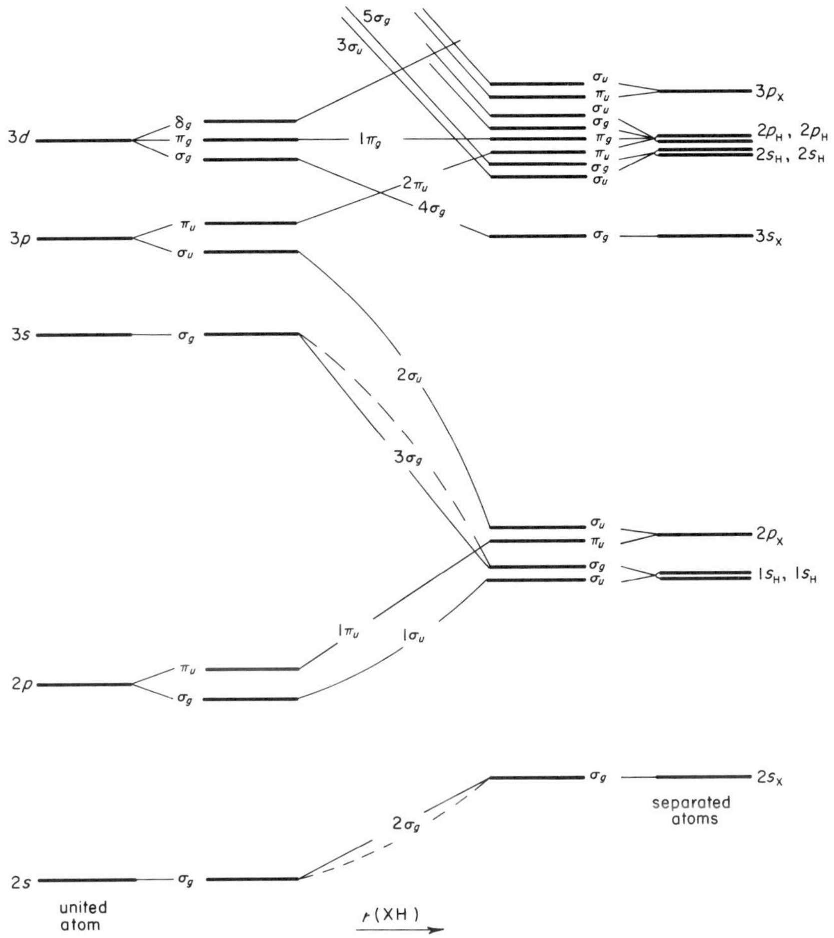


FIG. 120. **Correlation of orbitals between large and small internuclear distance in linear $XH_2$ molecules.** At the extreme left are the energies of the orbitals of the united atom, at the extreme right those of the separated atoms. It is assumed that the ionization potential of X from the $2p$ orbital is less than that of H from the $1s$ orbital.

etc. ($1\sigma_g$ corresponds to the $1s$ orbital of X which is not shown)[3]. The energetical order of the orbitals in the molecule is given by the intercepts of a vertical line (not necessarily a straight line) with the connecting lines between right and left. These connecting lines are drawn in such a way that there are as few intersections as possible. Orbitals of the same species cannot intersect.

It may be noted that, on account of this correlation, two of the orbitals, arising from $2p_X$ and $1s_H$ of the separated atoms, lower their energy on reducing

[3] Many authors leave out the $K$ electrons of atoms other than H in the counting, i.e., they would call the lowest $\sigma_g$ orbital shown in Fig. 120 $1\sigma_g$ rather than $2\sigma_g$ as is done here. Note that the numbers preceding the species symbols are not principal quantum numbers.

the internuclear separation, while the other two rapidly increase their energy. In addition the two $\sigma_u$ orbitals, $1\sigma_u$ and $2\sigma_u$, being of the same type "repel" each other. This is indicated by curvature (opposite in direction) of the two correlation lines. It must be emphasized, however, that this whole procedure is extremely crude and can give reasonable predictions only when checked as much as possible by experimental data.

If the two atoms attached to X are not H but some atom Y like X of the second period of the periodic system, the correlations are somewhat changed as shown in Fig. 121. Moreover the correlation to the united atom is less important since the vertical line corresponding to an actual molecule is still farther to the right. The orbitals as given at the right-hand side of Fig. 121 correspond very roughly to those in $CO_2$ and similar molecules. Again the pairs of orbitals of the same type ($3\sigma_g$ and $4\sigma_g$; $1\pi_u$ and $2\pi_u$; $3\sigma_u$ and $4\sigma_u$) arising from atomic orbitals of nearly the same

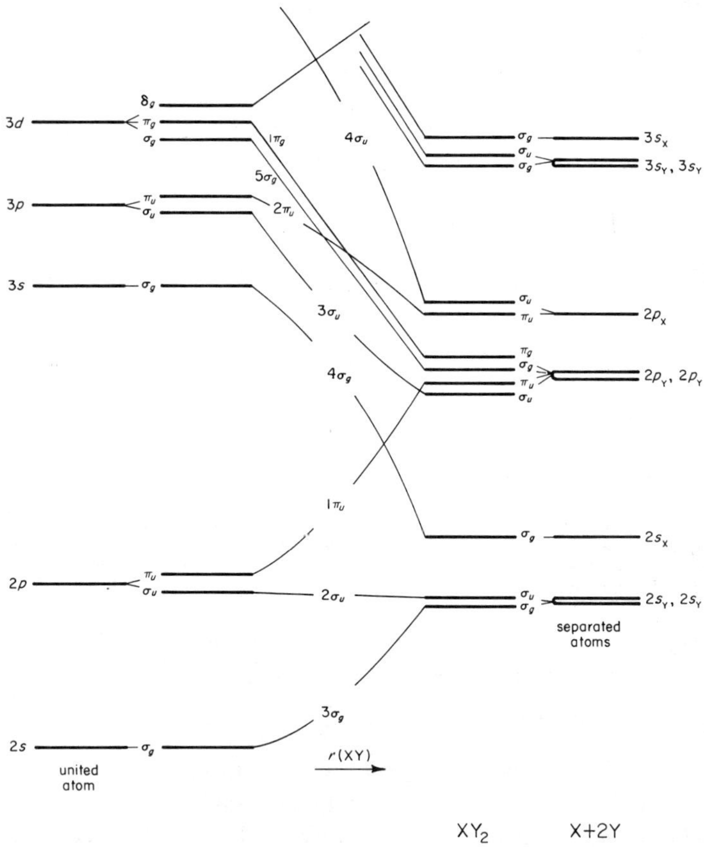


FIG. 121. **Correlation of orbitals between large and small internuclear distances in linear $XY_2$ molecules.** See caption of Fig. 120. It is assumed here that the ionization potentials of X from the $2p$ and $2s$ orbitals are smaller than those of Y.

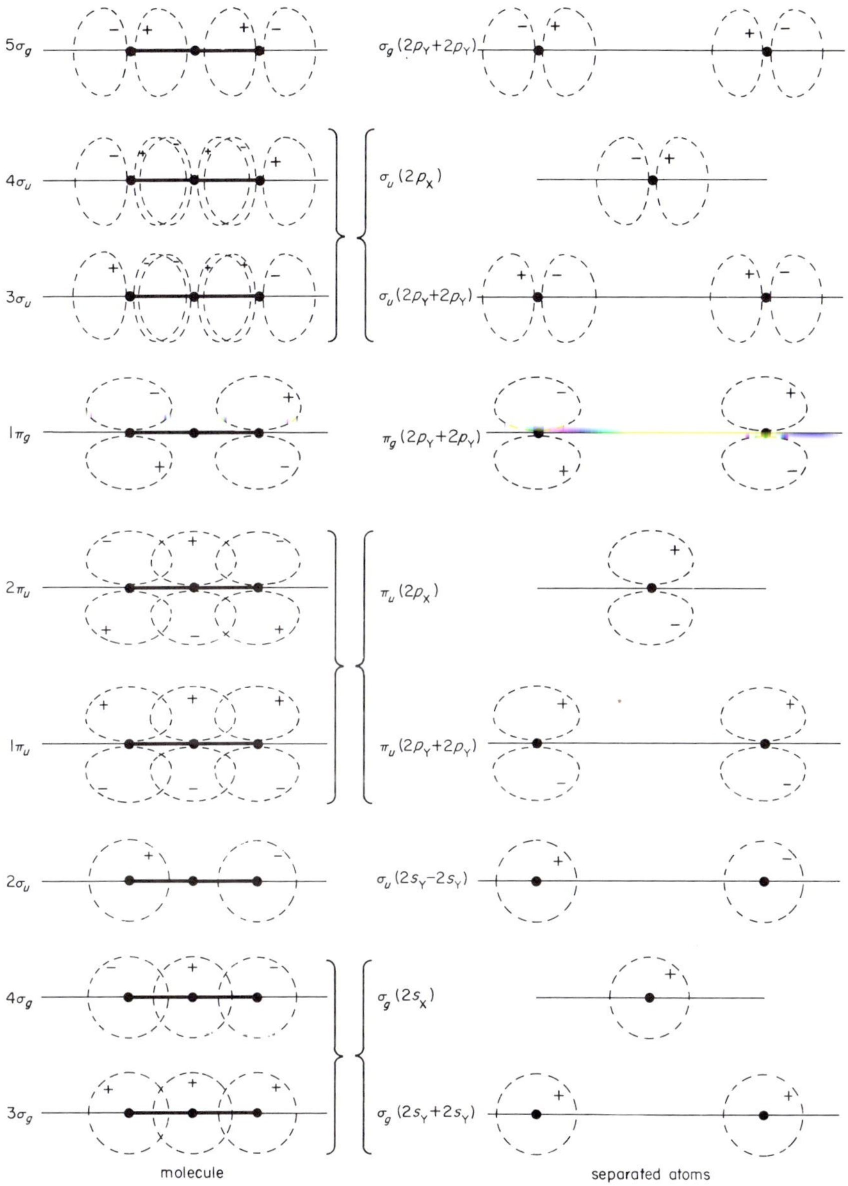


FIG. 122. **Shapes of the orbitals of the separated atoms and overlapping of these orbitals to form molecular orbitals for linear $XY_2$ (schematic).** The atomic orbitals in the form adapted to molecular orbitals are shown at right (corresponding to the right-hand part of Fig. 121). The combinations of these orbitals in the molecule are shown at left (corresponding to the center of Fig. 121). Only one component of the $\pi$ orbitals is shown.

energy "repel" each other as indicated by the curved connecting lines. On the other hand, orbitals like $2\sigma_u$, $1\pi_g$ are in a first rough approximation changed very little from $2s_Y - 2s_Y$ or $2p_Y - 2p_Y$ when the atoms are brought together, since the two Y atoms even in the molecule are far apart and since there is no orbital of X of the right symmetry to mix with them.

It is interesting to consider the shapes of the wave functions for the orbitals just discussed. They are shown schematically in Fig. 122. It is seen that the two $\sigma_g$ orbitals at very large separation (right-hand side of Fig. 122) can be overlapped in phase and out of phase. As we shall see in more detail later, the in-phase overlap corresponding to $\sigma_g(2s_X) + \sigma_g(2s_Y) \equiv 3\sigma_g$ gives rise to a lowering of the energy, the out-of-phase overlap corresponding to $\sigma_g(2s_X) - \sigma_g(2s_Y) \equiv 4\sigma_g$ gives rise to an increase of the energy. The same applies to the two $\sigma_u$ orbitals which give $\sigma_u(2p_X) + \sigma_u(2p_Y) \equiv 3\sigma_u$ and $\sigma_u(2p_X) - \sigma_u(2p_Y) \equiv 4\sigma_u$. For the two $\pi_u(2p)$ orbitals the overlap is, so to speak, sideways, again in-phase and out-of-phase overlap giving rise to the orbitals $\pi_u(2p_X) + \pi_u(2p_Y) \equiv 1\pi_u$ and $\pi_u(2p_X) - \pi_u(2p_Y) \equiv 2\pi_u$. In a first approximation no overlap occurs in the orbitals $2\sigma_u$, $1\pi_g$ and $5\sigma_g$, since each of them is formed entirely from orbitals of the Y atoms[3a].

**Bent $XY_2$ molecules.** The correlation of the orbitals of bent $XH_2$ between the united atom and the separate atoms is shown in Fig. 123. Here the $p$ atomic orbitals on both sides of the diagram split into three molecular orbitals (see Table 58 of Appendix IV), viz. $a_1$, $b_1$, $b_2$ corresponding to $p_z$, $p_x$, $p_y$. The two $1s$ orbitals of the two H atoms form $a_1$ and $b_2$ molecular orbitals in this point group corresponding to $1s_{H'} + 1s_{H''}$ and $1s_{H'} - 1s_{H''}$ (see p. 300). Again the connecting lines between the levels in the two limiting cases are drawn in such a way that orbitals of the same species, now in $C_{2v}$, do not cross. The two orbitals $3a_1$ and $4a_1$, and similarly $1b_2$ and $2b_2$, "repel" each other, as indicated by the opposite curvature of the connecting lines. In Fig. 124 the shapes of these orbitals are indicated showing the in-phase and out-of-phase overlap. An important difference between the bent and linear case (Figs. 123 and 120) should be noted: In the bent case both orbitals ($3a_1$ and $1b_2$) arising from the two H orbitals lower their energy (are "bonding" orbitals, see below) when the atoms are brought together, while in the linear case only one does ($1\sigma_u$). As a corollary, it will be remembered that for an angle of 90° in the bent molecule there is no $s$-$p$ hybridization, that is, the orbital $2a_1$ arising from $2s_X$ is little affected by the interaction with $3a_1$ and $4a_1$ while for a 180° angle (and intermediate values) $s$-$p$ hybridization is possible and as a consequence the $2\sigma_g$ orbital interacts strongly with the $3\sigma_g$ orbital and is pushed down somewhat (see the broken-line connecting curves in Fig. 120).

By combining the results of Figs. 120 and 123 we can now plot a rough diagram which shows the variation of the orbital energies in changing the angle from 180° to 90°. This is shown in Fig. 125. Such diagrams were first discussed by Walsh (1263a) and are now often called Walsh diagrams. The correlation of the orbitals between left and right follows of course the symmetry rules of Table 59 (p. 576).

[3a] See however footnote [10a] p. 393.

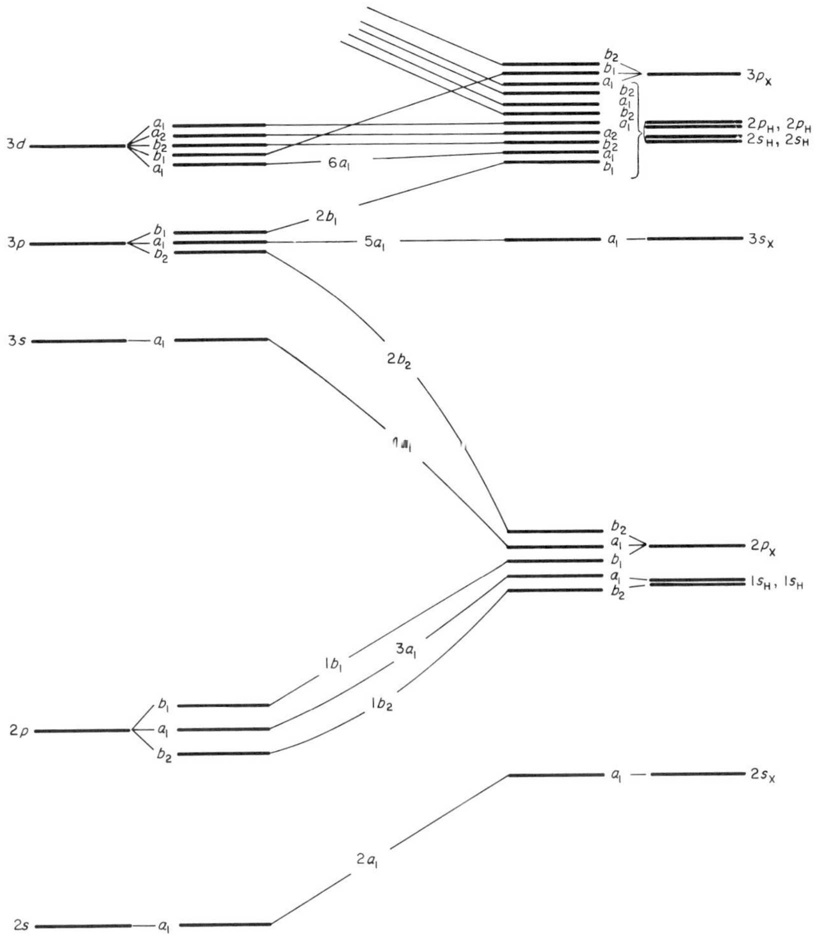


FIG. 123. **Correlation of orbitals between large and small internuclear distances in bent $XH_2$ molecules.** See caption of Fig. 120. It is assumed that the ionization potential of X from the $2p$ orbital is smaller than that of the H atom from the $1s$ orbital.

The lowering of the orbitals $2a_1 - 2\sigma_g$ and $1b_2 - 1\sigma_u$ from left to right is due to increasing hybridization. The orbital $1b_1 - 1\pi_u$ remains essentially unchanged since in both bent and linear $XY_2$ it is not subject to interaction with any other orbitals having $n = 2$; but $3a_1 - 1\pi_u$ changes strongly since in the bent form there is a strong interaction as pointed out above.

If the Y atoms of bent $XY_2$ are not H, the correlation diagram between united atom and separated atoms is of course considerably changed compared to Fig. 123; but since the relation to Fig. 123 is the same as that of Fig. 121 to Fig. 120 the reader can easily plot this diagram himself. Instead we give in Fig. 126 the resulting correlation of the orbital energies between the bent and linear form.

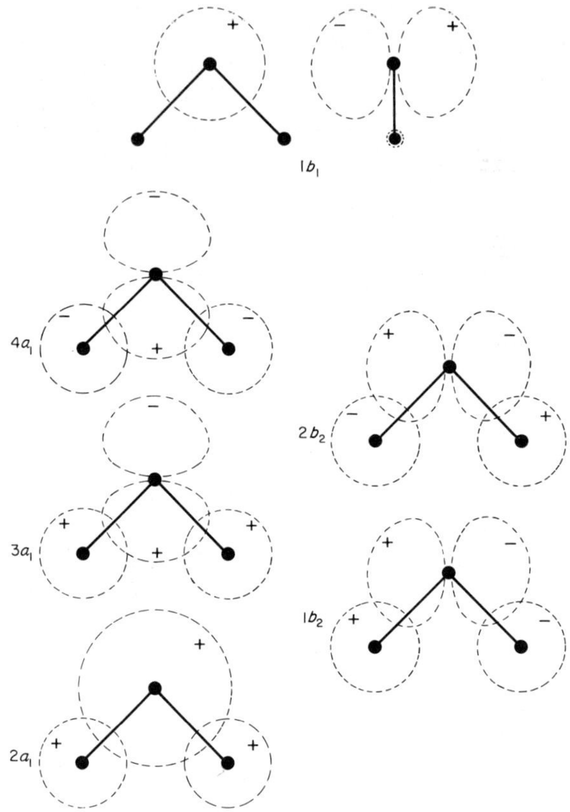


FIG. 124. **Shapes of molecular orbitals of bent $XH_2$ in terms of linear combinations of atomic orbitals (schematic).** For the $1b_1$ orbital two views are given. All other orbitals are symmetrical with respect to the molecular plane.

The original version of this diagram given by Walsh (1264) has been slightly modified as suggested by Fischer–Hjalmars (386) and Green and Linnett (446). It is seen that the $4\sigma_g$, $3\sigma_u$, $1\pi_g$, $5\sigma_g$ and $4\sigma_u$ orbitals favor the linear conformation while $1\pi_u - 1b_1$ and more strongly $2\pi_u - 6a_1$ favor the bent conformation. The $1s$ orbitals are not shown since they will remain essentially atomic orbitals.

A similar diagram for HXY molecules has been given by Walsh (1265). A more detailed theoretical justification of the basis of the Walsh diagrams has been attempted by Coulson and Nielsen (241).

**Planar and non-planar $XH_3$ molecules.** In Fig. 127a the correlation diagram for the orbitals of planar $XH_3$ (point group $\boldsymbol{D}_{3h}$) is given. The diagram is similar to Fig. 120 for linear $XH_2$, i.e. it shows the correlation between the orbitals at small and large XH distance. The three H atom $1s$ orbitals, as we have seen previously (p. 303f, see Table 61), form the molecular orbitals $a_1'$ and $e'$.

There is strong interaction (repulsion) between the orbitals $1e'$ and $2e'$ and to a smaller extent between $2a_1'$ and $3a_1'$.

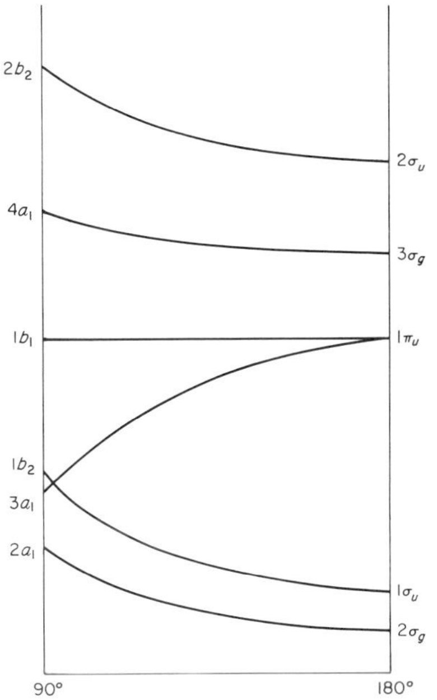


FIG. 125. **Walsh diagram for $XH_2$ molecules.** The variation of orbital energies in going from a bent (90°) to a linear conformation is shown. The 1*s* orbital of X is not included.

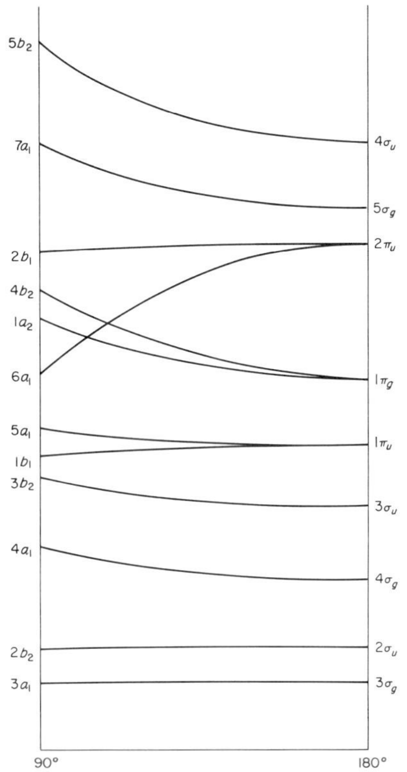


FIG. 126. **Walsh diagram for $XY_2$ molecules.** The order of the orbitals at right and left is not uniquely determined from theory. The 1*s* orbitals of X and Y are omitted but have been taken into account in the numbering of the orbitals.

For non-planar $XH_3$ (point group $C_{3v}$) the correlation is somewhat different, as shown in Fig. 127b, since the species $a_1'$ and $a_2''$ are now identical and are called $a_1$. Also the mutual interactions are somewhat different, since at a 90° angle between the X—H bonds there is no hybridization between $2p_X$ and $2s_X$ and, as a consequence, no strong "repulsion" between $3a_1$ and $2a_1$; rather, $3a_1$ interacts with $4a_1$ and on that account is pushed down.

In Fig. 128 is shown a Walsh diagram correlating the orbitals of planar $XH_3$ to those of non-planar (90°) $XH_3$ [see Walsh (1266)]. The energies of the orbitals at left and right are taken from appropriate vertical lines in Fig. 127b and a respectively. As before the correlation rules of Table 60 have been observed in connecting the levels at the left with those at the right. The most important point in this diagram is the sharp rise of the $3a_1 - 1a_2''$ orbital from left to right, already

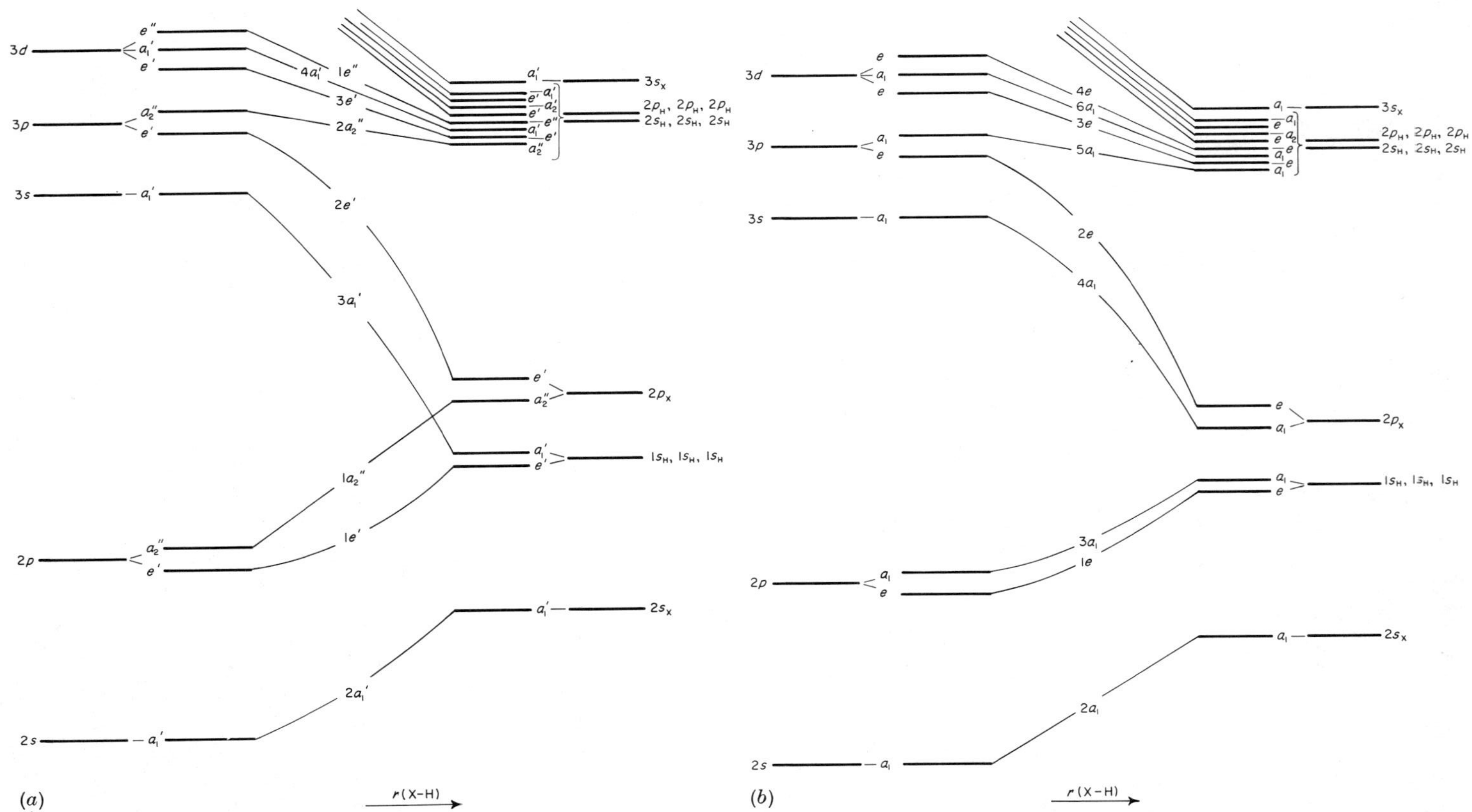


FIG. 127. **Correlation of orbitals between large and small internuclear distances** (*a*) **in planar** $XH_3$ **(point group** $D_{3h}$**),** (*b*) **in non-planar** $XH_3$ **(point group** $C_{3v}$**).** See caption of Fig. 120. It is assumed that the ionization potential of X from the 2*p* orbital is somewhat less than that of H from the 1*s* orbital.

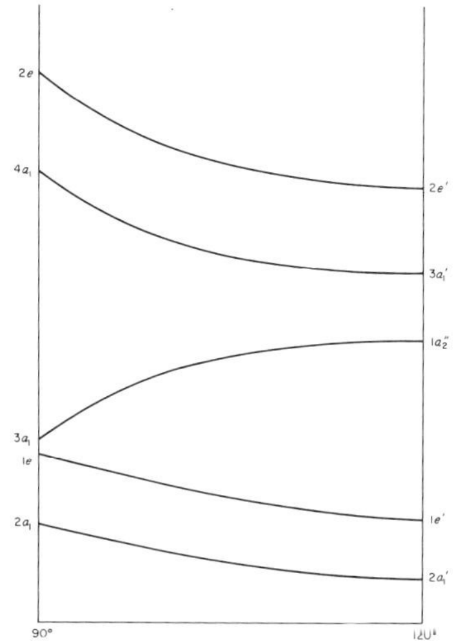


FIG. 128. **Walsh diagram for the correlation of orbitals between non-planar and planar $XH_3$.** The $1s$ orbital of X is not shown.

explained above. Closely related to this is the pronounced drop of the $4a_1-3a_1'$ orbital and the slight drop of $2a_1-2a_1'$. The similar drop of the $1e-1e'$ orbital follows according to Walsh from the H—H anti-bonding character (see section 3) of this orbital which is less pronounced in the planar conformation since the H atoms are farther away from one another.

A diagram similar to Fig. 128 for $XY_3$ (non-hydride) molecules has been given and discussed by Walsh (1267).

**Tetrahedral $XH_4$ and $XY_4$ molecules.** In Fig. 129 the correlation diagram of the orbitals of tetrahedral $XH_4$ is given. The molecular orbitals arising from the atomic orbitals of the central (X) atom are immediately read off Table 58, p. 574, as are the molecular orbitals arising from the united atom. The four $1s$ orbitals of the four H atoms (and similarly the four $2s$ orbitals) yield, according to a discussion entirely similar to that for the three H atoms in $CH_3$ on p. 303, the molecular orbitals $a_1$ and $f_2$ (see Table 61, p. 578). The determination of the orbitals derived from the four $2p_H$ orbitals is similar to that for three $2p_H$ discussed earlier (p. 304). As listed in Table 61 the molecular orbitals $a_1, f_2, e, f_1, f_2$ arise. In the $XH_4$ molecule we expect strong interaction (repulsion) between the orbitals $2a_1$ and $3a_1$ and similarly between $1f_2$ and $2f_2$.

In $XH_4$ molecules in general the orbitals formed from $2s_H$ and $2p_H$ need not be considered. However when H is replaced by Y they must be considered (for example for $CF_4$). In that case in Fig. 129 the central atom orbitals $2s_X$, $2p_X, \ldots$ have to be moved upwards appropriately such that the correct relation between the ionization potentials of X and Y is represented. For heavier central atoms a strong interaction between orbitals derived from $2s_Y$ or $2p_Y$ with those derived from $3d_X$ or $4d_X$ or even $4f_X$ may arise [see section 3(b)].

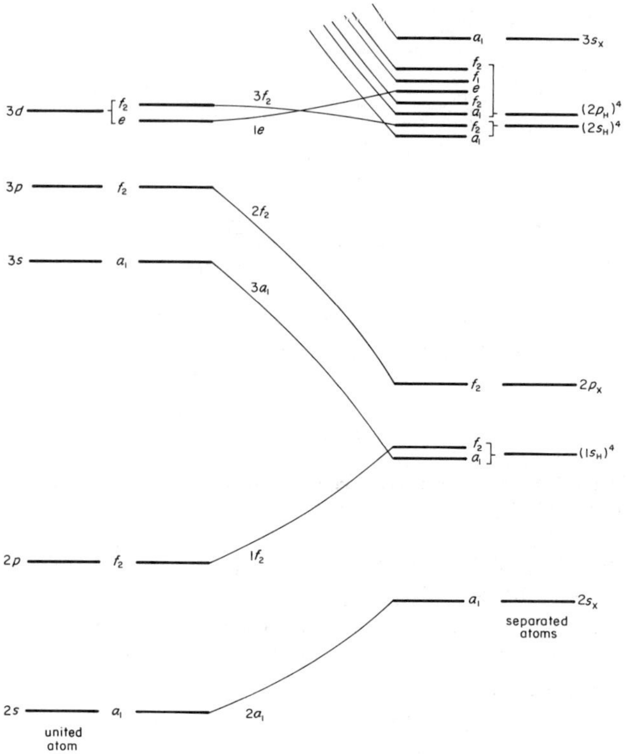


FIG. 129. **Correlation of orbitals between large and small internuclear distances for tetrahedral $XH_4$.** See caption of Fig. 120.

**Octahedral $XY_6$ molecules.** In octahedral $XY_6$ molecules we have six equivalent $2s_Y$ and $2p_Y$ orbitals. From group theoretical considerations similar to those given earlier (see Table 61) we find that the $2s_Y$ orbitals give rise to the molecular orbitals $a_{1g}$, $e_g$, $f_{1u}$ and the $2p_Y$ orbitals to $a_{1g}$, $e_g$, $f_{1g}$, $2f_{1u}$, $f_{2g}$, $f_{2u}$. Again the molecular orbitals arising from the central atom can be read off directly from Table 58, p. 574. Octahedral molecules do not arise for central atoms of the first period but only when $3d$ orbitals are partially occupied in the ground state.

In Fig. 130 a diagram of the orbital energies is given in a somewhat different manner from the preceding diagrams: omitting the correlation to the united atom (which is not of interest here) we give at left the orbitals of the central atom (X) up to $4p$ and the molecular orbitals that arise from them when the atoms are far apart; similarly at the right the orbitals of the "*ligand*" atoms (Y) and the molecular orbitals that arise from them according to Table 61. In the center the orbitals of the resulting molecule, when center and ligand atoms are brought together, are shown qualitatively. They arise by the mutual interaction of orbitals of the same species at left and right: the number and type of orbitals remain

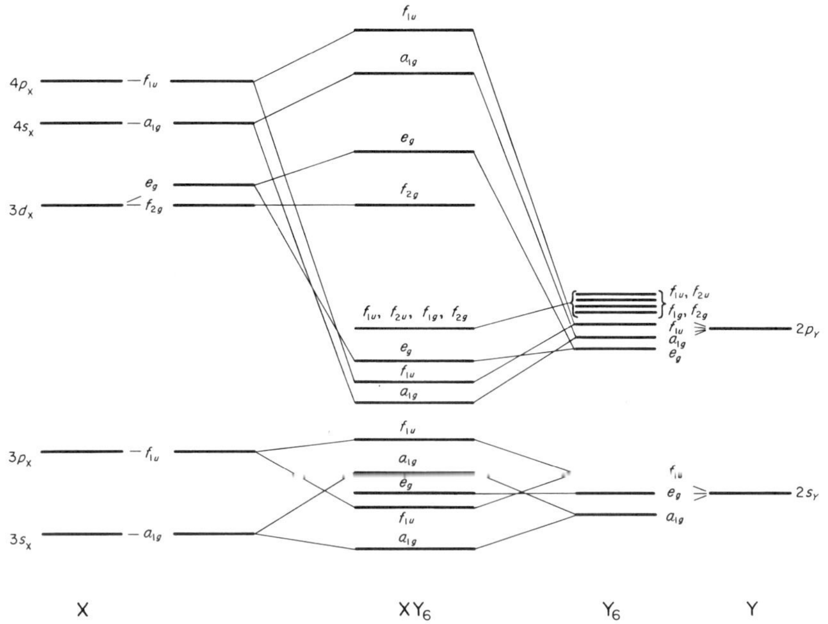


FIG. 130. **Correlation of orbitals between the separated atoms and the molecule for octahedral $XY_6$.** At left are the orbitals of X, at right those of Y. It is assumed that the ionization potential of X from the $3d$ orbital is somewhat less than that of Y from the $2p$ orbital. The $3s$, $3p$, $3d$ orbitals of Y are considered to be unoccupied in the ground state of Y and are not shown. The ligand orbitals $f_{1g}, f_{2g}, f_{1u}, f_{2u}$ arising from $2p_x$, $2p_y$ of $2p_Y$ are usually disregarded for reasons that are not entirely clear. They are shown here, assuming that they do not mix appreciably with the other orbitals of the same species.

unchanged in going from the sides (large separation) to the center. For example the $a_{1g}$ orbital from $2s_Y$ and the $a_{1g}$ orbital from $3s_X$ form two $a_{1g}$ orbitals of the molecule, one appreciably higher than the higher of the two and one appreciably lower than the lower of the two. In these two molecular orbitals the atomic orbitals are strongly mixed as indicated by the connecting lines to both atomic orbitals. There are also orbitals like $f_{2g}$ from $3d_X$ or $e_g$ from $2s_Y$ which even in the molecule are essentially orbitals belonging to the central atom alone or to the ligand atoms alone.

Diagrams like Fig. 130 are widely used in crystal field and ligand field theory. In these theories one often considers the effect of the "ligands" Y as a small perturbation and is then able to estimate the splitting of the atomic orbitals of X under the action of the field produced by the Y atoms (or ions).

The molecular orbitals for many other polyhedral molecules of various symmetries have been discussed by Hoffman and Gouterman (561).

**Planar $H_2XY$ molecules.** As mentioned before when more than one heavy atom is present in a molecule that contains in addition several H atoms it is more significant to consider the correlation to the united molecule than that to the united

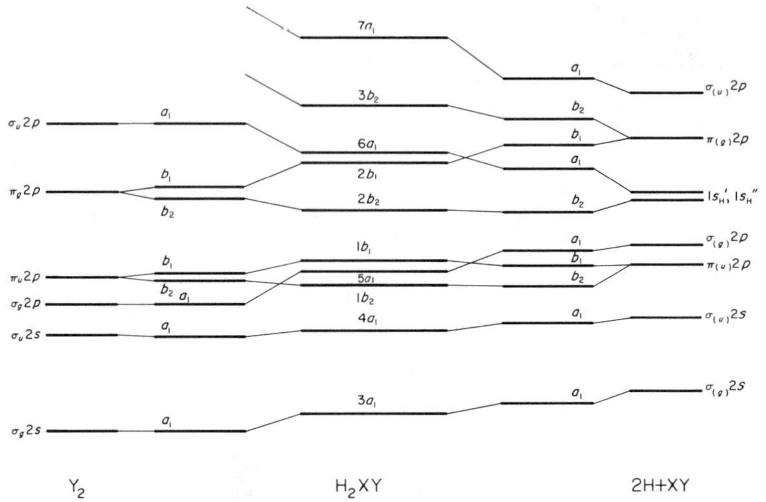


FIG. 131. **Correlation of the orbitals of planar $H_2XY$ to those of the united molecule $Y_2$ and to those of 2H + XY.** The variable along the abscissa is the XH distance. Note that at the left, since $Y_2$ is homonuclear, the orbitals are $\sigma_g$, $\sigma_u$, $\pi_g$, $\pi_u$ while at the right, since XY is heteronuclear, the $g$, $u$ characteristic does not strictly apply. However, just as at the left, the orbitals $\sigma 2s$ are mixtures (but not 50:50 mixtures) of the $2s$ orbitals of X and Y and similarly for the other XY orbitals. The order of $\pi_{(u)}2p$ and $\sigma_{(g)}2p$ is reversed at the right compared to the left in accordance with the situation in CO as compared to $O_2$ (see Vol. I, p. 346).

atom. As an example, in Fig. 131 the correlation diagram for planar $H_2XY$ is shown assuming that the united atom of $XH_2$ is Y and that therefore the united molecule is $Y_2$ (as in $H_2CO$ whose united molecule is $O_2$). At left the orbitals of $Y_2$ are shown as well as the orbitals arising from them in $C_{2v}$ symmetry. At the right the orbitals of the XY molecule and the two H atoms are given as well as the orbitals arising from them in $C_{2v}$ symmetry. If, as before, we connect the orbitals at the right with those of equal type at the left (avoiding all intersections of orbitals of the same species), we obtain an idea, even though an extremely rough one, of the relative order of the orbitals for intermediate X—H distances.

Very similar, even simpler considerations can be applied to HCN and to $C_2H_2$. The reader may easily plot diagrams similar to Fig. 131 for these cases. One can of course also plot diagrams similar to Figs. 120, 121, 123 giving the correlation to the united atoms.

**Planar $X_2H_4$ molecules.** For a discussion of the orbitals of planar $X_2H_4$ or more generally $X_2Y_4$ molecules, one may consider the correlation to a united molecule $Z_2$ (where Z is the united atom of $XH_2$ or $XY_2$) or to a united molecule $UY_4$ (where U is the united atom of $X_2$) or to the united atom of the whole molecule. Even without correlation to a united atom or molecule most of the qualitative information about the order of the orbitals can be obtained in this case from the interaction of orbitals of the same species. This is shown in Fig. 132 where at the extreme right the orbital energies of $XH_2$ are plotted. Each of the $XH_2$ orbitals splits into two when the $XH_2$ radicals are brought together; one of these two orbitals is symmetric, the other antisymmetric, with respect to the new plane of symmetry. As Table 60, p. 577, shows, $a_1$ gives $a_g$ and $b_{1u}$, $b_1$ gives $b_{2g}$ and $b_{3u}$, and $b_2$ gives $b_{2u}$ and $b_{3g}$. The two lowest pairs of $a_g$ and $b_{1u}$ orbitals interact

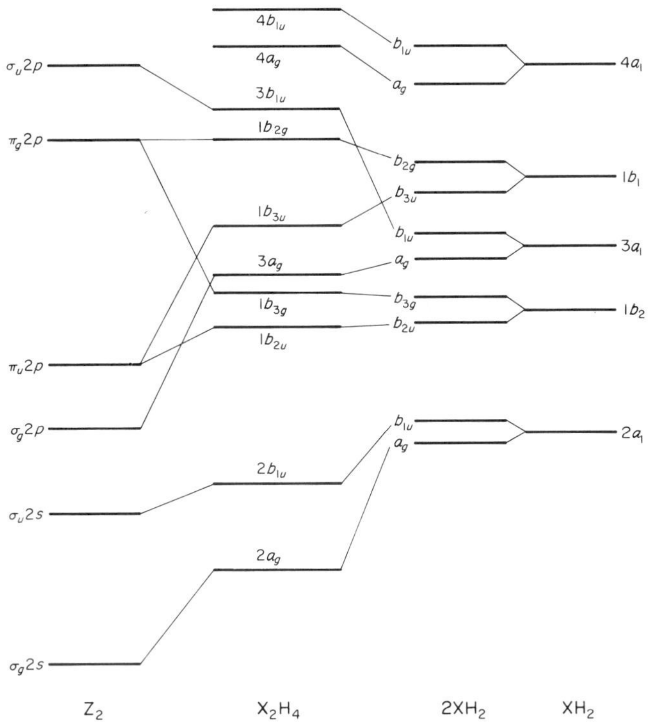


FIG. 132. **Correlation of orbitals of two $XH_2$ groups to those of an $X_2H_4$ molecule of point group $D_{2h}$ and to the united molecule $Z_2$.** In the right-hand part of the diagram the X–X distance, in the left-hand part the X–H distances increase from left to right.

strongly leading to the energies given in the center of the diagram. As we have seen above, in $XH_2$, of the two molecular orbitals arising from the two $1s_H$ orbitals the one without a newly formed nodal plane ($a_1$) lies lower than the one with such a plane ($b_2$); similarly here, in the pairs $a_g - b_{1u}$, $b_{2u} - b_{3g}$, $b_{3u} - b_{2g}$, the ones without a new nodal plane, i.e. $a_g$, $b_{2u}$, $b_{3u}$, lie lower than their partners formed from the same state of $XH_2$. Here as elsewhere this rule is important for an understanding of the order of the orbitals as was first pointed out by Mulliken (891). The correlation of the orbitals to the united molecule ($O_2$ for $C_2H_4$) has been added at the left in Fig. 132. It would be easy to draw a diagram similar to Fig. 132 for $X_2Y_4$, with $Y \neq H$ making use of Fig. 126 for $XY_2$.

The pairs of orbitals arising from a given orbital, say $2a_1$, of the $XH_2$ group, may be represented by wave functions which are in a first approximation the sum and difference of the two $XH_2$ orbitals, e.g.[4], $\psi(2a_1) + \psi(2a_1)$ and $\psi(2a_1) - \psi(2a_1)$ or, for short, $2a_1 + 2a_1$ and $2a_1 - 2a_1$.

**$X_2H_6$ molecules of $D_{3d}$ symmetry.** In Fig. 133 the correlation of orbitals for an $X_2H_6$ molecule of $D_{3d}$ symmetry is given when it is built up from two non-planar $XH_3$ groups of $C_{3v}$ symmetry. The energies of the orbitals of the $XH_3$ groups are taken from Fig. 127b. Again we have a doubling of all the orbitals of

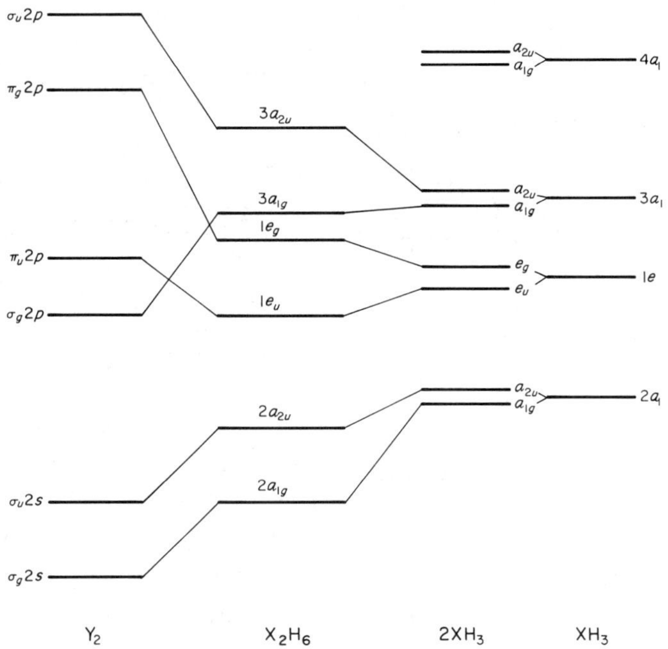


FIG. 133. **Correlation of orbitals of two $XH_3$ groups to those of an $X_2H_6$ molecule and to the united molecule $Y_2$.** It is assumed that the symmetry of the system is $D_{3d}$ throughout. See caption of Fig. 132.

$XH_3$ and again it is the interaction of orbitals of the same species (in particular $2a_{1g} - 3a_{1g}$ and $2a_{2u} - 3a_{2u}$) which causes the main readjustment of orbital energies. In addition, just as for $X_2H_4$ we must take account of the fact that the orbitals with a newly formed nodal plane lie higher than those without such a plane if they are derived from the same $XH_3$ orbital; for example $2a_{1g}$ is lower than $2a_{2u}$, $1e_u$ is lower than $1e_g$, and so on. The correlation to the united diatomic molecule $Y_2$ [that is for $r(X—H) \to 0$] is also shown in Fig. 133.

If the $X_2H_6$ molecule is of symmetry $D_{3h}$ instead of $D_{3d}$ all one has to do is change $a_{1g}$ to $a_1'$, $a_{2u}$ to $a_2''$, $e_u$ to $e'$ and $e_g$ to $e''$ (see Table 59, p. 576). If $XH_3$ were planar (point group $D_{3h}$) one would of course first have to correlate with the

[4] Here it is understood, but not explicitly noted in the designation, that the two $\psi$ functions have different centers and therefore $\psi(2a_1) - \psi(2a_1)$ is not zero.

corresponding species of $C_{3v}$ since these are the ones that matter in the formation of $X_2H_6$.

**$\pi$ orbitals in benzene and other unsaturated molecules.** Molecular orbitals that arise from $p$ atomic orbitals and are antisymmetric with respect to the plane of the molecule are often called $\pi$ orbitals[5]. For example the orbitals

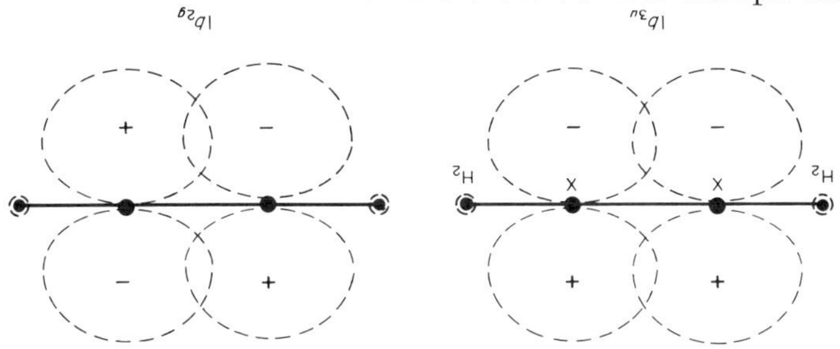


FIG. 134. **Form of the $1b_{3u}$ and $1b_{2g}$ orbital functions of $X_2H_4$ (schematic).** The diagram gives a side view from a point on the $y$-axis (the $z$-axis is the X–X axis, the $x$-axis is perpendicular to the plane of the molecule, i.e., in the plane of the paper). If filled by two electrons the first diagram represents a $\pi$-bond, the second a $\pi$ anti-bond (see p. 394).

$1b_{3u}$ and $1b_{2g}$ of $C_2H_4$ which arise from a $1b_1$ orbital in the $CH_2$ group (see Fig. 132) are antisymmetric with respect to the molecular plane, i.e. they are $\pi$ orbitals. Figure 134 shows schematically the form of the two wave functions: in one, $1b_{3u}$, the $p_x$ atomic orbitals (or $b_1$ molecular orbitals of $CH_2$) are overlapped in phase, in the other, $1b_{2g}$, out of phase. The former has the lower energy, the latter the higher energy. Since the $p_x$ orbitals are in general the most loosely bound orbitals still occupied by electrons in the separated groups, the corresponding $\pi$ orbitals of the complete molecule are very important for an understanding of the stability of the molecule, and, more specifically, of the nature of double-bonding; they also largely determine the lowest excited electronic states of these molecules.

In discussing the $\pi$ orbitals of the benzene molecule, in accordance with general custom, we choose the $z$-axis (rather than the $x$-axis, as in $C_2H_4$) perpendicular to the plane of the molecule, that is, the $\pi$ orbitals are derived from $p_z$ atomic orbitals of which there are six, one for each carbon atom. For the present discussion we disregard the molecular orbitals arising from $s$, $p_x$ and $p_y$ atomic orbitals.

It was first shown by Hückel (584) that when the six carbon atoms of $C_6H_6$, each with one $p_z$ atomic orbital, are brought symmetrically together four molecular orbitals of different energy arise, two of which are doubly degenerate (see Table 61, p. 578). This is shown schematically in Fig. 135. The orbitals have the species $a_{2u}$, $e_{1g}$, $e_{2u}$, $b_{2g}$ with energies $E_0 + 2\beta$, $E_0 + \beta$, $E_0 - \beta$, and $E_0 - 2\beta$ respectively

[5] This designation is not entirely consistent with other uses of $\pi$ but it is well established in the literature.

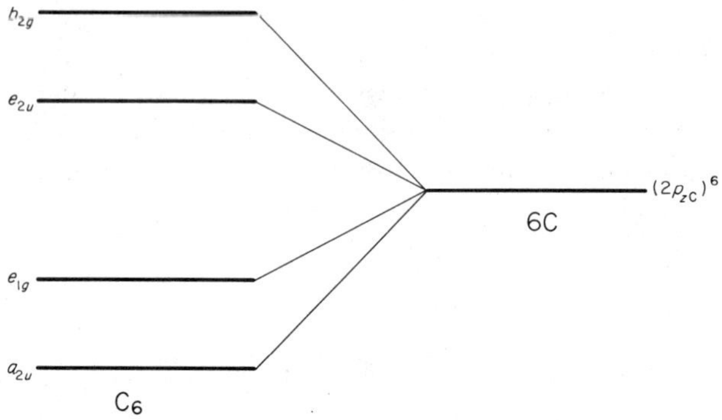


Fig. 135. **Splitting of orbitals derived from the six $2p_z$ orbitals of six carbon atoms in the formation of a $C_6$ ring of $D_{6h}$ symmetry.**

where $\beta$ is the so-called *resonance integral* giving the interaction between neighboring C atoms [see Coulson (7)]. $\beta$ is in general negative so that $a_{2u}$ with energy $E_0 + 2\beta$ is lowest. $E_0$ is the energy for zero interaction.

On the basis of the group-theoretical method previously described (p. 303f) it is easy to see how the four orbitals of Fig. 135 arise. Let $\varphi_i$ be the $2p_z$ orbital function of the $i$th carbon atom. It is antisymmetric with respect to the plane of the molecule. The transformation properties of the six functions $\varphi_i$ with respect to the symmetry elements of point group $\boldsymbol{D}_{6h}$ are easily derived. For example, a rotation by 60° about the six-fold axis changes $\varphi_1$ into $\varphi_2$, $\varphi_2$ into $\varphi_3$, etc., that is, the character of this symmetry operation is 0; reflection at the plane of the molecule ($xy$-plane) changes every $\varphi_i$ into its negative, that is, the character is $-6$; reflection at a plane of symmetry through $\varphi_1$ leaves $\varphi_1$ and $\varphi_4$ unchanged and changes $\varphi_2$ into $\varphi_6$ and $\varphi_3$ into $\varphi_5$, that is, the character is $+2$ and similarly for the other symmetry operations. In this way we obtain a (reducible) representation of the point group $\boldsymbol{D}_{6h}$ with characters

$$\begin{array}{cccccccccc} I & C_6 & C_3 & C_2'' & C_2 & C_2' & \sigma_h & \sigma_v & \sigma_d & \cdots \\ +6 & 0 & 0 & 0 & -2 & 0 & -6 & +2 & 0 & \cdots \end{array} \qquad \text{(III, 32)}$$

It is readily seen that this set of numbers is the sum of the sets of characters of the irreducible representations $A_{2u}$, $B_{2g}$, $E_{1g}$, $E_{2u}$ (see Table 51, p. 566) and this is the only way in which the representation (III, 32) can be reduced. No matter what the interactions, as long as the $\boldsymbol{D}_{6h}$ symmetry is preserved, the orbital wave function of any electron originally $2p_z$ of C must belong to one of the four species mentioned.

An alternative method of deriving the same result is to resolve the $p_z$ atomic orbital into the local symmetry of a C atom in $C_6H_6$, viz. $\boldsymbol{C}_{2v}$. This yields the species $b_1$. It is now only necessary to see which species of $\boldsymbol{D}_{6h}$ give on resolution to $\boldsymbol{C}_{2v}$ (with $z \rightarrow x$) the species $b_1$. According to Table 59, p. 576, these are precisely the species already given above.

The wave functions of the four $\pi$ orbitals can easily be expressed in terms of the functions $\varphi_i$ of the separated carbon atoms since they are entirely determined by symmetry. One finds

for the $a_{2u}$ orbital $\quad \psi_0 = \dfrac{1}{\sqrt{6}}(\varphi_1 + \varphi_2 + \varphi_3 + \varphi_4 + \varphi_5 + \varphi_6),$

$$\text{for the } e_{1g} \text{ orbital} \quad \psi_1{}^a = \frac{1}{\sqrt{12}}(\varphi_1 - \varphi_2 - 2\varphi_3 - \varphi_4 + \varphi_5 + 2\varphi_6)$$

$$\psi_1{}^b = \tfrac{1}{2}(\varphi_1 + \varphi_2 - \varphi_4 - \varphi_5), \qquad \text{(III, 33)}$$

$$\text{for the } e_{2u} \text{ orbital} \quad \psi_2{}^a = \frac{1}{\sqrt{12}}(-\varphi_1 - \varphi_2 + 2\varphi_3 - \varphi_4 - \varphi_5 + 2\varphi_6)$$

$$\psi_2{}^b = \tfrac{1}{2}(\varphi_1 - \varphi_2 + \varphi_4 - \varphi_5),$$

$$\text{for the } b_{2g} \text{ orbital} \quad \psi_3 = \frac{1}{\sqrt{6}}(-\varphi_1 + \varphi_2 - \varphi_3 + \varphi_4 - \varphi_5 + \varphi_6).$$

These orbitals are shown schematically in Fig. 136. All orbitals are antisymmetric with respect to the plane of the molecule. Only the part of the wave

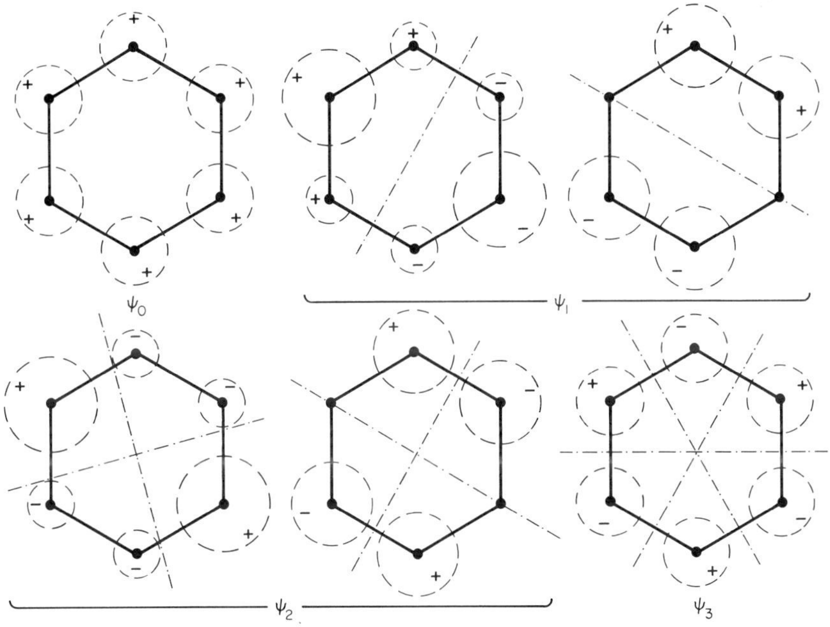


FIG. 136. **Shapes of orbital functions derived from six $2p_z$ orbitals of six carbon atoms in a hexagonal $C_6$ ring.** Only the shape and sign of the function above the $C_6$ plane are shown. Note the nodal planes through the $z$-axis indicated by dot-dash lines.

function above the plane is shown. It is seen that the orbital functions have 0, 1, 2 and 3 nodal planes through the symmetry axis. As usual in wave mechanical problems the energy increases with the number of nodal planes (compare Fig. 135 with Fig. 136).

The orbital wave functions can also be written in a more concise way as

$$\psi_l = c_1\varphi_1 + c_2\varphi_2 + \cdots + c_6\varphi_6 \qquad \text{(III, 34)}$$

with

$$c_k = \frac{1}{\sqrt{6}} e^{-(2\pi i/6)lk}.$$

Here $l = 0, \pm 1, \pm 2$ and 3 correspond to the $a_{2u}$, $e_{1g}$, $e_{2u}$ and $b_{2g}$ orbitals respectively. This form shows more clearly that all six orbitals of the separated carbon atoms enter symmetrically into all six $\pi$ orbital functions. But the forms (III, 33) are more easily visualized and the relation to the corresponding normal vibrations (see Vol. II) is made more obvious.

## (c) Molecular wave functions and the Pauli principle

Up to now in this section we have considered only the orbital wave functions of single electrons in the field of the nuclei and the average field of the other electrons. We must now consider the question as to how the *electronic wave function of the whole molecule* is related to those of the individual electrons. Or, in other words, knowing the possible orbitals of the individual electrons we can now build up the molecule by adding the electrons one by one to the molecule. We shall obtain the electronic ground state of the molecule if we add the electrons in the lowest possible orbitals. Here, just as for atoms and diatomic molecules, we meet immediately the restriction of the *Pauli principle*: At the most, two electrons can occupy a given non-degenerate orbital; at the most, four electrons can occupy a doubly degenerate orbital; at the most, six a triply degenerate orbital, and so on. One can verify readily that this form of the Pauli principle gives the same restrictions as an application of this principle in its original form (see Vol. I) to the united atom or to the separated atoms, since according to Ehrenfest's adiabatic principle the number of states is not altered by a change of the coupling conditions. Indeed we have used implicitly this principle for establishing the correlation between molecular orbitals and orbitals of the united atom or of the separated atoms.

**Electronic states resulting from non-equivalent electrons.** If two electrons are present in two different (non-equivalent) orbitals $\chi_1$ and $\chi_2$, the resulting wave function $\psi_e$ is in a first approximation the product of the two orbital functions

$$\psi_e = \chi_1\chi_2 \tag{III, 35}$$

and the species of this function is the product ("direct product", see Chap. I) of the species of the two orbitals. Similarly, if more than two non-equivalent orbitals are filled with one electron each, the resulting function is the product of the individual orbital functions, the resulting species is the product of the species of the individual orbitals. These resulting species are the same as the vibrational species resulting from the excitation of two or more different vibrations as discussed in detail in Volume II. There is, however, one important difference: the presence of the spin $s_i$ of each electron. The spins are added vectorially, just as for atoms and diatomic molecules, that is

$$\boldsymbol{S} = \sum \boldsymbol{s}_i \tag{III, 36}$$

where all quantum-mechanically possible orientations occur. For two electrons in non-equivalent orbitals we have for the quantum number of total spin (since $s_i = \frac{1}{2}$)

$$S = 1 \quad \text{or} \quad 0;$$

for three electrons

$$S = \tfrac{3}{2}, \quad \text{or } \tfrac{1}{2}, \quad \text{or } \tfrac{1}{2},$$

for four electrons

$$S = 2, 1, 1, 1, 0, 0.$$

There are rarely more than four non-equivalent electrons.

As for atoms and diatomic molecules, if an electron is in an orbital $x$, we call it for short an $x$ electron, using small letters for the species while capital letters refer to the resultant electronic state. If we have for example an $a_1$ and an $a_2$ electron in a molecule of symmetry $\boldsymbol{C}_{2v}$, since $A_1 \times A_2 = A_2$, as far as the orbital motions of the electrons are concerned only one state, $A_2$, arises, but since the spins can be parallel or antiparallel, we have $S = 1$ or 0, that is we have two states of different multiplicity $^3A_2$ and $^1A_2$. If there are two non-equivalent $e$ electrons in a molecule of point group $\boldsymbol{C}_{3v}$, since $E \times E = A_1 + A_2 + E$ (see Table 57, p. 570) we obtain the states

$$^1A_1, \quad ^1A_2, \quad ^1E, \quad ^3A_1, \quad ^3A_2, \quad ^3E.$$

Similarly, for a configuration $a_1''e'e''$ of a molecule of point group $\boldsymbol{D}_{3h}$ we obtain according to Table 57 the electronic states

$$^2A_1', \quad ^2A_2', \quad ^2E', \quad ^2A_1', \quad ^2A_2', \quad ^2E', \quad ^4A_1', \quad ^4A_2', \quad ^4E'.$$

The doublet states occur twice since there are two different ways of obtaining $S = \frac{1}{2}$. These examples together with a few others are collected in Table 30. But no attempt has been made to make this table complete since the determination of the resulting states of non-equivalent electrons is so simple with the aid of Table 57. In particular, linear molecules are not included since the resulting states for these molecules are the same as for diatomic molecules (see Vol. I, p. 333f).

TABLE 30. STATES RESULTING FROM NON-EQUIVALENT ELECTRONS

| Point group | Electron configuration | Resulting states |
|---|---|---|
| $\boldsymbol{C}_{2v}$ | $a_1$ | $^2A_1$ |
| | $a_1a_1$ | $^1A_1$, $^3A_1$ |
| | $a_1a_2$ | $^1A_2$, $^3A_2$ |
| | $b_1b_2$ | $^1A_2$, $^3A_2$ |
| $\boldsymbol{C}_{3v}$ | $e$ | $^2E$ |
| | $a_1e$ | $^1E$, $^3E$ |
| | $ee$ | $^1A_1$, $^1A_2$, $^1E$, $^3A_1$, $^3A_2$, $^3E$ |
| $\boldsymbol{D}_{3h}$ | $a_2''e'$ | $^1E''$, $^3E''$ |
| | $e'e''$ | $^1A_1''$, $^1A_2''$, $^1E''$, $^3A_1''$, $^3A_2''$, $^3E''$ |
| | $a_1''e'e''$ | $^2A_1'(2)$, $^2A_2'(2)$, $^2E'(2)$, $^4A_1'$, $^4A_2'$, $^4E'$ |
| $\boldsymbol{D}_{6h}$ | $a_{1u}e_{1g}e_{2u}e_{2u}$ | $^1B_{1u}(2)$, $^1B_{2u}(2)$, $^1E_{1u}(6)$, $^3B_{1u}(3)$, $^3B_{2u}(3)$, $^3E_{1u}(9)$, $^5B_{1u}$, $^5B_{2u}$, $^5E_{1u}(3)$ |

**Electronic states resulting from equivalent electrons.** According to the Pauli principle only two electrons can enter a non-degenerate orbital and, if they do, a single totally symmetric singlet state arises since the two electrons must have opposite spins and since any non-degenerate species multiplied by itself gives a totally symmetric species. No further discussion of equivalent electrons in non-degenerate orbitals is needed.

If there are two electrons in a doubly degenerate orbital, then several states, singlets and triplets, result but not all the states that can arise from two electrons in two non-equivalent degenerate orbitals occur. The singlet states are given by the symmetrical product of the species with itself since the spin function for two electrons with antiparallel spin is antisymmetric and the total eigenfunction (see below) must be antisymmetric. In contrast the triplet states (with symmetric spin function) are given by the antisymmetric product of the species with itself as listed in Table 57, p. 570. Thus two equivalent $e$ electrons of a molecule of point group $C_{3v}$ give the states $^1A_1$, $^1E$ and $^3A_2$ and similarly for other point groups.

TABLE 31. STATES OF EQUIVALENT ELECTRONS FOR THE MORE IMPORTANT POINT GROUPS

| Point group | Electron configuration | Resulting molecular states | Point group | Electron configuration | Resulting molecular states |
|---|---|---|---|---|---|
| $\boldsymbol{C}_{3v}$ | $a_2$ | $^2A_2$ | $\boldsymbol{C}_{\infty v}$, $[\boldsymbol{D}_{\infty h}]$[b] | $\pi$ | $^2\Pi$ |
| | $a_2{}^2$ | $^1A_1$ | | $\pi^2$ | $^1\Sigma^+$, $^1\Delta$, $^3\Sigma^-$ |
| | $e$ | $^2E$ | | $\pi^3$ | $^2\Pi$ |
| | $e^2$ | $^1A_1$, $^1E$, $^3A_2$ | | $\delta$ | $^2\Delta$ |
| | $e^3$ | $^2E$ | | $\delta^2$ | $^1\Sigma^+$, $^1\Gamma$, $^3\Sigma^-$ |
| | $e^4$ | $^1A_1$ | | $\delta^3$ | $^2\Delta$ |
| $\boldsymbol{D}_{3h}$,$[\boldsymbol{D}_3$, $\boldsymbol{D}_{3d}]$[a] | $e'$ | $^2E'$ | $\boldsymbol{O}$, $\boldsymbol{T}_d$,$[\boldsymbol{O}_h]$[b] | $e$ | $^2E$ |
| | $e'^2$ | $^1A_1'$, $^1E'$, $^3A_2'$ | | $e^2$ | $^1A_1$, $^1E$, $^3A_2$ |
| | $e'^3$ | $^2E'$ | | $e^3$ | $^2E$ |
| | | | | $e^4$ | $^1A_1$ |
| | $e'^4$ | $^1A_1'$ | | $f_1$ | $^2F_1$ |
| | $e''$ | $^2E''$ | | $f_1{}^2$ | $^1A_1$, $^1E$, $^1F_2$, $^3F_1$ |
| | $e''^2$ | $^1A_1'$, $^1E'$, $^3A_2'$ | | $f_1{}^3$ | $^2E$, $^2F_1$, $^2F_2$, $^4A_1$ |
| | $e''^3$ | $^2E''$ | | $f_1{}^4$ | $^1A_1$, $^1E$, $^1F_2$, $^3F_1$ |
| | $e''^4$ | $^1A_1'$ | | $f_1{}^5$ | $^2F_1$ |
| | | | | $f_1{}^6$ | $^1A_1$ |
| $\boldsymbol{D}_4$, $\boldsymbol{C}_{4v}$, $\boldsymbol{D}_{2d}$, $[\boldsymbol{D}_{4h}]$[b] | $e$ | $^2E$ | | $f_2$ | $^2F_2$ |
| | $e^2$ | $^1A_1$, $^1B_1$, $^1B_2$, $^3A_2$ | | $f_2{}^2$ | $^1A_1$, $^1E$, $^1F_2$, $^3F_1$ |
| | $e^3$ | $^2E$ | | $f_2{}^3$ | $^2E$, $^2F_1$, $^2F_2$, $^4A_2$ |
| | $e^4$ | $^1A_1$ | | $f_2{}^4$ | $^1A_1$, $^1E$, $^1F_2$, $^3F_1$ |
| | | | | $f_2{}^5$ | $^2F_2$ |
| $\boldsymbol{D}_6$, $\boldsymbol{C}_{6v}$,$[\boldsymbol{D}_{6h}]$[b] | $e_1$ | $^2E_1$ | | $f_2{}^6$ | $^1A_1$ |
| | $e_1{}^2$ | $^1A_1$, $^1E_2$, $^3A_2$ | | | |
| | $e_1{}^3$ | $^2E_1$ | | | |
| | $e_1{}^4$ | $^1A_1$ | | | |
| | $e_2$ | $^2E_2$ | | | |
| | $e_2{}^2$ | $^1A_1$, $^1E_2$, $^3A_2$ | | | |
| | $e_2{}^3$ | $^2E_2$ | | | |
| | $e_2{}^4$ | $^1A_1$ | | | |

[a] For $\boldsymbol{D}_3$ omit ′ and ″; for $\boldsymbol{D}_{3d}$ omit ′ and ″ and add $g$ and $u$ according to the $(g, u)$ rule.
[b] For $\boldsymbol{D}_{4h}$, $\boldsymbol{D}_{6h}$, $\boldsymbol{D}_{\infty h}$, $\boldsymbol{O}_h$ add $g$ and $u$ according to the $(g, u)$ rule.

If there are three electrons in one and the same doubly degenerate orbital only a single $^2E$ state arises since there are only two possible distributions of the three electrons over the two degenerate wave functions and these can occur only when two spins are antiparallel, that is, when the resulting state is a doublet state. In every case, necessarily the same state arises from three electrons as from a single electron in a doubly degenerate orbital. This is because four electrons in a doubly degenerate orbital always give a single totally symmetric singlet state and the states of one and three electrons in the same orbital must add up to this totally symmetric singlet state. Quite generally a hole in an otherwise filled orbital gives the same states as if only the hole were occupied.

For cubic point groups we must in addition consider triply degenerate orbitals. If two electrons are in one and the same $f$ orbital, we must again use the symmetric product of the species $F_1$ (or $F_2$) with itself to find the resulting singlet states and the antisymmetric product for the triplet states. Thus we obtain for both $f_1^2$ and $f_2^2$ the states $^1A_1$, $^1E$, $^1F_2$, $^3F_1$. The same states also result when two electrons are missing from a complete $f$ shell, that is, from $f_1^4$ and $f_2^4$. The determination of the states resulting from three equivalent $f$ electrons is less simple. We can, however, use the relation to the united atom in order to obtain the resulting states. The configuration $p^3$ gives the atomic states $^4S_u$, $^2D_u$, $^2P_u$. For the point group $\boldsymbol{T_d}$, $p^3$ goes over into $f_2^3$ and therefore the states resulting from $f_2^3$ are those species of $\boldsymbol{T_d}$ that correspond to $^4S_u$, $^2D_u$, $^2P_u$ (see Table 58), viz. $^4A_2$, $^2E$, $^2F_1$, $^2F_2$. Similarly the configuration $f_1^3$ gives states corresponding to $^4S_g$, $^2D_g$, $^2P_g$, viz. $^4A_1$, $^2E$, $^2F_1$, $^2F_2$.

In Table 31 these examples and several others are collected. For point groups with a center of symmetry the resulting states are $g$ or $u$ depending on whether there is an even or odd number of odd electrons. This is the same rule ("$g$, $u$ rule") that applies to the vibrational levels in which several vibrations are excited (see Vol. II, p. 124).

**Configurations with equivalent and non-equivalent electrons.** If equivalent as well as non-equivalent electrons are present the resulting states are found by first forming the resulting states of each group of equivalent electrons and then forming the direct product of the species so obtained. Since closed shells always give a single totally symmetric singlet state they can be entirely neglected in the determination of the resulting states.

As an example consider the configuration $a_1^2a_2e^4e^2f_1^6f_2^2$ of a tetrahedral molecule (point group $\boldsymbol{T_d}$). We can neglect the closed shells $a_1^2e^4f_1^6$ and need only consider the states resulting from $a_2e^2f_2^2$. The configurations $a_2$, $e^2$ and $f_2^2$ give respectively $^2A_2$; $^1A_1$, $^1E$, $^3A_2$; and $^1A_1$, $^1E$, $^1F_2$, $^3F_1$. Multiplying the first two sets yields $^2A_2$, $^2E$, $^2A_1$, $^4A_1$. Each of these must be multiplied by each of the species of $f_2^2$. This gives the following states of $a_2e^2f_2^2$ and therefore also of $a_1^2a_2e^4e^2f_1^6f_2^2$: $^2A_1(2)$, $^2A_2(2)$, $^2E(4)$, $^2F_1(5)$, $^2F_2(4)$, $^4A_1$, $^4E$, $^4F_1(3)$, $^4F_2(3)$, $^6F_1$ where the numbers in brackets indicate how many times each state occurs (if it is greater than 1). In the multiplication the spin rule has been taken into account. Usually even for fairly highly excited configurations fewer states result since in general fewer electrons are outside closed shells than in the example chosen.

In molecules of type $XY_n$ in which X is a transition element with $d$ (or $f$) electrons as outermost electrons and similarly in transition metal complexes in crystals and solutions it is often the $d$ (or $f$) electrons alone that determine the observed energy levels. It is easy, on the basis of the preceding discussion, to determine all the states that result from a given number of $d$ electrons in a given symmetry of the molecule or of the crystal field surrounding a complex ion. For example, let us determine the states resulting from a $d^3$ (or $d^7$) configuration in a molecule (or complex ion) of $\boldsymbol{D}_{3d}$ symmetry. According to Table 58, p. 574, a $d$ orbital gives the orbitals $a_{1g}$, $e_g$ and $\bar{e}_g$ (where the bar over the second $e_g$ is added merely to distinguish the two non-equivalent $e_g$ orbitals). Now there are nine different ways of distributing three electrons in the three orbitals in accordance with the Pauli principle, viz. $a_{1g}^2 e_g$, $a_{1g}^2 \bar{e}_g$, $a_{1g}e_g{}^2$, $a_{1g}\bar{e}_g{}^2$, $a_{1g}e_g\bar{e}_g$, $e_g{}^3$, $e_g{}^2\bar{e}_g$, $e_g\bar{e}_g{}^2$, $\bar{e}_g{}^3$ and these give, according to the rules outlined above (see, Table 31), the molecular states ${}^2A_{1g}(6)$, ${}^2A_{2g}(6)$, ${}^2E_g(14)$, ${}^4A_{1g}$, ${}^4A_{2g}(3)$, ${}^4E_g(3)$. The same states would of course have been obtained if we had first determined the atomic states resulting from a $d^3$ configuration and then resolved each of them according to the point group $\boldsymbol{D}_{3d}$ using Table 58. The resolutions of $d^n$ for various point groups have been given by Tanabe and Sugano (1185), Jørgensen (650) and Gilde and Bán (421).

**Antisymmetrized wave functions.** As has already been stated above, the molecular wave function corresponding to a certain electron configuration is in a first approximation simply the product of the orbital functions for each electron that is present. For example for a configuration $a_1{}^2 b_1{}^2 b_2$ of a $\boldsymbol{C}_{2v}$ molecule the molecular wave function is

$$\psi = \chi_{a_1}(q_1)\, \chi_{a_1}(q_2)\, \chi_{b_1}(q_3)\, \chi_{b_1}(q_4)\, \chi_{b_2}(q_5) \qquad \text{(III, 37)}$$

where $\chi_{a_1}, \chi_{b_1}, \chi_{b_2}$ are the orbital functions of the $a_1$, $b_1$ and $b_2$ electrons respectively and where $q_1, q_2, \ldots, q_5$ represent the coordinates $x_1, y_1, z_1, x_2, y_2, z_2, \ldots$ of the five electrons. If the spin is included, we must use the spin-orbital functions

$$\psi_i(j) = \chi_i(q_j)\, \varphi_i \qquad \text{(III, 38)}$$

where $j$ numbers the electrons and $i$ numbers the (spin) orbitals. Including spin the total eigenfunction of the configuration $a_1{}^2 b_1{}^2 b_2$ is

$$\psi_s = \psi_1(1)\psi_2(2)\psi_3(3)\psi_4(4)\psi_5(5). \qquad \text{(III, 39)}$$

The spin function $\varphi_2$ that goes with $\chi_{a_1}(q_2)$ must be different from $\varphi_1$ which goes with $\chi_{a_1}(q_1)$ and similarly for $\varphi_4$ and $\varphi_3$ since according to the Pauli principle two electrons can go into the same orbital only if they have opposite spin directions.

Because of the identity of the electrons the total electronic eigenfunction of any atomic system can only be symmetric or antisymmetric with respect to an exchange of any two electrons (see Vol. I, p. 24). The Pauli principle, in its wave mechanical form, postulates that only those states actually occur whose total eigenfunctions are antisymmetric with respect to exchange of any two electrons.

Now the product functions (III, 37) and (III, 39) are neither symmetric nor antisymmetric; e.g. if one exchanges electrons 1 and 2, $\psi_s$ is transformed into

$$\psi_1(2)\psi_2(1)\psi_3(3)\psi_4(4)\psi_5(5) \tag{III, 40}$$

which is neither the same as nor the negative of (III, 39). However, following Slater it is easy to form by combination of (III, 39), (III, 40) and all other functions obtained by permutations from (III, 39) a function that is antisymmetric and therefore obeys the Pauli principle, namely

$$\Phi = \frac{1}{\sqrt{5!}} \begin{vmatrix} \psi_1(1) & \psi_2(1) & \psi_3(1) & \psi_4(1) & \psi_5(1) \\ \psi_1(2) & \psi_2(2) & \psi_3(2) & \psi_4(2) & \psi_5(2) \\ \cdot & \cdot & \cdot & \cdot & \cdot \\ \psi_1(5) & \psi_2(5) & \psi_3(5) & \psi_4(5) & \psi_5(5) \end{vmatrix} \tag{III, 41}$$

and similarly in the general case with $n$ electrons.

If the determinant (III, 41) is developed it consists of $5! = 120$ terms of the form (III, 40) (or in general $n!$ terms) with positive or negative sign depending on whether the permutation is even or odd. From the rules for determinants it is immediately seen that if two electrons (i.e. two rows in the determinant) are exchanged the function $\Phi$ changes sign, i.e. is antisymmetric. $\Phi$ is called the antisymmetrized product of the spin orbitals. Symbolically one writes

$$\Phi = \mathfrak{A}\psi_1(1)\psi_2(2)\psi_3(3)\ldots\psi_n(n) \tag{III, 42}$$

where $\mathfrak{A}$ is the antisymmetrizing operator which may be expressed as

$$\mathfrak{A} = \frac{1}{\sqrt{n!}}\sum_P (-1)^P P. \tag{III, 43}$$

Here $P$ is a permutation of the electrons and the sum is to be extended over all $n!$ permutations.

The forms (III, 41) or (III, 42) of the total wave function make clear the limitations of the concept of an electron configuration. In the configuration $a_1^2\, b_1^2\, b_2$ for example there are always two electrons in the $a_1$ orbital, but by no means always the same electrons. In fact because of the indistinguishability of the electrons it has no meaning to ask whether it is the same or another electron. Precisely this lack of distinction between different electrons is taken into account in the form (III, 41).

If all electrons are in closed shells only one antisymmetrized function $\Phi$ arises corresponding to the one non-degenerate (singlet) state which is formed. If there is only one electron in one of the orbitals (as in the example $a_1^2b_1^2b_2$), there will be two functions $\Phi$ one with the spin variable in $\varphi_i$ having the value $+\frac{1}{2}$, the other with $-\frac{1}{2}$. For a single electron in a (doubly) degenerate orbital we have in addition the orbital degeneracy, i.e. two forms of $\chi_i(q_j)$. Therefore there are four functions $\Phi$. In general there will be as many functions $\Phi$ as indicated by the product of spin and orbital degeneracy. Any linear combination of these functions will be a proper wave function.

**Hartree–Fock self-consistent field molecular orbitals.** The self-consistent field method according to Hartree (see p. 303) uses the simple product representation (III, 37) in order to determine the best molecular orbitals. Actually, as first pointed out by Fock for atoms, we must use in this calculation the antisymmetrized product (III, 42). Both the exact form of the orbital wave functions and the energies of the resulting states are obtained by the solution of a variation problem: we must minimize the expression

$$E = \int \Phi^* H \Phi d\tau. \tag{III, 44}$$

Here $H$ is the Hamiltonian

$$H = \sum_j H_j + \frac{1}{2} e^2 \sum_{j \neq l} \frac{1}{r_{jl}} \tag{III, 45}$$

in which $H_j$ is the Hamiltonian of electron $j$ in the field of the bare nuclei and $r_{jl}$ is the distance between electrons $j$ and $l$; in other words the first term represents the interaction of the electrons with the nuclei, the second term their interaction with one another.

If (III, 42) and (III, 45) are introduced into (III, 44) one finds [see Mulliken (905)] for the energy of an $n$ electron system [assuming that all $\psi_i(j)$ are orthonormal]

$$E = \sum_{i=1}^{n} d_i^0 + \frac{1}{2} \sum_{i=1}^{n} \sum_{\substack{k=1 \\ i \neq k}}^{n} (J_{ik} - K_{ik}). \tag{III, 46}$$

Here $d_i^0$ depends only on spin orbital $i$. It is in fact the energy of the orbital in the field of the bare nuclei, that is

$$d_i^0 = \int \psi_i^*(j) H_j \psi_i(j) d\tau_j. \tag{III, 47}$$

The second sum in (III, 46) represents the effect of the electronic interactions; the integrals $J_{ik}$ and $K_{ik}$ depend on two orbitals $i$ and $k$, namely

$$J_{ik} = \iint \psi_i^*(j)\psi_i(j) \frac{e^2}{r_{jl}} \psi_k^*(l)\psi_k(l) d\tau_j d\tau_l \tag{III, 48}$$

$$K_{ik} = \iint \psi_k^*(j)\psi_i(j) \frac{e^2}{r_{jl}} \psi_i^*(l)\psi_k(l) d\tau_j d\tau_l. \tag{III, 49}$$

The summations in (III, 46) are over all filled orbitals. The integrals (III, 47) and (III, 48) are independent of the spin functions, that is, in them we can replace $\psi_i(j)$ by $\chi_i(q_j)$. But in (III, 49) the spin is essential: the integral vanishes if the spin function is different for $\psi_i(j)$ and $\psi_k(j)$ [and similarly for $\psi_i(l)$ and $\psi_k(l)$]; however, if the spin functions are the same, the (non-zero) value of the integral does not depend on the spin function.

The integrals $J_{ik}$ are called *Coulomb integrals* since they represent the mutual potential energy of the two charge distributions calculated in a classical way. The integrals $K_{ik}$ are called *exchange integrals* since in them the charge density $\psi_i^*(j)\psi_i(j)$ has been replaced by $\psi_k^*(j)\psi_i(j)$ in which both orbitals $i$ and $k$ enter. These contributions to the energy would not arise in the Hartree approximation. They are entirely due to the consideration of the possibility of exchange of electrons which is taken into account in the Hartree–Fock method. As already mentioned $K_{ik}$ vanishes if the two spin orbitals have different spin functions. This result may also be expressed by saying that only electrons with parallel spin can exchange.

We can rewrite the expression (III, 46) for the energy also in the following form

$$E = \sum_{i=1}^{n} \epsilon_i - \tfrac{1}{2} \sum_{i=1}^{n} \sum_{\substack{k=1 \\ k \neq i}}^{n} (J_{ik} - K_{ik}) \qquad \text{(III, 50)}$$

where

$$\epsilon_i = d_i^0 + \sum_{\substack{k=1 \\ k \neq i}}^{n} (J_{ik} - K_{ik})$$

is essentially the energy of an electron in the orbital $\psi_i$ when both the effect of the bare nuclei and of the other electrons (both Coulomb and exchange contributions) are taken into account. The total energy is not simply $\sum \epsilon_i$ since in this sum the mutual interaction of the electrons is contained twice. However in a first approximation $-\epsilon_i$ represents the energy required to remove an electron from the orbital $\psi_i$ when all the other $n - 1$ orbitals are filled. In particular the ionization potential of the molecule is equal to the smallest $-\epsilon_i$ value.

The method here described for the calculation of orbital energies was first developed by Mulliken (905) and Roothaan (1082) and applied by them to a few examples. More detailed calculations on this basis have been carried through for $CO_2$ by Mulligan (887), for COS by Clementi (205), for $C_3$ by Clementi and McLean (207), for $H_2O$ by Ellison and Shull (353) and McWeeny and Ohno (815), for $NH_3$ by Duncan (321) and Kaplan (656), for $CH_4$ by Woznick (1321), for $H_2O_2$ by Amako and Giguère (68), for HCN by McLean (814) and for $C_2H_2$ by Burnelle (166). In each case the linear combinations of atomic orbitals previously described (p. 300) were used as the starting point of the calculations. The coefficients in these linear combinations were determined by the condition of minimum energy and the orbital energies were obtained from eq. (III, 50).

**Large spin-orbit coupling.** Up to now in all the considerations of this subsection we have implicitly assumed that spin-orbit coupling is small. Many cases occur in which this assumption cannot be made. We must then combine the spin function with the orbital function before we determine the species of the resulting states. For linear molecules everything is similar to diatomic molecules: we have $(\omega, \omega)$ coupling or $(\Omega_c, \omega)$ coupling, also simply referred to as Hund's case (*c*) (see Vol. I, pp. 337f). The resulting states can be described as $\frac{1}{2}, \frac{3}{2}, \ldots$ states for an odd number of electrons and $0^+, 0^-, 1, 2, 3, \ldots$ states for an even number of electrons, as is done for diatomic molecules. We may, however, also follow the notation used for other polyatomic molecules and call them $E_{\frac{1}{2}}, E_{\frac{3}{2}}, \ldots$ states and $\Sigma^+, \Sigma^-, \Pi, \Delta, \ldots$ states respectively. This is done in Table 32.

For $C_{2v}$ molecules, for example for a single $b_1$ electron outside closed shells, we must multiply by the species of the spin function which according to Table 56 (Appendix II) is $E_{\frac{1}{2}}$. The result is an $E_{\frac{1}{2}}$ state, which corresponds to $^2B_1$ for small spin-orbit coupling. The $E_{\frac{1}{2}}$ state is doubly degenerate but since this is a Kramers degeneracy (see p. 17) it cannot be lifted by any distortion of the molecule but only by magnetic interactions such as occur in the rotating molecule (see p. 123) or in a magnetic field.

A somewhat different situation arises for a $C_{3v}$ molecule if there is a single $e$ electron outside closed shells. The spin function is again of species $E_{\frac{1}{2}}$ but combining now with the orbital function of species $E$ we obtain according to Table 57

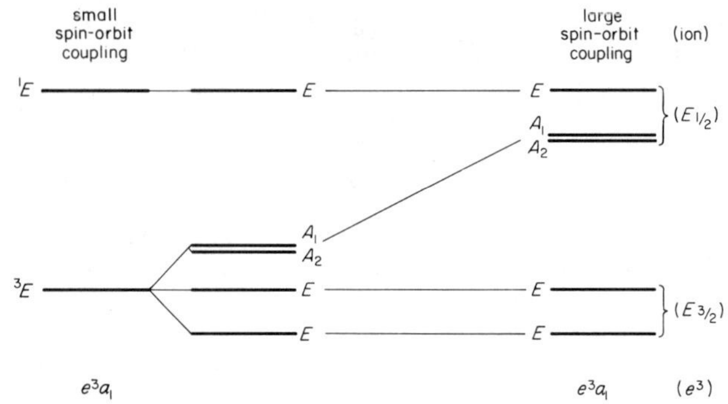


FIG. 137. **Energy levels of the configuration $e^3a_1$ (or $ea_1$) of a $C_{3v}$ molecule for small and for large spin-orbit coupling.** At the right the lower two $E$ states correspond to $E_{\frac{3}{2}}$ of the configuration $e^3$ of the ion while the upper group of states correspond to $E_{\frac{1}{2}}$ of this configuration

TABLE 32. MOLECULAR ELECTRONIC STATES RESULTING FROM VARIOUS ELECTRON CONFIGURATIONS FOR STRONG SPIN-ORBIT COUPLING

| Point group | Electron configuration | Resulting states[a] |
|---|---|---|
| $\boldsymbol{C_{2v}}$ | $a_1$ (or $a_2$ or $b_1$ or $b_2$) | $E_{\frac{1}{2}}$ |
| | $a_1a_1$ | $A_1$; $A_2$, $B_1$, $B_2$ |
| | $a_1b_1$ | $B_1$; $A_1$, $A_2$, $B_2$ |
| $\boldsymbol{C_{3v}}$ | $a_1$ (or $a_2$) | $E_{\frac{1}{2}}$ |
| | $e$ (or $e^3$) | $E_{\frac{1}{2}}$, $E_{\frac{3}{2}}$ |
| | $ea_1$ (or $e^3a_1$) | $E$; $A_1$, $A_2$, $E(2)$ |
| | $e^2$ | $A_1$, $E$; $A_1$, $E$ |
| | $e^2a_1$ (or $e^2a_2$) | $E_{\frac{1}{2}}(3)$, $E_{\frac{3}{2}}$; $E_{\frac{1}{2}}$, $E_{\frac{3}{2}}$ |
| | $e^2e$ | $E_{\frac{1}{2}}(5)$, $E_{\frac{3}{2}}(3)$; $E_{\frac{1}{2}}(3)$, $E_{\frac{3}{2}}$ |
| $\boldsymbol{D_{3h}}$,$(\boldsymbol{C_{3h}})$[b] | $a_1'$ (or $a_2'$) | $E_{\frac{1}{2}}$ |
| | $a_2''$ (or $a_1''$) | $E_{\frac{5}{2}}$ |
| | $e'$ (or $e'^3$) | $E_{\frac{3}{2}}$, $E_{\frac{5}{2}}$ |
| | $e''$ (or $e''^3$) | $E_{\frac{1}{2}}$, $E_{\frac{3}{2}}$ |
| | $e'a_2''$ (or $e'^3a_2''$) | $E''$; $A_1'$, $A_2'$, $E'$, $E''$ |
| | $e'^2$ (or $e''^2$) | $A_1'$, $E'$; $A_1'$, $E''$ |
| | $e'^2a_1'$ (or $e''^2a_1'$) | $E_{\frac{1}{2}}(2)$, $E_{\frac{3}{2}}$, $E_{\frac{5}{2}}$; $E_{\frac{1}{2}}$, $E_{\frac{3}{2}}$ |
| $\boldsymbol{D_6}$, $\boldsymbol{C_{6v}}$,$(\boldsymbol{D_{6h}})$[c] | $a_1$ (or $a_2$) | $E_{\frac{1}{2}}$ |
| | $b_1$ (or $b_2$) | $E_{\frac{5}{2}}$ |
| | $e_1$ (or $e_1{}^3$) | $E_{\frac{1}{2}}$, $E_{\frac{3}{2}}$ |
| | $e_2$ (or $e_2{}^3$) | $E_{\frac{3}{2}}$, $E_{\frac{5}{2}}$ |
| | $a_1e_1$ (or $a_2e_1$, $b_1e_2$, $b_2e_2$, $a_1e_1{}^3, \ldots$) | $E_1$; $A_1$, $A_2$, $E_1$, $E_2$ |
| | $a_1e_2$ (or $a_2e_2$, $b_1e_1$, $b_2e_1$, $a_1e_2{}^3, \ldots$) | $E_2$; $B_1$, $B_2$, $E_1$, $E_2$ |
| | $e_1{}^2$ (or $e_2{}^2$) | $A_1$, $E_2$; $A_1$, $E_1$ |
| | $e_1{}^2a_1$ (or $e_2{}^2a_1$ or $e_1{}^2a_2$ or $e_2{}^2a_2$) | $E_{\frac{1}{2}}(2)$, $E_{\frac{3}{2}}$, $E_{\frac{5}{2}}$; $E_{\frac{1}{2}}$, $E_{\frac{3}{2}}$ |
| | $e_1{}^2b_1$ (or $e_1{}^2b_2$ or $e_2{}^2b_1$ or $e_2{}^2b_2$) | $E_{\frac{1}{2}}$, $E_{\frac{3}{2}}$, $E_{\frac{5}{2}}(2)$; $E_{\frac{3}{2}}$, $E_{\frac{5}{2}}$ |

Table 32 (contd.)

| Point group | Electron configuration | Resulting states[a] |
|---|---|---|
| $\boldsymbol{C}_{\infty v}, (\boldsymbol{D}_{\infty h})$[c] | $\sigma$ | $E_{\frac{1}{2}}$[d] |
| | $\pi$ | $E_{\frac{1}{2}}, E_{\frac{3}{2}}$ |
| | $\delta$ | $E_{\frac{3}{2}}, E_{\frac{5}{2}}$ |
| | $\sigma\pi$ (or $\sigma\pi^3$) | $\Pi; \Sigma^+, \Sigma^-, \Pi, \Delta$ |
| | $\sigma\delta$ (or $\sigma\delta^3$) | $\Delta; \Pi, \Delta, \Phi$ |
| | $\pi^2$ | $\Sigma^+, \Delta; \Sigma^+, \Pi$ |
| | $\delta^2$ | $\Sigma^+, \Gamma; \Sigma^+, \Pi$ |
| | $\sigma\pi^2$ | $E_{\frac{1}{2}}(2), E_{\frac{3}{2}}, E_{\frac{5}{2}}; E_{\frac{1}{2}}, E_{\frac{3}{2}}$ |
| $\boldsymbol{O}, \boldsymbol{T}_d, (\boldsymbol{O}_h)$[c] | $a_1$ (or $a_2$) | $E_{\frac{1}{2}}$ |
| | $e$ (or $e^3$) | $G_{\frac{3}{2}}$ |
| | $f_1$ (or $f_1{}^5$) | $E_{\frac{1}{2}}, G_{\frac{3}{2}}$ |
| | $f_2$ (or $f_2{}^5$) | $E_{\frac{5}{2}}, G_{\frac{3}{2}}$ |
| | $e^2$ | $A_1, E; F_2$ |
| | $a_1e$ (or $a_2e$ or $a_1e^3$ or $a_2e^3$) | $E; F_1, F_2$ |
| | $a_1f_1$ (or $a_2f_2$ or $a_1f_1{}^5$ or $a_2f_2{}^5$) | $F_1; A_1, E, F_1, F_2$ |
| | $a_1f_2$ (or $a_2f_1$ or $a_1f_2{}^5$ or $a_2f_1{}^5$) | $F_2; A_2, E, F_1, F_2$ |
| | $f_1{}^2$ (or $f_2{}^2$ or $f_1{}^4$ or $f_2{}^4$) | $A_1, E, F_2; A_1, E, F_1, F_2$ |
| | $f_1f_2$ | $A_2, E, F_1, F_2; A_1, A_2, E(2), F_1(3), F_2(4)$ |
| | $a_1e^2$ | $E_{\frac{1}{2}}, E_{\frac{5}{2}}, G_{\frac{3}{2}}; G_{\frac{3}{2}}$ |
| | $f_1{}^3$ (or $f_2{}^3$) | $E_{\frac{1}{2}}, E_{\frac{5}{2}}, G_{\frac{3}{2}}(3); G_{\frac{3}{2}}$ |

[a] States derived from multiplets of different multiplicities are separated by a semicolon.
[b] For $\boldsymbol{C}_{3h}$ omit subscripts 1 and 2 of $a$.
[c] For $\boldsymbol{D}_{6h}$, $\boldsymbol{D}_{\infty h}$, $\boldsymbol{O}_h$ add the $g$, $u$ rule.
[d] In diatomic language the states given are $\frac{1}{2}$; $\frac{1}{2}$, $\frac{3}{2}$; $\frac{3}{2}$, $\frac{5}{2}$; 1, 0+, 0−, 1, 2, etc.

(Appendix III) the states $E_{\frac{1}{2}}$ and $E_{\frac{3}{2}}$ which may differ considerably in energy. Each of these states has a Kramers degeneracy similar to a $\frac{1}{2}$ and a $\frac{3}{2}$ state of a diatomic molecule. For small spin-orbit coupling the two states form a $^2E$ state.

A configuration that occurs frequently for $C_{3v}$ molecules is $e^3a_1$. The configuration $e^3$ gives the same states as $e$, that is $E_{\frac{1}{2}}$ and $E_{\frac{3}{2}}$, while $a_1$ gives $E_{\frac{1}{2}}$. Combining the former with the latter we obtain according to Table 57, $A_1$, $A_2$, $E$, $E$, $E$ which correspond to $^3E$ and $^1E$ for small spin-orbit coupling. Figure 137 shows qualitatively the energy levels for small and large coupling and their correlation. It is very probable that several Rydberg states of $CH_3I$ are of the type described by the right-hand side of Fig. 137 (see Chap. V, section 3b).

Other examples together with those just considered are collected in Table 32.

## (d) Term manifold of the molecule, examples

The procedure to determine the manifold of electronic states of a polyatomic molecule from the electron configuration is entirely analogous to that for diatomic molecules (Vol. I, p. 338f). To obtain the ground state we must put the electrons into the lowest possible orbitals as far as is compatible with the Pauli principle, choosing of course the most probable order of the orbitals as discussed in sub-section (b). Excited states of the molecule are then obtained by taking an electron from the highest occupied orbital (i.e. by taking the most loosely bound electron) to various unoccupied (higher) orbitals or, particularly when the highest orbital is not completely filled, by taking an electron from a lower orbital to the last partially filled or to higher orbitals. In each case the electronic states resulting from the particular electron configuration must be determined according to the methods described in the preceding sub-section.

For a molecule with an even number of electrons, if it has no partially filled degenerate orbitals, the lowest state is almost always one in which all electrons are paired in the occupied orbitals, that is, the ground state is a totally symmetric singlet state ($^1A$, $^1A_1$, $^1A_{1g}$, $^1A'$, etc.). However, it may happen that the highest occupied and the lowest unoccupied orbital lie close together. In that case the triplet state arising when an electron is taken from the last occupied to the first unoccupied orbital may be lower than the singlet state corresponding to both electrons in the highest occupied orbital and thus a triplet state may form the ground state. If the molecule has degenerate orbitals, the ground state will be triplet if the degenerate orbital is the last to be filled and contains only two electrons. This follows from Hund's rule which applies here as well as for atoms and diatomic molecules, and which indicates that of the three states arising from two electrons in a doubly degenerate orbital (see Table 31) the triplet state is always the lowest.

For a molecule with an odd number of electrons the ground state would usually be expected to be a doublet state with a species equal to that of the last (partially) filled orbital. Only for a cubic point group could a quartet state be the ground state, namely when a triply degenerate orbital is just half filled (see Table 31). For molecules of lower symmetry this could only happen if two orbitals at least one of which is degenerate had nearly the same energy and shared three electrons.

**Rydberg series of states.** More and more highly excited states are obtained by bringing an electron from one of the orbitals filled in the ground state to higher and higher orbitals. These higher orbitals are more and more like atomic orbitals and therefore give rise to Rydberg series of electronic states whose limit corresponds to the complete removal of the electron considered, i.e. to an ionization limit of the molecule.

The designation of the higher (Rydberg) orbitals usually follows the designation of the corresponding atomic orbitals, that is we have $nsa$, $npa_1$, $npb_1$, $npb_2$, $npe$, $nda_1$, $nde$, ... orbitals where $n$ is the principal quantum number and $s, p, d, \ldots$ indicate the $l$ value in the corresponding atom. For molecules built up from atoms of the first period, $n$ must at least be 3 to obtain a Rydberg orbital. Care must be taken to distinguish $n$ from the running number used in designating the lower orbitals (as for instance in $3a_1$, $1b_2$, etc.) and from the principal quantum number in the separated atoms.

The energies of Rydberg states, formed by excitation of a single electron to a Rydberg orbital can be represented in a good approximation by a Rydberg formula

$$E_n = A - \frac{R}{(n - \delta)^2} \tag{III, 51}$$

where $A$ is the ionization limit (corresponding to complete removal of the single electron), and where $\delta$ is the Rydberg correction (or quantum defect). For molecules built-up from atoms of the first period, $\delta$ is small ($\leq 0.1$) for states derived from $nd$ electrons, somewhat larger ($0.3-0.5$) for $np$ electrons and appreciably larger ($0.9-1.2$) for $ns$ electrons.

**Term manifold of non-linear $XH_2$ molecules.** In Table 33 the ground states and first excited states of a number of *non-linear* $XH_2$ molecules are given as derived in the way just described assuming the order of the orbitals given in the previous Fig. 123. In this Table $(1a_1)^2$ is the $K$ shell of the X atom. It is not repeated in the last column. Most of the predicted states in Table 33 have been observed.

TABLE 33. GROUND STATES AND FIRST EXCITED STATES OF SEVERAL NON-LINEAR $XH_2$ MOLECULES AS DERIVED FROM THE ELECTRON CONFIGURATIONS

| Molecule | Lowest electron configuration and ground state | | First excited electron configuration and resulting states | |
|---|---|---|---|---|
| $BH_2$ | $(1a_1)^2(2a_1)^2(1b_2)^2 3a_1$ | $^2A_1$ | $\{\ \ldots(2a_1)^2(1b_2)^2(1b_1)$ | $^2B_1$ |
| | | | $\ \ \ \ldots(2a_1)^2(1b_2)(3a_1)^2$ | $^2B_2$ |
| $CH_2$ | $(1a_1)^2(2a_1)^2(1b_2)^2(3a_1)^2$ | $^1A_1$ | $\ldots(2a_1)^2(1b_2)^2(3a_1)(1b_1)$ | $^3B_1$, $^1B_1$ |
| $NH_2$ | $(1a_1)^2(2a_1)^2(1b_2)^2(3a_1)^2(1b_1)$ | $^2B_1$ | $\ldots(2a_1)^2(1b_2)^2(3a_1)(1b_1)^2$ | $^2A_1$ |
| $H_2O$ | $(1a_1)^2(2a_1)^2(1b_2)^2(3a_1)^2(1b_1)^2$ | $^1A_1$ | $\ldots(2a_1)^2(1b_2)^2(3a_1)^2(1b_1)(4a_1)$ | $^3B_1$, $^1B_1$ |

Let us now briefly consider the more highly excited states of one of the molecules of Table 33, viz. $H_2O$ for which a fair number have actually been observed. Table 34 gives two sets of excited states, one set in which an electron from the $1b_1$ orbital (which is the last orbital filled in the ground state) is brought into the various higher (Rydberg) orbitals that arise from $ns$, $np$, $nd$, ... orbitals of the united atom and another in which the electron is taken from the $3a_1$ orbital. The first set approaches with increasing $n$ the first ionization limit of $H_2O$. Because of the splitting of all degeneracies in $H_2O$ there are in this set three Rydberg series of orbitals $np$, viz. $a_1$, $b_1$, $b_2$, five Rydberg series $nd$, viz. $a_1$, $a_1$, $a_2$, $b_1$, $b_2$ and so on, but only one $ns$, viz. $a_1$. Corresponding series arise in the second somewhat higher set. Still further sets of energy levels arise by excitation from the orbitals $1b_2$,

TABLE 34. PREDICTED RYDBERG STATES OF $H_2O$

| Electron configuration | States | Electron configuration | States |
|---|---|---|---|
| $\ldots(1b_2)^2(3a_1)^2(1b_1)(nsa_1)$ | $^3B_1$, $^1B_1$ | $\ldots(1b_2)^2(3a_1)(1b_1)^2(nsa_1)$ | $^3A_1$, $^1A_1$ |
| $\ldots(1b_2)^2(3a_1)^2(1b_1)(npa_1)$ | $^1B_1$, $^1B_1$ | $\ldots(1b_2)^2(3a_1)(1b_1)^2(npa_1)$ | $^3A_1$, $^1A_1$ |
| $\ldots(1b_2)^2(3a_1)^2(1b_1)(npb_1)$ | $^3A_1$, $^1A_1$ | $\ldots(1b_2)^2(3a_1)(1b_1)^2(npb_1)$ | $^3B_1$, $^1B_1$ |
| $\ldots(1b_2)^2(3a_1)^2(1b_1)(npb_2)$ | $^3A_2$, $^1A_2$ | $\ldots(1b_2)^2(3a_1)(1b_1)^2(npb_2)$ | $^3B_2$, $^1B_2$ |
| $\ldots(1b_2)^2(3a_1)^2(1b_1)(nda_1)$ | $^3B_1$, $^1B_1$ | $\ldots(1b_2)^2(3a_1)(1b_1)^2(nda_1)$ | $^3A_1$, $^1A_1$ |
| $\ldots(1b_2)^2(3a_1)^2(1b_1)(nda_1)$ | $^3B_1$, $^1B_1$ | $\ldots(1b_2)^2(3a_1)(1b_1)^2(nda_1)$ | $^3A_1$, $^1A_1$ |
| $\ldots(1b_2)^2(3a_1)^2(1b_1)(nda_2)$ | $^3B_2$, $^1B_2$ | $\ldots(1b_2)^2(3a_1)(1b_1)^2(nda_2)$ | $^3A_2$, $^1A_2$ |
| $\ldots(1b_2)^2(3a_1)^2(1b_1)(ndb_1)$ | $^3A_1$, $^1A_1$ | $\ldots(1b_2)^2(3a_1)(1b_1)^2(ndb_1)$ | $^3B_1$, $^1B_1$ |
| $\ldots(1b_2)^2(3a_1)^2(1b_1)(ndb_2)$ | $^3A_2$, $^1A_2$ | $\ldots(1b_2)^2(3a_1)(1b_1)^2(ndb_2)$ | $^3B_2$, $^1B_2$ |
| . . . . . . . . . . | . . . | . . . . . . . . . . | . . . |

$2a_1$ and $1a_1$ but these are not important except for an interpretation of very high energy levels much above the energy required to remove the $3a_1$ electron. Finally there are states corresponding to the excitation of two (or even more) electrons, such as the state $(1a_1)^2(2a_1)^2(1b_2)^2(3a_1)^2(3sa_1)^2\ {}^1A_1$ and others. These also, in the

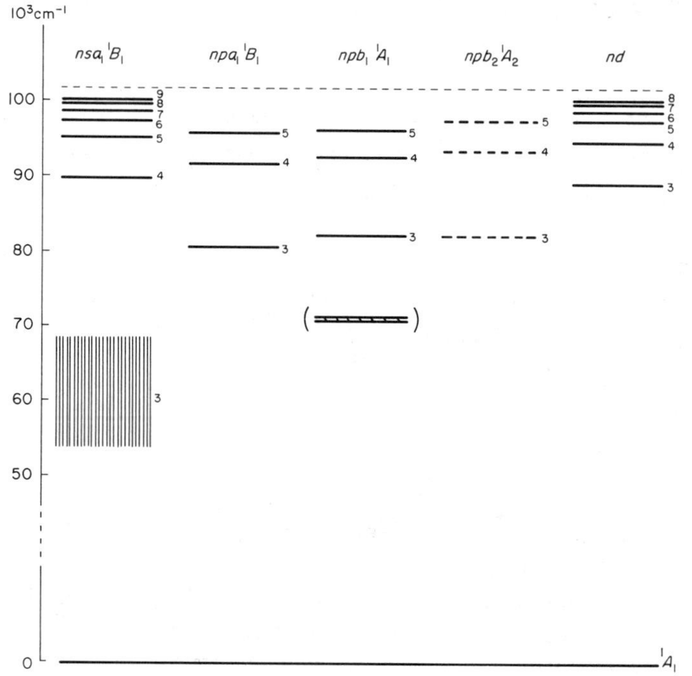


FIG. 138. **Observed electronic states of $H_2O$.** Apart from the ground state only the species of the $3pa_1\ {}^1B_1$ state has been established by a rotational fine structure analysis [Johns (631)]; the others have been assigned on the basis of the observed Rydberg corrections. The state $3sa_1\ {}^1B_1$ is continuous ($\tilde{A}$ of Table 62); the lowest excited $^1A_1$ state is diffuse and probably belongs to a series different from $npb_1\ {}^1A_1$ (see Table 62).

present case, lie very high and need not be considered further. Predictions similar to those for $H_2O$ in Table 34 can readily be made for the other molecules of Table 33.

The lowest Rydberg orbital of $H_2O$ is $3sa_1$ and this is identical with $4a_1$ of Table 33. In Fig. 138 the observed electronic states of $H_2O$ are plotted in an energy level diagram. It is seen that, as far as the evidence goes, there is good agreement with the predicted states of Table 34. The assignments have been made on the basis of the rule mentioned earlier for the magnitude of the Rydberg correction, and taking account of the observed species derived from the band structure. No $A_2$ states have been observed since they cannot combine with the ground state (see p. 132) and only absorption spectra from the ground state have as yet been studied (see Chap. V, section 1a).

The question of the relative order and the energy differences of the states arising from an *np* or *nd* electron has been discussed theoretically by Coulson and Stamper (242) and La Paglia (723)(724).

**Term manifold of linear $XH_2$ molecules.** If an $XH_2$ molecule is *linear* the expected ground and excited states are of course different from those given in Tables 33 and 34. In Table 35 the same molecules considered in Table 33 are dealt with under the assumption that they are linear. $CH_2$ has actually been observed to be linear in its ground state, and $BH_2$ as well as $NH_2$ have been observed to be linear (or nearly linear) in their first excited states. There is little question that for both these molecules there is strong Renner–Teller interaction in the $^2\Pi$ ground state of the linear conformation leading to a splitting into two states (as in Fig. 4b), one with a bent equilibrium position, the actual ground state, and one with a linear or nearly linear equilibrium position, the observed first excited state.

Whenever there are two electrons in a $\pi$ orbital several electronic states result, of which, according to the Hund rule, the state of highest multiplicity lies lowest, e.g. for $CH_2$ the $^3\Sigma_g^-$ state resulting from the lowest configuration. Table 36 gives the higher electronic states of $CH_2$ as predicted from the electron configurations. Again there are two sets of Rydberg states, one derived from $\ldots(1\sigma_u)^2(1\pi_u)$

TABLE 35. GROUND STATES AND FIRST EXCITED STATES OF LINEAR $XH_2$ MOLECULES AS DERIVED FROM THE ELECTRON CONFIGURATIONS

| Molecule | Lowest electron configuration and ground state | | First excited electron configuration and resulting states | |
|---|---|---|---|---|
| $H_2$, $CH_2^+$ | $(1\sigma_g)^2(2\sigma_g)^2(1\sigma_u)^2(1\pi_u)$ | $^2\Pi_u$ | $\ldots(2\sigma_g)^2(1\sigma_u)^2(3\sigma_g)$<br>$\ldots(2\sigma_g)^2(1\sigma_u)(1\pi_u)^2$ | $^2\Sigma_g^+$<br>$^4\Sigma_u^-, {}^2\Sigma_u^-, {}^2\Sigma_u^+, {}^2\Delta_u$ |
| $H_2$, $NH_2^+$ | $(1\sigma_g)^2(2\sigma_g)^2(1\sigma_u)^2(1\pi_u)^2$ | $^3\Sigma_g^-, {}^1\Delta_g, {}^1\Sigma_g^+$ | $\ldots(2\sigma_g)^2(1\sigma_u)^2(1\pi_u)(3\sigma_g)$<br>$\ldots(2\sigma_g)^2(1\sigma_u)(1\pi_u)^3$ | $^3\Pi_u, {}^1\Pi_u$<br>$^3\Pi_g, {}^1\Pi_g$ |
| $H_2$, $H_2O^+$ | $(1\sigma_g)^2(2\sigma_g)^2(1\sigma_u)^2(1\pi_u)^3$ | $^2\Pi_u$ | $\ldots(2\sigma_g)^2(1\sigma_u)^2(1\pi_u)^2(3\sigma_g)$<br>$\ldots(2\sigma_g)^2(1\sigma_u)(1\pi_u)^4$ | $^4\Sigma_g^-, {}^2\Sigma_g^-, {}^2\Sigma_g^+, {}^2\Delta_g$<br>$^2\Sigma_u^+$ |
| $_2O$ | $(1\sigma_g)^2(2\sigma_g)^2(1\sigma_u)^2(1\pi_u)^4$ | $^1\Sigma_g^+$ | $\ldots(2\sigma_g)^2(1\sigma_u)^2(1\pi_u)^3(3\sigma_g)$ | $^3\Pi_u, {}^1\Pi_u$ |

TABLE 36. PREDICTED RYDBERG STATES OF LINEAR $CH_2$

| Electron configuration | States | Electron configuration | States |
|---|---|---|---|
| $\ldots(2\sigma_g)^2(1\sigma_u)^2(1\pi_u)(ns\sigma_g)$ | $^3\Pi_u,\ ^1\Pi_u$ | $\ldots(2\sigma_g)^2(1\sigma_u)(1\pi_u)^2(ns\sigma_g)$ | $\left\{\begin{array}{l} ^5\Sigma_u^-,\ ^3\Sigma_u^+,\ ^3\Sigma_u^-(2) \\ ^3\Delta_u,\ ^1\Sigma_u^+,\ ^1\Sigma_u^-,\ ^1\Delta_u \end{array}\right.$ |
| $\ldots(2\sigma_g)^2(1\sigma_u)^2(1\pi_u)(np\sigma_u)$ | $^3\Pi_g,\ ^1\Pi_g$ | $\ldots(2\sigma_g)^2(1\sigma_u)(1\pi_u)^2(np\sigma_u)$ | $\left\{\begin{array}{l} ^5\Sigma_g^-,\ ^3\Sigma_g^+,\ ^3\Sigma_g^-(2) \\ ^3\Delta_g,\ ^1\Sigma_g^+,\ ^1\Sigma_g^-,\ ^1\Delta_g \end{array}\right.$ |
| $\ldots(2\sigma_g)^2(1\sigma_u)^2(1\pi_u)(np\pi_u)$ | $\left\{\begin{array}{l} ^3\Sigma_g^+,\ ^3\Sigma_g^-,\ ^3\Delta_g \\ ^1\Sigma_g^+,\ ^1\Sigma_g^-,\ ^1\Delta_g \end{array}\right.$ | $\ldots(2\sigma_g)^2(1\sigma_u)(1\pi_u)^2(np\pi_u)$ | $\left\{\begin{array}{l} ^5\Pi_g,\ ^3\Pi_g(4),\ ^3\Phi_g \\ ^1\Pi_g(3),\ ^1\Phi_g \end{array}\right.$ |
| $\ldots(2\sigma_g)^2(1\sigma_u)^2(1\pi_u)(nd\sigma_g)$ | $^3\Pi_u,\ ^1\Pi_u$ | $\ldots(2\sigma_g)^2(1\sigma_u)(1\pi_u)^2(nd\sigma_g)$ | $\left\{\begin{array}{l} ^5\Sigma_u^-,\ ^3\Sigma_u^+,\ ^3\Sigma_u^-(2) \\ ^3\Delta_u,\ ^1\Sigma_u^+,\ ^1\Sigma_u^-,\ ^1\Delta_u \end{array}\right.$ |
| $\ldots(2\sigma_g)^2(1\sigma_u)^2(1\pi_u)(nd\pi_g)$ | $\left\{\begin{array}{l} ^3\Sigma_u^+,\ ^3\Sigma_u^-,\ ^3\Delta_u \\ ^1\Sigma_u^+,\ ^1\Sigma_u^-,\ ^1\Delta_u \end{array}\right.$ | | |
| $\ldots(2\sigma_g)^2(1\sigma_u)^2(1\pi_u)(nd\delta_g)$ | $\left\{\begin{array}{l} ^3\Pi_u,\ ^3\Phi_u \\ ^1\Pi_u,\ ^1\Phi_u \end{array}\right.$ | | |
| . . . . . . . . . . | . . . . . | . . . . . . . . . . | . . . . . |

of the ion, the other from $\ldots(1\sigma_u)(1\pi_u)^2$. The configuration $\ldots(1\sigma_u)^2(1\pi_u)(3s\sigma_g)$ in Table 36 is the same as $\ldots(1\sigma_u)^2(1\pi_u)(3\sigma_g)$ of Table 35.

**Term manifold of linear and non-linear $XY_2$ molecules.** While the electron configurations of the $XH_2$ molecules can be based essentially on those of the united atom this is no longer possible when the hydrogen atoms are replaced by heavier atoms. We must then use in the correlation diagram of Fig. 121 for linear $XY_2$ a region which is closer to the separated atom orbitals. The resulting (very approximate) order of the orbitals is shown at the right in the previous Walsh diagram, Fig. 126, while the corresponding order for non-linear $XY_2$ is shown at the left of this diagram. In Table 37 the lowest and first excited electron configurations obtained from Fig. 126 as well as the resulting states are given for a number of linear molecules containing up to 16 valence electrons, and in Table 38 similar data for a number of non-linear molecules containing 17 to 20 valence electrons. In both tables the $K$ electrons have been omitted, but have been counted in the designation of the orbitals. It should be noted that between $C_3$ and $BO_2$ a reversal of the order of the $1\pi_u$ and $3\sigma_u$ orbitals occurs. This reversal is not obvious from Fig. 121 but is nevertheless very clear from the experimental evidence since the first observed excited state of $C_3$ is $^1\Pi_u$ while in $BO_2$ and $CO_2^+$ the excited $^2\Pi_u$ state is observed to be lower than $^2\Sigma_u^+$.

For linear $CO_2$ a detailed calculation on the basis of the self-consistent field method of Mulliken and Roothaan (see p. 336) has been carried through by Mulligan (887). The energies of the molecular orbitals so obtained are compared with the observed energies in Fig. 139. The observed values are obtained from the various observed Rydberg series of $CO_2$ which give the energies required to remove an electron from the $1\pi_g$, $1\pi_u$, $3\sigma_u$ and $4\sigma_g$ orbitals (see Chap. V, section 1c). It should be noted that Fig. 139 is not an energy level diagram of $CO_2$ but a diagram of the ionization energies of the various orbitals. As Figure 139 shows the theory gives the wrong order for the two orbitals $1\pi_u$ and $3\sigma_u$.

The order of orbitals in non-linear $XY_2$ is less certain than in linear $XY_2$. Note in Table 38 the difference between $NO_2$ and $CF_2$ which is caused by the smaller

TABLE 37. GROUND STATES AND FIRST EXCITED STATES OF LINEAR $XY_2$ MOLECULES AS DERIVED FROM THE ELECTRON CONFIGURATIONS[a]

| Molecule | Lowest electron configuration and ground state | | First excited electron configuration and resulting states | |
|---|---|---|---|---|
| $C_3$ | $(3\sigma_g)^2(2\sigma_u)^2(4\sigma_g)^2(1\pi_u)^4(3\sigma_u)^2$ | $\underline{^1\Sigma_g^+}$ | $\ldots(4\sigma_g)^2(1\pi_u)^4(3\sigma_u)(1\pi_g)$ | $^3\Pi_u$, $\underline{^1\Pi_u}$ |
| CNC, CCN[b] | $(3\sigma_g)^2(2\sigma_u)^2(4\sigma_g)^2(1\pi_u)^4(3\sigma_u)^2(1\pi_g)$ | $\underline{^2\Pi_g}$ | $\ldots(4\sigma_g)^2(1\pi_u)^4(3\sigma_u)(1\pi_g)^2$ | $^4\Sigma_u^-$, $\underline{^2\Sigma_u^-}$, $\underline{^2\Sigma_u^+}$, $\underline{^2\Delta_u}$ |
| NCN, NNC[b] | $(3\sigma_g)^2(2\sigma_u)^2(4\sigma_g)^2(1\pi_u)^4(3\sigma_u)^2(1\pi_g)^2$ | $\underline{^3\Sigma_g^-}$, $\underline{^1\Delta_g}$, $^1\Sigma_g^+$ | $\ldots(4\sigma_g)^2(1\pi_u)^4(3\sigma_u)(1\pi_g)^3$ | $\underline{^3\Pi_u}$, $\underline{^1\Pi_u}$ |
| | | | $\ldots(4\sigma_g)^2(1\pi_u)^3(3\sigma_u)^2(1\pi_g)^3$ | $^3\Sigma_u^-$, $^3\Sigma_u^+$, $^3\Delta_u$, $^1\Sigma_u^-$, $^1\Sigma_u^+$, $^1\Delta_u$ |
| $BO_2$, $CO_2^+$, $N_3$, NCO[b] | $(3\sigma_g)^2(2\sigma_u)^2(4\sigma_g)^2(3\sigma_u)^2(1\pi_u)^4(1\pi_g)^3$ | $\underline{^2\Pi_g}$ | $\ldots(4\sigma_g)^2(3\sigma_u)^2(1\pi_u)^3(1\pi_g)^4$ | $\underline{^2\Pi_u}$ |
| | | | $\ldots(4\sigma_g)^2(3\sigma_u)(1\pi_u)^4(1\pi_g)^4$ | $\underline{^2\Sigma_u^+}$ |
| $CO_2$ | $(3\sigma_g)^2(2\sigma_u)^2(4\sigma_g)^2(3\sigma_u)^2(1\pi_u)^4(1\pi_g)^4$ | $^1\Sigma_g^+$ | $\ldots(4\sigma_g)^2(3\sigma_u)^2(1\pi_u)^4(1\pi_g)^3(2\pi_u)$ | $^3\Sigma_u^-$, $^3\Sigma_u^+$, $^3\Delta_u$, $^1\Sigma_u^-$, $^1\Sigma_u^+$, $\underline{^1\Delta_u}$ |
| | | | $\ldots(4\sigma_g)^2(3\sigma_u)^2(1\pi_u)^4(1\pi_g)^3(5\sigma_g)$ | $^3\Pi_g$, $^1\Pi_g$ |

[a] Observed states are underlined.
[b] For CCN, NNC, NCO omit $g$ and $u$.

TABLE 38. GROUND STATES AND FIRST EXCITED STATES OF NON-LINEAR $XY_2$ MOLECULES AS DERIVED FROM THE ELECTRON CONFIGURATIONS

| Molecule | Lowest electron configuration and ground state | | First excited electron configuration and resulting states | |
|---|---|---|---|---|
| $CO_2$ | Linear, see Table 37 | | $\ldots(3b_2)^2(1b_1)^2(5a_1)^2(1a_2)^2(4b_2)(6a_1)$ | $^3B_2$, $^1B_2$ |
| $NO_2$, ($BF_2$) | $(3a_1)^2(2b_2)^2(4a_1)^2(3b_2)^2(1b_1)^2(5a_1)^2(1a_2)^2(4b_2)^2(6a_1)$ | $^2A_1$ | $\ldots(3b_2)^2(1b_1)^2(5a_1)^2(1a_2)^2(4b_2)^2(2b_1)$ | $^2B_1$ |
| | | | $\ldots(3b_2)^2(1b_1)^2(5a_1)^2(1a_2)^2(4b_2)(6a_1)^2$ | $^2B_2$ |
| $CF_2$, $O_3$ | $\ldots(3b_2)^2(1b_1)^2(5a_1)^2(4b_2)^2(1a_2)^2(6a_1)^2$ | $^1A_1$ | $\ldots(1b_1)^2(5a_1)^2(4b_2)^2(1a_2)^2(6a_1)(2b_1)$ | $^3B_1$ $^1B_1$ |
| $NF_2$ | $\ldots(3b_2)^2(1b_1)^2(5a_1)^2(4b_2)^2(6a_1)^2(1a_2)^2(2b_1)$ | $^2B_1$ | $\ldots(5a_1)^2(4b_2)^2(6a_1)^2(1a_2)(2b_1)^2$ | $^2A_2$ |
| | | | $\ldots(5a_1)^2(4b_2)^2(6a_1)(1a_2)^2(2b_1)^2$ | $^2A_1$ |
| $OF_2$ | $\ldots(5a_1)^2(6a_1)^2(4b_2)^2(1a_2)^2(2b_1)^2$ | $^1A_1$ | $\ldots(5a_1)^2(6a_1)^2(4b_2)^2(1a_2)^2(2b_1)(7a_1)$ | $^3B_1$, $^1B_1$ |

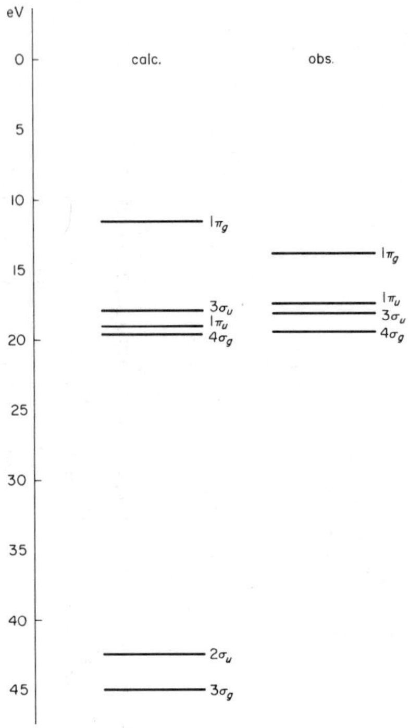


FIG. 139. **Observed and calculated orbital energies in $CO_2$.** The energies plotted are the energies required to remove an electron from the orbital considered. The calculated values are from Mulligan (887). For the observed values see p. 501f. Note that this is not an energy level diagram.

YXY angle in $CF_2$. Fischer–Hjalmars (386) and Gould and Linnett (444a) have made detailed orbital calculations for $O_3$. The order of the orbitals for $NO_2$ has been discussed recently by Green and Linnett (446). Mulliken (914) has given a general discussion of molecular orbitals in $XY_2$ molecules [see also Hijikata, Lin and Baird (548)].

**Term manifold of $XH_3$ molecules.** For a planar $XH_3$ molecule the order of the orbitals is given by Fig. 127a, for non-planar $XH_3$ by Fig. 127b. If we take

TABLE 39. GROUND STATES AND FIRST EXCITED STATES OF PLANAR $XH_3$ MOLECULES AS DERIVED FROM THE ELECTRON CONFIGURATIONS

| Molecule | Lowest electron configuration and ground state | | First excited electron configuration and resulting states | |
|---|---|---|---|---|
| $BH_3$ | $(1a_1')^2(2a_1')^2(1e')^4$ | $^1A_1'$ | $\ldots(2a_1')^2(1e')^3(1a_2'')$ | $^3E''$, $^1E''$ |
| $CH_3$ | $(1a_1')^2(2a_1')^2(1e')^4(1a_2'')$ | $^2A_2''$ | $\{\ldots(2a_1')^2(1e')^3(1a_2'')^2$ | $^2E'$ |
| | | | $\ldots(2a_1')^2(1e')^4(3a_1')$ | $^2A_1'$ |
| $NH_3$ | $(1a_1')^2(2a_1')^2(1e')^4(1a_2'')^2$ | $^1A_1'$ | $\ldots(2a_1')^2(1e')^4(1a_2'')(3a_1')$ | $^3A_2''$, $^1A_2''$ |

Table 40. Predicted Rydberg states of (planar) $CH_3$

| Electron configuration | State | Electron configuration | State |
|---|---|---|---|
| $\ldots(2a_1')^2(1e')^4(nsa_1')$ | $^2A_1'$ | $\ldots(2a_1')^2(1e')^4(nfe'')$ | $^2E''$ |
| $\ldots(2a_1')^2(1e')^4(npe')$ | $^2E'$ | $\ldots(2a_1')^2(1e')^4(nfe')$ | $^2E'$ |
| $\ldots(2a_1')^2(1e')^4(npa_2'')$ | $^2A_2''$ | $\ldots(2a_1')^2(1e')^4(nfa_2'')$ | $^2A_2''$ |
| $\ldots(2a_1')^2(1e')^4(nde'')$ | $^2E''$ | $\ldots(2a_1')^2(1e')^4(nfa_2')$ | $^2A_2'$ |
| $\ldots(2a_1')^2(1e')^4(nde')$ | $^2E'$ | $\ldots(2a_1')^2(1e')^4(nfa_1')$ | $^2A_1'$ |
| $\ldots(2a_1')^2(1e')^4(nda_1')$ | $^2A_1'$ | | |

account of the correlation rules of Table 60 (Appendix IV) we can apply any qualitative result derived for planar $XH_3$ immediately to non-planar $XH_3$, but must be prepared for considerable differences in the quantitative energy relations. Table 39 gives for the three molecules $BH_3$, $CH_3$ and $NH_3$ the ground states and first excited states derived from the electron configurations on the assumption of $\boldsymbol{D}_{3h}$ symmetry. The observed ground states and first excited states for $CH_3$ and

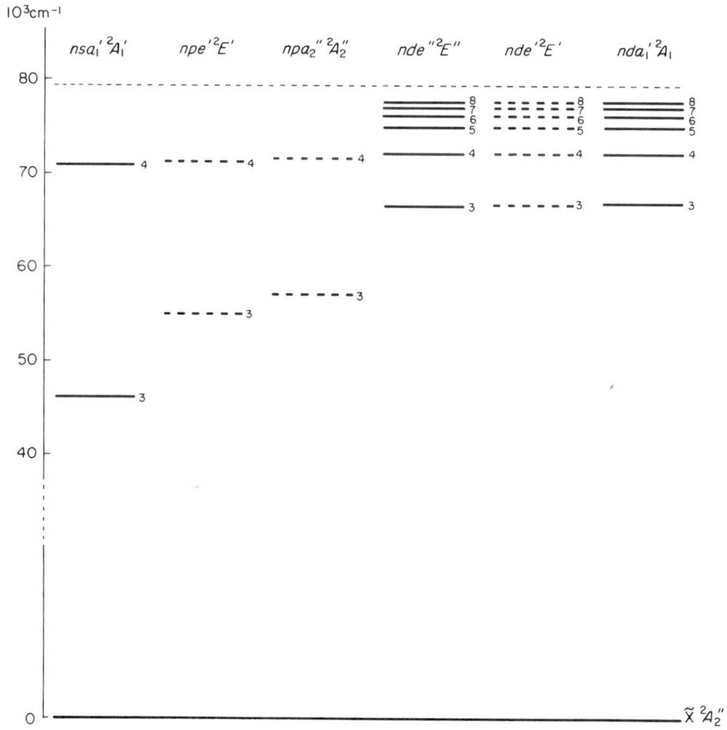


Fig. 140. **Observed electronic states of $CH_3$.** Several predicted states which have not been observed are indicated by broken lines. For the numerical data on which this diagram is based see Table 65 of Appendix VI. The position of the non-Rydberg state $(e')^3(a_2'')^2\ ^2E'$ is not known; it is presumably the lowest excited state.

$NH_3$ agree very well with the predicted ones except that for $NH_3$ since it is non-planar in the ground state $^1A_1'$ must be replaced by $^1A_1$.

In Table 40 the higher excited states of $CH_3$ predicted from the electron configurations are given, based on the ground state configuration $(1a_1')^2(2a_1')^2(1e')^4$ of the ion. In Fig. 140 the observed excited states of $CH_3$ are shown in an energy level diagram. They agree very well with the predicted states of Table 40. Here it must be remembered that only $^2A_1'$ and $^2E''$ states can be observed in absorption from the $^2A_2''$ ground state (see Chap. II, section 1a). The first $nsa_1'$ Rydberg state is identical with the $(1e')^4\ 3a_1'\ ^2A_1'$ state given in Table 39. Rydberg states based on the first excited state of the ion $\ldots(1e')^3(1a_2'')$ have not yet been observed.

**Term manifold of tetrahedral $XH_4$ molecules.** The manifold of electronic states of $CH_4$ has already been discussed on the basis of the correlation to the united atom (p. 278). The electron configuration gives the same results (see Fig. 129); we have for the ground state and the first excited states

$$(1a_1)^2(2a_1)^2(1f_2)^6\ ^1A_1 \quad \text{and} \quad (1a_1)^2(2a_1)^2(1f_2)^5(3a_1)\ ^3F_2,\ ^1F_2$$

where the $1f_2$ and $3a_1$ orbitals arise from $2p$ and $3s$ of the united atom.

It is interesting to consider the ground state and first excited states of $CH_4^+$ and $BH_4$ under the assumption that they have tetrahedral symmetry. One finds immediately $(1a_1)^2(2a_1)^2(1f_2)^5\ ^2F_2$ for the ground state and $(1a_1)^2(2a_1)(1f_2)^6\ ^2A_1$ and $(1a_1)^2(2a_1)^2(1f_2)^4\ 3a_1\ ^4F_1,\ ^2A_1,\ ^2E,\ ^2F_1,\ ^2F_2$ for the first excited states. The ground state is a triply degenerate state. According to the Jahn–Teller theorem it cannot have its energy minimum in the symmetrical conformation; but the potential function still has $\boldsymbol{T_d}$ symmetry (compare the discussion on p. 51). Unfortunately up to now no spectrum of either $CH_4^+$ or $BH_4$ has been observed.

**Term manifold of $H_2CO$.** The order of the orbitals in a planar $H_2XY$ molecule was given in Fig. 131. If we fill up the orbitals with 16 electrons, we find for the ground state of $H_2CO$

$$\ldots(3a_1)^2(4a_1)^2(1b_2)^2(5a_1)^2(1b_1)^2(2b_2)^2\ ^1A_1 \qquad \text{(III, 52)}$$

while the first excited states are

$$\ldots(3a_1)^2(4a_1)^2(1b_2)^2(5a_1)^2(1b_1)^2(2b_2)(2b_1)\ ^3A_2,\ ^1A_2. \qquad \text{(III, 52a)}$$

The first two observed excited states have indeed been found to be $^3A_2$ and $^1A_2$ (see Chap. V, section 2b). To be sure, the molecule is slightly non-planar in these states; but the deviation from planarity is small and the $\boldsymbol{C_{2v}}$ species can be applied (see p. 22f).

**Term manifold of $C_2H_4$.** Using the order of orbital energies given in Fig. 132 one finds immediately that the ground state of $C_2H_4$ (with 12 electrons outside the $K$ shells) must have the configuration

$$\ldots(2a_g)^2(2b_{1u})^2(1b_{2u})^2(1b_{3g})^2(3a_g)^2(1b_{3u})^2\ ^1A_g \qquad \text{(III, 53)}$$

and the first excited states

$$\ldots(2a_g)^2(2b_{1u})^2(1b_{2u})^2(1b_{3g})^2(3a_g)^2(1b_{3u})(1b_{2g})\ ^3B_{1u},\ ^1B_{1u}. \qquad \text{(III, 54)}$$

Here both the last filled and the first unfilled orbital in the ground state are orbitals that are antisymmetric with respect to the plane of the molecule; they are $\pi$ orbitals (see p. 327), which arise from the $1b_1$ orbital of $CH_2$. There are of course other excited states (as discussed below); it appears, however, that the first strong observed absorption beginning at 2000 Å corresponds to the transition from the ground state to the $(1b_{3u})(1b_{2g})\ ^1B_{1u}$ state. The much weaker transition to the $^3B_{1u}$ has also been found (see Chap. V, section 4a).

As can be seen from Fig. 132 the two configurations (III, 53) and (III, 54) of $C_2H_4$ correspond in the united molecule $O_2$ to $\sigma_g^2\sigma_u^2\sigma_g^2\pi_u^4\pi_g^2$ and $\sigma_g^2\sigma_u^2\sigma_g^2\pi_u^3\pi_g^3$ respectively. The strongest transition between these two configurations of $O_2$ is the well-known system of Schumann–Runge bands $^3\Sigma_u^- - {}^3\Sigma_g^-$, which is described by Mulliken as a $V \leftarrow N$ transition, i.e. a transition in which an electron goes from a bonding to a corresponding anti-bonding orbital (charge-transfer spectrum, see Vol. I, p. 384). As was first pointed out by Mulliken (895), the strong observed $^1B_{1u} - {}^1A_g$ transition in ethylene, if it is represented by the electron configurations given above, is a transition in which an electron goes from a bonding to the corresponding anti-bonding orbital, that is, is a $V \leftarrow N$ transition (see Fig. 132 and the discussion in section 3b($\beta$)). The more detailed analysis of the observed spectrum by Wilkinson and Mulliken (1303) has confirmed this interpretation.

Other excited states of $C_2H_4$ are obtained by taking an electron from the $1b_{3u}$ orbital to higher orbitals, e.g.

$$\ldots(1b_{2u})^2(1b_{3g})^2(3a_g)^2(1b_{3u})(3b_{1u})\ ^3B_{2g},\ ^1B_{2g} \qquad \text{(III, 55)}$$

$$\ldots(1b_{2u})^2(1b_{3g})^2(3a_g)^2(1b_{3u})(4a_g)\ ^3B_{3u},\ ^1B_{3u} \qquad \text{(III, 56)}$$

or by taking an electron from the next lower orbital $3a_g$ to higher orbitals, e.g.

$$\ldots(1b_{2u})^2(1b_{3g})^2(3a_g)(1b_{3u})^2(1b_{2g})\ ^3B_{2g},\ ^1B_{2g} \qquad \text{(III, 57)}$$

$$\ldots(1b_{2u})^2(1b_{3g})^2(3a_g)(1b_{3u})^2(3b_{1u})\ ^3B_{1u},\ ^1B_{1u} \qquad \text{(III, 58)}$$

and so on. Transitions to the states (III, 55) and (III, 57) from the ground state are forbidden. Transitions to (III, 58) are also of the $V \leftarrow N$ type and are probably mixed to some degree with the transition to (III, 54). The upper state of the observed strong absorption starting at 1740 Å must be identified with $^1B_{3u}$ of the configuration (III, 56). This state represents the first member of a Rydberg series.

Quantitative calculations of the excitation energies of some of the states just discussed on the basis of a consideration of the $\pi$ electrons only have been made by Parr and Pariser (966) with encouraging agreement with the experimental values.

**Term manifold of $C_6H_6$.** Just as for ethylene the $\pi$ orbitals of benzene are the most loosely bound orbitals and therefore the ground state and the lowest excited states are determined by them. Since there are six $\pi$ electrons in the lowest state, according to Fig. 135 they form the configuration $(a_{2u})^2(e_{1g})^4$ yielding

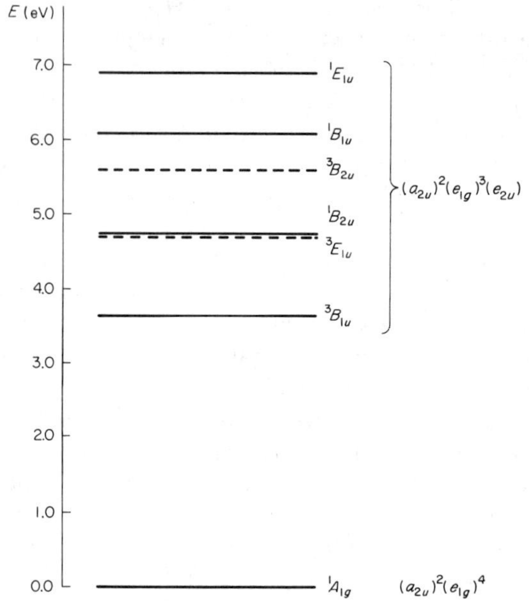


FIG. 141. **Observed low excited states of $C_6H_6$.** The predicted positions of the two states $^3E_{1u}$ and $^3B_{2u}$ which have not been observed are indicated by broken lines.

a $^1A_{1g}$ state. The first excited states are obtained by taking an electron from the $e_{1g}$ orbital and putting it in the next higher orbital $e_{2u}$. The configuration so obtained, $(a_{2u})^2(e_{1g})^3(e_{2u})$, yields six states

$$^3B_{1u},\ ^3B_{2u},\ ^3E_{1u},\ ^1B_{1u},\ ^1B_{2u},\ ^1E_{1u}.$$

Four states identified in the ultraviolet absorption spectrum of $C_6H_6$ are most probably to be ascribed to this configuration [see Mulliken (906)]. They are shown in Fig. 141. The spread in energy of these states is considerable and is due largely to inter-electronic repulsions which the simple orbital picture neglects.

Higher electronic states of $C_6H_6$ are obtained either by putting $\pi$ electrons into higher $\pi$ orbitals [e.g. yielding the configurations $(a_{2u})^2(e_{1g})^3(b_{2g})$ and $(a_{2u})(e_{1g})^4(e_{2u})$], or into Rydberg orbitals, or by exciting electrons from other orbitals which lie only slightly below the $\pi$ orbitals.

**Term manifold of the $Cr^{+++}$ ion in an octahedral complex.** The energy levels of complex ions in solutions and crystals can be determined in the same way as described in the preceding paragraphs for symmetrical molecules. For example, the orbitals of a transition metal ion in an octahedral field are given in the second column from the left of Fig. 130 if the surrounding ions or molecules causing the octahedral field have closed shells only, whose energies lie much below those of the $d$ orbitals here considered. In the $Cr^{+++}$ ion there are three $d$ electrons in addition to closed shells. Figure 130 shows that in an octahedral field these three electrons go into the $f_{2g}$ orbital derived from $3d$ and give according to Table 31 the states $^4A_{2g}$, $^2E_g$, $^2F_{1g}$, $^2F_{2g}$ of which $^4A_{2g}$ is expected to be the ground state. The first excited electron configuration is $(f_{2g})^2e_g$ which gives the states $^4F_{1g}$, $^4F_{2g}$, $^2F_{1g}(2)$, $^2F_{2g}(2)$, $^2E_g(2)$, $^2A_{1g}$, $^2A_{2g}$. As can be seen from the orbital energies given in the center of Fig. 130 the low-lying states just derived will not be greatly changed if the ligand interaction is strong (see also p. 407).

## 3. Stability of Molecular Electronic States: Valence

Up to now, in this chapter, we have limited ourselves to answering the question as to which electronic states of a polyatomic molecule arise by building up the molecule from its constituent atoms, or from its electrons, or from the united atom or molecule. Many of the numerous resulting states are unstable and only comparatively few are stable, that is, have a pronounced minimum of the potential surface. The question as to *which states are stable* must now be considered. It is intimately related to the problem of *valence* in polyatomic molecules. Closely connected with the question of stability is also the question of the *shape* of the molecule in its various electronic states.

In principle the stability of a molecular electronic state can be established theoretically by calculating the electronic energy for a number of internuclear distances and seeing whether (upon addition of the nuclear repulsion) a minimum in the potential energy results. In practice, however, such calculations are so difficult that even for the simplest polyatomic system, $H_3^+$, they have not yet been carried through with high accuracy. Thus far only the diatomic systems $H_2^+$ and $H_2$ have been treated with sufficient precision, and for them remarkable agreement with experiment has been obtained [see pp. 351 and 360 of Volume I and the more recent papers of Kolos and Roothaan (685) and Herzberg and Monfils (535)]. While a few other molecules have been treated in this way with much lower precision, for a qualitative understanding of molecule formation and the stability of molecular electronic states it is necessary to make certain greatly simplifying assumptions. As has already been discussed in Volume I two quite different methods have been developed and are widely used for an understanding of the stability of molecular electronic states: the *valence-bond method* (or electron-pair bond method) which is an extension of the original Heitler–London theory of the $H_2$ molecule, and the *molecular orbital method* which was first developed by Hund and Mulliken for diatomic molecules.

The basic assumptions of each of these two approaches represent quite drastic simplifications in the solution of the wave equation and one must therefore be prepared for disagreements between prediction and experiment. In some cases the valence-bond method leads to better agreement, in others the molecular orbital method. In either method higher approximations can be introduced which would eventually lead to the same result as a rigorous treatment. For other approaches to the problem of the nature of the chemical bond see for example

Hellman (19), Moffitt (867), Hurley (595), Preuss (1011), Hall (460), Arai (71), Ruedenberg (1088) and Bader and Jones (85).

In order to illuminate the situation it appears useful to quote a pertinent remark by Wigner (1297) with regard to the purpose of similar drastic simplifications made in the treatment of the related problem of cohesion in metals:

> "If I had a great calculating machine, I would perhaps apply it to the Schrödinger equation of each metal and obtain its cohesive energy, its lattice constant, etc. It is not clear, however, that I would gain a great deal by this. Presumably, all the results would agree with the experimental values and not much would be learned from the calculation. What would be preferable, instead, would be a vivid picture of the behavior of the wave function, a simple description of the essence of metallic cohesion and an understanding of the causes of its variation from element to element. Hence the task which is before us is not a purely scientific one; it is partly pedagogic. Nor can its solution be unique: the same wave function can be depicted in a variety of ways (just as a cubic close-packed lattice can), the same energy can be decomposed in a variety of ways into different basic constituents. Hence the value of any contribution to the problem will depend on the taste of the reader. In fact, from the point of view of the present article, the principal purpose of accurate calculations is to assure us that nothing truly significant has been overlooked."

## (a) The valence-bond (or electron-pair bond) method

**The elementary Heitler–London theory.** In the original Heitler–London theory of homopolar binding (see Vol. I, p. 350f) the attraction of two H atoms is related to the *exchange degeneracy* that exists for large internuclear distances. This degeneracy is split when the atoms are brought together resulting in two states, one of which is symmetric, the other antisymmetric with respect to an exchange of the two electrons. The symmetric state has a potential function with a minimum, i.e. corresponds to attraction; the antisymmetric state has no potential minimum, it corresponds to repulsion. According to the Pauli principle the former can only occur when the two electrons have anti-parallel spin directions ($S = 0$), the latter only for parallel spin directions ($S = 1$). It appears therefore that the formation of a chemical bond corresponds to the formation of an electron pair with anti-parallel spin directions from two unpaired electrons, one coming from one, the other from the other of the two atoms that enter the bond.

As was shown in Volume I, p. 353, the eigenfunctions of the two states arising from the two H atoms are

$$\begin{aligned}\psi_s &= N_s[\varphi_A(1)\varphi_B(2) + \varphi_A(2)\varphi_B(1)] \\ \psi_a &= N_a[\varphi_A(1)\varphi_B(2) - \varphi_A(2)\varphi_B(1)]\end{aligned} \qquad \text{(III, 59)}$$

where $A$ and $B$ refer to the two nuclei, 1 and 2 to the two electrons and $\varphi$ is a hydrogen eigenfunction centered on $A$ or $B$. The first function $\psi_s$ remains unchanged, the second $\psi_a$ changes sign for an exchange of the electrons. The normalization factors $N_s$ and $N_a$ are given by

$$N_s = \frac{1}{\sqrt{2 + 2S}}, \qquad N_a = \frac{1}{\sqrt{2 - 2S}}, \tag{III, 60}$$

where

$$S = \iint \varphi_A(1)\varphi_B(1)\varphi_A(2)\varphi_B(2)d\tau_1 d\tau_2. \tag{III, 61}$$

The energies in the two states, referred to the energy of the separated atoms as zero, are given by

$$E_s = \frac{J + K}{1 + S}, \qquad E_a = \frac{J - K}{1 - S} \tag{III, 62}$$

or, approximately, since usually $S \ll 1$

$$E_s = J + K, \qquad E_a = J - K.$$

Here $J$ is the Coulomb integral[6]

$$J = \iint \varphi_A(1)\varphi_B(2) W \varphi_A(1)\varphi_B(2)d\tau_1 d\tau_2 \tag{III, 63}$$

and $K$ is the exchange integral

$$K = \iint \varphi_A(1)\varphi_B(2) W \varphi_A(2)\varphi_B(1)d\tau_1 d\tau_2 \tag{III, 64}$$

where the perturbation function $W$ is the deviation of the potential energy of the system from that at large internuclear distance $R$. Since the exchange integral is strongly negative, the symmetric state with energy $E_s$ leads to strong attraction, the antisymmetric state with energy $E_a$ to strong repulsion. According to the Pauli principle the former must be combined with an antisymmetric spin function, i.e. $S = 0$ (antiparallel spins), the latter with a symmetric spin function, i.e. $S = 1$ (parallel spins). Thus the spin enters only in order to fulfil the Pauli principle but has nothing to do with the energy of attraction or repulsion.

Generalizing this result for atoms with several unpaired electrons Heitler and London concluded that the free valence of an atom is equal to the number of unpaired electrons, that is, equals $2S_i$ where $S_i$ is the total spin of the atom. Thus the alternation of even and odd valencies in successive columns of the periodic system is accounted for (see Table 35, p. 357, of Vol. I). In different states an atom may have different valencies corresponding to the different $S_i$ values, for example, carbon is divalent in the $^3P$ ground state but tetravalent in the excited $2s2p^3\ {}^5S$ state.

As has been stressed in Volume I the elementary Heitler–London theory is based on the assumption that the atoms are in $S$ states ($L = 0$) and that the energy

[6] In Volume I, $J$ was the exchange integral, $K$ the Coulomb integral. The change in designation corresponds to the present usage of the majority of (even though not all) authors in this field.

difference between the state considered and other states of the atom is large compared to the chemical binding energy.

For a diatomic molecule, if the two atoms forming it are in $S$ states, the resulting molecular states are $\Sigma$ states which occur with the spin values (see p. 282).

$$S = S_1+S_2, S_1+S_2-1, \ldots, |S_1-S_2|. \qquad \text{(III, 65)}$$

The energies of these states are according to Heitler (490)

$$E = J_{XY} + (p - n_X n_Y)K_{XY}. \qquad \text{(III, 66)}$$

Here the energy $E$ is referred to $E = 0$ at $R = \infty$; $J_{XY} = \sum J_{ij}$ is the Coulomb interaction energy summed over all pairs of electrons (see Vol. I, p. 353) and $K_{XY}$ is the exchange integral formed for one of the electron pairs newly formed in the bond; $p$ is the number of such pairs, $n_X$ and $n_Y$ are the number of electrons of X and Y that remain unpaired. Applying equation (III, 66) to two $N$ atoms in their $^4S$ ground states we obtain the energies of the four resulting molecular states $^1\Sigma_g^+$, $^3\Sigma_u^+$, $^5\Sigma_g^+$, $^7\Sigma_u^+$ as $J_{XY}+3K_{XY}$, $J_{XY}+1K_{XY}$, $J_{XY}-3K_{XY}$ and $J_{XY}-9K_{XY}$ respectively. Since the exchange integral $K_{XY}$ is negative (Vol. I, p. 354) and in general larger in magnitude than the Coulomb integral it follows that only the states $^1\Sigma_g^+$ and $^3\Sigma_u^+$ are stable while the other two states are unstable even though $^5\Sigma_g^+$ still has one new electron pair. Actually the observed states of $N_2$ are in agreement with this prediction: the $^1\Sigma_g^+$ is about three times as stable as the $^3\Sigma_u^+$ state, and the $^5\Sigma_g^+$ has only a very shallow minimum.

**Heitler and Rumer's extension of the Heitler–London theory.** Heitler and Rumer (491) and Heitler (489)(490) [see also Born (130)] have extended the preceding elementary considerations to polyatomic molecules. Again restricting the considerations to atoms in $S$ states we see that now, in general, several states of a given multiplicity (total spin) arise. For example for three atoms X, Y, Z with $S_1 = \frac{1}{2}$, $S_2 = \frac{3}{2}$, $S_3 = 2$ the resultant spin is $S = 0, 1, 1, 2, 2, 3, 3, 4$. The energy formulae for the molecular states are simple when only one state of a given multiplicity arises. This is the case for the state with the lowest $S$ value (in the example $S = 0$). For a molecule XYZ... with a central atom X of high multiplicity to which the atoms Y, Z, ... are bound, Heitler and Rumer found for the energy of the molecular state of lowest multiplicity

$$E = J_{XYZ\ldots} + p_1K_{XY} + p_2K_{XZ} + \cdots - p_1p_2K_{YZ} - \cdots \qquad \text{(III, 67)}$$

where as before $J_{XYZ\ldots}$ is the Coulomb energy, $K_{XY}$, $K_{XZ}$, ... are the exchange integrals between X and Y, X and Z, ... respectively and $p_1$, $p_2$, ... the number of newly formed electron pairs between X and Y, X and Z, ... (which in this case is equal to the number of unpaired electrons of the free atoms Y, Z, ...). According to (III, 67) the total interaction energy is made up additively of the interaction energies of the various pairs of atoms (note that $J_{XYZ\ldots}$ is the sum of contributions $J_{XY} + J_{XZ} + \cdots$).

Remembering again that the exchange integrals $K_{XY}$, $K_{XZ}, \ldots$ are negative we see from (III, 67) that each atom Y, Z, ... is attracted by X with an energy that is the same as that for the diatomic molecules XY, XZ, ... having the same number of bonds. On the other hand the atoms Y, Z, ... repel one another since the sign in front of $p_1 p_2 K_{YZ}, \ldots$ is negative. This repulsion is similar to the repulsion of two He atoms which also cannot form any new electron pairs.

If, for example, the X atom is the carbon atom in its $^5S$ state and Y, Z, ... are four H atoms in their $^2S$ ground states the lowest resulting molecular state is a singlet state of $CH_4$ ($^1A_1$). In this state each H atom is attracted by the C atom with an energy equal to the binding energy of the CH radical (formed from $^5S + {}^2S$). From this binding energy we have to subtract the mutual repulsion of the H atoms which as a consequence tend to take up positions as far removed from one another as possible, that is, they will assume a tetrahedral structure. The H—H distances are much greater than the C—H distances and therefore the repulsion energies are small compared to the C—H attractions, that is, a stable molecule is formed. The simplicity of this result is somewhat deceiving since only the $^5S$ state of carbon has been used and the states $^3P$, $^1D$, $^1S$ which lie below it have been neglected.

In a similar way the molecules $CH_3$ and $NH_3$ can be treated. Because of the repulsion of the H atoms one would predict on this basis a planar symmetrical structure. While this prediction is in agreement with experiment for $CH_3$ it is not for $NH_3$ which is well known to be a pyramidal molecule. This discrepancy must be ascribed to the neglect of nearby electronic states of the central atom (the $^2D$ and $^2P$ states of $N$). As we shall see the pyramidal structure of $NH_3$ follows in a natural way when all three low-lying states of the N atom, i.e. $^4S$, $^2D$, $^2P$, are treated together. Actually, observation shows that planar and non-planar $NH_3$ differ in energy by only 0.25 eV which is very small in comparison with the total binding energy.

As a further application of eq. (III, 67) consider the HCN molecule for which also a single state with $S = 0$ arises from the separate atoms H($^2S$) + C($^5S$) + N($^4S$). Equation (III, 67) gives for the energy

$$E = J_{HCN} + K_{CH} + 3K_{CN} - 3K_{NH}. \tag{III, 68}$$

The repulsion between the N and H atoms leads in this approximation to a linear configuration, in agreement with the observed structure of the ground state. Since $K_{NH}$, because of the large NH distance, is small compared to $K_{CH}$ and $K_{CN}$ the total binding energy $D$(HCN) is nearly the sum of $D$(CH) and $D$(CN). Here it must however be remembered that the dissociation energies are those involving the C atom in the $^5S$ state; the additivity is therefore difficult to establish experimentally.

If a fifth H atom is added to $CH_4$ or a fourth to $NH_3$ or a second one to HCN, in each case several states of lowest multiplicity are formed (doublet states) for which the formula (III, 67) no longer holds. It can however be shown that in all these cases the additional H atom is repelled [see Heitler (ref. (490), p. 558)], that is, a saturation of valencies exists. A central atom with spin $S$ can thus bind a maximum of $2S$ hydrogen atoms (or other univalent atoms) or a correspondingly smaller number of atoms with higher $S_i$ values.

We shall not consider further the general case of valence saturation on the basis of the Heitler–Rumer theory. It is however of special interest to consider the case of three univalent atoms X, Y, Z in $^2S$ states (i.e. $S_i = \frac{1}{2}$). According to the spin rule (III, 65) there are two states of this system with the lowest possible total spin $S = \frac{1}{2}$. On the other hand there are three "pure valence" states which may be described as follows

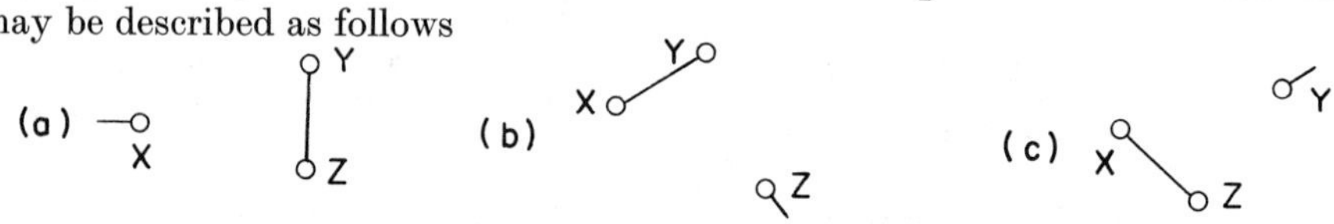


These three states are not independent of one another: each one may be expressed in terms of the two $S = \frac{1}{2}$ states. For these two states London (760) has derived the following energy formula:

$$E_{\pm} = J_{\mathrm{XYZ}} \pm \sqrt{K_{\mathrm{XY}}^2 + K_{\mathrm{YZ}}^2 + K_{\mathrm{XZ}}^2 - K_{\mathrm{XY}}K_{\mathrm{YZ}} - K_{\mathrm{XY}}K_{\mathrm{XZ}} - K_{\mathrm{YZ}}K_{\mathrm{XZ}}} \tag{III, 69}$$

where as before $J_{\mathrm{XYZ}}$ is the sum of the Coulomb integrals and $K_{\mathrm{XY}}$, $K_{\mathrm{YZ}}$, $K_{\mathrm{XZ}}$ are the exchange integrals, all of which depend on the separation of the respective atoms and vanish when this separation is very large.

Unlike the previous cases in which there was only one state of lowest multiplicity, now the energy is not simply the sum of contributions of the different pairs of atoms. If one atom, say X, is far removed from the two others the exchange integrals $K_{\mathrm{XY}}$ and $K_{\mathrm{XZ}}$ will be small compared to $K_{\mathrm{YZ}}$ and therefore (III, 69) becomes in a good approximation

$$E_{+} = J_{\mathrm{XYZ}} + K_{\mathrm{YZ}} - \tfrac{1}{2}(K_{\mathrm{XY}} + K_{\mathrm{XZ}}), \tag{III, 70}$$

$$E_{-} = J_{\mathrm{XYZ}} - K_{\mathrm{YZ}} + \tfrac{1}{2}(K_{\mathrm{XY}} + K_{\mathrm{XZ}}). \tag{III, 71}$$

Since the exchange integrals are negative, the first state ($E_+$) corresponds to a stable YZ molecule, the second state ($E_-$) to an unstable YZ molecule in a repulsive state. In the first state the X atom is repelled by YZ, in the second state it is attracted. This result corresponds well with the elementary idea of saturation of valencies: if YZ is a stable molecule a third atom X will be repelled (pure valence state (a) above). However there is repulsion only for fairly large distances between X and YZ. If X is close to Y while Z is at a larger distance we obtain from (III, 69)

$$E_{+} = J_{\mathrm{XYZ}} + K_{\mathrm{XY}} - \tfrac{1}{2}(K_{\mathrm{XZ}} + K_{\mathrm{YZ}})$$

instead of (III, 70), that is, now $E_+$ corresponds to a stable XY molecule repelled by Z (pure valence state (b)); and similarly, if X is close to Z while Y is at a larger

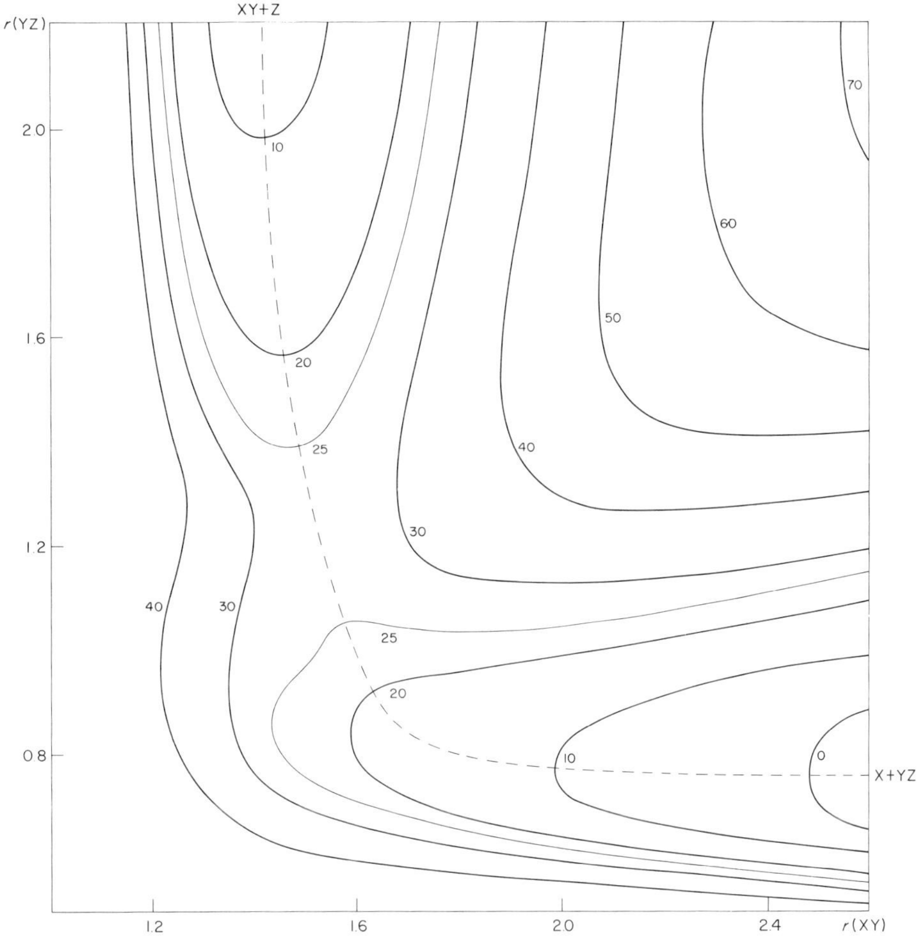


FIG. 142. **Two-dimensional cross section of the potential surface of three univalent atoms X, Y, Z assuming a linear conformation.** This figure is taken from Eyring and Polanyi (369) who calculated it for the system H + H + Br. The energy values in the diagram (in kcal/mol) refer to this system.

distance, $E^+$ corresponds to a stable XZ molecule repelled by Y (pure valence state (c)). Thus without a change of electronic state the transformations

$$X + YZ \rightarrow XY + Z \quad \text{and} \quad X + YZ \rightarrow XZ + Y$$

can take place, but there is a potential barrier between the three valence states since X is repelled by YZ (and similarly Z by XY and Y by XZ) for large separation of the two.

Figure 142 shows a two-dimensional cross section of the potential surface as a function of the XY and YZ distances assuming the system to be linear, that is, assuming $r(XZ) = r(XY) + r(YZ)$. The potential barrier between the valley corresponding to YZ and that corresponding to XY is clearly shown. If the complete potential function as a function of the three distances could be drawn it would show three valleys (in four-dimensional space) separated by potential barriers. Consideration of such potential surfaces, derived in a first approximation from eq. (III, 69), is of considerable importance for an understanding of chemical reactions (see also Chap. IV).

Potential barriers arise according to the Heitler–Rumer theory also in other cases in which there is more than one state of lowest multiplicity and in which therefore there is no additivity of the binding energies. For example, for $C_2H_2$ formed from $2H(^2S) + 2C(^5S)$ two singlet states result and the state $E_+$ in this case gives initially repulsion between the two H atoms and the central $C_2$ group. Further details may be found in Heitler's article (490).

Even though the conclusions of the Heitler–Rumer theory have been derived under the assumption that all atoms are in $S$ states, Heitler (490) has shown that the results remain valid if one atom is in a $P$ state as long as the resulting molecule is linear. If more than one atom is in a $P$ state or if, as frequently happens, there are several neighboring atomic states, the problem can be handled by a perturbation calculation, but the situation becomes then much less simple.

**Slater and Pauling's extension of the Heitler–London theory: *p* valence.** While Heitler, London, Rumer and Born considered the separation of states of the same electron configuration as large compared to the interaction energies between the atoms, Slater (1139) and Pauling (970)(971) have considered this separation as small compared to the chemical binding energies. In other words they started out from given electron configurations of the separated atoms rather than from given states of these electron configurations. They purposely neglected the finer interaction of the electrons which gives rise to the splitting of the states of one and the same electron configuration. This neglect is clearly a serious neglect since for example the three states arising from the ground configuration $p^3$ of the N atom actually have energies 0, 2.38 and 3.57 eV, while the binding energies of NH and $NH_3$ are 3.7 and 12 eV respectively. Nevertheless on the basis of this neglect we obtain in some way a much clearer and more direct picture of directional effects in valence, that is, of the observations that molecules like $H_2O$ are not linear, molecules like $NH_3$ are not planar.

As long as there is only one $s$ or one $p$ or one $d$ electron outside closed shells in the separate atoms the Slater–Pauling approach does not lead to any change compared to Heitler–London's since only one state arises ($^2S$ or $^2P$ or $^2D$). It does however lead to a change when there are two or more $p$ or $d$ electrons.

The form of atomic $p$ orbitals has been given in the previous Figs. 117 and 119a. In these figures the axes are of course entirely arbitrary. If an H atom with a $1s$ electron is brought up to an atom (X) with two $p$ electrons along the direction of one of the coordinate axes, say $x$, two states of the molecule result in the same way as in the $H_2$ molecule, one state with strong attraction, the other with strong repulsion. The magnitude of the attraction (or repulsion) is, as for

$H_2$, determined by the value of the exchange integral $K$ given by eq. (III, 64). We can write this also in the form

$$K = \iint \varphi_A(1)\varphi_B(1) W \varphi_A(2)\varphi_B(2) d\tau_1 d\tau_2 \qquad \text{(III, 72)}$$

which shows more clearly that $K$ is large if there are regions in which the products $\varphi_A(1)\varphi_B(1)$ and $\varphi_A(2)\varphi_B(2)$ are large or in other words if there are regions in which both orbitals $\varphi_A$ and $\varphi_B$ have fairly large values. Thus the *strength of the binding is determined by the magnitude of the overlap of the atomic orbitals from which the valence bond is formed.* This principle is used a great deal in valence bond theory.

It follows immediately from the *overlap principle* that if the H atom were brought up to the X atom with the two $p$ electrons along a line different from one of the axes $x$, $y$, $z$, the strength of the binding would be less than if it approaches the atom along one of the axes. While this difference has no effect if there is only one H atom, since the X atom can of course appropriately orient itself relative to the direction H—X, it does have a large effect when there are two H atoms. Clearly the two H atom orbitals will give maximum overlap with two of the $p$ orbitals if the two X—H bonds are at right angles to each other. Under this condition the maximum bond strength is obtained if for each bond an electron pair is available.

In this way the valence bond theory leads to a simple explanation of the fact that in the $H_2O$ molecule in its ground state the two O—H bonds are nearly at right angles to each other. Figure 143 shows the overlapping of the orbitals in this case. To be sure, the O atom has four $p$ electrons but two of these are already paired in the free atom, that is, only two are available and just sufficient for the two bonds.

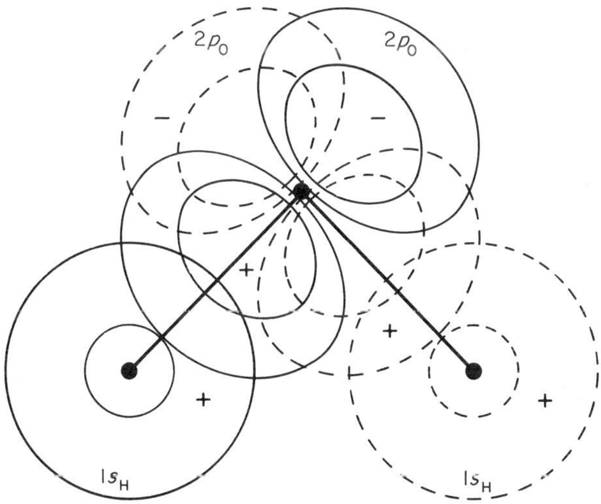


FIG. 143. **Overlapping of the 1s orbitals of two H atoms with two 2p orbitals of the O atom in $H_2O$ for a 90° valence angle.** For the sake of clarity one $p$ orbital of the O atom is shown by full-line, the other by broken-line curves. Two contour lines one corresponding to a high, the other to a low, value of $\psi$ are shown for each orbital.

The fact that the actual bond angle is slightly larger than 90° (viz. 104° 27′, see Vol. II, p. 480) may be attributed to the repulsion between the H atoms and to *s-p* hybridization in the O atom (see below). The same prediction as for $H_2O$ results of course for $CH_2$ and indeed the bond angle in the lowest singlet state has been found to be close to that of $H_2O$; but the actual ground state is a $^3\Sigma$ state of linear $CH_2$ (see p. 491). The explanation of this observation must also be sought in *s-p* hybridization (see below).

If there are *three* unpaired *p*-electrons as in the N atom, three new electron pairs can be formed with the three electrons of three H atoms and the greatest overlap of the atomic orbitals will arise if the three H nuclei are placed on the three coordinate axes $x$, $y$, $z$. In other words on this basis we expect the $NH_3$ molecule to have a pyramidal structure with an H—N—H angle of 90°. The actual structure of $NH_3$ (see Vol. II, p. 439) is not far from this prediction; again the angle is somewhat larger (106° 47′) than predicted.

For $H_2S$, $H_2Se$, ... and $PH_3$, $AsH_3$, ... to which similar considerations apply, the observed valence angles are even closer to the predicted value 90°.

Bond formation by the use of only $p$ electrons in one of the participating atoms is often referred to as *p-valence*. In the preceding examples the other atoms were H atoms with one $s$ electron each. However, the same considerations apply if the atoms to be attached are atoms like F, Cl, Br, ... with one hole in a complete $p$ shell (which is equivalent to having a single $p$ electron[7]). The unpaired $p$ electron can be used just like an $s$ electron for bond formation; in Fig. 143 we have simply to replace the spherically symmetrical $s$ orbital by a $p$ orbital with its two lobes (Fig. 119a) oriented in the direction of the bond to be formed. The exchange integrals are similar, and in consequence we have molecules like $F_2O$, $Cl_2O$, $NF_3$, $PCl_3$, ... all of which are characterized by valence angles near 90°.

It can be shown [see Van Vleck and Sherman (1240), and Coulson (7)] that the energy of a system consisting of a central atom to which a number of other atoms are attached, is in a first approximation given by

$$E = \sum J_{ij} + \sum K_{ik} - \tfrac{1}{2} \sum K_{il}. \qquad \text{(III, 73)}$$

This equation, while similar in form to eq. (III, 67) of the Heitler–Rumer theory, differs from it in that the Coulomb and exchange integrals $J_{ij}$, $K_{ik}$, $K_{il}$ are now formed separately for each pair of electrons, not for whole atoms. $J_{ij}$ are the Coulomb integrals between all pairs of electrons, $K_{ik}$ are the exchange integrals for all new electron pairs formed in the molecule and $K_{il}$ are the exchange integrals for electrons from different atoms not forming pairs.

If we apply (III, 73) to the $NH_3$ molecule we obtain, disregarding the $1s$ and $2s$ electrons of N,

$$\begin{aligned} E_{NH_3} = {} & J_{p_x s_a} + J_{p_y s_a} + J_{p_z s_a} + J_{p_x s_b} + \cdots + J_{p_x s_c} + \cdots \\ & + K_{p_x s_a} + K_{p_y s_b} + K_{p_z s_c} - \tfrac{1}{2}K_{p_x s_b} - \tfrac{1}{2}K_{p_x s_c} - \tfrac{1}{2}K_{p_y s_a} \qquad \text{(III, 74)} \\ & - \tfrac{1}{2}K_{p_y s_c} - \tfrac{1}{2}K_{p_z s_a} - \tfrac{1}{2}K_{p_z s_b} - \tfrac{1}{2}K_{s_a s_b} - \tfrac{1}{2}K_{s_a s_c} - \tfrac{1}{2}K_{s_b s_c}. \end{aligned}$$

[7] Actually atoms like B, Al, with a single $p$ electron do not readily attach themselves to a central O, N, ... atom.

Here the three H atom orbitals are designated $s_a$, $s_b$, $s_c$. The largest and most important terms in (III, 74) are $K_{p_x s_a} + K_{p_y s_b} + K_{p_z s_c}$: the exchange integrals referring to the three newly formed electron pairs. These integrals are negative and have their maximum magnitude when the three H atoms are placed in the three axes $x$, $y$, $z$, i.e. when the three bonds are at 90° to one another. The exchange integrals $K_{p_x s_b}$, $K_{p_x s_c}$, ... because of the negative sign in (III, 74) have an antibonding effect but this effect, because of the much larger separation between the corresponding orbitals, is small compared to the bonding effect of $K_{p_x s_a} + K_{p_y s_b} + K_{p_z s_c}$. It may be noted that the exchange integrals $K_{p_x s_b}$, etc., between non-bonding orbitals have a minimum value when the valence angles are 90° and therefore reinforce the directional effects.

If we consider that for equal N—H distances

$$K_{p_x s_a} = K_{p_y s_b} = K_{p_z s_c},$$
$$K_{p_x s_b} = K_{p_y s_a} = K_{p_z s_a} = \cdots$$
$$J_{p_x s_a} = J_{p_y s_b} = J_{p_z s_c}; J_{p_x s_b} = J_{p_y s_c} = \cdots.$$

then eq. (III, 74) may also be written

$$E_{\mathrm{NH_3}} = 3E_{\mathrm{NH}} + E_{\mathrm{HH}} \tag{III, 75}$$

where

$$E_{\mathrm{NH}} = J_{p_x s_a} + J_{p_y s_a} + J_{p_z s_a} + K_{p_x s_a} - \tfrac{1}{2}K_{p_y s_a} - \tfrac{1}{2}K_{p_z s_a}$$
$$E_{\mathrm{HH}} = -\tfrac{1}{2}K_{s_a s_b} - \tfrac{1}{2}K_{s_a s_c} - \tfrac{1}{2}K_{s_b s_c}.$$

In other words, except for the small repulsion term of the H atoms the total binding energy of $NH_3$ is three times that of the NH molecule. This additivity of bond energies is observed to hold approximately in many cases.

Formulae similar to (III, 74) and (III, 75) for $NH_3$ may be easily written down for $H_2O$ and other molecules. It is important to remember that the use of the basic formula (III, 73) implies the assumption that the wave function of the molecule can be approximated by a product of the atomic orbitals of the two atoms forming a bond and that the exchange integrals represent the main contribution to the bond energy. Moreover in this equation the normalizing integrals $S$ of p. 353 have been neglected.

***s*-*p* hybridization.** According to the first step in the Slater–Pauling theory thus far presented the carbon atom would have two free $p$-valencies, i.e. would be divalent. Just as in the elementary Heitler–London theory, for an understanding of the tetravalency of carbon it is necessary to make use of excited states. While Heitler and London ascribe the tetravalency to the $^5S$ state, Slater and Pauling, since they use electron configurations rather than states, ascribe it to the configuration $sp^3$. However this configuration would on the basis of the preceding considerations give three $p$ valencies at right angles to one another and a (weaker) $s$ valency without a fixed direction. The observed tetrahedral arrangement of bonds around a carbon atom led Pauling (969)(971) to inquire whether a suitable mixture of $s$ and $p$ orbitals would supply tetrahedral bonds.

As has already been shown in section 2 of this Chapter (p. 307) if the energy difference between the $2p$ and $2s$ orbitals of a C atom is neglected, the orbital functions can be mixed (hybridized) in such a way that four mutually orthogonal orbitals are obtained which point to the corners of a regular tetrahedron. They have the form (see eq. III, 18)

$$\psi_i = \tfrac{1}{2}\psi(2s) + \tfrac{1}{2}\sqrt{3}\,\psi(2p_i). \qquad \text{(III, 76)}$$

The functions $\psi(2p_i)$ and $\psi_i$ are illustrated in the previous Fig. 119 (p. 308). The four hybridized orbitals (III, 76) can now be used, in the same way as $p_x$, $p_y$, $p_z$, were used previously for the formation of electron pair bonds if four univalent atoms are brought up to the C atom. Not only does it follow immediately that the four bonds are directed to the corners of a tetrahedron but, because of the form of the hybridized orbitals with their much larger lobe on one side than on the other (Fig. 119b and c), these orbitals lead to a much larger overlapping, i.e. to a larger value of the exchange integral and therefore to much greater bond strength than do pure $p$ bonds. The gain in bond strength of hybridized tetrahedral orbitals compared to non-hybridized orbitals compensates for the need to excite the C atom to the $sp^3$ configuration. It is for this reason that compounds in which carbon is tetravalent are chemically so stable.

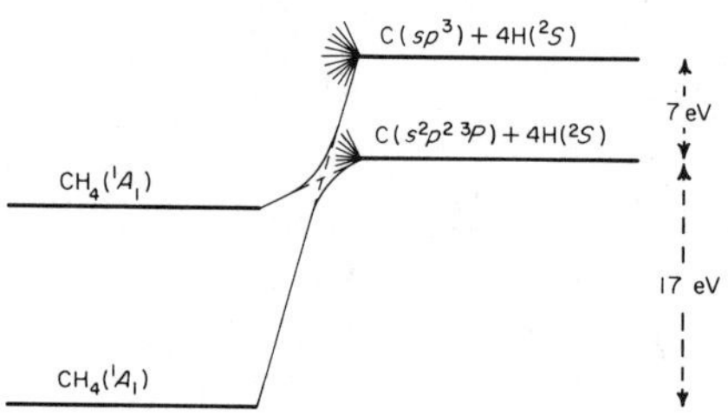


FIG. 144. **Correlation of the two lowest $^1A_1$ states of $CH_4$ with the states of the separated atoms.** Full-line correlation lines represent actual correlation; broken-line correlation lines correspond to zero-approximation valence-bond theory.

If in a $CH_4$ molecule the four H atoms are simultaneously and symmetrically removed in such a way that the four electron-pair bonds are maintained to large internuclear distances one obtains the so-called *valence state* of carbon which has the electron configuration $sp^3$. The valence state in the Slater–Pauling theory unlike the tetravalent $^5S$ state of the Heitler–London theory is not a spectroscopic state but is a mixture of the states $^5S$, $^3S$, $^3D$, $^1D$, $^3P$, $^1P$ which arise from $sp^3$ and are spread over a considerable range of energies, with an average energy of about 7 eV above the $^3P$ ground state [see Van Vleck (1236) who first developed the concept of valence state]. On the other hand, of the many molecular states that arise from the six states of C $sp^3$ by combining with four H atoms (see p. 291 and Table 26) the Slater–Pauling theory considers only the lowest one of species $^1A_1$. Such a state arises also from C $s^2p^2$ (the lowest state of which, $^3P$, is the ground state of carbon). Since states of the same species cannot intersect, if the H atoms are simultaneously and symmetrically removed from $CH_4$, actually, the ground state $^3P$ of C arises. The correlation is illustrated in Fig. 144. On the basis of this diagram one may be tempted to consider as the true strength of the CH bond in $CH_4$ one quarter of the energy to dissociate it into four H atoms and the C atom in the valence state, that is, a value that is higher by $\frac{1}{4}$ of 7 eV than the net bond strength which is $\frac{1}{4}$ of 17 eV (= 4.25 eV).

Strictly speaking, since not only the $^3P$ state but also the $^1D$ state of the $s^2p^2$ configuration as well as several of the states of the $sp^3$ and $p^4$ configurations (viz. $^5S$, $^3D$, $^1D$ and $^3P$, $^1D$ respectively), give $^1A_1$ states of $CH_4$, the ground state of the molecule will be a mixture of all these states. Conversely, the valence state of carbon is not simply an average of the states of the $sp^3$ configuration but the $^3P$ and $^1D$ states of the $s^2p^2$ and $p^4$ configurations must also be included [see Voge (1248)]. The true theoretical strength of the CH bond in $CH_4$, as opposed to the net thermochemical bond strength, is therefore a rather indefinite concept.

In view of the success of the assumption of *s-p* hybridization for the understanding of the tetravalency of carbon one is naturally inclined to ask whether its introduction in other atoms might be of advantage. If, for example, in the nitrogen atom one were to introduce complete *s-p* hybridization one would again have four tetrahedral orbitals, but since there are now five electrons, two will fill one of the four orbitals even before the three H atoms are brought up to the N atom. These two electrons form the so-called "lone" pair. Apart from it, there will be three new electron pairs formed corresponding to three NH bonds which would now have tetrahedral angles (109° 28′) between one another instead of 90° without hybridization. The observed angle in $NH_3$ (106° 47′) lies between these two extremes. Indeed the energy difference between the 2*s* and 2*p* orbitals in nitrogen is somewhat greater than in carbon while only three bonds are formed to compensate for the excitation of the N atom to the valence state. For these reasons it is reasonable to expect that hybridization is not as complete as in carbon. The same reasons apply even more strongly to oxygen atoms (and similarly S, Se, . . . atoms). Here only two of the hybridized tetrahedral orbitals are free for bond formation. Again the observed bond angle in $H_2O$ lies between 90° and 109° 28′.

Similar to $CH_4$ there are also for $H_2O$, $NH_3$, . . . several states of the central atom that give rise to $^1A_1$ states. They will mix in to some extent with the ground state of the molecule but the contribution of the $^1A_1$ states derived from the ground states of N and O is larger than for C.

The atoms of group II of the periodic system Be, Mg, . . . cannot form any electron pair bonds in their $s^2\ ^1S$ ground states but must be excited to the configuration *sp* for bond formation. As was shown in section 2, if we use only one component, say $p_z$, of the $p$ orbitals we obtain, by hybridization with $s$, two digonal hybrid orbitals pointing in opposite directions. Thus Be, Mg, . . . are expected to form stable linear molecules with two univalent atoms: $BeF_2$, $MgCl_2$, $CaCl_2$, . . . . This expectation is in general agreement with experiment, but in recent molecular beam work by Wharton, Berg and Klemperer (1292) some of the heavier dihalides ($BaF_2$, $BaCl_2$, . . .) have been found to have appreciable dipole moments that is, must be bent. If instead of digonal hybrids we had formed tetrahedral hybrids, using all three components of the $p$ orbital, only two of the four orbitals could be filled with an unpaired electron each and only two bonds with two other atoms could be formed which would subtend an angle of 109° 28′.

However, the observation of a linear conformation for the ground states of many of these molecules suggests that $sp$ rather than $sp^3$ hybridization gives the state of lowest energy. The reason for this state of affairs may be (1) that the exchange integrals involving non-bonding electrons leading to repulsion are smallest for the linear configuration and (2) that the unfilled non-hybridized $p_x$ and $p_y$ orbitals can accept more readily the lone pairs of the halogen atoms forming $\pi$ bonds (see below).

For B, Al, Ga, . . . atoms in their $s^2p\ ^2P$ ground states, there would be only a single $p$ valence, but the configuration $sp^2$ if hybridized gives three equivalent orbitals with their axes in one plane and with angles of 120° between them (see section 2, p. 309). In this way the trivalence of B, Al, Ga, . . . can be understood and the observed planar $\boldsymbol{D}_{3h}$ structure of molecules like $BF_3$, $BCl_3$, . . . is accounted for. The planar symmetrical structure is further stabilized by the fact that exchange integrals between non-bonding electrons have their minimum values and because the lone pair electrons of the halogens can make use of the $p_z$ orbital not used in the $sp^2$ hybridization.

If one had used complete hybridization of all three components of the $p$ orbitals one would again have had four equivalent tetrahedral orbitals, only three of which would have been occupied leading to three electron-pair bonds. It appears that in this case the advantage of $sp^2$ hybridization over $sp^3$ hybridization is not very strong and in certain cases the tetrahedral arrangement is preferred as for example in $H_3BCO$ where the two lone pair electrons of CO fill the fourth tetrahedral orbital. Possibly the chemical stability of $B_2H_6$, as opposed to the chemical instability of $BH_3$, may be related to the preference for a tetrahedral arrangement of bonds around each B atom, giving rise to the bridge structure of $B_2H_6$ which is now well established [see Coulson (7)].

**π bonds.** All the electron-pair bonds considered thus far were formed either from $s$ orbitals of the separated atoms (which are non-directional) or from those components of $p$ orbitals or hybridized orbitals that point in the direction of the bond. In the language of diatomic molecules these orbitals are $\sigma$ orbitals (either $s\sigma$ or $p\sigma$). Correspondingly the bonds are called *σ bonds*. Up to now in our discussion of the valence bond method we have considered only single bonds. There are in this method two ways of describing *double bonds*:

(a) If two carbon atoms with tetrahedral (hybridized) orbitals are brought together along a line bisecting the axes of two orbitals, we obtain an overlap between two pairs of orbitals as shown in Fig. 145a. This overlap is of course not as great as is obtained in two ordinary $\sigma$ bonds but nevertheless is sufficiently strong to yield a binding stronger than that of a single bond. The two electron-pair-bonds in this case could be described as "bent" bonds (of a banana shape). This representation corresponds closely to classical stereo-chemistry in which a double bond was considered as formed by two tetrahedral carbon atoms touching along a common edge of the two tetrahedrons. It is clear that on this model the four H atoms of $C_2H_4$ will all lie in the same plane (at right angles to the plane of

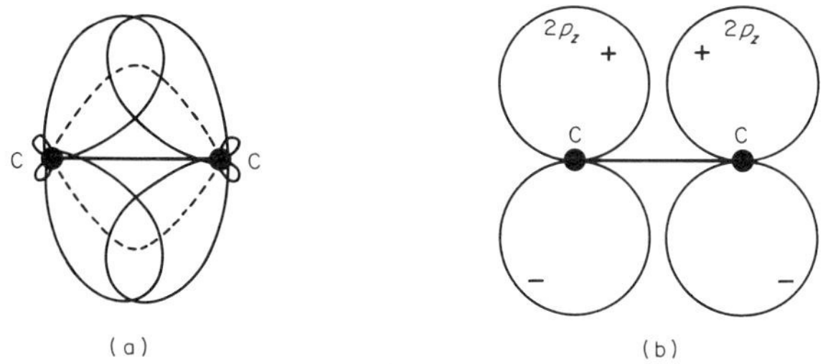


FIG. 145. **Two alternative representations of double bonds in valence-bond theory** (*a*) **by overlapping of tetrahedral orbitals;** (*b*) **by sideways overlap of $p_z$ orbitals.** Strictly speaking, orbital functions as in Fig. 119 should be used, but for simplicity, in conformity with common practice, greatly simplified orbitals are shown. In (*a*) both bonds of the double bond are equivalent, we have two "bent bonds" indicated by the broken lines. In (*b*) only the orbitals forming the $\pi$ bond are shown.

the paper in Fig. 145) and that the H—C—H angle in each $CH_2$ group would be the tetrahedral angle: 109° 28′.

(b) If instead, as was first done by Hückel (583), we start out from trigonally hybridized orbitals, that is, hybridize only $s$ with $p_x$ and $p_y$ but not with $p_z$ (if $xy$ is the plane of the $C_2H_4$ molecule) we obtain two CH bonds at each carbon atom which are at 120° to each other and to the (single) bond that connects the two carbon atoms with each other. The second bond between the two carbon atoms is then assumed to arise by the sideways overlap of the two $p_z$ orbitals in a way illustrated in Fig. 145b. These $p_z$ orbitals as already indicated (p. 327) are called $\pi$ orbitals: in $O_2$ (the united molecule) they correspond to the $\pi_u 2p$ orbitals. It is immediately clear that upon rotation of the two $CH_2$ groups with respect to each other about the C—C bond the overlap of the $\pi$ orbitals decreases and is zero when the $CH_2$ planes are at 90° to each other. In this way one can readily account for the stability of the planar form, just as for model (a). Because of its formation from two $\pi$ orbitals the second bond in $C_2H_4$ and other molecules is called a $\pi$ bond. The double bond in this representation consists of a $\sigma$ and a $\pi$ bond.

Penney (975) and Coulson and Moffitt (240) have shown that the second model (b) gives a lower energy than the first one (a) and is therefore preferable [see also Coulson (7)]. Indeed the HCH angle in $C_2H_4$ is observed to be 117.6° [Dowling and Stoicheff (305)], which is in better agreement with the second than with the first model. Similarly in $H_2CO$ this angle is found to be 118° [Lawrance and Strandberg (730).] On the other hand for substituents other than H (e.g. in $F_2C{=}CH_2$, $Cl_2C{=}CH_2$ and others) the XCX angle appears to be closer to 109° 28′. For this reason Pauling (30) still prefers the first model of the double bond. On either model there is no free rotation while for a single bond such free rotation exists, at least in the approximation of elementary valence theory.

Also for a linear molecule like $CO_2$, two models may be used for the double bonds: either (a) the orbitals of the central carbon atom are completely hybridized and two of the resulting tetrahedral orbitals overlap with the two unpaired $p$ orbitals of one O atom ($O_I$) giving rise to two bent electron-pair bonds while the

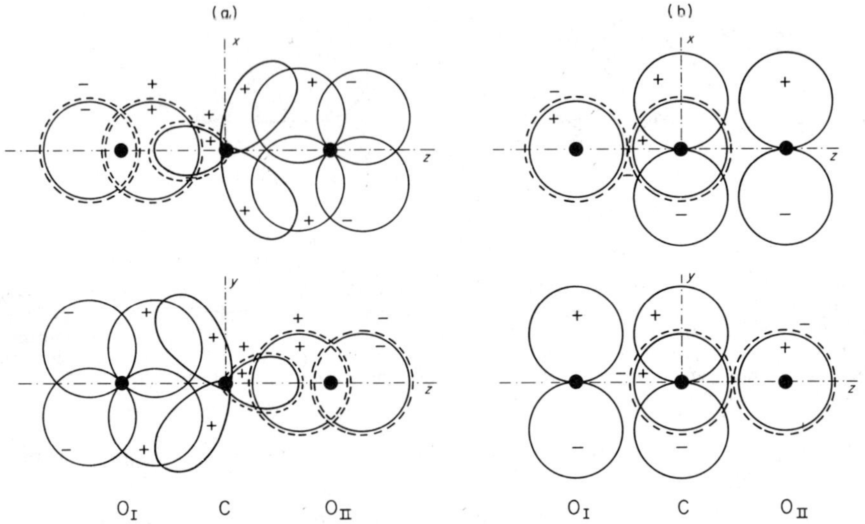


FIG. 146. **Valence-bond orbitals in $CO_2$; (*a*) using tetrahedral orbitals on C, (*b*) using $\pi$ bonds for the second bonds of each double bond.** Two projections in two perpendicular planes through the molecular axis are shown. In (*a*) the two unpaired $p$ orbitals of each O atom are oriented under 45° to the internuclear ($z$) axis; in (*b*) one of them is $p_z$ (not shown), the other is $p_y$ in $O_I$, $p_x$ in $O_{II}$.

other two tetrahedral orbitals of C overlap with the two unpaired $p$ orbitals of the other O atom ($O_{II}$) also giving rise to two bent bonds, but in a plane at right angles to the plane of the first double bond; or (b) only one of the $p$ orbitals of the C atom ($p_z$) is hybridized with the $s$ orbital forming two digonal orbitals in the $z$ direction (see p. 309) and forming two $\sigma$ bonds with the $p_z$ orbitals of the two O atoms while the $p_y$ and $p_x$ orbitals of C form $\pi$ bonds with the second unpaired orbitals ($p_y$ and $p_x$) of $O_I$ and $O_{II}$ respectively. The third $p$ orbital of each O atom of course does not take part in the binding since already in the free atom it contains two electrons. In one $\pi$ bond, say between C and $O_I$, the $xz$-plane is a nodal plane of the $\pi$ orbitals, in the other, between C and $O_{II}$, the $yz$-plane is a nodal plane. In Fig. 146 the orbitals in O + C + O which account for the binding are shown schematically according to the two models (but omitting the orbitals giving $\sigma$ bonds in the second model).

In a similar way triple bonds, as in HC≡CH, can either be considered as three bent bonds with three tetrahedral orbitals from one atom overlapping with three of another (the two tetrahedrons have one face in common) or they can be considered as the superposition of one $\sigma$ bond formed, as before, from digonal orbitals and two $\pi$ bonds formed by sideways overlap of the $p_x$ and $p_y$ orbitals of the two atoms.

**Resonance in the valence-bond method.** In the valence-bond method of Slater and Pauling just as in the Heitler–Rumer theory the concept of mixing of two or more pure valence states, here usually called resonance, is of importance. We must distinguish two types of resonance: *exact resonance* and *near-resonance*. The first is of the same type as that first described by Heisenberg in the discussion of the excited states of He [see (21), p. 66]: If $\psi_1$ is the eigenfunction when electron 1 is excited and $\psi_2$ when electron 2 is excited, the actual eigenfunctions, when the interaction of the electrons is taken into account, are found to be 50:50 mixtures of $\psi_1$ and $\psi_2$, in a first approximation simply

$$\psi_s = \psi_1 + \psi_2 \quad \text{and} \quad \psi_a = \psi_1 - \psi_2. \tag{III, 77}$$

These two eigenfunctions have opposite symmetry with regard to exchange of the two electrons and, quite generally, for exact resonance the two (or more) true eigenfunctions have different symmetries with respect to the element of symmetry that causes the resonance. The difference in energy between the two states is in a first approximation [see, e.g. Pauling and Wilson (31)]

$$E_s - E_a = 2 \int \psi_1 H_{12} \psi_2 d\tau \tag{III, 77a}$$

where for the He atom $H_{12} = e^2/r_{12}$. Exact resonance often arises when we start out from eigenfunctions that do not fulfill all the symmetry requirements of the system under consideration while suitable linear combinations of them do. Thus the localized orbitals used in valence-bond theory are in general not proper symmetry orbitals, but suitable linear combinations of them are. The energies of the latter differ from those of the former by the resonance energy. The introduction of resonance between localized orbitals leads to *de-localization.*

Resonance also arises between states of somewhat different energy. Unlike the situation in the preceding type of resonance, here the two zero-approximation eigenfunctions already have the correct symmetry and only states of the same symmetry (species) can "resonate" with each other. This resonance between different electronic states is the exact analogue of perturbations between rotational or vibrational levels treated previously (see also the discussion of perturbations in Volume I, pp. 282f, and that of Fermi resonance in Volume II, pp. 215f). Now the resulting eigenfunctions are not 50:50 mixtures but rather given by

$$\psi_a = c\psi_1 - d\psi_2, \qquad \psi_b = d\psi_1 + c\psi_2 \tag{III, 78}$$

where $c^2 + d^2 = 1$ and $c > d$.

The functions $\psi_1, \psi_2, \ldots$ are now the eigenfunctions of the molecule in different electronic states of (slightly) different energies corresponding, in the present discussion, to different valence configurations obtained from valence-bond theory. The actual states are mixtures of the two (or more) "original" states and their energies are changed by an amount which is called the resonance energy. The lowest state is always lowered by resonance. The magnitude of this lowering, i.e. of the resonance energy for two interacting states is in a first approximation

$$\Delta E = \frac{\left[\int \psi_1 W \psi_2 d\tau\right]^2}{|E_2{}^0 - E_1{}^0|} \tag{III, 79}$$

where $W$ represents the terms of the Hamiltonian neglected in the zero approximation. The upper state increases its energy by the same amount. It must be emphasized that the functions (III, 77 or III, 78) are only first approximations. The actual eigenfunctions are more complicated and in them the contributions of the two or more states can no longer be separated.

Both the concepts of exact resonance and of near resonance have already been used implicitly in the earlier discussion of linear combinations of atomic orbitals (p. 300) as well as of hybrid orbitals (p. 307). The concept of resonance just as that of perturbations is a somewhat artificial one since it is based on a certain zero approximation. If one could derive directly the energies from the wave equation by a computing machine of sufficiently high power one would not encounter the concept of resonance. In other words *resonance is not something that actually happens in the molecule* but is a mathematical construct that helps us to understand better the positions of energy levels and especially of the ground state without carrying the calculations to high orders.

After these preliminaries let us now consider a few examples of the application of resonance in valence-bond theory. One simple example already considered in Volume I is the HCl molecule. There are two low-lying states of this molecule, one derived from H + Cl, the other derived from $H^+ + Cl^-$; the first has a minimum because of homopolar attraction (formation of a $\sigma$ bond), the second because of the electrostatic attraction of the ions. A rough estimate shows that the first state forms the ground state since for separated nuclei the energy of $H^+ + Cl^-$ lies 10 eV above that of H + Cl. However, the ionic state will still "resonate" with the ground state since both states have the same symmetry ($^1\Sigma^+$). This resonance causes the energy of the ground state to be lower, that is, causes the molecule to be more stable than it otherwise would be. Thus HCl while predominantly homopolar in its ground state has to a certain small extent ionic (electrovalent) properties. This type of resonance is also called *ionic-homopolar resonance.* Actually the ionic states of HCl and HF have recently been observed as excited states of these molecules [Jacques and Barrow (615), Johns and Barrow (636)]. They have in turn some homopolar properties and lie higher than they would have without the interaction with the homopolar (ground) state.

A simple example of resonance in a polyatomic molecule is presented by $CO_2$. As we have seen the two double bonds (bent bonds or $\pi$ bonds) are in planes at right angles to each other. For a fixed configuration of the nuclei there are clearly two states, one in which the left double bond is in the $xz$-plane and one in which the right double bond is in that plane. These states are in exact resonance and therefore the ground state has equal contributions from each of these two valence states. In addition according to Pauling (30) two ionic states

$$^+O{\equiv}C{-}O^- \quad \text{and} \quad ^-O{-}C{\equiv}O^+$$

contribute to the ground state giving a considerable ionic-homopolar resonance energy and explaining the observed decrease of the C=O distance in $CO_2$ compared to that in molecules with only one C=O bond, for example $H_2CO$ (1.16 Å compared to 1.21 Å).

Similar considerations apply to $N_2O$ which is well known to be linear and unsymmetrical (see Vol. II, p. 277). The only valence-bond structure that does not require an excitation or ionization of one of the atoms is

$$\text{(a)} \qquad :\ddot{N}-\ddot{N}=\ddot{O}:$$

where the pairs of dots indicate "lone" pairs of electrons not used for bonding (including $2s$ electrons). In the structure (a) the outer N atom uses only one of its $p$ electrons for binding. The central N atom uses its three $p$ electrons only, resulting in three bonds at 90° to each other. Therefore the molecule would be expected to be strongly bent in this state. However, as first pointed out by Pauling (30), there are three valence states that probably lie much lower, viz.:

$$\text{(b)} \quad \overset{-}{.}\ddot{N}-N^{+}\equiv\underset{\cdot\cdot}{O}: \qquad \text{(c)} \quad \overset{-}{\cdot}\underset{\cdot\cdot}{N}=N^{+}=\ddot{O}: \qquad \text{(d)} \quad \dot{N}\equiv N^{+}-\ddot{O}\overset{-}{:}\ .$$

In each of these the central N atom is ionized and can therefore be in a tetravalent state (assuming $sp^3$ hybridization).

The structures (b) and (c) are similar to the two resonating structures of $CO_2$ considered above: they differ only by the orientation of the planes of the double bonds. In all three structures the larger number of bonds (four electron-pair bonds and one electrostatic bond) is assumed to cause the energy to be lower than that of structure (a) in spite of the energy needed to ionize the central N atom. A further lowering of the energy arises on account of resonance. On this basis the linearity of the molecule is readily understood: it follows from the same reasons as the linearity of molecules like $CO_2$ and FCN. While in each of the structures (b), (c), (d) the molecule would have a large dipole moment it is clear that in a state that is a mixture of the three the dipole moment would be quite small as is indeed observed for the ground state of $N_2O$.

There is a fifth structure that may be considered for $N_2O$, although it has been rejected by Pauling, viz.

$$\text{(e)} \qquad :N\equiv N=\ddot{O}:\ .$$

In this structure the central nitrogen atom is brought to a state $1s^22s2p^33s$ in which it is pentavalent. The energy required to bring the N atom to this state is clearly less than to bring it to the $2s2p^3$ state of $N^+$ which is required for (b), (c) and (d); on the other hand in the structures (b), (c) and (d) some energy is gained in the formation of the negative ions. Therefore these structures have probably about the same energy as (e). In fact it may be that (e) is only a different way of writing (b) or (c) in that the $3s$ electron in (c) forms an electron-pair bond with the third $p$ electron of the outer N atom while in (b) it is considered to have moved over completely to it forming an additional lone pair and giving rise to the negative charge. Just as (b), (c) and (d) the valence structure (e) yields a linear molecule.

The resonance concept has also been used by Pauling (30) to account for the chemical stability of molecules with an odd number of electrons, even though according to the elementary theory one would expect them to form stable dimers. The most striking case is that of NO which, on the basis of the valence structure

(a) $-\ddot{N}{=}\ddot{O}{:}$,

should form

(b) 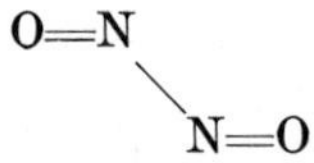

According to Pauling it is resonance with the structure

(c) $:\overset{-\,..}{N}{=}\underset{..}{\overset{+}{O}}-$

which stabilizes the ground state of NO and therefore reduces the energy gained in the formation of the dimer (b).

More recently the same type of argument has been used to account for the chemical stability of $XeF_4$ and other Xe and Kr compounds.

The most important example of resonance and complete electron delocalization is provided by the benzene molecule. The two Kékulé structures

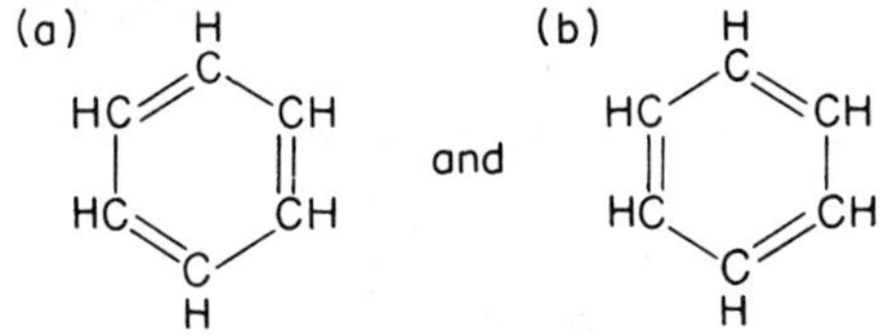

obviously have the same energy and the (exact) resonance between them must lead to a considerable lowering of the energy. The description of the electronic structure of each of the Kékulé forms, on the basis of the Slater–Pauling theory, is an extension of that already given for $C_2H_4$ (p. 365, model (b)) and was first proposed by Hückel (584): we hybridize only two of the $2p$ orbitals of each C atom with the $2s$ orbital obtaining three trigonal orbitals in the $xy$-plane (see p. 309) and one non-hybridized orbital $2p_z$ with its axis perpendicular to this plane. The three trigonal orbitals of each of the six carbon atoms combine in pairs with one another and with the $1s$ orbitals of the H atoms forming twelve localized valence-bond orbitals each of which is filled with two electrons. In this way the ring structure

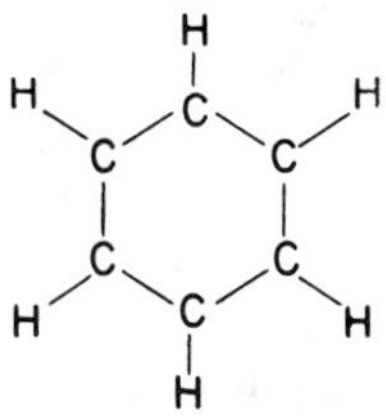

arises in which all bonds are $\sigma$ bonds and are at 120° to each other in the $xy$-plane.

There remain six electrons, which occupy the six $p_z$ orbitals centered on each carbon atom with their lobes above and below the $xy$-plane. The sideways overlap of two adjacent $p_z$ orbitals will lead to the formation of a $\pi$ bond just as in $C_2H_4$ (see Fig. 145b). Clearly only three such $\pi$ bonds can be formed (a given carbon atom can combine its $p_z$ orbital only with that of one or the other of its neighbors) and in this way we obtain one or the other of the two Kékulé forms. While in each Kékulé structure, because of the alternation of single and double bonds, one would expect an alternation of bond lengths the superposition of the two structures (i.e. the resonance) has the effect that all bond lengths are equal, in agreement with observation. Moreover, no difference has ever been found chemically between the different carbon—carbon bonds of benzene.

The magnitude of the resonance energy can be determined theoretically as follows: The energy of a Kékulé structure is readily obtained from eq. (III, 73) as

$$E_{\text{Kékulé}} = J + K_{12} + K_{34} + K_{56} - \tfrac{1}{2}(K_{23} + K_{45} + K_{61}) \qquad \text{(III, 80)}$$

where $J$ is the Coulomb energy and $K_{ij}$ the exchange energy between $p_z$ electrons on atoms $i$ and $j$ (numbered in order 1, 2, . . ., 6 around the ring). In (III, 80) we have neglected non-neighbor interactions and considered only the exchange of the $\pi$ electrons. The integrals $K_{12}$, $K_{34}$, $K_{56}$ are equal among themselves, say equal to $K_a$, and the integrals $K_{23}$, $K_{45}$, $K_{61}$ are equal among themselves, say equal to $K_b$; therefore

$$E_{\text{Kékulé}} = J + 3K_a - \tfrac{3}{2}K_b.$$

Usually the additional assumption is made[8] that $K_a = K_b = K$ so that

$$E_{\text{Kékulé}} = J + 1.5K. \qquad \text{(III, 81)}$$

If one now calculates the energy of the system in which the two Kékulé structures are mixed one finds from (III, 77a) [see Coulson (7)]

$$E_s = J + 2.4K. \qquad \text{(III, 82)}$$

Thus there is a resonance energy[9] of $-0.9K$. (Note that $K$ is negative.)

If the eigenfunctions of the two Kékulé structures are $\psi_1$ and $\psi_2$, the ground state, in the zero approximation, has the eigenfunction $\psi_1 + \psi_2$. It is totally symmetric, i.e. has the species $A_{1g}$ of point group $\boldsymbol{D}_{6h}$. The other state of the molecule arising from the two Kékulé structures has the eigenfunction $\psi_1 - \psi_2$ and belongs to the species $B_{2u}$. Its energy $E_a$ is higher than that of the Kékulé structure by the resonance energy, i.e. in the present approximation

$$E_a = J + 0.6K.$$

For a more complete description of the ground state of benzene, structures other than the Kékulé structures must also be considered. It can be shown [see

[8] This assumption is contrary to the essential characteristic of the Kékulé structure that single and double bonds alternate and that therefore alternate bonds have different lengths, i.e., that $K_a > K_b$. For this reason $E_{\text{Kékulé}} > J + 1.5K$; or in other words the value $J + 1.5K$ refers to a hypothetical Kékulé structure with equal bond lengths.

[9] Strictly speaking this is the resonance energy referred to a Kékulé structure of symmetrical form (see footnote 8).

Eyring, Walter and Kimball (13)] that all possible valence structures of benzene in the valence-bond theory can be expressed in terms of the two Kékulé structures and the three Dewar structures. The latter are

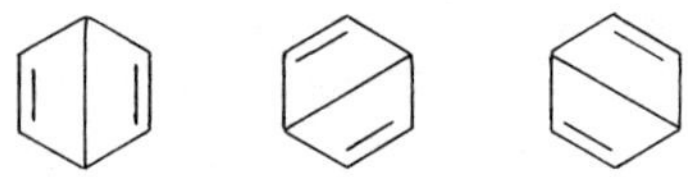

Let us designate the corresponding eigenfunctions by $\psi_3$, $\psi_4$, $\psi_5$ respectively. The energy of one of the Dewar structures is (similar to (III, 80))

$$\begin{aligned} E_{\text{Dewar}} &= J + K_{12} + K_{45} - \tfrac{1}{2}(K_{23} + K_{34} + K_{56} + K_{61}) \\ &= J + 2K_a - 2K_b \end{aligned} \qquad \text{(III, 83)}$$

which on the assumption that $K_a = K_b$ gives simply

$$E_{\text{Dewar}} = J.$$

Since this is higher by $-1.5K$ than the energy of the Kékulé structures, the Dewar structures have only a relatively small even though not negligible effect on the ground state.

There is, of course, exact resonance between the three Dewar structures. This resonance leads to a splitting into two states, one totally symmetric ($A_{1g}$) with eigenfunction

$$\psi_s(\text{Dewar}) = \psi_3 + \psi_4 + \psi_5$$

and one doubly degenerate ($E_{2g}$) with eigenfunctions

$$\psi_e(\text{Dewar}) = \psi_3 - \psi_4, \qquad \bar{\psi}_e(\text{Dewar}) = \psi_4 - \psi_5.$$

It is the first of these that can resonate with the ground state (since its species is the same). As was first shown by Pauling and Wheland (972) this resonance produces an additional lowering of the ground state by an amount of $-0.20_6K$ so that the total resonance energy is $-1.10_6K$. The percentage contributions of each of the two Kékulé and the three Dewar structures to the wave function of the ground state of benzene are found to be 39.0 and 7.3 per cent respectively.

All experimental determinations of the resonance energy have to be based on certain assumptions. The most direct determination seems to be the one based on a comparison of the heat of hydrogenation of cyclohexene ($C_6H_{10}$) with that of benzene [Kistiakowsky, Ruhoff, Smith and Vaughan (673)]. Assuming the same internuclear distances in the two molecules one would expect for a single Kékulé structure of benzene a heat of hydrogenation three times that of cyclohexene, that is $3 \times 28.6 = 85.8$ kcal/mole, since three instead of one double bond is broken. Actually only 49.8 kcal/mole is released showing that the energy of the ground state of $C_6H_6$ is 36.0 kcal/mole lower than expected. This is the observed resonance energy, which when compared to the theoretical value ($-1.10_6K$) leads to $K = -32.5$ kcal/mole. However, this value gives only moderate agreement of calculated with observed resonance energies in naphthalene and higher aromatic compounds [see Coulson (7), Wheland (43) and Streitwieser (39)].

It is interesting to note that the same four electronic states derived above from the two Kékulé and the three Dewar structures can also be obtained directly without any assumptions about $\pi$ bonds simply on a group theoretical basis: If six atoms in ${}^2S_g$ states are brought to the corners of a regular hexagon the resulting molecular states are those given at the bottom of the earlier Table 26, p. 290. It is easily seen that the same states arise from six atoms with a $p_z$ electron each. The singlet states that arise in this way are: two ${}^1A_{1g}$, one ${}^1B_{2u}$ and one ${}^1E_{2g}$ and these are precisely the states derived from the valence picture.

The ${}^1B_{2u}$ state, which corresponds to $\psi_1 - \psi_2$, has been observed as the upper state of the ultraviolet absorption bands of $C_6H_6$ near 2625 Å (see Chap. V, section 10). According to the rough approximation above, its excitation energy should be $-2.006K$. The observed energy yields thus $K = -54$ kcal/mole. Considering the crudeness of the approximation the agreement with the value derived from the "observed" resonance energy must be considered as satisfactory.

***d*-valence.** The $d$ orbitals of an atom have a five-fold degeneracy. Figure 147 shows in three projections a set of five mutually orthogonal (real) orbitals similar to Fig. 117 for $p$ orbitals. The corresponding mathematical expressions are

$$\begin{aligned}\psi(d_{z^2}) &= \frac{1}{\sqrt{3}}(3z^2 - r^2)f(r)\\ \psi(d_{x^2-y^2}) &= (x^2 - y^2)f(r)\\ \psi(d_{yz}) &= 2yzf(r)\\ \psi(d_{zx}) &= 2zxf(r)\\ \psi(d_{xy}) &= 2xyf(r),\end{aligned} \qquad \text{(III, 84)}$$

where $f(r)$ is given by (III, 15). Any five mutually orthogonal linear combinations of the functions (III, 84) would also form a possible set of orbitals[10].

If an atom had two $d$ electrons outside closed shells, according to valence-bond theory, two univalent atoms approaching the former would give, according to Fig. 147, maximum overlap if the two bond directions are at 45°, or 60°, or 90°, or 120°, or 135° to each other. If there are three $d$ electrons the same angles between the bond directions may occur. As a special case the three bonds may be in a plane at 120° to one another (see Table 29). The corresponding trigonal orbitals are linear combinations of $d_{z^2}$, $d_{x^2-y^2}$, $d_{xy}$.

The pure $d$ bonds just discussed are not important because in all cases in which $d$ electrons are the outermost electrons there are other electrons ($s$ or $p$) of similar energies with which the $d$ electrons can hybridize. Various possibilities of hybridization involving $d$ electrons were listed previously in Table 29, p. 310. Probably the most important of these is the one that leads to the formation of six equivalent octahedral orbitals from one $s$, three $p$ and two $d$ orbitals. As an example consider the formation of the $SF_6$ molecule which is known to have the form of a regular octahedron (see Vol. II, p. 336). The sulfur atom in its ground state has the configuration $3s^23p^4$ and is therefore divalent as in $H_2S$ and similar

[10] Note that the individual lobes except in $d_{z^2}$ are not axially symmetric.

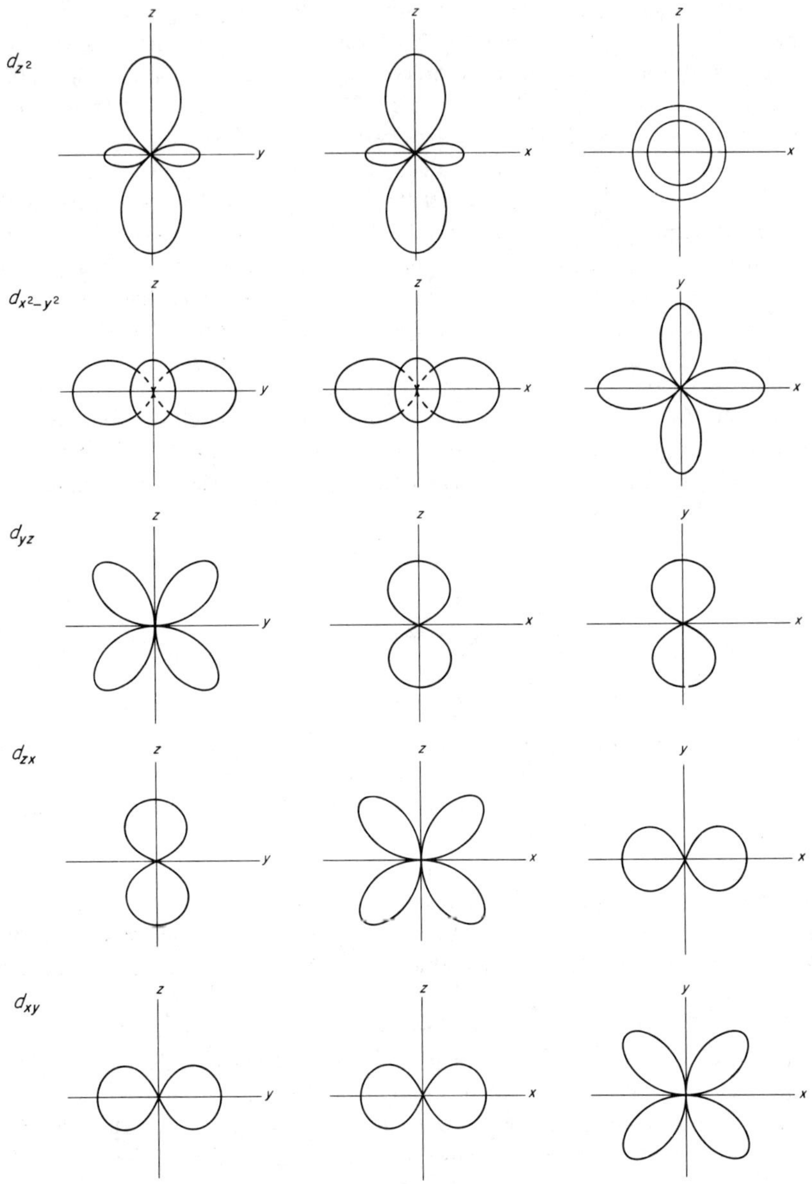


FIG. 147. **Five mutually orthogonal *d* orbitals of a free atom.** Only the directional dependence of the orbitals is shown in three projections. To obtain diagrams similar to Fig. 119 one would have to multiply the functions given by $r^2f(r)$.

compounds. A hexavalent state is obtained by taking one electron each from $3s$ and $3p$ to $3d$, that is, by using the configuration $3s3p^33d^2$. In this state six electron-pair bonds can be formed with the electrons of six univalent atoms and these bonds will be directed toward the corners of a regular octahedron. It is significant that for oxygen for which the $3d$ orbital lies fairly high no hexavalent compounds have been found while for S, Se, Te, the hexafluorides are well known. However, no hexahydrides have been found for these elements. The reason for the difference between F, H and other ligands has been discussed recently by Craig and Zauli (254) [see however Cruikshank, Webster and Mayers (257a)].

In the transition elements the $4s$ and $4p$ (or $5s$ and $5p$) orbitals are in general only slightly higher in energy than the $3d$ (or $4d$) orbitals and hybridization is readily achieved. Thus in Mo, which in the ground state has the configuration $4d^55s$, we obtain easily the configuration $4d^25s5p^3$ which can be hybridized as before to give six equivalent octahedral orbitals and in this way we can account for the stability of $MoF_6$. In a similar way $WF_6$ and $UF_6$ can be understood. The non-existence of $CrF_6$ in spite of an entirely similar configuration is less easily explained. If there are more electrons in the $d$ shell, as in Re, Os, Ir, Pt, we can still obtain octahedral hybridization by simply not using the additional $d$ electrons, and indeed hexahalides of these elements have been prepared.

For most transition elements the situation is complicated by electrostatic effects in the complex ions that are formed. For example in $[Cr(NH_3)_6]Cl_3$ we have a $Cr^{+++}$ ion with three $d$ electrons which would not be sufficient to form six electron-pair bonds; but if we assume that each $NH_3$ gives one of its lone pair electrons to the central atom we have a $Cr^{---}$ ion with nine outer electrons which may be in $3d$, $4s$ or $4p$ orbitals and may form six hybridized octahedral bond orbitals. To the effect of the electron-pair bonds thus formed is added the electrostatic attraction between the oppositely charged ions. We shall postpone a further discussion of transition elements until after a more detailed treatment of the molecular orbital approach.

**Excited states.** Valence-bond theory is mainly concerned with the question of the stability of the ground state of a molecule, but in principle it can also deal with the excited states of molecules. Indeed the concept of resonance implies the presence of at least one excited state of the same species as the ground state and information about this excited state is obtained automatically when the effect of resonance on the ground state is determined: the excited state is raised by the same amount that the ground state is lowered. It is more difficult to make precise predictions about excited states whose species and spin are different from that of the ground state. A very interesting and successful attempt has been made by Jordan and Longuet–Higgins (649) who have calculated the energies of the low-lying electronic states of $CH_2$, $CH_3$, $NH_2$, $BH_2$ and $BH_3$ as a function of the bond angles. The agreement of the calculations for $CH_2$ and $NH_2$ with the experimental data is very satisfactory. The prediction that $BH_2$ in its ground state would be linear was, however, not confirmed by subsequent experiments [Herzberg and

Johns (531)]. Similar calculations of low excited states have been made by Hutchinson (598) for $H_2O$, $BeH_2$, $CH_2$, $NH_3$, $CH_3$, $C_2H_2$, by Ellison (352) for $CH_2$, $CH_3$, $CH_4$, by Jordan (648) for $PH_2$ and by Dixon (286) for $CH_2$.

## (b) The molecular orbital method

### (α) *Basic idea*

There are two important aspects to molecular orbital theory: one is the determination of the order and energy of the various electronic states of a given molecule and the other is the prediction of the stability of each of these states, particularly of the ground state. The first aspect has already been dealt with fairly exhaustively in section 2 of this chapter; it remains now to discuss the second aspect, in other words the question of how on the basis of molecular orbitals the stability and the phenomenon of valence in polyatomic molecules can be understood.

**Bonding, anti-bonding and non-bonding electrons.** Just as in the first approximation of molecular orbital theory we consider independently the contributions of the individual electrons to the total energy and eigenfunction of each of the electronic states, so in a valence theory based on molecular orbitals we must consider independently the contributions of the individual electrons to the binding. This is done by means of the concepts of bonding, anti-bonding and non-bonding electrons (or orbitals). If in the formation of the molecule from the separated atoms (or groups of atoms) an electron occupies an orbital whose energy decreases as the interatomic distance $r$ decreases, then this electron gives a positive contribution to the binding energy, that is, it is a *bonding electron*; if the energy of the orbital increases as $r$ decreases the electron gives a negative contribution, it is an *anti-bonding electron*; if the energy of the orbital does not change appreciably the electron is called *non-bonding*. Thus in order to predict whether or not a certain molecular electronic state resulting from the separated atoms is stable one has to establish its electron configuration on the basis of one of the correlation schemes previously given (section 2b) and then see whether the contribution of the occupied bonding orbitals to the binding is larger in magnitude than the opposing contribution of the occupied anti-bonding orbitals (compare the corresponding rule for diatomic molecules, Vol. I, p. 362).

The difficulty in the molecular orbital method is that the magnitude of the bonding or anti-bonding action of a given orbital is not easy to assess without elaborate calculations. However, in many cases (see below) the bonding action of an electron in a bonding orbital is opposite and roughly equal to the anti-bonding action of an electron in the corresponding anti-bonding orbital, Therefore frequently the same *rule of thumb* holds as for diatomic molecules (Vol. I, p. 367), viz. *a stable bond is formed if the number* $n_b$ *of bonding electrons is greater than the number* $n_a$ *of anti-bonding electrons*, the binding energy being greater the greater the difference $n_b - n_a$.

**Resonance and the one-electron bond.** While the prototype for the electron-pair bond is the $H_2$ molecule, the prototype for bonding in the molecular orbital theory is the $H_2^+$ molecule. As we have seen in more detail in Volume I the reason for the strong attraction between an H atom and an $H^+$ ion (proton) is the *resonance degeneracy* that exists for the system, since the energy at large internuclear distance is the same whether the electron is near the one or the other nucleus. For smaller distances we have in consequence a splitting of the $1s$ orbital into two orbitals $\sigma_g$ and $\sigma_u$, one ($\sigma_g$) having an energy below, the other ($\sigma_u$) above that of the separated atoms. If the electron goes into the $\sigma_g$ orbital a stable molecule results, if it goes into the $\sigma_u$ orbital an unstable molecule results.

The magnitude of the decrease or increase of energy of a molecular orbital when the internuclear distance changes depends on the degree of overlapping of the two atomic orbitals from which it is formed. Here it must be noted that in molecular orbital theory it is simply the overlap of two atomic orbitals that matters, while in the valence-bond theory it is the overlap of orbitals each filled with an unpaired electron and the consequent "exchange" interaction. This difference is made clearer if one considers that in the valence-bond theory the binding depends on the exchange integrals (III, 72) which contain the six coordinates of the two electrons in the bond, while here the term that determines the strength of the binding is the resonance integral

$$R = \int \varphi_A(1)\mathrm{H}\varphi_B(1)d\tau \qquad \text{(III, 85)}$$

which depends on the three coordinates of one electron only. In other words the basic concept in the molecular orbital theory is the *one-electron bond*, not a two-electron bond as in the valence-bond theory. On the other hand each non-degenerate molecular orbital can accommodate only two electrons with anti-parallel spin directions (because of the Pauli principle) and in this way the electron-pair concept enters the molecular orbital theory quite naturally. In contrast, in the valence-bond theory spin pairing comes in rather indirectly because it is the symmetric space function that corresponds to binding and it must be combined with an antisymmetric spin function (i.e. anti-parallel spins) to obey the requirements of the Pauli principle.

The resonance in $H_2^+$ is exact: whether at large internuclear distance the electron is with the one or the other proton the energy is the same and we have in the molecule the splitting into $\sigma_g$ and $\sigma_u$ as described above, each of which contains equal amounts of the two atomic orbitals: the orbital function is in a first approximation the sum or difference of the two atomic orbitals. If the two nuclei are unequal the atomic orbitals have different energies but on bringing the nuclei together they will still influence each other even though less than in the case of exact resonance. On account of this mutual interaction the upper of the two levels will be "pushed" upwards the lower one downwards. The orbital wave function is no longer simply the sum or difference of the two atomic orbitals but in a first approximation is given by

$$\psi_{AB} = a\varphi_A + b\varphi_B. \qquad \text{(III, 86)}$$

In a higher approximation the effect of other orbitals must also be considered (see section 2, p. 300).

It will be clear from these considerations that resonance is introduced here at a very early stage, i.e. as resonance between the individual orbitals, while in the valence-bond theory it is introduced almost as an after-thought as a resonance between the valence states resulting from the theory.

**Contribution of ionic states.** Another important difference between the valence-bond and molecular orbital theories lies in the treatment of the contributions of ionic states. In the valence-bond theory the normal wave functions do not contain any ionic contributions but ionic states are introduced in the process of resonance between different valence states. In the molecular orbital treatment ionic states are taken into account from the beginning and, in fact, are over-emphasized. This can be seen immediately if two electrons are placed into an orbital such as (III, 86). The wave function of the system is then

$$\Psi = \psi_{AB}(1)\psi_{AB}(2) = a^2\varphi_A(1)\varphi_A(2) + b^2\varphi_B(1)\varphi_B(2) + ab[\varphi_A(1)\varphi_B(2) + \varphi_A(2)\varphi_B(1)]. \quad \text{(III, 87)}$$

Here the first term on the right implies that both electrons are with A, that is, we have $A^-B^+$, while the second term implies that both electrons are with nucleus B, that is, we have $A^+B^-$, and only the third term represents true homopolar binding.

According to (III, 87) if the two nuclei are separated there is a 50% probability that two ions $A^- + B^+$ or $A^+ + B^-$ arise while only the remaining probability of 50% leads to two neutral atoms $A + B$. In actual fact a dissociation of AB into ions requires in general much higher energy (for $H_2$ 13 eV higher) than a dissociation into neutral atoms and therefore the former will certainly not occur when the nuclei are adiabatically separated. On the other hand the valence-bond method leads to dissociation into neutral atoms only. It represents therefore the much better approximation at large internuclear distances. But it does not give in such a natural way as does the molecular orbital theory the behavior for small distances. Another way of describing the over-emphasis on ionic states by the molecular orbital method is by saying that *electron correlation* is neglected. Indeed much recent work on refinements of molecular orbital theory deals with the problem of properly introducing electron correlation, that is, the fact that the instantaneous field acting on a given electron deviates from the average (self-consistent) field used in the Hartree–Fock procedure [see, e.g. Löwdin (780), Kraus (692), Clementi (206)].

**Delocalization.** One of the principal characteristics of the molecular orbital method is that in general all orbitals extend over the whole molecule; they are *delocalized*. For example in the ground state of the $CH_4$ molecule, we have in molecular orbital theory, apart from the $1s$ electrons of C, two $2a_1$ and six $1f_2$ electrons all of which are spread over the whole molecule while in valence-bond theory there is a pair of electrons localized between each H atom and the central C atom and these pairs account for the four C—H bonds. On the other hand in molecular orbital theory the binding is produced collectively by the four electrons of the four H atoms and the four electrons of the C atom which fill together the $2a_1$ and $1f_2$ orbitals. Each of these bonding orbitals contains equal contributions from all four H atoms. Accordingly, it is not possible to say that the electron of a particular H atom together with one from the C atom binds that particular H atom to the C atom; rather each electron coming from the H atoms contributes equally to all four C—H bonds. When an electron is excited it does not come from a particular H atom but from an orbital that extends over all four H atoms. While this aspect of molecular orbital theory is an advantage for a discussion of excited electronic states it does not lend itself to a simple explanation of the observed constancy of C—H bond dissociation energies, distances, and force constants in a variety of molecules in their ground states. In contrast the valence-bond theory gives a simple and straightforward account of these facts.

**Charge distribution, population analysis.** Another way of obtaining information in molecular orbital theory about the strengths of bonds is by considering the electron density contributed by various orbitals throughout the molecule. Mulliken (911)(915) who developed this approach called it *electronic population analysis.* The basic consideration underlying this approach is the following. The reason for homopolar attraction between two atoms, both in valence-bond and molecular orbital theory, is the fact that there is a greater density of electrons between the atoms than there would be if the undeformed atoms were simply superposed. The greater concentration of electrons between the two nuclei leads to an attraction between them. Conversely a depletion of electrons in this region leads to a repulsion. London's well-known pictures for the electron distribution in the ${}^1\Sigma_g^+$ and ${}^3\Sigma_u^+$ states of $H_2$ (Fig. 162 of Vol. I) illustrate this point.

Let $\psi_i$ be one of the (normalized) molecular orbitals of a diatomic molecule which may be expressed [similar to (III, 86)] as a linear combination of (normalized) atomic orbitals $\varphi_{ar}$ and $\varphi_{bs}$

$$\psi_i = c^i_{ar}\varphi_{ar} + c^i_{bs}\varphi_{bs}. \tag{III, 88}$$

Here $a$ and $b$ stand for the two different centers (nuclei) and $r$ and $s$ for the quantum numbers of the two atomic orbitals which in general will be different. According to the general rules of quantum mechanics the square of the wave function is the electron density $\rho$ of an electron in the orbital $\psi_i$. If there are $N(i)$ electrons in the orbital $\psi_i$ we have therefore

$$\rho = N(i)\psi_i^2 = N(i)(c^i_{ar})^2(\varphi_{ar})^2 + 2N(i)c^i_{ar}c^i_{bs}\varphi_{ar}\varphi_{bs} + N(i)(c^i_{bs})^2(\varphi_{bs})^2. \tag{III, 89}$$

Integrating over the whole of space we obtain, since $\psi_i$, $\varphi_{ar}$ and $\varphi_{bs}$ are normalized

$$N(i) = N(i)(c^i_{ar})^2 + 2N(i)c^i_{ar}c^i_{bs}S_{arbs} + N(i)(c^i_{bs})^2 \tag{III, 90}$$

where

$$S_{arbs} = \int \varphi_{ar}\varphi_{bs}d\tau \tag{III, 91}$$

is the overlap integral of the two atomic orbitals.

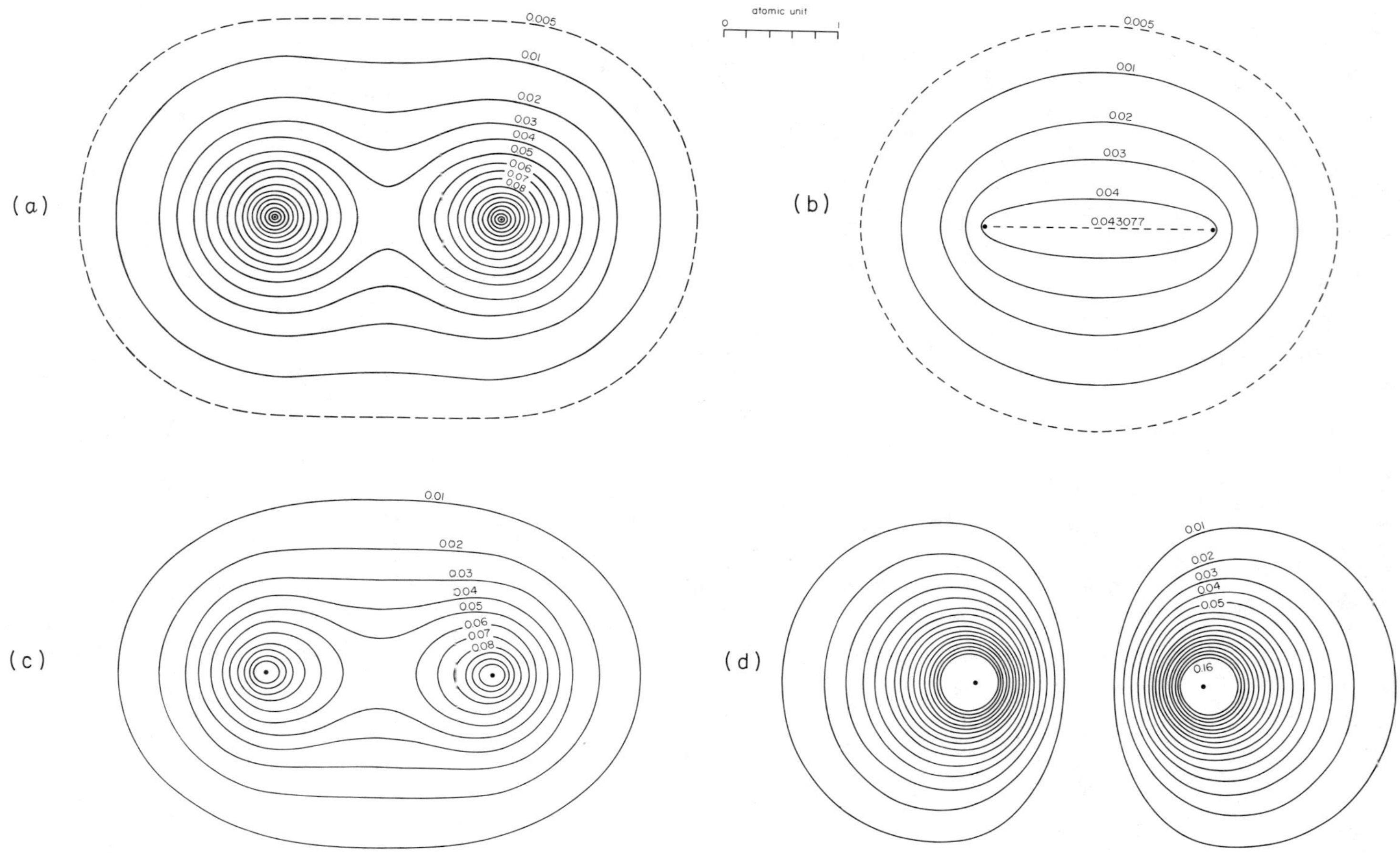


FIG. 148. **Electron distribution in the $1\sigma_g$ and $1\sigma_u$ orbitals of $H_2^+$ for $R = 2$ atomic units (1.06 Å).** (*a*) $\frac{1}{2}[(\varphi(1s_a))^2 + (\varphi(1s_b))^2]$ (*b*) $\varphi(1s_a)\varphi(1s_b)$ (*c*) $\rho(1\sigma_g)$ (*d*) $\rho(1\sigma_u)$. (*a*) and (*b*) represent the components from which $\rho(1\sigma_g)$ and $\rho(1\sigma_u)$ in (*c*) and (*d*) are formed according to eqs. (III, 95 and 96). The small black dots represent the positions of the nuclei. Successive contour lines correspond to equal

According to (III, 90) the total population $N(i)$ of the orbital $\psi_i$ (which for a non-degenerate orbital may be 0, 1 or 2) is resolved into three contributions: $N(i)(c^i_{ar})^2$ and $N(i)(c^i_{bs})^2$ called the *net atomic populations* on atoms $a$ and $b$ and $2N(i)c^i_{ar}c^i_{bs}S_{arbs}$ called the *overlap population*. It is clear that as the distance of the atoms increases, the overlap population decreases so that for large distances only the atomic populations remain. Conversely as the distance is decreased each atom gives up some of its electron population and contributes it to the overlap population thus giving rise to binding between the two atoms if the overlap population is positive or giving rise to repulsion if the overlap population is negative.

As an example let us consider the $H_2^+$ molecule in its two lowest molecular orbitals

$$\psi(1\sigma_g) = \frac{1}{\sqrt{2(1 + S)}}[\varphi(1s_a) + \varphi(1s_b)], \quad \text{(III, 92)}$$

$$\psi(1\sigma_u) = \frac{1}{\sqrt{2(1 - S)}}[\varphi(1s_a) - \varphi(1s_b)], \quad \text{(III, 93)}$$

$$S = \int \varphi(1s_a)\varphi(1s_b)d\tau. \quad \text{(III, 94)}$$

Here the normalizing factors have been obtained from (III, 90) by putting $c^i_{bs} = \pm c^i_{ar}$. With only one electron present the electron density according to (III, 89) is for the lowest (bonding) orbital

$$\rho(1\sigma_g) = \frac{1}{2(1 + S)}[(\varphi(1s_a))^2 + (\varphi(1s_b))^2] + \frac{1}{1 + S}\varphi(1s_a)\varphi(1s_b) \quad \text{(III, 95)}$$

and for the next lowest (anti-bonding) orbital

$$\rho(1\sigma_u) = \frac{1}{2(1 - S)}[(\varphi(1s_a))^2 + (\varphi(1s_b))^2] - \frac{1}{1 - S}\varphi(1s_a)\varphi(1s_b). \quad \text{(III, 96)}$$

In each case the first term represents the net atomic population density which, for large internuclear distance $R$ (i.e. $S \to 0$), approaches $\frac{1}{2}[(\varphi(1s_a))^2 + (\varphi(1s_b))^2]$ while the second term is the overlap population density.

In Fig. 148a and b the quantities $\frac{1}{2}[(\varphi(1s_a))^2 + (\varphi(1s_b))^2]$ and $\varphi(1s_a)\varphi(1s_b)$ for $R = 2$ a.u. ($= 1.06$ Å) are plotted in the form of contour diagrams for a plane going through the internuclear axis. The electron distribution (III, 95) for $1\sigma_g$ is obtained by adding the values in Fig. 148a and b and multiplying by $1/(1 + S)$ giving Fig. 148c, while the electron distribution (III, 96) for $1\sigma_u$ is obtained by subtracting the values in Fig. 148b from those in Fig. 144a and multiplying by $1/(1 - S)$ giving Fig. 148d. In Fig. 149 the variation of $\rho(1\sigma_g)$ and $\rho(1\sigma_u)$ and of their components along the internuclear axis is shown. Both figures clearly show that, in the bonding orbital, electron density near the nuclei is taken away and put in between them while in the anti-bonding orbital electron density between the nuclei is taken away and added near them.

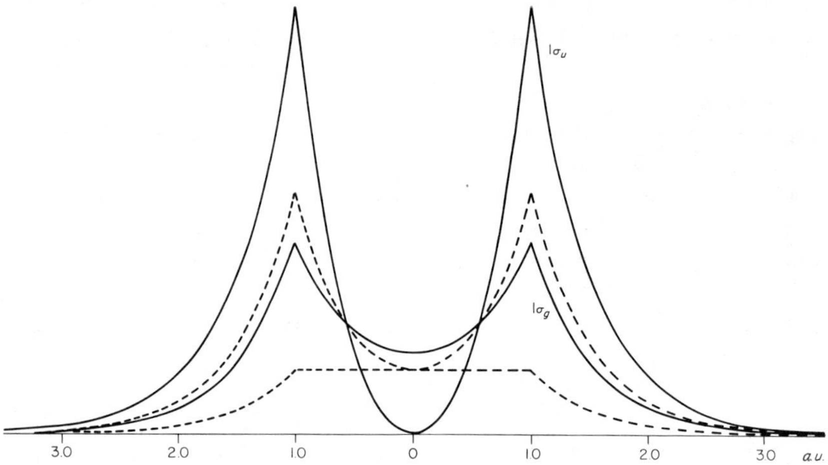


FIG. 149. **Variation of the electron distribution in the $1\sigma_g$ and $1\sigma_u$ orbitals of $H_2^+$ along the axis of the molecule.** The full-line curves give $\rho(1\sigma_g)$, and $\rho(1\sigma_u)$, the broken-line curves represent the component functions $\frac{1}{2}[(\varphi(1s_a))]^2 + (\varphi(1s_b))^2]$ and $\varphi(1s_a)\varphi(1s_b)$. The abscissa scale is in atomic units.

The total overlap population for $1\sigma_g$ and $1\sigma_u$ is

$$\frac{1}{1+S}\int \varphi_a\varphi_b d\tau = \frac{S}{1+S} \quad \text{and} \quad -\frac{1}{1-S}\int \varphi_a\varphi_b d\tau = -\frac{S}{1-S} \qquad \text{(III, 97)}$$

respectively. Since, as is easily seen,

$$\frac{1}{2(1 \pm S)}(\varphi_a^2 + \varphi_b^2) = \tfrac{1}{2}(\varphi_a^2 + \varphi_b^2) \mp \frac{S}{2(1 \pm S)}(\varphi_a^2 + \varphi_b^2)$$

we can immediately verify, if we integrate over the whole space, that the electron population taken away from the free atoms (that is, the amount by which the first term in (III, 95) is reduced compared to the case when $R = \infty$ and therefore $S = 0$) is exactly equal to the overlap population.

The importance of overlap populations for chemical binding becomes still clearer when we consider the formulae for the energies. From a simple perturbation calculation (see Vol. I, p. 362) one finds for the energy of $1\sigma_g$ and $1\sigma_u$ referred to the energy of free H + $H^+$ at large $R$:

$$\epsilon(1\sigma_g) = \frac{H_{aa} + H_{ab}}{1+S}, \qquad \epsilon(1\sigma_u) = \frac{H_{aa} - H_{ab}}{1-S} \qquad \text{(III, 98)}$$

where

$$H_{aa} = \int \varphi_a W \varphi_a d\tau, \qquad \text{(III, 99a)}$$

$$H_{ab} = \int \varphi_a W \varphi_b d\tau. \qquad \text{(III, 99b)}$$

Here $W$ is the potential energy of the two nuclei and the electron as a function of their mutual distances, taking the energy of the ground state of the system H + $H^+$ at large distance as zero. We can re-write (III, 98) as follows:

$$\epsilon(1\sigma_g) = H_{aa} + \frac{H_{ab} - SH_{aa}}{1 + S}, \qquad \epsilon(1\sigma_u) = H_{aa} - \frac{H_{ab} - SH_{aa}}{1 - S}. \qquad \text{(III, 100)}$$

If in (III, 99b) $W$ (which is always negative) is replaced by its average $W_{\text{av.}}$ we have

$$H_{ab} = W_{\text{av.}} \int \varphi_a \varphi_b d\tau = -\delta S \qquad \text{(III, 101)}$$

where $\delta$ is a positive constant whose value depends of course on $R$. Therefore

$$\epsilon(1\sigma_g) = H_{aa} - \alpha \frac{S}{1 + S}, \qquad \epsilon(1\sigma_u) = H_{aa} + \alpha \frac{S}{1 - S}, \qquad \text{(III, 102)}$$
$$\alpha = \delta + H_{aa}.$$

The energy is thus the sum of a (small) Coulomb term $H_{aa}$, which is the same for $1\sigma_g$ and $1\sigma_u$ and represents the ordinary Coulomb interaction between a proton and an H atom, and a resonance term which has opposite sign for the two states and depends on the degree of overlap of the two atomic orbitals. This latter term is the essential one in covalent binding according to molecular orbital theory. It is significant that this term is directly proportional to the overlap population in the two states $1\sigma_g$ and $1\sigma_u$, that is, the *energy of attraction or repulsion is in a good approximation proportional to the overlap population*. It is reasonable to suppose that this connection between overlap population and bond energy applies in more complicated molecules as well where the energy cannot be expressed by such simple formulae. Even in the simple case of $H_2^+$ the energy formulae used here are based on a very rough approximation since we have simply used linear combinations of atomic orbitals for the molecular orbitals.

In a *polyatomic molecule* the molecular orbitals are in general linear combinations of more than two atomic orbitals (see section 2). Therefore (III, 88) has to be replaced by

$$\psi_i = \sum_{a,r} c^i_{ar} \varphi_{ar}, \qquad \text{(III, 103)}$$

a sum that is extended over all atoms $a, b, c, \ldots$ of the molecule and over various atomic orbitals $r, s, \ldots$. Similar to (III, 90) the number of electrons in the orbital $\psi_i$ is now expressed by

$$N(i) = N(i) \sum_{a,r} (c^i_{ar})^2 + 2N(i) \sum_{\substack{a,r \\ b,s}} c^i_{ar} c^i_{bs} S_{arbs} \qquad \text{(III, 104)}$$

where as before $S_{arbs}$ is defined by (III, 91).

The first term in (III, 104) represents the net atomic population. It may be resolved into partial net atomic populations

$$n(i; a, r) = N(i)(c^i_{ar})^2 \qquad \text{(III, 105)}$$

contributed by each orbital $r, s, \ldots$ of each atom $a, b, c, \ldots$. The second term of

(III, 104) represents the overlap population, which gives a good indication of covalent bond strength in the molecule. It may be resolved into partial overlap populations

$$n(i; ar, bs) = 2N(i)c^i_{ar}c^i_{bs}S_{arbs} \quad \text{(III, 106)}$$

corresponding to contributions by pairs of orbitals of pairs of atoms. If we sum over all the orbitals of a given pair of atoms $(a, b)$ we get

$$n(i; a, b) = \sum_{r,s} n(i, ar, bs) \quad \text{(III, 107)}$$

The contribution of a given pair of atomic orbitals may be positive or negative giving a positive or negative contribution to the binding between the two atoms.

For some purposes a breakdown of the electron population of the molecule into *gross atomic populations* is useful even though somewhat arbitrary [Mulliken (911)]. In this approach all of the electron density is assigned either to one or to another atom. Since the overlap population is symmetrically related to the two atoms between which the overlap takes place, one assigns half of it to each of the two atoms, that is, one defines as the partial gross population of atom $a$, in a molecular orbital $\psi_i$, due to atomic orbital $\varphi_{ar}$,

$$N(i; a, r) = N(i)(c^i_{ar})^2 + N(i)c^i_{ar}c^i_{bs}S_{arbs} \quad \text{(III, 108)}$$

and correspondingly as the sub-total gross population of atom $a$ in molecular orbital $\psi_i$

$$N(i; a) = \sum_r N(i; a, r) \quad \text{(III, 109)}$$

and the total gross population on atom $a$

$$N(a) = \sum_i N(i; a) = \sum_i \sum_r N(i; a, r) \quad \text{(III, 110)}$$

The total gross population in orbital $\psi_i$

$$N(i) = \sum_a N(i; a) \quad \text{(III, 111)}$$

must be integral as long as configuration interaction is neglected. The total gross population in atomic orbital $\varphi_{ar}$ due to all molecular orbitals $\psi_i$

$$N(a, r) = \sum_i N(i; a, r) \quad \text{(III, 112)}$$

need not be integral. As an example we quote Mulliken's effective electron configuration for the two atoms in CO:

$$1s_{\mathrm{C}}{}^{2.00}2s_{\mathrm{C}}{}^{1.49}2p\sigma_{\mathrm{C}}{}^{1.40}2p\pi_{\mathrm{C}}{}^{1.02};1s_{\mathrm{O}}{}^{2.00}2s_{\mathrm{O}}{}^{1.86}2p\sigma_{\mathrm{O}}{}^{1.25}2p\pi_{\mathrm{O}}{}^{2.98}.$$

Here the exponents are the total gross atomic populations $N(a, r)$ for the respective atomic orbitals.

Another related application of electron population densities is in the calculation of the *forces* on individual nuclei of a molecule. These forces are the vector sum of the repulsion by the other nuclei and the attractive forces of the electronic charge distribution. With the help of the Hellmann–Feynman theorem Bader and Jones (84)(85)(86) have recently carried out detailed calculations of these forces using self-consistent wave functions and population densities for a number of diatomic and polyatomic molecules. They call binding or anti-binding molecular orbitals those orbitals that when occupied by two electrons give a positive or negative contribution to the force of attraction between two nuclei. The condition that in the equilibrium position the resultant force must be zero, gives a very stringent test of the quality of the orbital functions.

### (β) *Ground states*

**The ground state of $H_2O$.** In order to illustrate the way in which molecular orbital theory comes to conclusions about molecular stability let us now consider the ground states of a few individual molecules. The molecular orbitals of a linear and a bent $XH_2$ molecule have been discussed in section 2(b) and the order of the orbitals is given in Figs. 120 and 123. For a linear conformation the two H atom $1s$ orbitals form the non-localized orbitals $\sigma_g$ and $\sigma_u$ while the $2p$ orbital of the X atom forms a $\sigma_u$ and a $\pi_u$ molecular orbital. The two $\sigma_u$ orbitals (one from $1s_{\mathrm{H}} - 1s_{\mathrm{H}}$, the other from $2p_{\mathrm{X}}$), since they are close in energy, interact strongly ("resonate") and in the molecule form two orbitals of widely different energy: one, $1\sigma_u$, is bonding, the other, $2\sigma_u$, is anti-bonding. The $\pi_u$ orbital from $2p_{\mathrm{X}}$ has no other orbital in its neighborhood to interact with, and although eventually in the united atom according to Fig. 120 its energy goes down, it must for large separations of the nuclei be considered as a non-bonding orbital. Thus in the ground state of *linear* $H_2O$ with the electron configuration

$$K(2\sigma_g)^2(1\sigma_u)^2(1\pi_u)^4$$

there are only two bonding electrons. The $2\sigma_g$ orbital is only very slightly bonding since the $\sigma_g 2s_{\mathrm{O}}$ orbital interacts only very slightly with $\sigma_g$ $(1s_{\mathrm{H}} + 1s_{\mathrm{H}})$ because of their wide separation in energy; that is $2\sigma_g$ is, even in the molecule, essentially $\sigma_g 2s_{\mathrm{O}}$.

For a *non-linear*, but symmetrical conformation of the three atoms (point group $C_{2v}$) the two H atom $1s$ orbitals form the non-localized orbitals $a_1$ and $b_2$ while the $2p$ orbital of the O atom forms the orbitals $a_1$, $b_1$, $b_2$ (see Fig. 123). Now we have two resonating pairs of orbitals $a_1(1s_{\mathrm{H}} + 1s_{\mathrm{H}})$ and $a_1 2p_{\mathrm{O}}$, and $b_2(1s_{\mathrm{H}} - 1s_{\mathrm{H}})$ and $b_2 2p_{\mathrm{O}}$ giving rise to the two pairs of molecular orbitals $3a_1$, $4a_1$ and $1b_2$, $2b_2$. One of each pair is a bonding orbital ($3a_1$ and $1b_2$) while the other is an anti-bonding orbital ($4a_1$ and $2b_2$). The third orbital, $1b_1$, arising from the $2p_{\mathrm{O}}$ atomic orbital is non-bonding or only slightly bonding since there is no other orbital of the same species near-by with which it can interact. Thus in the ground state of *bent* $H_2O$ with the electron configuration

$$K(2a_1)^2(1b_2)^2(3a_1)^2(1b_1)^2$$

there are four bonding electrons. According to the rough rule of thumb previously given one would therefore expect bent $H_2O$ to be more stable than linear $H_2O$; here just as in the valence-bond method there are four electrons that produce the binding, but now the four electrons are spread over the whole molecule. Just as in the valence-bond method there is a lone pair of electrons, namely the two electrons in the $1b_1$ orbital.

One may of course argue that the bonding power of the two bonding electrons in linear $H_2O$ could be higher than the bonding power of the four electrons in bent $H_2O$. That this is not so may be seen from the following rough mathematical formulation of the preceding qualitative considerations. It must be kept in mind, however, that small neglects may have a considerable effect on the calculated energy differences of conformations with different angles since the energy differences are in general fairly small (of the order of 1 eV).

The mathematical form of the four orbital wave functions $3a_1$, $4a_1$, $1b_2$, $2b_2$ can immediately be written in the approximation that uses linear combinations of atomic orbitals; they are

$$\begin{aligned} \tilde{\psi}_3(3a_1) &= \psi_3 + \lambda\psi_4 \\ \tilde{\psi}_4(4a_1) &= -\lambda\psi_3 + \psi_4 \end{aligned} \qquad \text{(III, 113)}$$

$$\begin{aligned} \tilde{\psi}_5(1b_2) &= \psi_5 + \mu\psi_6 \\ \tilde{\psi}_6(2b_2) &= -\mu\psi_5 + \psi_6 \end{aligned} \qquad \text{(III, 114)}$$

where $\psi_3 = a_1 2p_z$, $\psi_4 = a_1(1s_{H'} + 1s_{H''})$, $\psi_5 = b_2 2p_y$ and $\psi_6 = b_2(1s_{H'} - 1s_{H''})$. These molecular orbitals are essentially the same as those in eq. (III, 9) except that the interaction with the orbitals $1s$ and $2s$ of the O atom has been neglected and that normalization factors have been omitted. The secular equation for the first pair of orbitals, $\tilde{\psi}_3$ and $\tilde{\psi}_4$, is

$$\begin{vmatrix} E_p - E & H_{34} \\ H_{34} & E_H - E \end{vmatrix} = 0 \qquad \text{(III, 115)}$$

and for the second pair, $\tilde{\psi}_5$ and $\tilde{\psi}_6$,

$$\begin{vmatrix} E_p - E & H_{56} \\ H_{56} & E_H - E \end{vmatrix} = 0 \qquad \text{(III, 116)}$$

Here $E_p$ and $E_H$ are the energies of the $2p_O$ and $1s_H$ electrons in the free atoms and

$$H_{34} = \int \psi_3 H \psi_4 d\tau, \qquad H_{56} = \int \psi_5 H \psi_6 d\tau, \qquad \text{(III, 117)}$$

$H$ being the interaction Hamiltonian of the three atoms at finite distances.

If the angle between the OH direction and the $z$-axis is $\alpha$, it is readily shown [see Coulson (7)] that

$$H_{34} = 2\beta_{OH} \cos\alpha, \qquad H_{56} = 2\beta_{OH} \sin\alpha \qquad \text{(III, 118)}$$

where $\beta_{OH}$ is the resonance integral between an O atom $2p$ orbital and an H atom

1$s$ orbital when the axis of the 2$p$ orbital is directed toward the H atom. Solving now the eqs. (III, 115) and (III, 116) after substitution of (III, 118) we find

$$\begin{aligned} E_{3,4} &= \tfrac{1}{2}(E_p + E_{\mathrm{H}}) \pm \tfrac{1}{2}\sqrt{(E_p - E_{\mathrm{H}})^2 + 16\beta_{\mathrm{OH}}^2 \cos^2 \alpha} \\ E_{5,6} &= \tfrac{1}{2}(E_p + E_{\mathrm{H}}) \pm \tfrac{1}{2}\sqrt{(E_p - E_{\mathrm{H}})^2 + 16\beta_{\mathrm{OH}}^2 \sin^2 \alpha}. \end{aligned} \qquad \text{(III, 119)}$$

If we fill the lower of each of these pairs of orbitals (i.e. $\tilde{\psi}_3$ and $\tilde{\psi}_5$) with four electrons we obtain for the energy of the ground state (relative to that of the separated atoms)

$$\begin{aligned} E = 2E_3 + 2E_5 = 2(E_p + E_{\mathrm{H}}) &- \sqrt{(E_p - E_{\mathrm{H}})^2 + 16\beta_{\mathrm{OH}}^2 \cos^2 \alpha} \\ &- \sqrt{(E_p - E_{\mathrm{H}})^2 + 16\beta_{\mathrm{OH}}^2 \sin^2 \alpha}. \end{aligned} \qquad \text{(III, 120)}$$

This expression has a minimum value for $\alpha = 45°$. Thus according to this approximate molecular orbital treatment the most stable configuration of $H_2O$ corresponds to an H—O—H angle of 90° in complete agreement with the elementary form of the valence-bond theory.

In Fig. 150 the variation of the energy of the orbitals $\tilde{\psi}_3$ and $\tilde{\psi}_5$ with angle $\alpha$ is shown graphically as well as the variation of the total energy $E = 2(E_3 + E_5)$. In this approximation $E_3 = E_5$ for $\alpha = 45°$. Figure 150 gives the justification for the way in which in the Walsh diagram, Fig. 125, the $3a_1$ and $1b_2$ orbitals have been drawn. Figure 125 corresponds to the right-hand part of Fig. 150 from 45° to 90°. Once the Walsh diagram is accepted the structure of other $XH_2$ molecules in their ground and low excited states is readily predicted.

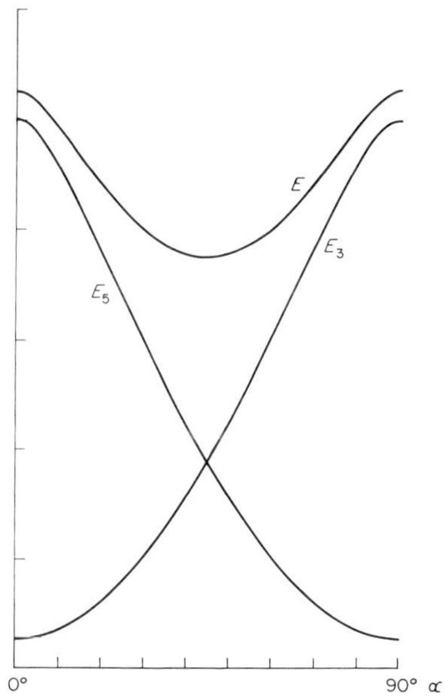


FIG. 150. **Orbital energies, $E_3$ and $E_5$, and molecular energy $E = 2(E_3 + E_5)$ of an $XH_2$ molecule as a function of the HXH angle 2α.** The scale for $E$ is twice that for $E_3$ and $E_5$.

Ellison and Shull (353) have carried out a detailed treatment of the ground state of $H_2O$ according to the LCAO self-consistent field molecular orbital method including all the orbitals given in eq. (III, 9) above. They have obtained the coefficients $c_{ik}$ in the linear combinations as well as the energies of the orbitals and of the molecule as a whole for various assumed H—O—H angles but a fixed OH distance equal to the observed value. In Table 41 we reproduce their results for the coefficients for an angle of 105°. It is not surprising to find that the contributions of $\psi_4(1s_{\mathrm{H'}} + 1s_{\mathrm{H''}})$ and $\psi_3(2p_{z\mathrm{O}})$ to $\tilde{\psi}_1(1a_1)$ are negligible. On the other hand the contribution of $\psi_2(2s_{\mathrm{O}})$ to the bonding orbital $\tilde{\psi}_3(3a_1)$ is comparable to that of $\psi_3(2p_{zO})$ and $\psi_4(1s_{H'} + 1s_{H''})$ contrary to expectation by the elementary treatment above. In other words, there is a considerable amount of $2s-2p$ mixing. The contributions of $\psi_1(1s_O)$, $\psi_3(2p_{zO})$ and $\psi_4(1s_{H'} + 1s_{H''})$ to $\tilde{\psi}_2(2a_1)$ and of $\psi_1(1s_O)$ to $\tilde{\psi}_3(3a_1)$ are small but not negligible. The calculated total energy has a minimum for an angle of 120° which deviates by 15° from the true value 105°.

TABLE 41. COEFFICIENTS $c_{ik}$ OF ATOMIC ORBITALS IN THE MOLECULAR ORBITALS OF $H_2O$ IN THE SELF-CONSISTENT FIELD TREATMENT OF ELLISON AND SHULL (353) FOR AN H—O—H BOND ANGLE OF 105° [COMPARE EQ. (III, 9)]

| Molecular orbital | Coefficient of atomic orbitals | | | | | | |
|---|---|---|---|---|---|---|---|
| | $\psi_1$ $a_1(1s_O)$ | $\psi_2$ $a_1(2s_O)$ | $\psi_3$ $a_1(2p_{zO})$ | $\psi_4$ $a_1$ $(1s_{H'}+1s_{H''})$ | $\psi_5$ $b_2(2p_{yO})$ | $\psi_6$ $b_2$ $(1s_{H'}-1s_{H''})$ | $\psi_7$ $b_1(2p_{xO})$ |
| $\tilde{\psi}_1(1a_1)$ | 1.0002 | 0.0163 | 0.0024 | −0.0033 | — | — | — |
| $\tilde{\psi}_2(2a_1)$ | −0.0286 | 0.8450 | 0.1328 | 0.1781 | — | — | — |
| $\tilde{\psi}_3(3a_1)$ | −0.0258 | −0.4601 | 0.8277 | 0.3341 | — | — | — |
| $*\tilde{\psi}_4(4a_1)$ | −0.086 | −0.833 | −0.642 | 1.061 | — | — | — |
| $\tilde{\psi}_5(1b_2)$ | — | — | — | — | 0.5428 | 0.7759 | — |
| $*\tilde{\psi}_6(2b_2)$ | — | — | — | — | −1.013 | 1.230 | — |
| $\tilde{\psi}_7(1b_1)$ | — | — | — | — | — | — | 1.000 |

* The orbitals marked by an asterisk are not occupied in the ground state. The numbering is the same as in eq. (III, 9).

This discrepancy must be ascribed to the fact that the total energy is a very large quantity and the change of energy with angle is only 0.13 per cent of the total energy. Therefore as already emphasized very minor neglects can have a large effect on the calculated angle.

For an appreciation of the way in which molecular orbital theory yields information about molecular stability and valence it is instructive to consider the overlap populations in $H_2O$ evaluated by Mulliken (911) from the orbital coefficients of Ellison and Shull. They are reproduced in Table 42. Considering first the total overlap population (last column of Table 42) one sees that the main contributions are from the orbitals $2a_1$ and $1b_2$. The partial overlap populations show that the overlap in $2a_1$ is mainly due to the overlap of $2s_O$ with the symmetrical combination of the two hydrogen $1s$ orbitals while the overlap population in $1b_2$ is due entirely to the overlap of $2p_{yO}$ with the antisymmetric combination of the $1s_H$ orbitals, which, incidentally, between themselves have a rather large negative overlap implying strong repulsion. It should however be noted that the positive contribution of the overlap of $2s_O$ and $1s_{H'} + 1s_{H''}$ to $2a_1$ is balanced by a very similar negative contribution to $3a_1$ so that the total contribution of the $(2s_O;$ $1s_{H'} + 1s_{H''})$ overlap is quite small (last line of Table 42). On the other hand the contribution of $(2p_{zO}; 1s_{H'} + 1s_{H''})$ is large as is the contribution of $(2p_{yO}; 1s_{H'} - 1s_{H''})$. Thus the preceding more elementary considerations in which only $2p_{zO}$ and $2p_{yO}$

TABLE 42. OVERLAP POPULATIONS IN THE GROUND STATE OF $H_2O$ ACCORDING TO MULLIKEN (911) BASED ON THE SELF-CONSISTENT FIELD TREATMENT OF ELLISON AND SHULL (353)

| Molecular orbital | Partial overlap population $n(i;\,ar,\,bs)$ | | | | | Total overlap population $n(i)$ |
|---|---|---|---|---|---|---|
| | $a_1(1s_O;\ 1s_{H'}+1s_{H''})$ | $a_1(2s_O;\ 1s_{H'}+1s_{H''})$ | $a_1(2p_{zO};\ 1s_{H'}+1s_{H''})$ | $b_2(2p_{yO};\ 1s_{H'}-1s_{H''})$ | $(1s_{H'};\ 1s_{H''})$ | |
| $\tilde{\psi}_1(1a_1)$ | −0.0012 | 0.000 | 0.000 | — | 0.000 | −0.001 |
| $\tilde{\psi}_2(2a_1)$ | −0.0018 | 0.419 | 0.028 | — | 0.024 | +0.469 |
| $\tilde{\psi}_3(3a_1)$ | −0.0030 | −0.432 | 0.332 | — | 0.084 | −0.019 |
| $\tilde{\psi}_5(1b_2)$ | — | — | — | 0.658 | −0.450 | +0.208 |
| $\tilde{\psi}_7(1b_1)$ | — | — | — | — | — | 0.000 |
| $n(ar,\,bs)$ | −0.0060 | −0.013 | 0.360 | 0.658 | −0.342 | $n = 0.657$ |

were used are not invalidated. For comparison with the total overlap population in $H_2O$, that is, $n = 0.657$, we note that the overlap population in $H_2$ is 0.858.

Finally in Table 43 the contributions of the individual atomic orbitals to the gross atomic populations are given for the orbitals of $H_2O$ that are occupied in the ground state. As it should be, the total gross population $N(i)$ in each molecular orbital comes out to be close to 2.0. The gross populations of the individual atomic orbitals $N(a, r)$ are not all 2.0. We can write, as above for CO, an effective electron configuration in the ground state of $H_2O$ as follows:

$$1s_O{}^{2.00}2s_O{}^{1.85}2p_O{}^{4.50};\quad 1s_{H'}{}^{0.82}1s_{H''}{}^{0.82}$$

According to this the total electronic charge on the O atom is 8.35 while on the two H atoms it is 1.65; in other words $H_2O$ is ionic to the extent of 0.35 electronic charges. In spite of the increased charge on O, the $2s$ orbital has a charge less than normal while $2p$ has appreciably more. This may be taken as an indication of $2s-2p$ hybridization in $H_2O$.

TABLE 43. GROSS ATOMIC POPULATIONS IN $H_2O$ ACCORDING TO MULLIKEN (911)

| olecular orbital | Partial populations N$(i;\,a,\,r)$ | | | | | | | Gross atomic populations | | |
|---|---|---|---|---|---|---|---|---|---|---|
| | $1s_O$ | $2s_O$ | $2p_{zO}$ | $2p_{yO}$ | $2p_{xO}$ | $1s_{H'}+1s_{H''}$ | $1s_{H'}-1s_{H''}$ | $N(i;\,O)$ | $N(i;\,H_2)$ | $N(i)$ |
| $_1(1a_1)$ | 2.0002 | 0.0005 | 0.0000 | — | — | −0.0005 | — | 2.0007 | −0.0005 | 2.000 |
| $_2(2a_1)$ | 0.0008 | 1.638 | 0.049 | — | — | 0.309 | — | 1.688 | 0.309 | 1.997 |
| $_3(3a_1)$ | −0.0001 | 0.209 | 1.534 | — | — | 0.257 | — | 1.743 | 0.257 | 2.000 |
| $_5(1b_2)$ | — | — | — | 0.918 | — | — | 1.080 | 0.918 | 1.080 | 1.998 |
| $_7(1b_1)$ | — | — | — | — | 2.000 | — | — | 2.000 | — | 2.000 |
| $N(a,r)$ | 2.0009 | 1.847 | 1.583 | 0.918 | 2.000 | 0.565 | 1.080 | N(O) = 8.349 | N($H_2$) = 1.645 | N = 9.995 |

**The ground state of $NH_3$.** The molecular orbitals of planar and non-planar $XH_3$ molecules have been discussed in section 2(b) and the order of the orbitals was given in Fig. 127a and b. In planar $XH_3$ the $e'$ orbital arising from the $1s$ orbitals of the three H atoms resonates with the $e'$ orbital from $2p_X$ and we obtain one strongly bonding and one strongly anti-bonding orbital ($1e'$ and $2e'$ respectively). The $a_1'$ orbital arising from $1s_H$ is slightly anti-bonding to the extent that it interacts with the $a_1'$ orbital from $2s_X$ which is slightly bonding; the $a_2''$ orbital from $2p_X$ is non-bonding. In the ground state of $NH_3$ assuming it to be planar the orbitals are filled up to (and including) $1a_2''$ (see Fig. 127a). Neglecting the slight bonding effect of the $2a_1'$ orbital we therefore have *four bonding electrons.*

On the other hand in non-planar $XH_3$ (Fig. 127b) both the species $a_2''$ and $a_1'$ go over into $a_1$ and we have therefore a pair of strongly resonating $a_1$ orbitals, one from $1s_H$ and one from $2p_X$ giving rise to one strongly bonding and one strongly anti-bonding orbital ($3a_1$ and $4a_1$). Compared to this strong interaction the effect of the $2a_1$ orbital from $2s_X$ can be neglected since its energy is rather different. The two $e$ orbitals from $1s_H$ and $2p_X$ behave in the same way as $e'$ in planar $XH_3$. Thus in the ground state of non-planar $NH_3$ we have the configuration $(2a_1)^2(1e)^4(3a_1)^2$ with *six bonding electrons.* Because of the greater number of bonding electrons in the non-planar as compared to the planar conformation we may expect $NH_3$ to be non-planar as is indeed observed. The Walsh diagram of Fig. 128 had already made this clear on the basis of a somewhat similar reasoning. A rough quantitative discussion similar to that for $H_2O$ (p. 387) would lead to the same result. The agreement of these predictions with experiment should however not lead one to overlook the roughness of the approximation used and the effect which small neglected terms may have on the small energy differences between conformations of different angles.

TABLE 44. COEFFICIENTS $c_{ik}$ OF ATOMIC ORBITALS IN THE MOLECULAR ORBITALS OF $NH_3$ IN THE SELF-CONSISTENT FIELD TREATMENT OF KAPLAN (656) FOR AN H—N—H ANGLE OF 106° 47′ [COMPARE EQ. (III, 12)]

| Molecular orbital[a] | Coefficients of atomic orbitals[b] | | | | | |
|---|---|---|---|---|---|---|
| | $\psi_1$ $a_1(1s_N)$ | $\psi_2$ $a_1(2s_N)$ | $\psi_3$ $a_1(1s_H)$ | $\psi_4$ $a_1(2p_{zN})$ | $\psi_5$ $e(2p_{xyN})$ | $\psi_6$ $e(1s_H)$ |
| $\tilde{\psi}_1(1a_1)$ | +1.0001 | −0.0033 | −0.0013 | −0.0020 | — | — |
| $\tilde{\psi}_2(2a_1)$ | +0.0286 | +0.7591 | −0.2711 | +0.1616 | — | — |
| $\tilde{\psi}_3(3a_1)$ | +0.0257 | −0.4418 | −0.2582 | +0.8956 | — | — |
| $*\tilde{\psi}_4(4a_1)$ | −0.1478 | +1.2773 | +1.5556 | +0.5527 | — | — |
| $\tilde{\psi}_5(1e)$ | — | — | — | — | +0.6195 | +0.4860 |
| $*\tilde{\psi}_6(2e)$ | — | — | — | — | −1.1303 | +1.1938 |

* The orbitals marked by an asterisk are not occupied in the ground state.

[a] The numbering of the molecular orbitals is the same as in eq. (III, 12) except that here $a_1' \to a_1$, $a_2'' \to a_1$, $e' \to e$.

[b] The numbering and form of the atomic orbitals is similar to that in eqs. (III, 10) and (III, 11). $a_1(1s_H)$ and $e(1s_H)$ are formed from all three H atomic orbitals as in eq. (III, 11).

Detailed molecular orbital calculations of $NH_3$ according to the self-consistent field method have been carried out by Kaplan (656) and Duncan (321). Peters (977) has used Kaplan and Duncan's wave functions for a population analysis. Table 44 gives the coefficients of the atomic orbitals (see eq. III, 12) obtained

TABLE 45. OVERLAP POPULATIONS IN THE GROUND STATE OF $NH_3$ ACCORDING TO PETERS (977) BASED ON THE SELF-CONSISTENT FIELD ORBITALS OF KAPLAN (656)

| Molecular orbital | Partial overlap population[a] $n(i; ar, bs)$ | | | | | Total overlap population $n(i)$ |
|---|---|---|---|---|---|---|
| | $a_1$ $(1s_N;1s_H)$ | $a_1$ $(2s_N;1s_H)$ | $a_1$ $(2p_{zN};1s_H)$ | $e$ $(2p_{xyN};1s_H)$ | $\left.\begin{matrix}a_1\\e\end{matrix}\right\}(1s_H;1s_H)$ | |
| $\tilde{\psi}_1(1a_1)$ | — | — | — | — | — | 0.000 |
| $\tilde{\psi}_2(2a_1)$ | −0.003 | +0.609 | +0.040 | — | +0.059 | +0.705 |
| $\tilde{\psi}_3(3a_1)$ | −0.003 | −0.338 | +0.212 | — | +0.053 | −0.076 |
| $\tilde{\psi}_5(1e)$ | — | — | — | +0.760 | −0.237 | +0.523 × 2 |
| $n(ar, bs)$ | −0.006 | +0.271 | +0.252 | +1.520 | −0.362 | $n$ = 1.675 |

[a] $1s_H$ stands for the three H atoms and implies, in the three $a_1$ orbitals, $\psi_3$ of (III, 11) and, in the $e$ orbital, $\psi_6$ or $\psi_8$ of (III, 11).

by Kaplan and Table 45 the overlap populations derived from them by Peters. The H—N—H angle has been assumed to be 106° 47′. It is seen from Table 44 that there is considerable mixing of $a_1(2s_N)$, $a_1(1s_H)$ and $a_1(2p_{zN})$. Correspondingly the $2a_1$ molecular orbital contributes much more to the overlap population (and therefore to the binding) than we expected above on the basis of qualitative considerations. There is a fairly strong contribution of $(2p_{zN};\ 1s_H)$ to the overlap population in $3a_1$ but a negative contribution of $(2s_N;\ 1s_H)$, resulting somewhat unexpectedly in a small negative overlap population of the $3a_1$ orbital as a whole. The $1e$ orbital, in agreement with expectation, has a large overlap population.

The total overlap population in $NH_3$ summed over all occupied orbitals is 1.675, i.e. more than twice as large as in $H_2O$. Considering that the experimentally determined atomic heat of formation of $NH_3$ is only 26 per cent larger than that of $H_2O$ we must conclude that the total overlap population does not correlate linearly with the total binding energy. Calculation of the gross atomic populations in $NH_3$ yields the following representation of the ground state in terms of atomic orbitals

$$1s_N{}^{2.00}2s_N{}^{1.68}2p_N{}^{4.08};\quad 1s_H{}^{2.24}.$$

Kaplan (656) has carried out self-consistent field calculations also for the planar form of $NH_3$. While he does find that the pyramidal form has the lower energy the energy difference comes out to be 16 times the observed value. This somewhat disappointing result reflects the difficulty already mentioned of obtaining with sufficient accuracy the difference of two very large numbers, the total energies of the planar and non-planar form when each of these is influenced by some of the neglects made in the calculations.

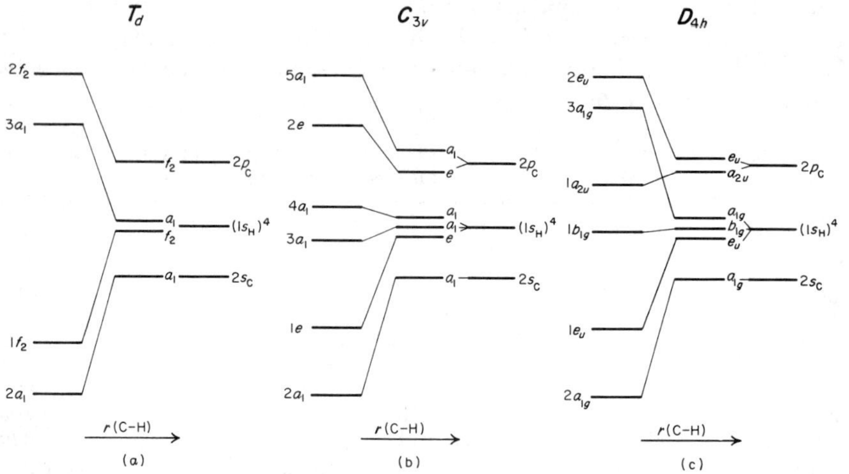


FIG. 151. **Correlation diagrams for the orbitals of $CH_4$ for three assumed structures of the nuclear frame** (*a*) **tetrahedron ($T_d$),** (*b*) **distorted tetrahedron ($C_{3v}$),** (*c*) **plane square ($D_{4h}$).** In each diagram the molecular orbital energies are at left, the energies of the atomic orbitals for large C–H distance at right.

**The ground state of $CH_4$.** From molecular orbital theory it is not as simple as from valence-bond theory to decide which conformation of the five atoms of methane would be the most stable one. Let us consider three possible structures: a regular tetrahedron ($T_d$), a non-regular tetrahedron ($C_{3v}$) and a square planar form ($D_{4h}$). In Fig. 151a, b, c the lowest orbitals for these three cases are shown. In the regular tetrahedral conformation the four $1s_H$ orbitals form an $f_2$ and $a_1$ orbital (see p. 321) while the $2s_C$ and $2p_C$ orbitals form $a_1$ and $f_2$ respectively. The resonance between the two $a_1$ and the two $f_2$ orbitals leads to two strongly bonding orbitals, $2a_1$ and $1f_2$, and two strongly anti-bonding orbitals, $3a_1$ and $2f_2$; the $1a_1$ orbital (not shown) is essentially $1s_C$ and therefore non-bonding. The eight outer electrons just fill the two bonding orbitals, i.e. we have eight bonding electrons all of which of course are distributed over the whole molecule.

If one H atom has a greater distance from the C atom than the others (point group $C_{3v}$), we have only $a_1$ and $e$ orbitals as shown in Fig. 151b. Of these the orbitals $2a_1$ and $1e$ are clearly bonding orbitals while $3a_1$ and $4a_1$ are non-bonding and $2e$ and $5a_1$ strongly anti-bonding. Thus there are in the ground state only six bonding electrons. The same applies to the square conformation ($D_{4h}$) as shown by Fig. 151c. Thus according to the previous rule of thumb (p. 376) the regular tetrahedron gives the most stable conformation.

It should be noted that the eight bonding electrons in tetrahedral $CH_4$ result here without the assumption of $sp^3$ hybridization. Indeed since $2s_C$ gives an $a_1$ orbital, $2p_C$ an $f_2$ orbital in the tetrahedral field they cannot interact for reasons of symmetry. Rather it is the fact that these carbon orbitals have the same symmetry as the $1s$ hydrogen orbitals which causes a particularly strong resonance and therefore the formation of two strongly bonding molecular orbitals. These bonding orbitals are just filled by the available electrons and thus give rise to a particularly stable conformation. A population analysis for $CH_4$ has not yet been made.

Recently a very detailed treatment using contributions of more than 25 atomic orbitals has been carried out by Moccia (864) [see also Kraus (692) and Woznick (1321)]. The tetrahedral structure of $CH_4$ is found to give the lowest energy. Moccia has also carried out similar calculations for $H_2O$, $NH_3$, $SiH_4$ and related molecules.

**The ground state of $CO_2$.** The order and correlation of the orbitals in linear $XY_2$ have been discussed earlier in connection with Fig. 121. The two orbitals $\sigma_g$ and $\sigma_u$ arising from the $2s$ orbitals of the two O atoms lie fairly low and may be assumed to interact only slightly with the other $\sigma_g$ and $\sigma_u$ orbitals [see Mulliken (892)]. But the $\sigma_g$ orbital arising from $2s_C$ lies higher and interacts strongly with $\sigma_g$ from $2p_{zO} + 2p_{zO}$ thus forming the bonding molecular orbital $4\sigma_g$ and the anti-bonding orbital $5\sigma_g$[10a]. Similarly the $\sigma_u$ orbital from $2p_{zO} - 2p_{zO}$ resonates with $\sigma_u$ from $2p_{zC}$ forming another bonding–anti-bonding pair: $3\sigma_u$ and $4\sigma_u$; and finally the two $\pi_u$ orbitals arising from $2p_O + 2p_O$ and $2p_C$ resonate giving rise to the bonding $1\pi_u$ and the anti-bonding $2\pi_u$. The $\pi_g$ orbital arising from $2p_O - 2p_O$, since it is the only one of its kind with $n = 2$, is non-bonding.

In the ground state of $CO_2$ the 16 electrons outside the $K$ shell fill the orbitals up to $1\pi_g$:

$$(K)(K)(K)(3\sigma_g)^2(2\sigma_u)^2(4\sigma_g)^2(3\sigma_u)^2(1\pi_u)^4(1\pi_g)^4. \qquad \text{(III, 121)}$$

Apart from non-bonding orbitals the bonding orbitals $4\sigma_g$, $3\sigma_u$, $1\pi_u$ are filled: we have eight bonding electrons, of which four are $\pi$ electrons. This is the same result as that of valence-bond theory (p. 365f). It is interesting to compare the form of the bonding orbitals in the two theories. For valence-bond theory they were given schematically in Fig. 146, for molecular orbital theory we may refer to the schematic representation in Fig. 122. The essential difference is that in molecular orbital theory all orbitals (both $\sigma$ and $\pi$) extend over the whole molecule. One component of the $1\pi_u$ orbital is in the plane of the paper (Fig. 122), the other in a plane perpendicular to it, and each is filled with two electrons. In contrast in valence-bond theory (Fig. 146b) the left $\pi$ bond is made up of two electrons, one from C and one from $O_I$ with a nodal plane in the $xz$ plane while the right $\pi$ bond

[10a] Note that in Figs. 121 and 122, which apply to linear $XY_2$, a strong interaction of the $2s_Y$ orbitals with $2s_X$ was assumed while the interaction of $2s_X$ with $2p_{zY}$ was considered as small, and therefore $3\sigma_g$ was considered as bonding and $4\sigma_g$ as anti-bonding. In fact, of course, all three orbitals are mixed and the relative contributions to $3\sigma_g$, $4\sigma_g$ and $5\sigma_g$ depend on the energy differences between $2s_X$, $2s_Y$ and $2p_Y$. For $CO_2$, according to Mulligan (887) and Mulliken (892)(914), $3\sigma_g$ has relatively little admixture of $2s_C$ and thus is non-bonding. Therefore in Fig. 122 the bottom line should be separated off, and $\sigma_g(2s_X)$ should be combined with $\sigma_g(2p_Y + 2p_Y)$ at the top. As a result in Fig. 121, if applied to $CO_2$, $4\sigma_g$ should show a downward bend corresponding to bonding and $5\sigma_g$ an upward one corresponding to anti-bonding.

arises from two other electrons, one from C and one from $O_{II}$ with a nodal plane in the $yz$ plane, that is, the bonds are localized. This localization is, however, lifted by the resonance between the two opposite valence structures (see p. 368), an effect that in molecular orbital theory is taken into account from the beginning since we use molecular symmetry orbitals.

If $CO_2$ were bent the $\sigma_g$ and $\sigma_u$ orbitals would go over into $a_1$ and $b_2$ respectively while $\pi_u$ would split into $a_1 + b_1$ and $\pi_g$ into $a_2 + b_2$. The $a_1(\pi_u)$ orbital can now interact with $4a_1$ (corresponding to $4\sigma_g$) and will become a non-bonding orbital. Thus there are only six instead of eight bonding electrons and in a very rough approximation we may say that for this reason $CO_2$ is linear. A more detailed discussion has been given by Walsh (1264).

**The ground state of $C_2H_4$.** The order of the orbitals in planar $C_2H_4$ and their correlation with those of $CH_2$ are given in the previous Fig. 132. Each of the orbitals of $CH_2$ splits in $C_2H_4$ into a bonding and an anti-bonding orbital (just as in $H_2^+$), the first without, the second with a node between the two C atoms. However, both members of the pair arising from $2a_1$ and of the pair arising from $1b_2$ of $CH_2$ are filled in the ground state of $C_2H_4$ and therefore the net bonding action of these four orbitals is small. In fact the bonding and anti-bonding actions of $1b_{2u}$ and $1b_{3g}$ are probably very small since they come from an orbital ($1b_2$) of $CH_2$ that is strongly C—H bonding and thus is likely to remain mainly concentrated in the two $CH_2$ groups. This is indicated in Fig. 132 by the fact that the energy of these orbitals changes very little when the C–C distance is reduced. On the other hand the pairs of orbitals derived from $3a_1$ and $1b_1$ of $CH_2$ are fairly widely split into a bonding and an anti-bonding orbital each, and only the bonding one in each pair, viz. $3a_g$ and $1b_{3u}$, is occupied in the ground state of $C_2H_4$.

The twelve electrons outside the $K$ shells of the C atoms fill all orbitals just up to $1b_{3u}$ (Fig. 132) and thus we have a net number of *four* C—C bonding electrons in agreement with elementary valence considerations. The $3a_g$ bonding orbital arises from the $2p_z$ orbital of carbon and the two electrons in this orbital therefore correspond to the $\sigma$ bond while the $1b_{3u}$ orbital which is antisymmetric with respect to the plane of the molecule arises from the $2p_x$ orbital of C and therefore the two electrons in this orbital represent the $\pi$ bond (sideways overlap of the two $1b_1$ electrons of the two $CH_2$ groups). The $3a_g$ orbital loses some of its bonding power on account of interaction with the $2a_g$ orbital; but to the extent that it does, the $2a_g$ orbital gains in bonding power which is then no longer compensated by the anti-bonding power of $2b_{1u}$ and thus the net bonding power remains essentially the same.

Thus far we have assumed the planar $\boldsymbol{D}_{2h}$ structure of $C_2H_4$. It is interesting to see what the stability would be if the two $CH_2$ groups were assumed to have their planes at right angles to each other (point group $\boldsymbol{D}_{2d}$). Figure 152 shows the orbitals of "perpendicular" $C_2H_4$ in a similar fashion to Fig. 132 for planar $C_2H_4$. Here the $a_1$ orbitals of $CH_2$ split into an $a_1$ (bonding) and $b_2$ (anti-bonding) orbital of $C_2H_4$ while the $b_1$ orbitals of $CH_2$ do not split but form $e$ orbitals of $C_2H_4$. This

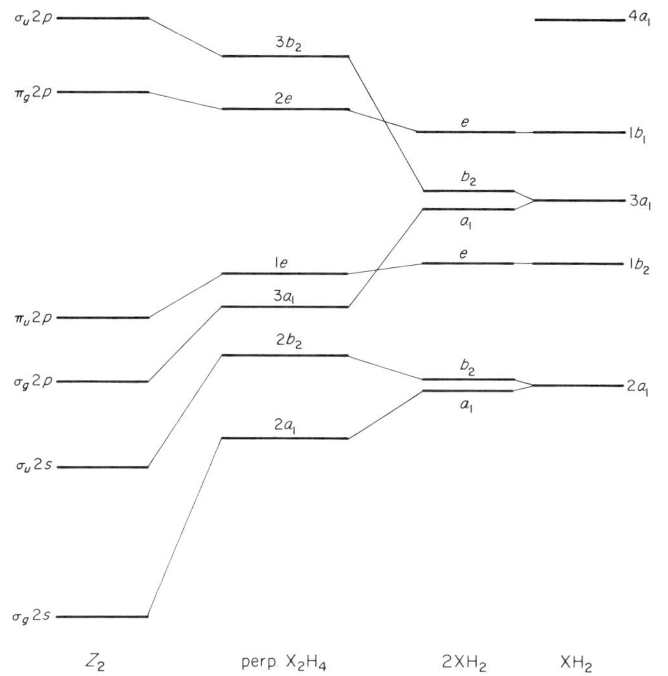


FIG. 152. **Correlation of orbitals of perpendicular $X_2H_4$ of point group $D_{2d}$ to those of $XH_2$ and to those of the united molecule $Z_2$.** In the right-hand part of the diagram the X–X distance, in the left-hand part the X–H distances increase from left to right.

is because in the point group $D_{2d}$ the orbitals $b_1(CH_2) + b_1(CH_2)$ and $b_1(CH_2) - b_1(CH_2)$ have the same energy, a conclusion that is immediately verified by considering Fig. 153, which gives an end-on view of the orbital lobes for these two $C_2H_4$ orbitals. The same conclusion applies to $b_2(CH_2)$. Therefore the orbitals $1e$ and $2e$ of perpendicular $C_2H_4$ are essentially non-bonding although their mutual interaction will result in a small degree of bonding and anti-bonding character respectively.

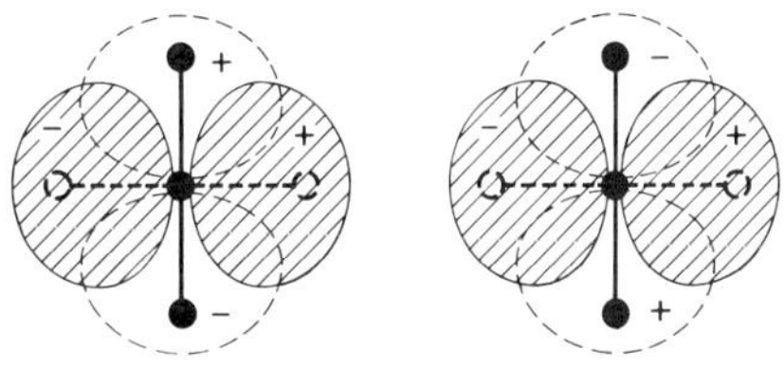

FIG. 153. **Schematic form of the two orbitals of perpendicular $C_2H_4$ derived from the $b_1$ orbitals of the two $CH_2$ groups.** The ($\pi$) orbitals are viewed along the C—C axis. The full lines refer to the lobes contributed by the $CH_2$ group in front, the broken lines to that behind it. One diagram goes over into the other by the operation $S_4$ showing that they are degenerate.

In the ground state of perpendicular $C_2H_4$ the orbitals are filled up to $2e$ which will contain two electrons yielding $^3A_2$ as the lowest state (see Table 31). Thus there are only two C—C bonding electrons ($3a_1$) and therefore the perpendicular form is less stable than the planar form. There is no second C—C bond ($\pi$ bond) formed from the $1b_1$ electrons of $CH_2$ as there is in the planar form; or, expressed in a different way, in the planar form we have maximum overlapping of the two $b_1(CH_2)$ orbitals while in the perpendicular form as shown by Fig. 153 there is an equal amount of overlapping of lobes of the same sign (bonding) as of lobes of opposite sign (anti-bonding) and therefore effectively no bonding action of the $b_1(CH_2)$ orbitals. Thus from molecular orbital theory we expect a strong potential barrier hindering the rotation of the two $CH_2$ groups away from the planar conformation; this is in agreement with experiment.

In spite of the strong similarities in the explanations of the planarity of $C_2H_4$ on the basis of valence-bond and of molecular orbital theory an important difference should not be overlooked. In valence-bond theory the $\pi$ bond is produced by the *exchange* between the two lone pair ($p_z$) electrons, one on each $CH_2$ group (see Fig. 145b); in molecular orbital theory it is produced by the *resonance* between the two $b_1$ orbitals (which occurs irrespective of whether or not they are occupied) and the fact that the lower of the two resultant orbitals, $b_{3u}$, is occupied by two electrons.

It should perhaps be emphasized that, except for the $1b_{3u}$ orbital (which is exclusively C—C bonding since it has the molecular plane as a nodal plane), all other occupied orbitals have both C—H and C—C bonding (or anti-bonding) properties; for some, like $2b_{1u}$, $1b_{2u}$, $1b_{3g}$, the CH bonding properties predominate while for others like $2a_g$ and $3a_g$ the C—C bonding predominates. But all are whole-molecule (non-localized) orbitals. A population analysis which would allow one to be more specific with regard to the distribution of bonding properties has not yet been attempted.

**Saturation of valencies.** The phenomenon of saturation of valencies is most easily visualized on the basis of the valence-bond theory. The understanding of this phenomenon on the basis of molecular orbital theory is not as clear-cut, but we shall try to illustrate it by a few examples.

If instead of two $CH_2$ groups as in the preceding discussion we bring together two $H_2O$ molecules with four more electrons it is clear from Fig. 132 that, if the symmetry of the resultant "molecule" is $\boldsymbol{D}_{2h}$, the four additional electrons will occupy in the lowest state the two anti-bonding orbitals $1b_{2g}$ and $3b_{1u}$. Thus there would be as many anti-bonding as there are bonding electrons and therefore $O_2H_4$ is not stable. The same result is obtained if the symmetry of the resulting molecule is $\boldsymbol{D}_{2d}$ (see Fig. 152): there would be two bonding and two anti-bonding electrons. For a lower assumed symmetry the result would be the same since bonding and anti-bonding orbitals are always formed in pairs and in the present case both would be occupied.

If two $NH_2$ groups are brought together symmetrically (either in point group $\boldsymbol{D}_{2h}$ or $\boldsymbol{D}_{2d}$), there will be two more bonding than anti-bonding electrons and so a stable molecule can arise as is indeed observed. That in the ground state the molecule has actually a lower symmetry is not so readily seen according to the molecular orbital theory as it is according to the valence-bond theory [see Penney and Sutherland (976)], but, as emphasized earlier, the energy differences between the different conformations are relatively small.

It is equally difficult to see in a simple way from molecular orbital theory why molecules like $H_2F$ or $H_3O$ or $NH_4$ either do not exist or have a relatively small energy of dissociation into H + HF, H + $H_2O$, H + $NH_3$; in other words, to see why the valency of F is saturated with one H atom, O with two, N with three H atoms. If in Fig. 123 nine electrons are put in the lowest orbitals the last electron has to go into the anti-bonding $4a_1$ orbital so that only a net number of three bonding electrons remains. On this basis stability of $H_2F$ would be possible; however its stability would be much less than that of $H_2O$, particularly since $4a_1$ is a very strongly anti-bonding orbital which correlates with a three-quantum state of the united atom. Therefore a small dissociation energy of $H_2F$ into H + HF is expected. No evidence for the existence of an $H_2F$ radical has yet been obtained. However, polymers of HF have been observed and their (moderate) stability is more readily accounted for on the basis of molecular orbital theory than on the basis of valence bond theory.

Similarly, if in Fig. 127 nine electrons are put in the lowest orbitals of $XH_3$ the last electron must go into the anti-bonding orbital $4a_1$ (or $3a_1'$) and therefore the net number of bonding electrons is reduced. Because of the strongly anti-bonding property of $4a_1$ the bonding is probably weakened more than proportionately to the lowering of the net number of bonding electrons. This, combined with the high stability of $H_2O$, makes the low stability of $H_3O$ understandable. Similar arguments would apply to $NH_4$ or $CH_5$.

In general, valence-bond theory gives a more straightforward account of the saturation of valencies than does molecular orbital theory, but the latter theory is more readily adaptable to an explanation of exceptions to the classical valence rules such as the chemical stability of molecules with an odd number of electrons or of the fluorides of Xe and Kr recently observed. Whether molecules like NO, $ClO_2$, $NO_2$ or whether their dimers $N_2O_2$, $Cl_2O_4$, $N_2O_4$ are chemically stable depends according to molecular orbital theory on the quantitative balance between the bonding and anti-bonding action of the molecular orbitals newly formed in the dimer. Without very detailed calculations it would be difficult to predict in a given case what the result of this balance would be. But if in the correlation diagrams the lines representing the various orbitals are adjusted to fit the observed facts in one or two cases they can be used for predictions in others. There is no necessary pairing of electrons from the component molecules of a dimer and therefore the chemical stability of molecules with an odd number of electrons is much more easily understandable than according to valence-bond theory.

**The ground state of $C_6H_6$.** In Fig. 154 the energies of the $2s$ and $2p$ orbitals of six carbon atoms and of the $1s$ orbitals of six hydrogen atoms are given at the left and the right for large separation and in the center for the actual conformation of the benzene molecule assuming $\boldsymbol{D}_{6h}$ symmetry. According to the methods previously explained (p. 328) the six $2s$ orbitals of the carbon atoms just as the six $1s$ orbitals of the hydrogen atoms give the orbitals $a_{1g}$, $e_{1u}$, $e_{2g}$ and $b_{1u}$ (in this order). Their wave functions are similar to those given in Fig. 136 except that they are

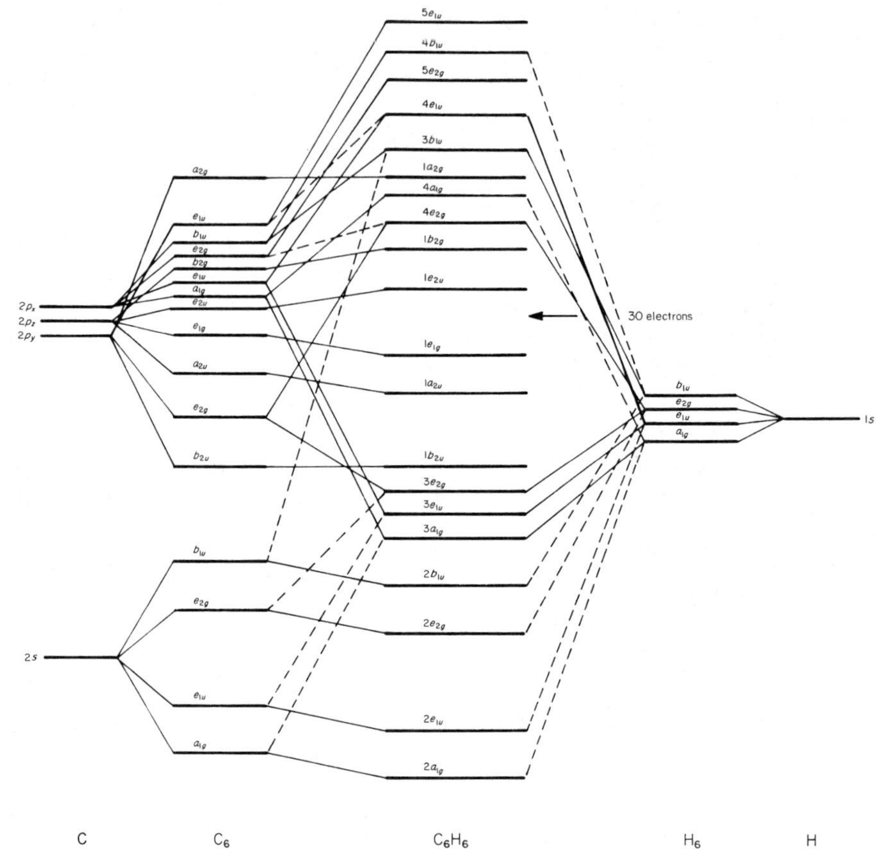


FIG. 154. **Correlation of orbitals of $C_6H_6$ to those of the separated atoms.** The orbital energies of six C atoms at large and medium distance are shown at left, those of six H atoms at right. The molecular orbitals of $C_6H_6$ are shown in the center (assuming $\boldsymbol{D}_{6h}$ symmetry). Those that are mixtures of both H atom and C atom orbitals have widely different energies from those of the $C_6$ and $H_6$ groups and are connected by leading lines to all those zero-approximation orbitals that contribute to them (broken lines refer to minor contributions).

symmetrical with respect to the plane of the molecule. The eighteen $2p$ orbitals of the carbon atoms give three groups of six orbitals, one derived from $2p_z$, one from $2p_y$ and one from $2p_x$ where, for the purpose of this discussion, the $x$, $y$, $z$ axes are local axes at each C atom, $x$ directed to the center of the molecule, $y$ in the plane and $z$ perpendicular to it. It is easily seen that the six $2p_z$ orbitals give (see p. 328)

$$a_{2u},\ e_{1g},\ e_{2u},\ b_{2g}$$

molecular orbitals, the six $2p_y$ orbitals give (see Table 61, Appendix V)

$$b_{2u},\ e_{2g},\ e_{1u},\ a_{2g}$$

and the six $2p_x$ orbitals give

$$a_{1g},\ e_{1u},\ e_{2g},\ b_{1u}.$$

As shown by Fig. 136 the $a_{2u}$ orbital wave function of the first group has the same sign for adjacent C atoms and therefore leads to a positive overlap, that is, to strong bonding for all six C—C bonds. The same applies to the $b_{2u}$ and $a_{1g}$ molecular orbitals derived from $2p_y$ and $2p_x$ respectively. On the other hand in the $b_{2g}$ orbital of $2p_z$ (see Fig. 136) and similarly $a_{2g}$ of $2p_y$ and $b_{1u}$ of $2p_x$ the wave function has opposite sign for adjacent C atoms leading to strong C—C anti-bonding action. The degenerate orbitals are less bonding or anti-bonding.

The group of orbitals arising from $2p_x$, since they have the same species as those from $2s$, is mixed with them in the $C_6$ molecule, that is, they are pushed upwards. The same applies to the $e_{1u}$ orbital from $2p_y$, while $e_{2g}$ being slightly C—C bonding is not much affected. The C—C bonding orbital $b_{2u}$ (from $2p_y$) is the only one of its kind in the valence shell and therefore not affected by any mixing. Rather it moves downward as the C atoms approach one another showing its bonding action. In $C_6$ it is in all probability much below the bonding orbitals from $2p_z$ (viz., $a_{2u}$ and $e_{1g}$) since the overlap is stronger for $p$ orbitals in the plane if they have the same sign on adjacent atoms (see Fig. 122).

In this way we obtain the order of the orbitals for $C_6$ in the column to the left of center in Fig. 154. If now the orbitals formed from $1s_H$ are introduced a further rearrangement of orbitals of type $a_{1g}$, $e_{1u}$, $e_{2g}$ and $b_{1u}$ takes place as indicated in Fig. 154 in the column designated $C_6H_6$: $a_{1g}$, $e_{1u}$ and $e_{2g}$ from $1s_H$ are pushed down because of mixing with the higher orbitals of the same species, which in turn are pushed up. The orbital $b_{1u}(1s_H)$ is pushed up because of mixing with $b_{1u}(2s_C)$ which lies comparatively high. The molecular orbitals derived from $2p_z$, since they have different species, are not affected by this rearrangement.

If now we fill the orbitals of $C_6H_6$ with the 30 electrons (other than $K$ electrons) present in the neutral molecule we see that in the ground state the orbitals are filled up to and including $1e_{1g}$ yielding a ${}^1A_{1g}$ state. The occupied orbitals mainly responsible for C—C bonding are $2a_{1g}$, $2e_{1u}$, $2e_{2g}$ and $1b_{2u}$ corresponding to normal $\sigma$ bonds and $1a_{2u}$ and $1e_{1g}$ corresponding to $\pi$ bonds, while the orbitals mainly responsible for C—H bonding are $3a_{1g}$, $3e_{1u}$, $3e_{2g}$ and $2b_{1u}$. However it must be realized that this distinction is not clear-cut since all orbitals are non-localized and most of them are mixtures of several atomic orbitals. Only the orbitals $a_{2u}$, $e_{1g}$, $e_{2u}$, $b_{2g}$ derived from $2p_z$, and $a_{2g}$, $b_{2u}$ from $2p_y$ definitely do not contribute to C—H bonding.

In the usual treatment of $C_6H_6$ only the orbitals derived from $2p_z$ are considered and the effect of the other orbitals is neglected (see p. 327f). Figure 154 shows that it is somewhat of an accident that the orbitals $3b_{1u}$ and $4e_{2g}$ are above the orbitals from $2p_z$ and not in the same energy region. This result has however been confirmed by the study of the ultraviolet spectrum of $C_6H_6$.

A more detailed discussion of the effect of the electrons derived from the $2p_z$ electrons of carbon, also called *unsaturation electrons* by Mulliken (896), shows that, since the orbitals are spread over the whole molecule, a lowering of the energy compared to the case of three localized orbitals takes place and as a consequence the internuclear distance is smaller than would correspond to the average of a single and a double bond. Coulson (237)(238) has defined a *fractional bond order* which when evaluated for $C_6H_6$ gives a value of 1.67 for every C—C bond in this molecule. If, using this bond order, one interpolates between the C—C distances for a pure single, double and triple bond ($C_2H_6$, $C_2H_4$ and $C_2H_2$) one predicts for $C_6H_6$ a bond length of 1.40 Å which agrees very well with the experimental value of 1.397 Å [Stoicheff (1164)].

We have seen previously (p. 327) that the two $\pi$ orbitals which are filled in the ground state of $C_6H_6$, i.e. $1a_{2u}$ and $1e_{1g}$, have energies $E_0 + 2\beta$ and $E_0 + \beta$. Therefore the energy of the six $\pi$ electrons is

$$2(E_0 + 2\beta) + 4(E_0 + \beta) = 6E_0 + 8\beta \qquad \text{(III, 122)}$$

where $\beta$ is the resonance integral giving the interaction between two neighboring C atoms produced by the $2p_z$ electrons (all interactions between non-neighbors are neglected). Now, for an electron localized between two atoms a and b the wave function of a bonding orbital is in a first approximation $\psi_a + \psi_b$ and according to (III, 98) (assuming $S = 0$ and putting $H_{ab} = \beta$) the energy is $E_0 + \beta$. Thus for six localized electrons the energy would be $6(E_0 + \beta)$. The effect of the delocalization is therefore a lowering of the energy by $-2\beta$ ($\beta$ is negative). It is this *delocalization energy* that corresponds to the resonance energy in valence-bond theory, that is, to $-1.106K$ (see p. 372). A rough theoretical value for $-2\beta$ is 40 kcal/mole [see Coulson (7)] which agrees quite satisfactorily with the observed resonance energy of 36.0 kcal/mole (see p. 372). In molecular orbital theory it is thus the fact that the $2p_z$ electrons become completely delocalized that causes the lowering of the energy and of the bond distance compared to the average of the values for single and double bonds. This state of affairs induced Lennard–Jones to call the $\pi$ electrons *mobile electrons*. In view of the fact that the $\sigma$ electrons are also delocalized we prefer the name *unsaturation electrons* suggested by Mulliken.

The study of $\pi$ electrons in other aromatic molecules has been of great value in the elucidation of their behavior. For a much more detailed discussion see the books of Coulson (7), Daudel (8), Hartmann (18), Streitwieser (39) and the review by Longuet–Higgins (765) as well as the recent papers by Hartmann (482) and Ruch (1087).

**Conjugation and hyperconjugation.** The effects of delocalization of the $\pi$ electrons in benzene represent an example of a more general phenomenon called *conjugation*. It occurs whenever in the elementary valence picture there are two or more double bonds separated by single bonds. In valence-bond theory the observed effects of conjugation are accounted for by the assumption of resonance between different valence structures, in benzene between the two Kékulé structures and the three Dewar structures (see p. 371). In molecular orbital theory we do not use a definite valence structure but distribute the $\pi$ electrons over the whole molecule and obtain a lowering of the energy on account of this *delocalization*. Of course, in molecular orbital theory all orbitals are whole-molecule orbitals, that is, all electrons are delocalized. However the $\sigma$ electrons in benzene and other conjugated systems form density distributions between the nuclei which can be interpreted in terms of a pair of electrons for each single bond while the $\pi$ electrons show quite a different distribution: in benzene, two more or less uniform rings of electronic charge above and below the plane of the six carbon nuclei. The average number of $\pi$ electrons between adjacent nuclei is one, not two. In this sense the $\pi$ electrons are more completely delocalized (or *mobile* in the sense of Lennard–Jones) than the $\sigma$ electrons.

The simplest example of conjugation is provided by butadiene

$$H_2C{=}CH{-}CH{=}CH_2$$

in which we have two double bonds. The four $\pi$ electrons of the two double bonds, as in benzene, are not restricted to the immediate neighborhood of these double bonds but are delocalized, i.e. spread over the whole molecule. If we assume for butadiene the planar trans structure of point group $\boldsymbol{C}_{2h}$, which is strongly supported by spectroscopic evidence [see Marais, Sheppard and Stoicheff (797)], we see immediately that the four $2p_z$ orbitals of the four C atoms will form the molecular orbitals $a_u$, $b_g$, $a_u$, $b_g$ since $a_u$ and $b_g$ are the only species antisymmetric to the plane of the molecule. According to Mulliken (896) the four orbital wave functions in the LCAO approximation are

$$\begin{aligned} \psi_1 &= 0.401(\varphi_1 + \varphi_4) + 0.582(\varphi_2 + \varphi_3) \\ \psi_2 &= 0.582(\varphi_1 - \varphi_4) + 0.401(\varphi_2 - \varphi_3) \\ \psi_3 &= 0.582(\varphi_1 + \varphi_4) - 0.401(\varphi_2 + \varphi_3) \\ \psi_4 &= 0.401(\varphi_1 - \varphi_4) - 0.582(\varphi_2 - \varphi_3) \end{aligned} \qquad \text{(III, 123)}$$

where $\varphi_1$, $\varphi_2$, $\varphi_3$, $\varphi_4$ are $2p_z$ atomic orbitals at the four carbon atoms. The corresponding energies are

$$E_0 + 1.422\beta, \quad E_0 + 0.674\beta, \quad E_0 - 0.674\beta, \quad E_0 - 1.422\beta \qquad \text{(III, 124)}$$

respectively while without delocalization the energies would be

$$E_0 + \beta, \quad E_0 + \beta, \quad E_0 - \beta, \quad E_0 - \beta.$$

In the ground state the four electrons go into the two lowest orbitals $\psi_1$ and $\psi_2$ and, in a first approximation, the energy is $4E_0 + 4.192\beta$ with, and $4E_0 + 4\beta$ without, the effect of delocalization. We see that the delocalization effect is much smaller in butadiene than in benzene. The effect increases with the number of conjugated double bonds and is particularly strong in cyclic polyenes [see Mulliken (897)]. In benzene as a consequence of conjugation, that is, of the delocalization of the $\pi$ electrons, all C—C bonds are equivalent [$r_0$(C—C) $= 1.397$ Å]; in butadiene one would expect at least a partial approach to equality of the distances. Indeed the C—C single bond distance is observed to be 1.476 Å [Marais, Sheppard and Stoicheff (797)] compared to the value 1.536 in $C_2H_6$; but no increase of the double bond distance has been found within 0.005 Å.

If triple bonds are conjugated as in diacetylene ($HC{\equiv}C{-}C{\equiv}CH$) similar effects of delocalization arise. Here there are eight unsaturation electrons and the difference from the elementary valence model is more pronounced than in butadiene. The C—C single bond distance is only 1.376 Å [Callomon and Stoicheff (179)].

It was first recognized by Mulliken (897) that some delocalization may occur when a $CH_3$ group is conjugated with a double or triple bond. This phenomenon is called *hyperconjugation*. The simplest case of this type is represented by methyl acetylene which may be written

$$H_3{\equiv}C{-}C{\equiv}CH$$

that is, the three C—H bonds of the $CH_3$ group are conjugated with the C≡C triple bond. The reason for delocalization, i.e. for a mild form of conjugation (called hyperconjugation), is easy to see: the three $1s$ orbitals of the three H atoms of $CH_3$ form the molecular orbitals $a_1$ and $e$ (see p. 302f). The $e$ orbital interacts with the $\pi$ orbitals of the C≡C bond since $\pi$ orbitals of linear molecules become $e$ orbitals in $C_{3v}$ molecules. Each component of $e$ has a nodal plane through the axis of the molecule just as $\pi$. As previously, the interaction of these orbitals of equal species leads to one orbital with lower and one with higher energy; to each of them both $H_3{\equiv}C$ and C≡C contribute, that is, the electrons in these two triple bonds are delocalized. Again, as a result of this delocalization the energy of the ground state is slightly lowered and the C—C single bond takes on partial triple bond characteristics. In agreement with this expectation the C—C single bond distance has been found to be only 1.459 Å [Herzberg, Patat and Verleger (536), Herzberg and Stoicheff (541), Costain and Stoicheff (236)] compared to 1.536 for $C_2H_6$. To be sure, there are additional causes for the bond shortening, other than hyperconjugation [see Coulson (7)].

In valence-bond theory hyperconjugation is accounted for by resonance of the normal valence structure with ionic structures and anomalous valence structures with one very long bond, for example, for methyl acetylene with the structures

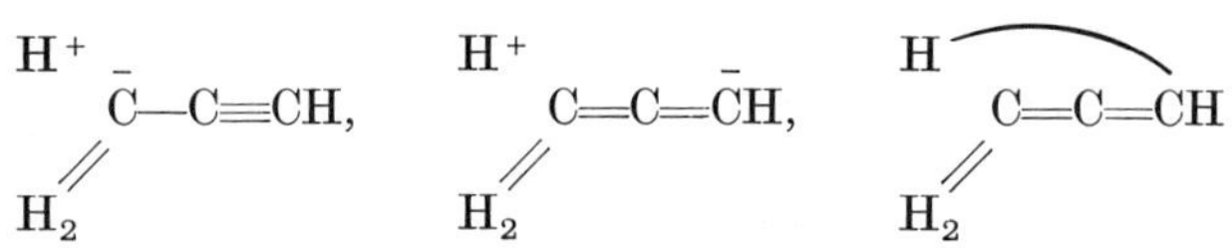

The explanation by molecular orbital theory seems more natural. Hyperconjugation effects have been investigated in a large number of compounds. More details may be found in the books by Baker (2) and Dewar (9) and the review by Ham (465).

**Configuration interaction.** Up to now we have assumed that the ground state of a molecule (or any of its excited states) can be described by a single electron configuration, e.g. the ground state of $H_2O$ by

$$(1a_1)^2(2a_1)^2(1b_2)^2(3a_1)^2(1b_1)^2. \tag{III, 125}$$

The corresponding electronic wave function of the molecule is a Slater determinant (see p. 335) made up of the orbital wave functions occupied in the ground state. Even with the best molecular orbital wave functions, viz., the self-consistent field functions, the Slater determinant is only a first approximation to the true electronic wave function since it neglects the finer interaction of the electrons. As mentioned before (p. 378), the electron correlation produced by the Coulomb repulsion between electrons is largely neglected in the molecular orbital method and therefore the probability of finding two (or more) electrons in the same element of volume is much larger than it would be if Coulomb repulsion were properly taken into account.

In principle it is possible to take electron correlation fully into account if the interaction of states of different electron configuration (but of the same species) is included in the calculations. This *configuration interaction*, similar to the phenomenon of resonance in valence-bond theory, leads to a lowering of the energy of the ground state and at the same time to a modification of the electronic wave function corresponding to it.

As an example consider the following excited configurations of $H_2O$ that will give ${}^1A_1$ states and will therefore interact with the ground state:

$$\begin{array}{l} (1a_1)^2(2a_1)^2(1b_2)^2(3a_1)(1b_1)^2(4a_1) \\ (1a_1)^2(2a_1)^2(1b_2)^2(3a_1)^2(4a_1)^2 \\ (1a_1)^2(2a_1)^2(1b_2)^2(1b_1)^2(4a_1)^2 \\ (1a_1)^2(2a_1)^2(1b_2)(3a_1)^2(1b_1)^2(2b_2) \\ (1a_1)^2(2a_1)^2(1b_2)^2(3a_1)^2(2b_2)^2 \\ \qquad \text{etc.} \qquad\qquad \text{etc.} \end{array} \tag{III, 126}$$

It is clear that configuration interaction with the lowest excited configurations (giving states of the correct symmetry) will be most important. However, certain low-lying configurations can be eliminated according to a theorem by Brillouin which states that, if the configuration considered differs from that of the ground state by a change of only one electron, then the configuration interaction between these two can be neglected to second order [see Kotani, Ohno and Kayama (690)]. Thus in the example of $H_2O$ the first and fourth configuration of the above list need not be considered for calculations up to second order.

Configuration interaction has been introduced in calculations for a considerable number of diatomic molecules but up to now only for a few simple polyatomic molecules, e.g. $H_2O$ [McWeeny and Ohno (815)] and $NH_3$ [Kaplan (656)]. In these two cases the energy is lowered by 1.12 and 0.3 eV respectively compared to single determinant electronic wave functions.

Configuration interaction is in general much more important for excited states because the interacting configurations are usually closer in energy. But the calculation of these effects is even more cumbersome.

**Free electron model.** A drastic simplification of the molecular orbital approach has been discussed by Bayliss (103), Kuhn (700), Platt (989) and others [see the review of Platt (991)]. They assume that the electrons, more particularly the $\pi$ electrons in unsaturated compounds, can move freely within the confines of the molecule similar to electrons in a metal. A more detailed mathematical form was given to this model by Ruedenberg and Scherr (1089)(1090) and Ham and Ruedenberg (466). The free electron model is especially useful for larger molecules which we are not including here. However, a brief discussion for a simple case may be in order.

Consider the molecule hexatriene which is similar to butadiene previously discussed, but with three conjugated double bonds:

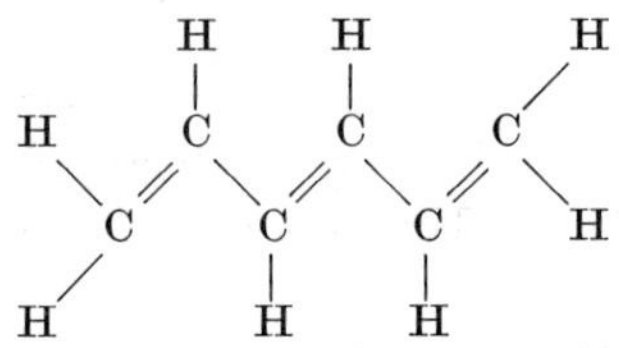

The $\pi$ orbitals in this molecule may be thought of as extending over the whole length of the carbon chain. In a very rough approximation we may consider simply the orbitals of a one-dimensional box as appropriate to the $\pi$ electrons. The first four

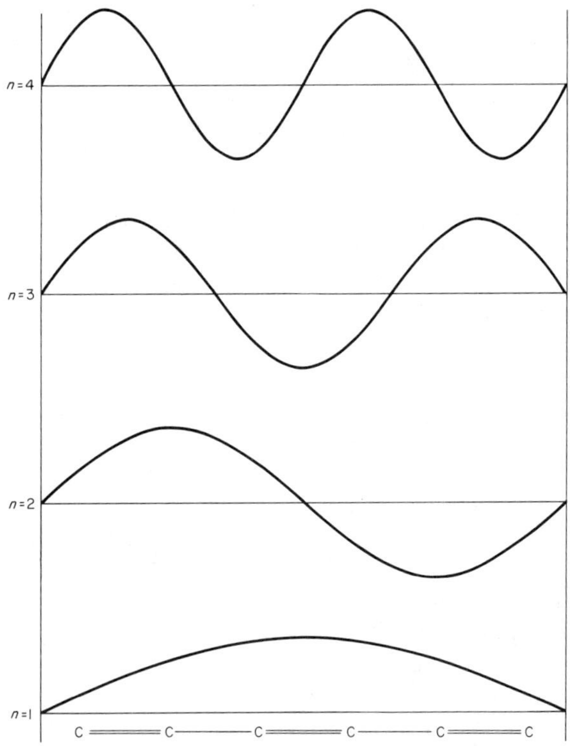


FIG. 155. **First four free electron orbitals in hexatriene after Platt (991).** The approximate positions of the C atoms relative to the orbital functions are shown at the bottom.

orbitals may be represented by the simple sinusoidal functions in Fig. 155. The corresponding energy values are[11]

$$E_n = \frac{h^2}{8ml^2} n^2 \qquad \text{(III, 127)}$$

where $n = 1, 2, 3, 4, \ldots$; $m$ is the mass of the electron and $l$ is the length of the box [for the effect of the width of the box see Ruedenberg and Scherr (1089)].

Since there are six $\pi$ electrons in the example, the orbitals $n = 1$, 2 and 3 are filled in the ground state. The first excited state corresponds to the transition of an electron from $n = 3$ to $n = 4$. It is immediately seen that the excitation energy would be

$$E_4 - E_3 = \frac{7h^2}{8ml^2} \qquad \text{(III, 128)}$$

without any adjustable parameter. Substituting $h$, $m$ and $l$ and converting to $\text{cm}^{-1}$, we find for this energy 30,000 $\text{cm}^{-1}$ which, considering the roughness of the approximation, agrees surprisingly well with the first observed absorption maximum at 39,750 $\text{cm}^{-1}$.

For a general polyene with $N$ carbon atoms and $l = Nr(\text{C—C})$ one obtains for the excitation energy

$$\Delta E = \frac{h^2}{8m[r(\text{C—C})]^2} \frac{N+1}{N^2}. \qquad \text{(III, 129)}$$

In other words the frequency of the longest-wavelength transition is inversely proportional to N in agreement with experiment.

**Molecules containing atoms of transition elements (so-called ligand-field theory).** When $d$ (or $f$) electrons are present in the central atom of an $XY_n$ molecule or complex, more orbitals arise and the situation becomes more complex. However, there are some simplifying features in that the perturbation of the $d$ (or $f$) orbitals by the molecule formation, i.e. by the presence of the ligands, is often slight and therefore the energies of the resultant molecular orbitals can be calculated more readily than is possible for $p$ or $s$ orbitals. This state of affairs has led, unfortunately in my opinion, to the introduction of a separate name, ligand-field theory, for molecular orbital theory as applied to atoms with $d$ (or $f$) electrons[12].

In a first approximation one considers the ligand atoms to be negative ions with closed shells only and simply determines the orbitals of the central atoms in the field of these charges which are assumed to be point-like. Inorganic chemists call this procedure crystal field theory but in fact it is precisely the same procedure as is applied in the early stages of molecular orbital theory (see section 2) when the possible molecular orbitals arising from the atomic orbitals of a central atom are

[11] The energy formula follows either from a straightforward solution of the Schrödinger equation or, more simply, by applying the de Broglie relation $\lambda_n = h/mv_n$ (where, from Fig. 155, $\lambda_n = 2l/n$) and forming $E_n = \frac{1}{2}mv_n^2$.

[12] Even if it were admitted that a separate name is desirable it appears to me that the term ligand-field theory is rather poor since it fails to make the intended distinction, e.g., between $CuCl_2$ and $CCl_4$. It is not at all clear why the effect of the Cl atoms should be different in principle in one case from that in the other. In both cases they are the "ligands" whose field causes a splitting of the states of the central atom. But "ligand-field theory" is never applied to $CCl_4$. The main feature of this theory, viz. that it is concerned with $d$ (or $f$) electrons of the central atoms, is not expressed by its name. In this connection see also the remarks by Jørgensen (651a).

determined. For $d$ orbitals, as long as the negative ions are not too close to the central atom, it is relatively easy to draw conclusions about the order and splitting of the resultant orbitals from the positions of the ions relative to the lobes of the $d$ orbital functions. In Fig. 156 are shown the results of such considerations (see below) for a few point groups. The symmetry types of the resulting orbitals (just

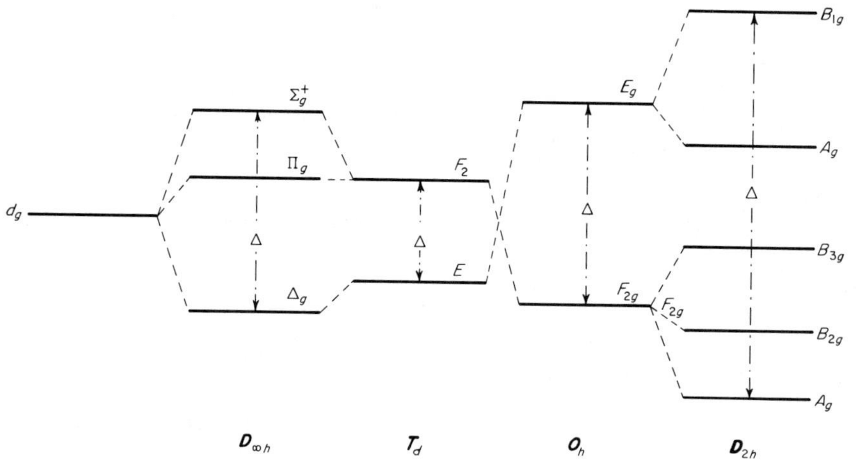


FIG. 156. **Splitting of a $d$ orbital in fields of $D_{\infty h}$, $T_d$, $O_h$ and $D_{2h}$ symmetry.**

as previously for $2p$ or $2s$ orbitals) follow immediately from Table 58, of Appendix IV.

It is easily seen that in an octahedral or tetrahedral field the component orbitals $d_{z^2}$ and $d_{x^2-y^2}$ of Fig. 147 [see eq. (III, 84)] form the $e_g$ (or $e$) orbital while $d_{xy}$, $d_{yz}$ and $d_{zx}$ form the $f_{2g}$ (or $f_2$) orbital. If the octahedral field is produced by six negative point charges on the $x$, $y$ and $z$ axes (which are the $C_4$ axes), it is immediately clear from Fig. 147 that the $e_g$ orbital must have the higher energy since in it the electrons are closer to the negative point charges than in $f_{2g}$. On the other hand in a tetrahedral field the four negative point charges producing it must be on the diagonal $C_3$ axes, not on the $S_4$ axes which form the $x$, $y$, $z$ axes, and therefore the $f_2$ orbital formed by $d_{xy}$, $d_{yz}$, $d_{zx}$ is higher than the $e$ orbital. In a similar way the order of the orbitals for other point groups as shown in Fig. 156 can be understood [for more details see Coulson (7), Orgel (28), McClure (805), Griffith (16) and Ballhausen (3)]. The over-all splitting is generally designated $\Delta$ (or, in the older literature, $10D_q$).

The ground state of the molecule (or ion) is obtained by filling the lowest possible orbitals of Fig. 156 with the available electrons. It is much more frequent here than for molecules without $d$ electrons, that the lowest electron configuration gives rise to several electronic states which spread over a certain energy interval. As in atoms the lowest of these is in general determined by Hund's rule, that is, the state of highest multiplicity lies lowest. When $\Delta$ is small it may even happen that a state with an excited electron configuration, since it can have higher multiplicity, may give rise to the ground state (see below).

Let us consider three simple examples. In linear $CuCl_2$, if following Hougen, Leroi and James (578), we assume it to have the ionic form $Cu^{++}Cl_2^-$, the $3d$ orbital of $Cu^{++}$ in the field of the two $Cl^-$ ions gives the molecular orbitals $\delta_g$, $\pi_g$, $\sigma_g$ (see Fig. 156). The nine electrons in the $3d$ shell fill $\delta_g$ and $\pi_g$ completely leaving one electron in the $\sigma_g$ orbital and giving rise to a $^2\Sigma_g^+$ ground state. In $VCl_4$, assuming it to be tetrahedral in structure and of the ionic form $V^{4+}Cl_4^-$, the $d$ orbital of $V^{4+}$ splits into $e$ and $f_2$. Since only one $3d$ electron is present the ground state would be $^2E$. In the $[Cr(H_2O)_6]^{++}$ complex ion, if it has octahedral symmetry and if the $H_2O$ groups are uncharged we have four $3d$ electrons on the central $Cr^{++}$ ion which can form the configuration $(f_{2g})^4$ or $(f_{2g})^3e_g$. The lowest state of the first configuration is $^3F_{1g}$ that of the second $^5E_g$ (see Table 31, p. 332). It is difficult to predict which of these is the ground state of the ion, but from the spectrum it appears to be $^3F_{1g}$ of the configuration $(f_{2g})^4$.

Up to this point we have considered the modifications and splittings of the *orbitals* of a central atom (or ion) in the field of point-like charges. If the modifying forces of the crystal field are small, it is better to consider the modifications of the various *states* of the central atom rather than that of the orbitals, applying the rules previously discussed for the correlation between molecule and united atom (section 1a). This *weak-field crystal field theory* in many cases reproduces the observed data in a very satisfactory way [see McClure (805) and Ballhausen (3)]. However, when the fields are so strong that the splittings are comparable to the energy differences of states of the same electron configuration, the orbital description is more appropriate (*strong-field crystal field theory*). Here the number of adjustable parameters is much larger and predictions are more difficult to make.

A third stage arises when interactions with the orbitals of the ligand atoms are no longer negligible. Just as before in ordinary molecular orbital theory for $2s$ and $2p$ orbitals, the ligand orbitals will mix with the X atom orbitals of the same species and, consequently, some will be shifted downwards, others upwards when the molecule is formed. This is the so-called ligand-field theory which is really nothing other than molecular orbital theory applied to molecules with a central atom and having $d$ or $f$ electrons.

Again let us illustrate the situation by a few examples. In Fig. 157 are given the energies of the orbitals of a linear $XY_2$ molecule (center) and their correlations with the orbitals of the separated X and Y atoms. The figure is similar to Fig. 121 except that we are now including $3d$ orbitals of X and Y, and have drawn the orbitals of X on the left side of the diagram. Also we are assuming here that the $3p$ orbitals of Y lie below the $3d$ orbitals of X as would be the case if Y were an S or Cl atom (but not if it were an Na or Mg atom). The $3d$ orbitals of the X atom give the molecular orbitals $\sigma_g$, $\pi_g$ and $\delta_g$ as before; but in the second approximation which we are now considering we must take account of the interaction of these orbitals with those derived from the Y orbitals. Thus the $\sigma_g$ and $\pi_g$ orbitals

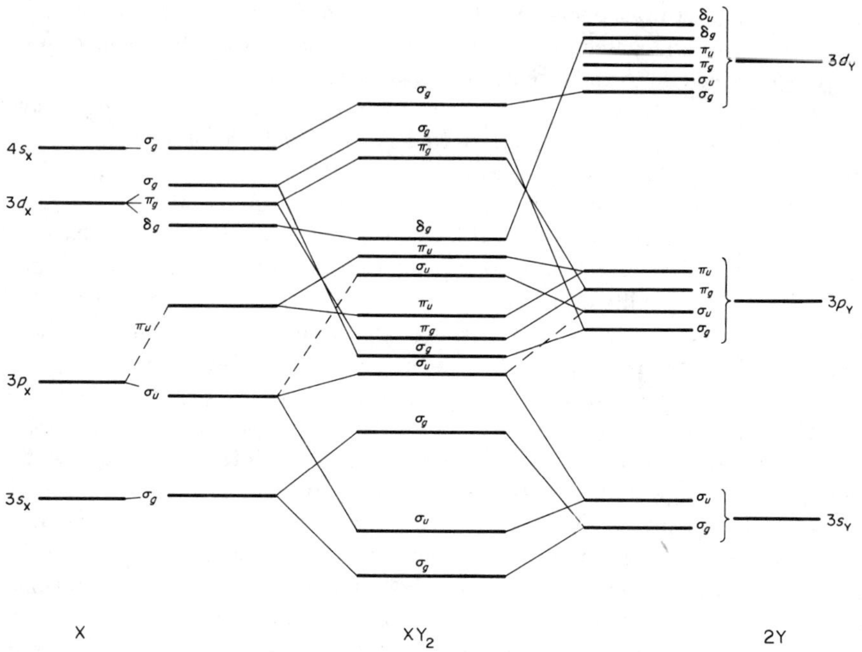


FIG. 157. **Correlation of orbitals of linear $XY_2$ molecules to those of the separated atoms, including *d* orbitals.** Next to the column at the extreme left are shown the orbital energies of the X atom in a field of $\boldsymbol{D}_{\infty h}$ symmetry. Similarly next to the column at the extreme right are shown the orbital energies of the two Y atoms in a field of $\boldsymbol{D}_{\infty h}$ symmetry. In the molecule (center) the orbital energies are further modified by mixing of orbitals of equal species.

from $3p_{\mathrm{Y}}$ push $\sigma_g$ and $\pi_g$ from $3d_{\mathrm{X}}$ up while $\delta_g$ is unaffected. The resulting orbitals are shown in the center of Fig. 157. Using these orbitals we see that the ground states of $CuCl_2$ and similar molecules as well as the first excited states are unchanged compared to the more elementary treatment given earlier since the same three orbitals $\delta_g$, $\pi_g$, $\sigma_g$ are the outermost orbitals, but the energies (i.e. the splitting constants $\Delta$) are changed considerably. Moreover the bonding in the molecule assumes some homopolar character: in $CuCl_2$ there is only one electron in the (slightly) anti-bonding $\sigma_g$ orbital while the corresponding bonding orbital $\sigma_g(3p_{\mathrm{Y}})$ is filled.

In $NiCl_2$ with one electron less than $CuCl_2$ the lowest electron configuration is $\delta_g^4\pi_g^4$ which gives a ${}^1\Sigma_g^+$ state, but there is a slightly excited configuration $\delta_g^4\pi_g^3\sigma_g$ which gives the states ${}^3\Pi_g$ and ${}^1\Pi_g$ of which the former will be the lower one and may indeed be lower than the ${}^1\Sigma_g^+$ state. Actually the spectrum of free $Ni^{++}$ shows that the ${}^3F$ state of the configuration $3d^8$ is the lowest state (in agreement with Hund's rule). This state can give rise to the ${}^3\Pi_g$ molecular state but not to ${}^1\Sigma_g^+$. Considering that singlet states of $Ni^{++}$ which can give ${}^1\Sigma_g^+$ lie fairly high and that the electrostatic interaction energies are small, we must conclude that the ${}^3\Pi_g$ state is the ground state of $NiCl_2$.

As a further example let us consider a tetrahedral $XY_4$ molecule. Figure 158 shows the energies of the orbitals in the separated atoms and their probable modification in the molecule. It is similar to Fig. 129 except that now it is assumed that $3d$ is the outermost occupied orbital of X and $3p$ of Y. The $3s$, $3p$, $3d$ and $4s$ orbitals of the X atom give $a_1$, $f_2$, $e + f_2$ and $a_1$ molecular orbitals respectively

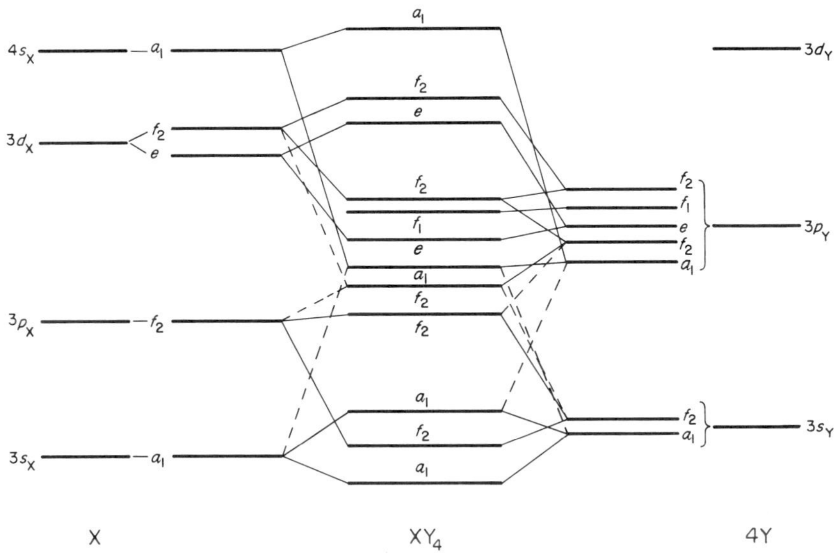


FIG. 158. **Correlation of orbitals of tetrahedral $XY_4$ molecules to those of the separated atoms, including *d* orbitals.** See caption of Fig. 157.

(see Table 58). The four $3s$ orbitals of Y give $a_1 + f_2$, and the four $3p$ orbitals give $a_1 + f_2 + e + f_1 + f_2$ molecular orbitals (see Table 61, Appendix V). Introducing now the interaction between orbitals of the same species we obtain the molecular orbitals given in the center of Fig. 158. Upon adding the electrons to these orbitals we see that the orbitals derived from $3p_Y$ will be filled up first, that is, the Y atoms will form somewhat distorted negative ions. Thus, for example, for $VCl_4$ in its ground state we obtain the same electron configuration as before. However, all orbitals except $f_1$ (derived from $3p_Y$) are somewhat shifted. If the interaction is strong the shifts may be much greater than shown in Fig. 158. At the same time the bonding and antibonding nature of the orbitals will make itself felt. In discussions of the binding usually only the $\sigma$ components of the $3p_Y$ orbitals are considered which give $a_1$ and $f_2$ molecular orbitals. These together with the $a_1$ and $f_2$ orbitals derived from $4s$ and $3d$ of the central atom give bonding—antibonding pairs of orbitals. Since in the ground state only the bonding $a_1$ and $f_2$ orbitals are filled, we have a net bonding action ($\sigma$-bonding). It is clear from Fig. 158 that, in addition, some of the $f_2$ and $e$ orbitals (but not $f_1$) arising from the $\pi$ components of $3p_Y$ can strongly affect the binding since they can interact with $f_2$ and $e$ of $3d_X$, resulting in $\pi$-bonding.

Figure 158 can also be applied to tetrahedral complexes in which the Y atoms are replaced by $H_2O$, $NH_3$, $CN^-$ and other groups.

A diagram showing the energies of the orbitals in octahedral $XY_6$ molecules or ions having $3d$ electrons has already been given in Fig. 130. The six $2s$ orbitals of the Y atoms (ligands) give the molecular orbitals $a_{1g} + e_g + f_{1u}$ as do the $\sigma$ components of the six $2p$ orbitals. The $\pi$ components of the latter give $f_{1g} + f_{1u} + f_{2g} + f_{2u}$ (see Table 61). The effect of these $\pi$ components has been neglected in Fig. 130 in agreement with common practice in determining the molecular orbitals of $XY_6$ (center of Fig. 130), but the reader can easily derive qualitatively what the effects of these orbitals would be. In agreement with Fig. 156 the order of the $e$ and $f_2$ orbitals arising from $3d_X$ is reversed in $XY_6$ compared to $XY_4$. When, as assumed in Fig. 130, the $2p_Y$ orbitals are lower than the $3d_X$ orbitals, the interaction between $e_g(2p_Y)$ and $e_g(3d_X)$ causes the quantity $\Delta$ (see Fig. 156) to be somewhat larger than in the approximation in which pointlike ions are assumed as ligands. Similarly the interaction of $f_{1u}(2p_Y)$ and $a_{1g}(2p_Y)$ with $f_{1u}(4p_X)$ and $a_{1g}(4s_X)$ causes the higher one of each pair to shift upwards, the lower one downwards. The $f_{2g}(3d_X)$ orbital is not shifted by molecule formation as long as interaction with the $\pi$ components of $3p_Y$ is neglected. An example is the $CoF_6^{3-}$ ion in which six electrons occupy the $f_{2g}(3d_X)$ and $e_g(3d_X)$ orbitals. The ground state is either $(f_{2g})^6\ {}^1A_{1g}$ or one of the triplet states of $(f_{2g})^5e_g$, i.e., ${}^3F_{1g}$ or ${}^3F_{2g}$.

Both for tetrahedral and octahedral molecules consisting of atoms with $d$ electrons, not infrequently an orbitally degenerate state may result as the ground state, for example ${}^2E$ for $VCl_4$, ${}^3F_{1g}$ for $CoF_6^{3-}$. In that case the Jahn–Teller (vibronic) interaction (Chapter I, section 2) causes a splitting of the potential function in such a way that the symmetrical position is no longer that of minimum energy. Van Vleck (1237) was the first to consider the effect of Jahn–Teller interactions in complex ions of the form $XY_6$ in which the $d$ shell is partially filled. Recent detailed discussions of this topic have been given by Öpik and Pryce (950) and Ballhausen and Liehr (92) [see also Ballhausen (3)] while Weinstock and Goodman (1286) have treated corresponding effects in free $XY_6$ molecules.

The magnitude of the energy difference $\Delta$ of the two orbitals, $e$ and $f_2$ or $e_g$ and $f_{2g}$, arising from the $3d$ orbital of the central atom depends of course on the nature of the attached ions or groups of atoms. Empirically it is found [Tsuchida (1227)] that $\Delta$ increases for a given central atom in the series (called spectrochemical series)

$$I^-,\ Br^-,\ Cl^-,\ F^-,\ C_2H_5OH,\ H_2O,\ NH_3,\ C_2H_4(NH_2)_2,\ NO_2^-,\ CN^-.$$

For large $\Delta$, as with $CN^-$, the ground states always correspond to the lowest electron configuration while for small $\Delta$, as with $I^-$, $Br^-$, . . . frequently the ground state arises from a slightly excited electron configuration if it can give a state of higher multiplicity (compare the example of $NiCl_2$ given above).

Considerations very similar to those for systems with $d$ electrons apply to those with $f$ electrons, that is, to molecules containing rare earth atoms or atoms of the actinium group. Without going into details we mention only one complicating factor: spin-orbit coupling is in general large for these elements. It is therefore necessary to consider the different multiplet components separately, i.e. to consider the species of spin-orbit functions which for even multiplicities are the representations of the extended point groups (see p. 337f). The modifications that arise compared to the treatment sketched above have been discussed by many authors [see, e.g. Griffith and Orgel (448), Eisenstein and Pryce (348), McClure (805), Ballhausen (3), Griffith (16), Dunn (325), Jørgensen (651)].

Another problem related to the preceding discussion is the existence of "sandwich" molecules made up of an atom with 3*d* electrons, like Fe or Cr, "sandwiched" between two aromatic rings, $C_5H_5$ in ferrocene or $C_6H_6$ in $Cr(C_6H_6)_2$ and similarly in other molecules. Again the 3*d* orbitals have to be resolved according to the point group of the molecule and their interaction with molecular orbitals of the same species formed from the orbitals (usually the $\pi$ orbitals) of the two aromatic groups considered. Here the molecular orbital method is definitely superior to the valence-bond method which cannot account for the existence of these molecules in a simple way. We shall, however, not consider the details of the molecular orbital treatment of these molecules [see Coulson (7), Orgel (28), Cotton (6)].

Somewhat similar considerations apply also to molecules like $B_2H_6$, $Al_2Cl_6$ and others which have a bridge structure. Here the dominating feature is a three-center bond which fits quite naturally into molecular orbital theory but can only poorly be represented in valence-bond theory.

Similarly the existence of stable molecules formed from Xe and fluorine is more readily accounted for on the basis of molecular orbital theory than by the valence-bond method [see Lohr and Lipscomb (759a) and Bilham and Linnett (118a)].

### ($\gamma$) *Excited States*

For an understanding of the (electronic) ground states of polyatomic molecules, valence-bond theory and molecular orbital theory are probably of equal value. Some features are more readily understood by one theory, others more readily by the other theory. On the whole the two theories supplement each other very well and, if higher and higher approximations are introduced, they lead even quantitatively to the same conclusions. However, for a treatment of excited states the molecular orbital theory is unquestionably much superior and very few attempts have been made to treat excited states by the valence-bond method (see p. 375).

In section 2 we have already considered the excited states of a number of molecules simply on the basis of the order of the orbitals without considering the problem of stability. We shall now consider this question of stability for excited states.

**Rydberg states.** Every molecule has Rydberg states, that is, states in which one electron has a principal quantum number $n$ at least one larger than in the ground state (see section 2). Since $n$ can take all values greater than a certain $n_0$, we always have a series (Rydberg series) of such states. The stability of these states depends on the stability of the ground state (or low excited state) of the molecular ion obtained by removal of the electron in question, since an electron in a Rydberg orbital is in general a non-bonding electron: its orbital is large and does not deviate much from that of an atom. The stability of the ion depends on whether an electron is removed from a bonding, anti-bonding or non-bonding orbital of the ground state of the neutral molecule. In addition the possibility exists of a change of conformation in going from the neutral to the ionized molecule.

Consider as an example the $H_2O$ molecule. Its ground state electron configuration was given in Table 33 (see also Table 41). The most loosely bound orbital occupied in the ground state is $1b_1$ which is non-bonding. Therefore the ground state of $H_2O^+$ with one electron removed from this orbital is expected to have nearly the same stability as $H_2O$ and the same would be expected to apply to those Rydberg states of $H_2O$ which converge to the ground state of $H_2O^+$. The observed states [Price (1015), Johns (631)] conform to this expectation: the OH distance and the H—O—H angle as well as the vibrational frequencies differ only very slightly from the ground state values. As a consequence the $0-0$ bands are by far the strongest bands in each of the band systems corresponding to transitions from the ground state to the Rydberg states. Similar remarks apply to the Rydberg states of $H_2S$ [Price (1015), Watanabe and Jursa (1274)].

In the $CH_3$ radical in its ground state the most loosely bound occupied orbital is $1a_2''$, but only one electron is present in it (see Fig. 127 and Table 39). It is a non-bonding electron since the molecule is planar. The orbital has a node in the molecular plane. In the ground state of $CH_3^+$ this orbital is no longer occupied but all bonding electrons are the same as in $CH_3$. Therefore again we expect Rydberg states (see Table 40) with vibrational frequencies and rotational constants very similar to those of the ground state and this has indeed been found [Herzberg (521)]. On the other hand, in $NH_3$ the situation is somewhat different. Here there are two electrons in the $1a_2''$ orbital (see Table 39) and, since in non-planar $NH_3$ they are mixed with other bonding electrons, they actually cause $NH_3$ to be non-planar in the ground state. However $NH_3^+$ has the same electron configuration as $CH_3$ and is therefore in all probability planar. Thus one expects $NH_3$ to be planar in the Rydberg states and this has indeed been found for several of these states (see Chap. V, section 2a). The NH distance is only slightly larger than in the ground state, but because of the change of valence angle the out-of-plane bending vibration is strongly excited in transitions between the ground state and the Rydberg states and consequently the Rydberg series are not very obvious in the spectrum. The stretching vibrations are apparently not excited in the observed Rydberg transitions and their frequencies are therefore presumably not very different from the ground state values. Thus it is not surprising that the dissociation energy $D(NH_2^+-H)$ of the ion is somewhat larger than $D(NH_2—H)$ of the neutral molecule [since I.P. $(NH_3) <$ I.P. $(NH_2)$, see Foner and Hudson (389)].

A very different situation arises for example in $CH_4$. Here the last occupied orbital in the ground state is the $1f_2$ orbital which is strongly bonding (see Fig. 129 and p. 392). If one of the electrons in this orbital is removed the bond strength is greatly reduced. This is confirmed by the observation that it requires only 1.3 eV more to produce $CH_3^+$ from $CH_4$ than to produce $CH_4^+$ [i.e. $D(CH_4^+ \rightarrow CH_3^+ + H) = 1.3$ eV]. The same small stability toward dissociation into $CH_3 + H$ must be expected for the Rydberg states of $CH_4$. As a consequence the equilibrium conformation in these states will deviate considerably from that of the ground state and will have much larger $r_e$ values; therefore in absorption from the ground state the unstable part of the potential surface is reached and only a continuous absorption spectrum is expected in agreement with observation. In fact no discrete electronic spectrum of $CH_4$ has been observed.

Similar conclusions may be drawn for other fully saturated hydrocarbons like $C_2H_6$ and $C_3H_8$ for which also no discrete electronic spectra are known. However, the situation is quite different in a molecule like $CH_3I$ (and similarly $C_2H_5I$). In the formation of $CH_3I$ from $CH_3$ and I, the outermost $a_2''$ orbital of $CH_3$ becomes $a_1$ in $C_{3v}$ symmetry and interacts with $a_1$ from $5p_z$ of I giving a C—I bonding and anti-bonding orbital. $5p_x$ and $5p_y$ of I remain unaffected (there is no $\pi$ bonding) forming a non-bonding $e$ orbital of $CH_3I$ which will be the highest filled orbital. If an electron is removed from this $e$ orbital to form $CH_3I^+$ the binding is not changed and therefore stable Rydberg states of neutral $CH_3I$ are expected with internuclear distances and vibrational frequencies similar to those of the ground state. The corresponding Rydberg series in the absorption spectrum should therefore be simple and easily recognizable and this is indeed observed to be the case [see Chap. V, section 3b]. Since in the iodine atom the spin splitting of the ground state is large (0.94 eV), it is expected to be large also in $CH_3I^+$ as well as in the Rydberg states of $CH_3I$, in agreement with observation.

In $C_2H_4$ the highest occupied orbital is $1b_{3u}$ (see Fig. 132). It corresponds to the $\pi$ bond between the two C atoms. Removal of one electron from this orbital will weaken the C—C bond (three instead of four bonding electrons). Therefore we expect in the Rydberg states the equilibrium distance to be increased and the C—C frequency decreased. Both predictions are borne out by the observations on $C_2H_4$ and $C_2D_4$ (see Chap. V, section 4a): each Rydberg transition is found to consist of a progression of bands in the C—C stretching vibration ($\nu_2'$), indicating a change in $r_0$(C—C). The value of $\nu_2$ in the Rydberg states is 1370 cm$^{-1}$ compared to 1623 cm$^{-1}$ in the ground state. As we have seen previously the $\pi$ electrons ($1b_{3u}$) are mainly responsible for the stability of the planar conformation of $C_2H_4$. The removal of one of them means a considerable loosening of this rigidity. It is therefore not surprising that the torsional vibration is very substantially reduced in frequency (from 1027 to 235 cm$^{-1}$ in $C_2H_4$) and that as a consequence this non-totally symmetric vibration is fairly strongly excited in the Rydberg transitions even though the molecule remains planar in the upper states.

The ground state of the $C_6H_6^+$ ion has one electron missing from the $1e_{1g}$ orbital (see Fig. 154) and is therefore a $^2E_{1g}$ state. Even though $e_{1g}$ is a bonding orbital ($\pi$ bonding) the removal of one electron from it does not change the total binding energy very much since there are still 17 bonding electrons (not counting those giving C—H bonding). We expect therefore, in benzene, normal Rydberg series without much vibrational structure and these have indeed been found both for $C_6H_6$ and $C_6D_6$ [see Chap. V, section 10]. Of the four observed series, three have small Rydberg corrections while one has a large one. Probably these series correspond to *nf* and *np* orbitals of the excited electron; *nd* and *ns* do not give rise to states that can combine with the ground state. From Table 58 (Appendix IV) we see that an *np* orbital gives $a_{2u} + e_{1u}$ molecular orbitals of $C_6H_6$, while an *nf* orbital gives $a_{2u} + b_{1u} + b_{2u} + e_{1u} + e_{2u}$. If each of these is combined in turn with the ground state of the ion ($^2E_{1g}$) one obtains, in addition to states that do not combine with the ground state of $C_6H_6$, three $^1E_{1u}$ and two $^1A_{2u}$ states which do combine with the ground state. The observed four Rydberg series fit in very well with this prediction but the observations are not sufficient to establish the species of the upper states. Some evidence for Jahn–Teller interaction in the upper states has been found but is not conclusive. Only three of the predicted states are $^1E_{1u}$ and only these can show Jahn–Teller effects.

As already emphasized, series of Rydberg states arise not only by taking an electron from the most loosely bound orbital in the ground state to Rydberg orbitals but also by taking it from any other orbital occupied in the ground state. As before the stability of these Rydberg states depends on whether the particular orbital from which the electron is removed is bonding, non-bonding or anti-bonding. Such higher Rydberg series have been observed for a number of molecules, among others $CO_2$, $CS_2$, COS, $N_2O$ and $C_6H_6$ (see Chap. V).

For $CO_2$, $CS_2$ and $N_2O$ the limits of the Rydberg series have been found to differ by amounts that are exactly equal to the known excitation energies of the ions $CO_2^+$, $CS_2^+$ and $N_2O^+$ respectively. Most of the observed series consist of single bands indicating that in the excited states the molecule has the same structure as in the ground state, that is, a non-bonding electron is excited. This conclusion applies in particular to the $CO_2$ series corresponding to removal of an electron from the $1\pi_g$ and $3\sigma_u$ orbitals (see Fig. 139). The $3\sigma_u$ orbital is usually considered to be slightly bonding but the evidence from the Rydberg series shows that its bonding power cannot be large. For $CO_2$ a Rydberg series has also been found [Tanaka and Ogawa (1191)] that corresponds to the removal of an electron from the $1\pi_u$ orbital, but here, in agreement with the strongly bonding character of $1\pi_u$, each member of the Rydberg series consists of a long progression in $\nu_1'$ showing that the equilibrium position is greatly changed compared to the ground state.

For $C_6H_6$ also, a second Rydberg limit has been found, namely at 11.49 eV, while the first one discussed earlier occurs at 9.25 eV. It is very probable that the second series corresponds to the removal of an $a_{2u}$ electron rather than an $e_{1g}$ electron (see Figs. 135 and 154). The $a_{2u}$ orbital is bonding just as $e_{1g}$, but in view of the large number of bonding electrons, there is no difficulty in understanding the observation of a simple Rydberg series in the spectrum. The energy difference 2.24 eV is of the order expected for the two $\pi$ electron orbitals $a_{2u}$ and $e_{1g}$. A third

Rydberg limit occurs at 16.84 eV. It corresponds probably to the removal of an electron from an $e_{1u}$ or $e_{2g}$ orbital which participates in $\sigma$ bonding.

**Sub-Rydberg states.** When an electron is taken from one low-lying orbital to another (without change of principal quantum number) the resulting states are called sub-Rydberg states. The distinction between Rydberg and sub-Rydberg states is not a sharp one since the principal quantum number is not well defined for the lower electronic states. Consequently some states may in one approximation be considered as Rydberg states, in another as sub-Rydberg states.

While in a Rydberg state the excited orbital to which the electron is transferred from the ground state is always non-bonding, for sub-Rydberg states it may be bonding, non-bonding or anti-bonding. An important case is that in which the electron goes from a bonding to the corresponding anti-bonding orbital. As already mentioned (p. 349), excited states arising in this way have been called $V$ states by Mulliken (893) if the multiplicity is the same as in the ground state. The prototype of such a $V$ state is the $B\ ^1\Sigma_u^+$ state of $H_2$ in which an electron has been transferred from the lowest orbital $\sigma_g 1s$ to the corresponding anti-bonding orbital $\sigma_u 1s$ (see Vol. I, p. 384). The corresponding triplet state, called $T$ by Mulliken, is the well-known repulsive state $^3\Sigma_u^+$ arising from normal atoms. As we have seen in Volume I the $B\ ^1\Sigma_u^+$ state of $H_2$ is an "ionic" state corresponding to $H^+H^-$ and a similar conclusion applies to many $V$ states[13]. It is for that reason that $V$ states are often stable in spite of the reduction in the number of bonding electrons; but usually the equilibrium internuclear distances are considerably larger than in the ground state.

As an example of an excited $V$ state in a polyatomic molecule we refer again (see p. 349) to the state of the $C_2H_4$ molecule in which one electron has been transferred from the $1b_{3u}$ orbital to the $1b_{2g}$ orbital (Fig. 132). These two orbitals are a bonding–anti-bonding pair. The configuration $(1b_{3u})(1b_{2g})$ gives the states $^1B_{1u}$ $^3B_{1u}$ of which the former has all the characteristics of a $V$ state. Mulliken (895) has identified the $V-N$ transition of $C_2H_4$ with the strong broad absorption beginning at 2000 Å and having a maximum at 1750 Å (see Chap. V, section 4a). As expected the observed excited state is much less tightly bound than the ground state. In addition in the excited state the energy minimum should correspond to the perpendicular form of $C_2H_4$ in which the $CH_2$ groups are rotated by 90° compared to the planar form (see p. 394f). While no direct spectroscopic evidence exists in confirmation of this theoretical conclusion, there is much indirect evidence to support it [see Mulliken (889)].

If a molecule has conjugated double bonds several $V$ states arise. For example in trans-butadiene (see p. 401) we have the four orbitals (III, 123) arising from $2p_z$ atomic orbitals. The ground state is

$$N: \quad (\psi_1)^2(\psi_2)^2. \tag{III, 130}$$

[13] It is interesting to note that in united atom approximation the $B^1\Sigma_u^+$ state has the configuration $1s\sigma 2p\sigma$ and might therefore be considered as a Rydberg state with $n = 2$.

There are four excited states ($V$ states) involving the $\pi$ orbitals

$$\begin{aligned} &V_1: \quad (\psi_1)^2(\psi_2)\,(\psi_3)\\ &V_2: \quad (\psi_1)^2(\psi_2)\,(\psi_4)\\ &V_3: \quad (\psi_1)\,(\psi_2)^2(\psi_3)\\ &V_4: \quad (\psi_1)\,(\psi_2)^2(\psi_4)\end{aligned} \qquad \text{(III, 131)}$$

which correspond to excitation of an electron from a bonding to an anti-bonding orbital. From the orbital energies (III, 124) we get immediately the transition energies ($\beta$ is negative)

$$\begin{aligned} &V_1-N: \quad -1.348\beta, \qquad V_2-N: \quad -2.096\beta,\\ &V_3-N: \quad -2.096\beta, \qquad V_4-N: \quad -2.844\beta \end{aligned} \qquad \text{(III, 132)}$$

while without conjugation (i.e. without delocalization) all four transition energies would have been $-2\beta$. Thus, on account of conjugation the *first* excited state is considerably stabilized, in fact relatively much more so than the ground state.

The magnitude of the stabilization energy of the first excited state increases with the number of double bonds that are conjugated. At the same time it can be shown that the intensity is more and more concentrated in the $V_1-N$ transition unless it is forbidden by symmetry. In this way the color of organic dyes containing conjugated polyenes and its variation with the number of double bonds can be understood [Mulliken (896)(900)(901)]. Similar considerations apply to cyclic dienes and benzene [Mulliken (897)(898)], although for benzene the transition to the lowest excited singlet state is forbidden on account of the high symmetry (see p. 556).

For molecules in which hyperconjugation can arise (see p. 402) also the first excited state is stabilized. In fact it appears that the observation of such a stabilization would represent a strong indication for the presence of hyperconjugation [see Mulliken (897)].

There are many cases in which excited states arise by transferring an electron from a *non-bonding* to an *anti-bonding* orbital. Such states are called *Q states* by Mulliken (893). Consider for example in $H_2O$ an electron taken from the non-bonding $1b_1$ orbital, which in the ground state contains the "lone pair", to the anti-bonding $4a_1$ orbital (see Fig. 123). The latter orbital forms a bonding–anti-bonding pair with $3a_1$ which like $4a_1$ is formed from the two $1s$ orbitals of H and the $2p$ orbital of O. The state $\ldots(1b_1)(4a_1)$ is therefore clearly much less stable than the ground state. It is probably the upper state of the continuous absorption of $H_2O$ which begins at 1800 Å. From the point of view of the united atom, the $4a_1$ orbital is a $3s$ Rydberg orbital (with a large Rydberg correction). Similarly in $NH_3$ the first excited state is one in which an electron is transferred from a slightly bonding orbital, $3a_1$ (see Fig. 127b), to an anti-bonding orbital $4a_1$. Because of mixing of the wave functions with that of $2a_1$ the orbitals $3a_1$ and $4a_1$ do not form a bonding–anti-bonding pair. Such a pair is formed by the $1e$ and $2e$ orbitals but the state corresponding to transfer of an electron from $1e$ to $2e$ has not yet been identified.

Excited states in which an electron has been transferred from a bonding to a non-bonding orbital or from a bonding to another bonding orbital occur frequently as low-lying states of free radicals. For example in $NH_2$ whose ground state is $\ldots(3a_1)^2(1b_1)\ ^2B_1$ (see Fig. 123 and Table 33) the first excited state ($^2A_1$) is formed by the transfer of an electron from the bonding $3a_1$ to the non-bonding $1b_1$ orbital. Since the energy of the latter is only slightly higher than that of the former the excitation energy is expected to be small and the stability not much less than that of the ground state; this is in agreement with observation (see Chap. V, section 1a). A similar conclusion applies to the first excited state $(1e')^3(1a_2'')^2\ ^2\mathrm{E}'$ of $CH_3$ (see Table 39) which has however not yet been observed.

The preceding discussion is perhaps sufficient to show how the general rules about stability on the basis of the molecular orbital method can be applied to excited states. The question of which geometrical configuration is the most stable in a given excited state can be answered in the same way as for the ground states, that is, by means of Walsh diagrams [see sub-section ($\beta$) p. 387]. In this way it is readily seen that, for example, HCO in the first excited state is linear while in the ground state it is bent or that HCN and $C_2H_2$ in their first excited states are bent. The reason why in $C_2H_2$ the trans-bent form and not the cis-bent form occurs is less easy to predict [see Ingold and King (600) and Mulliken (914)].

In many molecules electronic excitation occurs frequently between orbitals that are at least approximately localized (i.e., locally orthogonal). Transitions between such states are often described by a short-hand notation due to Kasha (659) and others, in which $\sigma$ or $\pi$ stands for electrons that are bonding according to local symmetry, $\sigma^*$ or $\pi^*$ for those that are anti-bonding, and $n$ for those that are non-bonding, i.e., "lone pairs". In this notation the $V-N$ transitions of the groupings —C═C—, —C≡C— become $\pi^*-\pi$ transitions; the far ultraviolet continua of aliphatic hydrocarbons are examples of $\sigma^*-\sigma$ transitions; and the near ultraviolet transitions of carbonyl compounds, e.g., the aldehydes and ketones, are of $\pi^*-n$ type.

**Intensities of electronic transitions.** In view of the large number of electronic states of polyatomic molecules it is important for the interpretation of the observed spectra to have some idea of the expected relative intensities of the predicted transitions. Such studies were initiated by an extensive series of papers by Mulliken (894)(895)(896)(897)(898)(899)(900)(901)(902)(903). In order to predict precise transition probabilities precise wave functions of the states involved are needed and since in most cases such wave functions are not yet available, we must be satisfied with very rough approximations.

If we sum over all the component vibrational and rotational transitions from a given level $m$, $v''$, $J''$ we find for the electronic transition probability between states $n$ and $m$ in absorption (see Vol. I, p. 382)

$$B_{mn} = \sum_{v'}\sum_{J'} B_{mn,v''v',J''J'} = \frac{8\pi^3}{3h^2c}\,|R_e^{nm}|^2, \tag{III, 133}$$

where $R_e^{nm}$ is the electronic transition moment

$$R_e^{nm} = \int \psi_e^{n*} M_e \psi_e^m d\tau_e. \tag{III, 134}$$

For a transition to a continuous range of energy levels (corresponding to dissociation, see Chap. IV) the summation must be replaced by appropriate integrals.

The experimentally determined quantity is usually the absorption coefficient $k_\nu$ and its integral over the whole band system. It can be expressed in terms of the transition probability $B_{mn}$ by the following relation (see Vol. I, p. 383)

$$\int k_\nu d\nu = N_m B_{mn} h\nu_{nm} = \frac{8\pi^3 \nu_{nm}}{3hc} N_m |R_e^{nm}|^2. \tag{III, 135}$$

Frequently one uses instead of the transition probability the so-called oscillator strength $f^{nm}$ which is related to $B_{mn}$ by

$$f^{nm} = \frac{\mu h c^2 \nu_{nm}}{\pi \epsilon^2} B_{mn} \tag{III, 136}$$

where $\mu$ and $\epsilon$ are the mass and charge of the electron. For strong transitions $f^{nm}$ is of the order of one (cf. the $f$ sum rule Vol. I, p. 383). The transition probability $B_{mn}$ is related to the mean life $\tau_n$ of the molecule in the excited state before spontaneous radiation of frequency $\nu_{nm}$ occurs. One finds (see Vol. I, p. 21)

$$B_{mn} = \frac{1}{8\pi h c \nu_{nm}{}^3} \frac{1}{\tau_n}. \tag{III, 137}$$

In this relation it is assumed that only one transition from the state $n$ to lower states is possible. If several transitions are possible $B_{mn}$ in (III, 137) must be replaced by $\sum_m B_{mn}$. Equations (III, 133–137) are written for non-degenerate states; for degenerate states the degeneracies $d_m$ and $d_n$ must be introduced in the way indicated in eqs. (I, 54–I, 56) of Vol. I (p. 21).

Genuine Rydberg transitions are usually very strong, provided that $\Delta S = 0$ and that the transition is allowed by the symmetry rules. For the first member of such a series $f^{nm}$ may be of order 1. Just as in an atomic Rydberg series the intensity drops with increasing $n$, but often at the series limit enough intensity is left that the adjoining continuous spectrum can be observed. If in the first member most of the intensity is concentrated in one band the peak absorption may have an absorption coefficient of the order 5000 cm$^{-1}$ atm.$^{-1}$, that is, at 1 atm. pressure a path length of only $\frac{1}{5000}$ cm will reduce the incident intensity by a factor $1/e$ or, in other words, at 1 m path a pressure of about 1.5 $\mu$ will have the same effect. An example is the first Rydberg transition of $CH_3I$ (see Fig. 102) whose strongest band at 2011.6 Å was found to appear at a pressure of only 10$\mu$ at a path of 10 cm [Herzberg and Scheibe (539)].

In contrast to the Rydberg transitions, for most sub-Rydberg transitions the intensity would vanish if the molecule were separated into its constituent atoms. We must therefore expect that sub-Rydberg transitions are in general quite weak. However, it was first shown by Mulliken (894)(895) that, as an exception to this rule, $V-N$ transitions can be expected to be quite strong, sometimes as strong as Rydberg transitions. This is because the wave functions of a bonding–antibonding pair of orbitals are given by equations like (III, 92–94) and the $z$ component of the transition moment ($z$-axis in the direction of the bond) is therefore given by

$$R_z^{nm} = \frac{\epsilon}{2\sqrt{1 - S^2}} \left[\int \varphi(1s_a)\, z\, \varphi(1s_a) d\tau - \int \varphi(1s_b)\, z\, \varphi(1s_b) d\tau\right] \qquad \text{(III, 138)}$$

which in a rough approximation can be simplified to

$$R_z^{nm} \approx \frac{\epsilon(z_a - z_b)}{2\sqrt{1 - S^2}} = \frac{\epsilon r}{2\sqrt{1 - S^2}}, \qquad \text{(III, 139)}$$

where $z_a$ and $z_b$ are the $z$-coordinates of the two nuclei and $r$ is the internuclear distance. Since $S$ is, in general, small compared to 1 (for not too small $r$), we see that for a $V-N$ transition $R^{nm}$ is of the order of $\frac{1}{2}\epsilon r$ and therefore the intensity of absorption of the order of that of a classically oscillating electron with amplitude $\frac{1}{2}r$ which is of the order of the intensity of a Rydberg transition. [Substituting $\frac{1}{2}\epsilon r$ for $R$ in (III, 133) and (III, 136) one finds with $\nu_{nm} \approx 100{,}000\ \mathrm{cm}^{-1}$ and $r \approx 10^{-8}$ cm, $f = 0.3$.] Since in a $V-N$ transition the system goes from an atomic to an ionic state (e.g. $HH \rightarrow H^+H^-$) Mulliken has designated these transitions as *charge-transfer spectra*.

When several $V-N$ transitions arise as in molecules with conjugated double-bonds it was found by Mulliken (896)(897)(900)(901) that the $V-N$ transition of smallest frequency has the highest intensity with an $f$ value of order 1 (see p. 416) provided it is not forbidden by symmetry.

### (c) Intermolecular forces

When all normal valencies in a molecule are saturated there still remain weak forces of attraction between it and other equal or unequal molecules (or atoms). These forces are responsible for the condensation of the gas consisting of the molecules considered as well as in many cases for the formation of dimers, polymers or molecular complexes. There is not always a sharp distinction between these weak forces of interaction and the strong valence forces because, as we have seen, the concept of saturation of valencies is not a very precise one. In fact some of the weak attractions may be considered as due to imperfect saturation of valencies or residual valencies. On the other hand there are other weak attractions which, for all practical purposes, must be considered as genuine forces between saturated molecules not related to imperfect saturations. These forces are usually referred to as *van der Waals forces*, since even in the absence of residual valence forces they cause deviations from the ideal gas laws and produce the additional terms in the van der Waals equation for a real gas.

**Van der Waals forces.** There are three contributions to van der Waals forces between saturated molecules. The first and usually most important contribution is made by the *dispersion forces* first recognized by London (761)(762). They have been discussed in some detail in Volume I (p. 377) for atoms at large distances, particularly inert gas atoms. The dispersion forces are due to the perturbation of the ground state of the compound system consisting of two (saturated) atoms or molecules by all the higher electronic states that can combine with the ground state (just as for dispersion).

A second contribution to van der Waals forces, associated with the names of Debye and Falkenhagen, is the *induction effect.* This contribution does not arise for the interaction between atoms but is important for the interaction between molecules. It is due to the electrical polarization of one molecule by the other which occurs when the molecules are not spherically symmetrical, i.e. have a dipole (or quadrupole) moment. The direction of the induced dipole moment is always such that an additional attraction results. Even for a large dipole moment this effect is small compared to the dispersion effect and it is almost always negligible when the dipole moment is zero.

The third contribution is the *orientation effect* (Debye and Keesom) which arises for molecules with dipole moments by the direct interaction of their dipole moments. Depending on their mutual orientation two dipoles may attract or repel each other. A quantum theoretical treatment shows that only for the lowest rotational levels is the average of all orientations such that an attraction (potential minimum) arises. Thus the orientation effect is strongly dependent on temperature.

To illustrate the relative magnitude of the three contributions to van der Waals interaction, Table 46 gives their relative values for a number of molecules as listed

TABLE 46. RELATIVE MAGNITUDE OF INDUCTION AND ORIENTATION EFFECTS COMPARED TO THE DISPERSION EFFECT IN SEVERAL SIMPLE MOLECULES AFTER HELLMANN (19)

| | HI | HBr | HCl | $NH_3$ | $H_2O$ | |
|---|---|---|---|---|---|---|
| dipole moment | 0.38 | 0.78 | 1.03 | 1.47 | 1.84 | debye |
| induction: dispersion | 0.0035 | 0.018 | 0.037 | 0.085 | 0.158 | |
| orientation: dispersion | 0.0007 | 0.028 | 0.13 | 0.72 | 3.0 | |

by Hellmann (19) [see also Hirschfelder, Curtiss and Bird (23a)]. For all three contributions the potential energy can be written in a first approximation

$$U(r) = Be^{-r/\rho} - \frac{C}{r^6}. \tag{III, 140}$$

The first term represents the repulsion due to valence forces at small distances $r$, the second represents the van der Waals interaction which predominates at large $r$. The constant $C$ is the sum of three contributions corresponding to the dispersion, induction and orientation effects.

Van der Waals interaction between two atoms or molecules arises in every electronic state at large distances, even when at smaller distances there is a strong valence attraction. However, for two like molecules in unlike states because of the resonance the term $r^{-6}$ in (III, 140) is replaced by $r^{-3}$, that is, the attraction extends to greater $r$ values (see Vol. I, p. 379).

Van der Waals forces, particularly in excited states, are responsible for the pressure broadening of molecular lines. This broadening just like that of atomic lines can be fairly well represented at high pressures by the statistical theory (see Vol. I, p. 396). According to this theory in order to obtain the intensity distribution in a pressure-broadened line one only needs to ascertain how much of the time, on the average, the molecules are at the various intermolecular distances and then apply the Franck–Condon principle to the transitions to the excited state from each of these positions.

There are a few molecules such as NO and $ClO_2$ for which elementary valence theory indicates incomplete saturation of valences, yet van der Waals forces of the type discussed above represent the main intermolecular interaction in their ground states.

**Hydrogen bonding.** During the last three decades it has been firmly established that under certain conditions a hydrogen atom bound to an atom in a given molecule can exert a weak attracting force on a second similar atom in another molecule. In other words, contrary to elementary valence concepts, a hydrogen atom can be bound to two atoms rather than to only one. Such a configuration is called a *hydrogen bond* or sometimes a *hydrogen bridge.*

In recent years hydrogen bonds have become extremely important for an understanding of the structure of proteins and the nature of certain biological processes. Pauling (30) believes "that as the methods of structural chemistry are further applied to physiological problems it will be found that the significance of the hydrogen bond for physiology is greater than that of any other single structural feature". A detailed summary of the present status of our knowledge of the hydrogen bond has been given by Pimentel and McClellan (33).

The dimers of the fatty acids provide well-known examples of hydrogen bonds: e.g., the dimer of formic acid has been shown by spectroscopic and electron diffraction studies to have the structure

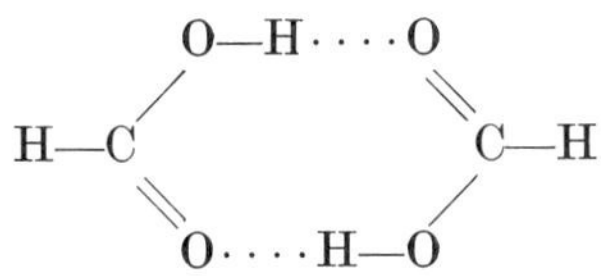


(III, 141)

There are two hydrogen bonds: O—H· · · ·O and O· · · ·H—O. The O· · · · O distance has been found to be about 2.69 Å, considerably more than twice the O—H distance in the monomer which is 0.97 Å. The binding energy of the dimer is about 13 kcal/mole [obtained both from the vapor density (Coolidge (225)) and from spectroscopically determined equilibrium data (Herman (510))], that is, 6.5 kcal/mole are required to split each hydrogen bond. This value is of course very small compared to the OH bond dissociation energy but large compared to the dissociation energies of van der Waals molecules. It is interesting to note that the deuterium bond is appreciably weaker than the hydrogen bond, the O···· O distance being 2.80 Å [Costain and Srivastava (235)].

Another important example is the hydrogen bond in liquid water. Here we have chains of hydrogen bonds for example as follows

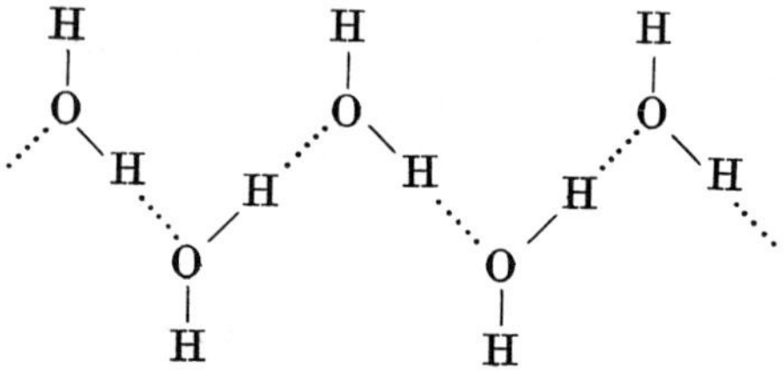

which account for the anomalously high boiling point of $H_2O$ compared to $H_2S$, $H_2Se$, $H_2Te$.

The simplest example of a hydrogen bond occurs in the $(HF_2)^-$ ion which is now generally assumed to have the structure

$$(F \cdots\cdot H \cdots\cdot F)^-$$

with a symmetrical position of the H atom. Here the F···· H distance is found to be 1.13 Å compared to 0.92 Å for free HF, and the dissociation energy into $HF + F^-$ is 27 kcal mole [see Coulson (7)]. The F····H bond in the $(HF_2)^-$ ion is the strongest known hydrogen bond.

It is clear that an understanding of the stability of the hydrogen bond is not possible on the basis of elementary valence theory. According to the valence-bond method a hydrogen atom, since it contributes only one electron, can take part in only one bond. Pauling (30) has therefore developed a purely electrostatic theory of hydrogen bonding. A proton, since its radius is vanishingly small, can attract not only one but several negative ions (e.g. $F^-$). It can approach two of these negative ions to a distance equal to their radius. But a third negative ion cannot approach the proton to the same distance, as is illustrated in Fig. 159. The stability of the $(FHF)^-$ ion may thus be understood as due to the stability of the system $F^-H^+F^-$. For large internuclear distances the energy of the state $F^- + H^+ + F^-$ lies very high above that of the state $F + H + F^-$, but as the internuclear distances are reduced and interaction sets in, since both states are of the same type $(^1\Sigma^+)$, the first state will be pushed up, the second will be pushed down by "resonance" in the Pauling sense, to a value even below the energy of

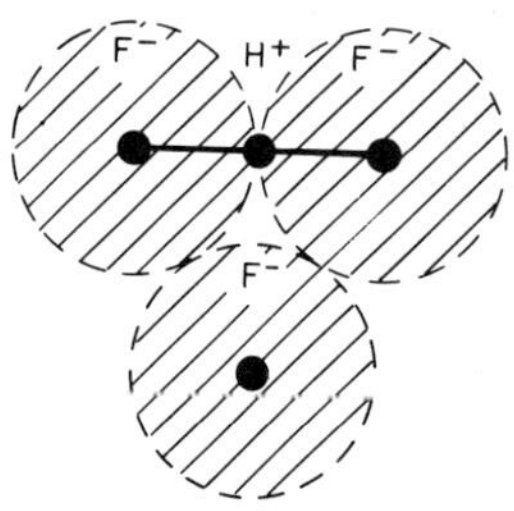


FIG. 159. **Electrostatic explanation of hydrogen bonding in $(FHF)^-$.** The circles indicate the size of the charge clouds of the $F^-$ ions; only two $F^-$ ions can approach $H^+$ to within the radius of this cloud.

$FH + F^-$. Thus in this interpretation the relatively large dissociation energy of $(FHF)^-$ into $FH + F^-$ is entirely the result of ionic-covalent resonance. It will be recalled that a similar resonance, according to Pauling, causes the dissociation energy of HF to be larger than it would be if it were a purely covalent compound. For $(FHF)^-$, the covalent contribution is assumed to be zero.

In a similar fashion the hydrogen bond between two O atoms, e.g., in the formic acid dimer, can be ascribed to ionic-covalent resonance. Here the resonance is between the states

$$O^{-}\cdots H^{+}\cdots O \quad \text{and} \quad O\cdots H\cdots O \qquad \text{(III, 142)}$$

The $H^+$ ion attracts a neutral O atom bound to other atoms because the lone pair electrons protrude toward it [see Lennard-Jones and Pople (739)]. Therefore the first state in (III, 142), which is much higher than the second at large distances, is lowered considerably at smaller distances and mixes with the second state pushing it downward and causing it to be stable.

In addition to the two states just considered, a third one

$$O^{-}\cdots H\cdots O^{+}\!<$$

has also been suggested as contributing to the bonding. In this state one O atom is positively charged (instead of H) and is therefore trivalent. However calculations by Coulson and Danielson (239) have shown that the contribution of this state to the ground state of the hydrogen-bonded system is only a few percent.

A purely electrostatic contribution to the hydrogen bond due to the dipole moment of the OH bond and its effect on the charge cloud of the other O atom must also be considered [see Coulson (7), Lennard-Jones and Pople (739)]. From a discussion of Coulson (7) it appears that the ionic-covalent resonance at least for the hydrogen-bond water ice gives the largest contribution.

For the formic acid dimer there are two equivalent structures, that given by (III, 141) and

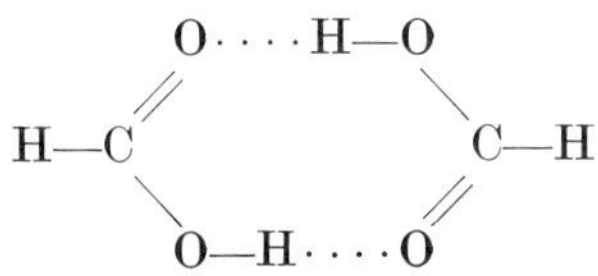

(III, 143)

These two structures will resonate in the same way as the two Kékulé structures of benzene. Indeed, a part of the stability of the dimer may be accounted for in this way as was shown by Gillette and Sherman (422); but even when the H atoms are in the most favorable positions (half-way between the O atoms) this resonance can at most account for half of the dissociation energy of the dimer [see Coulson (7)].

As already noted, the $(FHF)^-$ ion has the strongest hydrogen bond. On the basis of the electrostatic explanation one must expect that the strength decreases in the order F, O, N, since the electron affinity goes down, and this is indeed observed. However, no hydrogen bonds have been observed between atoms other than the three mentioned even though at least for Cl and S the electron affinities are of the same order. The difference is that, because of the much greater ion radius, the electrostatic interactions are much smaller and apparently not sufficient for hydrogen bond formation.

It is interesting to compare the preceding considerations with a qualitative molecular orbital treatment of hydrogen bonding. Let us simplify the hydrogen bond between two $H_2O$ molecules

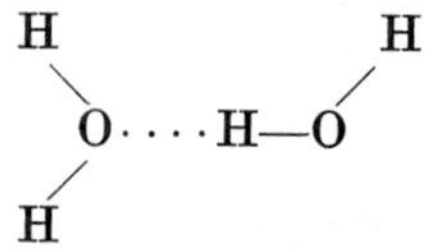


by replacing the left $H_2O$ molecule by its united atom Ne and by replacing the OH group on the other side of the H atom that forms the bond by its united atom F. In other words let us consider a hydrogen bond between Ne and HF

$$\text{Ne}\cdots\cdot\text{H—F}.$$

In Fig. 160 at the left the orbitals of Ne in an axial field, at the right those of HF are shown. If we disregard for a moment the excited orbital $3s\sigma$ of HF we see that the two $2p\sigma$ orbitals of Ne and HF will interact to give two molecular orbitals of Ne····HF (center of Fig. 158), one a little lower than $2p\sigma$ of Ne and the

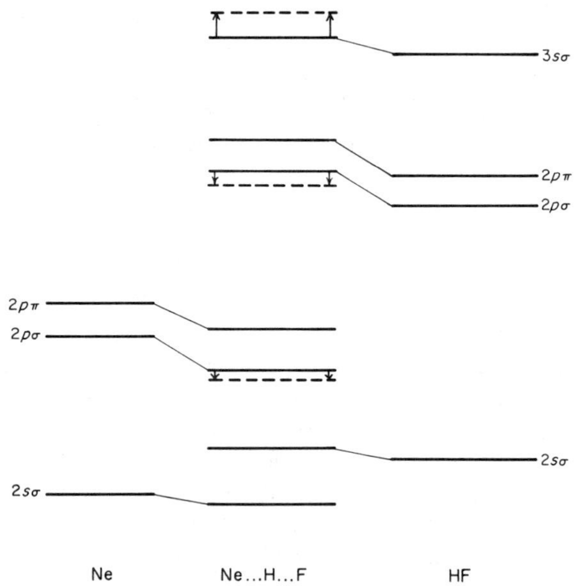


FIG. 160. **Molecular orbital explanation of hydrogen bonding between Ne and F.** The solid lines in the center give the orbital energies before introduction of configuration mixing. The arrows indicate the shifts produced by the mixing.

other a little higher (by the same amount) than $2p\sigma$ of HF. The same applies to the $2p\pi$ orbitals. If therefore we fill the two $\sigma$ and the two $\pi$ molecular orbitals with 12 electrons corresponding to the ground state of Ne + HF the bonding and anti-bonding actions compensate each other exactly and no stable system would be formed. However the rather low-lying $3s\sigma$ orbital of HF, which is unoccupied in the ground state, will mix with the two $\sigma$ orbitals arising from the $2p\sigma$ orbitals of the separated systems, push them down and thus reduce the anti-bonding action of one and increase the bonding action of the other (indicated by the short vertical arrows in Fig. 160). As a consequence of this configuration interaction, in the ground state bonding and anti-bonding actions no longer compensate and a small net binding energy arises.

When an electron in free HF is excited from the $2p\sigma$ to the $3s\sigma$ orbital the resulting $^1\Sigma^+$ state has strongly ionic properties, that is, it is essentially $H^+F^-$ [see Johns and Barrow (636)]. Thus, as in the valence-bond treatment it is also in molecular orbital theory in a sense the effect of the ionic state that causes the stability of the hydrogen bond. On the other hand, since the relation between the $3s\sigma$ orbital and the ionic state is not a simple one, it is perhaps better to consider the stability of the hydrogen bond as due to the effect of one (or several) low-lying excited orbital in much the same way as in Volume I (p. 357f) the stability of molecules like BeH was explained. This view emphasizes the contrast to the dispersion forces which are due to the collective action of all excited states of the right symmetry.

While a Ne—HF complex has not yet been observed complexes of HCl with Ar and Xe having a dissociation energy of 1.1 and 1.6 kcal/mole respectively have recently been found by Rank, Sitaram, Glickman and Wiggins (1047).

If the Ne atom in our model (Fig. 160) is replaced by the iso-electronic $F^-$ we obtain the $(FHF)^-$ ion. Here the energy separation of the $2p\sigma$ and $2p\pi$ orbitals at the left from those at the right is much smaller if not reversed. Therefore the bonding and anti-bonding action is stronger. At the same time the $3s\sigma$ orbital of $F^-$ must now be considered which is much lower than in Ne and even lower than $3s\sigma$ of HF. Its mixture with the two $2p\sigma$ orbitals is therefore stronger, that is, these orbitals are lowered much more than before: thus a much stronger hydrogen bond is obtained.

Returning now to the case of the $H_2O$ dimer we show in Fig. 161 a diagram similar to Fig. 160. The $\pi$ orbitals of Fig. 160 are now split into $b_1$ and $b_2$. If the two $H_2O$ molecules approach each other thus

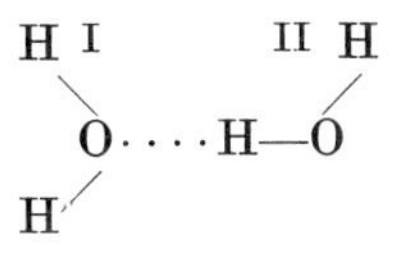


it is clear that the lone pair orbital $1b_1$ of I will be most affected while in II it will be the H—O bonding orbitals $1b_2$ and $3a_1$ (see p. 385). If the dimer has a planar conformation, $b_1$ becomes $a''$ while both $b_2$ and $a_1$ become $a'$. Therefore in a planar conformation those orbitals of one $H_2O$ which are localized between the two O atoms do not interact with those of the other $H_2O$. An interaction does take place when the molecules are not co-planar since then there are no symmetry restrictions. But even then just as in the Ne····H—F system the bonding and anti-bonding actions exactly compensate for the orbitals filled in the ground state.

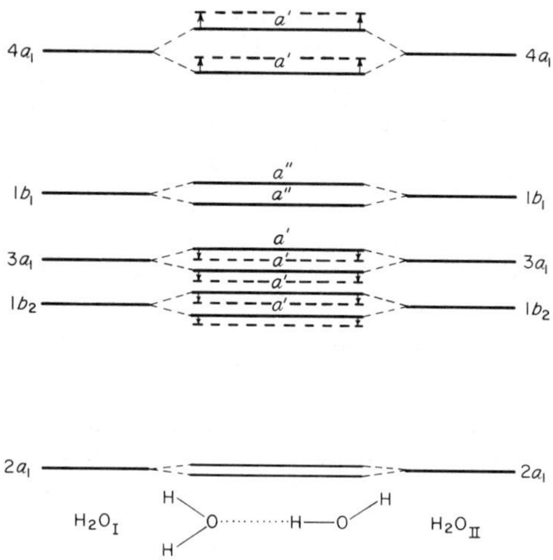


FIG. 161. **Correlation of orbitals in the hydrogen-bonded dimer of $H_2O$.** The symmetry of the dimer is assumed to be $\mathbf{C}_s$. The arrows as in Fig. 160 indicate the shifts produced by configuration mixing.

Only if we take account of the lowest excited orbital $4a_1$ (that is, take account of configuration interaction) does this compensation break down whether or not the dimer is planar. The $a'$ orbitals are pushed down and a positive binding energy results.

The preceding considerations, while entirely qualitative, do perhaps give a feeling for the way in which molecular orbital theory can deal with hydrogen bonded molecules. More detailed discussions have been given by Hofacker (559a) and Pimentel and McClellan (33).

**Charge-transfer forces.** Closely related to hydrogen-bonded systems are loosely bound molecular complexes which have been discussed by Mulliken and his collaborators in a series of papers (907)(908)(909)(1066)(910)(913)(951)(996)(1226)(920). The starting point of Mulliken's considerations was the observation that a solution of iodine in benzene shows a new strong absorption near 3000 Å not present in either iodine or benzene alone. Similar spectra have been found in many other combinations of halogens with aromatic compounds and have been ascribed to complex formation in the mixture. Mulliken recognized that the additional absorptions are particularly strong in complexes formed from a molecule that easily gives up an electron ("donor" molecule) and one that easily accepts it ("acceptor").

Therefore as in Pauling's explanation of hydrogen bonding we can also assume here an ionic-covalent resonance as the main reason for the stability of the complex.

Let X and Y be the two component molecules and let the ionization potential of X be appreciably smaller than that of Y. If the state $X^+Y^-$ is not too high above the non-bonded ground state $X \cdots Y$, the latter may acquire stability by resonance with the former. Again the difference from ordinary ionic-covalent resonance is that here the two molecules X and Y have all their homopolar valencies saturated and do not attract each other without this resonance. The resonance supplies the whole of the binding energy.

If the wave function of the non-bonded ground state is $\psi_0(XY)$ and that of the ionic state $\psi_1(X^+Y^-)$ then the wave function of the complex will be a mixture of the two:

$$\psi_N = a\psi_0(XY) + b\psi_1(X^+Y^-) \qquad \text{(III, 144)}$$

where $b$ is very small compared to $a$. The lowering of the energy of the ground state, according to second order perturbation theory, is given by

$$\Delta E = \frac{{W_{01}}^2}{E_1 - E_0} \qquad \text{(III, 145)}$$

where the matrix element $W_{01}$ is to be taken between the two eigenfunctions $\psi_0$ and $\psi_1$. The lowering of the energy depends inversely on the energy difference $E_1 - E_0$ of the ionic and covalent state. The resulting forces have been called *charge-transfer* forces since they depend on the transfer of an electron from X to Y [eq. (III, 144)].

The ionic state $X^+Y^-$ corresponds to an excited state of the complex whose wave function is

$$\psi_E = -b\psi_0(XY) + a\psi_1(X^+Y^-) \qquad \text{(III, 146)}$$

and whose energy is $\Delta E$ above where it would be without the ionic-covalent resonance (see the similar considerations p. 367). The energy relations are shown in Fig. 162. The broken lines give the energies of the unperturbed states. It is the transition between the perturbed states $N$ and $E$ which gives rise to the observed new spectrum of the complex. It is strong for the same reason that charge-transfer spectra of stable molecules are strong (see p. 419).

It would be difficult to predict quantitatively the magnitude of $\Delta E$, that is, the stability of the complex; but it is plausible that it should have a sufficient magnitude to account for the observed dissociation energies. For the case of iodine dissolved in benzene we know that the ionization potential of $C_6H_6$ is fairly small (9.24 eV) while the electron affinity of $I_2$ is about 2 eV [see Mulliken (907)]. Thus for large distances $E_1 - E_0 \approx 7$ eV. In the complex the energy difference between the states $E_1$ and $E_0$ will be much less because of the strong attraction between $X^+$ and $Y^-$ (see Fig. 162). Considering that the energy of attraction of two point charges at a distance of 4 Å is about 3.6 eV it seems reasonable that the center of the observed absorption occurs at about 33000 $cm^{-1}$ ($\approx$ 4 eV).

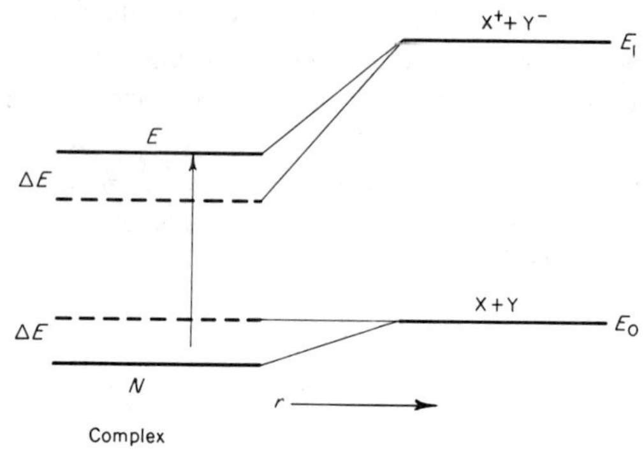


FIG. 162. **Energy level diagram explaining charge-transfer complexes.** The broken-line levels at left correspond to zero charge transfer. $\Delta E$ is the energy shift produced by the transfer. The vertical arrow represents the optical transition in the complex.

A large number of charge-transfer complexes have been studied both experimentally and theoretically during the last 15 years and in many cases conclusions have been drawn with regard to the geometric structure of the complexes. Mulliken (909) distinguishes a number of different types of charge-transfer complexes but these will not be discussed here. It appears that these complexes are of considerable importance for the understanding of many reactions in liquids and also for an understanding of the structure of crystals of mixtures of various compounds. For more details see for example the review by Orgel (950a) and the monograph by Briegleb (5).

Since the interaction that leads to the stability of charge-transfer complexes is of the same type as other covalent interactions, it is clear that the force of attraction just as for these other interactions decreases exponentially with the distance of the two components of the complex. The same behavior is exhibited by the forces giving rise to hydrogen bonds while van der Waals forces decrease much less rapidly with distance (see p. 420).

# CHAPTER IV

## DISSOCIATION, PREDISSOCIATION AND RECOMBINATION: CONTINUOUS AND DIFFUSE SPECTRA

Continuous and diffuse absorption and emission spectra are observed for polyatomic molecules even more frequently than for diatomic molecules. According to elementary considerations of quantum theory, these spectra must correspond to ionization or dissociation processes[1]. Continua corresponding to ionization join on to the Rydberg series previously considered, just as for atoms and diatomic molecules. These ionization continua, whose interpretation is usually straightforward and does not present any difficulty, will not be included in the present chapter.

On the other hand, the detailed interpretation of the continuous and diffuse spectra corresponding to dissociation and predissociation, respectively, often presents considerable difficulties. As a preparation for their discussion, we must consider first in somewhat greater detail than was necessary in Volume II the forms of the potential energy surfaces of polyatomic molecules in various electronic states.

### 1. Potential Surfaces

The potential energy of an $N$-atomic molecule is a function of $3N-6$ (or, for linear molecules, $3N-5$) internal coordinates, that is, three coordinates for triatomic, six for four-atomic, and so on. This function can be represented by a $3N-6$ dimensional hypersurface in a $3N-6+1$ dimensional space. Such hypersurfaces, unlike the simple potential curves of diatomic molecules, are difficult to visualize and still more difficult to represent graphically. For the discussion of the vibrational motion it is not necessary to have such a graphical representation, but for a discussion of dissociation and predissociation phenomena it is necessary to have it even if drastic simplifications have to be made.

**Symmetric linear triatomic molecules.** The potential energy of a linear triatomic molecule depends on four coordinates. It is customary, in order to visualize this function, to put the two bending coordinates equal to zero, that is, to consider the motion rigidly fixed to the symmetry axis ($z$-axis). With this

[1] It must be emphasized that in polyatomic molecules much more often than in diatomic molecules the line density in a discrete spectrum (that is, in a transition between two stable electronic states) may be so high that *apparent* continua or *apparently* diffuse spectra arise. Indeed, if the line spacing is smaller than the Doppler width of the lines, the spectrum would appear continuous or diffuse, even under the highest resolution. In this chapter we shall not consider such apparently continuous or apparently diffuse spectra.

simplification the potential energy depends on only two coordinates and can therefore be represented by a two-dimensional surface in ordinary (three-dimensional)

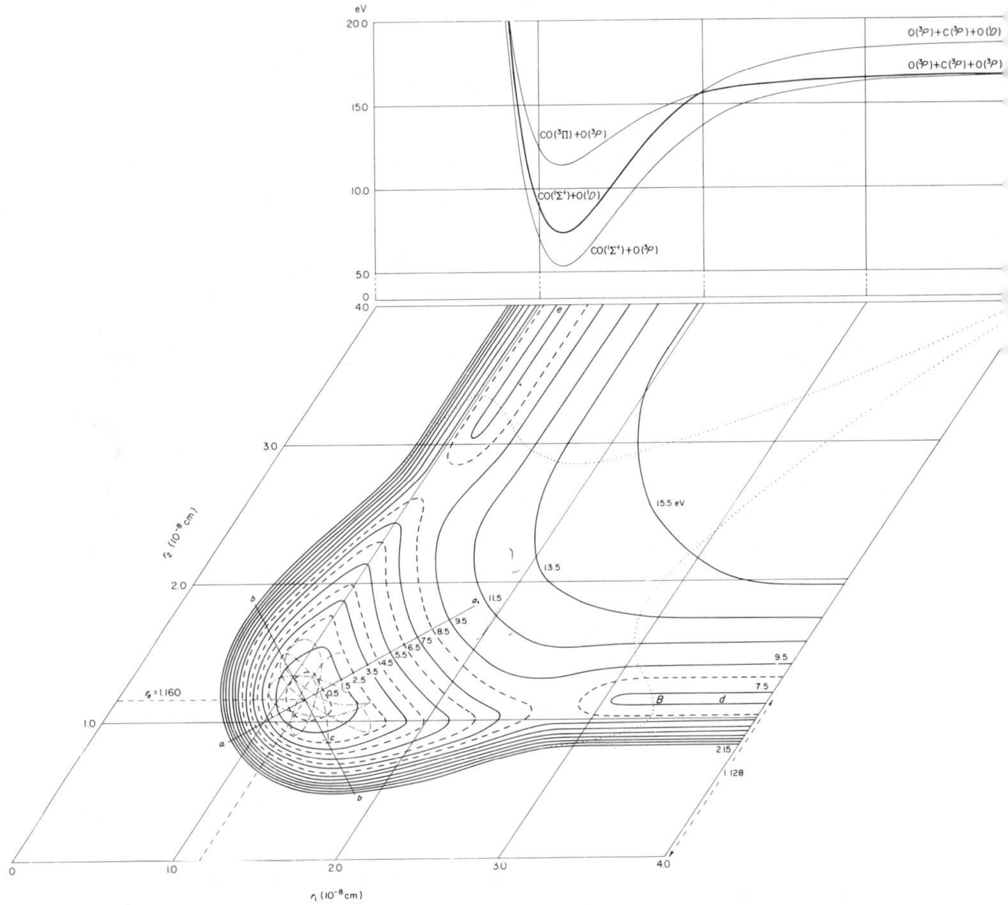


FIG. 163. **Potential surface of the lowest singlet state of $CO_2$ as a function of the two C–O distances $r_1$ and $r_2$.** The contour lines shown are lines of constant potential energy. The intervals between the full contour lines are 2 eV. The broken contour lines are intermediate. The oblique coordinate system has been chosen to correspond to the condition (IV, 2) for $CO_2$. At the top a cross section of the potential surface for large $r_2$ is shown by the heavy-curve. The light curves give similar cross sections for other nearby states including the one shown in Fig. 164 (see caption of Fig. 164). The dot-dash curve starting at $C$ represents qualitatively the Lissajous motion of the molecule when it starts with a large value of the normal coordinate corresponding to $\nu_3$.

space. This is done by means of contour lines in Fig. 163 for the electronic ground state of $CO_2$ as a function of the two C—O distances $r_1$ and $r_2$. Figure 163 is similar to Fig. 66(b) of Volume II (p. 203) except that the angle between the coordinate axes has been changed from 90° to 55° and that account has been taken of the fact that for large $r_1$ (or $r_2$) one obtains $CO(^1\Sigma) + O(^1D)$ not $CO(^1\Sigma) + O(^3P)$ according to the correlation rule for the electron spin (see p. 282).

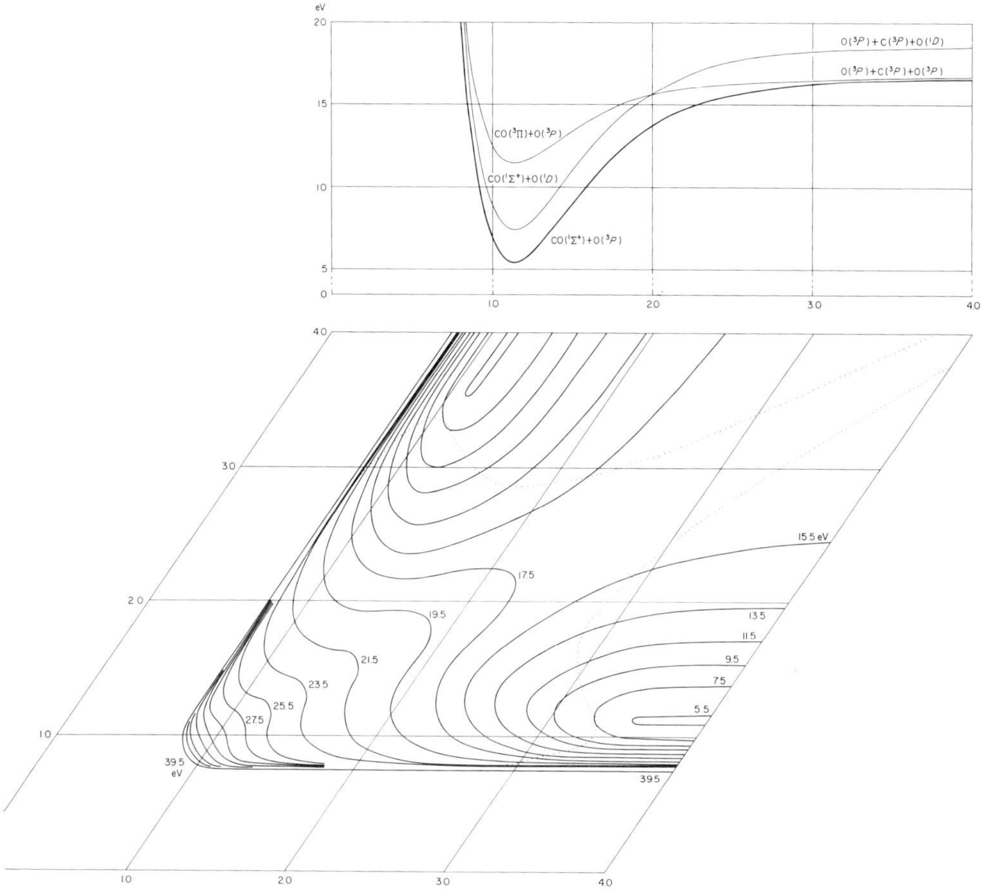


FIG. 164. **Potential surface of the lowest triplet state of $CO_2$ as a function of the two C–O distances $r_1$ and $r_2$.** See caption of Fig. 163. The two figures should be superimposed to get a picture of their mutual relation. The cross sections of the surfaces at the top are given in order to help visualize this mutual relation. The dotted lines in each of the main diagrams represent the approximate curve of intersection of the two surfaces.

In Fig. 164 the corresponding contour diagram is shown for the lowest triplet state arising from $CO(^1\Sigma) + O(^3P)$. Actually one should visualize the two potential surfaces, Figs. 163 and 164, one above the other and intersecting each other. As a guide for this purpose at the top in each case a cross section of the potential surfaces for large $r_1$ is shown and in the main diagrams the curve of intersection is roughly indicated by the dotted lines.

It is seen from Fig. 163 that even in this simple triatomic case there are three dissociation limits of the electronic ground state corresponding to O + CO, OC + O and O + C + O of which, of course, the first two have the same energy. It should also be noted that while the electronic ground state on dissociation to CO + O gives $^1\Sigma^+ + {}^1D$, i.e., an excited state, it does correlate with O + C + O in their lowest ($^3P$) states.

As we have seen in Chapter III, section 1c($\alpha$), there are in general a large number of molecular electronic states arising from a given combination of atomic states (there are 119 states from $^3P + {}^3P + {}^3P$, see Table 24), and there will be just as many potential surfaces. Only the two lowest of these are shown for $CO_2$ in Figs. 163 and 164.

For a representation of the potential energy function alone a rectangular coordinate system would, of course, be just as good as (and more convenient than) an oblique coordinate system. However, if one wants the motion of a small mass moving without friction on the surface to represent the motions of the atoms in the molecule, an oblique system of the axes must be used for the following reasons. In order to have a one-to-one correspondence, not only must the potential energy V, but also the kinetic energy $\frac{1}{2}m(\dot{x}^2 + \dot{y}^2)$ of the mass point be proportional to that of the molecule. For the general case of an XYZ molecule the kinetic energy of the molecule in terms of $r_1$ and $r_2$ is readily seen to be

$$T = \frac{1}{2(m_1 + m_2 + m_3)} [m_1(m_2 + m_3)\dot{r}_1{}^2 + 2m_1m_3\dot{r}_1\dot{r}_2 + m_3(m_1 + m_2)\dot{r}_2{}^2] \tag{IV, 1}$$

where $m_1$, $m_2$, $m_3$ are the masses of X, Y and Z, respectively. Clearly, if $r_1$ and $r_2$ were plotted as a rectangular coordinate system (i.e., if $x = r_1$, $y = r_2$) the kinetic energy of the mass point would not be of the form (IV, 1); but if we choose oblique coordinates, $\tilde{x}$, $\tilde{y}$, as shown in Fig. 165, such that

$$\tilde{x} = r_1 = x - y \cot \vartheta, \qquad \tilde{y} = \frac{r_2}{c} = \frac{y}{\sin \vartheta}$$

where $c$ is a factor by which $r_2$ is reduced as ordinate in the oblique coordinate system, we can, by a suitable choice of $\vartheta$ and $c$, make $T$ of the form $\frac{1}{2}m(\dot{x}^2 + \dot{y}^2)$. We find

$$\cos \vartheta = +\sqrt{\frac{m_1m_3}{(m_1 + m_2)(m_2 + m_3)}} \tag{IV, 2}$$

$$c = +\sqrt{\frac{m_1(m_2 + m_3)}{m_3(m_1 + m_2)}}. \tag{IV, 3}$$

For symmetrical $XY_2$ molecules ($m_1 = m_3$) these formulae yield $c = 1$ and $\cos \vartheta = m_1/(m_1 + m_2)$ which for $CO_2$ gives $\vartheta = 55° \, 9'$, for $CH_2$ $\vartheta = 85° \, 35'$. Thus we must plot directly $r_1$ and $r_2$ (without reduction) under the angle given, as has been done in Figs. 163 and 164.

Let us now consider the potential surface of the ground state of $CO_2$ (Fig. 163) in a little more detail. The symmetrical stretching vibration $\nu_1$ corresponds to the motion of the image point along the line $a - a$; if the amplitude of this vibration is increased, it would lead eventually to a dissociation into O + C + O. On the other hand, the antisymmetric stretching mode $\nu_3$ corresponds to motion of the image point along the line $b-b$ perpendicular to $a-a$ and if its amplitude were increased without deviating from the line $b-b$, no dissociation would result since

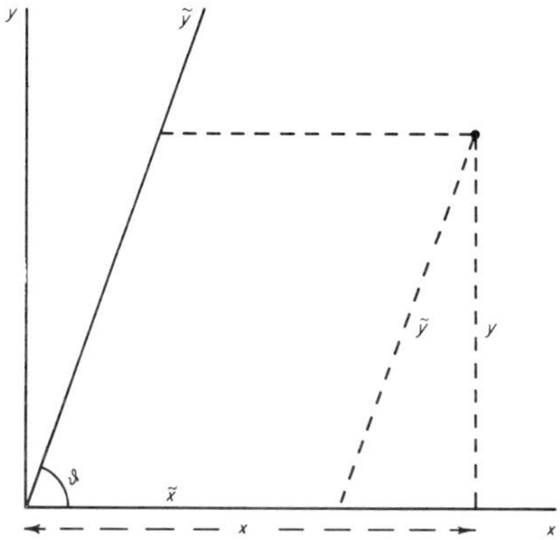


FIG. 165. **Relation between oblique and rectangular coordinate systems used for potential surfaces of linear XYZ molecules.**

the potential energy goes to infinity at the intersection of the line $b-b$ with the coordinate axes. Actually, however, classically, a simple oscillation along $b-b$ is not possible except for infinitesimal amplitudes because at the end point of the motion the direction of greatest slope does not coincide with the direction $b-b$. For this reason the image point, once it has reached a point on $b-b$ at some distance from the minimum, will start along the line of greatest slope and describe a complicated Lissajous motion, such as the one indicated by the dot-dash curve in Fig. 163 starting at the point marked $C$. This Lissajous motion is a superposition of the symmetric and antisymmetric vibration with different amplitudes and phases. Thus, classically, it is impossible to excite the antisymmetric vibration to higher amplitudes without simultaneously exciting the symmetric vibration.

If the amplitude of the antisymmetric vibration is sufficiently high, the Lissajous motion will lead the image point into the valley $d$ or $e$, that is, the molecule dissociates into CO + O. But this dissociation is not an instantaneous one. Only after carrying out an extended Lissajous motion such as the dot-dash curve in Fig. 163 would the dissociation occur, quite unlike the dissociation into O + C + O that would occur when the molecule starts out from a conformation corresponding to a point sufficiently high on the left-hand part of the line $a-a$. In that case we would have dissociation in half an "oscillation" (in one "swing"). The former case is the classical prototype of a unimolecular chemical reaction [see e.g. Slater (36) and Thiele and Wilson (1202)].

Translating these classical considerations into quantum-theoretical ones, it may be expected that the higher vibrational levels of $\nu_1$ converge to the dissociation energy $D(CO_2 \rightarrow O + C + O)$ in much the same way as the vibrational levels of a diatomic molecule in a given electronic state converge to its dissociation energy. On the other hand, the higher vibrational levels of $\nu_3$ will *not* converge to

$D(CO_2 \rightarrow O + CO)$ but will continue above this limit even though considerably broadened by predissociation (see section 3). To my knowledge no detailed calculations of the higher vibrational levels of a triatomic $XY_2$ molecule have been attempted. We know, however, that the eigenfunctions of the levels $v_3\nu_3$ are antisymmetric for odd, symmetric for even $v_3$.

Near the minimum the potential function can be represented by a power series in the displacement coordinates $q_i = r_i - r_i^e$ as discussed in Volume II, p. 204, but for a representation of the whole of the function more complicated expressions are required. The simplest assumption is that the potential function is a sum of Morse functions, one for each bond [see e.g. Thiele and Wilson (1202)]. Accordingly in the present case, we would have, still neglecting bending motions,

$$V = D[1 - e^{-a(r_1 - r_1{}^e)}]^2 + D[1 - e^{-a(r_2 - r_2{}^e)}]^2. \qquad \text{(IV, 4)}$$

This function approaches the energy $D$ when either $r_1$ or $r_2$ approaches $\infty$ while $r_2$ and $r_1$ respectively have their equilibrium values. The function approaches the energy $2D$ when both $r_1$ and $r_2$ go to $\infty$. Clearly this is an extremely rough approximation. A much more sophisticated potential function starting out from a Morse–Lippincott function for the stretching coordinates [see Lippincott (754a, b)] and a trigonometric function for the bending coordinates has been given by Plíva (993) and applied to the discussion of the anharmonic constants in the ground states of a number of triatomic molecules [see also Kuchitsu and Bartell (699a)]. However, he also has not considered the higher vibrational levels which would be of interest in connection with the problem of dissociation.

In quantum theory half a quantum of the antisymmetric vibration $\nu_3$ is always excited, and therefore when $\nu_1$ is excited to higher $v_1$ values, the corresponding motion is not entirely restricted to the line $a-a$. Thus a Lissajous-like motion starts also in this case and it may eventually bring the image point to one of the two valleys $d$ and $e$. The levels $v_1\nu_1$ above the lowest dissociation limit are therefore also subject to predissociation.

Let us now consider the motion of the nuclei at right angles to the internuclear axis which we have disregarded so far. One way to represent diagrammatically the dependence of the potential energy on the bending coordinate is to keep one CO distance fixed and plot the potential energy of the molecule as a function of the $x$ and $z$ coordinates of the other O atom in a plane through the C—O axis. This is done in Fig. 166 assuming that the approaching O atom is in the $^1D$ state while CO is in the $^1\Sigma^+$ ground state. It is likely that there is a small potential maximum (not shown) which the O atom has to overcome when it comes from infinity to enter the bowl corresponding to the ground state of $CO_2$. While actually the internuclear distance in the CO group does not remain fixed in the process of combining it with another O atom (as assumed in Fig. 166), the change (from $r_e = 1.13$ to 1.16) is relatively small. Diagrams similar to Fig. 166 were first used by Hirschfelder, Eyring and Topley (555) for the systems $H_2 + H$ and $H_2 + Cl$.

Another way of representing the dependence of the potential energy of $CO_2$ on the bending coordinate is to plot it as a function of the O – O distance ($x$) and

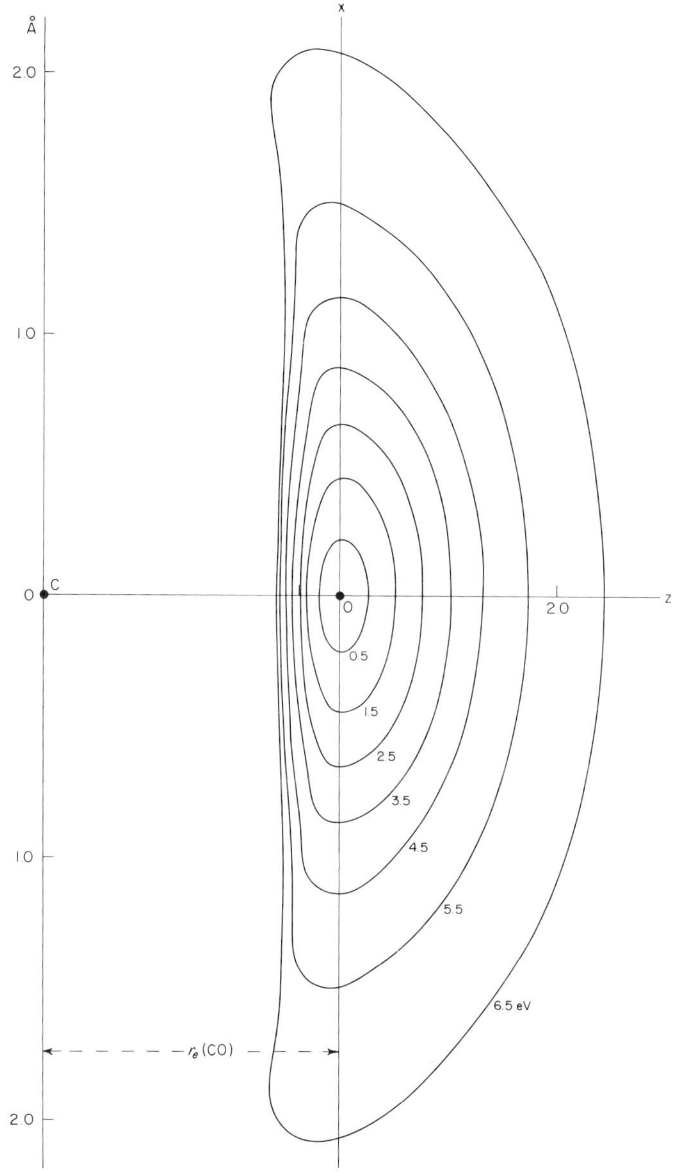


FIG. 166. **Potential surface of the lowest singlet state of $CO_2$ as a function of the *x* and *z* coordinates of one O nucleus, keeping the rest of the molecule fixed on the abscissa (*z*) axis.** The central C atom is fixed at the origin of the coordinate system while the second O atom (not shown) is fixed at $-1.16$Å of the abscissa axis. A quantitative calculation of this potential surface has recently been presented by Pariseau, Suzuki and Overend (960a).

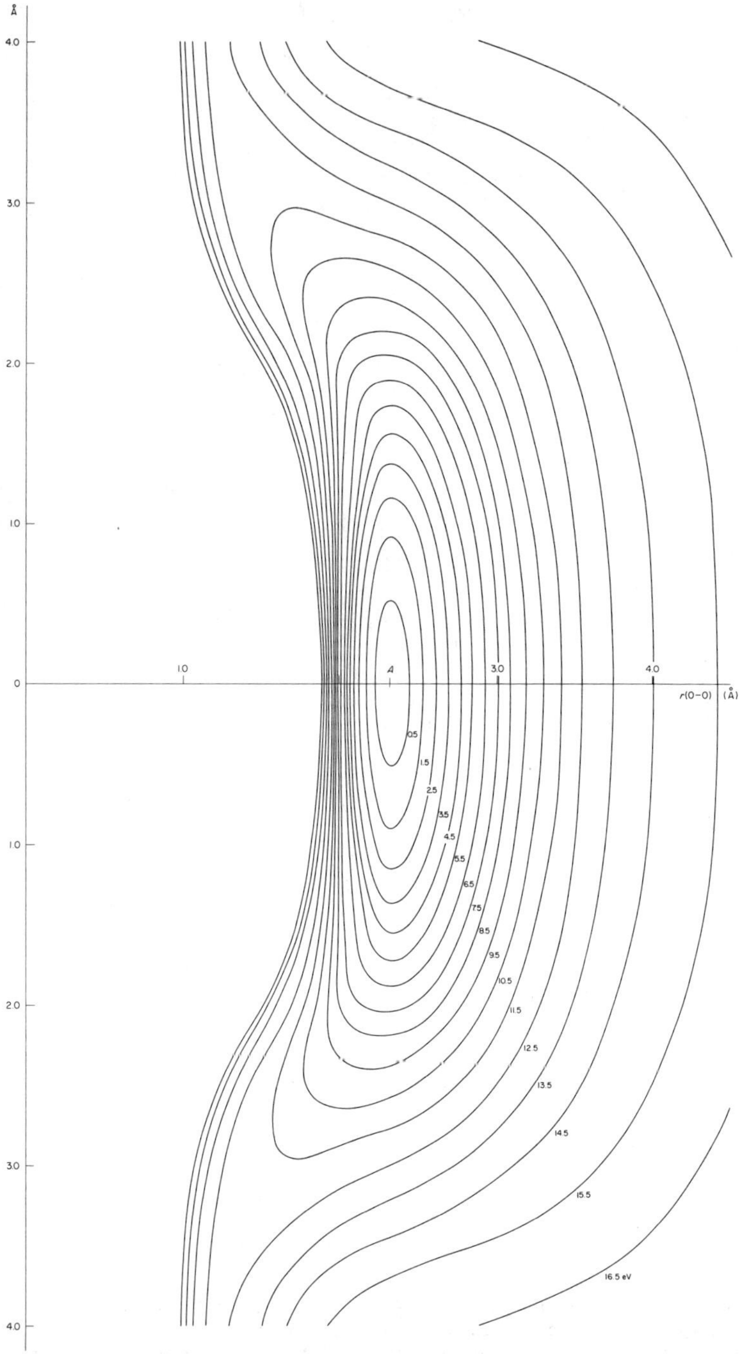


FIG. 167. **Potential surface of the lowest singlet state of $CO_2$ as a function of the O – O distance ($x$) and the perpendicular distance ($y$) of the C nucleus from the O – O line.** The vibrations $\nu_1$ and $\nu_2$ for small amplitudes correspond to motions of the image point parallel to the two coordinate axes through the bottom of the central bowl marked $A$.

the perpendicular distance ($y$) of the C atom from the O – O axis assuming it to be restricted to the perpendicular plane through the center of the O – O line. This is done in Fig. 167. For large $y$ the cross section of this surface gives the potential energy curve of $O_2$ in its $^3\Sigma_g^-$ ground state. The abscissa axis ($y = 0$) corresponds to the diagonal $a-a$ in Fig. 163. Again it is seen that classically the vibration $\nu_1$ can be excited to high amplitudes without losing its character while excitation of $\nu_2$ ($\perp$ to the abscissa axis at $A$) soon leads to Lissajous motion and a mixing of $\nu_1$ and $\nu_2$. If Fig. 167 were plotted for various positions of the C atom between the two O atoms, one would obtain an idea of the whole potential function as indeed one would if one plotted Fig. 166 for various values of the C—O distance.

**Unsymmetric linear triatomic molecules.** If we plot the potential surface of a linear XYZ molecule as a function of the XY and YZ distances ($r_1$ and $r_2$), assuming the molecule to remain linear throughout, we no longer obtain a symmetrical pattern as in Fig. 163 for $CO_2$ (or generally for $XY_2$). As we have seen earlier we must use a different scale on the two (oblique) coordinate axes if we want the motion of a mass point on the potential surface to represent the intramolecular motion. From the previous formulae (IV, 2) and (IV, 3) we obtain for example for HCN $\vartheta = 78° 15'$ and $c = 0.378$. In Fig. 168 the potential surface of HCN is plotted schematically for the ground state. Again we have two valleys, but now of different depth and slope, one leading to H($^2S$) + CN($^2\Sigma^+$), and the other to CH($^4\Sigma^-$) + N($^4S$). We do not show the intersecting surface leading to CH($^2\Pi$) + N($^4S$), which gives only triplet and quintet states, nor that leading to CH($^2\Pi$) + N($^2D$) which does yield singlet states, but for large $r_2$ lies in all probability higher[2] than CH($^4\Sigma^-$) + N($^4S$) and therefore does not correspond to the ground state of the molecule (see Fig. 170). The two normal vibrations $\nu_1$ and $\nu_3$, which are essentially a CH and a CN vibration, correspond to the motions in the two mutually perpendicular directions indicated in Fig. 168. It is clear from the figure that neither vibration remains a simple vibration when the amplitude is increased. Rather they give then Lissajous motions, and either Lissajous motion, if the energy is high enough, will lead to a dissociation of the molecule corresponding to the lowest dissociation limit, in the present case H + CN.

A plot of the potential energy as a function of the $x$- and $z$-coordinates of the H atom (or the N atom), keeping the CN group (or the CH group) fixed, would look very similar to Fig. 166 for $CO_2$, and would give an idea of the dependence on the bending coordinate.

**Non-linear symmetric triatomic molecules.** The potential function of a non-linear $XY_2$ molecule plotted as a function of the two XY distances, assuming the YXY angle to be kept constant, is of exactly the same type as Fig. 163 for a linear molecule. However, such a plot is much less significant since, in contrast to the linear case, in a collision Y + XY the angle YXY does certainly not remain

[2] While the $^4\Sigma^-$ state of CH has not yet been observed its excitation energy is almost certainly less than 1 eV.

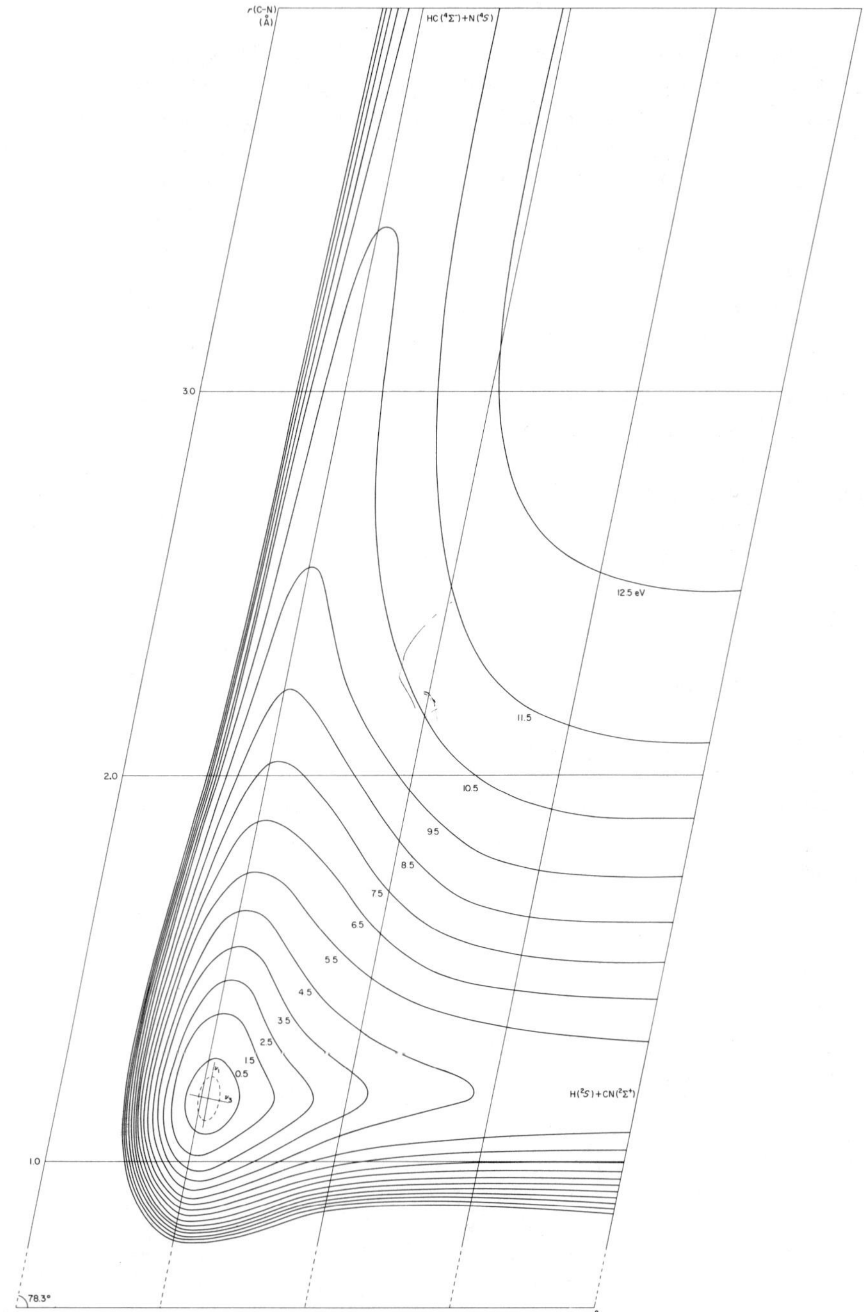


FIG. 168. **Potential surface of the lowest singlet state of HCN (ground state) as a function of the CH and CN distances (schematic).** In accordance with the previous formulae (IV, 3) the CN distance is plotted on a scale which is $\frac{1}{0.378}$ times as large as that of the CH distance. The angle of the oblique axis system corresponds to eq. (IV, 2). The directions of motion for the vibrations $\nu_1$ and $\nu_3$ near the minimum of the bowl are indicated.

fixed and in addition, even near the potential minimum, the dynamics of the motion would not be represented by the motion of a mass point on such a potential surface.

More useful representations are those similar to Figs. 166 and 167. As an example, in Fig. 169a the potential energy of $H_2O$ is shown by contour lines as a function of the H—H distance ($x$) and the distance ($y$) of the O nucleus from the H—H line, assuming that this nucleus is at all times symmetrically placed with respect to the two H atoms. The possibility of antisymmetric motion (vibration) is neglected, just as in Fig. 167 for a linear molecule. The minimum corresponding to the equilibrium position lies now, of course, above the $x$ axis, not on it as in the linear case. The two symmetrical normal modes near the equilibrium position are again represented by the motions of the image point in the directions of maximum and minimum curvature in the potential well (*aa* and *bb* in Fig. 169a).

For large $y$ the cross section of the potential surface approaches the potential function of an $H_2$ molecule in the ground state. As for $CO_2$ (Fig. 167) there is a potential valley coming from the top of the figure and ending in the potential well representing the equilibrium position. In all probability there is a substantial potential hump in this valley corresponding to the rearrangement of the bonds, from an H—H bond to two OH bonds. The O atom at the upper end of the valley must be in the $^1D$ state not the $^3P$ ground state, since the ground states of both $H_2O$ and $H_2$ are singlets. The potential surface corresponding to the combination $O(^3P) + H_2(^1\Sigma_g^+)$ is lower at large $y$ and intersects the surface shown at an intermediate $y$ value. Nevertheless, if we go from the minimum of the well to the plateau at the right, we obtain an O atom in the $^3P$ state (+ two H atoms). Such a correlation is possible because at large $x$ the $^1\Sigma_g^+$ and $^3\Sigma_u^+$ states of $H_2$ have the same energy.

In order to make this situation clearer, we have drawn in Fig. 169b and c two cross sections of the potential surface along two lines parallel to the $y$ axis. The combination of an $O(^3P)$ atom with $H_2(^1\Sigma_g^+)$ gives only triplet states ($^3B_2$, $^3B_1$, $^3A_2$) which, near the conformation of the $H_2O$ ground state, lie very high. But if the $O(^3P)$ atom is combined with $H_2$ in the (repulsive) $^3\Sigma_u^+$ state, singlet, triplet and quintet states arise (see Table 27). Of these, the $^1A_1$ state interacts with the two $^1A_1$ states from $O(^1D) + H_2(^1\Sigma_g^+)$. For intermediate $x$ the latter combination is lower and forms the ground state (as shown in Fig. 169b), for large $x$ the combination $O(^3P) + H_2(^3\Sigma_u^+)$ is lower at large $y$, and therefore, the $^1A_1$ state at small $y$ is derived from it (see Fig. 169c). The three $^1A_1$ surfaces intersect only at very large $y$ near $A$ at the upper edge of Fig. 169a (see the cross section for large $y$ above Fig. 169a). They separate as soon as $y$ is no longer very large since then, because of the interaction between the O atom and the H atoms, the non-crossing rule becomes operative.

The potential surface Fig. 169 can be used for qualitative (but not quantitative) discussions of the dynamics of collision processes. A collision between an O atom (in the $^1D$ state) and an $H_2$ molecule whose axis is perpendicular to the direction of motion of the O atom, is represented by the motion of the image point along the

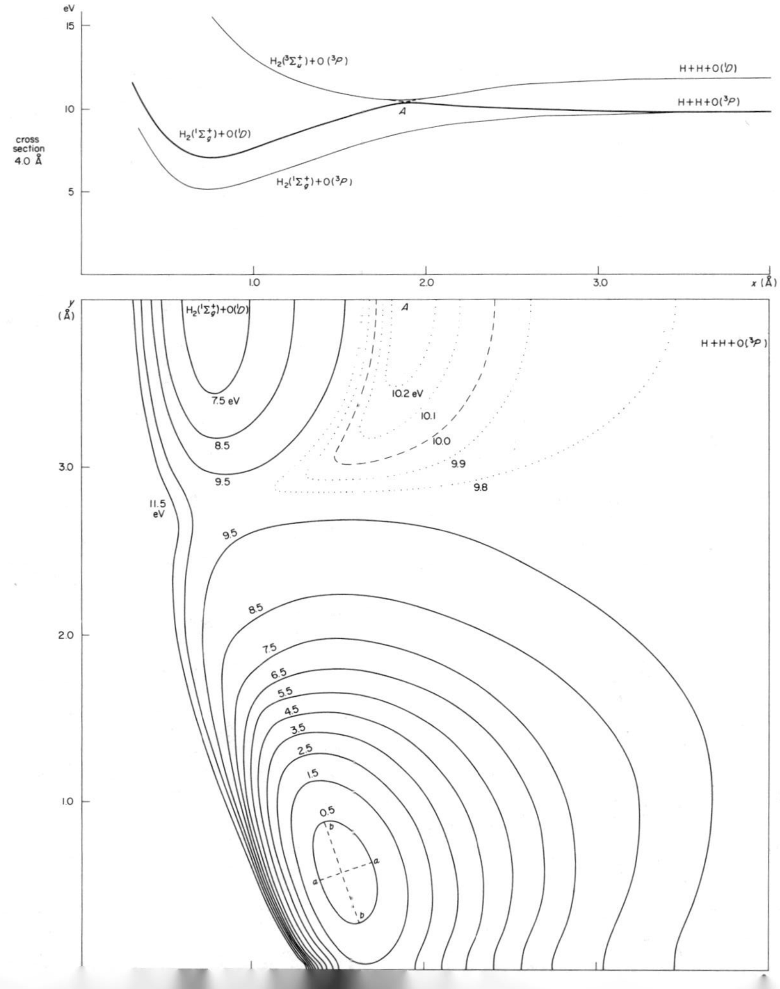


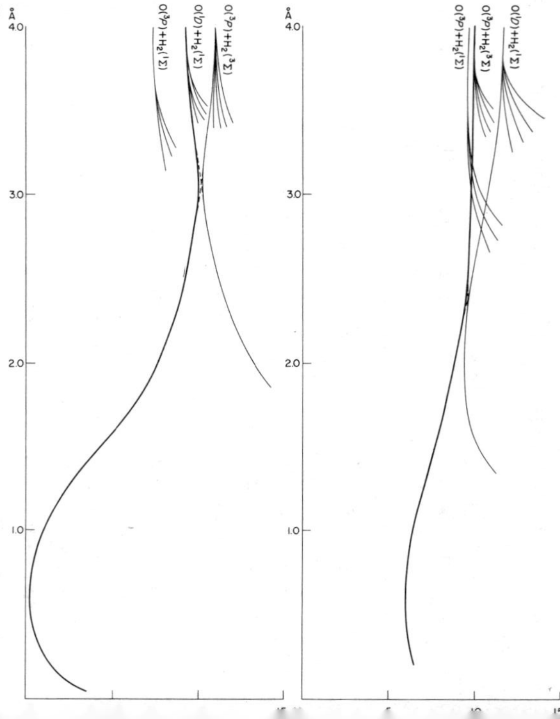


Fig. 169. **Potential surface of the lowest singlet state of $H_2O$ as a function of the H—H distance ($x$) and the distance ($y$) of the O nucleus from the H—H line.** At the right in ($b$) and ($c$) two cross sections of the potential surface along two lines parallel to the $y$-axis are shown; at the top a cross section along a line parallel to the $x$-axis for large $y$ is shown. These cross sectional diagrams include as thin-line curves some of the triplet states that arise from the separated atoms. The cross sections of the main (singlet) potential surface are shown by heavy curves.

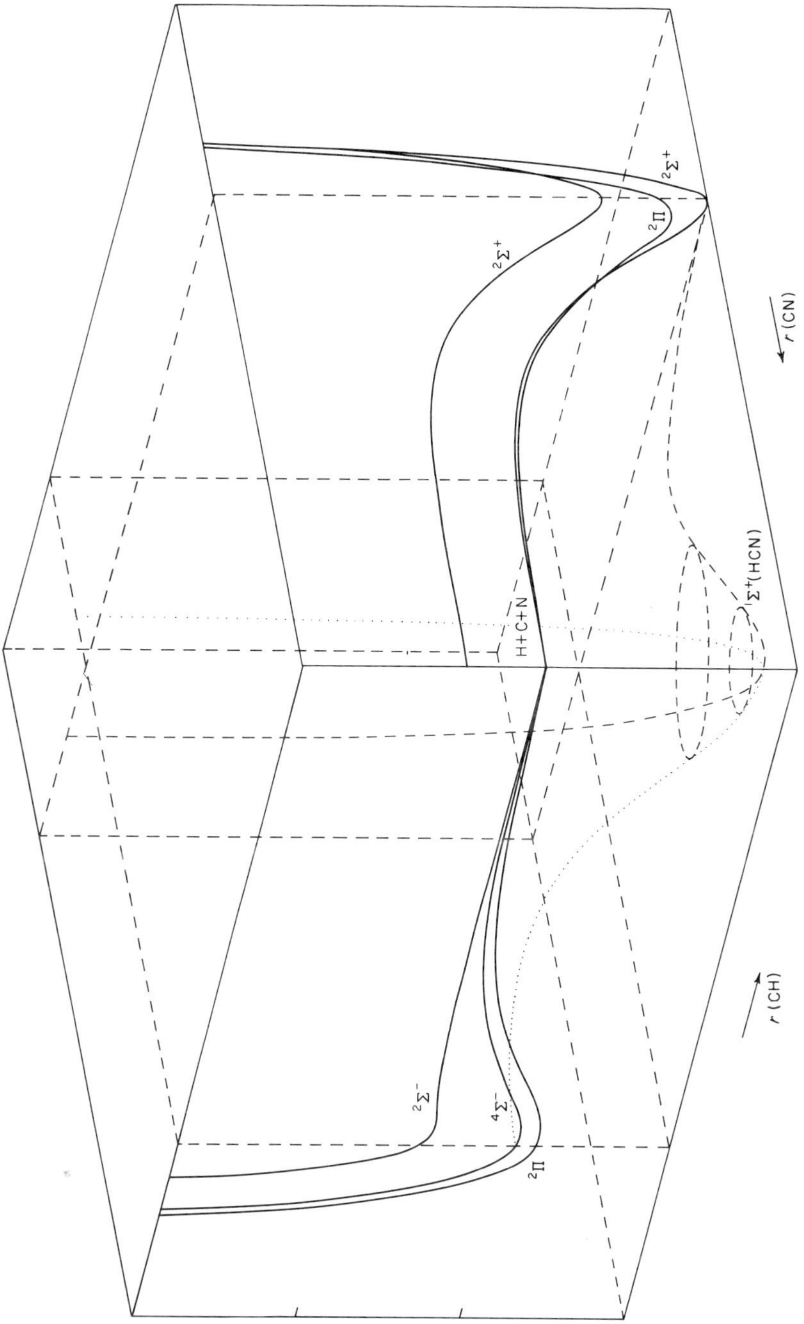


Fig. 170. **Simplified representation of the lowest potential surfaces of HCN.** The curves on the faces of the cube represent the potential energy for large $r$(CH) or $r$(CN). The potential bowl corresponding to the ground state of HCN is sketched qualitatively.

valley that enters the diagram at the top. It is clear that because of the curvature of the valley, the point will move up on the left slope until its momentum component in the direction of steepest descent is used up, after which it will roll down the slope, enter the bowl corresponding to the ground state of $H_2O$ (if it had sufficient energy to overcome the potential hump), rise on the other side, and so on, carrying out a complicated Lissajous motion. Eventually it will return and leave the valley again, but the lifetime of the collision complex is obviously much larger than the period of one vibration.

A simplified method of representing the potential functions of triatomic molecules has been used by Laidler (712) and others. It consists simply in drawing a cube on whose faces the limiting potential curves are drawn for large $r_1$, $r_2$, etc. This is shown for HCN in Fig. 170. Such figures are useful both for the discussion of dissociation processes and of reactions between three atoms.

**More complicated cases.** Even for triatomic molecules, two-dimensional potential surfaces represent the potential function only in a very incomplete way since they neglect one (or two) degrees of freedom. If one wanted to represent the complete potential function, one would have to draw a series of two-dimensional surfaces for various values of the third parameter, e.g. the angle between the two XY bonds. However, such a series of surfaces would still not allow one to visualize the motion of the image point and therefore of the atoms in the molecule. Recently, using electronic computers several groups of authors have calculated classical trajectories of atoms in simple reactions taking account of all degrees of freedom [e.g. Polanyi and Rosner (998)].

For molecules with more than three atoms, even a representation by a series of two-dimensional potential surfaces is quite incomplete; for a four-atomic molecule, one would need a four-dimensional series of such surfaces since there are six degrees of freedom. The best one can do is to draw one- or two-dimensional cross sections of the six-dimensional hypersurface, keeping five or four parameters fixed, and similarly for more atoms. For example, for $H_2CO$ we can plot the potential energy as a function of the H—H distance and the distance of the H—H line from the C nucleus, keeping everything else fixed and keeping the $CH_2$ configuration symmetrical. In this way we obtain a diagram very similar to Fig. 169. We might also consider $H_2CO$ as a triatomic molecule $(H_2)$—C—O and represent its potential energy by a diagram like Fig. 168.

**Non-crossing rule and conical intersections.** For diatomic molecules, potential curves of states of the same species do not cross (see Vol. I, p. 295). If in a certain approximation they do cross, the effect of the neglected terms in the wave equation will always be such that the levels "repel" each other and therefore that in fact they do not cross. It was first recognized by Teller (1197), that this non-crossing rule has to be somewhat modified for polyatomic molecules.

Mathematically, the condition for crossing of two states $n$ and $i$, i.e. for (accidental) degeneracy, is that for them

$$\text{both} \quad W_{nn} = W_{ii} \quad \text{and} \quad W_{ni} = 0, \qquad \text{(IV, 5)}$$

where $W_{nn}$, $W_{ii}$ and $W_{ni}$ are the matrix elements of the perturbation function $W$ (see Vol. I, p. 14). In a diatomic molecule, the two conditions (IV, 5) cannot be simultaneously fulfilled by variation of the internuclear distance $r$, unless because of symmetry $W_{ni} = 0$ for any $r$, i.e., when the two states $n$ and $i$ have different species. However, for a polyatomic molecule, since there are several internuclear distances, the two conditions can be fulfilled for certain $r_i$ values even if the two states have the same species, because the electronic eigenfunctions depend on the $r_i$ as parameters. Teller has shown that in a two-dimensional case near the point $r_1^0$, $r_2^0$ for which the two conditions (IV, 5) are fulfilled, the two potential surfaces form a double cone in $r_1$ and $r_2$. In other words, the two potential surfaces intersect in a single point if they are represented as functions of $r_1$ and $r_2$: we have a *conical intersection*. Actually, of course, the potential functions depend on several other variables and the intersection occurs for any values of these, i.e. in such a case the two $f$-dimensional surfaces would intersect along an $f-2$ dimensional "line".

If spin-orbit interaction is large (i.e., if magnetic forces have to be considered), the matrix element $W_{ni}$ is complex, and the condition $W_{ni} = 0$ implies two conditions: the real and imaginary part must separately vanish. Therefore, only for a set of three variables, $r_1$, $r_2$, $r_3$, can the conditions (IV, 5) be fulfilled, and thus the "line" of intersection has only $f-3$ dimensions.

It can be shown [Herzberg and Longuet–Higgins (534)] that the electronic eigenfunction changes sign when one goes once around a conical intersection. This is also the characteristic of the electronic eigenfunction for the self-intersecting potential functions arising in the Jahn–Teller effect (see p. 40 and Fig. 16). Indeed Jahn–Teller distorted potential functions of degenerate electronic states are typical examples of conical intersections.

Conical intersections also arise in molecules for which a linear conformation is possible and for which in this linear conformation an intersection of two degenerate states of different orbital species exists (for example between a $\Pi$ and a $\Delta$ state). As the molecule is bent each state splits into an $A'$ and an $A''$ state. The two resulting $A'$ states cannot intersect except in the linear conformation: we have a conical intersection. This is schematically illustrated by Fig. 171. A similar intersection results for the $A''$ components.

In all these cases the apex of the cone corresponds to a conformation of the molecule in which the two electronic states are non-accidentally degenerate with each other or have different species. One may ask whether a more general case is possible in which the apex does not correspond to a special (symmetrical) conformation. As was shown by Herzberg and Longuet–Higgins (534) such a situation arises for the two $^2A'$ states that result when three unlike atoms in $^2S$ states (e.g. H + Li + Na) are brought together [compare London's formula (III, 69); the apex corresponds to the point where the square root vanishes]. This is a genuine case in which the intersection contravenes the non-crossing rule in its "diatomic" form.

Several other types of intersection may arise in polyatomic molecules. As we have seen in Chapter I, section 2(b), vibronic interactions in linear molecules lead to a splitting of the potential function, but in such a way that near the linear conformation they are in contact (see Fig. 4), i.e. the angle of the cone is 180°. These intersections are more properly called *glancing intersections.* Here again the point of intersection is determined by symmetry and thus is in accord with the elementary non-crossing rule.

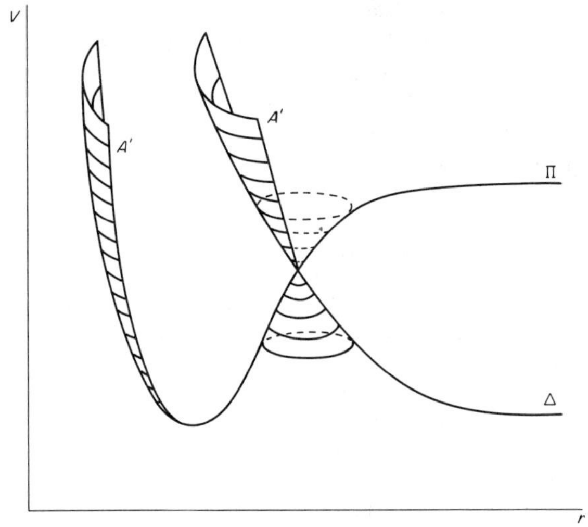


FIG. 171. **Conical intersection of the potential surfaces of the $A'$ (or $A''$) states derived from a Π and a Δ state of a linear molecule.**

Another type is exemplified by the potential surfaces of the $H_2O$ system partly represented in Fig. 169. Here for large $y$ the surfaces for the two $^1A_1$ states derived from $O(^1D) + H_2(^1\Sigma)$ and $O(^3P) + H_2(^3\Sigma)$ intersect since the matrix element $W_{ni}$ vanishes because the O atom is at a great distance from the $H_2$ group. As soon as the O atom is brought closer, however, this intersection is "avoided" and, as a consequence, the ground state of $H_2O$, while it yields $O(^1D) + H_2(^1\Sigma)$ upon removal of the O atom, does yield oxygen in the $^3P$ state when the two H atoms are removed separately without the formation of a $^1\Sigma$ $H_2$ molecule.

Near-intersections can arise just as for diatomic molecules: for example, an intersection between a state arising from two ions and one arising from neutral parts, or an intersection between a state in which one electron is highly excited and one in which no such excitation exists. Just as for diatomic molecules, such near intersections usually take place at fairly large internuclear distances where $W_{ni}$ is very small (see Vol. I, p. 375). Moreover, such intersections in polyatomic molecules would extend along $f - 1$ dimensional lines, or ridges (if there are $f$ degrees of freedom). They would not be conical intersections.

## 2. Continuous Spectra: Dissociation of Polyatomic Molecules

Just as for diatomic molecules, continuous spectra of polyatomic molecules can occur in absorption and emission. Very few of these continua correspond to ionization processes which we are not considering in this chapter; almost all of the observed continua correspond to dissociation processes. The interpretation of these continua is, however, much more difficult than for diatomic molecules because of the existence, in each electronic state, of several dissociation limits corresponding to different dissociation products. Consequently, only in relatively few cases is the interpretation of observed continua of polyatomic molecules as detailed and definitive as for many continua of diatomic molecules. Another reason for this uncertainty is that the possibilities of predissociation are much more numerous (see section 3) and that often predissociation, i.e. diffuseness, may be so strong that the resulting spectrum cannot easily be distinguished from a true continuum corresponding to a continuous range of energy levels.

In this section we shall consider only those continua for which the corresponding continuous range of energy levels represents dissociation in one single oscillation, without going through a complicated Lissajous motion ("direct dissociation"). In such a case, the molecule dissociates within a time of the order of $10^{-12}$ to $10^{-13}$ seconds after having been brought into the unstable state.

### (a) Absorption continua

**Upper state with stable equilibrium position.** If both lower and upper states have a stable equilibrium position, absorption may yet, on the basis of the Franck–Condon principle, lead to the continuous range of levels of the excited state, and a true continuous spectrum may result just as for diatomic molecules.

For example, consider a linear symmetric $XY_2$ molecule for which the potential surface of an excited state is similar to that of the ground state (shown in Fig. 163) except that the potential well is shifted to higher $r_1$ and $r_2$ values and is not as deep as for the ground state; then in absorption from the minimum of the ground state a point vertically above on the upper surface may be reached which is higher than the plateau corresponding to dissociation into three atoms. If that is the case, the image point representing the system will slide down to the minimum and up to the plateau; that is, the molecule will dissociate in one oscillation. Figure 172 shows a cross section of the two potential surfaces along the diagonal *a–a* of Fig. 163. It is entirely similar to the potential curves of $Cl_2$ shown in Fig. 175 of Volume I. As in the diatomic case, depending on the relative positions of the upper and lower surfaces, one may expect a progression of bands converging to a limit with an adjoining continuum, or only a continuum.

No clear-cut example of either of these two cases has yet been found. The reason for the lack of examples is two-fold: (1) the higher vibrational levels of the symmetric vibration are strongly predissociated (see section 3), and (2) the energy required for a dissociation into three atoms is fairly high, and the corresponding spectrum would lie far in the vacuum ultraviolet and could be obscured by many other discrete transitions or by strong continuous absorptions involving a change of shape (see below) which often lie at longer wavelengths.

For an unsymmetrical linear triatomic molecule, XYZ, again a continuous spectrum will only arise if position and depth of the potential well differ appreciably in upper and lower state. Here, in addition to the possibility of dissociation into three atoms (when both the equilibrium distances $r_1^e$ and $r_2^e$ change proportionately),

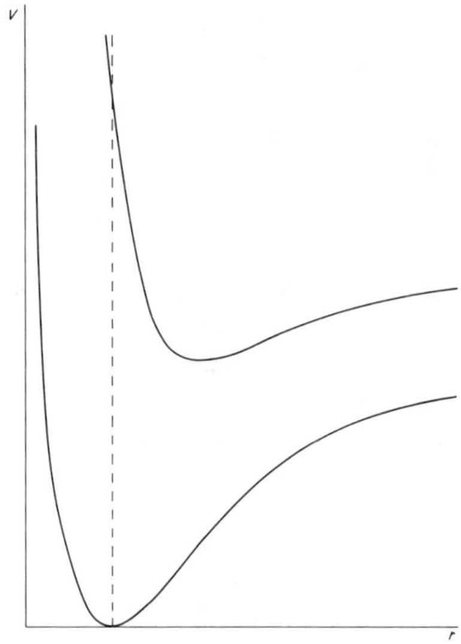


FIG. 172. **Cross sections of the potential surfaces in two electronic states of a linear $XY_2$ molecule explaining the origin of a continuous absorption spectrum.** The two potential surfaces are assumed to be of the type of Fig. 163 and the cross section is taken along the diagonal $r_1 = r_2$.

there is also the possibility of dissociation in one oscillation into X + YZ or XY + Z if only $r_1^e$ or only $r_2^e$ changes appreciably. This is because in such a case, after the photon absorption, the system may reach a point on the side of the well such that after going through the minimum, the image point flies out through one of the two valleys (see Fig. 168). The greater the change of the internuclear distance involved, the more likely is such a process to occur. However, if the change is not large even though the energy is sufficiently high, the image point may not immediately reach the valley corresponding to dissociation but carry out a Lissajous motion which only after some time leads it to this valley. Such a mechanism will be discussed more fully in the next section.

Unsymmetric dissociation processes similar to those in linear XYZ molecules may also occur in linear $XY_2$ molecules if in the excited state the equilibrium conformation is unsymmetric, that is, if in the excited state the potential well is unsymmetrically located with respect to that of the ground state. In that case the image point, which is initially (i.e. immediately after the "quantum jump") symmetrically located, may, after going through the new minimum, leave it on the opposite side through the valley corresponding to Y + XY. Conversely, one may say that a linear symmetric $XY_2$ molecule will show a genuine (quasi-diatomic) continuous spectrum corresponding to dissociation into Y + XY only if in the upper state the equilibrium conformation is unsymmetric.

If in a *non-linear* symmetric $XY_2$ molecule the X—Y distance is much greater in an excited state than in the ground state, while the Y—X—Y angle is little changed, a direct dissociation into Y + X + Y may take place similar to that for linear $XY_2$. Again the energy required is rather high, and no observed case has been clearly shown to be of this type. In order to obtain a direct dissociation into Y + XY, one would, as in the linear case, have to have an unsymmetrical equilibrium position in the excited state.

If the Y—X—Y angle changes appreciably in the transition, it is better to consider a potential diagram of the type of Fig. 169. It is clear that if in the excited state the angle is smaller than in the ground state, that is, if the potential well is at a greater $y$ value, the image point after going through the minimum may well fly out of the valley at the top of the diagram, that is, a dissociation into X + $Y_2$ may take place. The same conclusion applies for a molecule that is linear in the ground state (see Fig. 167) if it is bent in the excited state, as often happens. In either case a diatomic molecule is formed from two atoms (Y) not bound together in the original molecule. Such a dissociation process thus leads to *internal recombination.*

Possible examples of continua corresponding to dissociation into X + $Y_2$ are the continuous absorption spectra of $H_2O$, $H_2S$, $H_2Se$, . . . in the near ultraviolet (see Chap. V, section 1). In these molecules the highest orbital filled in the ground state is a non-bonding orbital ($1b_1$, in $H_2O$—see Fig. 123 and Table 33), while the lowest unfilled orbital is antibonding ($4a_1$ in $H_2O$). In the first excited state corresponding to removal of an electron from the former to the latter orbital, there are therefore only three net bonding electrons compared to four in the ground state; therefore, in this excited state the binding would be expected to be weaker and the equilibrium position at larger $r$ values than in the ground state. It is therefore possible that by absorption from the ground state according to the Franck–Condon principle, an energy range above one of the dissociation limits is reached. However, if the equilibrium conformation in the excited state is symmetrical, an unsymmetrical dissociation into HX + H is unlikely to occur while for the symmetrical dissociation into H + X + H the energy is not sufficient. Thus, only dissociation into X + $H_2$ would remain. Indeed, the $4a_1$ orbital favors bonding between the two H atoms (see Table 41), that is, the H—X—H angle is likely to be reduced in the excited state, and this, as we have seen, favors dissociation into X + $H_2$. On account of the spin conservation rule, the X atom must be in the $^1D$ state upon dissociation. The long wavelength limits of the observed continua for $H_2O$, $H_2S$, . . . are indeed in agreement with this condition as was pointed out already in 1931 by Goodeve and Stein (430) who were the first to suggest the possibility of a primary dissociation into X + $H_2$. However, for $H_2O$, photochemical evidence [see Ung and Back (1231)] shows that at 1849 Å dissociation into H + OH predominates, suggesting that the upper state of the continuum is either entirely repulsive or has an unsymmetrical equilibrium position.

For more-than-triatomic molecules the possibilities of direct photo-dissociation, when the upper state has a potential minimum, are even less easily visualized. We must try to approximate the molecule by a three-particle system and then apply the previous considerations. For example, a molecule like $CH_3Cl$ could be considered as a linear system $(H_3)\equiv C—Cl$ whose potential surface would be of the type of Fig. 168. It is then seen that if $r_e$(C—Cl) in the upper state is much larger than in the lower state, a dissociation into $CH_3$ + Cl in one oscillation may occur after light absorption. A dissociation into 3H + CCl would occur if the C—H distance is much larger in the excited than in the ground state, but the energy required for such a photo-dissociation would be very large. An unsymmetrical direct dissociation, for example into $H_2$ + HCCl, would require an upper state with an unsymmetrical equilibrium position.

In $CH_4$ photochemical investigations with different isotopes have given conclusive evidence that at least part of the continuous absorption spectrum corresponds to dissociation into $CH_2$ + $H_2$ [see Mahan and Mandal (791) and Magee (788)].

The general conclusion from the preceding considerations is entirely analogous to that for diatomic molecules: for an upper state with a stable equilibrium position a continuous absorption spectrum corresponding to direct dissociation (i.e. dissociation in a single oscillation; see p. 445) can arise only if the equilibrium internuclear distances and angles in the upper state are greatly different from those in the ground state. Such a large difference always arises when there is a change of symmetry, from linear to bent or from planar to non-planar, but can in principle occur also without change of symmetry.

**Upper state without stable equilibrium position.** If the upper state has no stable equilibrium position (or only a very shallow minimum), certain additional possibilities of dissociation may arise. As an example, let us consider a linear symmetric $XY_2$ molecule for which the upper state has a potential surface as shown in Fig. 173, consisting essentially of two valleys which rise toward the diagonal where there is either a ridge, as in Fig. 173, or a very shallow bowl at large $r_1 - r_2$. A mathematical expression for such a potential function, derived from Morse functions for the corresponding diatomic molecules, has recently been given by Wall and Porter (1258). If the upper state of an electronic transition has such a potential function, a direct dissociation into Y + XY will occur for almost any point reached by light absorption. To be sure, because of the curvature of each of the valleys, the image point will not follow a nearly straight line along the bottom of the valley, but an oscillation from one side to the other will take place while the point is leaving the valley, as indicated in Fig. 173. Nevertheless, this process must still be classified as a direct dissociation since the time required is essentially the same as if the "secondary" vibration did not take place. This "secondary" vibration of the image point will persist for large $r_1$ (or $r_2$) and corresponds to vibrational energy of the XY molecule which is produced in the dissociation process.

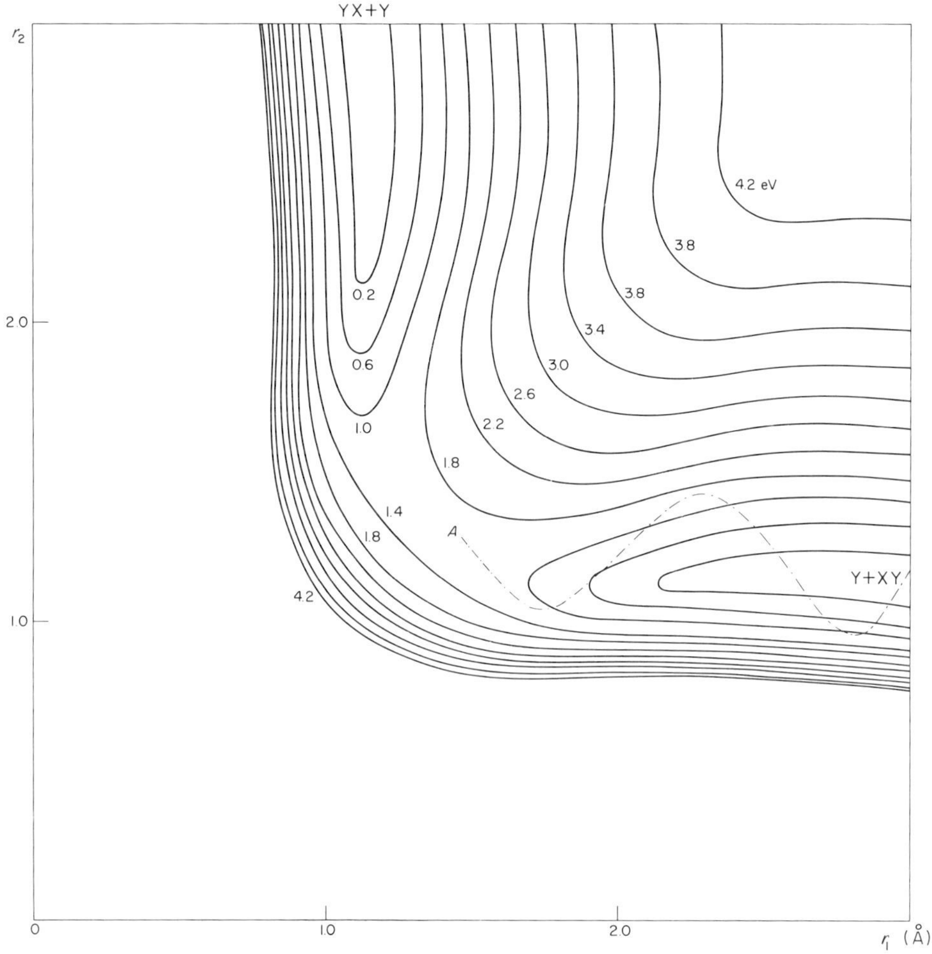


Fig. 173. **Potential surface of a repulsive state of a linear $XY_2$ molecule as a function of $r_1$ and $r_2$.** Although calculated on the basis of the formulae of Wall and Porter (1258) the figure must be considered as qualitative rather than quantitative. The dot-dash line gives the trajectory of the image point if it starts out from the point $A$. Since the diagram has not been plotted in an appropriate oblique coordinate system (see p. 432) the motion of the image point is not given directly by the dynamics of a mass point moving on the potential surface as plotted.

For bent $XY_2$ molecules, entirely similar considerations apply. Even if there is no asymmetry in the potential function of the upper state, as long as it is mainly repulsive, direct dissociation into Y + XY can take place; but in general, on the basis of the Franck–Condon principle, the continuum will start only at a wave number appreciably greater than corresponds to the net dissociation energy.

Fig. 174 shows the same repulsive potential surface as in Fig. 173 but now in terms of the coordinates $x$ and $y$ previously introduced (see Fig. 167). It is clear

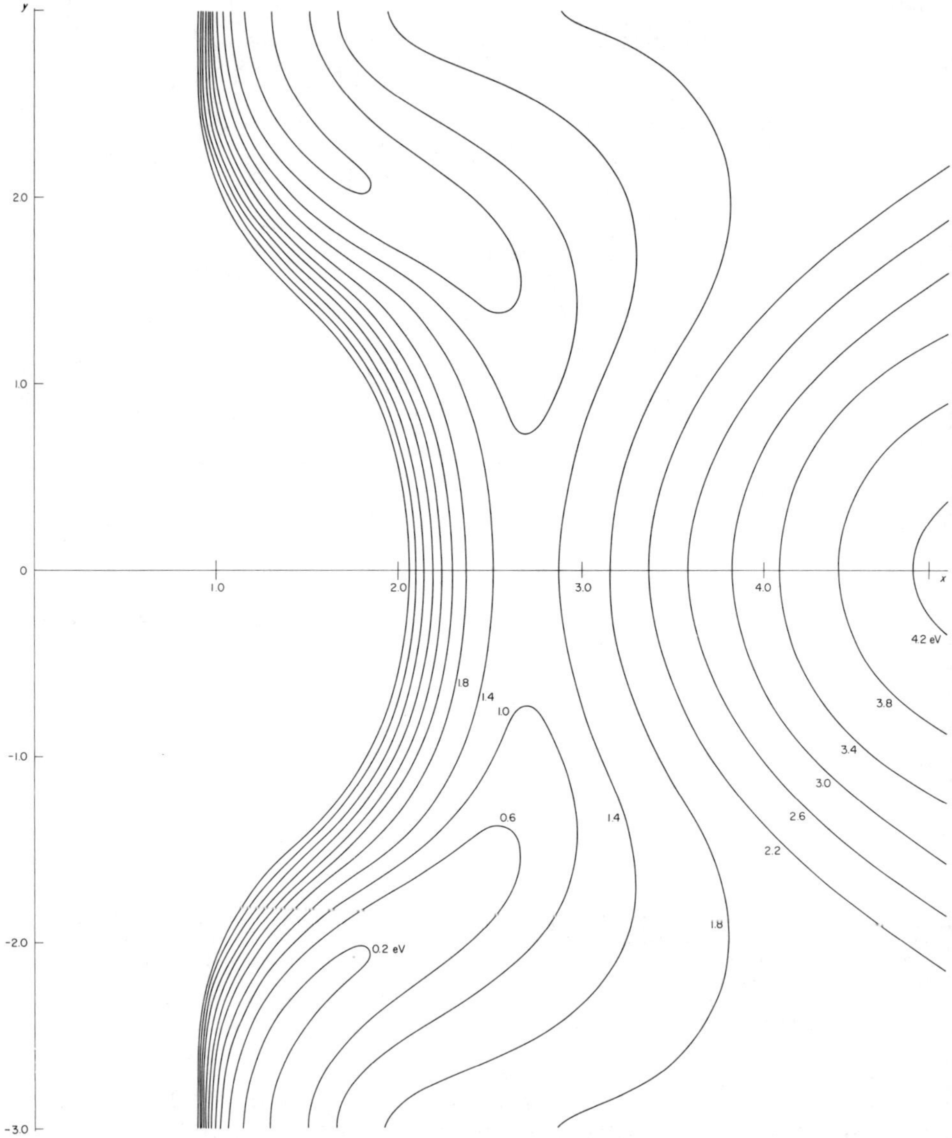


FIG. 174. **Potential surface of a repulsive state of a linear $XY_2$ molecule as a function of the Y - - - Y distance ($x$) and the displacement of X from the Y - - - Y axis ($y$).** This diagram represents the same surface as Fig. 173 (compare the energies written on the contour lines).

from this figure that if a transition from a stable ground state (as in Fig. 167) to such an excited state as represented in Fig. 174 takes place, the image point will immediately move to $y \to \infty$ along the valleys that extend toward the top or bottom of the diagram, that is, there will be direct dissociation into $X + Y_2$ (possibly with vibrational energy of the resulting $Y_2$ molecule).

Since both diagrams Fig. 174 and Fig. 173 represent one and the same (three-dimensional) potential surface, both dissociation processes $XY_2 \to Y + XY$ and $XY_2 \to X + Y_2$ may take place, and, if the dissociation energies are similar, even in the same region of the spectrum. Thus, one and the same continuous absorption spectrum may correspond to two different dissociation processes.

Similar conclusions may also be drawn for more complicated molecules: for example, the unsymmetrical dissociation of $CH_3Cl$ into $H_2$ + HCCl can now take place even if the potential function of the upper state is symmetrical, as long as it is mainly repulsive. The same consideration applies to other unsymmetrical dissociations. Under suitable conditions there may now be even more than two dissociation processes corresponding to one and the same continuous absorption spectrum. However, it must be realized that a certain dissociation process, although possible from the point of view of the available energy, may in fact not occur or occur at much higher energies because of intervening potential ridges or simply because of an unfavorable shape of the potential surface.

**Intensity distribution.** The distribution of intensity in a continuous absorption spectrum is, just as for diatomic molecules (see Vol. I, p. 391), given by

$$I_\nu \sim \nu\left[\int \psi_v' \psi_v'' d\tau_v\right]^2 \tag{IV, 6}$$

where, however, here the vibrational eigenfunctions $\psi_v'$ and $\psi_v''$ of the upper and lower states depend on the $3N - 6$ normal coordinates. $\psi_v''$ is in a first approximation a product of harmonic oscillator functions, but $\psi_v'$ is a more complicated function: In a very rough approximation it can be split into a product of harmonic oscillator functions of those normal coordinates which do not take part in the dissociation process and functions of those coordinates which are involved in the dissociation process. The latter functions depend on the coordinates in a way similar to the functions described in Volume I (p. 392) for diatomic molecules. These components of $\psi_v'$ vary in a regular way with the energy. No detailed discussion of the form of these functions has been given. One usually assumes that, as for diatomic molecules, a delta function will give a good approximation and therefore that a "reflection" of $\psi_v''^2$ at the potential surface of the upper state (after multiplication by $\nu$) gives the theoretical intensity distribution.

The first application of this method to a polyatomic molecule was carried out by Fink and Goodeve (382) for $CH_3Br$. Assuming that the observed continuum corresponds to a dissociation into $CH_3$ + Br and using only the eigenfunction of the C—Br vibration in the lower state, they obtained from the observed intensity distribution the variation of the potential energy with $r$(C—Br) in the upper state.

The fact that a reasonable curve was obtained in this way makes it probable that the interpretation of the absorption continuum by a direct dissociation process with a simple (repulsive) upper potential surface is correct. For $CH_3I$, an application of the same method led Porret and Goodeve (1005) to an upper potential curve with a point of inflexion. Since this seemed unlikely to them, they re-interpreted the data by resolving the observed intensity distribution into two curves of the type that give a single conventional upper potential function. The two upper states that they obtained in this way probably correspond to dissociation into $CH_3$ + I ($^2P_{\frac{1}{2}}$) and $CH_3$ + I ($^2P_{\frac{3}{2}}$). While this appears to be a very reasonable interpretation, one must remember that an apparently anomalous form of the upper one-dimensional potential curve could also be caused by an asymmetry of the potential surface of the excited state or other peculiarities of this state.

Let $r$ be the internuclear distance that will go to $\infty$ in the photodissociation process and let us assume that the potential function of the upper state depends linearly on $r$ near the $r_e$ value of the lower state. Then the quasi-diatomic method outlined above leads to the conclusion [see Gordus and Bernstein (439)] that the absorption coefficient $\varepsilon$ for absorption from the $v'' = 0$ state (for which $\psi'' = ae^{-\beta x^2}$) is given by

$$\varepsilon = \varepsilon_m e^{-\alpha(\nu_m - \nu)^2}$$

or

$$\log \varepsilon = \log \varepsilon_m - \alpha(\nu_m - \nu)^2. \qquad \text{(IV, 7)}$$

Here $\varepsilon_m$ and $\nu_m$ refer to the absorption maximum and $\alpha$ is a constant depending on $\beta$ and the slope of the upper state potential function.

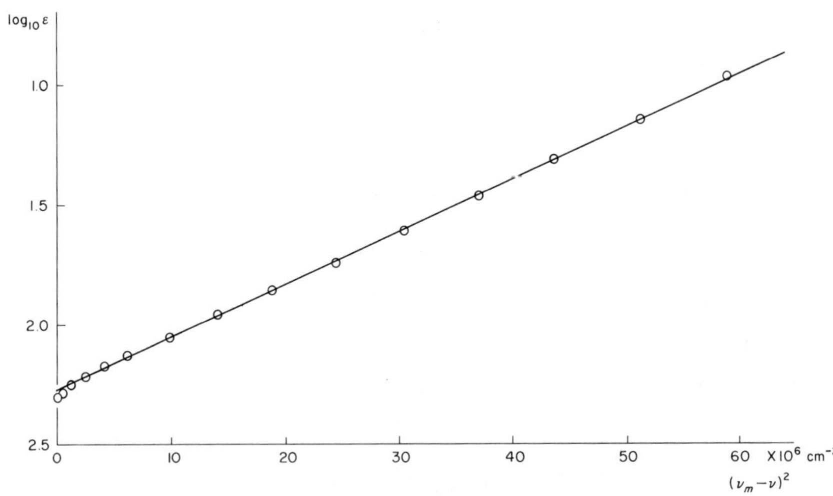


FIG. 175. **Dependence of $\log_{10}\varepsilon$ on $(\nu_m - \nu)^2$ for the absorption continuum of $CH_3Br$ after Gordus and Bernstein (439).**

It is indeed found in several cases that if log $\varepsilon$ is plotted against $(\nu_m - \nu)^2$, a straight line is obtained in agreement with (IV, 7). Figure 175 shows this for $CH_3Br$. Surprisingly such a straight-line plot is also obtained for the continuous absorption of $CCl_4$ even though this molecule can hardly be considered as a quasi-diatomic molecule. It seems probable, however, that a quadratic dependence of log $\varepsilon$ on $(\nu_m - \nu)$ will result in any molecule for which the repulsive part of the potential surface of the upper state directly above the equilibrium position of the lower state has its greatest slope approximately in the direction of one of the normal coordinates. Therefore, many and perhaps most continua corresponding to direct dissociations follow the relation (IV, 7). But this conjecture remains to be confirmed both theoretically and experimentally.

**Isotope shifts.** Gordus and Bernstein (439) have compared the intensity distribution in different isotopes and find, for example for $CH_3Br$ and $CD_3Br$, that except for a shift to the violet of 280 ($\pm 50$) $cm^{-1}$, the two absorption curves agree. If $CH_3$—Br were entirely a quasi-diatomic case, one would have expected a narrowing by about 10% (i.e. an increase of the slope in Fig. 175) corresponding to the change of the reduced mass and of the vibrational frequency. On the other hand, the shift should have been only 17 $cm^{-1}$ since that is the change of zero-point energy of the C—Br vibration. The difference between prediction and observation clearly shows the inadequacy of the quasi-diatomic approximation to represent the intensity distribution in continuous spectra of polyatomic molecules. The total change of zero-point energy in going from $CH_3Br$ to $CD_3Br$ is 1976 $cm^{-1}$. From this value we have to subtract the change of zero-point energy in the excited state near the conformation first reached in absorption. Unfortunately, this isotopic zero-point energy change in the excited state cannot be calculated since the frequencies of the remaining genuine vibrations are not known. The zero-point energy connected with the reaction coordinate is, of course, zero. We may conversely use the difference of the known zero-point change of the ground state and the observed shift as an observed value of the zero-point energy change in the excited state (in the present case 1700 $cm^{-1}$).

**Both upper and lower state without stable equilibrium position.** When neither for the upper nor for the lower state a stable equilibrium position exists, we obtain continuous spectra close to the absorption (or emission) lines or bands of the separated groups. They can only arise in the process of collision and therefore, in general, produce simply a broadening of the lines or bands of the separated groups. For example, the potential function of the system Ne + $O_2$ in its ground state has no minimum other than a van der Waals minimum. The same applies to many, though by no means all, of the excited states of the system. If a transition takes place from the unstable ground state in a conformation in which Ne and $O_2$ are close to each other (i.e. for a collision pair or quasi molecule) to an unstable excited state of the system a continuous spectrum arises which will be close to the absorption lines or bands of Ne or $O_2$. The intensity of the wings so produced will increase with the square of the pressure. Pressure broadening of Ne atomic lines by $O_2$, and similarly of other atomic lines by diatomic and polyatomic gases, or of molecular bands by foreign gases, may be understood in this way. We shall refrain from more detailed considerations since they would in most respects be similar to corresponding considerations for pairs of atoms (see Vol. I, p. 394).

### (b) Emission continua

There are not many cases of emission continua which can be definitely assigned to a polyatomic molecule while, as was discussed in Volume I, there are many emission continua of diatomic molecules.

**Upper state continuous.** Just as for diatomic molecules, a continuous emission spectrum may arise as a result of a radiative two-body recombination of two parts of the molecule. If the two parts, say X + YZ, come together on the potential surface of an excited state, an electronic transition to the ground state may take place during the collision time, and since the kinetic energy of the colliding partners is not quantized, a continuous spectrum is emitted corresponding to molecule formation (*recombination continuum*). The short wave length limit of this continuum is a lower limit to the dissociation energy if the two partners are in their ground states.

The well-known green air afterglow may represent an example of such a process. The spectrum of this glow appears to be continuous and is now generally ascribed to the recombination process [see Kaufman (663)]

$$O + NO \rightarrow NO_2 + h\nu. \tag{IV, 8}$$

The efficiency of this process has been determined by Kaufman (663) and Fontijn, Meyer and Schiff (392) who find that only one recombination in $10^6$ to $10^7$ collisions takes place. Such a low yield is to be expected for a two-body recombination if the radiative lifetime is $10^{-6}$ sec while the duration of a collision is of the order $10^{-13}$ sec.

The collision time $10^{-13}$ sec corresponds to a simple to and fro motion. Actually, as we have seen, in general, a collision X + YZ proceeds via a more or less complicated Lissajous motion in the bowl whose minimum represents the equilibrium position of the XYZ molecule in the electronic state formed from X + YZ. The Lissajous motion may appreciably lengthen the collision time and therefore increase the yield of the recombination reaction. The lengthening of the collision time becomes rapidly larger as the number of atoms in the molecule increases. At the same time this process of recombination changes from being the inverse of a direct dissociation to the inverse of a predissociation by vibration (see the next section).

Several continua observed in flames have been ascribed to radiative two-body recombinations. For example, the continua observed by James and Sugden (621) in hydrogen flame gases when traces of alkali elements (M) are present, have been assigned to

$$M + OH \rightarrow MOH + h\nu. \tag{IV, 9}$$

Similarly, the continuum observed in $NO + F_2$ and $NO + F_2O$ flames has been assigned by Johnston and Bertin (643) and Goodfriend and Woods (433) to the recombination

$$F + NO \rightarrow FNO + h\nu. \tag{IV, 9a}$$

Emission continua observed in the afterglow of discharges through $SO_2$ and $SeO_2$ have been interpreted by Herman, Grenat and Akriche (508)(509) as produced by the recombination reactions:

$$SO + O \rightarrow SO_2 + h\nu \tag{IV, 10}$$

$$SeO + O \rightarrow SeO_2 + h\nu. \tag{IV, 10a}$$

Without more detailed investigations it is difficult to be certain whether or not the observed spectra are true continua and if so, whether the interpretation given is the correct one. For more complicated molecules, such assignments would be extremely difficult to make with any degree of assurance.

**Lower state continuous.** In a transition from a stable upper state to the continuous range of a lower state, a continuous spectrum is emitted which in favorable cases could be very strong as shown by the diatomic examples of the $H_2$ and $He_2$ continua. However, for polyatomic molecules no clear-cut examples have as yet been found. One reason for this lack of examples is that in electric discharges polyatomic molecules are usually decomposed and only rarely and under very special conditions is an emission spectrum characteristic of the original molecule observed.

## 3. Diffuse Spectra: Predissociation of Polyatomic Molecules

If in any atomic system certain discrete energy levels, $a, b, c, \ldots$ of a series $A$ have the same energy as a continuous range $M$ of energy levels joining on to a series $B$ (see Fig. 176), the former assume to a small extent properties of the latter, that is, the levels $a, b, c, \ldots$ become diffuse, provided that certain selection rules are fulfilled. Since the continuous range $M$ corresponds to the system flying apart with various amounts of kinetic energy, the "mixing" of the eigenfunctions also means that once the system is in one of the discrete states $a$ or $b$ or $c, \ldots$ it will after a time find itself in the continuous range $M$, that is, it will fly apart. We have a radiationless decomposition process usually referred to as *Auger process*. When the continuous range of levels corresponds to dissociation, the process is called *predissociation*; when it corresponds to ionization, it is called *pre-ionization*, or by many authors, *auto-ionization*.

Predissociation occurs much more frequently in polyatomic than in diatomic

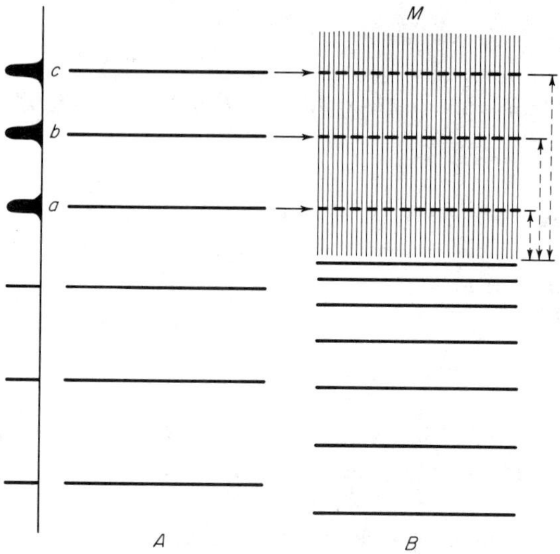


FIG. 176. **Energy level diagram for the Auger process.** The three uppermost levels ($a$, $b$, $c$) of the series $A$ are overlapped by the continuum $M$ of the series B. To the extreme left the width of the levels is indicated schematically. The radiationless transitions from the discrete to the continuous state are indicated by horizontal arrows. The broken vertical arrows give the amount of kinetic energy of the products.

molecules because there are many more continuous ranges of energy levels corresponding to the various dissociation processes and because, at least for molecules of low symmetry, the selection rules are much less restrictive.

**Criteria for predissociation.** The brief discussion just given of the nature of Auger processes makes it clear that predissociation can be recognized by the following three criteria:

(1) photochemical decomposition by absorption of light that brings the system to the levels which can undergo predissociation;
(2) broadening of the absorption lines or bands corresponding to these transitions;
(3) weakening of the emission lines that have the "diffuse" levels as upper levels.

The photochemical decomposition of the molecule occurs after a certain lifetime $\tau_l$ in the discrete state. The connection between diffuseness and photochemical decomposition was first established by Bonhoeffer and Farkas (129) when they showed that the photodecomposition of $NH_3$ produced by absorption in the diffuse bands near 2100 Å (see Fig. 179) occurs even at very low pressure, i.e. independent of collisions, with a constant yield.

According to the Heisenberg uncertainty relation, the product of the half-width $b$ of a level and its lifetime is $\sim h/2\pi$; and since the lifetime is the reciprocal of the radiationless transition probability, $\gamma$, we have

$$b = \frac{h}{2\pi}\frac{1}{\tau_l} = \frac{h}{2\pi}\gamma. \tag{IV, 11}$$

The radiationless decomposition will be readily observable only if the transition probability $\gamma$ is not much smaller than the probability $\beta$ of a radiative transition into lower states. Strictly speaking, in the Heisenberg relation, both radiative and radiationless transition probabilities must be included; the half-width of the level is then:

$$b = \frac{h}{2\pi}(\gamma + \beta) = \frac{h}{2\pi}\left(\frac{1}{\tau_l} + \frac{1}{\tau_r}\right) \tag{IV, 12}$$

where $\tau_r = 1/\beta$ is the mean life with respect to radiation. For allowed transitions, $\tau_r$ is of the order $10^{-8}$ sec and therefore, if $\gamma = 0$, the line width is about 0.0005 $\text{cm}^{-1}$. This is the so-called *natural line width*. The actual line width in a non-predissociated spectrum is much larger on account of the Doppler effect, e.g. at room temperature at 40,000 $\text{cm}^{-1}$ for a molecular weight of 40 this width is 0.08 $\text{cm}^{-1}$. Thus, only if $\gamma$ is more than one hundred times greater than $\beta$ (i.e. $\tau_l < 10^{-10}$ sec) will a broadening of the lines due to predissociation become noticeable. In other words, photochemical decomposition caused by predissociation may be observed long before a broadening of the lines becomes noticeable: broadening is a very insensitive criterion of predissociation. If diffuseness is observed, the yield of the photodecomposition (which is given by $\gamma/(\gamma + \beta)$) is always very close to 1.

The weakening of the emission lines (third criterion) is in general detectable only if a series of lines (a branch) or of bands (a progression) is observed. This series will break off suddenly or will show a sudden lowering of the intensity if at a

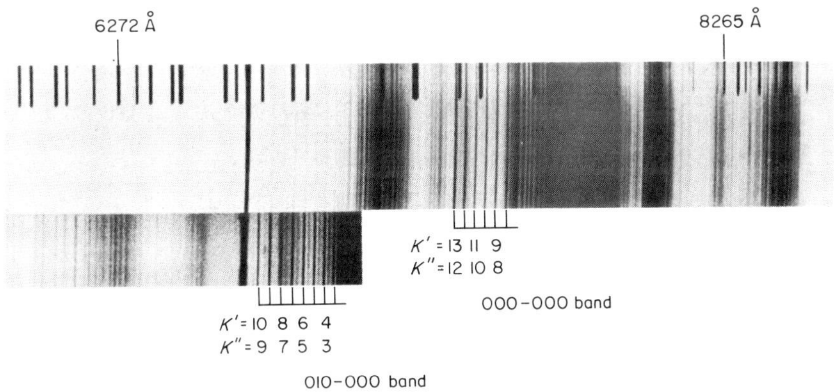


FIG. 177. **Breaking-off in the rotational ($K$) structure of the emission bands of HNO after Clement and Ramsay (204).** The $^rR$ heads of the 000 – 000 and 010 – 000 bands are marked. The last strong sub-band in the former band is 13 – 12, in the latter 10 – 9.

certain rotational or vibrational quantum number predissociation becomes possible with a sufficient probability. If, for example, at this point the radiationless transition probability $\gamma$ were about equal to the radiative transition probability $\beta$, there would be a sudden drop in the intensity of emission by 50 per cent which would be easy to observe. The breaking-off in emission is therefore also a much more sensitive criterion of predissociation than the diffuseness in absorption, but it is not as sensitive as the photochemical decomposition.

In diatomic molecules a breaking-off in emission has been observed in many instances, but for polyatomic molecules up to now only a single example is known, namely the breaking-off in the emission spectrum of HNO recently found by Clement and Ramsay (204). One of their spectra is shown in Fig. 177. The breaking-off at $K = 13$ in the $0-0$ band is clearly visible.

**Types of predissociation.** Three main cases of predissociation may be distinguished (see Vol. I, p. 413) which may be briefly referred to as (I) predissociation by electronic transition, (II) predissociation by vibration, and (III) predissociation by rotation.

In the first case (I), the dissociation continuum that causes the radiationless decomposition (see Fig. 176) belongs to another electronic state than that to which the discrete levels belong. Just as for diatomic molecules, we must distinguish two sub-cases, one in which the two electronic states have different species: *heterogeneous predissociation*, and one in which they have the same species: *homogeneous predissociation*.

In the second case (II) of predissociation, the dissociation continuum belongs to the same electronic state but joins on to a lower dissociation limit than that to which the series of "discrete" levels converges. Only the vibrational motion is changed in the predissociation process (predissociation by vibration). Unlike case I, here a decomposition can occur in a purely classical way as a result of the Lissajous motion of the image point on the multidimensional potential surface (see Fig. 163).

In the third case (III) of predissociation, the dissociation continuum belongs to the same electronic state and the same series of vibrational levels as the "discrete" higher rotational levels of a given stable vibrational level. In this predissociation process rotational energy is transformed into vibrational energy (predissociation by rotation). The higher rotational levels are mechanically unstable (see Vol. I, p. 425f), and a dissociation can occur in a purely classical way, as in case II.

**Selection rules for predissociation.** For an understanding of the selection rules for predissociation, it is important to realize that in a continuous range of energy levels the rotational quantum numbers $J$ and $K$ and the symmetry properties of the rotational levels are still well defined just as are the electronic species, total spin, etc. A non-zero value of $J$ or $K$ in the continuum means that the two parts of the molecule fly apart (or approach each other) with a certain angular momentum, that is, the asymptotic directions of motion of the two parts do not intersect. The energy differences corresponding to different $J$ or $K$ values are, of course, zero at infinite distance of the parts, that is, at every energy in the continuum all $J$ and $K$ values are represented.

According to Kronig (see Vol. I, p. 416), the following *rigorous selection rules* hold for predissociation of *linear molecules*:

$$\Delta J = 0, \qquad + \nleftrightarrow -, \qquad s \nleftrightarrow a. \tag{IV, 13}$$

Here + and − refer to the behavior of the total eigenfunction with respect to inversion, while $s$ and $a$ refer to the behavior with respect to exchange of identical nuclei (see Chap. I, p. 70f).

As a consequence of the first two selection rules (IV, 13), only one Λ-doubling component of a Π state of a linear molecule can be predissociated by a Σ state (see Vol. I, Fig. 183, p. 417); but if the molecule dissociates into a non-linear conformation, both components could predissociate since then there would be levels with either parity for each $J$.

For *non-linear molecules*, the selection rule for $J$ and for parity must obviously be the same while the rule $s \nleftrightarrow a$ is replaced by a corresponding rule for the over-all species according to the rotational subgroup (the same as for optical transitions, see pp. 223 and 246). However, these rules introduce even fewer restrictions than for linear molecules. No example in which the selection rules (IV, 13) exclude a certain predissociation has yet been found for a polyatomic molecule.

In addition to the rigorous rules (IV, 13) there are a number of *approximate selection rules*. For the spin quantum number $S$ we have, as for diatomic molecules,

$$\Delta S = 0 \tag{IV, 14}$$

which holds when spin-orbit coupling is small. For diatomic molecules a number of violations of this rule have been found, but all of them are based on observations of breaking-off in emission for which the sensitivity of detection is high. Rather strong spin-orbit coupling would be required to make a predissociation violating the rule (IV, 14) detectable by broadening of lines in absorption either in diatomic or polyatomic molecules.

For linear molecules, when Λ is defined, we have the additional (approximate) selection rules

$$\Delta\Lambda = 0, \quad \text{or} \quad \Delta\Lambda = \pm 1 \tag{IV, 15}$$

corresponding to homogeneous and heterogeneous predissociations respectively. For the predissociations with $\Delta\Lambda = 0$ and $\Lambda = 0$ we have

$$\Sigma^+ \nleftrightarrow \Sigma^-. \tag{IV, 16}$$

As long as vibronic interactions can be neglected (see below), heterogeneous predissociations (with $\Delta\Lambda = \pm 1$) occur only on account of the interaction of rotation and electronic motion and therefore for them the transition probability increases as $J(J + 1)$; it vanishes for $J = 0$. Since for polyatomic linear molecules, except for the hydrides $XH_2$, the rotational velocities are smaller than for diatomic molecules, the heterogeneous predissociations are in general much weaker than in diatomic molecules. Because of the rigorous rules (IV, 13) $\Sigma^+ - \Sigma^-$ predissociations cannot be made allowed by the interaction of rotation and electronic motion; they do, however, become possible for $\Sigma$ states with $S \neq 0$ when spin-orbit coupling is large.

If the quantum numbers $\Sigma$ or $N$ are defined in a linear molecule (Hund's case (*a*) and (*b*) respectively, see p. 19 and p. 73), we have the selection rules (just as in the diatomic case)

$$\Delta\Sigma = 0 \quad \text{or} \quad \Delta N = 0, \tag{IV, 17}$$

provided that $\Delta S = 0$. For intercombinations ($\Delta S \neq 0$) and in Hund's case (*c*) the rules (IV, 15), (IV, 16) and (IV, 17) would have to be replaced by

$$\Delta\Omega = 0, \quad \text{or} \quad \Delta\Omega = \pm 1 \tag{IV, 18}$$

and

$$0^+ \not\leftrightarrow 0^-. \tag{IV, 19}$$

The selection rule $\Delta N = 0$ is also valid for the predissociation of non-linear molecules as long as spin-orbit coupling is small.

In *symmetric top molecules* we have the analogue of the rule (IV, 15), viz:

$$\Delta K = 0, \pm 1. \tag{IV, 20}$$

Again the radiationless transitions with $\Delta K = 0$ are independent of $J$; those with $\Delta K = \pm 1$ have a probability proportional to $J(J + 1)$.

The selection rules discussed so far are readily derived from the expression for the radiationless transition probability first given by Wentzel (see Vol. I, p. 407), viz:

$$\gamma = \frac{4\pi^2}{h} |W_{ni}|^2, \qquad W_{ni} = \int \psi_n^* W \psi_i d\tau. \tag{IV, 21}$$

Here $\psi_n$ and $\psi_i$ are the wave functions of the discrete and the continuous state respectively, and $W$ is the perturbation function representing certain neglected terms in the Hamiltonian. Since $W$ is totally symmetric, it follows immediately that the states $n$ and $i$ must have the same species in order that a non-vanishing transition probability be obtained. In other words, we have the selection rule

$$\Gamma_n = \Gamma_i \tag{IV, 22}$$

if $\Gamma_n$ and $\Gamma_i$ represent the species of the discrete and the continuous state.

The selection rule (IV, 22) holds rigorously for the over-all species only, but if we can separate electronic, vibrational and rotational motion so that

$$\psi = \psi_e \psi_v \psi_r \quad \text{and} \quad W = W^e + W^v + W^r \tag{IV, 23}$$

we see immediately that (IV, 21), and therefore (IV, 22), can be applied to the electronic wave functions separately, that is, we have the selection rule

$$\Gamma_n^e = \Gamma_i^e. \tag{IV, 24}$$

In this approximation therefore only predissociations between electronic states of the same species, that is, homogeneous predissociations, are possible, e.g. for linear molecules only predissociations with $\Delta\Lambda = 0$. Similarly, in this approximation the vibrational species of the two states must be the same, i.e.

$$\Gamma_n^v = \Gamma_i^v. \tag{IV, 25}$$

If, as often happens, the equilibrium conformation in the two states is different, the selection rules (IV, 22, 24 and 25) refer to the species of the point group formed from the joint elements of symmetry[3]. Since, in addition, in the state $i$ in general all vibrational species are available for a given energy (because we are in a continuous range), it is clear that the vibrational selection rule (IV, 25) does not lead to any significant restriction of the possibilities of predissociation.

If now the *interaction of rotation and electronic motion* is no longer neglected heterogeneous predissociations become possible, that is, predissociations for which the electronic species $\Gamma^e$ does not remain the same. A more detailed consideration shows that a heterogeneous predissociation is possible on account of interaction of rotation and electronic motion if the species of the two states differ by the species $\Gamma^r$ of a rotation, that is,

$$\Gamma_n^e \times \Gamma^r = \Gamma_i^e. \tag{IV, 26}$$

Such a predissociation increases with the particular rotation and is rigorously forbidden for zero rotation. For example, an $A_2''$ state of a $\boldsymbol{D}_{3h}$ molecule could be predissociated by an $A_1''$ state if the molecule is rotating about the $z$-axis ($\Gamma_z^r = A_2'$) and by an $E'$ state if it is rotating about an axis perpendicular to $z$ ($\Gamma_{x,y}^r = E''$), but it could not be predissociated by $A_1'$, $A_2'$ or $E''$ states. Similarly, an $A_1$ state of a $\boldsymbol{C}_{2v}$ molecule can be predissociated by an $A_2$ or $B_1$ or $B_2$ state, depending on whether the molecule rotates about the $z$-, $y$- or $x$-axis. The rule (IV, 26) includes, of course, the linear case, i.e., the possibility of predissociations with $\Delta\Lambda = \pm 1$.

If finally we introduce the *interaction of vibration and electronic motion* (vibronic interaction) the resolution (IV, 23) of $\psi$ into a product of $\psi_e$ and $\psi_v$ and of $W$ into $W^e + W^v$ is no longer possible. We obtain in that case from the general rule (IV, 22) the selection rule

$$\Gamma_n^{ev} = \Gamma_i^{ev} \tag{IV, 27}$$

where $\Gamma^{ev}$ is the vibronic species. Thus *only states of the same vibronic species can predissociate into each other*. Here as before the point group is that of the joint elements of symmetry.

We see immediately from the selection rule (IV, 27) that radiationless transitions that are forbidden by the electronic selection rule (IV, 24) can be made

[3] If there is no potential minimum for one of the states, we must use the symmetry of the potential function.

allowed by vibronic interactions. For example, in a linear molecule a radiationless transition between a $\Pi$ and a $\Delta$ electronic state is forbidden for zero rotation. But if the bending vibration is singly excited in the $\Pi$ state, the vibronic states $\Sigma^+$, $\Sigma^-$, $\Delta$ result, of which the last can be predissociated by the $\Delta$ electronic state; or if the bending vibration is excited in the $\Delta$ state resulting in $\Pi$ and $\Phi$ vibronic states, the former can be predissociated by the $\Pi$ electronic state. Thus, $\Pi-\Delta$ predissociations can be caused not only by rotational-electronic interactions, but also by vibronic interactions. Unlike rotational-electronic interactions, vibronic interactions can in principle (see below) also cause $\Sigma-\Delta$ (or even $\Sigma-\Phi$, etc.) predissociations: if in the $\Sigma$ state the bending vibration is doubly excited, we have a $\Sigma^+$ and a $\Delta$ vibronic level, of which the latter can interact with the $\Delta$ electronic state; alternatively, if in both states the bending vibration is singly excited, a $\Pi$ vibronic state arises in each, and these $\Pi$ states can predissociate into each other.

Similarly in a non-linear molecule, say of point group $\boldsymbol{D}_{3h}$, we can expect predissociation of a ${}^1A_2''$ electronic state by a ${}^1E''$ electronic state if either in the one or the other state a degenerate vibration of type $E'$ is singly (or multiply) excited, since again we obtain then two states of the same vibronic symmetry ($E''$ if $E'$ is excited in ${}^1A_2''$, but $A_2''$ if it is excited in ${}^1E''$). In this case the predissociation cannot be produced by rotational-electronic interaction since there is no rotation of species $E'$.

In the preceding examples we have assumed that in the "discrete" and the continuous state the molecule has the same symmetry of the equilibrium position or that, if there is no equilibrium position of the continuous state, the symmetry of the potential function is the same in the two states. Often the symmetry will be different. For example, if we consider the predissociation of a planar symmetrical $XY_3$ molecule (point group $\boldsymbol{D}_{3h}$) into $XY_2 + Y$, the symmetry in the state causing the predissociation is likely to be $\boldsymbol{C}_{2v}$. If the dissociated state has the (electronic) species ${}^eA_1$, it could in a first approximation not cause the predissociation of a ${}^eA_2''$ state of $\boldsymbol{D}_{3h}$ since $A_2''$ goes into $B_1$ of $\boldsymbol{C}_{2v}$ (see Appendix IV) and the selection rule (IV, 24) is not fulfilled. However, if in the ${}^eA_2''$ state the out-of-plane vibration (species $A_2''$) is excited, alternate vibrational levels have vibronic species $A_1'$ which in $\boldsymbol{C}_{2v}$ goes into $A_1$ and can thus, according to (IV, 27), predissociate into ${}^eA_1$.

A more detailed consideration shows that several stages of vibronically produced predissociations may be distinguished. The strongest vibronically induced predissociations are those in which the two electronic species differ only by the species of one (non-totally symmetric) normal vibration. This was the case in most of the examples given above. If the excitation of two quanta of a vibration in one, or one quantum in each electronic state is required, as in the example of a $\Sigma-\Delta$ predissociation, then the radiationless transition probability is much smaller than in the first case, and if three quanta are required, the transition probability is again much smaller. In what follows we shall disregard the possibility of predissociations that require excitation of more than one quantum of vibration.

The vibration that induces an electronically forbidden predissociation can be excited either in the state $n$ or the state $i$. If for any reason it is excited in $n$ only, we shall find that none of the strong absorption bands (at low temperature) which involve totally symmetric vibrational levels (see Chap. II, section 2b) will show diffuseness, and only the weak bands corresponding to a forbidden component of the dipole moment will do so. On the other hand, the strong bands will be diffuse when the non-totally symmetric vibration is excited in the continuous ($i$) state. It is difficult to predict whether vibronic interaction is more effective in state $n$ or state $i$ [see Sponer and Teller (1155)].

If both vibronic and electronic-rotational interaction are taken into account we must again make use of the rule (IV, 26) except that in it we have to replace the electronic species $\Gamma^e$ by the vibronic species $\Gamma^{ev}$. As a consequence, vibronically heterogeneous predissociations may occur if the two vibronic species differ by the species of a rotation. Such predissociations will again increase in probability as the rotation about the axis corresponding to $\Gamma^r$ increases. However, practically, for non-linear molecules this rotation-induced predissociation is not very important since in most cases suitable vibronic interactions can cause the same predissociation more readily.

**Case I of predissociation.** For the occurrence of a strong predissociation of one electronic state by interaction with another one, it is not sufficient that all the preceding selection rules are obeyed: predissociation will be strong only if, in addition, the Franck–Condon principle is fulfilled for the radiationless transition, that is, if the two potential surfaces intersect or at least come very close to each other. The arguments for this conclusion are essentially the same as for diatomic molecules (see Vol. I, p. 420f) and will not be repeated here.

For a homogeneous predissociation an intersection of potential surfaces does in general not arise except for Teller's conical intersections (see p. 443). Rather, we may have the situation illustrated in one dimension in Fig. 178a. If the system is at the level $E$, it can dissociate without going from one potential surface to another; it simply remains on the lowest one shown. From this point of view this would be a predissociation of type II or III without an electronic transition. Just as for diatomic molecules one must, however, consider the degree of deviation from the "original" pair of intersecting potential surfaces (broken-line correlation). If, on approaching the point of intersection with non-zero velocity, there is a non-zero probability that the molecule will go to $F$ rather than to $H$ and also a finite probability of going back to $E$ rather than $G$, then we must clearly consider this as predissociation by electronic transition. However, if the interaction between the two states is very large so that in effect two new states result, as in Fig. 178b, then a transition from one to the other resulting state will no longer take place, and any diffuseness found must be ascribed to case II (or III). The dividing line between case I and II (or III) is, of course, not sharp: when the two resulting curves are separated by an energy of the order of a vibrational quantum, they can equally well be assigned to the one as to the other type.

The predissociation of $NH_3$ at 2170 Å, one of the first to be recognized as such, has recently been re-investigated by Douglas (294). He has found in $ND_3$ that

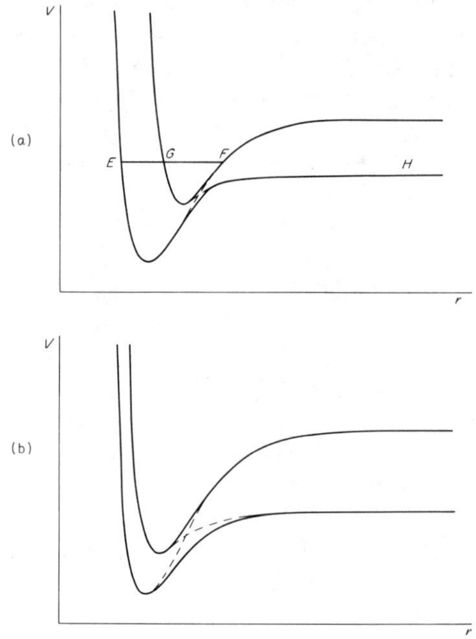


Fig. 178. **Weakly and strongly interacting potential surfaces.** In (a) there is a near-intersection, in (b) interaction is so strong that it is no longer possible to speak about an intersection.

the line width of the individual rotational lines decreases in going from the $0-0$ to the $1-0$ band but then increases markedly to a fairly constant value in the $2-0, 3-0, 4-0, \ldots$ bands. For the latter bands the line width is so large that individual rotational lines cannot be recognized, but the band width remains fairly constant and the bands are well separated as shown by Fig. 179. In view of the great width of the lines (2.5 $\text{cm}^{-1}$ in the $0-0$ band) and the low atomic number of the atoms involved, one must assume that the rule $\Delta S = 0$ is obeyed, that is, that the state causing the predissociation has the same multiplicity as the upper state of the bands, viz., singlet. Since the predissociation limit is at less than 5.7 eV above the ground state, the dissociation products cannot be N + 3H (or NH + 2H) but must be either $NH_2$ + H or $NH(^1\Delta)$ + $H_2(^1\Sigma)$.

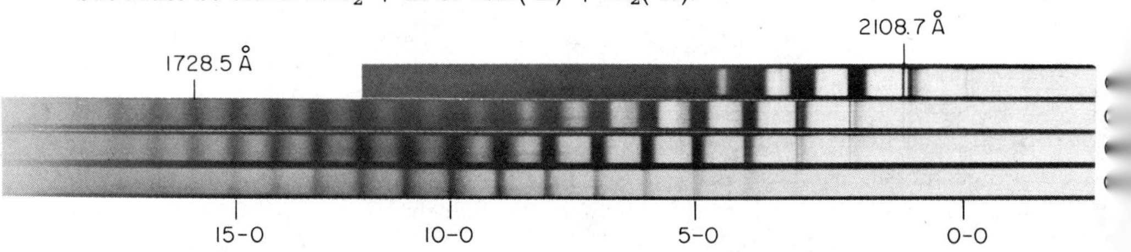


Fig. 179. **Predissociation in the $\tilde{A}-\tilde{X}$ bands of $ND_3$.** Spectrograms at four different pressures (given at the right) are shown. The path length was 40 cm.

Experimentally only a photodecomposition into $NH_2$ + H has been found and this is the only alternative that we shall consider here. Figure 180 shows a potential diagram as a function of the H—$NH_2$ distance ($r$). At large $r$ when the point group is $\boldsymbol{C}_{2v}$ there are two singlet states, $^1B_1$ and $^1A_1$, corresponding to the

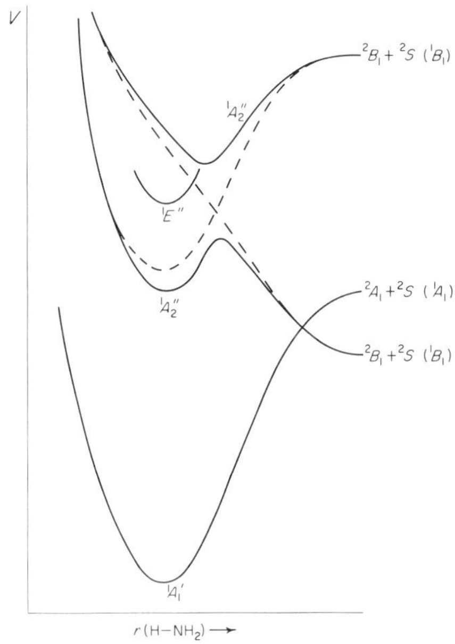


Fig. 180. **Potential functions of the lowest electronic states of $NH_3$ as a function of the H—$NH_2$ distance [after Douglas (294)].** The two broken-line curves are zero-approximation curves which give rise to the two $^1A_2''$ states shown.

two lowest states $^2B_1$ and $^2A_1$ of $NH_2$. Of these two states $^1A_1$ must correlate with the $^1A_1'$ ground state in the planar symmetrical ($\boldsymbol{D}_{3h}$) conformation of $NH_3$. The "discrete" upper state of the 2170 Å bands is $^1A_2''$ (see Chap. V, section 2a) which in point group $\boldsymbol{C}_{2v}$ becomes $^1B_1$. In view of the selection rule (IV, 24), an allowed predissociation is therefore possible only into the $^1B_1$ state at large $r$, that is into $NH_2(^2B_1)$ + H. Thus the $NH_2$ formed is in its ground state. The allowed nature of this predissociation is in agreement with the fact that no dependence on $J$ has been found. It is however not clear how to account for the presence of two $^1A_2''$ states in the molecule if they are as close together as seems to be required (see Fig. 180). The electron configuration gives only one $^1A_2''$ state in this energy range, derived from $3s$ of the united atom.

The strong diffuseness observed in the absorption bands of $CH_3$ near 2160 Å is apparently produced by a predissociation similar to that of $NH_3$ except that here from the ground state ($^2A_2''$) only the two lowest vibrational levels of the excited state ($^2A_1'$) are reached with any intensity. The state causing the predissociation is in all probability a $^2A_1$ state of the system $CH_2$ + H. Such a state can only arise from $CH_2$ in the lowest singlet state $^1A_1$ but not from the lowest triplet state $^3\Sigma_g^-$.

In diatomic molecules the probability of predissociation is often found to decrease in a progression of bands after a fairly pronounced maximum because the system goes through the point of intersection more and more rapidly and therefore has less and less time to switch over to the other electronic state. This is expected from the Franck–Condon principle (Vol. I, p. 423) and indeed, more detailed (independent) calculations by Zener (1331), Landau (719) and Stueckelberg (1172) have shown that such a dependence of the result of passing through the crossing point on the speed of the particles must necessarily occur[4].

For polyatomic molecules such a decrease of diffuseness in a series of absorption bands at shorter wave lengths has been observed only rarely. It does occur in a progression of bands in the far ultraviolet absorption spectrum of $H_2O$ first observed by Henning (494) and in the progression of visible absorption bands of $HNO_2$ described by Melvin and Wulf (820). In these two cases apparently the vibrational motion in the excited state is quasi-diatomic as indicated by the observation of a single progression, and therefore the predissociation behaves in the same way as for diatomic molecules. In most other cases there is a Lissajous motion in the excited state. In that event we may expect the radiationless transition probability to depend inversely on the component of the velocity at right angles to the ridge (which in the polyatomic case replaces the point of intersection in the diatomic or quasi-diatomic case). This component may be small even if the vibrational energy is large and therefore there is no reason for a decrease of the transition probability with increasing energy, that is, even at fairly high energy above the limit, predissociation can occur with undiminished intensity. The predissociation in the first excited state ${}^1A_2''$ of $NH_3$ (see Fig. 179) is an example: as mentioned before, the diffuseness of the bands does not change much from $v_2 = 2$ on.

The initial decrease of diffuseness from $v_2 = 0$ to $v_2 = 1$ observed in $ND_3$, as pointed out by Longuet–Higgins (private discussion), may be connected with the fact that the vibrational symmetry of the $v_2 = 1$ state is $A_2''$, that is, the wave function of this state has a nodal plane in the plane of the molecule.

In many diffuse spectra of polyatomic molecules the diffuseness sets in very gradually and continues to increase at shorter and shorter wave lengths until a continuum is produced. Striking examples of this type are found in the longest wave length absorption systems of HCN [from 1800 to 1570 Å, Herzberg and Innes (527)] and $ClO_2$ [from 3750 to 2700 Å; see Finkelnburg and Schumacher (384)]. The reasons for this behavior were first clearly recognized by Franck, Sponer and Teller (397):

(1) If a simple progression of bands occurs in absorption, that is, if the corresponding motion of the image point on the potential surface is one-dimensional, strong predissociation will start only (according to the Franck–Condon principle) when the one-dimensional motion reaches the line of intersection of the two potential surfaces. In general this line will be at the top of a ridge formed from

[4] The specific formula derived by them (known as the Landau–Zener formula) has, according to recent work by Bates (99) and Coulson and Zalewski (244), only very limited validity.

the two surfaces. No predissociation can occur for energies less than the minimum of the ridge. If the line representing the one-dimensional vibration intersects the ridge at a point other than the minimum only weak predissociation can occur for energies between that of the minimum and the point of intersection. This weak predissociation takes place because anharmonicity or zero-point motion in the other normal vibrations may lead the image point off the one-dimensional motion, and this effect increases with increasing energy above the minimum of the ridge, until the full strength of predissociation is reached at the point of intersection. An example would be represented by the two potential surfaces, Fig. 163 and Fig. 164, if the symmetric stretching vibration is excited in the first surface and the difference in multiplicity is disregarded. Predissociation could start very weakly at an energy corresponding to the point $B$ (except for the effect of tunneling). An actual example is probably $ClO_2$ for which, even though it is not linear, the potential surfaces are similar to Figs. 163 and 164. Since the potential surface has always more than two dimensions, a similar situation may also arise for a two- or higher-dimensional motion of the image point, that is, when two or more, but not all, vibrations are excited.

(2) If the motion of the image point has the same number of dimensions as the potential surface (as drawn), the Lissajous motion will fill every point of the potential surface which has an energy less than the energy of the system (and which is not separated by a barrier from the minimum). Therefore, as soon as the energy is higher than the lowest point of the ridge of intersection (assuming it to be above a dissociation limit), predissociation can take place. However, if the energy of the molecule is just sufficient to reach the lowest point of the ridge, predissociation is possible only in one particular conformation, and in general (classically) it takes considerable time before this conformation is reached during the Lissajous motion. As the energy is increased, a greater part of the ridge is accessible to the image point, and consequently, it takes less time for predissociation to occur. To the gradual decrease of the (classical) lifetime corresponds an increase in line width [see eq. (IV, 11)] and thus again the diffuseness of the absorption bands will increase gradually. Depending on the shape of the potential surface the increase of diffuseness may be very slow. An example is the first predissociation of HCN mentioned earlier, where, as in $ClO_2$, the beginning of the diffuseness is very gradual, but where (unlike $ClO_2$) at least two vibrations are prominent in the spectrum, and therefore the motion of the image point is more complicated.

An additional cause for lack of sharpness of the predissociation limit is provided by the "*tunnel effect*" in cases in which the lowest point of the ridge of intersection is above the dissociation limit. However, in general, just as for diatomic molecules, this effect is quite small except when hydrogen atoms are dissociating off the molecule. In the latter case the presence of the tunnel effect is often strikingly demonstrated by the widely different diffuseness in some of the absorption bands of isotopic molecules when H is replaced by D.

As an example, we refer to the previous Fig. 96 (p. 227) which shows the 2160 Å absorption band of $CH_3$ and for comparison, the corresponding band of $CD_3$ at

2140 Å. In the former the line width is so large that individual lines are not recognizable, while in the latter they are recognizable even though they are still broad. At least part of the difference in this and similar cases is due to the fact that D atoms go through a given barrier much less easily than H atoms, but another part is due to the fact that the width of the barrier through which the D atom has to "tunnel" is greater than for the H atom since the vibrational level of the upper state lies lower on account of the lower zero-point energy.

Similar differences in diffuseness between hydrides and deuterides have been found for $CH_2$ (see Fig. 185, p. 491), $NH_3$ [Douglas (294)], $H_2O$ (see Fig. 112, p. 265) and $H_2S$ [Johns (634)]. For $CH_2$ a change of diffuseness has even been observed in going from $^{12}CH_2$ to $^{13}CH_2$. For $H_2S$ one electronic transition shows the opposite effect, i.e. a widening of the lines upon deuteration. In this case clearly it cannot be a tunnel effect that is responsible for the diffuseness. Apparently here the ridge of intersection lies below the excited states of both $H_2S$ and $D_2S$ and for $H_2S$ which has the higher zero-point energy the discrete levels are above the region of maximum predissociation probability.

**Dependence on *J* and *K*.** As mentioned earlier, heterogeneous predissociations can occur on account of interaction of rotation and electronic motion. Such predissociations can be recognized by the dependence of the diffuseness on $J$ and $K$: no predissociation can occur for zero rotation. A clear dependence on $K$ has been observed for HCN in the $\tilde{A}$ state in which the molecule is non-linear. Bands of the $\tilde{A}-\tilde{X}$ system with higher $K'$ values, when they are sufficiently resolved, show a greater diffuseness than bands with lower $K'$ of about the same energy [Herzberg and Innes (527)]. To be sure, in this case at somewhat higher energy predissociation occurs also for $K' = 0$, that is, a homogeneous predissociation can also occur, but the heterogeneous one occurs at lower energy.

A somewhat similar case occurs in HCO [Herzberg and Ramsay (538)]. In the excited state of the red absorption bands, in which the molecule is linear, only the levels with $l' = 0$ are found to be relatively sharp, while those with $l' = 1, 2, \ldots$ are quite diffuse. The excited state ($^2A''$ or $^2\Sigma^-$) as well as the ground state ($^2A'$) is derived from $^2\Pi$ of the linear conformation (see p. 28) which in turn is derived from $CO(^3\Pi) + H(^2S)$ as shown schematically in Fig. 181 [see Johns, Priddle and Ramsay (638)]. The $^1\Sigma^+$ ground state of CO leads to a $^2\Sigma^+$ state of HCO in the linear conformation or $^2A'$ in the bent conformation. The intersection of this $^2A'$ state with the $^2A'$ ground state is avoided (see the full-line curves in Fig. 181) and therefore the ground state will dissociate over a potential barrier into $CO(^1\Sigma) + H(^2S)$. The predissociation of the excited ($^2A''$) state must take place into this same state since the energy of the first excited state of the products is much too high. Thus the predissociation is heterogeneous ($^2\Sigma^- \rightarrow {}^2\Sigma^+$ for the linear and $^2A'' \rightarrow {}^2A'$ for the bent conformation). It can arise either by interaction with the rotation about the $a$-axis or by interaction with the over-all rotation. The former interaction causes the very strong diffuseness for all levels with $l' > 0$, while the latter is expected to lead to an increasing diffuseness in the bands having $l' = 0$ with increasing $J$; such a variation has indeed been observed. The $l' > 0$ levels correspond to the $K' > 0$ levels of a non-linear molecule.

Finally, in the 1240 Å band of $H_2O$ shown in the previous Fig. 112, a diffuseness of the lines is seen which increases strongly for the higher $J$ (and $K$) values.

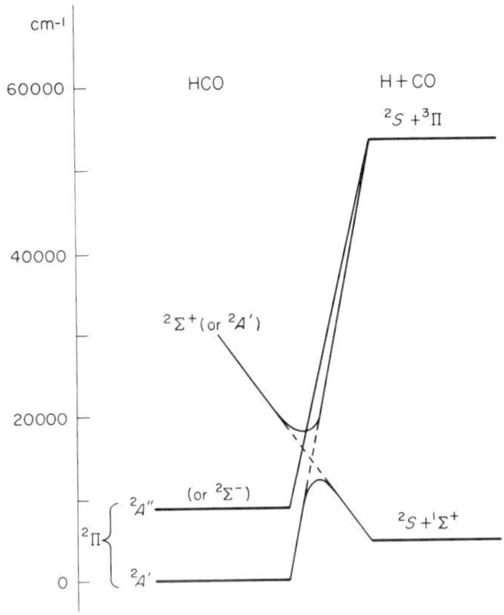


FIG. 181. **Correlation of electronic states of HCO to those of H + CO [after Johns, Priddle and Ramsay (638)].** The two observed states of HCO correspond to $^2\Pi$ of the linear conformation. In the upper one of the two states the molecule is linear; therefore this state is equivalent to a $^2\Sigma^-$-state.

Again we conclude that this is a heterogeneous predissociation. Since the discrete upper state of the bands has species $B_1$, and therefore $A''$ with respect to the point group $C_s$ of H + OH, the state causing the predissociation must be $A'$ if this is to be a heterogeneous predissociation. Such a state arises both from OH($^2\Pi$) and from OH($^2\Sigma$). The $A'$ state arising from OH($^2\Pi$) is in all probability the ground state of $H_2O$ which is not likely to intersect the $B_1$ state in such a way as to cause predissociation. But the $A'$ state from OH($^2\Sigma$) has its lowest dissociation limit at 8.4 eV above the ground state of $H_2O$ and is thus quite likely to cause predissociation in the $B_1$ state at 10.0 eV.

An extreme case of a dependence of the predissociation probability on $J$ or $K$ arises, of course, when the predissociation limit falls within a series of rotational levels, and as a consequence, a sudden increase of line width in absorption or a sudden breaking-off in emission is observed. Such a rapid change at a certain $K$ or $J$ value would be independent of whether the predissociation is homogeneous or heterogeneous. The only clear-cut case thus far known in polyatomic molecules is that of HNO [Clement and Ramsay (204)] where both breaking off in emission and diffuseness in absorption have been observed (Fig. 177).

**Case II of predissociation.** As mentioned before, the dividing line between case I and II of predissociation is far from sharp. In principle, the difference is that in case II dissociation can occur in a purely classical way whenever the image point representing the vibrational motion reaches an appropriate saddle point of the potential surface, while in case I in addition an electronic transition must occur whose probability is less than one. In other words, in case I predissociation, not every time the ridge of intersection (saddle) is reached does a dissociation occur but only for a fraction $\alpha$ of such favorable conformations[5]. If $\alpha$ is very small for a given case I predissociation, for example for a triplet-singlet intercombination, then there is no problem in distinguishing it from a case II predissociation. However, if $\alpha$ is larger than say 1/10, a distinction is difficult, particularly for excited electronic states. A few necessary conditions for identifying a given predissociation as case II are readily established, but none of them are sufficient:

(1) Except for very peculiar shapes of the potential surfaces (which might be brought about by interaction with other electronic states), one would expect that in case II the diffuseness increases fairly rapidly with increasing energy above the limit.

(2) Case II predissociation is more likely to happen when several vibrations are simultaneously excited, that is, when the spectrum does not simply consist of a single progression.

(3) Since case II is by definition a homogeneous predissociation, there can be no strong dependence on $J$ and $K$.

(4) The correlation rules must be fulfilled, that is, the state undergoing predissociation must be obtainable from the assumed dissociation products.

(5) The dissociation energy of the (excited) state necessary to account for the predissociation limit must be reasonable.

Even if all these conditions are fulfilled in a given case, they do not entirely exclude the possibility that the predissociation under consideration belongs to case I. For example, the predissociation found in the longest wavelength absorption system of $NO_2$ [see Henri (20) (498) and Fig. 192 below] does fulfill the first four of the above conditions. But it is difficult to say whether the small dissociation energy of the excited state, which would result from the assumption of case II, is reasonable [condition (5)]. Similarly, the predissociation of $H_2CO$ near 2750 Å fulfils the criteria (1) to (3), and criterion (4) can be made to fit if dissociation into H + HCO, with HCO in the excited $^2A''$ ($^2\Sigma^-$) state, is assumed, but the dissociation energy of the excited state would have to be very low. Unfortunately the present knowledge about the C—H stretching vibrations in the excited state is insufficient to say whether such a low $D$ value is reasonable. However, the low value for the dissociation energy $D$(H—HCO) in the ground state which would result makes this interpretation very doubtful. In all probability this is case I predissociation.

[5] Strictly speaking in quantum theory, even if no second electronic state is present, the crossing of a barrier does not proceed with the probability 1, but with a slightly smaller probability; however, in all cases in which the energy of the system is greater than that of the barrier, this probability is greater than 0.5 [see e.g., Bell (105)].

**Unimolecular decompositions.** One can be much more certain of having a case II predissociation if it is the ground state of the molecule that is predissociating. Spectroscopically such a predissociation can be investigated only in emission by observing transitions from a stable excited state to the higher vibrational levels of the ground state. No examples of this type have yet been found.

Chemically, case II predissociations of electronic ground states are well known as unimolecular decompositions. Such decompositions are observed when a given molecular gas is heated to a sufficiently high temperature. Thermal collisions bring the molecules into vibrational levels above the lowest dissociation limit, and the spontaneous decomposition process (predissociation by vibration) sets in. The lifetime before decomposition is greater than the time for a simple oscillation ($10^{-13}$ sec) and, therefore, also greater than the collision time. If the lifetime is greater than the time between successive collisions, deactivation may occur before the decomposition has taken place. When the pressure is sufficiently high, the number of molecules decomposing per second depends only on the number of activated molecules (which in turn depends only on the temperature), and we have an exponential decay as in radioactive decay, the decay constant (rate constant) being proportional to the probability $\gamma$ of the radiationless transition. Several cases of this type are known [see Slater (36), Rice (1070)] and must definitely be ascribed to case II predissociations even though in the literature on the subject the situation is usually not expressed in this way. At low pressures the reaction rate is determined by the rate of activation since every molecule once activated will decompose. Therefore in this low pressure region the decomposition reaction is of second order.

In addition to unimolecular reactions caused by case II predissociations, there are also unimolecular reactions for which case I predissociation is responsible. This happens when the lowest dissociation limit cannot be correlated with the ground electronic state. For example for $N_2O$ the $^1\Sigma^+$ ground state cannot be correlated with the lowest state of $N_2 + O$ which is a triplet state ($^1\Sigma + {}^3P$). Thus, when $N_2O$ is thermally activated to an energy above the lowest dissociation limit it has a much longer lifetime than expected on the basis of case II predissociation because a forbidden electronic transition (singlet $\rightarrow$ triplet) is required, i.e. not every time the image point reaches a favorable position, but only in a very small fraction of these favorable instants does a switch-over to the triplet state and therefore a dissociation occur [Herzberg (518), Lindars and Hinshelwood (754)]. On account of the longer lifetime, an exponential decay is observable to much lower pressures than would otherwise be the case. A similar situation would arise in $CO_2$, but its thermal decomposition requires much higher temperatures and has not yet been studied. (See, however, the discussion of the inverse reaction on p. 477f.)

Another example of case I and case II predissociations, closely related to unimolecular decompositions, is provided by the "metastable" ions observed in mass spectrometry [see, e.g., Field and Franklin (14)]. These ions form relatively diffuse peaks of small intensity at non-integral mass-numbers in the mass spectrum and have been shown to result from the spontaneous dissociation of ions after they have been accelerated but before they enter very far into the analyzing field. When the occurrence of these "metastable" ions can be shown to be independent of collisions (as is often the case), that is, if the dissociation is spontaneous, it must clearly be a radiationless phenomenon, that is, it must be predissociation.

The lifetime $\tau_l$ of the parent ions before decomposition must be of the order of the time for them to move from the accelerating field to the analyzing field, that is, of the order of $10^{-6}$ sec. Thus the predissociation is a very slow one. At the same time the radiative lifetime $\tau_r$ for transitions to lower stable states must also be of this order or larger, that is, they must be metastable in the usual optical sense. But metastability while necessary is not the principal characteristic of the "metastable" ions; the principal characteristic is their spontaneous decomposition and it would have been better to call them predissociating ions: by no means all optically metastable ions predissociate.

To account for metastability ($\tau_r \geqslant 10^{-6}$ sec) we must assume that the "metastable" ions are either in high vibrational levels of the ground state ($\tau_r > 10^{-3}$ sec) or in an excited electronic state which cannot easily combine with any lower stable state. To account for the long radiationless lifetime $\tau_l$ we have to assume that either we have case II predissociation (if the number of atoms is large enough so that the Lissajous motion is sufficiently complicated) or a case I predissociation that is strongly forbidden.

For large ions like the butane ion $C_4H_{10}^+$, for which this phenomenon was first observed [Hipple and Condon (554)], the lifetime $\tau_l$ in a case II predissociation is very probably long enough to account for the observation of "metastable" ions. By electron impact the molecule is brought to a highly vibrating level of the ground state of the parent ion which lies above the first dissociation limit and thus predissociation can occur. A similar case has recently been found by Dibeler and Rosenstock (275) for $CD_4^+$. Here a spontaneous dissociation into $CD_3^+ + D$ is observed which must be interpreted as a case II predissociation. For $CH_4^+$ a corresponding dissociation process has not been observed. Remembering the great difference in strength of predissociation of $CH_3$ and $CD_3$ (see Fig. 96) which is caused by the much slower rate of tunneling of D compared to H, it seems probable that a similar difference exists between $CH_4^+$ and $CD_4^+$. Only for the latter is the case II predissociation slow enough to bring the process into the range observable with the mass spectrometer.

Recently Begun and Landau (104) and Dibeler and Rosenstock (276) have observed "metastable" ions in $N_2O$ and $H_2S$ respectively, even though here the lifetime $\tau_l$ for a case II predissociation can hardly be long enough to make such predissociating ions observable in the mass spectrometer. However, in both these cases the ground state of the products of the decomposition process is a quartet state while that of the parent ion is a doublet state. We have

$$N_2O^+(^2\Pi) \rightarrow NO^+(^1\Sigma^+) + N(^4S) \qquad \text{(IV, 28a)}$$

$$H_2S^+(^2B_1) \rightarrow H_2(^1\Sigma_g^+) + S^+(^4S) \qquad \text{(IV, 28b)}$$

Thus in both cases the decomposition must be due to case I predissociation involving a violation of the spin rule $\Delta S = 0$. Because of this violation the lifetime $\tau_l$ is larger, by a factor of at least $10^3$, than if the rule $\Delta S = 0$ were obeyed and if we had in effect a case II predissociation. In order for the reactions (IV, 28) to proceed the parent ions must, of course, have a sufficient amount of vibrational energy. This amount of vibrational energy probably results from an initial electronic excitation of the parent ion followed by radiative transition to the ground state. Direct formation is unlikely on account of the Franck–Condon principle since the internuclear distances and angles in the ground states of the particular ions are so similar to those of the neutral molecules.

Newton and Sciamanna (931) have observed a "metastable" $CO_2^{++}$ ion which decomposes into $CO^+ + O^+$. Here again intercombination is likely to be the cause for the long life of the ion before decomposition.

**Case III of predissociation.** Predissociation by rotation (case III) would arise if the rotational energy is sufficiently excited in vibrational levels that are slightly below the lowest dissociation limit of a given electronic state. Just as for diatomic molecules there is a rotational barrier and therefore, in general, more rotational energy would be required for predissociation to occur than corresponds to the difference between the energy of the vibrational level considered and the dissociation limit. No such case has yet been observed for polyatomic molecules.

It should be noted that rotation will affect the predissociation limit also in case I predissociations in a way analogous to that for diatomic molecules. However, in the only case in which a breaking-off of the rotational structure has been observed (HNO), the effect of a rotational barrier has not been found.

**Inverse predissociation, recombination in a two-body collision.** The inverse of a predissociation process is a recombination process. We have previously discussed emission continua which correspond to a transition from an unstable (continuous) upper state to a stable ground state and represent the inverse of a continuous absorption spectrum. In an analogous way the inverse of a diffuse absorption spectrum would be a diffuse emission spectrum, and like the continuous emission spectrum, it would correspond to a two-body (*radiative*) recombination process.

The inverse of a case I predissociation would proceed as follows: the two radicals (or atoms) would approach each other on the potential surface of the state causing the predissociation; when the image point corresponding to this motion comes close to the ridge of intersection with the discrete excited state, and if the partners have the right amount of energy, a transition into the discrete state may take place, and from there a radiative transition to the ground state may occur leading to the emission of one particular frequency in the diffuse emission band. For the inverse of case II predissociation, the image point may simply pass over a barrier into the well representing the discrete state, carry out a Lissajous motion and during that time make the transition to the ground state.

In inverse predissociation (case I or II), different from the inverse of the simple dissociation, the lifetime can be much longer than the collision time ($10^{-13}$ sec), but this is so only for certain narrow energy ranges corresponding to the diffuse energy levels. If the energy is outside these diffuse ranges, we would have much the same situation as for the inverse of a simple dissociation, that is the two radicals would fly apart again after a time of the order of $10^{-13}$ sec.

As we have seen before, the inverse of a simple dissociation which leads to radiative two-body recombination is a very rare phenomenon. Only about $10^{-5}$ of all favorable two-body collisions lead to emission of radiation and therefore to recombination. Similarly the inverse of case I or case II predissociation is a very rare phenomenon, and no polyatomic example has as yet been unambiguously established by observation of a diffuse recombination spectrum[6]. It is possible that the reaction

$$NO + O \rightarrow NO_2 + h\nu,$$

previously discussed, actually belongs in this class, namely, if the observed continuum is assumed to be only a pseudo-continuum which really consists of a large number of diffuse bands. At any rate a further discussion seems warranted since in many instances inverse predissociations may be responsible for observed recombinations.

If the two collision partners have the right energy (corresponding to one of the diffuse energy levels of the system), the collision time is much longer than the time of a simple oscillation, we have what some authors have called a "sticky collision". In this case, the chance that a light quantum is emitted and the molecule is stabilized is much greater than $10^{-5}$. It would be equal to 1 if the lifetime $\tau_l$ for radiationless decomposition is much greater than the radiative lifetime $\tau_r$. However, this is true only for certain energy values of the collision partners (which are the more sharply defined the longer the lifetime is), and, since in general the colliding atoms or radicals have a continuous distribution of (translational) energies, the net recombination yield (per collision) is again very low unless the density of diffuse levels is very high.

Let us consider a system in which there are three diffuse levels slightly above the dissociation limit as shown in Fig. 182 at the left and let the thermal distribution of collision energies of the parts that are to recombine be given by the exponential (Boltzmann) curve at the right, i.e. let the abscissa of this curve represent the relative number of collision pairs of the energy given by the ordinate. The area under this curve represents the total number of collisions per unit time integrated over all energies.

The number of successful collisions (leading to recombinations) is obtained by multiplying the thermal distribution function by the capture probability. The result is represented by the area under the curves *a*, *b*, *c* at the right if the lifetime of the collision complex (i.e. the molecule in the diffuse level) is much longer than

[6] Inverse predissociation has been definitely observed for a diatomic example, AlH, not only in the laboratory (see Vol. I, p. 415) but also, very strikingly, in the atmospheres of certain low temperature stars [Herbig (507)].

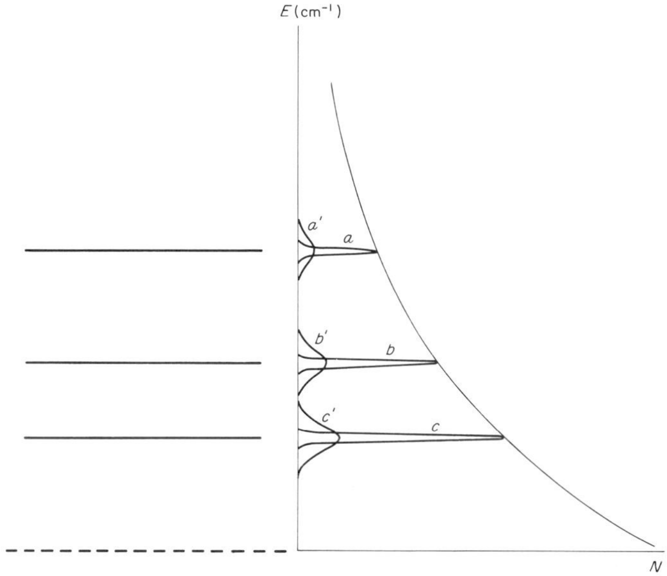


FIG. 182. **Thermal distribution of levels and capture probability for diffuse levels above a dissociation limit.** At the left are three levels of a system whose dissociation limit is indicated by the broken line at the bottom. At the right is a Boltzmann distribution curve (plotted sideways, to have the same energy scale as in the energy level diagram); two different widths of the levels are indicated showing the capture probability.

the radiative lifetime ($\tau_l \gg \tau_r$, that is, $\tau_l > 10^{-8}$ sec). In that event, as mentioned earlier the yield is close to 1 in the center of the diffuse level and therefore the peaks of the curves $a$, $b$, $c$ nearly reach the distribution curve. But the width of each of the curves $a$, $b$, $c$ is then extremely small, less than 0.001 $cm^{-1}$, while the thermal distribution curve at room temperature extends several hundred $cm^{-1}$ above the dissociation limit. Thus the total recombination yield will be very little greater than what it would be without the presence of the diffuse levels (viz. $10^{-5}$).

If the levels are broader, the lifetime is shorter and the capture probability is no longer 1 at the center of a diffuse level, that is, the curves $a'$, $b'$, $c'$ corresponding to $a$, $b$, $c$ no longer reach the thermal distribution curve but the area under these curves is approximately the same as before (since the height is roughly proportional to the lifetime $\tau_l$, that is, inversely proportional to the width). Thus the total recombination yield is in a first approximation independent of the width and, if there are only a few diffuse levels, is very little different from the value it would have if no diffuse levels were present.

However, as the *number of levels* in the range of the thermal distribution curve increases, the recombination yield will increase. If, for example, there are so many levels that the area under the curves $a, b, c, \ldots$ is $10^{-3}$ of the area under the thermal distribution curve, then the total recombination yield will be

$10^{-3}$, that is, 100 times what it would be for a simple inverse dissociation. If the density of vibrational levels were so large that the wings overlap, the yield would become of the order 1: every collision would lead to recombination.

It is clear that the density of vibrational levels above a dissociation limit is fairly large even for a triatomic molecule (cf. Fig. 28 of Vol. II) and will increase rapidly as the number of atoms in the molecule increases. Here it may be pointed out that the number of levels available for inverse predissociation is far greater than the number of levels reached in absorption from the ground state since the latter number is greatly restricted by the Franck–Condon principle.

In a rough approximation, neglecting anharmonicity, the density of levels (per $cm^{-1}$) at an energy $G_0(v_1, v_2, \ldots)$ above the lowest vibrational level is given by

$$P = \frac{G_0{}^{f-1}}{(f-1)!\prod_i \nu_i} \tag{IV, 29}$$

where $f$ is the number of vibrational degrees of freedom and $\nu_i$ are the fundamental frequencies (in $cm^{-1}$) [for more rigorous formulae see, e.g. Haarhoff (458)]. From (IV, 29) we find that for example for $H_2CO$ and $G_0 = 30{,}000\ cm^{-1}$ the level density is 6.3 per $cm^{-1}$, that is the average spacing of the levels is 0.16 $cm^{-1}$. For somewhat larger $G_0$ values, or for a larger number of atoms, the level spacing rapidly approaches, and may become even smaller than, the natural line width, that is, the recombination yield will be of the order of 1. Thus if molecules with more than four or five atoms are formed from their dissociation products via an inverse predissociation, in general every collision leads to radiative recombination provided that the radiative transition from the excited electronic state to the electronic ground state is allowed ($\tau_r \approx 10^{-8}$ sec). The restriction to a few narrow energy ranges no longer applies; every collision is a "sticky collision" quite unlike the collision of two atoms[7].

The relatively high yield of radiative recombination applies only to the inverse of case II predissociations, that is, of predissociations in which no third electronic state in addition to upper and lower state of the absorption bands is involved. Case I predissociations for which a third electronic state is involved occur in general for fairly low vibrational levels of the upper electronic state of the absorption bands and therefore the level density is low even for fairly complicated molecules. Thus, in general, inverse case I predissociations lead to a very low yield of radiative recombinations: not much higher than the yield of recombinations which are the inverse of a direct photodissociation.

[7] It has been pointed out by Callomon (173a) that a further lengthening of the collision time may be brought about by the process of "internal conversion" invoked by many authors for an understanding of energy transfer processes in large molecules. Here, the internal conversion would consist in a radiationless transition of the molecule from the state formed by inverse predissociation to high vibrational levels of a lower electronic state, possibly the ground state. The mechanism of this "conversion" is not too well understood as yet but is presumably connected with strong perturbations between the two states involved. Thus in the present case the over-all process would be closely related to, if not identical with, an inverse "accidental" predissociation (see Vol. I, p. 415).

In the preceding discussion we have implicitly assumed that only central collisions are important, that is that the total angular momentum $J$ of the whole system is zero. Actually, of course, most collisions are non-central, corresponding to various non-zero values of $J$, depending on the impact parameter (separation of the two linear momentum vectors of the colliding partners in the center of mass system) and the relative velocities of approach. For each group of collisions of given $J$ the preceding considerations must be applied. Thus, in effect, the rotational levels in each vibrational level must be counted when the total area $a + b + c + \cdots$ in Fig. 182 is evaluated, except that very high $J$ values should not be included in the evaluation, since there is, just as for diatomic molecules, an increasing rotational barrier with increasing $J$; or, in other words, collisions with very large $J$ are really not genuine collisions.

Up to now we have assumed that the two recombining groups (collision partners) approach each other always on one and the same potential surface belonging to an excited state of the resulting molecule. However, in general, several electronic states of the resulting molecule arise from a given combination of the separated groups (see Chap. III, section 1c) and these groups may approach each other on any of the corresponding potential surfaces. In general only one or two, if any, of these electronic states will combine with the ground state and thus be effective for radiative recombination. Therefore the recombination yield is reduced by a factor corresponding to the ratio of the statistical weights.

If none of the electronic states formed from the collision partners combines in an allowed transition with the ground state, the yield would be reduced by a considerable factor. Even if one or two states do combine with the ground state, their potential surfaces may not be favorable, they may be entirely repulsive, or only for certain relative orientations of the collision partners may they be favorable (i.e. there may be a very small steric factor), or transitions may have to occur in regions of the potential surface where the transition probability is low. Thus, there are many causes which for allowed transitions may lower the recombination yield. On the other hand, even if the radiative transition is forbidden for the most symmetrical conformation, the system may go to less symmetrical conformations on the potential surface from which the electronic transition is not forbidden.

If all electronic transitions to the ground state from the various excited states formed by the collision partners are forbidden, there remains the possibility of vibrational (infrared) transitions within the electronic ground state if this state is also formed from the collision partners. Thus we would have an inverse case II predissociation directly into the ground state followed by a radiative (infrared) transition from the diffuse level to a sharp level below the dissociation limit. However, infrared transition probabilities are low, of the order of $10^3$ (rather than $10^8$), and therefore, the same considerations as to the yield apply as for forbidden electronic transitions.

As an example, consider the recombination

$$O + CO \rightarrow CO_2 + h\nu.$$

Recently Mahan and Solo (792) have shown that at low pressure the reaction of O atoms with CO proceeds by such a radiative two-body recombination. Since the ground state of $O(^3P)$ + $CO(^1\Sigma^+)$ is a triplet state (see Fig. 164) while the ground state of $CO_2$ is singlet, a spin change must take place either in the inverse predissociation process or in the emission of radiation. Mahan and Solo have shown that the lifetime $\tau_l$ for the radiationless process is of the order $10^{-5}$ sec which is larger by a factor $10^3$ than the lifetime $\tau_l$ for a case II predissociation. Thus it appears probable that we have here an inverse case I predissociation violating the spin rule and leading the molecule to an excited singlet state (probably $\tilde{A}\ ^1B_2$) from which it goes over to the ground state with emission of radiation: a blue luminescence is observed.

As a second example, let us consider the collision of two $CH_3$ groups. In order to form $C_2H_6$ by an inverse predissociation, the two $CH_3$ groups must approach each other in at least approximately such a way that their symmetry axes coincide. This introduces a fairly small steric factor. According to Table 28, p. 295, two $CH_3$ groups in their ground states ($^2A_2''$) give the states $^1A_{1g}$, $^3A_{2u}$ of $C_2H_6$ of point group $\boldsymbol{D}_{3d}$ or $^1A_1'$, $^3A_2''$ for point group $\boldsymbol{D}_{3h}$. In either case, only one electronically excited state ($^3A_{2u}$ or $^3A_2''$) is formed which, on account of the spin rule ($\Delta S = 0$), cannot combine with the ground state. Without the spin rule and without the steric factor, the recombination yield might be as high as 1 because of the high density of levels above the dissociation limit for an eight-atomic molecule. But the addition of the spin rule and of the steric factor bring the yield of inverse predissociation down to perhaps $10^{-6}$, i.e., lower than for the radiative recombination of a diatomic molecule. The same applies for recombination of the two $CH_3$ groups in the ground state ($^1A_{1g}$) by infrared radiation. A very similar situation arises for the recombination of two $CH_2$ groups to form $C_2H_4$ (see Table 28).

As a third example, consider the formation of $H_2CO$ from $CH_2$ in the $^3\Sigma_g^-$ ground state and O in the $^3P_g$ ground state. According to Table 27, there are nine electronic states (i.e., nine potential surfaces) arising from $CH_2(^3\Sigma_g^-)$ + $O(^3P)$, viz., $^1A_1$, $^1A_2$, $^1B_2$, $^3A_1$, $^3A_2$, $^3B_2$, $^5A_1$, $^5A_2$, $^5B_2$. Of these, the first, $^1A_1$, forms the ground state of the molecule. Of the other states, only one, $^1B_2$, can combine with the ground state as an allowed transition. Statistically, this state is formed in only one in 27 collisions. If the $^1B_2$ state is not entirely repulsive, an inverse case II predissociation, followed by a radiative transition to the ground state, would give a radiative recombination yield much larger than $10^{-5}$ assuming that the steric factor is not too small and that the density of levels is large. An allowed inverse case I predissociation could arise from any one of the singlet states $^1A_1$, $^1A_2$, $^1B_2$ of $CH_2$ + O if there is a suitable excited state (belonging to a different dissociation limit), into which the radiationless transition could take place; but, as emphasized before, the yield of radiative recombination would be much smaller than for inverse case II predissociation since the density of levels in the discrete excited state is much smaller. Neither the predissociation $CH_2O \rightarrow CH_2 + O$ nor its inverse have as yet been observed. The predissociation of $H_2CO$ observed in the $\tilde{A}-\tilde{X}$ bands corresponds to dissociation into $H_2$ + CO or H + HCO.

Inverse predissociation that leads to an excited state of the molecule is associated with the emission of visible or ultraviolet light; its wave length is in general longer and often much longer than that of the corresponding absorption spectrum. The over-all intensity of this spectrum is a measure of the recombination yield, i.e., it is large when the radiationless transition probability is small and the level density high. If the latter condition is fulfilled, the spectrum will be more or less continuous and difficult to distinguish from that of an inverse dissociation. The luminescence observed in reaction (IV, 29) is an example but its spectrum has not yet been studied. No example has been established of an inverse predissociation in which the radiation emitted is in the infrared and corresponds to vibrational transitions.

As a summary of the types of radiative recombinations, it is perhaps useful to consider the following scheme:

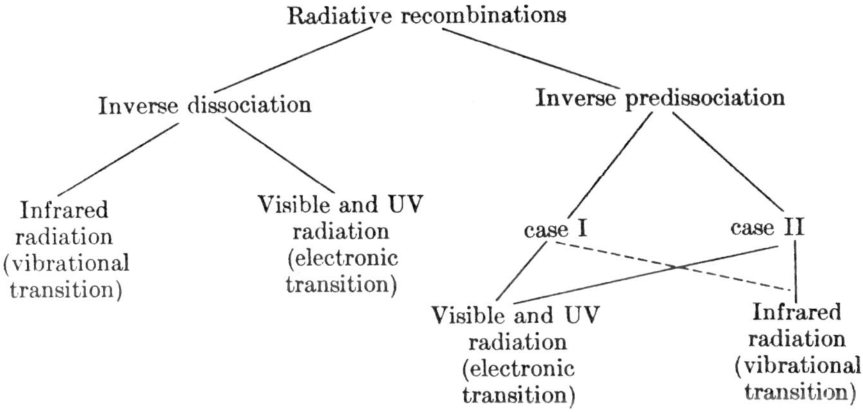


It is interesting to note that in nuclear physics radiative recombination plays a very important role [see, for example, Eisenbud and Wigner (11)]. Neutron and proton capture by nuclei proceeds via a direct analogue of inverse case II predissociation followed by an analogue of vibrational transitions. The frequencies of the radiative transitions from the diffuse states to lower states are of course very much higher and therefore, because of the $\nu^4$ factor in the transition probability, radiative capture is a fairly frequent phenomenon in nuclear physics. Many examples of resonant captures corresponding to diffuse levels in the compound nucleus have been found. The theory of this nuclear phenomenon was first given by Breit and Wigner (147) (one speaks of Breit–Wigner resonances) and has been developed extensively by many other investigators. It is in a far more advanced stage than the theory of the corresponding phenomenon in molecules.

**Three-body recombinations.** Radiative recombination of molecules is of importance only at very low pressures. In general, at normal laboratory pressures during the duration of the collision of two partners (i.e. the lifetime of the collision complex), collisions with third partners occur which can take away some of the excitation energy and thus stabilize the molecule formed, leaving it in a state slightly below the dissociation limit.

If $\tau$ is the duration of a two-body collision, and $Z$ the number of collisions suffered by a given atom or molecule per second, it is clear that $\tau Z$ is the fraction of two-body collisions which are also three-body collisions ($1/Z$ is the average time between two-body collisions). The fraction $\tau Z$ is proportional to the pressure for not too high pressures. At very high pressures the ratio of three- to two-body collisions reaches the limiting value 1, that is, every collision is then a three-body collision. The larger $\tau$ is, the smaller is the pressure at which this limit is reached. For diatomic molecules or for polyatomic molecules in repulsive states, $\tau$ is of the order $10^{-13}$ sec. At a pressure of 1 atm. the collision frequency $Z$ is of the order of $10^{10}$, and therefore the fraction of three-body collisions is $10^{-3}$. Since almost every three-body collision will lead to recombination, while only every $10^5-10^6$th two-body collision will lead to radiative recombination, we see that at atmospheric pressure three-body collisions are about 100 times more effective in producing recombinations than two-body collisions.

If an inverse predissociation can take place in a recombination process, it means that the duration ($\tau$) of the collision is lengthened, at least when the energy has certain values, and therefore, the yield is increased for both two-body and three-body recombinations. If the radiative transition involved in the recombination by inverse predissociation is an allowed one the ratio of the yields of two-body and three-body recombinations is the same as for the quasi-diatomic case; but if the radiative transition is forbidden or is a vibrational transition in the infrared, the yield of two-body recombinations at a given pressure is much reduced while that of three-body recombinations is unchanged. In any case, when at the same time the density of levels is high, the duration of *all* two-body collisions is increased so much that even at fairly low pressures collision with a third partner (or the wall of the reaction chamber), and therefore recombination, will almost invariably occur during this duration of the two-body collision, that is almost all two-body collisions are also three-body collisions.

It has indeed been found in many photochemical investigations that the yield of a recombination reaction is far greater than one might have expected from considering diatomic examples. A case in point is the recombination of $CH_3$ radicals to form $C_2H_6$. This recombination would be exceedingly slow by radiative recombination because of the difference in multiplicity of the two states involved (see p. 478), but it can readily take place by three-body collisions because of the long life of the collision complex. In this case recombination yields of about 0.1 have been found [see Steacie (38)].

It is clear that at sufficiently low pressures two-body recombinations predominate over three-body recombinations, but for molecules with eight or more atoms, such pressures are exceedingly low, less than $10^{-3}$ mm Hg. While at intermediate pressures both mechanisms occur simultaneously, at higher pressures two-body recombinations can be neglected in the study of reaction yields even though they do take place and may be responsible for some of the light that is emitted.

Clyne and Thrush (213)(214) have studied the light emission during the reactions

$$H + NO + M \rightarrow HNO + M \tag{IV, 30}$$

$$O + NO + M \rightarrow NO_2 + M \tag{IV, 31}$$

$$O + CO + M \rightarrow CO_2 + M \tag{IV, 32}$$

and its dependence on the partial pressures of the reactants. They found that the light yield depends linearly on the partial pressure of H, NO, O and CO but does not depend on the partial pressure of M although it does depend on the nature of M. From this result they conclude that the light emission is due to a three-body not a two-body mechanism. The yield does not depend on [M] because both the steps leading to formation of HNO, $NO_2$, $CO_2$ and those leading to deactivation (without radiation) depend on [M], and thus the [M] dependence cancels out assuming that light emission is slow compared to collisional deactivation. For radiative (two-body) recombination, an inverse dependence on [M] would result because of increasing collisional deactivation. The first step in each of the three reactions (before the arrival of M) is in all probability an inverse predissociation. More recently Reeves, Harteck and Chace (1063a) have studied the light emission in the O + NO reaction at very low pressure (3 to 20 $\mu$) and have established that under these conditions a third collision partner is not required, that is, that the light emission is due to transitions from the diffuse levels (formed by inverse predissociation) to the ground state.

For a more complete understanding of the three-body collision process, it would be necessary to consider the potential surface for a system including the third body. In general, such a system would have too many degrees of freedom for a graphical representation. We may, however, use a triatomic system as a model. Consider, for example, the recombination of a C and an O atom when the third partner is another O atom. If we restrict the motions to a straight line, we may use the previous Fig. 163. In a three-body collision the image point starts out from the plateau at the top right. Depending on the initial direction (and velocity), the image point will enter one of the valleys, oscillate rather violently about the bottom of the valley, and fly out of the valley. This type of trajectory clearly represents formation of a vibrating CO molecule. The same result is obtained if the image point first enters the bowl corresponding to the ground state of $CO_2$. Thus, classically, practically every three-body collision leads to recombination. Only those collisions for which the image point would move at large $r_2$ parallel to the $r_1$ axis (or at large $r_1$ parallel to the $r_2$ axis) would not lead to CO formation since the image point would then return to the plateau. Quantum-theoretically, of course, for a three-body collision to lead to recombination, at least one vibrational quantum must be removed by the third partner, and for that to happen there must be in the corresponding classical model a sufficient deviation of the motion of the image point in Fig. 163 from a linear motion.

It is seen from the model that because of the possibility of Lissajous motion the duration of the three-body collision may be much greater than if there were only repulsive potential regions. This is quite analogous to the lengthening of two-body collisions discussed earlier.

## 4. Determination of Dissociation Energies

In principle, the study of continuous and diffuse spectra of polyatomic molecules allows a determination of dissociation energies just as does the study of corresponding spectra of diatomic molecules, but in practice the situation is much less favorable.

**Determination of dissociation limits.** For diatomic molecules (see Vol. I, p. 438f) dissociation limits can be derived from band convergences, extrapolation of such convergences, long wave-length limits of absorption continua, predissociation limits, excitation of atomic fluorescence, photodissociation and chemiluminescence. For polyatomic molecules, in principle the same (or corresponding) methods are applicable.

Band convergences are extremely rare for polyatomic molecules and no clear-cut example has yet been observed. There are two reasons for this state of affairs. Only totally symmetric vibrations in molecules of high symmetry are likely to give rise to simple series of vibrational levels of diatomic type since only for them are the higher levels not mixed up with other vibrations. But for these vibrations the dissociation limit lies very high and thus it is in general not reached in absorption. A good example is the vibration $\nu_1$ of linear symmetric $XY_2$ (see Fig. 163). On the other hand, in an excited state in which the bond energies are much reduced (and this is necessary to observe a long progression), in general the shape of the molecule is less symmetrical, and therefore the higher vibrational levels can no longer be ascribed to a single vibration and no longer follow a simple quasi-diatomic relation.

Although it is possible to *extrapolate* the convergence limit of a progression of bands, this method is even less reliable here than for diatomic molecules since it is not clear how the interaction with other vibrations is to be taken into account in this extrapolation except possibly in cases of high symmetry. In the literature, an extrapolation of a dissociation limit has been seriously attempted only in one case, that of $ClO_2$ [Finkelnburg and Schumacher (384)] where a fairly long progression has been observed. The extrapolated limit was assigned to dissociation into ClO + O($^1D$) but does not agree well with the thermochemical value of the dissociation energy $D$(ClO − O) (see Table 47).

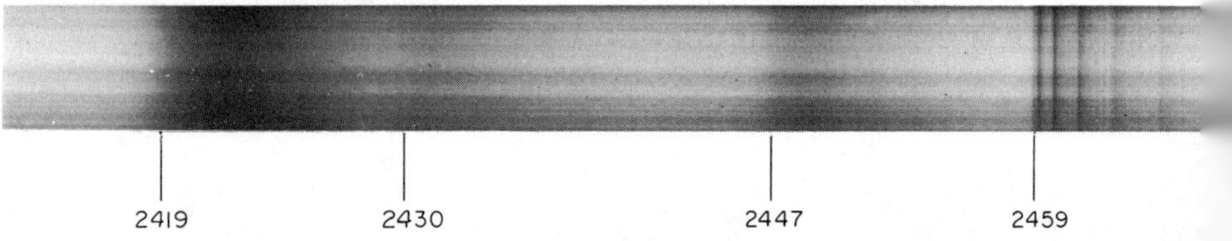

The longward limit of a continuum, as well as an observed predissociation limit, represent obviously upper bounds to the corresponding dissociation limits. However, for polyatomic molecules it is more difficult than for diatomic molecules to establish whether or not the upper bound is close to the actual value. Maxima in the potential surface which separate the discrete stable levels from the dissociated states are much more frequent. In addition, as we have seen, even without such a maximum, a continuum or a diffuseness may start rather sharply at a point appreciably above the dissociation limit because of the particular form of the vibrational (Lissajous) motion produced in the excited state. Therefore, even a sharp longward limit of a continuum or a sharp predissociation limit gives only an upper bound to the corresponding dissociation limit. Nevertheless such definite upper bounds are, of course, of considerable value in the discussion of dissociation energies.

As an example, we consider the two predissociation limits observed in the absorption spectrum of $NO_2$. In the visible and near ultraviolet absorption bands under medium resolution the lines begin to show a slight diffuseness at about 3950 Å, and this diffuseness increases gradually so that at 3700 Å it is noticeable even under fairly low resolution [Henri (496), Mecke (817)]. Recent studies under very high resolution by Douglas and Huber (296) have shown that the onset of diffuseness is really quite sharp, namely, at 3979 Å (see Fig. 192 below). This limit corresponds to 3.115 eV which must be considered as an upper bound to the dissociation energy $D(\mathrm{NO{-}O})$. Indeed an independent value, obtained from $D(O_2)$ and the heat formation of $NO_2$ from NO and $O_2$, is found to be 3.112 eV. It is therefore clear that the first predissociation limit (as obtained under high resolution) is a true dissociation limit. A second very clear predissociation limit is found in the ultraviolet band system near 2450 Å [Henri (497)]. As the spectrogram Fig. 183 shows, the lines of the band at 2491 Å are quite sharp, those of the band at 2459 Å are slightly broadened while the band at 2447 Å is completely diffuse. The predissociation limit is thus between 4.98 and 5.04 eV. The difference between the two limits lies between 1.86 and 1.92 eV which is smaller than the excitation energy 1.967 eV of the $^1D$ state of the O atom. Therefore, the second limit cannot be ascribed, as was at one time thought, to dissociation into $\mathrm{NO}(^2\Pi) + \mathrm{O}(^1D)$, but must either correspond to dissociation into $\mathrm{NO}(^2\Pi) + \mathrm{O}(^3P)$ or into $\mathrm{N}(^4S) + \mathrm{O}_2(^3\Sigma_g^-)$. For either assumption the predissociation limit, in spite of its sharpness, lies substantially above the dissociation limit, namely, by 1.9 eV and 0.5 eV, respectively.

Another example is supplied by the absorption spectrum of $SO_2$. At medium pressures a diffuseness with a very indistinct limit near 2700 Å has been reported by Henri (498), but at higher resolution it becomes clear that this diffuseness is only apparent and that there is no predissociation in this region [Douglas (294b)]. However, at shorter wave lengths at 1950 Å (6.36 eV) there is a fairly sharp limit of a genuine predissociation. The energy required to dissociate $SO_2$ into SO + O

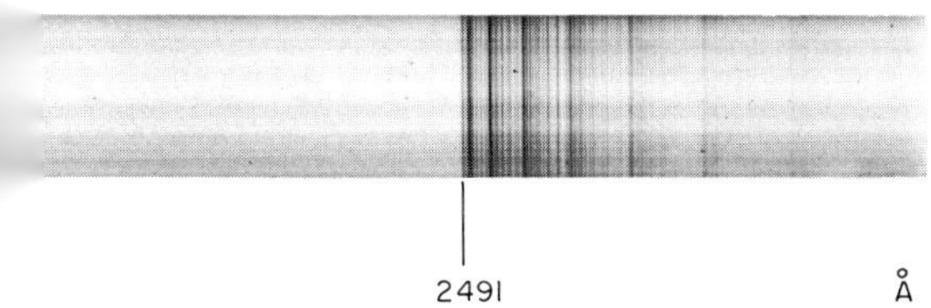


FIG. 183. **Predissociation in the ultraviolet ($\tilde{B}-\tilde{X}$) bands of $NO_2$ after Huber (unpublished).**

TABLE 47. OBSERVED LIMITS OF CONTINUA AND DIFFUSENESS AND THEIR RELATION TO DISSOCIATION LIMITS IN SIMPLE POLYATOMIC MOLECULES

| Molecule | Spectral feature | | Dissociation limit $cm^{-1}$ | Products | Calculated[a] limit |
|---|---|---|---|---|---|
| $CH_2$ | diffuse absorption | ≤ 1415 Å | < 70600 | $CH(^2\Sigma^-) + H(^2S)$ | ≥ 59800[b] |
| $H_2O$ | continuous absorption | < 1860 Å | < 53800 | $OH(^2\Pi) + H(^2S)$ | 41246 |
| | diffuseness of band lines | < 1240 Å | < 80700 | $O(^1D) + H_2(^1\Sigma_g^+)$ | 56452 |
| $H_2S$ | continuous absorption | < 2700 Å | < 37000 | $SH(^2\Pi) + H(^2S)$ | 26329[c] |
| | | | | $S(^1D) + H_2(^1\Sigma_g^+)$ | 27926[c] |
| $AlH_2$ | diffuse absorption | < 6470 | < 15450 | $AlH(^1\Sigma) + H(^2S)$ | |
| HCN | diffuseness of band lines | < 1810 Å[d] | < 55260 | $H(^2S) + CN(^2\Pi)$ | 54900[e] |
| HCO | diffuseness of bands with $l \neq 0$, $v_2' > 3$ | < 8062 Å | < 12400 | $H(^2S) + CO(^1\Sigma^+)$ | |
| HNO | Breaking off in emission above $K' = 13$ in 000 and above $K' = 10$ in 010 level; Diffuseness of absorption lines in 101 – 000 band for $K' > 0$ | | < 17000 | $H(^2S) + NO(^2\Pi)$ | |
| $N_2O$ | continuous absorption | < 3065 Å | < 32600 | $N_2(^1\Sigma_g^+) + O(^3P)$ | 13507 |
| | | | | $N_2(^1\Sigma_g^+) + O(^1D)$ | 29375 |
| $NO_2$ | Diffuse absorption in $\tilde{A} - \tilde{X}$ system | ≤ 3979 Å | 25125 | $NO(^2\Pi) + O(^3P)$ | 25105 |
| | Diffuse absorption in $\tilde{B} - \tilde{X}$ system | < 2490 Å | < 40150 | $N(^4S) + O_2(^3\Sigma_g^-)$ | 36320 |
| $SO_2$ | diffuse absorption bands in $\tilde{C} - \tilde{X}$ system | < 1950 Å | < 51300 | $SO(^3\Sigma^-) + O(^3P)$ | 45250[c] |
| | | | | $S(^3P) + O_2(^3\Sigma_g^-)$ | 47270 |
| $ClO_2$ | diffuse bands in $\tilde{A} - \tilde{X}$ system | < 3750 Å | < 26650 | $ClO(^2\Pi) + O(^3P)$ | 20205 |
| | convergence limit | ~ 2560 Å | ~ 39060 | $ClO(^2\Pi) + O(^1D)$ | 36073 |
| $CH_3$ | diffuse absorption | < 2164 Å | < 46205 | $CH_2(^3\Sigma_g^-) + H(^2S)$ | ≤ 39500 |
| | | | | $CH(^2\Pi) + H_2(^1\Sigma_g^+)$ | 35200 |
| $NH_3$ | diffuse absorption bands in $\tilde{A} - \tilde{X}$ system | < 2167 Å | < 46136 | $NH_2(^2B_1) + H(^2S)$ | 35300 |
| | | | | $NH_2(^2A_1) + H(^2S)$ | 45549 |
| $H_2CO$ | Diffuseness in $\tilde{A} - \tilde{X}$ system | < 2750 Å | < 36360 | $HCO(^2A') + H(^2S)$ | ~ 26050 |
| | Breaking off of vibrational structure in emission above $v_4 = 0$ | | ≤ 28736 | $CO(^1\Sigma^+) + 2H(^2S)$ | 36560 |
| $CH_4$ | continuous absorption | < 1455 Å | < 68730 | $CH_3(^2A_2'') + H(^2S)$ | 35500 |
| | | | | $CH_2(^3\Sigma_g^-) + H_2(^1\Sigma_g^+)$ | 36700 |

[a] For references see Tables 62–82.
[b] Using Prophet's (1033) atomic heat of formation the limit would be at 60,600 ± 1500 $cm^{-1}$.
[c] Assuming $D(S_2) = 98.5$ kcal.
[d] In the main progression.
[e] Assuming $D(CN) = 7.5$ eV.

is 5.61 eV [8]. Therefore, the predissociation limit would be 0.75 eV above the dissociation limit unless one wanted to consider that the SO formed is in the metastable $^1\Delta$ state (whose energy is not yet known) or that dissociation into $S + O_2$ takes place with either S or $O_2$ suitably excited. The minimum dissociation energy into $S + O_2$ is 5.86 eV. For either assumption the energy discrepancy may correspond to translational or vibrational energy of the dissociation products.

In Table 47 these and other examples of long-wave-limits of diffuseness as well as of continua are listed. It is seen that there are only a few cases in which these long-wave-limits coincide with the known dissociation limits.

Relatively little work has been done to use the analogue of the method of atomic fluorescence in the determination of dissociation limits of polyatomic molecules, that is, to ascertain the minimum photon energy required for the production of fluorescence of the dissociation products. Terenin (1199), and later Wieland (1293)(1294) have used this method for the interpretation of the continua of the mercury halides and the determination of their dissociation energies. They found that upon irradiation with light in one of the shorter wave length continua of $HgX_2$, a fluorescence spectrum due to diatomic HgX is emitted, and they have obtained in this way upper bounds for the corresponding dissociation limits. A number of other cases have been studied by Terenin and his collaborators [see the summary in Terenin (1200)], including such cases as

$$H_2O + h\nu \rightarrow OH(^2\Sigma) + H$$

$$NH_3 + h\nu \rightarrow NH_2(^2A_1) + H$$

$$CH_3OH + h\nu \rightarrow CH_3 + OH(^2\Sigma)$$

$$CH_3CN + h\nu \rightarrow CH_3 + CN(B^2\Sigma)$$

More recently, Style and his co-workers [Dyne and Style (334); Style and Ward (1173)(1174)] have used the fluorescence method to obtain the spectra of a number of free radicals. As strong sources of vacuum ultraviolet radiation become more readily available and free radical spectra become better understood, this method for the study of dissociation processes will gain in importance. It has the great advantage that it tells us immediately the degree of excitation of at least one of the dissociation products. However, just as the other methods, it gives us only upper bounds to the dissociation limits considered.

Photodissociation is in principle an excellent method to establish upper bounds of dissociation limits since it does not require a detailed interpretation of the absorption spectrum. In practice, however, it suffers from the occurrence of secondary reactions and the possibility of reactions of the parent molecule in an excited state below the dissociation limit. Similarly, the use of chemiluminescence is not of much help because side reactions may obscure the luminescence produced

[8] In a recent paper Warneck, Marmo and Sullivan (1271) give 5.43 eV for this energy on the basis of a continuum which, they believe, underlies the discrete bands below 2280 Å.

by a given reaction. The development of the method of crossed molecular beams may remedy this situation.

**Determination of the dissociation products.** In order to obtain reliable dissociation energies it is necessary, just as for diatomic molecules, to know the state of excitation of the products of dissociation at a given dissociation limit. The methods of determining them are again very similar to those for diatomic molecules.

(1) If several dissociation limits are known, their differences must be equal to possible energy differences of the dissociation products. Since usually one obtains only upper bounds of dissociation limits, the energy condition does not often lead to a clear-cut decision between various possible dissociation products. Moreover, the dissociation products, unlike those of diatomic molecules, may now have various amounts of vibrational energy leading to a much larger number of possibilities between which in general the energy differences of upper bounds of dissociation limits cannot decide.

(2) In determining the dissociation products, the analogues of the Wigner–Witmer correlation rules discussed in Chapter III, section 1c, must be fulfilled for the electronic states of the products into which the molecule dissociates. In other words, the upper state of a continuous spectrum or the state causing a predissociation must be obtainable from the assumed states of the dissociation products according to these correlation rules. For example, the state causing the predissociation in the 2160 Å band of $CH_3$ (Fig. 96) which must be a $^2A_1$ state (assuming a planar conformation) cannot arise from the ground state ($^3\Sigma_g^-$) of $CH_2$ but must arise from $CH_2(^1A_1) + H(^2S)$. Therefore the dissociation energy $D(CH_3 \rightarrow CH_2 + H)$ must be smaller than the predissociation limit by at least the excitation energy of the $^1A_1$ state of $CH_2$ (which is not yet known). Even that leads to an upper limit that is presumably much higher than the true value.

(3) The non-crossing rule does not necessarily apply to polyatomic molecules because of the possibility of conical intersections. On the other hand, since there are so many more degrees of freedom and since it is not necessary to preserve the symmetry of a molecule in a dissociation process, in effect the non-crossing rule does hold, that is, a crossing of states of the same species does in general not occur in a dissociation process.

(4) As already mentioned, the observation of fluorescence of the dissociation products gives direct information about their state of excitation. However, in general this is true only for one of the products; the other may or may not be excited to a state from which fluorescence either does not occur (metastable state) or is not readily observable.

(5) In some cases thermochemical data give information about the lowest dissociation energy as in the examples of $NO_2$ and $SO_2$ discussed above. Sometimes it is the combination of thermochemical data with spectroscopic data on other molecules that can lead to a decision.

(6) Frequently electron impact studies of a molecule give rough values for

certain dissociation energies and may be used for a decision on the dissociation products corresponding to a given continuous or diffuse spectrum. Unfortunately, the electron impact values also are in general only upper bounds to the dissociation energies considered.

As an example, consider the dissociation energy of $CH_3$ into $CH_2$ + H which may be determined from the ionization potential of $CH_2$ obtained spectroscopically [Herzberg (522)] and the appearance potential of $CH_2^+$ when $CH_3$ [produced by pyrolysis of $Hg(CH_3)_2$] is bombarded by electrons. We have

$$D(\mathrm{CH_2—H}) = \mathrm{A.P.}(\mathrm{CH_2^+ \text{ from } CH_3}) - \mathrm{I.P.}(\mathrm{CH_2}).$$

Two independent determinations of A.P.($CH_2^+$) have yielded 15.30 ± 0.10 eV [Langer, Hipple and Stevenson (720); Waldron (1255)] while I.P.($CH_2$) = 10.396 eV [9]. Therefore

$$D(\mathrm{CH_2—H}) = 4.90_4 \text{ eV} \quad (= 113.1 \text{ kcal/mole})$$

However, this value can only be considered as an upper limit since it is not known whether the $CH_2^+$ ions formed by electron impact have some internal energy. There is a good deal of evidence that indeed $D(CH_2—H)$ is smaller than the value given.

In the tables of Appendix VI spectroscopically obtained dissociation energies are included for a large number of molecules [see also Table 47, and the handbook by Vedeneiev, Gurvich, Kondratiev, Medvedev and Frankevich (41)].

[9] This value is based on the assumption that $CH_2^+$ in its ground state is linear. In view of the recent finding [Herzberg and Johns (530)] that the iso-electronic $BH_2$ is non-linear in its ground state, it is likely that the same applies to $CH_2^+$ and therefore that the value for I.P.($CH_2$) given in the text is only an upper limit, making the upper limit for $D(CH_2—H)$ still higher.

## CHAPTER V

# ELECTRONIC SPECTRA OF INDIVIDUAL MOLECULES AND THEIR INTERPRETATION

The theoretical considerations of the preceding four Chapters form the basis on which all observed electronic spectra of polyatomic molecules must be discussed. In this Chapter we shall briefly survey the observed spectra for all those molecules with up to 12 atoms which have been studied in some detail. In doing so we shall attempt to improve the interpretation of older data on the basis of the more recent theoretical developments and shall point out gaps in our knowledge of these spectra. In the tables of Appendix VI, the more important constants of all the known electronic states are summarized including those for molecules not specifically discussed in the text. For molecules for which only fragmentary data are available or only continuous absorption spectra have been found, references to the most recent work are given. Early summaries of the state of knowledge in this field were given by Sponer (37) and Sponer and Teller (1155). For a recent summary see Robinson (1077).

In the discussion of observed electronic states some designation is needed in addition to the group-theoretical species symbols to distinguish states of the same species and states whose species has not been established (or is not important for the particular discussion). For diatomic molecules the letters $X$, $A$, $B, \ldots,$ $a$, $b$, $c, \ldots,$ are used, the capital letters referring to states of the same multiplicity as the ground state ($X$), the small letters to states of different multiplicity. This system of designations cannot be taken over for non-linear polyatomic molecules because of the possible confusion with species symbols. Therefore we shall follow here the suggestion of Douglas (294) to add a tilde to the letters $X$, $A$, $B, \ldots,$ i.e., write $\tilde{X}$, $\tilde{A}$, $\tilde{B}, \ldots,$ $\tilde{a}$, $\tilde{b}, \ldots,$ but otherwise follow the rules well established for diatomic molecules.

## 1. Triatomic Molecules

It is convenient, because of the great differences of their spectra, to distinguish three groups of triatomic molecules: di-hydrides, mono-hydrides and non-hydrides.

### (a) Di-hydrides

The electronic spectra of both chemically stable and unstable triatomic di-hydrides have been studied. Stable di-hydrides that have been investigated are $H_2O$, $H_2S$, $H_2Se$ and $H_2Te$. Particular attention has been devoted in recent years to the six free radicals belonging to this class, $BH_2$, $CH_2$, $NH_2$, $AlH_2$, $SiH_2$ and $PH_2$, which have yielded very interesting spectra. Spectra of $BeH_2$ and $MgH_2$ have not yet been discovered. In Table 62 of Appendix VI the known states and transitions of the molecules mentioned are summarized.

**$H_2O$ and $H_2S$.** The ground state data for $H_2O$ and $H_2S$ given in Table 62 are based on infrared work. Since these are chemically stable molecules with all shells closed in the ground state, the first electronic transition occurs at fairly short wavelengths; it consists of a broad continuum, extending for $H_2O$ from 1860 to 1450 Å, for $H_2S$ from 2700 to 1900 Å. These transitions in all probability correspond in each case to the first member of a *ns* Rydberg series. The $3sa_1$ Rydberg orbital of $H_2O$ is identical with the anti-bonding orbital $4a_1$ of Fig. 123. (We use the description $3sa_1$ when we want to emphasize the relation to the higher members of the Rydberg series, while we use the description $4a_1$ when we want to consider the bonding properties of the orbitals and the stability of the resulting electronic states.) Higher members of the *ns* Rydberg series, as well as several other Rydberg series (probably *np* and *nd*), were first observed by Price (1015). Figure 184 shows an absorption curve of $H_2O$ in the region 1900–1200 Å according to Watanabe and Zelikoff (1278).

In $H_2O$ the second absorption consists of a long progression of diffuse bands extending from 1411 to 1256 Å with spacings of about 800 cm$^{-1}$. Such a low frequency can hardly correspond to a vibration other than a bending vibration. A long progression in a bending vibration implies that a large change in angle has taken place. Indeed, this transition does not fit in with the various Rydberg series going to the first ionization potential (removal of a $1b_1$ electron) but probably is the first member of a series corresponding to removal of a $3a_1$ electron (see p. 341f). The corresponding state of $H_2O^+$ is the analogue of the $\tilde{A}$ state of $NH_2$ (see below) and it is probable that $H_2O^+$ in this state is nearly linear like $NH_2$. If an electron in a Rydberg orbital is added to $H_2O^+$ in this state, the $H_2O$ molecule thus formed will have the same conformation as

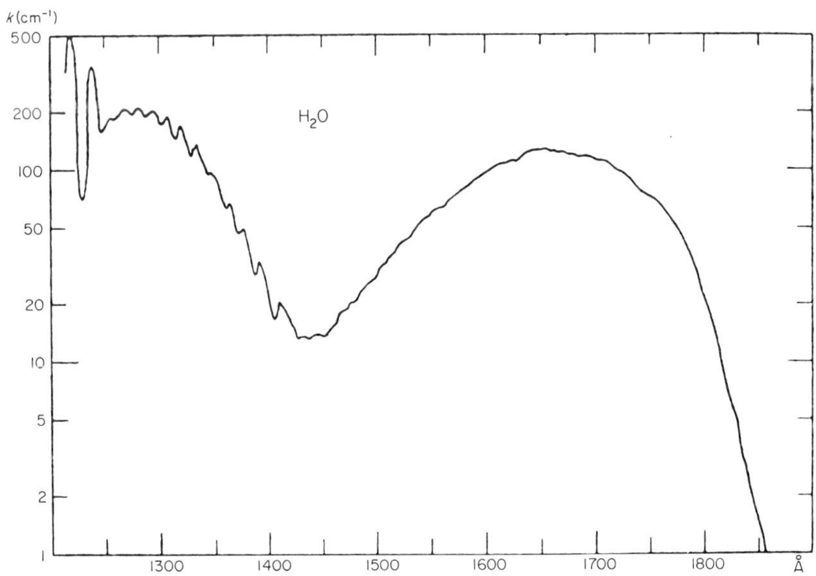


FIG. 184. **Absorption curve of $H_2O$ vapor in the region 1900 to 1200 Å after Watanabe and Zelikoff (1278).** The ordinate is the absorption coefficient in cm$^{-1}$.

$H_2O^+$ (or at least nearly so), and in this way the observed vibrational structure of the transition $\tilde{B} \leftarrow \tilde{X}$ can be understood.

Most of the other higher electronic transitions of both $H_2O$ and $H_2S$ have comparatively little vibrational structure. The $0-0$ band of the first member of the $npa_1\ {}^1B_1$ series of $H_2O$ was recently studied under high resolution and analysed in detail by Johns (631). Figure 112, p. 265, shows a spectrogram of this band. Johns' analysis confirms the conclusion derived from the observation of simple Rydberg series with little vibrational structure that in the Rydberg states the X—H distance and H—X—H angle are not very different from the values in the ground state; and that the symmetry is the same, i.e., $\boldsymbol{C_{2v}}$.

The $H_2O$ band at 1240 Å has a somewhat complicated structure which was readily analysed only because the rotational levels of the lower state were well known from infrared work. In contrast, the bands of $H_2S$ appear to have a fairly simple structure. This is because $H_2S$ in both upper and lower states is fairly close to a symmetric top, i.e., $A$ and $B$ are much more nearly alike than in $H_2O$. The observed predissociation phenomena in the two molecules have been discussed in Chapter IV. In Fig. 138 an energy level diagram showing the observed electronic states of $H_2O$ was given. The situation for $H_2S$ is very similar.

The ionization potentials obtained from the observed Rydberg series are $12.61_8$ and $10.47_2$ eV, for $H_2O$ and $H_2S$ respectively [see Price (1015) and Price, Teegan and Walsh (1024)]. Henning (494) has observed for $H_2O$ two progressions of (diffuse) bands occurring beyond the first ionization limit, starting at 116760 and 125820 cm$^{-1}$. These progressions probably represent members of a Rydberg series whose limit may be the sharp absorption limit at 16.5 eV. This interpretation is supported by electron impact experiments of Frost and McDowell (401) [confirming earlier work of Price and Sugden (1023)], who observed three appearance potentials of $H_2O^+$ at 12.60, 14.35 and 16.34 eV, of which the last very probably corresponds to Henning's absorption limit. The three appearance potentials must correspond to three states of $H_2O^+$, the ground state ${}^2B_1$, a first excited state ${}^2A_1$ at $1.7_5$ eV, and a second excited state ${}^2B_2$ at $3.7_4$ eV (see p. 341). The corresponding figures for $H_2S^+$ are $2.0_1$ and $3.7_3$ eV. Absorption coefficients and photoionization efficiencies of $H_2O$ and $H_2S$ have been studied by a number of authors, most recently by Watanabe and Jursa (1274). The absorption of $H_2O$ from 110,000 cm$^{-1}$ to 600,000 cm$^{-1}$ has been investigated by Astoin (78)(79) and Metzger and Cook (836).

The spectra of $H_2Se$ and $H_2Te$ have been studied by Price, Teegan and Walsh (1024) (see Table 62).

**$BH_2$ and $AlH_2$.** The six di-hydride free radicals, in contrast to $H_2O$, $H_2S$, ... do show absorption bands in the visible region which correspond to electronic transitions within the valence shell. The recently observed spectrum of $BH_2$ [Herzberg and Johns (531)] extends from 6000 Å to longer wavelengths and consists of a simple progression of bands alternately with even and odd $K$ values showing that the molecule is bent in the lower state and linear or nearly linear in the excited state in agreement with expectation from the Walsh diagram (p. 319). The two states arise by large Renner–Teller interaction from one and the same ${}^2\Pi_u$ state of the linear conformation in the way previously discussed (p. 27f). A small doublet splitting has been found. The transition is ${}^2B_1 - {}^2A_1$ as expected for the most longward transition from the electron configuration (Table 33). The angle in the ground state, since there is only one electron in the $3a_1$ orbital, is not as small as in the other molecules of this group, for which two electrons are in this orbital (131° as compared to 105°). In the excited state, with no electron in the $3a_1$ orbital, the molecule is apparently strictly linear unlike the corresponding states of the other di-hydride radicals. The vibrational levels show no sign of a minimum in the $\Delta G$ curve (see p. 121f).

The spectrum of $AlH_2$ is very similar to that of $BH_2$ except that the rotational constants are correspondingly smaller. As for $BH_2$ only one electronic transition has been observed. The analysis of the $AlH_2$ spectrum is as yet incomplete.

**$CH_2$.** Up to now, the absorption spectrum of $CH_2$ has been observed only by the flash photolysis technique (just as those of $BH_2$ and $AlH_2$). The strongest spectra have been obtained by using diazomethane ($CH_2N_2$) as the parent compound. Two quite different spectra are obtained at different times after the beginning of the photolysis flash. The first spectrum occurring earlier consists of two band systems, one, fairly strong in the red and near infrared region of the spectrum, and another, very weak, in the near ultraviolet. These systems have been assigned to singlet states of $CH_2$, the lowest being $\tilde{a}\,{}^1A_1$. The second spectrum, which appears somewhat later, occurs in the vacuum ultraviolet near 1415 Å and is reproduced in Fig. 185. It has been assigned to a triplet transition ${}^3\Sigma_u^- - {}^3\Sigma_g^-$ of $CH_2$. The lower state $\tilde{X}\,{}^3\Sigma_g^-$ is the ground state of the molecule. The band at 1415 Å is followed by a Rydberg series of similar transitions [Herzberg (522)].

The simple fine structure of the principal band at 1415Å (see Fig. 185) shows that the molecule is linear (or nearly linear) in both upper and lower state. On the other hand, the red (singlet) spectrum (see Fig. 93) shows an alternation of band type similar to $BH_2$, indicating that the molecule is strongly bent in the lower and linear or nearly linear in the upper state. The energy difference between the lowest singlet ($\tilde{a}\,{}^1A_1$) and the lowest triplet ($\tilde{X}\,{}^3\Sigma_g^-$) state is not yet known, but there is good evidence that the singlet is above the triplet state [Herzberg (521)], probably by less than 1 eV.

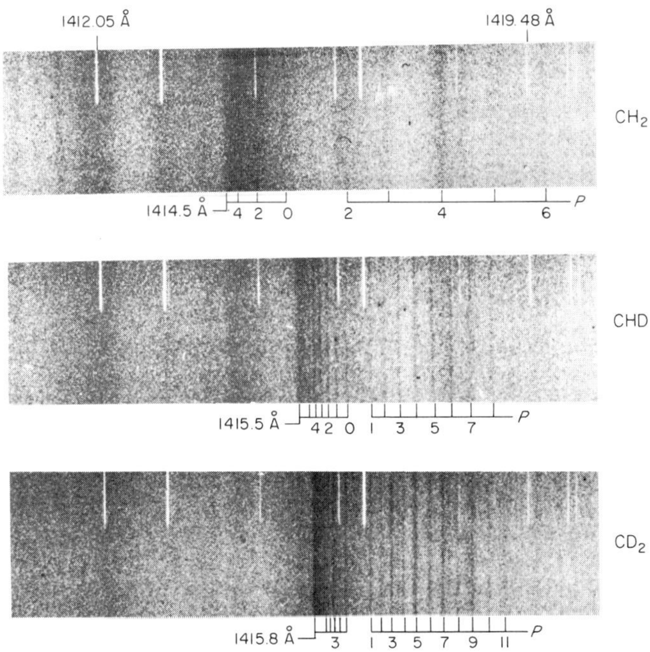


FIG. 185. **Ultraviolet absorption spectrum of $CH_2$, CHD and $CD_2$ in the region 1420–1410 Å after Herzberg (521).** Only a single band of each of the species has been clearly resolved. The lines of the $CH_2$ band are very diffuse and were recognized only after the spectrum of $C^{13}H_2$, which is much sharper, had been observed.

It should perhaps be emphasized that the triplet splitting in the $^3\Sigma_u^- - {}^3\Sigma_g^-$ transition has not yet been resolved, but the indirect evidence that it is a triplet transition appears conclusive.

The low-lying electronic states of $CH_2$ established in this way are in excellent accord with the predictions from molecular orbital theory (see Chap. III, section 2d), and the shape of the molecule in these states is in good agreement with expectation from the Walsh diagram (p. 319) as well as with more detailed valence-bond calculations of Jordan and Longuet–Higgins (649). The two states $\tilde{a}$ and $\tilde{b}$ arise by strong vibronic (Renner–Teller) interaction from the $^1\Delta_g$ state of the linear conformation (see p. 343). The upper state $\tilde{c}\ {}^1A_1$ of the very weak near ultraviolet absorption bands is readily accounted for as due to the electron configuration $\ldots(1b_2)^2(1b_1)^2$. The upper states of the Rydberg series have the electron configurations given in Table 36 [for details see Herzberg (521)]. In all these states the molecule must be linear (or nearly linear) since the transitions have very little vibrational structure.

Figure 186 shows an energy level diagram of all observed electronic states. The limit of the Rydberg series is at 10.396 eV [Herzberg (522)]. It corresponds to the ionization of $CH_2$ to a state of the $CH_2^+$ ion in which it is linear (since $CH_2$ is linear in the Rydberg states). In view of the non-linearity of the ground state of the iso-electronic $BH_2$ it is likely that $CH_2^+$ is non-linear in its ground state and that the ionization potential 10.396 eV corresponds to the first excited state, the analogue of the

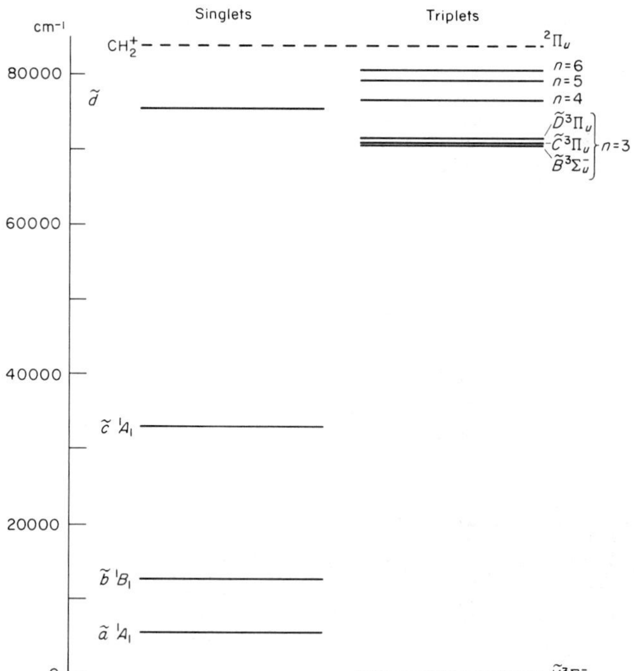


FIG. 186. **Observed electronic states of $CH_2$.** The separation between the triplet and singlet levels is as yet unknown. It seems certain, however, that the lowest triplet state is the ground state of the molecule. The broken line indicates the ionization limit corresponding to the $^2\Pi_u$ state of the $CH_2^+$ ion, assuming it to be linear.

excited state of $BH_2$. In that case the true ionization potential of $CH_2$ would be smaller by an amount of the order of 0.3 eV.

The low-lying $^3\Pi_g$ state of $CH_2$ arising from the configuration $\sigma_u\pi_u{}^3$ has not yet been observed; the transition to it from the ground state is forbidden but might be made allowed by vibronic interaction. Also the $^3\Pi_u$ and $^1\Pi_u$ states arising from $\sigma_u^2\pi_u 3s\sigma_g$ have not been observed.

**$NH_2$.** In $NH_2$ up to now only one electronic transition $^2A_1-{}^2B_1$ has been observed [Herzberg and Ramsay (537)] and analysed in considerable detail [Dressler and Ramsay (308)] both for $NH_2$ and $ND_2$. This transition consists of the so-called $\alpha$ bands of ammonia, known since 1864 in emission in oxy-ammonia flames, but identified as due to $NH_2$ only 12 years ago when they were first observed in absorption in the flash photolysis of $NH_3$.

In the ground state, $\tilde{X}\ {}^2B_1$, the molecule is strongly bent as in the ground state of $H_2O$ while in the excited state $\tilde{A}\ {}^2A_1$ it is nearly linear (see p. 218). Again the two states arise from a single $^2\Pi_u$ state of the linear conformation on account of a strong vibronic (Renner–Teller) interaction. Because of the large change of angle a long progression of bands is observed alternately with even and odd $K$ values (just as for the red bands of $BH_2$ and $CH_2$). The spacing $\Delta_2G$ of the levels with $l = 0$ in the upper state first increases and only toward the end of the progression starts to decrease. The doublet nature of the electronic transition is recognized by small splittings of almost all "lines" (see p. 221). Just as for the red bands of $BH_2$ and $CH_2$ the transition moment of the $NH_2$ bands is perpendicular to the plane of the molecule (type $C$ bands). A number of hot bands have been analysed by Johns and Ramsay (638a) yielding a value for the bending frequency $\nu_2$ in the ground state. The rotational and vibrational constants are summarized in Table 62.

**$SiH_2$ and $PH_2$.** For both $SiH_2$ and $PH_2$ a progression of bands has been observed very similar in character to the red bands of $CH_2$ and the $\alpha$-bands of $NH_2$, respectively, [Herzberg and Verma (546), Ramsay (1042)]. The bands appear to be somewhat more complex than those of $CH_2$ and $NH_2$, but an analysis of one of the $PH_2$ bands has recently been accomplished by Dixon, Duxbury and Ramsay (286b); none of the $SiH_2$ bands have as yet been fully analysed. Both spectra correspond in all probability to a transition between two states which in the linear conformation coincide, forming $^1\Delta$ and $^2\Pi$ states, respectively. For $PH_2$ and $PD_2$ the vibrational intervals of the upper state first decrease normally, but then increase appreciably indicating, as first pointed out by Dixon (285), an appreciable deviation from linearity in the excited state (see p. 121). The analysis of the rotational structure gives an angle of 123.1°, as compared to 91.5° in the ground state. For $SiH_2$, unlike $CH_2$, no evidence for the triplet states has as yet been found. Possibly here the lowest state is a singlet state, viz., the lower state of the observed bands.

## (b) Mono-hydrides

Only two stable triatomic mono-hydrides, HCN and HCP, have been studied in detail, but analyses have been made of the spectra of seven free radicals: HCO, HNO, HPO, HSiCl, HSiBr, HCF and HCCl.

**HCN.** The far ultraviolet absorption spectrum of HCN was first studied by Price (1012) and later by Price and Walsh (1029). They found strong diffuse bands below 1120 Å and somewhat weaker discrete bands between 1550 and 1350 Å. Hilgendorff (549) first observed a system of much weaker bands at longer wavelengths from 2000 to 1700 Å. This system was extended, and the fine structure of the bands was investigated in detail for both HCN and DCN by Herzberg and Innes (527) who also obtained a partial resolution and analysis of the 1550–1350 Å system.

In the excited state of the main system, $\tilde{A}-\tilde{X}$, near 1800 Å, the molecule is strongly bent, unlike the ground state, and this causes long progressions in the bending vibration $\nu_2$ to appear. The bands have a very simple structure since in the ground state the molecule is linear, and in the excited state it is very nearly a symmetric top. For such transitions the selection rule $K' - l'' = 0, \pm 1$ is applicable (see p. 193) and since in the lower state only the 000, 010 and 020 levels (with $l'' = 0, 1, 2$) contribute to the absorption, the number of sub-bands is greatly restricted. The previous Fig. 82 reproduces one of the absorption bands showing a simple $P$, $Q$ and $R$ branch. The large $K$-type doubling in the upper state, particularly for $K' = 1$ (see Fig. 80), confirms the non-linear structure. The sign of the $K$-type doubling, i.e., the fact that the upper levels of the $Q$ lines are the lower $K$-type doubling components, shows unambiguously that the upper state is $^1A''$, that is, the transition moment is perpendicular to the plane of the (bent) molecule. This conclusion is confirmed by the complete absence of sub-bands with $K' - l'' = 0$. Precise values for all three rotational constants $A, B, C$ in the excited state have been determined (see Table 63 of Appendix VI). The strong predissociation observed for the higher vibrational levels of the $\tilde{A}\ ^1A''$ state has already been discussed in Chapter IV.

In DCN for which predissociation is less strong, a fragment of a second system in the same region has been found, with a similar bending frequency $\nu_2'$. Apparently for HCN this system is either much more strongly predissociated or more strongly overlapped by the main system, so that it is not observed. The fine structure of this second system shows that here also the upper state is a $^1A''$ state. The two $^1A''$ states probably arise from the excited $^1\Delta$ and $^1\Sigma^-$ states of the linear conformation derived from the electron configuration $\sigma^2\pi^3\pi$. Thus, for linear HCN the two transitions corresponding to $\tilde{A}-\tilde{X}$ and $\tilde{B}-\tilde{X}$ would be forbidden. It is for this reason

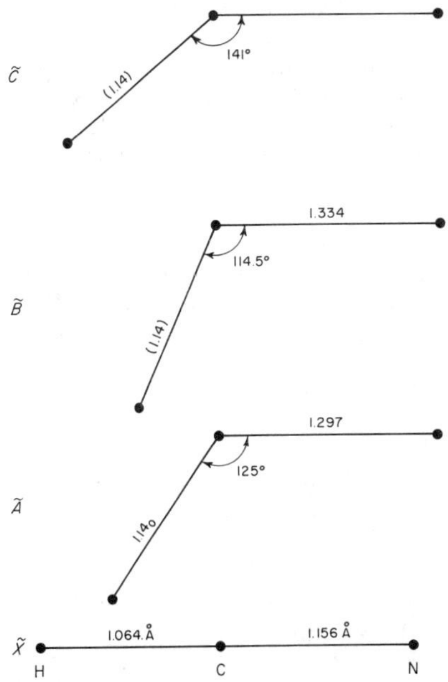


FIG. 187. **Geometrical structure of HCN in the four known electronic states.**

that in both band systems only transitions to upper levels below the potential maximum corresponding to the linear conformation are observed (see Fig. 68) and that the spacings $\Delta G$ in the observed progressions vary so regularly. Toward the end of the $0v_2'0-000$ progressions a slight bending up of the $\Delta G$ curve is noticeable [Johns (635)] in agreement with expectation from the discussion of quasi-linear molecules (Fig. 46).

As shown by Fig. 88 the bands of the strong progressions starting at 1550 Å are diffuse and the diffuseness rapidly increases at shorter wavelengths. For the first members of this transition ($\tilde{C}-\tilde{X}$) the $R$ and $Q$ heads of the $J$ structure are still clearly recognizable and the $K$ structure is well resolved, showing both sub-bands with $K' - l'' = 0$ and $K' - l'' = \pm 1$ (see p. 206f.). Thus the bands are hybrid bands having both $\parallel$ and $\perp$ components and therefore the transition moment is in the plane of the (bent) molecule, that is, the upper state is $^1A'$ (not $^1A''$). From the rotational constants $A$, an approximate value for the H—C—N angle can be obtained (see Table 63). In Fig. 187 the geometrical structures of HCN in the four electronic states discussed here are represented graphically. The $\tilde{C}\ ^1A'$ state is probably derived from the $\sigma\pi^4\pi\ ^1\Pi$ state of the linear conformation. If that is so, the transition $\tilde{C}-\tilde{X}$ would be allowed in the linear conformation. Indeed it is much stronger than the $\tilde{A}-\tilde{X}$ and $\tilde{B}-\tilde{X}$ transitions. Unfortunately, because of the increasing diffuseness the vibrational structure near the "vertical" transition could not be analysed; it should show strong effects of quasi-linearity (see Chap. I, section 3e).

The bands below 1120 Å have not yet been analysed, and no Rydberg series have been observed. Thus, no spectroscopic ionization potential is available.

**HCP.** The analogue of HCN with P replacing N is a compound that has only fairly recently become available. The ultraviolet absorption spectrum of HCP has been studied by Johns and Tyler (639). It is very similar to that of HCN but shifted to longer wavelengths. It begins with a weak extensive absorption system in the region 4100 to 3050 Å. The bands show a simple fine structure with two $P$ and two $R$ branches showing that the transition is $^3\Sigma^+ - {}^1\Sigma^+$. The analogue of this transition in HCN has not yet been found.

The next absorption consists of an extensive system of strong bands with well-resolved fine structure extending from 3000 to 2300 Å. The principal system is the exact analogue of the $\tilde{A}-\tilde{X}$ system of HCN. The molecule is strongly bent in the upper state which is found to be an $A''$ state, that is, the transition moment, as for HCN, is perpendicular to the plane of the molecule. Overlapping the main system, three other systems have been identified. In the upper states of at least one of these systems the molecule is linear. Another of these systems is probably the analogue of the $\tilde{B}-\tilde{X}$ system of HCN. The analogue of the $\tilde{C}-\tilde{X}$ system of HCN has not yet been found. It may lie in the vacuum ultraviolet region which has not yet been studied.

The molecular constants thus far obtained are summarized in Table 63.

**HCO.** For the HCO radical two electronic transitions are known, the hydrocarbon flame bands of Vaidya (1233) in the near ultraviolet and the red bands first observed by Ramsay (1040). In spite of much effort, the hydrocarbon flame bands have not yet been analyzed in detail [see, e.g., Hornbeck and Herman (567) and the recent paper by Vaidya (1235)], but the red bands have been fairly completely analyzed, both for HCO and DCO [Herzberg and Ramsay (538), Johns, Priddle and Ramsay (638)]. The principal bands consist each of a simple $P$, $Q$ and $R$ branch as shown by Fig. 92. The considerable combination defect which is even larger in DCO (see p. 213) shows conclusively that the molecule is strongly bent in the lower state but linear in the upper state of the bands. Assuming a CH distance of 1.08 ± 0.01 Å, one finds for the ground state an angle of 119.5° and a CO distance of 1.20 Å.

Even though the molecule is strongly bent in the ground state, it is still nearly a symmetric top. The presence of strong $Q$ branches in the observed bands shows that they must be classified as sub-bands of $\perp$ bands ($\Delta K = \pm 1$): the transition moment is perpendicular to the figure axis. Since the $Q$ lines are found to originate in the lower components of the $K$ doublets (see Fig. 81), the transition moment must be perpendicular to the plane of the molecule. Thus the electronic transition is either $^2A'' - {}^2A'$ or $^2A' - {}^2A''$. From the electron configuration there can be little doubt that the ground state of HCO is a $^2A'$ state and therefore that the observed transition is $^2A'' - {}^2A'$.

According to Chapter II, section 3a($\beta$), one would have expected each band to consist of a number of sub-bands corresponding to the different $K$ values of the lower and the different $l$ values of the upper state. The reason that only one sub-band is observed has been found to be the result of a strong predissociation which affects all but the sub-level with $l = 0$ of the upper state. Johns, Priddle and Ramsay (638) have given strong evidence that both electronic states of HCO arise by a strong Renner–Teller interaction from a $^2\Pi$ state of the linear conformation similar to the situation in $NH_2$. If the upper state is electronically $^2A''$, its observed sharp levels (with $l = 0$) must be vibronically $^2\Sigma^-$. Since the only state into which the molecule can predissociate at the observed energies is the $^2\Sigma^+$ (or $^2A'$) electronic state arising from $H(^2S) + CO(^1\Sigma^+)$ (which is also the ground state, see Fig. 181) and since this state has no $^2\Sigma^-$ (or $^2A''$) vibronic components it is clear that in the excited state the $^2\Sigma^-$ vibronic levels are not subject to predissociation except at higher $J$ values (see Chap. IV, section 3). $^2\Sigma^-$ vibronic levels occur in an electronic $\Pi$ state only for odd $v_2$ and they form the upper states of the sharp absorption bands. Halfway between the sharp bands diffuse bands are observed corresponding to even $v_2'$, that is odd $K'$. These levels are strongly predissociated. Sub-bands with $K' = 2$ which should be close to the sharp ($K' = 0$) bands are so strongly predissociated that they are not easily recognized; they have however been detected by Johns, Priddle and Ramsay by careful photometry of the background near the sharp bands.

The hydrocarbon flame bands have probably the same lower state as the red bands of HCO; but up to now no confirmation of this assumption by agreement of vibrational (or rotational) combination differences has been obtained.

**HNO.** The absorption spectrum of HNO (and DNO) was first observed in the flash photolysis of nitromethane and related compounds as well as of mixtures of $NH_3$ and NO by Dalby (264) who analyzed three bands of each isotope. More recently Bancroft, Hollas and Ramsay (93) have greatly extended the analysis. The spectrum consists of a strong band with numerous sub-bands near 7500 Å and a number of weaker bands of similar structure at shorter wavelengths extending to 5500 Å. The corresponding emission spectrum was first observed in the reaction of H atoms with NO by Cashion and Polanyi (183) and studied in detail by Clement and Ramsay (204). It extends from 6000 to 9600 Å. Figure 109 shows a part of the main band in absorption. The fine structure of the sub-bands is easily analyzed. $K$-type doubling is readily recognized in many of the sub-bands (see Fig. 109). The band is a typical $\perp$ band of a nearly symmetric top, that is, the molecule is bent in both upper and lower state. The rotational constants have been accurately determined and are listed together with the geometrical parameters derived from them in Table 63. The H—N—O angle is 108.6° in the ground state and 116.3° in the excited state.

As in HCO, the transition moment of the $\tilde{A} - \tilde{X}$ bands is perpendicular to the plane of the molecule, that is, the transition is $^1A'' - {}^1A'$, or $^1A' - {}^1A''$. Since from the electron configuration we expect the ground state to be $^1A'$, the upper state must be $^1A''$. For linear HNO, the ground electronic state would have the electron configuration $\ldots \sigma^2\pi^4\pi^2$ (like $O_2$) giving rise to the states $^3\Sigma^-$, $^1\Delta$, $^1\Sigma^+$ of which $^3\Sigma^-$

would be lowest. However, for bent HNO according to the Walsh diagram [see Walsh (1265)] the lowest state would probably be $\ldots (a')^2(a'')^2(a')^2\ {}^1A'$ which correlates with the $^1\Delta$ state of the linear conformation. The first excited singlet state would be expected to be $\ldots (a')^2(a'')^2(a')(a'')\ {}^1A''$ which also correlates with $^1\Delta$. Thus the two observed electronic states are in all probability derived from the $^1\Delta$ state of the linear conformation. The $^3\Sigma^-$ state or its analogue in the bent conformation has not yet been found. It must lie very low and might even be the true ground state.

Clement and Ramsay (204) have observed a breaking-off of the $K$ structure in the $000-000$ and $010-000$ emission bands of the $\tilde{A}-\tilde{X}$ system, and Bancroft, Hollas and Ramsay (93) have observed a diffuseness in the $101-000$ absorption band. The predissociation limit lies at 17,000 cm$^{-1}$ for HNO and 17,190 cm$^{-1}$ for DNO. These numbers represent upper limits to the corresponding dissociation energies into H + NO and D + NO respectively. The state causing the predissociation is probably the $^3A''$ state arising from normal H + NO, and correlating with $^3\Sigma^-$ in the linear conformation. This point as well as the mechanism of the emission of the HNO bands in the H + NO reaction has been discussed in more detail by Clyne and Thrush (213).

No other spectra associated with HNO have yet been found. Since the lifetime of HNO is about 0.1 sec, a further study of its ultraviolet spectrum would not be difficult.

**HPO.** A spectrum in the region 4500–6500 Å long known to occur in the flame of phosphorus burning in hydrogen [see Ludlam (782)] has recently been studied under improved conditions and at high resolution by Lam Thanh and Peyron (716). They have shown that this spectrum is the exact analogue in HPO of the HNO spectrum discussed above. It also represents a $^1A''-{}^1A'$ transition and the molecule is strongly bent in both electronic states. The molecular constants obtained by Lam Thanh and Peyron are included in Table 63.

**HSiCl and HSiBr.** In the flash photolysis of $SiH_3Cl$ and $SiH_3Br$, Herzberg and Verma (545) observed two extensive systems of absorption bands in the region 4000–5000 Å consisting each of widely spaced branches of "lines". Closer inspection shows that the "lines" are actually $Q$ branches of sub-bands each of which is accompanied by $P$ and $R$ branches. The widely spaced $K$ structure shows that the molecules are nearly symmetric tops with one very small moment of inertia in both upper and lower state. The deuterium isotope effect as well as the absence of an intensity alternation show that only one H atom is present in the molecule. Similarly the Cl and Br isotope shifts show that one Cl and one Br respectively are present in the molecules responsible for the two systems. These observations together with the detailed fine structure analysis leave no doubt that we are dealing with the spectra of (non-linear) HSiCl and HSiBr. The HSiX angle in the ground state is 103°, in the excited state 116°. From the shading of the unresolved $Q$ branches with $K' = 1$ and $K'' = 1$ it follows (see p. 253f.) that the transitions are of the type $A''-A'$.

The same band systems have also been observed in emission, in the fluorescence produced by the same flash that photolyses the parent compound. These fluorescence spectra extend to 6000 Å and yield some of the vibrational levels of the ground states. The resulting vibrational frequencies of both states as well as the rotational constants and geometrical data are included in Table 63.

An interesting feature of the HSiCl and HSiBr bands is the presence of branches with $\Delta K = \pm 2$ in addition to those with $\Delta K = \pm 1$ and 0. The appearance of these branches cannot be ascribed to deviations from the symmetric top since these deviations are extremely small ($b = -0.00052$ for HSiCl) and much more asymmetric molecules do not show any evidence of such branches in agreement with their predicted weakness. Herzberg and Verma (545) and Hougen (574) have suggested that the reason for this anomaly is spin-orbit interaction, in other words, that the electronic transition is $^3A''-{}^1A'$ (see p. 268). However, the absence of an observable triplet splitting is difficult to account for. An alternative explanation would be "axis switching" (see p. 208).

**HCF and HCCl.** Merer and Travis (825)(829) have recently obtained extensive spectra of HCF and HCCl in the flash photolysis of $CHFBr_2$ and $CHClBr_2$, respectively. The observed bands are clearly those of a nearly symmetric top, that is, just as for the iso-electronic HNO and HPO, the molecules are bent in upper and lower states. The electronic transition is of the type $^1A''-{}^1A'$. Preliminary values of the molecular constants are included in Table 63.

## (c) Non-hydrides

Electronic spectra of a large number of triatomic non-hydrides are known. We shall consider here only the more important ones, but include several others in Table 64.

**$C_3$.** The spectrum of $C_3$ near 4050 Å was first observed in comets by Huggins (586) and in the laboratory by Herzberg (519); it was first identified as belonging to $C_3$ by Douglas (292). The $0-0$ band of the system at 4050 Å is reproduced in Fig. 72; it consists of simple $P$, $Q$ and $R$ branches with alternate missing lines [with $C_3^{13}$ an intensity alternation is found: Clusius and Douglas (211)]. Recent work in absorption under very high resolution has established that this system is a $^1\Pi_u-{}^1\Sigma_g^+$ transition and that the molecule is linear in upper and lower state [Gausset, Herzberg, Lagerqvist and Rosen (410)(411)]. Even in absorption at low temperature the vibrational structure of the spectrum is fairly complex. The reason for this complexity has been found to lie in the fact that (1) the bending frequency of the ground state is extremely small ($\nu_2 = 63$ cm$^{-1}$) and (2) that the Renner–Teller splittings in the excited state are very large ($\epsilon = 0.537$). Thus even at room temperature $\nu_2''$ is excited by several quanta and the $1-1$, $2-2, \ldots$ bands in $\nu_2$ consist of several strong and widely spaced sub-bands (see p. 158f.).

There is some evidence for a $^3\Pi_u-{}^1\Sigma_g^+$ transition at longer wavelengths, and several perturbations in the main system [see Gausset *et al.* (411)] are probably caused by the $^3\Pi_u$ state.

As previously discussed in connection with Table 37, the states $^1\Sigma_g^+$, $^1\Pi_u$ and $^3\Pi_u$ are expected from the electron configurations as ground and first excited states. Possibly one or two of the states of the configuration

$$(4\sigma_g)^2\ (1\pi_u)^3(3\sigma_u)^2\ 1\pi_g\ \ {}^3\Sigma_u^-,\ {}^1\Sigma_u^-,\ {}^3\Sigma_u^+,\ {}^1\Sigma_u^+,\ {}^3\Delta_u,\ {}^1\Delta_u,$$

lie in the energy range of the $^1\Pi_u$ state but have not as yet been identified. A continuum with a peak at 4000 Å which occurs in conjunction with the 4050 group in emission in a King furnace [Phillips and Brewer (978)] has more recently been interpreted by Brewer and Engelke (148) as due to the large number of unresolved vibrational bands which must be expected in the $^1\Pi-{}^1\Sigma$ system at high temperature. The spectrum of $C_3$ at liquid helium temperature in inert gas matrices has been studied by Barger and Broida (95) and Weltner and his collaborators (1288)(1289)(1287).

**CCN and CNC.** Merer and Travis (826) have recently observed in the flash photolysis of diazoacetonitrile a spectrum in the region 4710 to 3480 Å, which they have shown to be due to the $C_2N$ molecule. Three electronic transitions, two $^2\Sigma-{}^2\Pi$ and one $^2\Delta - {}^2\Pi$ have been identified. The molecule is linear in both upper and lower states. No intensity alternation occurs in the branches showing that the molecule must have the unsymmetrical structure CCN. The agreement of the observed electronic states with those predicted from the electron configuration (Table 37), as well as the isotope shifts observed when an $N^{15}$ substituted parent compound was used, confirm the assignment of this spectrum to the CCN molecule. Renner–Teller splittings have been found for both the $^2\Pi$ ground state and the excited $^2\Delta$ state. Still more recently Merer and Travis (828) have identified two band systems of the isomeric CNC molecule in the region 3320 to 2830 Å. Alternate lines are clearly missing in the branches of these bands, thus establishing the symmetrical structure. The molecular constants of both isomers are included in Table 64.

**NCN.** A strong group of bands near 3285 Å was observed in emission by Jennings and Linnett (627) by introducing various hydrocarbons into a stream of active nitrogen and in absorption in the flash photolysis of diazomethane by Herzberg (521). Isotope investigations and the observed intensity alternation in the fine structure show that the carrier of this spectrum is symmetrical linear NCN [Herzberg and Travis (543)]. The fine structure analysis shows it to be a $^3\Pi_u - {}^3\Sigma_g^-$ transition in agreement with prediction (see Table 37). The C—N distance is so nearly the same in upper and lower state that only sequences have been observed and no vibrational frequencies could be determined. The spectrum exhibits the first clear example of a Renner–Teller splitting in a triplet state ($^3\Pi_u$). Milligan, Jacox and Bass (851a) have recently observed the infrared spectrum of NCN in a matrix at low temperature, and in addition a further ultraviolet absorption system in the region 3000 to 2400 Å. Quite recently Kroto (699aa) has observed a singlet transition ($^1\Pi_u$ – $^1\Delta_g$) of NCN.

**$BO_2$, $CO_2^+$, $CS_2^+$, $N_3$, NCO, $N_2O^+$.** For the symmetric molecules with 15 valence electrons $BO_2$, $CO_2^+$, $CS_2^+$, $N_3$, one or both of the predicted transitions (Table 37) $^2\Pi_u - {}^2\Pi_g$ and $^2\Sigma_u^+ - {}^2\Pi_g$ have been observed in the visible and near ultraviolet regions (see Fig. 78, p. 192). The linear and symmetrical structure of these molecules follows unambiguously from the fact that alternate lines in the branches are missing and from the absence of *K* structure. Similarly the unsymmetrical molecules NCO and $N_2O^+$ exhibit the corresponding transitions $^2\Pi - {}^2\Pi$ and $^2\Sigma^+ - {}^2\Pi$ (see Fig. 74 and 75). All observed $^2\Pi$ states show a Renner–Teller splitting whenever the bending vibration $\nu_2$ is excited. The magnitude of this splitting is largest in the ground state, the Renner parameter $\epsilon$ being about 0.20, while in the excited $^2\Pi$ state it is much smaller. The most detailed analyses are those of $BO_2$ [Johns (630)], NCO [Dixon (281) (282)] and $N_2O^+$ [Callomon and Creutzberg (174)]. As usual for linear molecules, the spin-doubling is well resolved. The geometrical structure is completely determined in the symmetrical molecules. For the unsymmetrical molecules isotope studies would be required to obtain the two internuclear distances separately. Such studies have so far been carried out only in $N_2O^+$ (174).

It is interesting to compare the bending frequencies $\nu_2$ in the ground states of these molecules with those in corresponding molecules with 16 valence electrons on the one hand and those with 12, 13 and 14 valence electrons on the other. As the number of valence electrons increases from 12 to 16 the bending frequency increases from a very small value ($C_3$: 63 cm$^{-1}$) to a value ten times as large ($CO_2$: 664 cm$^{-1}$). It appears that it is the number of electrons that are in the $1\pi_g$ orbital that determines the stability with respect to bending in accordance with the Walsh diagram Fig. 126 [see also Walsh (1269)]. The molecules with 12, 13, . . ., 16 valence electrons have 0, 1, 2, 3 and 4 electrons in this orbital (see Table 37). In all the low excited states, the bending frequency is higher than in the ground state since more $\pi_g$ electrons are present.

A diffuseness has been observed in the $^2\Pi - {}^2\Pi$ bands of NCO above 33700 cm$^{-1}$

[Dixon (282)]. This predissociation limit cannot correspond to a dissociation into $N(^4S) + CO(^1\Sigma^+)$ because the diffuseness can hardly be caused by a transition with $\Delta S = 1$. It must either correspond to dissociation into $N(^2D) + CO(^1\Sigma)$ or into $NC(^2\Sigma) + O(^3P)$ leading to an upper limit of 1.8 eV for $D$(N—CO) or 4.2 eV for $D$(NC—O).

All the absorption bands of $N_3$ except the $0-0$ band are diffuse, implying $D(N—N_2) < 4.55$ eV (again assuming that the predissociation cannot be an intercombination).

Only transitions involving the orbitals $1\pi_u$, $3\sigma_u$ and $1\pi_g$ have been observed for the molecules here considered but a more highly excited $^2\Sigma_g^+$ state of $CO_2^+$ lying 45320 cm$^{-1}$ above the $^2\Pi_{\frac{3}{2}}$ ground state has been derived from Rydberg series of $CO_2$ by Tanaka, Jursa and LeBlanc (1189). It probably corresponds to removal of a $4\sigma_g$ electron from $CO_2$ (see Table 37).

**$CO_2$.** Carbon dioxide does not show any absorption in the visible and near ultraviolet regions. Lyman (785) and Leifson (736) were the first to observe the vacuum ultraviolet absorption spectrum. This early work was extended to shorter wavelengths by Henning (494) and Rathenau (1058). More recently, Wilkinson and Johnston (1302), Inn, Watanabe and Zelikoff (604) have reinvestigated the longward part of the vacuum ultraviolet absorption spectrum while Tanaka, Jursa and LeBlanc (1189) and Tanaka and Ogawa (1191) have reinvestigated and extended the shortward part.

In emission in discharges through $CO_2$ the spectrum of $CO_2^+$ is most prominent (see p. 499), but in addition, under special conditions, for example, in a high voltage arc [Feast (373)], a group of bands is observed between 5000 and 3500 Å which also appear in emission in the carbon monoxide flame. They are usually referred to as the carbon monoxide flame bands [Weston (1291), Kondratiev (688), Fowler and Gaydon (394), Gaydon (413)]. These bands were first ascribed to $CO_2$ by Kondratiev (688) and while this assignment has been generally accepted following Gaydon's (413) work a detailed analysis and confirmation was lacking until the recent work of Dixon (284)(283). Dixon obtained high resolution spectra of the afterglow of a discharge through $CO_2$ which gives some simplification to the spectrum because of a low rotational temperature. It is probable that the upper state of the flame bands is the same as the upper state of the diffuse absorption bands in the region 1750–1400 Å. Because in the excited state the molecule is strongly bent (122° ± 2°) transitions from the minimum of the upper state lead to very highly excited vibrational levels of the ground state and occur in the visible and near ultraviolet regions. The coarse and fine structure of the bands fit with the predicted higher vibrational levels of the ground state. It is greatly affected by Fermi resonances between these vibrational levels. The excited state has been identified as a $^1B_2$ state derived from a $^1\Delta_u$ state of the linear conformation (see Table 37).

In absorption $CO_2$ is completely transparent at least to 2100 Å. This transparency has been confirmed by experiments in the liquid phase by Eiseman and Harris (347) as well as by unpublished investigations of the writer with a path of up to 250 m of $CO_2$ at atmospheric pressure.

The first known absorption of $CO_2$ starts very gradually at about 1750 Å, reaching a maximum at 1475 Å. It consists of rather diffuse bands which are overlapped by a continuum [see Price and Simpson (1021)]. The intensities and spacings of the bands are quite irregular. Figure 188a gives a recording of this region as observed by Inn, Watanabe and Zelikoff (604). The second absorption region is represented in Fig. 188b. Here the bands are somewhat sharper and more regular and have an intensity maximum near 1330 Å. There are two progressions in the frequency 1225 cm$^{-1}$, but the bands are too diffuse for a detailed analysis.

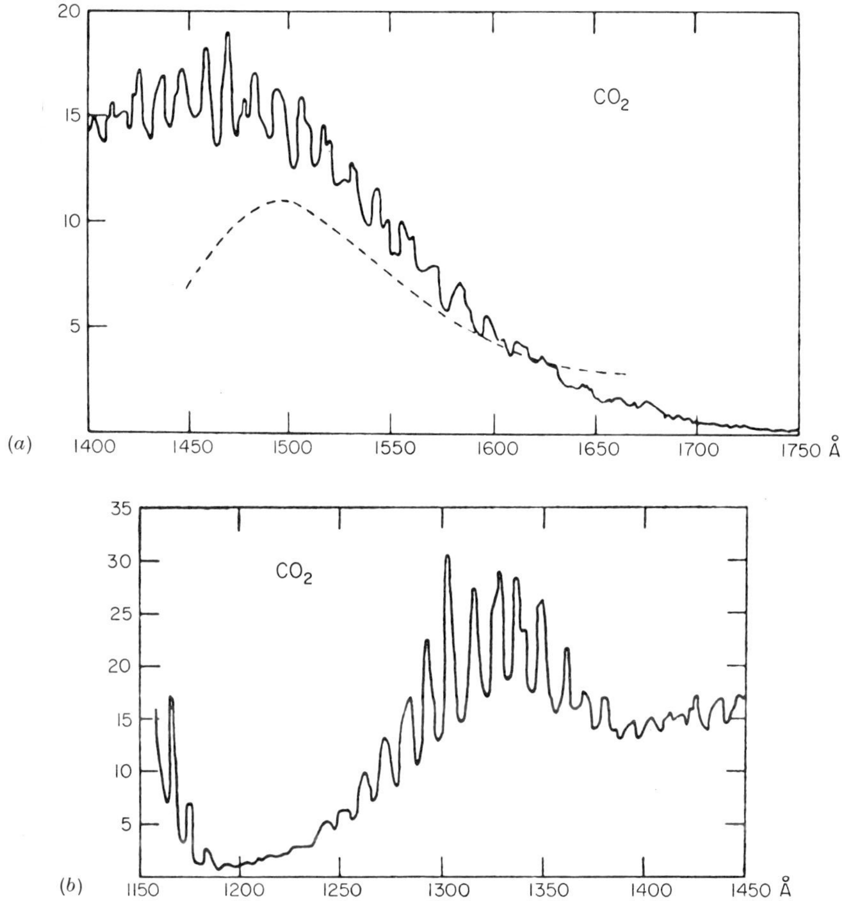


FIG. 188. **Absorption curves of $CO_2$ in the regions** (*a*) **1750 to 1400 Å and** (*b*) **1450 to 1150 Å, after Inn, Watanabe and Zelikoff (604).** The ordinate represents the absorption coefficient in $cm^{-1}$.

According to Walsh (1264) and Mulliken (914), the upper state of the first absorption system (which is probably also the upper state of the carbon monoxide flame bands) corresponds in all probability to the configuration $\ldots 1\pi_g^3 2\pi_u$ of the linear conformation; but in fact the molecule is bent in this state in accordance with the Walsh diagram Fig. 126. Here it is interesting to note that the ground state of $NO_2$ would in the linear conformation have the electron configuration $\ldots 1\pi_g^4 2\pi_u$, and again because of the presence of the $2\pi_u$ electron the molecule is actually bent. The second absorption system of $CO_2$ according to Mulliken corresponds to an excitation from the $1\pi_u$ orbital to $2\pi_u$. Such a transition would be strongly forbidden in linear $CO_2$ but may well occur weakly if the molecule is bent in the excited state.

Below 1150 Å there are extremely strong absorption bands most of which belong to four Rydberg series going to the first ionization limit of $CO_2$ (see Table 64). The two main Rydberg series with slightly different limits (111060 and 111240 $cm^{-1}$) correspond to the two components of the $^2\Pi_g$ ground state of $CO_2^+$ which according to the analysis of the $CO_2^+$ spectrum are separated by 160 cm$^{-1}$. The smaller of the two Rydberg limits gives an ionization potential of 13.769 eV. An ionization continuum joining onto the Rydberg limit has been observed by Tanaka, Jursa and LeBlanc (1189).

Even though the rotational fine structure has not been resolved for any of the Rydberg bands, it is clear that in all the upper states the molecule is linear since $CO_2{}^+$ is linear (see above). This conclusion is also in accord with the absence of long progressions in the bending vibration. No definite assignment of the observed Rydberg series to specific electron configurations has been made. Several progressions in $\nu_1$ not belonging to the Rydberg series but occurring in the same spectral region have been recognized. Further work is needed to establish their interpretation.

Below 899 Å, the first ionization limit of $CO_2$, further strong Rydberg series have been found by Henning (494), Tanaka, Jursa and LeBlanc (1189), and under much improved resolution by Tanaka and Ogawa (1191). We reproduce in Fig. 189 one of their spectra which shows most of these Rydberg series. One series with a fairly long progression in $\nu_1$ for each $n$ value converges to the first excited state $^2\Pi_u$ of $CO_2^+$ and thus corresponds to the removal of the $1\pi_u$ electron from the ground configuration of

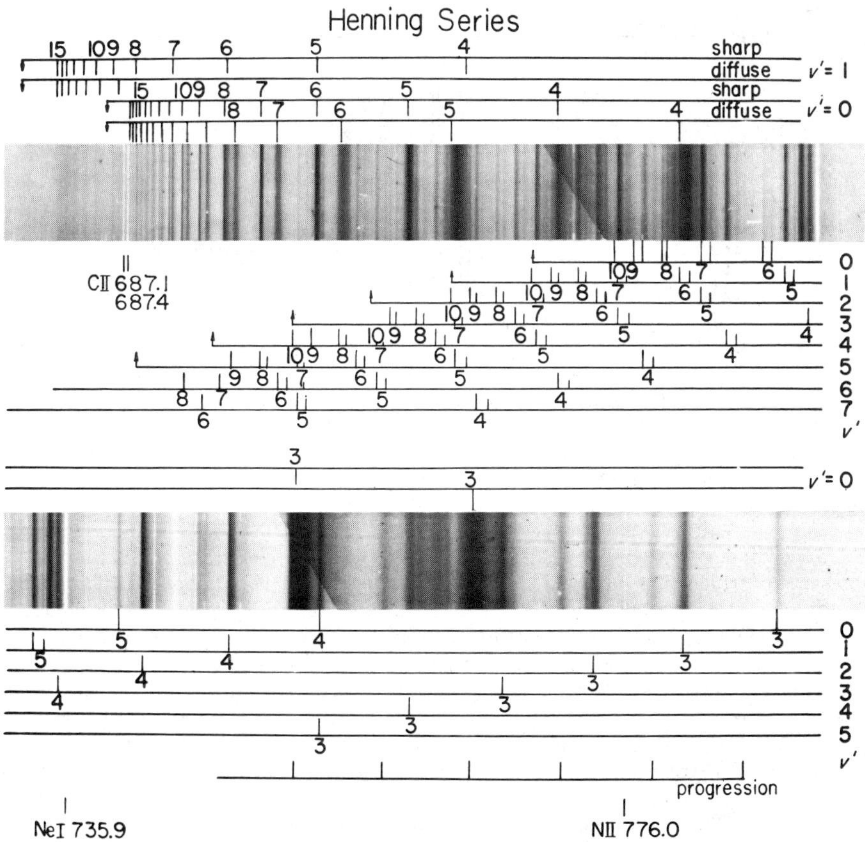


FIG. 189. **Absorption spectrum of $CO_2$ in the region 790 to 670 Å after Tanaka and Ogawa (1191).** At the top of each strip the bands of the Rydberg series converging to the $^2\Sigma_u^+$ state of $CO_2^+$ are indicated, at the bottom those converging to the $^2\Pi_u$ state.

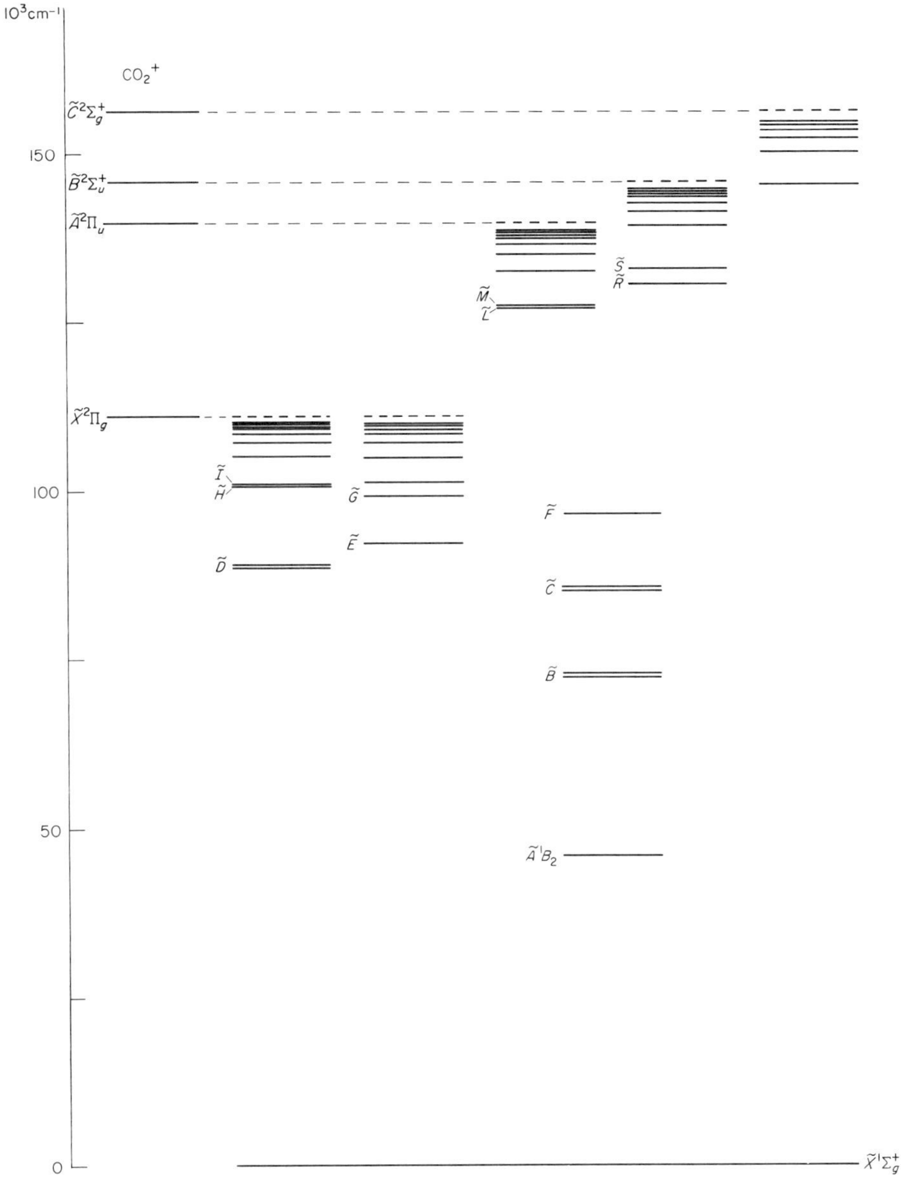


FIG. 190. **Observed electronic states of $CO_2$.** The levels are grouped according to the Rydberg series to which they belong, except the levels, $\tilde{A}$, $\tilde{B}$, $\tilde{C}$ and $\tilde{F}$, which, like the ground state $\tilde{X}$, cannot be assigned to any particular series. The three series going to the highest level of $CO_2^+$ are not separated in the diagram. Except for the lowest states of these series, they would coincide on the scale of this diagram.

$CO_2$ (see Table 37). The higher series members are split into two components about 90 $cm^{-1}$ apart, corresponding to the two components $^2\Pi_{\frac{3}{2}u}$ and $^2\Pi_{\frac{1}{2}u}$ of $^2\Pi_u(CO_2^+)$. The vibrational intervals converge within the accuracy of the measurements to those of $CO_2^+$ known from the near ultraviolet emission spectrum (see Table 64). The two series first found by Henning (see Fig. 189) converge to 685.9 Å, corresponding to the second excited state $^2\Sigma_u^+$ of $CO_2^+$. Bands with $v_1' = 1$ have also been observed. Finally, Tanaka, Jursa and LeBlanc (1189) have observed three Rydberg series below 686 Å which converge to 639.4 Å, corresponding to the predicted (but not yet directly observed) $^2\Sigma_g^+$ state of $CO_2^+$. The last two limits ($^2\Sigma_u^+$ and $^2\Sigma_g^+$) correspond to removal of an electron from the $3\sigma_u$ and $4\sigma_g$ orbital respectively.

The absorption spectrum of $CO_2$ at still smaller wavelengths, down to 160 Å, has been studied under low resolution by Astoin, Sanson and Bonnelle (80). They observe three distinct continua which they ascribe to various dissociative ionization processes. Sun and Weissler (1178) have measured absorption coefficients between 1300 and 375 Å. The strongest absorption occurs at 923 Å, with an absorption coefficient $k = 3000\ cm^{-1}$.

In Fig. 190 we present a diagram of the electronic energy levels of $CO_2$ based on the data just discussed (see also Table 64).

**$CS_2$.** Carbon disulfide with the same number of valence electrons as $CO_2$ shows a very characteristic absorption in the region 4000–3500 Å which extends to shorter and longer wavelengths when the pressure is increased. Even under medium resolution it appears like a line spectrum. It was first recognized by Jenkins (626) that each "line" is a head of a band with a simple fine structure. The first detailed rotational analysis of several bands was accomplished by Liebermann (746). In spite of the linearity of the molecule in the ground state, the vibrational analysis has presented considerable difficulties, but recently Kleman (680) has been able to obtain a fairly complete analysis of the longward part of the spectrum on the basis of high dispersion spectra which allowed a decision on the band type of each vibrational transition. A considerable extension of Liebermann's rotational analysis has recently been accomplished by Douglas and Milton (298). In the previous Fig. 84 several bands of this band system were reproduced which show clearly the effects of $K$-type doubling and therefore establish the fact that the molecule is bent in the excited state.

As one approaches the center of the absorption region, strong perturbations in the vibrational structure begin to appear which soon make further analysis impossible. It is likely that a second electronic transition is overlapping the first, and is the cause for the high degree of complication of the shortward end of the spectrum.

Mulliken (892)(914) has suggested that the upper states of the $CS_2$ bands must be analogues of those of the $CO_2$ absorption bands near 1475 and 1330 Å (see above), that is, are probably $^1B_2$ and $^1A_2$ derived from the configuration $\pi_g^3\pi_u$ of the linear conformation. The rotational structure of the first system is indeed compatible with a $^1B_2$ upper state. However, Douglas (293) has shown that the $CS_2$ bands under discussion show a noticeable Zeeman splitting, indicating that the excited state has a magnetic moment and therefore strongly suggesting that it is a triplet state. But no triplet splitting was found in any of the absorption bands.

More recent work by Douglas and Milton (298) has shown that the magnetic moment in the excited state, instead of being independent of $J$ as would be expected for a pure case (b), actually increases approximately linearly with $J$. Such an increase is expected for the $B_2$ component of a $^3A_2$ state if it is near case (c) (see Chap. II, section 4). The weakness of the other two triplet components ($A_1 + B_1$) can be understood by a more detailed theoretical discussion [Hougen (576)]. The $^3A_2$ excited state probably arises either from the $\ldots \pi_g^3\pi_u\ ^3\Delta_u$ or the $\pi_g^3\pi_u\ ^3\Sigma_u^-$ state of the linear conformation.

A second much stronger absorption of $CS_2$ occurs in the region from 2200 to 1800 Å with a maximum at about 1970 Å [Price and Simpson (1020), Hauptman (483), Ramasastry and Rao (1038)]. Recently Douglas and Zanon (302) have obtained a partial rotational analysis of a few bands in this system and have shown that, as for the near ultraviolet system, the molecule is strongly bent in the upper state and that the transition moment is parallel to the $a$-axis, i.e., that the upper state is $^1B_2$. The vibrational structure is still only very incompletely understood.

Between the 1970 Å system and the beginning of the strong Rydberg series at 1375 Å, the following features are observed: a single strong diffuse band with two intensity maxima at 1815 Å, a weak progression in a frequency 830 cm$^{-1}$, a very condensed group of sharp bands, probably a sequence $(0-0, 1-1, \ldots)$ in the bending vibration, at 1595 Å, and another fairly intense and complicated system between 1535 and 1450 Å. The two narrow groups very probably correspond to upper electronic states in which the molecule has very nearly the same structure as in the ground state, just as for the upper states of the Rydberg bands. More detailed investigation of these transitions under high resolution would be interesting.

At 1375 Å begin two strong Rydberg series which converge to 81299 and 81735 cm$^{-1}$ corresponding to the two components of the $^2\Pi_g$ ground state of $CS_2^+$. These series, first observed by Price and Simpson (1020), have recently been extended by Tanaka, Jursa and LeBlanc (1190). The separation of the two limits is in good agreement with the splitting found by Callomon (172) in the emission spectrum of $CS_2^+$, viz., 440 cm$^{-1}$. A number of further strong Rydberg series have been found at still shorter wavelengths by Tanaka, Jursa and LeBlanc (1190). Two of them converge to 116760 cm$^{-1}$, three others to 130600 cm$^{-1}$ and one fairly weak one to 157390 cm$^{-1}$. The three limits correspond to the three excited states $^2\Sigma_u^+$, $^2\Pi_u$, $^2\Sigma_g^+$ of $CS_2^+$ of which, however, up to now only the first has been established in detail [Callomon (172)]. The agreement of the energy difference of the two limits 116760 and 81299 cm$^{-1}$ with the energy of the (only known) excited state ($^2\Sigma_u^+$) of $CS_2^+$ is again very satisfactory.

**$N_2O$.** In contrast to carbon dioxide, nitrous oxide (which has the same number of electrons) does absorb in the near ultraviolet. Using an absorbing path of 33 m and a pressure of 5 atm., Sponer and Bonner (1150) observed a continuous absorption starting at 3065 Å with a flat maximum near 2900 Å. It is followed by a second and third continuum starting at 2820 and at 2600 Å, and with maxima at 2730 and 1820 Å, respectively. The tail of the third continuum near 2600 Å was first observed by Dutta (331) with an absorbing path of 1 m at 1 atm. pressure. There is a break at 2300 Å, after which there is a rapid rise so that the maximum at 1820 Å can be observed with a path of 5 cm and less than $\frac{1}{4}$ atm. pressure [Zelikoff, Watanabe and Inn (1330)]. Weak very diffuse bands are superimposed on this continuum. Both Sponer and Bonner, and Zelikoff, Watanabe and Inn have discussed the correlation of these three continua with the various dissociation products $N_2(^1\Sigma) + O(^3P)$, $N_2(^1\Sigma) + O(^1D)$, $NO(^2\Pi) + N(^4S)$, $N_2(^1\Sigma) + O(^1S), \ldots$. Zelikoff and Aschenbrand (1327)(1328) have eliminated some possible reactions by photochemical experiments at 1849, 1470 and 1236 Å.

The first discrete region of absorption of $N_2O$ occurs between 1600 and 1400 Å [Duncan (320), Zelikoff, Watanabe and Inn (1330)]. There is a progression of fairly rapidly converging diffuse bands with a spacing of 800 cm$^{-1}$ at the longward end and 400 cm$^{-1}$ at the shortward end. Figure 191a gives a photometer curve of this region.

The absorption just mentioned is followed by another continuum, representing the strongest absorption feature above 1000 Å. It is shown in Fig. 191b. The symmetrical shape of this absorption is remarkable. There is an additional narrow peak near the maximum. It is not clear whether the continuum or the peak is the first member of a Rydberg series. Beyond 1200 Å begin two Rydberg series of absorption bands which

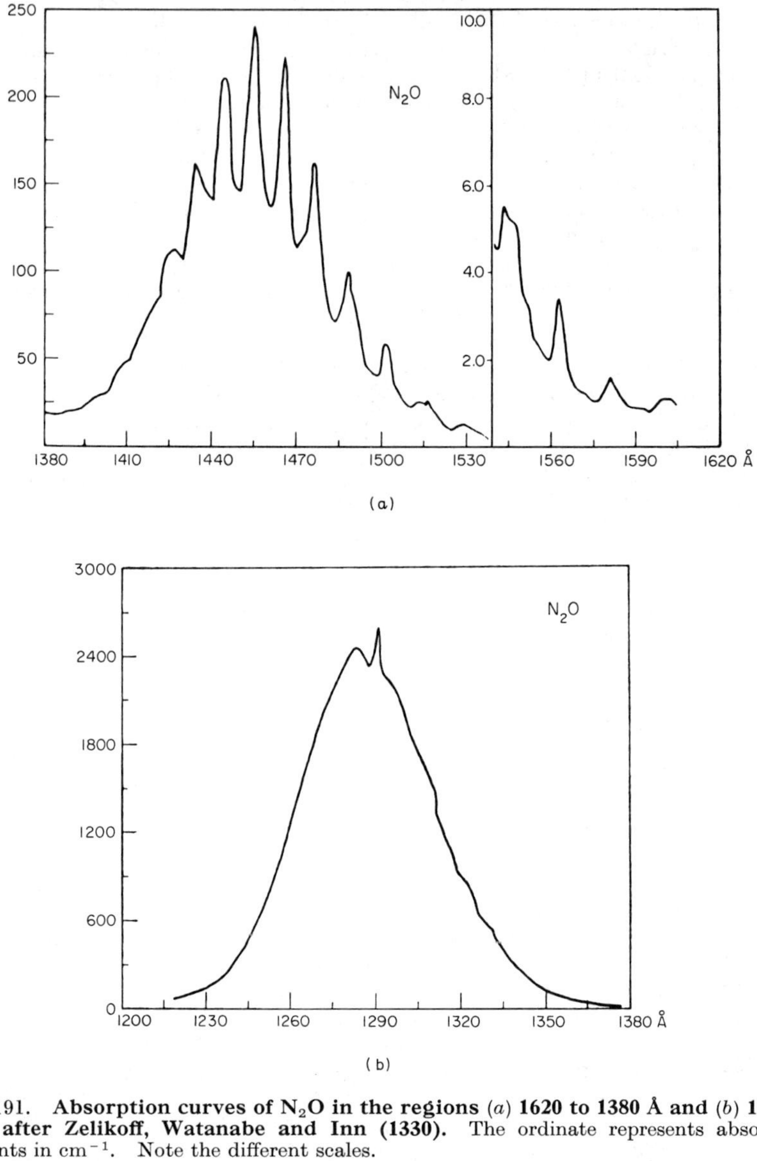


FIG. 191. **Absorption curves of $N_2O$ in the regions** (*a*) **1620 to 1380 Å and** (*b*) **1380 to 1200 Å after Zelikoff, Watanabe and Inn (1330).** The ordinate represents absorption coefficients in $cm^{-1}$. Note the different scales.

converge to 104000 and 104300 $cm^{-1}$ corresponding to the two components of the $^2\Pi$ ground state of $N_2O^+$ [Tanaka, Jursa and LeBlanc (1190)]. The splitting of the $^2\Pi$ state as obtained from the $N_2O^+$ spectrum by Callomon (173) is 133.1 $cm^{-1}$. The poor agreement is due to the fact that the second Rydberg series could not be followed to higher members. The formulae for the two series and the resulting ionization potential are given in Table 64. A few irregular bands are superimposed on the first members of the Rydberg series [see Zelikoff, Watanabe and Inn (1330)]. Whether they belong to them or represent other electronic transitions is not clear.

At wavelengths shorter than the first Rydberg limit Tanaka, Jursa and LeBlanc (1190) have observed two further Rydberg series converging to 132230 $cm^{-1}$, and at still shorter wavelengths, four Rydberg series converging to 162165 $cm^{-1}$. The difference of the limit 132230 $cm^{-1}$ from the lowest limit is 28230 $cm^{-1}$, which agrees remarkably well with the energy of the $^2\Sigma^+$ state of $N_2O^+$ above the ground state for which Callomon gives 28229.8 $cm^{-1}$. The limit 132230 $cm^{-1}$ corresponds, therefore, to the removal of a $7\sigma$ electron from the ground state configuration of $N_2O$, while the limit 162165 $cm^{-1}$ presumably corresponds to the removal of an electron from the next lower orbital. A number of vibrational bands have been found to accompany the Rydberg series with the intermediate limit. This is not surprising since the change of $B$ value in going from the ground state of $N_2O$ to the $^2\Sigma^+$ state of $N_2O^+$ is larger than in going to the $^2\Pi$ ground state of $N_2O^+$. There is, however, no question that in all Rydberg states the molecule is linear since the $N_2O^+$ ion is linear in both known states and very probably also in the state corresponding to the highest observed Rydberg limit of $N_2O$.

Astoin (79) has extended the absorption spectrum of $N_2O$ as far as 160 Å but with very low resolution. There appear to be strong discrete bands between 200000 and 375000 $cm^{-1}$, and a continuum between 400000 and 600000 $cm^{-1}$.

**$NO_2$.** Nitrogen dioxide with one more electron than $CO_2$ (or $N_2O$) does show absorption in the visible region causing its orange color. This absorption consists of a large number of mostly ill-defined bands each of which consists of a large number of irregularly spaced fine lines. Figure 192 shows three sections of this spectrum, one showing a rather distinct band (at 8370 Å), the second showing the more usual very irregular ill-defined structure between 4555 and 4542 Å, with a superimposed band of simple structure recently recognized by Douglas and Huber (296), and the third showing a similar band in the diffuse region near 3910 Å (see Chap. IV, p. 483). The spectrum extends weakly into the near infrared up to 10300 Å and possibly beyond [Douglas and Huber (296)], its intensity maximum is at about 4000 Å, but it extends at least to 3000 Å in the ultraviolet. No complete description of the spectrum has as yet been published [see, however, the curve for the absorption coefficient under low resolution published by Hall and Blacet (462)].

The absorption of $NO_2$ in a solid matrix at 4°K has been studied by Robinson, McCarty and Keelty (1080). Even at this temperature the spectrum remains surprisingly complex. Wood and Dieke (1318) and more recently Douglas (294a) have studied the magnetic rotation spectrum corresponding to the visible absorption and have obtained some simplification of the spectrum.

The fluorescence of $NO_2$ produced by exciting lines in the visible has been studied by Heil (488), Neuberger and Duncan (929), and recently by Douglas (294a). The life-time of the excited state was found to be 44 $\mu$sec. This life-time is longer by a factor 100 than the life-time expected from the integrated absorption coefficient. Douglas (294a) suggests that this difference is connected with strong perturbations of the excited state by the numerous vibrational levels of the ground state.

The visible $NO_2$ bands have also been observed as chemiluminescence in the reaction between NO and atomic oxygen [Kaufman (663), Broida, Schiff and Sugden (157)]; indeed, it is probable that the spectrum of the air afterglow is a pseudo-continuum produced by many overlapping bands of this transition.

Quite recently Douglas and Huber (296) have succeeded in analysing a simple progression of bands among the visible absorption bands. The spacing of the bands is 880 $cm^{-1}$. They consist of simple $P$, $Q$ and $R$ branches and must be assigned to the

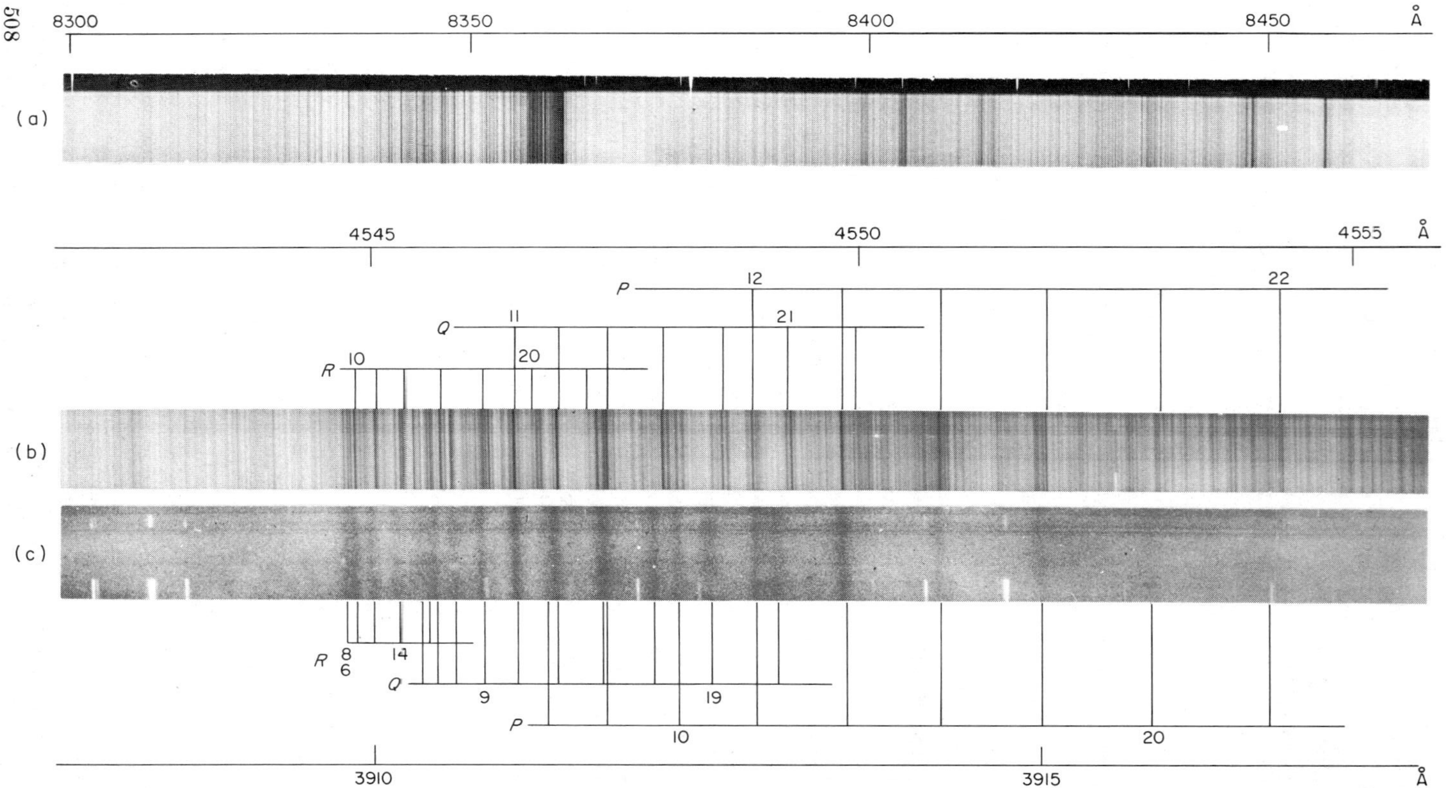


FIG. 192. **Sections of the visible and near ultraviolet absorption spectrum of $NO_2$ after Douglas and Huber (296).** (*a*) **Part of photographic infrared spectrum showing strong band head at 8361 Å.** (*b*) **Small section of absorption spectrum in the blue region of the spectrum showing three branches of a simple band.** (*c*) **Small section of the near ultraviolet spectrum showing diffuse band.**

$K' = 0 \leftarrow K'' = 1$ sub-bands of a $^2B_1 - {}^2A_1$ electronic transition. A low-lying $^2B_1$ state of $NO_2$ is predicted from the electron configuration (see Table 38), viz.,

$$\ldots(4b_2)^2(2b_1)\ {}^2B_1$$

It arises together with the $\ldots(4b_2)^2(6a_1)\ {}^2A_1$ ground state from the $\ldots 1\pi_g^4 2\pi_u\ {}^2\Pi_u$ state of the linear conformation. From the Walsh diagram the molecule is predicted to be much less bent in the $^2B_1$ state than in the ground state. Bands with higher $K'$ values have not been identified, probably because $K' \neq 0$ levels, in contrast to the $K' = 0$ levels, are strongly perturbed. At any rate, the $K' = 0$ levels of the $^2B_1$ state, which correspond to $\Sigma_g^-$ vibronic levels in the linear conformation, cannot be perturbed by any of the higher vibrational levels of the $^2A_1$ ground state.

A much more regular absorption of $NO_2$ extends from 2500 Å to shorter wavelengths with a few "hot" bands preceding it (see Fig. 183). It shows a very striking predissociation starting at about 2480 Å (see Chap. IV, section 3). The $0-0$ band at 2491 Å has recently been analyzed in detail by Ritchie, Walsh and Warsop (1072). As had already been recognized by Harris and King (476), the band is a parallel band of a nearly symmetric top, i.e., the transition moment lies in the top axis ($\perp$ to the symmetry axis). Thus, the upper state must be a $^2B_2$ state. There is a considerable change in the angle and in the N—O distance. It seems probable that the upper state $\tilde{B}\ {}^2B_2$ arises from the configuration $\ldots(4b_2)(6a_1)^2$.

The $\tilde{B}-\tilde{X}$ transition is followed by a progression of diffuse bands from 2350 to 2000 Å with a spacing of about 940 cm$^{-1}$. This progression may possibly be part of the $\tilde{B}-\tilde{X}$ system. A good reproduction is given by Nakayama, Kitamura and Watanabe (926). At still shorter wavelengths there is a continuous absorption of rapidly increasing intensity overlapped by a few diffuse bands [Mori (879), Nakayama, Kitamura and Watanabe (926)] until at 1650 Å an extensive system of fairly sharp bands starts which extends to 1350 Å. This system ($\tilde{E}-\tilde{X}$) was first observed by Price and Simpson (1022) [see also Mori (879)] and has recently been analyzed in detail by Ritchie and Walsh (1071). Progressions in two vibrational frequencies $\nu_1' = 1420$, $\nu_2' = 596._5$ have been found. There is an alternation in the sub-head structure of successive bands of the progression in $\nu_2'$ showing conclusively that the molecule in the upper state is linear (see p. 211). The sub-heads fit excellently with the assumption that the electronic transition is of the parallel type, i.e., that the upper state is $^2\Sigma_u^+$. A linear conformation is expected for a Rydberg state of $NO_2$ since the ion ($NO_2{}^+$) has $CO_2$-like structure.

The $\tilde{E}-\tilde{X}$ system is followed by three progressions of diffuse bands which extend from 1305 to 1055 Å and have spacings of 610, 560 and 624 cm$^{-1}$, respectively. Nakayama, Kitamura and Watanabe (926) have observed weak photoionization starting at 9.78 eV. This probably corresponds to the first ionization potential. A break in the photoionization curve at 10.83 eV is interpreted by them as corresponding to the dissociation process

$$NO_2 + h\nu \rightarrow NO^+ + O^- \qquad (V, 1)$$

No Rydberg series going to the first ionization limit has been identified. This was to be expected in view of the difference in geometrical structures of $NO_2$ and $NO_2^+$. Transitions to higher Rydberg states must necessarily have an extended structure like the $\tilde{E}-\tilde{X}$ transition and are therefore not readily identified without a detailed analysis. The Rydberg series given by Price and Simpson (1022) and Nakayama, Kitamura and Watanabe (926) with limits at 12.3 and 11.62 eV are not convincing. Even the more extended Rydberg series with a limit at 18.87 eV (see Table 64) recently found by Tanaka and Jursa (1187) are not entirely convincing. The limit would correspond to an excited state of $NO_2^+$, 9.09 eV above the ground state. From electron impact studies Collin and Lossing (222)(221) have obtained evidence for an excited state of $NO_2^+$ at about 4 eV.

**$CF_2$.** A spectrum of the $CF_2$ radical (with 18 valence electrons) was first observed and identified in emission in a discharge through $CF_4$ by Venkateswarlu (1243). A large number of bands occur in the region 3300–2300 Å. They can be assigned to progressions in two vibrations, $\nu_1$ and $\nu_2$, of the upper and lower state. Each band has a coarse fine-structure which must be interpreted as the $K$ structure of a nearly symmetric top with one fairly small moment of inertia. Several of these bands were observed in absorption in a discharge through a mixture of fluorocarbons by Laird, Andrews and Barrow (713) and more recently in the flash photolysis of $CF_2Br_2$ by Mann and Thrush (796). In absorption a single progression in $\nu_2'$ dominates the spectrum. The separation of the bands varies from 505 to 490 cm$^{-1}$. In addition, a few hot bands corresponding to excitation of one or two quanta of $\nu_2''$ in the ground state have been observed. Mann and Thrush have thrown some doubt on the assignment of a large number of emission bands as involving $\nu_1$.

Mulliken, in Venkateswarlu's paper, has interpreted the $CF_2$ bands as a $^1B_2-{}^1A_1$ transition. However, a recent detailed fine-structure analysis by Mathews (800b) shows the bands to belong to a $^1A_1$ (or $^1B_1$) $-$ $^1A_1$ transition. No other electronic transitions of $CF_2$ have yet been found. Observation of triplet transitions similar to those in $CH_2$ would be interesting. It is likely, but by no means certain, that the $^1A_1$ state of $CF_2$ is the ground state.

An emission spectrum similar to that of $CF_2$ has been found for $SiF_2$ by Johns, Chantry and Barrow (637) and analyzed in greater detail by Rao and Venkateswarlu (1049). The resulting data are included in Table 64.

**$O_3$.** The ozone molecule (with 18 valence electrons) shows weak absorption even in the photographic infrared as far as 10000 Å. Indeed, the longward limit of this absorption has not been established. According to Wulf (1324), the (diffuse) bands in the photographic infrared form a progression of ten members with a spacing of 567 cm$^{-1}$, the first one being at 10000 cm$^{-1}$. Wulf's reproduction does not clearly show this progression, but the wavelengths of Lefebvre (735) seem to confirm it. A second somewhat stronger absorption discovered by Chappuis (188) extends from 6100 to 5500 Å. The two strongest bands are at 6020 and 5730 Å [cf. the reproduction of Wulf (1324)]. As has been shown by Humphrey and Badger (591) by a study under high resolution, the bands are genuinely diffuse. The occurrence of predissociation at this long wavelength is not surprising since the dissociation energy (obtained from thermochemical data) is only $1.0_4$ eV. A satisfactory vibrational analysis of the Chappuis bands has not been given (Wulf's formula gives the wrong $\nu_2''$). At long paths the Chappuis bands extend to shorter wavelengths [see Wulf (1324) and Vigroux (1247)] and merge into a very weak continuum at about 4000 Å.

Beginning at 3740 Å, the continuum is followed by weak extensions of the so-called Huggins bands. The main bands of this system first observed by Huggins (587) extend from 3450 to 3000 Å. The shortward end of the Huggins bands is overlapped by the shoulder of the Hartley band, which represents the strongest absorption of ozone in the near ultraviolet. The Hartley band consists of a broad continuum between 3000 and 2200 Å with a very high and almost symmetrical peak near 2550 Å [for a comparison of various measurements of this continuum, see Vassy (1241)]. The continuum appears to be overlapped by weak diffuse bands; it is not certain whether these bands belong to the Huggins bands or, indeed, whether the Huggins bands and the Hartley band belong to the same electronic transition. However, there is an enormous difference in intensity: The peak of the Hartley band at 2550 Å has an absorption coefficient of 150 cm$^{-1}$, that is, a layer of 0.007 cm at atmospheric pressure absorbs 60 per cent of the incident light (and that is the reason why the atmosphere with an ozone layer of 0.3 cm-atm is so completely opaque to solar and stellar radiation in this spectral region). On the other hand, the absorption coefficient in the Huggins bands varies from about 5 cm$^{-1}$ at 3000 Å to about 0.01 at 3430 Å and 0.0003 at 3650 Å [see Vigroux (1247)]. For comparison, the absorption coefficient for the strongest of the Chappuis bands is 0.06 cm$^{-1}$.

Although the Huggins bands are still diffuse, they are much sharper than the Chappuis bands. Jakovleva and Kondratiev (619) have attempted a vibrational analysis which yields the frequencies 636.3 and 351.7 for the vibrations $\nu_1'$ and $\nu_2'$ of the upper state, while in the lower state only a single vibration, 1047 cm$^{-1}$, was observed. Melcher (819) and Eberhardt and Shand (341) have measured additional bands at the longward end and have found evidence for a second vibration, 710 cm$^{-1}$, in the ground state. According to the most recent study of the infrared spectrum [Kaplan, Migeotte and Neven (657)], the frequency $\nu_3$ is 1042.2 cm$^{-1}$ while $\nu_1$ and $\nu_2$ according to Wilson and Badger (1309) are 1110 and 705 cm$^{-1}$ respectively. If the presence of $\nu_3''$ in the Huggins bands were established it would imply that the molecule is asymmetric in the upper electronic state. However, in view of the large anharmonicity found by Jakovleva and Kondratiev, it is yet possible that the observed transitions belong to $\nu_1''$ (and $\nu_2''$) as expected when in both upper and lower state the molecule has $\boldsymbol{C}_{2v}$ symmetry.

Jakovleva and Kondratiev (619) have found several progressions in the diffuse bands overlapping (or forming part of) the Hartley band. The spacing of 300 cm$^{-1}$ in these progressions corresponds presumably to the bending vibration $\nu_2'$ in the upper state. The large extent of the Huggins and Hartley bands strongly suggests that there is a considerable change of angle between upper and lower state. The possibility that both band systems are really one electronic transition cannot be entirely eliminated on the basis of present data. There have been a number of investigations of the effect of temperature on the Huggins bands [Wulf and Melvin (1325), Vassy (1241), Eberhardt and Shand (341), Barbier and Chalonge (94) and Vigroux (1247)]. A striking effect at low temperature is the sharpening up of the bands. This may partly be due to shortening of the rotational branches of the bands but also, and perhaps more importantly, to the disappearance of $1-1$, $2-2$ bands in the bending fundamental which accompany every main band.

In the vacuum ultraviolet the absorption of $O_3$ is almost entirely continuous with a few very diffuse and not very distinct overlapping bands [Price and Simpson (1022), Tanaka, Inn and Watanabe (1186), Ogawa and Cook (945)]. There are at least six distinct continua whose maxima occur at 1725, 1450, 1330, 1215, 1120 and 750 Å. Tanaka, Inn and Watanabe suggest that the diffuse bands overlapping the continua at 1330, 1215 and 1120 Å form progressions with spacings of 600 and 800 cm$^{-1}$. The continuum with maximum at 750 Å may be an ionization continuum; it starts at 950 Å, a wavelength which corresponds to the electron impact value of the ionization potential.

The interpretation of some of the observed states of $O_3$ in terms of electron configurations has been discussed by Mulliken (914).

**$SO_2$.** In contrast to ozone, and in spite of the similarity of its electronic and geometrical structure, sulphur dioxide has almost exclusively discrete and sharp absorption bands. The most longward (and weakest) absorption observed so far lies in the near ultraviolet between 3900 and 3400 Å. Immediately adjoining it between 3400 and 2600 Å lies a much stronger absorption system. In the earlier work of Watson and Parker (1280), Clements (208) and Asundi and Samuel (83) it was assumed that the bands of both regions belong to one and the same electronic transition, the weaker bands at the longward end being assigned as hot bands. But Metropolis and Beutler (835) showed that there are in fact two electronic transitions, a weak one (3900–3400 Å) and a stronger one (3400–2600 Å). The second system has been analyzed in some detail by Metropolis (834) while for the first, Metropolis and Beutler gave a brief vibrational analysis. Both systems consist of fairly long progressions in both $\nu_1'$ and $\nu_2'$ as expected on the basis of the application of the Franck–Condon principle (see Chap. II, section 2) if the S—O distance, as well as the angle, changes in the transition. Gaydon (412) had earlier observed the first system in emission in the after-glow of a discharge through $SO_2$ and arrived at the same $0-0$ band for the system as Metropolis and Beutler (835). It has been suggested by Coon and Mulliken [cf. Mulliken (914)] that the molecule may be slightly asymmetric in the excited state because some of the weaker bands which are assigned as bands involving a change of $v_3$ ($\Delta v_3 = \pm 2, \pm 4$) seem to have a larger intensity than compatible with the Franck–Condon principle on the assumption of $\boldsymbol{C}_{2v}$ symmetry in the excited state.

Douglas (293) has shown that the bands of the first system exhibit an appreciable Zeeman effect thus demonstrating that the upper state must be a triplet state. For this reason we shall designate the transition $\tilde{a}-\tilde{X}$. Merer (822) has analyzed the rotational structure of some of these bands and has found clear evidence of triplet splitting although he did not identify some of the expected branches (see p. 268). He has, however, established that the transition is ${}^3B_1-{}^1A_1$ since only sub-bands with $\Delta K = \pm 1$ appear. It is probable but has not been confirmed by a detailed analysis of the rotational structure that the $\tilde{A}-\tilde{X}$ system is the corresponding ${}^1B_1-{}^1A_1$ transition. Another interesting feature of the 3900 Å system is the appearance of a $1-0$ band in $\nu_3$ (the antisymmetric stretching vibration) with an intensity comparable to that of the $0-0$ band. According to van der Waals (1248a), the occurrence of this forbidden component of the ${}^3B_1-{}^1A_1$ transition is not caused by simple vibronic interaction with another triplet state (of type ${}^3B_2$) but by vibrationally induced spin-orbit coupling. This coupling can mix the $A_2$ component of the ${}^3B_1$ state with the ${}^1B_1$ state if $\nu_3(b_2)$ is excited by an odd number of quanta and thus can enable the $1-0$, $3-0, \ldots$ bands of the ${}^3B_1-{}^1A_1$ transition to borrow intensity from the neighbouring $\tilde{A}\,{}^1B_1-\tilde{X}\,{}^1A_1$ transition.

A third region of absorption several times stronger than the $\tilde{A}-\tilde{X}$ system extends from 2350 to 1800 Å. This absorption, studied by Chow (195), Price and Simpson (1020), Duchesne and Rosen (315) and Dubois and Rosen (311)(310), probably consists of two and perhaps more electronic transitions. Corresponding emission spectra have been studied in an electric discharge by Chow and Smyth (196) and Chow (195), and in fluorescence by Lotmar (778) and Douglas and Zanon (303). Dubois (310) has recently obtained a fairly complete analysis of the second system ($\tilde{D}-\tilde{X}$) in absorption. The analysis of the emission spectrum is more certain since the vibrational frequencies of the ground state are known from infrared work. In both absorption and emission all three normal vibrations of upper and lower state occur with non-zero $\Delta v_i$. For the ground state frequencies this conclusion has been confirmed by a recent investigation of the resonance fluorescence at low pressure and high resolution by Douglas and Zanon (303). By excitation with the strong Zn lines at 2139 and 2100 Å they have observed resonance progressions entirely similar to those well known for diatomic molecules, each member of a progression consisting of only two or three lines; progressions in $\nu_3''$ with both even and odd $v_3''$ are quite strong. Such progressions are difficult to account for even if one assumes that the $SO_2$ molecule is unsymmetrical in the $\tilde{C}$ state.

Except for three fairly weak diffuse bands near 1550 Å, $SO_2$ is transparent between 1800 and 1350 Å [Price and Simpson (1020)]. At 1335 and 1140 Å there are two narrow groups of bands which in all probability are the first two members of a Rydberg series.

The narrowness of these groups implies that there is very little difference in the geometrical structure in these Rydberg states from that in the ground state. Two other much broader band groups appear at 1260 and 1040 Å and represent probably another Rydberg series with upper states of somewhat different conformation from the ground state.

The preceding discussions of triatomic non-hydrides are summarized in Table 64. The vibrational and rotational constants in the various electronic states are listed. Data for a number of other triatomic molecules not discussed in the text are included in Table 64.

## 2. Four-atomic molecules

We divide the four-atomic molecules into four classes according to the number of hydrogen atoms present: tri-hydrides, di-hydrides, mono-hydrides and non-hydrides.

### (a) Tri-hydrides

The only tri-hydrides for which electronic transitions are known are $NH_3$, $PH_3$, $AsH_3$ and the methyl radical $CH_3$. In spite of considerable effort, no spectra of the radicals $BH_3$ and $SiH_3$ have as yet been observed. We shall consider here only $CH_3$ and $NH_3$; for the other tri-hydrides see Table 65.

**$CH_3$.** A spectrum of $CH_3$ was first observed by Herzberg and Shoosmith (540) in the flash photolysis of $Hg(CH_3)_2$ and later in the photolysis of many other compounds. In spite of considerable searching, no absorption was found in the visible and near ultraviolet regions even though, if $CH_3$ were non-planar, a transition from the $\ldots(e)^4a_1\ ^2A_1$ ground state to the $\ldots(e)^3a_1^2\ ^2E$ excited state would be expected in this region. For planar $CH_3$, the absence of the corresponding transition [from $\ldots(e')^4a_2''\ ^2A_2''$ to $\ldots(e')^3(a_2'')^2\ ^2E'$] is easily understood since it would be an electronically forbidden transition (see p. 132). Even if, as seems likely, the molecule were non-planar in the excited $^2E$ state and therefore, strictly speaking, only the selection rules for the joint point group, $\boldsymbol{C}_{3v}$, would apply, the transition would still be very weak since the "vertical", i.e., Franck–Condon allowed, part of the transition would obey the $\boldsymbol{D}_{3h}$ selection rules. If in spite of its expected weakness, this transition could be observed, it would give a good deal more information about $CH_3$ than at present available because it would not be diffuse. Moreover, the excited $^2E$ state is expected to show an interesting Jahn–Teller effect.

The most longward observed absorption of $CH_3$ occurs at 2160 Å. It consists of two diffuse maxima. The corresponding band of $CD_3$ (at 2140 Å) is sharper and shows a partly resolved fine structure (see Fig. 96). In addition, three very much weaker bands appear at longer and shorter wavelengths. The presence of a single strong band in the system shows that the conformation of the molecule in the upper and lower state must be very nearly the same. The fine structure of the band can be understood completely as that of a || band of a symmetric top (see Fig. 97). The observed weak intensity alternation, and particularly the very low intensity of the $R(0)$ line, can only arise if at least in one of the two combining states the molecule is planar because only for $\boldsymbol{D}_{3h}$ symmetry is there an intensity alternation in the $K = 0$ sub-band (see Fig. 97). Thus, one must conclude, combining the evidence from the vibrational and rotational intensity distribution, that the molecule is planar in both states involved. It is not possible to exclude the possibility that $CH_3$ is *slightly* non-planar, but only to the extent that would allow a large inversion doubling to arise, so large that only one inversion-component is observed in absorption.

Further, much stronger absorption bands occur in the region 1500 to 1280 Å, as shown in Fig. 193. They form three Rydberg series, designated, $\beta$, $\gamma$ and $\delta$. Although the first (strongest) member of the $\gamma$ as well as that of the $\delta$ series is accompanied by a few much weaker vibrational bands, the dominating intensity of the main $(0-0)$ bands indicates that for all Rydberg states the conformation is the same as in the ground state, viz., the planar one. The first member of the $\beta$ series is apparently the 2160 Å band. The $\gamma$ series consists of fairly broad unresolved maxima (even in $CD_3$) which

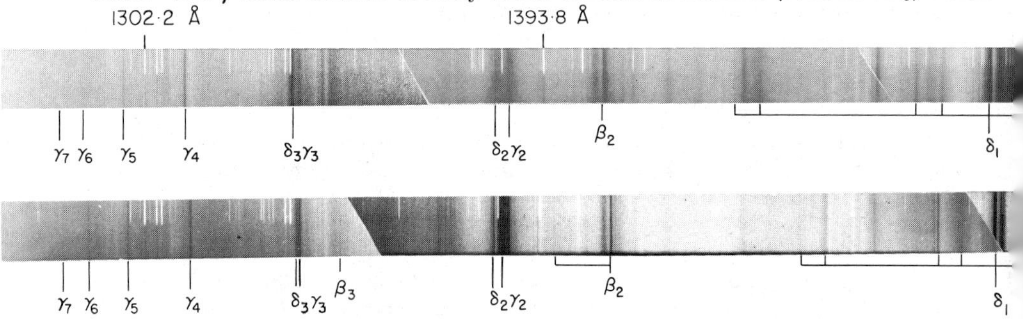


FIG. 193. **Absorption spectrum of $CH_3$ and $CD_3$ in the region 1550 to 1300 Å after Herzberg (521).** The upper states of the bands $\gamma_1$, $\delta_1$ and $\beta_2$ have been called $\tilde{C}$, $\tilde{D}$ and $\tilde{E}$, respectively, in Table 65. Note that the two spectrograms are composites taken from different exposures; thus the apparent intensities of different bands do not correspond well to the true intensity ratios.

are in all probability unresolved $\perp$ bands. The bands of the $\delta$ series have two maxima and in the first member some resolved fine structure, indicating that these bands are $\|$ bands. The Rydberg corrections for both $\gamma$ and $\delta$ series are very small ($\sim 0.08$–$0.09$), suggesting that these series correspond to excitation of the outermost electron into an *nd* orbital, while the large Rydberg correction of the $\beta$ series suggests that it corresponds to an *ns* orbital. The absence of a series corresponding to *np* is in striking agreement with expectation from molecular orbital theory since the two states $\ldots(e')^4(npe')\ ^2E'$ and $(e')^4(npa_2'')\ ^2A_2''$ cannot combine with the $^2A_2''$ ground state (see p. 132). Similarly, of the three states arising from an *nd* atomic orbital, $^2A_1'$, $^2E'$, $^2E''$, only two, $^2A_1'$ and $^2E''$, can combine with the ground state, and they correspond clearly to the observed $\delta$ and $\gamma$ series respectively. No third series with small Rydberg correction has been observed.

All three observed Rydberg series converge to 79392 cm$^{-1}$ for $CH_3$ and 79315 cm$^{-1}$ for $CD_3$, corresponding to the first ionization potential. The fact that the ionization potential of $CH_3$ is greater than that of $CD_3$ (by 77 cm$^{-1}$) indicates that the zero-point vibrational energy of $CH_3^+$ in its ground state is greater than that of $CH_3$ in its ground state, that is, that the sum of the frequencies of all normal vibrations is greater. The same conclusion applies for all the upper states of the $\gamma$ and $\delta$ Rydberg series, but not to those of the $\beta$ series. Unfortunately, the individual vibrational frequencies of both excited and ground states are still not known with any certainty. Some suggested frequencies are included in Table 65.

Since most of the absorption bands of $CH_3$ and $CD_3$ are to a greater or smaller extent diffuse, only provisional values of the rotational constants have as yet been obtained. But the analyses of both the $\tilde{D}-\tilde{X}$ band of $CH_3$ and the $\tilde{B}-\tilde{X}$ and $\tilde{D}-\tilde{X}$ bands of $CD_3$ agree in yielding a C—H distance $r_0'' = 1.079$ Å for the $^2A_2''$ ground state assuming the planar symmetrical structure. In the $\tilde{B}$ state $r_0'$ is larger, in the $\tilde{D}$ state it is smaller than $r_0''$. The only bands that show a fairly sharp fine structure are the second members of the $\gamma$ and $\delta$ series of $CD_3$ near 1385 Å, but they have not yet been analyzed. The striking difference in diffuseness between corresponding bands of $CH_3$ and $CD_3$ has been discussed in Chapter IV, section 3. Most of the information about the structure of free methyl in its ground state is derived from the spectrum of $CD_3$ (see Table 65).

A nearly planar structure of $CH_3$ in its ground state has more recently also been derived by a careful discussion of the observed hyperfine splitting in the electron-spin resonance spectrum by Karplus (658).

**$NH_3$.** Ammonia is transparent in the whole visible and near ultraviolet regions. As was first shown by Leifson (736), a long progression of diffuse absorption bands starts at 2168 Å and extends to about 1700 Å. This progression is shown for $ND_3$ in Fig. 179. At elevated temperatures additional bands appear particularly at the longward end. Dixon (280) has extended these "hot" bands to 2431 Å. The band system ($\tilde{A}-\tilde{X}$) has been further investigated by Duncan (317), and recently Walsh and Warsop (1270) have studied it in great detail and have obtained a complete vibrational analysis. The corresponding band system of $ND_3$, briefly mentioned by Benedict (106), Duncan (318), and Walsh and Warsop (1270), has been studied in detail by Douglas (294). None of the bands of $NH_3$ in this system show a resolved rotational fine structure, but the first four bands of the main progression, as well as most of the hot bands, show two peaks each, corresponding to $Q$ and $R$ heads (similar to the 2160 Å band of $CH_3$). The bands of $ND_3$ with $v_2' = 0$ and 1 show fine structure. The line width of the individual lines is smallest for the bands with $v_2' = 1$ although even for them it is still of the order of 0.5 $cm^{-1}$.

According to Walsh and Warsop, the $Q$ heads of the main progression of the $\tilde{A}-\tilde{X}$ system can be well represented by

$$\nu_{\tilde{A}-\tilde{X}} = 46136 + 874v_2' + 4.0v_2'^2 \qquad \text{(V, 2)}$$

The frequency 874 $cm^{-1}$ must be interpreted as that of the (symmetrical) bending vibration $\nu_2'$ in the upper state, and the existence of a long progression in this vibration with an intensity maximum at about $v_2' = 7$ implies a considerable change in angle of the $NH_3$ pyramid in going from the ground state to the excited state. Indeed, it was shown by Walsh and Warsop that the bands of the hot progression in which $\nu_2''$ is singly excited come alternately from the upper and lower inversion component of the lower state for odd and even $v_2'$ respectively, that is, are alternately shifted to longer and shorter wavelengths because the $v_2'' = 1$ level in $NH_3$ is split by 36 $cm^{-1}$ (see Vol. II, p. 223). Such a situation can only arise if the molecule is planar in the excited state since then alternate vibrational levels in the progression $v_2'\nu_2'$ are symmetric and antisymmetric with respect to the plane of the molecule, and can only combine with one or the other of the inversion doubling levels of the lower state as illustrated in Fig. 67b. (If the molecule were non-planar in the excited state, there would be, as shown by Fig. 67a, two bands for each $v_2'-v_2''$ transition which for $v_2'' = 1$ would be separated by 36 $cm^{-1}$ and thus easily resolved.)

The planarity of the $\tilde{A}$ state is nicely confirmed by the rotational structure of the $ND_3$ bands [Douglas (294)]: The bands with $v_2' = 0$ have a strong $R(0)$ line, those with $v_2' = 1$ a very weak $R(0)$ line as expected if the upper state is planar since only the $K = 0$ sub-band contributes to $R(0)$ and since for $K = 0$ of a planar molecule either the even or the odd levels (in $J$) are strong depending on whether the state is vibronically $A_1'$ or $A_2''$. Considering that the $R(0)$ line has $J' = 1$, it follows that the $v_2' = 0$ level is vibronically $A_2''$, and therefore, the $\tilde{A}$ electronic state is $A_2''$. In agreement with this conclusion Dressler (307) finds in absorption of $NH_3$ in an argon matrix at 4.2°K only alternate bands since only $J = 0$ is present in the lower state and since in the upper state the $J = 1$ level (or its analogue in the solid) is present only for even $v_2'$ values.

A second weaker system consisting of a long progression of fairly sharp bands occurs in the region 1690 to 1400 Å. It was first studied by Duncan (317) and Duncan and Harrison (322) and more recently, under much higher resolution by Douglas and Hollas (295) and Douglas (294). The latter authors obtained a complete rotational analysis of several bands of the progression which shows unambiguously that the bands are $\perp$ bands and that the upper electronic state is a $^1E''$ state of planar $NH_3$. The proof for the planarity of the excited ($\tilde{B}$) state rests again on the observation of an intensity alternation in the sub-bands with $K'' = 0$ (see p. 234f.), while the alternation of the sign of this intensity alternation proves that the progression is again in $v_2'$. The band origins of successive bands in the progression can be represented by

$$\nu_{\tilde{B}-\tilde{X}} = 59225.5 + 880.60v_2' + 18.437v_2'^2 - 0.71863v_2'^3 \qquad \text{(V, 3)}$$

The rotational analysis shows the presence of a large first order Coriolis splitting in the excited state with a $\zeta_e$ value (see p. 86) that varies from 0.873 for $v' = 1$ to 0.804 for $v' = 8$. In addition, a large $j$-type doubling is present in one Coriolis component [$(+j)$ levels] of the $K' = 1$ levels of the upper state (see p. 96). The electronic nature of $\zeta_e$ has been verified by Douglas by investigation of the Zeeman effect in these bands.

Another weaker progression with similar spacings has been found by Douglas and Hollas (295) and Douglas (294) to occur in the same region as the $\tilde{B}-\tilde{X}$ bands. The fine-structure analysis shows that these bands are || bands. They are therefore in all probability due to another electronic transition, $\tilde{C}-\tilde{X}$, but the possibility that they represent a forbidden component of the $\tilde{B}-\tilde{X}$ system made possible by vibronic interaction is difficult to exclude with certainty. In Table 65 we have chosen the former alternative. Since the vibrational numbering has not been established the $\tilde{C}$ state can be either $A_1'$ or $A_2''$ but on the basis of the electron configuration $A_1'$ is more likely to be correct.

At wavelengths below 1450Å further much stronger progressions of absorption bands start [Duncan (317)]. According to Walsh and Warsop (1270), there are at least three distinct and partially overlapping progressions in the region 1450 to 1220 Å corresponding to at least three different electronic states as listed in Table 65. Below 1220 Å Duncan (317)(319) has observed still another progression starting at 82857 cm$^{-1}$.

All observed electronic transitions show only progressions in the out-of-plane bending vibration $\nu_2'$, and in all of them the frequency of this vibration is of similar magnitude (and similar to its value in the ground state). It is striking that in each case the anharmonicity constant $x_{22}$ is positive, that is, the vibrational interval increases with $v_2'$.

Walsh and Warsop have arranged the states $\tilde{A}$, $\tilde{D}$ and $\tilde{E}$ into a Rydberg series with the formula

$$\nu = 82150 - \frac{R}{(n - 1.02_4)^2}, \quad n = 3, 4, 5 \qquad \text{(V, 4)}$$

The Rydberg correction obtained is in line with that expected for an $ns$ series, and the first electronic state ($\tilde{A}$) is certainly $\ldots(e')^4(a_2'')(3s\, a_1')\ ^1A_2''$. The ionization potential obtained from (V, 4) which is of low accuracy, agrees very well with the probably more accurate photoionization value of Inn (603) and Watanabe and Mottl (1276), viz., $10.15_4$ eV.

Thompson and Duncan (1215) have looked for the transition to the lowest triplet state with an absorbing path of 10 m atm. but have not found it.

Absorption coefficients of $NH_3$ in the region 2200 to 1050 Å have been measured by Watanabe (1272). Below 1400 Å the absorption coefficients reach 1000 cm$^{-1}$. A strong continuous absorption begins at 1150 Å [see also Duncan (319) and Metzger and Cook (836)]. Sun and Weissler (1177) have measured the absorption coefficients between 1300 and 374 Å using a discontinuous background. The total $f$ value in this region is 4.9.

## (b) Di-hydrides

Electronic absorption spectra of several four-atomic di-hydrides are known, but only for two of them have they been studied in some detail.

**$C_2H_2$.** The most longward absorption of acetylene consists of a large number of very weak absorption bands extending from 2400 to 2100 Å. At elevated temperatures, bands have been observed as far as 2500 Å. This band system (here called $\tilde{A}-\tilde{X}$) was first studied in some detail by Kistiakowsky (672), Göpfert (438) and Woo, Liu, Chu and Chih (1317) who identified a number of progressions. A detailed and convincing analysis of both the vibrational and rotational structure was first accomplished by Ingold and King (600) and Innes (605). They showed that in the excited state the $C_2H_2$ molecule is bent, having $\boldsymbol{C}_{2h}$ symmetry (trans form). In this way the long progressions in the bending vibration $\nu_3'$ ($\approx 1050$ cm$^{-1}$) of the upper state are accounted for. That this vibration is really a bending vibration (and not the C—C stretching vibration) is shown by the large isotope shift in $C_2D_2$, for which its frequency is observed to be about 830 cm$^{-1}$. The C—C stretching vibration is also excited and has the frequency $\nu_2' = 1385$ cm$^{-1}$ which changes much less (to 1299 cm$^{-1}$) in going to $C_2D_2$.

Each one of the main bands of the $\tilde{A}-\tilde{X}$ system is accompanied by a group of hot bands. The analysis of the $K$-structure of the hot bands is similar to that for HCN (see p. 203 and Figs. 65 and 79) and establishes that, in the excited state, the molecule is non-linear. The analysis of the $J$ structure has confirmed this conclusion by the observation of a combination defect caused by asymmetry ($K$-type) doubling in the excited state. The sign of this doubling shows that the excited state is an $A_u$ state (presumably $^1A_u$). This state is to be correlated with a $^1\Sigma_u^-$ or a $^1\Delta_u$ state of the linear conformation. The fact that the transition is forbidden for $\boldsymbol{D}_{\infty h}$ symmetry accounts for its relatively low intensity. The dimensions of the molecule in the excited state derived from the analysis of the rotational structure are given in Table 66. A few weak bands with $\Delta K = 0$ and $\pm 2$ have been observed. The suggestion by Herzberg (523) that the reason for the occurrence of these forbidden transitions is spin-orbit interaction (see p. 221f.) has not been confirmed by Zeeman investigations of Douglas (unpublished). But Hougen and Watson (580) have shown that these weak bands can be quantitatively accounted for by the mechanism of axis switching discussed earlier (p.208f.).

At shorter wavelengths, between 2000 and 1550 Å there is a rather complex spectrum of diffuse bands of gradually increasing intensity [Herzberg (517)] which has not yet been analyzed [see the remarks of Wilkinson (1301)]. It is possible but far from certain that these bands belong to the $\tilde{A}-\tilde{X}$ system.

An extremely strong discrete absorption system sets in at 1519 Å [Herzberg (517), Price (1013), Moe and Duncan (865)]. It has recently been studied under high resolution both for $C_2H_2$ and $C_2D_2$ by Wilkinson (1301). The system consists of a single short progression in the C—C stretching vibration $\nu_2' \approx 1849$ cm$^{-1}$ (for $C_2$,$D_2$, $\nu_2' \approx 1720$ cm$^{-1}$). The first band of the progression (0 – 0 band) is the strongest. It has an absorption coefficient of 1400 cm$^{-1}$. A second similar system of similar intensity starts at 1342 Å. Here the second band in the $v_2'$ progression is the strongest ($\nu_2' = 1781$ cm$^{-1}$ for $C_2H_2$ and 1559 cm$^{-1}$ for $C_2D_2$), and $\nu_1'$ is also (singly) excited in one of the bands. Even though the bands are diffuse on account of predissociation and their rotational structure can therefore not be studied, it is clear that the molecule is linear in the upper states since only totally symmetric vibrations are excited.

The two strong band systems just discussed form the first members of two fairly long Rydberg series observed by Price (1013). The second member (at 1250 Å) of the first series has recently been studied under high resolution by Herzberg (523) and has been found to be a $^1\Pi_u - {}^1\Sigma_g^+$ transition. The rotational structure confirms that in the excited state the molecule is linear. The hot bands accompanying the main bands show clear evidence of Renner–Teller splitting. It must be concluded that the other members of this Rydberg series also have $^1\Pi_u$ upper states. The limit of this series is at 91950 cm$^{-1}$ while the second Rydberg series converges to 92076 cm$^{-1}$. It is not possible to say whether or not the two limits are identical. The predicted ground state of $C_2H_2^+$ is $^2\Pi$ which might have a splitting of the order of 50 cm$^{-1}$.

Overlapping the first member, $\tilde{D}-\tilde{X}$, of the second Rydberg series are two other band systems of similar intensity $\tilde{E}-\tilde{X}$ and $\tilde{F}-\tilde{X}$ which show a more complicated vibrational structure, similar to the $\tilde{A}-\tilde{X}$ system. It appears from this structure that in both the $\tilde{E}$ and $\tilde{F}$ state the molecule is non-linear; indeed, Wilkinson suggests that in the $\tilde{E}$ state it is non-planar. For further data see Table 66.

Quantitative absorption measurements on acetylene have been made by Nakayama and Watanabe (927) who have also given very useful absorption curves in the region 2000 to 1050 Å. Similar low resolution data in the region 1000 to 600 Å have been obtained by Walker and Weissler (1257) and Metzger and Cook (836). There are two maxima of continuous absorption at 920 and 800 Å. [See also the recent mass-spectrometric photoionization work of Dibeler and Reese (273).] Ingold and King (601) were unsuccessful in finding any absorption longward of the $\tilde{A}-\tilde{X}$ system with an absorbing path of 20 m atm.

**$H_2CO$.** Formaldehyde exhibits a beautiful sharp absorption spectrum in the near ultraviolet which has been studied by a large number of investigators ever since the first detailed investigations by Henri and Schou (503) and Schou (1108). These authors recognized that there is a coarse and a fine rotational structure corresponding to a small and a large moment of inertia (about the C—O axis and an axis perpendicular to it: $K$ and $J$ structure respectively), but a conclusive rotational analysis of a number of the absorption bands was first accomplished by Dieke and Kistiakowsky (277), in a classic investigation which resulted in the first unambiguous interpretation of an electronic spectrum of a polyatomic molecule. They analyzed the sub-bands with $K \geq 3$ of the six main absorption bands at 3530, 3430, 3390, 3370, 3295 and 3260 Å, and obtained $A_v$ and $B_v + C_v$ values. For the $K$ values investigated, the band structure is almost exactly that of a $\perp$ band of a symmetric top. Only for the lowest $K$ values were small deviations observed which are due to the fact that the moments of inertia $I_B$ and $I_C$ are different and which show clearly that the transition moment is along the $b$ axis, that is, in the plane of the molecule, perpendicular to the C—O axis.

More recently, Robinson (1075) and in greater detail Parkin (962a) and Callomon and Innes (178) have analyzed sub-bands with lower $K$ values and have thus been able to obtain $B'_v$ and $C'_v$ separately. An important result of this work was that the inertial defect $\Delta = I_C - I_A - I_B$ for the upper state of the 3530 Å band, which is the apparent $0-0$ band of the system (see however below), is strongly negative amounting to $-0.265$ a.m.u. Å$^2$ while for the ground state it is well known, from microwave investigations [Lawrance and Strandberg (730)], to be positive, viz., $+0.057$ a.m.u. Å$^2$. This observation supplies unambiguous proof for the conclusion which was already strongly suggested by the vibrational analysis (see below) that the molecule is non-planar in the upper state of the near ultraviolet bands, a result that had been predicted on the basis of Walsh diagrams (see p. 316) by Walsh (1268).

The vibrational analysis of the main bands at first appeared to present no difficulties. The strongest bands form a fairly long progression in a frequency 1182

$cm^{-1}$, which must be interpreted as the CO stretching vibration $\nu_2'$ in the excited state. Each of the strongest bands is accompanied by short progressions with frequencies 824, 1322 and 2872 $cm^{-1}$ which have been assigned respectively to the molecule bending vibration $\nu_4'$ [corresponding to the out-of-plane bending vibration $\nu_4$ of the planar molecule (see Vol. II, p. 300)], to the $CH_2$ deformation vibration $\nu_3'$, and to the CH stretching vibration $\nu_1'$ [see Brand (138)]. Considerable difficulties arose, however, with the interpretation of a number of hot bands appearing at the longward side of the main bands and with the interpretation of the emission spectrum observed in fluorescence [Herzberg and Franz (526), Gradstein (443)], in electric discharges [Schüler and Woeldike (1115), Schüler and Reinebeck (1111), Dyne (332), Brand (138)], and in chemiluminescence [Emeléus (355), Pearse and Gaydon (32)].

The principal hot band is a band (designated $\alpha$ by Henri and Schou) which lies 1280 $cm^{-1}$ longward of the first strong absorption band (designated $A$). In the emission spectrum the $\alpha$ band forms the starting point of progressions in $\nu_2''$ (=1744) and $\nu_4''$ (=1167 $cm^{-1}$) extending toward longer wavelengths. Such a difference between the first absorption and the first fluorescence band arises as we have seen (p. 177 and Fig. 69) when the electronic transition is a forbidden one. In the present case, since the ground state is undoubtedly $^1A_1$, the excited state would have to be $^1A_2$, and the separation of the $A$ and $\alpha$ bands would have to be the sum $\nu_4'' + \nu_4'$ if one assumes that the molecule is planar in both states and that the transition moment for the main bands is in the plane of the molecule[1]. The rotational structure of the $\alpha$ band is in agreement with these assumptions: it is a $\perp$ band like the $A$ band, and the intensity alternation is opposite to that in the $A$ band [Dyne (332)], a reversal that is expected for a $\nu_0 - \nu_4''$ band (type $A_2 - B_1$) as compared with a $\nu_0 + \nu_4'$ band (type $B_2 - A_1$). However, since $\nu_4'' = 1167$ $cm^{-1}$ is definitely established from infrared work [see Vol. II, p. 300, and Callomon and Innes (178)], one would have to conclude that $\nu_4' = 120$ $cm^{-1}$, at first sight a rather strange result.

The way out of this difficulty was shown by Walsh (1268) and worked out in more detail by Brand (138). Walsh pointed out that if $H_2CO$ is non-planar in the excited state, but does not deviate too much from planarity, the lowest vibrational level in this state will be split on account of inversion doubling. If the electronic species is $A''$, then the lower inversion component will behave like an $A_2$, the upper like a $B_2$ vibronic level of a planar molecule in an $A_2$ electronic state. This is shown schematically in the energy level diagram Fig. 194 (compare Fig. 1a for an $A_1$ electronic state). Only the $B_2$ vibronic levels can be reached from the lowest vibrational level of the ground state while in fluorescence from the lowest level of the upper state ($A_2$) only the $B_1$ vibrational levels of the ground state can be reached. In the sum $\nu_4' + \nu_4''$ which would be the separation of $A$ and $\alpha$ for a planar upper state now $\nu_4'$ must be replaced by the inversion doubling in the lowest vibrational level ($v_4 = 0$) of the excited state which comes out to be 124 $cm^{-1}$. On this basis Brand (138) has given a fairly complete interpretation of the vibrational structure of both the absorption and emission spectrum of both $H_2CO$ and $D_2CO$.

It may be noted that the frequency difference 824 $cm^{-1}$ found among the main absorption bands is the difference between the two upper inversion doubling components of the levels $v_4' = 1$ and $v_4' = 0$ of the bending vibration $\nu_4'$ in the excited state (see Fig. 194). The lower components, observed in the emission spectrum, show that the doubling for $v_4 = 1$ is much larger (viz., 407 $cm^{-1}$) than for $v_4 = 0$ (124 $cm^{-1}$), and thus the average value of $\nu_4'$ is only 683 $cm^{-1}$. The emission progression in $\nu_4''$ shows a

[1] Note that in Volume II and in Brand's paper $\nu_4$, $\nu_5$ and $\nu_6$ were designated $\nu_6$, $\nu_4$ and $\nu_5$, respectively, because of the different choice of axes for $B_1$ and $B_2$ before and after Mulliken's report (912).

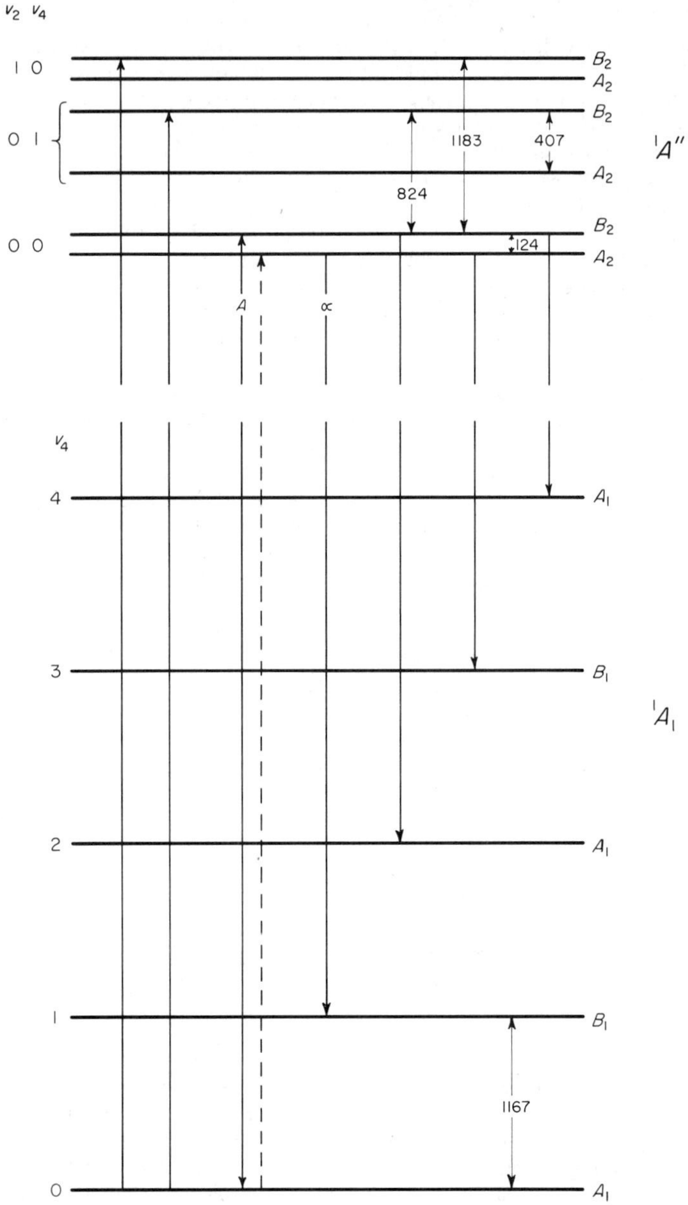


FIG. 194. **Energy level diagram for the main absorption and emission bands in the near ultraviolet spectrum of $H_2CO$.** In the upper state only the levels 0, $\nu_4'$ and $\nu_2'$ with their inversion components, in the lower state only the levels $v_4\nu_4$ are shown.

considerable staggering because the even levels are reached only from the upper, the odd levels from the lower inversion component of the upper state.

Summarizing the preceding discussion, we can say that the main bands of $H_2CO$ (in absorption or emission) are due to a $^1A'' - ^1A_1$ electronic transition. In the $^1A''$ upper electronic state the molecule is non-planar, but this state is derived from a $^1A_2$ state of the planar conformation. The individual inversion doublet components (which are split rather widely) must be described in terms of the vibronic species of point group $\boldsymbol{C}_{2v}$. The height of the barrier between the two non-planar equilibrium conformations has been estimated from the $0^+ - 0^-$ splitting under the assumption of a simple potential function to be 650 cm$^{-1}$ [Brand (138)], that is, the barrier is little higher than the zero-point energy of $\nu_4'$ (which is about 400 cm$^{-1}$). In Fig. 194 we have distinguished the inversion doubling components by $0^+, 0^-, 1^+, 1^-, \ldots$ (see also Fig. 1, p. 23). For some purposes it is more convenient to number the levels serially, $v_4 = 0, 1, 2, \ldots$; in that case the numbering remains unchanged in the transition to a planar conformation.

There are four further interesting points with regard to the near ultraviolet absorption spectrum of formaldehyde. In addition to the main absorption bands of $\perp$ structure, a number of much weaker bands with $\|$ structure have been found [Dieke and Kistiakowsky (277)]. It was recognized by Brand (138) that these $\|$ bands correspond to transitions to the opposite inversion doublet components as the main bands. In particular, one of them lies to the longward side of the $A$ band at a distance of 124 cm$^{-1}$, that is, corresponds to the transition from the lowest level of the ground state to the lowest vibrational level of the excited state (the lower inversion component of the $v_4' = 0$ level). These bands have been further studied by Callomon and Innes (178). The upper state is vibronically $A_2$. The transition to the $A_1$ lower state is rigorously forbidden by the general (electric dipole) selection rule (II, 19) even if interaction of vibrational and electronic motion is taken into account. An $A_2 - A_1$ vibronic transition can be made allowed by interaction with the rotation about the $a$-axis (see p. 265f) but such a transition should only show sub-bands with higher $K$ values while the $K = 0$ sub-band is definitely observed. Therefore rovibronic interaction is not responsible for the occurrence of these bands. Rather, Callomon and Innes (178) have shown that these bands are made possible by magnetic dipole radiation for which indeed $A_2 - A_1$ transitions are allowed (see p. 135). This conclusion is supported by much other evidence.

The strongest $\perp$ bands, representing the main progressions, are all of type $B$. Besides these, however, a number of weaker $\perp$ bands of type $C$ have been identified by Callomon and Innes (178). They must be assigned to $1 - 0$ transitions in $\nu_5'$ and $\nu_6'$, two vibrations that in the planar conformation would have the species $B_2$. The $v_5' = 1$ and $v_6' = 1$ levels have therefore $B_1$ vibronic symmetry. Transitions to these levels from the ground state arise by vibronic interaction with a $^1B_1$ electronic state while the main bands arise by interaction with a $^1B_2$ electronic state (both correspond to forbidden components of the dipole moment, the electronic transition being forbidden by symmetry in the planar conformation). The type $C$ bands may account for as much as a quarter of the total intensity of the system.

Among the hot bands at the longward side of the first main band (at 3530 Å) there are a number of very weak bands which are not temperature sensitive and persist at very low temperature. These were first observed by Cohen and Reid (217) and soon after, independently confirmed by Brand (138) and Robinson (1075). It was recognized immediately that these bands must belong to an excited electronic state different from that of the main bands and that the most likely identification for this state is $^3A_2$ of the same electron configuration as $^1A_2$. Robinson and Di Giorgio (1078)(278) have shown that in the $^3A_2$ state also the molecule is non-planar (the inertial defect is $-0.33$ a.m.u. Å$^2$). The inversion splitting is 30 cm$^{-1}$ for $v_4' = 0$ and 242 cm$^{-1}$ for $v_4' = 1$, that is, substantially smaller than in the $^1A_2$ state [see also Hodges, Henderson and Coon (559)]. Therefore, the angle of bend must be larger (about 35° for $^3A_2$, against 20° for $^1A_2$). The $^3A_2$ state splits by spin-orbit interaction into $A_1 + B_1 + B_2$ (see p. 17f) and can therefore mix with neighbouring $^1A_1$, $^1B_1$, $^1B_2$ states. Thus the vibrational transitions in $^3A_2 - {}^1A_1$ are those of a symmetry-allowed electronic transition, and the first main band corresponds to $0^+ - 0$ rather than $0^- - 0$ as in $^1A_2 - {}^1A_1$ (see Fig. 194).

In the main progression of the $H_2CO$ bands diffuseness sets in under medium resolution at about 2750 Å [Schou (1106), Herzberg (517)]. The progression continues with increasing diffuseness and rapidly decreasing intensity to 2300 Å [Everett and Minkoff (367)]. In emission, no bands with $v_4' > 0$ have been observed, suggesting that the upper states of these bands are predissociated [Brand and Reed (143)] even though in absorption the rotational lines for bands with $v_4' > 0$ are quite sharp. It is probable that this is a case of very gradual setting-in of predissociation (see p. 466); only at 2750 Å is the predissociation strong enough for a line broadening to be visible. Brand and Reed (143) assume the existence of two different predissociation limits. The first predissociation limit corresponds to an energy of 28736 cm$^{-1}$, which must be considered as an upper limit for $D$(H—CHO).

No absorption of $H_2CO$ is apparent between 2300 and 1750 Å. At shorter wavelengths four strong Rydberg series have been observed [Price (1014), Allison and Walsh (61)]. The first members are at 1750 Å, 1556, 1524 and 1397 Å respectively and the limit which is the same for all four series is at 87765 cm$^{-1}$ for $H_2CO$ and at 87905 cm$^{-1}$ for $D_2CO$ corresponding to the first ionization potential.

The four Rydberg series are easily accounted for in terms of electron configurations. The first, with the larger Rydberg correction, corresponds to a transfer of an electron from the last filled shell of the ground state, $(2b_2)^2$, to the $nsa_1$ orbitals. The next two series correspond to a transfer to two of the orbitals formed from $np$. There are three such orbitals $a_1$, $b_1$, $b_2$ (see Table 58) yielding, with one electron in $2b_2$, the states $B_2$, $A_2$, $A_1$ of which only $B_2$ and $A_1$ can combine with the ground state. The fourth Rydberg series probably corresponds to a transfer of an electron from $2b_2$ to $nd$ although the Rydberg correction (0.40) is rather large for this assignment. The splitting of the $nd$ orbitals has not been observed.

Some vibrational structure has been observed by Allison and Walsh in the first two members of the first and fourth Rydberg series. Provisionally this structure leads to the conclusion that in the upper Rydberg states also the molecule may be non-planar.

The absorption of $H_2CO$ beyond the ionization limit at 1139 Å has not yet been investigated. The Stark effect in the spectrum of $H_2CO$ has been discussed in Chapter II, section 4.

## (c) Mono-hydrides

Several four-atomic mono-hydrides have been found to have discrete electronic spectra (see Table 67). In almost all cases only a single electronic transition has been observed.

**HNCN.** The free HNCN radical and its spectrum were observed by Herzberg and Warsop (547) in the flash photolysis of diazomethane ($H_2CN_2$). A single band with the typical appearance of a ⊥ band of a (nearly) symmetric top appears at 3440 Å (see the previous Fig. 108). That this band is really due to the HNCN radical follows from the band structure and the study of the effect of isotopic substitution: The large spacing of the sub-bands and the halving of this spacing in the deuterated molecule can only be understood if the three heavier atoms are located nearly on a straight line and only the H atom is off that line. The absence of an intensity alternation (as well as the magnitude of the sub-band spacing) shows that only one H atom is present. The fact that only two sub-bands for every single one in the normal compound appear when a 60 per cent $C^{13}$, 40 per cent $C^{12}$ isotopic mixture is used, shows that only one carbon atom is present. The analysis of the $J$ structure gives a $B$ value of 0.3699 $cm^{-1}$ in the lower state which is of the same order as for other HXYZ molecules like $HN_3$ and HNCO in agreement with the assumption that three and only three heavier atoms are present. The conditions of production of the spectrum leave only N atoms as the remaining constituents even though isotope investigations with $N^{15}$ have not been carried out. The study of the $J$ structure of the $C^{13}$ substituted molecule gives nearly the same $B$ and $C$ values as for the ordinary ($C^{12}$) molecule. This result leads to the important conclusion that the carbon atom must be the central atom, that is, must be located between the two N atoms.

Even though the molecule HNCN is very nearly a symmetric top, the $K$-type doubling characteristic of an asymmetric top is clearly exhibited for $K' = 1$ and $K'' = 1$ as a doubling of all branches in the $2-1$ and $1-2$ sub-bands and by a combination defect between $P$, $R$ and $Q$ branches in the $0-1$ and $1-0$ sub-bands. The sign of the defect indicates that the band is a type C band, that is, the transition moment is perpendicular to the plane of the molecule. The planarity of the molecule in both electronic states follows from the positive sign and small magnitude of the inertial defect. Other geometrical data for the two states are given in Table 67. The observation of a single band for this electronic transition is in conformity with the Franck–Condon principle since the dimensions change so little in the transition.

From the band structure it follows that the electronic transition is either $A''-A'$ or $A'-A''$. The electron configuration of the ground state must be similar to that of the iso-electronic NCO, i.e., $\ldots \pi^4\sigma^2\pi^3$. The replacement of O by NH, pointing in a direction off the N—C—N axis, results in a splitting of each of the $\pi$ orbitals into an $a'$ and $a''$ orbital, of which the former might be expected to lie lower. The ground state of HNCN would therefore have the electron configuration

$$\ldots a'^2 a''^2 a'^2 a'^2 a''\ {}^2A''$$

and the observed transition would be ${}^2A' - {}^2A''$. The doublet splitting has not been resolved. It is expected to be small.

**HONO.** Melvin and Wulf (820) have observed in the absorption spectrum of mixtures of NO, $NO_2$ and $H_2O$, a system of simple bands in the region 3850 to 3150 Å which they have ascribed to HONO. This spectrum has been further studied by Tarte and D'Or (1193)(291), Porter (1008) and quite recently by King and Moule (669). Porter and King and Moule have also studied the spectrum after deuterium and $N^{15}$ substitution and have confirmed the identity of the carrier. The independent studies by D'Or and Tarte (290) and Jones, Badger and Moore (647) of the infrared spectrum have firmly established the presence of two tautomeric species *cis*- and *trans*-HONO of which the latter has the lower energy. King and Moule have interpreted the ultraviolet spectrum on this basis. Both isomers give rise to a progression in an N=O stretching vibration $\nu_2' \simeq 1110\ cm^{-1}$ thus accounting for the two strongest progressions. In addition a weak progression in which a lower state frequency 620 $cm^{-1}$ is excited has been found.

The electronic transition according to King and Moule is one in which an electron goes from a lone pair orbital to an antibonding $\pi$ orbital. All the absorption bands are diffuse presumably because of predissociation. Unlike most other cases of polyatomic molecules, here the diffuseness decreases with increasing energy.

**HFCO.** An extensive discrete spectrum of HFCO (and DFCO) has recently been discovered and studied in detail by Giddings and Innes (418)(419). The spectrum extends from 2700 to 2000 Å with an intensity maximum at 2100 Å. The most prominent frequency difference in the spectrum is 1105 $cm^{-1}$ which gives rise to long progressions with slowly decreasing spacing. In addition, shorter progressions in the frequencies 467 and 570 are observed. The frequencies 1105 and 467 change only slightly upon deuterium substitution and are therefore assigned to the C—O stretching and OCF bending vibrations respectively. The large change of $\nu$ (CO) in going from the ground state (where it is 1837 $cm^{-1}$) to the excited state corresponds well with the length of the progression. Clearly the CO distance is changed considerably in the electronic transition, even more than in $H_2CO$.

There can be no question that the electronic transition of HFCO is the analogue of that in $H_2CO$, and it is therefore probable that in the excited state the molecule is non-planar. This suggestion appears to be confirmed by the presence of progressions of three members in the vibration 570 $cm^{-1}$ which Giddings and Innes (419) interpret as the out-of-plane bending vibration. The remaining three fundamentals of the excited state of HFCO (see Table 67) are uncertain because they are based on the assumed assignment of very few bands.

Although the rotational structure is only incompletely resolved, even under the highest resolution used, Giddings and Innes (418) and in greater detail Parkin and Innes (964) have succeeded in obtaining reliable rotational constants for the upper state by the procedure of calculating the band structure with various trial values of the constants and modifying them until a complete fit is obtained. They have shown that the transition moment, at least for the bands that have been studied, is perpendicular to the plane of the molecule in the ground state. The resulting constants and geometrical parameters are given in Table 67.

## (d) Non-hydrides

The ultraviolet absorption spectra of only a few four-atomic molecules not containing hydrogen have been studied in detail. Several of them show only continuous or diffuse absorption, and not much information can be obtained. We shall consider here only those for which fairly extensive investigations have been carried out.

**$C_2N_2$.** With an absorbing path of about 6 m atm. $C_2N_2$ shows a system of absorption bands ($\tilde{a}-\tilde{X}$) in the region 3020 to 2400 Å [Woo and Liu (1316)]. The bands are sharp and slightly degraded to the red. They form two fairly long progressions in a frequency 2050 $cm^{-1}$ which are separated by 896 $cm^{-1}$. If $C_2N_2$ were linear in the upper state of this band system, as it is in the ground state, only two vibrations could be excited strongly, viz., the two totally symmetric vibrations $\nu_1$ and $\nu_2$, and the magnitude of the two observed intervals fits with this interpretation since they are not too different from the values in the ground state (see Table 68). To be sure, there are a number of weaker bands which cannot be readily assigned to any of the progressions in $\nu_1$ and $\nu_2$. They correspond probably to sequences in the bending vibrations $\nu_4$ and $\nu_5$. Recently Callomon and Davey (175) using much longer absorbing paths have been able to resolve the fine structure of the $0-0$ band of this system finding four branches. The form of these branches and the observed intensity alternation identify the transition unambiguously as ${}^3\Sigma_u^+ - {}^1\Sigma_g^+$ (see Chap. II, section 3a($\alpha$)) and confirm that the molecule is linear in the excited (${}^3\Sigma_u^+$) state just as in the ground state. A more detailed vibrational analysis by the same authors (175a) gives the frequencies listed in Table 68.

A somewhat stronger absorption system begins at 2300 Å and extends to 1800 Å

[Woo and Badger (1312), Mooney and Reid (875)]. From an examination of high-resolution plates taken by P. K. Carroll at Ottawa, it appears that all the bands are degraded to the red, and most, but by no means all, are double-headed. Some show resolved rotational structure indicating that the double heads correspond to $R$ and $Q$ branches and suggesting that the transition is vibronically $^1\Pi_u - {}^1\Sigma_g^+$ of a linear or near-linear conformation. A more detailed vibrational analysis by Callomon and Davey (175a) and Cartwright, Walsh and Warsop (182b) has shown that the main features of the system can be interpreted in terms of a forbidden electronic transition of the linear – linear type made vibronically allowed by $\nu_4(\pi_g)$. The $1-0$ and $0-1$ bands in $\nu_4$ are at 45675.5 and 44896.9 cm$^{-1}$. The upper electronic state, $\tilde{A}$, is probably $^1\Delta_u$ or $^1\Sigma_u^-$ of the same configuration... $(\pi_g)^3\pi_u$ as the state $\tilde{a}\ {}^3\Sigma_u^+$. Several progressions with spacings of about 2140 cm$^{-1}$ have been observed. Most likely this frequency corresponds to the $\nu_1$ vibration of the upper state. The main bands are accompanied by closely-spaced groups of bands which represent probably sequences in the low-frequency bending vibration.

Price and Walsh (1029) appear to be the only authors who have studied the absorption of $C_2N_2$ in the vacuum ultraviolet. A system of fairly sharp strong absorption bands begins at 1700 Å and extends to 1450 Å. On Price and Walsh's published spectrograms the strongest progression has a spacing of about 1900 cm$^{-1}$, again probably corresponding to $\nu_1$. A still stronger system of absorption bands starts at 1320 Å. These bands are diffuse and appear to come to a Rydberg limit at about 900 Å corresponding to an ionization potential of 13.8 eV. No detailed analysis of these bands has been given.

**$NO_3$.** When $NO_2$ is treated with ozone diluted with oxygen, the resulting gas mixture shows a blue color which, according to Jones and Wulf (646), is due to the absorption spectrum of the free $NO_3$ radical temporarily formed in the reaction. The spectrum of $NO_3$ has recently been studied under high resolution by Ramsay (1044). There are about 20 diffuse bands in the region 6650 to 5000 Å. A small isotope shift observed when $N^{15}O_2$ is used confirms that the molecule contains an N atom but is not sufficient to make the identification of the molecule and the vibrational assignments unambiguous. The $0-0$ band is at 6625 Å; it is followed by a short progression in $\nu_1' = 930$ cm$^{-1}$. The remaining bands must correspond to excitation of the vibrations $\nu_2'$, $\nu_3'$ and $\nu_4'$. If the molecule had $\boldsymbol{D}_{3h}$ symmetry in both upper and lower state, excitation of these non-totally symmetric vibrations should be very weak. That, in fact, at least some of them are quite strongly excited is in all probability due to Jahn–Teller interaction in the excited state. According to Walsh (1268), this state is a $^2E'$ state produced from the ground state $\ldots(e')^4a_2'\ {}^2A_2'$ by excitation of an electron from the $e'$ to the $a_2'$ orbital.

**$Cl_2CO$.** The near ultraviolet spectrum of phosgene was first studied by Henri and Howell (501) and more recently by Giddings and Innes (419). Starting at 3100 Å there are a large number of sharp absorption bands which become gradually stronger and more diffuse at shorter wavelengths. The strongest bands are near 2800 Å. Giddings and Innes have studied the effect of temperature on the bands and have ascertained that many of the bands assigned as hot bands by Henri and Howell do not increase in intensity with temperature. Their revised vibrational analysis yields long progressions in the two frequencies $\nu_4' = 581$ and $\nu_6' = 430$ cm$^{-1}$ of the excited state and short progressions in the corresponding frequencies $\nu_4'' = 580$ and $\nu_6'' = 440$ cm$^{-1}$ of the ground state. Since these vibrations are out-of-plane and antisymmetrical bending vibrations respectively in the ground state Giddings and Innes conclude that the molecule is non-planar and skewed in the excited state (i.e. has no plane of symmetry).

The absorption spectrum in the vacuum ultraviolet has been studied by LaPaglia and Duncan (725) who found a broad continuum with maximum at 1550 Å followed by five fairly well separated discrete band systems. Each of these appears to consist of a single progression in a vibration that can either be the symmetrical bending or C—Cl stretching vibration (see Table 68). The most intense feature of the absorption spectrum is the system at 1240 Å. There is strong continuous absorption beyond 1050 Å, a wavelength that corresponds to the ionization potential of 11.77 eV obtained from electron impact studies. While the upper states of the discrete band systems are clearly Rydberg states no Rydberg series have been found.

**$Cl_2CS$.** Four band systems of thiophosgene have been found in the visible and near ultraviolet regions. Burnelle (165), using a path length of 50 cm, found a very weak absorption system extending from 7000 to 5300 Å, which he ascribes to the first triplet – singlet transition $^3A_2 - {}^1A_1$, the analogue of the $\tilde{a} - \tilde{X}$ bands of $H_2CO$. At shorter wavelengths between 5950 and 3900 Å lies a much stronger system first observed by Henri (20) which corresponds to the $\tilde{A} - \tilde{X}$ system of $H_2CO$. The vibrational analysis of this system, first attempted by Henri and Duchesne (500) and extended by Duchesne (313) and Burnelle (165), has recently been completed in a detailed investigation by Brand, Callomon, Moule, Tyrrell and Goodwin (139). They have shown conclusively that the molecule is non-planar in the excited state with an out-of-plane angle of 32°. Just as for $H_2CO$, the inversion doubling, particularly for the higher levels of the out-of-plane bending vibration, is fairly large, and therefore the upper state and its vibrational levels can be classified according to point group $\boldsymbol{C}_{2v}$. All but one of the vibrational frequencies in the excited state have been determined (see Table 68). $Cl^{35}$—$Cl^{37}$ isotope shifts for some bands are considerable and have helped in the assignment. A third system extending from 2970 to 2690 Å has been found by Henri (20) and is assigned by Burnelle to an allowed $^1B_2 - {}^1A_1$ transition. A fourth system extending from 2770 to 2390 Å, and also found by Henri, consists of diffuse bands. It is the strongest of the four systems and is assigned to a $^1A_1 - {}^1A_1$ transition.

## 3. Five-Atomic Molecules

Again, we shall consider the five-atomic molecules in groups according to the number of hydrogen atoms they contain: tetra-hydrides, tri-hydrides, di-hydrides, mono-hydrides and non-hydrides.

### (a) Tetra-hydrides

**$CH_4$ (methane).** As far as is known at present, all known $XH_4$ molecules exhibit only continuous absorptions. We shall consider briefly only the $CH_4$ molecule because of its general importance.

Earlier workers did report discrete absorption bands of methane in the region 1800–1400 Å, but later workers have shown that these are due to impurities [see for example Duncan and Howe (323)]. Quantitative absorption measurements have been carried out by Wilkinson and Johnston (1302), Moe and Duncan (866), Sun and Weissler (1177) and Ditchburn (279). No absorption with an absorption coefficient greater than 0.3 $cm^{-1}$ has been observed above 1455 Å, and probably the absorption coefficient is much less. At 1455 Å absorption begins rather suddenly reaching $k = 70$ $cm^{-1}$ at 1370 Å and rising to a maximum of 1500 $cm^{-1}$ at about 930Å [2]. Figure 195 shows the absorption curve of Ditchburn (279). It is clear from the shape

[2] In solid $CH_4$ the limit is shifted to 1390 Å [see Dressler and Schnepp (309)].

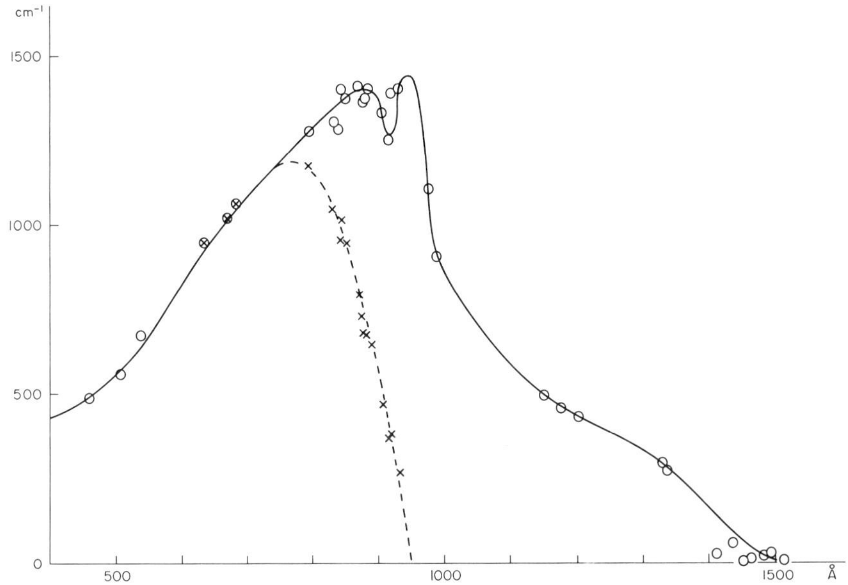


FIG. 195. **Absorption curve of $CH_4$ in the region 1500 to 400 Å after Ditchburn (279).** The broken line curve gives the photoionization cross section as determined by Ditchburn from the data of Wainfan, Walker and Weissler (1253).

of the absorption curve that it is composite in origin. The photoionization efficiency in $CH_4$ was measured by Wainfan, Walker and Weissler (1253). Following Ditchburn, their values, converted to ionization cross sections, have been included in Fig. 195. The ionization potential is $12.99 \pm 0.01$ eV [as given by Watanabe (1273)] corresponding to 946 Å. It is seen that below 770 Å all the absorption is due to photoionization; but at longer wavelengths part of and above 946 Å all of the absorption must be due to dissociation processes.

All excited (unstable) states of $CH_4$ are of the Rydberg type. Continua may arise because these Rydberg states are reached in a region of their potential surfaces above the lowest dissociation limit (here it must be noted that $CH_4^+$ has a relatively small dissociation energy, viz., 1.0 eV) or because they are intersected by one of the states arising from $CH_3 + H$ or $CH_2 + H_2$ or $CH_2 + H + H$ or $CH + H_2 + H$ and because as a consequence such a strong predissociation results that in effect the spectrum is continuous.

The first excited singlet (Rydberg) state (see p. 348) is $\ldots(1f_2)^5 3a_1\ ^1F_2$ where $3a_1$ is essentially a $3s$ atomic orbital[3]. In all probability this $^1F_2$ state is the upper state of the first absorption continuum starting at 1455 Å. It is not likely that the potential surface of this state is intersected by or goes over into the potential surface of the ground state of $CH_4$ which arises from $CH_3(^2A_2'') + H(^2S)$. Therefore, since only one singlet state arises from these products and since the spin rule must be observed for a strong

[3] It must be noted that the $3a_1$ orbital, while it is of the Rydberg type for small internuclear distances, goes over, according to Fig. 129, for larger internuclear distances into the strongly anti-bonding orbital that arises from the four H atom $1s$ orbitals (with some admixture of $2s_C$). Thus the first Rydberg transition may also be described as a $V-N$ or $\sigma^* - \sigma$ transition, that is, it would be of the charge-transfer type (see p. 419).

predissociation, it is most unlikely that the first continuum corresponds to dissociation into $CH_3 + H$ in their ground states. The same consideration applies to the other strong continua whose upper states are singlet states. However, a dissociation into $CH_3(^2E') + H(^2S)$ is definitely possible. Here the $^2E'$ state is the as yet unobserved first excited state of $CH_3$.

The energy at 1455 Å would also be sufficient for dissociation into $CH_2 + H_2$. However, on account of spin conservation (assuming the excited state of $CH_4$ to be a singlet state), the resulting $CH_2$ cannot be in the $^3\Sigma_g^-$ ground state but must be in a singlet state, probably the excited $^1B_1$ state; the lowest singlet state of $CH_2$, $^1A_1$, would result from the ground state of $CH_4$. The energy would also be just sufficient for a dissociation into $CH_2 + H + H$ or $CH + H_2 + H$. At 1236 Å photodetachment of $H_2$ has been shown to be an important but not exclusive primary photolytic step by Mahan and Mandel (791) in mixed isotope studies.

## (b) Tri-hydrides

**$CH_3I$.** Methyl iodide has a continuous absorption in the near ultraviolet which, with a path length of 6 m, extends to 3600 Å [Bayliss (102)] and has a maximum at 2600 Å. Quantitative measurements of the absorption coefficients to 2100 Å were made by Porret and Goodeve (1004). From their data, one finds $k \approx 30\ cm^{-1}$ near the maximum. As was first found by Herzberg and Scheibe (539), an extremely strong discrete absorption system starts at 2012 Å (with a few much weaker hot bands at longer wavelengths). For the strongest bands the absorption coefficient is about 5000 $cm^{-1}$. The previous Fig. 102 shows part of this band system. The main progression consists of four or five narrow bands with a spacing of 1090 $cm^{-1}$. This spacing is readily interpreted as corresponding to the totally symmetric $CH_3$ deformation vibration $\nu_2'$, which in the ground state is 1251 $cm^{-1}$. The change of frequency is in harmony, according to the Franck–Condon principle, with the length of the progression. Several of the weaker bands of similar appearance can be interpreted by the excitation of one quantum of the other two totally symmetric vibrations $\nu_1' = 2660$ and $\nu_3' = 499\ cm^{-1}$. All bands are slightly diffuse.

In addition to the narrow bands there are a number of weaker bands of quite different structure, first resolved by Scheibe, Povenz and Linström (1102) and Henrici and Grieneisen (505). Additional such bands have been found more recently by Polo (1000) and Dunn and Herzberg (326). Each of these bands consists of a series of slightly diffuse lines with the typical intensity alternation ..., strong, weak, weak, strong, ... showing that they represent the unresolved $Q$ branches of the sub-bands of $\perp$ bands while the main bands appear like $\parallel$ bands. The spacing of the line-like $Q$ branches in some of the $\perp$ bands is about 20 $cm^{-1}$, i.e., more than twice the spacing in the infrared $\perp$ bands. The separations of these bands from the main bands can be easily understood as corresponding to an excitation of the vibration $\nu_5' = 1243$ and $\nu_6' = 844\ cm^{-1}$; indeed, one hot band of the same type appears at 888 $cm^{-1}$ longward of the $0-0$ main band which agrees nicely with the infrared value $\nu_6'' = 882\ cm^{-1}$.

The occurrence of these weak $\perp$ bands in an allowed strong system of $\parallel$ bands could be interpreted as produced by a forbidden dipole component induced by perturbation by a nearby $E$ electronic state (see p. 241f). However, this interpretation would not account for the anomalous spacing in these bands. A satisfactory explanation of the wide spacing and, at the same time, of the occurrence of these bands, was first given by Mulliken and Teller (917) on the basis of a careful study of the interaction of electronic, vibrational and rotational angular momenta in this case. They showed that the upper electronic state of this band system is an $E$ state, that is, that the main bands are $\perp$ bands. As explained in more detail in Chapter II, section 3(b), if the $E$

state has a $\zeta_e$ value of approximately 1, then the spacing within the main bands, viz. $2[A(1 - \zeta_e) - B]$, will be small (of the order $2B$), and these bands will look like $\parallel$ bands as indeed they do.

When in the upper state one of the $\perp$ vibrations $\nu_4'$, $\nu_5'$ or $\nu_6'$ is singly excited, three vibronic states $E$, $A_1$, $A_2$ arise of which the $E$ vibronic state can combine with the $^1A_1$ ground state (see Fig. 61, p. 161). Since in the upper state the effective $\zeta$ is now approximately $-(\zeta_e + \zeta_v)$ (see p. 64) the spacing of the sub-bands ($Q$ branches) in the $1-0$ ($E - A_1$) band will be $2[A(1 + \zeta_e + \zeta_v) - B]$ which for $\zeta_e \approx 1$ and $\zeta_v$ positive is indeed greater than twice the spacing in the corresponding infrared band for which it is $2[A(1 - \zeta_v) - B]$. Similar results are found for the $0-1$ hot band. The intensity of the widely spaced bands relative to the main bands is determined by the strength of the vibronic interaction (see p. 236). Conversely we can conclude from the observed intensity ratio $I_{1-0}/I_{0-0}$ that Jahn–Teller interaction is weak and therefore that the equilibrium conformation in the excited state deviates only very slightly from the symmetrical ($\boldsymbol{C}_{3v}$) conformation (see, for example, Fig. 23a). Unfortunately, the vibronic splitting (Jahn–Teller splitting) between the three levels $E$, $A_1$, $A_2$ arising when one of the degenerate vibrations is singly excited has not been determined since the $1-1$ bands which should show transitions to all three levels (see Fig. 61) are too strongly overlapped by the neighboring much stronger $0-0$ band.

Even after all but one ($\nu_4'$) of the upper state fundamentals have been identified in this way there remain a number of bands of medium intensity which cannot be accounted for on the basis of a single $E-A_1$ electronic transition. In particular, the band at 2000.4 Å (Fig. 102) shows a much narrower $K$ structure than the bands of the main system yielding an effective $\zeta$ value of $-0.063$. Dunn and Herzberg (326) have recently shown that this band and several others can be well accounted for as produced by another $E-A_1$ electronic transition. Such a transition close to the main $E-A_1$ transition is indeed expected on the basis of a consideration of the electron configuration. In the ground state the last filled orbital must be an $e$ orbital [see Mulliken and Teller (917)] and therefore the first excited Rydberg state is $\ldots(e)^3(6sa_1)\ {}^3E$, $^1E$ corresponding to $^3\Pi$, $^1\Pi$ of a linear molecule. Because of the presence of the iodine atom, spin-orbit coupling is large and we expect the arrangement of levels at the right of Fig. 137, p. 338, that is, there are two close-lying $E$ states corresponding to $^2E_{\frac{3}{2}}$ of the $CH_3I^+$ ion. The higher one of the two $E$ states, corresponding to $^3\Pi_1$ of the linear case, combines strongly with the $^1A_1$ ground state and represents the main absorption system at 2012 Å; the lower one, corresponding to $^3\Pi_2$, gives rise to the weaker system. The $\zeta$ value observed for the 2000.4 Å band can be accounted for if it is assumed to be the $\nu_6'-0$ band of the weaker system.

Another strong band system $\tilde{D}-\tilde{X}$ starting at 1831 Å is similar in structure to the system $\tilde{C}-\tilde{X}$ starting at 2012 Å except that it is slightly more diffuse and no bands with $K$ structure are resolved. This system presumably corresponds to the third (highest) $E$ state in Fig. 137 (derived from $^1\Pi_1$) which is correlated with $^2E_{\frac{1}{2}}$ of the ion. The $\tilde{D}-\tilde{X}$ system is separated from the $\tilde{C}-\tilde{X}$ system by 4900 $cm^{-1}$ which is close to the energy difference $E_{\frac{1}{2}} - E_{\frac{3}{2}}$ of the ion.

At shorter wavelengths a large number of further discrete electronic transitions have been observed by Price (1016). They form several clear Rydberg series going to two limits, at 76930 and 81990 $cm^{-1}$, which correspond to the two components of the $^2E$ ground state of the ion. Some of the lower members of the Rydberg series are strongly predissociated. A more detailed analysis of the sharp members would be of interest. Unfortunately, at present there is no theoretical guide to the relative intensities of different Rydberg series, and therefore a detailed assignment of the different observed series is not possible.

Table 69 lists the data for the lower electronic states of $CH_3I$. It includes also corresponding data about $CH_3Br$ and $CH_3Cl$, whose absorption spectra are similar to that of $CH_3I$ except that the lower excited states are much more strongly predissociated. It is interesting to note that the separation of the series limits (as well as of the first two strong band systems) is smaller than for $CH_3I$ in agreement with the expected smaller spin-orbit interaction.

## (c) Di-hydrides

**$H_2CCO$ (ketene).** Even though ketene is iso-electronic with $CO_2$, its ultraviolet absorption spectrum shows little similarity to that of $CO_2$. A fairly weak but extensive system of diffuse bands in the region 3850 to 2600 Å was first observed by Lardy (726) and studied in more detail by Norrish, Crone and Saltmarsh (940). The bands are each about 12 Å wide and appear strongly with a path of 15 cm atm. No vibrational analysis has been attempted, but a vibrational interval of 400 $cm^{-1}$ occurs frequently. The system is generally ascribed to an absorption in the =CO group similar to that in $H_2CO$. The strong occurrence of a low frequency fundamental (bending vibration) suggests that the molecule is non-planar in the excited state just as is $H_2CO$. The diffuseness of the bands gives an upper limit to the dissociation energy, presumably $D(CH_2—CO)$, of 3.3 eV.

Another much stronger system of diffuse bands occurs between 2130 and 1930 Å [Price, Teegan and Walsh (1025)]. It appears to consist of a simple progression with a spacing of about 1100 $cm^{-1}$ which probably corresponds to the CCO stretching vibration in the upper state.

Several very strong discrete absorption systems occur in the region 1850 to 1300 Å. They are easily observable at a pressure of 0.05 mm in an absorption tube of 15 cm length. The first system, between 1850 and 1700 Å, is well separated from the next which starts at 1630 Å. At shorter wavelengths there is more and more mutual overlapping, but Price, Teegan and Walsh have recognized five further separate band systems. All of them have progressions in a frequency of about 1025 $cm^{-1}$. In the first system the main progression is in a frequency of 830 $cm^{-1}$. In the others, a second frequency that varies between 494 and 627 $cm^{-1}$ occurs with one quantum. Four of the band systems, together with two further bands at the shortward end, form a good Rydberg series which yields an ionization potential of $9.60_7$ eV.

All bands below 1850 Å are quite sharp, but no fine structure has been resolved. It is probable that the observed features represent $Q$ branches of unresolved $\|$ bands. For $\perp$ bands, since $A$ is large, the $K$ structure would have been easily resolved. While it may seem strange that all observed electronic transitions are of the $\|$ type (i.e., ${}^1A_1 - {}^1A_1$), it must be remembered that in equally strong $\perp$ bands the intensity is distributed over a much greater interval, and therefore, such bands will not be as prominent. The $\perp$ transitions which are expected are probably forming an unresolved background in the spectrum.

**$H_2CN_2$ (diazomethane).** It is not surprising that the spectrum of diazomethane is closely similar to that of ketene, since these two molecules are isoelectronic and have the same geometrical structure. However, the whole spectrum is shifted to longer wavelengths since the ionization potential is smaller (see below). As is well known, diazomethane is a yellow gas. The color is caused by a weak absorption below 4750 Å first studied by Kirkbride and Norrish (671) [see also Brinton and Volman (151)]. Starting at 4710 Å there are a number of very diffuse bands which at about 4200 Å merge into a continuous spectrum with a maximum at 3950 Å. The diffuse nature even of the first bands (they are about twice as wide as those of ketene) implies that the dissociation energy $D(CH_2—N_2)$ is less than 2.6 eV. A second much stronger continuous absorption starts at 2650 Å with a maximum at 2175 Å.

The absorption spectrum below 2100 Å was first studied by Herzberg (521) and in much greater detail by Merer (823). In this region only discrete absorption bands are found. Figure 196 shows a spectrogram in the region 1700 to 1350 Å. There is an extensive band system near 1900 Å consisting mainly of $\perp$ bands. Merer has resolved the $K$ structure of several bands and for $D_2CN_2$ also the $J$ structure, and

FIG. 196. **Absorption spectrum of diazomethane ($CH_2N_2$) in the region 1700 to 1350 Å.** The main feature is the extended Rydberg series which is marked. At longer wavelengths several non-Rydberg transitions have been observed [see Merer (823)].

has shown that there are three mutually overlapping (and interacting) electronic transitions, two of type ${}^1B_1-{}^1A_1$ and one of type ${}^1B_2-{}^1A_1$. The (weak) presence of $\|$ components in these transitions and other evidence suggest strongly that the molecule retains $\boldsymbol{C}_{2v}$ symmetry in the excited states.

As can be seen from Fig. 196 an extended and very clear Rydberg series begins with a narrow group of bands at 1670 Å and leads to a limit at 1378 Å corresponding to an ionization potential of $8.99_9$ eV. The small Rydberg correction (0.10) suggests that this series corresponds to an excitation of the most loosely bound electron to an *nd* orbital. From such a configuration five close-lying singlet states arise only one of which (${}^1A_2$) cannot combine with the ground state. It is probable that the main Rydberg series corresponds to one of the ${}^1A_1$ components since the Rydberg bands, just as for ketene, appear to be $\|$ bands.

There is, in addition, a perpendicular-type system at 1585 Å, similar in character to the 1900 Å group. These two systems represent probably the first two members of an *np* Rydberg series. The Rydberg correction is 0.67, an acceptable value for an *np* orbital. Finally there is a group of diffuse bands near 1750 Å which do not seem to belong to any Rydberg series.

## (d) Mono-hydrides

Until recently the haloforms $CHF_3$, $CHCl_3$, $CHBr_3$ and $CHI_3$ were the only five-atomic monohydrides that had been studied. Their absorption spectra are entirely continuous; references to the literature are given in Table 70.

**NC-$C_2$H (cyano-acetylene).** Quite recently Job and King (628b) have obtained interesting discrete spectra of cyano-acetylene. Two absorption systems have been found, one $\tilde{A}-\tilde{X}$ extending from 2700 to 2400 Å and consisting of a large number of sharp violet-degraded bands and a second somewhat stronger one, $\tilde{B}-\tilde{X}$, extending from 2300 Å to shorter wavelengths and also consisting of sharp bands.

The bands of the first system have both a $K$ and a $J$ structure. They appear to be the exact analogue of the $\tilde{A}-\tilde{X}$ system of $C_2H_2$. In the excited state the molecule is non-linear, the H atom lying well off the C≡C axis and the chain of heavy atoms being bent probably between C≡C and C≡N such that the molecule has a trans structure.

The second system, $\tilde{B}-\tilde{X}$, is much simpler in structure than the first. In the excited state the molecule appears to be linear. The 0–0 band is absent, that is, the electronic transition is forbidden. Its occurrence is induced by the bending vibrations; $1-0$ and $0-1$ bands in all three bending vibrations have been observed. The upper state is either $^1\Delta$ or $^1\Sigma^-$. The analogy with the $\tilde{A}-\tilde{X}$ system of $C_2N_2$ is striking. The vibrational frequencies in the $\tilde{B}$ state of $NCC_2H$ (see Table 70) are similar to those in the $\tilde{A}$ states of $C_2N_2$ and $C_2H_2$.

## (e) Non-hydrides

The absorption spectra of a large number of tetrahalides of C, Si, Sn, Se, Te have been investigated, but only continuous absorptions (analogous to those of $CH_4$, see sub-section (a)) have been found. We shall not discuss them here. References to earlier work may be found in Sponer and Teller (1154).

**$CF_3I$.** The spectrum of $CF_3I$ was first described by Sutcliffe and Walsh (1180). Independently it had also been studied under high resolution by Polo, Herzberg and Dunn (unpublished). In many ways, the spectrum is similar to that of $CH_3I$: there is a weak continuous absorption with a peak at 2650 Å, followed by a very strong absorption system consisting of sharp bands in the region 1740 to 1630 Å (see Fig. 62, p. 162). A second system of sharp bands begins at 1599 Å. The two systems clearly correspond to the strong $\tilde{C}-\tilde{X}$ and $\tilde{D}-\tilde{X}$ systems of $CH_3I$. Like the latter, they are separated by about 5000 cm$^{-1}$, but they are shifted appreciably to shorter wavelengths. There can be little question that the two systems correspond to the two doublet components of the ground state of $CF_3I^+$ for which, just as for $CH_3I^+$, the splitting is expected to be of the same order as in the ground state of the iodine atom. While a number of further bands and band systems have been found by Sutcliffe and Walsh at shorter wavelengths, these are all diffuse and have not been arranged in Rydberg series.

The structure of the 1740 Å system has been analyzed in considerable detail [Sutcliffe and Walsh (1180), Dunn and Herzberg (326)]. As discussed in Chapter II, section 2b($\alpha$), the principal bands are immediately assigned to progressions in the three totally symmetric vibrations $\nu_1'$, $\nu_2'$, $\nu_3'$. Weaker bands longward of the main bands are easily accounted for as hot bands in which the three totally symmetric vibrations of the ground state $\nu_1''$, $\nu_2''$, $\nu_3''$ are singly (or in one case doubly) excited, while the upper state is the same as that of the $0-0$ band at 1740 Å. Each of the main bands, and at higher pressures even each of the hot bands just mentioned, is accompanied by a number of satellites which clearly represent the first and possibly second members of sequences in all but the two highest fundamentals ($\nu_1$ and $\nu_4$). In the sequence members corresponding to the degenerate vibration $\nu_6$, there is evidence of Jahn–Teller splitting in the upper state (see p. 163 and Fig. 63) confirming the expectation that the excited state is an $E$ electronic state.

In addition to the bands mentioned so far which are pseudo-parallel since $\zeta_e \approx 1$ (see p. 235), two very weak bands have been found which consist each of a nearly equidistant series of lines with a characteristic intensity alternation . . ., strong, weak, weak, strong, . . . . These bands are exactly analogous to the similar $CH_3I$ bands previously discussed (see Fig. 102). They are typical $\perp$ bands, but, just as for $CH_3I$, the spacing of the $Q$ branches is much larger than corresponds to the $2(A - B)$ value. Again this large spacing must be ascribed to the fact that in the upper state an $e$ vibration is singly excited and that vibronic interaction causes the effective $\zeta$ value to be negative and large.

No analogue of the $\tilde{B}-\tilde{X}$ transition of $CH_3I$ (corresponding to $^3\Pi_2-{}^1\Sigma^+$ of the linear case) has been found for $CF_3I$. Its absence may only be apparent because of overlapping by the much stronger main system.

The molecular constants of $CF_3I$ in the electronic states discussed above are collected in Table 71.

**$OsO_4$.** One of the very few tetrahedral molecules for which discrete absorption systems have been observed and analyzed is osmium tetroxide. The spectrum was first observed by Lifschitz and Rosenbohm (749) and Kato (662) and studied in detail by Langseth and Qviller (722). It consists of a long progression of band groups extending from 3200 to 2145 Å. The intensity distribution in this progression is, however, not what would be expected for a single progression. There is a minimum at 2630 Å, and all bands below 2500 Å are much stronger than those above 2500 Å. Langseth and Qviller have arranged the bands into three band systems, with $0-0$ bands at 33388, 36225 and 38101 cm$^{-1}$. The spacings in the main progressions are 811, 835 and 832 cm$^{-1}$ in the first, second and third system respectively. Most probably these frequencies correspond to the ground state frequency 971 cm$^{-1}$, which is observed as by far the strongest line in the Raman spectrum and is therefore in all probability the symmetrical breathing vibration $\nu_1$. These assignments are in conformity with the vibrational selection rules for allowed electronic transitions (see Chap. II, section 2b($\alpha$)); but Langseth and Qviller have assigned some other bands to transitions involving changes of the non-totally symmetric vibrations $\nu_3$ by one quantum (i.e. $\Delta v_3 = \pm 1$)[4]. It would appear that these bands can be reassigned to members of $\Delta v_i = 0$ sequences in the non-totally symmetric vibrations.

## 4. Six-Atomic Molecules

### (a) Tetra-hydrides

**$C_2H_4$ and $C_2D_4$.** The history of the investigation of the absorption spectrum of ethylene has been well presented by Wilkinson and Mulliken (1303) and will not be repeated here. The longest wavelength absorption is a progression of extremely weak and diffuse bands extending with increasing intensity from 3400 to 2600 Å, first observed in a path of 2.5 m of liquid ethylene by Reid (1064). In view of the extremely low intensity of these bands, it seems very probable that they correspond to a transition from the ground state to the lowest predicted triplet state $^3B_{1u}$ [the $T$ state of Mulliken (904)(914a)]. Indeed, more recent work by Evans (365a) has shown that this absorption system can be observed in the gas if it is mixed with oxygen which, because of its triplet ground state, is known to bring out singlet – triplet transitions in other molecules (see p. 559). As we have seen in Chapter III, section 2d, in the triplet state as well as in the corresponding singlet state, the $C_2H_4$ molecule is predicted to have a $\boldsymbol{D}_{2d}$ equilibrium position, i.e., the two $CH_2$ groups are twisted by 90° with respect to each other. There is nothing in the observed spectrum to prove or disprove this structure. The average spacing, 995 cm$^{-1}$, of the bands in the observed spectrum probably corresponds to the C—C stretching vibration $\nu_2$ in the $^3B_{1u}$ state.

A much stronger absorption begins in the liquid at 2600 Å, but in the gas has only been observed from about 2100 Å on. As shown by Fig. 197 it also consists of a progression of diffuse bands with rapidly increasing intensity. They merge into a continuum at about 1750 Å where at the same time another strong electronic transition is starting. The continuum reaches a flat maximum at about 1620 Å.

It is now generally agreed that the diffuse bands and the continuum form a single electronic transition, namely, the transition to the first excited singlet state $^1B_{1u}$ corresponding to the triplet state $^3B_{1u}$ already mentioned. The transition may be designated $\tilde{A}\ ^1B_{1u} - \tilde{X}\ ^1A_g$. The upper state is Mulliken's $V$ state which corresponds to the upper state of the Schumann–Runge bands of the isoelectronic $O_2$ molecule. In $C_2D_4$ Wilkinson and Mulliken (1303) have found that each of the diffuse bands is resolved into a number of peaks which they consider as corresponding to the twisting

[4] From the more recent Raman work of Woodward and Roberts (1320), it appears that Langseth and Qviller's values for $\nu_3$ were erroneous.

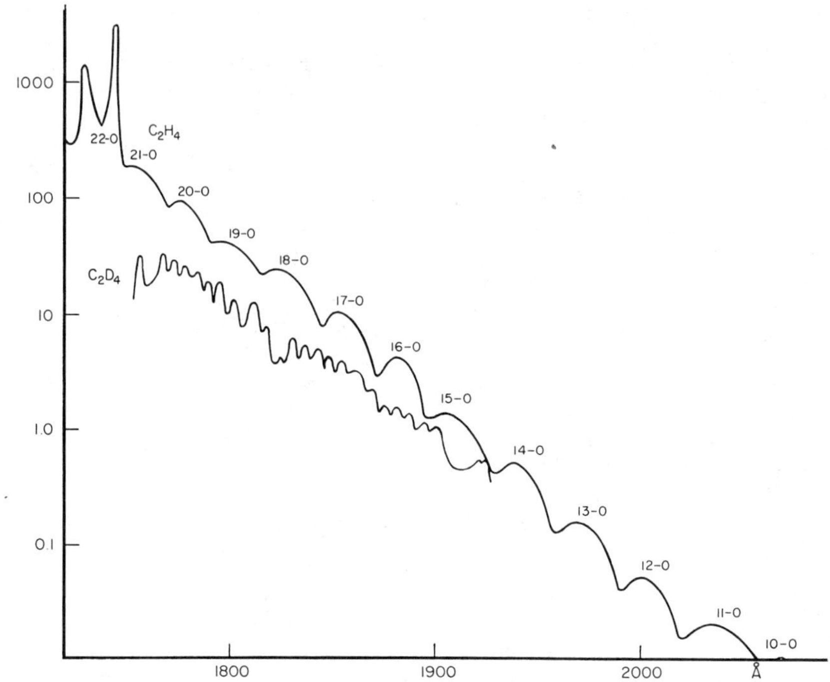


FIG. 197. **Absorption curve of $C_2H_4$ and $C_2D_4$ in the region 2100 to 1700 Å after Wilkinson and Mulliken (1303).**

vibration, strongly excited because of the change of equilibrium angle between the planes of the two $CH_2$ groups. The spacing of the main bands corresponds again to excitation of the C—C stretching vibration $\nu_2$. By assuming the theoretical ratio of this vibration in $C_2H_4$ and $C_2D_4$, Wilkinson and Mulliken (1303) have obtained the vibrational numbering of the bands (see Fig. 197). The extrapolated $0-0$ band lies outside the observed region in the gas but shortward of the beginning of the $\tilde{A}-\tilde{X}$ absorption in the liquid.

The system of fairly sharp bands which starts at 1744.1 Å (1735.5 Å in $C_2D_4$) was

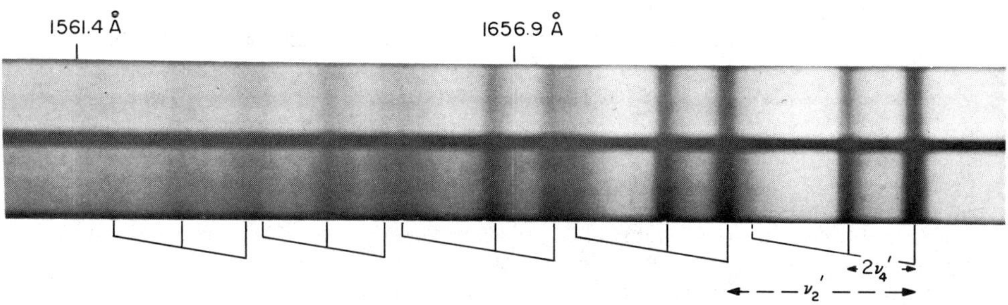


FIG. 198. **Spectrogram of the $\tilde{B}\leftarrow\tilde{X}$ absorption bands of $C_2H_4$ in the region 1750 to 1550 Å.** The band system consists mainly of a progression of "doublets" in $\nu_2 = 1370$ cm$^{-1}$. The spacing of the doublets corresponds to $2\nu_4$ ($=472$ cm$^{-1}$). A third member of the progression in $2\nu_4$ is visible in several members of the main progression as indicated. The pressures for the two spectrograms shown were 0.001 and 0.002 mm at a path length of 40 cm.

recognized by Price and Tutte (1026) as representing the first Rydberg transition. As shown in Fig. 198 it consists of a progression of doublets which corresponds again to the C—C stretching vibration but its frequency is much larger than for the $^3B_{1u}$ and $^1B_{1u}$ states. In $C_2D_4$ a third (weak) component of each member of this progression is observed and such a component can just be recognized for a few members in $C_2H_4$ (see Fig. 198). The interpretation of these two or three components is not obvious. They cannot be sequences since there is no vibration of sufficiently low frequency to produce such an intensity in the second member. But to interpret them as progressions in a vibration of the upper state leads to a very low frequency since the spacing is only 470 cm$^{-1}$ for $C_2H_4$ and 280 cm$^{-1}$ for $C_2D_4$. If we assume, as do Wilkinson and Mulliken (1303), that the molecule is planar and symmetrical in the excited state there can hardly be a totally symmetric fundamental of such small frequency. Therefore Wilkinson and Mulliken (1303) have assigned these bands to $2\nu_4$ and $4\nu_4$ where $\nu_4$ is the twisting vibration. Because of the large change of this frequency in going from the ground state to the Rydberg state (from 1027 to 236 in $C_2H_4$), the bands with $\Delta v_4 = +2$ are strong. The change of frequency in going from $C_2H_4$ to $C_2D_4$ is much larger than would follow from the isotope relations if the twisting vibration were harmonic. A very low barrier and a considerable deviation from either a harmonic or a cosine potential must be assumed to account for the observed isotope effect. Such a deviation does fit with the observed rapid increase of the vibrational intervals with $v_4$ (in $C_2H_4$: $2\nu_4' = 472$, $4\nu_4' = 1084$).

Wilkinson (1300) has studied in detail and under high resolution the absorption of both ethylenes from 1520 to 1280 Å and has found six Rydberg transitions which belong to four different Rydberg series. Three of these had already been recognized under much lower resolution by Price and Tutte (1026). Each Rydberg transition is accompanied by vibrational transitions similar to those discussed above for the first Rydberg transition. The vibrational frequencies are given in Table 72.

Further members of the strongest Rydberg series have been found by Price and Tutte (1026) between 1300 and 1200 Å and have yielded an ionization potential of $10.50_7$ eV. An absorption curve extending to 1060 Å has been given by Zelikoff and Watanabe (1329). The absorption coefficient of the first and strongest Rydberg transition at 1700 Å reaches a value of 1500 cm$^{-1}$. Some of the other members go up to $k \approx 1000$ cm$^{-1}$. These values refer to the total absorption: underlying continuum and Rydberg transition.

**$CH_3SH$.** While $CH_3OH$ shows only continuous absorption in the vacuum region[5] $CH_3SH$ does show both discrete and continuous absorption. From 2780 to about 1800 Å the absorption is continuous [Hukumoto (590)], but in the vacuum ultraviolet, as first shown by Price, Teegan and Walsh (1024), the absorption is almost entirely discrete to 1300 Å. There is a diffuse band at about 1840 Å followed by a system of sharp bands between 1760 and 1700 Å. Further systems follow below 1640. They have been arranged into three Rydberg series by Price, Teegan and Walsh as indicated in Table 72, leading to an ionization potential of $9.43_9$ eV. There is a striking similarity to the absorption spectrum of $H_2S$ (see p. 489) except that the latter is shifted by about 200 Å to shorter wavelengths. Strangely, such a similarity does not exist between $CH_3OH$ and $H_2O$.

## (b) Tri-hydrides

**$CH_3CN$ (acetonitrile).** Acetonitrile has no absorption in the visible and near ultraviolet regions. Its vacuum ultraviolet absorption spectrum has been studied by

[5] Beynon and Evans (117) gave discrete bands at 1607, 1588, 1565 and 1492, 1470, 1445 Å, but these have not been mentioned by any other investigator.

Herzberg and Scheibe (539) and Cutler (263). There is continuous absorption starting at 1820 Å, which is followed at 1600 Å by a short Rydberg series of discrete bands. The first two members of the series, at 1295 and 1150 Å, show vibrational structure. The C≡N stretching vibration is most prominent (see Table 73). The third member of the Rydberg series is quite diffuse and therefore the limit of the series, i.e., the ionization potential cannot be very precisely given.

**$C_2H_3Cl$ (vinyl chloride).** The absorption spectrum of vinyl chloride has been studied by Walsh (1259). The spectrum begins at about 2200 Å with a strong almost continuous absorption having a few very diffuse bands on the shortward side. The peak of the absorption is at about 1840 Å. The spacing of the diffuse bands is very similar in magnitude to the large spacing (1370 $cm^{-1}$) in the $\tilde{B}-\tilde{X}$ system of $C_2H_4$, that is, corresponds to excitation of the C—C stretching vibration. It seems probable, therefore, that the diffuse bands represent the first Rydberg transition as do the $\tilde{B}-\tilde{X}$ bands of $C_2H_4$, while the continuum represents the $V \leftarrow N$ transition of $C_2H_3Cl$ similar to the $\tilde{A}-\tilde{X}$ bands of $C_2H_4$.

At shorter wavelengths there are two progressions of sharp bands starting at 1585 and 1471 Å with spacings of 1372 and 1200 $cm^{-1}$ respectively, which are followed by two fairly long Rydberg series and one fragmentary one. They in turn are followed by an ionization continuum at 1240 Å. In addition, there is a single very strong band at 1462 Å which is ascribed by Walsh to a transition involving a non-bonding orbital near the Cl atom.

The ionization potential of $C_2H_3Cl$ comes out to be 0.51 eV lower than that of $C_2H_4$. Because of the high electron affinity of Cl, one might have expected the opposite change on the basis of a simple inductive effect. According to Walsh (1259) the explanation must be sought in a strong "resonance" between the $\pi$ electrons of the C=C bond and the non-bonding lone pair of Cl which has the same species (i.e., is antisymmetric with respect to the plane of the molecule).

## (c) Di-hydrides

**$C_4H_2$ and $C_4H_2^+$ (diacetylene).** The spectrum of diacetylene in the near ultraviolet was first studied by Woo and Chu (1313). They found an extensive system of weak, sharp bands in the region 2970–2650 Å. Callomon (171), who studied these bands under very high resolution, did not find any rotational structure, although the heads are extremely sharp, and concluded that the upper state is slightly predissociated. The vibrational structure is exceedingly complicated, and this suggests (but does not prove) that the molecule is slightly bent in the excited state, similar to acetylene in its first excited state. Woo and Chu (1313) found one very distinct progression of doublets with spacings between doublets of 690 $cm^{-1}$ and a doublet separation of about 50 $cm^{-1}$. The frequency 690 $cm^{-1}$ could correspond to the totally symmetric vibration $\nu_3'$ which is 874 $cm^{-1}$ in the ground state, or it could be a bending vibration if the molecule is bent in the excited state. Under the latter assumption, the "doublets" are readily accounted for as $\Sigma-\Pi$ and $\Delta-\Pi$ pairs similar to those observed in HCN and $C_2H_2$ (see p. 203f). This interpretation leads to a rotational constant of $A' \approx 12$ $cm^{-1}$, a reasonable value. The lower state of the doublets would be one in which the lowest bending vibration is singly excited.

At shorter wavelengths the bands of this system become more diffuse. At the same time, at about 2600 Å a new much stronger system starts ("low pressure bands" of Woo and Chu). All bands of this system appear diffuse even under the low resolution used by Woo and Chu. However, in this system regularities in the vibrational structure are more apparent. Woo and Chu give ten progressions in the frequency ~2100 $cm^{-1}$, which is probably the symmetric C≡C stretching vibration $\nu_2'$; in the ground state the corresponding frequency is 2184 cm$^{-1}$. One of the progressions is a "hot" progression corresponding to excitation of $\nu_6''$ or $\nu_8''$ in the ground state[6]. Such a progression with $\Delta v_{\text{bending}} = -1$ can occur only if the molecule is bent in the excited state (see p. 203). The mutual relation of the other progressions to one another is not clear. Woo and Chu have measured this system to 2000 Å. It is not certain how far it extends into the vacuum region since Price and Walsh (1029) in their work on the vacuum ultraviolet absorption used much lower pressures.

Price and Walsh did find two extremely strong and very short progressions at 1645 and 1447 Å which represent the first members of two Rydberg series having one and the same limit, viz., 87042 cm$^{-1}$. This limit must correspond to the first ionization potential (10.79$_1$ eV). The spacing in the short progressions (not given by Price and Walsh but measured by Callomon (171) on their reproductions) is 2150 ± 50 cm$^{-1}$, clearly corresponding to the vibration $\nu_2'$. The absence of other vibrational bands makes it practically certain that in all the Rydberg states the molecule is linear just as in the ground state.

At 1120 Å, i.e., just below the first ionization limit, a new set of strong absorption bands starts. These bands must be the first members of new Rydberg series going to the next ionization limit. While the higher members of these series have not been observed, one can estimate, comparing with the average of the two previous series, that the next ionization limit will be at about 111000 cm$^{-1}$, that is, there must be a fairly low-lying excited state of the $C_4H_2^+$ ion. Such a state is to be expected since there will be two $\pi$ orbitals corresponding to the two triple bonds. The lower one will be $\pi_u$, the higher $\pi_g$, the splitting being a measure of the "conjugation" between the two triple bonds. Both are filled in the neutral molecule. Removing an electron from $\pi_g^4$ results in the ground state ($^2\Pi_g$) of the ion, removing one from $\pi_u^4$ gives an excited state ($^2\Pi_u$) of the ion corresponding to the second Rydberg limit. Because of the resonance between the two $\pi$ orbitals, the first ionization limit of $C_4H_2$ is less, the second is greater than the single limit in $C_2H_2$.

The allowed transition $^2\Pi_u - {}^2\Pi_g$ of the ion has been identified by Callomon (171) as the so-called "*T*" spectrum discovered by Schüler and Reinebeck (1112) in discharges through acetylene and related compounds. It consists of a number of bands with very narrow fine structure in the region 5900–5000 Å. The $0-0$ band is at 19723 cm$^{-1}$, which compares well with the very rough estimate, 24000 cm$^{-1}$, from the two ionization limits of $C_4H_2$. Callomon (171) has succeeded in a full vibrational and rotational analysis of this spectrum. The doublet splittings in the upper and lower $^2\Pi$ states are 30.6 and 33.3 cm$^{-1}$ respectively. The rotational constants and vibrational frequencies in the ground state of the $C_4H_2^+$ ion are very similar to those of the ground state of the neutral molecule (see Table 74). The molecular ion is linear and symmetric in both $^2\Pi$ states.

**$HC_2CHO$ (propynal).** The violet and near ultraviolet absorption spectrum of propynal (propargyl aldehyde) was first studied under low resolution by Howe and Goldstein (581). High resolution studies of the main absorption system starting at 3830 Å were made by Brand, Callomon and Watson (140)(141) who have been able to obtain an exceptionally complete rotational and vibrational analysis of this interesting spectrum. The molecule is very nearly a prolate symmetric top and the main bands are perpendicular bands of this top with the transition moment perpendicular to the plane of the molecule (type C bands). Figure 199 shows the central part of the $0-0$

[6] Woo and Chu did not know the true fundamentals of $C_4H_2$ in the ground state. It is therefore significant that the value 635 cm$^{-1}$ found by them is close to two of the bending frequencies ($\nu_6$ and $\nu_8$) now known (see Table 74).

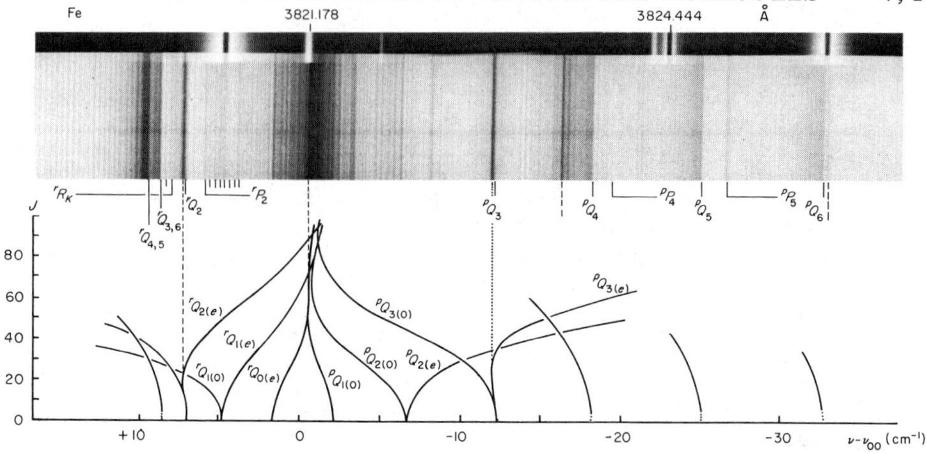


FIG. 199. **Central part of 0 − 0 band of the Ã–X̃ system of propynal after Brand, Callomon and Watson (141).** The $Q$ heads, $^pQ_6$, $^pQ_5, \ldots,$ $^pQ_3$ and $^rQ_2$, $^rQ_3, \ldots$ form a regular series, but because of large asymmetry doubling, the $Q$ heads of the sub-bands $1-2$, $0-1$, $1-0$ and $2-1$ are not obvious (unlike the HNCN band in Fig. 108 for which the asymmetry effects are much smaller). Below the spectrogram a Fortrat diagram of the $Q$ branches is shown. The higher asymmetry terms produce a concentration of lines near the origin of the band.

band. The slight deviations from the symmetric top cause deviations of the branches from the normal form at high $J$ as indicated below the spectrogram; in particular a concentration of lines results halfway between $^pQ_1$ and $^rQ_0$ which gives the band a less regular appearance than one might have expected.

The type $C$ character of the main bands shows that the electronic transition is $^1A''-{}^1A'$. In addition to these main bands there are also weaker bands of type $B$ and type $A$ corresponding to odd $\Delta v$ values of the out-of-plane vibrations. They are forbidden components in the $^1A''-{}^1A'$ electronic transition (see Chap. II, section 2b($\beta$)). Although this transition is analogous to the near ultraviolet bands of $H_2CO$ (see p. 518) unlike $H_2CO$ the inertial defect in the excited state is found to be very nearly zero, indicating that the molecule is at least nearly planar in this state. The levels of the out-of-plane vibrations show no irregularity which might suggest a potential maximum for the planar conformation. This is in contrast to the excited state of $H_2CO$ where the barrier is about 600 cm$^{-1}$.

All but one of the 12 vibrational frequencies of the excited state have been determined (see Table 74). At somewhat longer wavelengths beginning at about 4200 Å is an extremely weak band system which in all probability has as upper state the $^3A''$ state corresponding to the $^1A''$ upper state of the 3800 Å system, for it shows sub-bands with $\Delta K = 0, \pm 1$ and $\Delta N = 0, \pm 1, \pm 2$ (Callomon, unpublished). This system is also emitted in electrodeless discharges through propynal vapour. As in the case of formaldehyde, it probably owes its appearance mainly to mixing of the $^3A''$ state with some higher $^1A'$ state through spin-orbit coupling. Vibronic interaction which causes the forbidden components in the $^1A''-{}^1A'$ transition is not effective in the $^3A''-{}^1A'$ transition: bands with odd $\Delta v$ values in the out-of-plane vibrations are absent.

At shorter wavelengths there is a group of diffuse bands near 2100 Å and continuous absorption below 1550 Å.

**$C_2H_2O_2$ (glyoxal).** Glyoxal vapor exhibits a system of weak, sharp absorption bands in the region 5400 to 3900 Å first observed by Purvis and McCleland (1034), and first studied in detail under higher resolution by Thompson (1206). Brand (137), with the help of comparisons with $C_2D_2O_2$, recognized that there are actually two band systems, an extremely weak one with its $0-0$ band at 5208 Å, and another relatively much stronger one with its $0-0$ band at 4549 Å. Both band systems have been observed in fluorescence [Thompson (1206), Brand (137)] and in a high frequency discharge [Gaydon (414)]. While the $0-0$ bands have the greatest intensity, enough vibrational transitions are observed that a fairly complete analysis is possible. Brand showed on this basis, and considering the observed band contours, that the main electronic transition (4549 Å band) is $^1A_u-{}^1A_g$, the molecule having nearly the same ($\boldsymbol{C}_{2h}$) structure in both upper and lower state. This conclusion was fully confirmed by the analysis of the rotational structure by King (667) and Paldus and Ramsay (957). While King was able to resolve only the $K$ structure Paldus and Ramsay also resolved the $J$ structure and showed unambiguously that the $0-0$ band is a type $C$ band. Precise values of the rotational constants $A$, $B$ and $C$ were determined both for the upper and lower state (see Table 74) for both ordinary and deuterated glyoxal. In addition to the main type $C$ bands, there are also weaker bands of mixed type $A$ and $B$ (hybrid bands). These bands must be assigned as corresponding to odd changes of non-totally symmetric vibrations, i.e., they represent a forbidden component of vibronic type $B_u-A_g$, made possible by vibronic interactions (see p. 175f.). The very weak green bands are ascribed by Brand to an intercombination $^3A_u-{}^1A_g$. This assignment was confirmed by the observation of a strong magnetic rotation spectrum by Eberhardt and Renner (340). The $^3A_u$ state belongs presumably to the same electron configuration as the $^1A_u$ upper state of the $\tilde{A}-\tilde{X}$ system.

A second absorption region extends from 3200 to 2300 Å. According to Thompson (1206) it consists of diffuse bands, but no detailed description of this system has been given. The spectrum in the vacuum ultraviolet has been studied by Walsh (1262). There is a fairly weak progression of diffuse bands extending from 2050 to 1850 Å with a spacing of about 560 cm$^{-1}$. It is followed by a very strong fairly broad continuum with maximum at about 1667 Å, which broadens in both directions with increasing pressure and is overlapped by two diffuse bands at 1750 Å and 1601 Å. Finally, two strong moderately sharp bands at 1355 and 1324 Å have been found, but there is no clear evidence of a Rydberg series.

**$C_2H_2Cl_2$ (*cis*-, *trans*- and 1,1-dichloroethylene).** The absorption spectra of *cis*- and *trans*-dichloroethylene have been studied by Mahnke and Noyes (793) and Walsh (1259), those of 1:1 dichloroethylene by Teegan and Walsh (1195). These spectra are very similar to that of vinyl chloride discussed above. All have a strong broad continuum between 1800 and 2100 Å which shifts slightly to longer wavelengths in going from the 1:1 to the *cis* and the *trans* molecule[7]. These continua are followed in each case by one or more Rydberg series, each member of which shows some vibrational structure. The Rydberg series correspond in all probability to excitation of an electron from the most loosely bound orbital which is the analogue of $1b_{3u}$ of $C_2H_4$ (i.e., they are the $\pi$ electrons of the double bond). In addition, there are in each case strong single bands, as in $C_2H_3Cl$, which are ascribed by Walsh to excitation of the lone-pair electrons of the Cl atoms. For 1,1-dichloroethylene a tentative Rydberg series of these bands has been found yielding a second ionization limit which is higher by 0.91 eV than the minimum ionization potential.

From the main Rydberg series, the first ionization potentials of the three molecules are found to be 9.953 (*trans*), 9.657 (*cis*) and 9.459 eV (1,1). All of these values are lower than for vinyl chloride. The lowering is ascribed by Teegan and Walsh (1195) to the mutual repulsion of the lone-pair electrons of the two chlorine atoms combined

[7] At higher pressures, absorption extends to 2400 Å.

with their interaction with the $\pi$ electrons of the C=C bond. This effect gives the right order of the ionization potentials. Quantitative absorption measurements for trans $C_2H_2Cl_2$ in the whole region discussed here have recently been made by Goto (440).

It is well known that *cis*- or *trans*-dichloroethylene can be converted into a mixture of the two isomers by ultraviolet light in the quartz region. It was first suggested by Mulliken (889) that this interconversion (*cis-trans* isomerization) is made possible by the fact that in the upper state $\tilde{A}$ of the absorption (Mulliken's $V$ state), the equilibrium conformation is one in which the two groups are rotated by 90° compared to the planar conformation. In other words, the potential surface of this state has a maximum for the planar conformation, and this region of the potential surface is preferentially reached according to the Franck–Condon principle in absorption from the ground state. Thus, after the absorption act the molecule will oscillate violently (with an amplitude of nearly 90°) about the perpendicular equilibrium position, and the return to the ground state can take place equally well near the *cis* as near the *trans* conformation, i.e., a mixture of isomers is obtained. This explanation implies that the absorption continuum near 2200 Å is not a real continuum since it does not lead to dissociation. Rather, it must arise by the superposition of a large number of bands corresponding to the large number of vibrational levels near the maximum of the potential function of the upper state.

## (d) Mono-hydrides

**$C_2HCl_3$ (trichloroethylene).** The ultraviolet absorption spectrum of trichloroethylene described by Walsh (1259) is similar to those of the mono- and dichloroethylenes treated earlier. There is again a strong broad continuum extending, at the highest pressure used, from 2600 to 1800 Å with a maximum at 1960 Å. Strong discrete bands start at 1686 Å and extend with increasing intensity to about 1400 Å. They have been assigned by Walsh to two band systems with vibrational intervals of 1400, 660, 280 and 120 cm$^{-1}$ in the first, and 1400 and 426 cm$^{-1}$ in the second. The 1400 cm$^{-1}$ vibration is presumably the C=C stretching vibration in these excited states. Its value in the ground state is 1590 cm$^{-1}$. If the $0-0$ bands of these two band systems are considered as the first two members of a Rydberg series, several higher members can be found with a limit corresponding to 8.79 eV. However, the series must be considered as doubtful. Watanabe (1273) from photoionization measurements obtains an ionization potential of 9.47 eV. Below 1400 Å, there are a number of broad diffuse bands of great intensity. They are ascribed by Walsh to removal of an electron from one of the C—Cl bonding orbitals, or one of the non-bonding orbitals of Cl, to high Rydberg orbitals.

Trichloroethylene seems to be the only six-atomic mono-hydride that has as yet been studied in the ultraviolet in any detail (see Table 75).

## (e) Non-hydrides

**$C_2O_2Cl_2$ (oxalyl chloride).** Oxalyl chloride, according to Saksena and Kagarise (1095), has two regions of ultraviolet absorption from 3800 to 3000 and from 2900 to 2400. The second region consists of a few very broad bands. In the first region there is much more structure, which has been investigated in more detail by Sidman (1130). There is one fairly prominent progression of five members with a spacing decreasing from 1463 to 1413 cm$^{-1}$. This frequency is assigned to the totally symmetric C=O stretching vibration in the excited state, which is 1778 in the ground state. In addition, a frequency 397 cm$^{-1}$ occurs in several combinations. The sharp bands are assigned by Sidman to the *trans* isomer while an underlying continuum which becomes dominating at higher temperature is assigned to the *cis* isomer. These assignments to the *trans* and *cis* forms have, however, been questioned by Saksena and Jauhri (1093).

**$C_2Cl_4$.** The ultraviolet absorption spectrum of tetrachloroethylene, studied by Walsh (1259), is similar to that of trichloroethylene. Like the latter, it starts with a strong broad continuum with maximum at 1970 Å which extends to shorter and longer wavelengths as the pressure is increased. Indeed, at atmospheric pressure in a 10 cm path, Lacher, Hummel, Bohmfalk and Park (705) find strong absorption to 2700 Å. On the shortward side of the maximum and overlapping the continuum there is a fairly extended progression of diffuse bands with a spacing of about 1350 $cm^{-1}$. A second frequency of 450 $cm^{-1}$ is also indicated. It is probable that the continuous absorption corresponds to the $\tilde{A}-\tilde{X}$ system of $C_2H_4$ although for $C_2H_4$ the diffuse bands are on the longward side of the continuum.

In $C_2Cl_4$ a second broad continuum occurs at shorter wavelengths with a maximum at 1615 Å. At its shortward side it is overlapped by a progression of strong sharp bands starting at 1573 Å and having a spacing of 1343 $cm^{-1}$, with indications of two other frequencies (480 and 260 $cm^{-1}$). It appears that this band system is the first member of a Rydberg series of which two further members are also observed (see Table 76). However, the series is not definitely established, and therefore the resulting ionization potential of 9.5 eV is not reliable. Photoionization measurements [see Bralsford, Harris and Price (135)] yield an ionization potential of 9.32 eV.

Quantitative absorption measurements have recently been made by Goto (441). The absorption coefficient at the first maximum reaches 800 $cm^{-1}$; in the region below 1350 Å it reaches 3000 $cm^{-1}$.

## 5. Seven-Atomic Molecules

**$CH_3NH_2$.** The ultraviolet absorption spectrum of methylamine has been investigated to about 2000 Å by Herzberg and Kölsch (533), Henri and Lasareff (502), Emeléus and Jolley (357), Emeléus and Briscoe (356), and Förster and Jungers (393). The observed spectrum is very similar in character to the first absorption of $NH_3$ except that it is shifted to longer wavelengths. There is apparently a single progression of diffuse bands beginning at about 2450 Å whose intensity increases rapidly toward shorter wavelengths. From intensity irregularities at the longward end, it was soon recognized that the spacing of the bands, $\sim 350$ $cm^{-1}$, is not a vibrational frequency of the upper state, but that twice this separation is such a frequency. Indeed, Förster and Jungers (393) showed by comparison of the absorption of several deuterated species that a frequency of about three times the spacing is an even more important vibrational frequency of the upper state; in other words, we have progressions in the frequencies 650 and 1000 $cm^{-1}$ overlapping one another. These two frequencies correspond probably to the wagging vibration of the $NH_2$ group and the symmetrical deformation vibration of the $CH_3$ group respectively. The electronic transition is clearly similar to that in $NH_3$, i.e., corresponds to removal of an electron from the lone-pair orbital of the N atom into a 3*s* Rydberg orbital. The spectrum of $CH_3NH_2$ in the vacuum ultraviolet has not yet been studied.

**$C_3H_4$ (allene).** The ultraviolet absorption spectrum of allene has been studied by Sutcliffe and Walsh (1179). The first absorption is continuous with a maximum at 1710 Å but extending to 2030 Å and probably beyond as the pressure is increased. Overlapping this continuum is a progression of diffuse bands with a spacing of about 610 $cm^{-1}$. It seems likely that this frequency corresponds to the symmetrical C=C stretching vibration which in the ground state has the value 1073 $cm^{-1}$.

At wavelengths shorter than 1550 Å a large number of strong bands appears, but they are all fairly diffuse. Sutcliffe and Walsh (1179) have arranged these bands into nine Rydberg series. Some of them are rather fragmentary and probably represent simply vibrational members of the main Rydberg series. This conclusion is supported by the fact that among the nine Rydberg series of Sutcliffe and Walsh there are three with a Rydberg correction of about 1, while only one such series is expected from an *ns* electron in a molecule of $\boldsymbol{D}_{2d}$ symmetry (viz. $a_1$). On the other hand, an *np* electron becomes either a $b_2$ or an $e$ electron, and therefore, combined with the core of configuration $\ldots e^3$ gives rise to the five states $E$, $A_1$, $A_2$, $B_1$, $B_2$ of which only $E$ and $B_2$ can combine with the $^1A_1$ ground state. Thus, only two Rydberg series with a Rydberg correction of about 0.5 are expected unless one were to assume that the symmetry in the excited states is lower than $\boldsymbol{D}_{2d}$. Sutcliffe and Walsh find four such series. Similar considerations apply to the *nd* Rydberg orbitals.

**$CH_3C_2H$ (methyl acetylene).** The ultraviolet absorption spectrum of methyl acetylene has been studied by Price and Walsh (1029), by Watanabe and Namioka (1277) and by Nakayama and Watanabe (927). Methyl acetylene begins to absorb at about 2000 Å. There are two broad continua with maxima at 1920 and 1650 Å. Several very diffuse bands overlap these continua, or one may also say that these continua arise by the superposition of a number of diffuse bands. At about 1595 Å discrete absorption sets in, and a large number of fairly sharp bands appear in the region 1595 to 1100 Å. Price and Walsh (1029) have identified two fairly long Rydberg series with limits at 91100 and 91240 cm$^{-1}$ while Watanabe and his associates (1277)(927) have assigned three other Rydberg series with a common limit at 83590 cm$^{-1}$. The latter limit agrees well with the first ionization potential obtained by the electron impact and photoionization methods [Franklin and Field (398); Coates and Anderson (216); Watanabe and Namioka (1277)]. The first member of one of Price and Walsh's Rydberg series is identical with the first member of one of Nakayama and Watanabe's series. Such a coincidence throws perhaps some doubt on the reality of one of the series, but there are other bands which may replace the first member in one of them. If both Rydberg limits are real, it would indicate the existence of a low excited state of the $CH_3C_2H^+$ ion 7580 cm$^{-1}$ above its ground state. Since Price and Walsh consider the slight difference between their two limits as significant, the excited state of the ion would have a doublet splitting of 140 cm$^{-1}$. The ground state of the ion would correspond to removal of an electron from the $\pi$ orbital of the C≡C bond while the excited state of the ion would correspond to a removal of an electron from the $e$ orbital of the $CH_3$ group. The doublet splitting of the excited state is probably spin doubling.

The lowering of the ionization potential of $CH_3C_2H$ compared to $C_2H_2$ (viz. 10.36 compared to 11.41 eV) may be ascribed to hyperconjugation between the three C—H bonds of the $CH_3$ group and the C≡C bond. In other words, the reason for the lowering is similar to the lowering of the ionization potential of $C_4H_2$ compared to that of $C_2H_2$ (viz. 10.79 compared to 11.41 eV).

**$CH_3CHO$ (acetaldehyde).** Like formaldehyde (see p. 518) and other aldehydes, acetaldehyde has a weak absorption $\tilde{A}-\tilde{X}$ (presumably $\pi^*-n$) in the region 3500 to 2500 Å. At the longward side of this absorption there are a large number of bands with a complicated fine structure; at shorter wavelengths the bands become diffuse and merge into a continuous absorption [Henri and Schou (503), Schou (1107), Leighton and Blacet (737), Rao and Rao (1057) and Innes and Giddings (606)]. There can be no question that this absorption belongs to excitation in the >C=O group. The vibrational structure of the $\tilde{A}-\tilde{X}$ system is extremely complicated and contradictory values for the vibrational frequencies and even the position of the $0-0$ band may be found in the papers quoted. Innes and Giddings (606) on the basis of studies at two different temperatures have shown that many bands previously considered as hot bands actually originate in the vibrationless ground state. They suggest that the $0-0$ band is near 28700 $cm^{-1}$ or even at somewhat lower frequency and that the interval 1125 $cm^{-1}$ found by Rao and Rao (1057) in several progressions corresponds to the C—O stretching vibration in the excited state. In the ground state this vibration has the frequency 1743 $cm^{-1}$. The large change is very similar to that in $H_2CO$. A second more doubtful frequency of 480 $cm^{-1}$ is assigned by Innes and Giddings to the out-of-plane vibration of the C—C(H)=O frame.

While in static systems acetaldehyde fluoresces in the region 6100–5100 Å, in flow systems it fluoresces in the region 3550–4700 Å. The former fluorescence has been shown to be due to biacetyl formed as a result of irradiation [Matheson and Zabor (800)]; but the latter fluorescence, studied by Murad (918) and Longin (764), is a genuine $CH_3CHO$ fluorescence corresponding to the same transition ($\tilde{A}-\tilde{X}$) as the first absorption system. Figure 200 shows low resolution recordings of the absorption and fluorescence spectrum according to Innes, Giddings and Longin. The symmetrical arrangement of the two curves with respect to the $0-0$ band is nicely illustrated. It must be assumed that all molecules are brought to the lowest vibrational level of the excited state before emission. An emission spectrum very similar to the fluorescence

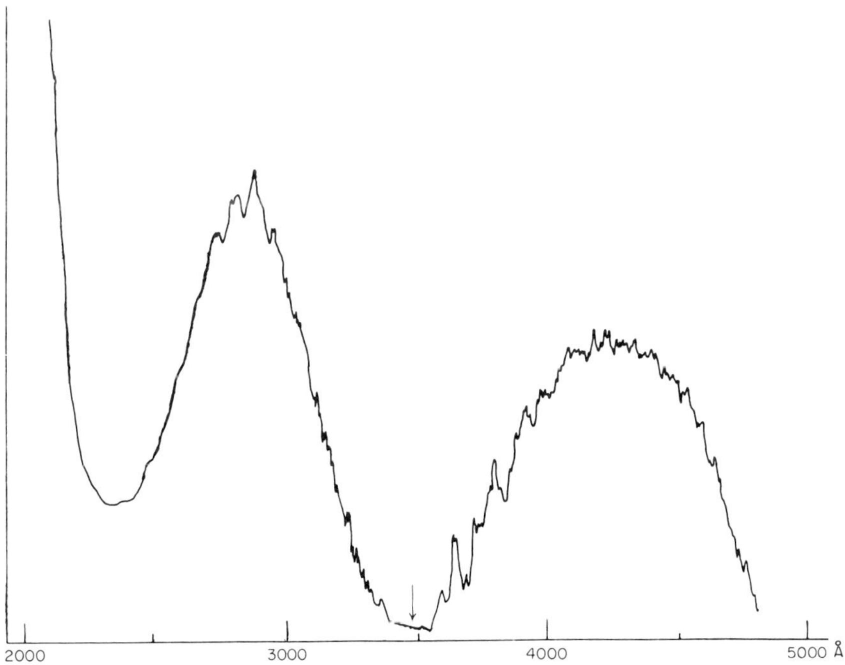


FIG. 200. **Low resolution absorption and fluorescence curves of acetaldehyde in the visible and near ultraviolet regions after Innes and Giddings (606).** The position of the $0-0$ band is indicated by the short arrow. The fluorescence curve (to the right of the arrow) is from Longin (764).

spectrum has been found by Robinson (1074) in a high frequency discharge in rapidly streaming acetaldehyde at relatively high pressure.

In the vacuum ultraviolet the absorption spectrum of acetaldehyde has been investigated by Walsh (1263) and Lake and Harrison (714). There are two distinct band systems with fairly extensive vibrational structure starting each with a very strong pair of bands, one at 1818 and the other at 1662 Å. In both systems a vibrational difference of 1200 $cm^{-1}$ is prominent, probably corresponding to the C=O stretching frequency in these excited states. At shorter wavelengths there are a large number of discrete bands which are readily assigned to three Rydberg series [Walsh (1263)]. One of these (see Table 77) is one of the longest Rydberg series thus far observed for a polyatomic molecule and leads to a very precise value for the ionization potential (10.229 eV). None of the members of the Rydberg series, other than the two systems at 1818 and 1662 Å, shows any indication of vibrational structure. The system starting at 1662 Å which represents the strongest absorption above 1200 Å has an $f$ value of 0.13 [Lake and Harrison (714)] and is assigned by Walsh as the $V \leftarrow N$ transition of the C=O bond (in the sense of Mulliken, see Chap. III), that is, as a transition in which one electron goes from a $\pi$ bonding to a corresponding anti-bonding orbital. At the same time it forms the first member of the long Rydberg series. This implies that the anti-bonding ($\pi^*$) orbital may also be considered as a ($3s$ or $3p$) Rydberg orbital.

**$C_2H_4O$ (ethylene oxide).** The ultraviolet absorption spectrum of ethylene oxide is very different from that of acetaldehyde even though these two molecules are isomers. There is no analogue of the long-wave absorption of acetaldehyde. In ethylene oxide the first absorption occurs at 2120 Å [Liu and Duncan (755)]. It is continuous and extends to 1600 Å. Overlapping this continuum are two or three strong diffuse bands starting at 1715 Å which in all probability represent an electronic transition different from the continuum. A second somewhat sharper band system starts at 1572 Å; it is also overlapped by a continuum. The vibrational structure suggests the frequencies 780 $cm^{-1}$ in the first and 724 and 1125 $cm^{-1}$ in the second band system.

Below 1435 Å there are very strong sharp absorption bands which appear to form Rydberg series. However, in the first assignment of these series by Liu and Duncan (755), an ionization potential of 10.81 eV was obtained while a later determination with the aid of the photoionization method by Watanabe (1273) led to a value of 10.565 eV. Lowrey and Watanabe (781) have subsequently reinvestigated and reassigned the Rydberg series (see Table 77). On the basis of the new assignment, a precise agreement between the Rydberg limit and the photoionization value of the ionization potential is obtained. Almost all strong members of the Rydberg series are accompanied by vibrational transitions. There is very strong continuous absorption below the Rydberg limit but only a small part of it corresponds to ionization.

The near equality of the ionization potentials of $C_2H_4O$ and $C_2H_4$ (10.565 and 10.507 eV) is striking. Also, the first absorption region of $C_2H_4O$ has great similarity to that of $C_2H_4$. Both absorptions have been interpreted as $V \leftarrow N$ transitions, but it is not clear why in $C_2H_4O$, in which the $\pi$ electrons of $C_2H_4$ are used to bind the O atom, the transition from a bonding to an anti-bonding orbital should take about the same amount of energy as in $C_2H_4$. It seems more likely that the first ionization limit of $C_2H_4O$ corresponds to removal of an oxygen "lone-pair" electron (as in $H_2O$, $H_2CO$ and others), that therefore the first strong absorption is of the $\sigma^* - n$ type and that the resemblance to $C_2H_4$ is coincidental.

**$SF_6$, $MoF_6$, $WF_6$, $UF_6$.** A large number of hexafluorides of various elements are now known. Most of them are believed to have octahedral symmetry ($\boldsymbol{O_h}$) in their ground states. The electronic spectra of these molecules are of interest since they might exhibit examples of the special selection rules and vibronic interactions in this type of molecule. Unfortunately, in the regions thus far investigated, none of these molecules exhibit sharp band systems. We shall consider only a few examples.

$SF_6$ shows only continuous absorption which at the highest pressures used (620 mm in a 1 m path) starts at 2170 Å [Liu, Moe and Duncan (756)]. At lower pressures (80 mm in a 1 m path) $SF_6$ is transparent to 1563 Å. At still lower pressures ($< 1$ mm) it is transparent to 1100 Å. Below this wavelength four very broad absorption maxima appear, at 1054, 936, 872 and 830 Å. These must correspond to at least four different allowed electronic transitions. The only allowed type of electronic transitions in an octahedral molecule is $^1F_{1u}-{}^1A_{1g}$ if the ground state is $^1A_{1g}$. Thus, the four upper states, at least near the equilibrium position of the ground state, must be $^1F_{1u}$ states. There is, of course, likely to be strong Jahn–Teller interaction in the excited states, that is, the potential minima in these states will not correspond to octahedral symmetry.

The ultraviolet absorption spectra of $MoF_6$ and $WF_6$ have been studied by Tanner and Duncan (1192). They find continuous absorption starting at 2815 and 2715 Å respectively in a 40 cm cell filled to 500 mm pressure. The centers of these continuous absorptions are estimated to lie at about 1850 and 1750 Å respectively. In $MoF_6$ preceding the maximum, between 2020 and 1870 Å, a progression of diffuse bands is observed with an average separation of 639 cm$^{-1}$. This frequency fits well as the totally symmetric $a_{1g}$ vibration which is 741 cm$^{-1}$ in the ground state. Its strong excitation, rather than that of a non-totally symmetric vibration, suggests that in this case Jahn–Teller interaction in the excited state is not strong.

The ultraviolet absorption spectrum of $UF_6$ has been reported by Dieke and Duncan (10). They find a system of diffuse bands in the region 4100 to 3500 Å and strong continuous absorption below 3300 Å. The most prominent frequency in the band system is 600 cm$^{-1}$. Other band separations are 240 and 100 cm$^{-1}$. The main frequency is similar to the totally symmetric ($a_{1g}$) frequency of the ground state which is 667 cm$^{-1}$. The presence of other frequencies might suggest that the molecule, at least in the excited state, is not octahedral. For the ground state, the spectroscopic evidence for the $\mathbf{O}_h$ structure appears to be quite conclusive [Bigeleisen, Mayer, Stevenson and Turkevich (118), Gaunt (409), Claassen, Weinstock and Malm (202)]. While at first such a structure seemed to be contradicted by electron diffraction [Bauer (100)], later refinements of the theory of electron diffraction by Glauber and Schomaker (425) have resolved this discrepancy in favor of the symmetrical (octahedral) structure.

## 6. Eight-Atomic Molecules

**$C_2H_6$ (ethane) and $B_2H_6$ (diborane).** Just like methane, ethane is quite transparent to fairly short wavelengths. Its absorption spectrum is almost entirely continuous. It has been briefly described by Scheibe and Grieneisen (1101) and Price (1013). According to the former authors, it begins at 1600 Å, and according to Price it becomes more intense at 1350 and continues beyond 1000 Å. In this last region there are extremely diffuse bands, but these have not been measured. Below 1200 Å there is a broad continuous absorption region with a maximum at 850 Å [Schoen (1105)].

In spite of the different geometrical structure of diborane, the ultraviolet absorption is similar to that of ethane except that it is shifted to longer wavelengths. It begins weakly at 2000 Å [Blum and Herzberg (126)]; a large increase of continuous absorption occurs at 1600 Å leading to several very diffuse maxima below 1400 Å [Price (1018)].

The continuous nature of the absorption of these molecules and their complete transparency in the near ultraviolet is connected with the fact that all valence electrons are used in single bonds (there are no lone-pair and $\pi$ electrons). Therefore, the only non-occupied orbitals arising from the valence electrons are anti-bonding orbitals. States in which one electron has gone from a bonding to an anti-bonding orbital are likely to lie fairly high and above the dissociation limit corresponding to the bond in question. Moreover, the ion produced by removal of the most loosely bound electron is not very stable since one bond is now only a one-electron bond. Indeed, from appearance potential measurements it is well known that it requires only about 1 eV to remove an H atom from $C_2H_6^+$ and even less from $B_2H_6^+$. Therefore, it is probable that Rydberg states of $C_2H_6$ and $B_2H_6$ also have low dissociation energies and that the equilibrium positions of the nuclei are greatly changed from those of the ground state. Thus in absorption from the ground state, the most likely parts of the potential surfaces to be reached are the repulsive parts, i.e., a continuous spectrum will be produced just as in $CH_4$ (see p. 526). The dissociation processes resulting from absorption in the continuous spectra of $C_2H_6$ have been studied in detail by Hampson, McNesby, Akimoto and Tanaka (470). The most important process seems to be the splitting off of an $H_2$ molecule.

**$C_2H_5I$ (ethyl iodide).** The ultraviolet absorption spectrum of ethyl iodide has been studied by Scheibe, Povenz and Linström (1102), Henrici and Milazzo (506) and Price (1017). It is in many respects similar to that of methyl iodide (see p. 528). The first absorption is a continuous one with a maximum at 2565 Å. It is followed by a number of discrete band systems, starting at 2020 Å and extending all the way to the two Rydberg limits at 1327 and 1249 Å. The first three discrete systems ($\tilde{B}-\tilde{X}$, $\tilde{C}-\tilde{X}$, $\tilde{D}-\tilde{X}$, see Table 78), with their first strong bands at 2020, 2005 and 1837 Å, are clearly the analogues of the methyl iodide systems $\tilde{B}-\tilde{X}$, $\tilde{C}-\tilde{X}$, $\tilde{D}-\tilde{X}$ (see Table 69). However, the $\tilde{B}-\tilde{X}$ and $\tilde{C}-\tilde{X}$ transitions have more nearly the same intensity and are closer together than the corresponding transitions in $CH_3I$. In $CH_3I$ the $\tilde{B}$, $\tilde{C}$ and $\tilde{D}$ states are degenerate ($E$) electronic states. It is probable that the corresponding states of $C_2H_5I$ are close-lying pairs of $A'$, $A''$ states.

Corresponding to the much larger number of normal vibrations, the $C_2H_5I$ band systems are much richer in bands than those of $CH_3I$, making a complete vibrational analysis extremely difficult. Some tentative assignments have been made for the $\tilde{B}-\tilde{X}$ and $\tilde{C}-\tilde{X}$ systems by Henrici and Milazzo (506). Each one of the main bands is accompanied by a group of bands which are clearly members of sequences in several of the low-frequency fundamentals.

Following the first three systems of sharp bands, there are, just as for $CH_3I$, several systems of diffuse bands in the region from 1740 to 1420 Å. They in turn are followed by two very clear Rydberg series of sharp bands. The two series limits differ by 4700 $cm^{-1}$, which is slightly less than the corresponding value (5060 $cm^{-1}$) for $CH_3I$. As for $CH_3I$, these two limits correspond to the two spin doublet components of the ground state of the ion. While in $CH_3I^+$ these two components differ in type ($E_{\frac{3}{2}}$ and $E_{\frac{1}{2}}$), in $C_2H_5I^+$ with its lower symmetry they are of the same type, viz., $E_{\frac{1}{2}}$. Combining an $ns$ electron (species $e_{\frac{1}{2}}$) with the lowest $E_{\frac{1}{2}}$ state of the ion gives two pairs of $A'$, $A''$ states (see Table 57). The states $\tilde{B}$ and $\tilde{C}$ are probably these two pairs for $n = 5$. While in $CH_3I$ the $\tilde{B}$ state has approximately $\Omega = 2$ and $\tilde{B}-\tilde{X}$ is thus very weak, $\Omega$ is not even approximately defined in $C_2H_5I$ and therefore $\tilde{B}-\tilde{X}$ and $\tilde{C}-\tilde{X}$ are of nearly the same intensity. The ionization potential of $C_2H_5I$ is smaller by 0.2 eV than that of $CH_3I$.

The spectra of $C_2H_5Br$ and $C_2H_5Cl$ have been briefly described by Price (1017); they differ from that of $C_2H_5I$ in much the same way as the spectra of $CH_3Br$ and $CH_3Cl$ differ from that of $CH_3I$: There is a shift to shorter wavelengths and all bands are diffuse, the diffuseness for the chloride being greater than for the bromide.

**$C_2H_3CHO$.** Acrolein (or propenal), like most aldehydes, has a characteristic though fairly weak absorption in the region 3870 to 3000 Å with a maximum at about 3300 Å. This absorption was first studied by Lüthy (783) and re-examined by Thompson and Linnett (1209), Blacet, Young and Roof (121) and Inuzuka (614). It was studied under fairly high resolution first by Eastwood and Snow (337) and recently by Brand and Williamson (145). The latter authors, with the help of a study of $C_2H_3CDO$, have obtained a fairly detailed vibrational analysis identifying nine of the eighteen fundamentals of the excited state (see Table 78). The most prominent progression is that in the CO stretching vibration $\nu_5' = 1265$ cm$^{-1}$ which in the ground state has the frequency 1723 cm$^{-1}$. Toward shorter wavelengths the intensity increases rapidly in the 1265 cm$^{-1}$ progression. At the same time the bands become more diffuse and eventually merge into a continuous absorption. Hollas (565) has independently come to substantially the same vibrational analysis and has added a partial rotational analysis.

Under high resolution the principal bands show a structure similar to that of the $C$ type bands of glyoxal (see p. 539). This structure must be interpreted as the $K$ structure of perpendicular bands of a nearly symmetric top. The molecule forms a nearly symmetric top only for the trans structure

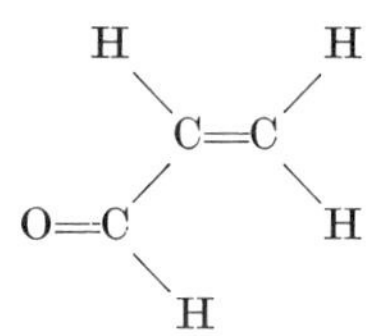


but not for the cis structure; the trans structure must therefore be considered as established. The transition moment is perpendicular to the plane of the molecule, that is, the transition is ${}^1A''-{}^1A'$. Since the $K$ structure is shaded to shorter wavelengths (while the unresolved $J$ structure is shaded to longer wavelengths), it follows that the rotational constant $A$ in the excited state is even greater than in the ground state and therefore that the molecule has the trans structure also in the excited state. From an assignment of the torsional vibrations in the excited state, Brand and Williamson derive the barrier heights 2700 and 5300 cm$^{-1}$ for the terminal and central C—C bonds respectively in the excited state, while the corresponding values in the ground state are 8700 and 2270 cm$^{-1}$. In other words, a reversal in the single and double bond character of these two bonds seems to take place on excitation.

At the longward side of the $\tilde{A}-\tilde{X}$ system is a very weak system of absorption bands starting at 4122 Å [Brand and Williamson (145)]. It is probable that the upper state of these bands is the triplet state corresponding to the singlet upper state of the 3865 Å system. As a confirmation of this assignment Eberhardt and Renner (340) observed the 4122 Å bands in magnetic rotation. Like the corresponding transition in $H_2CO$ and similar molecules, the two upper states (${}^1A''$ and ${}^3A''$) arise by taking an electron from the most loosely bound (lone pair) orbital predominantly centered on the O atom to an anti-bonding $\pi$ orbital of the C=O bond, that is, these transitions may be described as $\pi^*-n$ transitions.

At shorter wavelengths a second much stronger absorption, almost entirely continuous, starts at 2350 Å. Thompson and Linnett (1209) report a few diffuse bands at the longward end with a spacing of 300 cm$^{-1}$. It is probable that this continuous absorption is the same as that observed by Walsh (1260) at lower pressure, in the region 2050 and 1800 Å with a maximum at 1935 Å. This absorption probably represents a $V \leftarrow N$ transition of the C=C bond corresponding to a removal of an electron from a bonding ($\pi$) to an anti-bonding ($\pi^*$) orbital in this bond ($\pi^* - \pi$ transition).

At still shorter wavelengths there are a large number of slightly diffuse bands. Most of these bands have been assigned by Walsh to three Rydberg series (see Table 78). They all go to the same limit, which clearly corresponds to the first ionization potential (removal of the lone-pair electron of the oxygen atom). A very strong band at 1750 Å followed by some vibrational structure, as well as a strong doublet near 1600 Å, do not belong to these Rydberg series. They may represent the beginning of Rydberg series corresponding to removal of an electron from a $\pi$ orbital of the C=C bond.

There is a striking similarity of the spectrum of acrolein with that of acetaldehyde, except that in the latter there is no analogue of the strong continuum near 1900 Å. For a more detailed comparison see Walsh (1260).

## 7. Nine-Atomic Molecules

**$(CH_3)_2O$ and $(CH_3)_2S$.** Dimethyl ether shows only weak continuous absorption in the quartz ultraviolet, beginning at about 2350 Å [Thompson (1204)]. This continuum is followed by a progression of diffuse bands starting at 1880 Å [Scheibe and Grieneisen (1101), Harrison and Price (481)]. The spacing varies between 340 and 430 cm$^{-1}$ and probably corresponds to the C—O—C bending vibration $\nu_7$ in the excited state (in the ground state $\nu_7'' = 413$ cm$^{-1}$). After a gap in the absorption near 1790 Å similar groups of bands are found more and more frequently. Some of them have been arranged in a Rydberg series by Hernandez (511) yielding an ionization limit at 80330 cm$^{-1}$.

Dimethyl sulfide has been investigated by Thompson and Linnett (1210) and Price, Teegan and Walsh (1024). With an absorbing path of 10 cm and a pressure of 3 mm, a system of diffuse bands appears in the region 2300 to 2100 Å. Two bands, at 2278 and 2142 Å, are particularly prominent. Their separation of 2787 cm$^{-1}$ corresponds presumably to a CH stretching vibration in the excited state. The interpretation of the other observed intervals is much less certain. A much stronger very broad diffuse band appears at 2025 Å. It is followed by a system of strong sharp bands between 1960 and 1850 Å with frequency differences of 650 and 330 cm$^{-1}$. Another broad diffuse band (weaker than the one at 2025 Å) appears at 1795 Å. Finally, at 1645 Å several systems of very strong sharp bands begin which, however, rapidly merge into a continuous absorption in the region 1530 to 1400 Å. It is probable that the strong bands at 1645 Å represent the beginning of a Rydberg series. On this basis Price, Teegan and Walsh suggest an ionization potential of 8.7 eV.

**$C_4H_4O$ (furan).** The ultraviolet absorption spectrum of furan has been investigated by Pickett (979)(980), Price and Walsh (1028) and Watanabe and Nakayama (1275). Contrary to earlier observations of Menczel (821) and Fialkovskaja (380), Pickett has shown that furan vapor is transparent to 2300 Å, even at a pressure of 400 mm and a path of 60 cm. Near this wavelength there is a steep increase of continuous absorption leading to four broad maxima at 2110, 2043, 1996 and 1953 Å. These appear to form a progression in a frequency 1200 cm$^{-1}$. At somewhat shorter wavelength, 1915 Å, begins a system of sharp strong bands whose vibrational structure has been fairly fully analyzed by Pickett. Only short progressions appear, the 0 – 0 band at 1915 Å being by far the strongest. Three vibrational frequencies, 1395, 1068 and 848 cm$^{-1}$, have been fairly definitely identified; two others, 2965 and 465 cm$^{-1}$, are tentative. Price and Walsh interpret the most prominent frequency, 1395 cm$^{-1}$, as that of a stretching vibration in the C=C bonds; the corresponding ground state frequency is 1486 cm$^{-1}$. Since excitation in all probability corresponds to removal of an electron from one of the $\pi$ orbitals in the ground state, excitation of the C═C vibration was to be expected.

Furan is again fairly transparent between 1800 and 1700 Å. At 1680 Å strong sharp absorption bands are found. At shorter wavelengths they become more and more numerous. Price and Walsh (1028) picked out of this large number of bands two Rydberg series with a common limit at 73050 cm$^{-1}$. But this limit did not agree well with the later photoionization experiments of Watanabe and Nakayama (1275) who, with the help of intensity measurements, picked out two entirely different Rydberg series with a different limit (71683 cm$^{-1}$); this latter limit does agree very well with the ionization potential obtained by photoionization experiments. A number of vibrational bands have been assigned for several members of these Rydberg series. The frequency 1068 cm$^{-1}$ is most prominent.

Overlapping the ionization continuum of the first ionization limit is a considerable number of diffuse bands in the region 1400 to 1250 Å, of which a band at 1383 Å is particularly prominent. Watanabe and Nakayama (1275) have assigned this band with six others to a Rydberg series with a limit at 80229 cm$^{-1}$, i.e., 1.060 eV above the first limit. Since furan must have two orbitals corresponding to the four $\pi$ electrons, it seems reasonable to assign the second ionization limit to ionization from the lower of these two orbitals.

**$C_4H_4S$ (thiophene).** As has been known for a long time, thiophene has a discrete absorption starting at about 2400 Å (with weak extensions to 2600 Å). This absorption appears to be similar to the well-known near-ultraviolet absorption of benzene (see section 10). Such a similarity is not surprising in view of the aromatic behavior of thiophene. Spectrograms of the near-ultraviolet absorption system of thiophene have been published by Godart (426), Price and Walsh (1028) and Milazzo (841). They show clearly a progression in a frequency 965 cm$^{-1}$. Just as the 925 cm$^{-1}$ frequency of $C_6H_6$ it must be assigned as a totally symmetric vibration corresponding essentially to a "breathing" motion in the excited state. Several other vibrational frequencies have been identified by Milazzo (841) (see Table 79). At shorter wavelengths the bands of this system become diffuse. Milazzo (841) assigns these shorter wavelength bands to other electronic transitions, but it seems more probable that they belong to the same transition as the bands at longer wavelengths.

Much stronger absorption starts rather abruptly, after a transparent region, at 1880 Å. Only the first few bands of this system are sharp; the remaining ones become rapidly very diffuse. Vibrational frequencies of 540 and 1250 cm$^{-1}$ in the upper state are apparent. A large number of sharp absorption bands occur in the region 1600 to 1400 Å. Apparently they form a number of Rydberg series of which the most prominent, according to Price and Walsh, gives an ionization potential of 8.95 eV.

## 8. Ten-Atomic Molecules

**$(C_2H_3)_2$ (1,3 butadiene).** The ultraviolet absorption spectrum of butadiene was first studied by Scheibe and Grieneisen (1101) and later in more detail by Price and Walsh (1027). It begins with a short progression of very diffuse bands between 2170 and 1970 Å. Their spacing of 1440 cm$^{-1}$ very probably corresponds to the symmetrical C═C stretching vibration whose value in the ground state is 1643 cm$^{-1}$. The 2170–1970 Å system is probably the analogue of the first singlet transition ($\tilde{A} \leftarrow \tilde{X}$ of Table 72) of ethylene. In $C_2H_4$ the frequency of the C═C vibration drops from 1623 to

850 $cm^{-1}$ since one of the $\pi$ electrons goes from a bonding to an anti-bonding orbital ($V \leftarrow N$ transition of Mulliken). The reason that in butadiene it drops only from 1643 to 1440 $cm^{-1}$ is presumably that there are two pairs of $\pi$ electrons, and thus removal of one electron to an anti-bonding orbital does not lower the bond strength as much.

The diffuse bands are followed (and partly overlapped) by a number of strong single and double bands in the region 2000–1600 Å which represent apparently separate electronic transitions. The separation of the components of the pairs is about 350 $cm^{-1}$ which may be (as assumed by Price and Walsh) the analogue of the splitting 460 $cm^{-1}$ observed in $C_2H_4$ and there ascribed to twice the torsional frequency. A group of diffuse bands sets in at 1760 Å [assigned to the transition $V_{2,3} \leftarrow N$ by Price and Walsh] and below 1520 Å a large number of sharp bands are observed which represent separate electronic transitions. Price and Walsh (1027) have assigned two Rydberg series which lead to an ionization limit of 9.061 eV. There is a splitting of the first members of these series which rapidly diminishes and could be ascribed to the splitting between singlet and triplet levels arising for each $n$ by interaction between the outer electron and the core. A number of bands appear beyond the ionization limit at 1370 Å. These must correspond to excitation of an electron from the second $\pi$ orbital which lies 1.5 eV below the first (see p. 401).

Sugden and Walsh (1176) have assigned another fragmentary Rydberg series among the many bands below 1520 Å. This series yields an ionization potential of 8.75 eV and has been ascribed to the presence of cis-butadiene while the main Rydberg series belong to trans-butadiene. However, the photoionization experiments of Watanabe (1273) have not confirmed the existence of a lower ionization potential, and infrared and Raman spectra give little evidence of the presence of cis-butadiene [Marais, Sheppard and Stoicheff (797), see, however, Panchenko, Pentin, Tyulin and Tatevskii (959)].

**$(CH_3)_2CO$ (acetone).** Like all ketones, acetone has a moderately strong absorption in the near ultraviolet extending from 3300 to 2200 Å with a maximum at about 2750 Å. The older literature on this spectrum is well summarized by Noyes, Duncan and Manning (942). As was first recognized by Norrish, Crone and Saltmarsh (941) there is sharp fine structure at the longward end from 3300 to 3000 Å, but this becomes rapidly diffuse and merges into a continuum extending from 3000 to 2200 Å. Three or four indistinct maxima are observed in the continuum [see Norrish *et al.* (941)] which are separated by about 1100 $cm^{-1}$. By irradiation with light in the discrete region, acetone vapor can be made to fluoresce [Damon and Daniels (266) and Crone and Norrish (256)]. The presence of this fluorescence shows that the absorption in the longward region is really discrete; but there are some complications which we shall not discuss here [see, for example, Groh, Luckey and Noyes (449)]. Noyes, Duncan and Manning (942) have made a tentative vibrational analysis of the discrete part of the absorption spectrum and find long progressions in the frequency 210 and short progressions in the frequency 1198 $cm^{-1}$. The former probably corresponds to the ground state frequency 391 $cm^{-1}$ (C—C—C deformation vibration).

The absorption of acetone in the vacuum ultraviolet, first studied by Scheibe, Povenz and Linström (1102), has been investigated in detail by Lawson and Duncan (731) both for the ordinary and the deuterated compound. There are three or four strong band groups, starting at 1944 Å, with a spacing of 1194 $cm^{-1}$. The vibrational analysis by Lawson and Duncan gives, in addition, the upper state frequencies 1047, 714, 315 and 269 $cm^{-1}$. A quantitative absorption curve in this region is given by Lake and Harrison (714).

Following an absorption minimum near 1740 Å, a group of rather irregularly spaced fairly sharp bands starts at 1665 Å [Duncan (316)]. The strongest band in this region is at 1533 Å; it represents in all probability a separate electronic transition. Many other bands follow at shorter wavelengths. They represent probably a number of Rydberg series. Duncan (316) gave one Rydberg series, but its limit was not confirmed by the photoionization experiments of Watanabe (1272), who made a different assignment of Duncan's data and obtained a series leading to an ionization potential of $9.70_5$ eV. There is strong continuous absorption from 1280 Å on; it is probably the ionization continuum.

**$C_4H_4NH$ (pyrrole).** The ultraviolet absorption spectrum of pyrrole, studied by Scheibe and Grieneisen (1101), Price and Walsh (1028) and Milazzo (840), is similar to that of furan (see p. 548) which has the same number of electrons. The first absorption between 2200 and 1900 Å consists of a progression of very diffuse bands (spacing 1360 $cm^{-1}$). It is shifted somewhat to shorter wavelengths compared to furan [see Milazzo (838)] and is followed by a continuous absorption between 1900 and 1770 Å which seems to correspond to the discrete bands of furane in the same wavelength region (transition $\tilde{B} \leftarrow \tilde{X}$). At shorter wavelengths, between 1750 and 1590 Å, there is a weak system of diffuse bands probably corresponding to the $\tilde{C} \leftarrow \tilde{X}$ system of furan. Starting at 1530 Å, a number of single sharp bands have been observed which have not been classified but do seem to represent Rydberg series going to the first ionization potential.

In hexane solution Menczel (821) reports a weak longward extension of the absorption of pyrrole to 2800 Å with a number of very faint bands in the region 2743–2596 Å.

**$p$-$C_4H_4N_2$ (pyrazine).** Several absorption systems of pyrazine have been observed. Two systems in the near ultraviolet are of particular interest: an extremely weak system of sharp bands in the region 3760–3560 Å, and a moderately strong system of sharp bands in the region 3300–2900 Å. These two systems, first described by Hirt (556) and Ito, Shimada, Kuraishi and Mizushima (613), have recently been further studied by Innes and his co-workers who have shown that they supply interesting applications of the selection rules discussed in Chapter II.

The absorption system at 3300 Å consists of bands of two different types: some are exceedingly sharp under medium resolution, while others appear broad. Ito *et al.* (613) concluded that there are two different electronic transitions in close proximity, but Innes, Simmons and Tilford (610) showed conclusively, by comparing the spectra of $C_4H_4N_2$ and $C_4D_4N_2$, that there is only one electronic transition of which the sharp bands form the allowed $\parallel$ component, while the broad bands represent a forbidden $\perp$ component made possible by vibronic interaction. The sharp bands show under high resolution a remarkably simple appearance [Merritt and Innes (831)] as illustrated in Fig. 105. They have clearly the structure of $\parallel$ bands of a symmetric top with nearly the same $A$ and $B$ values in the upper and lower state. Thus, the electronic transition moment must be in the direction perpendicular to the plane of the molecule, i.e., we have a $^1B_{3u}-{}^1A_g$ electronic transition, assuming a planar $\boldsymbol{D}_{2h}$ structure for the molecule and using the choice of axes recommended by Mulliken (912).

On the other hand, according to Innes and Parkin (609) the broad bands show under high resolution a perpendicular type structure, that is, the transition moment is in the plane of the molecule; more particularly, the observation of a central peak in these bands shows that they are of $A$ (not $B$) type, that is, the transition moment lies in the $a$-axis which passes through the two nitrogen atoms. An $A$ type component, which is forbidden in an allowed $^1B_{3u}-{}^1A_g$ transition, may occur when a $b_{2g}$ vibration is excited by 1, 3, ... quanta. The vibrational analysis shows that the strongest broad bands are shifted by 383 $cm^{-1}$ compared to the main sharp bands. Moreover, there are hot bands[8] of the $\perp$ type separated by 919 cm$^{-1}$ toward longer wavelengths from the main $\|$ bands. The latter frequency does correspond to an observed $b_{2g}$ Raman frequency. Innes, Simmons and Tilford (610) therefore conclude that 383 cm$^{-1}$ is the excited state $b_{2g}$ frequency corresponding to 919 cm$^{-1}$ in the ground state (the corresponding frequencies for $C_4D_4N_2$ are 292 and 721 cm$^{-1}$ respectively). It is significant and in agreement with the vibronic selection rules that hot bands in which the frequency 919 cm$^{-1}$ (or 721 cm$^{-1}$ for $C_4D_4N_2$) is doubly excited have again the same sharp $\|$ structure as the corresponding main bands.

The intensity of the forbidden component of the $^1B_{3u}-{}^1A_g$ electronic transition is determined by the strength of the vibronic interaction. For the deuterated compound, because of the smaller amplitude of the vibrations, the vibronic interaction is less strong, and therefore the intensity of the forbidden component relative to the allowed one should be less strong. This is indeed observed [Innes, Simmons and Tilford (610)], and this observation represents a striking confirmation of the theory of forbidden transitions (see Chap. II, section 2b($\beta$)).

The extremely weak 3700 Å system has recently been studied in absorption under very high resolution by Innes and Giddings (607). They find that the band structure is very similar to that of the 3300 Å bands, i.e., is that of a $\|$ transition. However, there is a slight intensity alternation in the branches which suggests the presence of $\Delta K = \pm 2$ transitions in addition to the main $\Delta K = 0$ transitions. For a planar nearly symmetric top molecule the spacing $4(\bar{B} - C)$ of the $Q$ branches with $\Delta K = \pm 2$ is very nearly the same as that of the $P$ and $R$ branches (viz. $2\bar{B}$) in the $\Delta K = 0$ component; but the $\Delta K = +2$ component will have an intensity alternation in the ratio 13:11 as a function of $K$ since the top axis is a two-fold axis of symmetry. The presence of $\Delta K = \pm 2$ branches can be accounted for by the assumption that the transition is a triplet – singlet transition [Herzberg (523); see Chap. II, section 3b]. This triplet – singlet transition is most probably $^3B_{3u}-{}^1A_g$ corresponding to the $^1B_{3u}-{}^1A_g$ transition at 3300 Å. The proposed interpretation has been strikingly confirmed by Douglas and Milton's (299) observation of a large Zeeman splitting in the 3700 Å system.

The values of the rotational constant $\bar{B}_0''$ derived by Innes *et al.* from the two-band systems for the lower state agree in a very satisfactory way (0.2050 and 0.2048 cm$^{-1}$); those for the excited states differ slightly (0.2041 and 0.2036 cm$^{-1}$ respectively). These constants fit very well with reasonable geometric data of the molecule in both upper and lower states but are, of course, not sufficient to determine them.

Goodman and Kasha (437) and El-Sayed and Robinson (351) have studied the phosphorescence of pyrazine in various solid matrices. The observed spectrum clearly represents the $^3B_{3u}-{}^1A_g$ transition in emission. It consists mainly of a fairly long progression in a frequency 600 cm$^{-1}$ extending from the 0 – 0 band as obtained in the absorption spectrum to longer wavelengths and, in addition, several weaker combination bands. A frequency of 609 cm$^{-1}$ has been observed in the Raman spectrum of the liquid and has been assigned by Lord, Marston and Miller (770) to a totally symmetric ring vibration ($\nu_5$) in which the two N atoms move outward while the four CH groups move inward[9]. One would conclude from the observation of a long progression in this vibration that in the $^3B_{3u}-{}^1A_g$ transition the C—N—C angle changes appreciably. In the absorption spectrum a corresponding progression is more difficult to establish because of the overlapping by the $\tilde{A}-\tilde{X}$ system.

[8] El-Sayed and Robinson (351) have apparently observed some of these hot bands in a solid matrix at very low temperature and conclude that they are not hot bands but correspond to a different electronic transition.

[9] From the $^1B_{3u}-{}^1A_g$ transition the frequency of this vibration in the vapor has been found to be 596 cm$^{-1}$.

A third absorption system $\tilde{B}-\tilde{X}$, consisting of diffuse bands, has been observed and analyzed by Hirt (556). In the vacuum ultraviolet five band systems have been found and studied in detail by Parkin and Innes (963). One of them, $\tilde{E}-\tilde{X}$, consists of a large number of very sharp bands, the others of diffuse bands. The vibrational data on these band systems are included in Table 80.

The four transitions $\tilde{a} \leftarrow \tilde{X}$, $\tilde{A}-\tilde{X}$, $\tilde{B}-\tilde{X}$ and $\tilde{C}-\tilde{X}$ can be readily interpreted in terms of molecular orbitals: there are two orbitals $a_g$ and $b_{1u}$ derived from the lone-pair electrons (also called $n$) of the two N atoms. Of these, $b_{1u}(n)$ is probably the most loosely bound filled orbital in the ground state. In addition, there are the six $\pi$ orbitals of the three double bonds: three bonding ones of species $b_{1g}$, $b_{2g}$ and $b_{3u}$ (often simply called $\pi$), and three anti-bonding ones of species $a_u$, $b_{2g}$ and $b_{3u}$ (often called $\pi^*$); only the former are filled in the ground state [see Ito *et al.* (613)]. Transition of an electron from the $b_{1u}(n)$ orbital to $b_{2g}$ $(\pi^*)$ gives rise to ${}^3B_{3u}-{}^1A_g$ and ${}^1B_{3u}-{}^1A_g$ transitions, thus accounting for the two long wavelength systems $\tilde{a}-\tilde{X}$ and $\tilde{A}-\tilde{X}$. The $\tilde{B} \leftarrow \tilde{X}$ and $\tilde{C} \leftarrow \tilde{X}$ transitions are then most naturally interpreted as those in which an electron goes either from $b_{1g}$ $(\pi)$ orbital to $a_u$ $(\pi^*)$, or from $b_{3u}$ $(\pi)$ to $b_{2g}$ $(\pi^*)$ both of which give rise to ${}^1B_{1u}-{}^1A_g$ transitions. Experimentally, the band types of the $\tilde{B}-\tilde{X}$ and $\tilde{C}-\tilde{X}$ transitions have not been established.

The spectra of the less symmetrical diazines, viz., pyridazine (1,2-diazine) and pyrimidine (1,3-diazine) have been studied by Uber (1229), Halverson and Hirt (464) and Parkin and Innes (963). Their results are included in Table 80.

## 9. Eleven-Atomic Molecules

**$C_5H_6$ (1,3 cyclo-pentadiene).** Even though cyclopentadiene differs from furan only by the substitution of a $CH_2$ group for the O atom, its ultraviolet absorption spectrum is rather different from that of furan. Indeed, there is a greater similarity with thiophene. The spectrum has been investigated by Scheibe and Grieneisen (1101), Pickett, Paddock and Sackter (982) and Price and Walsh (1028). The first absorption is a broad diffuse region extending from 2600 to 2100 Å with an intensity maximum at 2320 Å. There is a progression of extremely diffuse bands with a spacing of about 770 $cm^{-1}$.

Very suddenly, at 1985 Å a strong discrete absorption begins extending to 1860 Å and consisting of several progressions of bands with vibrational intervals 1440, 1070, 780 and 470 $cm^{-1}$. The frequency 1440 clearly corresponds to the strong Raman frequency 1496 $cm^{-1}$ [Reitz (1067)], which must be considered as the symmetric vibration of the two C=C bonds in the ground state. Similarly the frequencies 1070 and 780 $cm^{-1}$ probably correspond to the totally symmetric vibrations 1105 and 911 $cm^{-1}$ of the ground state. The frequency 470 $cm^{-1}$ is not as readily interpreted. It cannot be the frequency of the $CH_2$ twisting vibration as assumed by Price and Walsh (1028) since this vibration is not totally symmetrical. It could, however, be twice this frequency, similar to the interpretation of the band pairs in ethylene (see p. 535). On this assumption the twisting vibration would be only 235 $cm^{-1}$ in the excited state. On the other hand the frequency difference 470 $cm^{-1}$ could also be interpreted as the difference of two electronic transitions as was assumed by Pickett, Paddock and Sackter.

Cyclo-pentadiene is fairly transparent between 1860 and 1670 Å. Below 1670 Å many strong individual absorption bands occur which appear to represent separate electronic transitions without much vibrational structure. Some of them are rather diffuse. Price and Walsh (1028) have identified tentatively one Rydberg series among them which leads to an ionization potential of $8.62_3$ eV.

Cyclo-pentadiene (like furan and thiophene) has four $\pi$ orbitals, two bonding and two anti-bonding, only the former being filled in the ground state. It seems likely that the first absorption ($\tilde{A}-\tilde{X}$) corresponds to removal of an electron from the highest bonding $\pi$ orbital to the lowest anti-bonding $\pi^*$ orbital ($\pi^* - \pi$ transition). Whether the second absorption ($\tilde{B} - \tilde{X}$) is also a transition between $\pi$ orbitals or corresponds to excitation into a Rydberg orbital is difficult to decide.

**$C_5H_5N$ (pyridine).** The near ultraviolet absorption spectrum of pyridine vapor in the region 3000–2500 Å was first investigated in detail by Henri and Angenot (499). Ten years later Sponer and Stücklen (1153) reinvestigated the spectrum and accomplished a fairly satisfactory vibrational analysis.

The spectrum consists of a large number of very sharp line-like bands which reach a maximum of intensity at about 2750 Å, where at the same time the bands become somewhat diffuse. The most prominent feature is a fairly long progression in a frequency 542 cm$^{-1}$ which starts with a band at 34769 cm$^{-1}$. This band must be considered as the $0-0$ band of the system. Preceding it at a distance of 601 cm$^{-1}$ is a much weaker but similar band which must be the $0-1$ band in the same vibration. Indeed, from the Raman spectrum the lowest totally symmetric vibration in the ground state is observed to be 605 cm$^{-1}$ (in the liquid). It seems certain that the interval 542 cm$^{-1}$ is the frequency of the corresponding vibration in the excited electronic state. Five other totally symmetric vibrations of the ground state (1482, 1218, 1068, 1029, 992) have been identified by Sponer and Stücklen among the hot bands on the longward side of the $0-0$ band. In the upper state, in addition to 542, only the frequencies 995 and 968 cm$^{-1}$, corresponding to 1029 and 992 cm$^{-1}$ respectively in the ground state, have been definitely identified. Tentative identifications of several other vibrational frequencies, including some non-totally symmetric ones, have been made by Sponer and Stucklen. A progression of weaker bands accompanying the main progression at a separation of 139 cm$^{-1}$ at the shortward side is puzzling and has not been interpreted.

The almost line-like structure of the near ultraviolet pyridine bands suggests that they are $\|$ bands ($\Delta K = 0$) and that the rotational constants are nearly the same in the upper and lower state. This suggestion implies that the electronic transition is $^1B_1 - {}^1A_1$. This assignment is in accordance with the interpretation of the excited state as one in which an electron is removed from a lone-pair orbital ($a_1$) to an anti-bonding $\pi$ orbital ($b_1$); in other words, the transition is of the type $\pi^* - n$ just as the near ultraviolet system of pyrazine (see p. 553). This interpretation first suggested by Kasha (659) has since been supported by a great deal of other evidence [see Goodman (436)]. For pyridine the triplet – singlet transition corresponding to the main bands has not yet been found[10].

The far ultraviolet spectrum of pyridine has been briefly described by El-Sayed, Kasha and Tanaka (350) and El-Sayed (349). There are four Rydberg series, one converging to $9.26_6$ eV, presumably the lowest ionization potential; another, in the region 1200–1266 Å, converging to 10.3 eV; and two series converging to 11.56 eV. The three ionization limits are interpreted as corresponding to removal of an electron from the outermost filled $\pi$ orbital (of species $a_2$), from the nitrogen lone-pair orbital (of species $a_1$), and from the second $\pi$ orbital (of species $b_1$) respectively.

**$C_5H_4O_2$ (2-furfuraldehyde).** The spectrum of the aldehyde of furan is interesting because it shows, in addition to absorptions very similar to those of furan, a strong absorption in the near ultraviolet (2780–2460 Å) which must be ascribed to the aldehyde group even though it is considerably shifted to the ultraviolet compared to other aldehydes[11]. This absorption has been studied by several investigators, most recently

[10] Reid (1065) mistakenly thought he had observed such a transition, but his bands were shown by Brealey (146) to be due to pyrazine present as an impurity.

[11] Walsh (1261) considers it to be the analogue of the extremely weak and doubtful absorption of furan in the region 2800–2500 Å. The enormous difference in intensity makes this interpretation rather unlikely.

by Santhamma (1097). The absorption bands are quite sharp to about 2600 Å but then become more and more diffuse. A tentative vibrational analysis has been given by Santhamma, but since there are 19 totally symmetric vibrations (i.e., vibrations that are symmetric with respect to the plane of the molecule, the only element of symmetry), a definite assignment is hardly possible.

At longer wavelengths there is an extremely weak system of diffuse bands extending from 3700 to 3300 Å. At shorter wavelengths, according to Walsh (1261), there is an almost continuous absorption from 2000 to 1700 Å with a maximum at 1800 Å and a number of overlapping diffuse bands on the longward side. This absorption probably corresponds to a combination of the $\tilde{A}-\tilde{X}$ and $\tilde{B}-\tilde{X}$ bands of furan. Two excited state frequencies, 887 and 230 cm$^{-1}$, have been recognized among the diffuse bands.

Following a short gap at 1700 Å, strong diffuse absorption bands begin at 1600 Å and continue up to the limit of observation (1100 Å). Presumably this absorption corresponds to the superposition of a large number of Rydberg series, but no specific ones have been identified. Probably the ionization potential is in the neighborhood of that of furan.

Data for a few other eleven-atomic molecules are included in Table 81.

## 10. Twelve-Atomic Molecules

**$CH_3C{\equiv}C{-}C{\equiv}CCH_3$ (dimethyldiacetylene).** The absorption spectrum of dimethyldiacetylene has been studied by Price and Walsh (1029). The vapor is transparent to about 2050 Å. Here a system of discrete absorption bands starts. The vibrational structure is not at all clear except that a few of the more prominent bands are separated by about 2500 cm$^{-1}$. Possibly there are two (or even three) electronic transitions superimposed. At 1670 Å an extremely strong progression of bands starts, again with a (somewhat irregular) spacing of about 2500 cm$^{-1}$. Such a frequency seems much too low for a C—H stretching vibration but is rather high for a C≡C stretching vibration which is 2264 cm$^{-1}$ in the ground state. The first two bands of the progression are the strongest bands of the spectrum. At somewhat higher pressure many additional bands of this band system appear which have not been classified. This system is at nearly the same wavelengths as the strongest system of diacetylene ($\tilde{C} \leftarrow \tilde{X}$ of Table 74). It is followed by other systems similar to the Rydberg series of diacetylene, but the series are not as well developed, and no precise value for the first ionization potential can be given; it is likely to be slightly smaller than that of diacetylene. There is, however, starting at 1270 Å, a second Rydberg series which is more clearly defined than in diacetylene and leads to a second ionization limit of 11.51 eV.

**$C_6H_6$ (benzene).** The strong ultraviolet absorption system of benzene vapor near 2600 Å has been known since the early days of spectroscopy. Because of the great sharpness of the absorption bands and the obvious regularity in their coarse structure, this system has stimulated the curiosity of many investigators, and in the course of time spectra of higher and higher resolution have been obtained and more and more details have been studied and explained. Historically, these bands presented the first extensive and clear-cut example of an electronic transition forbidden by the symmetry selection rules and of an application of the vibronic selection rules (see Chap. II, section 2b($\beta$)).

In the previous Fig. 70 spectrograms showing the absorption bands of $C_6H_6$ and $C_6D_6$ are reproduced. The most prominent feature is a progression of bands in a frequency 925 cm$^{-1}$ for $C_6H_6$ and in a frequency 879 cm$^{-1}$ for $C_6D_6$. The smallness of the isotope shift indicates that this frequency corresponds to a vibration of the $C_6$

ring, and according to the selection rules, unless the symmetry in the excited state were reduced, it must be a vibration symmetric with respect to all symmetry elements of point group $\boldsymbol{D}_{6h}$. The only vibration of this type is the symmetric breathing vibration $\nu_2$ which, in the ground state, has the frequencies 995 and 946 cm$^{-1}$ for $C_6H_6$ and $C_6D_6$ respectively[12]. The fact that the bands are shaded to the red shows that in the excited state the $C_6$ hexagon is slightly larger than in the ground state and therefore, on the basis of the Franck–Condon principle, we expect a fairly long progression in the vibration $\nu_2$ in agreement with observation.

At the longward side of the first strong group of bands, as shown by Fig. 70, there is a very weak group which is clearly a group of hot bands, as was indeed confirmed by investigations of the effect of temperature on these bands by Radle and Beck (1037). However, the separation of the principal hot band from the first main band, which is 1130 cm$^{-1}$ for $C_6H_6$, should be equal to the value of $\nu_2$ in the ground state, or, if benzene were not symmetrical in the excited state, to some other ground state fundamental. No such frequency exists in the ground state[13]. The explanation for this apparent discrepancy was first given by Sponer, Nordheim, Sklar and Teller (1152). It lies in the fact that the 2600 Å bands represent a forbidden electronic transition made allowed by interaction between the electronic motion in the excited state and a non-totally symmetric vibration, say $\nu_a$. In such a case, as we have seen in Chapter II, p. 177f, the $0-0$ band of the system is rigorously forbidden and the main bands go to levels in all of which $\nu_a$ is singly excited, in addition to the excitation of the totally symmetrical vibrations by 0, 1 or more quanta. As a consequence, the $0-1$ hot band in the anti-symmetric vibration which is now vibronically allowed is separated from the first main band $1-0$ by $\nu_a' + \nu_a''$ (see Fig. 69).

The question is now to identify $\nu_a$. We know that $\nu_a''$ must be appreciably less than 1130 cm$^{-1}$ (it would be about $\frac{1}{2}$(1138) if there were not much difference between the force constants in the upper and lower state). Since for $C_6D_6$ the corresponding figure is 1076 cm$^{-1}$, it follows that $\nu_a$ must be a ring vibration (not a vibration involving considerable motion of the H atoms). These arguments lead to the two vibrations $\nu_{18}(e_{2g})$ or $\nu_6(b_{1u})$ (see Vol. II, p. 364f) of which the former has the well-established value 608.0 cm$^{-1}$, while the latter (an inactive fundamental) has the somewhat uncertain value 1010 cm$^{-1}$. The most reasonable assignment is $\nu_a'' = \nu_{18}(e_{2g}) = 608.0$ cm$^{-1}$ and therefore $\nu_a' = 522$ cm$^{-1}$. If the latter value is correct, the $0-0$ band which is rigorously forbidden would be 522 cm$^{-1}$ longward from the first main band, i.e., at 38086 cm$^{-1}$. A striking confirmation of this conclusion is supplied by spectra of solid $C_6H_6$ obtained many years ago by Kronenberger (698) at the temperature of liquid hydrogen ($-259$°C). Here all hot bands are eliminated, but a very weak band appears longward of the first main band at a separation of about 520 cm$^{-1}$. This must be the $0-0$ band which because of the effect of crystal fields is no longer strictly forbidden.

Knowing now the species of the vibration which causes the forbidden transition to appear, viz., $e_{2g}$, we can use the previous relations (II, 37) and (II, 39) to determine the species of the excited electronic state. Since it is not immediately obvious from the band structure whether the bands are $\parallel$ or $\perp$ bands, we consider both alternatives. The species of the dipole moment is $A_{2u}$ or $E_{1u}$ respectively (see Table 51, p. 566, or Table 55 of Vol. II). Therefore, the excited electronic state must have species $E_{2u}$

[12] The only other totally symmetric vibration is the symmetrical C—H vibration $\nu_1$ which in the ground state has the frequencies 3073 and 2292 cm$^{-1}$ for $C_6H_6$ and $C_6D_6$ respectively. Even without a knowledge of the $C_6D_6$ spectrum, one would have been justified to conclude that 925 cm$^{-1}$ of the excited state corresponds to $\nu_2$, not $\nu_1$.

[13] This argument was not entirely cogent as long as not all fundamentals of $C_6H_6$ were known.

or $B_{1u}$ or $B_{2u}$. Either of the last two alternatives is compatible with observation if the main bands are ⊥ bands. It is easy to check that in forbidden $E_{2u}-A_{1g}$ or $B_{1u}-A_{1g}$ or $B_{2u}-A_{1g}$ electronic transitions, when an $e_{2g}$ vibration is singly excited in the upper or lower state, vibronic transitions with an allowed dipole component, either $A_{2u}$ or $E_{1u}$, are indeed obtained.

Recent work by Callomon, Dunn and Mills (176) on the partially resolved rotational structure has shown that the bands are ⊥ bands. This observation removes the possibility that the electronic transition is $^1E_{2u}-{}^1A_{1g}$, but it is almost impossible to decide between $^1B_{1u}-{}^1A_{1g}$ and $^1B_{2u}-{}^1A_{1g}$ from the rotational or vibrational structure alone[14]. However, a decision is possible on the basis of molecular orbital theory: As we have seen in Chapter III, p. 350, the first excited configuration $(a_{2u})^2(e_{1g})^3(e_{2u})$ of the $\pi$ electrons of benzene gives the three singlet states $^1B_{1u}$, $^1B_{2u}$ and $^1E_{1u}$. A more detailed calculation, first by Goeppert–Mayer and Sklar (427) and later refined by several authors, most recently Parr, Craig and Ross (965), has shown that of the three singlet states $^1B_{2u}$ lies lowest and $^1E_{1u}$ highest. Indeed, a theoretical energy value of 4.4 eV is obtained for $^1B_{2u}$ which compares well with the observed energy of 4.9 eV. On the other hand, the $^1B_{1u}$ state is predicted to lie at 9.0 eV. There can therefore be little doubt that the 2600 Å system of $C_6H_6$ represents the predicted (forbidden) $^1B_{2u}-{}^1A_{1g}$ electronic transition. The valence-bond method also gives directly $^1B_{2u}$ as the first excited singlet state. It corresponds essentially to the out-of-phase overlap of the two Kékulé structures (see p. 373).

Thus far we have only considered the main bands. Most of the remaining bands have been assigned, in accordance with the selection rules, by Sponer, Nordheim, Sklar and Teller (1152) and Garforth and Ingold (404) for $C_6H_6$ and by Sponer (1147) and Garforth and Ingold (405) for $C_6D_6$. Without going into details, it is perhaps of interest to mention a few of the more important assignments. As can be seen in Fig. 70 each one of the main bands is accompanied by a group of weaker bands at the longward side. It is clear that these bands must belong to sequences in the various vibrations which can be thermally excited at room temperature. The closest of these hot bands is separated by only 86 cm$^{-1}$ from the main band. This is just the difference between $\nu_{18}(e_{2g})$ in the upper and lower state ($\nu''_{18} - \nu'_{18} = 608 - 522 = 86$ cm$^{-1}$), and these bands are therefore $2-1$ bands in $\nu_{18}$, the main bands being $1-0$ bands in this vibration. The third band of the group, the strongest next to the main band, is separated by 162 cm$^{-1}$ from the main band. It was first shown by Kistiakowsky and Solomon (674) by a study of the effect of temperature on the relative intensities of these bands, that this hot band corresponds to the $1-1$ band in the lowest frequency fundamental $\nu_{20}(e_{2u})$ of $C_6H_6$ which in the ground state has the frequency 399 cm$^{-1}$ and therefore in the excited state 237 cm$^{-1}$. The $2-2$ and $3-3$ bands of this sequence have also been observed with the same drop in intensity as between $0-0$ and $1-1$. The rather large reduction of $\nu_{20}$ in going from the ground to the excited state is reasonable because this is a ring bending vibration (see Vol. II, p. 118) which would be very sensitive to an excitation of the $\pi$ electrons.

In principle, any one of the four fundamentals of species $e_{2g}$ could cause the forbidden $^1B_{2u}-{}^1A_{1g}$ electronic transition to appear. However, only the carbon ring

[14] Only if weak bands with a ∥ structure were found in addition to the main ⊥ bands would a decision be possible since a ∥ component of the dipole moment can arise only in a $^1B_{1u}-{}^1A_{1g}$ transition through vibronic interaction with a $b_{2g}$ vibration. A ∥ component of a $^1B_{2u}-{}^1A_{1g}$ transition would require vibronic interaction with a $b_{1g}$ vibration of which there are none in $C_6H_6$. Albrecht and Simpson (55) by polarized absorption in a crystal of para-dimethoxybenzene have obtained the direction of the transition moment in the analogue of the 2600 Å benzene absorption. It is in accord with the assumption that in benzene the transition is $^1B_{2u}-{}^1A_{1g}$.

vibrations are likely to do it strongly as does $\nu_{18}(e_{2g})$. The only other carbon ring vibration of this type is $\nu_{16}(e_{2g})$. $1-0$ bands in this vibration, if present at all, are at least 20 times weaker than the $1-0$ bands in $\nu_{18}$. [Bands previously assigned as $1-0$ in $\nu_{16}$ by Sponer *et al.* (1152) and Garforth and Ingold (404) were shown by Callomon, Dunn and Mills (176) to have rotational contours incompatible with such an assignment.] In the ground state the level $\nu_{16}''$ is in almost exact Fermi resonance with $\nu_2'' + \nu_{18}''$; transitions from these two levels have been observed but give little information about the importance of $\nu_{16}$ for the vibronic interaction between the $^1B_{2u}$ electronic state and higher $^1E_{1u}$ states. Theoretical considerations by Craig (251) predict that $\nu_{16}$ is much less important than $\nu_{18}$.

Excellent confirmation of the analysis comes from a study of the fluorescence spectrum of benzene (see Sponer, Nordheim, Sklar and Teller (1152) and Garforth and Ingold (404)(405)). Unless the fluorescence is observed at extremely low pressure one finds a spectrum that is essentially the same for all methods of excitation. Most of the spectrum is longward of the main absorption bands. Progressions in the frequency 995 cm$^{-1}$, the "breathing" vibration of the ground state, are most prominent, starting from the band $0-\nu_{18}''$ and from a few other bands observed in absorption, including one from the first main absorption band $\nu_{18}'-0$. Each of these bands is again accompanied by sequences. While collisions have a tendency to bring all molecules to thermal vibrational equilibrium in the excited state before emission, clearly some molecules radiate before all vibrational energy is removed. Up to two quanta of $\nu_{18}'$, five quanta of $\nu_{20}'$ and one quantum of $\nu_2'$ have been identified [Garforth and Ingold (404)].

In addition to the main system of absorption bands of benzene there is at longer wavelengths, near 3400 Å, an exceedingly weak absorption first observed by Sklar (1138) in an absorbing path of 20 cm of liquid $C_6H_6$. The ratio of the intensity of this absorption to that of the main system was of the order $10^{-5}$. From the first excited electron configuration $(a_{2u})^2(e_{1g})^3e_{2u}$ three triplet states $^3B_{1u}$, $^3B_{2u}$ and $^3E_{1u}$ arise of which a more detailed calculation [see Parr, Craig and Ross (965)] shows $^3B_{1u}$ to lie lowest, well below the $^1B_{2u}$ state[15]. Sklar (1138) had predicted this state from a valence-bond treatment, and this caused him to look for it and to identify the very weak absorption that he found with the $^3B_{1u}-{}^1A_{1g}$ transition. A more detailed discussion of the identification of this transition has recently been given by Albrecht (54). Absorption curves of liquid $C_6H_6$ in the 3400–2600 Å region have been published by Lewis and Kasha (742) and Pitts (987).

Lewis and Kasha (741) had earlier established that in benzene, as in many other molecules, the lowest triplet state is responsible for the phosphorescence observed in a rigid matrix at low temperature. Indeed, it appears that the most longward absorption band of the liquid agrees approximately with the most shortward phosphorescence band of the solid glass. The phosphorescence spectrum has a rather different appearance when observed in different matrices, presumably because in different crystal fields the symmetry selection rules are different. This is strikingly illustrated if one compares the spectrum obtained by Shull (1128) in EPA (a mixture of ether, isopentane and ethyl alcohol) with those obtained by Robinson (1076) in a krypton and an argon matrix. There can, however, be little doubt that the $0-0$ band (at 29470 cm$^{-1}$) occurs only because of crystal perturbations and that the transition is forbidden not only by the spin rule but also by symmetry, and that it appears because of vibronic interactions just like the 2600 Å system. A vibrational analysis of the phosphorescence spectrum has been given by Shull (1128), Robinson (1076) and Leach and Lopez-Delgado (731a). Considering the crystal perturbations, the observed vibrational

[15] The $^3B_{2u}$ state is predicted to lie well above the $^1B_{2u}$ state, contrary to the usual behavior of corresponding singlet and triplet states.

transitions are quite compatible with a ${}^3B_{1u}-{}^1A_{1g}$ transition [see Craig (251), Mizushima and Koide (863) and Albrecht (54)].

The $\tilde{a}\ {}^3B_{1u}-\tilde{X}\ {}^1A_{1g}$ bands of benzene have not yet been observed in the free molecule. Evans (363)(364) and Craig, Hollas and King (252) showed that even its observation in the liquid depends on the presence of traces of oxygen. Evans (365) did observe the transition in benzene vapor in the presence of high pressures of oxygen (130 atm) with a path length of only 5.2 cm, whereas Craig, Hollas and King (252) were unable to detect it in 22.5 m of carefully purified and de-oxygenated liquid benzene. Quite recently King and Pinnington (669a) have observed the $\tilde{a}-\tilde{X}$ transition in $C_6H_6$ vapor mixed with oxygen at pressures as low as 1 atm by using path lengths up to 440 m. Although, as already mentioned, the $\tilde{a}-\tilde{X}$ transition is generally considered to be both spin and symmetry forbidden, in the gas spectra induced by oxygen the $0-0$ band does occur. King and Pinnington on this basis suggest that in the $\tilde{a}$ state the symmetry is lowered to $\boldsymbol{D}_{2h}$ and that therefore the $0-0$ band is allowed. A distorted ($\boldsymbol{D}_{2h}$) equilibrium conformation in the $\tilde{a}$ state is also suggested by de Groot and van der Waals (450) on the basis of the electron-spin resonance spectrum in rigid matrices and has been predicted theoretically by Liehr (747a).

Three mechanisms have been invoked to account for the enhancement of the $\tilde{a}-\tilde{X}$ bands by oxygen. Evans attributed it to the magnetic field of the paramagnetic additive (NO is also effective) but Tsubomura and Mulliken (1226) showed that this effect was negligibly small. Indeed, the addition of oxygen has no enhancing effect on the $\tilde{a}-\tilde{X}$ transition of $C_2N_2$ which occurs in a similar region (Callomon, unpublished). Hoijting (562) instead invokes purely covalent exchange interactions and suggests that it is the mixing of the two triplet states of the $O_2-C_6H_6$ complex, one from ${}^3B_{1u}$ and one from ${}^1B_{2u}$, that causes the effect. This interpretation is in turn rejected by Murrell (920a) who, as did Tsubomura and Mulliken, prefers a charge-transfer mechanism involving ionic structures of the complex (see p. 426f.). Further studies are needed before this question can be definitely answered.

The life-time of the free $C_6H_6$ molecule in the ${}^3B_{1u}$ state has been determined indirectly by Lim (751) to be $28 \pm 4$ sec. It is 16 sec in an argon matrix [Wright, Frosch and Robinson (1322)][16]. On the other hand, the absence of the $\tilde{a}-\tilde{X}$ transition in the experiments of Craig, Hollas and King (252), mentioned earlier, indicates a lower limit of 300 sec. However, it must be noted that Craig, Hollas and King were unable to observe the spectrum beyond 29410 cm$^{-1}$, and therefore their limit to the life-time refers to the $0-0$ band which would be vibronically forbidden if the electronic transition is ${}^3B_{1u}-{}^1A_{1g}$ of $\boldsymbol{D}_{6h}$.

Shibata, Kushida and Mori (1123) have reported an absorption in liquid benzene at still longer wavelengths, but this has up to now not been confirmed. The problem of locating the second excited triplet state (${}^3E_{1u}$) has been discussed by Platt (992).

On the shortward side of the 2600 Å system, strong absorption (stronger by about a factor 10) starts at 2050 Å. First observed by Scheibe, Povenz and Linström (1102), it has later been studied by a number of investigators. We reproduce in Fig. 201 the absorption curve of Pickett, Muntz and McPherson (981). An instructive spectrogram has been given by Price and Walsh (1031). As can be seen from Fig. 201, there are in the region 2050–1700 Å at least two electronic transitions: a system of very diffuse bands between 2050 and 1850 Å and an extremely strong broad absorption with a few sharp peaks and an underlying continuum. While in the first system a progression in a frequency of about 900 cm$^{-1}$ can be clearly recognized, the diffuse nature of the bands prevents a more detailed analysis [see, however, Dunn and Ingold (327)]. The

[16] This life-time may be compared with 0.59 μsec determined by Donovan and Duncan (289a) as the life-time of the ${}^1B_{2u}$ state.

same applies to the second system at 1800 Å. One is therefore dependent on theoretical arguments for the correct assignment of these electronic transitions.

It is generally assumed that these two transitions must arise by simple excitation of one of the six $\pi$ electrons into one of the two (anti-bonding) $\pi$ orbitals that are vacant in the ground state, viz., $e_{2u}$ and $b_{2g}$. The only allowed transition arising in this way

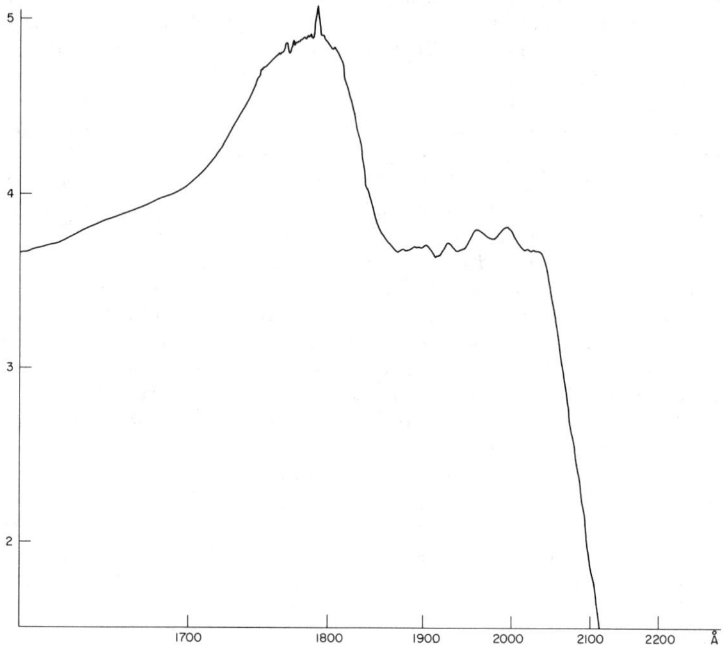


FIG. 201. **Absorption curve of $C_6H_6$ in the region 2050 to 1700 Å after Pickett, Muntz and McPherson (981).**

is that to the $^1E_{1u}$ state of the configuration $(a_{2u})^2(e_{1g})^3e_{2u}$ (see p. 350). The 1800 Å system is generally ascribed to this transition [see Nordheim, Sponer and Teller (939)]. It would then seem most plausible to assign the 2000 Å system to a transition to the $^1B_{1u}$ state which arises from the same configuration as $^1E_{1u}$ and $^1B_{2u}$ (see Fig. 141). However, calculations of the energies of the various states resulting from the $\pi$ electrons suggest the possibility that, instead, the 2000 Å band may be due to a transition to the $^1E_{2g}$ state of the configuration $(a_{2u})^2(e_{1g})^3b_{2g}$. Arguments aimed at a decision between the two assignments, $^1B_{1u}$ or $^1E_{2g}$, may be found in the papers by Dunn and Ingold (327), Parr, Craig and Ross (965) and Murrell and Pople (921).

Price and Walsh (1031) have emphasized that the strong 1800 Å system would fit in very well as the first member of one of the Rydberg series (see below) particularly since the observed vibrational structure is similar to that of the other members of this Rydberg series. It appears very probable that both transitions, the Rydberg transition and the $\pi$ electron transition, are overlapping at 1800 Å, and more particularly that the sharp features near 1800 Å belong to the Rydberg transition. This is also assumed by Wilkinson (1299) in his detailed analysis of the vacuum ultraviolet spectrum of $C_6H_6$ and $C_6D_6$.

Below 1700 Å there are several strong Rydberg series, first observed by Price and

Wood (1032), which converge to the first ionization potential 9.247 eV (see Table 82). Wilkinson (1299), who has reinvestigated the spectrum under high resolution, gives four such series. Many of the members of the Rydberg series show vibrational structure. Two progressions are prominent, one in a frequency of about 970, and another in a frequency of about 690 $cm^{-1}$, both with a large anharmonicity. The former frequency is clearly $\nu_2$, the breathing vibration (995 $cm^{-1}$ in the ground state), but the second frequency is not as readily identified since the only other totally symmetric vibration is the C—H stretching vibration which can hardly be as low as 690 $cm^{-1}$. Unless one wanted to assume that the symmetry in all the Rydberg states is lower than in the ground state, one must assume that the 690 $cm^{-1}$ progression is made possible by vibronic interactions in the excited states in a way very similar to the $1-0$ bands in the degenerate vibrations observed for $CH_3I$ and $CF_3I$ (see pp. 528 and 532). The upper electronic states of the Rydberg transitions must be either $A_{2u}$ or $E_{1u}$. Those that are $E_{1u}$ will have Jahn–Teller interaction, and therefore in them the $1-0$, as well as the $2-0$, and $3-0$ transitions in vibrations of species $e_{2g}$ can appear weakly. This explanation was suggested by Liehr and Moffitt (748). The $e_{2g}$ vibration is taken by them to be $\nu_{18}$ which in the ground state has the frequency 608 $cm^{-1}$. However, the vibration responsible may also be $\nu_{17}(e_{2g})$ which in the ground state is 1178 $cm^{-1}$. A fairly large Jahn–Teller interaction may also account for the large apparent anharmonicity in these progressions, in the way illustrated by Fig. 22.

Wilkinson (1299) had thought that the spacing 690 $cm^{-1}$ in the Rydberg transitions corresponds to $2\nu_{18}(e_{2g})$ and that observation of a progression in $2\nu_{18}$ establishes the presence of Jahn–Teller interaction. Actually, vibrational transitions with $\Delta v_{18} = \pm 2, \pm 4, \ldots$ could occur weakly even without vibronic interaction if there is a large change of vibrational frequency in going to the excited state (see p. 152). Such an explanation would be necessary if it could be shown that the Rydberg states showing these 690 $cm^{-1}$ progressions are $^1A_{2u}$ rather than $^1E_{1u}$ electronic states since $^1A_{2u}$ is not subject to Jahn–Teller interaction.

The example just discussed illustrates the difficulty of establishing the presence of Jahn–Teller interaction. A decision could be reached if the $0-1$ band in $\nu_{18}$ were observed, or if the Jahn–Teller splitting in the $1-1$ band could be found, or if the rotational structure could be resolved.

Near the first ionization limit there are two strong absorption bands (at 1342.5 and 1341.5 Å) which are followed by further bands at shorter wavelengths forming two Rydberg series with a common limit at $11.48_9$ eV. El-Sayed, Kasha and Tanaka (350) have assigned this limit to ionization from the $a_{2u}$ orbital of the $\pi$ electrons. A further Rydberg series was observed by the same authors in the region 850 to 700 Å with a limit at 16.84 eV. Since the first members of this series show fairly long progressions in a frequency 960 $cm^{-1}$, which is in all probability $\nu_2$, the limit is assigned to the removal of a $\sigma$ electron of the $C_6$ ring.

**$(CH_3CO)_2$ (biacetyl).** The absorption spectrum of biacetyl is similar to that of glyoxal (in which the two methyl groups are replaced by H atoms). The absorption spectrum in the visible and near ultraviolet was first studied by Henri (20) and his students Lardy (726) and Light (750). In the region 4700–4000 Å there is an extensive system of fairly sharp red-shaded absorption bands. This system is clearly the analogue of the $\tilde{A}-\tilde{X}$ system of glyoxal. There are six band groups separated by intervals 1060 and 620 $cm^{-1}$ which must be assigned as upper state vibrational frequencies. At higher pressure a continuous absorption joins on which extends to 3500 Å. A second region of absorption, entirely continuous, extends from 2800 to 2200 Å and corresponds presumably to the $\tilde{B}-\tilde{X}$ transition of glyoxal.

The vacuum ultraviolet absorption of biacetyl has been studied by Ells (354).

Two systems of discrete bands were found, one extending from 2000 to 1880 Å, the other from 1750 to 1650 Å. A frequency difference of 1220 $cm^{-1}$ is prominent in the first system and one of 1350 $cm^{-1}$ in the second. There appears to be continuous absorption at shorter wavelengths, but the spectrum has not been studied below 1550 Å.

Biacetyl vapor shows an interesting fluorescence studied by a number of investigators [see, for example, Almy, Fuller and Kinzer (64)]. The spectrum of the fluorescence is independent of the wavelength of the exciting radiation (λ4358, 4047 or 3650 Å). It extends from 5000 to 6200 Å consisting of a number of discrete bands. No vibrational analysis of this spectrum has yet been made. The life-time of the fluorescence measured directly by a phosphoroscope is $1.6 \times 10^{-3}$ sec [Almy and Anderson (63)] while the life-time of the $\tilde{A}$ state, as obtained from the integrated absorption coefficient, is $10^{-5}$ sec. It has therefore been concluded that the upper state of the fluorescence is not the $\tilde{A}$ state but another state, $\tilde{a}$, which is probably a triplet state.

The mechanism of the fluorescence has been studied by several authors [in addition to the papers quoted, see Noyes and Henriques (943)(944), Almy and Gillette (65), Kaskan and Duncan (660) and Coward and Noyes (247)]. It is noteworthy that the fluorescence can be excited not only by radiation in the discrete region (λ4358 and 4047), but also by radiation in the region of continuous absorption (λ3650 Å). However, while in the former region the quantum yield of fluorescence is independent of pressure and has a value of 0.145 at 27°C, in the latter region it is zero at zero pressure but rises quickly with pressure to almost the same value as at longer wavelengths. One must assume that the region of continuous absorption is really a region of strong predissociation and that at high pressure, collisions bring the molecules down to stable levels of the $\tilde{A}$ state before they have had time to undergo radiationless decomposition. In the discrete region of the $\tilde{A}$ state, a radiationless transition to the $\tilde{a}$ state (internal conversion) may take place for those vibrational levels which perturb each other in a way entirely similar to that responsible for extra lines in perturbations (see pp. 291 and 292 of Vol. I). Following such a "transition" the molecule is brought down by collisions to the lowest vibrational level of the $\tilde{a}$ state before fluorescence takes place. In addition to the green fluorescence discussed so far, a blue fluorescence is mentioned by Coward and Noyes (247) which may correspond to the direct transition $\tilde{A}-\tilde{X}$.

It may be noted that the fluorescence of acetone, acetaldehyde and other ketones is found to be identical with that of biacetyl [see Matheson and Zabor (800)], and is now generally ascribed to biacetyl formed in the photolysis of these compounds.

**Other twelve-atomic molecules.** The ultraviolet absorption spectra of a large number of halogen-substituted benzenes have been studied by various investigators, but very little work has been done in the vacuum ultraviolet; only $C_6H_5Cl$, $C_6H_5Br$, $C_6H_5I$ and $C_6H_4Cl_2$ have been studied in this region by Price and Walsh (1031). On the whole, the absorption spectra of these substituted benzenes are very similar to that of benzene with minor shifts and relatively small changes of intensity in spite of the much lower symmetry of most of these molecules, for which the 2600 Å transition is no longer forbidden. This observation confirms the conclusion that the absorption near 2600 Å is entirely determined by the $\pi$ electrons of the $C_6$ ring and that these are very little affected by the substitution. It is interesting to note that in the (unsymmetrically) substituted benzenes the $0-0$ band which is absent in benzene does appear. A general discussion of substituted benzenes may be found in Sponer and Teller (1155). In Table 82 references to work on individual halogen-substituted benzenes are included as well as for a few other twelve-atomic molecules.

# APPENDIX I

## SPECIES AND CHARACTERS OF THE EXTENDED POINT GROUPS

In the following Tables the symbols and numbers to the left of and above the broken lines refer to the standard point groups (see Vol. II, pp. 105–123). Below and to the right of the broken lines are the species and characters of the corresponding extended point groups (see p. 15f). For these point groups the order of classes of planes of symmetry and of two-fold axes is double that shown which refers to the ordinary point groups. The designation *sep.* indicates separably degenerate species (see Vol. II, p. 99). The symmetry element $R$ is the artificial one discussed on p. 15. To the right in each Table the translations and rotations belonging to the particular species are given. For some point groups (say $\boldsymbol{P}$) with a center of inversion, the character tables are not given explicitly but can be immediately derived from those of the corresponding groups ($\boldsymbol{Q}$) without a center of inversion by replacing each species by two: one symmetric ($g$), the other antisymmetric ($u$) with respect to the center. This relation is expressed by the symbolic equations $\boldsymbol{P} = \boldsymbol{Q} \times \boldsymbol{C}_i$.

TABLE 48

POINT GROUPS $\boldsymbol{C}_1$, $\boldsymbol{C}_2$, $\boldsymbol{C}_3$, $\boldsymbol{C}_s$, $\boldsymbol{C}_i$

| $\boldsymbol{C}_1$ | $I$ | $R$ | |
|---|---|---|---|
| $A$ | 1 | 1 | $T$, $R$ |
| $B_{\frac{1}{2}}$ | 1 | −1 | |

| $\boldsymbol{C}_2$ | $I$ | $C_2(z)$ | $R$ | |
|---|---|---|---|---|
| $A$ | 1 | 1 | 1 | $T_z$, $R_z$ |
| $B$ | 1 | −1 | 1 | $T_x$, $T_y$, $R_x$, $R_y$ |
| $E_{\frac{1}{2}}$ | 2 | 0 | −2 | sep. |

| $\boldsymbol{C}_3$ | $I$ | $2C_3$ | $R$ | $2C_3^2$ | |
|---|---|---|---|---|---|
| $A$ | 1 | 1 | 1 | 1 | $T_z$, $R_z$ |
| $E$ | 2 | −1 | 2 | −1 sep. | $T_x$, $T_y$, $R_x$, $R_y$ |
| $E_{\frac{1}{2}}$ | 2 | 1 | −2 | −1 sep. | |
| $B_{\frac{3}{2}}$ | 1 | −1 | −1 | 1 | |

| $\boldsymbol{C}_s$ | $I$ | $\sigma(xy)$ | $R$ | |
|---|---|---|---|---|
| $A'$ | 1 | 1 | 1 | $T_x$, $T_y$, $R_z$ |
| $A''$ | 1 | −1 | 1 | $T_z$, $R_x$, $R_y$ |
| $E_{\frac{1}{2}}$ | 2 | 0 | −2 | sep. |

| $\boldsymbol{C}_i$ | $I$ | $i$ | $R$ | $iR$ | |
|---|---|---|---|---|---|
| $A_g$ | 1 | 1 | 1 | 1 | $R_x$, $R_y$, $R_z$ |
| $A_u$ | 1 | −1 | 1 | −1 | $T_x$, $T_y$, $T_z$ |
| $B_{\frac{1}{2}g}$ | 1 | 1 | −1 | −1 | |
| $B_{\frac{1}{2}u}$ | 1 | −1 | −1 | 1 | |

TABLE 49

POINT GROUPS $\boldsymbol{C}_{2h}$ AND $\boldsymbol{C}_{3h}$

| $\boldsymbol{C}_{2h}$ | $I$ | $C_2$ | $\sigma_h$ | $i$ | $R$ | $iR$ | |
|---|---|---|---|---|---|---|---|
| $A_g$ | 1 | 1 | 1 | 1 | 1 | 1 | $R_z$ |
| $A_u$ | 1 | 1 | $-1$ | $-1$ | 1 | $-1$ | $T_z$ |
| $B_g$ | 1 | $-1$ | $-1$ | 1 | 1 | 1 | $R_x, R_y$ |
| $B_u$ | 1 | $-1$ | 1 | $-1$ | 1 | $-1$ | $T_x, T_y$ |
| $E_{\frac{1}{2}g}$ | 2 | 0 | 0 | 2 | $-2$ | $-2$ | sep. |
| $E_{\frac{1}{2}u}$ | 2 | 0 | 0 | $-2$ | $-2$ | 2 | sep. |

| $\boldsymbol{C}_{3h}$ | $I$ | $2C_3$ | $\sigma_h$ | $2S_3$ | $R$ | $2C_3^2$ | $2S_3^5$ | |
|---|---|---|---|---|---|---|---|---|
| $A'$ | 1 | 1 | 1 | 1 | 1 | 1 | 1 | $R_z$ |
| $A''$ | 1 | 1 | $-1$ | $-1$ | 1 | 1 | $-1$ | $T_z$ |
| $E'$ | 2 | $-1$ | 2 | $-1$ | 2 | $-1$ | $-1$ | sep. $T_x, T_y$ |
| $E''$ | 2 | $-1$ | $-2$ | 1 | 2 | $-1$ | 1 | sep. $R_x, R_y$ |
| $E_{\frac{1}{2}}$ | 2 | 1 | 0 | $\sqrt{3}$ | $-2$ | $-1$ | $-\sqrt{3}$ | sep. |
| $E_{\frac{3}{2}}$ | 2 | $-2$ | 0 | 0 | $-2$ | 2 | 0 | sep. |
| $E_{\frac{5}{2}}$ | 2 | 1 | 0 | $-\sqrt{3}$ | $-2$ | $-1$ | $\sqrt{3}$ | sep. |

TABLE 50

POINT GROUPS $\boldsymbol{D}_2$, $\boldsymbol{C}_{2v}$, $\boldsymbol{D}_{2h}(\equiv \boldsymbol{V}_h)$; $\boldsymbol{D}_3$, $\boldsymbol{C}_{3v}$, $\boldsymbol{D}_{3d}$; $\boldsymbol{D}_4$, $\boldsymbol{C}_{4v}$, $\boldsymbol{D}_{2d}(\equiv \boldsymbol{V}_d)$, $\boldsymbol{D}_{4h}$

| $\boldsymbol{D}_2$ / $\boldsymbol{C}_{2v}$ | $I$ / $I$ | $C_2(z)$ / $C_2(z)$ | $C_2(y)$ / $\sigma_v(xz)$ | $C_2(x)$ / $\sigma_v(yz)$ | $R$ / $R$ | |
|---|---|---|---|---|---|---|
| $A$ $A_1$ | 1 | 1 | 1 | 1 | 1 | $T_z$ for $\boldsymbol{C}_{2v}$ |
| $B_1$ $A_2$ | 1 | 1 | $-1$ | $-1$ | 1 | $T_z$ for $\boldsymbol{D}_2$; $R_z$ |
| $B_2$ $B_1$ | 1 | $-1$ | 1 | $-1$ | 1 | $T_y$, $R_y$ for $\boldsymbol{D}_2$; $T_x$, $R_y$ for $\boldsymbol{C}_{2v}$ |
| $B_3$ $B_2$ | 1 | $-1$ | $-1$ | 1 | 1 | $T_x$, $R_x$ for $\boldsymbol{D}_2$; $T_y$, $R_x$ for $\boldsymbol{C}_{2v}$ |
| $E_{\frac{1}{2}}$ $E_{\frac{1}{2}}$ | 2 | 0 | 0 | 0 | $-2$ | |

$\boldsymbol{D}_{2h} \equiv \boldsymbol{D}_2 \times \boldsymbol{C}_i$ (see Vol. II, p. 108)

| $\boldsymbol{D}_3$ / $\boldsymbol{C}_{3v}$ | $I$ / $I$ | $2C_3$ / $2C_3$ | $3C_2$ / $3\sigma_v$ | $R$ / $R$ | $2C_3^2$ / $2C_3^2$ | |
|---|---|---|---|---|---|---|
| $A_1$ | 1 | 1 | 1 | 1 | 1 | $T_z$ for $\boldsymbol{C}_{3v}$ |
| $A_2$ | 1 | 1 | $-1$ | 1 | 1 | $T_z$ for $\boldsymbol{D}_3$; $R_z$ |
| $E$ | 2 | $-1$ | 0 | 2 | $-1$ | $T_{x,y}$; $R_{x,y}$ |
| $E_{\frac{1}{2}}$ | 2 | 1 | 0 | $-2$ | $-1$ | |
| $E_{\frac{3}{2}}$ | 2 | $-2$ | 0 | $-2$ | 2 | sep. |

$\boldsymbol{D}_{3d} \equiv \boldsymbol{D}_3 \times \boldsymbol{C}_i$ (see Vol. II, p. 116)

Table 50 (*Continued*)

| $\boldsymbol{D}_4$<br>$\boldsymbol{C}_{4v}$<br>$\boldsymbol{D}_{2d}$ | $I$<br>$I$<br>$I$ | $2C_4(z)$<br>$2C_4(z)$<br>$2S_4(z)$ | $C_4{}^2 \equiv C_2''$<br>$C_4{}^2 \equiv C_2''$<br>$S_4{}^2 \equiv C_2''$ | $2C_2$<br>$2\sigma_v$<br>$2C_2$ | $2C_2'$<br>$2\sigma_d$<br>$2\sigma_d$ | $R$<br>$R$<br>$R$ | $2C_4{}^3$<br>$2C_4{}^3$<br>$2S_4{}^3$ | |
|---|---|---|---|---|---|---|---|---|
| $A_1$ | 1 | 1 | 1 | 1 | 1 | 1 | 1 | $T_z$ for $\boldsymbol{C}_{4v}$ |
| $A_2$ | 1 | 1 | 1 | $-1$ | $-1$ | 1 | 1 | $T_z$ for $\boldsymbol{D}_4$; $R_z$ |
| $B_1$ | 1 | $-1$ | 1 | 1 | $-1$ | 1 | $-1$ | |
| $B_2$ | 1 | $-1$ | 1 | $-1$ | 1 | 1 | $-1$ | $T_z$ for $\boldsymbol{D}_{2d}$ |
| $E$ | 2 | 0 | $-2$ | 0 | 0 | 2 | 0 | $T_{x,y}$; $R_{x,y}$ |
| $E_{\frac{1}{2}}$ | 2 | $+\sqrt{2}$ | 0 | 0 | 0 | $-2$ | $-\sqrt{2}$ | |
| $E_{\frac{3}{2}}$ | 2 | $-\sqrt{2}$ | 0 | 0 | 0 | $-2$ | $+\sqrt{2}$ | |

$\boldsymbol{D}_{4h} \equiv \boldsymbol{D}_4 \times \boldsymbol{C}_i$ (see Vol. II, p. 117)

TABLE 51

POINT GROUPS $\boldsymbol{D}_5$, $\boldsymbol{C}_{5v}$, $\boldsymbol{D}_{5d}$; $\boldsymbol{D}_6$, $\boldsymbol{C}_{6v}$, $\boldsymbol{D}_{3h}$, $\boldsymbol{D}_{6h}$; $\boldsymbol{D}_8$, $\boldsymbol{C}_{8v}$, $\boldsymbol{D}_{4d}(\equiv \boldsymbol{S}_{8v})$, $\boldsymbol{D}_{8h}$

| $\boldsymbol{D}_5$<br>$\boldsymbol{C}_{5v}$ | $I$<br>$I$ | $2C_5$<br>$2C_5$ | $2C_5{}^2$<br>$2C_5{}^2$ | $5C_2$<br>$5\sigma_v$ | $R$<br>$R$ | $2C_5{}^4$<br>$2C_5{}^4$ | $2C_5{}^3$<br>$2C_5{}^3$ | |
|---|---|---|---|---|---|---|---|---|
| $A_1$ | 1 | 1 | 1 | $+1$ | 1 | 1 | 1 | $T_z$ for $\boldsymbol{C}_{5v}$ |
| $A_2$ | 1 | 1 | 1 | $-1$ | 1 | 1 | 1 | $T_z$ for $\boldsymbol{D}_5$; $R_z$ |
| $E_1$ | 2 | $\frac{-1+\sqrt{5}}{2}$ | $\frac{-1-\sqrt{5}}{2}$ | 0 | 2 | $\frac{-1+\sqrt{5}}{2}$ | $\frac{-1-\sqrt{5}}{2}$ | $T_{x,y}$, $R_{x,y}$ |
| $E_2$ | 2 | $\frac{-1-\sqrt{5}}{2}$ | $\frac{-1+\sqrt{5}}{2}$ | 0 | 2 | $\frac{-1-\sqrt{5}}{2}$ | $\frac{-1+\sqrt{5}}{2}$ | |
| $E_{\frac{1}{2}}$ | 2 | $\frac{1+\sqrt{5}}{2}$ | $\frac{-1+\sqrt{5}}{2}$ | 0 | $-2$ | $\frac{-1-\sqrt{5}}{2}$ | $\frac{1-\sqrt{5}}{2}$ | |
| $E_{\frac{3}{2}}$ | 2 | $\frac{1-\sqrt{5}}{2}$ | $\frac{-1-\sqrt{5}}{2}$ | 0 | $-2$ | $\frac{-1+\sqrt{5}}{2}$ | $\frac{1+\sqrt{5}}{2}$ | |
| $E_{\frac{5}{2}}$ | 2 | $-2$ | 2 | 0 | $-2$ | $+2$ | $-2$ | sep. |

$\boldsymbol{D}_{5d} \equiv \boldsymbol{D}_5 \times \boldsymbol{C}_i$

Table 51 (*Continued*)

| | $\boldsymbol{D}_6$ | $I$ | $2C_6(z)$ | $2C_6^2 \equiv 2C_3$ | $C_6^3 \equiv C_2''$ | $3C_2$ | $3C_2'$ | $R$ | $2C_6^5$ | $2C_6^4 \equiv 2C_3^2$ | |
|---|---|---|---|---|---|---|---|---|---|---|---|
| | $\boldsymbol{C}_{6v}$ | $I$ | $2C_6(z)$ | $2C_6^2 \equiv 2C_3$ | $C_6^3 \equiv C_2''$ | $3\sigma_v$ | $3\sigma_d$ | $R$ | $2C_6^5$ | $2C_3^2$ | |
| $\boldsymbol{D}_{3h}$ | | $I$ | $2S_3(z)$ | $2C_3(z)$ | $\sigma_h$ | $3C_2$ | $3\sigma_v$ | $R$ | $2S_3^5$ | $2C_3^2$ | |
| $A_1'$ | $A_1$ | 1 | 1 | 1 | 1 | 1 | 1 | 1 | 1 | 1 | $T_z$ for $\boldsymbol{C}_{6v}$ |
| $A_2'$ | $A_2$ | 1 | 1 | 1 | 1 | −1 | −1 | 1 | 1 | 1 | $T_z$ for $\boldsymbol{D}_6$; $R_z$ |
| $A_1''$ | $B_1$ | 1 | −1 | 1 | −1 | 1 | −1 | 1 | −1 | 1 | |
| $A_2''$ | $B_2$ | 1 | −1 | 1 | −1 | −1 | 1 | 1 | −1 | 1 | $T_z$ for $\boldsymbol{D}_{3h}$ |
| $E''$ | $E_1$ | 2 | 1 | −1 | −2 | 0 | 0 | 2 | 1 | −1 | $T_{x,y}$ for $\boldsymbol{D}_6$, $\boldsymbol{C}_{6v}$; $R_{x,y}$ |
| $E'$ | $E_2$ | 2 | −1 | −1 | 2 | 0 | 0 | 2 | −1 | −1 | $T_{x,y}$ for $\boldsymbol{D}_{3h}$ |
| $E_{\frac{1}{2}}$ | $E_{\frac{1}{2}}$ | 2 | $\sqrt{3}$ | 1 | 0 | 0 | 0 | −2 | $-\sqrt{3}$ | −1 | |
| $E_{\frac{3}{2}}$ | $E_{\frac{3}{2}}$ | 2 | 0 | −2 | 0 | 0 | 0 | −2 | 0 | 2 | |
| $E_{\frac{5}{2}}$ | $E_{\frac{5}{2}}$ | 2 | $-\sqrt{3}$ | 1 | 0 | 0 | 0 | −2 | $\sqrt{3}$ | −1 | |

$\boldsymbol{D}_{6h} \equiv \boldsymbol{D}_6 \times \boldsymbol{C}_i$ (see Vol. II, p. 117)

| $\boldsymbol{D}_8$ | $I$ | $2C_8(z)$ | $2C_8^2 \equiv 2C_4$ | $2C_8^3$ | $C_8^4 \equiv C_2''$ | $4C_2$ | $4C_2'$ | $R$ | $2C_8^7$ | $2C_4^3$ | $2C_8^5$ | |
|---|---|---|---|---|---|---|---|---|---|---|---|---|
| $\boldsymbol{C}_{8v}$ | $I$ | $2C_8(z)$ | $2C_8^2 \equiv 2C_4$ | $2C_8^3$ | $C_8^4 \equiv C_2''$ | $4\sigma_v$ | $4\sigma_d$ | $R$ | $2C_8^7$ | $2C_4^3$ | $2C_8^5$ | |
| $\boldsymbol{D}_{4d}$ | $I$ | $2S_8(z)$ | $2S_8^2 \equiv 2C_4$ | $2S_8^3$ | $S_8^4 \equiv C_2''$ | $4C_2$ | $4\sigma_d$ | $R$ | $2S_8^7$ | $2C_4^3$ | $2S_8^5$ | |
| $A_1$ | 1 | 1 | 1 | 1 | 1 | +1 | +1 | 1 | 1 | 1 | 1 | $T_z$ for $\boldsymbol{C}_{8v}$ |
| $A_2$ | 1 | 1 | 1 | 1 | 1 | −1 | −1 | 1 | 1 | 1 | 1 | $T_z$ for $\boldsymbol{D}_8$; $R_z$ |
| $B_1$ | 1 | −1 | 1 | −1 | 1 | +1 | −1 | 1 | −1 | 1 | −1 | |
| $B_2$ | 1 | −1 | 1 | −1 | 1 | −1 | +1 | 1 | −1 | 1 | −1 | $T_z$ for $\boldsymbol{D}_{4d}$ |
| $E_1$ | 2 | $\sqrt{2}$ | 0 | $-\sqrt{2}$ | −2 | 0 | 0 | 2 | $\sqrt{2}$ | 0 | $-\sqrt{2}$ | $T_{x,y}$ |
| $E_2$ | 2 | 0 | −2 | 0 | +2 | 0 | 0 | 2 | 0 | −2 | 0 | |
| $E_3$ | 2 | $-\sqrt{2}$ | 0 | $\sqrt{2}$ | −2 | 0 | 0 | 2 | $-\sqrt{2}$ | 0 | $\sqrt{2}$ | $R_{x,y}$ |
| $E_{\frac{1}{2}}$ | 2 | $\sqrt{2+\sqrt{2}}$ | $\sqrt{2}$ | $\sqrt{2-\sqrt{2}}$ | 0 | 0 | 0 | −2 | $-\sqrt{2+\sqrt{2}}$ | $-\sqrt{2}$ | $-\sqrt{2-\sqrt{2}}$ | |
| $E_{\frac{3}{2}}$ | 2 | $\sqrt{2-\sqrt{2}}$ | $-\sqrt{2}$ | $-\sqrt{2+\sqrt{2}}$ | 0 | 0 | 0 | −2 | $-\sqrt{2-\sqrt{2}}$ | $+\sqrt{2}$ | $\sqrt{2+\sqrt{2}}$ | |
| $E_{\frac{5}{2}}$ | 2 | $-\sqrt{2-\sqrt{2}}$ | $-\sqrt{2}$ | $\sqrt{2+\sqrt{2}}$ | 0 | 0 | 0 | −2 | $\sqrt{2-\sqrt{2}}$ | $+\sqrt{2}$ | $-\sqrt{2+\sqrt{2}}$ | |
| $E_{\frac{7}{2}}$ | 2 | $-\sqrt{2+\sqrt{2}}$ | $\sqrt{2}$ | $-\sqrt{2-\sqrt{2}}$ | 0 | 0 | 0 | −2 | $\sqrt{2+\sqrt{2}}$ | $-\sqrt{2}$ | $\sqrt{2-\sqrt{2}}$ | |

$\boldsymbol{D}_{8h} \equiv \boldsymbol{D}_8 \times \boldsymbol{C}_i$

TABLE 52

POINT GROUPS $\boldsymbol{D}_\infty$, $\boldsymbol{C}_{\infty v}$; $\boldsymbol{D}_{\infty h}$

| $\boldsymbol{D}_\infty$<br>$\boldsymbol{C}_{\infty v}$ | $I$<br>$I$ | $2C_\infty^\varphi$<br>$2C_\infty^\varphi$ | $2C_\infty^{2\varphi}$<br>$2C_\infty^{2\varphi}$ | $2C_\infty^{3\varphi}$<br>$2C_\infty^{3\varphi}$ | ...<br>... | $\infty C_2$<br>$\infty\sigma_v$ | $R$<br>$R$ | $2C_\infty^\varphi R$<br>$2C_\infty^\varphi R$ | ...<br>... | |
|---|---|---|---|---|---|---|---|---|---|---|
| $\Sigma^+$ | 1 | 1 | 1 | 1 | | 1 | 1 | 1 | | $T_z$ |
| $\Sigma^-$ | 1 | 1 | 1 | 1 | | −1 | 1 | 1 | | $R_z$ |
| $\Pi$ | +2 | $2\cos\varphi$ | $2\cos 2\varphi$ | $2\cos 3\varphi$ | ... | 0 | +2 | $2\cos\varphi$ | ... | $T_{x,y}$; $R_{x,y}$ |
| $\Delta$ | +2 | $2\cos 2\varphi$ | $2\cos 2\cdot 2\varphi$ | $2\cos 3\cdot 2\varphi$ | ... | 0 | +2 | $2\cos 2\varphi$ | ... | |
| $\Phi$ | +2 | $2\cos 3\varphi$ | $2\cos 2\cdot 3\varphi$ | $2\cos 3\cdot 3\varphi$ | ... | 0 | +2 | $2\cos 3\varphi$ | ... | |
| . . . | . . . | . . . | . . . | . . . | . . . | . . . | . . . | . . . | . . . | |
| $E_{\frac{1}{2}}$ | +2 | $2\cos\frac{1}{2}\varphi$ | $2\cos\varphi$ | $2\cos 3\cdot\frac{1}{2}\varphi$ | ... | 0 | −2 | $-2\cos\frac{1}{2}\varphi$ | ... | |
| $E_{\frac{3}{2}}$ | +2 | $2\cos\frac{3}{2}\varphi$ | $2\cos 3\varphi$ | $2\cos 3\cdot\frac{3}{2}\varphi$ | ... | 0 | −2 | $-2\cos\frac{3}{2}\varphi$ | ... | |
| $E_{\frac{5}{2}}$ | +2 | $2\cos\frac{5}{2}\varphi$ | $2\cos 5\varphi$ | $2\cos 3\cdot\frac{5}{2}\varphi$ | ... | 0 | −2 | $-2\cos\frac{5}{2}\varphi$ | ... | |
| . . . | . . . | . . . | . . . | . . . | . . . | . . . | . . . | . . . | . . . | |

$\boldsymbol{D}_{\infty h} \equiv \boldsymbol{C}_{\infty v} \times \boldsymbol{C}_i$ (see Vol. II, p. 119)

TABLE 53

POINT GROUPS $\boldsymbol{O}$, $\boldsymbol{T}_d$; $\boldsymbol{O}_h$

| $\boldsymbol{O}$<br>$\boldsymbol{T}_d$ | a | $I$<br>$I$ | $8C_3$<br>$8C_3$ | $6C_2$<br>$6\sigma_d$ | $6C_4$<br>$6S_4$ | $3C_4{}^2 \equiv 3C_2''$<br>$3S_4{}^2 \equiv 3C_2$ | $R$<br>$R$ | $8C_3{}^2$<br>$8C_3{}^2$ | $6C_4{}^3$<br>$6S_4{}^3$ | |
|---|---|---|---|---|---|---|---|---|---|---|
| $A_1$ | $\Gamma_1$ | 1 | 1 | 1 | 1 | 1 | 1 | 1 | 1 | |
| $A_2$ | $\Gamma_2$ | 1 | 1 | −1 | −1 | 1 | 1 | 1 | −1 | |
| $E$ | $\Gamma_3$ | 2 | −1 | 0 | 0 | 2 | 2 | −1 | 0 | |
| $F_1$ ($T_1$) | $\Gamma_4$ | 3 | 0 | −1 | 1 | −1 | 3 | 0 | 1 | $T_{x,y,z}$ for $\boldsymbol{O}$; $R_{x,y,z}$ |
| $F_2$ ($T_2$) | $\Gamma_5$ | 3 | 0 | 1 | −1 | −1 | 3 | 0 | −1 | $T_{x,y,z}$ for $\boldsymbol{T}_d$ |
| $E_{\frac{1}{2}}$ | $\Gamma_6$ | 2 | 1 | 0 | $\sqrt{2}$ | 0 | −2 | −1 | $-\sqrt{2}$ | |
| $E_{\frac{5}{2}}$ | $\Gamma_7$ | 2 | 1 | 0 | $-\sqrt{2}$ | 0 | −2 | −1 | $\sqrt{2}$ | |
| $G_{\frac{3}{2}}$ | $\Gamma_8$ | 4 | −1 | 0 | 0 | 0 | −4 | 1 | 0 | |

$\boldsymbol{O}_h \equiv \boldsymbol{O} \times \boldsymbol{C}_i$ (see Vol. II, p. 123)

[a] Alternative species designations used by some authors.

TABLE 54
POINT GROUPS $\boldsymbol{I}$ AND $\boldsymbol{I}_h$

| $I$ | $I$ | $12C_5$ | $12C_5^2$ | $20C_3$ | $15C_2$ | $R$ | $12C_5^4$ | $12C_5^3$ | $20C_3^2$ | |
|---|---|---|---|---|---|---|---|---|---|---|
| $A$ | 1 | 1 | 1 | 1 | 1 | 1 | 1 | 1 | 1 | |
| $F_1$ | 3 | $\frac{1+\sqrt{5}}{2}$ | $\frac{1-\sqrt{5}}{2}$ | 0 | $-1$ | 3 | $\frac{1+\sqrt{5}}{2}$ | $\frac{1-\sqrt{5}}{2}$ | 0 | $T_{x,y,z}$; $R_{x,y,z}$ |
| $F_2$ | 3 | $\frac{1-\sqrt{5}}{2}$ | $\frac{1+\sqrt{5}}{2}$ | 0 | $-1$ | 3 | $\frac{1-\sqrt{5}}{2}$ | $\frac{1+\sqrt{5}}{2}$ | 0 | |
| $G$ | 4 | $-1$ | $-1$ | 1 | 0 | 4 | $-1$ | $-1$ | 1 | |
| $H$ | 5 | 0 | 0 | $-1$ | 1 | 5 | 0 | 0 | $-1$ | |
| $E_{\frac{1}{2}}$ | 2 | $\frac{1+\sqrt{5}}{2}$ | $\frac{-1+\sqrt{5}}{2}$ | 1 | 0 | $-2$ | $\frac{-1-\sqrt{5}}{2}$ | $\frac{1-\sqrt{5}}{2}$ | $-1$ | |
| $E_{\frac{7}{2}}$ | 2 | $\frac{1-\sqrt{5}}{2}$ | $\frac{-1-\sqrt{5}}{2}$ | 1 | 0 | $-2$ | $\frac{-1+\sqrt{5}}{2}$ | $\frac{1+\sqrt{5}}{2}$ | $-1$ | |
| $G_{\frac{3}{2}}$ | 4 | 1 | $-1$ | $-1$ | 0 | $-4$ | $-1$ | 1 | 1 | |
| $I_{\frac{5}{2}}$ | 6 | $-1$ | 1 | 0 | 0 | $-6$ | 1 | $-1$ | 0 | |

$$\boldsymbol{I}_h \equiv \boldsymbol{I} \times \boldsymbol{C}_i$$

TABLE 55
POINT GROUPS $\boldsymbol{K}$ AND $\boldsymbol{K}_h$

| $\boldsymbol{K}$ | $I$ | $\infty C_\infty^\varphi$ | $\infty C_\infty^{2\varphi}$ | . . . | $R$ | $\infty C_\infty^\varphi R$ | . . . |
|---|---|---|---|---|---|---|---|
| $D_0 \equiv S$ | 1 | 1 | 1 | . . . | 1 | 1 | . . . |
| $D_1 \equiv P$ | 3 | $1 + 2\cos\varphi$ | $1 + 2\cos 2\varphi$ | . . . | 3 | $1 + 2\cos\varphi$ | . . . |
| $D_2 \equiv D$ | 5 | $1 + 2\cos\varphi + 2\cos 2\varphi$ | $1 + 2\cos 2\varphi + 2\cos 4\varphi$ | . . . | 5 | $1 + 2\cos\varphi + 2\cos 2\varphi$ | . . . |
| $D_3 \equiv F$ | 7 | $1 + 2\cos\varphi + 2\cos 2\varphi + 2\cos 3\varphi$ | . . . | . . . | 7 | . . . | . . . |
| . . . | . . . | . . . | . . . | . . . | . . . | . . . | . . . |
| $D_{\frac{1}{2}}$ $(\equiv E_{\frac{1}{2}})$ | 2 | $2\cos\frac{1}{2}\varphi$ | $2\cos\varphi$ | . . . | $-2$ | $-2\cos\frac{1}{2}\varphi$ | . . . |
| $D_{\frac{3}{2}}$ $(\equiv G_{\frac{3}{2}})$ | 4 | $2\cos\frac{1}{2}\varphi + 2\cos\frac{3}{2}\varphi$ | $2\cos\varphi + 2\cos 3\varphi$ | . . . | $-4$ | $-2\cos\frac{1}{2}\varphi - 2\cos\frac{3}{2}\varphi$ | . . . |
| $D_{\frac{5}{2}}$ $(\equiv I_{\frac{5}{2}})$ | 6 | $2\cos\frac{1}{2}\varphi + 2\cos\frac{3}{2}\varphi + 2\cos\frac{5}{2}\varphi$ | . . . | . . . | $-6$ | . . . | . . . |

# APPENDIX II

TABLE 56

## SPECIES OF SPIN FUNCTIONS FOR THE MORE IMPORTANT POINT GROUPS

For the groups in ( ) the subscripts $g$ and $u$ should be omitted. The number in front of the species indicates the number of times the species occurs. Note that the species $B_{\frac{1}{2}}$ and $B_{\frac{3}{2}}$ do not occur singly but in pairs corresponding to Kramers doublets.

| $S$ | $\boldsymbol{C}_i(\boldsymbol{C}_1)$ | $\boldsymbol{C}_s$ | $\boldsymbol{C}_{2h}(\boldsymbol{C}_2)$ | $\boldsymbol{C}_{2v}$ | $\boldsymbol{D}_{2h}(\boldsymbol{D}_2)$ |
|---|---|---|---|---|---|
| 0 | $A_g$ | $A'$ | $A_g$ | $A_1$ | $A_g$ |
| $\frac{1}{2}$ | $2B_{\frac{1}{2}g}$ | $E_{\frac{1}{2}}$ | $E_{\frac{1}{2}g}$ | $E_{\frac{1}{2}}$ | $E_{\frac{1}{2}g}$ |
| 1 | $3A_g$ | $A'+2A''$ | $A_g+2B_g$ | $A_2+B_1+B_2$ | $B_{1g}+B_{2g}+B_{3g}$ |
| $\frac{3}{2}$ | $2B_{\frac{1}{2}g}+2B_{\frac{1}{2}g}$ | $2E_{\frac{1}{2}}$ | $2E_{\frac{1}{2}g}$ | $2E_{\frac{1}{2}}$ | $2E_{\frac{1}{2}g}$ |
| 2 | $A_g+2A_g+2A_g$ | $3A'+2A''$ | $3A_g+2B_g$ | $2A_1+A_2+B_1+B_2$ | $2A_g+B_{1g}+B_{2g}+B_{3g}$ |
| $\frac{5}{2}$ | $2B_{\frac{1}{2}g}+2B_{\frac{1}{2}g}+2B_{\frac{1}{2}g}$ | $3E_{\frac{1}{2}}$ | $3E_{\frac{1}{2}g}$ | $3E_{\frac{1}{2}}$ | $3E_{\frac{1}{2}g}$ |

| $S$ | $\boldsymbol{C}_3$ | $\boldsymbol{C}_{3h}$ | $\boldsymbol{C}_{3v}$ | $\boldsymbol{D}_{3d}(\boldsymbol{D}_3)$ | $\boldsymbol{D}_{3h}$ |
|---|---|---|---|---|---|
| 0 | $A$ | $A'$ | $A_1$ | $A_{1g}$ | $A_1'$ |
| $\frac{1}{2}$ | $E_{\frac{1}{2}}$ | $E_{\frac{1}{2}}$ | $E_{\frac{1}{2}}$ | $E_{\frac{1}{2}g}$ | $E_{\frac{1}{2}}$ |
| 1 | $A+E$ | $A'+E''$ | $A_2+E$ | $A_{2g}+E_g$ | $A_2'+E''$ |
| $\frac{3}{2}$ | $2B_{\frac{3}{2}}+E_{\frac{1}{2}}$ | $E_{\frac{1}{2}}+E_{\frac{3}{2}}$ | $E_{\frac{1}{2}}+E_{\frac{3}{2}}$ | $E_{\frac{1}{2}g}+E_{\frac{3}{2}g}$ | $E_{\frac{1}{2}}+E_{\frac{3}{2}}$ |
| 2 | $A+2E$ | $A'+E'+E''$ | $A_1+2E$ | $A_{1g}+2E_g$ | $A_1'+E'+E''$ |
| $\frac{5}{2}$ | $2B_{\frac{3}{2}}+2E_{\frac{1}{2}}$ | $E_{\frac{1}{2}}+E_{\frac{3}{2}}+E_{\frac{5}{2}}$ | $2E_{\frac{1}{2}}+E_{\frac{3}{2}}$ | $2E_{\frac{1}{2}g}+E_{\frac{3}{2}g}$ | $E_{\frac{1}{2}}+E_{\frac{3}{2}}+E_{\frac{5}{2}}$ |

| $S$ | $\boldsymbol{C}_{4v}$, $\boldsymbol{D}_{2d}$ | $\boldsymbol{D}_{4h}(\boldsymbol{D}_4)$ | $\boldsymbol{D}_{4d}$ | $\boldsymbol{D}_{5d}(\boldsymbol{D}_5, \boldsymbol{C}_{5v})$ | $\boldsymbol{D}_{5h}$ [a] |
|---|---|---|---|---|---|
| 0 | $A_1$ | $A_{1g}$ | $A_1$ | $A_{1g}$ | $A_1'$ |
| $\frac{1}{2}$ | $E_{\frac{1}{2}}$ | $E_{\frac{1}{2}g}$ | $E_{\frac{1}{2}}$ | $E_{\frac{1}{2}g}$ | $E_{\frac{1}{2}}$ |
| 1 | $A_2+E$ | $A_{2g}+E_g$ | $A_2+E_1$ | $A_{2g}+E_{1g}$ | $A_2'+E_1''$ |
| $\frac{3}{2}$ | $E_{\frac{1}{2}}+E_{\frac{3}{2}}$ | $E_{\frac{1}{2}g}+E_{\frac{3}{2}g}$ | $E_{\frac{1}{2}}+E_{\frac{3}{2}}$ | $E_{\frac{1}{2}g}+E_{\frac{3}{2}g}$ | $E_{\frac{1}{2}}+E_{\frac{3}{2}}$ |
| 2 | $A_1+B_1+B_2+E$ | $A_{1g}+B_{1g}+B_{2g}+E_g$ | $A_1+E_1+E_2$ | $A_{1g}+E_{1g}+E_{2g}$ | $A_1'+E_1''+E_2'$ |
| $\frac{5}{2}$ | $E_{\frac{1}{2}}+2E_{\frac{3}{2}}$ | $E_{\frac{1}{2}g}+2E_{\frac{3}{2}g}$ | $E_{\frac{1}{2}}+E_{\frac{3}{2}}+E_{\frac{5}{2}}$ | $E_{\frac{1}{2}g}+E_{\frac{3}{2}g}+E_{\frac{5}{2}g}$ | $E_{\frac{1}{2}}+E_{\frac{3}{2}}+E_{\frac{5}{2}}$ |

| $S$ | $\boldsymbol{D}_{6h}(\boldsymbol{D}_6, \boldsymbol{C}_{6v})$ | $\boldsymbol{D}_{\infty h}(\boldsymbol{C}_{\infty v})$ | $\boldsymbol{O}_h(\boldsymbol{T}_d)$ | $\boldsymbol{I}_h(\boldsymbol{I})$ | $\boldsymbol{K}_h(\boldsymbol{K})$ |
|---|---|---|---|---|---|
| 0 | $A_{1g}$ | $\Sigma_g^+$ | $A_{1g}$ | $A_g$ | $D_{0g}$ |
| $\frac{1}{2}$ | $E_{\frac{1}{2}g}$ | $E_{\frac{1}{2}g}$ | $E_{\frac{1}{2}g}$ | $E_{\frac{1}{2}g}$ | $D_{\frac{1}{2}g}$ |
| 1 | $A_{2g}+E_{1g}$ | $\Sigma_g^-+\Pi_g$ | $F_{1g}$ | $F_{1g}$ | $D_{1g}$ |
| $\frac{3}{2}$ | $E_{\frac{1}{2}g}+E_{\frac{3}{2}g}$ | $E_{\frac{1}{2}g}+E_{\frac{3}{2}g}$ | $G_{\frac{3}{2}g}$ | $G_{\frac{3}{2}g}$ | $D_{\frac{3}{2}g}$ |
| 2 | $A_{1g}+E_{1g}+E_{2g}$ | $\Sigma_g^++\Pi_g+\Delta_g$ | $E_g+F_{2g}$ | $H_g$ | $D_{2g}$ |
| $\frac{5}{2}$ | $E_{\frac{1}{2}g}+E_{\frac{3}{2}g}+E_{\frac{5}{2}g}$ | $E_{\frac{1}{2}g}+E_{\frac{3}{2}g}+E_{\frac{5}{2}g}$ | $E_{\frac{5}{2}g}+G_{\frac{3}{2}g}$ | $I_{\frac{5}{2}g}$ | $D_{\frac{5}{2}g}$ |

[a] The character table for $\boldsymbol{D}_{5h}$ (not given in Appendix I) is isomorphic with that of $\boldsymbol{D}_{10}$ and has five double-valued species. The single-valued species are given in Vol. II, p. 117.

# APPENDIX III

## TABLE 57

## DIRECT PRODUCTS OF REPRESENTATIONS (SPECIES) FOR ALL IMPORTANT POINT GROUPS

Species to be omitted in the symmetrical product of a degenerate species with itself are put in square brackets. They represent the antisymmetric product.

**$C_s$**

| | $A'$ | $A''$ | $E_{\frac{1}{2}}$ |
|---|---|---|---|
| $A'$ | $A'$ | $A''$ | $E_{\frac{1}{2}}$ |
| $A''$ | | $A'$ | $E_{\frac{1}{2}}$ |
| $E_{\frac{1}{2}}$ | | | $[A'], A', A'', A''$ |

**$C_i$, $(C_1)$** [a]

| | $A_g$ | $A_u$ | $B_{\frac{1}{2}g}$ | $B_{\frac{1}{2}u}$ |
|---|---|---|---|---|
| $A_g$ | $A_g$ | $A_u$ | $B_{\frac{1}{2}g}$ | $B_{\frac{1}{2}u}$ |
| $A_u$ | | $A_g$ | $B_{\frac{1}{2}u}$ | $B_{\frac{1}{2}g}$ |
| $B_{\frac{1}{2}g}$ | | | $A_g$ | $A_u$ |
| $B_{\frac{1}{2}u}$ | | | | $A_g$ |

**$C_{2v}$, $(C_2)$** [b]**, $(C_{2h})$** [b,c]

| | $A_1$ | $A_2$ | $B_1$ | $B_2$ | $E_{\frac{1}{2}}$ |
|---|---|---|---|---|---|
| $A_1$ | $A_1$ | $A_2$ | $B_1$ | $B_2$ | $E_{\frac{1}{2}}$ |
| $A_2$ | | $A_1$ | $B_2$ | $B_1$ | $E_{\frac{1}{2}}$ |
| $B_1$ | | | $A_1$ | $A_2$ | $E_{\frac{1}{2}}$ |
| $B_2$ | | | | $A_1$ | $E_{\frac{1}{2}}$ |
| $E_{\frac{1}{2}}$ | | | | | $[A_1], A_2, B_1, B_2$ |

**$C_3$**

| | $A$ | $E$ | $E_{\frac{1}{2}}$ | $B_{\frac{3}{2}}$ |
|---|---|---|---|---|
| $A$ | $A$ | $E$ | $E_{\frac{1}{2}}$ | $B_{\frac{3}{2}}$ |
| $E$ | | $[A], A, E$ | $E_{\frac{1}{2}}, 2B_{\frac{3}{2}}$ | $E_{\frac{1}{2}}$ |
| $E_{\frac{1}{2}}$ | | | $[A], A, E$ | $E$ |
| $B_{\frac{3}{2}}$ | | | | $A$ |

**$D_2$, $(D_{2h})$** [c]

| | $A$ | $B_1$ | $B_2$ | $B_3$ | $E_{\frac{1}{2}}$ |
|---|---|---|---|---|---|
| $A$ | $A$ | $B_1$ | $B_2$ | $B_3$ | $E_{\frac{1}{2}}$ |
| $B_1$ | | $A$ | $B_3$ | $B_2$ | $E_{\frac{1}{2}}$ |
| $B_2$ | | | $A$ | $B_1$ | $E_{\frac{1}{2}}$ |
| $B_3$ | | | | $A$ | $E_{\frac{1}{2}}$ |
| $E_{\frac{1}{2}}$ | | | | | $[A], B_1, B_2, B_3$ |

**$D_3$, $C_{3v}$, $(D_{3d})$** [c]

| | $A_1$ | $A_2$ | $E$ | $E_{\frac{1}{2}}$ | $E_{\frac{3}{2}}$ |
|---|---|---|---|---|---|
| $A_1$ | $A_1$ | $A_2$ | $E$ | $E_{\frac{1}{2}}$ | $E_{\frac{3}{2}}$ |
| $A_2$ | | $A_1$ | $E$ | $E_{\frac{1}{2}}$ | $E_{\frac{3}{2}}$ |
| $E$ | | | $A_1, [A_2], E$ | $E_{\frac{1}{2}}, E_{\frac{3}{2}}$ | $E_{\frac{1}{2}}, E_{\frac{1}{2}}$ |
| $E_{\frac{1}{2}}$ | | | | $[A_1], A_2, E$ | $E, E$ |
| $E_{\frac{3}{2}}$ | | | | | $[A_1], A_2, A_1, A$ |

[a] For this point group $g$ and $u$ should be omitted. [b] For these point groups the subscripts 1 and 2 should be droppe
[c] For these point groups the $(g, u)$ rule must be added, that is $g \times g = g, g \times u = u, u \times u = g$.

TABLE 57 (*Continued*)

$\boldsymbol{D}_{3h}$, $(\boldsymbol{C}_{3h})$ [b]

| | $A_1'$ | $A_2'$ | $A_1''$ | $A_2''$ | $E'$ | $E''$ | $E_{\frac{1}{2}}$ | $E_{\frac{3}{2}}$ | $E_{\frac{5}{2}}$ |
|---|---|---|---|---|---|---|---|---|---|
| $A_1'$ | $A_1'$ | $A_2'$ | $A_1''$ | $A_2''$ | $E'$ | $E''$ | $E_{\frac{1}{2}}$ | $E_{\frac{3}{2}}$ | $E_{\frac{5}{2}}$ |
| $A_2'$ | | $A_1'$ | $A_2''$ | $A_1''$ | $E'$ | $E''$ | $E_{\frac{1}{2}}$ | $E_{\frac{3}{2}}$ | $E_{\frac{5}{2}}$ |
| $A_1''$ | | | $A_1'$ | $A_2'$ | $E''$ | $E'$ | $E_{\frac{5}{2}}$ | $E_{\frac{3}{2}}$ | $E_{\frac{1}{2}}$ |
| $A_2''$ | | | | $A_1'$ | $E''$ | $E'$ | $E_{\frac{5}{2}}$ | $E_{\frac{3}{2}}$ | $E_{\frac{1}{2}}$ |
| $E'$ | | | | | $A_1', [A_2'], E'$ | $A_1'', A_2'', E''$ | $E_{\frac{3}{2}}, E_{\frac{5}{2}}$ | $E_{\frac{1}{2}}, E_{\frac{5}{2}}$ | $E_{\frac{1}{2}}, E_{\frac{3}{2}}$ |
| $E''$ | | | | | | $A_1', [A_2'], E'$ | $E_{\frac{1}{2}}, E_{\frac{3}{2}}$ | $E_{\frac{1}{2}}, E_{\frac{5}{2}}$ | $E_{\frac{3}{2}}, E_{\frac{5}{2}}$ |
| $E_{\frac{1}{2}}$ | | | | | | | $[A_1'], A_2', E''$ | $E', E''$ | $A_1'', A_2'', E'$ |
| $E_{\frac{3}{2}}$ | | | | | | | | $[A_1'], A_2', A_1'', A_2''$ | $E', E''$ |
| $E_{\frac{5}{2}}$ | | | | | | | | | $[A_1'], A_2', E''$ |

$\boldsymbol{D}_4$, $\boldsymbol{C}_{4v}$, $\boldsymbol{D}_{2d}$, $(\boldsymbol{D}_{4h})$ [c]

| | $A_1$ | $A_2$ | $B_1$ | $B_2$ | $E$ | $E_{\frac{1}{2}}$ | $E_{\frac{3}{2}}$ |
|---|---|---|---|---|---|---|---|
| $A_1$ | $A_1$ | $A_2$ | $B_1$ | $B_2$ | $E$ | $E_{\frac{1}{2}}$ | $E_{\frac{3}{2}}$ |
| $A_2$ | | $A_1$ | $B_2$ | $B_1$ | $E$ | $E_{\frac{1}{2}}$ | $E_{\frac{3}{2}}$ |
| $B_1$ | | | $A_1$ | $A_2$ | $E$ | $E_{\frac{3}{2}}$ | $E_{\frac{1}{2}}$ |
| $B_2$ | | | | $A_1$ | $E$ | $E_{\frac{3}{2}}$ | $E_{\frac{1}{2}}$ |
| $E$ | | | | | $A_1, [A_2], B_1, B_2$ | $E_{\frac{1}{2}}, E_{\frac{3}{2}}$ | $E_{\frac{1}{2}}, E_{\frac{3}{2}}$ |
| $E_{\frac{1}{2}}$ | | | | | | $[A_1], A_2, E$ | $B_1, B_2, E$ |
| $E_{\frac{3}{2}}$ | | | | | | | $[A_1], A_2, E$ |

$\boldsymbol{D}_5$, $\boldsymbol{C}_{5v}$, $(\boldsymbol{D}_{5d})$ [c]

| | $A_1$ | $A_2$ | $E_1$ | $E_2$ | $E_{\frac{1}{2}}$ | $E_{\frac{3}{2}}$ | $E_{\frac{5}{2}}$ |
|---|---|---|---|---|---|---|---|
| $A_1$ | $A_1$ | $A_2$ | $E_1$ | $E_2$ | $E_{\frac{1}{2}}$ | $E_{\frac{3}{2}}$ | $E_{\frac{5}{2}}$ |
| $A_2$ | | $A_1$ | $E_1$ | $E_2$ | $E_{\frac{1}{2}}$ | $E_{\frac{3}{2}}$ | $E_{\frac{5}{2}}$ |
| $E_1$ | | | $A_1, [A_2], E_2$ | $E_1, E_2$ | $E_{\frac{1}{2}}, E_{\frac{5}{2}}$ | $E_{\frac{1}{2}}, E_{\frac{5}{2}}$ | $E_{\frac{3}{2}}, E_{\frac{3}{2}}$ |
| $E_2$ | | | | $A_1, [A_2], E_1$ | $E_{\frac{3}{2}}, E_{\frac{3}{2}}$ | $E_{\frac{1}{2}}, E_{\frac{3}{2}}$ | $E_{\frac{3}{2}}, E_{\frac{3}{2}}$ |
| $E_{\frac{1}{2}}$ | | | | | $[A_1], A_2, E_1$ | $E_1, E_2$ | $E_2, E_2$ |
| $E_{\frac{3}{2}}$ | | | | | | $[A_1], A_2, E_2$ | $E_1, E_1$ |
| $E_{\frac{5}{2}}$ | | | | | | | $[A_1], A_2, A_1, A_2$ |

$\boldsymbol{D}_6$, $\boldsymbol{C}_{6v}$, $(\boldsymbol{D}_{6h})$ [c]

| | $A_1$ | $A_2$ | $B_1$ | $B_2$ | $E_1$ | $E_2$ | $E_{\frac{1}{2}}$ | $E_{\frac{3}{2}}$ | $E_{\frac{5}{2}}$ |
|---|---|---|---|---|---|---|---|---|---|
| | $A_1$ | $A_2$ | $B_1$ | $B_2$ | $E_1$ | $E_2$ | $E_{\frac{1}{2}}$ | $E_{\frac{3}{2}}$ | $E_{\frac{5}{2}}$ |
| | | $A_1$ | $B_2$ | $B_1$ | $E_1$ | $E_2$ | $E_{\frac{1}{2}}$ | $E_{\frac{3}{2}}$ | $E_{\frac{5}{2}}$ |
| | | | $A_1$ | $A_2$ | $E_2$ | $E_1$ | $E_{\frac{5}{2}}$ | $E_{\frac{3}{2}}$ | $E_{\frac{1}{2}}$ |
| | | | | $A_1$ | $E_2$ | $E_1$ | $E_{\frac{5}{2}}$ | $E_{\frac{3}{2}}$ | $E_{\frac{1}{2}}$ |
| | | | | | $A_1, [A_2], E_2$ | $B_1, B_2, E_1$ | $E_{\frac{1}{2}}, E_{\frac{3}{2}}$ | $E_{\frac{1}{2}}, E_{\frac{5}{2}}$ | $E_{\frac{3}{2}}, E_{\frac{5}{2}}$ |
| | | | | | | $A_1, [A_2], E_2$ | $E_{\frac{3}{2}}, E_{\frac{5}{2}}$ | $E_{\frac{1}{2}}, E_{\frac{5}{2}}$ | $E_{\frac{1}{2}}, E_{\frac{3}{2}}$ |
| | | | | | | | $[A_1], A_2, E_1$ | $E_1, E_2$ | $B_1, B_2, E_2$ |
| | | | | | | | | $[A_1], A_2, B_1, B_2$ | $E_1, E_2$ |
| | | | | | | | | | $[A_1], A_2, E_1$ |

Table 57 (*Continued*)

$\boldsymbol{D}_8$, $\boldsymbol{C}_{8v}$, $\boldsymbol{D}_{4d}$, $(\boldsymbol{D}_{8h})$[b]

| | $A_1$ | $A_2$ | $B_1$ | $B_2$ | $E_1$ | $E_2$ | $E_3$ | $E_{\frac{1}{2}}$ | $E_{\frac{3}{2}}$ | $E_{\frac{5}{2}}$ | $E_{\frac{7}{2}}$ |
|---|---|---|---|---|---|---|---|---|---|---|---|
| $A_1$ | $A_1$ | $A_2$ | $B_1$ | $B_2$ | $E_1$ | $E_2$ | $E_3$ | $E_{\frac{1}{2}}$ | $E_{\frac{3}{2}}$ | $E_{\frac{5}{2}}$ | $E_{\frac{7}{2}}$ |
| $A_2$ | | $A_1$ | $B_2$ | $B_1$ | $E_1$ | $E_2$ | $E_3$ | $E_{\frac{1}{2}}$ | $E_{\frac{3}{2}}$ | $E_{\frac{5}{2}}$ | $E_{\frac{7}{2}}$ |
| $B_1$ | | | $A_1$ | $A_2$ | $E_3$ | $E_2$ | $E_1$ | $E_{\frac{7}{2}}$ | $E_{\frac{5}{2}}$ | $E_{\frac{3}{2}}$ | $E_{\frac{1}{2}}$ |
| $B_2$ | | | | $A_1$ | $E_3$ | $E_2$ | $E_1$ | $E_{\frac{7}{2}}$ | $E_{\frac{5}{2}}$ | $E_{\frac{3}{2}}$ | $E_{\frac{1}{2}}$ |
| $E_1$ | | | | | $A_1, [A_2], E_2$ | $E_1, E_3$ | $B_1, B_2, E_2$ | $E_{\frac{1}{2}}, E_{\frac{3}{2}}$ | $E_{\frac{1}{2}}, E_{\frac{5}{2}}$ | $E_{\frac{3}{2}}, E_{\frac{7}{2}}$ | $E_{\frac{5}{2}}, E_{\frac{7}{2}}$ |
| $E_2$ | | | | | | $A_1, [A_2], B_1, B_2$ | $E_1, E_3$ | $E_{\frac{3}{2}}, E_{\frac{5}{2}}$ | $E_{\frac{1}{2}}, E_{\frac{7}{2}}$ | $E_{\frac{1}{2}}, E_{\frac{7}{2}}$ | $E_{\frac{3}{2}}, E_{\frac{5}{2}}$ |
| $E_3$ | | | | | | | $A_1, [A_2], E_2$ | $E_{\frac{5}{2}}, E_{\frac{7}{2}}$ | $E_{\frac{3}{2}}, E_{\frac{7}{2}}$ | $E_{\frac{1}{2}}, E_{\frac{5}{2}}$ | $E_{\frac{1}{2}}, E_{\frac{3}{2}}$ |
| $E_{\frac{1}{2}}$ | | | | | | | | $[A_1], A_2, E_1$ | $E_1, E_2$ | $E_2, E_3$ | $B_1, B_2, E_3$ |
| $E_{\frac{3}{2}}$ | | | | | | | | | $[A_1], A_2, E_3$ | $B_1, B_2, E_1$ | $E_2, E_3$ |
| $E_{\frac{5}{2}}$ | | | | | | | | | | $[A_1], A_2, E_3$ | $E_1, E_2$ |
| $E_{\frac{7}{2}}$ | | | | | | | | | | | $[A_1], A_2, E_1$ |

$\boldsymbol{D}_\infty$, $\boldsymbol{C}_{\infty v}$, $(\boldsymbol{D}_{\infty h})$[b]

| | $\Sigma^+$ | $\Sigma^-$ | $\Pi$ | $\Delta$ | $\Phi$ | ... | $E_{\frac{1}{2}}$ | $E_{\frac{3}{2}}$ | $E_{\frac{5}{2}}$ | |
|---|---|---|---|---|---|---|---|---|---|---|
| $\Sigma^+$ | $\Sigma^+$ | $\Sigma^-$ | $\Pi$ | $\Delta$ | $\Phi$ | ... | $E_{\frac{1}{2}}$ | $E_{\frac{3}{2}}$ | $E_{\frac{5}{2}}$ | ... |
| $\Sigma^-$ | | $\Sigma^+$ | $\Pi$ | $\Delta$ | $\Phi$ | ... | $E_{\frac{1}{2}}$ | $E_{\frac{3}{2}}$ | $E_{\frac{5}{2}}$ | ... |
| $\Pi$ | | | $\Sigma^+, [\Sigma^-], \Delta$ | $\Pi, \Phi$ | $\Delta, \Gamma$ | ... | $E_{\frac{1}{2}}, E_{\frac{3}{2}}$ | $E_{\frac{1}{2}}, E_{\frac{5}{2}}$ | $E_{\frac{3}{2}}, E_{\frac{7}{2}}$ | ... |
| $\Delta$ | | | | $\Sigma^+, [\Sigma^-], \Gamma$ | $\Pi, \mathrm{H}$ | ... | $E_{\frac{3}{2}}, E_{\frac{5}{2}}$ | $E_{\frac{3}{2}}, E_{\frac{7}{2}}$ | $E_{\frac{1}{2}}, E_{\frac{9}{2}}$ | ... |
| $\Phi$ | | | | | $\Sigma^+, [\Sigma^-], \mathrm{I}$ | ... | $E_{\frac{5}{2}}, E_{\frac{7}{2}}$ | $E_{\frac{3}{2}}, E_{\frac{9}{2}}$ | $E_{\frac{1}{2}}, E_{\frac{11}{2}}$ | ... |
| ... | ... | ... | ... | ... | ... | ... | ... | ... | ... | ... |
| $E_{\frac{1}{2}}$ | | | | | | | $[\Sigma^+], \Sigma^-, \Pi$ | $\Pi, \Delta$ | $\Delta, \Phi$ | ... |
| $E_{\frac{3}{2}}$ | | | | | | | | $[\Sigma^+], \Sigma^-, \Phi$ | $\Pi, \Gamma$ | ... |
| $E_{\frac{5}{2}}$ | | | | | | | | | $[\Sigma^+], \Sigma^-, \mathrm{H}$ | ... |
| ... | ... | ... | ... | ... | ... | ... | ... | ... | ... | ... |

Table 57 (*Continued*)

**$O$, $T_d$, $(O_h)$**[b]

| | $A_1$ | $A_2$ | $E$ | $F_1$ | $F_2$ | $E_{\frac{1}{2}}$ | $E_{\frac{5}{2}}$ | $G_{\frac{3}{2}}$ |
|---|---|---|---|---|---|---|---|---|
| $A_1$ | $A_1$ | $A_2$ | $E$ | $F_1$ | $F_2$ | $E_{\frac{1}{2}}$ | $E_{\frac{5}{2}}$ | $G_{\frac{3}{2}}$ |
| $A_2$ | | $A_1$ | $E$ | $F_2$ | $F_1$ | $E_{\frac{5}{2}}$ | $E_{\frac{1}{2}}$ | $G_{\frac{3}{2}}$ |
| $E$ | | | $A_1, [A_2], E$ | $F_1, F_2$ | $F_1, F_2$ | $G_{\frac{3}{2}}$ | $G_{\frac{3}{2}}$ | $E_{\frac{1}{2}}, E_{\frac{5}{2}}, G_{\frac{3}{2}}$ |
| $F_1$ | | | | $A_1, E, [F_1], F_2$ | $A_2, E, F_1, F_2$ | $E_{\frac{1}{2}}, G_{\frac{3}{2}}$ | $E_{\frac{5}{2}}, G_{\frac{3}{2}}$ | $E_{\frac{1}{2}}, E_{\frac{5}{2}}, 2G_{\frac{3}{2}}$ |
| $F_2$ | | | | | $A_1, E, [F_1], F_2$ | $E_{\frac{5}{2}}, G_{\frac{3}{2}}$ | $E_{\frac{1}{2}}, G_{\frac{3}{2}}$ | $E_{\frac{1}{2}}, E_{\frac{5}{2}}, 2G_{\frac{3}{2}}$ |
| $E_{\frac{1}{2}}$ | | | | | | $[A_1], F_1$ | $A_2, F_2$ | $E, F_1, F_2$ |
| $E_{\frac{5}{2}}$ | | | | | | | $[A_1], F_1$ | $E, F_1, F_2$ |
| $G_{\frac{3}{2}}$ | | | | | | | | $[A_1], A_2, [E], 2F_1, [F_2], F_2$ |

**$I$, $(I_h)$**[b]

| | $A$ | $F_1$ | $F_2$ | $G$ | $H$ | $E_{\frac{1}{2}}$ | $E_{\frac{7}{2}}$ | $G_{\frac{3}{2}}$ | $I_{\frac{5}{2}}$ |
|---|---|---|---|---|---|---|---|---|---|
| $A$ | $A$ | $F_1$ | $F_2$ | $G$ | $H$ | $E_{\frac{1}{2}}$ | $E_{\frac{7}{2}}$ | $G_{\frac{3}{2}}$ | $I_{\frac{5}{2}}$ |
| $F_1$ | | $A, [F_1], H$ | $G, H$ | $F_2, G, H$ | $F_1, F_2, G, H$ | $E_{\frac{1}{2}}, G_{\frac{3}{2}}$ | $I_{\frac{5}{2}}$ | $E_{\frac{1}{2}}, G_{\frac{3}{2}}, I_{\frac{5}{2}}$ | $E_{\frac{7}{2}}, G_{\frac{3}{2}}, 2I_{\frac{5}{2}}$ |
| $F_2$ | | | $A, [F_2], H$ | $F_1, G, H$ | $F_1, F_2, G, H$ | $I_{\frac{5}{2}}$ | $E_{\frac{7}{2}}, G_{\frac{3}{2}}$ | $E_{\frac{7}{2}}, G_{\frac{3}{2}}, I_{\frac{5}{2}}$ | $E_{\frac{1}{2}}, G_{\frac{3}{2}}, 2I_{\frac{5}{2}}$ |
| $G$ | | | | $A, [F_1, F_2], G, H$ | $F_1, F_2, G, 2H$ | $E_{\frac{7}{2}}, I_{\frac{5}{2}}$ | $E_{\frac{1}{2}}, I_{\frac{5}{2}}$ | $G_{\frac{3}{2}}, 2I_{\frac{5}{2}}$ | $E_{\frac{1}{2}}, E_{\frac{7}{2}}, 2G_{\frac{3}{2}}, 2I_{\frac{5}{2}}$ |
| $H$ | | | | | $A, [F_1, F_2, G], G, 2H$ | $G_{\frac{3}{2}}, I_{\frac{5}{2}}$ | $G_{\frac{3}{2}}, I_{\frac{5}{2}}$ | $E_{\frac{1}{2}}, E_{\frac{7}{2}}, G_{\frac{3}{2}}, 2I_{\frac{5}{2}}$ | $E_{\frac{1}{2}}, E_{\frac{7}{2}}, 2G_{\frac{3}{2}}, 3I_{\frac{5}{2}}$ |
| $E_{\frac{1}{2}}$ | | | | | | $[A], F_1$ | $G$ | $F_1, H$ | $F_2, G, H$ |
| $E_{\frac{7}{2}}$ | | | | | | | $[A], F_2$ | $F_2, H$ | $F_1, G, H$ |
| $G_{\frac{3}{2}}$ | | | | | | | | $[A], F_1, F_2, G, [H]$ | $F_1, F_2, 2G, 2H$ |
| $I_{\frac{5}{2}}$ | | | | | | | | | $[A], 2F_1, 2F_2, [G], G, [2H], H$ |

# APPENDIX IV

## RESOLUTION OF SPECIES OF SYMMETRIC POINT GROUPS INTO THOSE OF POINT GROUPS OF LOWER SYMMETRY

TABLE 58

RESOLUTION OF SPECIES OF ATOMS INTO THOSE OF MOLECULES OF VARIOUS POINT GROUPS

| Atomic species | Molecular species | | | | |
|---|---|---|---|---|---|
| | $\boldsymbol{O}_h$, ($\boldsymbol{O}$[a]) | $\boldsymbol{T}_d$ | $\boldsymbol{D}_{\infty h}$, ($\boldsymbol{C}_{\infty v}$[a]) | $\boldsymbol{D}_{6h}$, ($\boldsymbol{D}_6$[a], $\boldsymbol{C}_{6v}$[b]) | $\boldsymbol{D}_{4h}$, ($\boldsymbol{D}_4$[a], $\boldsymbol{C}_{4v}$[b]) |
| $S_g$ | $A_{1g}$ | $A_1$ | $\Sigma_g^+$ | $A_{1g}$ | $A_{1g}$ |
| $S_u$ | $A_{1u}$ | $A_2$ | $\Sigma_u^-$ | $A_{1u}$ | $A_{1u}$ |
| $P_g$ | $F_{1g}$ | $F_1$ | $\Sigma_g^- + \Pi_g$ | $A_{2g} + E_{1g}$ | $A_{2g} + E_g$ |
| $P_u$ | $F_{1u}$ | $F_2$ | $\Sigma_u^+ + \Pi_u$ | $A_{2u} + E_{1u}$ | $A_{2u} + E_u$ |
| $D_g$ | $E_g + F_{2g}$ | $E + F_2$ | $\Sigma_g^+ + \Pi_g + \Delta_g$ | $A_{1g} + E_{1g} + E_{2g}$ | $A_{1g} + B_{1g} + B_{2g} + E_g$ |
| $D_u$ | $E_u + F_{2u}$ | $E + F_1$ | $\Sigma_u^- + \Pi_u + \Delta_u$ | $A_{1u} + E_{1u} + E_{2u}$ | $A_{1u} + B_{1u} + B_{2u} + E_u$ |
| $F_g$ | $A_{2g} + F_{1g} + F_{2g}$ | $A_2 + F_1 + F_2$ | $\Sigma_g^- + \Pi_g + \Delta_g + \Phi_g$ | $A_{2g} + B_{1g} + B_{2g} + E_{1g} + E_{2g}$ | $A_{2g} + B_{1g} + B_{2g} + 2E_g$ |
| $F_u$ | $A_{2u} + F_{1u} + F_{2u}$ | $A_1 + F_1 + F_2$ | $\Sigma_u^+ + \Pi_u + \Delta_u + \Phi_u$ | $A_{2u} + B_{1u} + B_{2u} + E_{1u} + E_{2u}$ | $A_{2u} + B_{1u} + B_{2u} + 2E_u$ |
| $G_g$ | $A_{1g} + E_g + F_{1g} + F_{2g}$ | $A_1 + E + F_1 + F_2$ | $\Sigma_g^+ + \Pi_g + \Delta_g + \Phi_g + \Gamma_g$ | $A_{1g} + B_{1g} + B_{2g} + E_{1g} + 2E_{2g}$ | $2A_{1g} + A_{2g} + B_{1g} + B_{2g} + 2E_g$ |
| $G_u$ | $A_{1u} + E_u + F_{1u} + F_{2u}$ | $A_2 + E + F_1 + F_2$ | $\Sigma_u^- + \Pi_u + \Delta_u + \Phi_u + \Gamma_u$ | $A_{1u} + B_{1u} + B_{2u} + E_{1u} + 2E_{2u}$ | $2A_{1u} + A_{2u} + B_{1u} + B_{2u} + 2E_u$ |

TABLE 58 (*continued*)

Molecular Species

| Atomic species | $\boldsymbol{D}_{3h}$, ($\boldsymbol{D}_{3d}$[c], $\boldsymbol{D}_3$[d], $\boldsymbol{C}_{3h}$[e]) | $\boldsymbol{C}_{3v}$ | $\boldsymbol{D}_{2d}$ | $\boldsymbol{D}_{2h}$, ($\boldsymbol{D}_2$)[a] | $\boldsymbol{C}_{2v}$ | $\boldsymbol{C}_s$ |
|---|---|---|---|---|---|---|
| $S_g$ | $A_1'$ | $A_1$ | $A_1$ | $A_g$ | $A_1$ | $A'$ |
| $S_u$ | $A_1''$ | $A_2$ | $B_1$ | $A_u$ | $A_2$ | $A''$ |
| $P_g$ | $A_2'+E''$ | $A_2+E$ | $A_2+E$ | $B_{1g}+B_{2g}+B_{3g}$ | $A_2+B_1+B_2$ | $A'+2A''$ |
| $P_u$ | $A_2''+E'$ | $A_1+E$ | $B_2+E$ | $B_{1u}+B_{2u}+B_{3u}$ | $A_1+B_1+B_2$ | $2A'+A''$ |
| $D_g$ | $A_1'+E'+E''$ | $A_1+2E$ | $A_1+B_1+B_2+E$ | $2A_g+B_{1g}+B_{2g}+B_{3g}$ | $2A_1+A_2+B_1+B_2$ | $3A'+2A''$ |
| $D_u$ | $A_1''+E'+E''$ | $A_2+2E$ | $A_1+A_2+B_1+E$ | $2A_u+B_{1u}+B_{2u}+B_{3u}$ | $A_1+2A_2+B_1+B_2$ | $2A'+3A''$ |
| $F_g$ | $A_1''+A_2'+A_2''+E'+E''$ | $A_1+2A_2+2E$ | $A_2+B_1+B_2+2E$ | $A_g+2B_{1g}+2B_{2g}+2B_{3g}$ | $A_1+2A_2+2B_1+2B_2$ | $3A'+4A''$ |
| $F_u$ | $A_1'+A_2'+A_2''+E'+E''$ | $2A_1+A_2+2E$ | $A_1+A_2+B_2+2E$ | $A_u+2B_{1u}+2B_{2u}+2B_{3u}$ | $2A_1+A_2+2B_1+2B_2$ | $4A'+3A''$ |
| $G_g$ | $A_1'+A_1''+A_2''+2E'+E''$ | $2A_1+A_2+3E$ | $2A_1+A_2+B_1+B_2+2E$ | $3A_g+2B_{1g}+2B_{2g}+2B_{3g}$ | $3A_1+2A_2+2B_1+2B_2$ | $5A'+4A''$ |
| $G_u$ | $A_1'+A_1''+A_2'+E'+2E''$ | $A_1+2A_2+3E$ | $A_1+A_2+2B_1+B_2+2E$ | $3A_u+2B_{1u}+2B_{2u}+2B_{3u}$ | $2A_1+3A_2+2B_1+2B_2$ | $4A'+5A''$ |

[a] For these point groups disregard the subscripts $g$ and $u$.
[b] Substitute $A_1$ for $A_{1g}$ and $A_{2u}$, and $A_2$ for $A_{2g}$ and $A_{1u}$, and disregard the subscripts $g$ and $u$ in the remaining species.
[c] Disregard ′ and ″ and add $g$ and $u$ to agree with that in the corresponding atomic species.
[d] Disregard ′ and ″.
[e] Disregard the difference between $A_1$ and $A_2$.

TABLE 59

## RESOLUTION OF SPECIES OF LINEAR MOLECULES (POINT GROUPS $D_{\infty h}$, $C_{\infty v}$) INTO THOSE OF MOLECULES OF LOWER SYMMETRY

The $z$ axis in the linear case is assumed to lie in the internuclear axis. For the axial point groups it is assumed to coincide with the main symmetry axis, but for the point groups $D_{2h}$, $C_{2v}$, $C_{2h}$ and $C_s$ the correlation is given for several assumptions about the relative positions of the linear axis and the axis of the molecule to be formed.

| Linear[a] molecule | $D_{6h}$, $(D_6)$[a] | $C_{6v}$ | $D_{4h}$, $(D_4)$[a] | $D_{3h}$, $(D_3)$[bc] | $C_{3v}$ | $D_{2d}$ | $D_{2h}$ | | $C_{2v}$ | | | $C_{2h}$, $(C_2)$[a] | | $C_s$ | |
|---|---|---|---|---|---|---|---|---|---|---|---|---|---|---|---|
| | | | | | | | $z \to z$ | $z \to x$ | $z \to z$ | $z \to y$ | $z \to x$ | $z \to z$ | $z \to x,y$ | $\sigma_h \to \sigma$ | $\sigma_v \to \sigma$ |
| $\Sigma_g^+$ | $A_{1g}$ | $A_1$ | $A_{1g}$ | $A_1'$ | $A_1$ | $A_1$ | $A_g$ | $A_g$ | $A_1$ | $A_1$ | $A_1$ | $A_g$ | $A_g$ | $A'$ | $A'$ |
| $\Sigma_u^+$ | $A_{2u}$ | $A_1$ | $A_{2u}$ | $A_2''$ | $A_1$ | $B_2$ | $B_{1u}$ | $B_{3u}$ | $A_1$ | $B_2$ | $B_1$ | $A_u$ | $B_u$ | $A''$ | $A'$ |
| $\Sigma_g^-$ | $A_{2g}$ | $A_2$ | $A_{2g}$ | $A_2'$ | $A_2$ | $A_2$ | $B_{1g}$ | $B_{3g}$ | $A_2$ | $B_1$ | $B_2$ | $A_g$ | $B_g$ | $A'$ | $A''$ |
| $\Sigma_u^-$ | $A_{1u}$ | $A_2$ | $A_{1u}$ | $A_1''$ | $A_2$ | $B_1$ | $A_u$ | $A_u$ | $A_2$ | $A_2$ | $A_2$ | $A_u$ | $A_u$ | $A''$ | $A''$ |
| $\Pi_g$ | $E_{1g}$ | $E_1$ | $E_g$ | $E''$ | $E$ | $E$ | $B_{2g}+B_{3g}$ | $B_{1g}+B_{2g}$ | $B_1+B_2$ | $A_2+B_2$ | $A_2+B_1$ | $2B_g$ | $A_g+B_g$ | $2A''$ | $A'+A''$ |
| $\Pi_u$ | $E_{1u}$ | $E_1$ | $E_u$ | $E'$ | $E$ | $E$ | $B_{2u}+B_{3u}$ | $B_{1u}+B_{2u}$ | $B_1+B_2$ | $A_1+B_1$ | $A_1+B_2$ | $2B_u$ | $A_u+B_u$ | $2A'$ | $A'+A''$ |
| $\Delta_g$ | $E_{2g}$ | $E_2$ | $B_{1g}+B_{2g}$ | $E'$ | $E$ | $B_1+B_2$ | $A_g+B_{1g}$ | $A_g+B_{3g}$ | $A_1+A_2$ | $A_1+B_1$ | $A_1+B_2$ | $2A_g$ | $A_g+B_g$ | $2A'$ | $A'+A''$ |
| $\Delta_u$ | $E_{2u}$ | $E_2$ | $B_{1u}+B_{2u}$ | $E''$ | $E$ | $A_1+A_2$ | $A_u+B_{1u}$ | $A_u+B_{3u}$ | $A_1+A_2$ | $A_2+B_2$ | $A_2+B_1$ | $2A_u$ | $A_u+B_u$ | $2A''$ | $A'+A''$ |
| $\Phi_g$ | $B_{1g}+B_{2g}$ | $B_1+B_2$ | $E_g$ | $A_1''+A_2''$ | $A_1+A_2$ | $E$ | $B_{2g}+B_{3g}$ | $B_{1g}+B_{2g}$ | $B_1+B_2$ | $A_2+B_2$ | $A_2+B_1$ | $2B_g$ | $A_g+B_g$ | $2A''$ | $A'+A''$ |
| $\Phi_u$ | $B_{1u}+B_{2u}$ | $B_1+B_2$ | $E_u$ | $A_1'+A_2'$ | $A_1+A_2$ | $E$ | $B_{2u}+B_{3u}$ | $B_{1u}+B_{2u}$ | $B_1+B_2$ | $A_1+B_1$ | $A_1+B_2$ | $2B_u$ | $A_u+B_u$ | $2A'$ | $A'+A''$ |
| $\Gamma_g$ | $E_{2g}$ | $E_2$ | $A_{1g}+A_{2g}$ | $E'$ | $E$ | $A_1+A_2$ | $A_g+B_{1g}$ | $A_g+B_{3g}$ | $A_1+A_2$ | $A_1+B_1$ | $A_1+B_2$ | $2A_g$ | $A_g+B_g$ | $2A'$ | $A'+A''$ |
| $\Gamma_u$ | $E_{2u}$ | $E_2$ | $A_{1u}+A_{2u}$ | $E''$ | $E$ | $B_1+B_2$ | $A_u+B_{1u}$ | $A_u+B_{3u}$ | $A_1+A_2$ | $A_2+B_2$ | $A_2+B_1$ | $2A_u$ | $A_u+B_u$ | $2A''$ | $A'+A''$ |

[a] For $C_{\infty v}$, $D_6$, $D_4$, $C_2$ disregard the subscripts $g$ and $u$.
[b] For $D_3$ disregard $'$ and $''$.
[c] For $D_{3d}$ disregard $'$ and $''$ and add $g$ and $u$ to agree with linear case.

TABLE 60

RESOLUTION OF SPECIES OF POINT GROUPS $C_{2v}$, $D_{2h}$, $D_{3h}$, $D_{4h}$, $T_d$ INTO THOSE OF POINT GROUPS OF LOWER SYMMETRY

| $C_{2v}$ | $C_s$ $\sigma(xz)\rightarrow\sigma$ | $C_s$ $\sigma(yz)\rightarrow\sigma$ |
|---|---|---|
| $A_1$ | $A'$ | $A'$ |
| $A_2$ | $A''$ | $A''$ |
| $B_1$ | $A'$ | $A''$ |
| $B_2$ | $A''$ | $A'$ |

| $D_{2h}$ | $C_{2h}$ $z\rightarrow z$ | $C_{2v}$ $y\rightarrow y$, $z\rightarrow z$ | $C_s$ $\sigma(yz)\rightarrow\sigma$ |
|---|---|---|---|
| $A_g$ | $A_g$ | $A_1$ | $A'$ |
| $A_u$ | $A_u$ | $A_2$ | $A''$ |
| $B_{1g}$ | $A_g$ | $A_2$ | $A''$ |
| $B_{1u}$ | $A_u$ | $A_1$ | $A'$ |
| $B_{2g}$ | $B_g$ | $B_1$ | $A''$ |
| $B_{2u}$ | $B_u$ | $B_2$ | $A'$ |
| $B_{3g}$ | $B_g$ | $B_2$ | $A'$ |
| $B_{3u}$ | $B_u$ | $B_1$ | $A''$ |

| $D_{3h}$ | $C_{3v}$ | $C_s$ $\sigma_v\rightarrow\sigma$ | $C_{2v}$ $\sigma_h\rightarrow\sigma_v(yz)$ |
|---|---|---|---|
| $A_1'$ | $A_1$ | $A'$ | $A_1$ |
| $A_1''$ | $A_2$ | $A''$ | $A_2$ |
| $A_2'$ | $A_2$ | $A''$ | $B_2$ |
| $A_2''$ | $A_1$ | $A'$ | $B_1$ |
| $E'$ | $E$ | $A' + A''$ | $A_1 + B_2$ |
| $E''$ | $E$ | $A' + A''$ | $A_2 + B_1$ |

| $D_{4h}$ | $C_{4v}$ | $D_{2d}$ | $D_{2h}$ $z\rightarrow z$ |
|---|---|---|---|
| $A_{1g}$ | $A_1$ | $A_1$ | $A_g$ |
| $A_{1u}$ | $A_2$ | $B_1$ | $A_u$ |
| $A_{2g}$ | $A_2$ | $A_2$ | $B_{1g}$ |
| $A_{2u}$ | $A_1$ | $B_2$ | $B_{1u}$ |
| $B_{1g}$ | $B_1$ | $B_1$ | $A_g$ |
| $B_{1u}$ | $B_2$ | $A_1$ | $A_u$ |
| $B_{2g}$ | $B_2$ | $B_2$ | $B_{1g}$ |
| $B_{2u}$ | $B_1$ | $A_2$ | $B_{1u}$ |
| $E_g$ | $E$ | $E$ | $B_{2g} + B_{3g}$ |
| $E_u$ | $E$ | $E$ | $B_{2u} + B_{3u}$ |

| $T_d$ | $D_{2d}$ | $C_{3v}$ | $C_{2v}$ |
|---|---|---|---|
| $A_1$ | $A_1$ | $A_1$ | $A_1$ |
| $A_2$ | $B_1$ | $A_2$ | $A_2$ |
| $E$ | $A_1 + B_1$ | $E$ | $A_1 + A_2$ |
| $F_1$ | $A_2 + E$ | $A_2 + E$ | $A_2 + B_1 + B_2$ |
| $F_2$ | $B_2 + E$ | $A_1 + E$ | $A_1 + B_1 + B_2$ |

# APPENDIX V

TABLE 61

MOLECULAR ORBITALS (GROUP ORBITALS) FORMED FROM ATOMIC ORBITALS OF EQUIVALENT ATOMS

| Point group | Atomic orbitals | Molecular orbitals[a] |
|---|---|---|
| $\boldsymbol{C}_{2v}(\boldsymbol{C}_2)$ | $(ns) \times 2$[b] | $a_1 + b_2$ |
| | $(np) \times 2$[b] | $2a_1 + \underline{a_2} + \underline{b_1} + 2b_2$ |
| | $(nd) \times 2$[b] | $3a_1 + 2a_2 + 2b_1 + 3b_2$ |
| $\boldsymbol{C}_{2h}(\boldsymbol{C}_s, \boldsymbol{C}_i)$ | $(ns) \times 2$ | $a_g + b_u$ |
| | $(np) \times 2$ | $2a_g + \underline{a_u} + \underline{b_g} + 2b_u$ |
| | $(nd) \times 2$ | $3a_g + 2a_u + 2b_g + 3b_u$ |
| $\boldsymbol{D}_{2h}(\boldsymbol{D}_2)$ | $(ns) \times 2$[c] | $a_g + b_{1u}$ |
| | $(ns) \times 4$[b] | $a_g + b_{1u} + b_{2u} + b_{3g}$ |
| | $(np) \times 2$[c] | $a_g + b_{1u} + \underline{b_{2g}} + b_{2u} + b_{3g} + \underline{b_{3u}}$ |
| | $(np) \times 4$[b] | $2a_g + \underline{a_u} + \underline{b_{1g}} + 2b_{1u} + \underline{b_{2g}} + 2b_{2u} + 2b_{3g} + \underline{b_{3u}}$ |
| | $(nd) \times 2$[c] | $2a_g + a_u + b_{1g} + 2b_{1u} + b_{2g} + b_{2u} + b_{3g} + b_{3u}$ |
| | $(nd) \times 4$[b] | $3a_g + 2a_u + 2b_{1g} + 3b_{1u} + 2b_{2g} + 3b_{2u} + 3b_{3g} + 2b_{3u}$ |
| $\boldsymbol{C}_{3v}$ | $(ns) \times 3$ | $a_1 + e$ |
| | $(np) \times 3$ | $\underline{a_1} + a_1 + a_2 + \underline{e} + 2e$ |
| | $(nd) \times 3$ | $3a_1 + 2a_2 + 5e$ |
| $\boldsymbol{C}_{3h}$ | $(ns) \times 3$ | $a' + e'$ |
| | $(ns) \times 6$ | $a' + a'' + e' + e''$ |
| | $(np) \times 3$ | $2a' + 2e' + \underline{a''} + \underline{e''}$ |
| | $(nd) \times 3$ | $3a' + 2a'' + 3e' + 2e''$ |
| $\boldsymbol{D}_{3h}$ | $(ns) \times 2$ | $a_1' + a_2''$ |
| | $(ns) \times 3$ | $a_1' + e'$ |
| | $(ns) \times 6$ | $a_1' + a_2'' + e' + e''$ |
| | $(np) \times 2$ | $\underline{a_1'} + \underline{a_2''} + e' + e''$ |
| | $(np) \times 3$ | $a_1' + a_2' + \underline{a_2''} + 2e' + \underline{e''}$ |
| | $(np) \times 6$ | $\underline{a_1'} + a_1' + a_1'' + a_2' + \underline{a_2''} + a_2'' + \underline{e'} + 2e' + \underline{e''} + 2e''$ |
| $\boldsymbol{D}_{3d}$ | $(ns) \times 2$ | $a_{1g} + a_{2u}$ |
| | $(ns) \times 6$ | $a_{1g} + a_{2u} + e_g + e_u$ |
| | $(np) \times 2$ | $\underline{a_{1g}} + \underline{a_{2u}} + e_g + e_u$ |
| | $(np) \times 6$ | $\underline{a_{1g}} + a_{1g} + a_{1u} + a_{2g} + \underline{a_{2u}} + a_{2u} + a_{2u} + \underline{e_g} + 2e_g + \underline{e_u} + 2e_u$ |
| $\boldsymbol{D}_{2d}$ | $(ns) \times 2$ | $a_1 + b_2$ |
| | $(ns) \times 4$ | $a_1 + b_2 + e$ |
| | $(np) \times 2$ | $\underline{a_1} + \underline{b_2} + 2e$ |
| | $(np) \times 4$ | $\underline{a_1} + a_1 + a_2 + b_1 + 2b_2 + \underline{e} + 2e$ |
| | $(nd) \times 2$ | $2a_1 + a_2 + b_1 + 2b_2 + 2e$ |
| | $(nd) \times 4$ | $3a_1 + 2a_2 + 2b_1 + 3b_2 + 5e$ |

Table 61 (*Continued*)

| Point group | Atomic orbitals | Molecular orbitals[a] |
|---|---|---|
| $\boldsymbol{D}_{4h}$ | $(ns) \times 2$ | $a_{1g} + a_{2u}$ |
| | $(ns) \times 4$ | $a_{1g} + b_{1g} + e_u$ |
| | $(ns) \times 8$ | $a_{1g} + a_{2u} + b_{1g} + b_{2u} + e_g + e_u$ |
| | $(np) \times 2$ | $\underline{a_{1g}} + \underline{a_{2u}} + e_g + e_u$ |
| | $(np) \times 4$ | $a_{1g} + a_{2g} + \underline{a_{2u}} + \underline{b_{1g}} + b_{2g} + \underline{b_{2u}} + \underline{e_g} + 2e_u$ |
| | $(nd) \times 2$ | $a_{1g} + a_{2u} + b_{1g} + b_{1u} + b_{2g} + b_{2u} + e_g + e_u$ |
| | $(nd) \times 4$ | $2a_{1g} + a_{1u} + a_{2g} + a_{2u} + 2b_{1g} + b_{1u} + b_{2g} + b_{2u} + 2e_g + 3e_u$ |
| $\boldsymbol{D}_{5h}$ | $(ns) \times 2$ | $a_1' + a_2''$ |
| | $(ns) \times 5$ | $a_1' + e_1' + e_2'$ |
| | $(ns) \times 10$ | $a_1' + a_2'' + e_1' + e_1'' + e_2' + e_2''$ |
| | $(np) \times 2$ | $\underline{a_1'} + \underline{a_2''} + e_1' + e_1''$ |
| | $(np) \times 5$ | $a_1' + a_2' + \underline{a_2''} + 2e_1' + \underline{e_1''} + 2e_2' + \underline{e_2''}$ |
| | $(nd) \times 2$ | $a_1' + a_2'' + e_1' + e_1'' + e_2' + e_2''$ |
| | $(nd) \times 5$ | $2a_1' + a_1'' + a_2' + a_2'' + 3e_1' + 2e_1'' + 3e_2' + 2e_2''$ |
| $\boldsymbol{D}_{6h}$ | $(ns) \times 2$ | $a_{1g} + a_{2u}$ |
| | $(ns) \times 6$ | $a_{1g} + b_{1u} + e_{1u} + e_{2g}$ |
| | $(np) \times 2$ | $\underline{a_{1g}} + \underline{a_{2u}} + e_{1g} + e_{1u}$ |
| | $(np) \times 6$ | $a_{1g} + a_{2g} + \underline{a_{2u}} + \underline{b_{1u}} + \underline{b_{2g}} + b_{2u} + \underline{e_{1g}} + 2e_{1u} + 2e_{2g} + \underline{e_{2u}}$ |
| | $(nd) \times 2$ | $a_{1g} + a_{2u} + e_{1g} + e_{1u} + e_{2g} + e_{2u}$ |
| | $(nd) \times 6$ | $2a_{1g} + a_{1u} + a_{2g} + a_{2u} + b_{1g} + 2b_{1u} + b_{2g} + b_{2u} + 2e_{1g}$ $+ 3e_{1u} + 3e_{2g} + 2e_{2u}$ |
| $\boldsymbol{D}_{\infty h}$ | $(ns) \times 2$ | $\sigma_g + \sigma_u$ |
| | $(np) \times 2$ | $\underline{\sigma_g} + \underline{\sigma_u} + \pi_g + \pi_u$ |
| | $(nd) \times 2$ | $\sigma_g + \sigma_u + \pi_g + \pi_u + \delta_g + \delta_u$ |
| $\boldsymbol{T}_d$ | $(ns) \times 4$ | $a_1 + f_2$ |
| | $(np) \times 4$ | $\underline{a_1} + \underline{f_2} + e + f_1 + f_2$ |
| | $(nd) \times 4$ | $a_1 + 2e + 2f_1 + 3f_2$ |
| $\boldsymbol{O}_h$ | $(ns) \times 6$ | $a_{1g} + e_g + f_{1u}$ |
| | $(ns) \times 8$ | $a_{1g} + a_{2u} + f_{1u} + f_{2g}$ |
| | $(np) \times 6$ | $\underline{a_{1g}} + \underline{e_g} + \underline{f_{1u}} + f_{1g} + f_{1u} + f_{2g} + f_{2u}$ |
| | $(np) \times 8$ | $a_{1g} + a_{2u} + e_g + e_u + f_{1g} + 2f_{1u} + 2f_{2g} + f_{2u}$ |
| | $(nd) \times 6$ | $a_{1g} + a_{2g} + a_{2u} + 2e_g + e_u + f_{1g} + 2f_{1u} + 2f_{2g} + 2f_{2u}$ |

[a] Molecular orbitals arising from $2p_z$ atomic orbitals are underlined.
[b] In the $yz$-plane.
[c] In the $z$-axis.

# APPENDIX VI

## MOLECULAR CONSTANTS OF THE ELECTRONIC STATES OF MOLECULES WITH THREE TO TWELVE ATOMS

The following tables are similar to the table of molecular constants of diatomic molecules presented in Volume I (Table 39, p. 502). It was found convenient to prepare separate tables for each group of molecules consisting of a given number of atoms; in addition, within such a group, for molecules consisting of six or less atoms a further subdivision was made according to the number of hydrogen atoms present. Thus, the following tables were set up:

| | | | |
|---|---|---|---|
| Table 62 | Triatomic dihydrides | Table 73 | Six-atomic trihydrides |
| Table 63 | Triatomic monohydrides | Table 74 | Six-atomic dihydrides |
| Table 64 | Triatomic non-hydrides | Table 75 | Six-atomic monohydrides |
| Table 65 | Four-atomic trihydrides | Table 76 | Six-atomic non-hydrides |
| Table 66 | Four-atomic dihydrides | Table 77 | Seven-atomic molecules |
| Table 67 | Four-atomic monohydrides | Table 78 | Eight-atomic molecules |
| Table 68 | Four-atomic non-hydrides | Table 79 | Nine-atomic molecules |
| Table 69 | Five-atomic tetra- and trihydrides | Table 80 | Ten-atomic molecules |
| Table 70 | Five-atomic dihydrides and monohydrides | Table 81 | Eleven-atomic molecules |
| Table 71 | Five-atomic non-hydrides | Table 82 | Twelve-atomic molecules |
| Table 72 | Six-atomic tetrahydrides | | |

For more-than-six-atomic molecules for which the tables are not subdivided according to the number of H atoms, the molecules are listed according to the number of H atoms, those with maximum number first. Within each such sub-group the molecules are listed according to increasing number of valence electrons. Molecules with the same number of H atoms and the same number of valence electrons are listed according to the total number of electrons, e.g., $CO_2$ comes before COS which in turn comes before $CS_2$. In order to find the data for a given molecule in the tables, it is therefore necessary to ascertain the number of atoms, the number of H atoms and the number of valence electrons in it, and sometimes the total number of electrons.

Only molecules for which spectra have been observed in the gaseous phase have been included in the tables. For molecules showing only continuous absorptions, in general, no detailed list of states is given but only a reference to one or two recent papers. The same applies to several other molecules for which the data are not very complete. In all other cases all known electronic states (designated as indicated in the text, p. 488) are listed except higher Rydberg states for which instead the series formula is given. For each state the point group of the molecule, the energy $T_0$ above the lowest state (not $T_e$ as in Vol. I for diatomic molecules), the vibrational frequencies $\nu_i$, the rotational constants $A_0$, $B_0$, $C_0$ and the resulting geometrical data (internuclear distances and angles) are given. For triatomic and four-atomic molecules in addition, where possible, the electron configuration is included. Finally, the observed transitions and their wavelength regions as well as references to the relevant literature are included. In listing the transitions, in accordance with the internationally agreed rule, the upper state is always put first irrespective of whether the transition is observed in absorption ($\leftarrow$) or emission ($\rightarrow$).

In many cases not all the data mentioned are available, but where additional data are known which cannot be accommodated in the body of the tables, they are presented in the footnotes (similar to Vol. I). Since only for a few states of a few molecules all the normal frequencies are known, it was decided to limit the number listed in the body of the table to six. Wherever more frequencies are known, they are listed in the footnotes. This applies particularly to the ground states of more-than-four-atomic molecules.

An attempt has been made to collect the best available ground state data and include them in the tables, thus bringing up to date the information in Volume II on the molecules considered. The order of the fundamentals is the one used in Volume II, which agrees with Mulliken's report[1]. The only difference compared to Volume II arises for $\boldsymbol{C_{2v}}$ molecules for which in most cases, because of the different choice of axes in Mulliken's report, the meaning of $B_1$ and $B_2$ is interchanged. Care has been taken to make the numbering of the vibrations consistent (and therefore in some cases different from that of the original authors).

[1] It should be noted that according to Mulliken's report, as an exception to the general rule, the bending vibration of linear triatomic molecules is always designated $\nu_2$. In accordance with the practice of most authors, we have extended this exception to include non-linear triatomic molecules.

The $\nu_i$ values given are the observed first vibrational quanta (i.e., the fundamentals) without applying any correction for anharmonicity or Fermi resonance[2]. For bending vibrations of linear molecules the band center without correction by the term $-Bl^2$ or for non-linear molecules without correction by the term $-A\zeta^2$ are given; thus, the numbers listed in the tables for $\nu_i$ give directly the energy of the $J = 0$ level or the $J = 0$, $K = 0$ level of the state in which the vibration considered is singly excited.

For doublet and triplet states the $T_0$ values in the third column refer to the center of the multiplet. The actual components differ by $\pm\frac{1}{2}A$ or 0, $\pm A$ from this value; the spin coupling constants $A$ are given in the footnotes.

In the first line of each section referring to a given molecule, whenever available, the ionization potential and one or more dissociation energies are included. These data are given in eV (conversion factors used: 1 eV = $8066.0_3$ cm$^{-1}$ = $23.062_3$ kcal/mole); the internuclear distances $r_0$ are in Å (= $10^{-8}$ cm); all other numbers are in cm$^{-1}$ both in the body of the tables and the footnotes.

In order to save space in these tables, as well as in the footnotes, reference numbers (referring to the bibliography, p. 671) without the names of the authors are used.

Much effort has been devoted to making the data given as reliable as possible. They depend, of course, on the original papers quoted. In many ways these tables are similar to those of Sponer and Teller (1155) published twenty-five years ago. Comparison of the new with the old tables will show what enormous advances have been made in the detailed analysis of electronic spectra of polyatomic molecules.

Abbreviations:

( ) Constants and symbols in parentheses are uncertain.
repr. ( ) refers to a reference where a good reproduction of the particular spectrum considered may be found.
sp. = spectrum
ass. = assumed
max. = maximum
pred. = predissociation

[2] Even where the anharmonic constant $x_{ii}$ is given in a footnote, the value given in the body of the table is $\nu_i$, not $\omega_i$.

Table 62

## MOLECULAR CONSTANTS OF THE ELECTRONIC STATES OF TRIATOMIC DIHYDRIDES

| State | Point Group | $T_0$ | Vibrational Frequencies $\nu_1$ | $\nu_2$ | $\nu_3$ | Rotational Constants $A_0$ | $B_0$ | $C_0$ | $r_0$ (Å) | $\alpha$ | Electron Configuration | Observed Transitions | References | Remarks |
|---|---|---|---|---|---|---|---|---|---|---|---|---|---|---|
| **$BH_2$** | | I.P. = 9.8 eV [a] | | | | | | | | | | | | |
| $\tilde{A}\ ^2B_1(\Pi)$ | $D_{\infty h}$ | $\sim 5150$ [d] | | 954 [d] | | — | $6.1_2$ | — | 1.17 | 180° | $\ldots(2a_1)^2(1b_2)^2(1b_1)$ [b] | $\tilde{A} \leftarrow \tilde{X}$ 8650–6400 Å | (531) | Widely spaced bands with open fine structure |
| $\tilde{X}\ ^2A_1$ | $C_{2v}$ | 0 | | (1030) [c] | | 41.64 | $7.24_8$ | $6.00_8$ | 1.18 | 131° | $\ldots(2a_1)^2(1b_2)^2(3a_1)$ [b] | | (531) | |
| **$AlH_2$** | | | | | | | | | | | | | | |
| $\tilde{A}\ ^2B_1(\Pi)$ | $D_{\infty h}$ | [a] | | | | — | $3.57_2$ | — | 1.53 | 180° | $\ldots(b_2)^2(b_1)$ | $\tilde{A} \leftarrow \tilde{X}$ 6584 Å | (532) | Widely spaced bands with open fine structure [b] |
| $\tilde{X}\ ^2A_1$ | $C_{2v}$ | 0 | | | | 13.6 | $4.4_1$ | $3.3_3$ | 1.59 | 119° | $\ldots(b_2)^2(a_1)$ | | | |
| **$CH_2$** | | I.P. = 10.396 eV [a]; $D_0$(HC—H) $\geq 4.2_3$ eV [b] (see energy level diagram Fig. 186) | | | | | | | | | | | | |
| | | Fragments of singlet absorption systems near 70000 cm$^{-1}$ | | | | | | | | | | | (530) | |
| $\tilde{c}\ ^1A_1$ | $(C_{2v})$ | $\leq x + 27700$ | | | | | | | | | $\ldots(2a_1)^2(1b_2)^2(1b_1)^2$ | $\tilde{c} \leftarrow \tilde{a}$ 3620–3300 Å | (530) | Three fragmentary bands with open structure |
| $\tilde{b}\ ^1B_1$ | $(D_{\infty h})$ | $x + 7100$ | $\sim 3000$ | 557 | | — | $7.57$ [c] | — | $1.05_6$ | $\approx 180°$ [c] | $\ldots(2a_1)^2(1b_2)^2(3a_1)(1b_1)$ [d] | $\tilde{b} \leftarrow \tilde{a}$ 9000–5000 Å | (521)(529) (530) | Widely spaced bands with open fine structure See Fig. 93 and (521)(530) |
| $\tilde{a}\ ^1A_1$ | $C_{2v}$ | $x$ [e] | | | | $20.1_4$ | $11.1_6$ | $7.0_6$ | 1.11 | 102.4° | $\ldots(2a_1)^2(1b_2)^2(3a_1)^2$ [d] | | | |
| | | Rydberg series joining on to $\tilde{C}$ and $\tilde{D}$ | | | | | | | | | | | (522) | |
| | | Rydberg series joining on to $\tilde{B}$: $\nu = 83851 - R/(n - 0.12)^2$; $n = 3, 4, 5, 6$ | | | | | | | | | | | (522) | |
| $\tilde{D}(^3\Pi_u)$ | $(D_{\infty h})$ | 71592 | | | | | | | | | $\ldots(1\sigma_u)^2(1\pi_u)(3d\delta_g)$ | $\tilde{D} \leftarrow \tilde{X}$ 1397 Å | (521) | Diffuse band |
| $\tilde{C}(^3\Pi_u)$ | $(D_{\infty h})$ | 70917 | | | | | | | | | $\ldots(1\sigma_u)^2(1\pi_u)(3d\sigma_g)$ | $\tilde{C} \leftarrow \tilde{X}$ 1410 Å | (521) | Diffuse band |

$BH_2$: [a] From electron impact data of (374). [b] These two states are derived from $^2\Pi$ of the configuration $2\sigma_g^2 1\sigma_u^2 1\pi_u$ of linear $BH_2$ (see p. 343). [c] Estimated from inertial defect.
[d] Uncertain, based on long extrapolation; $x_{22} = 1.0$.

$AlH_2$: [a] Only one band (at 6584 Å) has been analyzed as yet, but the spectrum extends to longer wavelengths.
[b] The band at 6584 Å breaks off above $J' = 8$, presumably on account of predissociation; limit at 15450 cm$^{-1}$.

TABLE 62 (*continued*)

| State | Point Group | $T_0$ | Vibrational Frequencies $\nu_1$ | $\nu_2$ | $\nu_3$ | Rotational Constants $A_0$ | $B_0$ | $C_0$ | $r_0$ (Å) | $\alpha$ | Electron Configuration | Observed Transitions | References | Remarks |
|---|---|---|---|---|---|---|---|---|---|---|---|---|---|---|
| | **$CH_2$** (*continued*) | | | | | | | | | | | | | |
| $\tilde{B}\ {}^3\Sigma_u^-$ | $D_{\infty h}$ | 70634 | | | | — | $[3.59_5]^f$ | — | $[1.079]^f$ | 180° | $\ldots(1\sigma_u)^2(1\pi_u)(3d\pi_g)$ | $\tilde{B}\leftarrow\tilde{X}$ 1415 Å | (521) | Diffuse for $CH_2$, sharp fine structure for $CD_2$, see Fig. 185 |
| $\tilde{X}\ {}^3\Sigma_g^-$ | $D_{\infty h}$ | 0 | | | | — | $[3.95_0]^f$ | — | $[1.029]^f$ | 180° | $\ldots(2\sigma_g)^2(1\sigma_u)^2(1\pi_u)^2$ | | | |
| | **$SiH_2$** | | | | | | | | | | | | | |
| $\tilde{A}\ {}^1B_1$ | | | | | | | | | | | $\ldots(b_2)^2(a_1)(b_1)$ | $\tilde{A}\leftarrow\tilde{X}$ 6000–5000 Å | (546) | Widely spaced bands with extensive fine structure |
| $(\tilde{X})\ {}^1A_1$ [a] | $C_{2v}$ | 0 | | | | | | | | | $\ldots(b_2)^2(a_1)^2$ | | | |
| | **$NH_2$** | I.P. = 11.4 eV [a] | | | | | | | | | | | | |
| $\tilde{A}\ {}^2A_1(\Pi)$ | $C_{2v}$ | 10249 [b] | 3325 | 633 [c] | | — | $8.7_8$ [d] | — | $1.00_4$ | (144°) [e] | $\ldots(1b_2)^2(3a_1)(1b_1)^2$ [f] | $\tilde{A}\leftrightarrow\tilde{X}$ 9000–4300 Å | (308) | Widely spaced bands with open fine structure rep. (308) [g] |
| $\tilde{X}\ {}^2B_1$ | $C_{2v}$ | 0 | | 1497.2 | | $23.72_8$ | $12.94_2$ | $8.16_9$ | 1.024 | 103.4° | $\ldots(1b_2)^2(3a_1)^2(1b_1)$ [f] | | (308)(338)(638a) | |

$CH_2$: [a] From (522); if $CH_2^+$ is non-linear in its ground state, as is $BH_2$, the value given for I.P. is only an upper limit. [b] (1033) gives $D_0 = 4.3_3$ eV.
[c] $\alpha_1 = 0.1_1$, $\alpha_2 = -0.01_9$. There is evidence for a slight potential hump in the linear conformation (530).
[d] Derived from ${}^1\Delta_g$ of the configuration $2\sigma_g^2 1\sigma_u^2 1\pi_u^2$ of linear $CH_2$.
[e] There is strong evidence that $\tilde{a}\ {}^1A_1$ is above $\tilde{X}\ {}^3\Sigma_g^-$ but the magnitude of $x$ is quite uncertain. It is unlikely to be larger than 8000 $cm^{-1}$.
[f] Refers to $CD_2$. The $B$ value of $CH_2$ is very nearly twice as large but has not been accurately determined because of the diffuseness of the lines.
$SiH_2$: [a] Not certain whether this is the ground state.
$NH_2$: [a] From mass spectrometric data of (389). [b] This (extrapolated) number refers to $v = 0$, a level that does not actually occur here.
[c] $x_{22}^0 = +11._4$, at higher $v_2$ higher order terms become important. [d] $\alpha_1 = 0.12$, $\alpha_2 = -0.024$. [e] The evidence for a slight potential hump in the linear conformation is now quite conclusive (286a).
[f] Derived from ${}^2\Pi$ of the configuration $2\sigma_g^2 1\sigma_u^2 1\pi_u^3$ of linear $NH_2$.
[g] So-called $\alpha$ bands of ammonia; appear in emission in flames and discharges involving $NH_3$ and in absorption in the flash photolysis of $NH_3$, $N_2H_4$ and other parent compounds [for references see (1043)]; observed in a solid matrix at 40°K by (1079).

TABLE 62 (*continued*)

| State | Point Group | $T_0$ | Vibrational Frequencies $\nu_1$ | $\nu_2$ | $\nu_3$ | Rotational Constants $A_0$ | $B_0$ | $C_0$ | $r_0$ (Å) | $\alpha$ | Electron Configuration | Observed Transitions | References | Remarks |
|---|---|---|---|---|---|---|---|---|---|---|---|---|---|---|
| **PH$_2$** | | | | | | | | | | | | | | |
| $\tilde{A}\,^2A_1$ | $C_{2v}$ | 18276.6 | | 951.3 | | 20.340 | 5.606 | 4.311 | 1.403 | 123.1° | $\ldots(b_2)^2(a_1)(b_1)^2$ | $\tilde{A}\leftrightarrow\tilde{X}$ 8520–[a] 3600 Å | (1042)(452) (286b) | Widely spaced bands with complicated fine structure |
| $\tilde{X}\,^2B_1$ | $C_{2v}$ | 0 | | 1102 | | 9.120 | 8.087 | 4.225 | 1.428 | 91.5° | $\ldots(b_2)^2(a_1)^2(b_1)$ | | | |
| **H$_2$O** | | I.P. = $12.61_8$ eV; $D$(H—OH) = $5.113_6$ eV (see Fig. 138) | | | | | | | | | | | | |
| | | Ionization continuum starting at 745 Å (134200 cm$^{-1}$) | | | | | | | | | | | (494)(836) | |
| $\tilde{S}$ | | 125820 | | (1700) | | | | | | | | $\tilde{S}\leftarrow\tilde{X}$ 795–750Å | (494)(1274) | Progressions of diffuse bands |
| $\tilde{R}$ | | 116760 | | (900) | | | | | | | | $\tilde{R}\leftarrow\tilde{X}$ 857–815Å | (494)(1274) | |
| | | Two merging strong Rydberg series joining on to $\tilde{E}$ and $\tilde{F}$; $\nu = 101780 - R/(n-0.05)^2$; $n = 3, 4, \ldots$ [interpreted by (631) as the $nd$ and $(n+1)s$ series] | | | | | | | | | | | (1015) | |
| | | Two merging weaker Rydberg series joining on to $\tilde{C}$ and $\tilde{D}$; $\nu = 101780 - R/(n-0.7)^2$; $n = 3, 4, \ldots$ | | | | | | | | | | | (1015) | |
| $\tilde{F}$ | $C_{2v}$ | 89680 | | | | | | | | | $\ldots(3a_1)^2(1b_1)(3d)$ | $\tilde{F}\leftarrow\tilde{X}$ 1115 Å | (1015)(631) | Single band |
| $\tilde{E}(^1B_1)$ | $C_{2v}$ | 88660 | | | | | | | | | $\ldots(3a_1)^2(1b_1)(4sa_1)$ | $\tilde{E}\leftarrow\tilde{X}$ 1128 Å | (1015)(631) | Single band |
| $\tilde{D}\,^1A_1$ | $C_{2v}$ | 82038 | 3268 | 1636 | (3335) | | | | | | $\ldots(3a_1)^2(1b_1)(3pb_1)$ | $\tilde{D}\leftarrow\tilde{X}$ 1219–1172 Å | (1015)(105a) | Diffuse bands |
| $\tilde{C}\,^1B_1$ | $C_{2v}$ | 80624.8 | 3170 | (1422)[a] | (3224)[a] | $25.6_7$ | $12.5_5$ | $8.5_5$ | $1.01_6$ | $106._9$° | $\ldots(3a_1)^2(1b_1)(3pa_1)$ | $\tilde{C}\leftarrow\tilde{X}$ 1241–1194 Å | (1015)(631) (1278)(105a) | See Fig. 112. Lines of higher $J$ are diffuse |
| $\tilde{B}(^1A_1)$ | $(D_{\infty h})$ | 70870[b] | | (800) | | | | | | | $\ldots(3a_1)(1b_1)^2(3sa_1)$ | $\tilde{B}\leftarrow\tilde{X}$ 1411–1256 Å | (1278) | Progression of very diffuse bands[c] |
| $\tilde{A}(^1B_1)$ | $C_{2v}$ | (53800) | continuous absorption with broad maximum at 1655 Å[d] | | | | | | | | $\ldots(1b_2)^2(3a_1)^2(1b_1)(3sa_1)$ | $\tilde{A}\leftarrow\tilde{X}$ 1860–1450 Å | (1302) (1278) | $k_{max} = 124$cm$^{-1}$ |
| $\tilde{X}\,^1A_1$ | $C_{2v}$ | 0 | $3657.0_5$ | $1594.7_8$ | $3755.7_9$ | 27.877[e] | 14.512[e] | 9.285[e] | 0.956[f] | 105.2°[f] | $\ldots(1b_2)^2(3a_1)^2(1b_1)^2$ | infrared sp. | (107)(265) | |
| **H$_2$S** | | I.P. = $10.47_2$ eV[a]; $D$(H—SH) = 3.26 eV[b] | | | | | | | | | | | | |
| | | Rydberg series joining on to $\tilde{A}$: $\nu = 84420 - R/(n-1.04)^2$; $n = 4, 5, \ldots, 13$ | | | | | | | | | | | (1015)(1024) | No vibrational structure |
| | | Rydberg series of three members joining on to $\tilde{F}$ and $\tilde{G}$ for which no formula is given | | | | | | | | | | | (1015) | |
| | | Rydberg series joining on to $\tilde{D}$: $\nu = 84520 - R/(n-1.57)^2$; $n = 4, 5, \ldots, 10$ | | | | | | | | | | | (1015)(1024) | |

PH$_2$: [a] In absorption 5500–3600 Å, in emission 8520–4540 Å.

H$_2$O: [a] Estimated from $\nu_2' = 1038._4$ for D$_2$O assuming valence force system (631). (105a) gives $\nu_1 = 3179$, $\nu_2 = 1407$, $\nu_3 = (3238)_{calc}$.

[b] Wave number of first diffuse band; origin of band system may be much lower. [c] Overlapped by continuum.

[d] (1302) give three maxima at 1718, 1648 and 1608 Å, but these are not observed in the photoelectric work of (1278) [see also (629)].

[e] $\alpha_1^A = 0.74_7$, $\alpha_2^A = -3.32_3$, $\alpha_3^A = 1.241$; $\alpha_1^B = 0.22_2$, $\alpha_2^B = -0.16_7$, $\alpha_3^B = 0.11_2$; $\alpha_1^C = 0.18_0$, $\alpha_2^C = 0.13_5$, $\alpha_3^C = 0.12_9$ cm$^{-1}$. Slightly different $A_0$, $B_0$, $C_0$ and $\alpha_i$ values are given by (106a).

[f] These are $r_0$ and $\alpha_0$ values. The equilibrium values, given by (106a), are $0.957_2$ Å and $104.5_2$°.

TABLE 62 (*continued*)

| State | Point Group | $T_0$ | Vibrational Frequencies $\nu_1$ | $\nu_2$ | $\nu_3$ | Rotational Constants $A_0$ | $B_0$ | $C_0$ | $r_0$ (Å) | $\alpha$ | Electron Configuration | Observed Transitions | References | Remarks |
|---|---|---|---|---|---|---|---|---|---|---|---|---|---|---|
| **$H_2S$** (*continued*) | | | | | | | | | | | | | | |
| $\tilde{H}\ ^1B_1$ | $(C_{2v})$ | 71895 | | | | | | | | | | $\tilde{H} \leftarrow \tilde{X}$ 1391 Å | (1015)(1274) | Sharp band with $P$, $Q$, $R$ branch |
| $\tilde{G}$ | $(C_{2v})$ | 71060 | | | | | | | | | | $\tilde{G} \leftarrow \tilde{X}$ 1407 Å | (1015)(1274) | Diffuse band, $D_1$ of (1015) |
| $\tilde{F}$ | $(C_{2v})$ | 69850 | | | | | | | | | | $\tilde{F} \leftarrow \tilde{X}$ 1432 Å | (1015)(1274) | $C_1$ of (1015)[c] |
| $\tilde{E}$ | $(C_{2v})$ | (66600)[d] | | | | | | | | | | $\tilde{E} \leftarrow \tilde{X}$ 1500 Å | (1015)(1274) | $B_1$ of (1015) |
| $\tilde{D}$ | $(C_{2v})$ | 65963 | | | | | | | | | | $\tilde{D} \leftarrow \tilde{X}$ 1516 Å | (1015)(1274) | $A_1$ of (1015) |
| $\tilde{C}$ | $(C_{2v})$ | 64711 | | | | | | | | | | $\tilde{C} \leftarrow \tilde{X}$ 1545 Å | (1015)(1274) (1024) | Diffuse band |
| $\tilde{B}$ | $(C_{2v})$ | 63330 | | | | | | | | | | $\tilde{B} \leftarrow \tilde{X}$ 1579 Å | (1015)(1024) (1274) | Diffuse band |
| $\tilde{A}$ | | 37000 | continuum with maximum at 1950 Å | | | | | | | | | $\tilde{A} \leftarrow \tilde{X}$ 2700–1900 Å | (430)(1274) | [h] |
| $\tilde{X}\ ^1A_1$ | $C_{2v}$ | 0 | 2614.6 | 1182.7 | (2627.5)[e] | 10.374[f] | 8.991[f] | 4.732[f] | 1.328[g] | 92.2°[g] | $\ldots(a_1)^2(b_1)^2$ | infrared sp. | (58) | |
| **$H_2Se$** | | I.P. = 9.881 eV | | | | | | | | | | | | |
| | | Long Rydberg series joining on to $\tilde{E}$: $\nu = 79703 - R/(n - 2.05)^2$; $n = 5, 7, \ldots, 19$ | | | | | | | | | | | (1024) | |
| | | Rydberg series joining on to $\tilde{D}$: $\nu = 79677 - R/(n - 2.15)^2$; $n = 5, 6, \ldots, 9$ | | | | | | | | | | | (1024) | |
| | | Rydberg series joining on to $\tilde{C}$, not well developed | | | | | | | | | | | (1024) | |
| $\tilde{E}$ | $(C_{2v})$ | 67098 | | | | | | | | | | $\tilde{E} \leftarrow \tilde{X}$ 1491 Å | (1024) | |
| $\tilde{D}$ | $(C_{2v})$ | 66179 | | | | | | | | | | $\tilde{D} \leftarrow \tilde{X}$ 1511 Å | (1024) | |

$H_2S$: [a] From Rydberg series (1015); from photoionization measurements (1272) obtains 10.458 eV. [b] From thermochemical data assuming $D(S_2) = 4.27$ eV.

[c] Band has line-like $Q$ branch, strong $R$ but no $P$ branch.

[d] Wave number not explicitly given, read from reproduction of (1015). [e] Calculated from $\nu_2 + \nu_3$, $2\nu_2 + \nu_3$, $\nu_1 + \nu_3$, ....

[f] $\alpha_1^A = 0.124_3$, $\alpha_2^A = -0.346_8$, $\alpha_3^A = 0.177_3$, $\alpha_1^B = 0.159_5$, $\alpha_2^B = -0.2198$, $\alpha_3^B = 0.1266$; $\alpha_1^C = 0.069_0$, $\alpha_2^C = 0.0620$, $\alpha_3^C = 0.056_7$.

[g] These values refer to the equilibrium position, i.e. are $r_e$ and $\alpha_e$ values.

[h] Overlapped by a few very diffuse bands.

TABLE 62 (*continued*)

| State | Point Group | $T_0$ | Vibrational Frequencies $\nu_1$ | $\nu_2$ | $\nu_3$ | Rotational Constants $A_0$ | $B_0$ | $C_0$ | $r_0$ (Å) | $\alpha$ | Electron Configuration | Observed Transitions | References | Remarks |
|---|---|---|---|---|---|---|---|---|---|---|---|---|---|---|
| **$H_2Se$** (*continued*) | | | | | | | | | | | | | | |
| $\tilde{C}$ | | 60600 | | | | | | | | | | $\tilde{C} \leftarrow \tilde{X}$ 1650 Å | (1024) | Strong discrete bands[a] |
| $\tilde{B}$ | | 59170 | | | | | | | | | | $\tilde{B} \leftarrow \tilde{X}$ 1690 Å | (1024) | Strong discrete bands[a] |
| $\tilde{A}$ | | | Diffuse absorption with maximum at 1970 Å | | | | | | | | | $\tilde{A} \leftarrow \tilde{X}$ | (1024) | No details available |
| $\tilde{X}\ ^1A_1$ | $C_{2v}$ | 0 | 2344.5 | 1034.2 | 2357.8 | 8.1703[b] | 7.7272[b] | 3.9013[b] | 1.460 | 91° | | infrared and microwave sp. | (552)(958) (614a) | |
| **$H_2Te$** | | I.P. = 9.138 eV | | | | | | | | | | | | |
| | | Long Rydberg series joining on to $\tilde{E}$: $\nu = 73705 - R/(n - 2.95)^2$; $n = 6, 7, \ldots, 13$ | | | | | | | | | | | (1024) | |
| | | Rydberg series joining on to $\tilde{B}$, not analyzed | | | | | | | | | | | (1024) | |
| $\tilde{E}$ | ($C_{2v}$) | 61559 | | | | | | | | | | $\tilde{E} \leftarrow \tilde{X}$ 1624 Å | (1024) | |
| $\tilde{D}$ | | 59207 | | | | | | | | | | $\tilde{D} \leftarrow \tilde{X}$ 1689 Å | (1024) | |
| $\tilde{C}$ | | (55000) | many unclassified bands | | | | | | | | | $\tilde{C} \leftarrow \tilde{X}$ 1830–1700 Å | (1024) | |
| $\tilde{B}$ | | 52190 | | | | | | | | | | $\tilde{B} \leftarrow \tilde{X}$ 1916–1830 Å | (1024) | [a] |
| $\tilde{A}$ | | | Diffuse absorption near 2000 Å | | | | | | | | | $\tilde{A} \leftarrow \tilde{X}$ | (1024) | |
| $\tilde{X}\ ^1A_1$ | $C_{2v}$ | 0 | (2000) | 860.79 | (2000) | $6.248_6$[b] | $6.097_0$[b] | $3.036_1$[b] | 1.653 | $90.2_5$° | | infrared sp. | (1086)(553) | |

$H_2Se$: [a] Vibrational structure observed but not analyzed.
[b] $\alpha_1^A = 0.12_5$, $\alpha_2^A = -0.23_0$, $\alpha_3^A = 0.14_5$, $\alpha_1^B = 0.09_2$, $\alpha_2^B = -0.19_0$, $\alpha_3^B = 0.06_2$, $\alpha_1^C = 0.05_5$, $\alpha_2^C = 0.04_0$, $\alpha_3^C = 0.04_5$ cm$^{-1}$.

$H_2Te$: [a] Vibrational structure observed but not analyzed.
[b] $\alpha_2^A = -0.182$, $\alpha_2^B = -0.129$, $\alpha_2^C = 0.030$.

TABLE 63

MOLECULAR CONSTANTS OF THE ELECTRONIC STATES OF TRIATOMIC MONOHYDRIDES

| State | Point Group | $T_0$ | Vibrational Frequencies $\nu_1$ | $\nu_2$ | $\nu_3$ | Rotational Constants $A_0$ | $B_0$ | $C_0$ | $r_0$(HX) | $r_0$(XY) | $\alpha$ | Electron Configuration | Observed Transitions | References | Remarks |
|---|---|---|---|---|---|---|---|---|---|---|---|---|---|---|---|
| **HCN** | | I.P. = 13.91 eV[a]; $D$(H—CN) = 5.6$_5$ eV[b]; $D$(HC—N) = 9.69 eV | | | | | | | | | | | | | |
| | | Very strong unclassified diffuse bands below 1120 Å | | | | | | | | | | | | | |
| $\tilde{D}$ | | 71629 | | (1038) | | | | | | | | | | (1029) | |
| $\tilde{C}\ {}^1A'$ | $C_s$ | 65644 | 2273 | 869 | (1530) | 35.$_8$ | | | (1.14) | | 141° | ...$(a')(a')^2(a'')^2(a')$[c] | $\tilde{C}\leftarrow\tilde{X}$ 1550–1350 Å | (528) (528) | Diffuse bands; $\gamma - X$ system of (527) see Fig. 88 |
| $\tilde{B}\ {}^1A''$ | $C_s$ | (54620)[d] | | [728.6][d] | | [12.0][d] | [1.157][d] | [1.043][d] | (1.14) | 1.334 | 114.5° | ...$(a')^2(a')(a'')^2(a'')$ | $\tilde{B}\leftarrow\tilde{X}$ 1830–1535 Å | (527) | $\beta - X$ system of (527) predissociation |
| $\tilde{A}\ {}^1A''$ | $C_s$ | 52256.4[e] | | 940.6 | 1495.$_9$ | 22.8 | 1.332 | 1.251 | 1.14$_0$ | 1.297 | 125.0° | ...$(a')^2(a')^2(a'')(a')$[f] | $\tilde{A}\leftarrow\tilde{X}$ 1915–1600 Å | (527) | $\alpha - X$ system of (527), see Figs. 82 and 87, predissociation[g] |
| $\tilde{X}\ {}^1\Sigma^+$ | $C_{\infty v}$ | 0 | 3311.47 | 713.46 | 2096.7[h] | — | 1.47822[i] | — | 1.064[j] | 1.156[j] | 180° | ...$\sigma^2\pi^4$ | infrared and microwave sp. | (1048) (1136) (301) | |
| **HCP** | | | | | | | | | | | | | | | |
| $\tilde{D}\ {}^1\Sigma^+$ | $C_{\infty v}$ | 40255 | | (615) | (970) | — | (0.61) | — | | | (180°) | | $\tilde{D}\leftarrow\tilde{X}$ 2490–2360 Å | (639) | |
| $\tilde{C}$ ? | | 35980 | | | 950 | | | | | | | | $\tilde{C}\leftarrow\tilde{X}$ 2780–2650 Å | (639) | |
| $\tilde{B}\ {}^1A'$ | $C_{\infty v}$ | 35926.3 | | | 964 | — | 0.600$_5$ | — | | | 180° | ...$\sigma^2\pi^3\pi^*$ | $\tilde{B}\leftarrow\tilde{X}$ 2780–2580 Å | (639) | |
| $\tilde{A}\ {}^1A''$ | $C_s$ | 34769.9 | | 566.6 | 950.9 | ~24 | 0.589 | 0.577 | (1.14) | 1.69 | 128° | ...$(a')^2(a')^2(a'')(a')$ | $\tilde{A}\leftarrow\tilde{X}$ 3000–2300 Å | (639) | Extensive system of discrete bands |
| $\tilde{a}\ {}^3\Sigma^+$ | ($C_{\infty v}$) | 24400 | 2720 | 440 | 950 | — | (0.576) | | | | (180°) | | $\tilde{a}\leftarrow\tilde{X}$ 4100–3050 Å | (639) | $\lambda + \frac{1}{2}\gamma = -1.1_0\ \mathrm{cm}^{-1}$ |
| $\tilde{X}\ {}^1\Sigma^+$ | $C_{\infty v}$ | 0 | 3216.9$_0$ | 674.2$_5$ | 1278.2$_3$ | — | 0.66625[a] | — | 1.067 | 1.542 | 180° | ...$\sigma^2\pi^4$ | microwave and infrared sp. | (1228) (639) | |

HCN: [a] From electron impact experiments of (881). [b] Using $D$(CN) = 7.5$_0$ eV after (112). [c] Derived from $\sigma\pi^4\pi^*$.
[d] Refers to DCN since only fragments of this system have been found for HCN. [e] This is $\nu_0$ of the 000 – 000 band less $(A - \bar{B})$ since the upper state of this band has $K = 1$.
[f] Derived from $\sigma^2\pi^3\pi^*$. [g] See Table 47. [h] Quoted without explanation by (60). (301) give from Raman measurements: 2095.5.
[i] $\alpha_1 = 0.0104_2$, $\alpha_2 = -0.0036_1$, $\alpha_3 = 0.0095$, $D_{000} = 2.914 \times 10^{-6}$. [j] For $r_e$ (301) give 1.0657 and 1.1530 Å.

HCP: [a] $\alpha_1 = 0.0031$, $\alpha_2 = -0.0003_4$, $\alpha_3 = 0.003_4$.

TABLE 63 (*continued*)

| State | Point Group | $T_0$ | Vibrational Frequencies $\nu_1$ | $\nu_2$ | $\nu_3$ | Rotational Constant₁ $A_0$ | $B_0$ | $C_0$ | $r_0$(HX) | $r_0$(XY) | $\alpha$ | Electron Configuration | Observed Transitions | References | Remarks |
|---|---|---|---|---|---|---|---|---|---|---|---|---|---|---|---|
| **HCO** | | I.P. = 9.88 eV[a]; 1.54 eV[b] > $D$(H—CO) ≥ $1.0_4$ eV[c] | | | | | | | | | | | | | |
| $\tilde{C}$ | $C_s$ | (35000) | Not yet fully analyzed | | | | | | | | | | $\tilde{C} \rightarrow (\tilde{X})$ 4100–2600 Å | (1233) (567) (1234) | Hydrocarbon flame bands[d] |
| $\tilde{B}\ ^2A'$ | | | Continuum underlying $\tilde{A} \leftarrow \tilde{X}$ bands | | | | | | | | | | $\tilde{B} \leftarrow \tilde{X}$ | (638) | Possibly formed by diffuse bands of $\tilde{A} \leftarrow \tilde{X}$ |
| $\tilde{A}\ ^2A''(\Pi)$ | $C_{\infty v}$ | $9294._0$ | $3316._2$ | 802.3 | 1813.4 | — | $1.337_7$ | — | $1.04_4$[g] | 1.187[g] | 180° | $\ldots\sigma^2\pi^4\pi^*$[e] | $\tilde{A} \leftarrow \tilde{X}$ 8600–4600 Å | (538) (638) | See Fig. 92. All bands with $l' > 0$ are diffuse |
| $\tilde{X}\ ^2A'$ | $C_s$ | 0 | (2700)[f] | $1083._0$ | $1820._2$ | $22.36_5$ | $1.494_4$ | $1.400_8$ | (1.08) | $1.19_8$ | 119.5° | $\ldots(a')^2(a'')^2(a')$[e] | | | |
| **HNO** | | $D$(H—NO) ≤ 2.11 eV | | | | | | | | | | | | | |
| $\tilde{A}\ ^1A''$ | $C_s$ | 13154.4 | 2854.2 | 1420.8 | 981.2 | 22.164[a] | 1.3255[a] | 1.2426[a] | 1.036 | 1.241 | 116.3° | $\ldots(a')^2(a'')^2(a')(a'')$[b] | $\tilde{A} \leftrightarrow \tilde{X}$ 7700–5500 Å | (264) (93) (204) | Extensive fine structure (see Fig. 109); prediss. above 16994 $cm^{-1}$[c] |
| $(\tilde{X})$[d] $^1A'$ | $C_s$ | 0 | 3596 | 1562 | 1110 | $18.479_2$ | 1.4115 | 1.3071 | 1.063 | 1.212 | 108.6° | $\ldots(a')^2(a'')^2(a')^2$[b] | | | |
| **HPO** | | | | | | | | | | | | | | | |
| $\tilde{A}\ ^1A''$ | $C_s$ | 19032.8 | (2308) | 1309 | 865 | 8.273 | 0.6418 | 0.5925 | | | | | $\tilde{A} \rightarrow \tilde{X}$ 6800–4650 Å | (716) (715a) | Widely spaced bands with extensive fine structure |
| $\tilde{X}\ ^1A'$ | $C_s$ | 0 | | 1187 | 985 | 8.855 | 0.7024 | 0.6488 | (1.433) | 1.512 | 104.7° | | | | |

HCO: [a] From mass spectrometric studies by (1062). [b] From predissociation. [c] From photochemical experiments on $H_2CO$ by (678)(1246).
[d] Complicated fine structure; not certain that lower state is ground state of HCO. [e] These two states probably arise from the same $\sigma^2\pi^4\pi^*\ ^2\Pi$ state of linear HCO.
[f] (851) give 2488 $cm^{-1}$ in a CO matrix at 14°K. [g] These are $r_e$ values.

HNO: [a] $A_e$ = 22.158, $B_e$ = 1.3300, $C_e$ = 1.2540 are given by (93). [b] The two observed states are probably derived from the $\sigma^2\pi^4\pi^{*2}\ ^1\Delta$ state of the linear conformation.
[c] Observed in emission in the reaction of H with NO by (183) and (204) (see Fig. 177).
[d] Not certain that this is the ground state. The triplet state $\sigma^2\pi^4\pi^{*2}\ ^3\Sigma^-$ or its analogue in the bent conformation may lie lower but has not been observed.

TABLE 63 (*continued*)

| State | Point Group | $T_0$ | Vibrational Frequencies $\nu_1$ | $\nu_2$ | $\nu_3$ | Rotational Constants $A_0$ | $B_0$ | $C_0$ | $r_0$(HX) | $r_0$(XY) | $\alpha$ | Electron Configuration | Observed Transitions | References | Remarks |
|---|---|---|---|---|---|---|---|---|---|---|---|---|---|---|---|
| **HCF** | | | | | | | | | | | | | | | |
| $\tilde{A}\ ^1A''$ | $C_s$ | 17287 | | 1021 | | $25.4_5$[a] | 1.160 | 1.104 | (1.121)[b] | 1.297 | 127.2° | | $\tilde{A} \leftarrow \tilde{X}$ 6000–4500 Å | (825) | |
| $\tilde{X}\ ^1A'$ | $C_s$ | 0 | | 1403 | | $15.5_5$[c] | 1.221 | 1.126 | (1.121)[b] | 1.314 | 101.6° | | | | |
| **HCCl$^{35}$** | | | | | | | | | | | | | | | |
| $\tilde{A}\ ^1A''$ | $C_s$ | (12288)[a] | | 870 | | | $0.609_3$[b] | | | | 134° | | $\tilde{A} \leftarrow \tilde{X}$ 8200–5500 Å | (829) | |
| $\tilde{X}\ ^1A'$ | $C_s$ | 0 | | | | 15.75 | 0.6054 | 0.5882 | 1.12 | $1.68_9$ | 103.4° | | | | |
| **HSiCl$^{35}$** | | | | | | | | | | | | | | | |
| $\tilde{a}(^3A'')$[a] | $C_s$ | 20717.7 | (1250) | $563._7$ | $532._6$ | 9.857 | [0.2464][b] | [0.2404][b] | $1.49_9$ | $2.04_7$ | 116.1° | | $\tilde{a} \leftrightarrow \tilde{X}$ 6100–4100 Å[c] | (545) | Bands with widely spaced $K$ structure |
| $\tilde{X}\ ^1A'$ | $C_s$ | 0 | | $805._5$ | $522._4$ | 7.587 | [0.2461][b] | [0.2383][b] | 1.561 | 2.064 | 102.8° | | | | |
| **HSiBr$^{79}$** | | | | | | | | | | | | | | | |
| $\tilde{a}(^3A'')$[a] | $C_s$ | 19903.1 | (1270) | $535._1$ | $412._2$ | 9.906 | [0.1589][b] | [0.1563][b] | $(1.49_9)$ | 2.208 | 116.6° | | $\tilde{a} \leftrightarrow \tilde{X}$ 6200–4300 Å[c] | (545) | Bands with widely spaced $K$ structure |
| $\tilde{X}(^1A')$ | $C_s$ | 0 | $1547._8$ | $771._4$ | $408._0$ | 7.580 | [0.1578][b] | [0.1546][b] | $(1.56_1)$ | 2.231 | 102.9° | | | | |
| **$HO_2$** | see (215)[a] I.P. = $11.5_3$ eV[b] $D(\text{H—O}_2) = 1.99$ eV[b]; $D(\text{HO—O}) = 2.7_4$ eV[c] | | | | | | | | | | | | | | |
| **$HS_2$** | see (1007) | | | | | | | | | | | | | | |
| **HClO** | see (378) | | | | | | | | | | | | | | |

HCF: [a] $D_K = 0.031\ \text{cm}^{-1}$. [b] Assumed. [c] $D_K = 0.0024\ \text{cm}^{-1}$.

HCCl: [a] Extrapolated. [b] For $K = 0$, $v_2 = 2$.

HSiCl$^{35}$: [a] It is possible that this state is $^1A''$ (see p. 270). [b] From $A$ and $\bar{B}$ assuming zero inertial defect. [c] Absorption 4800–4100 Å, fluorescence 6100–4800 Å.

HSiBr$^{79}$: [a] It is possible that this state is $^1A''$ (see p. 270). [b] From $A$ and $\bar{B}$ assuming zero inertial defect. [c] Absorption 5000–4300 Å, fluorescence 6200–5000 Å.

$HO_2$: [a] Infrared bands of $HO_2$ have been observed in an Ar matrix by (850). The electronic bands reported in (288) have turned out to be an artefact. [b] From mass spectrometric studies of (390). [c] From (444a).

TABLE 64

MOLECULAR CONSTANTS OF THE ELECTRONIC STATES OF TRIATOMIC NON-HYDRIDES

| State | Point Group | $T_0$ | Vibrational Frequencies $\nu_1$ | $\nu_2$ | $\nu_3$ | Rotational Constants $A_0$ | $B_0$ | $C_0$ | $r_0$(XY) | $r_0$(YZ) | $\alpha$ | Electron Configuration | Observed Transitions | References | Remarks |
|---|---|---|---|---|---|---|---|---|---|---|---|---|---|---|---|
| **$C_3$** | | | | | | | | | | | | | | | |
| $\tilde{A}\ ^1\Pi_u$ | $D_{\infty h}$ | 24675.5 | 1085.9 | $307._9$ [a] | | — | 0.4124 [b] | — | 1.305 | | 180° | $\ldots(1\pi_u)^4(3\sigma_u)(1\pi_g)$ | $\tilde{A}\leftrightarrow\tilde{X}$ 4100–3400 Å | (519)(292)(211)(411) | 4050 Å group of comets [c]; Fig. 72 |
| $\tilde{X}\ ^1\Sigma_g^+$ | $D_{\infty h}$ | 0 | (1230) [d] | 63.1 | 2040 [e] | — | 0.4305 [f] | — | 1.277 | | 180° | $\ldots(1\pi_u)^4(3\sigma_u)^2$ | | | |
| **SiCC** | | | | | | | | | | | | | | | |
| $\tilde{A}(^1\Pi)$ | $(C_{\infty v})$ | 20085.1 | 1461 | [a] | 456 [a] | | | | | | | | $\tilde{A}\leftrightarrow\tilde{X}$ 5635–4260 Å | (813)(679) | Merrill–Sanford bands [b]; repr. (679) |
| $\tilde{X}\ ^1\Sigma^+$ | $C_{\infty v}$ | 0 | 1742 | [a] | 591 [a] | | | | | | | | | | |
| **CCN** | $D$(CCN) < 3.61 eV | | | | | | | | | | | | | | |
| $\tilde{C}\ ^2\Sigma^+$ | $C_{\infty v}$ | 26661.73 | 1859.20 | (465) | | — | 0.4129 [a] | — | | | 180° | $\ldots\sigma^2\pi^4\sigma\pi^2$ | $\tilde{C}\leftarrow\tilde{X}$ 3750–3480 Å | (826) | Predissociation above 29100 cm$^{-1}$ |
| $\tilde{B}(^2\Sigma^-)$ | $C_{\infty v}$ | 22413.25 | | | | — | 0.4051 | — | | | 180° | $\ldots\sigma^2\pi^4\sigma\pi^2$ | $\tilde{B}\leftarrow\tilde{X}$ 4465–4450 Å | (826) | Single band |
| $\tilde{A}\ ^2\Delta_i$ | $C_{\infty v}$ | 21259.15 [b] | 1770.77 | ~475 | 1241.64 | — | 0.4137 [c] | — | | | 180° | $\ldots\sigma^2\pi^4\sigma\pi^2$ | $\tilde{A}\leftarrow\tilde{X}$ 4710–3770 Å | (826) | |
| $\tilde{X}\ ^2\Pi_r$ | $C_{\infty v}$ | 0 [d] | | (325) [e] | | — | 0.3981 [f] | — | | | 180° | $\ldots\sigma^2\pi^4\sigma^2\pi$ | | (826) | |

$C_3$: [a] Because of Renner–Teller interaction (with $\epsilon = 0.537$) the 010 level is split into three components at 136.3, 259.2 and 480.8 cm$^{-1}$ above the 000 level. [b] $q = -0.0004_2$ cm$^{-1}$.
[c] A continuum observed in emission and absorption at high temperature in the same region is, according to (148), produced by overlapping bands of this system.
[d] From the fluorescence spectrum in an argon matrix by (1287). [e] From matrix infrared absorption by (1289).
[f] $\alpha_2 = -0.0123$; strong $l$-uncoupling in the higher $0v_20$ levels; $D_0 = 1.2 \times 10^{-6}$ cm$^{-1}$.

SiCC: [a] On the basis of matrix studies (1287a) have reassigned the gas spectra and give in the $\tilde{A}$ state $\nu_2 = 230$, $\nu_3 = 1015$; in the $\tilde{X}$ state $\nu_2 = 300$, $\nu_3 = 853$ cm$^{-1}$.
[b] First observed in low temperature carbon stars; see (813).

CCN: [a] $\alpha_1 = 0.0035$, $\alpha_2 = -0.0005$. [b] $A = -0.8$ cm$^{-1}$. [c] $\alpha_1 = 0.0034$, $\alpha_2 = -0.0005$, $\alpha_3 = 0.0015$. [d] $A = 40.34$ cm$^{-1}$. [e] $\epsilon\omega_2 \sim 144$ cm$^{-1}$; $\epsilon \sim +0.44$. [f] $\alpha_2 = -0.0023$.

TABLE 64 (*continued*)

| State | Point Group | $T_0$ | Vibrational Frequencies $\nu_1$ | $\nu_2$ | $\nu_3$ | Rotational Constants $A_0$ | $B_0$ | $C_0$ | $r_0$(XY) | $r_0$(YZ) | $\alpha$ | Electron Configuration | Observed Transitions | References | Remarks |
|---|---|---|---|---|---|---|---|---|---|---|---|---|---|---|---|
| **CNC** | | | | | | | | | | | | | | | |
| $\tilde{B}\ ^2\Sigma_u^-$ | $D_{\infty h}$ | 34802.3 | | 398 | | — | 0.4430[a] | — | 1.259 | | 180° | $\ldots(1\pi_u)^4(3\sigma_u)(1\pi_g)^2$ | $\tilde{B}\leftarrow\tilde{X}$ 2880–2830 Å | (828) | |
| $\tilde{A}\ ^2\Delta_u$ | $D_{\infty h}$ | 30338.5[b] | | 440 | | — | 0.4504[c] | — | 1.249 | | 180° | $\ldots(1\pi_u)^4(3\sigma_u)(1\pi_g)^2$ | $\tilde{A}\leftarrow\tilde{X}$ 3320–3250 Å | (828) | |
| $\tilde{X}\ ^2\Pi_g$ | $D_{\infty h}$ | 0[d] | | 321[e] | | — | 0.4535[f] | — | 1.245 | | 180° | $\ldots(1\pi_u)^4(3\sigma_u)^2(1\pi_g)$ | | | |
| **NCN** | | | | | | | | | | | | | | | |
| $\tilde{A}\ ^3\Pi_u$ | $D_{\infty h}$ | 30383.7[a] | [b] | (510)[c] | [b] | — | 0.3962 | — | 1.233 | | 180° | $\ldots(1\pi_u)^4(3\sigma_u)(1\pi_g)^3$ | $\tilde{A}\leftrightarrow\tilde{X}$ 3295–3260 Å | (543) | Emission and absorption bands with narrow fine structure[d] |
| $\tilde{X}\ ^3\Sigma_g^-$ | $D_{\infty h}$ | 0 | | (423)[f] | (1475)[f] | — | 0.3968[e] | — | 1.232 | | 180° | $\ldots(1\pi_u)^4(3\sigma_u)^2(1\pi_g)^2$ | | | |
| **NCO** | $D$(NC—O) $\leq$ 4.2 eV | | | | | | | | | | | | | | |
| $\tilde{B}\ ^2\Pi$ | $C_{\infty v}$ | 31753.1[a] | 2303 | | 1047 | — | [0.3765][b] | — | $r$(NC)+$r$(CO) $\leq$2.45 | | 180° | $\ldots\pi^3\sigma^2\pi^4$ | $\tilde{B}\leftarrow\tilde{X}$ 3200–2650 Å | (282) | Red degraded bands; diffuse above 33700 cm$^{-1}$ |
| $\tilde{A}\ ^2\Sigma^+$ | $C_{\infty v}$ | 22754.0 | 2338.0 | 680.8 | 1289.3[c] | — | $0.4021_1$[d] | — | $\leq$2.369 | | 180° | $\ldots\pi^4\sigma\pi^4$ | $\tilde{A}\leftrightarrow\tilde{X}$ 4500–3600 Å | (564)(281) | Violet degraded bands with narrow fine structure. See Figs. 74 and 75 |
| $\tilde{X}\ ^2\Pi_i$ | $C_{\infty v}$ | 0[e] | | (539)[f] | | — | $0.3894_0$[g] | — | $\leq$2.408 | | 180° | $\ldots\pi^4\sigma^2\pi^3$ | | | |

CNC: [a] $\gamma_0 = 0.030$ cm$^{-1}$. [b] $A = +0.33$. [c] $\alpha_2 = -0.0016$. [d] $A = +26.41$. [e] $\epsilon = 0.549$. [f] $\alpha_2 = -0.0037$.

NCN: [a] $A = -37.56$ cm$^{-1}$. [b] $\nu_1' - \nu_1'' = -5$, $\nu_3' - \nu_3'' = 58$ cm$^{-1}$. [c] Renner–Teller splitting with $\epsilon\omega_2 = -85.7$.
[d] First observed by (627); considered to be due to CNO by (812); see, however (544). A singlet transition ($^1\Pi_u - {}^1\Delta_g$) has recently been observed by Kroto (699aa). [e] $\lambda = 0.784$, $\gamma = -0.001$.
[f] Observed in Ar matrix by (851a).

NCO: [a] $A = -30.8$ cm$^{-1}$. [b] For the 1,0,0 vibrational level. [c] In Fermi resonance with $2\nu_2$ at 1385.3; (281) gives $\omega_1^0 = 1324.3$. [d] $\alpha_1 = 0.00308$, $\alpha_2 = -0.00057$, $\alpha_3 = 0.00150$.
[e] $A = -95.59$ cm$^{-1}$. [f] Fairly large Renner–Teller splitting with $\epsilon = -0.182$; the 010 level is split into three components at 441.4, 533.6 and 637.4 cm$^{-1}$. [g] $\alpha_2 = -0.0018_0$.

TABLE 64 (*continued*)

| State | Point Group | $T_0$ | Vibrational Frequencies $\nu_1$ | $\nu_2$ | $\nu_3$ | Rotational Constants $A_0$ | $B_0$ | $C_0$ | $r_0(XY)$ | $r_0(YZ)$ | $\alpha$ | Electron Configuration | Observed Transitions | References | Remarks |
|---|---|---|---|---|---|---|---|---|---|---|---|---|---|---|---|
| **NCS** | | see (564)(288a) | | | | | | | | | | | | | |
| **$N_2O^+$** | | | | | | | | | | | | | | | |
| $\tilde{C}$ | | 58020 | | | | | | | | | | | | (1190) | From Rydberg series of $N_2O$. Repr. (566)(173)[b] |
| $\tilde{A}\ ^2\Sigma^+$ | $C_{\infty v}$ | $28163.3_5$ | $2451._7$ | 614 | $1345._5$ | — | 0.43300[a] | — | 1.140 | 1.141 | 180° | $\ldots\pi^4\sigma\pi^4$ | $\tilde{A}\rightarrow\tilde{X}$ 4210–3260 Å | (152)(173) (566)(174) | |
| $\tilde{X}\ ^2\Pi_i$ | $C_{\infty v}$ | 0[c] | 1736.6 | 461.2[d] | 1126.4 | — | 0.41159[e] | — | 1.155 | 1.185 | 180° | $\ldots\pi^4\sigma^2\pi^3$ | | | |
| **$N_3$** | | $D(N—N_2) < 4.55$ eV | | | | | | | | | | | | | |
| $\tilde{B}\ ^2\Sigma_u^+$ | $(D_{\infty h})$ | 36739.1 | | | | — | $0.4323_8$[a] | — | $1.179_9$ | | 180° | $\ldots(\sigma_u)(\pi_u)^4(\pi_g)^4$ | $\tilde{B}\leftarrow\tilde{X}$ 2725–2600 Å | (1217)(297) | Only 0–0 band is sharp |
| $\tilde{X}\ ^2\Pi_g$ | $D_{\infty h}$ | 0[b] | | c | | — | 0.43117[d] | — | $1.181_5$ | | 180° | $\ldots(\sigma_u)^2(\pi_u)^4(\pi_g)^3$ | | | |
| **$BO_2$** | | | | | | | | | | | | | | | |
| $\tilde{B}\ ^2\Sigma_u^+$ | $D_{\infty h}$ | $24507.9_9$ | | 505 | (1410)[a] | — | 0.3250 | — | $1.273_3$ | | 180° | $\ldots(3\sigma_u)(1\pi_u)^4(1\pi_g)^4$ | $\tilde{B}\leftarrow\tilde{X}$ 4092–4050 Å | (630) | |
| $\tilde{A}\ ^2\Pi_u$ | $D_{\infty h}$ | $18291.5_9$[b] | 994 | 502[c] | 2357[d] | — | 0.3106 | — | $1.302_5$ | | 180° | $\ldots(3\sigma_u)^2(1\pi_u)^3(1\pi_g)^4$ | $\tilde{A}\leftrightarrow\tilde{X}$ 6450–3965 Å | (630) | Red-degraded bands with narrow fine structure.[e] See Fig. 78 |
| $\tilde{X}\ ^2\Pi_g$ | $D_{\infty h}$ | 0[f] | 1070 | 464[g] | 1322[d] | — | 0.3292 | — | $1.265_2$ | | 180° | $\ldots(3\sigma_u)^2(1\pi_u)^4(1\pi_g)^3$ | | (630) | |

$N_2O^+$: [a] $\alpha_1 = 0.00348$, $\alpha_3 = 0.00202$, $D_0 = 1.7_3 \times 10^{-7}$. [b] According to (269a) the $f$ value of this transition is 0.00365. [c] $A_0 = -132.36$. [d] $\epsilon = -0.19_3$. [e] $\alpha_1 = 0.00264$, $\alpha_3 = 0.00351$; $D_0 = 2.0 \times 10^{-7}$.

$N_3$: [a] $D_0 = 1.6 \times 10^{-7}$. [b] $A_{eff.} = -71.3$. [c] $\epsilon\omega_2 = -94.38$. [d] $D_0 = 1.5 \times 10^{-7}$.

$BO_2$: [a] Estimated from isotope effect. [b] $A = -101.3_5$. [c] $\epsilon\omega_2 = -13.1$. [d] These values are $\frac{1}{2}(2\nu_3)$. [e] Observed in emission as the "boric acid fluctuation bands" in flames and electric arcs, extending to 6450 Å. [f] $A = -148.5_8$. [g] $\epsilon\omega_2 \simeq -92._2$.

TABLE 64 (*continued*)

| State | Point Group | $T_0$ | Vibrational Frequencies $\nu_1$ | $\nu_2$ | $\nu_3$ | Rotational Constants $A_0$ | $B_0$ | $C_0$ | $r_0(XY)$ | $r_0(YZ)$ | $\alpha$ | Electron Configuration | Observed Transitions | References | Remarks |
|---|---|---|---|---|---|---|---|---|---|---|---|---|---|---|---|
| **$CO_2^+$** | | | | | | | | | | | | | | | |
| $\tilde{C}\ ^2\Sigma_g^+$ | $D_{\infty h}$ | 45320 | | | | | | | | | | | | (1189) | From Rydberg series of $CO_2$ |
| $\tilde{B}\ ^2\Sigma_u^+$ | $D_{\infty h}$ | 34597.9 | | | | — | 0.3784[a] | — | $1.180_1$ | | 180° | $\ldots(3\sigma_u)(1\pi_u)^4(1\pi_g)^4$ | $\tilde{B}\leftarrow\tilde{X}$ 2905–2875 Å | (160)(883) | Single strong band |
| $\tilde{A}\ ^2\Pi_u$ | $D_{\infty h}$ | 28500.6[b] | 1131 | | 2731[e] | — | 0.3493 | — | $1.228_3$ | | 180° | $\ldots(3\sigma_u)^2(1\pi_u)^3(1\pi_g)^4$ | $\tilde{A}\leftrightarrow\tilde{X}$ 4900–2900 Å | (882)(632) | Extensive band system |
| $\tilde{X}\ ^2\Pi_g$ | $D_{\infty h}$ | 0[c] | 1280 | [d] | 1469[e] | — | 0.3804 | — | $1.177_0$ | | 180° | $\ldots(3\sigma_u)^2(1\pi_u)^4(1\pi_g)^3$ | | (882)(632) | |
| **$OCS^+$** | | | | | | | | | | | | | | | |
| $\tilde{A}\ ^2\Pi$ | $C_{\infty v}$ | 31281[a] | | | | | | | | | | $\ldots\sigma^2\pi^3\pi^4$ | $\tilde{A}\rightarrow\tilde{X}$ 4320–3180 Å | (731aa) (566a) | |
| $\tilde{X}\ ^2\Pi_i$ | $C_{\infty v}$ | 0[b] | 2069 | | | | | | | | | $\ldots\sigma^2\pi^4\pi^3$ | | | |
| **$CS_2^+$** | | | | | | | | | | | | | | | |
| $\tilde{C}(^2\Sigma_g^+)$ | $(D_{\infty h})$ | 49068 | | | | | | | | | | | | (1190) | From Rydberg series of $CS_2$ |
| $\tilde{B}\ ^2\Sigma_u^+$ | $D_{\infty h}$ | 35238.2 | | | | — | 0.10776[a] | — | $1.554_2$ | | 180° | $\ldots(\pi_u)^4(\sigma_u)(\pi_g)^4$ | $\tilde{B}\rightarrow\tilde{X}$ 2910–2808 Å | (172) | |
| $\tilde{X}\ ^2\Pi_g$ | $D_{\infty h}$ | 0[b] | (624) | [204.8][c] | | — | 0.10914[d] | — | $1.564_1$ | | 180° | $\ldots(\pi_u)^4(\sigma_u)^2(\pi_g)^3$ | | | |

$CO_2^+$: [a] $\gamma = 0.019$. [b] $A = -95.5$. [c] $A = -159.5$. [d] Only the product $\epsilon\omega_2 = -93\ cm^{-1}$ is known. [e] These values are $\frac{1}{2}(2\nu_3)$.
$OCS^+$: [a] $A = -121$. [b] $A = -372$.
$CS_2^+$: [a] $D = 1.4 \times 10^{-8}$, $\gamma = +0.0043$. [b] $A = -440.39\ cm^{-1}$. [c] $\frac{1}{2}(2\nu_2)$. [d] $\alpha_2 = -0.00010$.

TABLE 64 (*continued*)

| State | Point Group | $T_0$ | Vibrational Frequencies $\nu_1$ $\nu_2$ $\nu_3$ | Rotational Constants $A_0$ $B_0$ $C_0$ $r_0$(XY) $r_0$(YZ) $\alpha$ | Electron Configuration | Observed Transitions | References | Remarks |
|---|---|---|---|---|---|---|---|---|
| | **$N_2O$** | I.P. = $12.89_3$ eV; $D(N_2—O) = 1.677_1$ eV; $D(N—NO) = 4.930_3$ eV | | | | | | |
| | | Rydberg series starting at 149600 $cm^{-1}$: | | $\nu = 162200 - R/(n - 0.06)^2$; | $n = 3, 4, \ldots, 11$ | | (1190) | |
| | | Rydberg series starting at 147340 $cm^{-1}$: | | $\nu = 162130 - R/(n - 0.31)^2$; | $n = 3, 4, \ldots, 11$ | | (1190) | |
| | | Rydberg series starting at 142900 $cm^{-1}$: | | $\nu = 162160 - R/(n - 0.58)^2$; | $n = 3, 4, 5, 6$ | | (1190) | apparent emission |
| | | Rydberg series starting at 141580 $cm^{-1}$: | | $\nu = 162200 - R/(n - 0.68)^2$; | $n = 3, 4, 5, 6$ | | (1190) | apparent emission |
| | | Rydberg series joining on to $\tilde{N}$: | | $\nu = 132250 - R/(n - 0.22)^2$; | $n = 3, 4, \ldots, 8$ | | (1190) } | Converging to |
| | | Rydberg series joining on to $\tilde{K}$: | | $\nu = 132210 - R/(n - 1.0)^2$; | $n = 3, 4, \ldots, 13$ | | (1190) } | $\tilde{A}\ ^2\Sigma^+$ of $N_2O^+$ |
| $\tilde{N}$ | $C_{\infty v}$ | 117930 | | | | $\tilde{N} \leftarrow \tilde{X}$ 848 Å | (1190) | |
| | | Rydberg series joining on to $\tilde{F}$: | | $\nu = 104300 - R/(n - 0.68)^2$; | $n = 3, 4, \ldots, 9$ | | (1190) | Converging to $\tilde{X}\ ^2\Pi_{\frac{1}{2}}$ of $N_2O^+$ |
| | | Rydberg series joining on to $\tilde{E}$: | | $\nu = 104000 - R/(n - 0.60)^2$; | $n = 3, 4, \ldots, 13$ | | (1190) | Converging to $\tilde{X}\ ^2\Pi_{\frac{3}{2}}$ of $N_2O^+$ |
| $\tilde{K}$ | $C_{\infty v}$ | 99710 | (1280)[a] | | | $\tilde{K} \leftarrow \tilde{X}$ 1003 Å | (1190) | |
| $\tilde{F}$ | $C_{\infty v}$ | 84960 | accompanied by weaker diffuse bands at shorter $\lambda$ | | | $\tilde{F} \leftarrow \tilde{X}$ 1177–1080 Å | (1330)(320) | $k_{max} = 3010$ |
| $\tilde{E}$ | $C_{\infty v}$ | 84633 | accompanied by weaker diffuse bands at shorter $\lambda$ | | | $\tilde{E} \leftarrow \tilde{X}$ 1182–1080 Å | (1330) | |

$N_2O$: [a] Only observed for higher members of Rydberg series.

TABLE 64 (*continued*)

| State | Point Group | $T_0$ | Vibrational Frequencies $\nu_1$ | $\nu_2$ | $\nu_3$ | Rotational Constants $A_0$ | $B_0$ | $C_0$ | $r_0(XY)$ | $r_0(YZ)$ | $\alpha$ | Electron Configuration | Observed Transitions | References | Remarks |
|---|---|---|---|---|---|---|---|---|---|---|---|---|---|---|---|
| **$N_2O$** (*continued*) | | | | | | | | | | | | | | | |
| $\tilde{D}$ | $C_{\infty v}$ | 77400 | | | | | | | | | | | $\tilde{D} \leftarrow \tilde{X}$ 1292 Å | (1330) | Single sharp band |
| $\tilde{C}$ | | 74000 | strong continuum with maximum at 1284 Å and, superimposed, very diffuse bands | | | | | | | | | | $\tilde{C} \leftarrow \tilde{X}$ 1350–1230 Å | (320)(1330) | $k_{max} = 2465$ |
| $\tilde{B}$ | $(C_s)$ | 59590 | | 975[b] | | | | | | | | | $\tilde{B} \leftarrow \tilde{X}$ 1680–1380 Å | (1330)(320) | Max. of intensity at 68730 $cm^{-1}$[c] |
| $\tilde{A}(^1\Sigma^+)$ | | (38500) | broad continuum with maximum at 1820 Å | | | | | | | | | | $\tilde{A} \leftarrow \tilde{X}$ 2600–1700 Å | (1150)(1330) | $k_{max} = 3.8\ cm^{-1}$[d] |
| $\tilde{b}(^3\Pi)$ | | (35460) | weak continuum[e] with maximum at 2730 Å | | | | | | | | | | $\tilde{b} \leftarrow \tilde{X}$ 2820–2600 Å | (1150) | Observed only with a path length of 33 m at 5 atm. |
| $\tilde{a}(^3\Pi)$ | | (32630) | very weak continuum[f] with flat maximum at 2900 Å | | | | | | | | | | $\tilde{a} \leftarrow \tilde{X}$ 3065–2800 Å | (1150) | |
| $\tilde{X}\ ^1\Sigma^+$ | $C_{\infty v}$ | 0 | 2223.76 | 588.78 | $1284.9_1$[g] | — | 0.419011[h] | — | $[1.128_2]$[i] | $[1.184_2]$[i] | 180° | $\ldots\sigma^2\sigma^2\pi^4\pi^4$ | infrared and microwave sp. | (640)(300) (715)(994) (1052)(396) (997) | |

$N_2O$: [b] Very rapid convergence ($x_{22} = -30\ cm^{-1}$). [c] The underlying continuum has $k_{max} = 147\ cm^{-1}$.
[d] Very diffuse bands are superimposed; perhaps two electronic transitions. Longward limit refers to path of 1 m atm.
[e] Probably corresponds to dissociation into $NO(^2\Pi) + N(^4S)$. [f] Probably corresponds to dissociation into $N_2(^1\Sigma) + O(^3P)$.
[g] In Fermi resonance with $2\nu_2$ which appears at $1168.2_0\ cm^{-1}$.
[h] $\alpha_1 = +0.00345$, $\alpha_2 = -0.00056$, $\alpha_3 = +0.00179$, $q = 0.000792\ cm^{-1}$, $D_J = 1.76 \times 10^{-7}\ cm^{-1}$. [i] These are the $r_e$ values of (994); $r_0$ values have not been given.

TABLE 64 (*continued*)

| State | Point Group | $T_0$ | Vibrational Frequencies $\nu_1$ | $\nu_2$ | $\nu_3$ | Rotational Constants $A_0$ | $B_0$ | $C_0$ | $r_0(XY)$ | $r_0(YZ)$ | $\alpha$ | Electron Configuration | Observed Transitions | References | Remarks |
|---|---|---|---|---|---|---|---|---|---|---|---|---|---|---|---|
| **$CO_2$** | | I.P. = 13.769 eV; $D(O—CO)$ = 5.453 eV (see energy level diagram Fig. 190) | | | | | | | | | | | | | |
| | | Three broad continua with maxima at 570, 350 and 280 Å | | | | | | | | | | | | (80) | |
| | | Three Rydberg series: $\nu = 156390 - \begin{cases} R/(n+0.29)^2; & n = 3, \ldots, 8 \\ R/(n+0.44)^2; & n = 3, \ldots, 8 \\ R/(n-0.05)^2; & n = 4, \ldots, 8 \end{cases}$ with first members at 146490, 147280 and 149500 respectively, converging to a new state of $CO_2^+$ viz. the predicted $^2\Sigma_g^+$ | | | | | | | | | | | | (1189) | |
| | | Two Rydberg series joining on to $\tilde{R}$ and $\tilde{S}$ $\nu = 145800 - \begin{cases} R/(n+0.068+3.25/n^3)^2; & n = 3, \ldots, 18 \\ R/(n-0.305)^2; & n = 3, \ldots, 18 \end{cases}$ [a] converging to the $^2\Sigma_u^+$ state of $CO_2^+$ | | | | | | | | | | | | (1191) | Henning series; see Fig. 189, p. 502 |
| $\tilde{S}$ | $D_{\infty h}$ | 132977 | 1273 | | | | | | | | | | $\tilde{S} \leftarrow \tilde{X}$ 752 Å | (1191) | Sharp bands; repr. (1191) |
| $\tilde{R}$ | $D_{\infty h}$ | 130774 | 1275 | | | | | | | | | | $\tilde{R} \leftarrow \tilde{X}$ 765 Å | (1191) | Diffuse bands; repr. (1191) |
| | | Rydberg series joining on to $\tilde{L}$: $\nu = 139726 - R/(n - 0.063 - 0.0069/n)^2$; $n = 3, \ldots, 10$, | | | | | | | | | | | | (1191) | Converging to the $^2\Pi_{\frac{1}{2}u}$ state of $CO_2^+$ [b] |
| | | Rydberg series parallel to above: $\nu = 139634 - R/(n - 0.044 - 0.34/n)^2$; $n = 4, \ldots, 9$ | | | | | | | | | | | | (1191) | Converging to the $^2\Pi_{\frac{3}{2}u}$ state of $CO_2^+$ [b] |
| $\tilde{M}$ | | 127443 | 1120 | | | | | | | | | | $\tilde{M} \leftarrow \tilde{X}$ 784–758 Å | (1191) | Progression of five members |
| $\tilde{L}$ | $D_{\infty h}$ | 127065 | 1127 | | | | | | | | | | $\tilde{L} \leftarrow \tilde{X}$ 787–712 Å | (1189) (1191) | Progression of six members; repr. (1191) |
| | | Two weak Rydberg series joining on to $\tilde{E}$ and $\tilde{G}$: $\nu = 111250 - \begin{cases} R/(n-0.57)^2; & n = 3, 4, \ldots, 11 \\ R/(n-0.97)^2; & n = 4, 5, \ldots, 11 \end{cases}$ | | | | | | | | | | | | (1189) | |
| | | Two strong Rydberg series joining on to $\tilde{D}$, $\tilde{H}$ and $\tilde{I}$: $\nu = \begin{cases} 111240 \\ 111060 \end{cases} - R/(n - 0.65)^2$; $n = 3, 4, \ldots, 15$ [c] | | | | | | | | | | | | (1189) | Converging to $^2\Pi_{\frac{1}{2}g}$ and $^2\Pi_{\frac{3}{2}g}$ of $CO_2^+$ |

$CO_2$: [a] Each series is accompanied by a vibrational series separated from the main series by 1274 $cm^{-1}$ which corresponds to $\Delta G_{\frac{1}{2}}$ in the $^2\Sigma_u^+$ state of $CO_2^+$.

[b] Each member consists of a long progression in $\nu_1$ with spacings similar to those in $^2\Pi_u$ of $CO_2$.

[c] In addition a vibrational series, 1410–1260 $cm^{-1}$ to shorter wavelengths, has been observed corresponding to the first $\Delta G$ in the ground state of $CO_2^+$.

TABLE 64 (*continued*)

| State | Point Group | $T_0$ | Vibrational Frequencies $\nu_1$ | $\nu_2$ | $\nu_3$ | Rotational Constants $A_0$ | $B_0$ | $C_0$ | $r_0$(XY) | $r_0$(YZ) | $\alpha$ | Electron Configuration | Observed Transitions | References | Remarks |
|---|---|---|---|---|---|---|---|---|---|---|---|---|---|---|---|
| **$CO_2$** (*continued*) | | | | | | | | | | | | | | | |
| $\tilde{I}$ | $\boldsymbol{D}_{\infty h}$ | 100940 | | | | | | | | | | | $\tilde{I}\leftarrow\tilde{X}$ 991 Å | (1189) | |
| $\tilde{H}$ | $\boldsymbol{D}_{\infty h}$ | $\begin{cases}100650\\100570\end{cases}$ | | | | | | | | | | | $\tilde{H}\leftarrow\tilde{X}$ 994 Å | (1189) | |
| $\tilde{G}$ | | 99331 | | | | | | | | | | | $\tilde{G}\leftarrow\tilde{X}$ 1007 Å | (1189) | |
| $\tilde{F}$ | $(\boldsymbol{D}_{\infty h})$ | 96600 | 1320 | | | | | | | | | | $\tilde{F}\leftarrow\tilde{X}$ 1035–969 Å | (1058)(1189) | |
| $\tilde{E}$ | $(\boldsymbol{D}_{\infty h})$ | $\begin{cases}92360\\91830\end{cases}$ | $\begin{cases}1240\\1172\end{cases}$ | | | | | | | | | | $\tilde{E}\leftarrow\tilde{X}$ 1070–1010 Å | (1058)(1189) | A weak and a strong progression |
| $\tilde{D}$ | $\boldsymbol{D}_{\infty h}$ | $\begin{cases}89111\\88535\end{cases}$ | $\begin{cases}1567\\1458\end{cases}$ | | | | | | | | | | $\tilde{D}\leftarrow\tilde{X}\begin{cases}1129–\\1122\text{ Å}\end{cases}$ | (1189) | Very strong, violet shaded bands |
| $\tilde{C}$ | $\boldsymbol{D}_{\infty h}$ | $\begin{cases}85840\\85160\end{cases}$ | $\begin{cases}1165\\1270\end{cases}$ | | | | | | | | | | $\tilde{C}\leftarrow\tilde{X}$ 1170–1130Å | (1058)(1189) | A weak and a strong progression |
| $\tilde{B}\begin{cases}{}^1B_1(\Sigma_g^-)\\{}^1A_1(\Delta_g)\end{cases}$ | $(\boldsymbol{C}_{2v})$ | $\begin{cases}73100\\72480\end{cases}$ | (1225) | | | | | | | | | $\begin{cases}\ldots b_1a_1^2a_2^2b_2^2a_1\\\ldots b_1^2a_1a_2^2b_2^2a_1\end{cases}$ | $\tilde{B}\leftarrow\tilde{X}$ 1390–1220 Å | (1021)(604) | Distinct red shaded bands, possibly two systems |
| $\tilde{A}\,{}^1B_2(\Delta_u)$ | $\boldsymbol{C}_{2v}$ | 46000 | | | | 5.3 | $\bar{B}=0.426$ | | 1.246 | | $122°\pm2°$ | $\ldots 1\pi_u^4a_2^2b_2a_1$ | $\tilde{A}\leftrightarrow\tilde{X}$ [d] | (1021)(604) (283) | In absorption weak diffuse bands[e] |
| $\tilde{X}\,{}^1\Sigma_g^+$ | $\boldsymbol{D}_{\infty h}$ | 0 | 1388.17[f g] | 667.40[f] | 2349.16[f] | — | 0.39021[h] | — | 1.1621 | | 180° | $\ldots 4\sigma_g^2 3\sigma_u^2 1\pi_u^4 1\pi_g^4$ | Raman and infrared sp. | (1166)(1085) (995)(245) | |

$CO_2$: [d] In absorption 1750–1400 Å, in emission 3800–3100 Å. [e] Overlapped by continuum; in emission carbon monoxide flame bands; see (283).
[f] For the anharmonic constants $x_{ik}$ see (246). [g] Shifted by Fermi resonance with $2\nu_2^0$ which is at 1285.40 cm$^{-1}$.
[h] $\alpha_1 = 0.00121$, $\alpha_2 = -0.00072$, $\alpha_3 = 0.00309$; $D = 13.5 \times 10^{-8}$.

TABLE 64 (*continued*)

| State | Point Group | $T_0$ | Vibrational Frequencies $\nu_1$ | $\nu_2$ | $\nu_3$ | Rotational Constants $A_0$ | $B_0$ | $C_0$ | $r_0(XY)$ | $r_0(YZ)$ | $\alpha$ | Electron Configuration | Observed Transitions | References | Remarks |
|---|---|---|---|---|---|---|---|---|---|---|---|---|---|---|---|
| | **OCS** | I.P. = (11.24 eV) $D(OC—S) \leq 3.71$ eV | | | | | | | | | | | | | |
| | | Rydberg series of apparent emission bands: $\nu = 144680 - R/(n + 0.10)^2$; | | | | | | | | | | $n = 4, 5, \ldots, 9$ | | (1190) | |
| | | Rydberg series of sharp red-degraded bands: $\nu = 129330 - R/(n + 0.04)^2$; | | | | | | | | | | $n = 3, 4, 5, 6$ | | (1190) | |
| | | Rydberg series of sharp bands: $\nu = 129390 - R/(n - 0.12)^2$; | | | | | | | | | | $n = 3, 4, \ldots, 9$ | | (1190) | |
| | | Rydberg series of violet-shaded bands: $\nu = 129400 - R/(n - 0.49)^2$; | | | | | | | | | | $n = 3, 4, \ldots, 8$ | | (1190) | |
| | | Rydberg series of diffuse bands: $\nu = 129400 - R/(n - 0.68)^2$; | | | | | | | | | | $n = 3, 4, \ldots, 13$ | | (1190) | |
| | | Two fragmentary Rydberg series with a limit of 90675 $cm^{-1}$ | | | | | | | | | | | | (1190) | |
| | | Several fragmentary progressions between 77200 and 82700 $cm^{-1}$ | | | | | | | | | | | | (1190) | Sharp bands |
| $\tilde{M}$ | | 77434 | | | 722 | | | | | | | | $\tilde{M} \leftarrow \tilde{X}$ 1290–1245 Å | (1190) | Sharp bands |
| $\tilde{L}$ | | 77109 | | | 688 | | | | | | | | $\tilde{L} \leftarrow \tilde{X}$ 1300–1230 Å | (1190) | Sharp bands |
| $\tilde{K}$ | | 72774 | | | 732 | | | | | | | | $\tilde{K} \leftarrow \tilde{X}$ 1375–1310 Å | (1190) | Sharp bands |
| $\tilde{J}$ | | 71887 | | | 744 | | | | | | | | $\tilde{J} \leftarrow \tilde{X}$ 1390–1315 Å | (1190) | Sharp bands[a] |
| $\tilde{I}$ | | 71785 | | | 719 | | | | | | | | $\tilde{I} \leftarrow \tilde{X}$ 1395–1310 Å | (1190) | Sharp bands[a] |
| $\tilde{H}$ | | 71696 | | | (723) | | | | | | | | $\tilde{H} \leftarrow \tilde{X}$ 1395–1320 Å | (1190) | Sharp bands[b] |
| $\tilde{G}$ | | 71531 | | | 703 | | | | | | | | $\tilde{G} \leftarrow \tilde{X}$ 1400–1340 Å | (1190) | Sharp bands[c] |
| $\tilde{F}$ | | 71435 | | | 727 | | | | | | | | $\tilde{F} \leftarrow \tilde{X}$ 1400–1310 Å | (1190) | Sharp bands[d] |
| $\tilde{E}$ | | 70907 | | | 724 | | | | | | | | $\tilde{E} \leftarrow \tilde{X}$ 1410–1320 Å | (1190) | Sharp bands |

OCS: [a] May belong to one of the other systems. [b] May belong to the $\tilde{E}-\tilde{X}$, or $\tilde{F}-\tilde{X}$, or $\tilde{G}-\tilde{X}$ systems. [c] May belong to the $\tilde{E}-\tilde{X}$ or the $\tilde{F}-\tilde{X}$ system. [d] May belong to the $\tilde{E}-\tilde{X}$ system.

TABLE 64 (*continued*)

| State | Point Group | $T_0$ | Vibrational Frequencies $\nu_1$ | $\nu_2$ | $\nu_3$ | Rotational Constants $A_0$ | $B_0$ | $C_0$ | $r_0(XY)$ | $r_0(YZ)$ | $\alpha$ | Electron Configuration | Observed Transitions | References | Remarks |
|---|---|---|---|---|---|---|---|---|---|---|---|---|---|---|---|
| **OCS** (*continued*) | | | | | | | | | | | | | | | |
| $\tilde{D}$ | | {63800, 63720} | | | 920[e] | | | | | | | | $\tilde{D} \leftarrow \tilde{X}$ 1570–1440 Å | (1190) | Diffuse bands |
| $\tilde{A}$ | | Continuous absorption beginning at 2550 Å | | | | | | | | | | | $\tilde{A} \leftarrow \tilde{X}$ 2550–2000 Å[f] | (757) | Sharp limit[g] |
| $\tilde{X}\,^1\Sigma^+$ | $C_{\infty v}$ | 0 | 2062.22 | $520.4_1$ | $858.9_5$ | — | $0.20285_7$[h] | — | 1.160 | 1.560 | 180° | $\ldots\sigma^2\sigma^2\pi^4\pi^4$ | infrared and microwave sp. | (795)(40) (179a)(233) | |
| **$CS_2$** | | I.P. = 10.079 eV | | | | | | | | | | | | | |
| | | Rydberg series starting at 144350 $cm^{-1}$: | | | | $\nu = 157390 - R/(n-0.95)^2$; | | | | | | $n = 4, 5, \ldots, 10$ | | (1190) | |
| $\tilde{P}$ | | 140410 | 590 | | | | | | | | | | $\tilde{P} \leftarrow \tilde{X}$ 712–688 Å | (1190) | Progression of weak bands |
| | | Rydberg series starting at 113420 $cm^{-1}$: | | | | $\nu = 130610 - R/(n-0.48)^2$; | | | | | | $n = 3, 4, \ldots, 7$ | } 882–766 Å | (1190) | |
| | | Rydberg series starting at 117780 $cm^{-1}$: | | | | $\nu = 130600 - R/(n-1.10)^2$; | | | | | | $n = 4, 5, \ldots, 10$ | } | (1190) | |
| | | Rydberg series starting at 120580 $cm^{-1}$: | | | | $\nu = 130540 - R/(n-0.81)^2$; | | | | | | $n = 4, 5, \ldots, 13$ | } | (1190) | |
| $\tilde{N}$ | | 117772 | (680) | | | | | | | | | | $\tilde{N} \leftarrow \tilde{X}$ 849–836 Å | (1190) | Irregular progression of strong bands |
| | | Rydberg series joining on to $\tilde{L}$: | | | | $\nu = 116760 - R/(n-0.58)^2$; | | | | | | $n = 3, 4, \ldots, 12$ | | (1190) } | Converge to $\tilde{B}\,^2\Sigma_u^+$ state of $CS_2^+$ |
| | | Rydberg series joining on to $\tilde{K}$: | | | | $\nu = 116760 - R/(n-0.94)^2$; | | | | | | $n = 3, 4, \ldots, 14$ | | (1190) } | |
| $\tilde{M}$ | | 100300 | 610 | | | | | | | | | | $\tilde{M} \leftarrow \tilde{X}$ 997–968 Å | (1190) | Progression of sharp bands |
| $\tilde{L}$ | | 96570 | (550) | | | | | | | | | | $\tilde{L} \leftarrow \tilde{X}$ 1035 Å | (1190) | |
| $\tilde{K}$ | | 89550 | (370) | | | | | | | | | | $\tilde{K} \leftarrow \tilde{X}$ 1117 Å | (1190) | $k > 4000$ |
| $\tilde{J}$ | | 88834 | (490) | | | | | | | | | | $\tilde{J} \leftarrow \tilde{X}$ 1126–1072 Å | (1190) | Long progression of diffuse bands |
| | | Long Rydberg series joining on to $\tilde{G}$: | | | | $\nu = 81735 - R/(n-0.46)^2$; | | | | | | $n = 4, 5, \ldots, 19$ | | (1020)(1190) | Converges to $\tilde{X}\,^2\Pi_{\frac{1}{2}}$ of $CS_2^+$ |
| | | Long Rydberg series joining on to $\tilde{F}$: | | | | $\nu = 81299 - R/(n-0.44)^2$; | | | | | | $n = 4, 5, \ldots, 19$ | | (1020)(1190) | Converges to $\tilde{X}^2\Pi_{\frac{3}{2}}$ of $CS_2^+$ |

OCS: [e] Very large anharmonicity. [f] The absorption in the region 2000–1570 Å has apparently not been studied.
[g] This limit probably corresponds to dissociation into $CO(^1\Sigma) + S(^1D)$ leading to the $D(OC—S)$ value quoted. [h] $\alpha_1 = 0.00121$, $\alpha_2 = -0.00034_8$, $\alpha_3 = 0.000683$.

TABLE 64 (*continued*)

| State | Point Group | $T_0$ | Vibrational Frequencies $\nu_1$ | $\nu_2$ | $\nu_3$ | Rotational Constants $A_0$ | $B_0$ | $C_0$ | $r_0(XY)$ | $r_0(YZ)$ | $\alpha$ | Electron Configuration | Observed Transitions | References | Remarks |
|---|---|---|---|---|---|---|---|---|---|---|---|---|---|---|---|
| **$CS_2$** (*continued*) | | | | | | | | | | | | | | | |
| $\tilde{G}$ | $D_{\infty h}$ | 73048 | some vibrational structure observed | | | | | | | | | | $\tilde{G} \leftarrow \tilde{X}$ 1369–1317 Å | (1020)(1190) | |
| $\tilde{F}$ | $D_{\infty h}$ | 72564 | some vibrational structure observed | | | | | | | | | | $\tilde{F} \leftarrow \tilde{X}$ 1378–1326 Å | (1020)(1190) | |
| $\tilde{E}$ | | 65150 | | | | | | | | | | | $\tilde{E} \leftarrow \tilde{X}$ 1535–1360 Å | (1020)(1190) | Complex system of sharp bands |
| $\tilde{D}$ | $D_{\infty h}$ | 62754 | (606) | | (1644) | | | | | | | | $\tilde{D} \leftarrow \tilde{X}$ 1612–1553 Å | (1020) | |
| $\tilde{C}$ | | 56490 | 830 | | | | | | | | | | $\tilde{C} \leftarrow \tilde{X}$ 1780–1620 Å | (1020) | Single progression, diffuse above 59000 |
| $\tilde{B}$ | $D_{\infty h}$ | {55198<br>55002 | | | | | | | | | | | $\tilde{B} \leftarrow \tilde{X}$ 1818–1811 Å | (1020) | Strong doublet |
| $\tilde{A}\ ^1B_2$ | $(C_{2v})$ | (45950) | (∼400) | (∼400) | | 11 | $\bar{B} = 0.10_0$ | | 1.66 | | 153° | | $\tilde{A} \leftarrow \tilde{X}$ 2200–1800 Å | (1020)(483)(302) | Extensive system of discrete bands; Repr. (302)[a] |
| | | Fragments of two or more systems in the region 29000–34000 $cm^{-1}$ | | | | | | | | | | | 3400–2900 Å | (680) | Stronger than $\tilde{a}-\tilde{X}$ |
| $\tilde{a}\ ^3A_2$[b] | $C_{2v}$ | 26187 | $691._5$ | 310.8 | (940) | 4.3 | 0.1129 | 0.1101 | 1.64 | | $135._8$° | | $\tilde{a} \leftarrow \tilde{X}$ 4300–3300 Å | (746)(680)(293)(298) | Extensive system of sharp bands; Repr. (680), see Fig. 84 |
| $\tilde{X}\ ^1\Sigma_g^+$ | $D_{\infty h}$ | 0 | 657.98[c] | 396.7 | 1532.5[d] | — | 0.10910[e] | — | $1.554_5$ | | 180° | | Infrared and Raman sp. | (1166)(403)(51)(452a) | |
| **$HgCl_2$** | | | | | | | | | | | | | | | |
| $\tilde{A}$ | $D_{\infty h}$ | 59016 | 289 | 60 | | | | | | | 180° | | $\tilde{A} \leftarrow \tilde{X}$ 1740–1660 Å | (1285) | see Fig. 58 |
| $\tilde{X}\ ^1\Sigma_g^+$ | $D_{\infty h}$ | 0 | 363 | 75 | 413 | | | | | | 180° | | Infrared emission | (681) | |

$CS_2$: [a] Also obtained with $C^{13}$ by (302). [b] Only the $B_2$ component of this state has been observed. [c] Shifted by Fermi resonance with $2\nu_2$ which is at 802.11 $cm^{-1}$.
[d] (51) give 1535.35 from unpublished work of G. R. Wilkinson. [e] $\alpha_1 = 0.00015_5$, $\alpha_2 = -0.00026$, $\alpha_3 = 0.000711$; $D = 1.0 \times 10^{-8}$ $cm^{-1}$.

TABLE 64 (*continued*)

| State | Point Group | $T_0$ | Vibrational Frequencies $\nu_1$ | $\nu_2$ | $\nu_3$ | Rotational Constants $A_0$ | $B_0$ | $C_0$ | $r_0$(Å) | $\alpha$ | Electron Configuration | Observed Transitions | References | Remarks |
|---|---|---|---|---|---|---|---|---|---|---|---|---|---|---|
| **HgClBr, HgBrI, HgClI** see (885) | | | | | | | | | | | | | | |
| **$HgBr_2$, $HgI_2$** see (1285) | | | | | | | | | | | | | | |
| **$NO_2$** I.P. = 9.78 eV[a]; $D$(ON—O) = $3.114_9$[b]; $D$(N—$O_2$) = $4.505_6$[b] | | | | | | | | | | | | | | |
| | | Fragments of two Rydberg series going to the limit at 152200 $cm^{-1}$ | | | | | | | | | | | (1187) | |
| | | Rydberg series joining on to $\tilde{J}$: $\nu = 152290 - R/(n-0.67)^2$; $n = 3, 4, \ldots, 7$ | | | | | | | | | | | (1187) | Repr. (1187) |
| | | Rydberg series joining on to $\tilde{I}$: $\nu = 152100 - R/(n-1.15)^2$; $n = 3, 4, \ldots, 9$ | | | | | | | | | | | (1187) | Repr. (1187) |
| $\tilde{J}$ | | 131680 | | | | | | | | | | $\tilde{J}-\tilde{X}$ 759 Å | (1187) | |
| $\tilde{I}$ | | 111480 | | | | | | | | | | $\tilde{I}-\tilde{X}$ 897 Å | (1187) | |
| | | Doubtful Rydberg series converging to 99500 and 93695 $cm^{-1}$ | | | | | | | | | | | (1022)(926) (1187) | |
| | | Fragments of a band system 89500–91500 $cm^{-1}$ | | | | | | | | | | | (926) | |
| $\tilde{H}$ | | 91508 | | 624 | | | | | | | | $\tilde{H}\leftarrow\tilde{X}$ 1095–1055 Å | (1187) | Progression of six bands |
| $\tilde{G}$ | | 87481 | | 560 | | | | | | | | $\tilde{G}\leftarrow\tilde{X}$ 1145–1115 Å | (926)(1187) | Repr. (926); somewhat doubtful |
| $\tilde{F}$ | | 76711 | | 610 | | | | | | | | $\tilde{F}\leftarrow\tilde{X}$ 1305–1250 Å | (926)(1187) | Progression of six bands; repr. (926) |
| $\tilde{E}\ ^2\Sigma_u^+$ | $D_{\infty h}$ | (58309) | 1420 | 596.5[c] | | | | | $1.1_3$[d] | 180° | $\ldots(1\pi_u)^4(1\pi_g)^4(3p\sigma)$ | $\tilde{E}\leftarrow\tilde{X}$ 1650–1350 Å | (1071)(879) | |
| $\tilde{D}$ | | (50000) | continuum with maximum at 1700 Å and diffuse bands | | | | | | | | | $\tilde{D}\leftarrow\tilde{X}$ 2000–1650 Å | (926)(879) | |
| $\tilde{C}$ | $C_{2v}$ | (42500) | (940) | | | | | | | | | $\tilde{C}\leftarrow\tilde{X}$ 2350–2000 Å | (926) | Repr. (926) |
| $\tilde{B}\ ^2B_2$ | $C_{2v}$ | 40125.9 | 1184 | | | 4.132 | 0.402 | 0.366 | 1.314 | 121.0° | $\ldots(5a_1)^2(1a_2)^2(4b_2)(6a_1)^2$ | $\tilde{B}\leftarrow\tilde{X}$ 2580–2350 Å | (477)(476) (1072) | Prediss. limit at 2480 Å; see Fig. 183 repr. (477)(1072) |
| $\tilde{A}(^2B_1)$ | $C_{2v}$ | < 15000 | | (896)[e] | | | $\bar{B}=0.370$[f] | | | | $\ldots(5a_1)^2(1a_2)^2(4b_2)^2(2b_1)$ | $\tilde{A}\leftrightarrow\tilde{X}$ 10000–3200 Å | (296) | Prediss. limit at 3979 Å; see Fig. 192 repr. (462) |
| $\tilde{X}\ ^2A_1$ | $C_{2v}$ | 0 | (1319.7) | 749.8 | 1617.75 | 8.0012 | 0.43364 | 0.41040 | 1.1934 | 134.1° | $\ldots(5a_1)^2(1a_2)^2(4b_2)^2(6a_1)$ | infrared and microwave sp. | (664)(72) (120) | |

$NO_2$: [a] By the photoionization method (926)(400); from electron impact 9.83 eV (223).
[b] From predissociation data of (296); thermochemical data give $3.112_7$ and $4.503_3$ respectively. [c] $g_{22} = 2._3$ $cm^{-1}$.
[d] From the fact that the $\tilde{E}-\tilde{X}$ bands are not shaded and thus $B' \approx B''$.
[e] It is not improbable that in the $\tilde{A}$ state the molecule is linear. In that case the number given represents $2\nu_2$.

TABLE 64 (*continued*)

| State | Point Group | $T_0$ | Vibrational Frequencies $\nu_1$ | $\nu_2$ | $\nu_3$ | Rotational Constants $A_0$ | $B_0$ | $C_0$ | $r_0$(Å) | $\alpha$ | Electron Configuration | Observed Transitions | References | Remarks |
|---|---|---|---|---|---|---|---|---|---|---|---|---|---|---|
| **FCO** | | see (852) | | | | | | | | | | | | |
| **$InCl_2$, $InBr_2$, $InI_2$** | | Continuous absorption only; see (1289a) | | | | | | | | | | | | |
| **$GaCl_2$** | | Continuous absorption only; see (1289a) | | | | | | | | | | | | |
| **$CF_2$** | | I.P. = (12.1)[a]; $D$(FC—F) ≤ 5.2 eV[b] | | | | | | | | | | | | |
| $\tilde{A}\ (^1A_1)$ | $C_{2v}$ | 37705 | | 495 | | 5.07 | 0.331 | 0.311 | | | $\ldots(4b_2)^2(1a_2)^2(6a_1)(2b_1)$ | $\tilde{A}\leftrightarrow\tilde{X}$ 3250–2200 Å | (1243)(713) (796)(1219) (800b) | Repr. (1243)(713) Prediss. above 41680 $cm^{-1}$ |
| $\tilde{X}\ ^1A_1$ | $C_{2v}$ | 0 | (1102)[c] | 667 | (1222)[c] | 2.95 | 0.417 | 0.365 | | | $\ldots(4b_2)^2(1a_2)^2(6a_1)^2$ | | | |
| **$SiF_2$** | | | | | | | | | | | | | | |
| $\tilde{A}\ (^1B_1)$ | $C_{2v}$ | 43078 | 598 | 342 | | | | | | | $\ldots(a_2)^2(b_2)^2(a_1)(b_1)$ | $\tilde{A}\rightarrow\tilde{X}$ 2755–2179 Å | (637)(1049) | Bands with coarse and narrow fine structure; repr. (637)(1049) (1055a) |
| $\tilde{X}\ ^1A_1$ | $C_{2v}$ | 0 | 776[a] | 427 | 947[a] | 1.02076 | 0.29433 | 0.22784 | 1.591 | 101.0° | $\ldots(a_2)^2(b_2)^2(a_1)^2$ | microwave sp. | | |

$CF_2$: [a] Indirectly from appearance potentials by (794); (1063) give 13.3 eV. [b] From predissociation (1219); according to (346) the heat of formation is $-2.00_8$ eV.
[c] From infrared spectrum of $CF_2$ in inert solid matrix (853).

$SiF_2$: [a] These are the values of (1049); (637) give $\nu_1$ = 937 and do not find $\nu_3$, which according to the selection rules should indeed not be prominent in the spectrum.

TABLE 64 (*continued*)

| State | Point Group | $T_0$ | Vibrational Frequencies $\nu_1$ | $\nu_2$ | $\nu_3$ | Rotational Constants $A_0$ | $B_0$ | $C_0$ | $r_0$(Å) | $\alpha$ | Electron Configuration | Observed Transitions | References | Remarks |
|---|---|---|---|---|---|---|---|---|---|---|---|---|---|---|
| **$SiCl_2$** | | see (82) | | | | | | | | | | | | |
| **$SnCl_2$** | | see (82) | | | | | | | | | | | | |
| **$O_3$** | | I.P. = $12.8_0$ eV[a]; $D(O_2—O) = 1.0_4$ eV | | | | | | | | | | | | |
| | | Six continua with maxima at 1725, 1450, 1330, 1215, 1120 and 750 Å | | | | | | | | | | | (1022)(1186)(945) | |
| $\tilde{D}$ | | ~33000 | | (300) | | | | | | | | $\tilde{D}-\tilde{X}$ 3000–2200 Å | (619)(1241) | Hartley band, extremely intense diffuse bands[b] |
| $\tilde{C}$ | | 28447 | (636)[c] | (352)[c] | | | | | | | | $\tilde{C}-\tilde{X}$ 3740–3000 Å | (619)(620) (271)(819) | Huggins bands; diffuse |
| $\tilde{B}$ | | 16625 | (1099)[d] | | | | | | | | | $\tilde{B}-\tilde{X}$ 6100–5500 Å | (1324)(1247) | Chappuis bands; diffuse; repr. (1324) |
| $\tilde{A}$ | | 10000 | | 566.7 | | | | | | | | $\tilde{A}-\tilde{X}$ 10000–7000 Å | (1324)(735) | Diffuse bands; repr. (1324) |
| $\tilde{X}\,^1A_1$ | $C_{2v}$ | 0 | 1110 | 705 | $1042.1_6$ | $3.5534_8$ | 0.44525 | 0.39479 | 1.278 | 116.8° | $\ldots(5a_1)^2(4b_2)^2(6a_1)^2(1a_2)^2$ | infrared and microwave sp. | (50)(1309) (657)(589) | |

$O_3$: [a] From electron impact data (515). [b] Intensity maximum at 2550 Å. [c] Anharmonicities are given by (620). [d] Very doubtful since assumed $\nu_2''$ is incorrect.

TABLE 64 (*continued*)

| State | Point Group | $T_0$ | Vibrational Frequencies $\nu_1$ | $\nu_2$ | $\nu_3$ | Rotational Constants $A_0$ | $B_0$ | $C_0$ | $r_0$(Å) | $\alpha$ | Electron Configuration | Observed Transitions | References | Remarks |
|---|---|---|---|---|---|---|---|---|---|---|---|---|---|---|
| **$SO_2$** | | I.P. = 12.34 eV[a]; $D(OS—O) = 5.61_3$[b] | | | | | | | | | | | | |
| | | Rydberg series joining on to $\tilde{G}$ and $\tilde{H}$: $\nu = 99500 - R/(n-1.02)^2$; $n = 3, 4, 5, 6$ | | | | | | | | | | | (429) | |
| $\tilde{H}$ | | 87110 | 1000 | | | | | | | | | $\tilde{H} \leftarrow \tilde{X}$ 1148–1123 Å | (429) | Repr. (429) |
| $\tilde{G}$ | | (80000) | broad absorption peak with some structure | | | | | | | | | $\tilde{G} \leftarrow \tilde{X}$ 1300–1200 Å | (429) | Repr. (429) |
| $\tilde{F}$ | | 74100 | | | | | | | | | | $\tilde{F} \leftarrow \tilde{X}$ 1350–1310 Å | (1020) | Sharp bands |
| $\tilde{E}$ | | (63570) | three diffuse bands at 63570, 64185 and 65400 cm$^{-1}$ | | | | | | | | | $\tilde{E} \leftarrow \tilde{X}$ 1600–1400 Å | (1020) | Absorption coeff. and additional bands in (429) |
| $\tilde{D}$ | $(C_s)$ | 42264 | 756.8 | 393.1 | 890.5 | | | | | | | $\tilde{D} \leftrightarrow \tilde{X}$ } 2350–1800 Å | (1020)(778) (315)(311) | Red shaded bands repr. in |
| $\tilde{C}$ | $(C_s)$ | 41413[c] | 963 | 377 | | $A - \bar{B} = 1.05$[c] | | | 1.53[c] | 109.1°[c] | [d] | $\tilde{C} \leftrightarrow \tilde{X}$ } | (1083)(310) | (1020)(1271) (778)[e] |
| $\tilde{A}(^1B_1)$ | $(C_{2v})$[f] | 29622 | 764 | 317.5 | 813 | $A - \bar{B} = 0.81$ | | | | | $\ldots(a_2)^2(b_2)^2(a_1)(b_1)$ | $\tilde{A} \leftrightarrow \tilde{X}$ 3400–2600 Å | (834)(447) | Red shaded bands repr. in (834)(1280) |
| $\tilde{a}(^3B_1)$ | $(C_{2v})$[f] | 25766.9 | 905.7 | 359.8 | | 2.296 | $\bar{B} = 0.2799$ | | 1.494 | 126.1° | $\ldots(a_2)^2(b_2)^2(a_1)(b_1)$ | $\tilde{a} \leftrightarrow \tilde{X}$ 3900–[g] 3400 Å | (835)(412) (822)(447) | Both violet and red shaded bands; repr. (822) |
| $\tilde{X}\,^1A_1$ | $C_{2v}$ | 0 | $1151.3_8$ | $517.6_9$ | $1361.7_6$ | 2.02736 | 0.34417 | 0.293535 | 1.4321[h] | 119.5°[h] | $\ldots(a_2)^2(b_2)^2(a_1)^2$ | infrared and microwave sp. | (675)(1120) (880) | |

$SO_2$: [a] From the Rydberg series of (429); identical with photoionization value of (1273).
[b] From (41). However (1271) derive $D(OS—O) = 5.3_5$ eV from the longward end of a continuum which they believe overlaps the $\tilde{D}-\tilde{X}$ system.; see also (310a).
[c] From (310a). [d] Electron configurations in (1264).
[e] Absorption coefficients are given by (429). According to (1083) there may be more than two electronic transitions in this region. According to (508) and (1271) there is in addition continuous absorption in this region; (508) also find in phosphorescence at longer wavelengths a corresponding emission continuum.
[f] According to (914) it is probable that the symmetry of the equilibrium position is only $C_s$. [g] Fluorescence spectrum 4460–3840 Å.
[h] According to (880) the equilibrium values are 1.4308 Å and $119.3_1$°.

TABLE 64 (*continued*)

| State | Point Group | $T_0$ | Vibrational Frequencies $\nu_1$ | $\nu_2$ | $\nu_3$ | Rotational Constants $A_0$ | $B_0$ | $C_0$ | $r_0$(XY) | $r_0$(YZ) | $\alpha$ | | Observed Transitions | References | Remarks |
|---|---|---|---|---|---|---|---|---|---|---|---|---|---|---|---|
| **$S_2O$** [a] | | $D(S_2—O) \leq 3.93$ eV | | | | | | | | | | | | | |
| $\tilde{A}$ | $C_s$ | $30099._8$ | | | $406._5$ | | | | | | | | $\tilde{A} \leftarrow \tilde{X}$ 3400–2500 Å | (230)(644) | Extensive system of bands with narrow fine structure [b] |
| $\tilde{X}\,{}^1A'$ | $C_s$ | 0 | 1165 | (388)[c] | 679 | 1.39811 | 0.16875 | 0.15034 | 1.884 | 1.465 | $118._0°$ | | Microwave and infrared sp. | (644)(832) | |
| **$SeO_2$** | | see (81)(314)(315)(1126)(472) | | | | | | | | | | | | | |
| **$TeO_2$** | | see (1126) | | | | | | | | | | | | | |
| **ONF** | | | | | | | | | | | | | | | |
| $\tilde{A}$ | $C_s$ | 30011 | (1450) | 343 | 1079[a] | | | | | | | | $\tilde{A} \leftarrow \tilde{X}$ 3350–2600 Å | (643) | [b] |
| $\tilde{X}\,{}^1A'$ | $C_s$ | 0 | $1844.0_3$ | 521 | $765.8_5$ | 3.17525 | 0.39507 | 0.35052 | 1.13 | 1.52 | 110° | | infrared and microwave sp. | (1311)(790) (330)(789) | [c] |
| **ONCl** | | see (1022) | | | | | | | | | | | | | |
| **$NF_2$** | | I.P. = 11.8 eV[a] | | | | | | | | | | | | | |
| $\tilde{A}$ | | (36000)[b] | | (380) | | | | | | | | | $\tilde{A} \leftarrow \tilde{X}$ 2800–2350 Å | (703)(434) | Indistinct bands overlying a continuum |
| $\tilde{X}(^2B_1)$ | $(C_{2v})$ | 0 | 1074.3 | (573.4)[c] | (930.7)[c] | $A - \bar{B} = 1.98$ | | | (1.37) ass. | | 104.2° | $\ldots 6a_1^2\, 1a_2^2\, 2b_1$ | infrared sp. | (473)(474) | |

$S_2O$: [a] Neither (230) nor (644) realized that the molecule they studied was SSO as now established by the microwave work of (832).
[b] Diffuseness sets in abruptly at $v_3' = 4$. [c] Measured in a solid film (125).

ONF: [a] The anharmonicity $x_{33} = -9.5$ is unusually large.
[b] A continuous emission spectrum was found to extend from 6400–5100 Å with a maximum at 6095 Å. It may belong to a different electronic transition; (643) consider it as produced by radiative two-body recombination of NO and F. [c] The microwave data have only been published in abstract form.

$NF_2$: [a] From mass spectrometric studies of (218). [b] This is the first band observed by (703); (434) give a band at 37739 cm$^{-1}$ as the first one. [c] Measured in a solid $N_2$ matrix.

TABLE 64 (*continued*)

| State | Point Group | $T_0$ | Vibrational Frequencies $\nu_1$ | $\nu_2$ | $\nu_3$ | Rotational Constants $A_0$ | $B_0$ | $C_0$ | $r_0(XY)$ | $r_0(XZ)$ | $\alpha$ | Electron Configuration | Observed Transitions | References | Remarks |
|---|---|---|---|---|---|---|---|---|---|---|---|---|---|---|---|
| **$Cl^{35}O_2$** | | I.P. = 11.1 eV[a]; $D(O—ClO) = 2.50_5$ eV[b] | | | | | | | | | | | | | |
| $\tilde{E}$ | | 63774 | (1000) | 508 | | | | | | | | $\ldots(a_2)^2(b_2)^2(a_1)(b_1)(a_1)$ | $\tilde{E} \leftarrow \tilde{X}$ 1570–1490 Å | (592) | |
| $\tilde{D}$ | | 61430 | 1051 | 521 | | | | | | | | $\ldots(a_2)^2(b_2)^2(a_1)^2(pa_1)$ | $\tilde{D} \leftarrow \tilde{X}$ 1630–1550 Å | (592) | |
| $\tilde{C}(^2A_1)$ | $(C_{2v})$ | 54389 | 1020 | | | | | | | | | $\ldots(a_2)^2(b_2)^2(a_1)^2(sa_1)$ | $\tilde{C} \leftarrow \tilde{X}$ 1830–1760 Å | (592) | Short progression |
| $\tilde{B}$ | | Strong continuous absorption below 2000 Å | | | | | | | | | | | | (592) | |
| $\tilde{A}(^2A_2)$ | $(C_{2v})$[c] | $21016._4$ | $707._1$ | 289.6 | $769._3$[d] | | | | | | | $\ldots(a_2)(b_2)^2(a_1)^2(b_1)^2$ | $\tilde{A} \leftarrow \tilde{X}$ 5100–2700 Å | (384)(228) | Repr. in (384)(228)[e] |
| $\tilde{X}\ ^2B_1$ | $C_{2v}$ | 0 | 945.5 | 447.4 | 1110.5 | $1.7371_8$ | $0.33197_1$ | $0.27799_2$ | 1.473 | | 117.6° | $\ldots(a_2)^2(b_2)^2(a_1)^2(b_1)$ | microwave sp. | (986) | [f] |
| **$S_2Cl$** | | see (810) | | | | | | | | | | | | | |
| **$Cl_2O$** | | see (385) | | | | | | | | | | | | | |
| **$SeCl_2$** | | see (1284) | | | | | | | | | | | | | |

$ClO_2$: [a] From mass spectrometric data of (274). [b] From thermochemical data.
[c] It is possible that the molecule is not symmetrical in this state; see (226).
[d] The observed vibrational interval, 1559 cm$^{-1}$, has more recently been interpreted by (226) as the interval $1^+ - 0^+$ in a double minimum potential for the antisymmetric normal coordinate.
[e] Predissociation starts at 3750 Å; (384) give extrapolated dissociation limit as 39000 cm$^{-1}$.
[f] The species $^2B_1$ of the ground state is definitely established by the observed hyperfine structure according to (259). The vibrational frequencies are from the analysis of the electronic spectrum by (228).

TABLE 64 (*continued*)

| State | Point Group | $T_0$ | Vibrational Frequencies $\nu_1$ | $\nu_2$ | $\nu_3$ | Rotational Constants $A_0$ | $B_0$ | $C_0$ | $r_0$(XY) | $r_0$(YZ) | $\alpha$ | Electron Configuration | Observed Transitions | References | Remarks |
|---|---|---|---|---|---|---|---|---|---|---|---|---|---|---|---|
| **$TeCl_2$** | | | | | | | | | | | | | | | |
| | | Two unclassified systems in the regions 2200–2150 and 1840–1685 Å | | | | | | | | | | | | | |
| $\tilde{A}$ | | 17142 | 313 | 58 | | | | | | | | | $\tilde{A} \leftarrow \tilde{X}$ 6500–5000 Å | (886) (1146) | Red shaded bands[a] |
| $\tilde{X}(^1A_1)$ | $C_{2v}$ | 0 | 386 | 71 | | | | | | | | | | | |
| **$SeBr_2$** | | see (1283) | | | | | | | | | | | | | |
| **$TeBr_2$** | | see (1282) and (597) | | | | | | | | | | | | | |
| **$XeF_2$** | | see (1308)(1035)(696a) | | | | | | | | | | | | | |

$TeCl_2$: [a] (1146) assumes that all observed bands have $\nu_3$ singly excited in the upper state, in order to account for the observed isotope shifts. On this basis, which seems doubtful, the transition would have to be classified as a forbidden transition, viz. $A_2 - A_1$.

Table 65

## MOLECULAR CONSTANTS OF THE ELECTRONIC STATES OF FOUR-ATOMIC TRI-HYDRIDES

| State | Point Group | $T_0$ | Vibrational Frequencies $\nu_1$ | $\nu_2$ | $\nu_3$ | $\nu_4$ | Rotational Constants $A_0$ | $B_0$ | $C_0$ | $r_0$(Å) | $\alpha$ | Electron Configuration | Observed Transitions | References | Remarks |
|---|---|---|---|---|---|---|---|---|---|---|---|---|---|---|---|
| | **$CH_3$** | I.P. = 9.843 eV[a]; $D(H—CH_2) \leq 4.9_0$ eV[b] | | | | | | | | | | | | | |
| | | Rydberg series joining on to $\tilde{D}$: $\nu = 79392 - R/(n - 0.083)^2$; $n = 3, 4, \ldots, 8$ | | | | | | | | | | | | (521) | δ series of (521), Fig. 193 |
| | | Rydberg series joining on to $\tilde{C}$: $\nu = 79392 - R/(n - 0.090)^2$; $n = 3, 4 \ldots, 8$ | | | | | | | | | | | | (521) | γ series of (521), Fig. 193 |
| $\tilde{E}\ ^2A_1'$ | $D_{3h}$ | 71042 | | | | | | | | | | $\ldots(e')^4 4sa_1'$ | $\tilde{E} \leftarrow \tilde{X}$ 1408 Å | (521) | $\beta_2$ of (521) |
| $\tilde{D}\ ^2A_1'$ | $D_{3h}$ | 66799 | | | | | — | (10.72) | | 1.020 | 120° | $\ldots(e')^4 3da_1'$ | $\tilde{D} \leftarrow \tilde{X}$ 1497 Å | (521) | $\delta_1$ of (521) |
| $\tilde{C}\ ^2E''$ | $D_{3h}$ | 66536 | | (1360) | | | | | | | | $\ldots(e')^4 3de''$ | $\tilde{C} \leftarrow \tilde{X}$ 1503 Å | (521) | $\gamma_1$ of (521) |
| $\tilde{B}\ ^2A_1'$ | $D_{3h}$ | 46205 | | (1360) | | | | | | $1.12_4$[c] | 120° | $\ldots(e')^4 3sa_1'$ | $\tilde{B} \leftarrow \tilde{X}$ 2160 Å | (521) | $\beta_1$ of (521), see Fig. 96 |
| $\tilde{X}\ ^2A_2''$ | $D_{3h}$ | 0 | | (580)[d] | | | — | (9.57) | | 1.079[c] | 120° | $\ldots(e')^4 a_2''$ | | (521) | |
| | **$NH_3$** | I.P. = $10.15_4$ eV[a]; $D(H—NH_2) = 4.3_8$ eV[b] | | | | | | | | | | | | | |
| | | Absorption coefficients between 1300 and 400 Å: see(1177)(1256) and (836) | | | | | | | | | | | | | |
| $\tilde{G}$ | | 82857 | | 988[c] | | | | | | | | | $\tilde{G} \leftarrow \tilde{X}$ 1210–1150 Å | (317)(1272)(319) | above first I.P. |
| $\tilde{F}$ | ($D_{3h}$) | (77739) | | (990) | | | | | | | | | $\tilde{F} \leftarrow \tilde{X}$ 1290–1220 Å | (1270)(1272) | |
| $\tilde{E}\ ^1A_2''$ | $D_{3h}$ | 75205 | | 927[d] | | | | | | | | | $\tilde{E} \leftarrow \tilde{X}$ 1330–1270 Å | (1270)(1272) | |
| $\tilde{D}\ ^1A_2''$ | $D_{3h}$ | 69731 | | 910.6[e] | | | | | | | | | $\tilde{D} \leftarrow \tilde{X}$ 1435–1270 Å | (1270)(1272) | |
| $\tilde{C}\ ^1A_1'$ | $D_{3h}$ | (63771)[f] | | (952) | | | — | 9.17[g] | 5.03[g] | | | $\ldots(1e')^4(1a_2'')(3pa_2'')$ | $\tilde{C} \leftarrow \tilde{X}$ 1570–1480 Å | (295)(294)(1272) | may be part of $\tilde{B}$ |
| $\tilde{B}\ ^1E''$ | $D_{3h}$ | 59225.5 | | $898._3$[h] | | | — | 10.29[i] | 5.21[i] | 1.027 | 120° | $\ldots(1e')^4(1a_2'')(3pe')$ | $\tilde{B} \leftarrow \tilde{X}$ 1690–1400 Å | (295)(1272) | see Fig. 101 |
| $\tilde{A}\ ^1A_2''$ | $D_{3h}$ | 46136 | | 878[j] | | | | | | 1.08[k] | 120° | $\ldots(1e')^4(1a_2'')(3sa_1')$ | $\tilde{A} \leftarrow \tilde{X}$ 2170–1700 Å | (1270)(294)(1272) | all bands diffuse[l] |
| $\tilde{X}\ ^1A_1$ | $C_{3v}$ | 0.793[m]<br>0 | 3337.2[m]<br>3336.2 | 968.3[m]<br>932.5 | 3443.9[m]<br>3443.6 | 1627.4[m]<br>1626.1 | — | 9.4443[n] | 6.196[p] | 1.0173[q] | 107.8°[q] | $\ldots(1e)^4(3a_1)^2$ | infrared bands | (108)(109) | |

$CH_3$: [a] I.P. ($CD_3$) = 9.833 eV. [b] From the appearance potential of $CH_2^+$ from $CH_3$ (720)(1255) combined with the spectroscopic I.P. of $CH_2$. [c] From $B_0$ of $CD_3$. [d] (69a) find 730 cm$^{-1}$ in the infra-red spectrum of $CH_3$ in an argon matrix.

$NH_3$: [a] From photoionization experiments (1276). The states $\tilde{A}$, $\tilde{D}$, $\tilde{E}$ form a Rydberg series with this limit. [b] See (956)(67) and (389). [c] $x_{22} = 17.2$. [d] $x_{22} = 10.0$. [e] $x_{22} = 10.04$.
[f] First observed band, probably not 0 – 0 band. [g] The vibrational numbering is not established; the $B$ and $C$ values given refer to the lowest observed level. $B$ decreases rapidly with $v_2'$.
[h] $x_{22} = 18.44$. [i] $\alpha_2^B = 0.577$, $\alpha_2^C = -0.043$, $\zeta_0 = 0.89$, $q_0 = 0.9_0$. [j] $x_{22} = 4.0$. [k] From $ND_3$ spectrum.
[l] Diffuseness is much less for the first bands of $ND_3$ and (294) has obtained a rotational analysis. [m] These are the two inversion components.
[n] $\alpha_1^B = 0.177$, $\alpha_2^B = 0.015$, $\alpha_3^B = 0.176$, $\alpha_4^B = -0.230$. [p] $\alpha_1^C = 0.037$, $\alpha_2^C = 0.098$, $\alpha_3^C = -0.009$, $\alpha_4^C = 0.066$. [q] The equilibrium values are 1.0124 Å and 106.67°.

TABLE 65 (*continued*)

| State | Point Group | $T_0$ | Vibrational Frequencies $\nu_1$ | $\nu_2$ | $\nu_3$ | $\nu_4$ | Rotational Constants $A_0$ | $B_0$ | $C_0$ | $r_0$(Å) | $\alpha$ | Electron Configuration | Observed Transitions | References | Remarks |
|---|---|---|---|---|---|---|---|---|---|---|---|---|---|---|---|
| | **$PH_3$** | | | | | | | | | | | | | | |
| $\tilde{D}$ | | (75000) | | | | | | | | | | | $\tilde{D} \leftarrow \tilde{X}$ 1330–1260 Å | (593) | [a] |
| $\tilde{C}$ | | (66660) | | 420 | | | | | | | | | $\tilde{C} \leftarrow \tilde{X}$ 1500–1360 Å | (593) | diffuse bands |
| $\tilde{B}$ | $C_{3v}$ | 62801 | | 495.8[b] | | | | | | | | | $\tilde{B} \leftarrow \tilde{X}$ 1590–1490 Å | (593) | [c] |
| $\tilde{A}$ | | (43000) | Continuum with maximum at 1800 Å | | | | | | | | | | $\tilde{A} \leftarrow \tilde{X}$ 2300–1600 Å | (189)(803) (593)(463) | [d] |
| $\tilde{X}\ {}^1A_1$ | $C_{3v}$ | 0 | 2322.9 | 992.0 | 2327.7 | 1122.4 | — | 4.45236[e] | 3.93 | 1.421 | 93.3°[f] | | infrared and microwave sp. | (806)(1171) (168)(560) | |
| | **$AsH_3$** | see (593) | | | | | | | | | | | | | |

$PH_3$: [a] Observed only for $PD_3$. [b] $x_{22} = 7.8_4$. [c] (593) conclude that in this state the barrier toward inversion is low.
[d] (463) gives maximum at 1910 Å; extent of continuum depends on pressure and path length.
[e] $D_J = 0.000147$ cm$^{-1}$ [(1171) give 0.000105]; $D_{JK} = -0.000136$ cm$^{-1}$; $\alpha_2^B = 0.175$, $\alpha_4^B = -0.109$, $\zeta_3 = 0.01$, $\zeta_4 = -0.456$. [f] From microwave data on $PH_2D$ and $PHD_2$ of (1137).

TABLE 66

MOLECULAR CONSTANTS OF THE ELECTRONIC STATES OF FOUR-ATOMIC DI-HYDRIDES

| State | Point Group | $T_0$ | Vibrational Frequencies $\nu_1$ | $\nu_2$ | $\nu_3$ | $\nu_4$ | $\nu_5$ | Rotational Constants $A_0$ | $B_0$ | $C_0$ | $r_0$(Å) | $\alpha$ | Electron Configuration | Observed Transitions | References | Remarks |
|---|---|---|---|---|---|---|---|---|---|---|---|---|---|---|---|---|
| $C_2H_2$ | | I.P. = 11.41 eV; $D$(H—$C_2$H) = 4.9 eV from (41); $D$(HC—CH) = 9.88$_6$ eV | | | | | | | | | | | | | | |
| | | Continuous absorption | | | | | | | | | | | | 1000–600 Å | (1257)(836) (1105) | |
| | | Rydberg series joining on to $\tilde{D}$: $\nu = 92076 - R/(n - 0.50)^2$; $n = 3, 4, \ldots, 10$ | | | | | | | | | | | | | (1013) | |
| | | Rydberg series joining on to $\tilde{C}$: $\nu = 91950 - R/(n - 0.95)^2$; $n = 3, 4, \ldots, 10$ | | | | | | | | | | | | | (1013) | |
| $\tilde{G}\ ^1\Pi_u$ | $D_{\infty h}$ | 80109.9 | | 1819 | | | | — | 1.1013[a] | — | | 180° | | $\tilde{G} \leftarrow \tilde{X}$ 1250–1160 Å | (523)(927) | $D$ bands of (927), fine structure resolved |
| $\tilde{F}$ | $C_{2h}$ | 74747[b] | | 1751 | 434[c] | | | | | | | | | $\tilde{F} \leftarrow \tilde{X}$ 1338–1307 Å | (1301) | $C-A$ system of (1301)[d] |
| $\tilde{E}$ | ($C_{2h}$) | 74622[b] | 2808 | 1500 | 759[c] | (630)[e] | | (8) | | | | | | $\tilde{E} \leftarrow \tilde{X}$ 1340–1278 Å | (1301) | $B-A$ system of (1301)[d] |
| $\tilde{D}$ | $D_{\infty h}$ | 74498 | (2748) | 1781 | | (393) | | | | | | 180° | | $\tilde{D} \leftarrow \tilde{X}$ 1342–1294 Å | (1013)(1301) | $3R'$ system of (1301)[d] |
| $\tilde{C}(^1\Pi_u)$ | $D_{\infty h}$ | 65814 | | 1849 | | | | | | | | 180° | | $\tilde{C} \leftarrow \tilde{X}$ 1519–1403 Å | (1013)(1301) | $3R$ system of (1301)[f] |
| $\tilde{B}$ | | | diffuse unassigned bands | | | | | | | | | | | $\tilde{B} \leftarrow \tilde{X}$ 2000–1550 Å | (517)(1301) (927) | repr. (927) |
| $\tilde{A}\ ^1A_u$ | $C_{2h}$ | 42197.7[g] | | 1385 | 1047.6[h] | | | 12.94[i] | 1.1243 | 1.0297 | $r_0$(CC) = 1.38$_8$ | 120° | $\ldots(1\pi_u)^3 1\pi_g$ | $\tilde{A} \leftarrow \tilde{X}$ 2369–2100 Å | (600)(605) | see Fig. 83[j] |
| $\tilde{X}\ ^1\Sigma_g^+$ | $D_{\infty h}$ | 0 | 3372.5 | 1973.5 | 3294.85 | 611.70 | 729.15 | — | 1.17660[k] | — | $r_0$(CH) = 1.058; $r_0$(CC) = 1.208[l] | 180° | $\ldots(1\pi_u)^4$ | infra-red and Raman sp. | (1296)(198) (377)(710) | |

$C_2H_2$: [a] $q = 0.0073$. [b] Corrected to $K = 0$ in upper state.

[c] The absence of a long progression in $\nu_3'$ throws doubt on the non-linearity of the molecule in this state, but the presence of hot bands with $\nu_4''$ is in its favor.

[d] All bands of these systems are slightly diffuse so that the fine structure cannot be resolved although band heads are pronounced.

[e] The identification of the twisting vibration seems doubtful. It it were confirmed it would prove that the molecule is non-planar in this state.

[f] Broad diffuse bands; peak absorption coefficient $k = 11500$ cm$^{-1}$. [g] 0 – 0 band from (605) corrected to $K = 0$ in upper state. The 0 – 1 hot band of (605) with $\nu_4''$ from infrared gives 42198.67

[h] $x_{33} = -8.71$. [i] $D_K = 0.03$, $\alpha_3^A$ is strongly negative.

[j] A system of emission bands has been found by (333) in a discharge through $C_2H_2$ in the region 2400–2900 Å. They may belong to the $\tilde{A} - \tilde{X}$ system.

[k] $\alpha_1 = 0.008_5$, $\alpha_2 = 0.0063_0$, $\alpha_3 = 0.0056_0$, $\alpha_4 = -0.0013_0$, $\alpha_5 = -0.0022_6$; $D_0 = 1.51 \times 10^{-6}$; $q_4 = 0.0052_7$, $q_5 = 0.0046_3$, cm$^{-1}$.

[l] Equilibrium values: $r_e$(CH) = 1.060, $r_e$(CC) = 1.203 Å [from (710)].

TABLE 66 (*continued*)

| State | Point Group | $T_0$ | Vibrational Frequencies $\nu_1$ | $\nu_2$ | $\nu_3$ | $\nu_4$ | $\nu_5$ | Rotational Constants $A_0$ | $B_0$ | $C_0$ | $r_0$(Å) | $\alpha$ | Electron Configuration | Observed Transitions | References | Remarks |
|---|---|---|---|---|---|---|---|---|---|---|---|---|---|---|---|---|
| $\mathbf{H_2CO}$ | I.P. = 10.88 eV; 3.02 eV ≤ $D$(H—HCO) ≤ 3.56 eV[a]; $D$(2H+CO) = 4.53$_3$ eV | | | | | | | | | | | | | | | |
| | | Rydberg series joining on to $\tilde{E}$: $\nu = 87830 - R/(n - 0.40)^2$; $n = 3, 4, \ldots, 8$ | | | | | | | | | | | | | (1014)(61) | See text |
| | | Two Rydberg series starting with $\tilde{C}$, $\tilde{G}$ and $\tilde{D}$, $\tilde{H}$ respectively but merging into a single series for the higher members: $\nu = 87710 - R/(n - 0.70)^2$; $n = 3, 4, \ldots, 9$ | | | | | | | | | | | | | (1014)(61) | See text |
| | | Rydberg series joining on to $\tilde{B}$, $\tilde{F}$: $\nu = 87809 - R/(n - 1.04)^2$; $n = 3, 4, \ldots, 8$ | | | | | | | | | | | | | (61) | See text |
| $\tilde{H}(^1B_2)$ | $(C_{2v})$ | 77643 | | | | | | | | | | | $\ldots(1b_1)^2(2b_2)(4pa_1)$ | $\tilde{H} \leftarrow \tilde{X}$ 1288 Å | (61) | Some vibrational structure |
| $\tilde{G}(^1A_1)$ | $(C_{2v})$ | 77287 | | | | | | | | | | | $\ldots(1b_1)^2(2b_2)(4pb_2)$ | $\tilde{G} \leftarrow \tilde{X}$ 1294 Å | (61) | Single band |
| $\tilde{F}$ | | 74648 | | | | | | | | | | | $\ldots(1b_1)^2(2b_2)(4sa_1)$ | $\tilde{F} \leftarrow \tilde{X}$ 1340 Å | (61) | Some vibrational structure |
| $\tilde{E}$ | $(C_{2v})$ | 71588 | | | | | | | | | | | $\ldots(1b_1)^2(2b_2)(3d)$ | $\tilde{E} \leftarrow \tilde{X}$ 1397 Å | (1014)(61) | Single band |
| $\tilde{D}\ ^1B_2$ | $(C_{2v})$ | 65634 | | | | | | | | | | | $\ldots(1b_1)^2(2b_2)(3pa_1)$ | $\tilde{D} \leftarrow \tilde{X}$ 1524 Å | (61) | Some vibrational structure |
| $\tilde{C}\ ^1A_1$ | $(C_{2v})$ | 64264 | | | | | | | | | | | $\ldots(1b_1)^2(2b_2)(3pb_2)$ | $\tilde{C} \leftarrow \tilde{X}$ 1556 Å | (1014)(61) | Repr. (1014) |
| $\tilde{B}\ ^1A''(B_2)$ | $(C_s)$ | 57133 | | | | (425) | | | | | | | $\ldots(1b_1)^2(2b_2)(3sa_1)$ | $\tilde{B} \leftarrow \tilde{X}$ 1750–1650 Å | (1014)(61) | Repr. (1014) diffuse |
| $\tilde{A}\ ^1A''(A_2)$ | $C_s$ | 28188.0 | | 1182.6[c] | | 683[d] | 2968[k] | 8.7517[e] | 1.1245[e] | 1.0123[e] | $r$(CH) = 1.09$_3$ $r$(CO) = 1.32$_3$ | ∠HCH = 119° out-of-plane angle 31° | $\ldots(1b_1)^2(2b_2)(2b_1)$ | $\tilde{A} \leftrightarrow \tilde{X}$ 3530–[f] 2300 Å | (277)(138) (178) (754aa)(962a) (964a) | Repr. in (1078)(367)[g] |
| $\tilde{a}\ ^3A''(A_2)$ | $C_s$ | 25194[h] | | 1251 | | 643[i] | | 8.6$_9$ | 1.156 | 1.041 | $r$(CO) = 1.31$_2$ | out-of-plane angle 35° | $\ldots(1b_1)^2(2b_2)(2b_1)$ | $\tilde{a} \leftarrow \tilde{X}$ 3967–3600 Å | (1078)(278) (559) | Repr. in (1078) |
| $\tilde{X}\ ^1A_1$ | $C_{2v}$ | 0 | 2766.4 | 1746.1 | 1500.6 | 1167.$_3$ | 2843.4[j] | 9.4053 | 1.2953$_6$ | 1.1342$_5$ | $r$(CH) = 1.102 $r$(CO) = 1.210 | ∠HCH = 121.1° | $\ldots(1b_1)^2(2b_2)^2$ | infrared and microwave sp. | (730) (342)(343) (123a)(962a) (754aa) | |
| $\mathbf{H_2O_2}$ | $D$(HO—OH) = 2.12 eV[a]; $D$(HOO—H) = 3.88 eV[b]; Continuous absorption only, see (1232), (1119) and (378) | | | | | | | | | | | | | | | |

$H_2CO$: [a] From weak predissociation. However, (678) find weak photodissociation by lines at 3650 Å corresponding to $D$(H—HCO) ≤ 3.40 eV. (1061) gives 3.23 eV from electron impact data. The lower limit given is from $D$(H—CO) (Table 63) and $D$(2H + CO) obtained thermochemically.

[b] This is the energy of the lower inversion level; the upper level is at 28312.6 corresponding to the first main band [see (277)]. [c] Refers to upper inversion level.

[d] This is an average frequency; the observed levels are 124.5, 542.3 and 948.6 cm$^{-1}$ above 0+.

[e] Refers to upper inversion level 0−. (1078) gives $A_0$ = 8.965, $B_0$ = 1.1253, $C_0$ = 1.0057 for the 0+ level. [f] In emission the system extends to 6000 Å.

[g] Absorption bands become diffuse at about 2750 Å. Breaking off in emission above $v_4' = 0$ suggests weak predissociation above 28736 cm$^{-1}$ [see (143)]. (628) have determined the lifetime of the $\tilde{A}$ state to be 2.3 × 10$^{-7}$ sec, i.e., much smaller than expected from the absorption intensity. This discrepancy may be due to weak predissociation.

[h] This is the number given by (1077); (1078) give 25200.2 cm$^{-1}$ for the transition to 0+.

[i] This is an average frequency; according to (1078) the observed levels are 30, 537, 779 and 1171 cm$^{-1}$ above 0+. [j] $\nu_6$ = 1251.2. [k] $\nu_6$ = 1445.

Table 67

MOLECULAR CONSTANTS OF THE ELECTRONIC STATES OF FOUR-ATOMIC MONO-HYDRIDES

| State | Point Group | $T_0$ | Vibrational Frequencies | | | | | | Rotational Constants | | | | | Observed Transitions | References | Remarks |
|---|---|---|---|---|---|---|---|---|---|---|---|---|---|---|---|---|
| | | | $\nu_1$ | $\nu_2$ | $\nu_3$ | $\nu_4$ | $\nu_5$ | $\nu_6$ | $A_0$ | $B_0$ | $C_0$ | $r_0$(Å) | $\alpha$ | | | |
| **HCCN** see (827)[a] | | | | | | | | | | | | | | | | |
| **HNCN** | | | | | | | | | | | | | | | | |
| $\tilde{A}\ ^2A'$ | $C_s$ | 28994.1 | | | | | | | 22.438[a] | 0.3759 | 0.3690 | $r$(NH) = $1.03_5$<br>$r$(N...N) = 2.443 | ∠HNC<br>= $120._6$° | $\tilde{A} \leftarrow \tilde{X}$ 3440 Å | (547) | See Fig. 108 |
| $\tilde{X}\ ^2A''$ | $C_s$ | 0 | | | | | | | 21.220[b] | 0.3699 | 0.3624 | $r$(NH) = $1.03_4$<br>$r$(N...N) = 2.470 | ∠HNC<br>= $116._5$° | | | |
| **$HN_3$** see (128)[a] | | | | | | | | | | | | | | | | |
| **HNCO** see (1291) | | | | | | | | | | | | | | | | |
| **$HCO_2$** | | | | | | | | | | | | | | | | |
| $\tilde{A}(^2B_1)$ | $(C_{2v})$ | (30400) | | | | | | | | | | | | $\tilde{A} \rightarrow \tilde{X}$ 3300–4400 Å | (334)(335)<br>(1173)(973) | Extensive band system repr. (335)<br>(1173)[a] |
| $\tilde{X}(^2A_1)$ | $(C_{2v})$ | 0 | 1120 | | | | | | | | | | | | | |

HCCN: [a] Discrete absorption bands in the region 3400–28006 Å obtained in the flash photolysis of N≡C—CH=$N_2$, not yet analyzed (827); main band at 3180 Å.
HNCN: [a] $D_K' = 0.0254$. [b] $D_K'' = 0.0131$ cm$^{-1}$.
$HN_3$: [a] Incidental spectrograms taken in this laboratory show discrete bands in the region 3150–2950 Å.
$HCO_2$: [a] Observed in fluorescence of formic acid excited by light in the vacuum ultraviolet region and in electric discharges; (334) gives schematic spectrum. No wavelengths of bands have beeen published.

TABLE 67 (*continued*)

| State | Point Group | $T_0$ | $\nu_1$ | $\nu_2$ | $\nu_3$ | $\nu_4$ | $\nu_5$ | $\nu_6$ | $A_0$ | $B_0$ | $C_0$ | $r_0$(Å) | $\alpha$ | Observed Transitions | References | Remarks |
|---|---|---|---|---|---|---|---|---|---|---|---|---|---|---|---|---|
| **HONO** | | | | | | | | | | | | | | | | |
| $\tilde{A}_{cis}(^1A'')$ | $C_s$ | 26500 | | 1107[a] | | | | | | | | | | $\tilde{A} \leftarrow \tilde{X}$ 3850–3150 Å | (820)(669) | Progressions of diffuse bands[c] |
| $\tilde{A}_{trans}(^1A'')$ | $(C_s)$ | 26034 | | 1117[b] | | | | | | | | | | | | |
| $\tilde{X}_{cis}\ ^1A''$ | $C_s$ | 180[d] | 3426 | 1639 | (1370) | 856 | 620 | 638 | $A-\bar{B}=2.40$ | | | $r(ON)=1.46$ | $\angle ONO = 114°$ | infrared spectrum | (647)(290) | |
| $\tilde{X}_{trans}\ ^1A''$ | $C_s$ | 0 | 3590 | 1698 | 1264 | 793 | 598 | 544 | $A-\bar{B}=3.2$ | | | $r(NO)=1.20$ | $\angle ONO = 118°$ | | | |
| **HFCO** | | | | | | | | | | | | | | | | |
| $\tilde{A}\ ^1A$ | $C_1$ | (36360)[a] | (2941) | 1105[b] | (1185) | (1107) | 467[c] | (570)[d] | [2.166][e] | [0.390$_4$][e] | [0.328$_2$][e] | $r(CO)=1.36$ | $\alpha_1=109°$[f] $\beta=\sim 20°$ | $\tilde{A} \leftarrow \tilde{X}$ 2700–2000 Å | (418)(419) (964) | [g]Predissociation suggested for $\lambda \lesssim 2200$ Å |
| $\tilde{X}\ ^1A'$ | $C_s$ | 0 | 2981.0 | 1836.9 | 1342.5 | 1064.8 | 662.5 | (1175)[h] | 3.04056 | 0.39227 | 0.34680 | $r(CF)=1.338$ $r(CO)=1.181$ $r(CH)=1.095$ | $\alpha_1=122.8$[f] $\alpha_2=109.9°$ $\alpha_3=127.3°$ | infrared and microwave sp. | (732)(372) (1170)(848) | |

HONO: [a] $x_{22} = -19.75\ cm^{-1}$. [b] $x_{22} = -22.3\ cm^{-1}$. [c] Diffuseness decreases to shorter wavelengths.
[d] The barrier separating *cis* and *trans* forms is estimated by (461) to be 3040 $cm^{-1}$.

HFCO: [a] Not certain, see (419). [b] $x_{22} = -6.5$. [c] $x_{55} = +16$. [d] See (419).
[e] Vibrational assignment of analyzed band (40281.5 $cm^{-1}$) is not certain, but its upper state is not the 00 . . . level. [f] $\alpha_1 = \angle FCO$, $\alpha_2 = \angle HCF$, $\alpha_3 = \angle HCO$, $\beta$ = pyramidal angle.
[g] The transition moment of the band that has been analyzed is in the *c*-axis. [h] From hot bands of the ultraviolet spectrum (418). (1245) calculate 1007 $cm^{-1}$ from $\nu_6$ of DFCO.

TABLE 68

MOLECULAR CONSTANTS OF THE ELECTRONIC STATES OF FOUR-ATOMIC NON-HYDRIDES

| State | Point Group | $T_0$ | Vibrational Frequencies $\nu_1$ | $\nu_2$ | $\nu_3$ | $\nu_4$ | $\nu_5$ | $\nu_6$ | Rotational Constants $A_0$ | $B_0$ | $C_0$ | $r_0$(Å) | $\alpha$ | Observed Transitions | References | Remarks |
|---|---|---|---|---|---|---|---|---|---|---|---|---|---|---|---|---|
| **$C_2N_2$** | | I.P. = 13.57 eV[a]; $D$(NC—CN) = $6.26_1$ eV[b] | | | | | | | | | | | | | | |
| | | Higher members of Rydberg series, not individually assigned; limit near 900 Å | | | | | | | | | | | | | (1029) | |
| $\tilde{C}$ | | (76000) | | | | | | | | | | | | $\tilde{C} \leftarrow \tilde{X}$ 1320–1250 Å | (1029) | Repr. (1029) |
| $\tilde{B}$ | | (60000) | 1900 | | | | | | | | | | | $\tilde{B} \leftarrow \tilde{X}$ 1680–1450 Å | (1029) | Repr. (1029) |
| $\tilde{A}(^1\Delta_u)$ | $D_{\infty h}$ | 44903 | 2140 | | | 274 | | — | | 0.1536[f] | — | | — | $\tilde{A} \leftarrow \tilde{X}$ 2260–1820 Å | (1312)(182aa) | Extensive band system |
| $\tilde{a}\ ^3\Sigma_u^+$ | $D_{\infty h}$ | 33289.9 | 2049.6[c] | 899.1 | | 264 | 211 | — | | 0.1532[d] | — | | — | $\tilde{a} \leftarrow \tilde{X}$ 3020–2400 Å | (1316)(175) (175a) | Red shaded double headed bands |
| $\tilde{X}\ ^1\Sigma_g^+$ | $D_{\infty h}$ | 0 | 2329.9 | 854.2[e] | 2157.8 | 507 | 233.1 | — | — | 0.1571 | — | $r$(CC) = 1.389<br>$r$(CN) = $1.15_4$ | 180° | infrared and Raman sp. | (872)(721) (860)(794a) | |
| **$NO_3$** | | | | | | | | | | | | | | | | |
| $\tilde{A}(^2E')$ | ($D_{3h}$) | 15089 | 930 | | | | | | | | | | | $\tilde{A} \leftarrow \tilde{X}$ 6650–5000 Å | (646)(1044) (596) | Diffuse bands, repr. (1044) |
| $\tilde{X}^2A_2'$ | $D_{3h}$ | 0 | | | | | | | | | | | | | | |
| **$F_2B^{11}O$** | | [c] | | | | | | | | | | | | | | |
| $\tilde{A}(^2A_2)$ | $C_{2v}$ | 17171.0 | | 875 | 480.6 | | | | | | | | | $\tilde{A} \rightarrow \tilde{X}$ 6330–5540 Å | (697)(801) | ∥ bands of oblate top[a] |
| $\tilde{X}(^2B_2)$[b] | $C_{2v}$ | 0 | 1369 | 856.0 | 491.0 | | | | | | | | | | | |

$C_2N_2$: [a] From electron impact experiments of (808). The VUV spectrum suggests 13.8 eV (1029).
[b] From thermochemical data assuming $D$(CN) = $7.5_0$ eV. From shock-tube experiments (1225) derive $D$(NC—CN) = 5.42 eV. [c] $x_{11} = -15.1$. [d] $\lambda = -0.22$, $\gamma = -0.018$.
[e] From the $\tilde{a} - \tilde{X}$ system (175a). [f] From $\nu(R_{\text{head}}) - \nu_{\text{origin}}$ of the band $0 - \nu_{4d}$ and $\alpha''_{4d}$ of (794a).

$F_2BO$: [a] Emission bands in discharge through $BF_3$. Emitter shown to contain one B and one O atom by isotope studies. [b] This is species of ground state expected from a Walsh diagram (1267).
[c] A second band system near 4465Å has recently been described by (800a).

TABLE 68 (*continued*)

| State | Point Group | $T_0$ | Vibrational Frequencies | | | | | | Rotational Constants | | | | | Observed Transitions | References | Remarks |
|---|---|---|---|---|---|---|---|---|---|---|---|---|---|---|---|---|
| | | | $\nu_1$ | $\nu_2$ | $\nu_3$ | $\nu_4$ | $\nu_5$ | $\nu_6$ | $A_0$ | $B_0$ | $C_0$ | $r_0$(Å) | $\alpha$ | | | |
| **$F_2CO$** | | | | | | | | | | | | | | | | |
| $\tilde{B}$ | | 48495 | 1705 | | | | | | | | | | | $\tilde{B}\leftarrow\tilde{X}$ 2065–1800 Å | (852) | Progression of red shaded bands |
| $\tilde{X}\,^1A_1$ | $C_{2v}$ | 0 | 1942[a] | 965 | 584 | 774 | 1249 | 626 | 0.394054 | 0.392037 | 0.196166 | $r$(CF) = 1.312 $r$(CO) = 1.174 | ∠FCF = 108.0° | Raman and microwave sp. | (932)(729) (728) | |
| **FClCO** | | see (1326) | | | | | | | | | | | | | | |
| **$Cl^{35}_2CO$** | | I.P. = $11.7_7$ eV[a] | | | | | | | | | | | | | | |
| $\tilde{G}$ | ($C_{2v}$) | 88253 | | 278 | | | | | | | | | | $\tilde{G}\leftarrow\tilde{X}$ 1150–1070 Å | (725) | Single long progression |
| $\tilde{F}$ | ($C_{2v}$) | 80816 | | 370 | | | | | | | | | | $\tilde{F}\leftarrow\tilde{X}$ 1240–1220 Å | (725) | |
| $\tilde{E}$ | ($C_{2v}$) | 74810 | | 582 | | | | | | | | | | $\tilde{E}\leftarrow\tilde{X}$ 1340–1290 Å | (725) | |
| $\tilde{D}$ | ($C_{2v}$) | 72538 | | 313 | | | | | | | | | | $\tilde{D}\leftarrow\tilde{X}$ 1380–1350 Å | (725) | |
| $\tilde{C}$ | ($C_{2v}$) | 66707 | | 270 | | | | | | | | | | $\tilde{C}\leftarrow\tilde{X}$ 1500–1410 Å | (725) | Single long progression |
| $\tilde{B}$ | | (61000) | continuum with max. at 64700 $cm^{-1}$ | | | | | | | | | | | $\tilde{B}\leftarrow\tilde{X}$ 1630–1470 Å | (725) | |
| $\tilde{A}(^1A)$ | ($C_1$) | 32730 | | | | 581 | | 430 | | | | | | $\tilde{A}\leftarrow\tilde{X}$ 3050–2380 Å | (501)(419) | Repr. (501); predissociation for $\lambda$ < 2750 Å |
| $\tilde{X}\,^1A_1$ | $C_{2v}$ | 0 | 1827 | 567 | 285 | 580 | 849 | 440 | 0.26414 | 0.11591 | 0.080464 | $r$(CCl) = 1.746 $r$(CO) = 1.166 | ∠ClCCl = 111.3° | infrared and microwave sp. | (953)(1073) | |

$F_2CO$: [a] Shifted by Fermi resonance with $2\nu_2$ which is at 1907 $cm^{-1}$.
$Cl_2CO$: [a] From electron impact data (881).

TABLE 68 (*continued*)

| State | Point Group | $T_0$ | Vibrational Frequencies $\nu_1$ | $\nu_2$ | $\nu_3$ | $\nu_4$ | $\nu_5$ | $\nu_6$ | Rotational Constants $A_0$ | $B_0$ | $C_0$ | $r_0$(Å) | $\alpha$ | Observed Transitions | References | Remarks |
|---|---|---|---|---|---|---|---|---|---|---|---|---|---|---|---|---|
| **$Cl^{35}_2CS$** | | | | | | | | | | | | | | | | |
| $\tilde{C}(^1A_1)$ | $C_{2v}$ | 36045 | | 442 | 208 | | | | | | | | | $\tilde{C} \leftarrow \tilde{X}$ 2770–2390 Å | (20)(165) | Diffuse bands |
| $\tilde{B}(^1B_2)$ | $C_{2v}$ | (33380) | | 468 | 310 | | | | | | | | | $\tilde{B} \leftarrow \tilde{X}$ 2970–2690 Å | (20)(165) | Sharp bands |
| $\tilde{A}(^1A_2)$ | $C_s$ | 18715.9 | 907.$_4$ | 480.0 | 245.0 | 285.8[a] | | 366.6 | | | | | out-of-plane angle = 32° | $\tilde{A} \leftarrow \tilde{X}$ 5950–3900 Å | (500)(313)(165)(139) | Sharp bands[b] |
| $\tilde{a}(^3A_2)$ | $C_s$ | (15000) | | (400) | | | | | | | | | | $\tilde{a} \leftarrow \tilde{X}$ 7000–5300 Å | (165) | Sharp bands[b] |
| $\tilde{X}\ ^1A_1$ | $C_{2v}$ | 0 | 1139.$_0$ | 503.$_5$ | 288.$_5$ | 471.$_0$ | 818 | (292) | | | | | | infrared sp. | (779)(306)(1205)(139) | c |
| **$O_4$** | see (383)(520)(76)(664a) | | | | | | | | | | | | | | | |
| **$SO_3$** | see (370) | | | | | | | | | | | | | | | |

$Cl_2CS$: [a] Average of two inversion doublets: inversion doubling for $v_4 = 0$ is 0.42, for $v_4 = 1$ it is 12.9 cm$^{-1}$.
[b] Although the molecule is definitely non-planar in these states the classification according to $C_{2v}$ is meaningful since the inversion doubling is large (see p. 22f). The barrier height is about 610 cm$^{-1}$. [c] The frequencies given are mainly the gas frequencies obtained from the electronic spectrum (139) which except for $\nu_3$ agree well with the infrared data of (779) and (306).

TABLE 68 (*continued*)

| State | Point Group | $T_0$ | Vibrational Frequencies $\nu_1$ | $\nu_2$ | $\nu_3$ | $\nu_4$ | $\nu_5$ | $\nu_6$ | Rotational Constants $A_0$ | $B_0$ | $C_0$ | $r_0$(Å) | $\alpha$ | Observed Transitions | References | Remarks |
|---|---|---|---|---|---|---|---|---|---|---|---|---|---|---|---|---|
| **$ClO_3$** | | see (432) continuous absorption only | | | | | | | | | | | | | | |
| **$F_2O_2$** | | see (153) continuous absorption only | | | | | | | | | | | | | | |
| **$PF_3$** | | | | | | | | | | | | | | | | |
| $\tilde{E}$ | | strong continuous absorption below 1130 Å | | | | | | | | | | | | $\tilde{E} \leftarrow \tilde{X}$ 1130–1000 Å | (593) | |
| $\tilde{D}$ | $C_{3v}$ | 82511[a] | | 461 | | | | | | | | | (110°)[b] | $\tilde{D} \leftarrow \tilde{X}$ 1215–1130 Å | (593) | Single long progression |
| $\tilde{C}$ | $C_{3v}$ | 71174[a] | | 460 | | | | | | | | | (110°)[b] | $\tilde{C} \leftarrow \tilde{X}$ 1405–1260 Å | (593) | Single long progression |
| $\tilde{B}$ | | (60000) | continuous absorption with two peaks at 1564 and 1515 Å | | | | | | | | | | | $\tilde{B} \leftarrow \tilde{X}$ 1650–1450 Å | (593) | |
| $\tilde{X}\,{}^1A_1$ | $C_{3v}$ | 0 | 892 | 487 | 860 | 344 | — | — | — | 0.26084 | | $1.53_5$ | (100°)[b] ass. | infrared and microwave sp. | (1310)(423) | |
| **$PCl_3$** | | see (593) and (463) | | | | | | | | | | | | | | |
| **$PBr_3$** | | see (622) | | | | | | | | | | | | | | |

$PF_3$: [a] Not certain that this is 0 – 0 band. [b] Angle of PF with $C_3$ axis.

TABLE 69

MOLECULAR CONSTANTS OF THE ELECTRONIC STATES OF FIVE-ATOMIC TETRA- AND TRI-HYDRIDES

| State | Point Group | $T_0$ | Vibrational Frequencies $\nu_1$ | $\nu_2$ | $\nu_3$ | $\nu_4$ | $\nu_5$ | $\nu_6$ | Rotational Constants $A_0$ | $B_0$ | $C_0$ | $r_0$(Å) | $\alpha$ | Observed Transitions | References | Remarks |
|---|---|---|---|---|---|---|---|---|---|---|---|---|---|---|---|---|
| **$CH_4$** | I.P. = 12.99 eV[a]; $D(CH_3—H) = 4.40_6$ eV[a] | | | | | | | | | | | | | | | |
| $\tilde{A}$ ($^1F_2$) | ($T_d$) | (68730) | Continuous absorption with peak at 930 Å probably consisting of several electronic transitions | | | | | | | | | | | $\tilde{A} \leftarrow \tilde{X}$ 1455–500 Å | (279)(1092) | |
| $\tilde{X}\,^1A_1$ | $T_d$ | 0 | 2916.5 | 1533.6 | 3019.49 | 1306.2 | — | — | — | 5.2412[b] | — | 1.0940 | | Infrared and Raman sp. | (1168)(514) (375) | |
| **$CH_3O$** (methoxy) see (1174)[a] | | | | | | | | | | | | | | | | |
| **$CH_3F$** see (1162) | | | | | | | | | | | | | | | | |
| **$CH_3Cl^{35}$** | I.P. = 11.22 eV; $D(CH_3—Cl)$ = 3.5 eV[a] | | | | | | | | | | | | | | | |
| | | Two parallel Rydberg series joining on to $\tilde{E}$ and $\tilde{F}$: $\nu = \left.\begin{matrix}91180\\90500\end{matrix}\right\} - R/(n-0.50)^2$; $n = 3, 4, \ldots$ | | | | | | | | | | | | | (1016) | [b] |
| $\tilde{F}$ | | (71760) | | (1320) | | | | | | | | | | $\tilde{F} \leftarrow \tilde{X}$ 1400–1320 Å | (1016) | Diffuse |
| $\tilde{E}$ | | (71110) | | (1320) | | | | | | | | | | $\tilde{E} \leftarrow \tilde{X}$ 1400–1320 Å | | Diffuse |
| $\tilde{C}$ | | 62500 | | (1040) | | | | | | | | | | $\tilde{C} \leftarrow \tilde{X}$ 1600–1565 Å | (539)(504) | Very diffuse |
| $\tilde{A}$ | | (50150) | Continuous absorption with maximum at 57900 cm$^{-1}$ | | | | | | | | | | | $\tilde{A} \leftarrow \tilde{X}$ 2000–1600 Å | (539)(504) | |
| $\tilde{X}\,^1A_1$ | $C_{3v}$ | 0 | 2967.8 | 1355.0 | 732.8 | 3043.6 | 1488.2 | 1017.5 | $5.09_7$[c] | 0.443401 | — | $r$(CCl) = 1.781 $r$(CH) = 1.113 | $110.5_2$° | infrared and microwave sp. | (40)(984) (563) | [d] |

$CH_4$: [a] From (1273) and (41). [b] $\alpha_2 = -0.122$, $\alpha_3 = 0.0416$, $\zeta_3 = 0.0552$.

$CH_3O$: [a] The spectrum assigned by (811) to $CH_3O$ has been shown by (927a) to be due to CHONO.

$CH_3Cl$: [a] From electron impact work (777). [b] The two limits correspond to the components $E_{\frac{1}{2}}$ and $E_{\frac{3}{2}}$ of the $^2E$ ground state of the $CH_3Cl^+$ ion. [c] $\zeta_4 = 0.100$, $\zeta_5 = -0.273$, $\zeta_6 = 0.222$.
[d] The values of $\nu_4$, $\nu_5$, $\nu_6$ differ slightly from those quoted in Volume II because of an error in Table 85 of that volume. They refer to the energy difference between the $J = 0$, $K = 0$ levels of upper and lower state (excluding the $\zeta^2$ term; see p. 404 of Vol. II).

TABLE 69 (*continued*)

| State | Point Group | $T_0$ | Vibrational Frequencies $\nu_1$ | $\nu_2$ | $\nu_3$ | $\nu_4$ | $\nu_5$ | $\nu_6$ | Rotational Constants $A_0$ | $B_0$ | $C_0$ | $r_0$(Å) | $\alpha$ | Observed Transitions | References | Remarks |
|---|---|---|---|---|---|---|---|---|---|---|---|---|---|---|---|---|
| | **$CH_3Br^{79}$** | I.P. = $10.54_1$ eV; $D(CH_3—Br) = 2.9$ eV [a] | | | | | | | | | | | | | | |
| | | Two parallel Rydberg series joining on to $\tilde{G}$ and $\tilde{H}$: $\nu = \left.\begin{matrix}87560\\85020\end{matrix}\right\} - R/(n - 0.10)^2$; $n = 4, 5, \ldots, 11$ | | | | | | | | | | | | 1282–1142 Å | (1016) | [b] |
| | | Rydberg series joining on to $\tilde{F}$: $\nu = 87560 - R/(n - 0.55)^2$; $n = 4, 5, \ldots$ | | | | | | | | | | | | 1456–1142 Å | (1016) | |
| | | Rydberg series joining on to $\tilde{E}$: $\nu = 85020 - R/(n - 0.58)^2$; $n = 4, 5, \ldots, 11$ | | | | | | | | | | | | 1515–1176 Å | (1016) | |
| $\tilde{H}$ | | 80360 | | | | | | | | | | | | $\tilde{H} \leftarrow \tilde{X}$ 1244 Å | (1016) | |
| $\tilde{G}$ | | 77980 | | | | | | | | | | | | $\tilde{G} \leftarrow \tilde{X}$ 1282 Å | (1016) | |
| $\tilde{F}$ | | 68680 | | | | | | | | | | | | $\tilde{F} \leftarrow \tilde{X}$ 1456 Å | (1016) | |
| $\tilde{E}$ | | 66020 | | | | | | | | | | | | $\tilde{E} \leftarrow \tilde{X}$ 1515 Å | (1016) | |
| $\tilde{D}$ | | 59135 | | 1080 | | | | | | | | | | $\tilde{D} \leftarrow \tilde{X}$ 1690–1660 Å | (539)(504) | Diffuse bands; *C* system of (539) |
| $\tilde{C}$ | | 55990 | | 1285 | | | | | | | | | | $\tilde{C} \leftarrow \tilde{X}$ 1790–1745 Å | (539)(504) | Diffuse bands; *B* system of (539) |
| $\tilde{A}$ | | (35000) | Continuous absorption with maximum at 49000 $cm^{-1}$ | | | | | | | | | | | $\tilde{A} \leftarrow \tilde{X}$ 2850–1800 Å | (382)(539) (439) | |
| $\tilde{X}\,^1A_1$ | $C_{3v}$ | 0 | 2972 | 1305.1 | 611 | 3056.3 | 1443.1 | 953.8 | $5.08_2$ [c] | 0.319160 | — | $r(CBr) = 1.939$ $r(CH) = 1.113$ | $111.2_3$° | infrared and microwave sp. | (23)(40) | [d] |

$CH_3Br$: [a] From electron impact work (777). [b] See footnote [b] of $CH_3Cl$. [c] $\zeta_4 = 0.049$, $\zeta_5 = -0.229$, $\zeta_6 = 0.169$ [from (159)]. [d] See footnote [d] of $CH_3Cl$.

TABLE 69 (*continued*)

| State | Point Group | $T_0$ | $\nu_1$ | $\nu_2$ | $\nu_3$ | $\nu_4$ | $\nu_5$ | $\nu_6$ | $A_0$ | $B_0$ | $C_0$ | $r_0$(Å) | $\alpha$ | Observed Transitions | References | Remarks |
|---|---|---|---|---|---|---|---|---|---|---|---|---|---|---|---|---|
| | | | Vibrational Frequencies | | | | | | Rotational Constants | | | | | | | |
| **$CH_3I$** | | I.P. = 9.538 eV; $D(CH_3—I)$ = 2.33 eV[a] | | | | | | | | | | | | | | |
| | | Two Rydberg series parallel to that starting with $\tilde{G}$, with Rydberg corrections $\delta = 0.45$ and 0.05 | | | | | | | | | | | | | (1016) | Members with $n = 5$ and 6 are present but diffuse; $\nu$ not given |
| | | Rydberg series joining on to $\tilde{G}$: $\nu = 81990 - R/(n - 0.20)^2$; $n = 7, 8, \ldots, 14$ | | | | | | | | | | | | | (1016) | |
| | | Two Rydberg series parallel to that starting with $\tilde{F}$, with Rydberg corrections $\delta = 0.50$ and 0.0 | | | | | | | | | | | | | (1016) | |
| | | Rydberg series joining on to $\tilde{F}$: $\nu = 76930 - R/(n - 0.25)^2$; $n = 7, 8, \ldots, 14$ | | | | | | | | | | | | | (1016) | |
| $\tilde{G}$ | | 79570 | | | | | | | | | | | | $\tilde{G} \leftarrow \tilde{X}$ 1257 Å | (1016) | |
| $\tilde{F}$ | | 74590 | | | | | | | | | | | | $\tilde{F} \leftarrow \tilde{X}$ 1341 Å | (1016) | |
| $\tilde{E}$ | | 58900 | | | | | | | | | | | | $\tilde{E} \leftarrow \tilde{X}$ 1700–1650 Å | (539) | *D* system of (539) |
| $\tilde{D}$ *E* | $C_{3v}$ | 54625 | | 1080 | | | | | | | | | | $\tilde{D} \leftarrow \tilde{X}$ 1870–1700 Å | (539) | *C* system of (539) |
| $\tilde{C}$ *E* | $C_{3v}$ | 49721 | 2660 | 1090 | 499 | | 1243 | 844 | 4.78 | | | | | $\tilde{C} \leftarrow \tilde{X}$ 2070–1880 Å | (539)(1102) (505) | *B* system of (539); see Fig. 102[b] |
| $\tilde{B}$ *E* | $C_{3v}$ | (49208) | | | | | | | | | | | | $\tilde{B} \leftarrow \tilde{X}$ 2070–1900 Å | (505)(326) | |
| $\tilde{A}$ | | (28000) | Continuous absorption with maximum at 38600 $cm^{-1}$ | | | | | | | | | | | $\tilde{A} \leftarrow \tilde{X}$ 3600–2000 Å | (1004)(539) (611) | |
| $\tilde{X}\,^1A_1$ | $C_{3v}$ | 0 | 2953.2 | 1250.8 | 533.2 | 3060.3 | 1438.2 | 882.7 | $5.11_9$[c] | 0.250217 | — | r(CI) = 2.1396 r(CH) = 1.106 | $111.1_7°$ | infrared and microwave sp. | (23)(40)(646a) | [d] |

$CH_3I$: [a] From thermochemical data (182a). [b] All bands are slightly diffuse. [c] $\zeta_4 = 0.059$, $\zeta_5 = -0.240$, $\zeta_6 = 0.206$ [from (646a)]. [d] See footnote [d] of $CH_3Cl$.

## TABLE 70

## MOLECULAR CONSTANTS OF THE ELECTRONIC STATES OF FIVE-ATOMIC DI-HYDRIDES AND MONO-HYDRIDES

| State | Point Group | $T_0$ | Vibrational Frequencies $\nu_1$ | $\nu_2$ | $\nu_3$ | $\nu_4$ | $\nu_5$ | $\nu_6$ | Rotational Constants $A_0$ | $B_0$ | $C_0$ | $r_0$(Å) | $\alpha$ | Observed Transitions | References | Remarks |
|---|---|---|---|---|---|---|---|---|---|---|---|---|---|---|---|---|
| **HCCCH** | | see (824)[a] | | | | | | | | | | | | | | |
| **$H_2CCO$** (ketene) | | I.P. = $9.60_7$ eV; $D(H_2C—CO) \leq 3.3$ eV | | | | | | | | | | | | | | |
| | | $\tilde{C}$, $\tilde{E}$, $\tilde{G}$, $\tilde{H}$ with two additional members form the Rydberg series: $\nu = 77491 - R/(n - 1.07)^2$, $n = 3, 4, \ldots, 8$ | | | | | | | | | | | | | (1025) | |
| $\tilde{H}\ ^1A_1$ | $C_{2v}$ | 72983 | | | 1010 | 611 | | | | | | | | $\tilde{H} \leftarrow \tilde{X}$ 1370–1340 Å | (1025) | Sharp bands |
| $\tilde{G}\ ^1A_1$ | $C_{2v}$ | 70371 | | | 1023 | 592 | | | | | | | | $\tilde{G} \leftarrow \tilde{X}$ 1421–1380 Å | (1025) | Sharp bands |
| $\tilde{F}\ ^1A_1$ | $C_{2v}$ | 66984 | | | 1025 | 627 | | | | | | | | $\tilde{F} \leftarrow \tilde{X}$ 1493–1430 Å | (1025) | Sharp bands |
| $\tilde{E}\ ^1A_1$ | $C_{2v}$ | 64760 | | | 1025 | 608 | | | | | | | | $\tilde{E} \leftarrow \tilde{X}$ 1545–1500 Å | (1025) | Sharp bands |
| $\tilde{D}\ ^1A_1$ | $C_{2v}$ | 61550 | | | 1020 | 494 | | | | | | | | $\tilde{D} \leftarrow \tilde{X}$ 1630–1550 Å | (1025) | Sharp bands |
| $\tilde{C}\ ^1A_1$ | $C_{2v}$ | 54680 | | | (1200)[a] | 830 | | | | | | | | $\tilde{C} \leftarrow \tilde{X}$ 1850–1700 Å | (1025) | Sharp bands at longer $\lambda$, diffuse bands at shorter $\lambda$ |
| $\tilde{B}$ | $C_{2v}$ | 46990[b] | | | | 1100 | | | | | | | | $\tilde{B} \leftarrow \tilde{X}$ 2130–1930 Å | (1025) | Diffuse bands |
| $\tilde{A}$ | ($C_s$) | (25975) | | | | | ($\nu_9 \approx 400$) | | | | | | | $\tilde{A} \leftarrow \tilde{X}$ 3850–2600 Å | (726)(940) | Diffuse bands |
| $\tilde{X}\ ^1A_1$ | $C_{2v}$ | 0 | 3070.4 | 2151.8 | 1388 | 1117.8 | 588 | 528[c] | 9.37[d] | 0.343347 | 0.330758 | $r$(CO) = 1.16[e] $r$(CC) = 1.315 | ∠HCH = 122.3° | infrared and microwave sp. | (74)(642) (73)(248) (169)(876) | |

HCCCH: [a] Discrete bands in the region 3650–3100 Å not yet analyzed.

$H_2CCO$: [a] (1025) give different assignment, viz. $\nu_6 = 380$ cm$^{-1}$. [b] From new plates of the writer. [c] $\nu_7(b_2) = 3166.1$, $\nu_8 = 976.7$, $\nu_9 = 433$ cm$^{-1}$ ($\nu_6$ and $\nu_8$ may have to be exchanged). [d] (169) give $A_0 = 9.34$ from $\nu_3 + \nu_5$. [e] Only the sum of $r$(CO) + $r$(CC) is accurately known: 2.475 Å. According to (876) $r$(CH) = 1.079 Å.

TABLE 70 (*continued*)

| State | Point Group | $T_0$ | Vibrational Frequencies $\nu_1$ | $\nu_2$ | $\nu_3$ | $\nu_4$ | $\nu_5$ | $\nu_6$ | Rotational Constants $A_0$ | $B_0$ | $C_0$ | $r_0$(Å) | $\alpha$ | Observed Transitions | References | Remarks |
|---|---|---|---|---|---|---|---|---|---|---|---|---|---|---|---|---|
| **$H_2CN_2$** (diazomethane) I.P. = 8.99₉ eV; $D(H_2C—N_2) \leq 2.6$ eV[a] | | | | | | | | | | | | | | | | |
| | | Rydberg series joining on to $\tilde{H}$: $\nu = 72585 - R/(n - 0.10)^2$; $n = 3, 4, \ldots, 10$ | | | | | | | | | | | | | (823) | Several components for each member, see Fig. 196 |
| $\tilde{I}$ | | 62831 | | | | | | | | | | | | $\tilde{I} \leftarrow \tilde{X}$ 1590–1540 Å | (823) | Group of ⊥ bands |
| $\tilde{H}$ | | 59670 | | | | | | | | | | | | $\tilde{H} \leftarrow \tilde{X}$ 1680–1610 Å | (823) | Group of ∥ bands |
| $\tilde{G}$ | | 56896 | | | | | | | | | | | | $\tilde{G} \leftarrow \tilde{X}$ 1770–1690 Å | (823) | Diffuse bands |
| $\tilde{F}\ ^1B_1$ | $C_{2v}$ | 52688.7 | 2908.8 | 2069.7 | 1174.1 | 792.0 | | 726.4 | 10.34[b] | $\bar{B} = 0.359$ | | | | $\tilde{F} \leftarrow \tilde{X}$ 1930–1780 Å | (823) | ⊥ bands with resolved $K$ structure |
| $\tilde{E}(^1B_2)$ | $C_{2v}$ | 52649 | | | | | | | | | | | | $\tilde{E} \leftarrow \tilde{X}$ 1930–1780 Å | (823) | ⊥ bands with resolved $K$ structure |
| $\tilde{D}\ ^1B_1$ | $C_{2v}$ | 52520.9 | | | | | | | 7.16[b] | $\bar{B} = 0.362$ | | | | $\tilde{D} \leftarrow \tilde{X}$ 1930–1780 Å | (823) | ⊥ bands with resolved $K$ structure see Fig. 196 |
| $\tilde{B}$ | | (37740) | continuum with peak at 2175 Å | | | | | | | | | | | $\tilde{B} \leftarrow \tilde{X}$ 2650–2000 Å | (671)(525) | |
| $\tilde{A}$ | | (21230) | | | | | | | | | | | | $\tilde{A} \leftarrow \tilde{X}$ 4750–3200 Å | (671) | Very diffuse bands merging into a continuum |
| $\tilde{X}\ ^1A_1$ | $C_{2v}$ | 0 | $3077._1$ | 2102.2 | 1414.4 | 1170 | $564._0$ | $406._0$[c] | 9.112 | 0.377109 | 0.361759 | $r$(NN) = 1.12, $r$(CN) = 1.32 [d]; $r$(CH) = 1.08 | ∠HCH = 127° | infrared and microwave sp. | (249)(877) (857) | |

$H_2CN_2$: [a] (968), from electron impact data, give 1.9 eV. [b] Modified by Coriolis interaction. [c] $\nu_7(b_2) = 3184.5$, $\nu_8 = 1109.0$, $\nu_9 = 421.2$.
[d] Only the sum $r(NN) + r(CN) = 2.442$ is accurately known.

TABLE 70 (*continued*)

| State | Point Group | $T_0$ | Vibrational Frequencies $\nu_1$ | $\nu_2$ | $\nu_3$ | $\nu_4$ | $\nu_5$ | $\nu_6$ | Rotational Constants $A_0$ | $B_0$ | $C_0$ | $r_0$(Å) | $\alpha$ | Observed Transitions | References | Remarks |
|---|---|---|---|---|---|---|---|---|---|---|---|---|---|---|---|---|
| $H_2CN_2$(diazirine) see (830) | | | | | | | | | | | | | | | | |
| **HCOOH** (formic acid) I.P. = 11.33 eV[a] | | | | | | | | | | | | | | | | |
| | | | Rydberg series joining on to $\tilde{C}$ and $\tilde{F}$: $\nu = 91370 - R/(n - 0.60)^2$; $n = 3, 4, \ldots$[b] | | | | | | | | | | | | (1019) | |
| $\tilde{F}$ | | 81880 | | | 1440 | | | | | | | | | $\tilde{F} \leftarrow \tilde{X}$ 1225–1180 Å | (1019) | Repr. (1019) |
| $\tilde{D}$ | | (78100)[c] | | | | | | | | | | | | $\tilde{D} \leftarrow \tilde{X}$ 1280 Å | (1019) | Narrow group |
| $\tilde{C}$ | | 72210 | | | 1480 | | | | | | | | | $\tilde{C} \leftarrow \tilde{X}$ 1410–1320 Å | (1019) | Repr. (1019) |
| $\tilde{B}$ | | 64500 | | | (1490) | | | | | | | | | $\tilde{B} \leftarrow \tilde{X}$ 1550–1450 Å | (1019)(924) | Diffuse bands according to (1019)[d] |
| $\tilde{A}$ | | (40000) | | | | | 1033 | 385 | | | | | | $\tilde{A} \leftarrow \tilde{X}$ 2600–2250 Å | (1175)(475) | Indistinct bands merging into a continuum at 2250 Å |
| $\tilde{X}\,{}^1A'$ | $C_s$ | 0 | 3570 | 2943 | 1770 | 1387 | 1229 | 1105[e] | 2.58548 | 0.402112 | 0.347447 | $r$(CO) = {1.202, 1.343}; $r$(CH) = 1.097; $r$(OH) = 0.972 | ∠OCO = 124.9°; ∠HCO = 124.1°; ∠COH = 106.3° | infrared and microwave sp. | (1224)(704) (854)(861) | |

HCOOH: [a] From Rydberg series; photoionization yields 11.05 eV; see (1273). [b] Not a very convincing series. [c] Taken from the spectrum reproduced by (1019). [d] Continuum with maximum at 1590 Å according to (924). [e] $\nu_7 = 625$, $\nu_8(a'') = 1033$, $\nu_9 = 638$ cm$^{-1}$.

TABLE 70 (*continued*)

| State | Point Group | $T_0$ | Vibrational Frequencies $\nu_1$ | $\nu_2$ | $\nu_3$ | $\nu_4$ | $\nu_5$ | $\nu_6$ | Rotational Constants $A_0$ | $B_0$ | $C_0$ | $r_0$(Å) | $\alpha$ | Observed Transitions | References | Remarks |
|---|---|---|---|---|---|---|---|---|---|---|---|---|---|---|---|---|
| **$CH_2F_2$** | | | see (1251) | | | | | | | | | | | | | |
| **$CH_2Cl_2$** | | | see (1333) I.P. = 11.35 eV[a] | | | | | | | | | | | | | |
| **$CH_2I_2$** | | | see (967)(611) continuous absorption only | | | | | | | | | | | | | |
| **NC—$C_2H$** (cyanoacetylene) | | | | | | | | | | | | | | | | |
| $\tilde{B}(^1\Delta$ or $^1\Sigma)$ | $C_{\infty v}$ | 44221 | | 2120 | | 953 | 290 | 414[a] | | | | | | $\tilde{B} \leftarrow \tilde{X}$ 2300–2100 Å | (628b) | Sharp bands[b] |
| $\tilde{A}(^1A')$ | $C_s$ [c] | 38485 | | | | | 884.5 | 160.3 | 7·00[d] | $\bar{B} = 0.1576$ | | | | $\tilde{A} \leftarrow \tilde{X}$ 2715–2300 Å | (628b) | Sharp double-headed bands[e] |
| $X(^1\Sigma^+)$ | $C_{\infty v}$ | 0 | 3326.6 | 2271.0 | 2076.9 | 876[f] | 663.3 | 499.7[g] | — | 1.151740[h] | — | $r$(CH) = 1.058<br>$r$(CC) = {1.205, 1.378}<br>$r$(CN) = 1.159 | 180° | infrared and microwave sp. | (1227a)<br>(628a,b)<br>(1228a) | |
| **CHONO** (nitroso formaldehyde), see (927)(811) | | | | | | | | | | | | | | | | |
| **$CHF_3$** | | | see (1162) | | | | | | | | | | | | | |
| **$CHCl_3$** | | | see (1333)(504) I.P. = 11.42 eV[a]; continuous absorption only | | | | | | | | | | | | | |
| **$CHBr_3$** | | | see (967)(504) continuous absorption only | | | | | | | | | | | | | |
| **$CHI_3$** | | | see (967)(504) continuous absorption only | | | | | | | | | | | | | |

$CH_2Cl_2$: [a] From photoionization (1273).

NC—$C_2H$: [a] $\nu_7(\pi) = 226$. [b] Forbidden transition; vibronic bands based on $\nu_5$, $\nu_6$ and $\nu_7$. [c] Probably *trans*-bent. [d] $\alpha_6^A$ very large. [e] Degraded to the violet; diffuse beyond 2400 Å.
[f] Measured in the liquid phase. [g] $\nu_7(\pi) = 229.7$ from $\tilde{A} \leftarrow \tilde{X}$; $g_{55} = 2.95$, $g_{66} = 1.00$, $g_{77} = 2.80$.
[h] $\alpha_5 = -0.0000576$, $\alpha_6 = -0.000308$, $\alpha_7 = -0.000482$; $q_5 = 0.0000853$, $q_6 = 0.000119$, $q_7 = 0.000218$, $D_0 = 1.8 \times 10^{-8}$.

$CHCl_3$: [a] From photoionization (1273).

TABLE 71

MOLECULAR CONSTANTS OF THE ELECTRONIC STATES OF FIVE-ATOMIC NON-HYDRIDES

| State | Point Group | $T_0$ | Vibrational Frequencies $\nu_1$ | $\nu_2$ | $\nu_3$ | $\nu_4$ | $\nu_5$ | $\nu_6$ | Rotational Constants $A_0$ | $B_0$ | $C_0$ | $r_0$(Å) | $\alpha$ | Observed Transitions | References | Remarks |
|---|---|---|---|---|---|---|---|---|---|---|---|---|---|---|---|---|
| **$C_3O_2$ (carbon suboxide)** | | | | | | | | | | | | | | | | |
| $\tilde{B}$ | | Continuous absorption extending from 2300 Å to shorter wavelengths | | | | | | | | | | | | | (1208) | [a] |
| $\tilde{A}$ | | 30697 | 2160 | 840 | | | | | | | | | | $\tilde{A} \leftarrow \tilde{X}$ 3300–2400 Å | (1208) | Progressions of pairs of bands[b] |
| $\tilde{X}\ {}^1\Sigma_g^+$ | $D_{\infty h}$ | 0 | 2200 | 830 | 2258 | 1573 | 577 | 550[c] | — | 0.07321 | — | | 180° | infrared and Raman sp. | (763)(845) (948)(707) | |
| **$CF_2N_2$** | | | see (853)(1133a) | | | | | | | | | | | | | |
| **$CF_4$** | | | see (1333) and (224) continuous absorption only | | | | | | | | | | | | | |
| **$CCl_4$** | | | see (1333) continuous absorption only | | | | | | | | | | | | | |
| **$CBr_4$, $CI_4$** | | | see (967)(611) continuous absorption only | | | | | | | | | | | | | |
| **$CF_3I$** | | | I.P. = 10.40 eV[a]; $D(CF_3—I)$ = 2.4$_7$ eV[b] | | | | | | | | | | | | | |
| | | Further absorption bands at 1357 Å and below, not analyzed in detail | | | | | | | | | | | | | (1180) | |
| $\tilde{E}$ | | 70623 | 950 | 720 | 225 | | | | | | | | | $\tilde{E} \leftarrow \tilde{X}$ 1455–1370 Å | (1180) | Diffuse bands |
| $\tilde{D}$ | | 62529 | 922 | 592 | 204 | | | | | | | | | $\tilde{D} \leftarrow \tilde{X}$ 1600–1540 Å | (1180) | |
| $\tilde{C}$ $E$ | $C_{3v}$ | 57494[c] | 969 | 682 | 231 | | | | | | | | | $\tilde{C} \leftarrow \tilde{X}$ 1750–1630 Å | (1180)(326) (1000) | Sharp bands, see Fig. 62 |
| $\tilde{A}$ | | 35500 | continuum with maximum at 37700 cm$^{-1}$ | | | | | | | | | | | $\tilde{A} \leftarrow \tilde{X}$ 2815–2480 Å | (326) | |
| $\tilde{X}\ {}^1A_1$ | $C_{3v}$ | 0 | 1073 | 741 | 286[d] | 1185 | 540 | 265[d] | 0.1910[e] | 0.050808 | — | $r$(CI) = 2.134, $r$(CF) = 1.332 ass. | ∠FCF = 108° ass. | infrared and microwave sp. | (345)(1122) (1161) | |

$C_3O_2$: [a] At higher pressures this absorption merges with $\tilde{A} - \tilde{X}$. [b] The separation of the members of a pair varies between 50 and 100 cm$^{-1}$. [c] $\nu_7$ < 70 cm$^{-1}$.

$CF_3I$: [a] From photoionization (135). [b] From electron collision data (371). [c] (1180) give 57468. [d] From combination bands and Raman spectrum of liquid. [e] Calculated from $B_0$ and the assumed parameters.

TABLE 71 (*continued*)

| State | Point Group | $T_0$ | Vibrational Frequencies $\nu_1$ | $\nu_2$ | $\nu_3$ | $\nu_4$ | $\nu_5$ | $\nu_6$ | Rotational Constants $A_0$ | $B_0$ | $C_0$ | $r_0$ | $\alpha$ | Observed Transitions | References | Remarks |
|---|---|---|---|---|---|---|---|---|---|---|---|---|---|---|---|---|
| **$CF_2Cl_2$** | | | see (1333) | | | | | | | | | | | | | |
| **$POCl_3$** | | | see (622) continuous absorption only | | | | | | | | | | | | | |
| **$CrO_2F_2$** | | | | | | | | | | | | | | | | |
| $\tilde{A}$ | | (17085) | | | | 214 | | | | | | | | $\tilde{A}-\tilde{X}$ 5650–4700 Å | (492)(56) | |
| $\tilde{X}\,^1A_1$ | $C_{2v}$ | 0 | 1006 | 727 | 364 | 208 | 259 | 1016[a] | $0.1471_0$ | $0.1430_0$ | $0.1412_7$ | | | infrared, Raman and microwave sp. | (558)(445) (1159) | |
| **$CrO_2Cl_2^{35}$** | | | | | | | | | | | | | | | | |
| $\tilde{C}$ | | Continuous absorption with maximum at 2950 Å | | | | | | | | | | | | $\tilde{C}\leftarrow\tilde{X}$ 3400–2500 Å | (56) | |
| $\tilde{B}$ | | Continuous absorption with maximum at 4100 Å | | | | | | | | | | | | $\tilde{B}\leftarrow\tilde{X}$ 4800–3600 Å | (56) | |
| $\tilde{A}$ | | 17098.6 | | | | 136.5 | | | | | | | | $\tilde{A}\leftarrow\tilde{X}$ 6000–5300 (Å) | (699)(56) | Sharp bands with fine structure |
| $\tilde{X}\,^1A_1$ | $C_{2v}$ | 0 | 981 | 465 | 356 | 140 | 224 | 995[a] | | | | | | infrared and Raman sp. | (844)(1159) | |

$CrO_2F_2$: [a] $\nu_7(b_1) = 274$, $\nu_8(b_2) = 789$, $\nu_9 = 304$.
$CrO_2Cl_2$: [a] $\nu_7(b_1) = 211$, $\nu_8(b_2) = 496$, $\nu_9 = 257$ (all in liquid or solution). (558) gives a few frequencies for the gaseous state.

TABLE 71 (*continued*)

| State | Point Group | $T_0$ | Vibrational Frequencies | | | | | | Rotational Constants | | | | | Observed Transitions | References | Remarks |
|---|---|---|---|---|---|---|---|---|---|---|---|---|---|---|---|---|
| | | | $\nu_1$ | $\nu_2$ | $\nu_3$ | $\nu_4$ | $\nu_5$ | $\nu_6$ | $A_0$ | $B_0$ | $C_0$ | $r_0$ | $\alpha$ | | | |
| **$RuO_4$** | | | see (691)(1036)(97)(1084) | | | | | | | | | | | | | |
| **$OsO_4$** | | | | | | | | | | | | | | | | |
| $\tilde{C}$ | | 38101 | 832 | | | | | | | | | | | $\tilde{C} \leftarrow \tilde{X}$ 2700–2270 Å | (722) | [a] |
| $\tilde{B}$ | | 36225 | 835 | | | | | | | | | | | $\tilde{B} \leftarrow \tilde{X}$ 2790–2240 Å | (722) | [a] |
| $\tilde{A}$ | | 33388 | 811 [b] | | | | | | | | | | | $\tilde{A} \leftarrow \tilde{X}$ 3200–2760 Å | (722)(1084) | [c] |
| $\tilde{X}\ {}^1A_1$ | $\boldsymbol{T_d}$ | 0 | 971 | (335) [d] | 960 | 328 | | | — | (0.114) | — | (1.85 Å) | | infrared and Raman sp. | (722)(1320) (289)(484) | |
| **$TiCl_4$, $TiBr_4$** | | | see (55a) | | | | | | | | | | | | | |
| **$VCl_4$** | | | see (122), (203), (974) and (55a) [a] | | | | | | | | | | | | | |
| **$XeF_4$** | | | see (652)(1035) | | | | | | | | | | | | | |

$OsO_4$: [a] Possibly the states $\tilde{C}$ and $\tilde{B}$ form a single state [see (1084)]. [b] (1084) give a mean separation of 827 $cm^{-1}$.
[c] Two very weak systems at longer wavelengths (4000–3200 Å) have been observed by (1084) but not definitely identified.
[d] In liquid (1320). (722) give 568 and 688 $cm^{-1}$ as doubtful Raman lines of the gas, but these are considered spurious by (1320).

$VCl_4$: [a] Far-reaching conclusions about Jahn–Teller splittings in $VCl_4$ are drawn by (122) from a few diffuse absorption maxima. From theory the ground state is ${}^2E$, but no definite experimental confirmation has yet been obtained. A careful reinvestigation of the infrared spectrum by (451) has not revealed any evidence for Jahn–Teller splitting in the ground state.

TABLE 72

MOLECULAR CONSTANTS OF THE ELECTRONIC STATES OF SIX-ATOMIC TETRA-HYDRIDES

| State | Point Group | $T_0$ | Vibrational Frequencies $\nu_1$ | $\nu_2$ | $\nu_3$ | $\nu_4$ | $\nu_5$ | $\nu_6$ | Rotational Constants $A_0$ | $B_0$ | $C_0$ | $r_0$(Å) | $\alpha$ | Observed Transitions | References | Remarks |
|---|---|---|---|---|---|---|---|---|---|---|---|---|---|---|---|---|
| $\mathbf{C_2H_4}$ | | I.P. = $10.50_7$ eV [a]; $D(H_2C—CH_2) = 7.2_6 \pm 0.3$ eV [b] | | | | | | | | | | | | | | |
| | | Continuous absorption with broad maxima at 820 and 690 Å | | | | | | | | | | | | 1000–600 Å | (1105)(836) (1257) | |
| | | Rydberg series joining on to $\tilde{D}$: $\nu = 84750 - R/(n - 0.4)^2$; $n = 3, 4, \ldots$ | | | | | | | | | | | | | (1026)(1300) | $nR''$ of (1300) |
| | | Rydberg series joining on to $\tilde{C}$: $\nu = 84750 - R/(n - 0.6)^2$; $n = 3, 4, \ldots$ | | | | | | | | | | | | | (1026)(1300) | $nR'$ of (1300) |
| | | Rydberg series joining on to $\tilde{B}$ and $\tilde{E}$: $\nu = 84750 - R/(n - 1.09)^2$; $n = 3, 4, 5, 6, 7$ | | | | | | | | | | | | | (1026)(1300) | $nR$ of (1300) [c] |
| $\tilde{F}$ | | 73011 | | 1460 | | 170 [d] | | | | | | | | $\tilde{F} \leftarrow \tilde{X}$ 1370–1335 Å | (1300) | $4R'''$ of (1300) [e] |
| $\tilde{E}$ | | 71788 | | 1451 | | 202 [d] | | | | | | | | $\tilde{E} \leftarrow \tilde{X}$ 1394–1340 Å | (1300) | $3R$ of (1300) |
| $\tilde{D}$ | | 69516 | | 1372 | | 260 [d] | | | | | | | | $\tilde{D} \leftarrow \tilde{X}$ 1440–1340 Å | (1300) | $3R''$ of (1300) |
| $\tilde{C}$ | | 66607 | | 1311 | | 235 [d] | | | | | | | | $\tilde{C} \leftarrow \tilde{X}$ 1502–1470 Å | (1300) | $3R'$ of (1300) |
| $\tilde{B}$ | $D_{2h}$ | 57336.0 | | 1370.3 [g] | | 236 [d] | | | | | | | | $\tilde{B} \leftarrow \tilde{X}$ 1745–1560 Å | (1303)(1329) | $2R$ of (1303) and (1300) [f] |
| $\tilde{A}\ ^1B_{1u}$ [m] | $D_{2d}$ | 40015 | | 850 [h] | | | | | | | | | | $\tilde{A} \leftarrow \tilde{X}$ 2100–1600 Å | (1303)(1329) | So-called $V$ state [i] |
| $\tilde{a}\ ^3B_{1u}$ [m] | ($D_{2d}$) | < 28700 | | 995 | | | | | | | | | | $\tilde{a} \leftarrow \tilde{X}$ 3400–2600 Å | (1064) | Very diffuse, extremely weak |
| $\tilde{X}\ ^1A_g$ | $D_{2h}$ | 0 | 3026.4 | 1622.6 | 1342.2 | 1027 | 3102.5 | 1236 [k] | 4.828 | 1.0012 [l] | 0.8282 [l] | $r$(CH) = 1.086; $r$(CC) = 1.339 | ∠HCH = 117.6° | infrared and Raman sp. | (75)(1163) (376)(305)(59) | |

$C_2H_4$: [a] From Rydberg series (1026); by photoionization (1272) finds 10.51 eV while (1257) give 10.46 eV. For $C_2D_4$ Rydberg series (1026) give I.P. = $10.51_9$ eV.

[b] From (1303); state of excitation of $CH_2$ on dissociation not known.

[c] Note that (1300) uses $n$ one lower and therefore uses $n - 0.09$ in the Rydberg denominator. This leads to the result (avoided here) that the lowest Rydberg state has $n = 2$.

[d] Half the observed $2\nu_4'$. [e] Considered to be the first member of a Rydberg series not further specified. [f] Extensive vibrational structure, see Fig. 198. [g] $x_{22} = -11.2$.

[h] $x_{22} = -1.9$. [i] Diffuse bands followed by continuum; see Fig. 197. [j] First observed in liquid $C_2H_4$ by (1064), more recently in gaseous $C_2H_4$ of high pressure mixed with $O_2$ by (365a).

[k] $\nu_7(b_{1u}) = 949.2$, $\nu_8(b_{2g}) = 950$, $\nu_9(b_{2u}) = 3105.3$, $\nu_{10} = 810.3$, $\nu_{11}(b_{3u}) = 2988.6_6$, $\nu_{12} = 1443.5$. [l] These are (305)'s values. (59) give 0.9998 and 0.8294 resp.

[m] These designations correspond to the conformation ($D_{2h}$) reached by vertical transitions from the ground state. In the equilibrium conformation the symmetry is presumably $D_{2d}$ and the states must be classified as $^1E$ and $^3E$.

TABLE 72 (*continued*)

| State | Point Group | $T_0$ | Vibrational Frequencies $\nu_1$ | $\nu_2$ | $\nu_3$ | $\nu_4$ | $\nu_5$ | $\nu_6$ | Rotational Constants $A_0$ | $B_0$ | $C_0$ | $r_0$(Å) | $\alpha$ | Observed Transitions | References | Remarks |
|---|---|---|---|---|---|---|---|---|---|---|---|---|---|---|---|---|
| **$N_2H_4$** | see (599); $D(H_2N—NH_2) = 2.5$ eV [a]; $D(H—N_2H_3) = 3.3$ eV | | | | | | | | | | | | | | | |
| **$CH_3OH$** | see (1014), (539), (946), and (430) [a]; I.P. = 10.85 eV [b] | | | | | | | | | | | | | | | |
| **$CH_3SH$** | I.P. = $9.43_9$ eV; $D(CH_3—SH) = 3.1_7$ eV [a] | | | | | | | | | | | | | | | |
| | | Weak Rydberg series joining on to $\tilde{D}$: $\nu = 76170 - R/(n - 1.30)^2$; $n = 4, 5, \ldots, 9$ | | | | | | | | | | | | | (1024) | |
| | | Strong Rydberg series joining on to $\tilde{E}$: $\nu = 76138 - R/(n - 1.05)^2$; $n = 4, 5, \ldots, 12$ | | | | | | | | | | | | | (1024) | |
| | | Strong Rydberg series joining on to $\tilde{C}$: $\nu = 76197 - R/(n - 1.55)^2$; $n = 4, 5, \ldots, 9$ | | | | | | | | | | | | | (1024) | |
| $\tilde{E}$ | | 61800 | | | | | | | | | | | | $\tilde{E} \leftarrow \tilde{X}$ 1620–1560 Å | (1024) | Vibrational structure, not analysed |
| $\tilde{D}$ | | 61144 | | | | | | | | | | | | $\tilde{D} \leftarrow \tilde{X}$ 1640 Å | (1024) | |
| $\tilde{C}$ | | 57250 | | | | | | | | | | | | $\tilde{C} \leftarrow \tilde{X}$ 1760–1700 Å | (1024) | Vibrational structure, not analysed |
| $\tilde{B}$ | | 54350 | | | | | | | | | | | | $\tilde{B} \leftarrow \tilde{X}$ 1840 Å | (1024) | Single diffuse band |
| $\tilde{A}$ | | (36000) | continuous absorption | | | | | | | | | | | $\tilde{A} \leftarrow \tilde{X}$ 2780–1800 Å | (590) | |
| $\tilde{X}\ ^1A'$ | $C_s$ | 0 | 2869 | 2607 | 1335 | 704 | 3010 | 1475 [b]<br>1430 | (5.68) | 0.43053 | 0.41306 | $r$(CH) = 1.104<br>$r$(SH) = 1.329<br>$r$(SC) = 1.818 | ∠HCH = 110.3°<br>∠CSH = 100.3° | infrared and microwave sp. | (1143)(665)<br>(1211)(1213) | Barrier height 247 $cm^{-1}$ |

$N_2H_4$: [a] See (389).
$CH_3OH$: [a] Continuous absorption only. [b] From photoionization, see (1273).
$CH_3SH$: [a] From thermochemical data of (787). [b] $\nu_7(e) = \left\{\begin{matrix}1060\\957\end{matrix}\right.$, $\nu_8 = 803$ [same numbering as for $CH_3OH$ in Vol. II, p. 335].

## Table 73

## MOLECULAR CONSTANTS OF THE ELECTRONIC STATES OF SIX-ATOMIC TRI-HYDRIDES

| State | Point Group | $T_0$ | Vibrational Frequencies | | | | | | Rotational Constants | | | | | Observed Transitions | References | Remarks |
|---|---|---|---|---|---|---|---|---|---|---|---|---|---|---|---|---|
| | | | $\nu_1$ | $\nu_2$ | $\nu_3$ | $\nu_4$ | $\nu_5$ | $\nu_6$ | $A_0$ | $B_0$ | $C_0$ | $r_0$(Å) | $\alpha$ | | | |
| **$CH_3CN$** (acetonitrile) I.P. = $11.9_5$ eV | | | | | | | | | | | | | | | | |
| $\tilde{E}$ | | 90853 | | | | | | | | | | | | $\tilde{E} \leftarrow \tilde{X}$ 1101 Å | (263) | |
| $\tilde{D}$ | | 86953 | | 2009 | (1107) | | | | | | | | | $\tilde{D} \leftarrow \tilde{X}$ 1150 Å | (263) | |
| $\tilde{C}$ | | 77374 | 2724 | 2001 | 1239 | | | | | | | | | $\tilde{C} \leftarrow \tilde{X}$ 1295–1250 Å | (263) | |
| $\tilde{B}$ | | (63000) | continuous absorption | | | | | | | | | | | $\tilde{B} \leftarrow \tilde{X}$ 1600–1300 Å | (263)(539) | |
| $\tilde{A}$ | | (55000) | continuous absorption with maximum at 59000 $cm^{-1}$ | | | | | | | | | | | $\tilde{A} \leftarrow \tilde{X}$ 1820–1600 Å | (539) | |
| $\tilde{X}\ ^1A_1$ | $C_{3v}$ | 0 | 2954.2 | 2267.7 | 1389 | 920.25 | 3009.16 | 1453.53[a] | (5.28)[b] | 0.306842 | — | r(CH) = 1.104<br>r(CC) = 1.458<br>r(CN) = 1.157 | ∠HCH = 109.5° | infrared and microwave sp. | (961)(1244)<br>(624)(1203)<br>(233)(925) | |
| **$C_2H_3O$** (vinoxy) see (261) | | | | | | | | | | | | | | | | |
| **$CH_3NO$** (nitrosomethane) | | | | | | | | | | | | | | | | |
| $\tilde{A}$ | $C_s$ | 14830 | (1400) | 1040 | 390 | | | | | | | | | $\tilde{A} \leftarrow \tilde{X}$ 7100–5900 Å | (287) | Sharp bands with fine structure |
| $\tilde{X}\ ^1A'$ | $C_s$ | 0 | 1564 | 842 | 400 | | | | | | | | | infrared sp. | (287)(784) | |

$CH_3CN$: [a] $\nu_7\ (e) = 1040.79$, $\nu_8 = 362.0$. [b] Calculated from microwave structure obtained from $B_0$ values. $\zeta_5 = 0.062$, $\zeta_6 = -0.384$, $\zeta_7 = 0.422$, $\zeta_8 = (0.945)$ calc.

TABLE 73 (*continued*)

| State | Point Group | $T_0$ | Vibrational Frequencies $\nu_1$ | $\nu_2$ | $\nu_3$ | $\nu_4$ | $\nu_5$ | $\nu_6$ | Rotational Constants $A_0$ | $B_0$ | $C_0$ | $r_0$(Å) | $\alpha$ | Observed Transitions | References | Remarks |
|---|---|---|---|---|---|---|---|---|---|---|---|---|---|---|---|---|
| **$HCONH_2$** (formamide) | | see (594)[a] | | | | | | | | | | | | | | |
| **$C_2H_3Cl$** | | I.P. = $9.99_8$ eV | | | | | | | | | | | | | | |
| | | Long Rydberg series joining on to $\tilde{D}$: | | | | | | | $\nu = 80645 - R/(n - 0.05)^2$; $n = 3, 4, \ldots, 11$ | | | | | | (1259) | |
| | | Fragment of Rydberg series joining on to $\tilde{C}$: | | | | | | | $\nu = 80645 - R/(n - 0.65)^2$; $n = 3, 4, \ldots$ | | | | | | (1259) | |
| | | Long Rydberg series joining on to $\tilde{B}$: | | | | | | | $\nu = 80700 - R/(n - 0.85)^2$; $n = 3, 4, \ldots, 9$ | | | | | | (1259) | |
| $\tilde{E}$ | | 68400 | | | | | | | | | | | | $\tilde{E} \leftarrow \tilde{X}$ 1462 Å | (1259) | Single strong band |
| $\tilde{D}$ | | 67979 | 1200 | | | | | | | | | | | $\tilde{D} \leftarrow \tilde{X}$ 1471–1420 Å | (1259) | Sharp bands |
| $\tilde{C}$ | | 63048 | 1372 | | | | | | | | | | | $\tilde{C} \leftarrow \tilde{X}$ 1585–1430 Å | (1259) | Sharp bands |
| $\tilde{B}$ | | 57330 | 1380 | very diffuse bands | | | | | | | | | | $\tilde{B} \leftarrow \tilde{X}$ 1750–1650 Å | (1259) | $\tilde{B} \leftarrow \tilde{X}$ and $\tilde{A} \leftarrow \tilde{X}$ are not well separated |
| $\tilde{A}$ | | (46000) | | continuum | | | | | | | | | | $\tilde{A} \leftarrow \tilde{X}$ 2200–1750 Å | (1259)(705) | |
| $\tilde{X}\ ^1A'$ | $C_s$ | 0 | 3121 | 3086 | 3030 | 1603 | 1369 | 1279[a] | | 0.201138 | 0.181635 | $r$(CH) = 1.079 and 1.090; $r$(CC) = 1.332; $r$(CCl) = 1.726 | ∠CCCl = 122.3°; ∠HCC = {123.8°, 119.5°, 121.0°} | infrared, Raman and microwave sp. | (453)(677) | [b] |

$HCONH_2$: [a] A fragmentary Rydberg series with a limit at 82566 cm$^{-1}$ has been observed.

$C_2H_3Cl$: [a] $\nu_7(a') = 1030$, $\nu_8 = 720$, $\nu_9 = 395$, $\nu_{10}(a'') = 941$, $\nu_{11} = 896$, $\nu_{12} = 620$. [b] The fundamentals of $C_2D_3Cl$ are given by (928).

TABLE 74

MOLECULAR CONSTANTS OF THE ELECTRONIC STATES OF SIX-ATOMIC DI-HYDRIDES

| State | Point Group | $T_0$ | Vibrational Frequencies $\nu_1$ | $\nu_2$ | $\nu_3$ | $\nu_4$ | $\nu_5$ | $\nu_6$ | Rotational Constants $A_0$ | $B_0$ | $C_0$ | $r_0$(Å) | $\alpha$ | Observed Transitions | References | Remarks |
|---|---|---|---|---|---|---|---|---|---|---|---|---|---|---|---|---|
| **$C_4H_2^+$** | | | | | | | | | | | | | | | | |
| $\tilde{A}\ ^2\Pi_u$ | $\boldsymbol{D}_{\infty h}$ | 19722.8[a] | (2858) | | | | | | — | $0.1333_2$ | — | | 180° | $\tilde{A}\leftarrow\tilde{X}$ 5900–5000 Å | (171) | $T$-spectrum of (1112)(1113) |
| $\tilde{X}\ ^2\Pi_g$ | $\boldsymbol{D}_{\infty h}$ | 0[b] | 3136.9 | 2176.6 | 860.6 | | | [c] | — | $0.1401_3$ | — | | 180° | | | Repr. (171) |
| **$C_4H_2$** (diacetylene) | | I.P. = $10.79_1$ eV; $D(HC_2—C_2H) \leq 4.2$ eV[a] | | | | | | | | | | | | | | |
| $\tilde{E}$ | $\boldsymbol{D}_{\infty h}$ | ~89290 | Beginning of second set of Rydberg series | | | | | | | | | | | $\tilde{E}\leftarrow\tilde{X}$ 1120–940 Å | (1029) | Not analyzed in detail |
| | | Rydberg series joining on to $\tilde{D}$: $\nu = 87042 - R/(n - 0.52)^2$; $n = 3, 4, \ldots, 8$ | | | | | | | | | | | | | (1029) | |
| | | Rydberg series joining on to $\tilde{C}$: $\nu = 87042 - R/(n - 0.95)^2$; $n = 3, 4, \ldots, 9$ | | | | | | | | | | | | | (1029) | |
| $\tilde{D}$ | $\boldsymbol{D}_{\infty h}$ | 69126 | | (2150) | | | | | | | | | | $\tilde{D}\leftarrow\tilde{X}$ 1447–1350 Å | (1029) | Progression of strong bands |
| $\tilde{C}$ | $\boldsymbol{D}_{\infty h}$ | 60790 | | (2150) | | | | | | | | | | $\tilde{C}\leftarrow\tilde{X}$ 1645–1550 Å | (1029) | Progression of very strong bands |
| $\tilde{B}$ | $(\boldsymbol{C}_{2h})$ | 39048 | | 2090 | | | | | | | | | | $\tilde{B}\leftarrow\tilde{X}$ 2650–2000 Å | (1313) | Diffuse bands |
| $\tilde{A}$ | $(\boldsymbol{C}_{2h})$ | (34000) | | | 690[b] | | | | | | | | | $\tilde{A}\leftarrow\tilde{X}$ 2970–2650 Å | (1313) | Extensive system of sharp bands[c] |
| $\tilde{X}\ ^1\Sigma_g^+$ | $\boldsymbol{D}_{\infty h}$ | 0 | (3329) | 2184 | 874 | 3329 | 2020 | 627[d] | — | 0.1469 | — | $r(C\equiv C) = 1.205$ ass. $r(C—C) = 1.376$ $r(C—H) = 1.446$ | 180° | infrared and Raman sp. | (645)(179) | |

$C_4H_2^+$: [a] $A = 30.6$ cm$^{-1}$. [b] $A = 33.3$ cm$^{-1}$. [c] $2\nu_7 = 971.5$.

$C_4H_2$: [a] From the predissociation in the $\tilde{A}\leftarrow\tilde{X}$ system. [b] May be $\nu_6$.
[c] Complicated vibrational structure; very slightly predissociated making resolution of rotational structure impossible. [d] $\nu_7(\pi_g) = 482$, $\nu_8(\pi_u) = 630$, $\nu_9 = (220)$.

TABLE 74 (*continued*)

| State | Point Group | $T_0$ | $\nu_1$ | $\nu_2$ | $\nu_3$ | $\nu_4$ | $\nu_5$ | $\nu_6$ | $A_0$ | $B_0$ | $C_0$ | $r_0$(Å) | $\alpha$ | Observed Transitions | References | Remarks |
|---|---|---|---|---|---|---|---|---|---|---|---|---|---|---|---|---|
| | | | Vibrational Frequencies | | | | | | Rotational Constants | | | | | | | |
| **$HC_2CHO$** (propynal) | | | | | | | | | | | | | | | | |
| $\tilde{D}$ | | Continuous absorption below 1600 Å | | | | | | | | | | | | $\tilde{D}\leftarrow\tilde{X}$ 1600–1300 Å | (525) | |
| $\tilde{C}$ | | 46800 | ($\Delta\nu \approx 1700$) | | | | | | | | | | | $\tilde{C}\leftarrow\tilde{X}$ 2150–1900 Å | (581) | Diffuse bands |
| $\tilde{B}(^1A'')$ | ($C_s$) | (38900) | $\Delta\nu \approx 670$ | | | | | | | | | | | $\tilde{B}\leftarrow\tilde{X}$ 2570–2150 Å | (1278a) | Diffuse bands |
| $\tilde{A}\ ^1A''$ | $C_s$ | 26162.9 | | 2952.5 | 1945.8 | 1304.0 | 1119.5 | 951.6[a] | $1.890_6$[b] | 0.16336 | 0.14987 | r(C=O) = 1.325<br>r(C≡C) = 1.238<br>r(CH) = 1.091 | | $\tilde{A}\leftarrow\tilde{X}$ 3830–3000 Å | (140)(141) | Sharp, well-resolved bands; see Fig. 199 |
| $\tilde{a}\ ^3A''$ | $C_s$ | 24127.2 | | | | 1323.1 | | [f] | | | | | | $\tilde{a}\leftarrow\tilde{X}$ 4200–3900 Å | (140)<br>(173a) | Sharp bands |
| $\tilde{X}\ ^1A'$ | $C_s$ | 0 | 3326 | $2858._2$ | 2106 | 1696.9 | 1389 | 943.7[c] | 2.2694[d] | $0.16098_5$ | 0.150091 | r(C=O) = 1.215<br>r(C≡C) = 1.209<br>r(C—C) = 1.445<br>r(CH) = 1.106, 1.055 | ∠CCO = 123.7°<br>∠CCC = 178.4°[e] | microwave and infrared sp. | (234)(668)<br>(144) | |

$HC_2CHO$: [a] $\nu_7(a') = (650)$, $\nu_8 = 506._9$, $\nu_9 = 189.4$, $\nu_{10}(a'') = 462.1$, $\nu_{11} = 389.7$, $\nu_{12} = 345.9$. [b] $D_K = 2.7_3 \times 10^{-4}$.
[c] $\nu_7(a') = 650.0$, $\nu_8 = 613.7$, $\nu_9 = 205.3$, $\nu_{10}(a'') = 981.2$, $\nu_{11} = 692.7$, $\nu_{12} = 260.6$. [d] $D_K = 2.90 \times 10^{-4}$; the effective microwave $A_0$ has been corrected for the effect of $D_K$.
[e] ∠C—CH = 113.9. [f] $\nu_9 = 181.0$, $\nu_{12} = 341.3°$. [g] Sub-bands with $\Delta K = 0, \pm 1$ and branches with $\Delta N = 0, \pm 1, \pm 2$ have been observed.

TABLE 74 (continued)

| State | Point Group | $T_0$ | Vibrational Frequencies $\nu_1$ | $\nu_2$ | $\nu_3$ | $\nu_4$ | $\nu_5$ | $\nu_6$ | Rotational Constants $A_0$ | $B_0$ | $C_0$ | $r_0$(Å) | $\alpha$ | Observed Transitions | References | Remarks |
|---|---|---|---|---|---|---|---|---|---|---|---|---|---|---|---|---|
| **$H_2C_2O_2$** (glyoxal) | | | I.P. $\leq$ 9.48 eV[a]; $D$(HCO—HCO) $\leq$ 2.84 eV[b] | | | | | | | | | | | | | |
| $\tilde{E}$ | | 73800 | | 1733 | | | | | | | | | | $\tilde{E} \leftarrow \tilde{X}$ 1355–1324 Å | (1262) | Two strong bands[c] |
| $\tilde{D}$ | | (57000) | broad continuum with several diffuse maxima | | | | | | | | | | | $\tilde{D} \leftarrow \tilde{X}$ 1770–1470 Å | (1262) | Repr. (1262) |
| $\tilde{C}$ | | 48974 | | | | 560 | | | | | | | | $\tilde{C} \leftarrow \tilde{X}$ 2050–1850 Å | (1262) | Progression of diffuse bands Repr. (1262) |
| $\tilde{B}$ | | (31000) | diffuse bands | | | | | | | | | | | $\tilde{B} \leftarrow \tilde{X}$ 3200–2300 Å | (1206) | |
| $\tilde{A}\ ^1A_u$ | $C_{2h}$ | 21973.4 | | 1391.4 | | 947 | 509 | [d] | 1.9635[e] | 0.1549 | 0.1435 | | | $\tilde{A} \leftrightarrow \tilde{X}$ 5400–3900 Å | (137)(667) (957) | Well-developed fine structure |
| $\tilde{a}\ ^3A_u$ | $C_{2h}$ | 19197 | | | (1457) | (465) | (106) | | | | | | | $\tilde{a} \leftrightarrow \tilde{X}$ 5750–5150 Å | (137)(340) | Overlapped by $\tilde{A}-\tilde{X}$ |
| $\tilde{X}\ ^1A_g$ | $C_{2h}$ | 0 | (2844) | 1742 | 1338 | 1060[f] | 553 | 801.5[g] | 1.8454[h] | 0.1599 | 0.1472 | | | infrared sp. | (220)(1206) (137)(478) (219) | |
| **$C_2H_2Cl_2$** *trans* | | | I.P. = 9.95$_3$ eV[a] | | | | | | | | | | | | | |
| | | | Rydberg series joining on to $\tilde{B}$ and $\tilde{D}$: $\nu = 80285 - R/(n - 0.72)^2$; $n = 3, 4, \ldots, 9$ | | | | | | | | | | | | (1259) | |
| $\tilde{D}$ | | 70081 | | | | (800)[b] | | | | | | | | $\tilde{D} \leftarrow \tilde{X}$ 1427 Å | (1259) | Very strong band |
| $\tilde{C}$ | | 65419 | | 1444 | | 732 | 184 | [c] | | | | | | $\tilde{C} \leftarrow \tilde{X}$ 1530–1440 Å | (1259) | Sharp bands, extensive system |
| $\tilde{B}$ | | (55500) | | | | | | | | | | | | $\tilde{B} \leftarrow \tilde{X}$ 1800–1600 Å | (1259) | Very diffuse bands |
| $\tilde{A}$ | | (41700) | continuum with maximum at 1950 Å | | | | | | | | | | | $\tilde{A} \leftarrow \tilde{X}$ 2100–1800 Å | (1259)(705) | |
| $\tilde{X}\ ^1A_g$ | $C_{2h}$ | 0 | 3071 | 1576 | 1270 | 844 | 349 | 898[d] | | | | | | infrared and Raman sp. | (115) | |

$H_2C_2O_2$: [a] Appearance potential of $H_2C_2O_2^+$ in mass spectrometer (1062). [b] From (1062). [c] May be two different electronic transitions. [d] $\nu_7(a_u) = 232$, $\nu_8(b_g) = 734$, $\nu_{12}(b_u) = 282$.
[e] $D_K = 2.7 \times 10^{-5}$. [f] (142) gives 1201 cm$^{-1}$ for this frequency. [g] $\nu_7(a_u) = 127$, $\nu_8(b_g) = 1048$, $\nu_9(b_u) = 2836.2$, $\nu_{10} = 1730$, $\nu_{11} = 1311.5$, $\nu_{12} = 342$.
[h] $D_K = 1.9 \times 10^{-5}$. The rotational constants are from the analysis of the $\tilde{A}-\tilde{X}$ electronic transition.

$C_2H_2Cl_2$ *trans*: [a] From the Rydberg series. (873) finds a photoionization limit at 9.64 eV. [b] Not given by (1259) but read from his spectrogram.
[c] Another observed frequency is 370 cm$^{-1}$, but it is not assigned.
[d] $\nu_7(a_u) = (192)$, $\nu_8(b_g) = 758$, $\nu_9(b_u) = 3090$, $\nu_{10} = 1200$, $\nu_{11} = 827$, $\nu_{12} = (265)$. The frequencies $\nu_1$, $\nu_2$, $\nu_3$, $\nu_4$, $\nu_5$, $\nu_8$ refer to the liquid.

TABLE 74 (*continued*)

| State | Point Group | $T_0$ | $\nu_1$ | $\nu_2$ | $\nu_3$ | $\nu_4$ | $\nu_5$ | $\nu_6$ | $A_0$ | $B_0$ | $C_0$ | $r_0$(Å) | $\alpha$ | Observed Transitions | References | Remarks |
|---|---|---|---|---|---|---|---|---|---|---|---|---|---|---|---|---|
| **$C_2H_2Cl_2$** *cis* | | I.P. = $9.65_7$ eV[a] | | | | | | | | | | | | | | |
| | | | Strong Rydberg series joining on to $\tilde{D}$ and $\tilde{G}$: | | | | | | $\nu = 77850 - R/(n - 0.07)^2$; $n = 3, 4, \ldots, 7$ | | | | | | (1259) | |
| | | | Strong Rydberg series joining on to $\tilde{A}$ and $\tilde{E}$: | | | | | | $\nu = 77937 - R/(n - 0.95)^2$; $n = 3, 4, \ldots, 7$ | | | | | | (1259) | |
| $\tilde{G}$ | | 70725 | | 1416 | | | | $\nu_7 = 260$ | | | | | | $\tilde{G} \leftarrow \tilde{X}$ 1414–1360 Å | (1259) | |
| $\tilde{F}$ | | 68166 | | | | | | | | | | | | $\tilde{F} \leftarrow \tilde{X}$ 1469–1464 Å | (1259) | Very strong sharp doublet |
| $\tilde{E}$ | | 65840 | | 1426 | | | | $\nu_7 = 250$ | | | | | | $\tilde{E} \leftarrow \tilde{X}$ 1535–1430 Å | (1259) | Sharp bands |
| $\tilde{D}$ | | 65182 | | | | | | | | | | | | $\tilde{D} \leftarrow \tilde{X}$ 1534 Å | (1259) | Sharp bands |
| $\tilde{C}$ | | 59749 | | 1423 | | | | $\nu_{10} = 805$ | | | | | | $\tilde{C} \leftarrow \tilde{X}$ 1700–1600 Å | (1259) | Strong sharp bands |
| $\tilde{A}$ | | (42300) | continuum with maximum at 1890 Å | | | | | | | | | | | $\tilde{A} \leftarrow \tilde{X}$ 2050–1770 Å | (1259)(705) | [b] |
| $\tilde{X}\ ^1A_1$ | $C_{2v}$ | 0 | 3086 | 1591 | 1179 | 711 | 173 | 876[c] | 0.384210 | 0.084897 | $0.069467_0$ | | | infrared, Raman and microwave sp. | (115)(388) | |
| **$C_2H_2Cl_2$** 1:1 | | I.P. = $9.45_9$ eV | | | | | | | | | | | | | | |
| | | | Doubtful Rydberg series joining on to $\tilde{E}$: | | | | | | $\nu = 83630 - R/(n - 0.60)^2$; $n = 3, 4, \ldots, 6$ | | | | | | (1195) | |
| | | | Rydberg series joining on to $\tilde{B}$, $\tilde{D}$: | | | | | | $\nu = 76293 - R/(n - 0.93)^2$; $n = 3, 4, \ldots, 9$ | | | | | | (1195) | |
| $\tilde{E}$ | | 67594 | | | | | | | | | | | | $\tilde{E} \leftarrow \tilde{X}$ 1479 Å | (1195) | Single band |
| $\tilde{D}$ | | 64650 | | | | | | | | | | | | $\tilde{D} \leftarrow \tilde{X}$ 1550–1500 Å | (1195) | Sharp bands |
| $\tilde{C}$ | | (61000)[a] | | | | | | | | | | | | $\tilde{C} \leftarrow \tilde{X}$ 1650–1550 Å | (1195) | Sharp bands |
| $\tilde{B}$ | | (54000) | | | | | | | | | | | | $\tilde{B} \leftarrow \tilde{X}$ 1870–1720 Å | (1195) | Diffuse bands, may belong to $\tilde{A} \leftarrow \tilde{X}$ |
| $\tilde{A}$ | | (41900) | continuum with maximum at 1920 Å | | | | | | | | | | | $\tilde{A} \leftarrow \tilde{X}$ 2400–1750 Å | (1195)(705) | |
| $\tilde{X}\ ^1A_1$ | $C_{2v}$ | 0 | 3035 | 1616 | 1391 | 601 | 299 | 686[b] | 0.24907 | 0.11379 | 0.07802 | $r$(CCl) = 1.727; $r$(CC) = 1.32 ass. | $\angle$ClCC = 123.2° | Raman, infrared and microwave sp. | (653) (1116) | |

$C_2H_2Cl_2$ *cis*: [a] From the Rydberg series. From photoionization the value 9.66 eV has been found (873). [b] The analogue of the $\tilde{B}-\tilde{X}$ transition of $C_2H_3Cl$ is not obvious but may well exist. [c] $\nu_7(a_2) = 406$, $\nu_8(b_1) = 697$, $\nu_9(b_2) = 3072$, $\nu_{10} = 1303$, $\nu_{11} = 857$, $\nu_{12} = 571$. The frequencies $\nu_3$, $\nu_4$, $\nu_5$, $\nu_6$, $\nu_7$ refer to the liquid.

$C_2H_2Cl_2$ 1:1: [a] Not given as separate electronic transition by (1195), but these bands cannot be hot bands of $\tilde{D}-\tilde{X}$. [b] $\nu_7(b_1) = 874$, $\nu_8 = 458$, $\nu_9(b_2) = 3130$, $\nu_{10} = 1088$, $\nu_{11} = 788$, $\nu_{12} = 375$ (Raman sp. of liquid and infrared sp. of vapor).

TABLE 75

MOLECULAR CONSTANTS OF THE ELECTRONIC STATES OF SIX-ATOMIC MONO-HYDRIDES

| State | Point Group | $T_0$ | Vibrational Frequencies $\nu_1$ | $\nu_2$ | $\nu_3$ | $\nu_4$ | $\nu_5$ | $\nu_6$ | Rotational Constants $A_0$ | $B_0$ | $C_0$ | $r_0$(Å) | $\alpha$ | Observed Transitions | References | Remarks |
|---|---|---|---|---|---|---|---|---|---|---|---|---|---|---|---|---|
| $C_2HCl_3$ | I.P. = 9.47 eV[a] | | | | | | | | | | | | | | | |
| $\tilde{H}$ | | 71000 | Very strong diffuse absorption bands | | | | | | | | | | | $\tilde{H} \leftarrow \tilde{X}$ 1400–1200 Å | (1259) | |
| | | Fragmentary Rydberg series joining on to $\tilde{B}$ and $\tilde{C}$: | | | | | | | $\nu = 70890 - R/(n - 0.90)^2$; $n = 4, \ldots, 9$ | | | | | | (1259) | |
| $\tilde{C}$ | | 64373 | | 1400 | | | 426 | | | | | | | $\tilde{C} \leftarrow \tilde{X}$ 1553–1450 Å | (1259) | Strong sharp bands, overlapped by continuum |
| $\tilde{B}$ | | 59295 | | 1400 | | 600 | 280[b] | 120[b] | | | | | | $\tilde{B} \leftarrow \tilde{X}$ 1686–1550 Å | (1259) | Sharp bands |
| $\tilde{A}$ | | (39500) | Broad continuum with maximum at 1960 Å | | | | | | | | | | | $\tilde{A} \leftarrow \tilde{X}$ 2600–1800 Å | (1259)(705) | |
| $\tilde{X}\ ^1A'$ | $C_s$ | 0 | 3096 | 1590 | 1250 | 850 | 633 | 452[c] | | | | | | infrared and Raman sp. | (1323)(113) | |

$C_2HCl_3$: [a] This is the photoionization value (1273). The Rydberg series gives 8.79 eV. [b] Assignment uncertain.
[c] $\nu_7(a') = 381$(liq.), $\nu_8 = 272$(liq.), $\nu_9 = 169$(liq.), $\nu_{10}(a'') = 940$, $\nu_{11} = 784$, $\nu_{12} = 212$(liq.).

TABLE 76

MOLECULAR CONSTANTS OF THE ELECTRONIC STATES OF SIX-ATOMIC NON-HYDRIDES

| State | Point Group | $T_0$ | Vibrational Frequencies $\nu_1$ | $\nu_2$ | $\nu_3$ | $\nu_4$ | $\nu_5$ | $\nu_6$ | Rotational Constants $A_0$ | $B_0$ | $C_0$ | $r_0$(Å) | $\alpha$ | Observed Transitions | References | Remarks |
|---|---|---|---|---|---|---|---|---|---|---|---|---|---|---|---|---|
| **$C_4N_2$ (dicyanoacetylene)** | | | | | | | | | | | | | | | | |
| $\tilde{B}(^1\Delta_u)$ | | 37814 | 2192 | | (591) | | | 458[a] | | | | | | $\tilde{B} \leftarrow \tilde{X}$ 2720–2250 Å | (846) | Slightly diffuse bands |
| $\tilde{A}(^1\Sigma_u^-)$ | | (35500) | | | | | | | | | | | | $\tilde{A} \leftarrow \tilde{X}$ 2880–2780 Å | (846)(641) | Extensive system of double-headed bands |
| $\tilde{X}\ ^1\Sigma_g^+$ | $\boldsymbol{D}_{\infty h}$ | 0 | 2267[b] | 2119 | 692 | 2241 | 1154 | 504[c] | | | | | | Raman and infrared sp. | (847) | |
| **$N_2O_4$** | $D(O_2N—NO_2) = 0.5937$ eV[a] | | Continuous absorption only, see (462) | | | | | | | | | | | | | |
| **$C_2O_2Cl_2$ (oxalyl chloride)** | | | | | | | | | | | | | | | | |
| $\tilde{B}$ | | (34600) | Very broad diffuse bands | | | | | | | | | | | 2900–2400 Å | (1095) | |
| $\tilde{A}(^1A_u)$ | $(\boldsymbol{C}_{2h})$ | 27189 | 1460 | | 397 | | | | | | | | | $\tilde{A} \leftarrow \tilde{X}$ 3800–3000 Å | (1095)(1130) | a |
| $\tilde{X}\ ^1A_g$ | $\boldsymbol{C}_{2h}$ | 0 | 1778 | 1078 | 619 | 465 | 276 | 360[b] | | | | | | infrared and Raman sp. | (1094)(492a) | c |

$C_4N_2$: [a] $\nu_7(\pi_g) = 260$, $\nu_8(\pi_u) = (443)$, $\nu_9 = 99$. [b] Shifted by Fermi resonance with $2\nu_5$ which occurs at 2333 cm$^{-1}$. [c] $\nu_7(\pi_g) = 263$, $\nu_8(\pi_u) = 472$, $\nu_9 = (107)$.

$N_2O_4$: [a] From (417); confirmed by (457).

$C_2O_2Cl_2$: [a] It is likely that both *cis* and *trans* forms are present, although (492a) consider that the *trans* form is sufficient to account for the entire spectrum. The vibrational frequencies and $T_0$ given here refer to the *trans* form.

[b] $\nu_7(a_u) = (159)$, $\nu_8(b_g) = 201$, $\nu_9(b_u) = 1790$, $\nu_{10} = 778$, $\nu_{11} = 520$, $\nu_{12} = (210)$. These frequencies are those given by (492a) for the *trans* form.

[c] The energy difference between the *cis* and *trans* form is given by (1094) as 980 cm$^{-1}$. See, however (492a).

TABLE 76 (*continued*)

| State | Point Group | $T_0$ | Vibrational Frequencies $\nu_1$ | $\nu_2$ | $\nu_3$ | $\nu_4$ | $\nu_5$ | $\nu_6$ | Rotational Constants $A_0$ | $B_0$ | $C_0$ | $r_0$(Å) | $\alpha$ | Observed Transitions | References | Remarks |
|---|---|---|---|---|---|---|---|---|---|---|---|---|---|---|---|---|
| | **$CF_3NO$** | see (798) | | | | | | | | | | | | | | |
| | **$C_2Cl_4$** | I.P. = $9.3_2$ eV | | | | | | | | | | | | | | |
| | | Very strong diffuse absorption bands below 1360 Å | | | | | | | | | | | | | (1259)(441) | |
| $\tilde{F}$ | | 72410 | | | | | | | | | | | | $\tilde{F} \leftarrow \tilde{X}$ 1381–1360 Å | (1259) | |
| $\tilde{E}$ | | 70097 | 1333 | 460 | 245 | | | | | | | | | $\tilde{E} \leftarrow \tilde{X}$ 1427–1360 Å | (1259)(441) | Sharp bands |
| $\tilde{D}$ | | 63557 | 1343 | 480 | 260 | | | | | | | | | $\tilde{D} \leftarrow \tilde{X}$ 1573–1440 Å | (1259)(441) | Sharp bands |
| $\tilde{C}$ | | (59000) | Broad continuum with maximum at 1615 Å | | | | | | | | | | | $\tilde{C} \leftarrow \tilde{X}$ 1700–1500 Å | (1259) | |
| $\tilde{B}$ | | (50350) | 1350 | 450 | | | | | | | | | | $\tilde{B} \leftarrow \tilde{X}$ 2000–1670 Å | (1259)(441) | Diffuse bands |
| $\tilde{A}$ | | (37700) | Broad continuum with maximum at 1970 Å | | | | | | | | | | | $\tilde{A} \leftarrow \tilde{X}$ 2700–1800 Å | (705)(1259)(441) | |
| $\tilde{X}\ ^1A_g$ | $D_{2h}$ | 0 | 1571 | 447 | 237 | | 1000 | 347[a] | | | | | | infrared and Raman sp. | (114) | |

$C_2Cl_4$: [a] $\nu_7(b_{1u}) = 288$, $\nu_8(b_{2g}) = 512$, $\nu_9(b_{2u}) = 782$, $\nu_{10} = 194$, $\nu_{11}(b_{3u}) = 913$, $\nu_{12} = 318$.

TABLE 77

MOLECULAR CONSTANTS IN THE ELECTRONIC STATES OF SEVEN-ATOMIC MOLECULES

| State | Point Group | $T_0$ | Vibrational Frequencies $\nu_1$ | $\nu_2$ | $\nu_3$ | $\nu_4$ | $\nu_5$ | $\nu_6$ | Rotational Constants $A_0$ | $B_0$ | $C_0$ | $r_0$(Å) | $\alpha$ | Observed Transitions | References | Remarks |
|---|---|---|---|---|---|---|---|---|---|---|---|---|---|---|---|---|
| **$C_2H_5$ (ethyl)** | | Diffuse bands at 2229 and 2242 Å: see (415) | | | | | | | | | | | | | | |
| **$CH_3NH_2$** | | I.P. = 8.97 eV[a]; $D(CH_3—NH_2)$ = (3.5) eV | | | | | | | | | | | | | | |
| $\tilde{A}$ | | 41680 | | | | | | 1000[b] | | | | | | $\tilde{A} \leftarrow \tilde{X}$ 2450–2000 Å | (393)(357) | Diffuse bands |
| $\tilde{X}\ ^1A'$ | $C_s$ | 0 | 3361 | 2961 | 2820 | 1623 | 1473 | 1430[c] | 3.440[d] | $0.7546_6$[d] | $0.7237_9$[d] | $r$(CN) = 1.474<br>$r$(NH) = 1.011 | ∠HNH = 105.9°<br>∠CNH = 112.1°[e] | infrared, Raman and microwave sp. | (444)(743)<br>(938) | [f] |
| **$H_2C{=}C{=}CH_2$ (allene)** | | I.P. = $10.19_2$ eV | | | | | | | | | | | | | | |
| | | Fragments of four Rydberg series (see text) | | | | | | | | | | | | | (1179) | |
| | | Rydberg series joining on to $\tilde{F}$: $\nu = 82210 - R/(n - 0.30)^2$; $n = 3, 4, 5, 6$ | | | | | | | | | | | | | (1179) | |
| | | Rydberg series joining on to $\tilde{E}$: $\nu = 82200 - R/(n - 0.40)^2$; $n = 3, 4, \ldots, 7$ | | | | | | | | | | | | | (1179) | |
| | | Rydberg series joining on to $\tilde{D}$: $\nu = 82190 - R/(n - 0.55)^2$; $n = 3, 4, \ldots, 7$ | | | | | | | | | | | | | (1179) | |
| | | Rydberg series joining on to $\tilde{C}$: $\nu = 82190 - R/(n - 0.70)^2$; $n = 3, 4, \ldots, 7$ | | | | | | | | | | | | | (1179) | |
| | | Rydberg series joining on to $\tilde{B}$: $\nu = 82210 - R/(n - 1.06)^2$; $n = 3, 4, \ldots, 8$ | | | | | | | | | | | | | (1179) | |
| $\tilde{F}$ | | 67726 | | | | | | | | | | | | $\tilde{F} \leftarrow \tilde{X}$ 1477 Å | (1179) | |
| $\tilde{E}$ | | 67108 | | | | | | | | | | | | $\tilde{E} \leftarrow \tilde{X}$ 1490 Å | (1179) | |
| $\tilde{D}$ | | {66188<br>65808 | | | | | | | | | | | | $\tilde{D} \leftarrow \tilde{X}$ 1520–1511 Å | (1179) | |
| $\tilde{C}$ | | 64705 | | | | | | | | | | | | $\tilde{C} \leftarrow \tilde{X}$ 1545–1535 Å | (1179) | Two strong bands |
| $\tilde{B}$ | | 57800 | | | 610 | | | | | | | | | $\tilde{B} \leftarrow \tilde{X}$ 1750–1580 Å | (1179) | Diffuse bands |
| $\tilde{A}$ | | < 49300 | broad continuum | | | | | | | | | | | $\tilde{A} \leftarrow \tilde{X}$ 2100–1600 Å | (1179) | |
| $\tilde{X}\ ^1A_1$ | $D_{2d}$ | 0 | 3015.0 | 1442.6 | 1072.6 | 865 | 3006.8 | 1957[a] | 4.81 | $0.2963_2$[b] | — | $r$(C=C) = $1.308_4$<br>$r$(C—H) = 1.087 | ∠HCH = 118.2° | infrared and Raman sp. | (775)(1165)<br>(954)(156)<br>(339)(952)<br>(795a) | |

$CH_3NH_2$: [a] From photoionization measurements of (1276). [b] $\nu_9$ = 650. [c] $\nu_7$ = 1130, $\nu_8$ = 1044, $\nu_9$ = 780, $\nu_{10}(a'')$ = 3427, $\nu_{11}$ = 2985, $\nu_{12}$ = 1485, $\nu_{13}$ = 1455, $\nu_{14}$ = 1195, $\nu_{15}$ = 264.
[d] Calculated from the moments of inertia given by (743).
[e] Distance from N atom to $CH_3$ symm. axis = 0.091 Å. The HCH angle has been assumed to be tetrahedral and the C—H distance to be 1.093 Å.
[f] Potential barrier for internal rotation: 691 $cm^{-1}$ [see (938) and (743)].

$H_2C{=}C{=}CH_2$: [a] $\nu_7(b_2)$ = 1398, $\nu_8(e)$ = 3085, $\nu_9$ = 1015, $\nu_{10}$ = 842, $\nu_{11}$ = 354 $cm^{-1}$. [b] $D_J = 11 \times 10^{-8}$, $D_{JK} = 5 \times 10^{-6}$ $cm^{-1}$.

TABLE 77 (*continued*)

| State | Point Group | $T_0$ | $\nu_1$ | $\nu_2$ | $\nu_3$ | $\nu_4$ | $\nu_5$ | $\nu_6$ | $A_0$ | $B_0$ | $C_0$ | $r_0$(Å) | $\alpha$ | Observed Transitions | References | Remarks |
|---|---|---|---|---|---|---|---|---|---|---|---|---|---|---|---|---|
| **$CH_3C{\equiv}CH$** (propyne) I.P. = $10.36_3$ eV[a] | | | | | | | | | | | | | | | | |
| | | Rydberg series joining on to $\tilde{G}$: $\nu = 91240 - R/(n - 0.52)^2$; $n = 3, 4, \ldots, 13$ | | | | | | | | | | | | | (1029) | |
| | | Rydberg series joining on to $\tilde{C}$: $\nu = 91100 - R/(n - 0.96)^2$; $n = 3, 4, \ldots, 12$ | | | | | | | | | | | | | (1029) | [b] |
| | | Rydberg series joining on to $\tilde{F}$: $\nu = 83570 - R/(n - 0.98)^2$; $n = 4, 5, 6, 7$ | | | | | | | | | | | | | (927) | |
| | | Rydberg series joining on to $\tilde{E}$: $\nu = 83600 - R/(n - 0.33)^2$; $n = 3, 4, \ldots, 10$ | | | | | | | | | | | | | (927) | |
| | | Rydberg series joining on to $\tilde{D}$: $\nu = 83580 - R/(n - 0.57)^2$; $n = 3, 4, \ldots, 9$ | | | | | | | | | | | | | (927) | |
| $\tilde{G}$ | | 73597 | | | | | | | | | | | | $\tilde{G} \leftarrow \tilde{X}$ 1359 Å | (1029) | |
| $\tilde{F}$ | | 71230 | | | 1960 | | | | | | | | | $\tilde{F} \leftarrow \tilde{X}$ 1405–1296 Å | (927) | $4R$ of (927) |
| $\tilde{E}$ | | 68140 | | | 1940 | | | | | | | | | $\tilde{E} \leftarrow \tilde{X}$ 1468–1388 Å | (927) | $3R''$ of (927) |
| $\tilde{D}$ | | 64913 | | | 1970 | | 990 | | | | | | | $\tilde{D} \leftarrow \tilde{X}$ 1540 Å | (1029)(927) | $3R'$ of (927) repr. (927) |
| $\tilde{C}$ | | 63370 | | | | | | | | | | | | $\tilde{C} \leftarrow \tilde{X}$ 1596–1578 Å | (1029)(927) | Repr. (1029)[c] (927) |
| $\tilde{B}$ | | (57000) | broad continuum with overlapping diffuse bands; maximum at 1650 Å | | | | | | | | | | | $\tilde{B} \leftarrow \tilde{X}$ 1750–1600 Å | (1029) | |
| $\tilde{A}$ | | (51000) | broad continuum with overlapping diffuse bands; maximum at 1920 Å | | | | | | | | | | | $\tilde{A} \leftarrow \tilde{X}$ 2000–1880 Å | (1029) | |
| $\tilde{X}\ {}^1A_1$ | $C_{3v}$ | 0 | 3334.6 | 2941.0 | 2142.0 | 1382 | 930.7 | 3008.3[d] | (5.21)[e] | 0.285058 | — | $r(CH) = \begin{cases}1.105\\1.056\end{cases}$ $r(C{-}C) = 1.459$ $r(C{\equiv}C) = 1.206$ | $\angle HCH = 110.2°$ | infrared, Raman, and microwave sp. | (199)(40) (233)(133) (101) | |

$CH_3C{\equiv}CH$: [a] From Rydberg series and photoionization (1277). [b] According to (1029) the first member is $D$; see text. [c] Shortward component of a doublet on (1029)'s spectrum as measured by (927). [d] $\nu_7(e) = 1452.0$, $\nu_8 = 1052.5$, $\nu_9 = 633.2$, $\nu_{10} = 328.0$. [e] Indirect, from (133).

TABLE 77 (*continued*)

| State | Point Group | $T_0$ | Vibrational Frequencies $\nu_1$ | $\nu_2$ | $\nu_3$ | $\nu_4$ | $\nu_5$ | $\nu_6$ | Rotational Constants $A_0$ | $B_0$ | $C_0$ | $r_0$(Å) | $\alpha$ | Observed Transitions | References | Remarks |
|---|---|---|---|---|---|---|---|---|---|---|---|---|---|---|---|---|
| **$C_2H_4O$** (ethylene oxide) I.P. = $10.56_5$ eV[a] | | | | | | | | | | | | | | | | |
| | | Strong continuous absorption below 1180 Å | | | | | | | | | | | | | (781) | |
| | | Rydberg series joining on to $\tilde{E}$: $\nu = 85220 - R/(n - 0.04)^2$; $n = 3, 4, \ldots, 8$ | | | | | | | | | | | | | (781) | |
| $\tilde{F}$ | | 77089 | | | | | | | | | | | | $\tilde{F} \leftarrow \tilde{X}$ 1297 Å | (781) | Sharp band[b] |
| $\tilde{E}$ | | 72307 | | 1456 | | 1141 | 686 | | | | | | | $\tilde{E} \leftarrow \tilde{X}$ 1385–1320 Å | (781)(755) | Sharp bands |
| $\tilde{D}$ | | 69682 | | | | 1114 | | | | | | | | $\tilde{D} \leftarrow \tilde{X}$ 1435–1390 Å | (781) | Sharp bands |
| $\tilde{C}$ | | 63605 | | | | 1125 | 724 | | | | | | | $\tilde{C} \leftarrow \tilde{X}$ 1575–1500 Å | (755)(781) | Sharp bands |
| $\tilde{B}$ | | 58377 | | | | | 780 | | | | | | | $\tilde{B} \leftarrow \tilde{X}$ 1715–1660 Å | (755)(781) (387) | Diffuse bands |
| $\tilde{A}$ | | (47000) | continuous absorption with maximum at 1680 Å | | | | | | | | | | | $\tilde{A} \leftarrow \tilde{X}$ 2120–1600 Å | (755)(781) | |
| $\tilde{X}\ ^1A_1$ | $C_{2v}$ | 0 | 3005 | 1490 | 1266 | 1120 | 877 | 3063[c] | 0.85004 | 0.73787 | 0.47026 | r(C—C) = 1.472<br>r(C—O) = 1.436<br>r(C—H) = 1.082 | ∠COC = 61.4°<br>∠HCH = 116.7°<br>∠$H_2$CC = 159.4° | infrared, Raman and microwave sp. | (774)(258) | |

$C_2H_4O$: [a] Confirmed by photoionization experiments of (1273); (see also (781)). [b] This system and $\tilde{D} - \tilde{X}$, according to (781), form the first two members of a second Rydberg series.
[c] $\nu_7(a_2) = (1345)$, $\nu_8 = 807$, $\nu_9(b_1) = 3079$, $\nu_{10} = 1143$, $\nu_{11} = 821$, $\nu_{12}(b_2) = 3019$, $\nu_{13} = 1470$, $\nu_{14} = 1153$, $\nu_{15} = 892$.

TABLE 77 (*continued*)

| State | Point Group | $T_0$ | $\nu_1$ | $\nu_2$ | $\nu_3$ | $\nu_4$ | $\nu_5$ | $\nu_6$ | $A_0$ | $B_0$ | $C_0$ | $r_0$(Å) | $\alpha$ | Observed Transitions | References | Remarks |
|---|---|---|---|---|---|---|---|---|---|---|---|---|---|---|---|---|
| **$CH_3CHO$** (acetaldehyde) I.P. = 10.229 eV; $D(CH_3$—$CHO) = (2.9_5)$ eV; $D(CH_3CO$—$H) = 3.69$ eV[a] | | | | | | | | | | | | | | | | |
| | | Rydberg series joining on to $\tilde{E}$: $\nu = 82475 - R/(n-0.20)^2$; $n = 3, 4, \ldots, 10$ | | | | | | | | | | | | | (1263) | |
| | | Rydberg series joining on to $\tilde{C}$: $\nu = 82505 - R/(n-0.70)^2$; $n = 3, 4, \ldots, 18$ | | | | | | | | | | | | | (1263) | |
| | | Rydberg series joining on to $\tilde{B}$: $\nu = 82504 - R/(n-0.90)^2$; $n = 3, 4, \ldots, 11$ | | | | | | | | | | | | | (1263) | |
| $\tilde{E}$ | | 68030 | | | | | | | | | | | | $\tilde{E} \leftarrow \tilde{X}$ 1470 Å | (1263) | Sharp band |
| $\tilde{D}$ | | 62500[b] | | | | | | | | | | | | $\tilde{D} \leftarrow \tilde{X}$ 1600 Å | (1263) | Sharp doublet |
| $\tilde{C}$ | | 60170 | | | | 1200 | | | | | | | | $\tilde{C} \leftarrow \tilde{X}$ 1665–1550 Å | (1263)(714) | Diffuse bands, doublets |
| $\tilde{B}$ | | 54996 | | | | 1200 | (750) | (350) | | | | | | $\tilde{B} \leftarrow \tilde{X}$ 1820–1670 Å | (1263)(714) | Sharp bands |
| $\tilde{A}$ | | (28700) | | | | 1125 | (480) | | | | | | | $\tilde{A} \leftrightarrow \tilde{X}$ 3500–2500 Å[c] | (1107)(800) (1074)(1057) (606)(764) (366)(666) (706) | Sharp bands followed by diffuse bands[d] |
| $\tilde{X}\ ^1A'$ | $C_s$ | 0 | 2967 | (2840) | 2736[e] | 1743 | 1441 | 1390[f] | 1.8877[g] | 0.33901[g] | 0.30354[g] | $r$(C—C) = 1.501<br>$r$(C=O) = 1.216<br>$r$(C—H) = $\begin{cases}1.086\\1.114\end{cases}$ | ∠CCO = 123.9°<br>∠HCH = 108.3°<br>∠CCH = 117.5° | infrared, Raman and microwave sp. | | [h] |
| **$CH_3NCS$, $CH_3SCN$** see (1314) | | | | | | | | | | | | | | | | |
| **$CH_3NO_2$** (nitromethane) see (1212) and (923) | | | | | | | | | | | | | | | | |
| **$CH_3ONO$** (methyl nitrite) see (1212) | | | | | | | | | | | | | | | | |
| **$CH_3COCl$** see (1242) diffuse bands | | | | | | | | | | | | | | | | |
| **$N_2O_5$** see (646) continuous absorption only | | | | | | | | | | | | | | | | |

$CH_3CHO$: [a] From mass spectrometric data of (919). [b] Only rough $\lambda$ quoted by (1263). Apparently does not belong to $\tilde{C} \leftarrow \tilde{X}$. [c] In emission to 6100 Å. [d] and continuum.
[e] Probably shifted by Fermi resonance. [f] $\nu_7 = 1352$, $\nu_8 = 1113$, $\nu_9 = 919$, $\nu_{10} = 509$, $\nu_{11}(a'') = 3024$, $\nu_{12} = (1420)$, $\nu_{13} = 867$, $\nu_{14} = 763.9$, $\nu_{15} = ?$
[g] From the moments of inertia cf (666). [h] Potential barrier for internal rotation: 406 $cm^{-1}$.

TABLE 77 (*continued*)

| State | Point Group | $T_0$ | Vibrational Frequencies $\nu_1$ | $\nu_2$ | $\nu_3$ | $\nu_4$ | $\nu_5$ | $\nu_6$ | Rotational Constants $A_0$ | $B_0$ | $C_0$ | $r_0$(Å) | $\alpha$ | Observed Transitions | References | Remarks |
|---|---|---|---|---|---|---|---|---|---|---|---|---|---|---|---|---|
| **$SF_6$** | | | | | | | | | | | | | | | | |
| $\tilde{F}$ | | (118250) | continuum with maximum at 120450 $cm^{-1}$ and width 4400 $cm^{-1}$ | | | | | | | | | | | $\tilde{F}\leftarrow\tilde{X}$ 846–815 Å | (756) | |
| $\tilde{E}$ | | (112000) | continuum with maximum at 114680 $cm^{-1}$ and width 5300 $cm^{-1}$ | | | | | | | | | | | $\tilde{E}\leftarrow\tilde{X}$ 893–852 Å | (756) | |
| $\tilde{D}$ | | (106000) | continuum with maximum at 106826 $cm^{-1}$ and width 1660 $cm^{-1}$ | | | | | | | | | | | $\tilde{D}\leftarrow\tilde{X}$ 944–929 Å | (756) | |
| $\tilde{C}$ | | (90400) | continuum with maximum at 94890 $cm^{-1}$ and width 9000 $cm^{-1}$ | | | | | | | | | | | $\tilde{C}\leftarrow\tilde{X}$ 1120–1000 Å | (756) | |
| $\tilde{B}$ | | (63980) | continuous absorption | | | | | | | | | | | $\tilde{B}\leftarrow\tilde{X}$ 1563– | (756) | [a] |
| $\tilde{A}$ | | (46000) | continuous absorption | | | | | | | | | | | $\tilde{A}\leftarrow\tilde{X}$ 2170– | (756) | [a] |
| $\tilde{X}\ ^1A_{1g}$ | $\boldsymbol{O_h}$ | 0 | 769.4 | 639.5 | 940[b] | 614[c] | 522 | 344 | | | | $1.56_4$[d] | | Raman and infrared sp. | (454)(711) (344)(409) | |
| **$MoF_6$** | | | | | | | | | | | | | | | | |
| $\tilde{B}$ | | (49555) | 639 | | | | | | | | | | | $\tilde{B}\leftarrow\tilde{X}$ 2020–1870 Å | (1192) | Diffuse bands |
| $\tilde{A}$ | | (35520) | continuum with maximum at 54000 $cm^{-1}$ | | | | | | | | | | | $\tilde{A}\leftarrow\tilde{X}$ 2815–1600 Å | (1192) | |
| $\tilde{X}\ ^1A_{1g}$ | $\boldsymbol{O_h}$ | 0 | 741 | 643 | 741 | 264 | 306 | 190 | | | | | | infrared and Raman sp. | (1192)(163) (409)(201) | |

$SF_6$: [a] Not certain that $\tilde{A}$ and $\tilde{B}$ are different. [b] This is the value of (711). (409) gives 932 while (344) give 947 $cm^{-1}$. [c] Average of the values of (711) and (409). [d] From electron diffraction (368a).

TABLE 77 (*continued*)

| State | Point Group | $T_0$ | Vibrational Frequencies $\nu_1$ | $\nu_2$ | $\nu_3$ | $\nu_4$ | $\nu_5$ | $\nu_6$ | Rotational Constants $A_0$ | $B_0$ | $C_0$ | $r_0$(Å) | $\alpha$ | Observed Transitions | References | Remarks |
|---|---|---|---|---|---|---|---|---|---|---|---|---|---|---|---|---|
| **$WF_6$** | | | | | | | | | | | | | | | | |
| $\tilde{A}$ | | (36800) | continuum with maximum at 57140 $cm^{-1}$ | | | | | | | | | | | $\tilde{A} \leftarrow \tilde{X}$ 2715–1600 Å | (1192) | |
| $\tilde{X}\ ^1A_{1g}$ | $O_h$ | 0 | 771 | 671 | 712 | 258 | 319 | 215 | | | | | | infrared and Raman sp. | (163)(409) (1192) | |
| **$UF_6$** | | | | | | | | | | | | | | | | |
| $\tilde{B}$ | | (30000) | continuous absorption | | | | | | | | | | | $\tilde{B} \leftarrow \tilde{X}$ 3300– | (10) | |
| $\tilde{A}$ | | (24400) | 600 | (240) | | (100) | | | | | | | | $\tilde{A} \leftarrow \tilde{X}$ 4100–3500 Å | (10) | Diffuse bands overlapped by continuum |
| $\tilde{X}\ ^1A_{1g}$ | $O_h$ | 0 | 666.6 | 535 | 626 | 189 | 200 | 144 | | | | | | Raman and infrared sp. | (202) (409)(118) | |
| **$ReF_6$, $OsF_6$, $IrF_6$, $PtF_6$** see (868a) and (347a) | | | | | | | | | | | | | | | | |
| **$PuF_6$** see (1160) | | | | | | | | | | | | | | | | |
| **$NpF_6$** see (435) and (348) | | | | | | | | | | | | | | | | |

TABLE 78

MOLECULAR CONSTANTS OF THE ELECTRONIC STATES OF EIGHT-ATOMIC MOLECULES

| State | Point Group | $T_0$ | Vibrational Frequencies | | | | | | Rotational Constants | | | | | Observed Transitions | References | Remarks |
|---|---|---|---|---|---|---|---|---|---|---|---|---|---|---|---|---|
| | | | $\nu_1$ | $\nu_2$ | $\nu_3$ | $\nu_4$ | $\nu_5$ | $\nu_6$ | $A_0$ | $B_0$ | $C_0$ | $r_0$(Å) | $\alpha$ | | | |
| **$B_2H_6$** | $D(H_3B—BH_3) = 1.61$ eV[a] | | | | | | | | | | | | | | | |
| $\tilde{B}$ | | (60000) | continuous absorption with diffuse maxima at 1350 Å and beyond | | | | | | | | | | | $\tilde{B} \leftarrow \tilde{X}$ 1650–1100 Å | (1018) | |
| $\tilde{A}$ | | (50000) | continuous absorption with maximum | | | | | | 1830 Å | | | | | $\tilde{A} \leftarrow \tilde{X}$ 2000–1700 Å | (126)(1018) | |
| $\tilde{X}\ ^1A_g$ | $D_{2h}$ | 0 | 2524 | 2104 | 1180 | 794 | 829 | 1745[b] | $2.66_2$ | $\bar{B}_0 = 0.582$ | | | | infrared and Raman sp. | (773)(69) | |
| **$C_2H_6$** | I.P. = 11.65 eV[a]; $D(H_3C—CH_3) = 3.64$ eV[b]; $D(C_2H_5—H) = 4.2_1$ eV[c] | | | | | | | | | | | | | | | |
| $\tilde{C}$ | | (83000) | broad continuum with maximum at 850 Å | | | | | | | | | | | $\tilde{C} \leftarrow \tilde{X}$ 1200–500 Å | (1105)(836) | |
| $\tilde{B}$ | | 74000 | continuous absorption with diffuse maxima | | | | | | | | | | | $\tilde{B} \leftarrow \tilde{X}$ 1350–1000 Å | (1013) | not certain that $\tilde{A}$ and $\tilde{B}$ are separate electronic states |
| $\tilde{A}$ | | 62000 | continuous absorption | | | | | | | | | | | $\tilde{A} \leftarrow \tilde{X}$ 1600–1350 Å | (1101) | |
| $\tilde{X}\ ^1A_{1g}$ | $D_{3d}$ | 0 | 2953.8 | 1388.4 | 944.8 | (278)[d] | 2895.7[e] | 1379.14[f] | 2.681 | $0.6621_5$[g] | — | $r$(C—C) = 1.536<br>$r$(C—H) = 1.091[h] | ∠HCH = 108.0[h] | infrared and Raman sp. | (1140)(1081)<br>(1167)(740) | [i] |
| **$Si_2H_6$** | | | see (358) | continuous absorption only | | | | | | | | | | | | |
| **$CH_2CHCH_2$** (allyl) | | | see (261) | | | | | | | | | | | | | |
| **$C_2H_5Cl$** | | | see (1017) | I.P. = 10.97 eV[a] | | | | | | | | | | | | |

$B_2H_6$: [a] From (374a). According to (455) the atomic heat of formation is 24.56 eV.
[b] $\nu_7 = (1035)$, $\nu_8(b_{1u}) = 2612$, $\nu_9 = (950)$, $\nu_{10} = 368$, $\nu_{11}(b_{2g}) = 2591$, $\nu_{12} = (920)$, $\nu_{13}(b_{2u}) = 1882$, $\nu_{14} = 973$, $\nu_{15}(b_{3g}) = 1012$, $\nu_{16}(b_{3u}) = 2525$, $\nu_{17} = 1602$, $\nu_{18} = 1177$. All frequencies refer to $B_2^{11}H_6$. (158) suggest a somewhat different assignment. The values for $\nu_6$ and $\nu_{13}$ are shifted by Fermi resonance.

$C_2H_6$: [a] From photoionization (1273). [b] Average of the values given by (343a) and (1160a). [c] From (41); (1160a) from mass spectrometric data give $4.34_5$ eV. [d] From combination bands
[e] Shifted by Fermi resonance with $\nu_8 + \nu_{11}$. [f] $\nu_7(e_u) = 2995.5$, $\nu_8 = 1472.2$, $\nu_9 = 821.52$, $\nu_{10}(e_g) = 2950.0$, $\nu_{11} = 1468.7$, $\nu_{12} = (1190)$.
[g] $D_J = 9.4 \times 10^{-7}$ from the Raman spectrum (1167); from the infrared spectrum (708) give $B_0 = 0.66310$, $D_J = 7 \times 10^{-7}$.
[h] These are the values of (740). Slightly different values are given by (1167) and (708). [i] Data for $C_2D_6$: (471) and (740).

$C_2H_5Cl$: [a] From photoionization (1273).

TABLE 78 (*continued*)

| State | Point Group | $T_0$ | $\nu_1$ | $\nu_2$ | $\nu_3$ | $\nu_4$ | $\nu_5$ | $\nu_6$ | $A_0$ | $B_0$ | $C_0$ | $r_0$(Å) | $\alpha$ | Observed Transitions | References | Remarks |
|---|---|---|---|---|---|---|---|---|---|---|---|---|---|---|---|---|
| **$C_2H_5Br$** | | see (1017) I.P. = 10.29 eV[a] | | | | | | | | | | | | | | |
| **$C_2H_5I$** | | I.P. = 9.345 eV | | | | | | | | | | | | | | |
| | | Rydberg series joining on to $\tilde{H}$: $\nu = 80080 - R/(n - 0.20)^2$; $n = 5, 6, \ldots, 10$ | | | | | | | | | | | | | (1017) | |
| $\tilde{H}$ | | 75380 | | | | | | | | | | | | $\tilde{H} \leftarrow \tilde{X}$ 1327 Å | (1017) | |
| | | Rydberg series joining on to $\tilde{G}$: $\nu = 75380 - R/(n - 0.20)^2$; $n = 5, 6, \ldots, 10$ | | | | | | | | | | | | | (1017) | |
| $\tilde{G}$ | | 70620 | | | | | | | | | | | | $\tilde{G} \leftarrow \tilde{X}$ 1416 Å | (1017) | |
| $\tilde{E}$ | | Many diffuse bands probably forming several band systems | | | | | | | | | | | | $\tilde{E} \leftarrow \tilde{X}$ 1740–1420 Å | (1017) | |
| $\tilde{D}$ | | 54430 | | 1100 | | | | | | | | | | $\tilde{D} \leftarrow \tilde{X}$ 1840–1770 Å | (1102) | |
| $\tilde{C}$ | | 49970 | | 1161 | 637 | 424 | 217 | | | | | | | $\tilde{C} \leftarrow \tilde{X}$ 2005–1840 Å | (506)(1102) | |
| $\tilde{B}$ | | 49710 | | 1137 | | | | | | | | | | $\tilde{B} \leftarrow \tilde{X}$ 2020–1840 Å | (506) | |
| $\tilde{A}$ | | Continuous absorption with maximum at 39000 $cm^{-1}$ | | | | | | | | | | | | $\tilde{A} \leftarrow \tilde{X}$ | (1102) | Longward limit not given |
| $\tilde{X}\ ^1A'$ | $C_s$ | 0 | 2914 | 1197 | 951 | 500 | 262 | | $0.9708_8$[a] | $0.09937_5$ | $0.09330_1$ | r(CI) = 2.139<br>r(CC) = 1.54 ass. | ∠CCI = 112.2° | Raman, infrared and microwave sp. | (25)(1121)<br>(734)(661) | [b] |

$C_2H_5Br$: [a] From photoionization (1273).

$C_2H_5I$: [a] $D_K = 1.2 \times 10^{-5}$, $D_{JK} = -0.50 \times 10^{-5}$.

[b] The frequencies given are the strongest Raman lines. They have not been definitely assigned to $\nu_1, \nu_2, \ldots$. (661) give for the torsional barrier 1390 $cm^{-1}$, and the torsional frequency 230 $cm^{-1}$.

TABLE 78 (*continued*)

| State | Point Group | $T_0$ | Vibrational Frequencies $\nu_1$ | $\nu_2$ | $\nu_3$ | $\nu_4$ | $\nu_5$ | $\nu_6$ | Rotational Constants $A_0$ | $B_0$ | $C_0$ | $r_0$(Å) | $\alpha$ | Observed Transitions | References | Remarks |
|---|---|---|---|---|---|---|---|---|---|---|---|---|---|---|---|---|
| **$C_2H_3CHO$** (acrolein) I.P. = $10.10_3$ eV | | | | | | | | | | | | | | | | |
| | | Rydberg series joining on to $\tilde{G}$: $\nu = 81460 - R/(n + 0.05)^2$; $n = 3, 4, \ldots, 7$ | | | | | | | | | | | | | (1260) | |
| | | Rydberg series joining on to $\tilde{F}$: $\nu = 81500 - R/(n - 0.15)^2$; $n = 3, 4, \ldots, 12$ | | | | | | | | | | | | | (1260) | |
| | | Rydberg series joining on to $\tilde{D}$: $\nu = 81516 - R/(n - 0.68)^2$; $n = 3, 4, \ldots, 10$ | | | | | | | | | | | | | (1260) | |
| $\tilde{G}$ | | 68513 | | | | | | | | | | | | $\tilde{G} \leftarrow \tilde{X}$ 1460 Å | (1260) | |
| $\tilde{F}$ | | 67511 | | | | | | | | | | | | $\tilde{F} \leftarrow \tilde{X}$ 1481 Å | (1260) | |
| $\tilde{E}$ | | (62500) | | | | | | | | | | | | $\tilde{E} \leftarrow \tilde{X}$ 1600 Å | (1260) | Doublet, not precisely specified |
| $\tilde{D}$ | | 60585 | | | | | | | | | | | | $\tilde{D} \leftarrow \tilde{X}$ 1650 Å | (1260) | |
| $\tilde{C}$ | | 57140 | | | | | 1259 | | | | | | | $\tilde{C} \leftarrow \tilde{X}$ 1750–1680 Å | (1260) | Very strong single band with two vibrational satellites |
| $\tilde{B}$ | | (42550) | (300) | continuous spectrum with max. at 1935 Å; a few diffuse bands at longward end | | | | | | | | | | $\tilde{B} \leftarrow \tilde{X}$ 2350–1800 Å | (1209)(1260) | |
| $\tilde{A}\ ^1A''$ | $C_s$ | 25858.1 | | | | | $1265._4$ | $1409._8$[a] | | | | $r$(C—C) = (1.35)<br>$r$(C=C) = (1.46)<br>$r$(C=O) = (1.32) | ∠CCC = (125°)<br>∠CCO = (125°) | $\tilde{A} \leftarrow \tilde{X}$ 3870–3000 Å[b] | (337)(783)<br>(614)(145)<br>(565) | Bands with resolved $K$-structure at longward end[c] |
| $\tilde{a}(^3A'')$ | $C_s$ | 24247 | | | | | $\nu_{18} = 372$ | | | | | | | $\tilde{a} \leftarrow \tilde{X}$ 4122–4025 Å | (145)(565) | Diffuse bands |
| $\tilde{X}\ ^1A'$ | $C_s$ | 0 | 3102 | | 3000 | 2800 | 1723 | 1625[d] | $1.5794_5$ | 0.155422 | 0.141522 | $r$(C—C) = 1.45<br>$r$(C=C) = 1.36<br>$r$(C=O) = 1.22 | ∠CCC = 122°<br>∠CCO = 125° | microwave, infrared and Raman sp. | (1252)(145)<br>(614) | [e] |

$C_2H_3CHO$: [a] $\nu_8 = 1133$, $\nu_{12} = 488$, $\nu_{14}(a'') = 335._9$, $\nu_{15} = 644$, $\nu_{16} = 582$, $\nu_{17} = 911._9$, $\nu_{18} = 254._4$. For $\nu_{12}$ (565) gives $508._7$ cm$^{-1}$. The frequencies given here for $\nu_{15}$ and $\nu_{16}$ are considered by (565) as $a'$ vibrations and an alternative $a''$ vibration at 812.1 cm$^{-1}$ is given.

[b] Hot bands extend to 3960 Å (337).

[c] The bands become diffuse at shorter wavelengths and merge into a continuum below 3600 Å. (1209) looked for fluorescence in the sharp region but found none. In agreement with this observation (565) under high resolution finds the $J$ structure blurred on account of diffuseness of the lines. The $K$ structure of the main bands is of ⊥ type ($\Delta K = \pm 1$) but some of the weaker bands have hybrid structure corresponding to a forbidden vibronic component. Torsional barriers: 2700 and 5300 cm$^{-1}$ for the terminal and central C—C bonds respectively.

[d] $\nu_7 = 1422$, $\nu_8 = 1361$, $\nu_9 = 1276$, $\nu_{10} = 1159$, $\nu_{11} = 913$, $\nu_{12} = 560._3$, $\nu_{13} = 340$, $\nu_{14}(a'') = 993$, $\nu_{15} = (980)$, $\nu_{16} = 959$, $\nu_{17} = 589$, $\nu_{18} = 152._9$. Very slightly different fundamental frequencies are given by (478). The values for $\nu_{12}$ and $\nu_{18}$ are from the electronic spectrum (565).

[e] Torsional barriers: 8700 and 2270 cm$^{-1}$ for the C=C and C—C bonds respectively (145).

TABLE 78 (*continued*)

| State | Point Group | $T_0$ | Vibrational Frequencies $\nu_1$ | $\nu_2$ | $\nu_3$ | $\nu_4$ | $\nu_5$ | $\nu_6$ | Rotational Constants $A_0$ | $B_0$ | $C_0$ | $r_0$(Å) | $\alpha$ | Observed Transitions | References | Remarks |
|---|---|---|---|---|---|---|---|---|---|---|---|---|---|---|---|---|
| **$C_2H_4N_2$** (diazoethane) | | | see (151) | | | | | | | | | | | | | |
| **$HCOOCH_3$** (methyl formate) | | | see (96) | | | | | | | | | | | | | |
| **$CH_3COOH$** (acetic acid) | | | see (96) | I.P. = 10.35[a] | | | | | | | | | | | | |
| **$CH_3ONO_2$** (methyl nitrate) | | | see (1212) | | | | | | | | | | | | | |
| **$C_2N_4H_2$** (*s*-tetrazine) | | | | | | | | | | | | | | | | |
| $\tilde{B}$ | | (36000) | continuum with maximum at 2470 Å | | | | | | | | | | | | (1144) | |
| $\tilde{A}\ ^1B_{3u}$ | $(D_{2h})$ | 18129[a] | | | | 700 | | [b] | $\frac{1}{2}(A+B) = 0.2234$ | | | | | $\tilde{A}\leftrightarrow\tilde{X}$ 5600–4700 Å[c] | (1144)(197) (799) | |
| $\tilde{X}\ ^1A_g$ | $D_{2h}$ | 0 | | | | 737 | | [d] | $\frac{1}{2}(A+B) = 0.2215$ | | | | | infrared sp. | (1145) | |

$CH_3COOH$: [a] From photoionization (1273).

$C_2N_4H_2$: [a] (1144) give, $\nu_{air}$ = 18134.

[b] A tentative frequency 149 cm$^{-1}$ corresponding to 232 in the lower state, but not identified, is given by (1144). (799) gives in addition the frequencies 1205 and 1520 cm$^{-1}$ in solution at low temperature.

[c] In fluorescence to 6500 Å.

[d] $\nu_9(b_{1u}) = 890$, $\nu_{11}(b_{2u}) = (3070)$, $\nu_{12} = 1200$, $\nu_{13} = 1090$, $\nu_{16}(b_{3u}) = 1448$, $\nu_{17} = 1106$, $\nu_{18} = 925$.

TABLE 79

MOLECULAR CONSTANTS IN THE ELECTRONIC STATES OF NINE-ATOMIC MOLECULES

| State | Point Group | $T_0$ | Vibrational Frequencies $\nu_1$ | $\nu_2$ | $\nu_3$ | $\nu_4$ | $\nu_5$ | $\nu_6$ | Rotational Constants $A_0$ | $B_0$ | $C_0$ | $r_0$(Å) | $\alpha$ | Observed Transitions | References | Remarks |
|---|---|---|---|---|---|---|---|---|---|---|---|---|---|---|---|---|
| **$Zn(CH_3)_2$** | | | continuous absorption only, see (1210) | | | | | | | | | | | | | |
| **$Cd(CH_3)_2$** | | | continuous absorption only, see (1210) | | | | | | | | | | | | | |
| **$Hg(CH_3)_2$** | | | | | | | | | | | | | | | | |
| $\tilde{D}$ | | (65000) | continuum extending from 64950 to 69700 $cm^{-1}$ | | | | | | | | | | | $\tilde{D} \leftarrow \tilde{X}$ 1540–1435 Å | (525) | |
| $\tilde{C}$ | | (59000) | | | (370) | | | | | | | | | $\tilde{C} \leftarrow \tilde{X}$ 1700–1550 Å | (525) | Diffuse bands |
| $\tilde{B}$ | | 47059 | | (1100) | 370 | | | | | | | | | $\tilde{B} \leftarrow \tilde{X}$ 2150–1850 Å | (1210)(1201) (525) | Diffuse bands |
| $\tilde{A}$ | | (40617) | | | 330 | | | | | | | | | $\tilde{A} \leftarrow \tilde{X}$ 2465–2175 Å | (1210) | Diffuse bands |
| $\tilde{X}\ ^1A_1'$ | $D_{3h}$[a] | 0 | 2910 | 1182 | 515 | ? | 2924 | 1205[b] | | 0.1162 | | r(Hg—C) = 2.094 | | infrared and Raman sp. | (456)(134) (1055)(1319) | |
| **$CH_3C_2H_3$** (propylene) I.P. = 9.74 eV | | | | | | | | | | | | | | | | |
| | | | Rydberg series joining on to $\tilde{C}$: $\nu = 78586 - R/(n - 0.15)^2$; $n = 4, 5, \ldots, 9$ | | | | | | | | | | | | (1096) | |
| $\tilde{C}$ | | 71120 | | | (310) | | | | | | | | | $\tilde{C} \leftarrow \tilde{X}$ 1406–1400 Å | (1096) | Diffuse bands |
| $\tilde{B}$ | | (66500) | | | | | | | | | | | | $\tilde{B} \leftarrow \tilde{X}$ 1500–1420 Å | (1026)(1096) | Diffuse bands, not measured |
| $\tilde{A}$ | | | Continuous absorption with maximum at 1730 Å; diffuse bands with spacing 1370 $cm^{-1}$ at shortward end | | | | | | | | | | | $\tilde{A} \leftarrow \tilde{X}$ 1900–1600 Å | (1026)(1096) | |
| $\tilde{X}\ ^1A'$ | $C_s$ | 0 | 3089.7 | (3013) | 2991.5 | 2932.7 | 2869.8 | 1651.6[a] | 1.537 | $0.31039_1$ | $0.27132_6$ | r(C=C) = 1.353; r(C—C) = 1.488 | ∠CCC = 124.8° | infrared, Raman and microwave sp. | (776)(745) | [b] |

$Hg(CH_3)_2$: [a] In all probability there is free rotation and $D_{3h}$ is not an exact description of the point group to which the molecule belongs.
[b] $\nu_7(a_2'') = 550$, $\nu_8(e') = 2980$, $\nu_9 = 1475$, $\nu_{10} = 788$, $\nu_{11} = 153$, $\nu_{12}(e'') = 2869$, $\nu_{13} = 1443$, $\nu_{14} = 700$.

$CH_3C_2H_3$: [a] $\nu_7(a') = 1473.6$, $\nu_8 = 1419.2$, $\nu_9 = 1377.6$, $\nu_{10} = 1298$, $\nu_{11} = 1229$, $\nu_{12} = 1171.9$, $\nu_{13} = 920.4$, $\nu_{14} = 427.5$, $\nu_{15}(a'') = 2954.0$, $\nu_{16} = 1442.6$, $\nu_{17} = 1044.7$, $\nu_{18} = 990.6$, $\nu_{19} = 912.4$, $\nu_{20} = 577.6$, $\nu_{21} = (174)$.
[b] Torsional barrier: 692 $cm^{-1}$. The equilibrium conformation is one in which one C—H bond of the $CH_3$ group lies in the plane of the vinyl group, eclipsed with respect to the double bond (516).

TABLE 79 (*continued*)

| State | Point Group | $T_0$ | Vibrational Frequencies $\nu_1$ | $\nu_2$ | $\nu_3$ | $\nu_4$ | $\nu_5$ | $\nu_6$ | Rotational Constants $A_0$ | $B_0$ | $C_0$ | $r_0$(Å) | $\alpha$ | Observed Transitions | References | Remarks |
|---|---|---|---|---|---|---|---|---|---|---|---|---|---|---|---|---|
| **$C_2H_5OH$** | see (539), (946) and (480) | | I.P. = 10.50 eV[a] | | | | | | | | | | | | | |
| **$(CH_3)_2O$** | | | I.P. = 9.96 eV[a] | | | | | | | | | | | | | |
| | | Rydberg series joining on to $\tilde{E}$: $\nu = 80330 - R/(n - 0.02)^2$; $n = 3, 4, 5, 6$ | | | | | | | | | | | | | (511) | First members have progressions in $\nu_7$ |
| $\tilde{F}$ | | 71170 | | | | | $\nu_7 = (410)$ | | | | | | | $\tilde{F} \leftarrow \tilde{X}$ 1405–1381 Å | (511) | Diffuse bands |
| $\tilde{E}$ | | 68120 | | | | | $v_7 = (400)$ | | | | | | | $\tilde{E} \leftarrow \tilde{X}$ 1468–1442 Å | (511) | Diffuse bands |
| $\tilde{D}$ | | 61390 | | | | | $\nu_7 = 410$ | | | | | | | $\tilde{D} \leftarrow \tilde{X}$ 1630–1595 Å | (511) | Diffuse bands |
| $\tilde{C}$ | | 58820 | | | | | $\nu_7 = 340$ | | | | | | | $\tilde{C} \leftarrow \tilde{X}$ 1700–1680 Å | (511)(481) | Diffuse bands |
| $\tilde{B}$ | | 53140 | | | | | $\nu_7 = 340$ | | | | | | | $\tilde{B} \leftarrow \tilde{X}$ 1880–1840 Å | (511)(481) | Diffuse bands |
| $\tilde{A}$ | | (42500) | continuum with rapidly increasing intensity toward shorter wavelengths | | | | | | | | | | | $\tilde{A} \leftarrow \tilde{X}$ 2350–1900 Å | (481)(1204) | |
| $\tilde{X}\ ^1A_1$ | $C_{2v}$ | 0 | 2997 | 2821 | 1448 | 1242 | 1053 | 929[b] | 1.29384 | 0.33545 | 0.29644 | $r$(CO) = 1.410<br>$r$(CH) = 1.096[c] | ∠COC = 111.7°<br>∠HCH = 109.5°[c] | infrared, Raman and microwave sp. | (23)(1194)(124) | |

$C_2H_5OH$: [a] From photoionization data of (1273).

$(CH_3)_2O$: [a] From photoionization experiments (1273) obtains 10.00 eV.

[b] $\nu_7(a_1) = 413$, $\nu_8(a_2) = (2889)$, $\nu_9 = (1456)$, $\nu_{10} = (1291)$, $\nu_{11} = (160)$, $\nu_{12}(b_1) = (2889)$, $\nu_{13} = (1440)$, $\nu_{14} = (1291)$, $\nu_{15} = 270$, $\nu_{16}(b_2) = (2889) = \nu_{17}$, $\nu_{18} = (1440) = \nu_{19}$, $\nu_{20} = (1291)$, $\nu_{21} = 1122$.

[c] CH distances and angles are averages. The actual $CH_3$ groups are slightly asymmetric; see (124).

TABLE 79 (*continued*)

| State | Point Group | $T_0$ | Vibrational Frequencies $\nu_1$ | $\nu_2$ | $\nu_3$ | $\nu_4$ | $\nu_5$ | $\nu_6$ | Rotational Constants $A_0$ | $B_0$ | $C_0$ | $r_0$(Å) | $\alpha$ | Observed Transitions | References | Remarks |
|---|---|---|---|---|---|---|---|---|---|---|---|---|---|---|---|---|
| **$(CH_3)_2S$** | | I.P. ≈ (8.7) eV[a] | | | | | | | | | | | | | | |
| $\tilde{F}$ | | (65360) | continuous absorption | | | | | | | | | | | $\tilde{F} \leftarrow \tilde{X}$ 1530–1400 Å | (1024) | |
| $\tilde{E}$ | | 60790 | | | | | | | | | | | | $\tilde{E} \leftarrow \tilde{X}$ 1645–1530 Å | (1024) | Sharp bands |
| $\tilde{D}$ | | 55710 | | | | | | | | | | | | $\tilde{D} \leftarrow \tilde{X}$ 1795 Å | (1024) | Broad diffuse band |
| $\tilde{C}$ | | 51000 | | 650[e] | 330[e] | | | | | | | | | $\tilde{C} \leftarrow \tilde{X}$ 1960–1850 Å | (1024) | Sharp bands |
| $\tilde{B}$ | | 49380 | | | | | | | | | | | | $\tilde{B} \leftarrow \tilde{X}$ 2025 Å | (1024) | Broad diffuse band |
| $\tilde{A}$ | | 43898 | 2787 | (1086)[e] | | | | | | | | | | $\tilde{A} \leftarrow \tilde{X}$ 2300–2100 Å | (1210) | Diffuse bands |
| $\tilde{X}\ ^1A_1$ | $C_{2v}$ | 0 | 2911 | 2832 | 1445 | 1325 | 1041 | 685[b] | 0.59406 | 0.25421 | 0.19073 | $r$(CS) = 1.802; $r$(CH) = 1.091 | ∠CSC = 98.9°; ∠HCH = 109.6°[c] | infrared, Raman and microwave sp. | (391)(1207) (1091)(985) | [d] |
| **$CH_3$—C≡C—CHO** (tetrolaldehyde) | | | | | | | | | | | | | | | | |
| $\tilde{C}$ | | 45440 | | | | 1890 | | | | | | | | $\tilde{C} \leftarrow \tilde{X}$ 2250–2000 Å | (884) | Strong diffuse peaks |
| $\tilde{B}$ | | (39000) | | | | | | | | | | | | $\tilde{B} \leftarrow \tilde{X}$ 2560–2480 Å | (884) | Diffuse bands |
| $\tilde{A}$ | | 26620 | | | | 1058 | 289 | | | | | | | $\tilde{A} \leftarrow \tilde{X}$ 3775–3115 Å | (884) | Sharp bands[a] |
| $\tilde{X}\ ^1A'$ | $C_s$ | 0 | | | | | | | | | | | | | | [b] |

$(CH_3)_2S$: [a] Based on first member of a Rydberg series.
[b] $\nu_7(a_1) = 285$, $\nu_8(a_2) = 2980$, $\nu_9 = 1426$, $\nu_{10} = (919)$, $\nu_{11} = (240)$, $\nu_{12}(b_1) = (2852)$, $\nu_{13} = (1426)$, $\nu_{14} = (1230)$, $\nu_{15} = ?$, $\nu_{16}(b_2) = (2980)$, $\nu_{17} = (2852)$, $\nu_{18} = (1426)$, $\nu_{19} = 1323$, $\nu_{20} = (1274)$, $\nu_{21} = 742$.
[c] Angle between axes of $CH_3$ groups: 104.4°. [d] Barrier toward internal rotation: $V_3 = 746\ cm^{-1}$. [e] Assignment doubtful.

$CH_3$—C≡C—CHO: [a] The numbering of the vibrations is uncertain. [b] Infrared, Raman and microwave spectra have apparently not yet been studied.

TABLE 79 (*continued*)

| State | Point Group | $T_0$ | $\nu_1$ | $\nu_2$ | $\nu_3$ | $\nu_4$ | $\nu_5$ | $\nu_6$ | $A_0$ | $B_0$ | $C_0$ | $r_0$(Å) | $\alpha$ | Observed Transitions | References | Remarks |
|---|---|---|---|---|---|---|---|---|---|---|---|---|---|---|---|---|
| **$C_4H_4O$** (furan) I.P. = 8.887 eV | | | | | | | | | | | | | | | | |
| | | Rydberg series joining on to $\tilde{F}$: $\nu = 80229 - R/(n - 0.28)^2$; $n = 4, 5, \ldots, 10$ | | | | | | | | | | | | 1330–1260 Å | (1275) | Diffuse bands |
| $\tilde{F}$ | | 72307 | | | | 980 | | | | | | | | $\tilde{F} \leftarrow \tilde{X}$ 1383 Å | (1275) | Diffuse |
| | | Rydberg series joining on to $\tilde{D}$: $\nu = 71688 - R/(n + 0.18)^2$; $n = 3, 4, \ldots, 7$ | | | | | | | | | | | | 1530–1435 Å | (1275) | |
| | | Rydberg series joining on to $\tilde{C}$: $\nu = 71678 - R/(n - 0.06)^2$; $n = 3, 4, \ldots, 11$ | | | | | | | | | | | | 1550–1408 Å | (1275) | |
| $\tilde{D}$ | | 60901 | | | 1068 | | | | | | | | | $\tilde{D} \leftarrow \tilde{X}$ 1645–1555 Å | (1275) | Sharp bands |
| $\tilde{C}$ | | 59524 | | | 1068 | | | | | | | | | $\tilde{C} \leftarrow \tilde{X}$ 1700–1565 Å | (1028)(1275) | Sharp bands[a] |
| $\tilde{B}$ | | 52230 | (2965) | | 1395 | 1068 | 848 | (465) | | | | | | $\tilde{B} \leftarrow \tilde{X}$ 1915–1750 Å | (979)(980) (1028) | Sharp bands |
| $\tilde{A}$ | | (47400) | | | 1200 | | | | | | | | | $\tilde{A} \leftarrow \tilde{X}$ 2150–1950 Å | (979)(980) (1028) | [b] |
| $\tilde{X}\ ^1A_1$ | $C_{2v}$ | 0 | (3120) | (3090) | 1486 | 1381 | 1137 | 1067[c] | 0.315117 | 0.308434 | 0.155804 | $r$(CO) = 1.362<br>$r$(CC) = $\begin{cases}1.361\\1.431\end{cases}$<br>$r$(CH) = 1.076 | ∠COC = 106.6°<br>∠OCC = 110.7°<br>∠CCC = 106.1°<br>∠CCH = 127.9°<br>∠OCH = 115.9° | microwave, infrared, and Raman sp. | (88)(1214) | |

$C_4H_4O$: [a] A band (at 58754 cm$^{-1}$) preceding the main band is assigned as a hot band in the frequency 770 cm$^{-1}$ of the ground state.
[b] Very diffuse bands; at high pressure continuous absorption to 2300 Å.
[c] $\nu_7(a_1) = 994$, $\nu_8 = 724$, $\nu_{12}(b_1) = 837$, $\nu_{13} = 744$, $\nu_{14} = 605$, $\nu_{15}(b_2) = (3120)$, $\nu_{16} = (3090)$, $\nu_{17} = 1579$, $\nu_{19} = 1270$, $\nu_{20} = 872$.

TABLE 79 (*continued*)

| State | Point Group | $T_0$ | Vibrational Frequencies $\nu_1$ | $\nu_2$ | $\nu_3$ | $\nu_4$ | $\nu_5$ | $\nu_6$ | Rotational Constants $A_0$ | $B_0$ | $C_0$ | $r_0$(Å) | $\alpha$ | Observed Transitions | References | Remarks |
|---|---|---|---|---|---|---|---|---|---|---|---|---|---|---|---|---|
| **$C_4H_4S$** (thiophene) I.P. = 8.95 eV | | | | | | | | | | | | | | | | |
| | | Several Rydberg series, the strongest of which is $\nu = 72170 - R/(n - 0.10)^2$; $n = 5, 6, \ldots, 8$ | | | | | | | | | | | | 1520–1400 Å | (1028) | |
| $\tilde{C}$ | | (63000)[a] | | | | | | | | | | | | $\tilde{C} \leftarrow \tilde{X}$ 1588 Å | (1028) | Sharp band |
| $\tilde{B}$ | | 53200 | | | 1250 | | | 540 | | | | | | $\tilde{B} \leftarrow \tilde{X}$ 1880–1650 Å | (1028) | Sharp bands, becoming diffuse at shorter $\lambda$ |
| $\tilde{A}$ | | 41595[b] | | | (1193) | (1128) | 965 | 709 | | | | | | $\tilde{A} \leftarrow \tilde{X}$ 2450–2100 Å | (841)(426) | Sharp bands becoming diffuse at shorter $\lambda$ |
| $\tilde{a}(^3B_1)$ | $(C_{2v})$ | (31400) | Very weak absorption observed in solution | | | | | | | | | | | $\tilde{a} \leftarrow \tilde{X}$ 3180–3130 Å | (955) | Two absorption maxima |
| $\tilde{X}\ ^1A_1$ | $C_{2v}$ | 0 | (3110) | (3080) | 1404 | 1358 | 1077 | 1032[c] | 0.268245 | 0.180729 | 0.107934 | $r$(CS) = 1.714<br>$r$(CC) = 1.370, 1.423<br>$r$(CH) = 1.079 | ∠CSC = 92.2°<br>∠SCC = 111.5°<br>∠CCC = 112.4°<br>∠CCH = 124.3°<br>∠SCH = 119.9° | infrared, Raman and microwave sp. | (1214)(90) | |
| **$C_4H_4Se$** (selenophene) | | | | | | | | | | | | | | | | |
| | | Continuous absorption 1865–1700 Å with three maxima | | | | | | | | | | | | | (842) | |
| $\tilde{C}$ | | 48839 | | | 1355 | 734 | | | | | | | | $\tilde{C} \leftarrow \tilde{X}$ 2095–1970 Å | (843) | |
| $\tilde{B}$ | | 40148 | | | 992 | 600 | 155 | | | | | | | $\tilde{B} \leftarrow \tilde{X}$ 2520–2420 Å | (843) | |
| $\tilde{A}$ | | (37300) | Diffuse bands, no analysis | | | | | | | | | | | $\tilde{A} \leftarrow \tilde{X}$ 2700–2650 Å | (843) | |
| $\tilde{X}\ ^1A_1$ | $(C_{2v})$[a] | 0 | 3096 | 3062 | 1428 | 1349 | 1019 | 923[b] | | | | | | Raman and infrared sp. | (416)(200) | |

$C_4H_4S$:

[a] Measured on (1028)'s published spectrogram.

[b] This value for the 0 – 0 band was given by (841). Because of the gradual setting in of the band system the true 0 – 0 band may be at longer wavelengths. (841) assumes two other electronic states at 42992 and 45378 cm$^{-1}$.

[c] $\nu_7(a_1) = 832$, $\nu_8 = 604$, $\nu_{13}(b_1) = (710)$, $\nu_{15}(b_2) = 3110$, $\nu_{16} = (3080)$, $\nu_{17} = 1590$, $\nu_{19} = 1252$, $\nu_{20} = 872$. $\nu_9$ to $\nu_{12}$, $\nu_{14}$, $\nu_{18}$, $\nu_{21}$ have not been identified.

$C_4H_4Se$:

[a] According to (416) selenophene is non-planar, i.e., belongs to $C_s$ rather than to $C_{2v}$.

[b] $\nu_7 = 761$, $\nu_8 = 459$; the assignment of the other frequencies is uncertain.

TABLE 79 (*continued*)

| State | Point Group | $T_0$ | Vibrational Frequencies | | | | | | Rotational Constants | | | | | Observed Transitions | References | Remarks |
|---|---|---|---|---|---|---|---|---|---|---|---|---|---|---|---|---|
| | | | $\nu_1$ | $\nu_2$ | $\nu_3$ | $\nu_4$ | $\nu_5$ | $\nu_6$ | $A_0$ | $B_0$ | $C_0$ | $r_0$(Å) | $\alpha$ | | | |
| **$C_3N_3H_3$** (*s*-triazine) | | | | | | | | | | | | | | | | |
| $\tilde{D}(^1A_2'')$ | $\boldsymbol{D_{3h}}$ | 55782 | | 952 | 738 | 661 | | | | | | | | $\tilde{D}\leftarrow\tilde{X}$ 1793–1695 Å | (150) | Sharp bands[a] |
| $\tilde{C}(^1A_2')$ | $(\boldsymbol{D_{3h}})$ | 44000 | | | | | | | | | | | | $\tilde{C}\leftarrow\tilde{X}$ 2275–2000 Å | (557) | Diffuse absorption |
| $\tilde{B}(^1A_1'')$ | $(\boldsymbol{D_{3h}})$ | 32500 | | | | | | | | | | | | $\tilde{B}\leftarrow\tilde{X}$ 3070– | (149) | Diffuse absorption under-lying $\tilde{A}-\tilde{X}$ |
| $\tilde{A}\ ^1A_2''$ | $(\boldsymbol{D_{3h}})$[b] | 31574 | | 1140 | 862 | 684 | 488 | | | | | | | $\tilde{A}\leftarrow\tilde{X}$ 3170–2700 Å | (149) | Sharp bands with ∥ structure[a] |
| $\tilde{a}\ ^3A_2''$ | $\boldsymbol{D_{3h}}$ | 26400 | | | | | | | | | | | | $\tilde{a}\leftrightarrow\tilde{X}$ 5000–3800 Å | (960) | Phosphorescence in rigid glass |
| $\tilde{X}\ ^1A_1'$ | $\boldsymbol{D_{3h}}$ | 0 | 3042 | 1132 | 991 | | | 837[c] | — | 0.21460 | | $r$(CN) = 1.338 | | infrared and Raman sp. | (717)(1158) (442)(718) | |
| **$Cl_2O_7$** | see (431) | | | | | | | | | | | | | | | |

$C_3N_3H_3$:

[a] The numbering of the observed vibrations is not certain.

[b] (149) conclude from the appearance of non-totally symmetric vibrations that the molecule is non-planar in the $\tilde{A}$ state.

[c] $\nu_7(a_2'') = 735$, $\nu_8(e') = 3056$, $\nu_9 = 1556$, $\nu_{10} = 1410$, $\nu_{11} = 1176$, $\nu_{12} = 675$, $\nu_{13}(e'') = ?$, $\nu_{14} = 341$.

TABLE 80

MOLECULAR CONSTANTS IN THE ELECTRONIC STATES OF TEN-ATOMIC MOLECULES

| State | Point Group | $T_0$ | Vibrational Frequencies $\nu_1$ | $\nu_2$ | $\nu_3$ | $\nu_4$ | $\nu_5$ | $\nu_6$ | Rotational Constants $A_0$ | $B_0$ | $C_0$ | $r_0$(Å) | $\alpha$ | Observed Transitions | References | Remarks |
|---|---|---|---|---|---|---|---|---|---|---|---|---|---|---|---|---|
| **$C_2H_5NH_2$** see (357) | | | | | | | | | | | | | | | | |
| **$(C_2H_3)_2$** (1,3 butadiene) I.P. = 9.061 eV[a] | | | | | | | | | | | | | | | | |
| | | Rydberg series joining on to $\tilde{H}$: $\nu = 73115 - R/(n - 0.10)^2$; $n = 4, 5, \ldots, 8$[b] | | | | | | | | | | | | | (1027) | |
| | | Rydberg series joining on to $\tilde{G}$: $\nu = 73066 - R/(n - 0.50)^2$; $n = 4, 5, \ldots, 8$ | | | | | | | | | | | | | (1027) | |
| $\tilde{H}$ | | 65900 | | | | | | | | | | | | $\tilde{H} \leftarrow \tilde{X}$ 1517 Å | (1027) | Doublet with spacing 200 $cm^{-1}$ |
| $\tilde{G}$ | | 64500 | | | | | | | | | | | | $\tilde{G} \leftarrow \tilde{X}$ 1550 Å | (1027) | Single sharp band |
| $\tilde{F}$ | | 56990 | | | | | | | | | | | | $\tilde{F} \leftarrow \tilde{X}$ 1755 Å | (1101)(1027) | Single fairly sharp band |
| $\tilde{E}$ | | 56820 | diffuse bands | | | | | | | | | | | $\tilde{E} \leftarrow \tilde{X}$ 1760–1600 Å | (1027) | |
| $\tilde{D}$ | | 54960 | | | | | | (350) | | | | | | $\tilde{D} \leftarrow \tilde{X}$ 1820–1808 Å | (1101)(1027) | Doublet |
| $\tilde{C}$ | | 53700[c] | | | | | | | | | | | | $\tilde{C} \leftarrow \tilde{X}$ 1862 Å | (1101)(1027) | Single fairly sharp band |
| $\tilde{B}$ | | 50500 | strong band followed by unassigned vibrational structure | | | | | | | | | | | $\tilde{B} \leftarrow \tilde{X}$ 1980–1870 Å | (1101)(1027) | |
| $\tilde{A}$ | | (46250)[d] | | | | 1440 | | | | | | | | $\tilde{A} \leftarrow \tilde{X}$ 2170–1970 Å | (1027)(1101) | Very diffuse bands[e] |
| $\tilde{X}\ ^1A_g$ | $C_{2h}$ | 0 | 3101 | 3014 | 3014 | 1643 | 1442 | 1279[f] | 1.370 | $\frac{1}{2}(B+C) = 0.1413$ | | $r$(C—C) = 1.476; $r$(C=C) = 1.337[g] | ∠CCC = 122.9° | infrared and Raman sp. | (797)(1183) (478)(959) | |

$(C_2H_3)_2$: [a] Confirmed by photoionization experiments of (1273). [b] (1027) give one further longward member at 60112 $cm^{-1}$, but this band seems to be part of $\tilde{E} \leftarrow \tilde{X}$.
[c] From (1101). On the spectrogram reproduced by (1027) it seems to appear at 53500 $cm^{-1}$. [d] Measured on the spectrogram reproduced by (1027).
[e] With overlapping continuum extending to longer wavelengths at higher pressure.
[f] $\nu_7(a_g) = 1205$, $\nu_8 = (890)$, $\nu_9 = 513$, $\nu_{10}(a_u) = 1013$, $\nu_{11} = 907.8$, $\nu_{12} = 520$, $\nu_{13} = 163$, $\nu_{14}(b_g) = 967$, $\nu_{15} = 911$, $\nu_{16} = 686$, $\nu_{17}(b_u) = 3102$, $\nu_{18} = 3056$, $\nu_{19} = 2985$, $\nu_{20} = 1599$, $\nu_{21} = 1385$, $\nu_{22} = 1285$, $\nu_{23} = 987$, $\nu_{24} = 301$.
[g] From electron diffraction (62).

TABLE 80 (*continued*)

| State | Point Group | $T_0$ | Vibrational Frequencies $\nu_1$ | $\nu_2$ | $\nu_3$ | $\nu_4$ | $\nu_5$ | $\nu_6$ | Rotational Constants $A_0$ | $B_0$ | $C_0$ | $r_0$(Å) | $\alpha$ | Observed Transitions | References | Remarks |
|---|---|---|---|---|---|---|---|---|---|---|---|---|---|---|---|---|
| **$C_4H_6$** (cyclo-butene) see (758) | | | | | | | | | | | | | | | | |
| **$C_2H_5CCH$** (ethyl acetylene) I.P. = $10.1_7$ eV see (927) | | | | | | | | | | | | | | | | |
| **$(CH_2)_3O$** (*c*-trimethylene oxide) I.P. = $9.66_8$ eV | | | | | | | | | | | | | | | | |
| | | | Rydberg series joining on to $\tilde{E}$: | | | | | | $\nu = 79330 - R/(n - 0.05)^2$; $n = 3, 4, 5, 6$ | | | | | | (512) | |
| | | | Rydberg series of doublets joining on to $\tilde{D}$: | | | | | | $\nu = 77980 - R/(n - 0.05)^2$; $n = 3, 4, \ldots, 10$ | | | | | | (512) | |
| $\tilde{E}$ | | 66300 | | | | | 591 | 322 | | | | | | $\tilde{E} \leftarrow \tilde{X}$ 1508–1413 Å | (512) | |
| $\tilde{D}$ | | 64910 | | | | | | | | | | | | $\tilde{D} \leftarrow \tilde{X}$ 1540 Å | (512) | Double band, separation 110 $cm^{-1}$ |
| $\tilde{C}$ | | 61727 | | 1188 | | | | | | | | | | $\tilde{C} \leftarrow \tilde{X}$ 1620–1525 Å | (512) | |
| $\tilde{B}$ | | 57312 | | 1209 | | | | | | | | | | $\tilde{B} \leftarrow \tilde{X}$ 1745–1635 Å | (512)(387) | Diffuse bands |
| $\tilde{A}$ | | 53276 | | 1280 | 1042 | | | | | | | | | $\tilde{A} \leftarrow \tilde{X}$ 1880–1830 Å | (387)(512) | Diffuse bands[a] |
| $\tilde{X}\ ^1A_1$ | $(C_{2v})$ | 0 | 2959 | 2930 | 1473 | 1461 | 1342 | 1134[b] | $0.40178_4$ | $0.39140_4$ | $0.22451_2$ | $r$(CC) = 1.549<br>$r$(CO) = 1.449<br>$r$(CH) = {1.091, 1.100} | ∠COC = 92.0°<br>∠CCC = 84.5°<br>∠HCH = {110.3°, 110.7°} | infrared, Raman and microwave sp. | (186)(187)<br>(1334)(267) | c |

$(CH_2)_3O$: [a] With underlying continuum, which for higher pressures extends to longer wavelengths.

[b] $\nu_7(a_1) = 1018$, $\nu_8 = 908$, $\nu_9(a_2) = (3000)$, $\nu_{10} = 1283$, $\nu_{11} = 1185$, $\nu_{12} = 986$, $\nu_{13}(b_1) = 3007$, $\nu_{14} = 2940$, $\nu_{15} = 1225$, $\nu_{16} = 1142$, $\nu_{17} = 836$, $\nu_{18} = 89.8$, $\nu_{19}(b_2) = 2966$, $\nu_{20} = 1508$, $\nu_{21} = 1458$, $\nu_{22} = 1363$, $\nu_{23} = 1228$, $\nu_{24} = 936$.

[c] The $C_3O$ ring is effectively planar, but there is a very slight puckering with a barrier of 35 $cm^{-1}$. The lowest level is 8 $cm^{-1}$ above the top of the barrier [see (186)].

TABLE 80 (*continued*)

| State | Point Group | $T_0$ | Vibrational Frequencies $\nu_1$ | $\nu_2$ | $\nu_3$ | $\nu_4$ | $\nu_5$ | $\nu_6$ | Rotational Constants $A_0$ | $B_0$ | $C_0$ | $r_0$(Å) | $\alpha$ | Observed Transitions | References | Remarks |
|---|---|---|---|---|---|---|---|---|---|---|---|---|---|---|---|---|
| **$CH_3N{=}NCH_3$** (azomethane) see (1045) | | | | | | | | | | | | | | | | |
| **$(CH_3)_2CO$** (acetone) I.P. = $9.70_5$ eV[a]; $D(CH_3CO—CH_3)$ = 3.34 eV[b] | | | | | | | | | | | | | | | | |
| | | Several overlapping Rydberg series; Rydberg series joining on to $\tilde{D}$: $\nu = 78280 - R/(n - 0.03)^2$; $n = 3, \ldots, 10$ | | | | | | | | | | | | | (316)(1272) | |
| $\tilde{D}$ | | 65213 | | | | | | | | | | | | $\tilde{D} \leftarrow \tilde{X}$ 1533 Å | (316)(96) | Strongest band |
| $\tilde{C}$ | | 60086 | extensive band system without clear progressions | | | | | | | | | | | $\tilde{C} \leftarrow \tilde{X}$ 1665–1550 Å | (316)(714) | Sharp bands |
| $\tilde{B}$ | | 51181 | | 1194 | 1047 | 714 | 315 | 269[c] | | | | | | $\tilde{B} \leftarrow \tilde{X}$ 1945–1790 Å | (731)(714) (96) | Sharp bands |
| $\tilde{A}$ | | (30924)[d] | | | | 1198 | $\nu_8 = 210$ | | | | | | | $\tilde{A} \leftrightarrow \tilde{X}$ 3300–2200 Å | (942) | Sharp bands to 3000 Å[e] |
| $\tilde{X}\ ^1A_1$ | $C_{2v}$ | 0 | 2922 | 2871 | 1710 | 1356 | 1066 | 787[f] | 0.33907 | 0.28404 | 0.16378 | r(CO) = 1.215[g]<br>r(CC) = 1.515<br>r(CH) = 1.086 | ∠CCC = 116.1°<br>∠HCC = 110.3° | microwave, infrared and Raman sp. | (1184)(209) | |
| **$C_2H_5CHO$** (propionaldehyde) see (96) | | | | | | | | | | | | | | | | |
| **$C_5H_5$** (cyclopentadienyl) see (1218) | | | | | | | | | | | | | | | | |

$(CH_3)_2CO$: [a] Agrees well with the photoionization value 9.69 eV. [b] From mass spectrometric data of (919). [c] Numbering is arbitrary. [d] Not certain that this is 0 – 0 band.
[e] Continuum adjoining discrete absorption may not belong to the same electronic transition.
[f] $\nu_7 = 530$, $\nu_8 = 391$. No detailed assignment has been made. Only the totally symmetric vibrations (corresponding to polarized Raman lines) are given here. [g] Assumed.

TABLE 80 (*continued*)

| State | Point Group | $T_0$ | Vibrational Frequencies $\nu_1$ | $\nu_2$ | $\nu_3$ | $\nu_4$ | $\nu_5$ | $\nu_6$ | Rotational Constants $A_0$ | $B_0$ | $C_0$ | $r_0$(Å) | $\alpha$ | Observed Transitions | References | Remarks |
|---|---|---|---|---|---|---|---|---|---|---|---|---|---|---|---|---|
| **$C_4H_4NH$** (pyrrole) | | | | | | | | | | | | | | | | |
| | | Many unclassified discrete bands below 1500 Å | | | | | | | | | | | | | (1028) | |
| $\tilde{D}$ | | 65880[a] | | | | | | | | | | | | $\tilde{D} \leftarrow \tilde{X}$ 1518 Å | (1028) | Single sharp band |
| $\tilde{C}$ | | 57250 | | | | | | | | | | | | $\tilde{C} \leftarrow \tilde{X}$ 1750–1590 Å | (1101)(1028) | Weak diffuse bands |
| $\tilde{B}$ | | (52500) | continuum with maximum at 1850 Å | | | | | | | | | | | $\tilde{B} \leftarrow \tilde{X}$ 1900–1770 Å | (1101)(1028) | |
| $\tilde{A}$ | | (45960) | | | | | 1360 | | | | | | | $\tilde{A} \leftarrow \tilde{X}$ 2200–1900 Å | (1101) | Very diffuse bands[b] |
| $\tilde{X}\ ^1A_1$ | $C_{2v}$ | 0 | 3400 | 3133 | 3100 | 1467 | 1384 | 1237[c] | $0.30456_1$ | $0.30025_1$ | $0.15117_4$ | | | microwave, infrared and Raman sp. | (89)(772) | [d] |
| **$C_2H_5NO_2$** (nitroethane) see (1212) | | | | | | | | | | | | | | | | |
| **$H_3C_2$—$C_2H_2Cl$** (chloroprene) see (1027) | | | | | | | | | | | | | | | | |
| **$C_6H_4$** (benzyne) see (115a) | | | | | | | | | | | | | | | | |

$C_4H_4NH$: [a] Measured on the spectrogram reproduced by (1028). [b] For an extension to much longer wavelengths see (821). (840) gives for the 0 – 0 band 47320 cm$^{-1}$.
[c] $\nu_7(a_1) = 1144$, $\nu_8 = 1076$, $\nu_9 = 711$, $\nu_{10}(a_2) = 868$, $\nu_{11} = (711)$, $\nu_{12} = (510)$, $\nu_{13}(b_1) = (1046)$, $\nu_{14} = 838$, $\nu_{15} = 768$, $\nu_{16} = 565$, $\nu_{17}(b_2) = 3133$, $\nu_{18} = 3111$, $\nu_{19} = 1530$, $\nu_{20} = 1418$, $\nu_{21} = (1146)$, $\nu_{22} = 1046$, $\nu_{23} = 1015$, $\nu_{24} = 647$. [d] No unique solution to the geometrical structure of pyrrole has as yet been obtained.

TABLE 80 (*continued*)

| State | Point Group | $T_0$ | Vibrational Frequencies $\nu_1$ | $\nu_2$ | $\nu_3$ | $\nu_4$ | $\nu_5$ | $\nu_6$ | Rotational Constants $A_0$ | $B_0$ | $C_0$ | $r_0$(Å) | $\alpha$ | Observed Transitions | References | Remarks |
|---|---|---|---|---|---|---|---|---|---|---|---|---|---|---|---|---|
| **1,4 $C_4H_4N_2$** (pyrazine) I.P. = 9.29 eV | | | | | | | | | | | | | | | | |
| | | One other system not further specified forming Rydberg series with $\tilde{E}-\tilde{X}$ and $\tilde{F}-\tilde{X}$ | | | | | | | | | | | | | (963) | |
| $\tilde{F}(^1B_{1u} + ^1B_{2u})$ | $\boldsymbol{D_{2h}}$ | 60700 | | | | 710 | | | | | | | | $\tilde{F}\leftarrow\tilde{X}$ 1646–1570 Å | (963)(349) | Very diffuse[a] |
| $\tilde{E}\ ^1B_{2u}$ | $\boldsymbol{D_{2h}}$ | 55154 | 2960 | 1443 | 1182 | 991 | 632 | | | | | | | $\tilde{E}\leftarrow\tilde{X}$ 1846–1640 Å | (963) | Sharp bands |
| $\tilde{D}(^1B_{3u})$ | ($\boldsymbol{D_{2h}}$) | (54000) | fragment of a system of sharp bands | | | | | | | | | | | $\tilde{D}\leftarrow\tilde{X}$ 1850–1795 Å | (963) | |
| $\tilde{C}(^1B_{1u})$ | ($\boldsymbol{D_{2h}}$) | 50880 | (2930) | | | 829 | | | | | | | | $\tilde{C}\leftarrow\tilde{X}$ 1967–1813 Å | (349)(963) | Diffuse bands |
| $\tilde{B}(^1B_{1u})$ | ($\boldsymbol{D_{2h}}$) | 38763 | | (1250) | (957) | | (585) | | | | | | | $\tilde{B}\leftarrow\tilde{X}$ 2700–2450 Å | (556) | Diffuse bands |
| $\tilde{A}\ ^1B_{3u}$ | $\boldsymbol{D_{2h}}$ | 30875.8 | | 1244 | 1155 | 882 | 582.6 | [b] | $\frac{1}{2}(A+B)=0.2036$ | | 0.1008 | | | $\tilde{A}\leftarrow\tilde{X}$ 3300–2900 Å | (831)(610) (609) | [c] |
| $\tilde{a}\ ^3B_{3u}$ | $\boldsymbol{D_{2h}}$ | 26820.2 | 3088 | 1222 | 1149 | 973 | 621 | | $\frac{1}{2}(A+B)=0.2041$ | | | | | $\tilde{a}\leftrightarrow\tilde{X}$ 3760–3560 Å | (607)(523) (1124) | Sharp bands, ‖ structure, $\Delta K = 0, \pm 2$[d] |
| $\tilde{X}\ ^1A_g$ | $\boldsymbol{D_{2h}}$ | 0 | 3054 | 1570 | $1230._0$ | $1015._0$ | 596.1[e] | (950)[f] | $\frac{1}{2}(A+B)=0.2049$ | | | | | infrared and Raman sp. | (770)(609) (1135)(170) | |

1,4 $C_4H_4N_2$: [a] Overlapped by continuum. [b] $\nu_{13}$ = 383 cm$^{-1}$; see (610). [c] Sharp bands with ‖ structure (see Fig. 105); broader bands with ⊥ structure.

[d] In emission (3760–4400 Å) observed in phosphorescence in solid matrices at low temperature by (437) and (351), spectrogram in (351).

[e] This is the value obtained by (831) from the UV spectrum.

[f] $\nu_7(a_u)$ = 363.0 [from (831)], $\nu_8(b_{1g})$ = 757, $\nu_9(b_{1u})$ = 3066, $\nu_{10}$ = 1484, $\nu_{11}$ = 1144, $\nu_{12}$ = 1110, $\nu_{13}(b_{2g})$ = 918.6, $\nu_{14}$ = 703, $\nu_{15}(b_{2u})$ = 3066, $\nu_{16}$ = 1418, $\nu_{17}$ = 1067, $\nu_{18}$ = 1022 [from (1135); (770) give 1148], $\nu_{19}(b_{3g})$ = 3041, $\nu_{20}$ = 1524, $\nu_{21}$ = 1118, $\nu_{22}$ = 516, $\nu_{23}(b_{3u})$ = 804, $\nu_{24}$ = 415.6 [from (831)].

TABLE 80 (*continued*)

| State | Point Group | $T_0$ | $\nu_1$ | $\nu_2$ | $\nu_3$ | $\nu_4$ | $\nu_5$ | $\nu_6$ | $A_0$ | $B_0$ | $C_0$ | $r_0$(Å) | $\alpha$ | Observed Transitions | References | Remarks |
|---|---|---|---|---|---|---|---|---|---|---|---|---|---|---|---|---|
| **1,3 $C_4H_4N_2$** (pyrimidine) | | | | | | | | | | | | | | | | |
| $\tilde{F}$ | $(C_{2v})$ | 58440 | | | | | | $\nu_8 = 910$ | | | | | | $\tilde{F} \leftarrow \tilde{X}$ 1710–1590 Å | (963)(349) | Strong diffuse bands |
| $\tilde{E}(^1A_1)$ | $(C_{2v})$ | 56271 | 3155 | 1165 | 950 | 324 | | | | | | | | $\tilde{E} \leftarrow \tilde{X}$ 1810–1620 Å | (963)(349) | Sharp bands |
| $\tilde{D}(^1A_1)$ | $(C_{2v})$ | 52340 | | | | | | $\nu_8 = 770$ | | | | | | $\tilde{D} \leftarrow \tilde{X}$ 1910–1800 Å | (963)(349) | Diffuse bands |
| $\tilde{C}(^1B_1)$ | $(C_{2v})$ | 51759 | 2991 | 1106 | 950 | 333 | | | | | | | | $\tilde{C} \leftarrow \tilde{X}$ 1932–1800 Å | (963)(349) | Sharp bands superposed on $\tilde{D}-\tilde{X}$ |
| $\tilde{B}(^1B_2)$ | $(C_{2v})$ | 40310 | | | | | | $\nu_8 = 950$ | | | | | | $\tilde{B} \leftarrow \tilde{X}$ 2450–2200 Å | (1230)(963) | Diffuse bands |
| $\tilde{A}\ ^1B_1$ | $(C_{2v})$[a] | 31072.5 | | | $\nu_7=1291$, $\nu_8=1015$, $\nu_9=530$ | | | | $\frac{1}{2}(A+B)=0.2047$ | | | | | $\tilde{A} \leftarrow \tilde{X}$ 3300–2700 Å | (608)(1229) | ‖ bands, extensive vibrational structure |
| $\tilde{a}\ ^3B_1$ | $(C_{2v})$ | 28300 | | (1080) | | | | | | | | | | $\tilde{a} \rightarrow \tilde{X}$ 4400–3530 Å | (1125)(694) (693) | Phosphorescence in rigid glass |
| $\tilde{X}\ ^1A_1$ | $C_{2v}$ | 0 | 3083 | 3048 | 3001 | 1570 | 1463 | 1146[b] | 0.20937 | 0.20238 | 0.10288 | | | infrared, Raman and microwave sp. | (770)(1104) (1134) | |

1,3 $C_4H_4N_2$: [a] The molecule is considered to be slightly non-planar in this state by (608).
[b] $\nu_7(a_1) = 1065$, $\nu_8 = 991$, $\nu_9 = 677$ [from UV spectrum, see (1134)], $\nu_{10}(a_2) = 870$, $\nu_{11} = 394$, $\nu_{12}(b_1) = 980$, $\nu_{13} = 806$, $\nu_{14} = 722$, $\nu_{15} = 679$, $\nu_{16} = 344$, $\nu_{17}(b_2) = 3095$, $\nu_{18} = 1570$, $\nu_{19} = 1402$, $\nu_{20} = 1371$, $\nu_{21} = 1227$, $\nu_{22} = 1161$, $\nu_{23} = 1021$, $\nu_{24} = 567$.

TABLE 80 (*continued*)

| State | Point Group | $T_0$ | Vibrational Frequencies $\nu_1$ | $\nu_2$ | $\nu_3$ | $\nu_4$ | $\nu_5$ | $\nu_6$ | Rotational Constants $A_0$ | $B_0$ | $C_0$ | $r_0$(Å) | $\alpha$ | Observed Transitions | References | Remarks |
|---|---|---|---|---|---|---|---|---|---|---|---|---|---|---|---|---|
| **1,2 $C_4H_4N_2$** (pyridazine) | | | | | | | | | | | | | | | | |
| $\tilde{E}(^1A_1)$ | | 57300 | | | | | | $\nu_8 = 850$ | | | | | | $\tilde{E} \leftarrow \tilde{X}$ 1745–1645 Å | (963) | Strong diffuse absorption |
| $\tilde{D}(^1B_1)$ | | 51503 | 2696 | 1038 | 644 | | | | | | | | | $\tilde{D} \leftarrow \tilde{X}$ 1942–1770 Å | (963) | Fairly sharp bands superposed on $\tilde{C} - \tilde{X}$ |
| $\tilde{C}(^1B_2)$ | | 50000 | continuous absorption | | | | | | | | | | | $\tilde{C} \leftarrow \tilde{X}$ 2050–1800 Å | (963) | |
| $\tilde{B}(^1A_1)$ | | 39500 | | | | | | $\nu_8 = 950$ | | | | | | $\tilde{B} \leftarrow \tilde{X}$ 2560–2320 Å | (963) | Diffuse bands |
| $\tilde{A}\ ^1B_1$ | $C_{2v}$ | 26648.8 | | | | | | | $\frac{1}{2}(A+B) = 0.2030$ | | | | | $\tilde{A} \leftarrow \tilde{X}$ 3750 Å | (608) | Sharp ‖ bands[a] |
| $\tilde{X}\ ^1A_1$ | $C_{2v}$ | 0 | 3063 | 3043 | 1572 | 1414 | 1283 | 1160[b] | 0.2083 | 0.1989 | 0.1018 | | | infrared, Raman and microwave sp. | (770)(1098) | |
| **$C_4H_4O_2$** (dioxadiene) | | | | | | | | | | | | | | | | |
| $\tilde{E}$ | | 57000 | (1700) | | (1110) | (550) | (220) | | | | | | | $\tilde{E} \leftarrow \tilde{X}$ 1760–1670 Å | (983) | |
| $\tilde{D}$ | | 50320 | (1720) | | (1190) | (520) | (220) | | | | | | | $\tilde{D} \leftarrow \tilde{X}$ 2000–1800 Å | (983) | |
| $\tilde{C}$ | | 43450 | (1640) | | (1290) | (500) | (200) | | | | | | | $\tilde{C} \leftarrow \tilde{X}$ 2300–2160 Å | (983) | |
| $\tilde{B}$ | | 38300 | (1690) | (1460) | (1070) | (525) | (230) | | | | | | | $\tilde{B} \leftarrow \tilde{X}$ 2640–2440 Å | (983) | |
| $\tilde{A}$[a] | | 30860 | (1800) | | (1160) | (510) | | | | | | | | $\tilde{A} \leftarrow \tilde{X}$ 3250–2950 Å | (983) | Indistinct but narrow bands |
| $\tilde{X}\ ^1A_g$ | $D_{2h}$ | 0 | | | | | | | | | | | | | | [b] |

**$C_4H_3N_2Cl$** (chloropyrazine) see (556)

**$C_4H_2N_2Cl_2$, $C_4HN_2Cl_3$, $C_4N_2Cl_4$** (di, tri, tetrachloro-pyrimidine) see (1230)

1,2 $C_4H_4N_2$: [a] Thus far only the $0-0$ band has been identified and analyzed.
[b] $\nu_7(a_1) = 1063$, $\nu_8 = 964$, $\nu_9 = 619$ [the ultraviolet spectrum gives 664 cm$^{-1}$, see (963)], $\nu_{10}(a_2) = 936$, $\nu_{11} = 863$, $\nu_{12} = 751$, $\nu_{13} = 421$, $\nu_{14}(b_1) = 760$, $\nu_{15} = 696$, $\nu_{16} = 370$, $\nu_{17}(b_2) = 3075$, $\nu_{18} = 3043$, $\nu_{19} = 1565$, $\nu_{20} = 1444$, $\nu_{21} = 1239$, $\nu_{22} = 1052$, $\nu_{23} = 1009$, $\nu_{24} = 664$.

$C_4H_4O_2$: [a] Possibly a triplet state. [b] Infrared and Raman spectra have not yet been studied.

TABLE 81

MOLECULAR CONSTANTS IN THE ELECTRONIC STATES OF ELEVEN-ATOMIC MOLECULES

| State | Point Group | $T_0$ | Vibrational Frequencies | | | | | | Rotational Constants | | | | | Observed Transitions | References | Remarks |
|---|---|---|---|---|---|---|---|---|---|---|---|---|---|---|---|---|
| | | | $\nu_1$ | $\nu_2$ | $\nu_3$ | $\nu_4$ | $\nu_5$ | $\nu_6$ | $A_0$ | $B_0$ | $C_0$ | $r_0$(Å) | $\alpha$ | | | |
| **$C_3H_7I$** | see (837), (839), (1017) | | | | | | | | | | | | | | | |
| **$C_5H_6$** (1,3 cyclopentadiene) I.P. = 8.62₃ eV | | | | | | | | | | | | | | | | |
| | | Rydberg series joining on to $\tilde{D}$: $\nu = 69550 - R/(n + 0.28)^2$; $n = 4, 5, 6, 7$ | | | | | | | | | | | | | (1028) | Tentative |
| $\tilde{D}$ | | 63540 | | | | | | | | | | | | $\tilde{D}\leftarrow\tilde{X}$ 1574 Å | (1028) | Single diffuse band[a] |
| $\tilde{C}$ | | 60140 | | | | | | | | | | | | $\tilde{C}\leftarrow\tilde{X}$ 1663 Å | (1101)(1028) | Single sharp band |
| $\tilde{B}$ | | 50400[b] | | | | 1440 | | 1070[c] | | | | | | $\tilde{B}\leftarrow\tilde{X}$ 1985–1860 Å | (1101)(982) (1028) | Sharp bands |
| $\tilde{A}$ | | 38880 | | | | $(\nu_8) = 770$ | | | very diffuse bands, forming nearly a continuum with maximum at 2320 Å | | | | | $\tilde{A}\leftarrow\tilde{X}$ 2600–2100 Å | (1101)(982) | |
| $\tilde{X}\ ^1A_1$ | $C_{2v}$ | 0 | 3088 | 2973 | 2880 | 1496 | 1368 | 1105[d] | 0.281064 | 0.274374 | 0.142483 | $r(C—C) = \begin{cases}1.46\\1.54^e\end{cases}$ $\angle CCC = 101°$ $r(C=C) = 1.34^e$ | | Raman, microwave and infrared sp. | (727)(1068) (733) | |
| **$CH_3CH=CHCHO$** (croton aldehyde) I.P. = 9.73 eV[a] | | | | | | | | | | | | | | | | |
| | | Discrete bands below 1800 Å not analysed | | | | | | | | | | | | | (1260) | |
| $\tilde{B}$ | | Continuous absorption with maximum at 2030 Å | | | | | | | | | | | | $\tilde{B}\leftarrow\tilde{X}$ 2100–1900 Å | (1260) | |
| $\tilde{A}$ | | 26518 | 1360 | 1237 | 450 | | | | | | | | | $\tilde{A}\leftarrow\tilde{X}$ 4000–3100 Å | (614)(1050) | Repr. (337) |
| $\tilde{X}\ ^1A'$ | $C_s$ | 0 | | | | | | | | | | | | Raman sp. | | [b] |
| **$C_6H_5$** (phenyl) | | | | | | | | | | | | | | | | |
| $\tilde{A}(^2B_1)$ | | 18908[b] | 896 | 722 | 571 | | | | | | | | | $\tilde{A}\leftarrow\tilde{X}$ 5290–4320 Å | (1009)(1010) | Sharp bands[a] |
| $\tilde{X}\ ^2A_1$ | $C_{2v}$ | 0 | | | | | | | | | | | | | | |

$C_5H_6$: [a] There are several bands between $\tilde{C}-\tilde{X}$ and $\tilde{D}-\tilde{X}$ on (1028)'s spectrogram which have not been assigned and whose wave numbers are not given.
[b] Average; the values by (1101), (982) and (1028) vary by $\pm 60$ cm$^{-1}$. [c] $\nu_8 = 780$, $\nu_{14} = (235)$.
[d] $\nu_7 = 1085$, $\nu_8 = 911$; no consistent and complete set of fundamentals is available. [e] Assumed.

$CH_3C_2H_2CHO$: [a] From photoionization (1273). [b] Raman spectrum observed by (682) but no assignment of fundamentals.

$C_6H_5$: [a] There is a doublet splitting of 10 cm$^{-1}$ in all bands. The transition is assumed to correspond to a removal of an electron from a $\pi$ orbital to a lone pair orbital. [b] At two places in (1010) the value 19908 is given for $T_0$, presumably by mistake.

TABLE 81 (*continued*)

| State | Point Group | $T_0$ | Vibrational Frequencies $\nu_1$ | $\nu_2$ | $\nu_3$ | $\nu_4$ | $\nu_5$ | $\nu_6$ | Rotational Constants $A_0$ | $B_0$ | $C_0$ | $r_0$(Å) | $\alpha$ | Observed Transitions | References | Remarks |
|---|---|---|---|---|---|---|---|---|---|---|---|---|---|---|---|---|
| **$C_5H_5N$ (pyridine)** I.P. = $9.26_6$ eV[a] | | | | | | | | | | | | | | | | |
| | | Two Rydberg series going to a limit at 93240 $cm^{-1}$ | | | | | | | | | | | | | (350) | No details |
| | | Rydberg series going to a limit at 83000 $cm^{-1}$ | | | | | | | | | | | | 1266–1200 Å | (350) | No details |
| | | Rydberg series going to a limit at 74740 $cm^{-1}$ | | | | | | | | | | | | | (350) | No details |
| $\tilde{F}$ | | Continuum with maximum at 1716 Å | | | | | | | | | | | | $\tilde{F} \leftarrow \tilde{X}$ | (349) | |
| $\tilde{E}$ | | 56405 | | (950) | | | | | | | | | | $\tilde{E} \leftarrow \tilde{X}$ 1775–(1600) Å | (349) | Diffuse [b] |
| $\tilde{C}(^1A_1)$ | | 49770 | | | | | | | | | | | | $\tilde{C} \leftarrow \tilde{X}$ 2010–? | (349) | No details |
| $\tilde{A}\ ^1B_1$ | $(C_{2v})$ | 34769 | $\nu_8 = 995$, $\nu_9 = 968$, $\nu_{10} = 542$ [c] | | | | | | | | | | | $\tilde{A} \leftarrow \tilde{X}$ 3000–2500 Å | (499)(1153) | Sharp bands, repr. (499) |
| $\tilde{X}\ ^1A_1$ | $C_{2v}$ | 0 | 3054 | (3054) | 3036 | 1583 | 1482 | 1218[d] | 0.201444 | 0.193624 | 0.098710 | r(CN) = 1.340<br>r(CC) = {1.390, 1.400} | ∠CNC = 116.7°<br>∠NCC = 124.0°<br>∠CCC = {118.6°, 118.1°} | infrared, Raman and microwave sp. | (231)(1304)<br>(91) | |
| **$C_5H_4O_2$ (2-furfuraldehyde)** | | | | | | | | | | | | | | | | |
| | | Large number of unassigned bands (probably Rydberg series) below 1600 Å | | | | | | | | | | | | | (1261) | |
| $\tilde{D}$ | | Continuous absorption with maximum at 1800 Å | | | | | | | | | | | | $\tilde{D} \leftarrow \tilde{X}$ 1950–1700 Å | (1261) | Diffuse [a] |
| $\tilde{C}$ | | (50055) | 887 | | | | | 230 | | | | | | $\tilde{C} \leftarrow \tilde{X}$ 2000–1940 Å | (1261) | Diffuse bands |
| $\tilde{B}$ | | 37107 | 883 | 742 | 561 | 446 | 380 | 202[b] | | | | | | $\tilde{B} \leftarrow \tilde{X}$ 2780–2460 Å | (1097) | Strong sharp bands followed by diffuse bands |
| $\tilde{A}$ | | (28385) | | | | | | | | | | | | $\tilde{A} \leftarrow \tilde{X}$ 3700–3300 Å | (1097) | Very weak, diffuse bands |
| $\tilde{X}\ ^1A'$ | $C_s$ | 0 | 3140 | 3120 | 1689 | 1666 | 1564 | 1466[c] | | | | | | Raman sp. | (127) | |

$C_5H_5N$: [a] From the first Rydberg series of (350). From photoionization (1273) obtains 9.23 eV. [b] A system of sharp bands is mentioned in the 1800–1600 Å region by (349).
[c] $\nu_{10}$ is the most prominent. Assignments other than $\nu_8$, $\nu_9$, $\nu_{10}$ are uncertain.
[d] $\nu_7(a_1) = 1068$, $\nu_8 = 1029$, $\nu_9 = 992$, $\nu_{10} = 605$, $\nu_{11}(a_2) = 981$, $\nu_{12} = 886$, $\nu_{13} = 374$, $\nu_{14}(b_1) = 942$, $\nu_{15} = 886$, $\nu_{16} = 749$, $\nu_{17} = 700$, $\nu_{18} = 405$, $\nu_{19}(b_2) = 3083$, $\nu_{20} = (3036)$, $\nu_{21} = 1572$, $\nu_{22} = 1439$, $\nu_{23} = 1375$, $\nu_{24} = (1217)$, $\nu_{25} = 1148$, $\nu_{26} = (1085)$, $\nu_{27} = 652$. This is the assignment of (1304) which differs only slightly, in the $b_1$ vibrations, from the original assignment of (231). For the in-plane vibrations ($a_1$, $b_2$) the assignment by (231) has been supported by calculations of (1332), but (1249) and (111) are led by similar calculations to slightly different assignments.

$C_5H_4O_2$: [a] $\tilde{D}$ and $\tilde{C}$ may be one and the same electronic state. [b] Several other vibrations are very doubtful. The numbering of the frequencies is not established.
[c] 1387, 1362, 1213, 1153, 1074, 1014, 922, 876, 497. These frequencies are the observed Raman frequencies. They may not all be fundamentals. The infrared spectrum has not been studied in detail.

TABLE 82

MOLECULAR CONSTANTS IN THE ELECTRONIC STATES OF TWELVE-ATOMIC MOLECULES

| State | Point Group | $T_0$ | Vibrational Frequencies $\nu_1$ | $\nu_2$ | $\nu_3$ | $\nu_4$ | $\nu_5$ | $\nu_6$ | Rotational Constants $A_0$ | $B_0$ | $C_0$ | $r_0$(Å) | $\alpha$ | Observed Transitions | References | Remarks |
|---|---|---|---|---|---|---|---|---|---|---|---|---|---|---|---|---|
| **$CH_3CH{=}CHCH_3$** (2-butene) see (181), (1026) and (408) | | | | | | | | | | | | | | | | |
| **$C_2H_5C_2H_3$** (1-butene) see (408) I.P. = 9.58 eV[a] | | | | | | | | | | | | | | | | |
| **$H_2C{=}C(CH_3)_2$** (2-methylpropene) see (408) | | | | | | | | | | | | | | | | |
| **$HCON(CH_3)_2$** (dimethylformamide) see (594) | | | | | | | | | | | | | | | | |
| **$C_6H_6$** (benzene) I.P. = 9.247 eV[a] | | | | | | | | | | | | | | | | |
| | | Rydberg series converging to 16.84 eV with fairly long progressions in $\nu_2 \approx 960\ cm^{-1}$ | | | | | | | | | | | | 850–700 Å | (350) | No details given[b] |
| | | Two Rydberg series joining on to $\tilde{P}$ and converging to $11.48_9$ eV | | | | | | | | | | | | 1342–1050 Å | (1032)(350) | No details given[b] |
| $\tilde{P}$ | $D_{6h}$ | 74543<br>74488 | | (640) | | | | | | | | | | $\tilde{P}\leftarrow\tilde{X}$ 1342 Å | (1299)(1032)<br>(350) | |
| | | Rydberg series joining on to $\tilde{G}$ and $\tilde{K}$: $\nu = 74587 - R/(n - 0.04)^2$; $n = 3, 4, \ldots, 9$ | | | | | | | | | | | | 1555–1341 Å | (1299) | Vibrational structure observed in all four series |
| | | Rydberg series joining on to $\tilde{F}$ and $\tilde{J}$: $\nu = 74587 - R/(n - 0.11)^2$; $n = 3, 4, \ldots, 9$ | | | | | | | | | | | | 1630–1341 Å | (1299) | |
| | | Rydberg series joining on to $\tilde{E}$ and $\tilde{I}$: $\nu = 74587 - R/(n - 0.16)^2$; $n = 3, 4, \ldots, 11$ | | | | | | | | | | | | 1675–1341 Å | (1299)(1032) | |
| | | Rydberg series joining on to $\tilde{D}$ and $\tilde{H}$: $\nu = 74587 - R/(n - 0.46)^2$; $n = 3, 4, \ldots, 10$ | | | | | | | | | | | | 1800–1341 Å | (1299)(1032) | |
| $\tilde{K}$ | $D_{6h}$ | 67683 | | 969 | $\nu_{18} = 670$ | | | | | | | | | $\tilde{K}\leftarrow\tilde{X}$ 1478–1455 Å | (1299) | $3R'''$ of (1299) |
| $\tilde{J}$ | $D_{6h}$ | 67607 | | 960 | $\nu_{18} = 676$ | | | | | | | | | $\tilde{J}\leftarrow\tilde{X}$ 1479–1458 Å | (1299) | $3R''$ of (1299) |
| $\tilde{I}$ | $D_{6h}$ | 67531 | | 971 | $\nu_{18} = 682$, ($\nu_{20} = 309$) | | | | | | | | | $\tilde{I}\leftarrow\tilde{X}$ 1493–1439 Å | (1299)(1032) | $3R'$ of (1299) |
| $\tilde{H}$ | $D_{6h}$ | 65718 | | 974 | $\nu_{18} = 696$, ($\nu_{20} = 314$) | | | | | | | | | $\tilde{H}\leftarrow\tilde{X}$ 1535–1490 Å | (1299)(1032) | $3R$ of (1299) |
| $\tilde{G}$ | $D_{6h}$ | 64387 | | | $\nu_{18} = 676$, ($\nu_{20} = 334$) | | | | | | | | | $\tilde{G}\leftarrow\tilde{X}$ 1555–1535 Å | (1299) | $2R'''$ of (1299) |

$C_2H_5C_2H_3$: [a] From photoionization (1273).
$C_6H_6$: [a] Photoionization experiments of (1272) give $9.24_5 \pm 0.01$ eV. [b] Absorption coefficients in the region 1335 to 430 Å given by (161).

TABLE 82 (*continued*)

| State | Point Group | $T_0$ | Vibrational Frequencies $\nu_1$ | $\nu_2$ | $\nu_3$ | $\nu_4$ | $\nu_5$ | $\nu_6$ | Rotational Constants $A_0$ | $B_0$ | $C_0$ | $r_0$(Å) | $\alpha$ | Observed Transitions | References | Remarks |
|---|---|---|---|---|---|---|---|---|---|---|---|---|---|---|---|---|
| **$C_6H_6$** (*continued*) | | | | | | | | | | | | | | | | |
| $\tilde{F}$ | $D_{6h}$ | 61421 | | | $\nu_{18} = 674$ | | | | | | | | | $\tilde{F} \leftarrow \tilde{X}$ 1630–1600 Å | (1299) | 2$R''$ of (1299) |
| $\tilde{E}$ | $D_{6h}$ | 59795 | | | $\nu_{18} = 744$ | | | | | | | | | $\tilde{E} \leftarrow \tilde{X}$ 1675–1615 Å | (1299) | 2$R'$ of (1299) |
| $\tilde{D}\ ^1E_{1u}$ or $^1A_{2u}$ | $D_{6h}$ | 55881 | | 974 | $\nu_{18} = 722$, ($\nu_{20} = 316$) | | | | | | | | | $\tilde{D} \leftarrow \tilde{X}$ 1800–1720 Å | (1299) | 2$R$ of (1299), possibly identical with $\tilde{C}$ |
| $\tilde{C}\ ^1E_{1u}$ | ($D_{6h}$) | (55000) | nearly a continuum unless $\tilde{D}$ and $\tilde{C}$ are identical | | | | | | | | | | | $\tilde{C} \leftarrow \tilde{X}$ 1820–1700 Å | (981)(468) | Very diffuse, $V - N$ of Mulliken |
| $\tilde{B}\ ^1B_{1u}$ or $^1E_{2g}$ | ($D_{6h}$) | 49100 | | 900 | | | | | | | | | | $\tilde{B} \leftarrow \tilde{X}$ 2050–1850 Å | (981)(327) (1102)(182) | Diffuse bands |
| $\tilde{A}\ ^1B_{2u}$ | $D_{6h}$ | $38086._1$ | $3134._9$ | $925._1$ | ? | 515 | ? | ?[c] | — | 0.1810[d] | | $r_0$(CC) = 1.434; $r_0$(CH) = 1.07 | | $\tilde{A} \leftrightarrow \tilde{X}$ 2670–2270 Å | (404)(1152) (1037)(176) (602)(405a) | Sharp bands, in absorption and fluorescence; see Fig. 70 [g] |
| $\tilde{a}\ ^3B_{1u}$ | ($D_{6h}$)[e] | 29510[f] | | (900) | | | | | | | | | | $\tilde{a} \leftrightarrow \tilde{X}$ 3400–3000 Å | (987)(1128) (1076)(365) (669a) | |
| $\tilde{X}\ ^1A_{1g}$ | $D_{6h}$ | 0 | 3073 | $995._4$ | 1350 | $674._0$ | 3057 | 1010[h] | — | $0.1896_0$[i] | | $r_0$(CC) = 1.397; $r_0$(CH) = 1.084 | | infrared and Raman sp. | (1164)(154) (268) | |

$C_6H_6$:

[c] $\nu_7(b_{2g}) = 775$, $\nu_8 = 365$, $\nu_{11}(e_{1g}) = 585$, $\nu_{15}(e_{2g}) = 3077._2$, $\nu_{17} = ?$ $\nu_{18} = 522._4$, $\nu_{19}(e_{2u}) = 706$, $\nu_{20} = 237._3$.

[d] $\zeta_{18} = 0.60$, $\zeta_{15} \sim 0$ [see (176)].

[e] (450) have studied electron spin resonance of $C_6H_6$ in the $\tilde{a}\ ^3B_{1u}$ state and conclude that the equilibrium conformation is not a regular hexagon; see however (731a).

[f] As observed by (365) in the vapor with high pressure $O_2$. (669a) give 29516 ± 9 in a similar mixture at atmospheric pressure. (987) finds 29410 in liquid. In an argon matrix (1076) observes 29575 cm$^{-1}$.

[g] In absorption in the vapor mixed with $O_2$ (365) (669a) and in liquid $C_6H_6$ (987); in fluorescence in a solid matrix (731a)(1128)(1076).

[h] $\nu_7(b_{2g}) = 990$, $\nu_8 = 707$, $\nu_9(b_{2u}) = 1309$, $\nu_{10} = 1146$, $\nu_{11}(e_{1g}) = 846$, $\nu_{12}(e_{1u}) = 3047, 3083$ [Fermi doublet; (154) give the unperturbed frequency as 3064 cm$^{-1}$], $\nu_{13} = 1482$, $\nu_{14} = 1037$, $\nu_{15}(e_{2g}) = 3056$, $\nu_{16} = 1585, 1606$ [Fermi doublet; (154) give the unperturbed frequency as 1599 cm$^{-1}$], $\nu_{17} = 1178$, $\nu_{18} = 608._0$, $\nu_{19}(e_{2u}) = 967$, $\nu_{20} = 398._6$; [$\nu_2$, $\nu_{18}$ and $\nu_{20}$ are from the ultraviolet spectrum (176)].

[i] $\alpha_4^B = 0.00012_2$ cm$^{-1}$ [see 268)]; $\zeta_{18} = +0.62$ [see (176)].

TABLE 82 (*continued*)

| State | Point Group | $T_0$ | Vibrational Frequencies $\nu_1$ | $\nu_2$ | $\nu_3$ | $\nu_4$ | $\nu_5$ | $\nu_6$ | Rotational Constants $A_0$ | $B_0$ | $C_0$ | $r_0$(Å) | $\alpha$ | Observed Transitions | References | Remarks |
|---|---|---|---|---|---|---|---|---|---|---|---|---|---|---|---|---|
| **$H_3CC{\equiv}C{-}C{\equiv}CCH_3$** (dimethyldiacetylene) I.P. < 11.5 eV | | | | | | | | | | | | | | | | |
| | | Rydberg series joining on to $\tilde{E}$: $\nu = 92810 - R/(n - 0.20)^2$; $n = 3, 4, \ldots, 7$ | | | | | | | | | | | | | (1029) | Second ionization limit |
| $\tilde{E}$ | | 78727 | | | | | | | | | | | | $\tilde{E} \leftarrow \tilde{X}$ 1270 Å | (1029) | Single band |
| $\tilde{C}$ | | 59984 | | 2500 | | | | | | | | | | $\tilde{C} \leftarrow \tilde{X}$ 1670–1380 Å | (1029) | Very strong discrete bands[a] |
| $\tilde{B}$ | | (48900)[b] | | 2500 | | | | | | | | | | $\tilde{B} \leftarrow \tilde{X}$ 2045–1700 Å | (1029) | Irregular vibrational structure[c] |
| $\tilde{X}\ ^1A_1'$ | $D_{3h}'$ | 0 | 2914 | 2264 | 1381 | 1228 | 554 | ?[d] | | | | | | infrared and Raman sp. | (379) | [e] |
| **$C_2H_3C{\equiv}CC_2H_3$** (divinyl-acetylene) see (1030) | | | | | | | | | | | | | | | | |
| **$(CH_3CO)_2$** (biacetyl) $D(CH_3CO{-}COCH_3) = 2.84$ eV[a] | | | | | | | | | | | | | | | | |
| $\tilde{D}$ | | 57295 | | | | | (1350) | (475) | | | | | | $\tilde{D} \leftarrow \tilde{X}$ 1750–1650 Å | (354) | Sharp bands |
| $\tilde{C}$ | | (50647) | | | | | 1220 | 220 | | | | | | $\tilde{C} \leftarrow \tilde{X}$ 2000–1880 Å | (354) | Sharp violet-shaded bands |
| $\tilde{B}\ ^1A_u$ | | (35500) | continuous absorption | | | | | | | | | | | $\tilde{B} \leftarrow \tilde{X}$ 2800–2200 Å | (750) | [b] |
| $\tilde{A}\ ^1A_u$ | $C_{2h}$ | (21983)[c] | | | | 1060 | 620 | 84 | | | | | | $\tilde{A} \leftrightarrow \tilde{X}$ 4700–3500 Å | (750)(726) | Sharp red-shaded bands[d] |
| $\tilde{a}\ ^3B_g$ | $C_{2h}$ | (19900) | fluorescence bands | | | | | | | | | | | $\tilde{a} \rightarrow \tilde{X}$ 5014–6135 Å | (64) | Excited by Hg **4358**, 4047, 3650[e] |
| $\tilde{X}\ ^1A_g$ | $C_{2h}$ | 0 | 3023 | 2940 | 1725 | 1444 | 1275 | 690[f] | | | | | | Raman sp. | (683) | |

$H_3CC{\equiv}C{-}C{\equiv}CCH_3$: [a] First member of a Rydberg series; higher members not clearly recognizable. [b] Measured on the reproduction of (1029).
[c] Strongest band at 1800 Å.
[d] $\nu_7(\bar{a}_2) = 2947$, $\nu_8 = 2163$, $\nu_9 = 1379$, $\nu_{10} = 941$, $\nu_{11}(e) = 2947$, $\nu_{12} = 1458$, $\nu_{13} = 1030$, $\nu_{14} = 353$, $\nu_{15} \sim (100)$, $\nu_{16}(\bar{e}) = 2953$, $\nu_{17} = 1426$, $\nu_{18} = 1022$, $\nu_{19} = 475$, $\nu_{20} = 247$.
[e] For the effect of free internal rotation on the point group and its representations, as well as on the vibrations and vibrational selection rules, see (575)(577a)(161a) and (162).

$(CH_3CO)_2$: [a] From mass spectrometric data of (919). [b] In the crystal, diffuse bands precede continuum beginning at 31475 cm$^{-1}$ (1133).
[c] Not certain that this is 0 – 0 band. (750) assumes 22594 cm$^{-1}$. In the crystal, according to (1133), the 0 – 0 band is at 22873 cm$^{-1}$.
[d] Extending to 4000 Å followed by a continuum. In fluorescence: bands extending from 4600 to 6000 Å.
[e] In the crystal discrete absorption starts at 20421 cm$^{-1}$ (1133) and is ascribed to an electronic transition ($^3A_u - {}^1A_g$) different from fluorescence which is ascribed to $^3B_g - {}^1A_g$. $^3B_g$ is below $^3A_u$. Lifetime of triplet biacetyl: $\tau = 1.2 \times 10^{-2}$ (312). [f] $\nu_7 = 375$ cm$^{-1}$, from Raman spectrum of (683); infrared spectra (not assigned) in (859)(1133) and (479).

TABLE 82 (*continued*)

| State | Point Group | $T_0$ | Vibrational Frequencies $\nu_1$ | $\nu_2$ | $\nu_3$ | $\nu_4$ | $\nu_5$ | $\nu_6$ | Rotational Constants $A_0$ | $B_0$ | $C_0$ | $r_0$(Å) | $\alpha$ | Observed Transitions | References | Remarks |
|---|---|---|---|---|---|---|---|---|---|---|---|---|---|---|---|---|
| **$C_5H_5NO$** (pyridine-N-oxide) | | see (612) | | | | | | | | | | | | | | |
| **$C_6H_5F$** | | see (98), (469) and (655) | | | | I.P. = 9.199[a] | | | | | | | | | | |
| **$C_6H_5Cl$** | | see (98), (1031) and (1110) | | | | I.P. = 9.07 eV[a] | | | | | | | | | | |
| **$C_6H_5Br$** | | see (1156), (77) and (1031) | | | | I.P. = 8.98 eV[a] | | | | | | | | | | |
| **$C_6H_5I$** | | see (329), (1031) and (328); | | | | I.P. = 8.73 eV[a] | | | | | | | | | | |
| **$C_6H_4O_2$** (benzoquinone) | | see (83a)(1136a)(70a)(624a) | | | | | | | | | | | | | | |
| **$C_6H_4F_2$** | | see (1056), (272), (229) and (469) | | | | | | | | | | | | | | |
| **$C_6H_4FCl$** | | see (695) | | | | | | | | | | | | | | |
| **$C_6H_4FBr$** | | see (696) | | | | | | | | | | | | | | |
| **$C_6H_4Cl_2$** | | see (1148), (1031) and (70) | | | | | | | | | | | | | | |
| **$C_6H_4Br_2$** | | see (1157) | | | | | | | | | | | | | | |
| **$C_6H_3F_3$** | | see (1054), (1149) and (469); | | | | I.P. = $9.68_5$ eV[a] | | | | | | | | | | |
| **$C_6H_3Cl_3$** | | see (1151), (684) and (858) | | | | | | | | | | | | | | |
| **$C_6H_2F_4$** | | see (1149) | | | | | | | | | | | | | | |
| **$C_6H_2Cl_4$** | | see (802) | | | | | | | | | | | | | | |

$C_6H_5F$: [a] From Rydberg series of (469).
$C_6H_5Cl$: [a] From photoionization (1273).
$C_6H_5Br$: [a] From photoionization (1273).
$C_6H_5I$: [a] From photoionization (1273).
$C_6H_3F_3$: [a] From a Rydberg series of (469).

# APPENDIX VII

## PHYSICAL CONSTANTS AND CONVERSION FACTORS

Since the publication of Volumes I and II, the values of the general physical constants have been very slightly changed and their accuracy has been improved. In addition, the International Unions of both Physics and Chemistry have agreed to adopt the scale of atomic weights based on the value 12.0000 for the carbon isotope of mass 12. A consistent set of constants, based on the discussion of many new measurements, has recently been proposed by Cohen and DuMond (217b) and is recommended by a committee of the U.S. National Academy of Sciences. Unfortunately this set became known only when this book was already in press. All numerical data in this book are based on an earlier report by Cohen and DuMond (217a), modified by the adoption of a very slightly improved value of the velocity of light. However, in almost all cases the numerical data obtained with the new set are, within the accuracy of the determinations, the same as those given in this book. In particular, the $B$ values of Appendix VI that were obtained from microwave measurements are based on the new value of $c$.

For the convenience of the reader we give in Table 83 the newly recommended values of those constants that are frequently used in spectroscopy and in Table 84 the corresponding values of the conversion factors of energy units. These tables replace Tables 1 and 2 of Volume I and Tables 147 and 148 of Volume II. Finally, Table 85 gives the numerical factors in the equations for moments of inertia, force constants, and Boltzmann factor, based on the numerical values of the physical constants of Table 83.

The conversion factors from eV to $cm^{-1}$ and cal actually used in the text and tables of this book were $8066.0_3$ and 23062.3, respectively, not those of Table 84. Thus, the ionization potentials and dissociation energies given in eV in Chapter V and in Appendix VI must be corrected by about one part in 20,000 in the few cases in which the accuracy of the determination warrants it.

TABLE 83. PHYSICAL CONSTANTS

| Constant | Symbol | Value | Error Limit[1] |
|---|---|---|---|
| Velocity of light in vacuum | $c$ | $2.997925 \times 10^{10}$ cm/sec | ± 3 |
| Planck's constant | $h$ | $6.6256 \times 10^{-27}$ erg sec | ± 5 |
| Electronic charge | $e$ | $4.80298 \times 10^{-10}$ e.s.u. | ± 20 |
| Electron rest mass | $m_e$ | $9.1091 \times 10^{-28}$ gm | ± 4 |
| Rydberg constant | $R_\infty$ | $109737.31$ cm$^{-1}$ | ± 3 |
| Bohr radius | $a_0$ | $0.529167 \times 10^{-8}$ cm | ± 6 |
| $\frac{1}{12}$ mass of $C^{12}$ atom | $M_1$ | $1.66043 \times 10^{-24}$ gm | ± 6 |
| Avogadro number (of molecules in a mol) | $N_A$ | $6.02252 \times 10^{23}$ | ± 28 |
| Boltzmann's constant | $k$ | $1.38054 \times 10^{-16}$ erg/degree | ± 18 |
| Gas constant | $R$ | $8.3143 \times 10^{7}$ erg/degree/mol | ± 12 |
| | | $1.98717$ cal/degree/mol | ± 29 |
| 1 thermochemical caloric | cal | $4.18400 \times 10^{7}$ ergs | |

[1] This is three times the standard deviation in units of the last digit quoted for the constant.

TABLE 84. CONVERSION FACTORS FOR ENERGY UNITS

| Unit | cm$^{-1}$ | erg/molecule | cal/mol | electron-volt |
|---|---|---|---|---|
| 1 cm$^{-1}$ | 1 | $1.9863_1 \times 10^{-16}$ | $2.8591_2$ | $1.23981_3 \times 10^{-4}$ |
| 1 erg/molecule | $5.0344_7 \times 10^{15}$ | 1 | $1.43942 \times 10^{16}$ | $6.2418_1 \times 10^{11}$ |
| 1 cal/mol | $0.34975_8$ | $6.9472_6 \times 10^{-17}$ | 1 | $4.3363_4 \times 10^{-5}$ |
| 1 eV (electron-volt) | $8065.7_3$ | $1.60210 \times 10^{-12}$ | 23060.9 | 1 |

TABLE 85. SOME NUMERICAL FACTORS

| Equation | Factor | Numerical value |
|---|---|---|
| $I = \dfrac{h}{8\pi^2 cB\ (\text{cm}^{-1})}$ | $\dfrac{h}{8\pi^2 c}$ | $27.990_8 \times 10^{-40}$ gm cm$^2$ × cm$^{-1}$<br>16.8575 a.m.u. Å$^2$ × cm$^{-1}$ |
| $I = \dfrac{h}{8\pi^2\ B\ (\text{Mc/sec})}$ | $\dfrac{h}{8\pi^2}$ | $8.39142 \times 10^{-35}$ gm cm$^2$ × (Mc/sec)<br>$50.5375 \times 10^4$ a.m.u. Å$^2$ × (Mc/sec) |
| $k = 4\pi^2 c^2 M_1 \mu_A \omega^2$ | $4\pi^2 c^2 M_1$ | $5.89145 \times 10^{-2}$ dynes/cm/(cm$^{-1}$)$^2$ |
| $N = N_0\, e^{-\frac{T(\text{cm}^{-1})hc}{kT}}$ | $\dfrac{k}{hc}$ | $0.69503$ cm$^{-1}$/degree |

# BIBLIOGRAPHY

## I. Handbooks, Monographs, Textbooks, Tables


1. H. C. Allen, Jr. and P. C. Cross, Molecular Vib-rotors (Wiley, New York, 1963).
2. J. W. Baker, Hyperconjugation (Oxford University Press, 1952).
3. C. J. Ballhausen, Introduction to Ligand Field Theory (McGraw-Hill, New York, 1962).
4. G. M. Barrow, Introduction to Molecular Spectroscopy (McGraw-Hill, New York, 1962).
5. G. Briegleb, Elektronen-Donator-Acceptor-Komplexe (Springer-Verlag, Berlin, 1961).
6. F. A. Cotton, Chemical Applications of Group Theory (Interscience, New York, 1963).

6a. T. L. Cottrell, The Strengths of Chemical Bonds (Butterworth, London, 1958).

7. C. A. Coulson, Valence, 2nd edition (Oxford University Press, 1961).
8. R. Daudel, Structure Electronique des Molécules (Gauthier-Villars, Paris, 1962).
9. M. J. S. Dewar, Hyperconjugation (Ronald Press Co., New York, 1962).
10. G. H. Dieke and A. B. F. Duncan, Spectroscopic Properties of Uranium Compounds (McGraw-Hill, New York, 1949).

10a. R. N. Dixon, Spectroscopy and Structure (Methuen, London, 1965).

11. L. Eisenbud and E. P. Wigner, Nuclear Structure (Princeton University Press, 1958).
12. M. A. Elyashevich, Atomnaya i Molekulyarnaya Spektroskopiya (Moscow, 1962).
13. H. Eyring, J. Walter and G. E. Kimball, Quantum Chemistry (Wiley, New York, 1944).
14. F. H. Field and J. L. Franklin, Electron Impact Phenomena (Academic Press, New York, 1957).
15. T. Förster, Fluoreszenz organischer Verbindungen (Vandenhoeck and Ruprecht, Göttingen, 1951).
16. J. S. Griffith, The Theory of Transition-Metal Ions (Cambridge University Press, 1961).
17. L. V. Gurvich, V. P. Glushko, G. A. Khachkuruzov, I. V. Veits and V. A. Medvedev, Termodinamicheskie Svoistva Individualnykh Veshchestv (Acad. Sci. USSR, Moscow, 1962).
18. H. Hartmann, Theorie der chemischen Bindung auf quanten-theoretischer Grundlage (Springer, Berlin, 1954).
19. H. Hellmann, Einführung in die Quantenchemie (Deuticke, Leipzig, 1937).
20. V. Henri, Structure des Molécules (Hermann, Paris, 1925).
21. G. Herzberg, Atomic Spectra and Atomic Structure (Dover Publications, New York, 1944).
22. ———, Molecular Spectra and Molecular Structure I, Spectra of Diatomic Molecules, 2nd edition (Van Nostrand, Princeton, N.J., 1950).
23. ———, Molecular Spectra and Molecular Structure II, Infrared and Raman Spectra of Polyatomic Molecules (Van Nostrand, Princeton, N.J., 1945).

23a. J. O. Hirschfelder, C. F. Curtiss and R. B. Bird, Molecular Theory of Gases and Liquids (Wiley, New York, 1964).

24. H. H. Jaffe and M. Orchin, Theory and Applications of Ultraviolet Spectroscopy (Wiley, New York, 1962).

24a. G. W. King, Spectroscopy and Molecular Structure (Holt, Rinehart and Winston, New York, 1964).

25. K. W. F. Kohlrausch, Ramanspektren, Hand- und Jahrbuch der chemischen Physik, vol. 9, part VI (Akademische Verlagsgesellschaft, Leipzig, 1943).

26. L. D. Landau and E. M. Lifshitz, Quantum Mechanics (Pergamon, London, 1958).
27. J. N. Murrell, The Theory of the Electronic Spectra of Organic Molecules (Methuen, London, 1963).
28. L. E. Orgel, An Introduction to Transition-Metal Chemistry: Ligand-Field Theory (Methuen, London, 1960).
29. R. G. Parr, The Quantum Theory of Molecular Electronic Structure (Benjamin, New York, 1963).
30. L. Pauling, The Nature of the Chemical Bond, 3rd edition (Cornell University Press, 1960).
31. ——— and E. B. Wilson, Jr., Introduction to Quantum Mechanics (McGraw-Hill, New York, 1935).
32. R. W. B. Pearse and A. G. Gaydon, The Identification of Molecular Spectra, 3rd edition (Chapman–Hall, London, 1963).
33. G. C. Pimentel and A. L. McClellan, The Hydrogen Bond (Freeman, San Francisco, 1960).
34. C. Sandorfy, Electronic Spectra and Quantum Chemistry (Prentice-Hall, Englewood Cliffs, N.J., 1964).
35. W. T. Simpson, Theories of Electrons in Molecules (Prentice-Hall, Englewood Cliffs, N.J. 1962).
36. N. B. Slater, Theory of Unimolecular Reactions (Methuen, London, 1959).
37. H. Sponer, Molekülspektren, Vols. I and II (Julius Springer, Berlin, 1935).
38. E. W. R. Steacie, Atomic and Free Radical Reactions, 2nd edition (Reinhold, New York, 1954).
39. A. Streitwieser, Jr., Molecular Orbital Theory for Organic Chemists (Wiley, New York, 1961).
39a. L. E. Sutton, Tables of Interatomic Distances and Configuration in Molecules and Ions, Spec. Publ. No. 11, Chem. Soc. London, 1958, Supplement Spec. Publ. No. 18 (1965).
40. C. H. Townes and A. L. Schawlow, Microwave Spectroscopy (McGraw-Hill, New York, 1955).
41. V. I. Vedeneiev, L. V. Gurvich, V. N. Kondratiev, V. A. Medvedev and E. L. Frankevich, Energii Razryva Khimicheskikh Svyazei (Acad. Sci. USSR, Moscow, 1962).
42. W. West, Chemical Applications of Spectroscopy, Vol. IX of Technique of Organic Chemistry (Interscience, New York, 1956).
43. G. W. Wheland, Resonance in Organic Chemistry (Wiley, New York, 1955).
44. E. P. Wigner, Group Theory and Its Application to the Quantum Mechanics of Atomic Spectra (Academic Press, New York, 1959).

## II. References to Individual Papers

50. A. Adel and D. M. Dennison, J. Chem. Phys. **14**, 379 (1946).
51. D. Agar, E. K. Plyler and E. D. Tidwell, J. Res. Natl. Bur. Std. **66**A, 259 (1962).
52. A. C. Albrecht, J. Chem. Phys. **33**, 156 (1960).
53. ———, J. Chem. Phys. **33**, 169 (1960).
54. ———, J. Chem. Phys. **38**, 354 (1963).
55. ——— and W. T. Simpson, J. Chem. Phys. **21**, 940 (1953).
55a. D. S. Alderdice, J. Mol. Spec. **15**, 509 (1965).
56. ——— and T. M. Dunn (to be published).
57. H. C. Allen, Jr. and W. B. Olson, J. Chem. Phys. **37**, 212 (1962).
58. ——— and E. K. Plyler, J. Chem. Phys. **25**, 1132 (1956).
59. ——— and ———, J. Amer. Chem. Soc. **80**, 2673 (1958).
60. ———, E. D. Tidwell and E. K. Plyler, J. Chem. Phys. **25**, 302 (1956).
61. K. Allison and A. D. Walsh, Chem. Inst. of Canada Symposium, Ottawa (1957).
62. A. Almenningen, O. Bastiansen and M. Traetteberg, Acta Chem. Scand. **12**, 1221 (1958).
63. G. M. Almy and S. Anderson, J. Chem. Phys. **8**, 805 (1940).

64. G. M. Almy, H. Q. Fuller and G. D. Kinzer, J. Chem. Phys. **8**, 37 (1940).
65. ——— and P. R. Gillette, J. Chem. Phys. **11**, 188 (1943).
66. ——— and R. B. Horsfall, Jr., Phys. Rev. **51**, 491 (1937).
67. A. P. Altshuller, J. Chem. Phys. **22**, 1947 (1954).
68. Y. Amako and P. A. Giguère, Can. J. Chem. **40**, 765 (1962).
69. A. B. Anderson and E. F. Barker, J. Chem. Phys. **18**, 698 (1950).
69a. W. L. S. Andrews and G. C. Pimentel (to be published).
70. T. Anno and I. Matubara, J. Chem. Phys. **23**, 796 (1955).
70a. ——— and A. Sadô, J. Chem. Phys. **32**, 1602 (1960).
71. T. Arai, Rev. Mod Phys. **32**, 370 (1960).
72. E. T. Arakawa and A. H. Nielsen, J. Mol. Spec. **2**, 413 (1958).
73. W. F. Arendale and W. H. Fletcher, J. Chem. Phys. **21**, 1898 (1953).
74. ——— and ———, J. Chem. Phys. **26**, 793 (1957).
75. R. L. Arnett and B. L. Crawford, Jr., J. Chem. Phys. **18**, 118 (1950).
76. S. J. Arnold, E. A. Ogryzlo and H. Witzke, J. Chem. Phys. **40**, 1769 (1964).
77. K. Asagoe and Y. Ikemoto, Phys. Math. Soc. Japan **22**, 685 (1940).
78. N. Astoin, C.R. Acad. Sci. (Paris) **242**, 2327 (1956).
79. ———, J. Rech. Cent. Nat. Rech. Sci., No. 38, 1 (1957).
80. ———, L. Sanson and M. C. Bonnelle, C.R. Acad. Sci. (Paris) **250**, 1824 (1960).
81. R. K. Asundi, M. Jan-Khan and R. Samuel, Proc. Roy. Soc. **157**A, 28 (1936).
82. ———, M. Karim and R. Samuel, Proc. Phys. Soc. **50**, 581 (1938).
83. ——— and R. Samuel, Proc. Ind. Acad. Sci. **2**A, 30 (1935).
83a. ——— and R. S. Singh, Nature **176**, 1223 (1955).

84. R. F. W. Bader, Can. J. Chem. **41**, 2303 (1963).
85. ——— and G. A. Jones, Can. J. Chem. **39**, 1253 (1961), **41**, 586, 2251 (1963).
86. ——— and ———, J. Chem. Phys. **38**, 2791 (1963).
87. B. Bak and F. A. Andersen, J. Chem. Phys. **22**, 1050 (1954).
88. ———, D. Christensen, W. B. Dixon, L. Hansen-Nygaard, J. Rastrup-Andersen and M. Schottländer, J. Mol. Spec. **9**, 124 (1962).
89. ———, ———, L. Hansen and J. Rastrup-Andersen, J. Chem. Phys. **24**, 720 (1956).
90. ———, ———, L. Hansen-Nygaard and J. Rastrup-Andersen, J. Mol. Spec. **7**, 58 (1961).
91. ———, L. Hansen and J. Rastrup-Andersen, J. Chem. Phys. **22**, 2013 (1954).
92. C. J. Ballhausen and A. D. Liehr, Acta Chem. Scand. **15**, 775 (1961).
93. J. L. Bancroft, J. M. Hollas and D. A. Ramsay, Can. J. Phys. **40**, 322 (1962).
94. D. Barbier and D. Chalonge, Ann. de Phys. (11) **17**, 272 (1942).
95. R. L. Barger and H. P. Broida, J. Chem. Phys. **37**, 1152 (1962).
96. E. E. Barnes and W. T. Simpson, J. Chem. Phys. **39**, 670 (1963).
97. G. B. Barton, Spectrochim. Acta **19**, 1619 (1963).
98. A. M. Bass and H. Sponer, J. Opt. Soc. Amer. **40**, 389 (1950).
99. D. R. Bates, Proc. Roy. Soc. **257**A, 22 (1960).
100. S. H. Bauer, J. Chem. Phys. **18**, 27 (1950).
101. R. W. Bayer and W. F. Edgell, J. Chem. Phys. **37**, 2502 (1962).
102. N. S. Bayliss, Nature **136**, 264 (1935).
103. ———, J. Chem. Phys. **16**, 287 (1948).
104. G. M. Begun and L. Landau, J. Chem. Phys. **35**, 547 (1961), **36**, 1083 (1962).
105. R. P. Bell, Trans. Far. Soc. **55**, 1 (1959).
105a. S. Bell, J. Mol. Spec. **16**, 205 (1965).
106. W. S. Benedict, Phys. Rev. **47**, 641 (1935).
106a. ———, N. Gailar and E. K. Plyler, J. Chem. Phys. **24**, 1139 (1956).
107. ——— and E. K. Plyler, J. Res. Natl. Bur. Std. **46**, 246 (1951).
108. ——— and ———, Can. J. Phys. **35**, 1235 (1957).
109. ———, ——— and E. D. Tidwell, J. Chem. Phys. **32**, 32 (1960).
110. J. M. Bennett, I. G. Ross and E. J. Wells, J. Mol. Spec. **4**, 342 (1960).
111. V. I. Berezin, Opt. and Spec. **15**, 167 (1963).
112. J. Berkowitz, J. Chem. Phys. **36**, 2533 (1963).

113. H. J. Bernstein, Can. J. Res. **28**B, 132 (1950).
114. ———, J. Chem. Phys. **18**, 478 (1950).
115. ——— and D. A. Ramsay, J. Chem. Phys. **17**, 556 (1949).
115a. R. S. Berry, G. N. Spokes and M. Stiles, J. Amer. Chem. Soc. **84**, 3570 (1962).
116. H. Bethe, Ann. Physik **3**, 133 (1929).
117. W. J. G. Beynon and E. J. Evans, Phil. Mag. **25**, 476 (1938).
118. J. Bigeleisen, M. G. Mayer, P. C. Stevenson and J. Turkevich, J. Chem. Phys. **16**, 442 (1948).
118a. J. Bilham and J. W. Linnett, Nature **201**, 1323 (1964).
119. G. R. Bird, J. Chem. Phys. **25**, 1040 (1956).
120. ———, J. C. Baird, A. W. Jache, J. A. Hodgeson, R. F. Curl, Jr., A. C. Kunkle. J. W. Bransford, J. Rastrup-Andersen and J. Rosenthal, J. Chem. Phys. **40**, 3378 (1964).
120a. F. W. Birss and D. A. Ramsay (to be published).
121. F. E. Blacet, W. G. Young and J. G. Roof, J. Amer. Chem. Soc. **59**, 608 (1937).
122. F. A. Blankenship and R. L. Belford, J. Chem. Phys. **36**, 633 (1962).
123. J. W. Blaker, M. Sidran and A. Kaercher, J. Chem. Phys. **37**, 684 (1962); J. Mol. Spec. **11**, 79 (1963).
123a. H. H. Blau, Jr. and H. H. Nielsen, J. Mol. Spec. **1**, 124 (1957).
124. U. Blukis, P. H. Kasai and R. J. Myers, J. Chem. Phys. **38**, 2753 (1963).
125. ——— and R. J. Myers, J. Phys. Chem. **69**, 1154 (1965).
126. E. Blum and G. Herzberg, J. Phys. Chem. **41**, 91 (1937).
127. G. B. Bonino and R. Manzoni-Ansidei, Z. physik. Chem. B**25**, 327 (1934).
128. M. Bonnemay and E. T. Verdier, J. chim. phys. **41**, 113 (1944).
129. K. F. Bonnhoeffer and L. Farkas, Z. physik. Chem. A**134**, 337 (1928).
130. M. Born, Erg. d. exakt. Naturw. **10**, 387 (1931).
131. ——— and R. Oppenheimer, Ann. Physik **84**, 457 (1927).
132. D. R. J. Boyd and H. C. Longuet-Higgins, Proc. Roy. Soc. **213**A, 55 (1952).
133. ——— and H. W. Thompson, Trans. Far. Soc. **48**, 493 (1952).
134. ———, ——— and R. L. Williams, Disc. Far. Soc. **9**, 154 (1950).
135. R. Bralsford, P. V. Harris and W. C. Price, Proc. Roy. Soc. **258**A, 459 (1960).
136. J. C. D. Brand, Trans. Far. Soc. **46**, 805 (1950).
137. ———, Trans. Far. Soc. **50**, 431 (1954).
138. ———, J. Chem. Soc., p. 858 (1956).
139. ———, J. H. Callomon, D. H. Moule, J. Tyrrell and T. H. Goodwin, Trans. Far. Soc. **61**, 2365 (1965).
140. ———, ——— and J. K. G. Watson, Can. J. Phys. **39**, 1508 (1961).
141. ———, ——— and ———, Disc. Far. Soc. **35**, 175 (1963).
142. ——— and G. J. Minkoff, J. Chem. Soc., p. 2970 (1954).
143. ——— and R. I. Reed, J. Chem. Soc., p. 2386 (1957).
144. ——— and J. K. G. Watson, Trans. Far. Soc. **56**, 1582 (1960).
145. ——— and D. G. Williamson, Disc. Far. Soc. **35**, 184 (1963).
146. G. J. Brealey, J. Chem. Phys. **24**, 571 (1956).
147. G. Breit and E. P. Wigner, Phys. Rev. **49**, 519 (1936).
148. L. Brewer and J. L. Engelke, J. Chem. Phys. **36**, 992 (1962).
149. J. S. Brinen and L. Goodman, J. Chem. Phys. **35**, 1219 (1961).
150. ———, R. C. Hirt and R. G. Schmitt, Spectrochim. Acta **18**, 863 (1962).
151. R. K. Brinton and D. H. Volman, J. Chem. Phys. **19**, 1394 (1951).
152. B. Brocklehurst, Nature **182**, 1366 (1958).
153. P. H. Brodersen, P. Frisch and H. J. Schumacher, Z. physik. Chem. B**37**, 25 (1937).
154. S. Brodersen and A. Langseth, Mat. Fys. Skr. Dan. Vid. Selsk. **1**, 1 (1956).
155. ——— and ———, J. Mol. Spec. **3**, 114, 450 (1959).
156. ——— and E. H. Richardson, J. Mol. Spec. **4**, 439 (1960).
157. H. P. Broida, H. I. Schiff and T. M. Sugden, Trans. Far. Soc. **57**, 259 (1961).
158. D. A. Brown and H. C. Longuet-Higgins, J. Inorg. and Nuclear Chem. **1**, 352 (1955).
159. R. G. Brown and T. H. Edwards, J. Chem. Phys. **37**, 1029, 1035 (1962).
159a. A. D. Buckingham and D. A. Ramsay, J. Chem. Phys. **42**, 3721 (1965).
160. F. Bueso-Sanllehi, Phys. Rev. **60**, 556 (1941).

161. S. M. Bunch, G. R. Cook, M. Ogawa and A. W. Ehler, J. Chem. Phys. **28**, 740 (1958).
161a. P. R. Bunker, J. Chem. Phys. **42**, 2991 (1965).
162. ——— and H. C. Longuet-Higgins, Proc. Roy. Soc. **280**A, 340 (1964).
163. T. G. Burke, D. F. Smith and A. H. Nielsen, J. Chem. Phys. **20**, 447 (1952).
164. D. G. Burkhard and J. C. Irvin, J. Chem. Phys. **23**, 1405 (1955).
165. L. Burnelle, J. Chem. Phys. **24**, 620 (1956).
166. ———, J. Chem. Phys. **35**, 311 (1961).
167. C. A. Burrus, J. Chem. Phys. **30**, 976 (1959).
168. ———, A. Jache and W. Gordy, Phys. Rev. **95**, 706 (1954).
169. P. E. B. Butler, D. R. Eaton and H. W. Thompson, Spectrochim. Acta **13**, 223 (1958).

170. S. Califano, G. Adembri and G. Sbrana, Spectrochim. Acta **20**, 385 (1964).
171. J. H. Callomon, Can. J. Phys. **34**, 1046 (1956).
172. ———, Proc. Roy. Soc. **244**A, 220 (1958).
173. ———, Proc. Chem. Soc., p. 313 (1959).
173a. ———, private discussion.
174. ——— and F. Creutzberg (to be published).
175. ——— and A. B. Davey, Proc. Phys. Soc. **82**, 335 (1963).
175a. ——— and ——— (to be published).
176. ———, T. M. Dunn and I. M. Mills, Phil. Trans. Roy. Soc. **259**A, 499 (1966).
177. ——— and A. C. Gilby, J. Chem. Soc., p. 1471 (1963).
178. ——— and K. K. Innes, J. Mol. Spec. **10**, 166 (1963).
179. ——— and B. P. Stoicheff, Can. J. Phys. **35**, 373 (1957).
179a. ——— and H. W. Thompson, Proc. Roy. Soc. **222**A, 431 (1954).
180. D. M. Cameron, W. C. Sears and H. H. Nielsen, J. Chem. Phys. **7**, 994 (1939).
181. E. P. Carr and H. Stücklen, J. Amer. Chem. Soc. **59**, 2138 (1937).
182. ——— and ———, J. Chem. Phys. **6**, 55 (1938).
182aa. P. K. Carroll (unpublished).
182a. A. S. Carson, W. Carter and J. B. Pedley, Proc. Roy. Soc. **260**A, 550 (1961).
182b. G. Cartwright, A. D. Walsh and P. A. Warsop, 8th European Congr. on Mol. Spec. Abstract 334, Copenhagen, 1965.
183. J. K. Cashion and J. C. Polanyi, J. Chem. Phys. **30**, 317 (1959).
184. J. W. Cederberg, C. H. Anderson and N. F. Ramsey, Phys. Rev. **136**, A960 (1964).
185. S. I. Chan and D. Stelman, J. Chem. Phys. **39**, 545 (1963).
186. ———, J. Zinn, J. Fernandez and W. D. Gwinn, J. Chem. Phys. **33**, 1643 (1960).
187. ———, ——— and W. D. Gwinn, J. Chem. Phys. **34**, 1319 (1961).
188. J. Chappuis, C.R. Acad. Sci. (Paris) **91**, 985 (1880).
189. G. H. Cheesman and H. J. Emeléus, J. Chem. Soc., p. 2837 (1932).
190. M. S. Child, Mol. Phys. **3**, 601 (1960).
191. ———, Mol. Phys. **5**, 391 (1962).
192. ———, Phil. Trans. Roy. Soc. **255**A, 31 (1962).
193. ———, J. Mol. Spec. **10**, 357 (1963).
194. ——— and H. C. Longuet-Higgins, Phil. Trans. Roy. Soc. **254**A, 259 (1961).
195. T. C. Chow, Phys. Rev. **44**, 638 (1933).
196. ——— and H. D. Smyth, Phys. Rev. **38**, 838 (1931).
197. M. Chowdhury and L. Goodman, J. Chem. Phys. **38**, 2979 (1963).
198. M. T. Christensen, D. R. Eaton, B. A. Green and H. W. Thompson, Proc. Roy. Soc. **238**A, 15 (1956).
199. ——— and H. W. Thompson, Trans. Far. Soc. **52**, 1439 (1956).
200. N. A. Chumaevskii, V. M. Tatevskii and Iu. K. Iurev, Opt. and Spec. **6**, 25 (1959).
201. H. H. Claassen, H. Selig and J. G. Malm, J. Chem. Phys. **36**, 2888 (1962).
202. ———, B. Weinstock and J. G. Malm, J. Chem. Phys. **25**, 426 (1956).
203. R. J. H. Clark and D. J. Machin, J. Chem. Soc., p. 4430 (1963).
204. M. J. Y. Clement and D. A. Ramsay, Can. J. Phys. **39**, 205 (1961).
205. E. Clementi, J. Chem. Phys. **36**, 750 (1962).
206. ———, J. Chem. Phys. **38**, 2248, **39**, 487 (1963).
207. E. Clementi and A. D. McLean, J. Chem. Phys. **36**, 45 (1962).

208. J. H. Clements, Phys. Rev. **47**, 224 (1935).
209. F. F. Cleveland, M. J. Murray, J. R. Coley and V. I. Komarewsky, J. Chem. Phys. **10**, 18 (1942).
210. W. L. Clinton and B. Rice, J. Chem. Phys. **30**, 542 (1959).
211. K. Clusius and A. E. Douglas, Can. J. Phys. **32**, 319 (1954).
212. M. A. A. Clyne and B. A. Thrush, Trans. Far. Soc. **57**, 69 (1961).
213. ——— and ———, Disc. Far. Soc. **33**, 139 (1962).
214. ——— and ———, Proc. Roy. Soc. **269**A, 404 (1962).
215. ——— and ———, Proc. Roy. Soc. **275**A, 559 (1963).
216. F. H. Coates and R. C. Anderson, J. Amer. Chem. Soc. **77**, 895 (1955).
217. A. D. Cohen and C. Reid, J. Chem. Phys. **24**, 85 (1956).
217a. E. R. Cohen and J. W. M. DuMond, Encyclop. Phys. **35**, 1 (1957).
217b. ——— and ———, Rev. Mod. Phys. **37**, 537 (1965).
218. C. B. Colburn and F. A. Johnson, J. Chem. Phys. **33**, 1869 (1960).
219. A. R. H. Cole and G. A. Osborne, J. Chem. Soc., p. 1532 (1964).
220. ——— and H. W. Thompson, Proc. Roy. Soc. **200**A, 10 (1949).
221. J. E. Collin, J. Chem. Phys. **30**, 1621 (1959).
222. ——— and F. P. Lossing, J. Chem. Phys. **28**, 900 (1958).
223. ———, Bull. Soc. Roy. Liège **32**, 133 (1963).
224. G. R. Cook and B. K. Ching, unpublished report.
225. A. S. Coolidge, J. Amer. Chem. Soc. **50**, 2166 (1928).
226. J. B. Coon, F. A. Cesani and C. M. Loyd, Disc. Far. Soc. **35**, 118 (1963).
227. ———, R. E. DeWames and C. M. Loyd, J. Mol. Spec. **8**, 285 (1962).
228. ——— and E. Ortiz, J. Mol. Spec. **1**, 81 (1957).
229. C. D. Cooper, J. Chem. Phys. **22**, 503 (1954).
230. H. Cordes, Z. Physik **105**, 251 (1937).
231. L. Corrsin, B. J. Fax and R. C. Lord, J. Chem. Phys. **21**, 1170 (1953).
232. C. C. Costain, Phys. Rev. **82**, 108 (1951).
233. ———, J. Chem. Phys. **29**, 864 (1958).
234. ——— and J. R. Morton, J. Chem. Phys. **31**, 389 (1959).
235. ——— and G. P. Srivastava, J. Chem. Phys. **35**, 1903 (1961), **41**, 1620 (1964).
236. ——— and B. P. Stoicheff, J. Chem. Phys. **30**, 777 (1959).
237. C. A. Coulson, Proc. Roy. Soc. **169**A, 413 (1939).
238. ———, Proc. Roy. Soc. **207**A, 91 (1951).
239. ——— and O. Danielsson, Ark. Fys. **8**, 239, 245 (1954).
240. ——— and W. E. Moffitt, Phil. Mag. **40**, 1 (1949).
241. ——— and A. H. Neilson, Disc. Far. Soc. **35**, 71 (1963).
242. ——— and J. G. Stamper, Mol. Phys. **6**, 609 (1963).
243. ——— and H. L. Strauss, Proc. Roy. Soc. **269**A, 443 (1962).
244. ——— and K. Zalewski, Proc. Roy. Soc. **268**A, 437 (1962).
245. C. P. Courtoy, Can. J. Phys. **35**, 608 (1957).
246. ———, Ann. Soc. Sci. Brux. **73**, 5 (1959).
247. N. A. Coward and W. A. Noyes, Jr., J. Chem. Phys. **22**, 1207 (1954).
248. A. P. Cox and A. S. Esbitt, J. Chem. Phys. **38**, 1636 (1963).
249. ———, L. F. Thomas and J. Sheridan, Nature **181**, 1000 (1958).
250. J. T. Cox, P. B. Peyton and W. Gordy, Phys. Rev. **91**, 222 (1953).
251. D. P. Craig, J. Chem. Phys. **18**, 236 (1950).
251a. ——— and R. D. Gordon, Proc. Roy. Soc. **288**A, 69 (1965).
252. ———, J. M. Hollas and G. W. King, J. Chem. Phys. **29**, 974 (1958).
253. ———, ———, M. F. Redies and S. C. Wait, Jr., Phil. Trans. Roy. Soc. **253**A, 543 (1961).
254. ——— and C. Zauli, J. Chem. Phys. **37**, 601, 609 (1962).
255. G. D. Craine and H. W. Thompson, Trans. Far. Soc. **49**, 1273 (1953).
256. H. G. Crone and R. G. W. Norrish, Nature **132**, 241 (1933).
257. P. C. Cross, R. M. Hainer and G. W. King, J. Chem. Phys. **12**, 210 (1944).
257a. D. W. J. Cruickshank, B. C. Webster and D. F. Mayers, J. Chem. Phys. **40**, 3733 (1964).

258. G. L. Cunningham, A. W. Boyd, R. J. Myers, W. D. Gwinn and W. I. Le Van, J. Chem. Phys. **19**, 676 (1951).
259. R. F. Curl, Jr., J. Chem. Phys. **37**, 779 (1962).
260. ——— and J. L. Kinsey, J. Chem. Phys. **35**, 1758 (1961).
261. C. L. Currie and D. A. Ramsay (to be published).
262. J. Curry and G. Herzberg, Nature **131**, 842 (1933).
263. J. A. Cutler, J. Chem. Phys. **16**, 136 (1948).

264. F. W. Dalby, Can. J. Phys. **36**, 1336 (1958).
265. ——— and H. H. Nielsen, J. Chem. Phys. **25**, 934 (1955).
266. G. H. Damon and F. Daniels, J. Amer. Chem. Soc. **55**, 2363 (1933).
267. A. Danti, W. J. Lafferty and R. C. Lord, J. Chem. Phys. **33**, 294 (1960).
268. ——— and R. C. Lord, Spectrochim. Acta **13**, 180 (1958).
269. H. L. Davis and J. E. Beam, J. Mol. Spec. **6**, 312 (1961).
269a. L. E. Dayton, F. W. Dalby and R. G. Bennett, J. Chem. Phys. **33**, 179 (1960).
270. J. C. Decius and E. B. Wilson, Jr., J. Chem. Phys. **19**, 1409 (1951).
271. G. Déjardin and A. Arnulf, C.R. Acad. Sci. (Paris) **205**, 1000 (1937).
272. A. H. Delsemme and J. Duchesne, C.R. Acad. Sci. (Paris) **234**, 612 (1952).
272a. D. M. Dennison, Rev. Mod Phys. **12**, 175 (1940).
273. V. H. Dibeler and R. M. Reese, J. Chem. Phys. **40**, 2034 (1964).
274. ———, ——— and D. E. Mann, J. Chem. Phys. **27**, 176 (1957).
275. ——— and H. M. Rosenstock, J. Chem. Phys. **39**, 1326 (1963).
276. ——— and ———, J. Chem. Phys. **39**, 3106 (1963).
277. G. H. Dieke and G. B. Kistiakowsky, Phys. Rev. **45**, 4 (1934).
278. V. E. Di Giorgio and G. W. Robinson, J. Chem. Phys. **31**, 1678 (1959).
279. R. W. Ditchburn, Proc. Roy. Soc. **229**A, 44 (1955).
280. J. K. Dixon, Phys. Rev. **43**, 711 (1933).
281. R. N. Dixon, Phil. Trans. Roy. Soc. **252**A, 165 (1960).
282. ———, Can. J. Phys. **38**, 10 (1960).
283. ———, Proc. Roy. Soc. **275**A, 431 (1963).
284. ———, Disc. Far. Soc. **35**, 105 (1963).
285. ———, Trans. Far. Soc. **60**, 1363 (1964).
286. ———, Mol. Phys. **8**, 201 (1964).
286a. ———, Mol. Phys. **9**, 357 (1965).
286b. ———, G. Duxbury and D. A. Ramsay, Proc. Roy. Soc., in press.
287. ——— and H. W. Kroto, Proc. Roy. Soc. **283**A, 423 (1965).
288. ——— and B. F. Mason, Nature **197**, 1198 (1963).
288a. ——— and D. A. Ramsay (to be published).
289. R. E. Dodd, Trans. Far. Soc. **55**, 1480 (1959).
289a. J. W. Donovan and A. B. F. Duncan, J. Chem. Phys. **35**, 1389 (1961).
290. L. D'Or and P. Tarte, Bull. Soc. Roy. Sci. Liège **20**, 478 (1951).
291. ——— and ———, Bull. Soc. Roy. Sci. Liège **20**, 685 (1951).
292. A. E. Douglas, Astrophys. J. **114**, 466 (1951).
293. ———, Can. J. Phys. **36**, 147 (1958).
294. ———, Disc. Far. Soc. **35**, 158 (1963).
294a. ———, (to be published).
294b. ———, private communication.
295. ——— and J. M. Hollas, Can. J. Phys. **39**, 479 (1961).
296. ——— and K. P. Huber, Can. J. Phys. **43**, 74 (1965).
297. ——— and W. J. Jones, Can. J. Phys. **43**, 2216 (1965).
298. ——— and E. R. V. Milton, J. Chem. Phys. **41**, 357 (1964).
299. ——— and ——— (to be published).
300. ——— and C. K. Møller, J. Chem. Phys. **22**, 275 (1954).
301. ——— and D. Sharma, J. Chem. Phys. **21**, 448 (1953).
302. ——— and I. Zanon, Can. J. Phys. **42**, 627 (1964).
303. ——— and ——— (to be published).
304. J. M. Dowling, J. Mol. Spec. **6**, 550 (1961).
305. ——— and B. P. Stoicheff, Can. J. Phys. **37**, 703 (1959).

306. A. J. Downs, Spectrochim. Acta **19**, 1165 (1963).
307. K. Dressler, J. Chem. Phys. **35**, 165 (1961).
308. ——— and D. A. Ramsay, Phil. Trans. Roy. Soc. **251**A, 553 (1959).
309. ——— and O. Schnepp, J. Chem. Phys. **33**, 270 (1960).
310. I. Dubois, Bull. Soc. Roy. Sci. Liège **32**, 777 (1963).
310a. ——— and A. Grötsch, Bull. Soc. Roy. Sci. Liège **33**, 833 (1964).
311. ——— and B. Rosen, Disc. Far. Soc. **35**, 124 (1963).
312. J. T. Dubois and F. Wilkinson, J. Chem. Phys. **39**, 899 (1963).
313. J. Duchesne, Mém. Acad. Roy. Belg. **28**, No. 8 (1955).
314. ——— and B. Rosen, Physica **8**, 540 (1941).
315. ——— and ———, J. Chem. Phys. **15**, 631 (1947).
316. A. B. F. Duncan, J. Chem. Phys. **3**, 131 (1935).
317. ———, Phys. Rev. **47**, 822 (1935).
318. ———, Phys. Rev. **47**, 886 (1935).
319. ———, Phys. Rev. **50**, 700 (1936).
320. ———, J. Chem. Phys. **4**, 638 (1936).
321. ———, J. Chem. Phys. **27**, 423 (1957).
322. ——— and G. R. Harrison, Phys. Rev. **49**, 211 (1936).
323. ——— and J. P. Howe, J. Chem. Phys. **2**, 851 (1934).
324. W. E. Duncanson and C. A. Coulson, Proc. Roy. Soc. Edinburgh **62**, 37 (1944).
325. T. M. Dunn, Pure and App. Chem. **6**, 1 (1963).
326. ——— and G. Herzberg (to be published).
327. ——— and C. K. Ingold, Nature **176**, 65 (1955).
328. ——— and T. Iredale, J. Chem. Soc., p. 1592 (1952).
329. R. A. Durie, T. Iredale and J. M. S. Jarvie, J. Chem. Soc., p. 1181 (1950).
330. J. R. Durig and R. C. Lord, Spectrochim. Acta **19**, 421 (1963).
331. A. K. Dutta, Proc. Roy. Soc. **138**A, 84 (1932).
332. P. J. Dyne, J. Chem. Phys. **20**, 811 (1952).
333. ———, Can. J. Phys. **30**, 79 (1952).
334. ——— and D. W. G. Style, Disc. Far. Soc. **2**, 159 (1947).
335. ——— and ———, J. Chem. Soc., p. 2122 (1952).

336. E. Eastwood and C. P. Snow, Proc. Roy. Soc. **149**A, 434 (1935).
337. ——— and ———, Proc. Roy. Soc. **149**A, 446 (1935).
338. D. R. Eaton, J. W. C. Johns and D. A. Ramsay (to be published).
339. ——— and H. W. Thompson, Proc. Roy. Soc. **250**A, 39 (1959).
340. W. H. Eberhardt and H. Renner, J. Mol. Spec. **6**, 483 (1961).
341. ——— and W. Shand, Jr., J. Chem. Phys. **14**, 525 (1946).
342. E. S. Ebers and H. H. Nielsen, J. Chem. Phys. **5**, 822 (1937).
343. ——— and ———, J. Chem. Phys. **6**, 311 (1938).
343a. B. H. Eckstein, H. A. Scheraga and E. R. Van Artsdalen, J. Chem. Phys. **22**, 28 (1954).
344. D. Edelson and K. B. McAfee, J. Chem. Phys. **19**, 1311 (1951).
345. W. F. Edgell and C. E. May, J. Chem. Phys. **22**, 1808 (1954).
346. J. W. Edwards and P. A. Small, Nature **202**, 1329 (1964).
347. B. J. Eiseman, Jr., and L. Harris, J. Amer. Chem. Soc. **54**, 1782 (1932).
347a. J. C. Eisenstein, J. Chem. Phys. **34**, 310 (1961).
348. ——— and M. H. L. Pryce, Proc. Roy. Soc. **255**A, 181 (1960).
349. M. A. El-Sayed, J. Chem. Phys. **36**, 552 (1962).
350. ———, M. Kasha and Y. Tanaka, J. Chem. Phys. **34**, 334 (1961).
351. ——— and G. W. Robinson, Mol. Phys. **4**, 273 (1961).
352. F. O. Ellison, J. Chem. Phys. **36**, 3112 (1962).
353. ——— and H. Shull, J. Chem. Phys. **23**, 2348 (1955).
354. V. R. Ells, J. Amer. Chem. Soc. **60**, 1864 (1938).
355. H. J. Emeléus, J. Chem. Soc., p. 2948 (1926).
356. ——— and H. V. A. Briscoe, J. Chem. Soc., p. 127 (1937).
357. ——— and L. J. Jolley, J. Chem. Soc., p. 1612 (1935).
358. H. J. Emeléus and K. Stewart, Trans. Far. Soc. **32**, 1577 (1936).

359. G. Erlandsson, Ark. Fys. **10**, 65 (1956).
360. ———, J. Chem. Phys. **28**, 71 (1958).
361. ———, Ark. Fys. **16**, 181 (1959).
362. J. R. Eshbach and M. W. P. Strandberg, Phys. Rev. **85**, 24 (1952).
363. D. F. Evans, Nature **178**, 534 (1956).
364. ———, J. Chem. Soc., p. 1351 (1957).
365. ———, J. Chem. Soc., p. 3885 (1957).
365a. ———, J. Chem. Soc., p. 1735 (1960).
366. J. C. Evans and H. J. Bernstein, Can. J. Chem. **34**, 1083 (1956).
367. A. J. Everett and G. J. Minkoff, Trans. Far. Soc. **44**, 816 (1948).
368. A. A. Evett, J. Chem. Phys. **31**, 565, 1419 (1959).
368a. V. C. Ewing and L. E. Sutton, Trans. Far. Soc. **59**, 1241 (1963).
369. H. Eyring and M. Polanyi, Z. physik. Chem. B**12**, 279 (1931).

370. E. Fajans and C. F. Goodeve, Trans. Far. Soc. **32**, 511 (1936).
371. J. B. Farmer, I. H. S. Henderson, F. P. Lossing and D. G. H. Marsden, J. Chem. Phys. **24**, 348 (1956).
372. P. Favero, A. M. Mirri and J. G. Baker, J. Chem. Phys. **31**, 566 (1954).
373. M. W. Feast, Proc. Phys. Soc. **63**A, 772 (1950).
374. T. P. Fehlner and W. S. Koski, J. Amer. Chem. Soc. **86**, 2733 (1964).
374a. ——— and ———, J. Amer. Chem. Soc. **87**, 409 (1965).
375. T. Feldman, J. Romanko and H. L. Welsh, Can. J. Phys. **33**, 138 (1955).
376. ———, ——— and ———, Can. J. Phys. **34**, 737 (1956).
377. ———, G. G. Shepherd and H. L. Welsh, Can. J. Phys. **34**, 1425 (1956).
378. W. C. Fergusson, L. Slotin and D. W. G. Style, Trans. Far. Soc. **32**, 956 (1936).
379. S. M. Ferigle, F. F. Cleveland and A. G. Meister, J. Chem. Phys. **20**, 1928 (1952).
380. O. V. Fialkovskaja, Acta Physicochim. USSR **9**, 215 (1938).
381. C. T. Fike, J. Chem. Phys. **31**, 568 (1959).
382. P. Fink and C. F. Goodeve, Proc. Roy. Soc. **163**A, 592 (1937).
383. W. Finkelnburg, Z. Phys. **90**, 1 (1934).
384. ——— and H. J. Schumacher, Z. phys. Chem. Bodensteinfestband, p. 704 (1931).
385. ———, ——— and G. Stieger, Z. phys. Chem. B**15**, 127 (1931).
386. I. Fischer-Hjalmars, Ark. Fys. **11**, 529 (1956).
386a. I. P. Fisher and G. A. Heath, Nature **208**, 1199 (1965).
387. G. Fleming, M. M. Anderson, A. J. Harrison and L. W. Pickett, J. Chem. Phys. **30**, 351 (1959).
388. W. H. Flygare and J. A. Howe, J. Chem. Phys. **36**, 440 (1962).
389. S. N. Foner and R. L. Hudson, J. Chem. Phys. **29**, 442 (1958).
390. ——— and ———, J. Chem. Phys. **36**, 2681 (1962).
391. R. Fonteyne, J. Chem. Phys. **8**, 60 (1940).
392. A. Fontijn, C. B. Meyer and H. I. Schiff, J. Chem. Phys. **40**, 64 (1964).
393. T. Förster and J. C. Jungers, Z. phys. Chem. B**36**, 387 (1937).
394. A. Fowler and A. G. Gaydon, Proc. Roy. Soc. **142**A, 362 (1933).
395. K. Fox, J. Mol. Spec. **9**, 381 (1962).
396. P. E. Fraley, W. W. Brim and K. N. Rao, J. Mol. Spec. **9**, 487 (1962).
397. J. Franck, H. Sponer and E. Teller, Z. phys. Chem. B**18**, 88 (1932).
398. J. L. Franklin and F. H. Field, J. Amer. Chem. Soc. **76**, 1994 (1954).
399. D. E. Freeman and W. Klemperer, J. Chem. Phys. **40**, 604 (1964).
400. D. C. Frost, D. Mak and C. A. McDowell, Can. J. Chem. **40**, 1064 (1962).
401. ——— and C. A. McDowell, Can. J. Chem. **36**, 39 (1958).
402. R. L. Fulton and M. Gouterman, J. Chem. Phys. **35**, 1059 (1961).

403. N. M. Gailar and E. K. Plyler, J. Res. Nat. Bur. Std. **48**, 392 (1952).
404. F. M. Garforth and C. K. Ingold, J. Chem. Soc., pp. 417, 427 (1948).
405. ——— and ———, J. Chem. Soc., pp. 433, 440 (1948).
405a. ———, ——— and H. G. Poole, J. Chem. Soc., p. 491 (1948).
406. J. S. Garing, H. H. Nielsen and K. N. Rao, J. Mol. Spec. **3**, 496 (1959).
407. W. R. S. Garton, Proc. Phys. Soc. **66**A, 848 (1953).

408. J. T. Gary and L. W. Pickett, J. Chem. Phys. **22**, 599 (1954).
409. J. Gaunt, Trans. Far. Soc. **49**, 1122 (1953).
410. L. Gausset, G. Herzberg, A. Lagerqvist and B. Rosen, Disc. Far. Soc. **35**, 113 (1963).
411. ———, ———, ——— and ———, Astrophys. J. **142**, 45 (1965).
412. A. G. Gaydon, Proc. Roy. Soc. **146**A, 901 (1934).
413. ———, Proc. Roy. Soc. **176**A, 505 (1940).
414. ———, Trans. Far. Soc. **43**, 36 (1947).
415. ———, G. N. Spokes and J. van Suchtelen, Proc. Roy. Soc. **256**A, 323 (1960).
416. H. Gerding, G. Milazzo and H. H. K. Rossmark, Rec. Trav. Chim. Pays-Bas **72**, 957 (1953).
417. W. F. Giauque and J. D. Kemp, J. Chem. Phys. **6**, 40 (1938).
418. L. E. Giddings, Jr., and K. K. Innes, J. Mol. Spec. **6**, 528 (1961).
419. ——— and ———, J. Mol. Spec. **8**, 328 (1962).
420. P. A. Giguère, J. Chem. Phys. **30**, 322 (1959).
421. F. J. Gilde and M. I. Bán, Acta Phys. Hung. **12**, 13 (1960).
422. R. H. Gillette and A. Sherman, J. Amer. Chem. Soc. **58**, 1135 (1936).
423. O. R. Gilliam, H. D. Edwards and W. Gordy, Phys. Rev. **75**, 1014 (1949).
424. N. Ginsburg, Phys. Rev. **74**, 1052 (1948).
425. R. Glauber and V. Schomaker, Phys. Rev. **89**, 667 (1953).
426. J. Godart, J. chim. phys. **34**, 70 (1937).
427. M. Goeppert-Mayer and A. L. Sklar, J. Chem. Phys. **6**, 645 (1938).
428. S. Golden, J. Chem. Phys. **16**, 78 (1948).
429. D. Golomb, K. Watanabe and F. F. Marmo, J. Chem. Phys. **36**, 958 (1962).
430. C. F. Goodeve and N. O. Stein, Trans. Far. Soc. **27**, 393 (1931).
431. ——— and B. A. M. Windsor, Trans. Far. Soc. **32**, 1518 (1936).
432. ——— and F. D. Richardson, Trans. Far. Soc. **33**, 453 (1937).
433. P. L. Goodfriend and H. P. Woods, J. Chem. Phys. **39**, 2379 (1963).
434. ——— and ———, J. Mol. Spec. **13**, 63 (1964).
435. G. L. Goodman and M. Fred, J. Chem. Phys. **30**, 849 (1959).
436. L. Goodman, J. Mol. Spec. **6**, 109 (1961).
437. ——— and M. Kasha, J. Mol. Spec. **2**, 58 (1958).
438. H. Göpfert, Z. wiss. Phot. **34**, 156 (1935).
438a. E. K. Gora, J. Mol. Spec. **16**, 378 (1965).
439. A. A. Gordus and R. B. Bernstein, J. Chem. Phys. **22**, 790 (1954).
440. K. Goto, Sci. Light **9**, 104 (1960).
441. ———, Sci. Light **11**, 119 (1962).
442. J. Goubeau, E. L. Jahn, A. Kreutzberger and C. Grundmann, J. Phys. Chem. **58**, 1078 (1954).
442a. R. D. Gould and J. W. Linnett, Trans. Far. Soc. **59**, 1001 (1963).
443. S. Gradstein, Z. physik. Chem. B**22**, 384 (1933).
444. A. P. Gray and R. C. Lord, J. Chem. Phys. **26**, 690 (1967).
444a. P. Gray, Trans. Far. Soc. **55**, 408 (1959).
445. J. D. Graybeal and D. W. Roe, Proc. West Virginia Acad. Sci. **33**, 37 (1961).
446. M. Green and J. W. Linnett, Trans. Far. Soc. **57**, 1 (1961).
447. K. F. Greenough and A. B. F. Duncan, J. Amer. Chem. Soc. **83**, 555 (1961).
448. J. S. Griffith and L. E. Orgel, J. Chem. Phys. **26**, 988 (1957).
449. H. J. Groh, G. W. Luckey and W. A. Noyes, Jr., J. Chem. Phys. **21**, 115 (1953).
450. M. S. de Groot and J. H. van der Waals, Mol. Phys. **6**, 545 (1963).
451. E. L. Grubb and R. L. Belford, J. Chem. Phys. **39**, 244 (1963).
452. H. Guenebaut and B. Pascat, C.R. Acad. Sci. (Paris) **259**, 2412 (1964).
452a. A. H. Guenther, T. A. Wiggins and D. H. Rank, J. Chem. Phys. **28**, 682 (1958).
453. C. W. Gullikson and J. R. Nielsen, J. Mol. Spec. **1**, 158 (1957).
454. ———, ——— and A. T. Stair, Jr., J. Mol. Spec. **1**, 151 (1957).
455. S. R. Gunn and L. G. Green, J. Chem. Phys. **36**, 1118 (1962).
456. H. S. Gutowsky, J. Chem. Phys. **17**, 128 (1949).

457. A. Guttman and S. S. Penner, J. Chem. Phys. **36**, 98 (1962).
458. P. C. Haarhoff, Mol. Phys. **7**, 101 (1963).
459. A. Hadni, C.R. Acad. Sci. (Paris) **239**, 349 (1954).
460. G. G. Hall, Rep. Progr. Phys. **22**, 1 (1959).
461. R. T. Hall and G. C. Pimentel, J. Chem. Phys. **38**, 1889 (1963).
462. T. C. Hall, Jr., and F. E. Blacet, J. Chem. Phys. **20**, 1745 (1952).
463. M. Halmann, J. Chem. Soc., p. 2853 (1963).
464. F. Halverson and R. C. Hirt, J. Chem. Phys. **19**, 711 (1951).
465. N. S. Ham, Rev. Pure and Appl. Chem. **11**, 159 (1961).
466. ——— and K. Ruedenberg, J. Chem. Phys. **29**, 1199 (1958).
467. H. F. Hameka, J. Chem. Phys. **36**, 2540, **37**, 2209 (1962).
467a. ——— and L. J. Oosterhoff, Mol. Phys. **1**, 358 (1958).
468. V. J. Hammond and W. C. Price, Trans. Far. Soc. **51**, 605 (1955).
469. ———, ———, J. P. Teegan and A. D. Walsh, Disc. Far. Soc. **9**, 53 (1950).
470. R. F. Hampson, Jr., J. R. McNesby, H. Akimoto and I. Tanaka, J. Chem. Phys. **40**, 1099 (1964).
471. G. E. Hansen and D. M. Dennison, J. Chem. Phys. **20**, 313 (1952).
472. P. B. Haranath and V. Sivaramamurty, Ind. J. Phys. **35**, 599 (1961).
473. M. D. Harmony and R. J. Myers, J. Chem. Phys. **37**, 636 (1962).
474. ———, ———, L. J. Schoen, D. R. Lide, Jr., and D. E. Mann, J. Chem. Phys. **35**, 1129 (1961).
475. L. Harris, Nature **118**, 482 (1926).
476. ——— and Gilbert W. King, J. Chem. Phys. **8**, 775 (1940).
477. ———, ———, W. S. Benedict and R. W. B. Pearse, J. Chem. Phys. **8**, 765 (1940).
478. R. K. Harris, Spectrochim. Acta **20**, 1129 (1964).
479. ——— and R. E. Witkowski, Spectrochim. Acta **20**, 1651 (1964).
480. A. J. Harrison, B. J. Cederholm and M. A. Terwilliger, J. Chem. Phys. **30**, 355 (1959).
481. ——— and D. R. W. Price, J. Chem. Phys. **30**, 357 (1959).
482. H. Hartmann, Z. Naturf. **15a**, 993 (1960).
483. E. Hauptman, Acta Phys. Pol. **7**, 86 (1938).
484. N. J. Hawkins and W. W. Sabol, J. Chem. Phys. **25**, 775 (1956).
485. K. T. Hecht, J. Mol. Spec. **5**, 355, 390 (1960).
486. ——— and D. M. Dennison, J. Chem. Phys. **26**, 31 (1957).
487. J. Heicklen, J. Chem. Phys. **36**, 721 (1962).
488. O. Heil, Z. Physik **77**, 563 (1932).
489. W. Heitler, Phys. Rev. **38**, 243 (1931).
490. ———, in Marx's Handb. d. Radiologie, II, 485 (1934).
491. ——— and G. Rumer, Z. Physik **68**, 12 (1931).
492. K. H. Hellwege, Z. Physik **117**, 596 (1940).
492a. J. L. Hencher and Gerald W. King, J. Mol. Spec. **16**, 158 (1965).
493. R. S. Henderson, Phys. Rev. **100**, 723 (1955).
494. H. J. Henning, Ann. Phys. **13**, 599 (1932).
496. V. Henri, Nature **125**, 202 (1930).
497. ———, Trans. Far. Soc. **25**, 765 (1929).
498. ———, in Leipziger Vorträge, p. 131 (Hirzel, Leipzig, 1931; see English translation in P. Debye, Structure of Molecules, p. 121, Blackie 1932).
499. ——— and P. Angenot, J. chim. phys. **33**, 641 (1936).
500. ——— and J. Duchesne, Nature **143**, 28 (1939).
501. ——— and O. R. Howell, Proc. Roy. Soc. **128**A, 192 (1930).
502. ——— and W. Lasareff, J. chim. phys. **32**, 353 (1935).
503. ——— and S. A. Schou, Z. Physik **49**, 774 (1928).
504. A. Henrici, Z. Physik **77**, 35 (1932).
505. ——— and H. Grieneisen, Z. physik. Chem. B**30**, 1 (1935).
506. ——— and G. Milazzo, Z. physik. Chem. B**33**, 201 (1936).
507. G. H. Herbig, Pub. Astron. Soc. Pac. **68**, 204 (1956); Mém. Soc. Roy. Sci. Liège (4) **18**, 288 (1957).

508. L. Herman, J. Akriche and H. Grenat, J. Quant. Spec. **2**, 215 (1962).
509. ———, H. Grenat and J. Akriche, Nature **194**, 468 (1962).
510. R. C. Herman, J. Chem. Phys. **8**, 252 (1940).
511. G. J. Hernandez, J. Chem. Phys. **38**, 1644 (1963).
512. ———, J. Chem. Phys. **38**, 2233 (1963).
513. J. Herranz, J. Mol. Spec. **6**, 343 (1961).
514. ——— and B. P. Stoicheff, J. Mol. Spec. **10**, 448 (1963).
515. J. T. Herron and H. I. Schiff, J. Chem. Phys. **24**, 1266 (1956).
516. D. R. Herschbach and L. C. Krisher, J. Chem. Phys. **28**, 728 (1958).
517. G. Herzberg, Trans. Far. Soc. **27**, 378 (1931).
518. ———, Z. physik. Chem. B**17**, 68 (1932).
519. ———, Astrophys. J. **96**, 314 (1942).
520. ———, Can. J. Phys. **31**, 657 (1955).
521. ———, Proc. Roy. Soc. **262**A, 291 (1961).
522. ———, Can. J. Phys. **39**, 1511 (1961).
523. ———, Disc. Far. Soc. No. **35**, 7 (1963).
524. ———, (to be published).
525. ———, (unpublished).
526. ——— and K. Franz, Z. Physik **76**, 720 (1932).
527. ——— and K. K. Innes, Can. J. Phys. **35**, 842 (1957).
528. ——— and ——— (to be published).
529. ——— and J. W. C. Johns, Mém. Soc. Roy. Sci. Liège (5) **7**, 117 (1963).
530. ——— and ——— Proc. Roy. Soc., in press.
531. ——— and ——— (to be published).
532. ——— and ——— (to be published).
533. ——— and R. Kölsch, Z. Elektrochem. **39**, 572 (1933).
534. ——— and H. C. Longuet-Higgins, Disc. Far. Soc. **35**, 77 (1963).
535. ——— and A. Monfils, J. Mol. Spec. **5**, 482 (1960).
536. ———, F. Patat and H. Verleger, J. Phys. Chem. **41**, 123 (1937).
537. ——— and D. A. Ramsay, J. Chem. Phys. **20**, 347 (1952).
538. ——— and ———, Proc. Roy. Soc. **233**A, 34 (1955).
539. ——— and G. Scheibe, Z. physik. Chem. B**7**, 390 (1930).
540. ——— and J. Shoosmith, Can. J. Phys. **34**, 523 (1956).
541. ——— and B. P. Stoicheff, Nature **175**, 79 (1955).
542. ——— and E. Teller, Z. physik. Chem. B**21**, 410 (1933).
543. ——— and D. N. Travis, Can. J. Phys. **42**, 1658 (1964).
544. ——— and ———, Nature **204**, 988 (1964).
545. ——— and R. D. Verma, Can. J. Phys. **42**, 395 (1964).
546. ——— and ——— (to be published).
547. ——— and P. A. Warsop, Can. J. Phys. **41**, 286 (1963).
548. K. Hijikata, C. C. Lin and J. C. Baird, J. Chem. Phys. **36**, 1183 (1962).
549. H. J. Hilgendorff, Z. Physik **95**, 781 (1935).
550. R. A. Hill and T. H. Edwards, J. Mol. Spec. **9**, 494 (1962).
551. ——— and ———, J. Mol. Spec. **11**, 433 (1963).
552. ——— and ———, J. Mol. Spec. **14**, 203 (1964).
553. ———, ———, K. Rossmann, K. N. Rao and H. H. Nielsen, J. Mol. Spec. **14**, 221 (1964).
554. J. A. Hipple and E. U. Condon, Phys. Rev. **68**, 54 (1945).
555. J. Hirschfelder, H. Eyring and B. Topley, J. Chem. Phys. **4**, 170 (1936).
556. R. C. Hirt, Spectrochim. Acta **12**, 114 (1958).
557. ———, F. Halverson and R. G. Schmitt, J. Chem. Phys. **22**, 1148 (1954).
558. W. E. Hobbs, J. Chem. Phys. **28**, 1220 (1958).
559. S. E. Hodges, J. R. Henderson and J. B. Coon, J. Mol. Spec. **2**, 99 (1958).
559a. L. Hofacker, in D. Hadzi, Symposium on Hydrogen Bonding, p. 375 (Pergamon, London, 1959).
560. J. M. Hoffman, H. H. Nielsen and K. N. Rao, Z. Elektrochem. **64**, 606 (1960).
561. R. Hoffmann and M. Gouterman, J. Chem. Phys. **36**, 2189 (1962).
562. G. J. Hoijtink, Mol. Phys. **3**, 67 (1960).

563. T. M. Holladay and A. H. Nielsen, J. Mol. Spec. **14**, 371 (1964).
564. R. Holland, D. W. G. Style, R. N. Dixon and D. A. Ramsay, Nature **182**, 336 (1958).
565. J. M. Hollas, Spectrochim. Acta **19**, 1425 (1963).
566. M. Horani and S. Leach, C.R. Acad. Sci. (Paris) **248**, 2196 (1959).
566a. ———, ———, J. Rostas and G. Berthier, J. chim. phys. (in press).
567. G. A. Hornbeck and R. C. Herman, Ind. Eng. Chem. **43**, 2739 (1951).
568. J. T. Hougen, J. Chem. Phys. **36**, 519 (1962).
569. ———, J. Chem. Phys. **36**, 1874 (1962).
570. ———, J. Chem. Phys. **37**, 403 (1962).
571. ———, J. Chem. Phys. **37**, 1433 (1962).
572. ———, J. Chem. Phys. **38**, 1167 (1963).
573. ———, J. Chem. Phys. **39**, 358 (1963).
574. ———, Can. J. Phys. **42**, 433 (1964).
575. ———, Can. J. Phys. **42**, 1920 (1964).
576. ———, J. Chem. Phys. **41**, 363 (1964).
577. ———, J. Mol. Spec. **13**, 149 (1964).
577a. ———, Can. J. Phys. **43**, 935 (1965).
578. ———, G. E. Leroi and T. C. James, J. Chem. Phys. **34**, 1670 (1961).
579. ——— and J. P. Jesson, J. Chem. Phys. **38**, 1524 (1963).
580. ——— and J. K. G. Watson, Can. J. Phys. **43**, 298 (1965).
581. J. A. Howe and J. H. Goldstein, J. Amer. Chem. Soc. **80**, 4846 (1958).
582. H. J. Hrostowski and R. J. Myers, J. Chem. Phys. **22**, 262 (1954).
583. E. Hückel, Z. Physik **60**, 423 (1930).
584. ———, Z. Physik **70**, 204 (1931).
586. W. Huggins, Proc. Roy. Soc. **33**, 1 (1882).
587. ———, Proc. Roy. Soc. **48**, 216 (1890).
588. E. C. Hughes and J. R. Johnson, J. Amer. Chem. Soc. **53**, 737 (1931).
589. R. H. Hughes, J. Chem. Phys. **24**, 131 (1956).
590. Y. Hukumoto, Sci. Rep. Tohoku Imp. Univ. **25**, 1162 (1936).
591. G. L. Humphrey and R. M. Badger, J. Chem. Phys. **15**, 794 (1947).
592. C. M. Humphries, A. D. Walsh and P. A. Warsop, Disc. Far. Soc. **35**, 137 (1963).
593. ———, ——— and ———, Disc. Far. Soc. **35**, 148 (1963).
594. H. D. Hunt and W. T. Simpson, J. Amer. Chem. Soc. **75**, 4540 (1953).
595. A. C. Hurley, Proc. Roy. Soc. **235**A, 224 (1956).
596. D. Husain and R. G. W. Norrish, Proc. Roy. Soc. **273**A, 165 (1963).
597. S. L. Hussain and R. Samuel, Proc. Phys. Soc. **49**, 679 (1937).
598. D. A. Hutchinson, Trans. Far. Soc. **58**, 1669 (1962).

599. S. Imanishi, Nature **127**, 782 (1931); Sci. Pap. Inst. Phys. Chem. Res. **15**, 166 (1931).
600. C. K. Ingold and Gerald W. King, J. Chem. Soc., p. 2702 (1953).
601. ——— and ———, J. Chem. Soc., p. 2725 (1953).
602. ——— and C. L. Wilson, J. Chem. Soc., p. 941, 1210 (1936).
603. E. C. Y. Inn, Phys. Rev. **91**, 1194 (1953).
604. ———, K. Watanabe and M. Zelikoff, J. Chem. Phys. **21**, 1648 (1953).
605. K. K. Innes, J. Chem. Phys. **22**, 863 (1954).
606. ——— and L. E. Giddings, Jr., J. Mol. Spec. **7**, 435 (1961).
607. ——— and ———, Disc. Far. Soc. **35**, 192 (1963).
608. ———, J. A. Merritt, W. C. Tincher and S. G. Tilford, Nature **187**, 500 (1960).
609. ——— and J. E. Parkin (to be published).
610. ———, J. D. Simmons and S. G. Tilford, J. Mol. Spec. **11**, 257 (1963).
611. M. Ito, P. C. Huang and E. M. Kosower, Trans. Far. Soc. **57**, 1662 (1961).
612. ——— and W. Mizushima, J. Chem. Phys. **24**, 495 (1956).
613. ———, R. Shimada, T. Kuraishi and W. Mizushima, J. Chem. Phys. **26**, 1508 (1957).
614. K. Inuzuka, Bull. Chem. Soc. Japan **33**, 678 (1960), **34**, 729 (1961).

614a. A. W. Jache, P. W. Moser and W. Gordy, J. Chem. Phys. **25**, 209 (1956).
615. J. K. Jacques and R. F. Barrow, Proc. Phys. Soc. **73**, 538 (1959).
616. H. A. Jahn, Proc. Roy. Soc. **164**A, 117 (1938).
617. ———, Proc. Roy. Soc. **168**A, 469, 495 (1938).
618. ——— and E. Teller, Proc. Roy. Soc. **161**A, 220 (1937).
619. A. J. Jakovleva and V. Kondratiev, Phys. Z. d. Sow. Union **1**, 47 (1932).
620. ——— and ———, Phys. Z. d. Sow. Union **9**, 106 (1936).
621. C. G. James and T. M. Sugden, Proc. Roy. Soc. **248**A, 238 (1958).
622. M. Jan-Khan and R. Samuel, Proc. Phys. Soc. **48**, 626 (1936).
623. N. Jannuzzi and S. P. S. Porto, J. Mol. Spec. **4**, 459 (1960).
624. T. S. Jaseja, Proc. Ind. Acad. Sci. A**50**, 108 (1959).
624a. M. G. Jayswal and R. S. Singh, Proc. Symp. Spec. Banaras, 1963, p. 134.
625. C. K. Jen, Phys. Rev. **81**, 197 (1951).
626. F. A. Jenkins, Astrophys. J. **70**, 191 (1929).
627. K. R. Jennings and J. W. Linnett, Trans. Far. Soc. **56**, 1737 (1960).
628. M. Jeunehomme and A. B. F. Duncan, J. Chem. Phys. **41**, 1692 (1964).
628a. V. A. Job and Gerald W. King, Can. J. Chem. **41**, 3132 (1963).
628b. ——— and ———, J. Mol. Spec. **19**, 155, 178 (1966).
629. A. Johannin-Gilles, J. Rech. C.N.R.S. No. 31, 205 (1955).
630. J. W. C. Johns, Can. J. Phys. **39**, 1738 (1961).
631. ———, Can. J. Phys. **41**, 209 (1963).
632. ———, Can. J. Phys. **42**, 1004 (1964).
633. ———, J. Mol. Spec. **15**, 473 (1965).
634. ———, (to be published).
635. ———, private communication.
636. ——— and R. F. Barrow, Proc. Roy. Soc. **251**A, 504 (1959).
637. ———, G. W. Chantry and R. F. Barrow, Trans. Far. Soc. **54**, 1589 (1958).
638. ———, S. H. Priddle and D. A. Ramsay, Disc. Far. Soc. **35**, 90 (1963).
638a. ——— and D. A. Ramsay (to be published).
639. ——— and J. K. Tyler (to be published).
640. C. M. Johnson, R. Trombarulo and W. Gordy, Phys. Rev. **84**, 1178 (1951).
641. G. D. Johnson, D. A. Ramsay and I. G. Ross (to be published).
642. H. R. Johnson and M. W. P. Strandberg, J. Chem. Phys. **20**, 687 (1952).
643. H. S. Johnston and H. J. Bertin, J. Mol. Spec. **3**, 683 (1959).
644. A. V. Jones, J. Chem. Phys. **18**, 1263 (1950).
645. ———, Proc. Roy. Soc. **211**A, 285 (1952).
646. E. J. Jones and O. R. Wulf, J. Chem. Phys. **5**, 873 (1937).
646a. E. W. Jones and H. W. Thompson, Proc. Roy. Soc. **288**A, 50 (1965).
647. L. H. Jones, R. M. Badger and G. E. Moore, J. Chem. Phys. **19**, 1599 (1951).
648. P. C. Jordan, J. Chem. Phys. **41**, 1442 (1964).
649. P. C. H. Jordan and H. C. Longuet-Higgins, Mol. Phys. **5**, 121 (1962).
650. C. K. Jørgenson, Acta Chem. Scand. **9**, 116 (1955).
651. ———, Sol. State Phys. **13**, 375 (1962).
651a. ———, Mol. Phys. **7**, 417 (1964).
652. J. Jortner, E. G. Wilson and S. A. Rice, J. Amer. Chem. Soc. **85**, 815 (1963).
653. P. Joyner and G. Glockler, J. Chem. Phys. **20**, 302 (1952).
654. B. R. Judd, Proc. Roy. Soc. **241**A, 122 (1957).

655. J. Kahane-Paillous, J. chim. phys. **55**, 815 (1958).
655a. ——— and S. Leach, in Advances of Molecular Spectroscopy, p. 398 (Pergamon, London, 1962).
656. H. Kaplan, J. Chem. Phys. **26**, 1704 (1957).
657. L. D. Kaplan, M. V. Migeotte and L. Neven, J. Chem. Phys. **24**, 1183 (1956).
658. M. Karplus, J. Chem. Phys. **30**, 15 (1959).
659. M. Kasha, Disc. Far. Soc. **9**, 14 (1950).
660. W. E. Kaskan and A. B. F. Duncan, J. Chem. Phys. **18**, 427 (1950).
661. T. Kasuya and T. Oka, J. Phys. Soc. Japan **15**, 296 (1960).
662. S. Kato, Sci. Pap. Inst. Phys. Chem. Res. **13**, 248 (1930).

663. F. Kaufman, Proc. Roy. Soc. **247**A, 123 (1958).
664. F. L. Keller and A. H. Nielsen, J. Chem. Phys. **24**, 636 (1956).
664a. A. U. Khan and M. Kasha, Nature **204**, 241 (1964).
665. R. W. Kilb, J. Chem. Phys. **23**, 1736 (1955).
666. ———, C. C. Lin and E. B. Wilson, Jr., J. Chem. Phys. **26**, 1695 (1957).
666a. G. E. Kimball, J. Chem. Phys. **8**, 188 (1940).
667. Gerald W. King, J. Chem. Soc., p. 5054 (1957).
668. ——— and D. Moule, Spectrochim. Acta **17**, 286 (1961).
669. ——— and ———, Can. J. Chem. **40**, 2057 (1962).
669a. ——— and E. H. Pinnington, J. Mol. Spec. **15**, 394 (1965).
670. Gilbert W. King, R. M. Hainer and P. C. Cross, J. Chem. Phys. **11**, 27 (1943).
671. F. W. Kirkbride and R. G. W. Norrish, J. Chem. Soc., p. 119 (1933).
672. G. B. Kistiakowsky, Phys. Rev. **37**, 276 (1931).
673. ———, J. R. Ruhoff, H. A. Smith and W. E. Vaughan, J. Amer. Chem. Soc. **58**, 137, 146 (1936).
674. ——— and A. K. Solomon, J. Chem. Phys. **5**, 609 (1937).
675. D. Kivelson, J. Chem. Phys. **22**, 904 (1954).
676. ——— and E. B. Wilson, Jr., J. Chem. Phys. **20**, 1575 (1952).
677. ———, ——— and D. R. Lide, Jr., J. Chem. Phys. **32**, 205 (1960).
678. R. Klein and L. J. Schoen, J. Chem. Phys. **24**, 1094 (1956), **29**, 953 (1958).
679. B. Kleman, Astrophys. J. **123**, 162 (1956).
680. ———, Can. J. Phys. **41**, 2034 (1963).
681. W. Klemperer and L. Lindeman, J. Chem. Phys. **25**, 397 (1956).
682. K. W. F. Kohlrausch and A. Pongratz, Z. physik. Chem. B**27**, 176 (1934).
683. ——— and ———, Chem. Ber. **67**, 976 (1934).
684. H. Kohn and H. Sponer, J. Opt. Soc. Amer. **39**, 75 (1949).
685. W. Kołos and C. C. J. Roothaan, Rev. Mod. Phys. **32**, 219 (1960).
686. K. Kondo, H. Hirakawa, A. Miyahara, T. Oka and K. Shimoda, J. Phys. Soc. Japan **15**, 303 (1960).
687. ——— and K. Shimoda, Proc. Int. Symp. Mol. Str. and Spec. Tokyo, p. C318–1 (1962).
688. V. Kondratiev, Z. Physik **63**, 322 (1930).
689. M. Kotani, Proc. Phys. Math. Soc. Japan **19**, 460 (1937).
690. ———, K. Ohno and K. Kayama, in Encyclop. Phys. **37** (2), 1 (1961).
691. S. E. Krasikov, A. N. Filippov and J. J. Chernyaev, Ann. secteur platine, Inst. chim. gén. (USSR) **13**, 19 (1936).
692. M. Krauss, J. Chem. Phys. **38**, 564 (1963).
693. V. G. Krishna and L. Goodman, J. Amer. Chem. Soc. **83**, 2042 (1961).
694. ——— and ———, J. Chem. Phys. **36**, 2217 (1962).
695. S. L. N. G. Krishnamachari, Ind. J. Phys. **29**, 603 (1955), **30**, 151, 319 (1956).
696. ———, Curr. Sci. **25**, 355 (1956).
696a. ———, N. A. Narasimham and M. Singh, Curr. Sci. **34**, 75 (1965).
697. ——— and B. R. Vengsarkar, Proc. Symp. Spec. Banaras, 1963, p. 87; Proc. Ind. Acad. Sci. **61**, 172 (1965).
698. A. Kronenberger, Z. Physik **63**, 494 (1930).
699. R. de L. Kronig, A. Schaafsma and P. K. Peerlkamp, Z. physik. Chem. B**22**, 323 (1933).
699aa. H. W. Kroto, J. Chem. Phys. **44**, 831 (1966).
699a. K. Kuchitsu and L. S. Bartell, J. Chem. Phys. **36**, 2460, 2470 (1962).
700. H. Kuhn, J. Chem. Phys. **16**, 840 (1948).
701. P. Kusch, J. Mol. Spec. **11**, 385 (1963).
702. ——— and F. W. Loomis, Phys. Rev. **55**, 850 (1939).
703. L. A. Kuznetsova, Y. Y. Kuzyakov and V. M. Tatevskii, Opt. and Spec. **16**, 295 (1964).
704. G. H. Kwei and R. F. Curl, Jr., J. Chem. Phys. **32**, 1592 (1960).

705. J. R. Lacher, L. E. Hummel, E. F. Bohmfalk and J. D. Park, J. Amer. Chem. Soc. **72**, 5486 (1950).

706. J. A. Ladd and W. J. Orville-Thomas, J. Chem. Soc., p. 2689 (1964).
707. W. J. Lafferty, A. G. Maki and E. K. Plyler, J. Chem. Phys. **40**, 224 (1964).
708. ——— and E. K. Plyler, J. Chem. Phys. **37**, 2688 (1962).
709. ———, ——— and E. D. Tidwell, J. Chem. Phys. **37**, 1981 (1962).
710. ——— and R. J. Thibault, J. Mol. Spec. **14**, 79 (1964).
711. R. T. Lagemann and E. A. Jones, J. Chem. Phys. **19**, 534 (1951).
712. K. J. Laidler, J. Chem. Phys. **10**, 34, 43 (1942); **22**, 1740 (1954).
713. R. K. Laird, E. B. Andrews and R. F. Barrow, Trans. Far. Soc. **46**, 803 (1950).
714. J. S. Lake and A. J. Harrison, J. Chem. Phys. **30**, 361 (1954).
715. K. Lakshmi, K. N. Rao and H. H. Nielsen, J. Chem. Phys. **24**, 811 (1956).
715a. M. Lam Thanh, Thesis, Univ. de Lyon, 1965.
716. ——— and M. Peyron, J. chim. phys. **59**, 688 (1962); **60**, 1289 (1963); **61**, 1531 (1964).
717. J. E. Lancaster and N. B. Colthup, J. Chem. Phys. **22**, 1149 (1954).
718. ——— and B. P. Stoicheff, Can. J. Phys. **34**, 1016 (1956).
719. L. Landau, Phys. Z. d. Sow. Union **1**, 88, **2**, 46 (1932).
720. A. Langer, J. A. Hipple and D. P. Stevenson, J. Chem. Phys. **22**, 1836 (1954).
721. A. Langseth and C. K. Møller, Acta Chem. Scand. **4**, 725 (1950).
722. ——— and B. Qviller, Z. physik. Chem. B**27**, 79 (1934).
723. S. R. La Paglia, J. Mol. Spec. **10**, 240 (1963).
724. ———, J. Chem. Phys. **41**, 1427 (1964).
725. ——— and A. B. F. Duncan, J. Chem. Phys. **34**, 125 (1961).
726. G. C. Lardy, J. chim. phys. **21**, 353 (1924).
727. V. W. Laurie, J. Chem. Phys. **24**, 635 (1956).
728. ——— and D. T. Pence, J. Mol. Spec. **10**, 155 (1963).
729. ———, ——— and R. H. Jackson, J. Chem. Phys. **37**, 2995 (1962).
730. R. B. Lawrance and M. W. P. Strandberg, Phys. Rev. **83**, 363 (1951).
731. M. Lawson and A. B. F. Duncan, J. Chem. Phys. **12**, 329 (1944).
731aa. S. Leach, J. chim. phys. **61**, 1493 (1964).
731a. ——— and R. Lopez-Delgado, J. chim. phys. **61**, 1636 (1964).
732. O. H. LeBlanc, Jr., V. W. Laurie and W. D. Gwinn, J. Chem. Phys. **33**, 598 (1960).
733. J. Lecomte, Bull. soc. chim., p. 415 (1946).
734. ———, in Encyclop. Phys. **26**, 244 (1958).
735. L. Lefebvre, C.R. Acad. Sci. (Paris) **200**, 1743 (1935).
736. S. W. Leifson, Astrophys. J. **63**, 73 (1926).
737. P. A. Leighton and F. E. Blacet, J. Amer. Chem. Soc. **55**, 1766 (1933).
738. J. E. Lennard-Jones, Proc. Roy. Soc. **198**A, 1, 14 (1949).
739. ——— and J. A. Pople, Proc. Roy. Soc. **205**A, 155, 163 (1951).
740. D. W. Lepard, D. M. C. Sweeney and H. L. Welsh, Can. J. Phys. **40**, 1567 (1962).
741. G. N. Lewis and M. Kasha, J. Amer. Chem. Soc. **66**, 2100 (1944).
742. ——— and ———, J. Amer. Chem. Soc. **67**, 994 (1945).
743. D. R. Lide, Jr., J. Chem. Phys. **27**, 343 (1957).
744. ——— and D. Christensen, J. Chem. Phys. **35**, 1374 (1961).
745. ——— and D. E. Mann, J. Chem. Phys. **27**, 868 (1957).
746. L. N. Liebermann, Phys. Rev. **60**, 496 (1941).
747. A. D. Liehr, J. Phys. Chem. **67**, 389 (1963).
747a. ———, Z. Naturf. **16**A, 641 (1961); Adv. Chem. Phys. **5**, 241 (1963).
748. ——— and W. Moffitt, J. Chem. Phys. **25**, 1074 (1956).
749. J. Lifschitz and E. Rosenbohm, Z. physik. Chem. **97**, 1 (1921).
750. L. Light, Z. physik. Chem. **122**, 414 (1926).
751. E. C. Lim, J. Chem. Phys. **36**, 3497 (1962).
752. C. C. Lin, Phys. Rev. **116**, 903 (1959).
753. ——— and J. D. Swalen, Rev. Mod. Phys. **31**, 841 (1959).
754. F. J. Lindars and C. Hinshelwood, Proc. Roy. Soc. **231**A, 162, 178 (1955).
754aa. D. C. Lindsey, Thesis, University of London, 1965.
754a. E. R. Lippincott, J. Chem. Phys. **21**, 2070 (1953).
754b. ——— and R. Schroeder, J. Chem. Phys. **23**, 1131 (1955).
755. T. K. Liu and A. B. F. Duncan, J. Chem. Phys. **17**, 241 (1949).

756. T. K. Liu, G. Moe and A. B. F. Duncan, J. Chem. Phys. **19**, 71 (1951).
757. W. Lochte-Holtgreven, C. E. H. Bawn and E. Eastwood, Nature **129**, 869 (1932).
758. B. B. Loeffler, E. Eberlin and L. W. Pickett, J. Chem. Phys. **28**, 345 (1958).
759. L. L. Lohr, Jr., and W. N. Lipscomb, J. Chem. Phys. **36**, 2225 (1962).
759a. ——— and ———, J. Amer. Chem. Soc. **85**, 240 (1963).
760. F. London, Z. Elektrochem. **35**, 552 (1929).
761. ———, Z. Physik **63**, 245 (1930).
762. ———, Z. physik. Chem. B**11**, 222 (1930).
763. D. A. Long, F. S. Murfin and R. L. Williams, Proc. Roy. Soc. **223**A, 251 (1954).
764. P. Longin, C.R. Acad. Sci. (Paris) **251**, 2499 (1960).
765. H. C. Longuet-Higgins, Proc. Chem. Soc., p. 157 (1957).
766. ———, Adv. in Spec. **2**, 429 (1961).
767. ———, Mol. Phys. **6**, 445 (1963).
768. ———, private communication.
769. ———, U. Öpik, M. H. L. Pryce and R. A. Sack, Proc. Roy. Soc. **244**A, 1 (1958).
770. R. C. Lord, A. L. Marston and F. A. Miller, Spectrochim. Acta **9**, 113 (1957).
771. ——— and R. E. Merrifield, J. Chem. Phys. **20**, 1348 (1952).
772. ——— and F. A. Miller, J. Chem. Phys. **10**, 328 (1942).
773. ——— and E. Nielsen, J. Chem. Phys. **19**, 1 (1951).
774. ——— and B. Nolin, J. Chem. Phys. **24**, 656 (1956).
775. ——— and P. Venkateswarlu, J. Chem. Phys. **20**, 1237 (1952).
776. ——— and ———, J. Opt. Soc. Amer. **43**, 1079 (1953).
777. F. P. Lossing, K. U. Ingold and I. H. S. Henderson, J. Chem. Phys. **22**, 1489 (1954).
778. W. Lotmar, Z. Physik **83**, 765 (1933).
779. R. J. Lovell and E. A. Jones, J. Mol. Spec. **4**, 173 (1960).
780. P. O. Löwdin, Adv. Chem. Phys. **2**, 207 (1959).
781. A. Lowrey and K. Watanabe, J. Chem. Phys. **28**, 208 (1958).
782. E. B. Ludlam, J. Chem. Phys. **3**, 617 (1935).
783. A. Lüthy, Z. physik. Chem. **107**, 285 (1923).
784. W. Lüttke, Z. Elektrochem. **61**, 302 (1957).
785. T. Lyman, Astrophys. J. **27**, 87 (1908).

786. J. H. Macek and G. H. Duffey, J. Chem. Phys. **34**, 288 (1961).
787. H. Mackle and R. T. B. McClean, Trans. Far. Soc. **58**, 895 (1962).
788. E. M. Magee, J. Chem. Phys. **39**, 855 (1963).
789. D. W. Magnuson, Phys. Rev. **83**, 485A (1951).
790. ———, J. Chem. Phys. **20**, 380 (1952).
791. B. H. Mahan and R. Mandal, J. Chem. Phys. **37**, 207 (1962).
792. ——— and R. B. Solo, J. Chem. Phys. **37**, 2669 (1962).
793. H. E. Mahncke and W. A. Noyes, Jr., J. Chem. Phys. **3**, 536 (1935).
794. J. R. Majer and C. R. Patrick, Nature **201**, 1022 (1964).
794a. A. G. Maki, J. Chem. Phys. **43**, 3193 (1965).
795. ———, E. K. Plyler and E. D. Tidwell, J. Res. Nat. Bur. Std. **66**A, 163 (1962).
795a. ——— and R. A. Toth, J. Mol. Spec. **17**, 136 (1965).
796. D. E. Mann and B. A. Thrush, J. Chem. Phys. **33**, 1732 (1960).
797. D. J. Marais, N. Sheppard and B. P. Stoicheff, Tetrahedron **17**, 163 (1962).
798. J. Mason, J. Chem. Soc., p. 3904 (1957).
799. S. F. Mason, J. Chem. Soc., p. 1263, 1269 (1959).
799a. ———, Quart. Rev. Chem. Soc. (London) **15**, 287 (1961).
799b. ———, Quart. Rev. Chem. Soc. (London) **17**, 20 (1963).
800. M. S. Matheson and J. W. Zabor, J. Chem. Phys. **7**, 536 (1939).
800a. C. W. Mathews, J. Mol. Spec. **19**, 203 (1966).
800b. ——— J. Chem. Phys. **45**, in press.
801. C. W. Mathews and K. K. Innes, J. Mol. Spec. **15**, 199 (1965).
802. I. Matubara and T. Anno, J. Chem. Phys. **24**, 595 (1956).
803. L. Mayor, A. D. Walsh and P. A. Warsop, J. Mol. Spec. **10**, 320 (1963).
804. D. S. McClure, J. Chem. Phys. **20**, 682 (1951).

805. D. S. McClure, Sol. State Phys. **9**, 399 (1959).
806. V. M. McConaghie and H. H. Nielsen, J. Chem. Phys. **21**, 1836 (1953).
807. H. M. McConnell, J. Chem. Phys. **34**, 13 (1961).
808. C. A. McDowell and J. W. Warren, Trans. Far. Soc. **48**, 1084 (1952).
809. K. L. McEwen, J. Chem. Phys. **32**, 1801 (1960).
810. W. D. McGrath, J. Chem. Phys. **33**, 297 (1960).
811. ——— and J. J. McGarvey, Nature **201**, 991 (1964); Trans. Far. Soc. **60**, 2196 (1964).
812. ——— and T. Morrow, Nature **203**, 619, **204**, 988 (1964).
813. A. McKellar, J. Roy. Astr. Soc. Canada **41**, 147 (1947).
814. A. D. McLean, J. Chem. Phys. **37**, 627 (1962).
815. R. McWeeny and K. A. Ohno, Proc. Roy. Soc. **255**A, 367 (1960).
816. J. H. Meal and S. R. Polo, J. Chem. Phys. **24**, 1119 (1956).
817. R. Mecke, Nature **125**, 526 (1930).
818. ———, Z. Physik **81**, 313 (1933).
819. D. Melcher, Helv. Phys. Acta **18**, 72 (1945).
820. E. H. Melvin and O. R. Wulf, J. Chem. Phys. **3**, 755 (1935).
821. S. Menczel, Z. physik. Chem. **125**, 161 (1927).
822. A. J. Merer, Disc. Far. Soc. **35**, 127 (1963).
823. ———, Can. J. Phys. **42**, 1242 (1964).
824. ——— (to be published).
825. ——— and D. N. Travis, Can. J. Phys. **44**, 1541 (1966).
826. ——— and ———, Can. J. Phys. **43**, 1795 (1965).
827. ——— and ——— (to be published).
828. ——— and ———, Can. J. Phys. **44**, 353 (1966).
829. ——— and ———, Can. J. Phys. **44**, 525 (1966).
830. J. A. Merritt, Can. J. Phys. **40**, 1683 (1962).
831. ——— and K. K. Innes, Spectrochim. Acta **16**, 945 (1960).
832. D. J. Meschi and R. J. Myers, J. Mol. Spec. **3**, 405 (1959).
833. N. Metropolis, Phys. Rev. **60**, 283 (1941).
834. ———, Phys. Rev. **60**, 295 (1941).
835. ——— and H. Beutler, Phys. Rev. **57**, 1078 (1940).
836. P. H. Metzger and G. R. Cook, J. Chem. Phys. **41**, 642 (1964).
837. G. Milazzo, Z. physik. Chem. B**33**, 109 (1936).
838. ———, R. Acad. d'Italia **4**, 87 (1942).
839. ———, Gazz. Chim. Ital. **68**, 747, 763 (1938).
840. ———, Spectrochim. Acta **2**, 245 (1942).
841. ———, Gazz. Chim. Ital. **78**, 835 (1948).
842. ——— and E. Miescher, Gazz. Chim. Ital. **83**, 782 (1953).
843. ——— and L. Paoloni, Gazz. Chim. Ital. **82**, 576 (1952).
844. F. A. Miller, G. L. Carlson and W. B. White, Spectrochim. Acta **15**, 709 (1959).
845. ——— and W. G. Fateley, Spectrochim. Acta **20**, 253 (1964).
846. ——— and R. B. Hannan, Jr., Spectrochim. Acta **12**, 321 (1958).
847. ———, ——— and L. R. Cousins, J. Chem. Phys. **23**, 2127 (1955).
848. R. F. Miller and R. F. Curl, Jr., J. Chem. Phys. **34**, 1847 (1961).
849. S. L. Miller and C. H. Townes, Phys. Rev. **90**, 537 (1953).
850. D. E. Milligan and M. E. Jacox, J. Chem. Phys. **38**, 2627 (1963).
851. ——— and ———, J. Chem. Phys. **41**, 3032 (1964).
851a. ———, ——— and A. M. Bass, J. Chem. Phys. **43**, 3149 (1965).
852. ———, ———, ———, J. J. Comeford and D. E. Mann, J. Chem. Phys. **42**, 3187 (1965).
853. ———, D. E. Mann, M. E. Jacox and R. A. Mitsch, J. Chem. Phys. **41**, 1199 (1964).
854. R. C. Millikan and K. S. Pitzer, J. Chem. Phys. **27**, 1305 (1957).
855. I. M. Mills, Mol. Phys. **7**, 549 (1964).
856. ——— and J. L. Duncan, J. Mol. Spec. **9**, 244 (1962).
857. ——— and H. W. Thompson, Trans. Far. Soc. **50**, 1270 (1954).
858. T. N. Misra, Ind. J. Phys. **37**, 173 (1963).
859. T. Miyazawa, J. Chem. Soc. Japan **74**, 743 (1953).
860. ———, J. Chem. Phys. **29**, 421 (1958).

861. T. Miyazawa and K. S. Pitzer, J. Chem. Phys. **30**, 1076 (1959).
862. M. Mizushima, J. Chem. Phys. **21**, 1222 (1953).
863. ——— and S. Koide, J. Chem. Phys. **20**, 765 (1952).
864. R. Moccia, J. Chem. Phys. **40**, 2164, 2176, 2186 (1964).
865. G. Moe and A. B. F. Duncan, J. Amer. Chem. Soc. **74**, 3136 (1952).
866. ——— and ———, J. Amer. Chem. Soc. **74**, 3140 (1952).
867. W. Moffitt, Proc. Roy. Soc. **210**A, 245 (1951).
868. ——— and C. A. Coulson, Phil. Mag. **38**, 634 (1947).
868a. ———, G. L. Goodman, M. Fred and B. Weinstock, Mol. Phys. **2**, 109 (1959).
869. ——— and A. D. Liehr, Phys. Rev. **106**, 1195 (1957).
870. ——— and W. Thorson, Phys. Rev. **108**, 1251 (1957).
871. ——— and ———, Coll. Int. C.N.R.S. **82**, 141 (1958).
872. C. K. Møller and B. P. Stoicheff, Can. J. Phys. **32**, 635 (1954).
873. J. Momigny, Nature **199**, 1179 (1963).
874. A. Monfils and B. Rosen, Nature **164**, 713 (1949).
875. R. B. Mooney and H. G. Reid, Proc. Roy. Soc. Edin. **52**, 152 (1932).
876. C. B. Moore and G. C. Pimentel, J. Chem. Phys. **38**, 2816 (1963).
877 ——— and ———, J. Chem. Phys. **40**, 329, 342 (1964).
878. J. Moret-Bailly, Cahiers de Phys. **15**, 237 (1961).
879. K. Mori, Sci. of Light **3**, 62 (1954); **4**, 130 (1955).
880. Y. Morino, Y. Kikuchi, S. Saito and E. Hirota, J. Mol. Spec. **13**, 95 (1964).
881. J. D. Morrison and A. J. C. Nicholson, J. Chem. Phys. **20**, 1021 (1952).
882. S. Mrozowski, Phys. Rev. **60**, 730 (1941); **62**, 270 (1942); **72**, 682, 691 (1947).
883. ———, Rev. Mod. Phys. **14**, 216 (1942).
884. J. S. Muirhead and J. A. Howe, J. Chem. Phys. **36**, 2316 (1962).
885. P. Müller, Helv. Phys. Acta **15**, 233 (1942).
886. ——— and M. Wehrli, Helv. Phys. Acta **15**, 307 (1942).
887. J. F. Mulligan, J. Chem. Phys. **19**, 347 (1951).
888. R. S. Mulliken, Phys. Rev. **41**, 49 (1932).
889. ———, Phys. Rev. **41**, 751 (1932).
890. ———, Phys. Rev. **43**, 279 (1933).
891. ———, J. Chem. Phys. **3**, 517 (1935).
892. ———, J. Chem. Phys. **3**, 720 (1935).
893. ———, Phys. Rev. **50**, 1017 (1936).
894. ———, J. Chem. Phys. **7**, 14 (1939).
895. ———, J. Chem. Phys. **7**, 20 (1939).
896. ———, J. Chem. Phys. **7**, 121 (1939).
897. ———, J. Chem. Phys. **7**, 339 (1939).
898. ———, J. Chem. Phys. **7**, 353 (1939).
899. ———, J. Chem. Phys. **7**, 356 (1939).
900. ———, J. Chem. Phys. **7**, 364 (1939).
901. ———, J. Chem. Phys. **7**, 570 (1939).
902. ———, J. Chem. Phys. **8**, 234 (1940).
903. ———, J. Chem. Phys. **8**, 382 (1940).
904. ———, Rev. Mod. Phys. **14**, 265 (1942).
905. ———, J. chim. phys. **46**, 497 (1949).
906. ———, J. chim. phys. **46**, 675 (1949).
907. ———, J. Amer. Chem. Soc. **72**, 600 (1950).
908. ———, J. Amer. Chem. Soc. **74**, 811 (1952).
909. ———, J. Phys. Chem. **56**, 801 (1952).
910. ———, J. Chem. Phys. **23**, 397 (1955).
911. ———, J. Chem. Phys. **23**, 1833, 1841 (1955).
912. ———, J. Chem. Phys. **23**, 1997 (1955).
913. ———, Rec. trav. chim. Pays-Bas **75**, 845 (1956).
914. ———, Can. J. Chem. **36**, 10 (1958).
914a. ———, J. Chem. Phys. **33**, 1596 (1960).
915. ———, J. Chem. Phys. **36**, 3428 (1962).
916. ———, C. A. Rieke and W. G. Brown, J. Amer. Chem. Soc. **63**, 41 (1941).

917. R. S. Mulliken, and E. Teller, Phys. Rev. **61**, 283 (1942).
918. E. Murad, J. Phys. Chem. **64**, 942 (1960).
919. ——— and M. G. Inghram, J. Chem. Phys. **41**, 404 (1964).
920. J. N. Murrell, J. Amer. Chem. Soc. **81**, 5037 (1959).
920a. ———, Mol. Phys. **3**, 319 (1960).
921. ——— and J. A. Pople, Proc. Phys. Soc. **69**A, 245 (1956).
922. R. J. Myers and E. B. Wilson, Jr., J. Chem. Phys. **33**, 186 (1960).

923. S. Nagakura, Mol. Phys. **3**, 152 (1960).
924. ———, K. Kaya and H. Tsubomura, J. Mol. Spec. **13**, 1 (1964).
925. I. Nakagawa and S. Shimanouchi, Spectrochim. Acta **18**, 513 (1962).
926. T. Nakayama, M. Y. Kitamura and K. Watanabe, J. Chem. Phys. **30**, 1180 (1959).
927. ——— and K. Watanabe, J. Chem. Phys. **40**, 558 (1964).
927a. I. M. Napier and R. G. W. Norrish, Nature **208**, 1090 (1965).
928. S. Narita, S. Ichinohe and S. Enomoto, J. Chem. Phys. **31**, 1151 (1959).
929. D. Neuberger and A. B. F. Duncan, J. Chem. Phys. **22**, 1693 (1954).
930. H. Neuïmin and A. Terenin, Acta Physicochim. **5**, 465 (1936).
931. A. S. Newton and A. F. Sciamanna, J. Chem. Phys. **40**, 718 (1964).
932. A. H. Nielsen, T. G. Burke, P. J. H. Woltz and E. A. Jones, J. Chem. Phys. **20**, 596 (1952).
933. H. H. Nielsen, Phys. Rev. **38**, 1432 (1931).
933a. ———, J. Chem. Phys. **5**, 818 (1937).
934. ———, Phys. Rev. **77**, 130 (1950).
935. ———, Rev. Mod. Phys. **23**, 90 (1951).
936. ———, in Encyclop. Phys. **37**/1, 173 (1959).
937. ——— and D. M. Dennison, Phys. Rev. **72**, 1101 (1947).
938. T. Nishikawa, T. Itoh and K. Shimoda, J. Chem. Phys. **23**, 1735 (1955).
939. G. Nordheim, H. Sponer and E. Teller, J. Chem. Phys. **8**, 455 (1940).
940. R. G. W. Norrish, H. G. Crone and O. D. Saltmarsh, J. Chem. Soc., p. 1533 (1933).
941. ———, ——— and ———, J. Chem. Soc., p. 1456 (1934).
942. W. A. Noyes, Jr., A. B. F. Duncan and W. M. Manning, J. Chem. Phys. **2**, 717 (1934).
943. ——— and F. C. Henriques, Jr., J. Chem. Phys. **7**, 767 (1939).
944. ——— and ———, J. Amer. Chem. Soc. **62**, 1038 (1940).

945. M. Ogawa and G. R. Cook, J. Chem. Phys. **28**, 173 (1958).
946. ——— and ———, J. Chem. Phys. **28**, 747 (1958).
947. T. Oka and Y. Morino, J. Mol. Spec. **8**, 300 (1962).
948. J. K. O'Loane, J. Chem. Phys. **21**, 669 (1953).
949. W. Opechowski, Physica **7**, 552 (1940).
950. U. Öpik and M. H. L. Pryce, Proc. Roy. Soc. **238**A, 425 (1957).
950a. L. E. Orgel, Quart. Rev. Chem. Soc. London **8**, 422 (1954).
951. ——— and R. S. Mulliken, J. Amer. Chem. Soc. **79**, 4839 (1957).
952. J. Overend and B. Crawford, Jr., J. Chem. Phys. **29**, 1002 (1958).
953. ——— and J. C. Evans, Trans. Far. Soc. **55**, 1817 (1959).
954. ——— and H. W. Thompson, J. Opt. Soc. Amer. **43**, 1065 (1953).

955. M. R. Padhye and S. R. Desai, Proc. Phys. Soc. **65**A, 298 (1952).
956. F. M. Page, Trans. Far. Soc. **57**, 1254 (1961).
957. J. Paldus and D. A. Ramsay (to be published).
958. E. D. Palik, J. Mol. Spec. **3**, 259 (1959).
959. Yu. N. Panchenko, Yu. A. Pentin, V. I. Tyulin and V. M. Tatevskii, Opt. and Spec. **13**, 488 (1962).
960. J. P. Paris, R. C. Hirt and R. G. Schmitt, J. Chem. Phys. **34**, 1851 (1961).
960a. M. A. Pariseau, I. Suzuki and J. Overend, J. Chem. Phys. **42**, 2335 (1965).
961. F. W. Parker, A. H. Nielsen and W. H. Fletcher, J. Mol. Spec. **1**, 107 (1957).
962. P. M. Parker, J. Chem. Phys. **37**, 1596 (1962).
962a. J. E. Parkin, Thesis, University of London, 1962.

962b. J. E. Parkin, J. Mol. Spec. **15**, 483 (1965).
963. ——— and K. K. Innes, J. Mol. Spec. **15**, 407 (1965).
964. ——— and ———, J. Mol. Spec. **16**, 93 (1965).
964a. ———, H. G. Poole and W. T. Raynes, Proc. Chem. Soc. (London), p. 248 (1962).
965. R. G. Parr, D. P. Craig and I. G. Ross, J. Chem. Phys. **18**, 1561 (1950).
966. ——— and R. Pariser, J. Chem. Phys. **23**, 711 (1955).
967. Y. P. Parti and R. Samuel, Proc. Phys. Soc. **49**, 568 (1937).
968. G. S. Paulett and R. Ettinger, J. Chem. Phys. **39**, 825 (1963).
969. L. Pauling, Proc. Nat. Acad. Sci. **14**, 359 (1928).
970. ———, J. Amer. Chem. Soc. **53**, 1367 (1931).
971. ———, Phys. Rev. **37**, 1185 (1931).
972. ——— and G. W. Wheland, J. Chem. Phys. **1**, 362 (1933).
973. T. E. Peacock, R. Rahman, D. H. Sleeman and E. S. G. Tuckley, Disc. Far. Soc. **35**, 144 (1963).
974. F. Pennella and W. J. Taylor, J. Mol. Spec. **11**, 321 (1963).
975. W. G. Penney, Proc. Roy. Soc. **144**A, 166; **146**A, 223 (1934).
976. ——— and G. B. B. M. Sutherland, J. Chem. Phys. **2**, 492 (1934).
977. D. Peters, J. Chem. Phys. **36**, 2743 (1962).
978. J. G. Phillips and L. Brewer, Mém. Soc. Roy. Sci. Liège (4), **15**, 341 (1955).
979. L. W. Pickett, J. Chem. Phys. **8**, 293 (1940).
980. ———, N. J. Hoeflich and T. C. Liu, J. Amer. Chem. Soc. **73**, 4865 (1951).
981. ———, M. Muntz and E. M. McPherson, J. Amer. Chem. Soc. **73**, 4862 (1951).
982. ———, E. Paddock and E. Sackter, J. Amer. Chem. Soc. **63**, 1073 (1941).
983. ——— and E. Sheffield, J. Amer. Chem. Soc. **68**, 216 (1946).
984. J. Pickworth and H. W. Thompson, Trans. Far. Soc. **50**, 218 (1954).
985. L. Pierce and M. Hayashi, J. Chem. Phys. **35**, 479 (1961).
986. M. G. K. Pillai and R. F. Curl, Jr., J. Chem. Phys. **37**, 2921 (1962).
987. A. C. Pitts, J. Chem. Phys. **18**, 1416 (1950).
988. G. Placzek and E. Teller, Z. Physik **81**, 209 (1933).
989. J. R. Platt, J. Chem. Phys. **17**, 484 (1949).
990. ———, in A. Hollaender, Radiation Biology Vol. III, p. 71 (1956).
991. ———, in Encyclop. Phys. **37**/2, 173 (1961).
992. ———, J. Mol. Spec. **9**, 288 (1962).
993. J. Plíva, Coll. Czech. Chem. Comm. **23**, 777 (1958).
994. ———, J. Mol. Spec. **12**, 360 (1964).
995. E. K. Plyler, L. R. Blaine and E. D. Tidwell, J. Res. Nat. Bur. Std. **55**, 183 (1955).
996. ——— and R. S. Mulliken, J. Amer. Chem. Soc. **81**, 823 (1959).
997. ———, E. D. Tidwell and A. G. Maki, J. Res. Nat. Bur. Std. **68**A, 79 (1964).
998. J. C. Polanyi and S. D. Rosner, J. Chem. Phys. **38**, 1028 (1963).
999. S. R. Polo, Can. J. Phys. **35**, 880 (1957).
1000. ——— (unpublished).
1001. J. A. Pople, Mol. Phys. **3**, 16 (1960).
1002. ——— and H. C. Longuet-Higgins, Mol. Phys. **1**, 372 (1958).
1003. ——— and J. W. Sidman, J. Chem. Phys. **27**, 1270 (1957).
1004. D. Porret and C. F. Goodeve, Trans. Far. Soc. **33**, 690 (1937).
1005. ——— and ———, Proc. Roy. Soc. **165**A, 31 (1938).
1006. G. Porter, Proc. Roy. Soc. **200**A, 284 (1950).
1007. ———, Disc. Far. Soc. **9**, 60 (1950).
1008. ———, J. Chem. Phys. **19**, 1278 (1951).
1009. ——— and B. Ward, Proc. Chem. Soc., p. 288 (1964).
1010. ——— and ———, Proc. Roy. Soc. **287**A, 457 (1965).
1011. H. Preuss, Z. Naturf. **12**a, 599 (1957).
1012. W. C. Price, Phys. Rev. **46**, 529 (1934).
1013. ———, Phys. Rev. **47**, 444 (1935).
1014. ———, J. Chem. Phys. **3**, 256 (1935).
1015. ———, J. Chem. Phys. **4**, 147 (1936).
1016. ———, J. Chem. Phys. **4**, 539 (1936).
1017. ———, J. Chem. Phys. **4**, 547 (1936).

1018. W. C. Price, J. Chem. Phys. **16**, 894 (1948).
1019. ——— and W. M. Evans, Proc. Roy. Soc. **162A**, 110 (1937).
1020. ——— and D. M. Simpson, Proc. Roy. Soc. **165A**, 272 (1938).
1021. ——— and ———, Proc. Roy. Soc. **169A**, 501 (1939).
1022. ——— and ———, Trans. Far. Soc. **37**, 106 (1941).
1023. ——— and T. M. Sugden, Trans. Far. Soc. **44**, 108 (1948).
1024. ———, J. P. Teegan and A. D. Walsh, Proc. Roy. Soc. **201A**, 600 (1950).
1025. ———, ——— and ———, J. Chem. Soc., p. 920 (1951).
1026. ——— and W. T. Tutte, Proc. Roy. Soc. **174A**, 207 (1940).
1027. ——— and A. D. Walsh, Proc. Roy. Soc. **174A**, 220 (1940).
1028. ——— and ———, Proc. Roy. Soc. **179A**, 201 (1941).
1029. ——— and ———, Trans. Far. Soc. **41**, 381 (1945).
1030. ——— and ———, Proc. Roy. Soc. **185A**, 182 (1946).
1031. ——— and ———, Proc. Roy. Soc. **191A**, 22 (1947).
1032. ——— and R. W. Wood, J. Chem. Phys. **3**, 439 (1935).
1033. H. Prophet, J. Chem. Phys. **38**, 2345 (1963).
1034. J. E. Purvis and N. P. McCleland, J. Chem. Soc., p. 1810 (1912).
1035. E. S. Pysh, J. Jortner and S. A. Rice, J. Chem. Phys. **40**, 2018 (1964).

1036. B. Qviller, Tids. Kjemi og Bergwesen **17**, 127 (1937).

1037. W. F. Radle and C. A. Beck, J. Chem. Phys. **8**, 507 (1940).
1038. C. Ramasastry and K. R. Rao, Ind. J. Phys. **21**, 313 (1947).
1039. D. A. Ramsay, J. Chem. Phys. **20**, 1920 (1952).
1040. ———, J. Chem. Phys. **21**, 960 (1953).
1041. ———, J. Chem. Phys. **25**, 188 (1956).
1042. ———, Nature **178**, 374 (1956).
1043. ———, Adv. in Spectroscopy **1**, 1 (1959).
1044. ———, Proc. Xth Coll. Spec. Int., p. 583 (1962).
1045. H. C. Ramsperger, J. Amer. Chem. Soc. **50**, 123 (1928).
1046. H. M. Randall, D. M. Dennison, N. Ginsburg and L. R. Weber, Phys. Rev. **52**, 160 (1937).
1047. D. H. Rank, P. Sitaram, W. A. Glickman and T. A. Wiggins, J. Chem. Phys. **39**, 2673 (1963).
1048. ———, G. Skorinko, D. P. Eastman and T. A. Wiggins, J. Mol. Spec. **4**, 518 (1960).
1049. D. R. Rao and P. Venkateswarlu, J. Mol. Spec. **7**, 287 (1961).
1050. I. A. Rao and V. R. Rao, J. Sci. Ind. Res. (India) **19**B, 121 (1960).
1051. K. N. Rao, W. W. Brim, J. M. Hoffman, L. H. Jones and R. S. McDowell, J. Mol. Spec. **7**, 362 (1961).
1052. ——— and H. H. Nielsen, Can. J. Phys. **34**, 1147 (1956).
1053. ——— and E. D. Palik, J. Mol. Spec. **1**, 24 (1957).
1054. ——— and H. Sponer, Can. J. Phys. **35**, 332 (1957).
1055. K. S. Rao, B. P. Stoicheff and R. Turner, Can. J. Phys. **38**, 1516 (1960).
1055a. V. M. Rao, R. F. Curl, Jr., P. L. Timms and J. L. Margrave, J. Chem. Phys. **43**, 2557 (1965).
1056. V. R. Rao and H. Sponer, Phys. Rev. **87**, 213 (1952).
1057. ——— and I. A. Rao, Ind. J. Phys. **28**, 491 (1954).
1058. G. Rathenau, Z. Physik **87**, 32 (1934).
1059. W. T. Raynes, J. Chem. Phys. **41**, 3020 (1964).
1060. ——— , J. Chem. Phys. **44**, 2755 (1966).
1061. R. I. Reed, Trans. Far. Soc. **52**, 1195 (1956).
1062. ——— and J. C. D. Brand, Trans. Far. Soc. **54**, 478 (1958).
1063. ——— and W. Snedden, Trans. Far. Soc. **54**, 301 (1958).
1063a. R. R. Reeves, P. Harteck and W. H. Chace, J. Chem. Phys. **41**, 764 (1964).
1064. C. Reid, J. Chem. Phys. **18**, 1299 (1950).
1065. ———, J. Chem. Phys. **18**, 1673 (1950).
1066. ——— and R. S. Mulliken, J. Amer. Chem. Soc. **76**, 3869 (1954).
1067. A. W. Reitz, Z. physik. Chem. B**33**, 179 (1936).

1068. A. W. Reitz, Z. physik. Chem. B**38**, 381 (1937).
1069. R. Renner, Z. Physik **92**, 172 (1934).
1070. O. K. Rice, Inst. Intern. Chim. Solvay, 12ème Conseil de Chimie (Interscience), p. 17 (1962).
1071. R. W. Ritchie and A. D. Walsh, Proc. Roy. Soc. **267**A, 395 (1962).
1072. ———, ——— and P. A. Warsop, Proc. Roy. Soc. **266**A, 257 (1962).
1073. G. W. Robinson, J. Chem. Phys. **21**, 1741 (1953).
1074. ———, J. Chem. Phys. **22**, 1147 (1954).
1075. ———, Can. J. Phys. **34**, 699 (1956).
1076. ———, J. Mol. Spec. **6**, 58 (1961).
1077. ———, in L. Marton: Methods of Experimental Physics Vol. **3**, p. 155 (1962).
1078. ——— and V. E. Di Giorgio, Can. J. Chem. **36**, 31 (1958).
1079. ——— and M. McCarty, Jr., Can. J. Phys. **36**, 1590 (1958).
1080. ———, ——— and M. C. Keelty, J. Chem. Phys. **27**, 972 (1957).
1081. J. Romanko, T. Feldman and H. L. Welsh, Can. J. Phys. **33**, 588 (1955).
1082. C. C. J. Roothaan, Rev. Mod. Phys. **23**, 69 (1951).
1083. B. Rosen, J. de physique (Radium) (8), **9**, 155 (1948).
1084. I. G. Ross and E. J. Wells (to be published).
1085. K. Rossmann, K. N. Rao and H. H. Nielsen, J. Chem. Phys. **24**, 103 (1956).
1086. ——— and J. W. Straley, J. Chem. Phys. **24**, 1276 (1956).
1087. E. Ruch, Z. Naturf. **16**a, 808 (1961).
1088. K. Ruedenberg, Rev. Mod. Phys. **34**, 326 (1962).
1089. ——— and C. W. Scherr, J. Chem. Phys. **21**, 1565, 1582 (1953).
1090. ——— and ———, J. Chem. Phys. **22**, 151, 1878 (1954).
1091. H. D. Rudolph, H. Dreizler and W. Maier, Z. Naturf. **15**a, 742 (1960).
1092. O. P. Rustgi, J. Opt. Soc. Amer. **54**, 464 (1964).

1093. B. D. Saksena and G. S. Jauhri, J. Chem. Phys. **36**, 2233 (1962).
1094. ——— and R. E. Kagarise, J. Chem. Phys. **19**, 987 (1951).
1095. ——— and ———, J. Chem. Phys. **19**, 999 (1951).
1096. J. A. R. Samson, F. F. Marmo and K. Watanabe, J. Chem. Phys. **36**, 783 (1962).
1097. V. Santhamma, Proc. Nat. Inst. Sci. India **22**, 256 (1956).
1098. T. N. Sarachman, Ph.D. Thesis, Harvard (1961).
1099. R. A. Satten, J. Chem. Phys. **27**, 286 (1957); **29**, 658 (1958).
1100. K. Schäfer, Z. physik. Chem. B**40**, 357 (1938).
1101. G. Scheibe and H. Grieneisen, Z. physik. Chem. B**25**, 52 (1934).
1102. ———, F. Povenz and C. Linström, Z. physik. Chem. B**20**, 283 (1933).
1103. R. Schlapp, Phys. Rev. **51**, 342 (1937).
1104. R. F. Schneider, Ph.D. Thesis, Columbia Univ. (1959).
1105. R. I. Schoen, J. Chem. Phys. **37**, 2032 (1962).
1106. S. A. Schou, J. chim. phys. **26**, 1 (1929).
1107. ———, J. chim. phys. **26**, 27 (1929).
1108. ———, J. chim. phys. **26**, 665 (1929).
1109. H. Schüler, Spectrochim. Acta **4**, 85 (1950).
1110. ——— and E. Lutz, Z. Naturf. **12**a, 334 (1957).
1111. ——— and L. Reinebeck, Z. Naturf. **5**a, 604 (1950).
1112. ——— and ———, Z. Naturf. **6**a, 160, 270 (1951); **7**a, 285 (1952).
1113. ——— and ———, Z. Naturf. **9**a, 350 (1954).
1114. ——— and ———, Spectrochim. Acta **6**, 288 (1954).
1115. ——— and A. Woeldike, Physik. Z. **43**, 415 (1942); **45**, 61 (1944).
1116. S. Sekino and T. Nishikawa, J. Phys. Soc. Japan **12**, 43 (1957).
1117. W. H. Shaffer, J. Chem. Phys. **9**, 607 (1941).
1118. ———, J. Chem. Phys. **10**, 1 (1942).
1119. R. S. Sharma, Proc. Acad. Sci. U.P. India, **4**, 51 (1934).
1120. R. D. Shelton, A. H. Nielsen and W. H. Fletcher, J. Chem. Phys. **21**, 2178 (1953).
1121. N. Sheppard, J. Chem. Phys. **17**, 79 (1949).
1122. J. Sheridan and W. Gordy, J. Chem. Phys. **20**, 591 (1952).
1123. K. Shibata, H. Kushida and S. Mori, Bull. Chem. Soc. Japan **24**, 188 (1951).

1124. R. Shimada, Spectrochim. Acta **17**, 14 (1961).
1125. ———, Spectrochim. Acta **17**, 30 (1961).
1126. C. Shin-Piaw, Ann. de Phys. (11) **10**, 173 (1938).
1127. J. H. Shirley, J. Chem. Phys. **38**, 2896 (1963).
1128. H. Shull, J. Chem. Phys. **17**, 295 (1949).
1129. S. Silver and W. H. Shaffer, J. Chem. Phys. **9**, 599 (1941).
1130. J. W. Sidman, J. Amer. Chem. Soc. **78**, 1527 (1956).
1131. ———, J. Chem. Phys. **29**, 644 (1958).
1132. ———, Chem. Rev. **58**, 689 (1958).
1133. ——— and D. S. McClure, J. Amer. Chem. Soc. **77**, 6461, 6471 (1955).
1133a. J. D. Simmons, I. R. Bartky and A. M. Bass, J. Mol. Spec. **17**, 48 (1965).
1134. ——— and K. K. Innes, J. Mol. Spec. **13**, 435 (1964).
1135. ———, ——— and G. M. Begun, J. Mol. Spec. **14**, 190 (1964).
1136. J. W. Simmons, W. E. Anderson and W. Gordy, Phys. Rev. **86**, 1055 (1952).
1136a. R. S. Singh, Ind. J. Phys. **33**, 376 (1959).
1137. M. H. Sirvetz and R. E. Weston, Jr., J. Chem. Phys. **21**, 898 (1953).
1138. A. L. Sklar, J. Chem. Phys. **5**, 669 (1937).
1139. J. C. Slater, Phys. Rev. **37**, 481; **38**, 1109 (1931).
1140. L. G. Smith, J. Chem. Phys. **17**, 139 (1949).
1141. H. D. Smyth, Phys. Rev. **44**, 690 (1933).
1142. R. L. Somerjai and D. F. Hornig, J. Chem. Phys. **36**, 1980 (1962).
1143. N. Solimene and B. P. Dailey, J. Chem. Phys. **23**, 124 (1955).
1144. G. H. Spencer, P. C. Cross and K. B. Wiberg, J. Chem. Phys. **35**, 1925 (1961).
1145. ———, ——— and ———, J. Chem. Phys. **35**, 1939 (1961).
1146. W. Spinnler, Helv. Phys. Acta **18**, 297 (1945).
1147. H. Sponer, J. Chem. Phys. **8**, 705 (1940).
1148. ———, Rev. Mod. Phys. **14**, 224 (1942).
1149. ———, J. Chem. Phys. **22**, 234 (1954).
1150. ——— and L. G. Bonner, J. Chem. Phys. **8**, 33 (1940).
1151. ——— and M. B. Hall, Phys. Rev. **61**, 107 (1942).
1152. ———, G. Nordheim, A. L. Sklar and E. Teller, J. Chem. Phys. **7**, 207 (1939).
1153. ——— and H. Stücklen, J. Chem. Phys. **14**, 101 (1936).
1154. ——— and E. Teller, J. Chem. Phys. **7**, 382 (1939).
1155. ——— and ———, Rev. Mod. Phys. **13**, 75 (1941).
1156. K. Sreeramamurty, Curr. Sci. **18**, 437 (1949).
1157. ———, Curr. Sci. **20**, 176 (1951).
1158. R. F. Stamm and J. E. Lancaster, J. Chem. Phys. **22**, 1280 (1954).
1159. H. Stammreich, K. Kawai and Y. Tavares, Spectrochim. Acta **15**, 438 (1959).
1160. M. J. Steindler and W. H. Gunther, Spectrochim. Acta **20**, 1319 (1964).
1160a. B. Steiner, C. F. Giese and M. G. Inghram, J. Chem. Phys. **34**, 189 (1961).
1161. F. Sterzer, J. Chem. Phys. **22**, 2094 (1954).
1162. S. Stokes and A. B. F. Duncan, J. Amer. Chem. Soc. **80**, 6177 (1958).
1163. B. P. Stoicheff, J. Chem. Phys. **21**, 755 (1953).
1164. ———, Can. J. Phys. **32**, 339 (1954).
1165. ———, Can. J. Phys. **33**, 811 (1955).
1166. ———, Can. J. Phys. **36**, 218 (1958).
1167. ———, Can. J. Phys. **40**, 358 (1962).
1168. ———, C. Cumming, G. E. St. John and H. L. Welsh, J. Chem. Phys. **20**, 498 (1952).
1169. A. J. Stone, J. Chem. Phys. **41**, 1568 (1964).
1170. R. F. Stratton and A. H. Nielsen, J. Mol. Spec. **4**, 373 (1960).
1171. R. E. Stroup, R. A. Oetjen and E. E. Bell, J. Opt. Soc. Amer. **43**, 1096 (1953).
1172. E. C. G. Stueckelberg, Helv. Phys. Acta **5**, 369 (1932).
1173. D. W. G. Style and J. C. Ward, J. Chem. Soc., p. 2125 (1952).
1174. ——— and ———, Trans. Far. Soc. **49**, 999 (1953).
1175. B. Sugarman, Proc. Phys. Soc. **55**, 429 (1943).
1176. T. M. Sugden and A. D. Walsh, Trans. Far. Soc. **41**, 76 (1945).
1177. H. Sun and G. L. Weissler, J. Chem. Phys. **23**, 1160 (1955).

1178. H. Sun and G. L. Weissler, J. Chem. Phys. **23**, 1625 (1955).
1179. L. H. Sutcliffe and A. D. Walsh, J. Chem. Soc., p. 899 (1952).
1180. ——— and ———, Trans. Far. Soc. **57**, 873 (1961).
1181. L. M. Sverdlov, Opt. and Spec. **8**, 17 (1960).
1182. ———, Opt. and Spec. **13**, 29 (1962).
1183. ——— and N. V. Tarasova, Opt. and Spec. **9**, 159 (1960).
1184. J. D. Swalen and C. C. Costain, J. Chem. Phys. **31**, 1562 (1959).

1185. Y. Tanabe and S. Sugano, J. Phys. Soc. Japan **9**, 753, 766 (1954).
1186. Y. Tanaka, E. C. Y. Inn and K. Watanabe, J. Chem. Phys. **21**, 1651 (1953).
1187. ——— and A. S. Jursa, J. Chem. Phys. **36**, 2493 (1962).
1188. ———, ——— and F. J. LeBlanc, J. Chem. Phys. **28**, 350 (1958).
1189. ———, ——— and ———, J. Chem. Phys. **32**, 1199 (1960).
1190. ———, ——— and ———, J. Chem. Phys. **32**, 1205 (1960).
1191. ——— and M. Ogawa, Can. J. Phys. **40**, 879 (1962).
1192. K. N. Tanner and A. B. F. Duncan, J. Amer. Chem. Soc. **73**, 1164 (1951).
1193. P. Tarte, Bull. Soc. Chim. Belg. **59**, 365 (1950).
1194. R. C. Taylor and G. L. Vidale, J. Chem. Phys. **26**, 122 (1957).
1195. J. P. Teegan and A. D. Walsh, Trans. Far. Soc. **47**, 1 (1951).
1196. E. Teller, in Hand- und Jahrbuch d. Chem. Phys. Vol. **9**, II, 43 (1934).
1197. ———, J. Phys. Chem. **41**, 109 (1937).
1198. ——— and L. Tisza, Z. Physik **73**, 791 (1932).
1199. A. Terenin, Z. Physik **44**, 713 (1927).
1200. ———, Usp. Fiz. Nauk **36**, 292 (1948).
1201. ——— and N. Prileshajewa, Acta Physicochim. **1**, 759 (1935).
1202. E. Thiele and D. J. Wilson, J. Chem. Phys. **35**, 1256 (1961).
1203. L. F. Thomas, E. I. Sherrard and J. Sheridan, Trans. Far. Soc. **51**, 619 (1955).
1204. H. W. Thompson, Proc. Roy. Soc. **150**A, 603 (1935).
1205. ———, J. Chem. Phys. **6**, 748 (1938).
1206. ———, Trans. Far. Soc. **36**, 988 (1940).
1207. ———, Trans. Far. Soc. **37**, 38 (1941).
1208. ——— and N. Healey, Proc. Roy. Soc. **157**A, 331 (1936).
1209. ——— and J. W. Linnett, J. Chem. Soc., p. 1452 (1935).
1210. ——— and ———, Proc. Roy. Soc. **156**A, 108 (1936).
1211. ——— and C. H. Miller, Trans. Far. Soc. **46**, 22 (1950).
1212. ——— and C. H. Purkis, Trans. Far. Soc. **32**, 674 (1936).
1213. ——— and N. P. Skerrett, Trans. Far. Soc. **36**, 812 (1940).
1214. ——— and R. B. Temple, Trans. Far. Soc. **41**, 27 (1945).
1215. R. J. Thompson and A. B. F. Duncan, J. Chem. Phys. **14**, 573 (1946).
1216. W. R. Thorson and I. Nakagawa, J. Chem. Phys. **33**, 994 (1960).
1217. B. A. Thrush, Proc. Roy. Soc. **235**A, 143 (1956).
1218. ———, Nature **178**, 155 (1956).
1219. ——— and J. J. Zwolenik, Trans. Far. Soc. **59**, 582 (1963).
1220. E. D. Tidwell, E. K. Plyler and W. S. Benedict, J. Opt. Soc. Amer. **50**, 1243 (1960).
1221. M. Tinkham and M. W. P. Strandberg, Phys. Rev. **97**, 937 (1955).
1222. ——— and ———, Phys. Rev. **97**, 951 (1955).
1223. W. M. Tolles, J. L. Kinsey, R. F. Curl and R. F. Heidelberg, J. Chem. Phys. **37**, 927 (1962).
1224. R. Trambarulo, A. Clark and C. Hearns, J. Chem. Phys. **28**, 736 (1958).
1225. W. Tsang, S. H. Bauer and M. Cowperthwaite, J. Chem. Phys. **36**, 1768 (1962).
1226. H. Tsubomura and R. S. Mulliken, J. Amer. Chem. Soc. **82**, 5966 (1960).
1227. R. Tsuchida, Bull. Chem. Soc. Japan **13**, 388 (1938).
1227a. G. C. Turrell, W. D. Jones and A. Maki, J. Chem. Phys. **26**, 1544 (1957).
1228. J. K. Tyler, J. Chem. Phys. **40**, 1170 (1964).
1228a. ——— and J. Sheridan, Trans. Far. Soc. **59**, 2661 (1963).

1229. F. M. Uber, J. Chem. Phys. **9**, 777 (1941).
1230. ——— and R. Winters, J. Amer. Chem. Soc. **63**, 137 (1941).

1231. A. Y. M. Ung and R. A. Back, Can. J. Chem. **42**, 753 (1964).
1232. H. C. Urey, L. H. Dawsey and F. O. Rice, J. Amer. Chem. Soc. **51**, 1371 (1929).

1233. W. M. Vaidya, Proc. Roy. Soc. **147**A, 513 (1934).
1234. ———, Proc. Phys. Soc. **64**A, 428 (1951).
1235. ———, Proc. Roy. Soc. **279**A, 572 (1964).
1236. J. H. Van Vleck, J. Chem. Phys. **2**, 20 (1934).
1237. ———, J. Chem. Phys. **7**, 72 (1939).
1238. ———, Rev. Mod. Phys. **23**, 213 (1951).
1239. ———, Physica **26**, 544 (1960).
1240. ——— and A. Sherman, Rev. Mod. Phys. **7**, 167 (1935).
1241. E. Vassy, Ann. de Phys. (11) **8**, 679 (1937).
1242. S. Vencov, C.R. Acad. Sci. (Paris) **208**, 801 (1939).
1243. P. Venkateswarlu, Phys. Rev. **77**, 676 (1950).
1244. ———, J. G. Baker and W. Gordy, J. Mol. Spec. **6**, 215 (1961).
1245. ———, S. Jagatheesan and K. V. Rajalakshmi, Proc. Ind. Acad. Sci. **58**, 373 (1963).
1246. M. Venugopalan and K. O. Kutschke, Can. J. Chem. **42**, 2451 (1964).
1247. E. Vigroux, Ann. de Phys. (12) **8**, 709 (1953).
1248. H. H. Voge, J. Chem. Phys. **4**, 581 (1936).

1248a. J. H. van der Waals, Disc. Far. Soc. **35**, 227 (1963).
1249. E. Wachsmann and E. W. Schmid, Z. physik. Chem. (Frankfurt) **27**, 145 (1961).
1250. E. L. Wagner, J. Chem. Phys. **37**, 751 (1962).
1251. P. Wagner and A. B. F. Duncan, J. Amer. Chem. Soc. **77**, 2609 (1955).
1252. R. Wagner, J. Fine, J. W. Simmons and J. H. Goldstein, J. Chem. Phys. **26**, 634 (1957).
1253. N. Wainfan, W. C. Walker and G. L. Weissler, Phys. Rev. **99**, 542 (1955).
1254. S. C. Wait, Jr., and M. P. Barnett, J. Mol. Spec. **4**, 93 (1960).
1255. J. D. Waldron, Trans. Far. Soc. **50**, 102 (1954).
1256. W. C. Walker and G. L. Weissler, J. Chem. Phys. **23**, 1540 (1955).
1257. ——— and ———, J. Chem. Phys. **23**, 1547 (1955).
1258. F. T. Wall and R. N. Porter, J. Chem. Phys. **36**, 3256 (1962).
1259. A. D. Walsh, Trans. Far. Soc. **41**, 35 (1945).
1260. ———, Trans. Far. Soc. **41**, 498 (1945).
1261. ———, Trans. Far. Soc. **42**, 62 (1946).
1262. ———, Trans. Far. Soc. **42**, 66 (1946).
1263. ———, Proc. Roy. Soc. **185**A, 176 (1946).
1263a. ———, J. Chem. Soc., p. 2260 (1953).
1264. ———, J. Chem. Soc., p. 2266 (1953).
1265. ———, J. Chem. Soc., p. 2288 (1953).
1266. ———, J. Chem. Soc., p. 2296 (1953).
1267. ———, J. Chem. Soc., p. 2301 (1953).
1268. ———, J. Chem. Soc., p. 2306 (1953).
1269. ———, Disc. Far. Soc. **35**, 223 (1963).
1270. ——— and P. A. Warsop, Trans. Far. Soc. **57**, 345 (1961).
1271. P. Warneck, F. F. Marmo and J. O. Sullivan, J. Chem. Phys. **40**, 1132 (1964).
1272. K. Watanabe, J. Chem. Phys. **22**, 1564 (1954).
1273. ———, J. Chem. Phys. **26**, 542 (1957).
1274. ——— and A. S. Jursa, J. Chem. Phys. **41**, 1650 (1964).
1275. ——— and T. Nakayama, J. Chem. Phys. **29**, 48 (1958).
1276. ——— and J. R. Mottl, J. Chem. Phys. **26**, 1773 (1957).
1277. ——— and T. Namioka, J. Chem. Phys. **24**, 915 (1956).
1278. ——— and M. Zelikoff, J. Opt. Soc. Amer. **43**, 753 (1953).
1278a. J. K. G. Watson, Thesis, University of London (1962).
1278b. ———, Can. J. Phys. **43**, 1996 (1965).
1279. ——— (to be published).
1280. W. W. Watson and A. E. Parker, Phys. Rev. **37**, 1484 (1931).
1281. G. G. Weber, J. Mol. Spec. **10**, 321 (1963).

1282. M. Wehrli, Helv. Phys. Acta **9**, 208 (1936).
1283. ———, Helv. Phys. Acta **9**, 329 (1936).
1284. ———, Helv. Phys. Acta **9**, 637 (1936).
1285. ———, Helv. Phys. Acta **11**, 339 (1938); **13**, 153 (1940).
1286. B. Weinstock and G. L. Goodman, Adv. Chem. Phys. **9**, 169 (1965).
1287. W. Weltner, Jr., and D. McLeod, Jr., J. Chem. Phys. **40**, 1305 (1964).
1287a. ——— and ———, J. Chem. Phys. **41**, 235 (1964).
1288. ——— and P. N. Walsh, J. Chem. Phys. **37**, 1153 (1962).
1289. ———, ——— and C. L. Angell, J. Chem. Phys. **40**, 1299 (1964).
1289a. W. Wenk, Helv. Phys. Acta **14**, 355 (1941).
1290. B. G. West and M. Mizushima, J. Chem. Phys. **38**, 251 (1963).
1291. F. R. Weston, Proc. Roy. Soc. **109**A, 176, 523 (1925).
1292. L. Wharton, R. A. Berg and W. Klemperer, J. Chem. Phys. **39**, 2023 (1963).
1293. K. Wieland, Z. Physik **76**, 801; **77**, 157 (1932).
1294. ———, Helv. Chim. Acta **24**, 1285 (1941).
1295. ———, Öster. Chem.-Ztg. **55**, 329 (1954).
1296. T. A. Wiggins, E. K. Plyler and E. D. Tidwell, J. Opt. Soc. Amer. **51**, 1219 (1961); **53**, 589 (1963).
1297. E. P. Wigner, Proc. Int. Conf. Theoret. Phys.; Science Council of Japan, p. 650 (1953).
1298. ——— and E. E. Witmer, Z. Physik **51**, 859 (1928).
1299. P. G. Wilkinson, Can. J. Phys. **34**, 596 (1956).
1300. ———, Can. J. Phys. **34**, 643 (1956).
1301. ———, J. Mol. Spec. **2**, 387 (1958).
1302. ——— and H. L. Johnston, J. Chem. Phys. **18**, 190 (1950).
1303. ——— and R. S. Mulliken, J. Chem. Phys. **23**, 1895 (1955).
1304. J. K. Wilmshurst and H. J. Bernstein, Can. J. Chem. **35**, 1183 (1957).
1305. E. B. Wilson, Jr., J. Chem. Phys. **3**, 276 (1935).
1306. ———, J. Chem. Phys. **6**, 740 (1938).
1307. ———, C. C. Lin and D. R. Lide, Jr., J. Chem. Phys. **23**, 136 (1955).
1308. E. G. Wilson, J. Jortner and S. A. Rice, J. Amer. Chem. Soc. **85**, 813 (1963).
1309. M. K. Wilson and R. M. Badger, J. Chem. Phys. **16**, 741 (1948).
1310. ——— and S. R. Polo, J. Chem. Phys. **20**, 1716 (1952).
1311. P. J. H. Woltz, E. A. Jones and A. H. Nielsen, J. Chem. Phys. **20**, 378 (1952).
1312. S. C. Woo and R. M. Badger, Phys. Rev. **39**, 932 (1932).
1313. ——— and T. C. Chu, J. Chem. Phys. **3**, 541 (1935); **5**, 786 (1937).
1314. ——— and ——— J. Chin. Chem. Soc. **5**, 162 (1937).
1315. ——— and T. K. Liu, J. Chem. Phys. **3**, 544 (1935).
1316. ——— and ———, J. Chem. Phys. **5**, 161 (1939).
1317. ———, ———, T. C. Chu and W. Chih, J. Chem. Phys. **6**, 240 (1938).
1318. R. W. Wood and G. H. Dieke, Nature **128**, 545 (1931).
1319. L. A. Woodward, Spectrochim. Acta **19**, 1963 (1963).
1320. ——— and H. L. Roberts, Trans. Far. Soc. **52**, 615 (1956).
1321. B. J. Woznick, J. Chem. Phys. **40**, 2860 (1964).
1322. M. R. Wright, R. P. Frosch and G. W. Robinson, J. Chem. Phys. **33**, 934 (1960)
1323. T. Y. Wu, Phys. Rev. **46**, 465 (1934).
1324. O. R. Wulf, Proc. Nat. Acad. Sci. **16**, 507 (1930).
1325. ——— and E. H. Melvin, Phys. Rev. **38**, 330 (1931).

1326. I. Zanon, G. Giacometti and D. Picciol, Spectrochim. Acta **19**, 301 (1963).
1327. M. Zelikoff and L. M. Aschenbrand, J. Chem. Phys. **22**, 1680, 1685 (1954).
1328. ——— and ———, J. Chem. Phys. **27**, 123 (1957).
1329. ——— and K. Watanabe, J. Opt. Soc. Amer. **43**, 756 (1953).
1330. ———, ——— and E. C. Y. Inn, J. Chem. Phys. **21**, 1643 (1953).
1331. C. Zener, Proc. Roy. Soc. **137**A, 696 (1932).
1332. G. Zerbi, B. Crawford, Jr., and J. Overend, J. Chem. Phys. **38**, 127 (1963).
1333. C. R. Zobel and A. B. F. Duncan, J. Amer. Chem. Soc. **77**, 2611 (1955).
1334. R. F. Zürcher and H. H. Günthard, Helv. Chim. Acta **38**, 849 (1955); **40**, 89 (1957).

# SUBJECT INDEX

As in the previous volumes, this index includes in addition to the usual material of an index also all *symbols* used in this volume and all *molecules* discussed or mentioned.

Italicized page numbers refer to more detailed discussions, ordinary page numbers to brief or casual references to the item listed; boldface page numbers refer to figures.

Mathematical symbols and symbols for species, point groups, molecular constants, etc., are listed at the beginning of the section devoted to the corresponding letter. Greek letter symbols are given under the letter with which they begin when they are written in English (for example, $\varphi$, $\pi$, $\psi$ under P, and in this order). Symbols to which a word is joined are listed under the corresponding symbol: for example, *R branch* is under *R*, not under *Rb*. But in all other cases the alphabeting is based on the part before the comma, for example, *electron impact* after *electronic angular momentum*.

All individual molecules are listed under their chemical formulae considered as words; for example, $CH_3Cl$ under *Chcl*; $HNO_2$ under *Hno*. If there are several molecules giving the same "word" they are listed in the order of increasing numbers of the first, second, . . . atom; for example, $CHCl_3$, $CH_3Cl$, $C_2H_2Cl_4$, $C_2H_4Cl_2$ in this order, but $C_2H_4$ is ahead of $CHCl_3$ since the corresponding "word" is *Ch*. It should be noted that this order is somewhat different from that in the formula index of Chemical Abstracts where the number of atoms has "priority" over the alphabet. This change appears to be necessary in a combined formula and subject index. For the benefit of the hurried reader, for all molecules discussed in detail, cross references are given under their chemical names.

The order of symbols in a chemical formula is the usual one for all organic compounds, including metal-organic compounds, that is, C comes first, then H, while the remaining atoms are in alphabetical order. For inorganic compounds the central atom, if any, is put first, then H and then the other atoms in alphabetical order, except in the case of acids and $H_2O$ for which H is put first. Thus we have $H_2S$, $HNO_2$, and so forth, that is, substantially the conventional order. Cross references are given in all ambiguous cases.

All material referring to a particular molecule, such as its electronic structure, dissociation, predissociation, molecular constants, spectrum, etc., is listed under the chemical formula of the molecule. Molecular types such as linear $XY_2$, planar $XY_3$, . . . molecules will be found under $XY_2$, $XY_3$, . . . where also various items relating to these types are given.

**B**

## F

## H

I

## J



## K

## O

## P

## U



## V

## W

# AUTHOR INDEX